Engineering Analysis with Pro/Mechanica and ANSYS

GUANGMING ZHANG

*Department of Mechanical Engineering
and the Institute for Systems Research*

The University of Maryland @ College Park

COLLEGE HOUSE ENTERPRISES, LLC
Knoxville, TN

THANKS FOR PERMISSIONS

The author and the publisher thank the copyright and trademark owners providing permission for use of copyrighted materials and trademarks. We apologize for any errors or omissions in obtaining permissions. Where appropriate, we referenced similar treatments and gave credit for prior work. Errors or omissions in obtaining permissions or in giving proper references are not intentional. We will correct them at the earliest opportunity after the error or omission is brought to our attention. Please contact the publisher at the address given below.

College House Enterprises, LLC.
5713 Glen Cove Drive
Knoxville, TN 37919-8611, U.S.A.
Phone: (865) 558 6111
FAX (865) 558 6111

ISBN 978-935673-03-3

**To my wife Jinyu
my children Zumei and Haowei, and my
grandchildren Qianqian, Jiajia and Lele**

who have always been with me especially at those difficult times. Their love and
understanding have supported me in a journey for survival and success.

ABOUT THE AUTHOR

Guangming Zhang obtained a bachelor degree and a master degree both in mechanical engineering from Tianjin University, the People's Republic of China. He obtained a master degree and a Ph. D. degree in mechanical engineering from the University of Illinois at Urbana-Champaign. He is currently an associate professor in the Department of Mechanical Engineering and holds a joint appointment with the Institute for Systems Research.

Professor Zhang worked at the Northwest Medical Surgical Instruments Factory in China where he served as a principal engineer to design surgical instruments and dental equipment. He also taught at the Beijing Institute of Printing and received the National Award for Outstanding Teaching from the Press and Publication Administration of the People's Republic of China in 1987. In 1992 he was selected by his peers at the University of Maryland to receive the Outstanding Systems Engineering Faculty Award. He was the recipient of the E. Robert Kent Outstanding Teaching Award of the College of Engineering in 1993, and the Poole and Kent Company Senior Faculty Teaching Award in 2004. He received the Professional Master of Engineering Program Award from the A. James Clark School of Engineering in 2006. He received the Pi Tau Sigma 2006 Faculty/Staff Appreciation Award. He was a recipient of the 1992 Blackall Machine Tool & Gage Award of the American Society of Mechanical Engineering. In 1995, 1997, 1998 and 1999, he received the Award of Commendation, Award of Member of the Year, and the Award of Appreciation from the Society of Manufacturing Engineers, Region 3, for his outstanding service as the faculty advisor to the SME Student Chapter at College Park. In 2006, he was named as one of the 6 Keystone Professors, the Clark School of Academy of Distinguished Professors.

Professor Zhang actively participated in the NSF sponsored ECSEL grant between 1990 and 2000. He served as the principal investigator for this grant at the University of Maryland between 1997 and 2000, and coordinated the ECSEL sponsored projects on integration of design, on active learning and hands-on experiences, and on developing methods for team learning. He is currently participating in the NSF sponsored ENGAGE program. He has written 4 textbooks, co-authored 3 textbooks, about 70 technical papers, and holds one patent.

PREFACE

This book presents an introduction to the fundamentals of finite element methods with computer applications. Finite element methods are numerical methods for solving problems of engineering and science. Those problems involve geometry, material properties, loading conditions and boundary conditions where seeking analytical based solutions is generally difficult, if not impossible. With capabilities of computers expanding at an unthinkable pace, the importance of using unbeatable computing power to perform finite element analysis has been well recognized. The book is written as an introductory text for undergraduate students in engineering. The book should also be useful to those engaged in the engineering design and engineering analysis.

The book is organized into 10 chapters. The first chapter provides a fundamental coverage of finite element methods. It stresses the mathematical and mechanical fundamentals, which support the development of finite element methods. Chapter 2 introduces the Pro/ENGINEER Wildfire design system and Pro/Mechanica, which is engineering simulation software to carry out finite element analysis. Chapter 3 focused on modeling a variety of load conditions. Chapter 4 focuses on modeling the boundary conditions of mechanical structures in general; illustrating the procedure of dealing the physical entities both at the component level and at the assembly level. Chapter 5 presents engineering analysis in the thermal domain. Chapter 6 presents an important branch of finite element methods. The branch is shell modeling through idealization. Chapter 6 also covers a review of the stress conditions so as to understand the mechanism of the engineering simulation software in the computing process. In Chapter 7, we focus on applications. Several case studies deal with engineering vibration analyses and buckling analyses. Chapter 8 introduces a specific concept called design optimization. Chapter 9 deals with the beam elements. Several examples related to mechanics of materials are presented. The engineering simulation system of ANSYS is introduced in Chapter 10. ANSYS is the FEA software most widely used in industry, not only nationally, but worldwide as well.

The material assembled in this book is an outgrowth of a senior-level design elective course taught by the author at the University of Maryland at College Park for the past ten years. Moreover, this book is written in a style adaptable for self-study and reference. Suggestions for improving the contents are welcome and the author deeply appreciate the efforts made by the readers in this regard.

It is a pleasure to extend special thanks to Dr. James W. Dally, President of the College House Enterprises, LLC. His support has been invaluable to the author, not only in the academic area, but also in many aspects of the author's life and career development.

Guangming Zhang
College Park, MD
July 2011

CHAPTER 1

INTRODUCTION TO FINITE ELEMENT METHODS

1.1 Introduction

Engineering design is an activity which demands the responsibility of the designer or the design team to meet the customer needs in terms of the design specifications. When an initial design is completed, a critical step to follow is to evaluate the performance of the designed component under a specified working environment. Results obtained from the evaluation lead to accepting the initial design or directing to an improvement through a redesign process. Finite element analysis is such a process, which has been used to carry out those evaluations of the initial designs through simulation under a computer based environment.

Finite element methods originated from an intuitive thinking, in which people viewed an entire body or an entire structure or a system (we consider the structure, the materials, the boundary conditions and the load condition as an integrated system) as composed of elements. As a result, the response of a system when it was subjected to an external load could be viewed as an integration of the responses of those elements under the loading conditions. To carry out such an integration process, numerical methods were introduced to facilitate the evaluation process, which was element based. It was clear that the results obtained from such an evaluation process were in the nature of approximation. The accuracy of the obtained results from the finite element method based evaluation was dependent on the number of elements used, the algorithm applied and the computing accuracy which could be achieved.

Finite element methods made their progress due to a great number of engineers and scientists devoted their efforts. Engineering analyses using finite element methods were gradually accepted in engineering applications because the methods had its unique capability of dealing with complex geometry and complex loading conditions by offering approximate solutions. Under those circumstances, obtaining a solution solely based on an analytical approach seemed very difficult, if not impossible. On the other hand, directly using an experimental method to evaluate the performance of a structural design would be extremely costly in general. In 1970s and 1980s, a significant amount of progress had been made in developing new computation algorithms, which significantly improved the computing efficiency and accuracy. While those progresses were being made, there were some difficulties in application implementations. All those difficulties were clustered at the limitation of the computation capability. As the number of elements was increased significantly in problem formulation, the requirement for data storage and distribution in computing was also increased accordingly. It was not uncommon that the computer time demanded to obtain a solution through the use of FEM was with day's time even weeks to complete.

Thanks to the rapid advancement of computer technology in past two decades, the capability of data storage and computing speed a computer system can offer not only has reached the needs of design engineers, but also has opened a new dimension to the design engineers for seeking solutions to those structures and systems, which are beyond their imagination to deal with during the past. Nowadays, finite element methods have already gained their incomputable popularity in engineering applications from the field of continuum mechanics, to heat transfer, flow dynamics, and even to medical fields where approximate solutions are needed for treatments and disease control. Finite element methods have already revolutionized to finite element analyses, which are a branch of engineering and science. Finite element analyses focus on obtaining approximate solutions to a wide variety of engineering and scientific problems, based on the integration of scientific fundamentals and computer based numerical analyses. Such a unique integration creates a favorable, stimulating environment for design evaluations. Today, software systems based on finite element methods to carry out finite element analyses for seeking solutions to real life challenges have been available. These software systems provide design engineers and scientists with a common platform for fast, efficient and cost-conscious product development, from design concept to final-stage testing and validation.

In this book, we will use two (2) of those FEA software systems available in the market to demonstrate the procedure(s) to perform FEA. These 2 software systems are Pro/Mechanica and ANSYS. We introduce Pro/Mechanica first in this book because we want to utilize the Pro/ENGINEER design system in the process of creating 3D solid models, and Pro/Mechanica and Pro/ENGINEER are an integrated system. We hope that the learning process related to the applications of Pro/Mechanica forms a basis for readers to understand the fundamentals related to the general framework embedded in the FEA software systems commercially available in the market.

In section 1.2, we introduce the finite element method in a one-dimensional space. In section 1.3, we introduce the finite element method in a two-dimensional space. In section 1.4, we introduce the finite element method in a three-dimensional space. In section 1.5, we emphasize the importance of setting a unit system when using a FEA software system.

1.2   Finite Element Method in a One-Dimensional Space

In this section, we use the concept of idealization applied in modeling a mechanical structure in one-dimensional space to introduce readers the basic concepts involved in using finite element methods. In the field of engineering, components such as springs are used in structural design. For the purpose of analyzing the response of a mechanical structure when subjected to loading, springs are widely used in so-called idealization. For example, when dealing with the displacement of a cylindrical bar, which is subjected to an axial load, the stress and strain can be evaluated based on the following constitution equations:

$$\sigma = \frac{F}{A}$$

$$A = \pi\, r^2$$

$$\varepsilon = E\sigma$$

$$F \longleftarrow \qquad \longrightarrow F$$

$$l_o$$

Where E = Young's modulus of elasticity, or elastic modulus, and ε is evaluated by

$$\varepsilon = \frac{\Delta l}{l_o} = E\sigma$$

By substitution, we will have the following relation:

$$F = \sigma A = (E\varepsilon)A = (E\frac{\Delta l}{l_o})A = (\frac{AE}{l_o})\Delta l$$

In the field of theoretical mechanics, we have the following constitution relation for a spring component where parameter k is the spring constant or the stiffness of the spring.

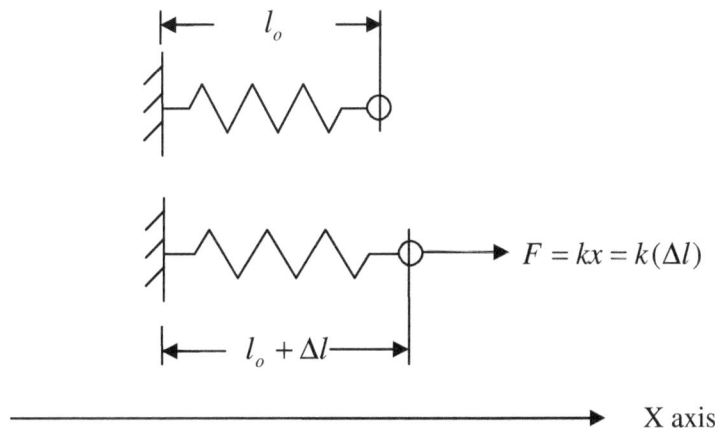

When we use a spring component to represent the cylindrical bar for evaluating the displacement, we may use a concept called an equivalent stiffness of the cylindrical bar, which is given by

$$k = \frac{AE}{l_o}$$

Now let us introduce a concept of shape function. The need to have a shape function comes from quantitatively and systematically representing the displacement at each node.

For example, when using FEA, a spring is modeled as an element. In a 2D space, each element has 2 nodes, as shown below.

Node 1   Node 2

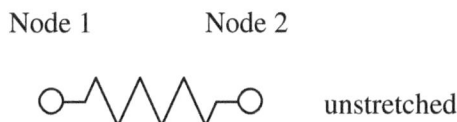   unstretched

Under a loading condition, the spring element can be stretched or compressed. The displacement at each node is used to quantify the deformation of the spring component. Let us assume that the displacement at node 1 is d_{1x} and the displacement at node 2 is d_{2x}.

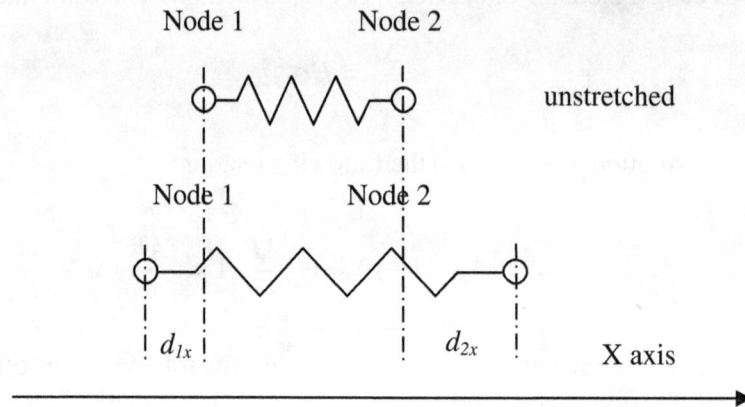

Mathematically, we are able to use the following relation to represent the displacement:

$$u(x) = \begin{cases} a_1 + a_2 x & 0 =< x <= l \\ d_{1x} & x = 0 \\ d_{2x} & x = l \end{cases}$$

where $a_1 = d_{1x}$ and $a_2 = \dfrac{d_{2x} - d_{1x}}{l}$.

To evaluate the displacement at a location within the element, the following relation applies:

$$u(x) = d_{1x} + \frac{d_{2x} - d_{1x}}{l} x \qquad 0 =< x <= l$$

In preparing a computer code to evaluate the displacement of an element, a matrix format is always used. To represent the displacement at a particular location within an element, the following matrix notation is introduced:

$$u(x) = \left[(1 - \frac{x}{l}) \quad \frac{x}{l} \right] \begin{Bmatrix} d_{1x} \\ d_{2x} \end{Bmatrix} = \left[N_1 \quad N_2 \right] \begin{Bmatrix} d_{1x} \\ d_{2x} \end{Bmatrix}$$

where $N_1 = 1 - \dfrac{x}{l}$ and $N_2 = \dfrac{x}{l}$ are called the shape function of the element to characterize the displacement at specific location within the element based on the known nodal displacements.

The shape function at node 1 is given by $N_1 = 1$ and $N_2 = 0$.
The shape function at node 2 is given by $N_1 = 0$ and $N_2 = 1$.

It is evident that N1 and N2 function as weighting factors. These weighting factors can be a zero-order polynomial function, such as the one in the current case. They can also be a first-order or second-order or higher-order polynomial functions, depending on the need in evaluation.

Now let us introduce a concept of stiffness matrix of an element. To characterize the system response of a spring component when subjected to loading, we have the following relations:

$$f_{1x} = k(d_{1x} - d_{2x})$$

$$f_{2x} = k(d_{2x} - d_{1x})$$

Pay a special attention to the sign of each entity with respect to the position direction of the X-axis defined, as shown below:

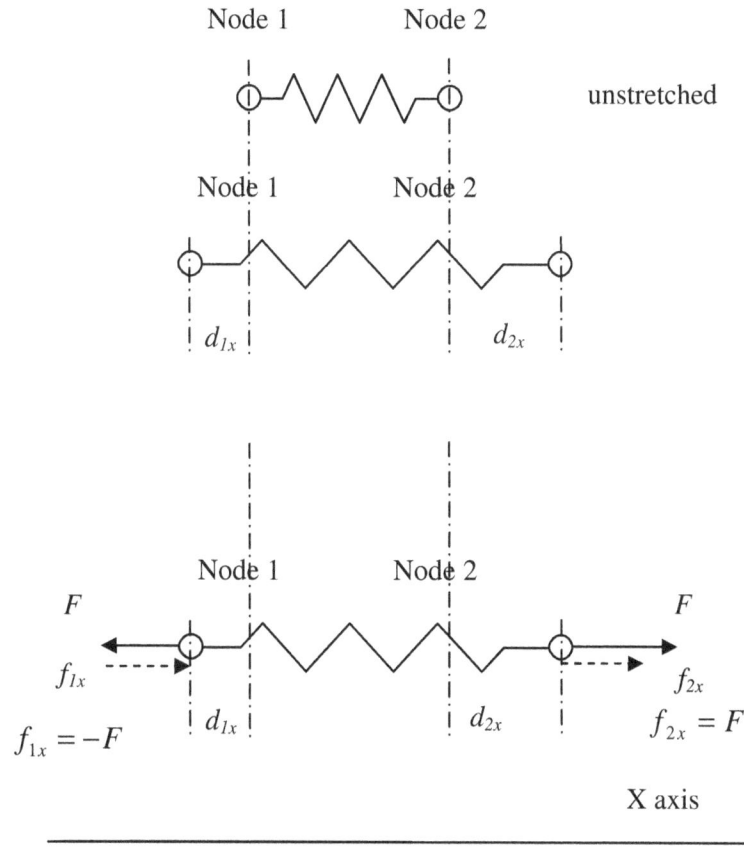

In preparing a computer code to evaluate the force acting on each node, a matrix format is always used. To represent the forces at the 2 nodes, the following matrix notation is introduced:

$$\begin{Bmatrix} f_{1x} \\ f_{2x} \end{Bmatrix} = \begin{bmatrix} k & -k \\ -k & k \end{bmatrix} \begin{Bmatrix} d_{1x} \\ d_{2x} \end{Bmatrix} = \begin{bmatrix} k_{11} & k_{12} \\ k_{21} & k_{22} \end{bmatrix} \begin{Bmatrix} d_{1x} \\ d_{2x} \end{Bmatrix} = \begin{bmatrix} N_1 & N_2 \end{bmatrix} \begin{Bmatrix} d_{1x} \\ d_{2x} \end{Bmatrix}$$

Where f_{1x} and f_{2x} are local nodal forces, and d_{1x} and d_{2x} are local nodal displacements. The 4 parameters, k_{11}, k_{12}, k_{21} and k_{22} are coefficients of the spring stiffness. These coefficients can be identified as

$$k_{11} = k$$

$$k_{12} = -k$$

$$k_{21} = -k$$

$$k_{22} = k$$

$$f_{1x} = k_{11}d_{1x} + k_{12}d_{2x}$$

$$f_{2x} = k_{21}d_{1x} + k_{22}d_{2x}$$

Node 1   Node 2

f_{1x}, d_{1x} ———→⊙〜〜〜⊙———→ f_{2x}, d_{2x}

X axis
—————————————————→

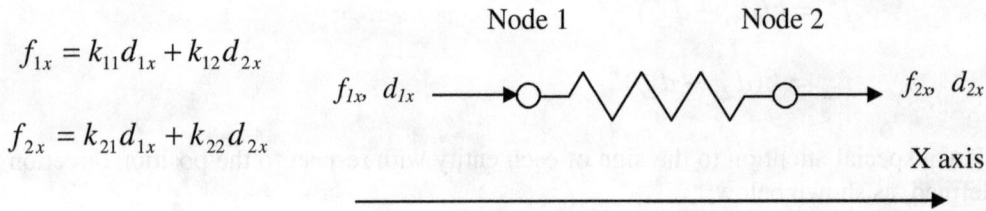

When a stiffness matrix for each element is constructed, we need to assemble all of them so that we will have a global stiffness matrix, which characterizes the entire system under study. We demonstrate the procedure to perform FEA in a one-dimensional space through a case study. The case study is a lumped mass system with 3 springs. An external force acts on the mass, as shown below. In this case, we view the lumped mass system consisting of three (3) elements. Each element is treated as a spring component. The system is in a static equilibrium status. Therefore, we model this system, using 3 elements and 4 nodes. Note that node 2 is shared by all 3 elements, as show below:

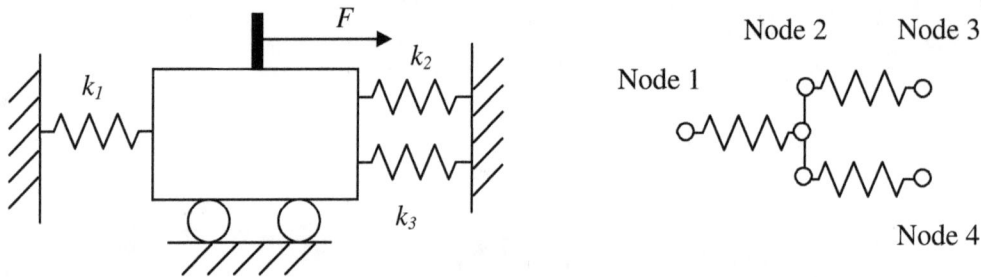

Secondly, we formulate a stiffness matrix for each of the 3 elements. The 3 stiffness matrices for the 3 elements are shown below:

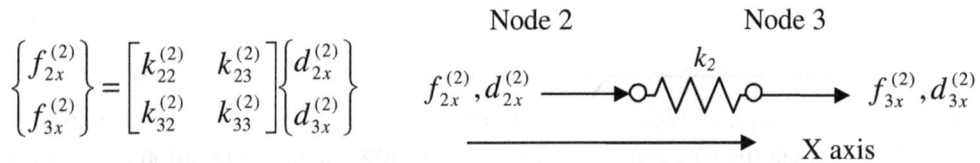

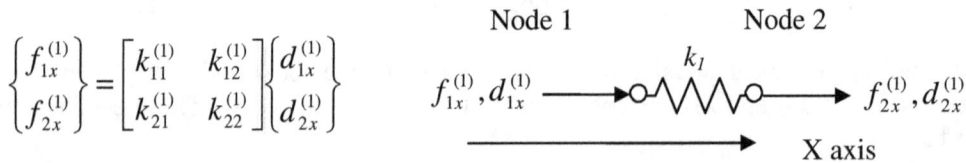

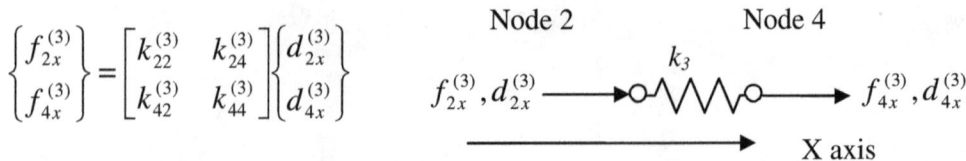

Node 2   Node 3

$$\left\{ \begin{matrix} f_{2x}^{(2)} \\ f_{3x}^{(2)} \end{matrix} \right\} = \begin{bmatrix} k_{22}^{(2)} & k_{23}^{(2)} \\ k_{32}^{(2)} & k_{33}^{(2)} \end{bmatrix} \left\{ \begin{matrix} d_{2x}^{(2)} \\ d_{3x}^{(2)} \end{matrix} \right\}$$

$f_{2x}^{(2)}, d_{2x}^{(2)}$ ———→⊙〜〜〜⊙———→ $f_{3x}^{(2)}, d_{3x}^{(2)}$

X axis
—————————————————→

Node 1   Node 2

$$\left\{ \begin{matrix} f_{1x}^{(1)} \\ f_{2x}^{(1)} \end{matrix} \right\} = \begin{bmatrix} k_{11}^{(1)} & k_{12}^{(1)} \\ k_{21}^{(1)} & k_{22}^{(1)} \end{bmatrix} \left\{ \begin{matrix} d_{1x}^{(1)} \\ d_{2x}^{(1)} \end{matrix} \right\}$$

$f_{1x}^{(1)}, d_{1x}^{(1)}$ ———→⊙〜〜〜⊙———→ $f_{2x}^{(1)}, d_{2x}^{(1)}$

X axis
—————————————————→

Node 2   Node 4

$$\left\{ \begin{matrix} f_{2x}^{(3)} \\ f_{4x}^{(3)} \end{matrix} \right\} = \begin{bmatrix} k_{22}^{(3)} & k_{24}^{(3)} \\ k_{42}^{(3)} & k_{44}^{(3)} \end{bmatrix} \left\{ \begin{matrix} d_{2x}^{(3)} \\ d_{4x}^{(3)} \end{matrix} \right\}$$

$f_{2x}^{(3)}, d_{2x}^{(3)}$ ———→⊙〜〜〜⊙———→ $f_{4x}^{(3)}, d_{4x}^{(3)}$

X axis
—————————————————→

Now let us assemble these 3 stiffness matrices to obtain the global stiffness matrix.

$$\left\{ \begin{array}{c} f_{1x}^{(1)} \\ f_{2x}^{(1)} + f_{2x}^{(2)} + f_{2x}^{(3)} \\ f_{3x}^{(2)} \\ f_{4x}^{(3)} \end{array} \right\} = \left[\begin{array}{cccc} k_{11}^{(1)} & k_{12}^{(1)} & 0 & 0 \\ k_{21}^{(1)} & k_{22}^{(1)} + k_{22}^{(2)} + k_{22}^{(3)} & k_{23}^{(2)} & k_{24}^{(3)} \\ 0 & k_{32}^{(2)} & k_{33}^{(2)} & 0 \\ 0 & k_{42}^{(3)} & 0 & k_{44}^{(3)} \end{array} \right] \left\{ \begin{array}{c} d_{1x}^{(1)} \\ d_{2x} \\ d_{3x}^{(2)} \\ d_{4x}^{(3)} \end{array} \right\}$$

where $d_{2x} = d_{2x}^{(1)} = d_{2x}^{(2)} = d_{2x}^{(3)}$.

It is important that we need to have a specific set of boundary conditions for the Assembled System. Those boundary conditions are also called constraint conditions, widely adopted in the FEA software systems.

Referring this specific case study, we have 2 constraint conditions. The first constraint condition has already been used in the process of assembling the 3 stiffness matrices of the 3 elements. The first constraint condition is "node 2 is shared by all 3 elements". The second constraint condition is $d_{1x}^{(1)} = d_{3x}^{(2)} = d_{4x}^{(3)} = 0$, or

$$\left\{ \begin{array}{c} d_{1x}^{(1)} \\ d_{2x} \\ d_{3x}^{(2)} \\ d_{4x}^{(3)} \end{array} \right\} = \left\{ \begin{array}{c} 0 \\ d_{2x} \\ 0 \\ 0 \end{array} \right\}$$

To determine the displacement of each node under a static equilibrium condition, we need a specific load condition, which is acting on the assembled system.

In general, we call $f_{1x}^{(1)}$ the reaction force at the joint between the spring of k_1 and the wall structure, $f_{3x}^{(2)}$ the reaction force at the joint between the spring of k_2 and the wall structure, and $f_{4x}^{(3)}$ the reaction force at the joint between the spring of k_1 and the wall structure. Note that we use UPPER CASE to represent the load condition or the reaction force because they characterize the load condition acting on the assembly system, not on a node of a single element. Certainly, the load acting on the system could be the nodal force under some circumstances.

$$\left\{ \begin{array}{c} F_{1x} \\ F_{2x} \\ F_{3x} \\ F_{4x} \end{array} \right\} = \left\{ \begin{array}{c} f_{1x}^{(1)} \\ f_{2x}^{(1)} + f_{2x}^{(2)} + f_{2x}^{(3)} \\ f_{3x}^{(2)} \\ f_{4x}^{(3)} \end{array} \right\}$$

After applying the boundary conditions and the load conditions, the mathematical form of FEA for this system is represented by:

$$\begin{Bmatrix} F_{1x} \\ F_{2x} \\ F_{3x} \\ F_{4x} \end{Bmatrix} = \begin{bmatrix} k_{11}^{(1)} & k_{12}^{(1)} & 0 & 0 \\ k_{21}^{(1)} & k_{22}^{(1)} + k_{22}^{(2)} + k_{22}^{(3)} & k_{23}^{(2)} & k_{24}^{(3)} \\ 0 & k_{32}^{(2)} & k_{33}^{(2)} & 0 \\ 0 & k_{42}^{(3)} & 0 & k_{44}^{(3)} \end{bmatrix} \begin{Bmatrix} 0 \\ d_{2x} \\ 0 \\ 0 \end{Bmatrix}$$

$$\begin{Bmatrix} F_{1x} \\ F_{2x} \\ F_{3x} \\ F_{4x} \end{Bmatrix} = \begin{bmatrix} k_1 & -k_1 & 0 & 0 \\ -k_1 & k_1 + k_2 + k_3 & -k_2 & -k_3 \\ 0 & -k_2 & k_2 & 0 \\ 0 & -k_3 & 0 & k_3 \end{bmatrix} \begin{Bmatrix} 0 \\ d_{2x} \\ 0 \\ 0 \end{Bmatrix}$$

Now let us the FEA formula to obtain solutions to 2 questions, namely, Example 1-1 and Example 1-2.

Example 1-1: Assume that $k_1 = 10$ Newton/mm, $k_2 = 8$ Newton/mm and $k_3 = 6$ Newton/mm. If we need to place the body 0.5 mm away from its original position, what is the load condition to keep the body a distance of 0.5 mm stationary? What are the reaction forces at the 3 joints?

$$\begin{Bmatrix} F_{1x} \\ F_{2x} \\ F_{3x} \\ F_{4x} \end{Bmatrix} = \begin{bmatrix} k_1 & -k_1 & 0 & 0 \\ -k_1 & k_1 + k_2 + k_3 & -k_2 & -k_3 \\ 0 & -k_2 & k_2 & 0 \\ 0 & -k_3 & 0 & k_3 \end{bmatrix} \begin{Bmatrix} 0 \\ d_{2x} \\ 0 \\ 0 \end{Bmatrix}$$

$$\begin{Bmatrix} F_{1x} \\ F_{2x} \\ F_{3x} \\ F_{4x} \end{Bmatrix} = \begin{bmatrix} 10 & -10 & 0 & 0 \\ -10 & 24 & -8 & -6 \\ 0 & -8 & 8 & 0 \\ 0 & -6 & 0 & 6 \end{bmatrix} \begin{Bmatrix} 0 \\ 0.5 \\ 0 \\ 0 \end{Bmatrix} = \begin{Bmatrix} -5 \\ 12 \\ -4 \\ -3 \end{Bmatrix}$$

Example 1-2: Assume that $k_1 = 10$ Newton/mm, $k_2 = 8$ Newton/mm and $k_3 = 6$ Newton/mm. There is lumped force acting on the body. Its magnitude is 48 Newton. What is the displacement of the body with respect to its original equilibrium condition? What are the reaction forces at the 3 joints?

$$\begin{Bmatrix} F_{1x} \\ F_{2x} \\ F_{3x} \\ F_{4x} \end{Bmatrix} = \begin{bmatrix} k_1 & -k_1 & 0 & 0 \\ -k_1 & k_1 + k_2 + k_3 & -k_2 & -k_3 \\ 0 & -k_2 & k_2 & 0 \\ 0 & -k_3 & 0 & k_3 \end{bmatrix} \begin{Bmatrix} 0 \\ d_{2x} \\ 0 \\ 0 \end{Bmatrix}$$

$$\begin{Bmatrix} F_{1x} \\ 48 \\ F_{3x} \\ F_{4x} \end{Bmatrix} = \begin{bmatrix} 10 & -10 & 0 & 0 \\ -10 & 24 & -8 & -6 \\ 0 & -8 & 8 & 0 \\ 0 & -6 & 0 & 6 \end{bmatrix} \begin{Bmatrix} 0 \\ d_{2x} \\ 0 \\ 0 \end{Bmatrix}$$

We have $48 = 24*d_{2x}$, and $d_{2x} = 2$ mm. Using the matrix multiplication, we have the 3 reaction forces listed below:

$$\begin{Bmatrix} F_{1x} \\ 48 \\ F_{3x} \\ F_{4x} \end{Bmatrix} = \begin{bmatrix} 10 & -10 & 0 & 0 \\ -10 & 24 & -8 & -6 \\ 0 & -8 & 8 & 0 \\ 0 & -6 & 0 & 6 \end{bmatrix} \begin{Bmatrix} 0 \\ 2 \\ 0 \\ 0 \end{Bmatrix} = \begin{Bmatrix} -20 \\ 48 \\ 16 \\ 12 \end{Bmatrix}$$

1.3 Finite Element Method in a Two-Dimensional Space

In the previous section, we dealt with structures which can be represented in a one-dimensional space. In this section, we will deal with structures, which have to be represented in a two-dimensional space. For example, a truss structure in real life is generally described in a two-dimensional space. In the following, we us a plane pin-jointed truss structure to demonstrate the procedure to obtain the stiffness matrix of an element in the 2-dimensional space. We assume that the loads act only at the joints, the axial force in each member is constant throughout the length of the member.

In FEA, we view each rod within a truss structure as an element. As illustrated, a rod is in an inclined position. The inclined angle with respect to the horizontal axis is 30 degrees. Under a load condition, there is a displacement along the x-axis, and there is no displacement along the y-axis because the force acting on the rod is along the x-axis. Assume that $d_{1x} = 1$.

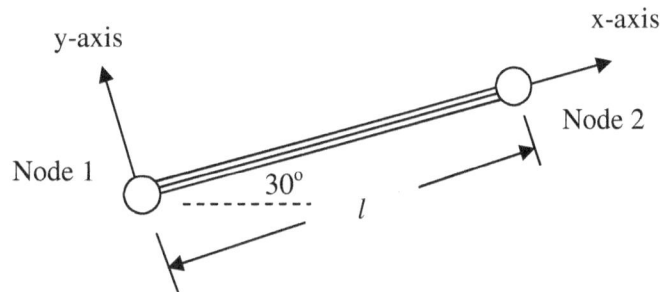

In general, a truss structure has more than 1 rod, and each rod has its own orientation with respect to the horizontal direction and/or the vertical direction. In order to effectively represent the system performance indices, such as displacement or load, we generally use a global coordinate system to define the boundary condition(s) and the load condition(s). This is extremely important when the assemblage of stiffness matrices is performed.

A displacement associated with an individual rod is also reflected into the displacement observed in the global coordinate system. The relationship between the displacements evaluated based on a local coordinate system and the displacements observed from the global coordinate system are established through the following transformation.

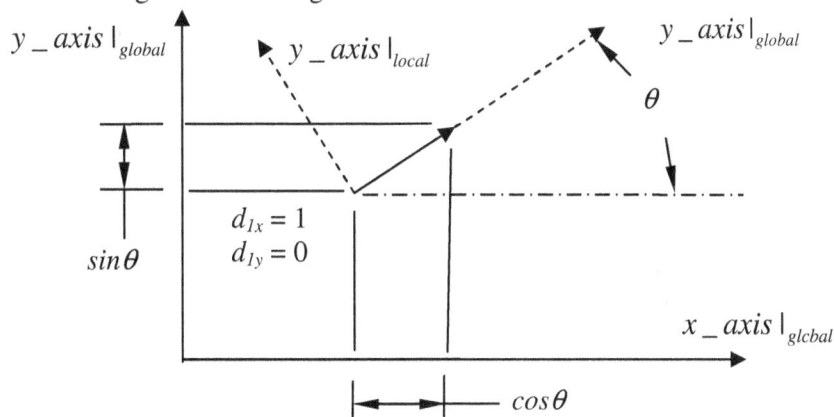

$$d_{1x\_global} = d_{1x\_local} \cos \theta \vec{i} - d_{1y\_local} \sin \theta \vec{j}$$

$$d_{1y\_global} = d_{1x\_local} \sin \theta \vec{i} + d_{1y\_local} \cos \theta \vec{j}$$

$$\left\{ \begin{array}{c} d_{1x} \\ d_{1y} \end{array} \right\}_{global} = \begin{bmatrix} \cos \theta & -\sin \theta \\ \sin \theta & \cos \theta \end{bmatrix} \left\{ \begin{array}{c} d_{1x} \\ d_{1y} \end{array} \right\}_{local} = \begin{bmatrix} C & -S \\ S & C \end{bmatrix} \left\{ \begin{array}{c} d_{1x} \\ d_{1y} \end{array} \right\}_{local}$$

$$\left\{ \begin{array}{c} d_{1x} \\ d_{1y} \end{array} \right\}_{local} = \begin{bmatrix} \cos \theta & \sin \theta \\ -\sin \theta & \cos \theta \end{bmatrix} \left\{ \begin{array}{c} d_{1x} \\ d_{1y} \end{array} \right\}_{global} = \begin{bmatrix} C & S \\ -S & C \end{bmatrix} \left\{ \begin{array}{c} d_{1x} \\ d_{1y} \end{array} \right\}_{global}$$

When we consider both nodes, or both Node 1 and Node 2, we have the following relations:

$$\left\{ \begin{array}{c} d_{1x} \\ d_{1y} \\ d_{2x} \\ d_{2y} \end{array} \right\}_{global} = \begin{bmatrix} \cos \theta & -\sin \theta & 0 & 0 \\ \sin \theta & \cos \theta & 0 & 0 \\ 0 & 0 & \cos \theta & -\sin \theta \\ 0 & 0 & \sin \theta & \cos \theta \end{bmatrix} \left\{ \begin{array}{c} d_{1x} \\ d_{1y} \\ d_{2x} \\ d_{2y} \end{array} \right\}_{local}$$

$$\left\{ \begin{array}{c} d_{1x} \\ d_{1y} \\ d_{2x} \\ d_{2y} \end{array} \right\}_{local} = \begin{bmatrix} \cos \theta & \sin \theta & 0 & 0 \\ -\sin \theta & \cos \theta & 0 & 0 \\ 0 & 0 & \cos \theta & \sin \theta \\ 0 & 0 & -\sin \theta & \cos \theta \end{bmatrix} \left\{ \begin{array}{c} d_{1x} \\ d_{1y} \\ d_{2x} \\ d_{2y} \end{array} \right\}_{global}$$

When considering the load conditions, it is natural to have the following relationship to interpret the mapping of a vector representing a force component from the global coordinate system to a local coordinate system and vice versa.

$$\left\{ \begin{array}{c} F_{1x} \\ F_{1y} \\ F_{2x} \\ F_{2y} \end{array} \right\}_{global} = \begin{bmatrix} \cos \theta & -\sin \theta & 0 & 0 \\ \sin \theta & \cos \theta & 0 & 0 \\ 0 & 0 & \cos \theta & -\sin \theta \\ 0 & 0 & \sin \theta & \cos \theta \end{bmatrix} \left\{ \begin{array}{c} f_{1x} \\ f_{1y} \\ f_{2x} \\ f_{2y} \end{array} \right\}_{local}$$

$$\left\{ \begin{array}{c} f_{1x} \\ f_{1y} \\ f_{2x} \\ f_{2y} \end{array} \right\}_{local} = \begin{bmatrix} \cos \theta & \sin \theta & 0 & 0 \\ -\sin \theta & \cos \theta & 0 & 0 \\ 0 & 0 & \cos \theta & \sin \theta \\ 0 & 0 & -\sin \theta & \cos \theta \end{bmatrix} \left\{ \begin{array}{c} F_{1x} \\ F_{1y} \\ F_{2x} \\ F_{2y} \end{array} \right\}_{global}$$

Now let us consider the stiffness matrix of a rod in the 2-dimensional space. We start from the stiffness matrix defined locally.

$$\begin{Bmatrix} f_{1x} \\ f_{1y} \\ f_{2x} \\ f_{2y} \end{Bmatrix}_{local} = \begin{bmatrix} k_1 & 0 & -k_1 & 0 \\ 0 & 0 & 0 & 0 \\ -k_1 & 0 & k_1 & 0 \\ 0 & 0 & 0 & 0 \end{bmatrix} \begin{Bmatrix} d_{1x} \\ d_{1y} \\ d_{2x} \\ d_{2y} \end{Bmatrix}_{local}$$

We substitute the local displacement vector by the global displacement vector. We have

$$\begin{Bmatrix} f_{1x} \\ f_{1y} \\ f_{2x} \\ f_{2y} \end{Bmatrix}_{local} = \begin{bmatrix} k_1 & 0 & -k_1 & 0 \\ 0 & 0 & 0 & 0 \\ -k_1 & 0 & k_1 & 0 \\ 0 & 0 & 0 & 0 \end{bmatrix} \begin{Bmatrix} d_{1x} \\ d_{1y} \\ d_{2x} \\ d_{2y} \end{Bmatrix}_{local} = \begin{bmatrix} k_1 & 0 & -k_1 & 0 \\ 0 & 0 & 0 & 0 \\ -k_1 & 0 & k_1 & 0 \\ 0 & 0 & 0 & 0 \end{bmatrix} \begin{bmatrix} C & S & 0 & 0 \\ -S & C & 0 & 0 \\ 0 & 0 & C & S \\ 0 & 0 & -S & C \end{bmatrix} \begin{Bmatrix} d_{1x} \\ d_{1y} \\ d_{2x} \\ d_{2y} \end{Bmatrix}_{global}$$

$$\begin{Bmatrix} f_{1x} \\ f_{1y} \\ f_{2x} \\ f_{2y} \end{Bmatrix}_{local} = \begin{bmatrix} k_1 & 0 & -k_1 & 0 \\ 0 & 0 & 0 & 0 \\ -k_1 & 0 & k_1 & 0 \\ 0 & 0 & 0 & 0 \end{bmatrix} [T] \begin{Bmatrix} d_{1x} \\ d_{1y} \\ d_{2x} \\ d_{2y} \end{Bmatrix}_{global}$$

$$[T] = \begin{bmatrix} C & S & 0 & 0 \\ -S & C & 0 & 0 \\ 0 & 0 & C & S \\ 0 & 0 & -S & C \end{bmatrix}$$

We substitute the local load vector by the global load vector. We have

$$\begin{bmatrix} C & S & 0 & 0 \\ -S & C & 0 & 0 \\ 0 & 0 & C & S \\ 0 & 0 & -S & C \end{bmatrix} \begin{Bmatrix} F_{1x} \\ F_{1y} \\ F_{2x} \\ F_{2y} \end{Bmatrix}_{local} = \begin{bmatrix} k_1 & 0 & -k_1 & 0 \\ 0 & 0 & 0 & 0 \\ -k_1 & 0 & k_1 & 0 \\ 0 & 0 & 0 & 0 \end{bmatrix} [T] \begin{Bmatrix} d_{1x} \\ d_{1y} \\ d_{2x} \\ d_{2y} \end{Bmatrix}_{global}$$

$$\begin{Bmatrix} F_{1x} \\ F_{1y} \\ F_{2x} \\ F_{2y} \end{Bmatrix}_{local} = [T]^T \begin{bmatrix} k_1 & 0 & -k_1 & 0 \\ 0 & 0 & 0 & 0 \\ -k_1 & 0 & k_1 & 0 \\ 0 & 0 & 0 & 0 \end{bmatrix} [T] \begin{Bmatrix} d_{1x} \\ d_{1y} \\ d_{2x} \\ d_{2y} \end{Bmatrix}_{global}$$

$$[T]^T = \begin{bmatrix} C & -S & 0 & 0 \\ S & C & 0 & 0 \\ 0 & 0 & C & -S \\ 0 & 0 & S & C \end{bmatrix}$$

$$\begin{Bmatrix} F_{1x} \\ F_{1y} \\ F_{2x} \\ F_{2y} \end{Bmatrix}_{local} = [K] \begin{Bmatrix} d_{1x} \\ d_{1y} \\ d_{2x} \\ d_{2y} \end{Bmatrix}_{global}$$

$$[K] = [T]^T \begin{bmatrix} k_1 & 0 & -k_1 & 0 \\ 0 & 0 & 0 & 0 \\ -k_1 & 0 & k_1 & 0 \\ 0 & 0 & 0 & 0 \end{bmatrix} [T]$$

Now let us the FEM formula to obtain the required solutions to the following truss structure. As illustrated, the truss structure has 3 bars. Their lengths are 300 mm, 400 mm and 500 mm, as shown. If we assign three nodes to the three joints, we will have 3 elements. Among the 3 joints, 2 joints are pin connections. The other is a roller support. Each element represents each bar. The load condition is a horizontal force acts on node 2.

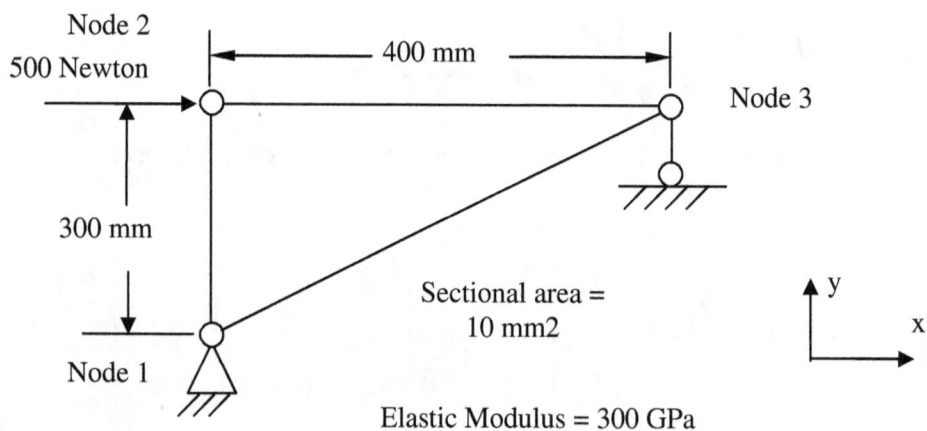

Node 2
500 Newton

400 mm

Node 3

300 mm

Sectional area =
10 mm2

y

x

Node 1

Elastic Modulus = 300 GPa

Our objectives of this study are
(1)   Obtain the stiffness matrix for each of the 3 elements.
(2)   Obtain the global stiffness matrix of the truss structure.
(3)   Determine the displacements of Node 2 and Node 3.
(4)   Determine the reaction forces on Node 1 and Node 3.

First let us examine the characteristics of this truss structure. There is no normal displacement of node 1. There is no normal displacement in node 3. As a result, there is no normal displacement in node 2. Therefore, under the global coordinate system, we have

$$
\left\{
\begin{array}{c}
d_{1x} \\
d_{1y} \\
d_{2x} \\
d_{2y} \\
d_{3x} \\
d_{3y}
\end{array}
\right\}_{global}
=
\left\{
\begin{array}{c}
d_{1x} \\
0 \\
d_{2x} \\
0 \\
d_{3x} \\
0
\end{array}
\right\}_{global}
$$

It is important to emphasize that in the local coordinate system the x direction is along the length of the element and the y direction is normal to the x direction.

Let us consider the first element, which will be the vertical bar. In this orientation $\theta = 90°$. The local stiffness matrix has the following form:

$$
\left[
\begin{array}{c}
f_{1x} \\
f_{1y} \\
f_{2x} \\
f_{2y}
\end{array}
\right]_{local}
=
\left[
\begin{array}{cccc}
k_1 & 0 & -k_1 & 0 \\
0 & 0 & 0 & 0 \\
-k_1 & 0 & k_1 & 0 \\
0 & 0 & 0 & 0
\end{array}
\right]
\left[
\begin{array}{c}
d_{1x} \\
d_{1y} \\
d_{2x} \\
d_{2y}
\end{array}
\right]_{local}
$$

In order to get the local coordinate system into the global coordinate system the force and displacement matrices must be multiplied by a matrix T.

$$
[T] =
\left[
\begin{array}{cccc}
\cos\theta & \sin\theta & 0 & 0 \\
-\sin\theta & \cos\theta & 0 & 0 \\
0 & 0 & \cos\theta & \sin\theta \\
0 & 0 & -\sin\theta & \cos\theta
\end{array}
\right]
$$

Because the force matrix is on the other side of the equation, the inverse of the T matrix is also required.

$$
[T]^T =
\left[
\begin{array}{cccc}
\cos\theta & -\sin\theta & 0 & 0 \\
\sin\theta & \cos\theta & 0 & 0 \\
0 & 0 & \cos\theta & -\sin\theta \\
0 & 0 & \sin\theta & \cos\theta
\end{array}
\right]
$$

When the T matrix is used to convert the local coordinate system into the global coordinate system the stiffness matrix takes the form

$$
\begin{bmatrix} F_{1x} \\ F_{1y} \\ F_{2x} \\ F_{2y} \end{bmatrix}_{Global} = [T]^T \begin{bmatrix} k_1 & 0 & -k_1 & 0 \\ 0 & 0 & 0 & 0 \\ -k_1 & 0 & k_1 & 0 \\ 0 & 0 & 0 & 0 \end{bmatrix} [T] \begin{bmatrix} d_{1x} \\ d_{1y} \\ d_{2x} \\ d_{2y} \end{bmatrix}_{global}
$$

K_1 can be determined from the equation

$$
K_1 = \frac{AE}{l_1} = \frac{(1*10^{-5})(300*10^9)}{.3} = 1*10^7 \, N/m
$$

The values can be inserted into the stiffness matrix

$$
\begin{bmatrix} F_{1x} \\ F_{1y} \\ F_{2x} \\ F_{2y} \end{bmatrix}_{Global} = \begin{bmatrix} 0 & -1 & 0 & 0 \\ 1 & 0 & 0 & 0 \\ 0 & 0 & 0 & -1 \\ 0 & 0 & 1 & 0 \end{bmatrix} \begin{bmatrix} k_1 & 0 & -k_1 & 0 \\ 0 & 0 & 0 & 0 \\ -k_1 & 0 & k_1 & 0 \\ 0 & 0 & 0 & 0 \end{bmatrix} \begin{bmatrix} 0 & 1 & 0 & 0 \\ -1 & 0 & 0 & 0 \\ 0 & 0 & 0 & 1 \\ 0 & 0 & -1 & 0 \end{bmatrix} \begin{bmatrix} d_{1x} \\ d_{1y} \\ d_{2x} \\ d_{2y} \end{bmatrix}_{global}
$$

The matrix multiplication can be evaluated through the use of a software system, such as Matlab.

$$
\begin{bmatrix} F_{1x} \\ F_{1y} \\ F_{2x} \\ F_{2y} \end{bmatrix}_{Global} = \begin{bmatrix} 0 & 0 & 0 & 0 \\ 0 & 1e7 & 0 & -1e7 \\ 0 & 0 & 0 & 0 \\ 0 & -1e7 & 0 & 1e7 \end{bmatrix} \begin{bmatrix} d_{1x} \\ d_{1y} \\ d_{2x} \\ d_{2y} \end{bmatrix}_{global}
$$

Now let us work with the next element, which is the horizontal bar. Let us name element 2. The orientation is $\theta=0°$. The local stiffness matrix takes the form

$$
\begin{bmatrix} f_{2x} \\ f_{2y} \\ f_{3x} \\ f_{3y} \end{bmatrix}_{local} = \begin{bmatrix} k_2 & 0 & -k_2 & 0 \\ 0 & 0 & 0 & 0 \\ -k_2 & 0 & k_2 & 0 \\ 0 & 0 & 0 & 0 \end{bmatrix} \begin{bmatrix} d_{2x} \\ d_{2y} \\ d_{3x} \\ d_{3y} \end{bmatrix}_{local}
$$

The matrix is the same form as the matrix for element one because it is again assumed that there is no motion normal to the local x coordinate. The T matrix is needed to transform the local coordinate system into the global coordinate system. The T matrix was already presented so will not be defined again. The result of the local to global transformation is

$$
\begin{bmatrix} f_{2x} \\ f_{2y} \\ f_{3x} \\ f_{3y} \end{bmatrix}_{global} = \begin{bmatrix} T^T \end{bmatrix} \begin{bmatrix} k_2 & 0 & -k_2 & 0 \\ 0 & 0 & 0 & 0 \\ -k_2 & 0 & k_2 & 0 \\ 0 & 0 & 0 & 0 \end{bmatrix} [T] \begin{bmatrix} d_{2x} \\ d_{2y} \\ d_{3x} \\ d_{3y} \end{bmatrix}_{global}
$$

The value of k_2 can be determined the same way as K_1

$$
[k_2] = \frac{AE}{l_2} = \frac{(1*10^{-5})(300*10^9)}{.4} = 7.5*10^6 \, N/m
$$

The values can now be used in the matrix

$$
\begin{bmatrix} f_{2x} \\ f_{2y} \\ f_{3x} \\ f_{3y} \end{bmatrix}_{global} = \begin{bmatrix} 1 & 0 & 0 & 0 \\ 0 & 1 & 0 & 0 \\ 0 & 0 & 1 & 0 \\ 0 & 0 & 0 & 1 \end{bmatrix} \begin{bmatrix} 7.5e6 & 0 & -7.5e6 & 0 \\ 0 & 0 & 0 & 0 \\ -7.5e6 & 0 & 7.5e6 & 0 \\ 0 & 0 & 0 & 0 \end{bmatrix} \begin{bmatrix} 1 & 0 & 0 & 0 \\ 0 & 1 & 0 & 0 \\ 0 & 0 & 1 & 0 \\ 0 & 0 & 0 & 1 \end{bmatrix} \begin{bmatrix} d_{2x} \\ d_{2y} \\ d_{3x} \\ d_{3y} \end{bmatrix}_{global}
$$

After the matrix multiplication, we have

$$
\begin{bmatrix} f_{2x} \\ f_{2y} \\ f_{3x} \\ f_{3y} \end{bmatrix}_{global} = \begin{bmatrix} 7.5e6 & 0 & -7.5e6 & 0 \\ 0 & 0 & 0 & 0 \\ -7.5e6 & 0 & 7.5e6 & 0 \\ 0 & 0 & 0 & 0 \end{bmatrix} \begin{bmatrix} d_{2x} \\ d_{2y} \\ d_{3x} \\ d_{3y} \end{bmatrix}_{global}
$$

The last element is the diagonal element. This is element 3. The orientation angle can be determined by the following calculation.

$$
\theta = \sin^{-1} \frac{300}{500} = 36.87°
$$

The local coordinate system for element 3 is the x direction along the beam and the y direction normal to the x direction. The local stiffness matrix is

$$
\begin{bmatrix} f_{3x} \\ f_{3y} \\ f_{1x} \\ f_{1y} \end{bmatrix}_{local} = \begin{bmatrix} k_3 & 0 & -k_3 & 0 \\ 0 & 0 & 0 & 0 \\ -k_3 & 0 & k_3 & 0 \\ 0 & 0 & 0 & 0 \end{bmatrix} \begin{bmatrix} d_{3x} \\ d_{3y} \\ d_{1x} \\ d_{1y} \end{bmatrix}_{local}
$$

By transforming the local matrix into the global matrix

$$
\begin{bmatrix} f_{3x} \\ f_{3y} \\ f_{1x} \\ f_{1y} \end{bmatrix}_{global} = [T]^T \begin{bmatrix} k_3 & 0 & -k_3 & 0 \\ 0 & 0 & 0 & 0 \\ -k_3 & 0 & k_3 & 0 \\ 0 & 0 & 0 & 0 \end{bmatrix} [T] \begin{bmatrix} d_{3x} \\ d_{3y} \\ d_{1x} \\ d_{1y} \end{bmatrix}_{global}
$$

K_3 can be determined from the equation

$$
k_3 = \frac{AE}{l_3} = \frac{(1*10^{-5})(300*10^9)}{.5} = 6*10^6 \, N/m
$$

The values can be used in the global coordinate stiffness matrix to get

$$
\begin{bmatrix} F_{3x} \\ F_{3y} \\ F_{1x} \\ F_{1y} \end{bmatrix}_{global} = \begin{bmatrix} .8 & .6 & 0 & 0 \\ -.6 & .8 & 0 & 0 \\ 0 & 0 & .8 & .6 \\ 0 & 0 & -.6 & .8 \end{bmatrix} \begin{bmatrix} 6e6 & 0 & -6e6 & 0 \\ 0 & 0 & 0 & 0 \\ -6e6 & 0 & 6e6 & 0 \\ 0 & 0 & 0 & 0 \end{bmatrix} \begin{bmatrix} .8 & -.6 & 0 & 0 \\ -.6 & .8 & 0 & 0 \\ 0 & 0 & .8 & -.6 \\ 0 & 0 & .6 & .8 \end{bmatrix} \begin{bmatrix} d_{3x} \\ d_{3y} \\ d_{1x} \\ d_{1y} \end{bmatrix}_{global}
$$

This matrix can be evaluated through matrix multiplication.

$$
\begin{bmatrix} F_{3x} \\ F_{3y} \\ F_{1x} \\ F_{1y} \end{bmatrix}_{global} = \begin{bmatrix} 3.84e6 & -.288e6 & -3.84e6 & 2.88e6 \\ -2.88e6 & 2.16e6 & 2.88e6 & -2.16e6 \\ -3.84e6 & 2.88e6 & 3.84e6 & -2.88e6 \\ 2.88e6 & -2.16e6 & -2.88e6 & 2.16e6 \end{bmatrix} \begin{bmatrix} d_{3x} \\ d_{3y} \\ d_{1x} \\ d_{1y} \end{bmatrix}_{global}
$$

The global stiffness matrix can be assembled by summing the parts of the individual stiffness matrices. The global stiffness matrix is

$$
\begin{bmatrix} F_{1x} \\ F_{1y} \\ 500 \\ F_{2y} \\ F_{3x} \\ F_{3y} \end{bmatrix} = \begin{bmatrix} 3.84 & -2.88 & 0 & 0 & 3.84 & -2.88 \\ -2.88 & 12.16 & 0 & -10 & 2.88 & -2.16 \\ 0 & 0 & 7.5 & 0 & -7.5 & 0 \\ 0 & -10 & 0 & 10 & 0 & 0 \\ -3.84 & 2.88 & -7.5 & 0 & 11.34 & -2.88 \\ 2.88 & -2.16 & 0 & 0 & -2.88 & 2.16 \end{bmatrix} *10^6 \begin{bmatrix} 0 \\ 0 \\ d_{2x} \\ 0 \\ d_{3x} \\ 0 \end{bmatrix}
$$

Let us apply the load condition, which is given by

$$\begin{bmatrix} F_{1x} \\ F_{1y} \\ 500 \\ F_{2y} \\ F_{3x} \\ F_{3y} \end{bmatrix}$$

Let us apply the boundary condition or the constraint condition, which is given by

$$\begin{bmatrix} 0 \\ 0 \\ d_{2x} \\ 0 \\ d_{3x} \\ 0 \end{bmatrix}$$

The results obtained from the calculations are listed below:

$$\begin{bmatrix} F_{1x} \\ F_{1y} \\ 500 \\ F_{2y} \\ F_{3x} \\ F_{3y} \end{bmatrix} = \begin{bmatrix} -500 \\ -375 \\ 500 \\ 0 \\ 0 \\ 375 \end{bmatrix} (Newton)$$

$$\begin{bmatrix} d_{1x} \\ d_{1y} \\ d_{2x} \\ d_{2y} \\ d_{3x} \\ d_{3y} \end{bmatrix} = \begin{bmatrix} 0 \\ 0 \\ 0.137 \\ 0 \\ 0.130 \\ 0 \end{bmatrix} (mm)$$

1.4   Finite Element Method in a Three-Dimensional Space

In this section, we deal with a plate structure. The plate structure is shown below. Its width or height varies along the x axis. The height at the left is 200 mm, and the height at the right is 60 mm.

The thickness of the plate is uniform and the thickness dimension is 20 mm. We assume that the material of the plate is aluminum. The modulus of elasticity is 70 GPa. The Poisson's ratio is 0.33.

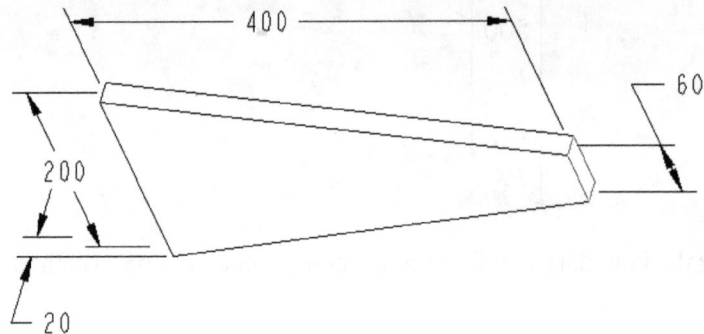

Assume that the left side of the plate is fixed, and the end on the right side is free to move. A tensile force is acting on the right side. The magnitude of the force is 1000 Newton.

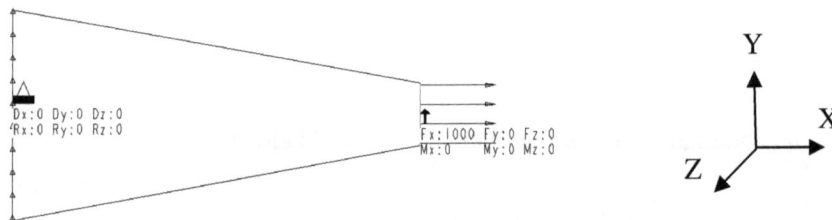

Step 1: divide this plate into four elements with an equal length of 100 mm, as shown below:

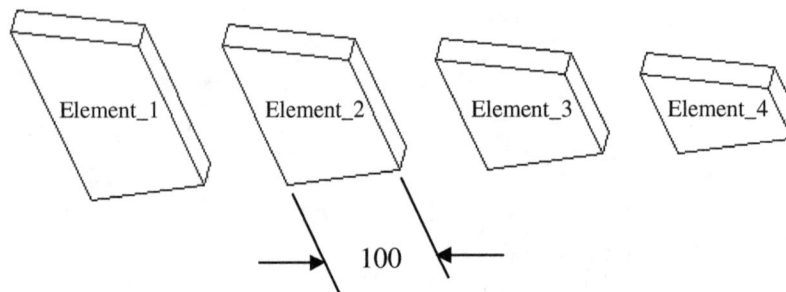

We assume that, under the loading, the stress and/or deformation are uniformly distributed on the cross sections normal to the horizontal axis. We define five nodal points, as shown below:

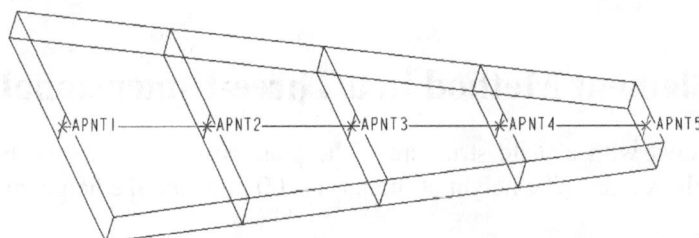

Step 2: We use a shape function to represent the physical behavior of an element so that we are able to interpolate entities of interest quantitatively. The interpolation can be linear, or quadratic or a specific form as needed. The following case illustrates a linear interpolation

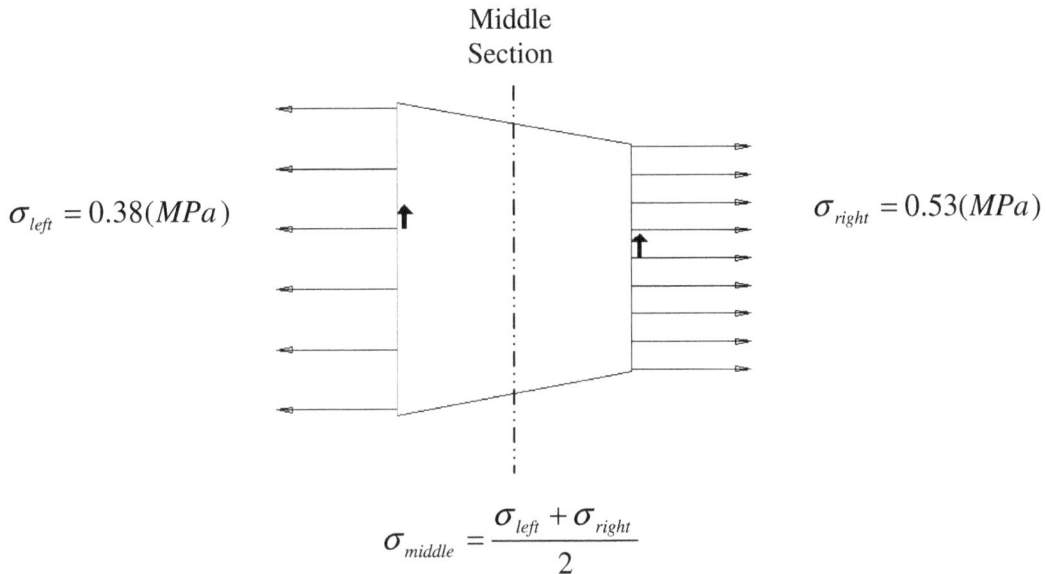

Middle
Section

$\sigma_{left} = 0.38(MPa)$

$\sigma_{right} = 0.53(MPa)$

$$\sigma_{middle} = \frac{\sigma_{left} + \sigma_{right}}{2}$$

Step 3: Develop an equation for an element so that an interpolation defined by the shape function can be carried out.

In the solid mechanics, we have the relations among stress, strain and elongation, as follows:

$$\sigma = \frac{F}{A} \qquad \sigma = E\varepsilon \qquad \varepsilon = \frac{\Delta l}{l_0}$$

When combining these relations, we may come up with

$$F = \left(\frac{AE}{l}\right)\Delta l$$

We may call the term AE/l as an equivalent stiffness of the element if we view the element is a spring component.

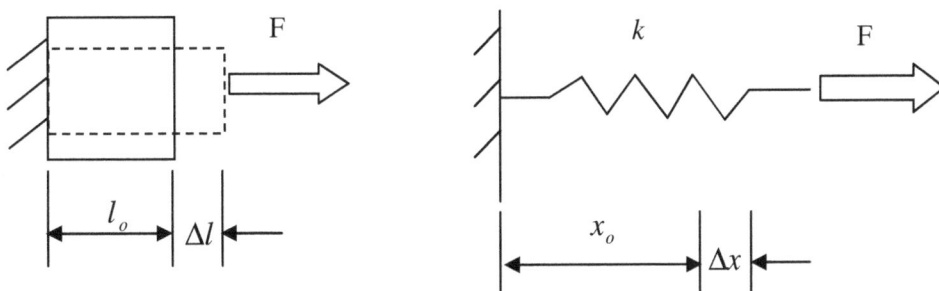

$$k_{equivalent} = \frac{A_{average}E}{l_o} = \frac{(A_{i+1} + A_i)E}{2l_o}$$

$$\Delta x_{equivalent} = u_{i+1} - u_i$$

Step 4: Derive the constituent equation for each element, using a free-body diagram.

Element 1

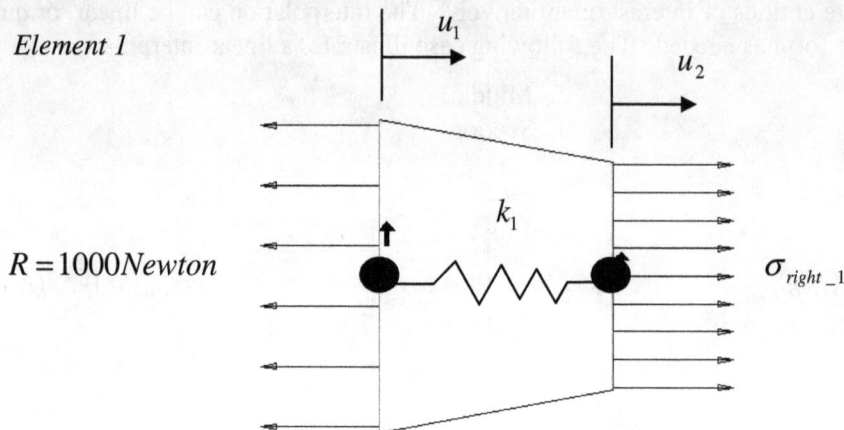

$$node1 : R - k_1(u_2 - u_1) = 0$$

Element 2

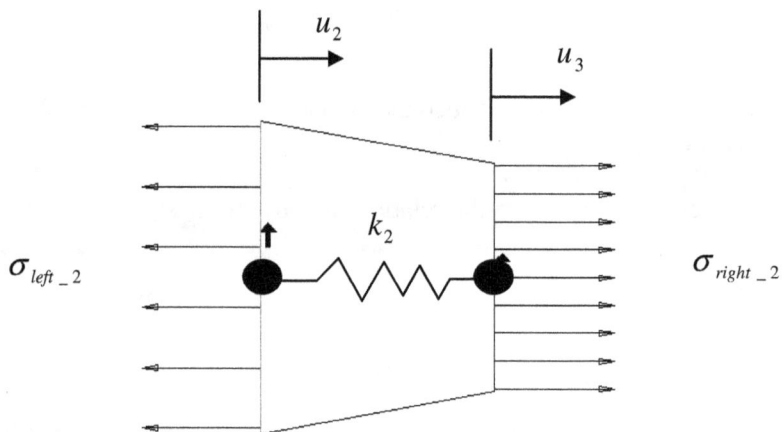

$$node2 : k_1(u_2 - u_1) - k_2(u_3 - u_2) = 0$$

Element 3

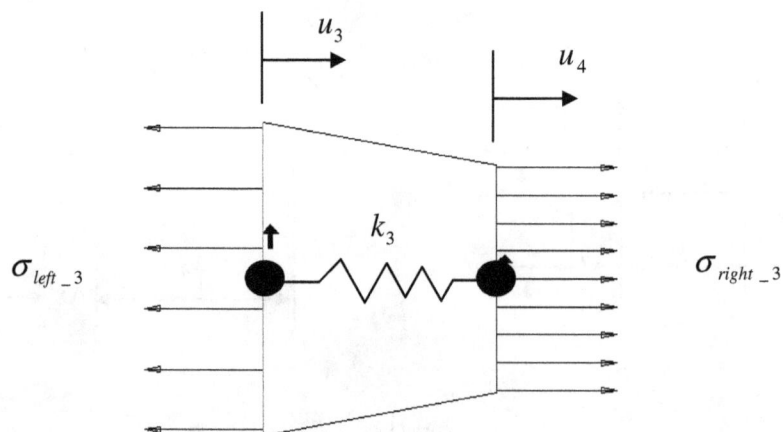

$$node3 : k_2(u_3 - u_2) - k_3(u_4 - u_3) = 0$$

Element 4

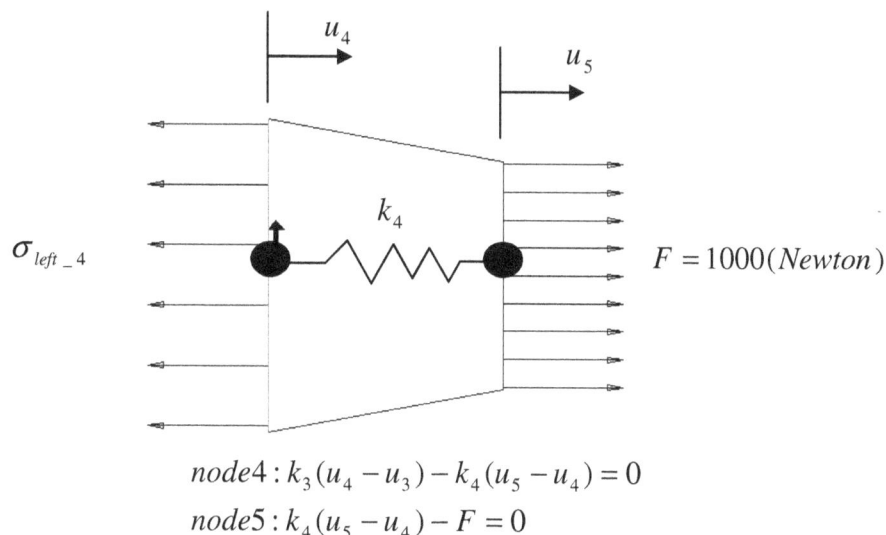

$$node4 : k_3(u_4 - u_3) - k_4(u_5 - u_4) = 0$$
$$node5 : k_4(u_5 - u_4) - F = 0$$

We arrange the above 5 equilibrium equations in a matrix form as follows:

$$
\begin{bmatrix}
k_1 & -k_1 & 0 & 0 & 0 \\
-k_1 & k_1 + k_2 & -k_2 & 0 & 0 \\
0 & -k_2 & k_2 + k_3 & -k_3 & 0 \\
0 & 0 & -k_3 & k_3 + k_4 & -k_4 \\
0 & 0 & 0 & -k_4 & k_4
\end{bmatrix}
\begin{Bmatrix}
u_1 \\ u_2 \\ u_3 \\ u_4 \\ u_5
\end{Bmatrix}
=
\begin{Bmatrix}
-R \\ 0 \\ 0 \\ 0 \\ F
\end{Bmatrix}
$$

If we decompose the force vector to a load vector and a reaction vector, we have

$$
\begin{bmatrix}
k_1 & -k_1 & 0 & 0 & 0 \\
-k_1 & k_1 + k_2 & -k_2 & 0 & 0 \\
0 & -k_2 & k_2 + k_3 & -k_3 & 0 \\
0 & 0 & -k_3 & k_3 + k_4 & -k_4 \\
0 & 0 & 0 & -k_4 & k_4
\end{bmatrix}
\begin{Bmatrix}
u_1 \\ u_2 \\ u_3 \\ u_4 \\ u_5
\end{Bmatrix}
=
\begin{Bmatrix}
0 \\ 0 \\ 0 \\ 0 \\ F
\end{Bmatrix}
+
\begin{Bmatrix}
-R \\ 0 \\ 0 \\ 0 \\ 0
\end{Bmatrix}
$$

Recognizing that the left side of the plate is fixed, the displacement of the left side is constrained to be zero, or

$$u_1 = 0$$
$$node1 : R - k_1(u_2 - u_1) = 0$$
$$node1 : R - k_1 u_2 = 0$$

When we move the reaction force vector to the left side of the above matrix equation, we have

$$
\begin{bmatrix}
k_1 & -k_1 & 0 & 0 & 0 \\
-k_1 & k_1+k_2 & -k_2 & 0 & 0 \\
0 & -k_2 & k_2+k_3 & -k_3 & 0 \\
0 & 0 & -k_3 & k_3+k_4 & -k_4 \\
0 & 0 & 0 & -k_4 & k_4
\end{bmatrix}
\begin{Bmatrix}
u_1 \\ u_2 \\ u_3 \\ u_4 \\ u_5
\end{Bmatrix}
+
\begin{Bmatrix}
R \\ 0 \\ 0 \\ 0 \\ 0
\end{Bmatrix}
=
\begin{Bmatrix}
0 \\ 0 \\ 0 \\ 0 \\ F
\end{Bmatrix}
$$

To eliminate the reaction vector, in the first row of the matrix, we replace k_1 by 1, and $-k_1$ by 0, and we have an explicit form of the stiffness matrix:

$$[K]\{u\} = \{F\}$$

$$
\begin{bmatrix}
1 & 0 & 0 & 0 & 0 \\
-k_1 & k_1+k_2 & -k_2 & 0 & 0 \\
0 & -k_2 & k_2+k_3 & -k_3 & 0 \\
0 & 0 & -k_3 & k_3+k_4 & -k_4 \\
0 & 0 & 0 & -k_4 & k_4
\end{bmatrix}
\begin{Bmatrix}
u_1 \\ u_2 \\ u_3 \\ u_4 \\ u_5
\end{Bmatrix}
=
\begin{Bmatrix}
0 \\ 0 \\ 0 \\ 0 \\ F
\end{Bmatrix}
$$

The stiffness matrix [K] represents the characteristics of the plate system when it is subjected to a load or loads.

Step 5: Determine the deformation of the plate when subjected to the uniform load acting on the free end of the plate.

Calculations of the equivalent stiffness values of the four (4) elements

(1) Calculate the areas of the cross sections at the five (5) nodal points.

$A_1 = 200 \times 20 = 4000 \ (\text{mm}^2)$

$A_2 = [\, 200 - \left(\dfrac{200-60}{400} \right) * 100 \,] \times 20 = 3300 \ (\text{mm}^2)$

$A_3 = [\, 200 - \left(\dfrac{200-60}{400} \right) * 200 \,] \times 20 = 2600 \ (\text{mm}^2)$

$A_4 = [\, 200 - \left(\dfrac{200-60}{400} \right) * 300 \,] \times 20 = 1900 \ (\text{mm}^2)$

$A_5 = [\, 200 - \left(\dfrac{200-60}{400} \right) * 400 \,] \times 20 = 1200 \ (\text{mm}^2)$

(2) Calculate the equivalent stiffness for the four elements, using the formula

$$k_{equivalent} = \frac{A_{average} E}{l_o} = \frac{(A_{i+1} + A_i)E}{2l_o}$$

$$k_1 = \frac{10^{-6}*(4000+3300)*70*10^9}{2*100*10^{-3}} = 1.28x10^9\,(pascal)$$

$$k_2 = \frac{10^{-6}*(3300+2600)*70*10^9}{2*100*10^{-3}} = 1.03x10^9\,(pascal)$$

$$k_3 = \frac{10^{-6}*(2600+1900)*70*10^9}{2*100*10^{-3}} = 0.788x10^9\,(pascal)$$

$$k_4 = \frac{10^{-6}*(1900+1200)*70*10^9}{2*100*10^{-3}} = 0.543x10^9\,(pascal)$$

Stiffness matrix for the 4 elements

Element 1

$$[K]^1 = \begin{bmatrix} k_1 & -k_1 \\ -k_1 & k_1 \end{bmatrix} = 10^9 \begin{bmatrix} 1.28 & -1.28 \\ -1.28 & 1.28 \end{bmatrix}$$

Element 2

$$[K]^2 = \begin{bmatrix} k_2 & -k_2 \\ -k_2 & k_2 \end{bmatrix} = 10^9 \begin{bmatrix} 1.03 & -1.03 \\ -1.03 & 1.03 \end{bmatrix}$$

Element 3

$$[K]^3 = \begin{bmatrix} k_3 & -k_3 \\ -k_3 & k_3 \end{bmatrix} = 10^9 \begin{bmatrix} 0.788 & -0.788 \\ -0.788 & 0.788 \end{bmatrix}$$

Element 4

$$[K]^4 = \begin{bmatrix} k_4 & -k_4 \\ -k_4 & k_4 \end{bmatrix} = 10^9 \begin{bmatrix} 0.543 & -0.543 \\ -0.543 & 0.543 \end{bmatrix}$$

Assemble the 4 stiffness matrices into a global stiffness matrix

$$10^9 \begin{bmatrix} 1.28 & -1.28 & 0 & 0 & 0 \\ -1.28 & 1.28+1.03 & -1.03 & 0 & 0 \\ 0 & -1.03 & 1.03+0.788 & -0.788 & 0 \\ 0 & 0 & -0.788 & 0.788+0.543 & -0.543 \\ 0 & 0 & 0 & -0.543 & 0.543 \end{bmatrix}$$

Apply the load condition (F=1000 Newton) and the boundary condition (u$_1$=0), we have the static equilibrium equation of the plate system in its matrix form, as follows:

$$10^9 \begin{bmatrix} 1.00 & 0 & 0 & 0 & 0 \\ -1.28 & 1.28+1.03 & -1.03 & 0 & 0 \\ 0 & -1.03 & 1.03+0.788 & -0.788 & 0 \\ 0 & 0 & -0.788 & 0.788+0.543 & -0.543 \\ 0 & 0 & 0 & -0.543 & 0.543 \end{bmatrix} \begin{Bmatrix} u_1 \\ u_2 \\ u_3 \\ u_4 \\ u_5 \end{Bmatrix} = \begin{Bmatrix} 0 \\ 0 \\ 0 \\ 0 \\ 1000 \end{Bmatrix}$$

Solving the above equation (for example, using Matlab), we have

$$\begin{Bmatrix} u_1 \\ u_2 \\ u_3 \\ u_4 \\ u_5 \end{Bmatrix} = 10^{-9} \begin{bmatrix} 1.00 & 0 & 0 & 0 & 0 \\ -1.28 & 1.28+1.03 & -1.03 & 0 & 0 \\ 0 & -1.03 & 1.03+0.788 & -0.788 & 0 \\ 0 & 0 & -0.788 & 0.788+0.543 & -0.543 \\ 0 & 0 & 0 & -0.543 & 0.543 \end{bmatrix}^{-1} \begin{Bmatrix} 0 \\ 0 \\ 0 \\ 0 \\ 1000 \end{Bmatrix}$$

$$\begin{Bmatrix} u_1 \\ u_2 \\ u_3 \\ u_4 \\ u_5 \end{Bmatrix} = 10^{-9} \begin{bmatrix} 1.00 & 0 & 0 & 0 & 0 \\ 1.00 & 0.7813 & 0.7813 & 0.7813 & 0.7813 \\ 1.00 & 0.7813 & 1.7521 & 1.7521 & 1.7521 \\ 1.00 & 0.7813 & 1.7521 & 3.0212 & 3.0212 \\ 1.00 & 0.7813 & 1.7521 & 3.0212 & 4.8628 \end{bmatrix} \begin{Bmatrix} 0 \\ 0 \\ 0 \\ 0 \\ 1000 \end{Bmatrix} = 10^{-6} \begin{Bmatrix} 0 \\ 0.7813 \\ 1.7521 \\ 3.0212 \\ 4.8628 \end{Bmatrix} (meter)$$

$$\begin{Bmatrix} u_1 \\ u_2 \\ u_3 \\ u_4 \\ u_5 \end{Bmatrix} = \begin{Bmatrix} 0 \\ 0.7813 \\ 1.7521 \\ 3.0212 \\ 4.8628 \end{Bmatrix} (\mu m)$$

Step 6: Post-processing information

Determine the normal strain in each of the four (4) elements, using the following formula

$$\varepsilon_i = \frac{u_{i+1} - u_i}{l_i}$$

Element 1

$$\varepsilon_1 = \frac{u_2 - u_1}{l_1} = \frac{0.7813 - 0}{100 * 10^{-3}} x10^{-6} = 7.813 x10^{-6}$$

Element 2

$$\varepsilon_2 = \frac{u_3 - u_2}{l_2} = \frac{1.7521 - 0.7813}{100 * 10^{-3}} x10^{-6} = 9.708 x10^{-6}$$

Element 3

$$\varepsilon_3 = \frac{u_4 - u_3}{l_3} = \frac{3.0212 - 1.7521}{100 * 10^{-3}} x10^{-6} = 12.691 x10^{-6}$$

Element 4

$$\varepsilon_4 = \frac{u_5 - u_4}{l_4} = \frac{4.8628 - 3.0212}{100 * 10^{-3}} x10^{-6} = 18.416 x10^{-6}$$

<u>Determine the normal stress in each of the four (4) elements, using the following formula</u>
$$\sigma_i = E\varepsilon_i$$

Element 1

$$\sigma_1 = E\varepsilon_1 = (70x10^9)(7.813x10^{-6}) = 0.546x10^6 (pascal) = 0.546(MPa)$$

Element 2

$$\sigma_2 = E\varepsilon_2 = (70x10^9)(9.708x10^{-6}) = 0.680x10^6 (pascal) = 0.680(MPa)$$

Element 3

$$\sigma_3 = E\varepsilon_3 = (70x10^9)(12.691x10^{-6}) = 0.888x10^6 (pascal) = 0.888(MPa)$$

Element 4

$$\sigma_4 = E\varepsilon_4 = (70x10^9)(18.416x10^{-6}) = 1.289x10^6 (pascal) = 1.289(MPa)$$

<u>Determine the reaction force R</u>

$$R = k_1(u_2 - u_1) = (1.28x10^9)(0.7813x10^6 - 0) = 1000(Newton)$$

<u>Determine the strain in the Y-direction and the strain in the Z-direction</u>

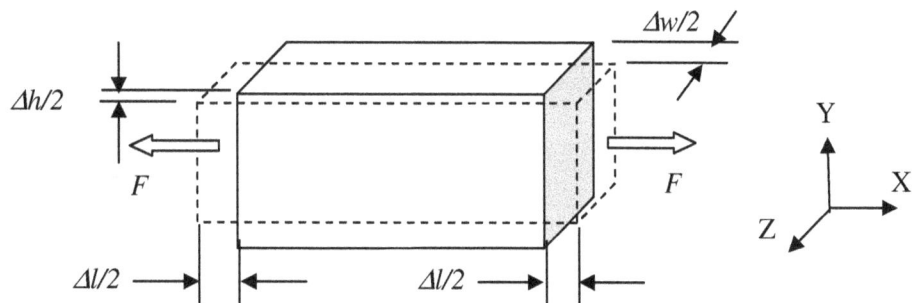

A parameter is used to characterize the relation between the axial elongation and the lateral contraction, either in the Y-direction or in the Z-direction. The parameter is called Poisson's ratio. It is defined as follows:

$$v = -\frac{lateral\_strain}{axial\_strain}$$

In this example, we assume that the material has the same properties in all directions, or the material is of an isotropic type.

Element 1
$$\varepsilon_{y1} = v\varepsilon_{x1} = 0.33 * 7.813x10^{-6} = 2.58x10^{-6}$$
$$\varepsilon_{z1} = v\varepsilon_{x1} = 0.33 * 7.813x10^{-6} = 2.58x10^{-6}$$

Element 2
$$\varepsilon_{y2} = v\varepsilon_{x2} = 0.33 * 9.708x10^{-6} = 3.20x10^{-6}$$
$$\varepsilon_{z2} = v\varepsilon_{x2} = 0.33 * 9.708x10^{-6} = 3.20x10^{-6}$$

Element 3
$$\varepsilon_{y3} = v\varepsilon_{x3} = 0.33 * 12.691x10^{-6} = 4.19x10^{-6}$$
$$\varepsilon_{z3} = v\varepsilon_{x3} = 0.33 * 12.691x10^{-6} = 4.19x10^{-6}$$

Element 4
$$\varepsilon_{y4} = v\varepsilon_{x4} = 0.33 * 18.416x10^{-6} = 6.08x10^{-6}$$
$$\varepsilon_{z4} = v\varepsilon_{x4} = 0.33 * 18.416x10^{-6} = 6.08x10^{-6}$$

Determine the stress in the Y-direction and the stress in the Z-direction

$$\sigma_{yi} = E\varepsilon_{yi} \qquad\qquad \sigma_{zi} = E\varepsilon_{zi}$$

Element 1

$$\sigma_{y1} = E\varepsilon_{y1} = (70x10^{9})(2.58x10^{-6}) = 0.181x10^{6}(pascal) = 0.181(MPa)$$
$$\sigma_{z1} = E\varepsilon_{z1} = (70x10^{9})(2.58x10^{-6}) = 0.181x10^{6}(pascal) = 0.181(MPa)$$

Element 2
$$\sigma_{y2} = E\varepsilon_{y2} = (70x10^{9})(3.20x10^{-6}) = 0.224x10^{6}(pascal) = 0.224(MPa)$$
$$\sigma_{z2} = E\varepsilon_{z2} = (70x10^{9})(3.20x10^{-6}) = 0.224x10^{6}(pascal) = 0.224(MPa)$$

Element 3
$$\sigma_{y3} = E\varepsilon_{y3} = (70x10^{9})(4.19x10^{-6}) = 0.293x10^{6}(pascal) = 0.293(MPa)$$
$$\sigma_{z3} = E\varepsilon_{z3} = (70x10^{9})(4.19x10^{-6}) = 0.293x10^{6}(pascal) = 0.293(MPa)$$

Element 4
$$\sigma_{y4} = E\varepsilon_{y4} = (70x10^{9})(6.08x10^{-6}) = 0.426x10^{6}(pascal) = 0.426(MPa)$$
$$\sigma_{z4} = E\varepsilon_{z4} = (70x10^{9})(6.08x10^{-6}) = 0.426x10^{6}(pascal) = 0.426(MPa)$$

Step 7: Graphical Representations
Use a set of 4 colors and characterize the strain distribution in the Y-direction or in the Z-direction. It is important to note that the numerical values represent the strains in the 4 middle cross sections.

0.181	
0.224	
0.293	
0.426	

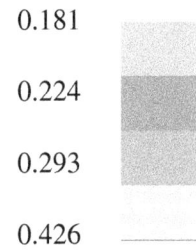

To verify the results obtained from this evaluation, we run **Pro/MECHANICA**, an FEA software system you will learn in Chapter 2.

Step 1: Setting the material properties

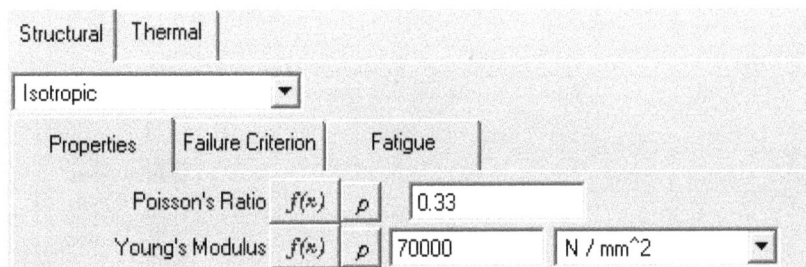

Step 2: Setting the constraint condition and the load condition

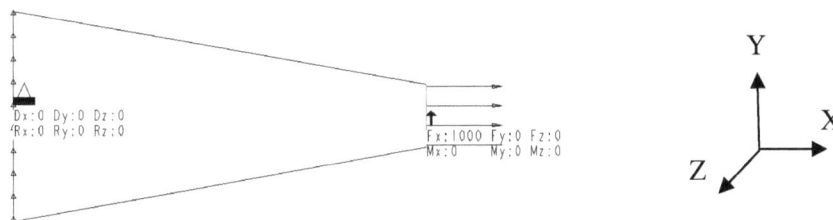

Step 3: Results obtained from running Pro/MECHANICA

Type and number of elements

```
Model Type: Three Dimensional

Points:              16
Edges:               47
Faces:               50

Springs:              0
Masses:               0
Beams:                0
Shells:               0
Solids:              18

Elements:            18
```

The maximum stress and displacement in the X direction

```
max_beam_bending:    0.000000e+00
max_beam_tensile:    0.000000e+00
max_beam_torsion:    0.000000e+00          Compare to
max_beam_total:      0.000000e+00
max_disp_mag:        2.534972e-03          4.8628 µm
max_disp_x:          2.530680e-03
max_disp_y:          1.282444e-04
max_disp_z:         -1.501406e-04
max_prin_mag:        8.830912e-01
max_rot_mag:         0.000000e+00
max_rot_x:           0.000000e+00
max_rot_y:           0.000000e+00          Compare to
max_rot_z:           0.000000e+00
max_stress_prin:     8.830912e-01          1.289 MPa
max_stress_vm:       8.430616e-01
max_stress_xx:       8.825928e-01
max_stress_xy:       1.486066e-01
max_stress_xz:      -8.724412e-02
max_stress_yy:       1.833535e-01
max_stress_yz:      -8.150371e-02
max_stress_zz:       2.279374e-01
```

Plot of the distribution of strain in the Y direction

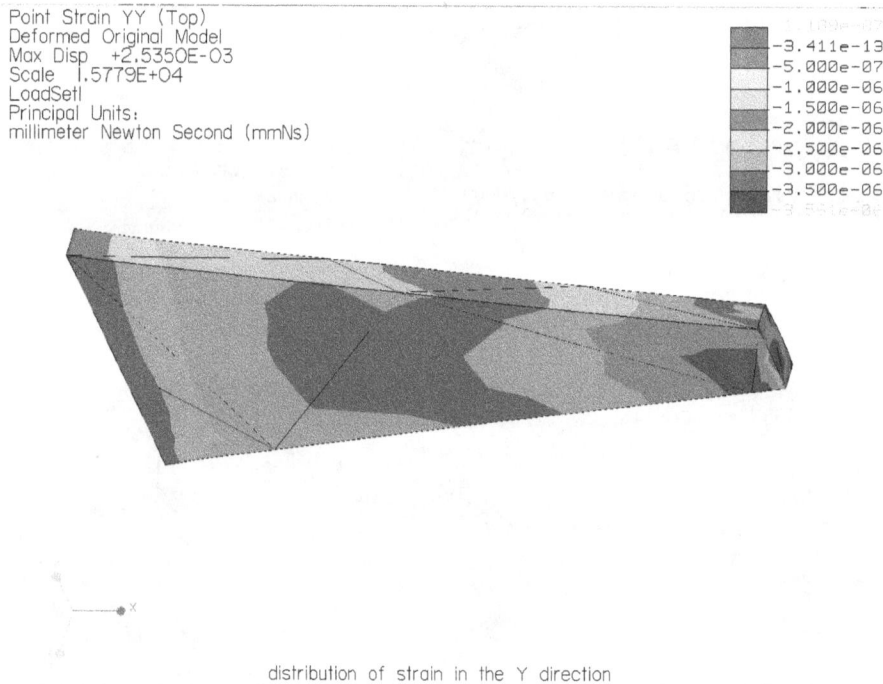

```
Point Strain YY (Top)                        -3.411e-13
Deformed Original Model                      -5.000e-07
Max Disp   +2.5350E-03                       -1.000e-06
Scale  1.5779E+04                            -1.500e-06
LoadSetl                                     -2.000e-06
Principal Units:                             -2.500e-06
millimeter Newton Second (mmNs)              -3.000e-06
                                             -3.500e-06
```

distribution of strain in the Y direction

It is evident that the results obtained from the use of Pro/MECHANICA are better in terms of clarification and clearness to represent the deformation and stress distributions. This is one of the reasons why people are using those software systems commercially available to perform engineering analyses, such as FEA.

1.5   Unit Systems Used in the FEA Software Systems

Although there are about 10 FEA software systems commercially available in the market, the unit systems used in those software systems are almost identical among them. In the following, we illustrate the unit systems used by Pro/Mechanica, ALGOR and ANSYS. It is important that users have a full understanding the basics of each of them so that the numerical values displayed on computer screen, or saved into the data files can be correctly interpreted.

Let us first review the unit systems used by Pro/Mechanica. From the set up menu of the Pro/Mechanica system, the 7 unit systems available under Pro/Mechanica are listed. Let us first examine Meter Kilogram Second (MKS) system, which is the SI unit system. There are 5 basic quantities, namely, length, mass, force, time and temperature. Besides, the gravity constant is set at 9.80665 m/second^2.

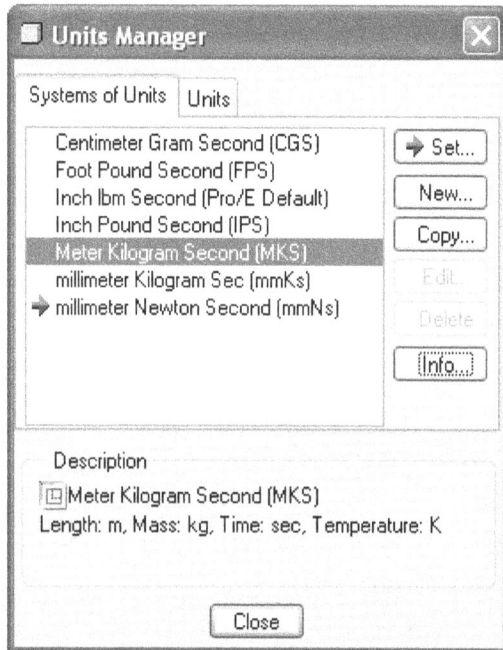

```
Basic quantities:

                        Length      m
                          Mass      kg
                         Force      N
                          Time      sec
                   Temperature      K

Gravity   9.80665   m / sec^2
```

Based on these 5 basic quantities, the units for other parameters can be derived. The following table shows a set of parameters, such as area, volume, velocity, acceleration, angular velocity, etc.

```
Derived quantities:


                      Area   m^2
                    Volume   m^3
                  Velocity   m / sec
              Acceleration   m / sec^2
          Angular Velocity   rad / sec
      Angular Acceleration   rad / sec^2
                 Frequency   1 / sec
                   Density   kg / m^3
             Torque/Moment   m N
         Distributed Force   N / m
        Distributed Moment   N
          Areal Distr Force   N / m^2
```

In the later chapters, we will use Pro/Mechanica to perform FEA. The unit system used in those chapters is the mmNs system. The 5 basic quantities are mm for the unit of length, tonne for the unit of mass, Newton for the unit of force, second for the unit of time and Celsius (deg C) for the unit of temperature. The gravity constant is set at 9806.65 mm/second^2.

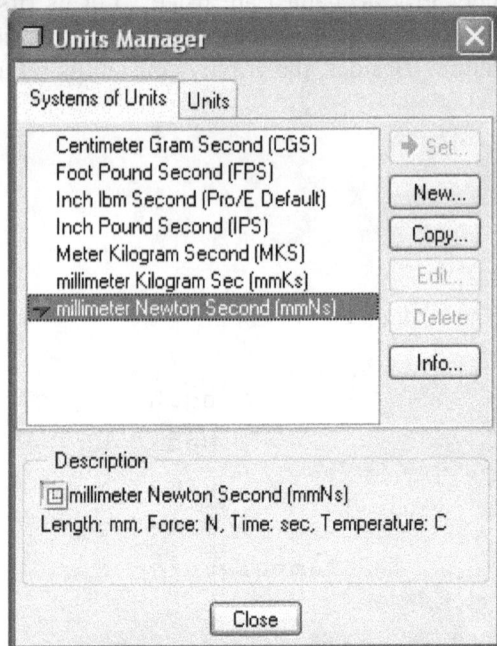

It is extremely important to remember that the unit of stress is

$$unit\_of\_stress = \frac{Newton}{mm^2} = \frac{Newton}{m^2} = 10^6(pascal) = 1(MPa)$$

It is extremely important to remember that the unit of displacement is

$$unit\_of\_deformation = mm$$

Now let us review the unit systems used by ANSYS. From the menu of Material Library > Select Unit. There are 4 unit systems used by ANSYS. For example, the currently selected unit system, as illustrated, is MKS or the SI unit system.

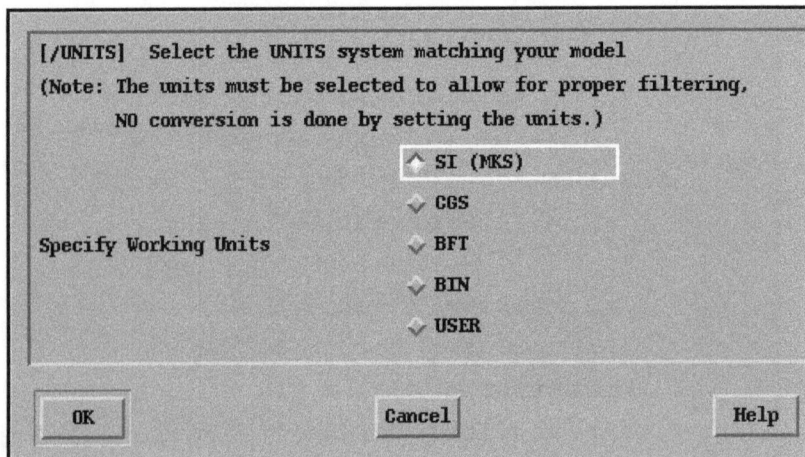

1.6 References

1. K. J. Bathe, Finite Element Procedure, Prentice Hall, Upper Saddle River, New Jersey, 1996.
2. J. W. Dally, Design Analysis of Structural Elements, Goodington, New York, NY, 2003.
3. J. H. Earle, Graphics for Engineers, AutoCAD Release 13, Addision-Wesley, Reading, Massachusetts, 1996.
4. J. M. Gere and S. P. Timoshenko, Mechanics of Materials, 3rd edition, PWS-KENT Publishing Company, Boston, 1990.
5. W. W. Hager, Applied Numerical Linear Algebra, Prentice-Hall, Englewood Cliffs, NJ, 1988.
6. M. A. N. Hendriks, H Jongedijk, J. G. Rots and W. J. E. van Spange (eds), Finite Elements in Engineering and Science, DIANA Computational Mechanics, 1997.
7. K. H. Huebner, D. L. Dewhirst, D. E. Smith, and T. G. Byrom, The Finite Element Method for Engineers, 4th edition, John Wiley & Sons, Inc., 2001.
8. D. L. Logan, A First Course in the Finite Element Method Using ALGOR, PWS Publishing, Boston, 1997.
9. R. L. Norton, Machine Design: An Integrated Approach, Prentice Hall, Upper Saddle River, New Jersey, 1996.
10. M. Petyl, Introduction to Finite Element Vibration Analysis, Cambridge University Press, 1990.
11. T. Ross, Advanced Applied Finite Element Methods, Horwood Publishing Limited, Chichester, England, 1998.
12. Y. Saad, Iterative Methods for Sparse Linear Systems, PWS Publishing Company, Boston, 1996.
13. P. Silvester, Higher-Order Polynomial Triangular Finite Element for Potential Problems, Int. J. Eng. Soc., Vol. 7, No. 8, pp. 849-861.

1.7 Exercises

1. Use your personal experience to describe the essential steps involved in engineering design and engineering analysis. Make sure that the following information is presented.
   (1) What was the product you designed?
   (2) Was the design process a team effort or your own effort?
   (3) What was the design objective(s)?
   (4) How did you and/or your design team perform an engineering analysis?

2. Present any experiences you have had using CAD and computer aided engineering software systems. Have you used Pro/Mechanica and/or ANSYS? What type or types of the problem(s) did you deal with at those times?

3. What is your expectation from taking a course in computer aided engineering analyses?

4 A truss structure is shown below.

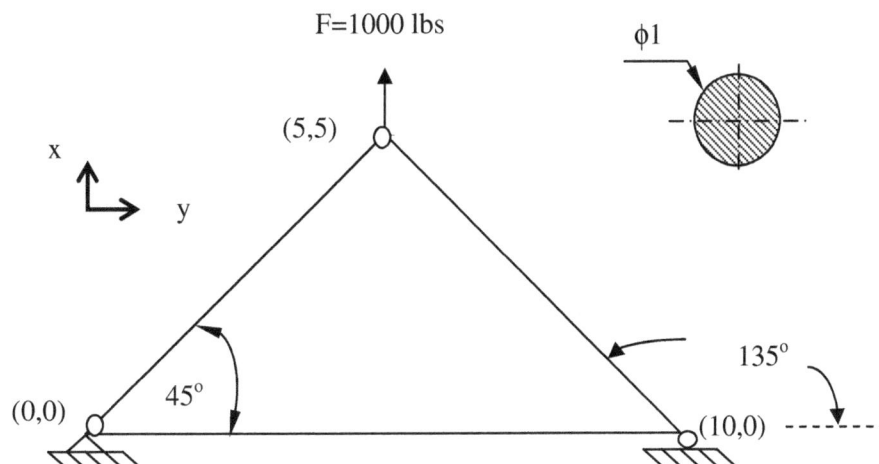

(1)   As illustrated, 2 elements associated with 3 nodes are needed. Calculate k_1 and k_2 using $\dfrac{AE}{l}$ where E = 10.1 x 10^6 lbf/in².

(2)   Determine the 2 stiffness matrices individually.

(3)   Assemble the 2 stiffness matrices and obtain the global stiffness matrix.

(4)   Assume that a vertical force is acting on the upper node, as shown. Determine the displacements at each of the 3 joints.

5.   A truss structure is shown below.

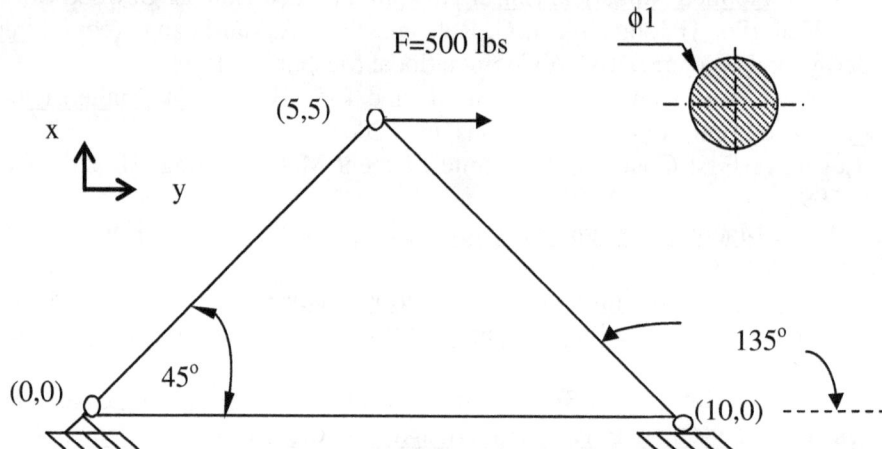

(1)   As illustrated, 2 elements associated with 3 nodes are needed. Calculate k_1 and k_2 using $\dfrac{AE}{l}$ where E = 10.1 x 10^6 lbf/in².

(2)   Determine the 2 stiffness matrices individually.

(3)   Assemble the 2 stiffness matrices and obtain the global stiffness matrix.

(4)   Assume that a vertical force is acting on the upper node, as shown. Determine the displacements at each of the 3 joints.

6.   The following figure illustrates a truss structure, which consists of 3 bars. The section area of each bar is circular. The diameter is 100 mm. The geometrical information of the truss structure is shown in the figure. The material type is determined by the student.

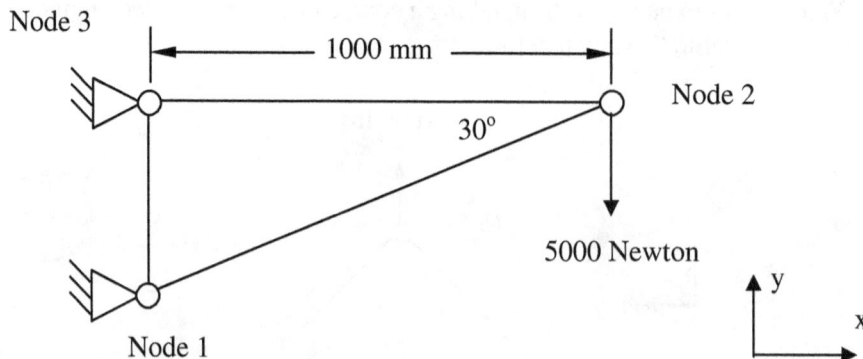

(1)   Obtain the stiffness matrix for each of the 3 elements.

(2)   Obtain the global stiffness matrix of the truss structure.

(3)   Under the shown load condition, determine the displacement of Node 2 in the horizontal and vertical directions.

(4)   Determine the reaction forces on Node 1 and Node 3.

CHAPTER 2

FEA USING Pro/Mechanica

2.1 Introduction

As discussed in Chapter 1, in order to perform finite element analysis, we need a geometrical model which is a representation of the physical object under study. When the geometrical model is available, the finite element analysis can be carried out. In this chapter, section 2.1 is a brief introduction to the Pro/ENGINEER Wildfire design system. We create a block component to demonstrate the essential steps needed to create a geometrical model. Section 2.2 is an introduction to the Pro/Mechanica system. We use a column structure subjected to a compressive load to demonstrate the essential steps needed for performing finite element analysis. In section 2.3, we present a case study. The case study deals with a cantilever beam subjected to a uniformly distributed load. The general procedure to perform finite element analysis is summarized at the end of this chapter.

2.2 Introduction to the Pro/ENGINEER Wildfire Design System

Before performing finite element analysis, we need a 3D solid model in general. In this chapter, we focus on using the Pro/ENGINEER Wildfire design system to create 3D solid models. Figure 2-1 presents the window on display when launching Pro/ENGINEER Wildfire 5.0. The window offers the following functions:

1. Provide a Main Toolbar, from which a new file can be opened for initiating the design process, such as creating a new object.
2. Provide tutorial materials on Pro/ENGINEER Wildfire.
3. Provide the guidelines for getting help from the PTC website.

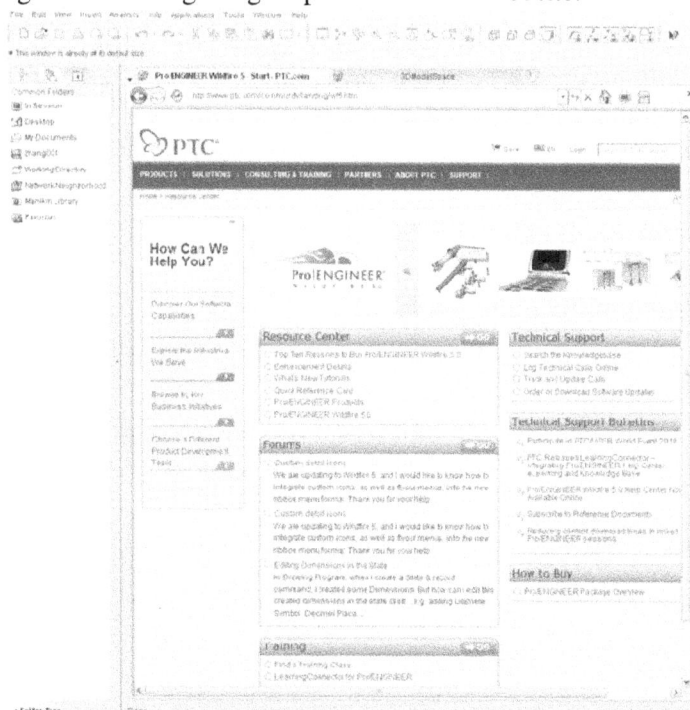

Figure 2-1 Window that Appears when Launching Pro/ENGINEER Wildfire 5.0

In the following, we use an example of creating a block component to demonstrate the process of creating a 3D solid model for an object. The three dimensions of the block are 190 x 60 x 40 mm.

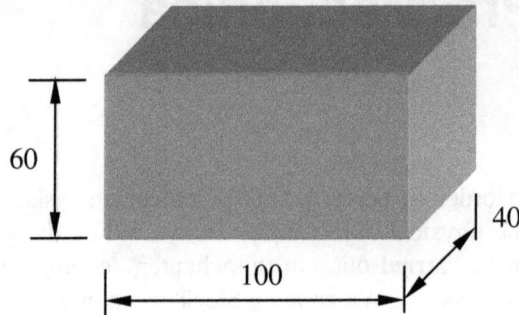

To start a design process using Pro/ENGINEER Wildfire, select the icon, called "**Create a new object**". The icon is displayed on the menu toolbar or the main toolbar, as illustrated below:

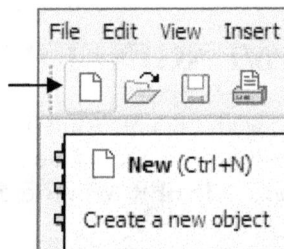

Make sure "**Part**" is selected from the window called **New**. Type the name of the file, or accept the default name, such as *prt0001* shown on display. Clear the icon of **Use default template > OK**.

The default unit system in Pro/ENGINEER is inch_pound_second. To change the unit system to mm_Newton_second, in the Select **mmns_part_solid** (units: Millimeter, Newton, Second) and type *block* in **DESCRIPTION**, and *student* in **MODELED_BY**, then **OK.**

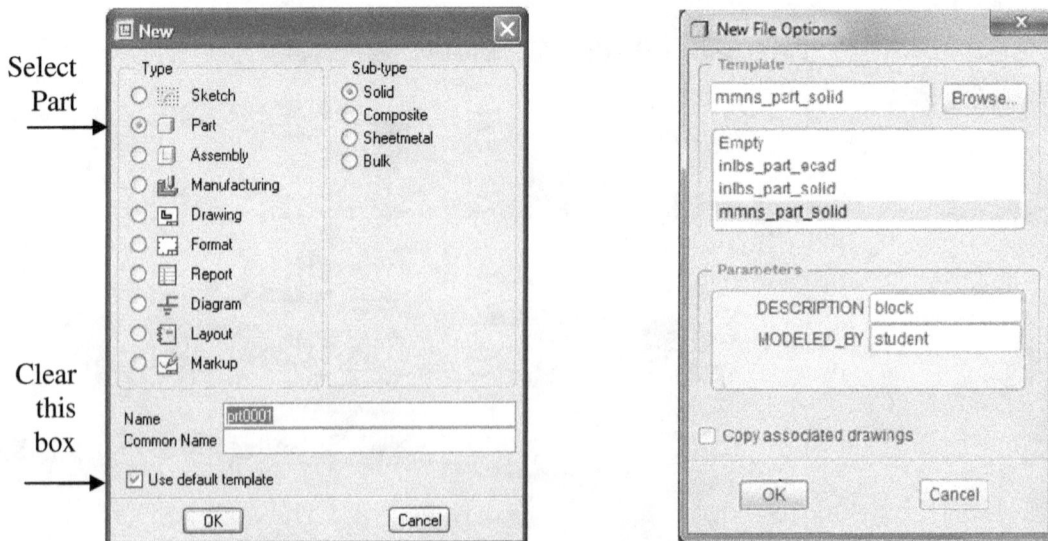

A design screen will be on display, as shown in Figure 2-2.

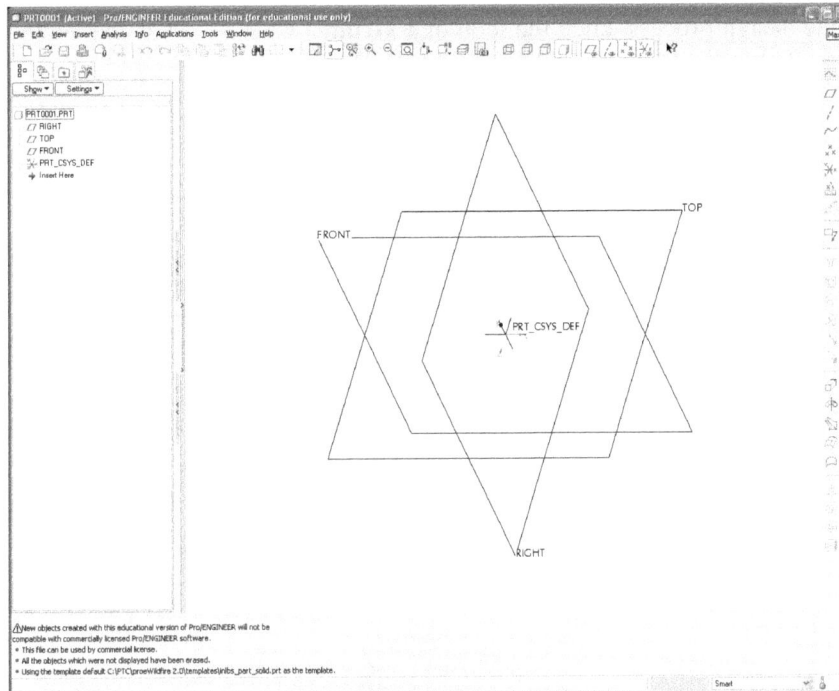

Figure 2-2 Window used for the Design Process

There are three datum planes displayed on screen. They are marked as **FRONT, TOP** and **RIGHT**. These three datum planes are orthogonal to each other. The intersection of these three planes is a point. This point represents the location of the origin of the default coordinate system. The orientation of the default coordinate system, called **PRT_CSYS_DEF**, is set in the following way, as illustrated in Figure 2-3:

The **FRONT** datum plane represents the X-Y plane.
The **TOP** datum plane represents the X-Z plane.
The **RIGHT** datum plane represents the Y-Z plane.

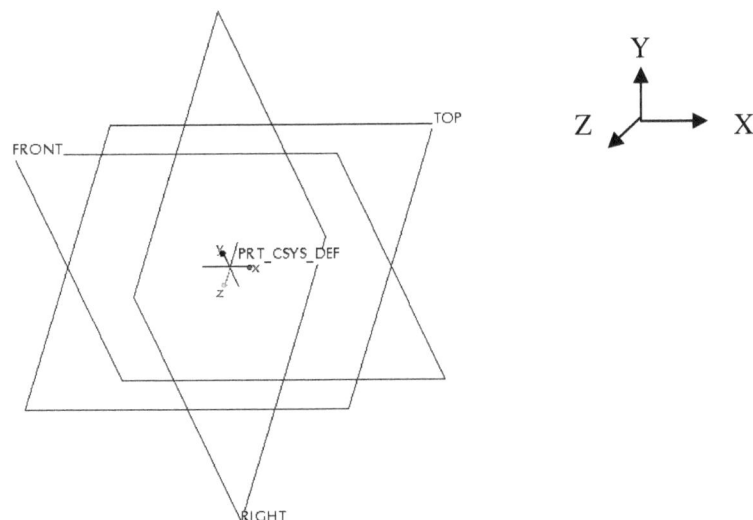

Figure 2-3 The Three Default Datum Planes and the Default Reference System

To start the design process, click the icon of **Extrude.** Specify 40 as the depth of extrusion. To initiate the process of sketching, activate **Placement > Define.**

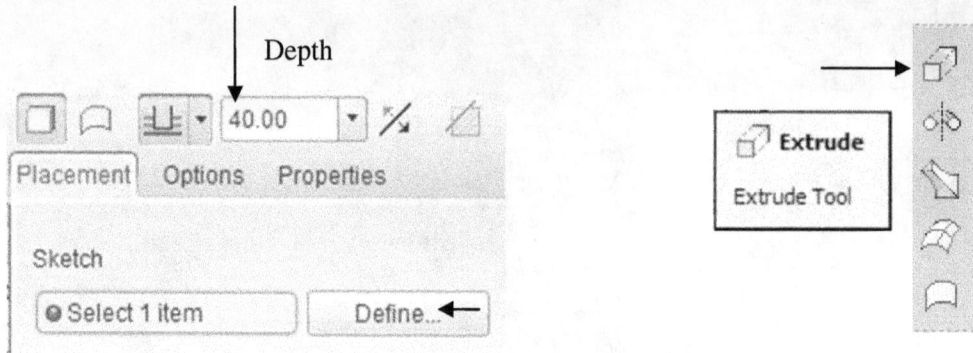

Select the **FRONT** datum plane displayed on screen as the sketch plane. Pick the box of **Sketch** to accept the **RIGHT** datum plane as the default reference to orient the sketch plane, as illustrated below:

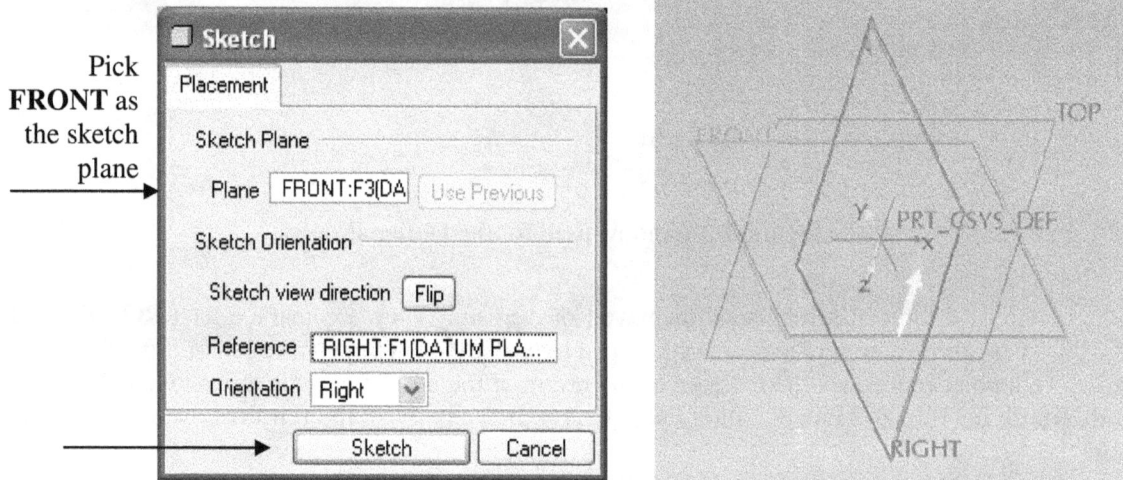

Let us create a 2D sketch. Because we have selected the **FRONT** datum plane as the sketch plane, the 2D sketch will be created in the x-y plane. From the toolbar of sketcher, click the icon of **Rectangle**, and start the sketch by picking one point on the Y axis and the other point on the X axis, as shown below. To change the horizontal dimension to 10, double click the numerical number, type *100*, and press the **Enter Key**. To change the vertical dimension to 6, double click the numerical number, type *60*, and press the **Enter Key**.

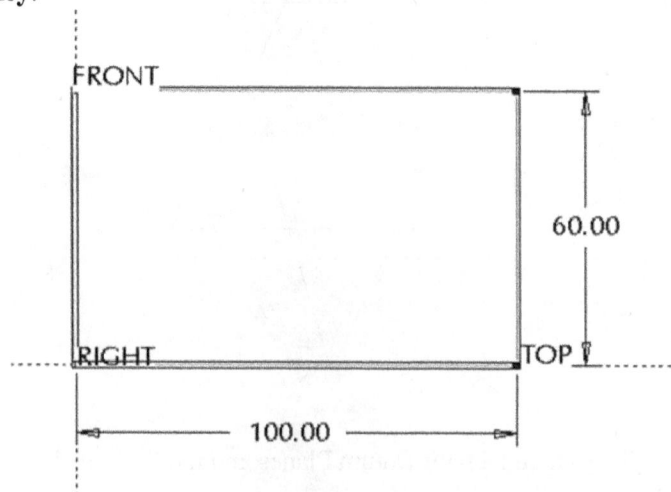

We have now completed the 2D sketch. Pick the icon of **Done,** and click the icon of Apply and Save.

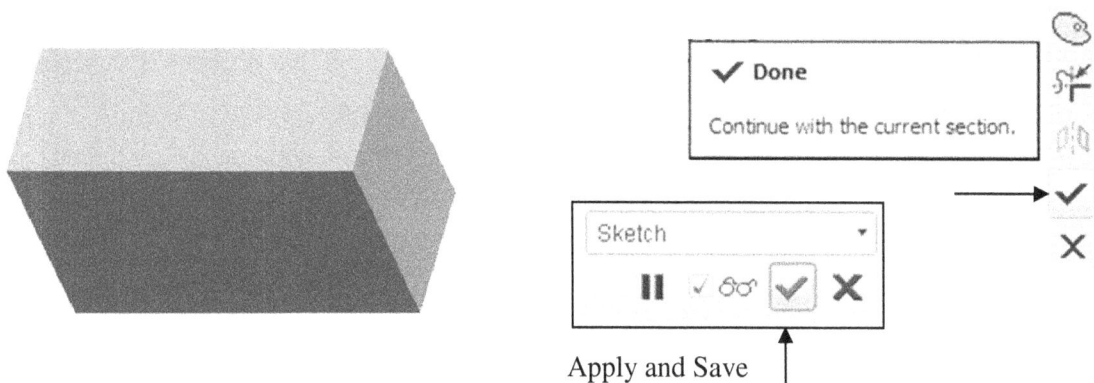

Apply and Save

When the creation of a 3D solid model is completed, users are able to view the 3D solid model from different viewing angle. Select the icon of **Saved view** from the main toolbar > select **Standard Orientation**. A 3D view of the created 3D model is on display.

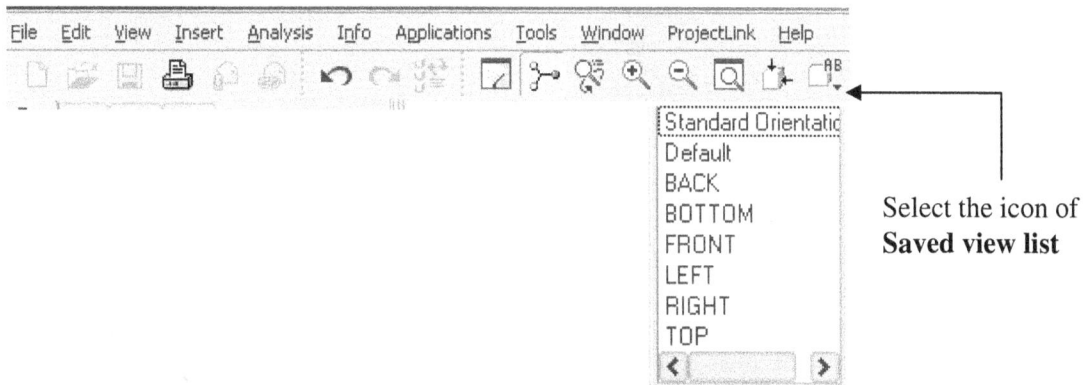

Select the icon of **Saved view list**

Users may also use the buttons of mouse to spin, turn, zoom and pan the 3D solid model. For example, holding down the middle button and moving the pointer will spin the 3D solid model. Information on the mouse controls is available on the Pro/ENGINEER Wildfire Web Tools, and the following information is copied from the Pro/ENGINEER Wildfire Web Tools:

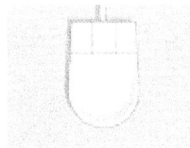

Spin the model by holding down the middle mouse button and moving the pointer.

Turn the model by moving your mouse from left to right while holding down the CTRL key and the middle mouse button.

Zooming takes advantage of the roll wheel, which pulls in the model or pushes it away. You can also zoom by moving your mouse up and down while holding down the CTRL key and middle mouse button.

Panning also complies with the new roll wheel standard by using the SHIFT key and the middle mouse button.

2.3   A Column Structure Subjected to Compressive Load

In section 2.1, we presented the process of creating a 3D solid model of a block component. In this section, we will get on the process of performing FEA. The example to be used in this section is a column structure, which is subjected to compressive loading. The column structure is shown below:

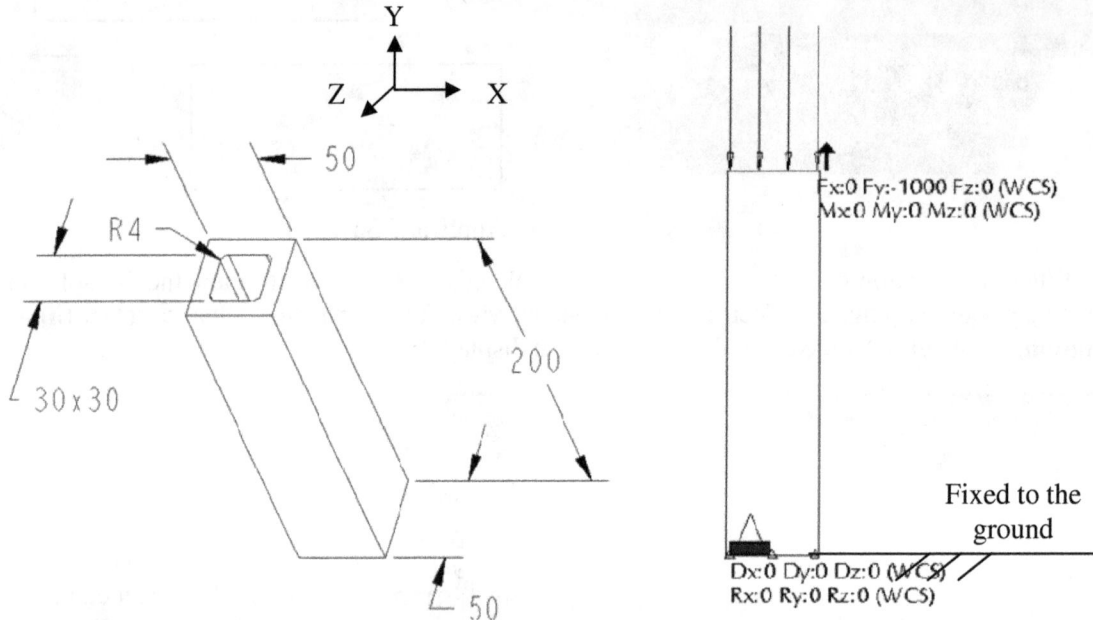

As illustrated, the height of the column is 200 mm. The section area is a square. The two dimensions of the section area are 50 x 50 mm, as shown. It is assumed that the bottom part of the column is fixed to the ground. As illustrated, a compressive load is uniformly acting on the top surface of the column. The magnitude of the load is 1000 Newton. There are 2 objectives to be expected from performing FEA:

(1) Find the location of the maximum displacement on the column when subjected to this compressive load and specify its magnitude.

(2) Find the location of the maximum principal stress and the location of the minimum principal stress on the column when subjected to this compressive load and specify their magnitudes.

To achieve these 2 objectives, in section 2.2.1, we work on the creation of a 3D solid model of the column structure under study. In section 2.2.2, we work on the process of performing FEA.

2.3.1   Creation of a 3D Solid Model of a Column Structure

To begin with the process of geometrical model, we launch the Pro/ENGINEER Wildfire design system. After launching the Pro/ENGINEER Wildfire, click the icon of **New** to initiate the creation of a 3D solid model.

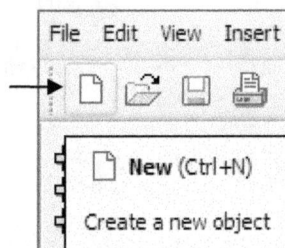

Type *column* as the file name, clear the box of **Use default template** > **OK**. Select the unit of **mmns_part_solid**, type *compression load* under the description of the model, and type *student* or *your name* under the **modeled_by** > **OK**. It is important to note the necessity of selecting the unit system. The default unit system used in Pro/ENGINEER Wildfire is the inch_lb_second system. In this book, we use the unit system is the mm_Newton_second system. If a unit system other than the mm_Newton_second system is used, a specific note will be made.

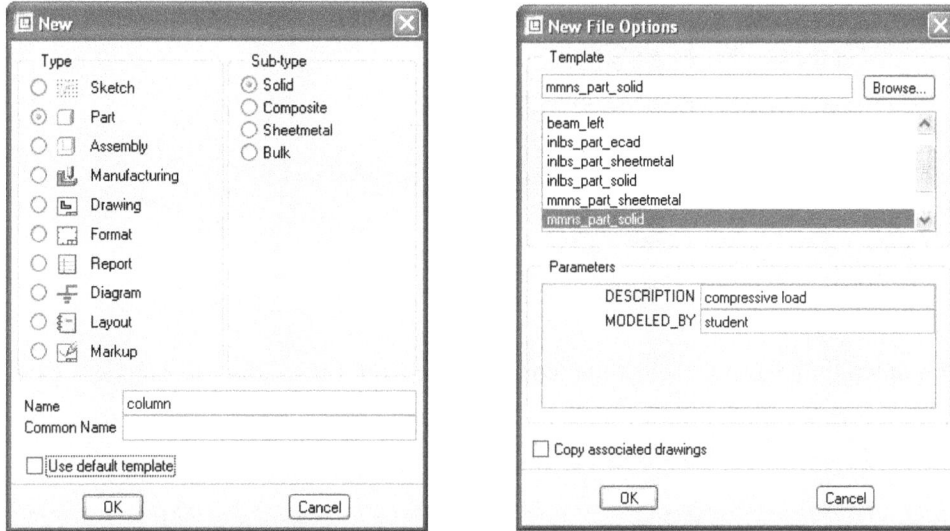

Pick up the icon of **Extrude** displayed on the toolbar of feature creation. Specify 50 as the depth value. To initiate the process of sketching, activate **Placement > Define.**

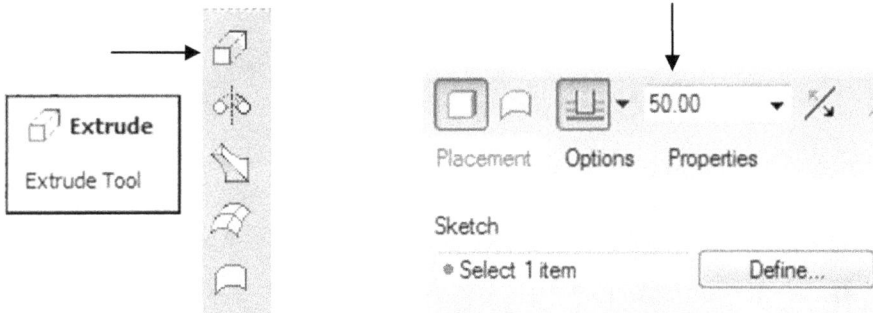

Select the **FRONT** datum plane displayed on screen as the sketch plane. Pick the box of **Sketch** to accept the **RIGHT** datum plane as the default reference to orient the sketch plane, as illustrated below:

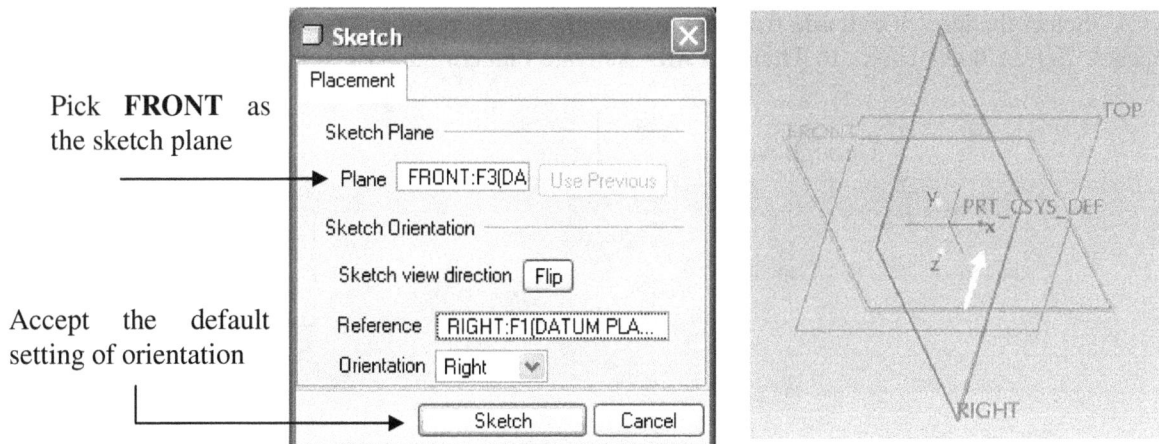

Pick **FRONT** as the sketch plane

Accept the default setting of orientation

Select the icon of **rectangle** from the toolbar of sketcher and sketch a rectangle. The 2 dimensions are 200 and 50, respectively.

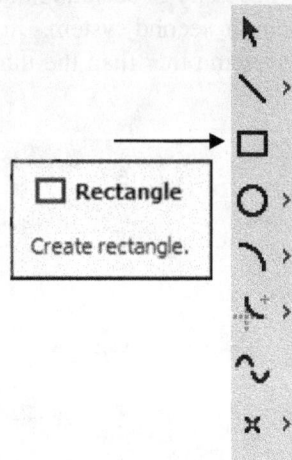

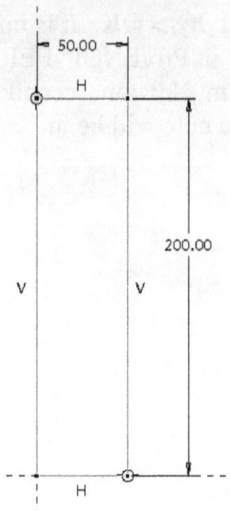

Upon completing the sketch, select the icon of **Done** from the toolbar of sketcher. Click the icon of **Apply and Save**.

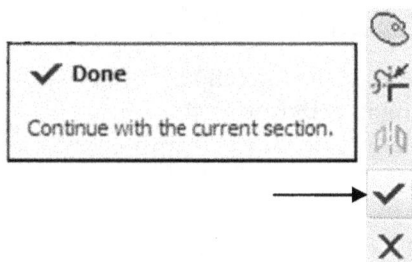

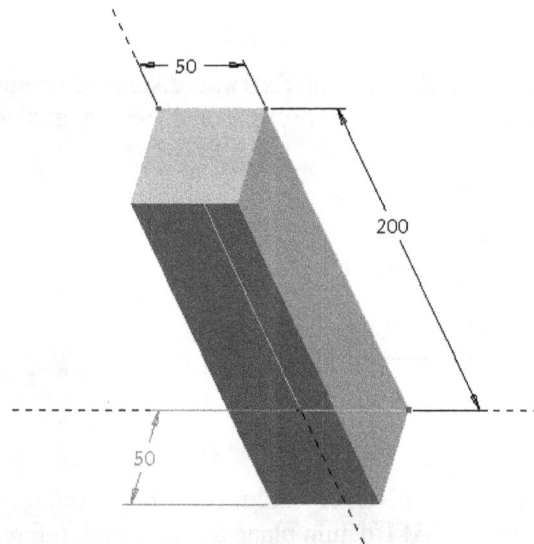

Apply and Save

Now let us create a cut feature so that the wall thickness becomes 10 mm.

Select the icon of **extrude** from the toolbar of feature creation. Make sure that the icon of **Cut** is selected. Set the depth choice to **Through All**. Activate **Placement > Define**.

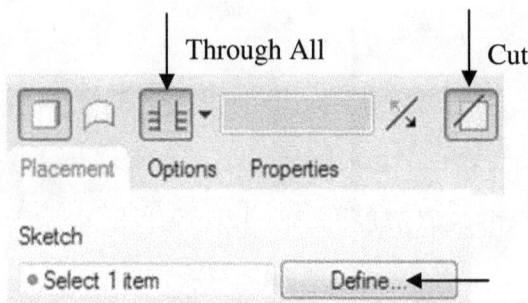

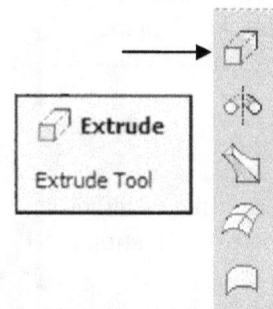

Through All   Cut

Select the top surface of the column feature as the sketch plane and click **Sketch** to accept the default setting for orientation, namely, use the right-sided surface to be at right to orient the sketch plane.

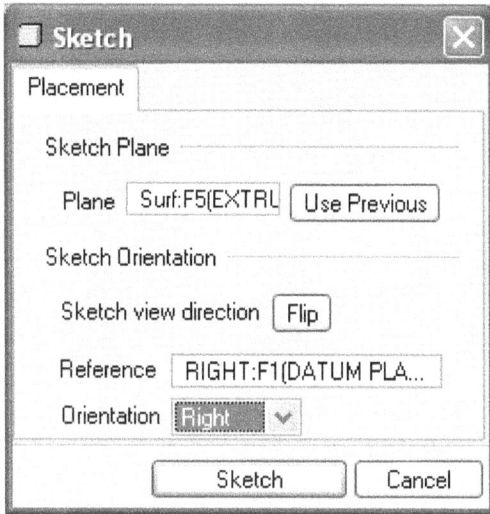

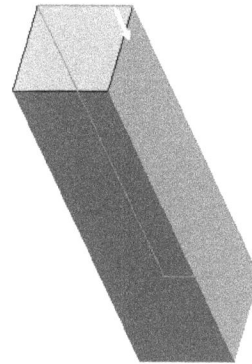

Select the icon of **Rectangle** from the toolbar of sketcher and sketch a rectangle. The 4 dimensions shown below are 10, 10, 30 and 30, respectively.

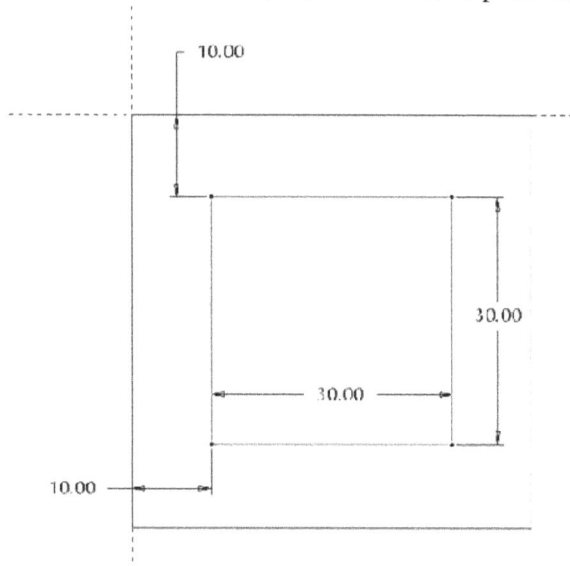

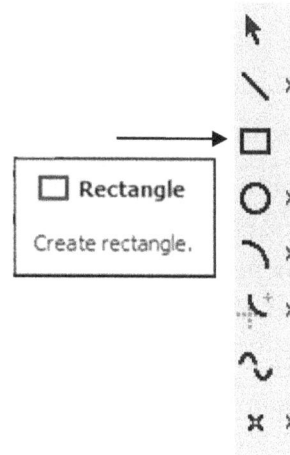

Upon completing the sketch, select the icon of **Done,** and click the icon of **Apply and Save**.

Apply and Save

Now let us add rounds to the 4 inner corners to avoid high stress concentration. Click the icon of **Round**. Specify 4 as the radius value. While holding down the **Ctrl** key, pick the 4 edges, as shown.

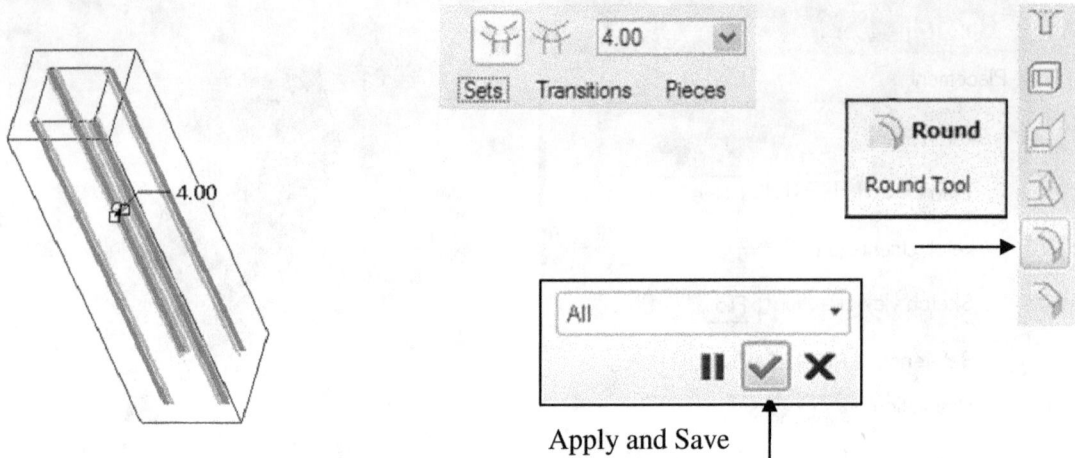

Apply and Save

Up to this point, the process of creating a 3D solid model for the column structure is completed. Click the icon of Save to save the 3D solid model.

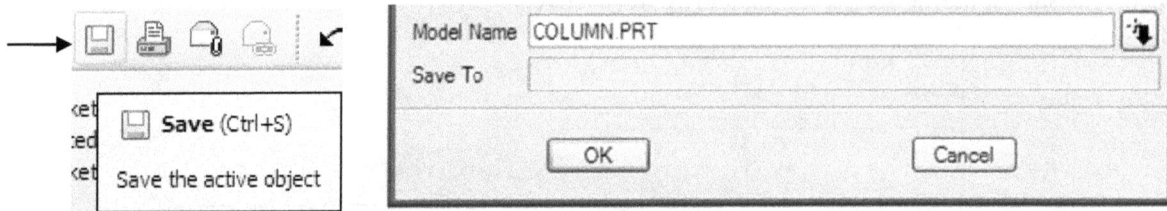

Now let us get on to the process of performing FEA.

2.3.2 Procedure of Performing FEA Using Pro/Mechanica

When the 3D solid model is available, we are able to use Pro/Mechanica to perform FEA for the column structure. There are 5 basic steps involved in such a process.

Step 1: Shift from the Pro/ENGINEER design system to Pro/Mechanica

From the main toolbar or the menu toolbar, select **Applications > Mechanica.**

A window called **Unit Info** appears. The displayed information is to verify the unit system used in the process of creating the 3D solid model. As illustrated, the unit system used to create the column model is mm_Newton_second system where the unit of mass is tonne and the unit of temperature is set to Celsius (deg C).

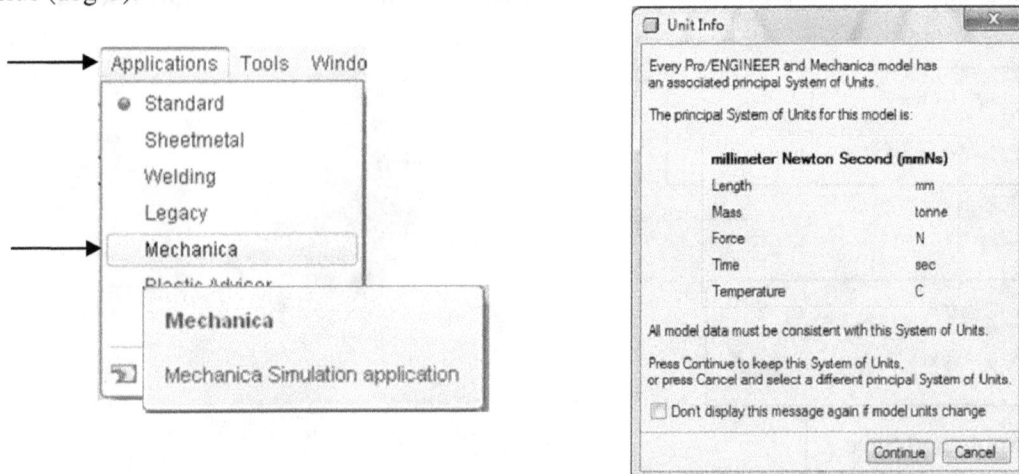

Click the box of **Continue** after checking the unit system.   Select **Structure** from the **Model Type** window > OK. To specify the material properties, click the icon of **Materials**.

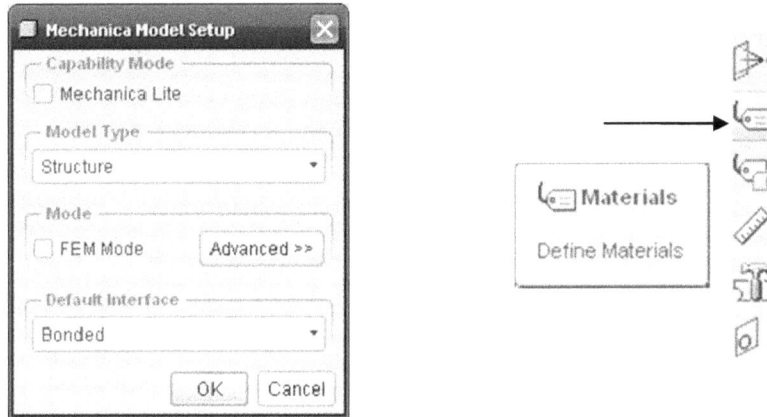

In the **Materials** window, select **STEEL** from the left side called Materials in Library > click the directional arrow so that the selected material type goes to the right side called Materials in Model.

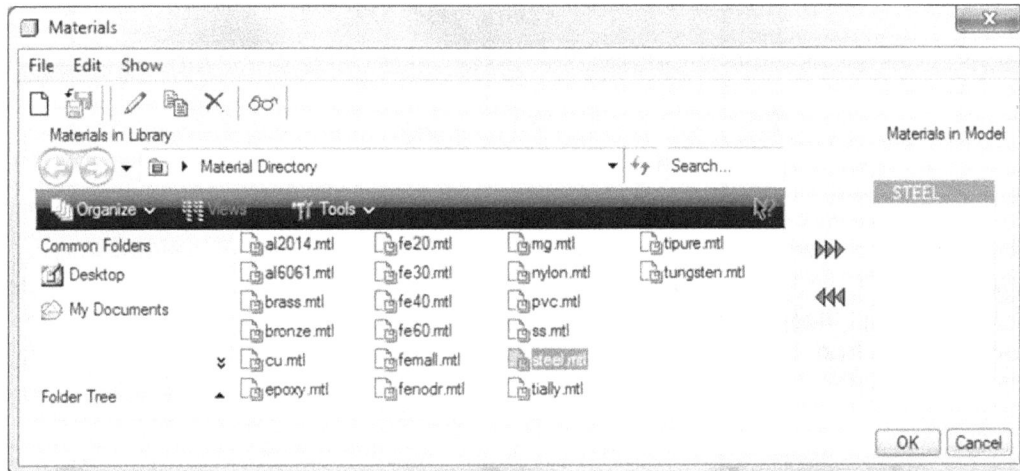

Click the icon of **Material Assignment**, the pop-up window indicates there is one component and the material type is STEEL. Just click **OK**.

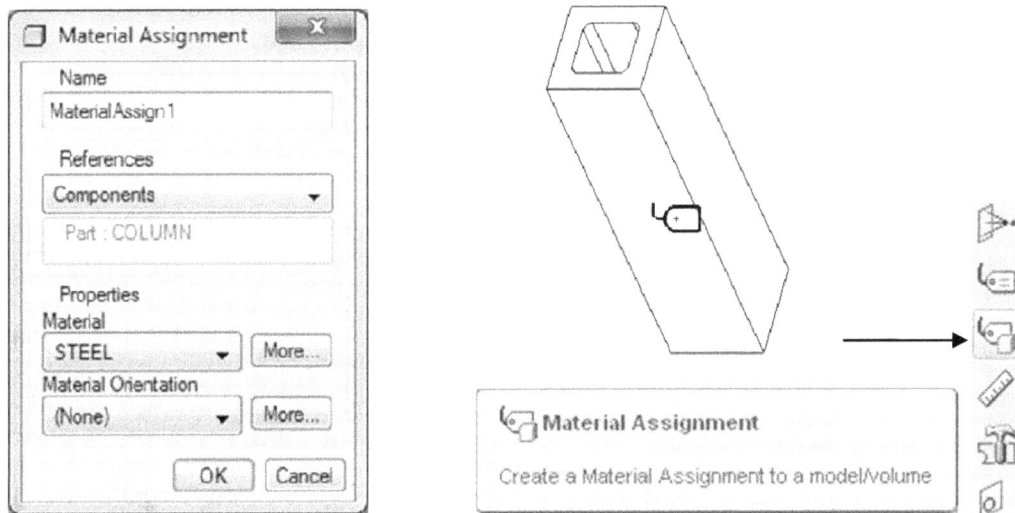

To define the fixed end constraint on the bottom of the column, select the icon of **New Displacement Constraint** from the toolbar of functions > **Surface(s)** > pick the surface at the bottom of the column > **OK** to accept the surface selection.

Because the constraint condition is "a fixed end", we set all 6 degrees of freedom equal to zero, which is the default setting > **OK**.

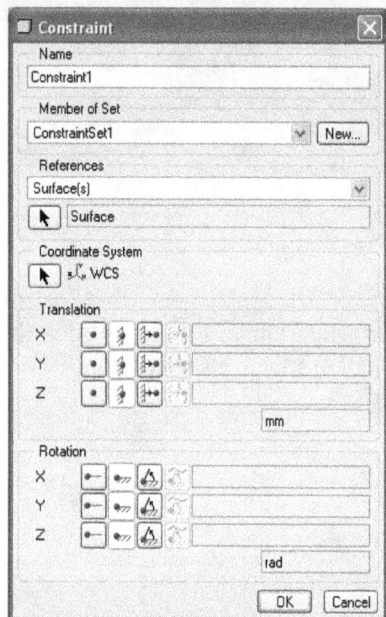

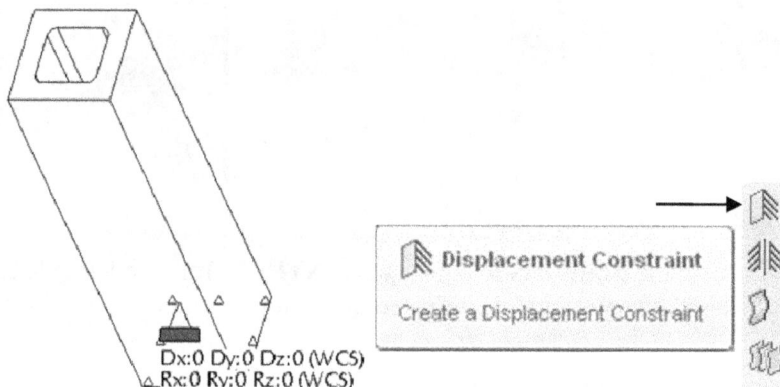

To define the load condition on the top surface of the column, select the icon of New Force/Moment Load from the toolbar of functions > **Surface(s)** > pick the surface at the top of the column > **OK** to accept the surface selection.

Because the load is acting along the negative Y direction, we type -5000 in the Y force component box > **OK**.

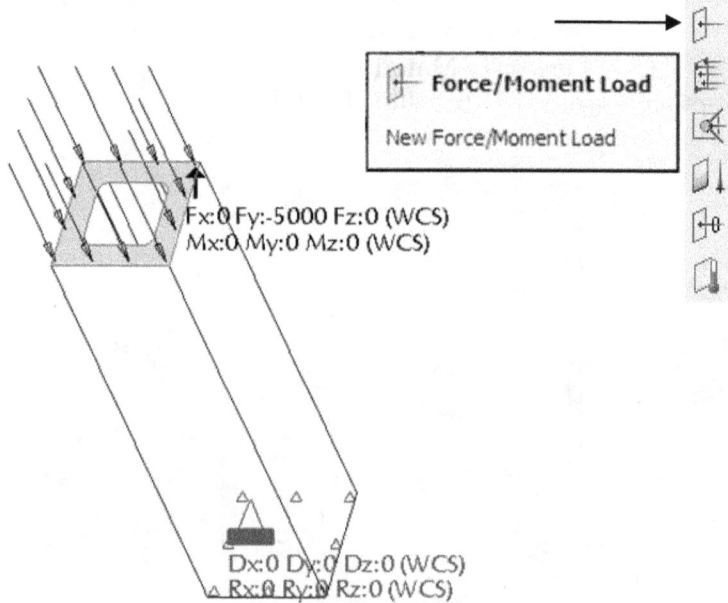

Under the Pro/MECHANICA environment, set up **Analyses** and **Run** it

Select the icon of **Run an Analysis** from the main toolbar > **File** > **New Static** > type *column* as the name of the analysis folder > **OK.** Click the box of **Run.**

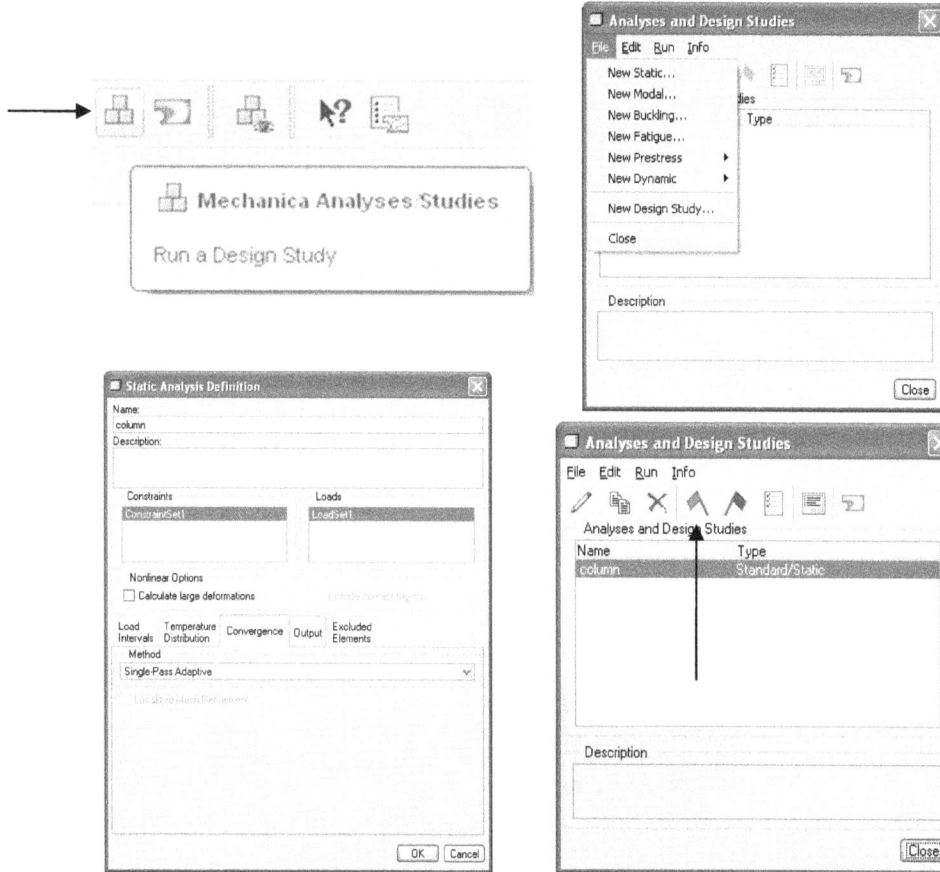

There is always a message displayed on screen. The message is asking the user "Do you want to run interactive diagnostics?" Click **Yes**. In this way a warning message will appear if an abnormal operation occurs.

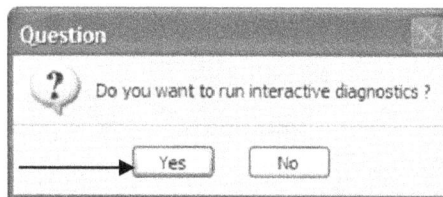

To monitor the computing process, users may click the icon of **Display study status**.

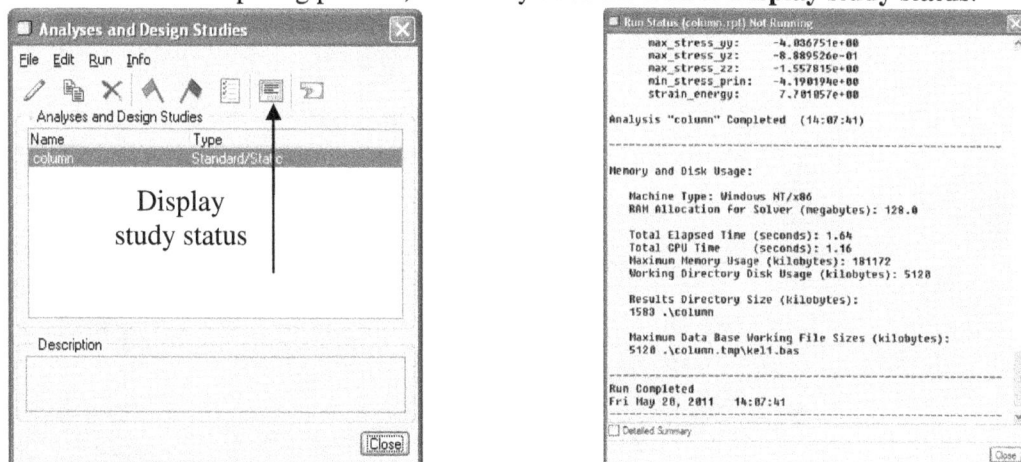

To study the obtained result, say the displacement of the column under the static loading condition, click the icon of **Review results** > **Displacement** > **Magnitude** > **OK and Show.** Note that if the folder called column does not appear in the Result Window Definition, click the icon of folder under Design Study, locate it and open it.

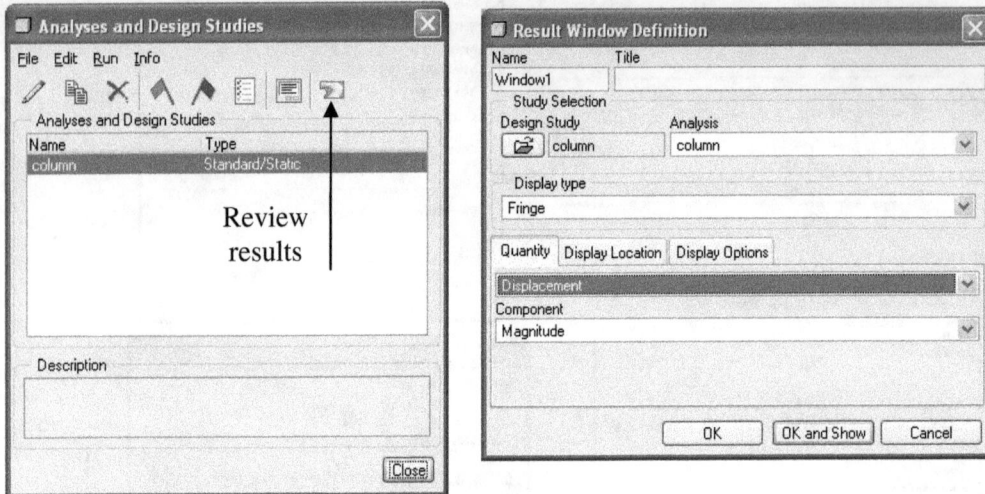

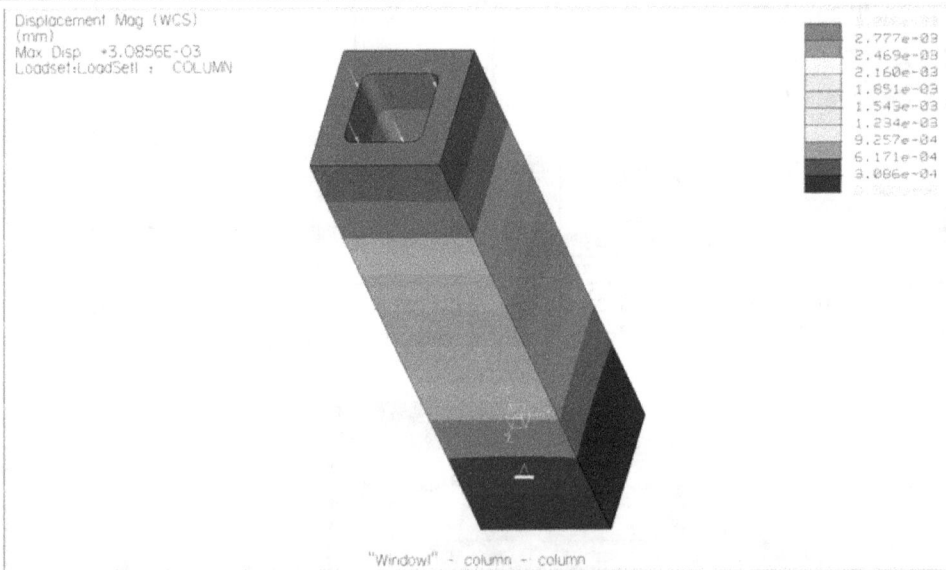

To study the distribution of stress, say the maximum principal stress distribution, select **Edit** from the main toolbar > **Copy** and a new Result Window appears > **Stress** > **Max Principal** > **OK and Show**.

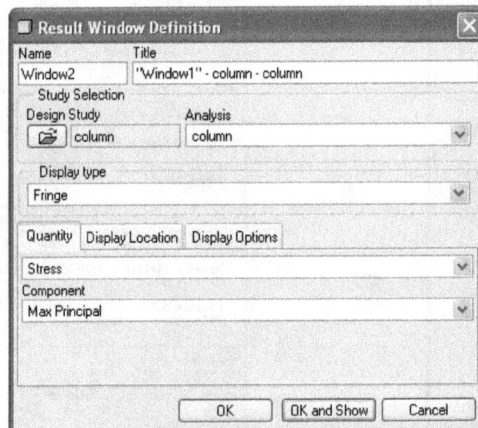

Stress Max Prin (WCS)
(N / mm^2)
Loadset:LoadSet1 : COLUMN

"Window1" - column - column

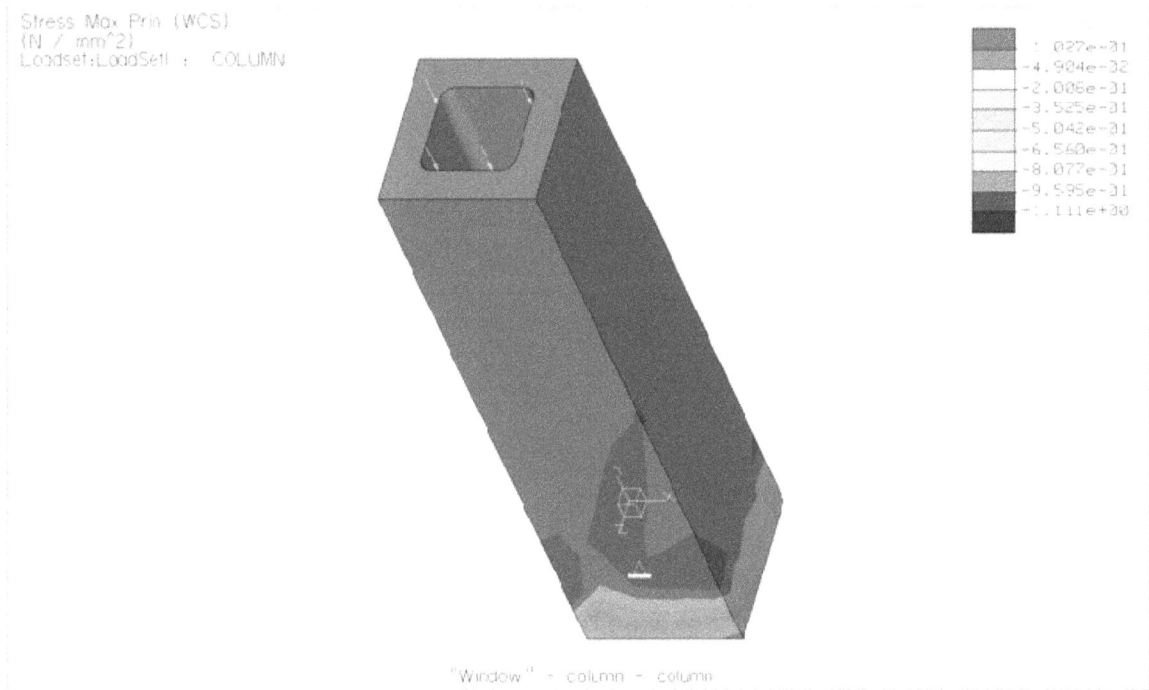

Now let us locate the position of the maximum displacement occurred on the column structure. We need to go back to the window illustrating the displacement distribution. From the main toolbar, select the icon of **Display definition**(s). A new window appears, on which both Window1 and Window2 are highlighted. De-highlighting Window2 and leaving Window1 highlighted will allow the distribution of displacement on display only.

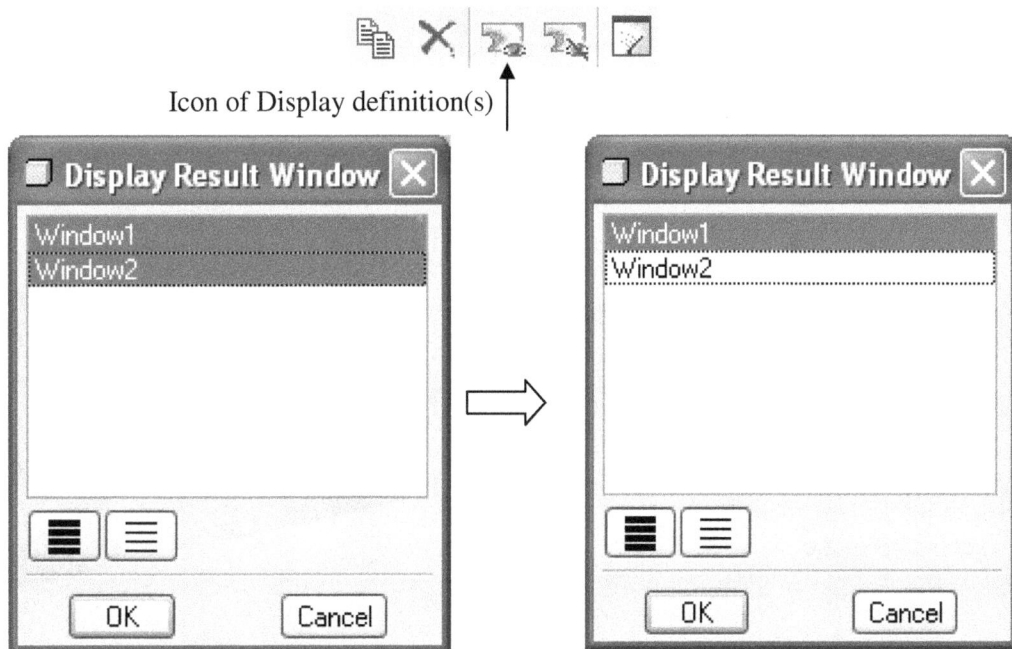

Icon of Display definition(s)

Click **Info** from the main toolbar > **Location in Dynamic Query**. Click **Info** > **Model Max,** and the magnitude and the location of the maximum displacement appear on the display. As illustrated, the maximum displacement is located at the corner of the top surface (x=50, y=200 and z=0) with the magnitude equal to 0.003086 mm.

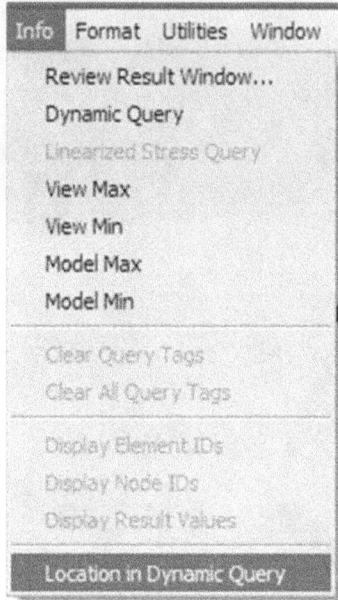

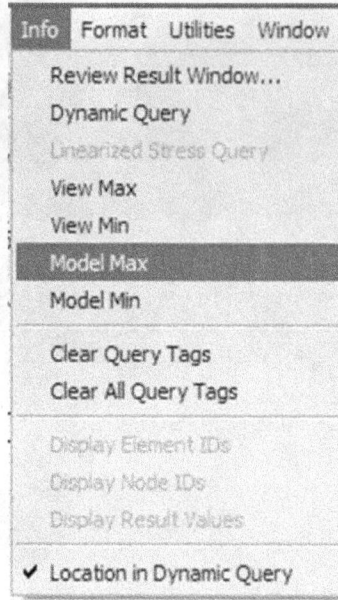

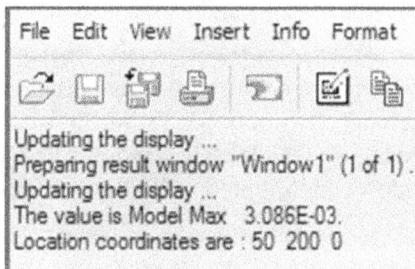

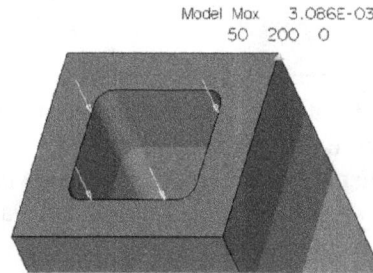

Click **Info > Model Min**. The minimum displacement is located at the bottom surface, which is fixed to the ground. Users can read the numerical values of those coordinates indicating the model max in the message window. All these values are zero, as shown below.

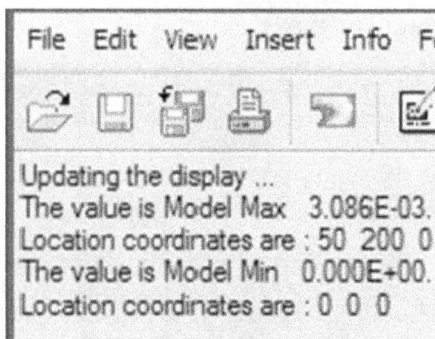

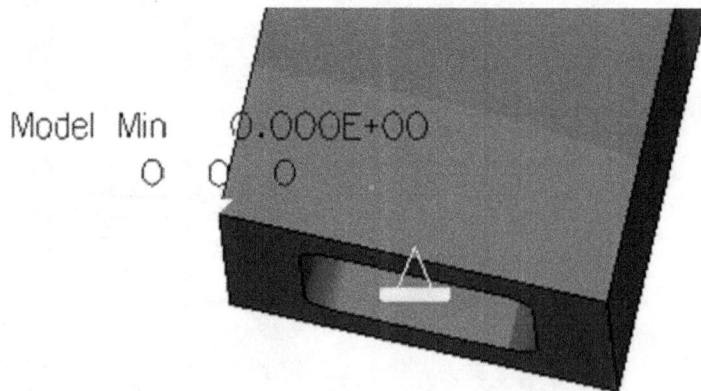

The maximum value of the maximum principal stress is 0.48 MPa. And the maximum value of the minimum principal stress is -1.16 MPa. Note that a negative value indicates compressive stress and a positive value indicates tensile stress. To specify their positions, click **Info > Location in Dynamic Query.** Click Info > **Model Max**. Afterwards, select **Info** from the main toolbar again > **Model Min**. As illustrated, these 2 locations are (x=0, y=11.2 z=50) and (x=46.5, y=0. Z=47.5), respectively.

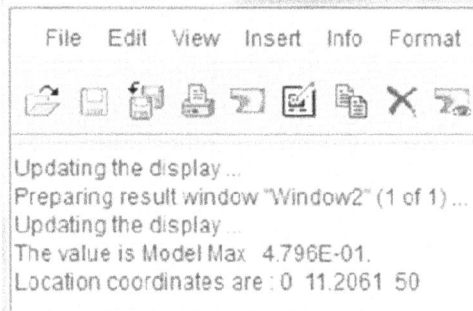

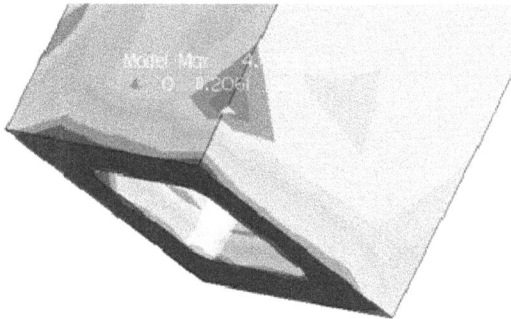

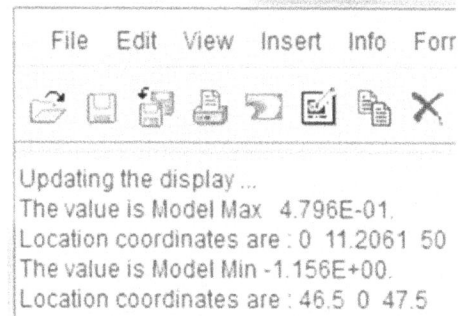

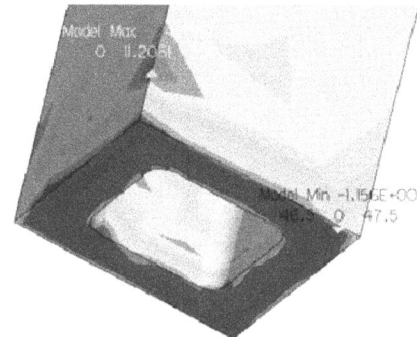

The maximum value of the minimum principal stress is -2.44 MPa. And the minimum value of the minimum principal stress is -5.17 MPa. Note that negative values indicate compressive stresses. To specify their positions, click **Info > Location in Dynamic Query.**   Click Info > **Model Max**. Afterwards, select **Info** from the main toolbar again > **Model Min**. As illustrated, these 2 locations are (x=11.2, y=0 z=38.8) and (x=0, y=0. Z=50), respectively.

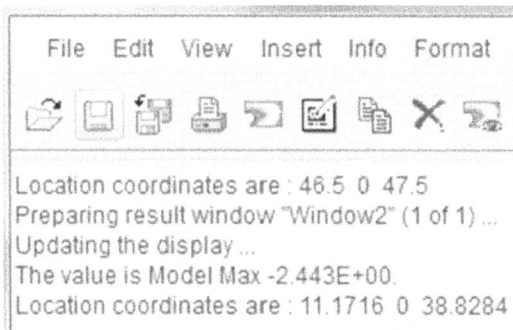

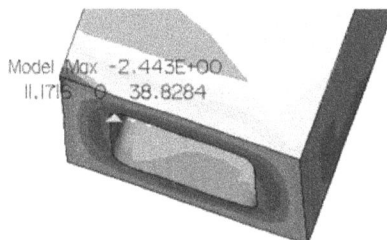

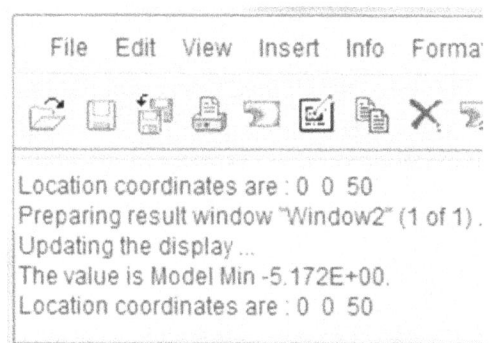

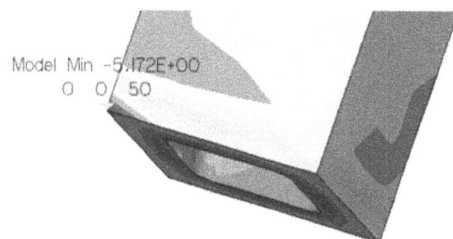

As a summary, we present the general procedure to perform FEA used in the software system. As illustrated in the flowchart, we need a 3D solid model first because the FEA software system assumes the availability of such a 3D solid model. The FEA software system requires users to define the material properties, constraint condition(s), load condition(s) and specification for mesh generation. After running FEA, the software system is able to provide the information on displacement, stress, strain, etc.

```
┌─────────────────────────────────────────────┐
│                                               │
│      Create a 3D solid model using a CAD system │
│                                               │
└─────────────────────────────────────────────┘
                        │
                        ▼
┌─────────────────────────────────────────────┐
│                                               │
│      Open the 3D solid model under a FEA system │
│                                               │
└─────────────────────────────────────────────┘
                        │
                        ▼
```

Material Properties	Constraint Condition(s)	Load Condition(s)	Mesh Generation

```
┌─────────────────────────────────────────────┐
│                                               │
│        Define a design study and run FEA        │
│                                               │
└─────────────────────────────────────────────┘
                        │
                        ▼
┌─────────────────────────────────────────────┐
│                                               │
│             Study the obtained results          │
│                                               │
└─────────────────────────────────────────────┘
```

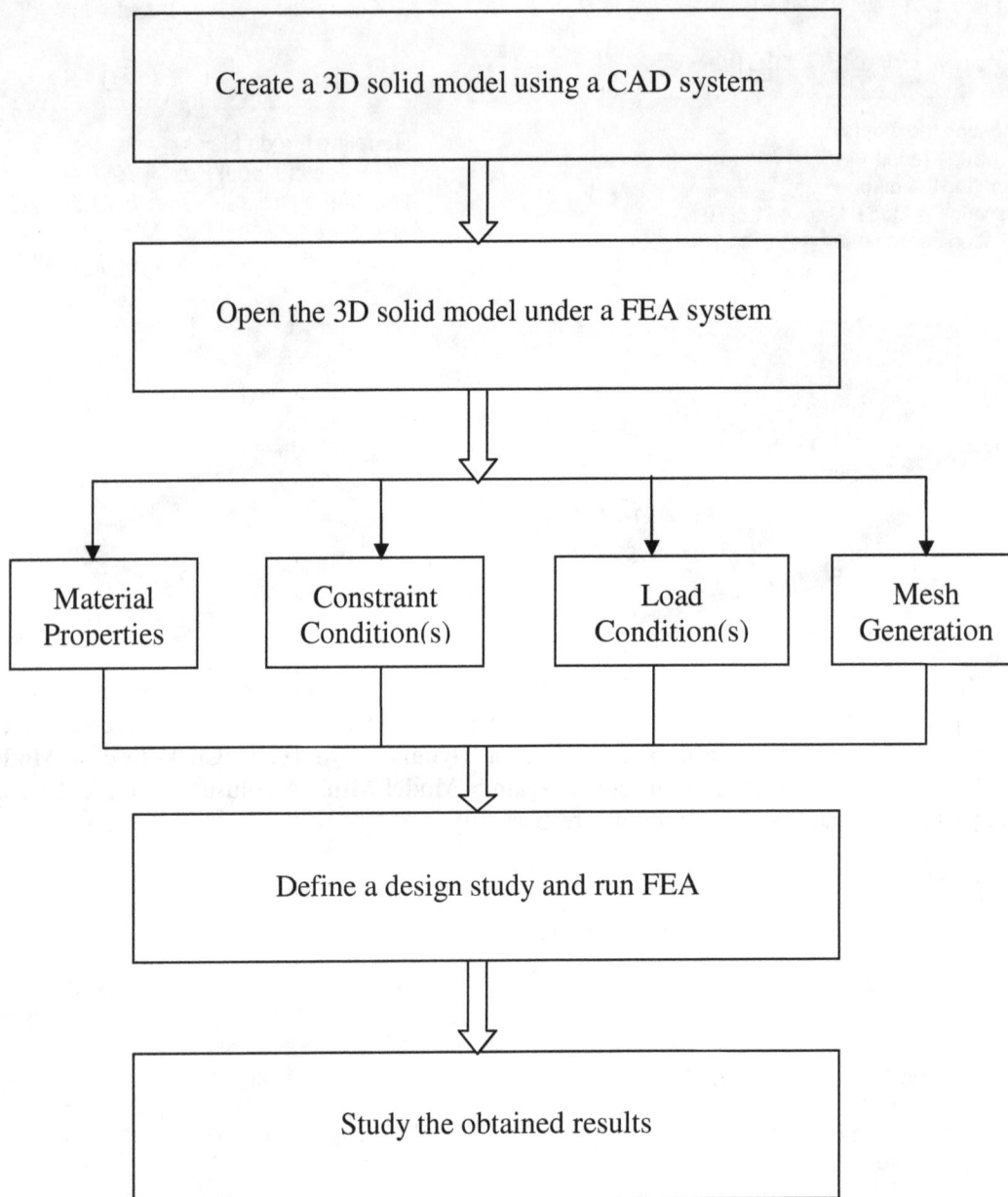

2.4   Beam Structure Subjected to a Uniformly Distributed Load

In section 2.1 and section 2.2, we presented the two examples to demonstrate the procedure of creating a 3D solid model for the physical object under investigation and perform FEA using Pro/Mechanica afterwards. In this section, we will perform a finite element analysis for a beam structure. The geometrical shape of the beam is shown below. The beam by its self is a block. The 3 dimensions are 400 x 80 x 20 mm. There is a hole cut through the beam. The diameter of the hole is 60 mm. The 2 dimensions to locate the hole positions are 200 and 40, respectively.

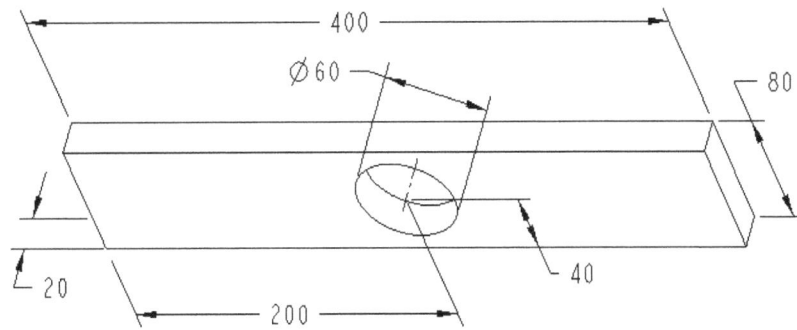

As illustrated below, the left end of the beam is fixed to the wall. Therefore, the structure is a cantilever beam. The load acts on the top surface of the beam. The load is a uniformly distributed load with a total value equal to 1000 Newton. Note the direction of the load is downward. Assume that the material type of the beam is steel.

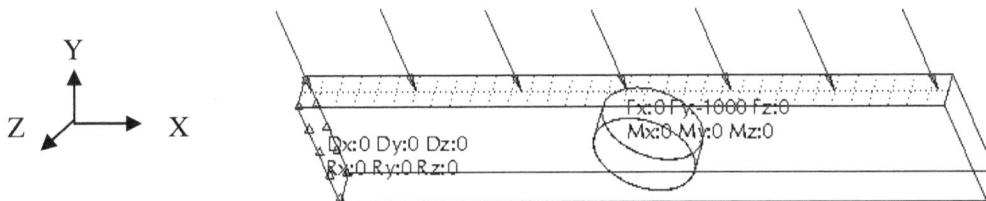

There are 2 objectives for performing FEA:

(1) Find the location of the maximum displacement on the beam when subjected to this uniformly distributed load and specify its magnitude.
(2) Find the location of the maximum principal stress and specify this magnitude.

To achieve these 2 objectives, we follow the procedure discussed in section 2.2. We first work on the creation of a 3D solid model of the beam structure, and afterwards work on performing FEA.

We click icon of **Create a new object** from the menu toolbar to initiate the creation of a 3D model.

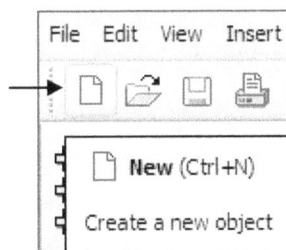

Type *beam_1* as the file name, clear the box of **Use default template** > **OK**. Select the unit of **mmns_part_solid**, type *cantilever beam* under the description of the model, and type *student* or *your name* under the **modeled_by** > **OK**.

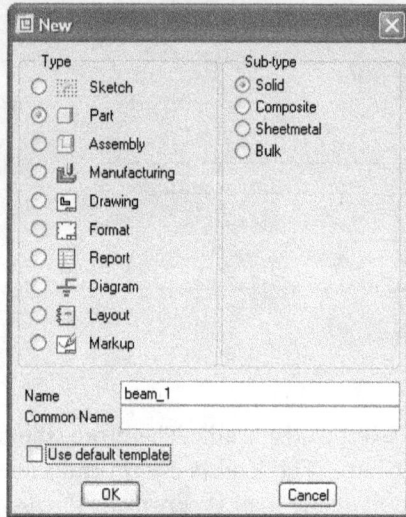

Click the icon of **Extrude.** Specify 20 as the depth value. To initiate the process of sketching, activate **Placement > Define.**

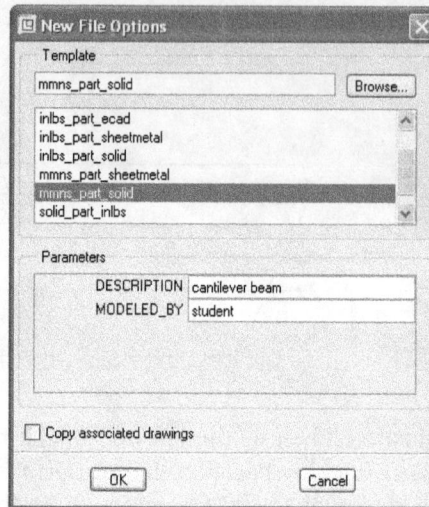

Depth = 20

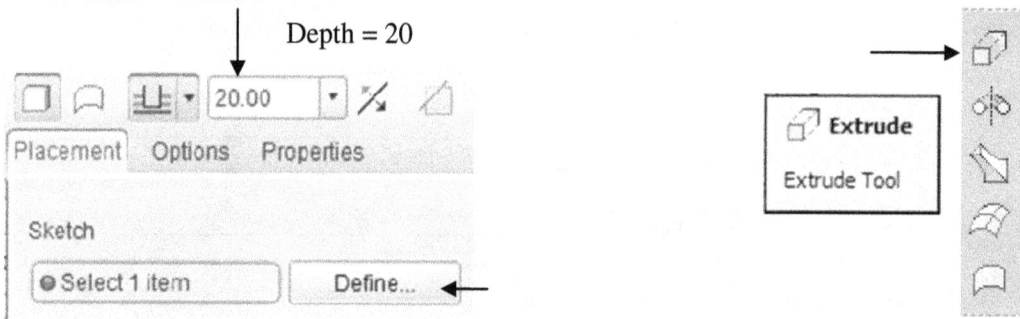

Select the **FRONT** datum plane displayed on screen as the sketch plane. Pick the box of **Sketch** to accept the **RIGHT** datum plane as the default reference to orient the sketch plane, as illustrated below:

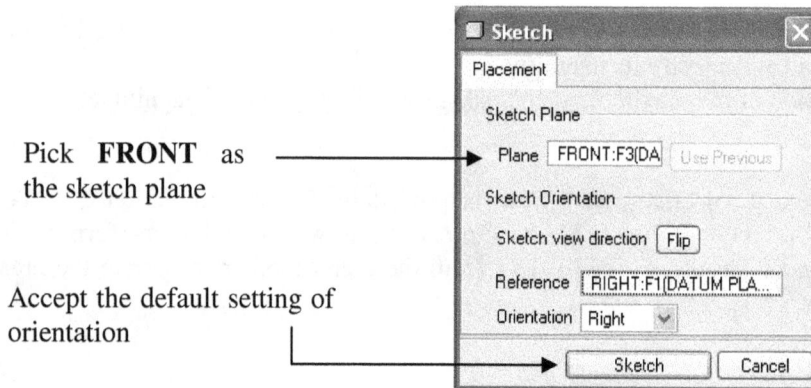

Pick **FRONT** as
the sketch plane

Accept the default setting of
orientation

Click the icon of **rectangle** from the toolbar of sketcher and sketch a rectangle. The 2 dimensions are 400 and 80, respectively.

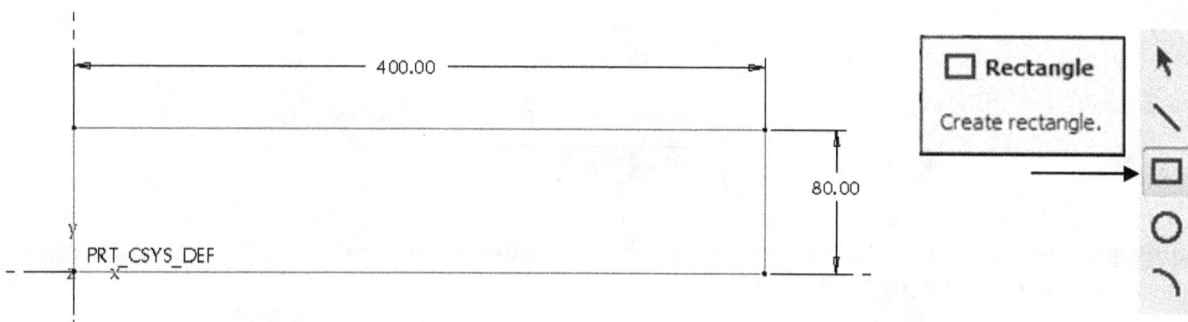

To create the hole feature, click the icon of **Circle.** Specify 60 as the diameter dimension. Specify 200 and 40 as the two position dimensions, as shown.

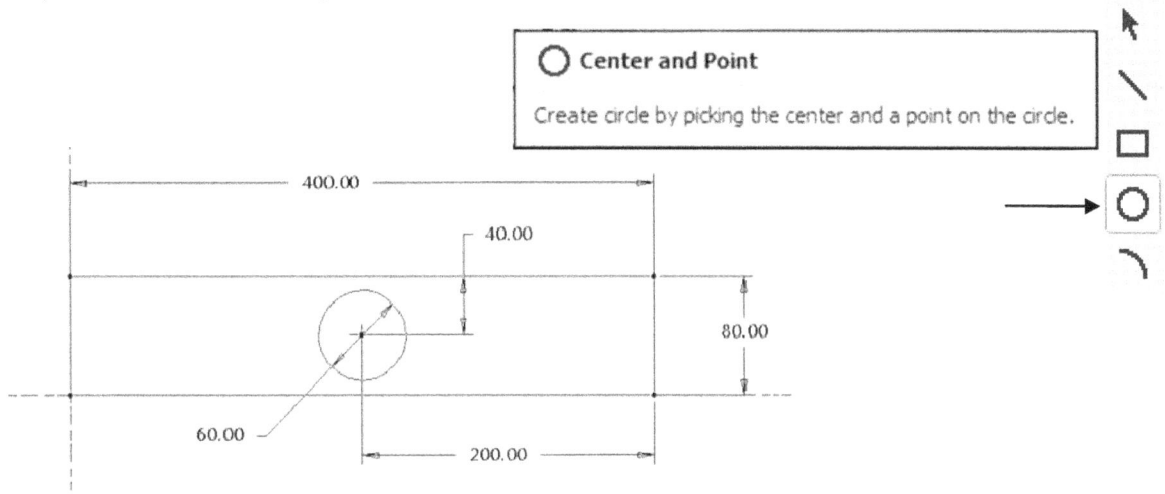

Upon completing the sketch, select the icon of **Done** and the icon of **Apply and Save**.

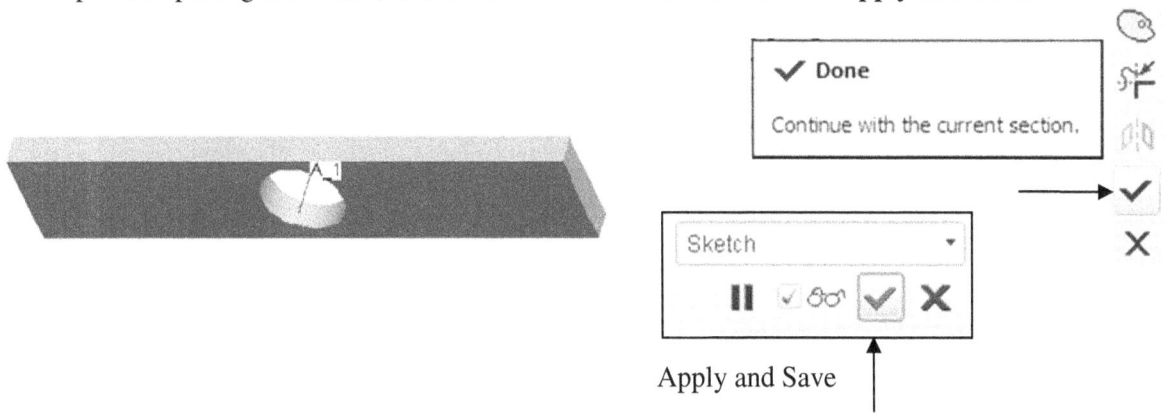

Apply and Save

Up to this point, the process of creating a 3D solid model for the beam structure is completed. Now let us get on to the process of performing FEA.

From the main toolbar or the menu toolbar, select **Applications > Mechanica.**

A window called **Unit Info** appears. The displayed information is to verify the unit system used in the process of creating the 3D solid model. As illustrated, the unit system used to create the column model is mm_Newton_second system where the unit of mass is tonne and the unit of temperature is set to Celsius (deg C).

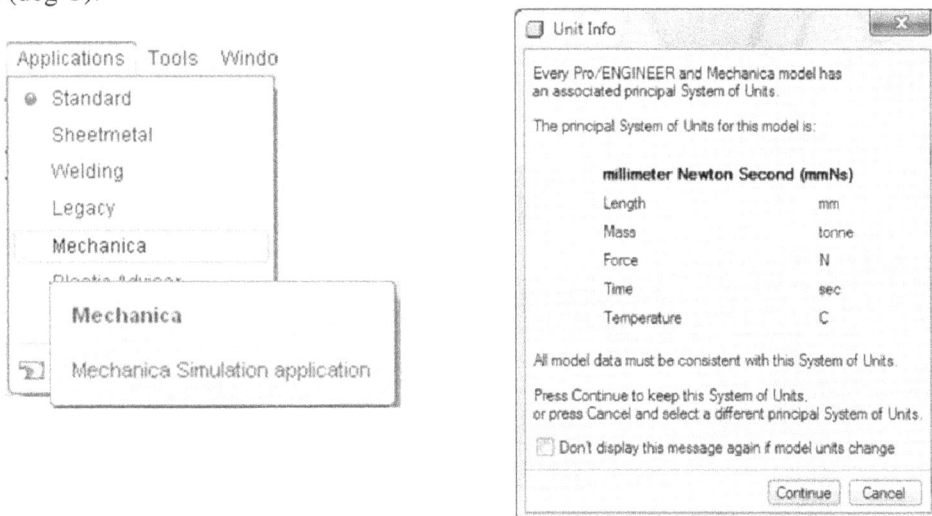

Click the box of **Continue** after checking the unit system. Select **Structure** from the **Model Type** window > OK. To specify the material properties, click the icon of **Materials**.

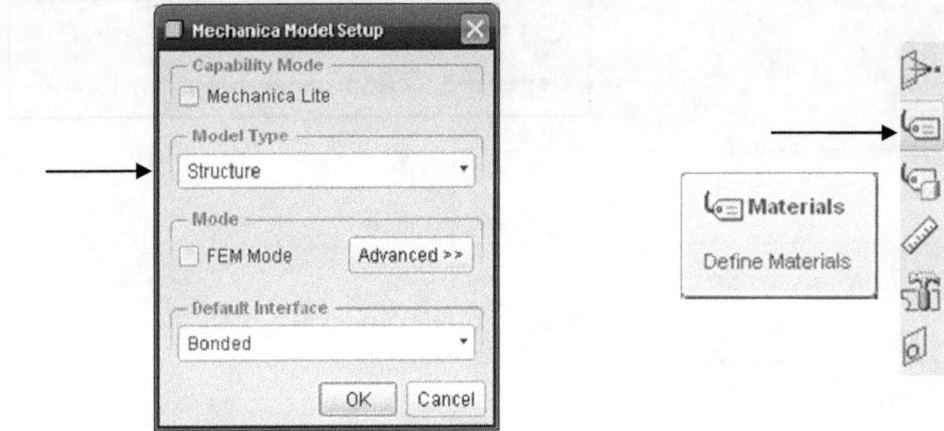

In the **Materials** window, select **STEEL** from the left side called Materials in Library > click the directional arrow so that the selected material type goes to the right side called Materials in Model.

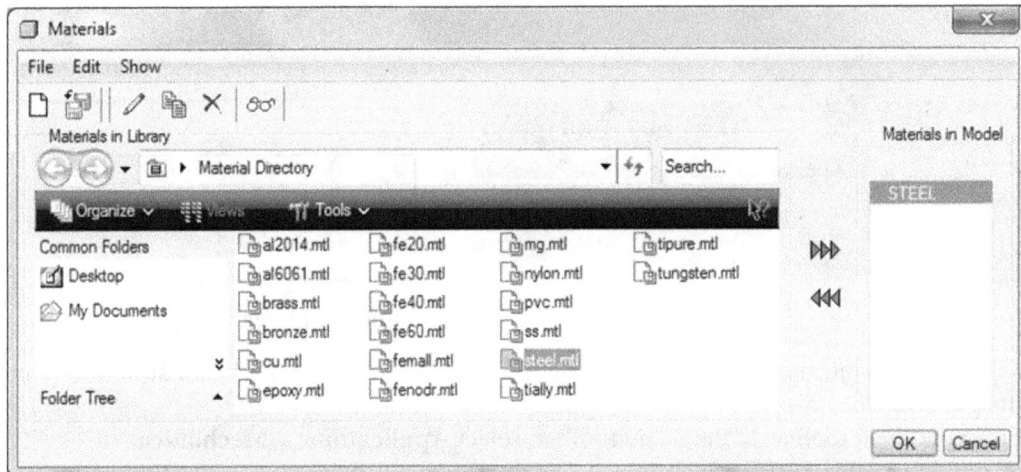

Click the icon of **Material Assignment**, the pop-up window indicates there is one component and the material type is STEEL. Just click **OK**.

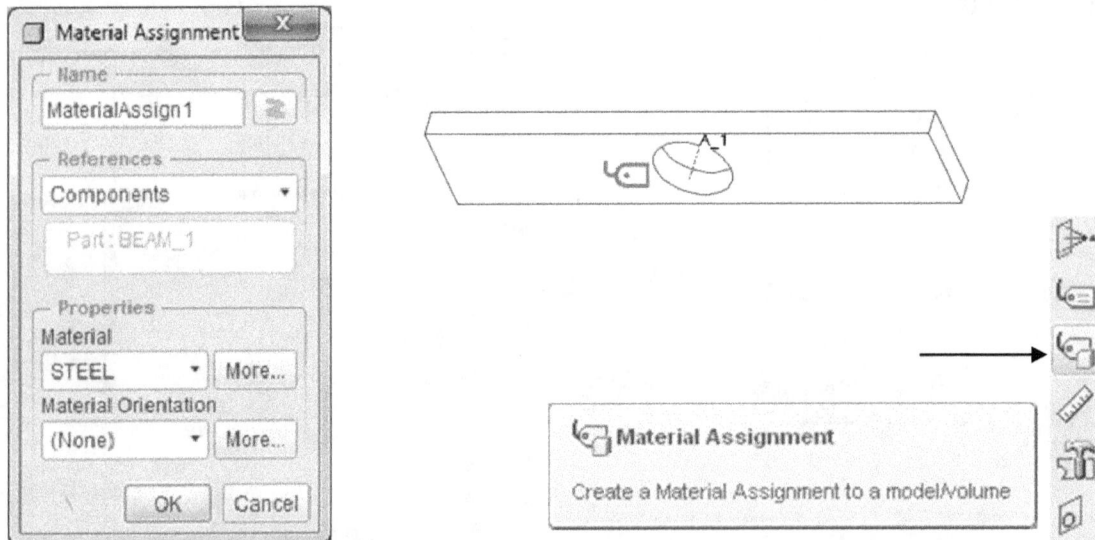

To define the fixed end constraint on the left side of the beam, select the icon of **New Displacement Constraint** from the toolbar of functions > **Surface(s)** > pick the surface at the left side of the beam > **OK** to accept the surface selection.

Because the constraint condition is "a fixed end", we set all 6 degrees of freedom equal to zero, which is the default setting > **OK**.

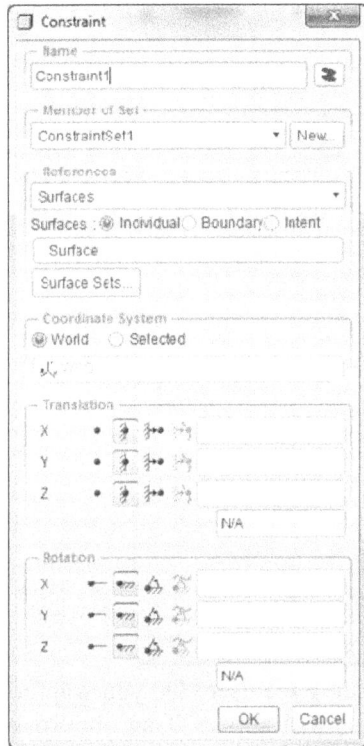

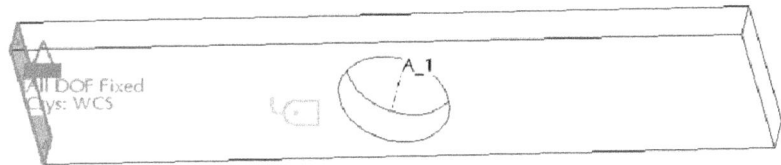

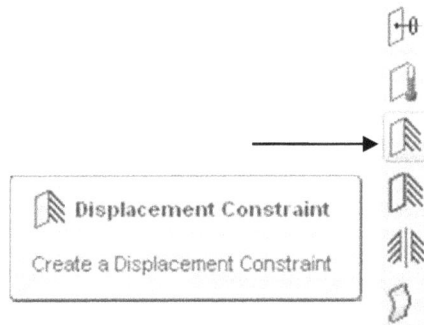

To define the load condition on the top surface of the beam, click the icon of **New Force/Moment Load** > **Surface(s)** > pick the surface at the top of the beam. Because the load is acting along the negative Y direction, we type -1000 in the Y force component box > **OK**.

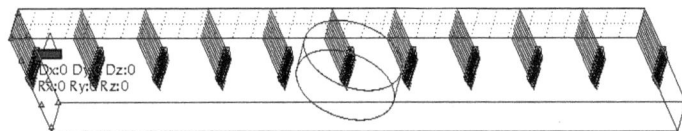

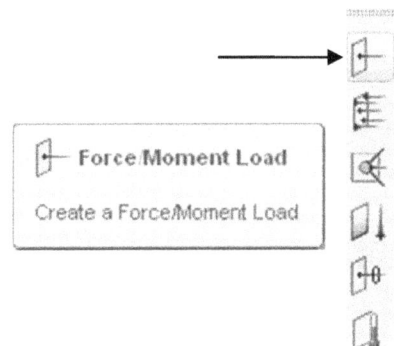

Under the Pro/MECHANICA, set up **Analyses** and **Run** it

Select the icon of **Mechanica Analyses/Studies** > **File** > **New Static** > type *beam_1* as the name of the analysis folder > **OK.** Click the box of **Run.**

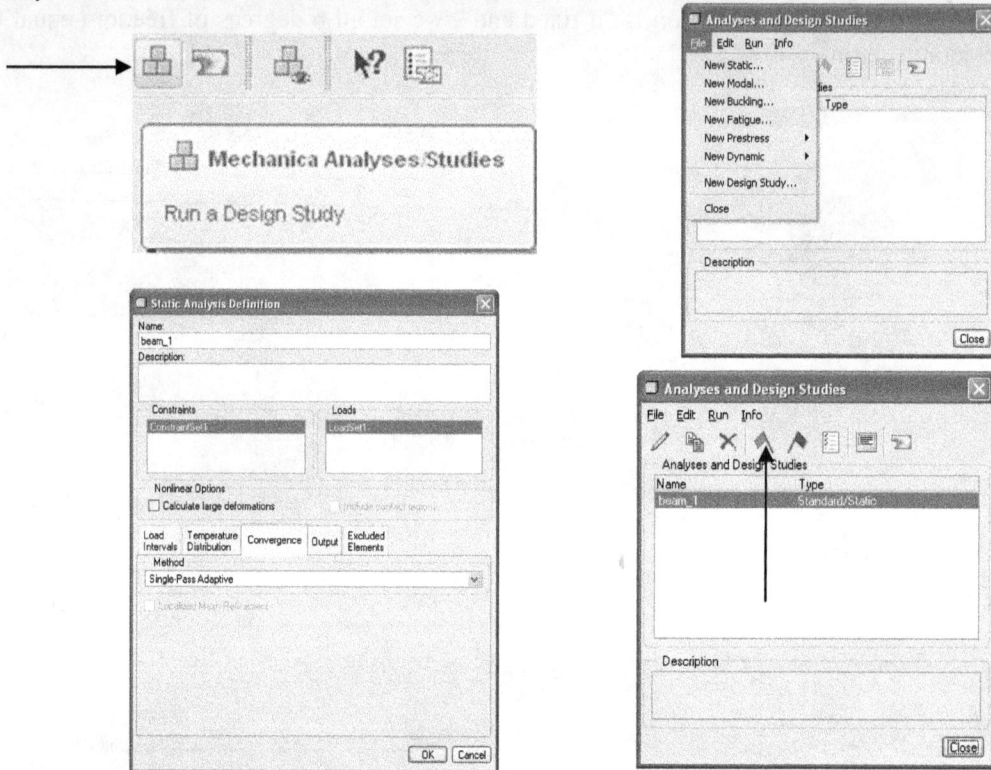

There is always a message displayed on screen. The message is asking the user "Do you want to run interactive diagnostics?" Click **Yes**. In this way a warning message will appear if an abnormal operation occurs.

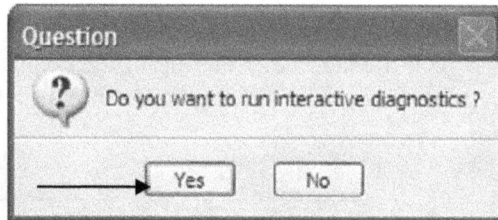

To monitor the computing process, users may click the icon of **Display study status**.

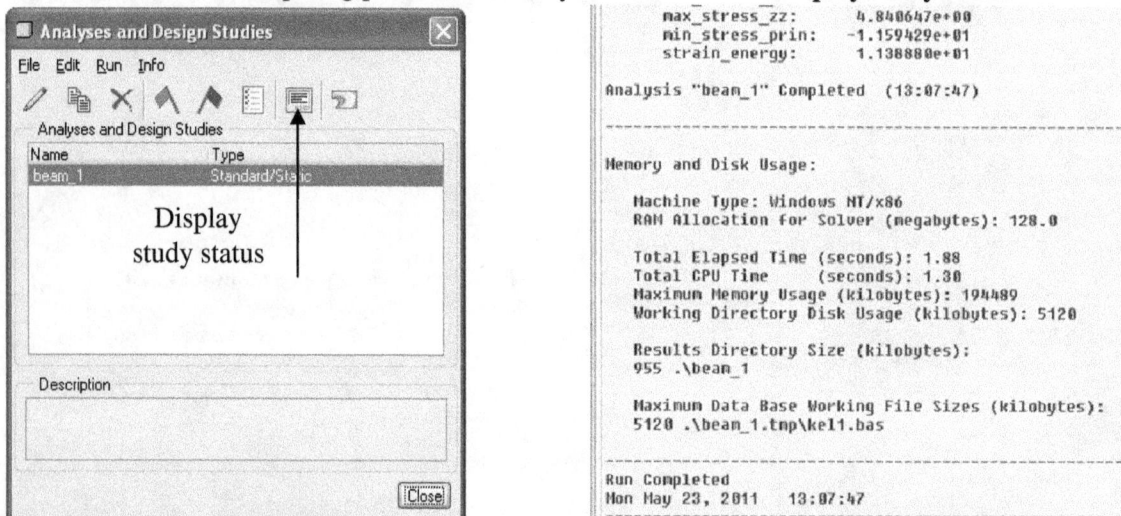

To study the obtained result, say the displacement of the cantilever beam under the static loading condition, click the icon of **Review results > Displacement > Magnitude > OK and Show**. Note that if the folder called beam_1 does not appear in the Result Window Definition, click the icon of folder under Design Study, locate it and open it.

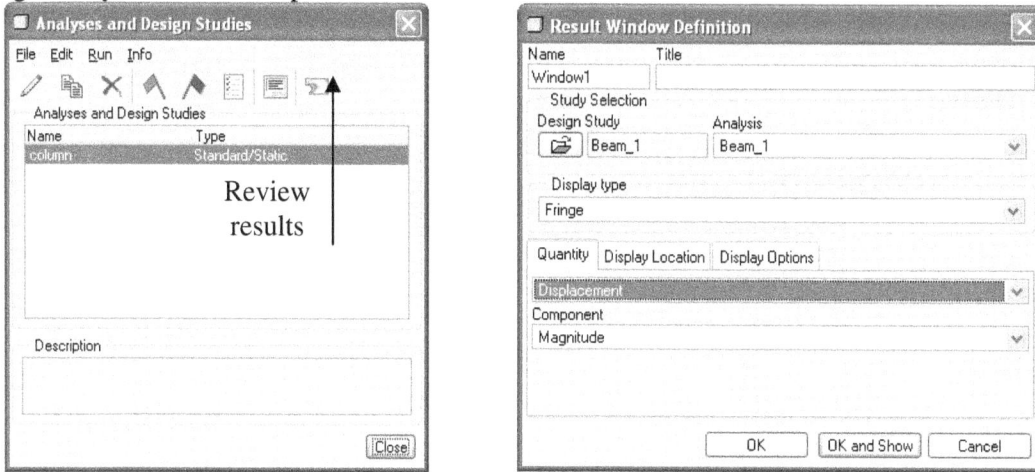

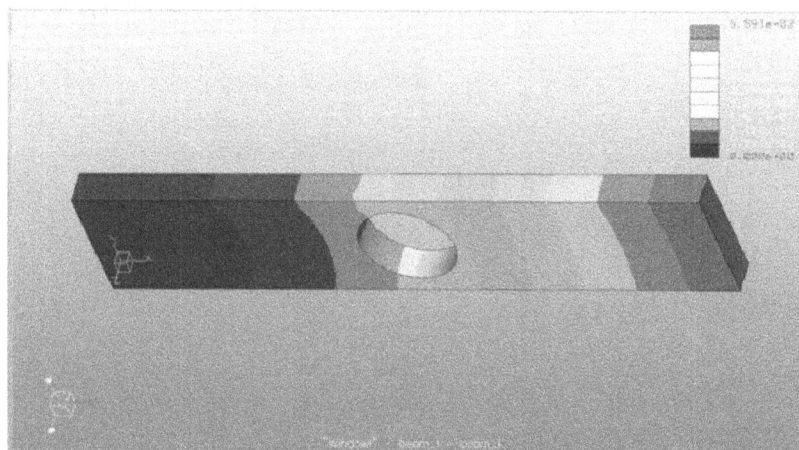

Click **Info** from the main toolbar > **Location in Dynamic Query**. Click **Info > Model Max,** and the magnitude and the location of the maximum displacement appear on the display. As illustrated, the maximum displacement is located at the corner of the top surface (x=50, y=200 and z=0) with the magnitude equal to 0.003086 mm.

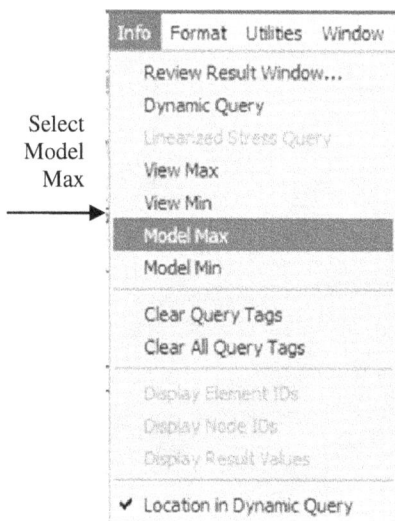

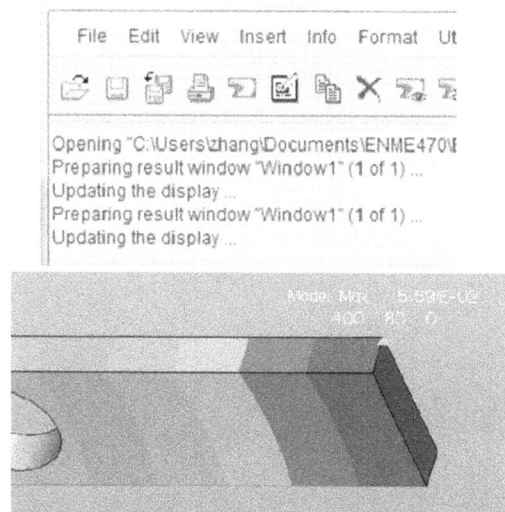

To change the background color to white, select **Format** from the main toolbar > **Result Window** > **White** > **OK**.

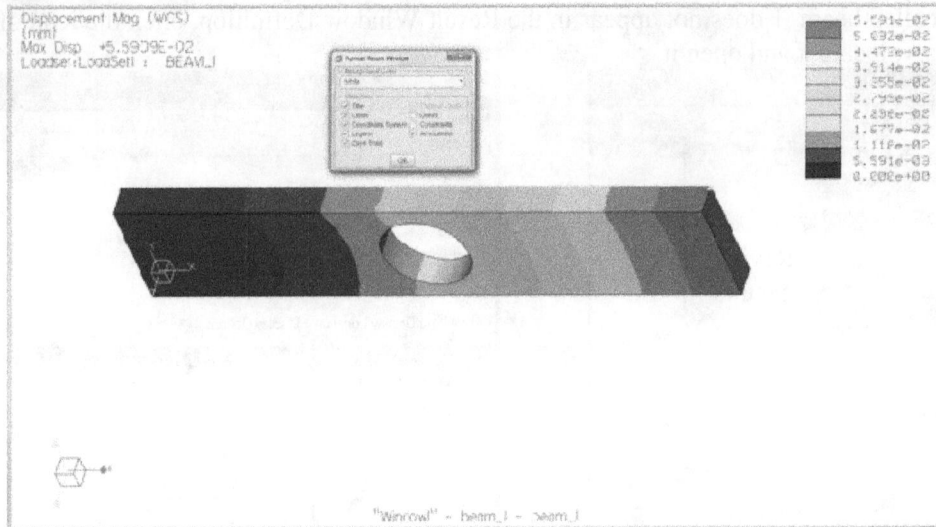

To study the distribution of the maximum principal stress distribution, select **Edit** from the main toolbar > **Copy** and a new Result Window appears > **Stress** > **Max Principal** > **OK and Show**.

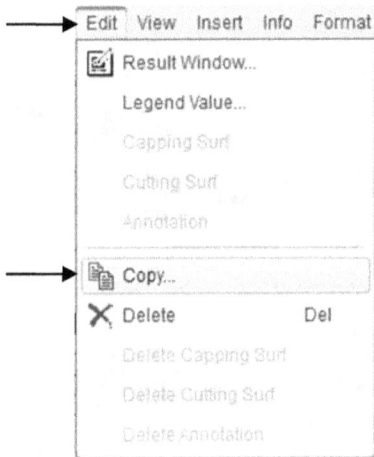

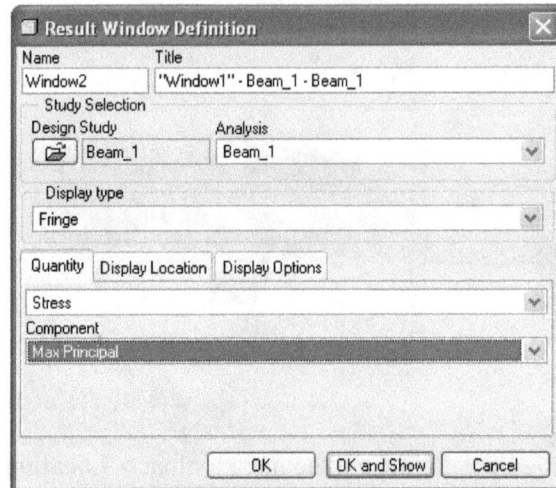

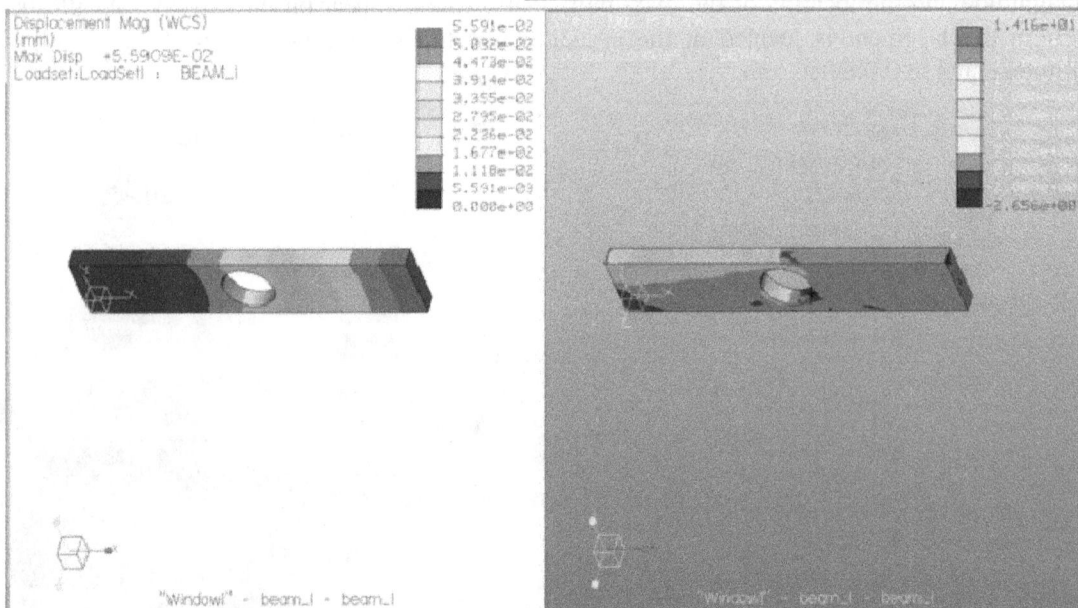

At this moment, we may have two windows or two definitions on display. One window (definition) is showing the displacement distribution and the other window is showing the distribution of the maximum principal stress. To have one window on display, click the icon of **Display definition(s)** from the main toolbar, a new window appears, on which both Window1 and Window2 are highlighted. De-highlighting Window1 and leaving Window2 highlighted will allow the distribution of maximum principal stress on display only.

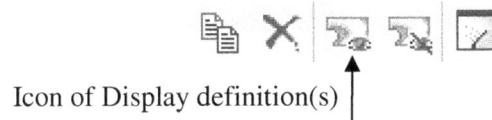

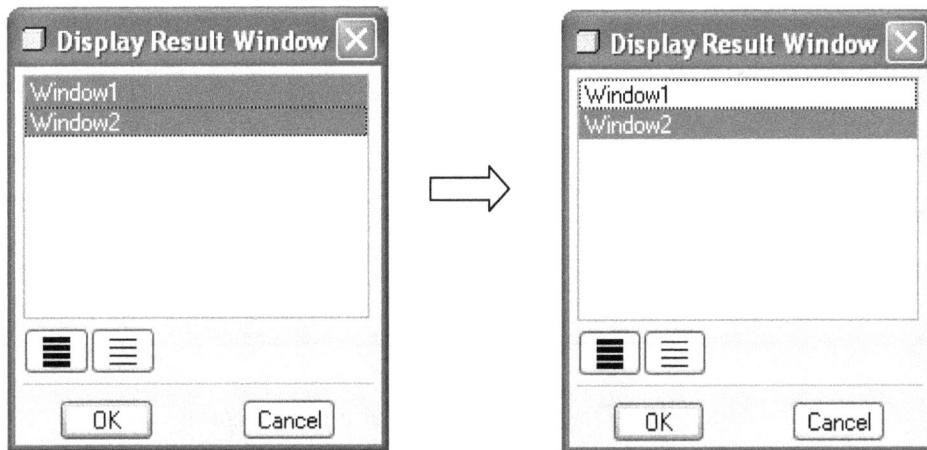

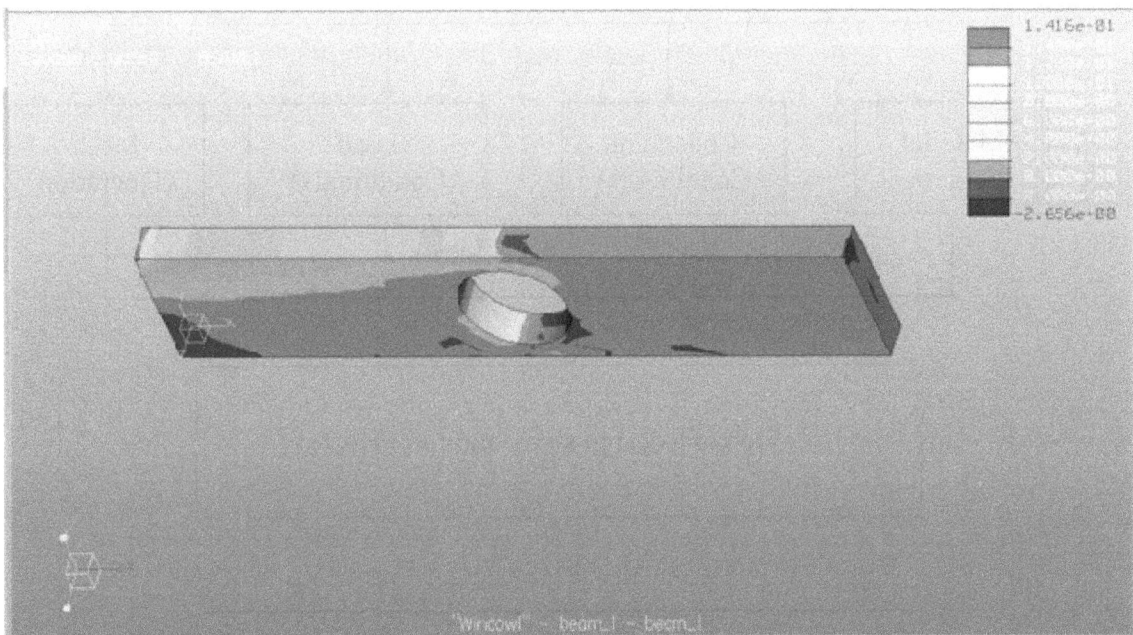

Icon of Display definition(s)

The maximum value of the maximum principal stress is 14.16 Map. To specify its location, click **Info > Model Max**. As illustrated, these 2 locations are (x=0, y=80 z=0).

File Edit View Insert Info Forma

Updating the display ...
Preparing result window "Window2" (1 of 1).
Updating the display ...
The value is Model Max 1.416E+01.
Location coordinates are : 0 80 0

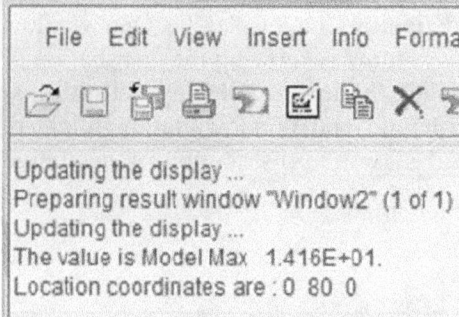

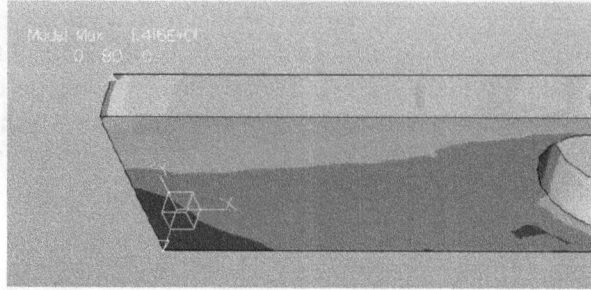

At the end of Chapter 2, let us summarize the basic steps required in the process of performing finite element analysis. As illustrate in the following flow chart, we start with the creation of a 3D solid model. Afterwards, we need to get into a FEA software program. Under the FEA software program, we specify the material properties, define the constraint and load condition(s), and generate the mesh (a subject will be discussed in Chapter 3). Finally, we perform finite element analysis and review the obtained results.

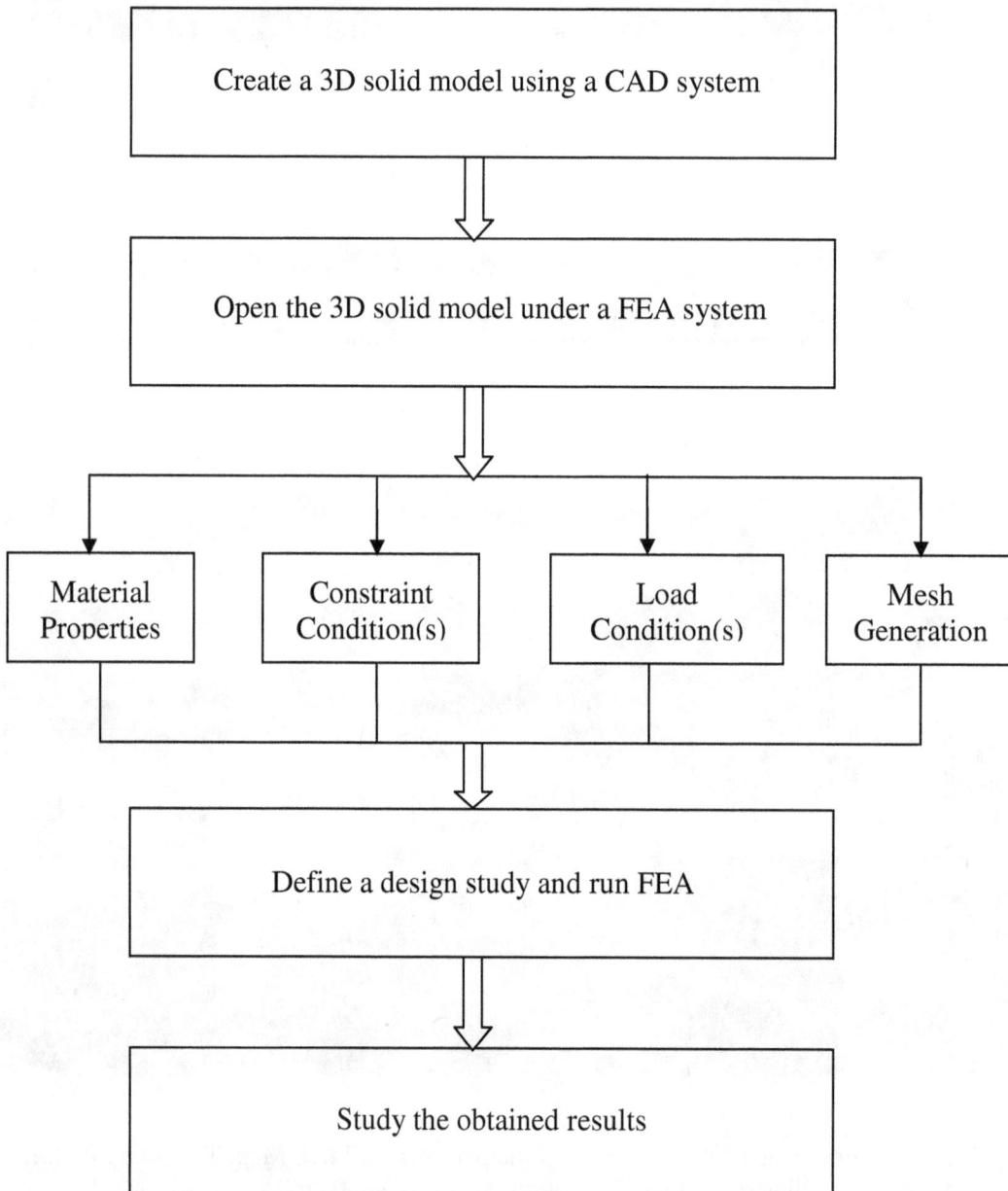

Create a 3D solid model using a CAD system

Open the 3D solid model under a FEA system

| Material Properties | Constraint Condition(s) | Load Condition(s) | Mesh Generation |

Define a design study and run FEA

Study the obtained results

2.5 References

1. K. J. Bathe, <u>Finite Element Procedure,</u> Prentice Hall, Upper Saddle River, New Jersey, 1996.
2. J. W. Dally and W. F. Riley, Experimental Stress Analysis, McGraw Hill, 1965.
3. J. W. Dally, Design Analysis of Structural Elements, Goodington, New York, NY, 2003.
4. Y. Y. Hsieh, <u>Elementary Theory of Structure,</u> 2nd edition, Prentice-Hall, Englewood Cliffs, NJ, 1982.
5. M. A. N. Hendriks, H Jongedijk, J. G. Rots and W. J. E. van Spange (eds), <u>Finite Elements in Engineering and Science,</u> DIANA Computational Mechanics, 1997.
6. K. H. Huebner, D. L. Dewhirst, D. E. Smith, and T. G. Byrom, <u>The Finite Element Method for Engineers</u>, 4th edition, John Wiley & Sons, Inc., 2001.
7. C.S. Krishnamoorthy, <u>Finite Element Analysis, Theory and Programming</u>, 2nd Ed., 1995.
8. R. L. Norton, <u>Machine Design: An Integrated Approach</u>, Prentice Hall, Upper Saddle River, New Jersey, 1996.
9. T. H. H. Pian and P. Tong, <u>Basis of Finite Element Methods for Solid Continua,</u> Int. J. Numer. Methods Eng., Vol. 2, No. 7, April 1964, pp. 1333-1336.
10. T. Ross, <u>Advanced Applied Finite Element Methods,</u> Horwood Publishing Limited, Chichester, England, 1998.
11. S. P. Timoshenko and D. H. Young, <u>Theory of Structure</u>, McGraw-Hill, Kogaksha Limited, 1965.
12. O. C. Zienkiewicz, <u>The Finite Element Method</u>, third edition, McGraw-Hill, New York, NY, 1977.

2.6 Exercises

1. The geometry of the cantilever beam is shown below:

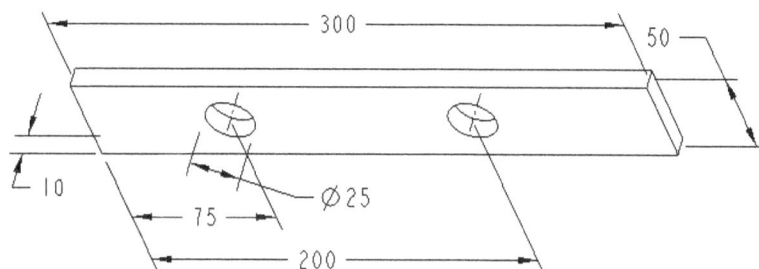

The material of the beam is steel. Analyze the stress and displacement distributions when subjected to

(1). Uniformly distributed pressure load. The magnitude is 0.2 N/mm^2.

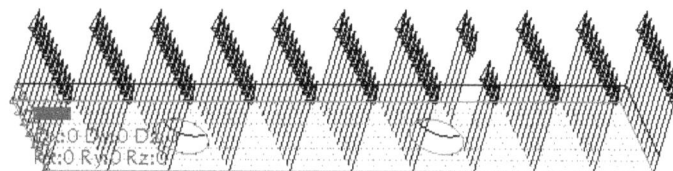

(2). Uniformly distributed load acting on the top surface. The magnitude is 1000 Newton.

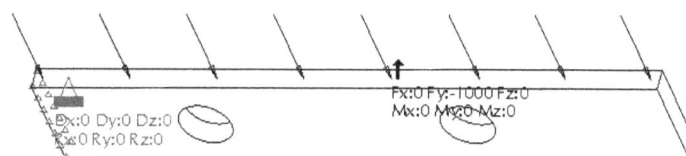

2. The geometry and dimensions of a plate component are shown below. Note that the geometry is symmetric with respect to the horizontal axis. The plate is subjected to an axial load during operation. The material of the plate is aluminum. The modulus of elasticity is 70 GPa. The Poisson's ratio is 0.33.

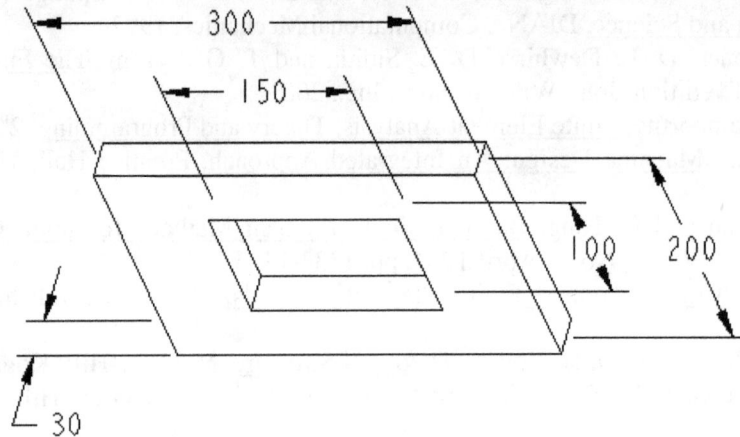

Assume that the left side of the plate is fixed, and the end on the right side is free to move. A tensile force is acting on the right side. The magnitude of the force is 500 Newton.

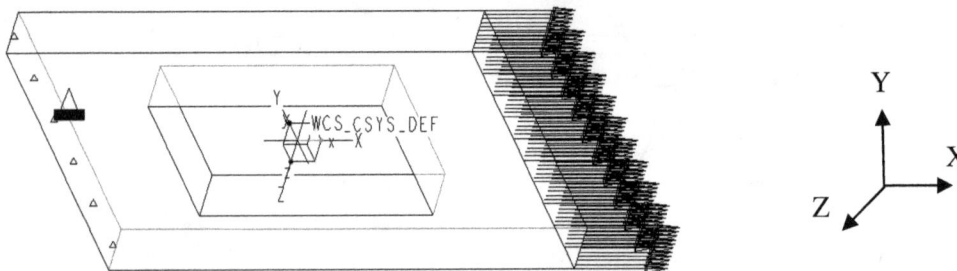

Analyze the stress and displacement distributions when subjected to the load of 500 Newton. The student has the choice to select the type of stress and the displacement in one of the three directions (x, y, z) or the total magnitude of the displacement.

3. Referring to the column structure, do the following:
    (1) Find the location (specify the values of the x, y and z coordinates of the location) of the maximum displacement on the column when subjected to this compressive load and specify its magnitude.
    (2) Find the location of the maximum value of the maximum principal stress and the location of the minimum value of the maximum principal stress on the column when subjected to this compressive load and specify their magnitudes. Make sure that you specify the values of the x, y and z coordinates of the location in each of those two cases.

4. A cantilever beam is shown below. Assume that the material type is steel. The left end of the beam is a fixed end and the right end is a free end. The over hanged length is 200 mm, the height is 30 mm and the width is 10 mm. The beam is subjected to a uniformly distributed load. The magnitude of the unit load or load per area is given by 0.5 Newton/mm², or the total load is 1000 Newton acting on the top surface of the beam.

Load intensity: q = 0.5 N/mm²

Fixed end

(200, 30, 10)

Free end

(0, 0, 0)

Y

X

Z

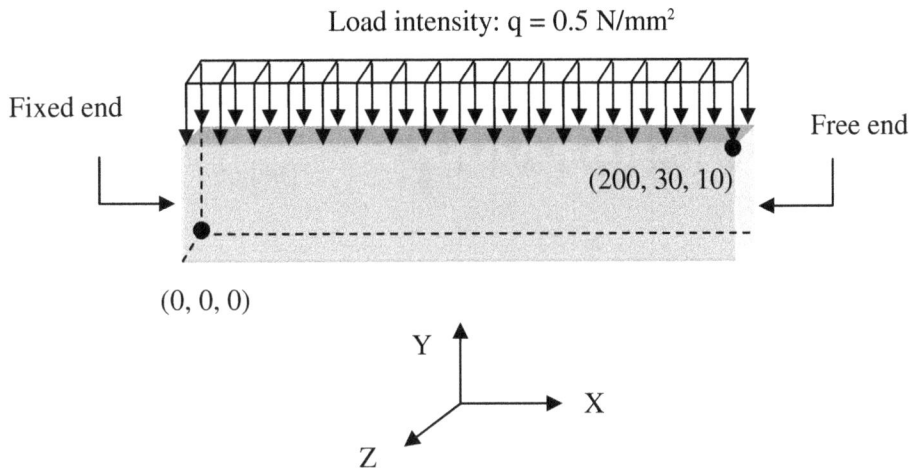

Find the location of the maximum displacement in the y-direction on the beam when subjected to this uniformly distributed load and specify its value with a positive or negative sign. Also find the maximum displacement in magnitude. Compare these 2 values. Are they the same in numerical value?

(1) Find the location of the maximum stress in the y-direction when subjected to this uniformly distributed load and specify their magnitude. Also Find the location of the maximum stress in the x-direction when subjected to this uniformly distributed load and specify their magnitude.

(2) Verify the obtained displacement value and the stress value as compared with the 2 values given by the following two formulas:

$$y_{max} = \frac{qL^4}{8EI}$$

$$\sigma_{max} = \frac{Mc}{I} = \frac{(\frac{1}{2}qL^2b)c}{I}$$

Where E = 200 GPa, L = 200 mm, c = 15 mm and I is given by

$$I = \frac{1}{12}bh^3 = \frac{1}{12}(10)(30)^3 \qquad (mm^4)$$

Are there any differences between the results obtained from FEA and the results obtained from the calculations using the formulas? List the possible reasons causing the observed differences.

When evaluating the magnitude of the uniform load. Pay attention to the unit used in the evaluation process. The units used in the textbook of Mechanics of Materials are shown below:

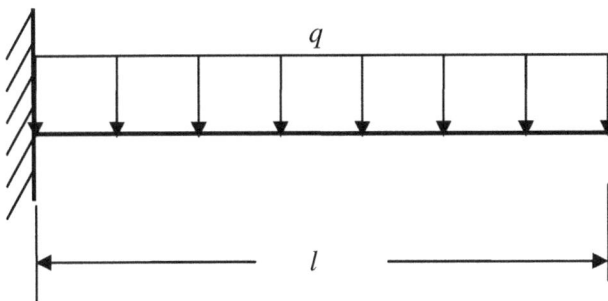

q

l

Unit of l: mm

Unit of q: Newton/mm

The units used in Pro/Mechanica in the process of evaluating the uniform load are listed below:

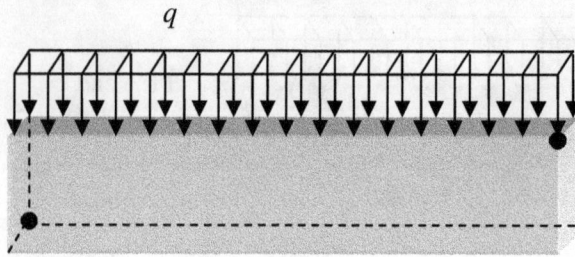

Unit of A: mm^2

Unit of q: Newton/mm^2

5.   Run the program in question 3 for 2 more times. Before running the program, modify the Poisson's ratio value listed in the window of material properties. Change the value to 0.1 and run it. Also change the value to 0.05 and run it. Examine the obtained results. Are you observing the difference(s)? Are you learning something new?

	Poisson's Ratio	σ_{xx} (Mpa)
case 1	0.27 (steel in the library)	
case 2	0.1	
case 3	0.05	

CHAPTER 3

MODELING OF LOAD CONDITIONS

3.1   Introduction

In Chapter 2, two examples were used to demonstrate the basic steps in the process of performing finite element analysis. Readers may notice the loads were uniformly distributed. In the engineering field, quite a few load conditions are not uniformly distributed. Furthermore, it is not uncommon to observe that a load acts on part of a surface, such as the one shown in Figure 3-1, namely, uniformly distributed load acting on part of the top surface of a beam structure.

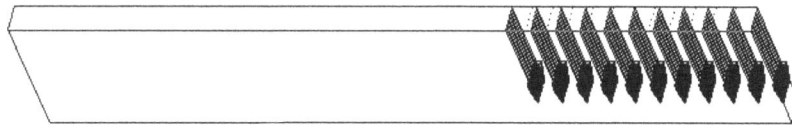

Figure 3-1 A Uniformly Distributed Load Acting on Part of the Top Surface

To model such a load condition, there is need to split the top surface of the beam structure so that the uniformly distributed load only acting on the right portion of the top surface. Sometimes the magnitude of a load varies on the area on which it is acting. Figure 3-2 presents 2 load cases as just described.

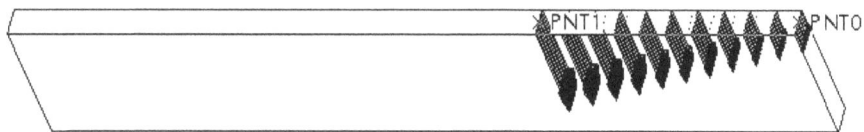

(a) A Non-Uniformly Distributed Load Acting on Part of the Top Surface

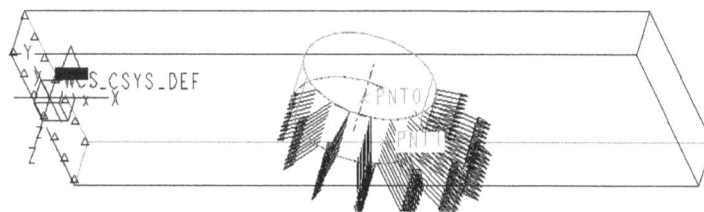

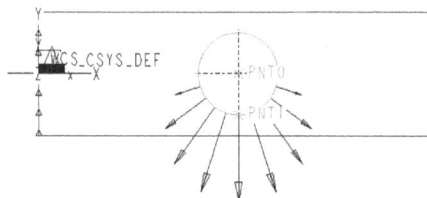

(b) Bearing Load

Figure 3-2 A Variety of Loads Observed in Engineering

3.2 Modeling a Uniformly Distributed Load with a Surface Region

A cantilever beam is shown below. As illustrated, a load is acting on part of the top surface. The load is a uniformly distributed load and the magnitude of the total load is 200 Newton, acting downward.

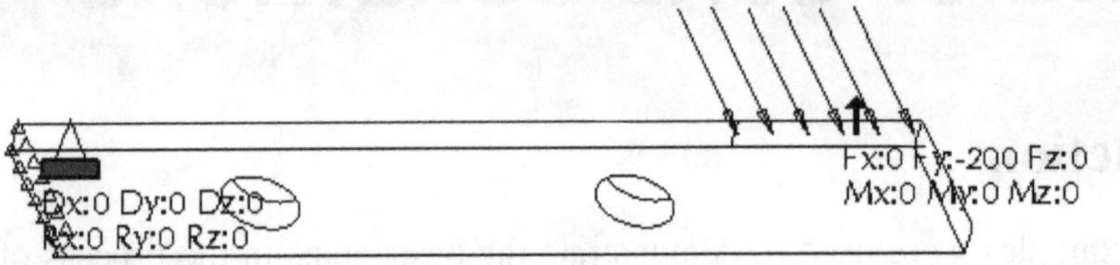

The geometrical shape and dimensions are shown in the following figure. The three dimensions of the beam are 300, 50 and 10 mm, respectively. The diameters of the two through holes are equal. The diameter dimension is 25 mm. The 2 dimensions of the surface area, on which the load is acting, are 60 mm and 10 mm, respectively.

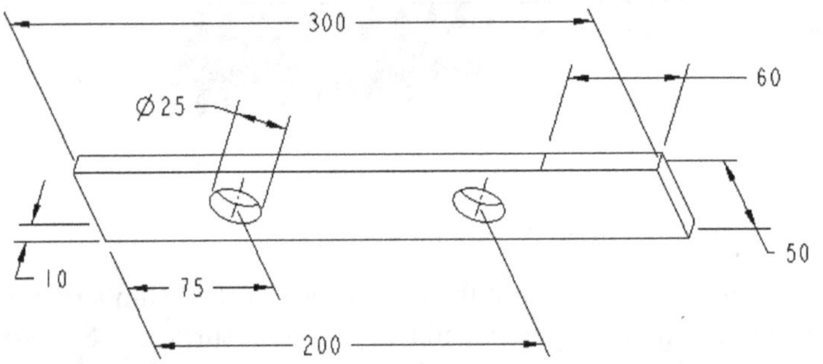

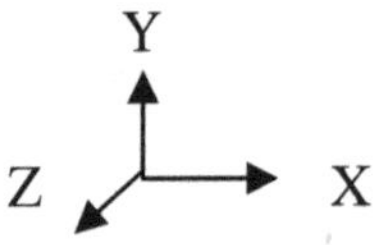

It is assumed that the material of the beam is steel. There are 2 objectives to be expected from performing FEA: the deflection pattern and the tensile stress distribution is the x direction.

Step 1: Create a file for the 3D solid mode.
File > New > Part > type *beam_2* as the file name and clear the icon of **Use default template.** Select **mmns_part_solid** (units: Millimeter, Newton, Second) and type *partial surface* in **DESCRIPTION**, and *student* in **MODELED_BY**, then **OK.**

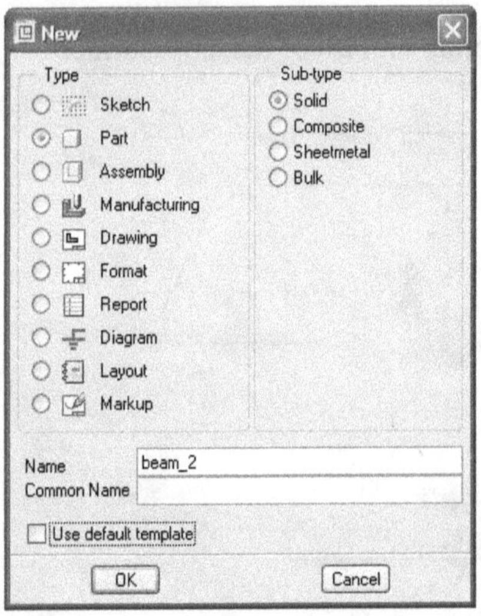

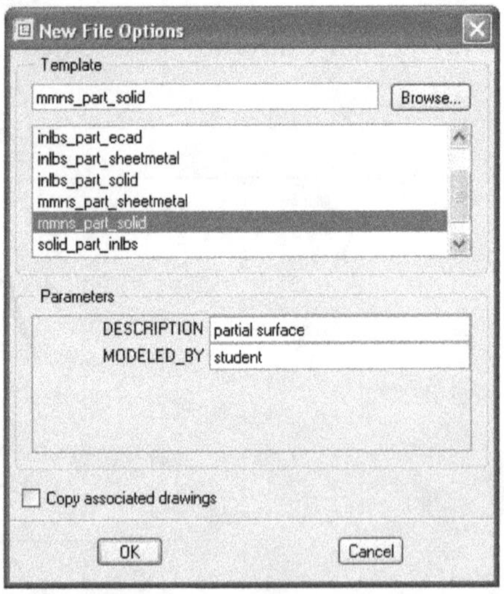

Click the icon of **Extrude**. Select Symmetry when specifying the depth value. Specify 10 as the depth value. Click the box of **Placement > Define**.

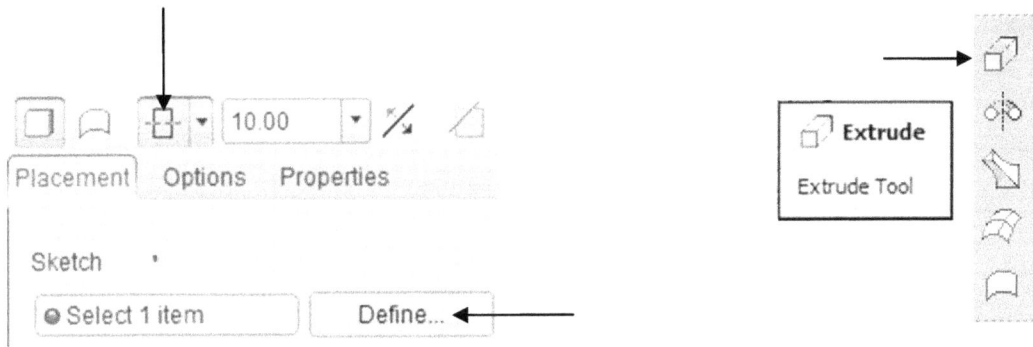

Select the **FRONT** datum plane displayed on screen as the sketch plane. Pick the box of **Sketch** to accept the **RIGHT** datum plane as the default reference to orient the sketch plane, as illustrated below:

Pick **FRONT** as the sketch plane

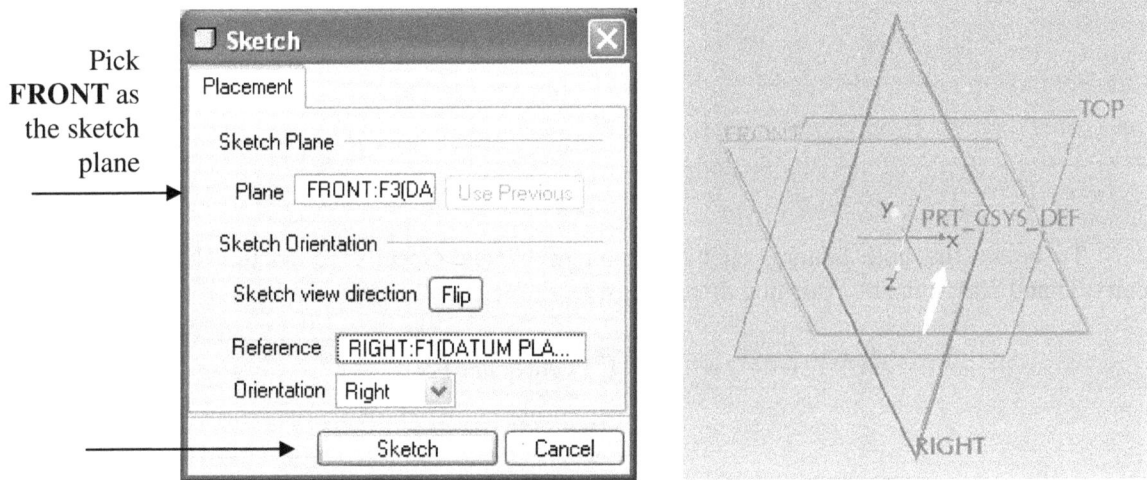

Click the icon of **Centerline**. Sketch a horizontal centerline along the axis, as shown below.

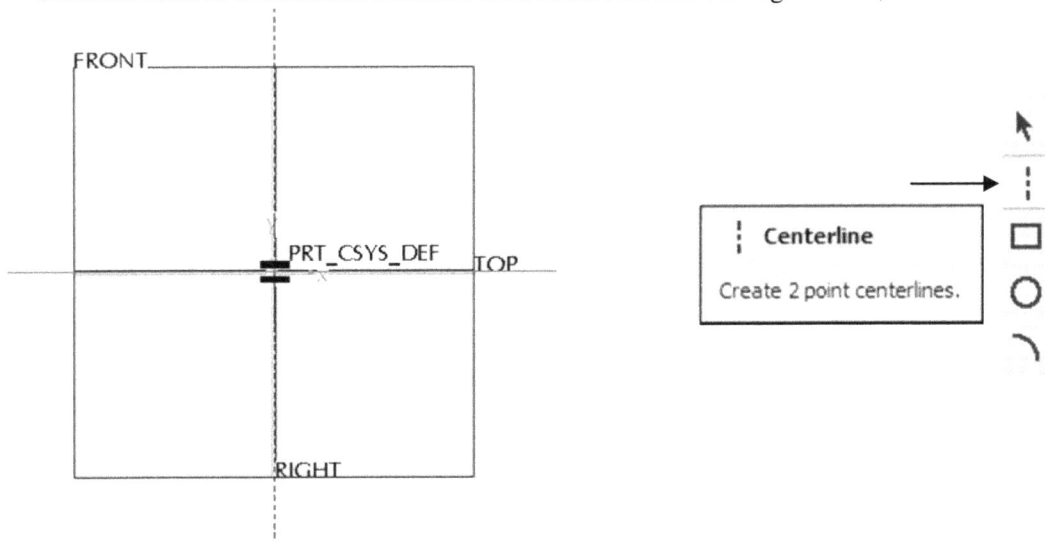

Click the icon of **Rectangle** and sketch a rectangle with the dimensions equal to 300 and 50, respectively. Keep the rectangle symmetric about the horizontal centerline.

□ **Rectangle**

Create rectangle.

300.00

H

50.00

V

V

H

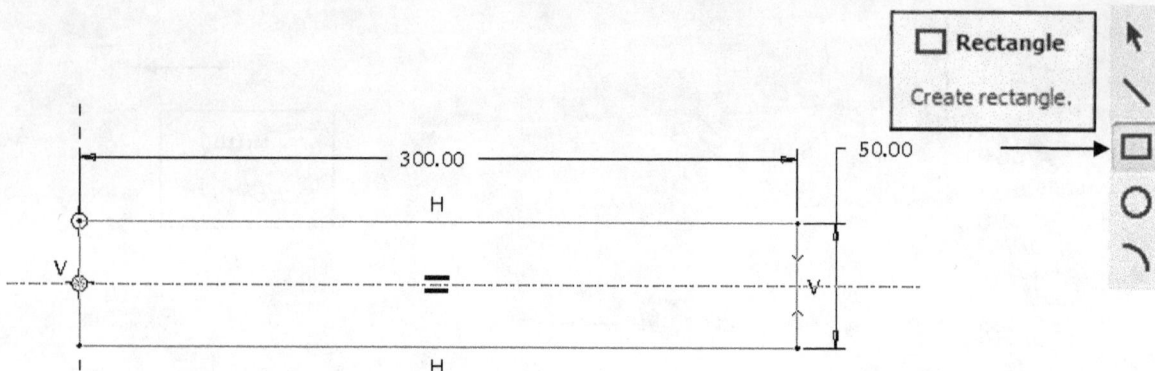

If your sketch is not symmetric about the horizontal centerline, pick the icon of **impose sketcher constraints** from the toolbar of sketcher, and select the icon of **vertices symmetric** from the pop up window called **Constraints** > click the horizontal centerline and click the 2 vertices so that the 2 vertices are symmetric about the horizontal centerline.

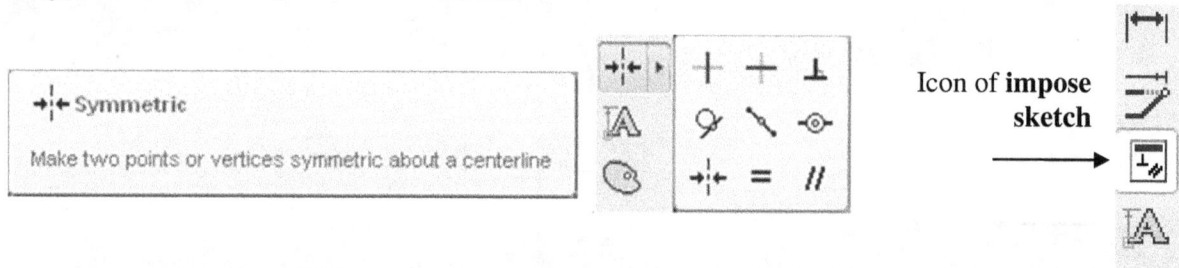

Symmetric

Make two points or vertices symmetric about a centerline

Icon of **impose sketch**

To create the hole feature, click the icon of **Circle.** Specify 25 as the diameter dimension. Specify 75 and 200 as the two position dimensions, as shown.

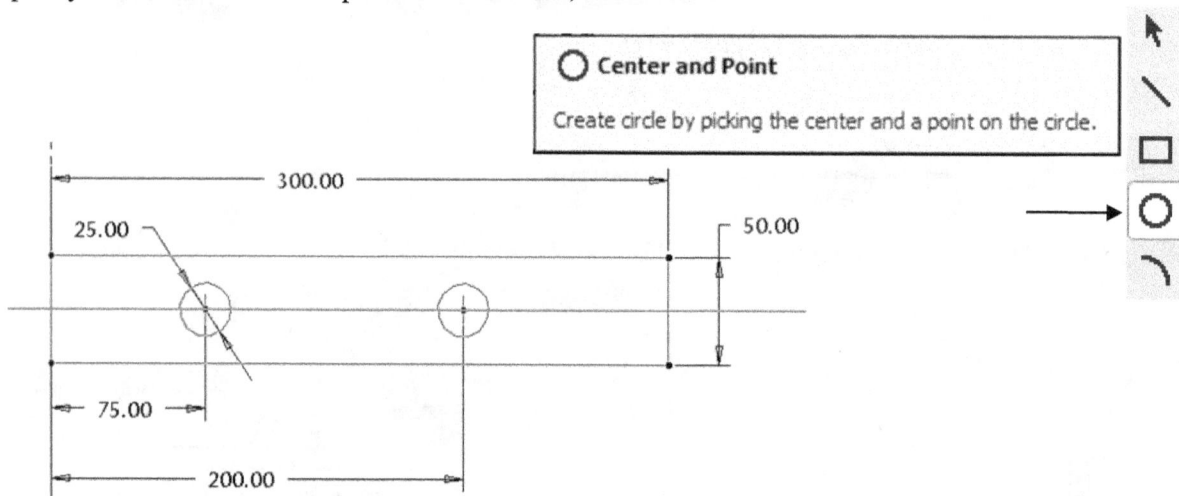

○ **Center and Point**

Create circle by picking the center and a point on the circle.

300.00

25.00

50.00

75.00

200.00

Upon completing the sketch, select the icon of **Done** and the icon of **Apply and Save**.

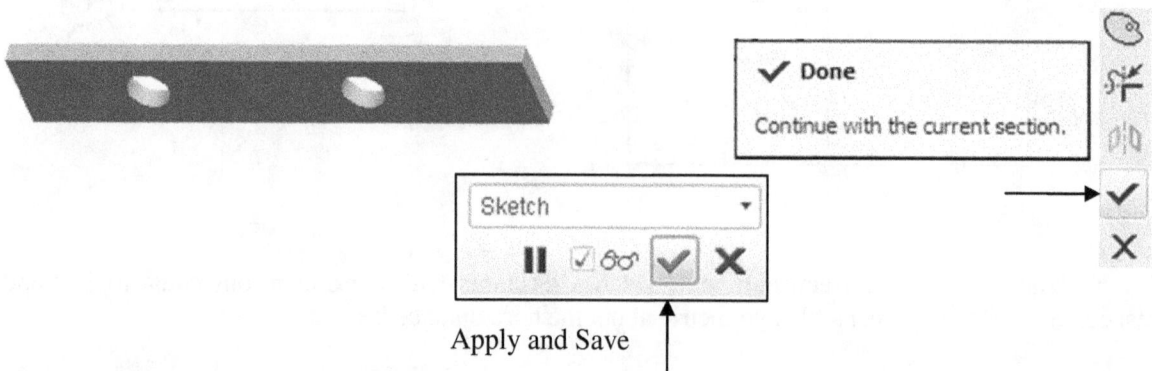

✓ **Done**

Continue with the current section.

Sketch

Apply and Save

Step 2: Create a datum curve so that a surface region can be specified for loading.

Let us first introduce the concept of a surface region. Assume that the surface is a rectangle shown below. The right portion of the rectangle is the area where a load acts. The left portion of the rectangle is the area where no load acts. As a result, a split operation is needed. After splitting, the right portion becomes a surface region.

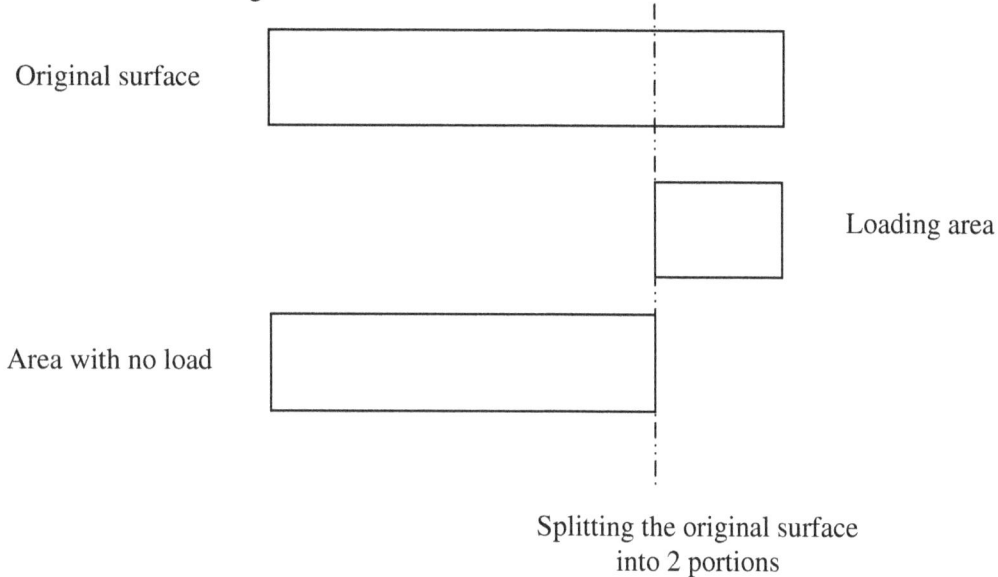

Original surface

Loading area

Area with no load

Splitting the original surface
into 2 portions

There are two steps in creating a surface region. The first step is to sketch the boundary of the surface region on the original surface. The second step is to split the original surface into 2 regions based on the sketched boundary.

To sketch the boundary, select the icon of **Sketch Tool** > Pick the top surface of the block as the sketch plane > accept the default orientation of the sketch plan.

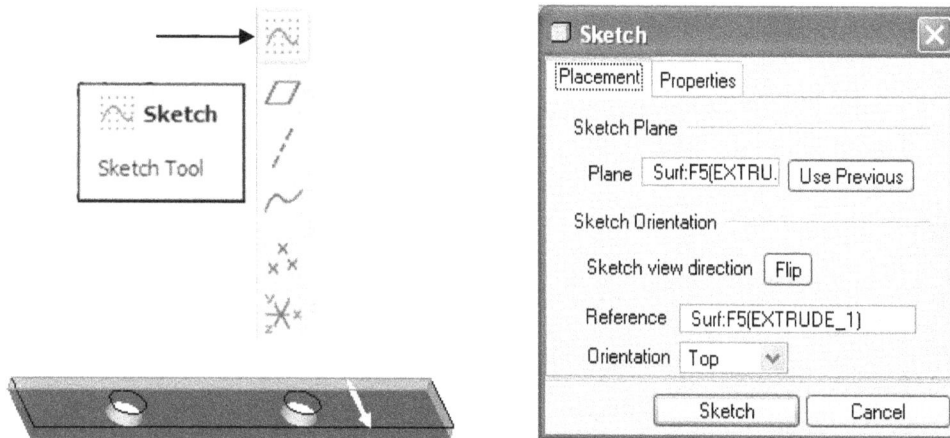

Before making a sketch, let us first add 2 new references. Click **Sketch**. These 2 new references are the surface at the right end and the surface in the front of the block, as shown. Afterwards, close the reference window.

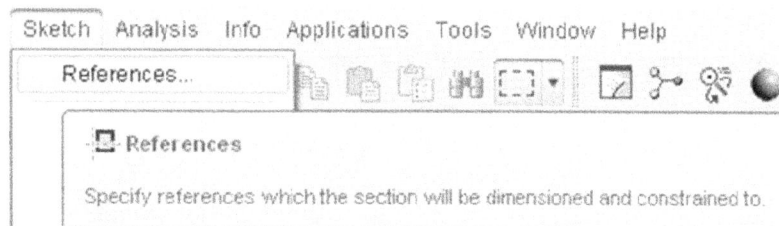

References

Surf:F5(EXTRUDE_1)
F1(RIGHT)
Surf:F5(EXTRUDE_1)
Surf:F5(EXTRUDE_1)

Select Use Edge/Offset   Delete

Reference status

Fully Placed

Close

Select this surface
as a new reference

PRT_CSYS_DEF

Select this surface
as a new reference

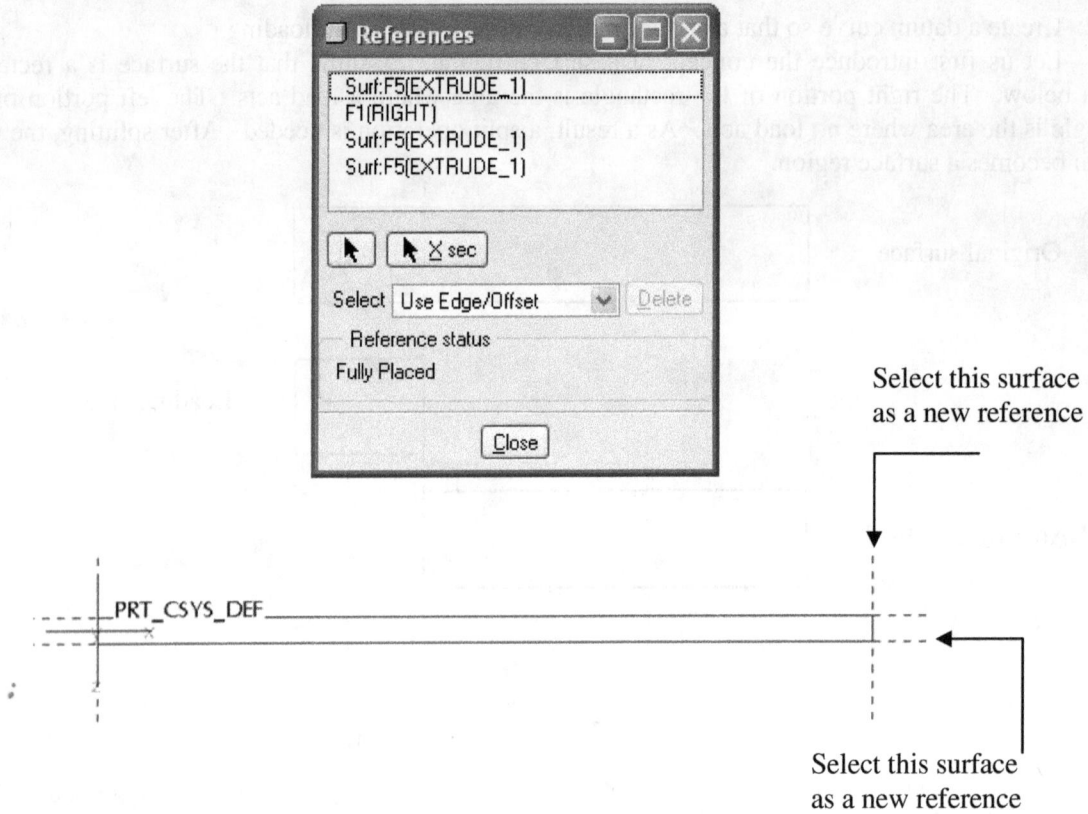

Select the icon of **Rectangle** from the toolbar of sketcher and draw a rectangle with the dimension of 60. Only one dimension is needed because the other sides of the sketched rectangle are aligned with the selected references. Click the icon of **Done** from the toolbar of sketcher.

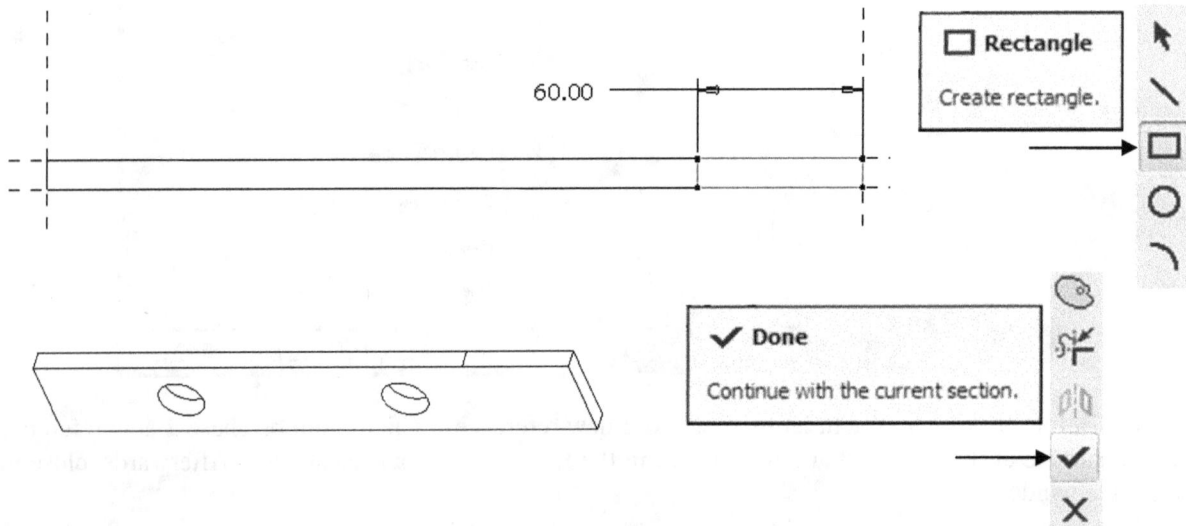

60.00

☐ **Rectangle**

Create rectangle.

✓ **Done**

Continue with the current section.

Step 3: Under the Pro/MECHANICA, perform FEA.

From the main toolbar or the menu toolbar, select **Applications > Mechanica.**

Continue after checking the unit system > **Structure > OK**.

In the **Materials** window, select **STEEL** from the left side called Materials in Library > click the directional arrow so that the selected material type goes to the right side called Materials in Model.

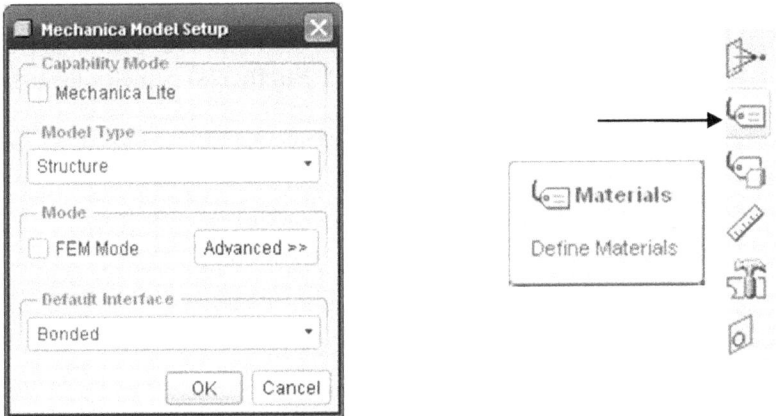

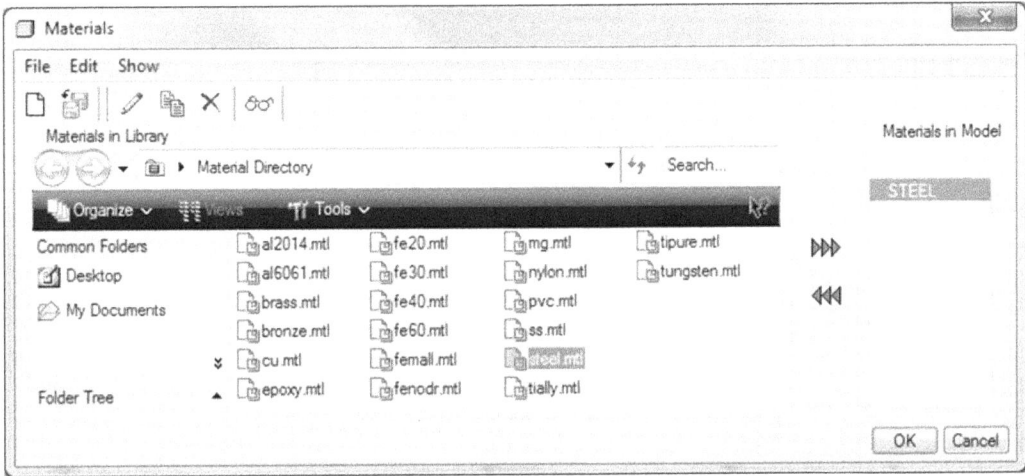

Click the icon of **Material Assignment**, the pop-up window indicates there is one component and the material type is STEEL. Just click **OK**.

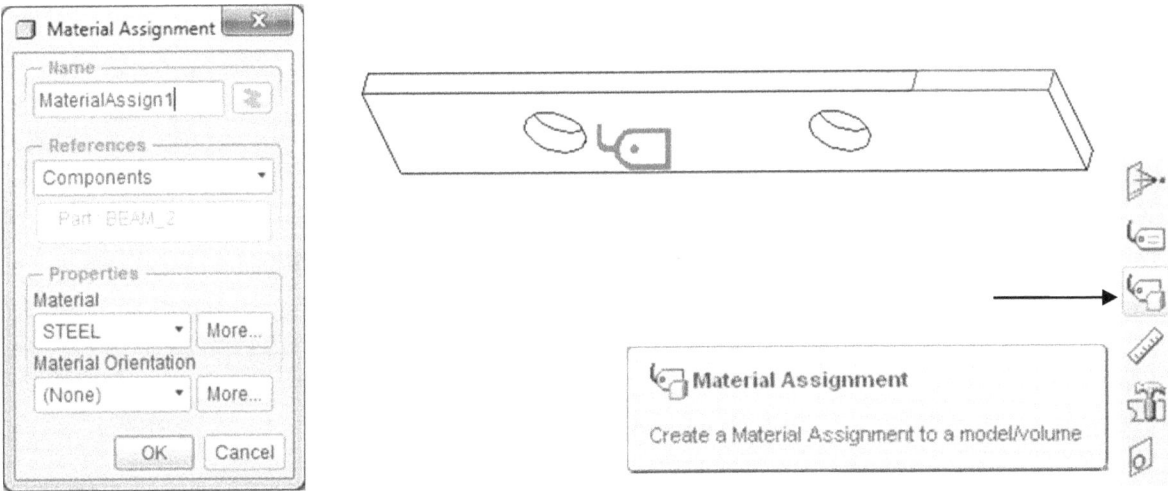

Step4:   Under the Pro/MECHANICA environment, define the fixed end constraint condition.
From the toolbar of functions, select the icon of **Constraints >** select the surface at the left end of the block as the fixed end, and accept the default settings, namely, fixing all six degrees of freedom.

To define the fixed end constraint on the left side of the beam, select the icon of **New Displacement Constraint** from the toolbar of functions > **Surface(s)** > pick the surface at the left side of the beam > **OK** to accept the surface selection.

Because the constraint condition is "a fixed end", we set all 6 degrees of freedom equal to zero, which is the default setting > **OK**.

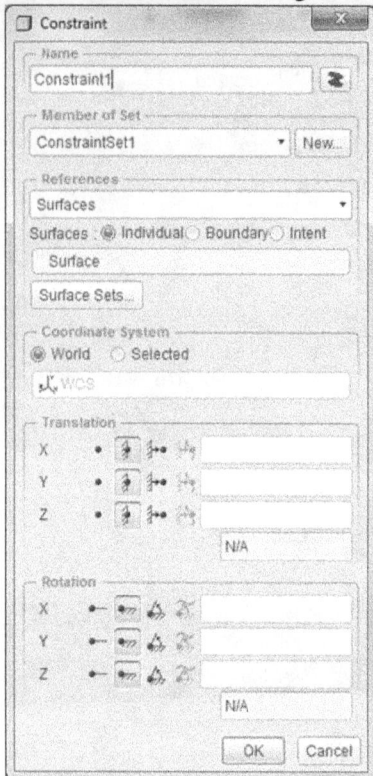

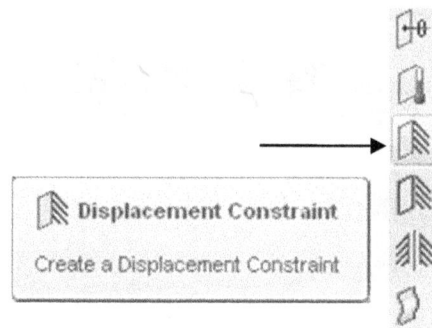

Step 5: Under the Pro/MECHANICA, define the load condition.

We first define a surface region, on which the uniformly distributed load will be defined.

Click the icon of **Surface Region**.

Click the datum curve of rectangle > click the surface being split.

Click the icon of **Apply and Save**.

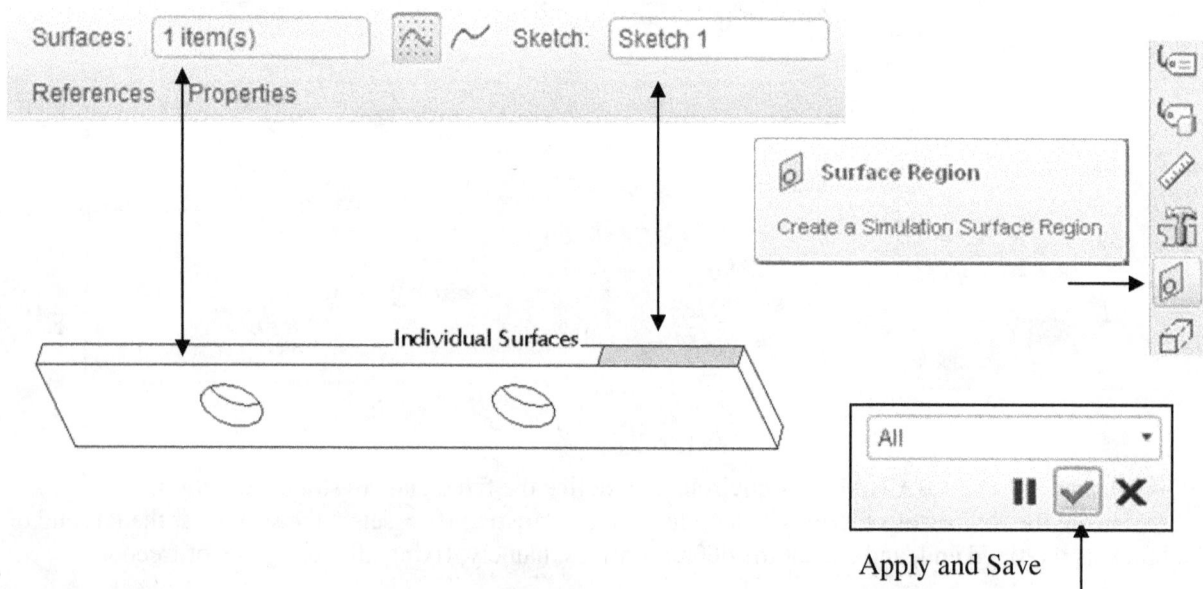

Apply and Save

To define the load condition, click the icon of **Force/Moment Loads > Surfaces >** pick the defined surface region.

Type -200 as the Y Force Component > **OK**.

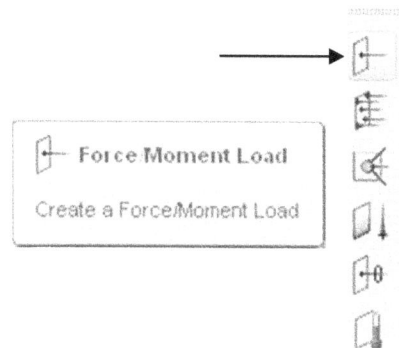

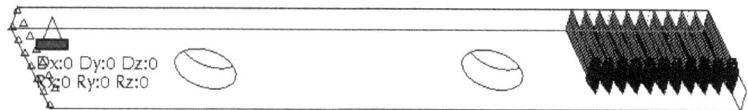

Step 6: Under the Pro/MECHANICA, set up **Analyses** and **Run.**

Select the icon of **Run a Design Study > File > New Static >** type in beam_2 > **OK**.

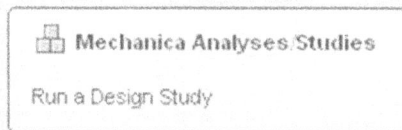

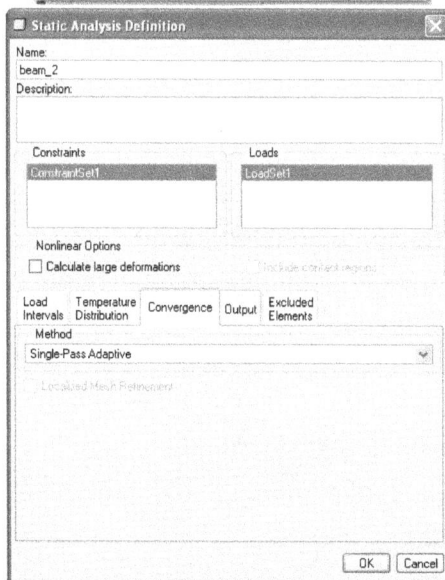

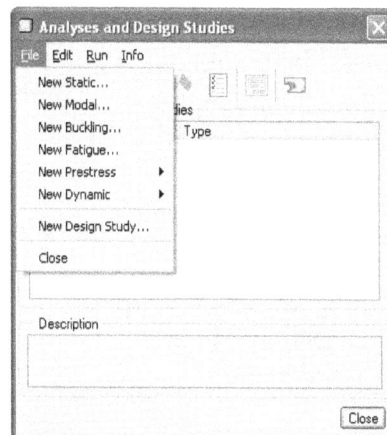

Before running the program, select Settings and specify a sufficient memory space, say 128 Megabytes for RAM Allocation. Run the program

There is always a message displayed on screen. The message is asking the user "Do you want to run interactive diagnostics?" Click **Yes**. In this way a warning message will appear if an abnormal operation occurs.

To monitor the computing process, users may click the icon of **Display study status**.

```
        min_stress_prin:    -1.680435e+01
        strain_energy:       6.668429e+00

Analysis "beam_2" Completed   (19:55:12)

------------------------------------------------

Memory and Disk Usage:

    Machine Type: Windows NT/x86
    RAM Allocation for Solver (megabytes):

    Total Elapsed Time (seconds): 1.43
    Total CPU Time      (seconds): 0.92
    Maximum Memory Usage (kilobytes): 24549
    Working Directory Disk Usage (kilobytes

    Results Directory Size (kilobytes):
    1161 .\beam_2

    Maximum Data Base Working File Sizes (k
    2048 .\beam_2.tmp\kel1.bas

------------------------------------------------
Run Completed
Mon May 23, 2011    19:55:12
```

Step 7: Under the Pro/MECHANICA, study the **Results**
Select the icon of **Results**
Type in Beam Deflection as the title > locate the Design Study folder called beam_2.
Select **Fringe** > choose **Displacement** > **Magnitude.**

Click Displace Options > Deformed > Show Element Edges > **OK and Show**.

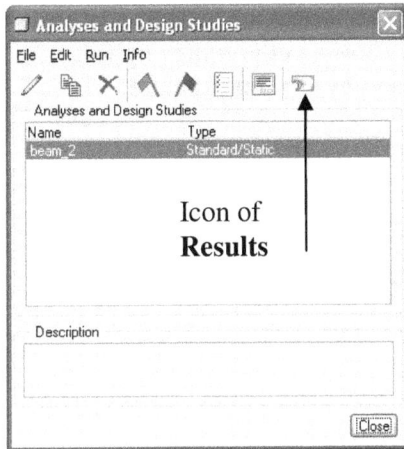

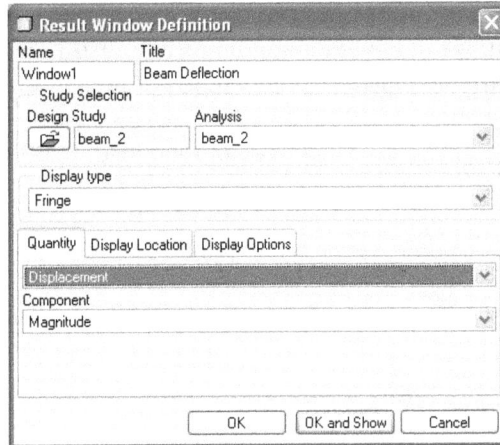

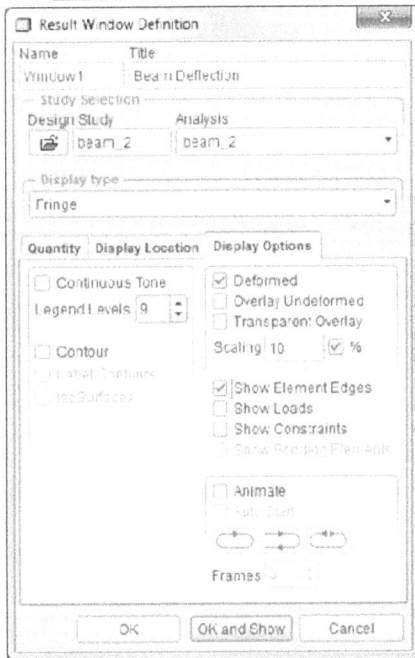

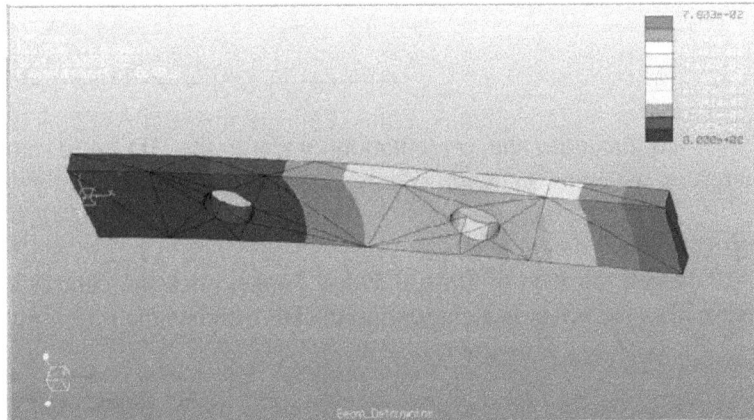

To study the distribution of stress, say the stress distribution in the x direction, select **Edit** from the main toolbar > **Copy** and a new Result Window appears > **Stress** > **XX** > **OK and Show**.

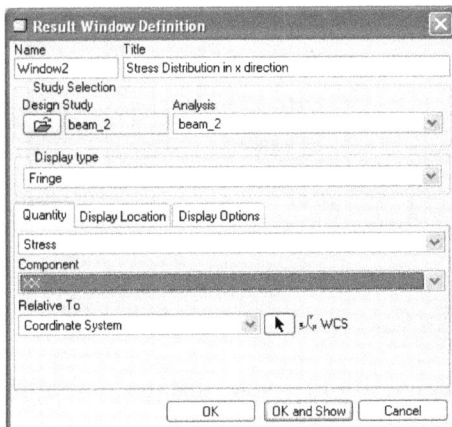

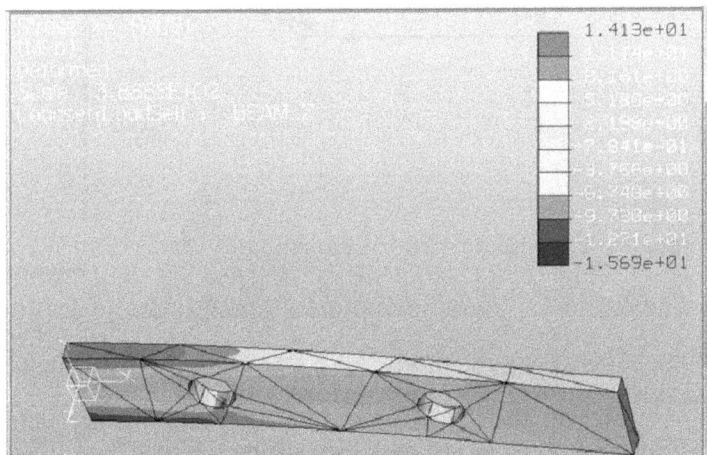

3.3   Modeling a Non-uniformly Distributed Load

In this section, we deal with a non-uniformly distributed load. As shown in the following figure, a non-uniformly distributed load is acting on part of the top surface of a cantilever beam structure.

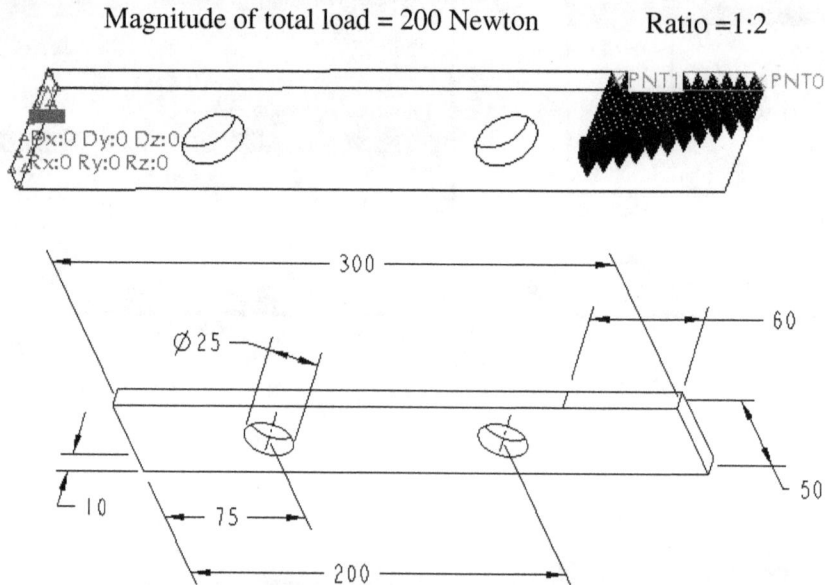

Magnitude of total load = 200 Newton   Ratio =1:2

To facilitate the presentation, we use the 3D solid model created in the previous section. Therefore, we focus on the procedure to define the non-uniformly distributed load.

Step 1: Create 2 datum points
       Click the icon of **Datum Point Tool** > click the central location on the right edge, as shown > type 0.50 as the offset value to define **PNT0**, which is the mid-point of the picked edge > **OK**.

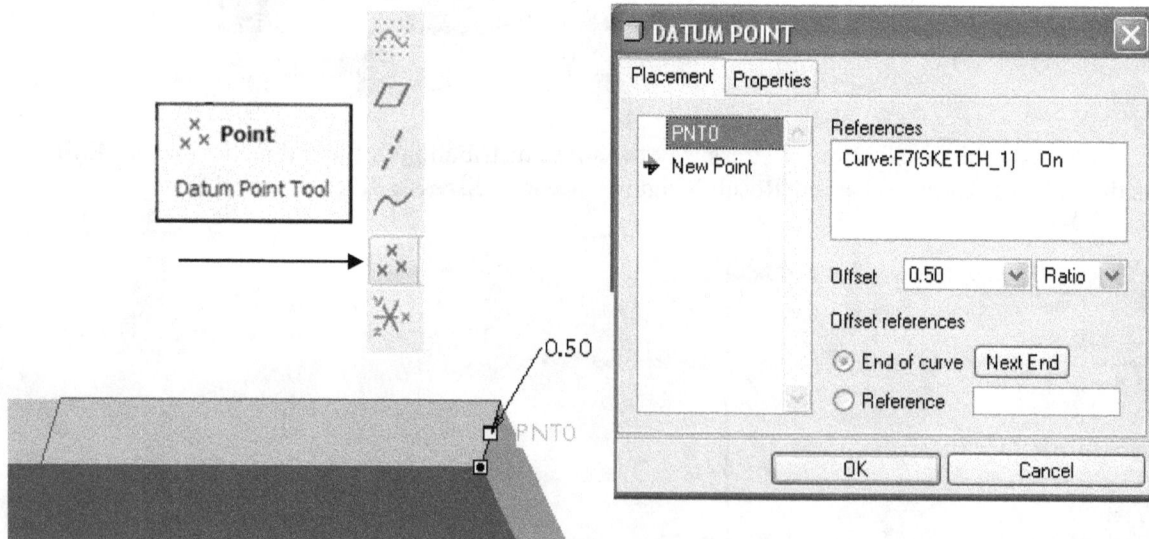

       Repeat the above procedure to create another datum point, PNT1. Note that the uniformly distributed load previously defined is still presented.

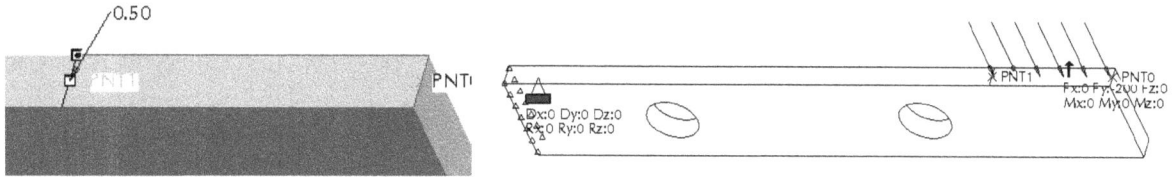

Step 2: Define a non-uniformly distributed load

Now let us define the non-uniformly distributed load. Click the icon of Force/Moment Load. To distinguish this load condition, click **New** under Member of Set > a new window called **Load Set Definition** appears and **LoadSet2** is automatically specified > type *non-uniformly distributed load* > **OK** to close the **Load Set Definition** window.

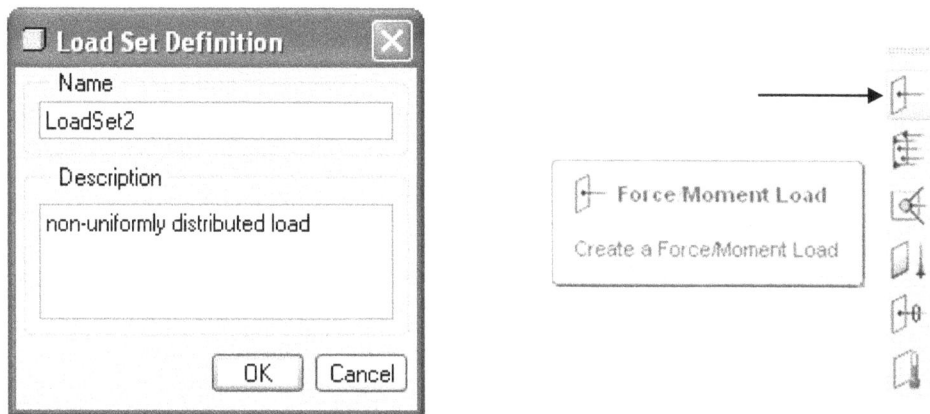

In the **Force/Moment Load** window, select **Surface(s) >** select the surface region just defined > **OK**.

Click **Advanced** > change Uniform to **Interpolated Over Entity** > **Define** and a new window called **Interpolation Over Entity** appears.

Add > pick **PNT0**, while holding down the **Ctrl** key, pick **PNT1** > **OK** > **Done/Return.**

Type the ratio numbers, namely, 1 and 2, and **OK.** Type -200 as the Y force component > **Preview** > **OK**.

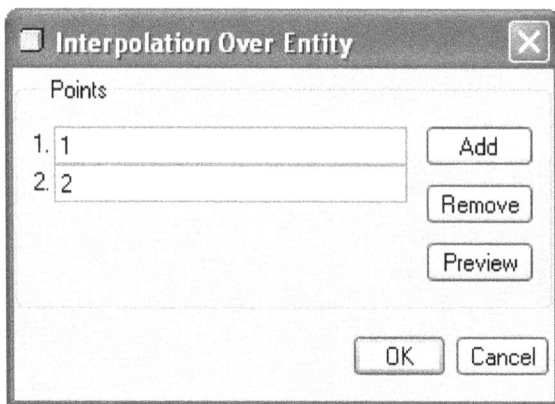

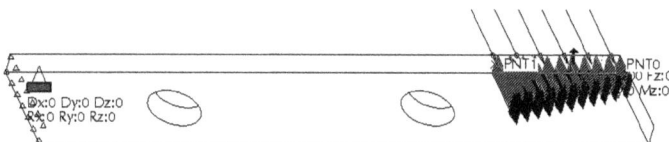

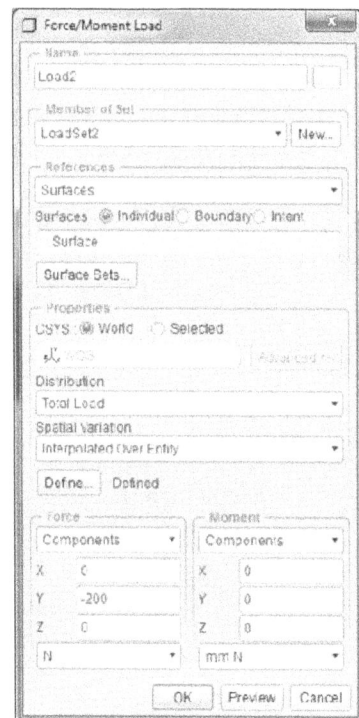

Step 3: Under the Pro/MECHANICA, set up **Analyses** and **Run** it
Select the icon of **Run an Analysis > File > New Static >** type *beam_3*. It is very important that you select LoadSet2 alone (do not select LoadSet1) > **OK**.

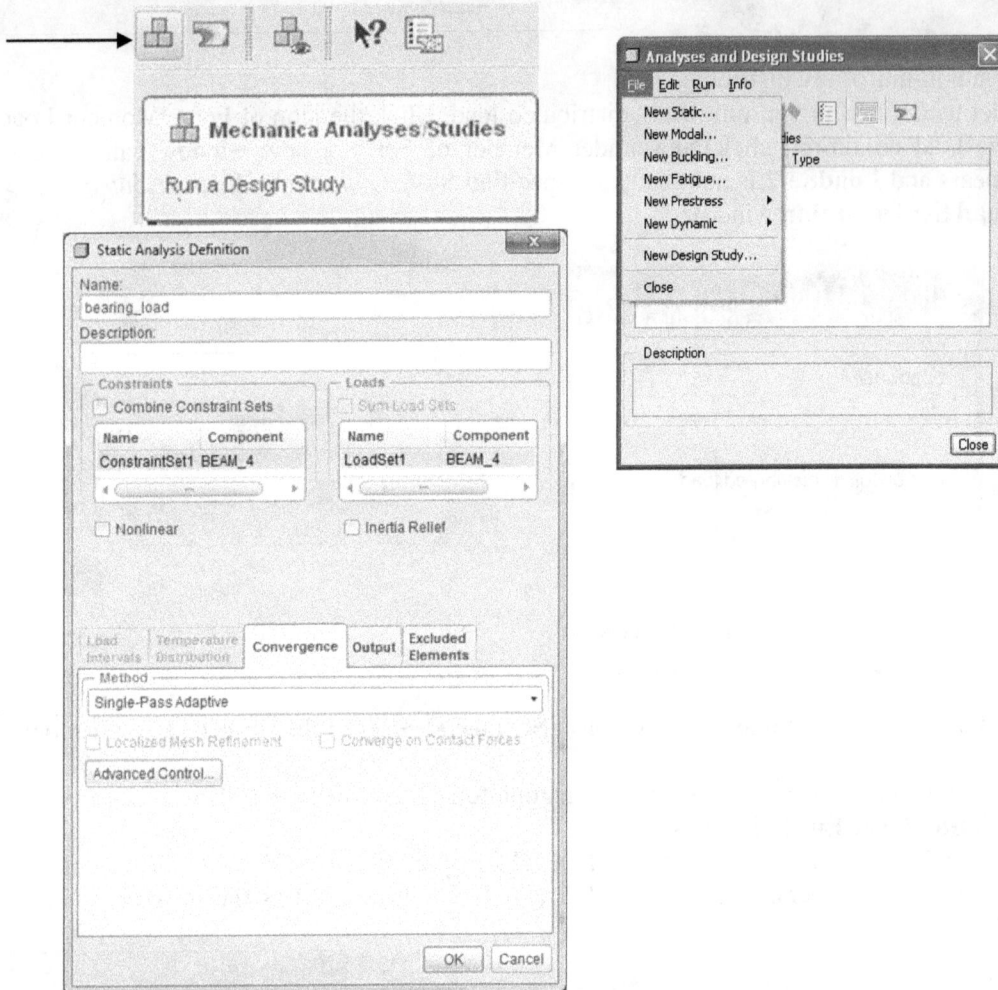

Run the program and afterwards, select the icon of **Results.** Type in *Beam Deflection* as the title > locate the Design Study folder called beam_3.
Select **Fringe >** choose **Displacement > OK and Show**.

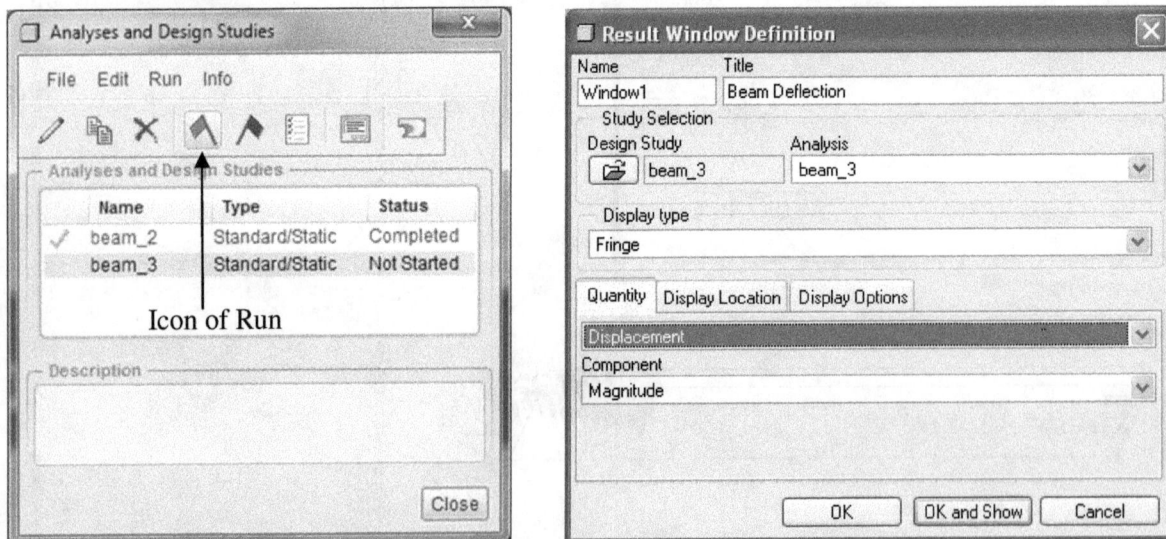

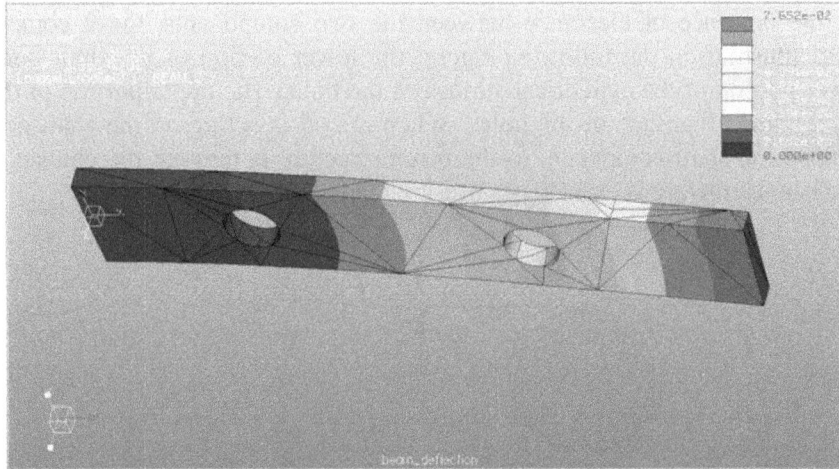

To study the distribution of stress, say the stress distribution in the x direction, select **Edit** from the main toolbar > **Copy** and a new Result Window appears > **Stress** > **XX** > **OK and Show**.

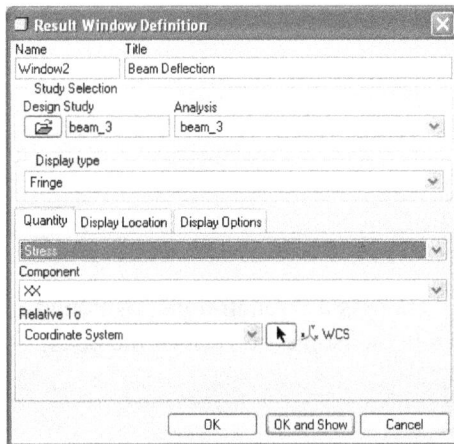

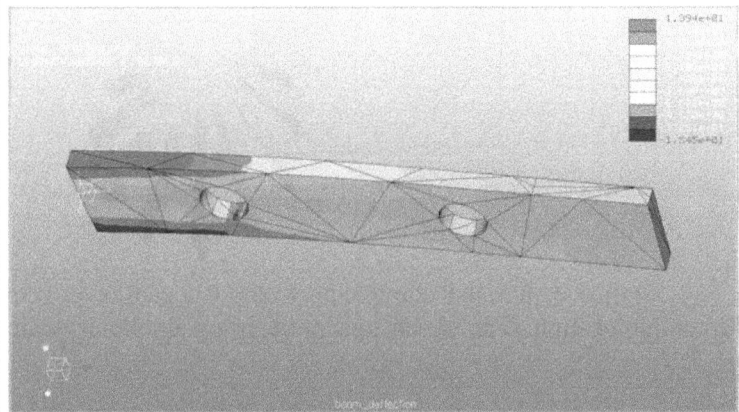

3.4   Modeling of a Bearing Load Condition

Figure 3-4 illustrates that a shaft component is being assembled with a beam structure. The beam structure has a through hole. The diameter dimension of the hole is slightly larger than the diameter dimension of the shaft so that a clearance fit can be assured during the assembly process.

(a) two components before assembling   (b) two components after assembling

Figure 3-4 Clearance Fit of 2 Components during Assembling

Because of the presence of clearance between the two components, the 2 components are in contact partially. As illustrate in the following figure, the lower portion of the shaft component is in contact with the lower portion of the cylindrical surface of the hole. The upper portion of the shaft is not in contact with the cylindrical surface of the hole. When a load is acting on the shaft component, the action transferring from the shaft component to the beam structure is through the contact area. Such a load condition to the beam structure is termed as a bearing load.

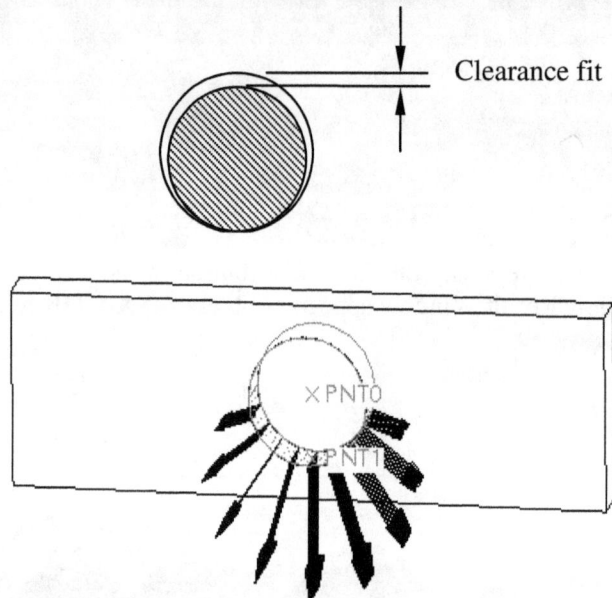

Clearance fit

In this section, the example presented is a cantilever beam. There is a through-hole positioned at the central location. The dimensions of the beam and hole are shown in the isometric view of the beam.

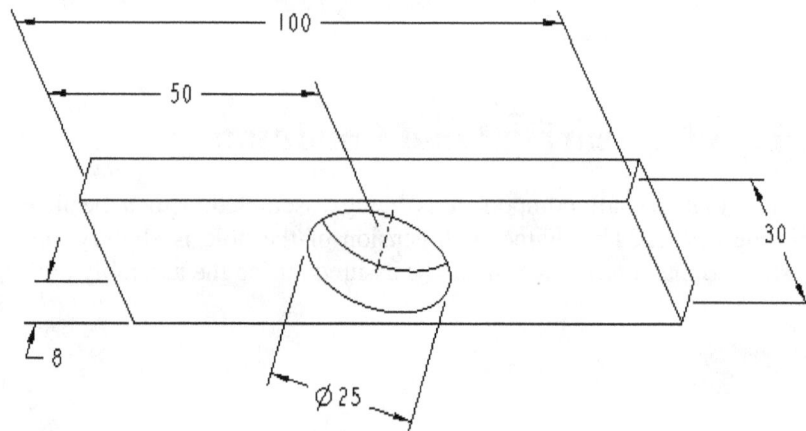

The following constraint and load conditions are assumed:
- The left side of the beam is fixed.
- The right side of the beam is a free end.
- The material of the cantilever is steel.
- The magnitude of the bearing load is 500 Newton, as shown.

The objectives of this study are
(1)   Plot the distribution of the deformation of the entire beam.
(2)   Plot the distribution of the maximum principal stress of the 3D solid model.
(3)   What is the highest value of the maximum principal stress? Where does it occur?

Step 1: Create a 3D solid model to represent the cantilever beam.

File > New > Part > type *beam_4* as the file name. Clear the box called **Use default template**, and select the **mmns_part_solid**. Fill in **DESCRIPTION** by typing in *bearing load*. Fill in **MODELED_BY** by typing your name, say *student* > **OK**.

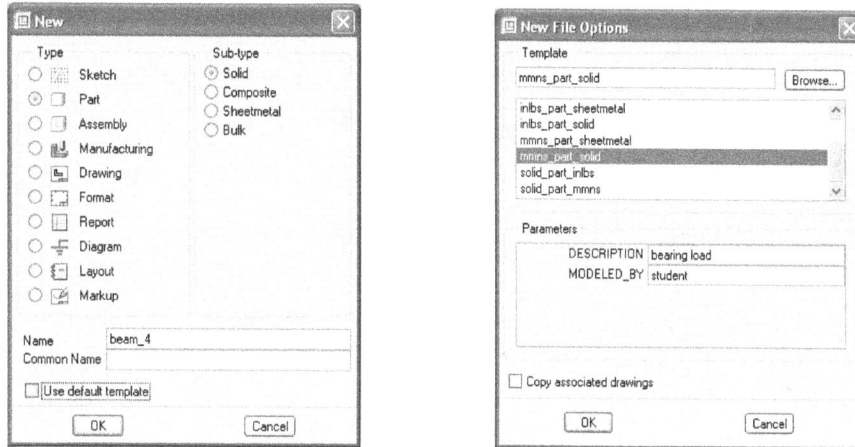

Click the icon of **Extrude**. Specify 8 as the depth of extrusion (the thickness of the beam). To initiate the process of sketching, activate **Placement > Define.**

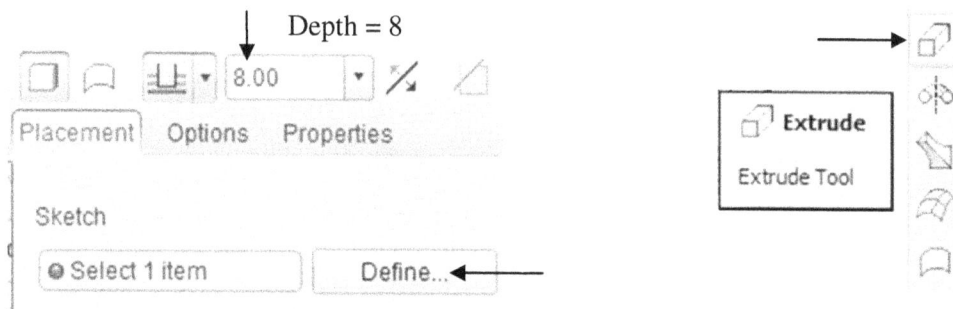

Select the **FRONT** datum plane displayed on screen as the sketch plane. Pick the box of **Sketch** to accept the **RIGHT** datum plane as the default reference to orient the sketch plane, as illustrated below:

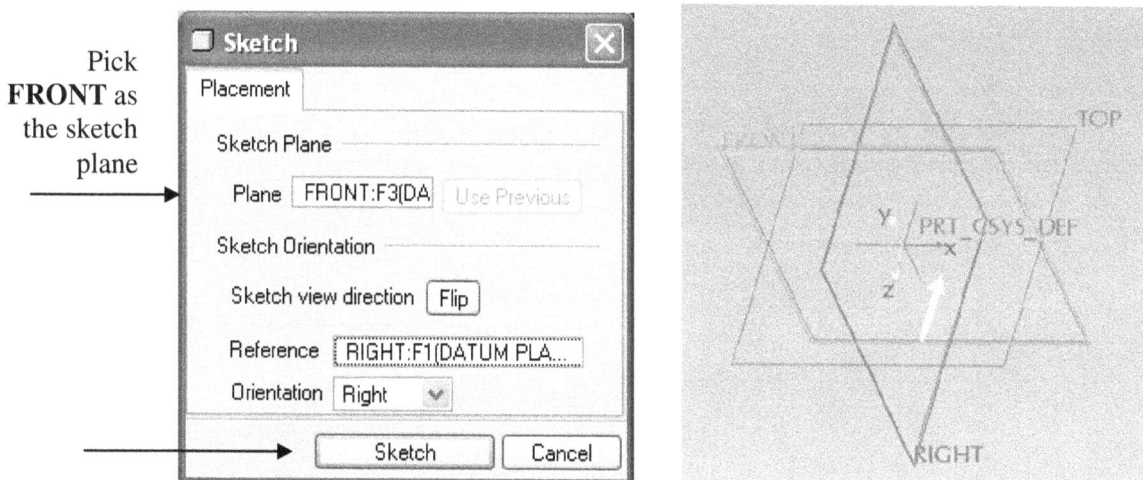

Click the icon of **Centerline**. Sketch a horizontal centerline along the axis, as shown below.

FRONT

PRT_CSYS_DEF   TOP

RIGHT

⋮ **Centerline**	□
Create 2 point centerlines.	○
	⌒

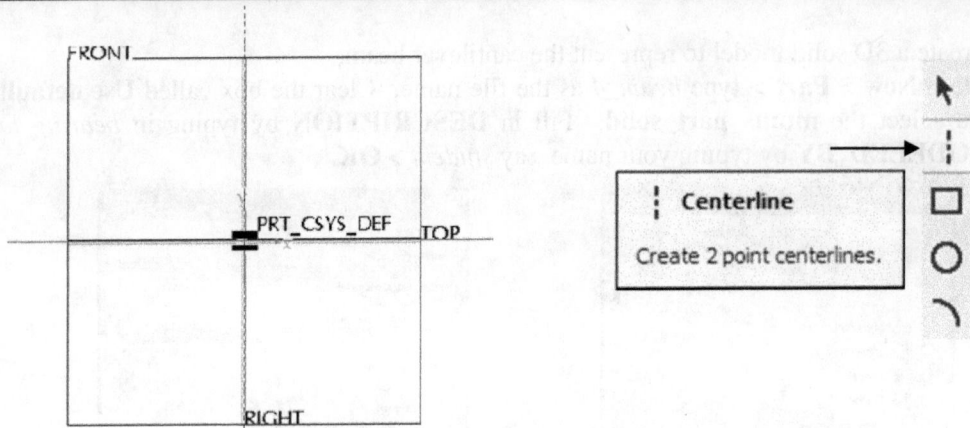

Click the icon of **Rectangle** and sketch a rectangle with the dimensions equal to 100 and 30, respectively. Keep the rectangle symmetric about the horizontal centerline.

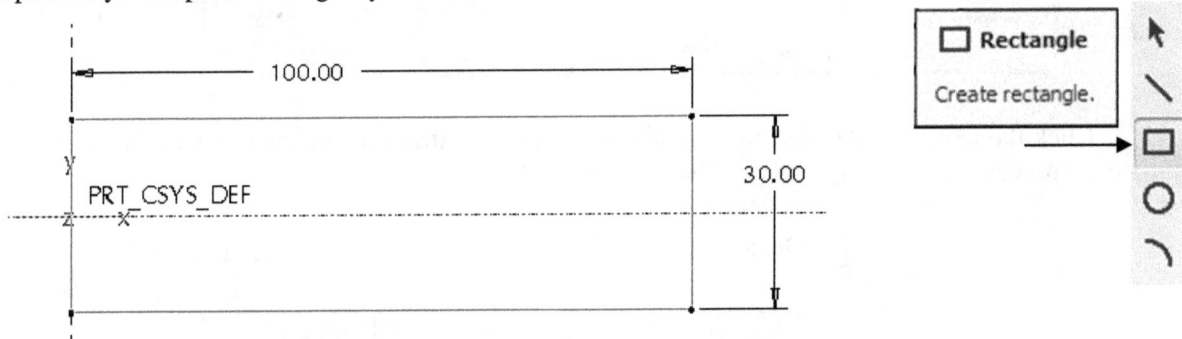

100.00

PRT_CSYS_DEF

30.00

□ **Rectangle**
Create rectangle.

If your sketch is not symmetric about the horizontal centerline, pick the icon of **impose sketcher constraints** from the toolbar of sketcher, and select the icon of **vertices symmetric** from the pop up window called **Constraints** > click the horizontal centerline and click the 2 vertices so that the 2 vertices are symmetric about the horizontal centerline.

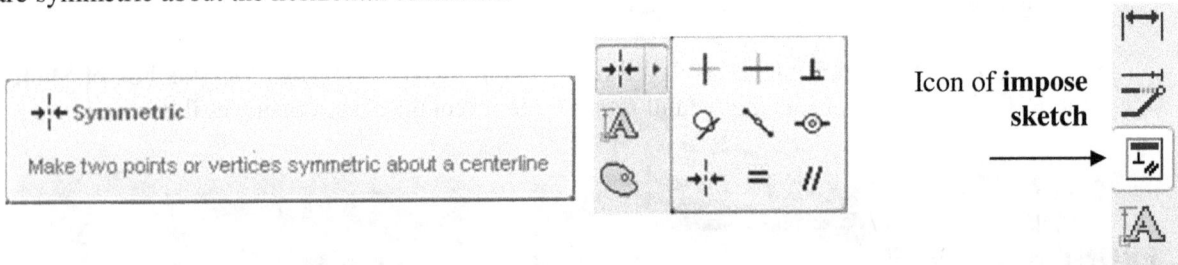

→⋮← **Symmetric**
Make two points or vertices symmetric about a centerline

Icon of **impose sketch**

To create the hole feature, click the icon of **Circle.** Specify 25 as the diameter dimension. Specify 50 as the position dimension, as shown.

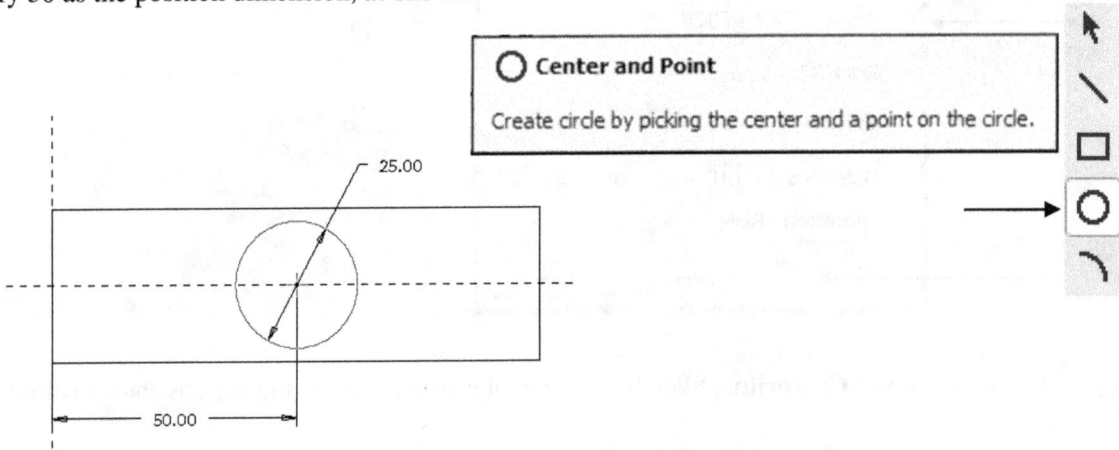

25.00

○ **Center and Point**
Create circle by picking the center and a point on the circle.

50.00

Upon completing the sketch, select the icon of **Done** and the icon of **Apply and Save**.

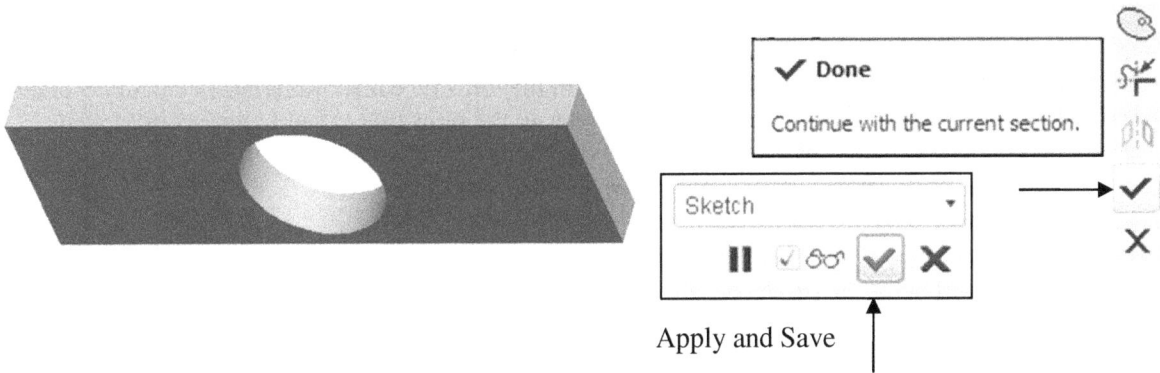

Apply and Save

Up to this point, the process of creating a 3D solid model for the beam structure is completed. Now let us get on to the process of performing FEA.

Step 2: Define two datum points to be used to define the bearing load condition.
To create **PNT0**, click the icon of **Datum Point Tool** from the toolbar of datum creation.
Pick the middle location from the arc or the upper half circle, as shown > **OK**.

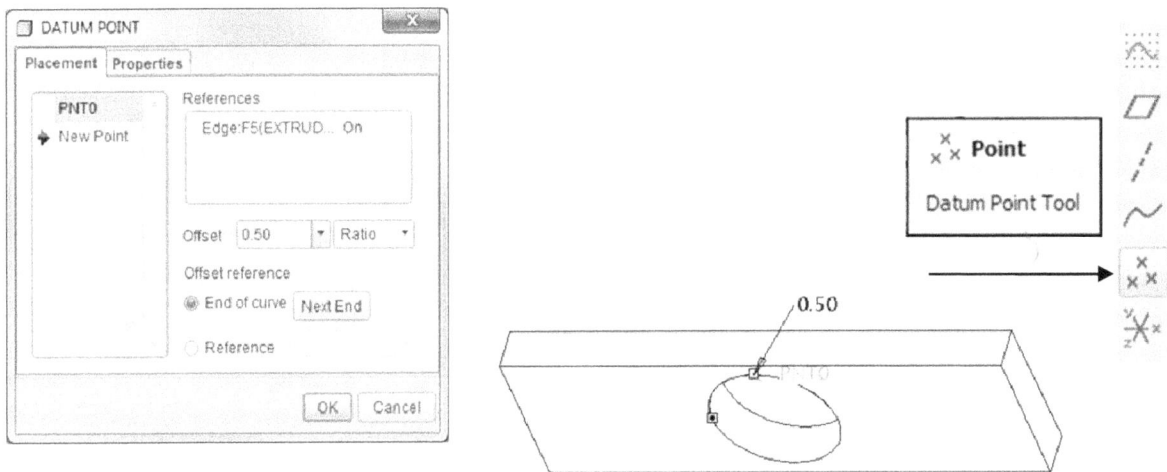

Click the icon of **Datum Point Tool**, again. Pick the arc or the lower half circle, as shown > **OK**.

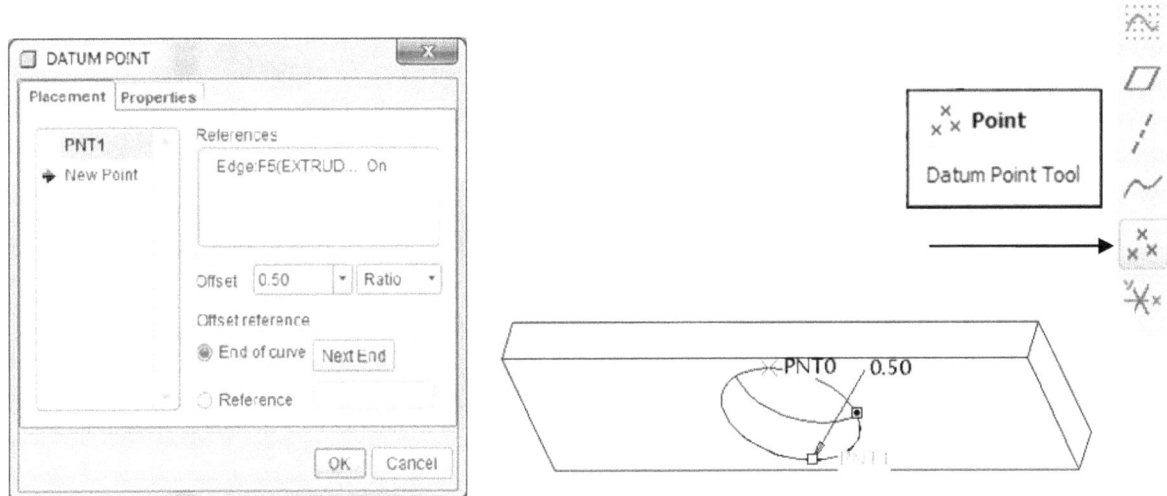

We will use PNT0 and PNT1 to form a directional vector, which is along the vertical direction downward.

Let us shift from the Pro/ENGINEER design system to Pro/Mechanica. From the main toolbar or the menu toolbar, select **Applications > Mechanica > Continue > Structure > OK.**

In the **Materials** window, select **STEEL** from the left side called Materials in Library > click the directional arrow so that the selected material type goes to the right side called Materials in Model.

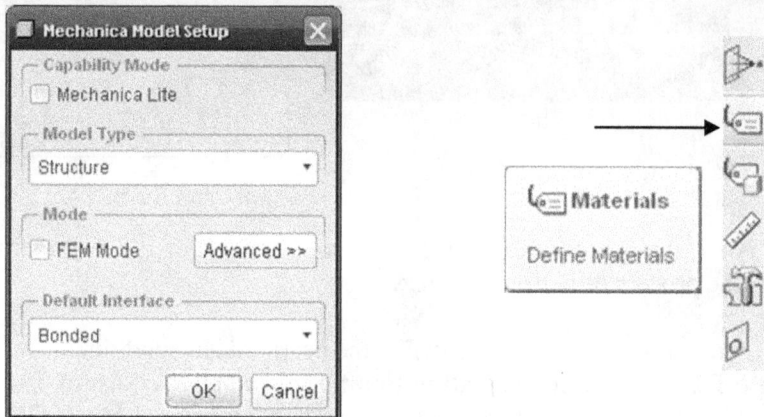

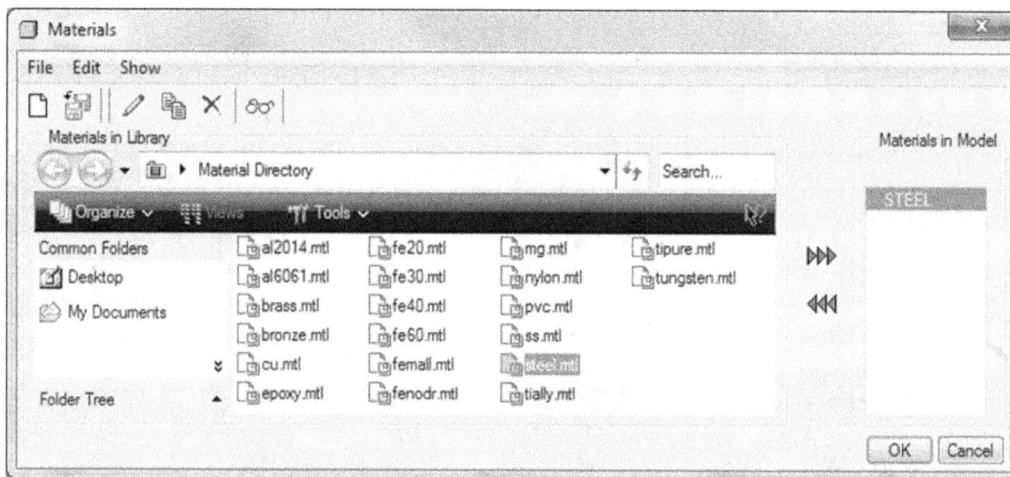

Click the icon of **Material Assignment**, the pop-up window indicates there is one component and the material type is STEEL. Just click **OK**.

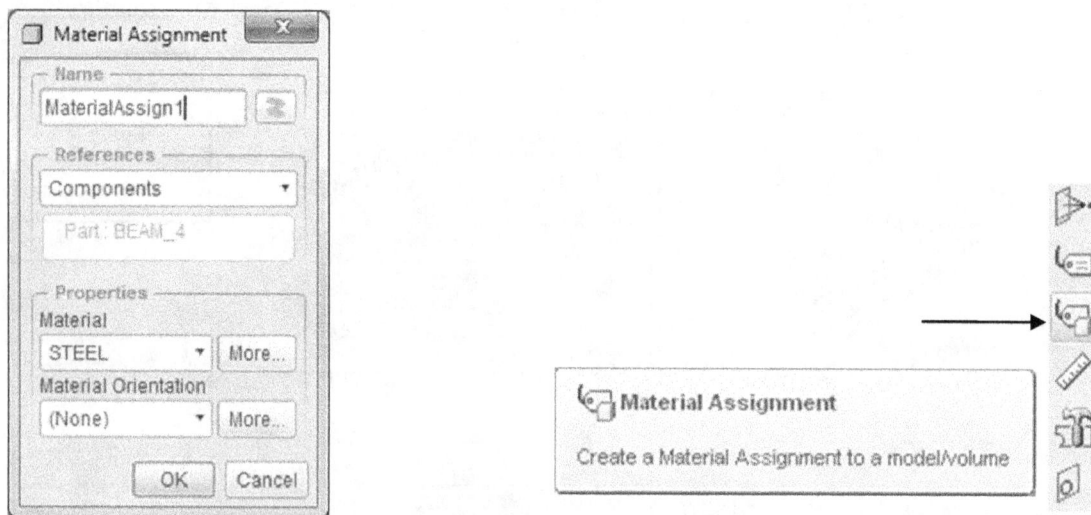

To define the fixed end constraint on the left side of the beam, click the icon of **Constraints >** select the surface at the left end of the block as the fixed end, and accept the default settings, namely, fixing all six degrees of freedom > **OK**.

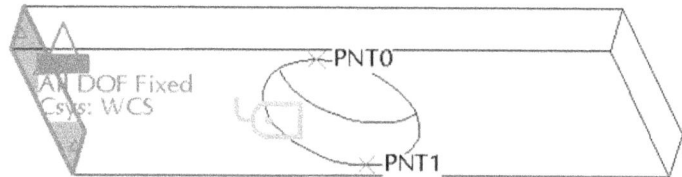

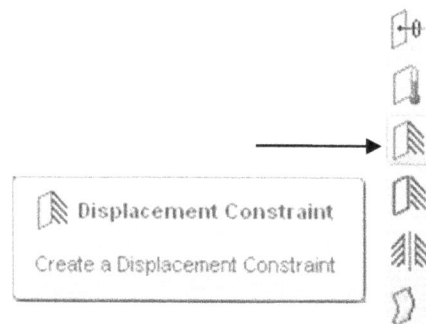

To define the bearing load condition, select the icon of **Bearing Load** > pick the low half surface of the hole > **OK** to complete the bearing hole selection. Select Dir **Points & Mag** from the box called **Force.** Pick PNT0 as the starting point or **From**. Pick PNT1 as the ending point or **To.**
Type 500 as the magnitude value > **OK**.

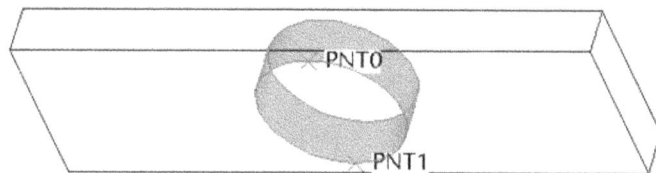

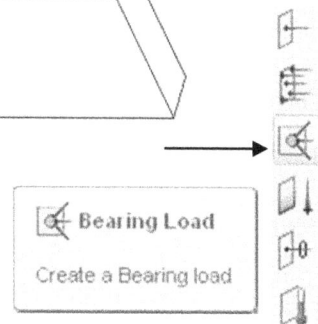

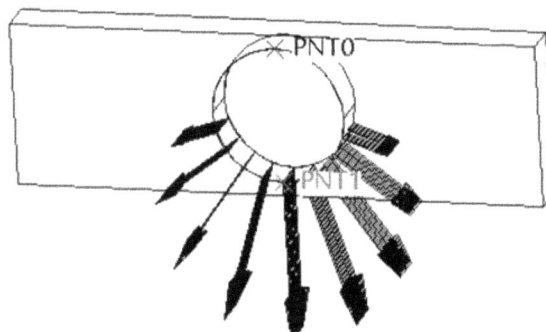

Under the Pro/MECHANICA, set up **Analyses** and **Run** it

Click the icon of **Run an Analysis** from the main toolbar > **File** > **New Static** > type *bearing_load* as the name of the analysis folder > **OK.** Click the box of **Run.**

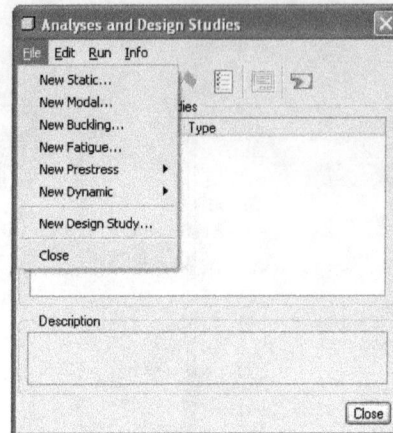

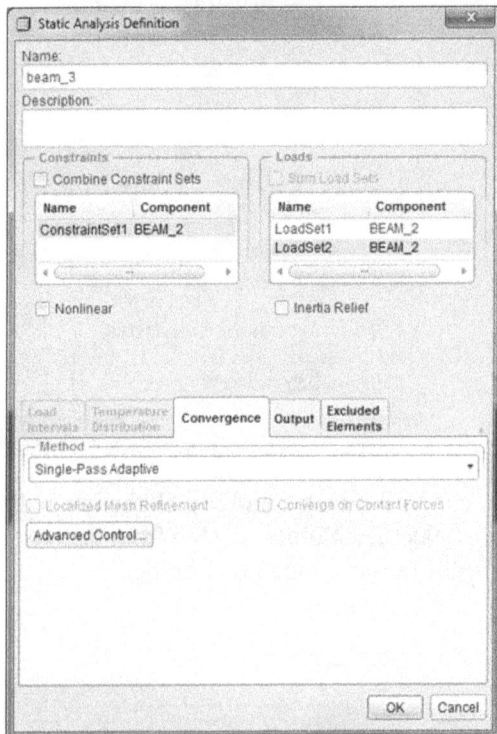

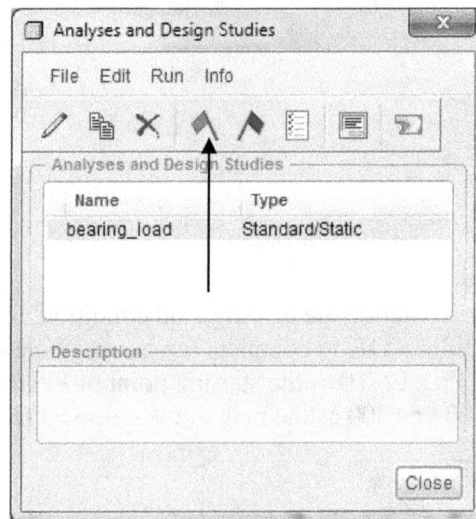

Run the program and afterwards, select the icon of **Results.** To study the obtained result, say the displacement of the canterliver beam under the static loading condition, select **Displacement** > **Magnitude.** Let us click Display Options. In the **Display Options** window, select Deformed and > **OK and Show.**

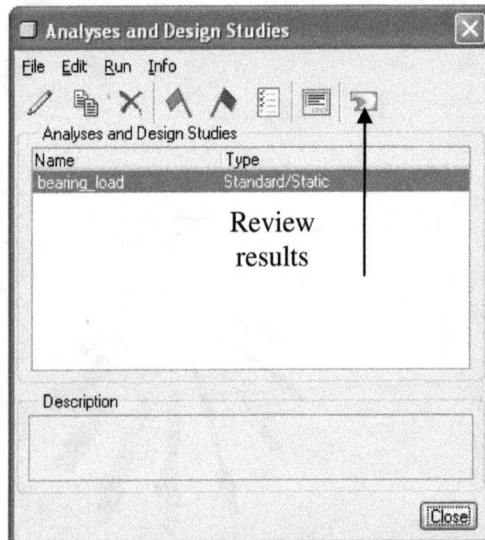

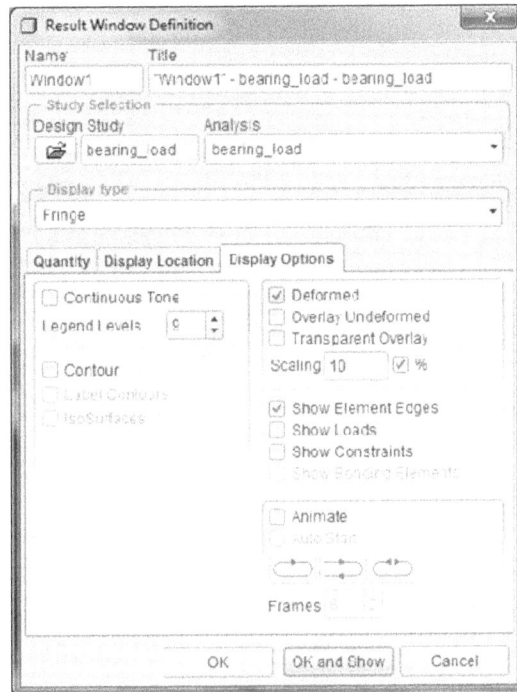

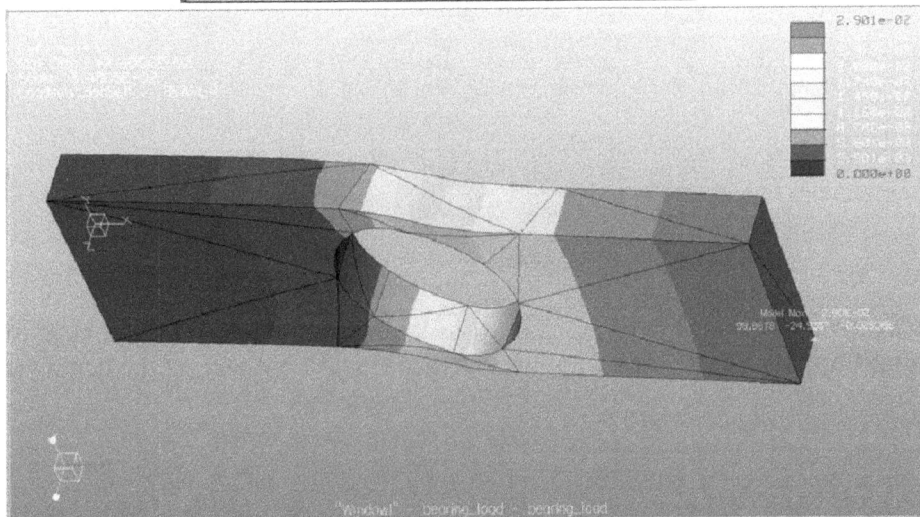

To study the distribution of the maximum principal stress distribution, select **Edit** from the main toolbar > **Copy** and a new Result Window appears > **Stress** > **von Mises** > **OK and Show**. Make sure that we do not use "Deformed". Let us use "Undeformed" because we want to identify the location of maximum value of von Mises stress.

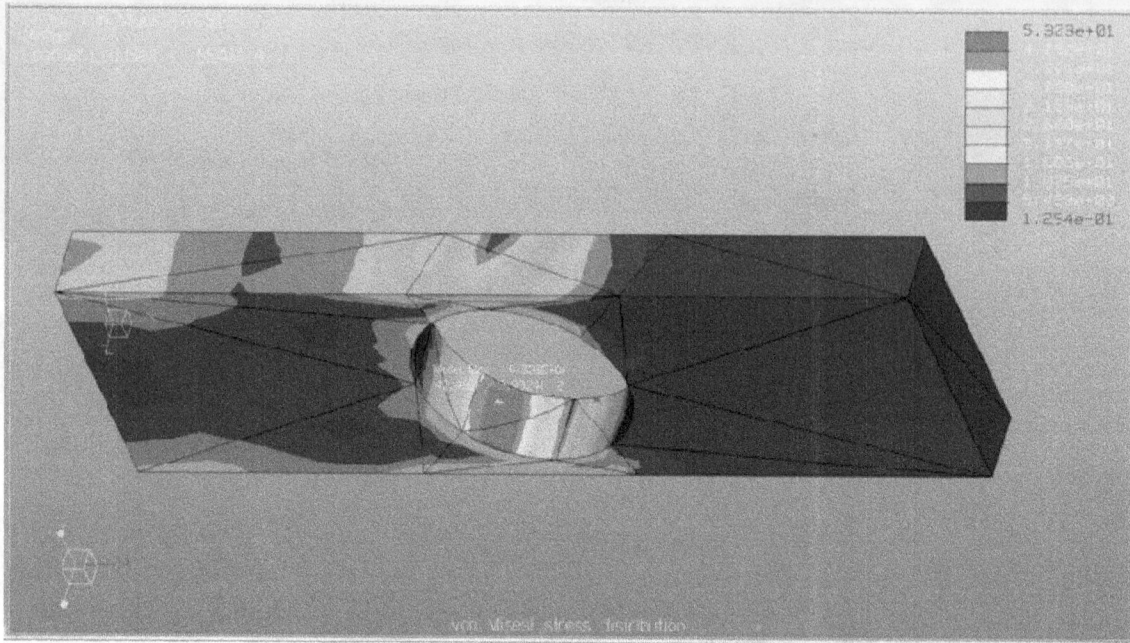

As indicated, the maximum value of von Mises stress is 53.23 MPa and the location x=42.5, y=-10.0 and z=2.

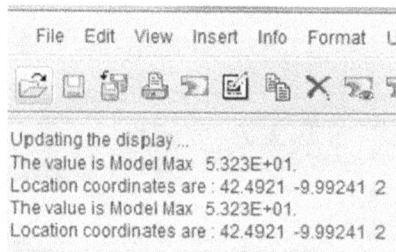

To facilitate the examination of the von Mises stress distribution, from the main toolbar, select **Insert > Cutting/Capping Surfs.** Pick **Capping Surface** > select **YZ** and set the depth value to 42.5 > **Apply > OK**.

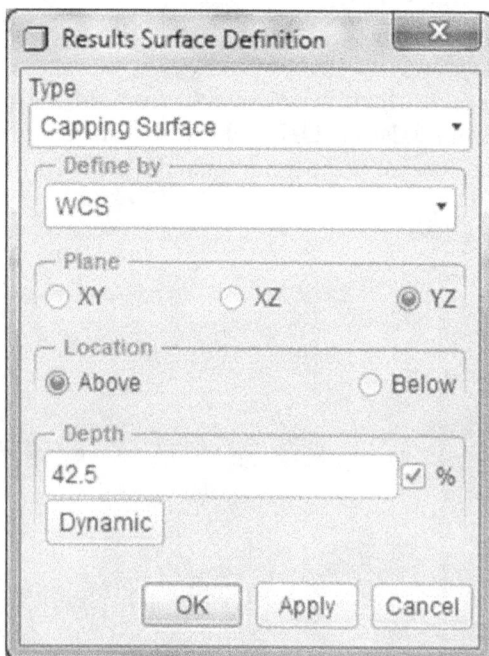

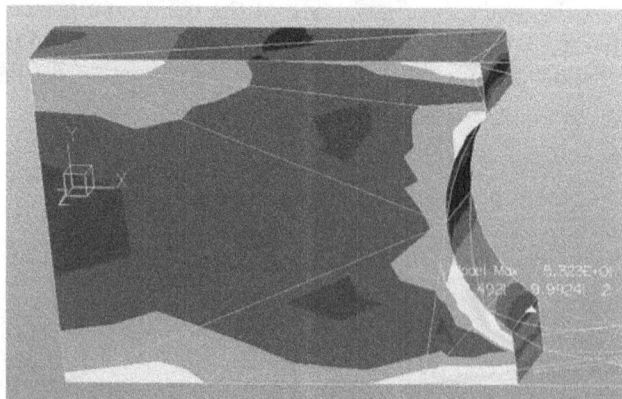

3.5   Modeling of a Pressure Load Condition

A support structure shown below is subjected to pressure loads. There are two pairs of pressure loads, as illustrated. The pressure acting on the walls of the 2 holes at the lower position is 20 N/mm^2. The pressure acting on the walls of the 2 holes at the upper position is 10 N/mm^2.

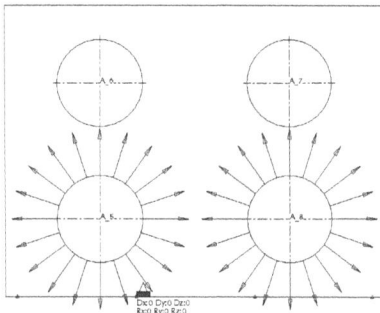
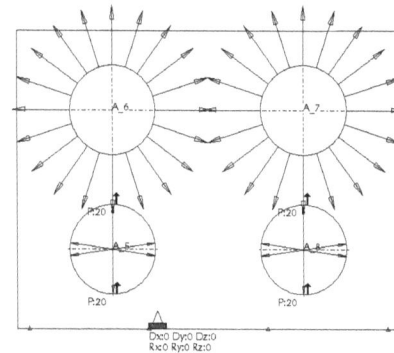

The geometrical shape and dimensions of the support structure are shown below.

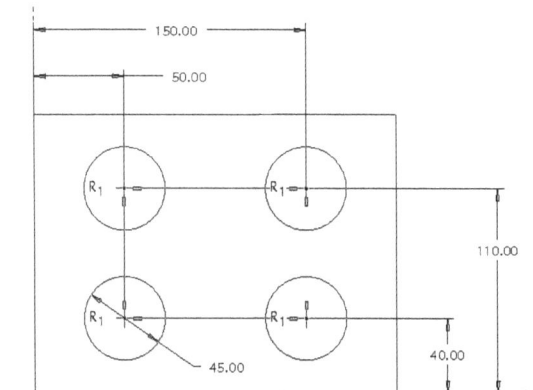
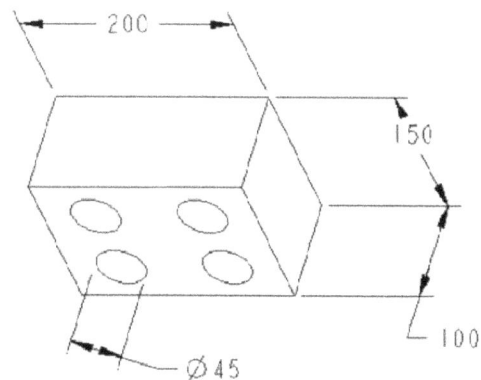

Assume that the bottom surface of the support structure is fixed to the ground. The material type of the support is steel. Our objectives of this study are "investigate the deformation pattern under the pressure loads and the maximum principal stress distribution when subjected to the pressure loads.

Step 1: Create a 3D solid model to represent the support structure.

File > New > Part > type *pipe_base* as the file name. Clear the box called **Use default template**, and select the **mmns_part_solid**. Fill in **DESCRIPTION** by typing *pressure loads*. Fill in **MODELED_BY** by typing your name, say *student.*> **OK.**

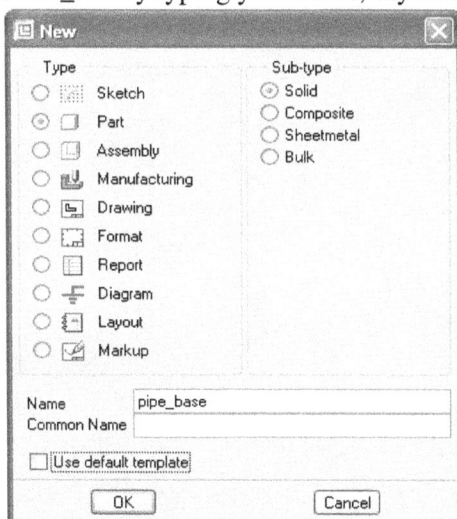
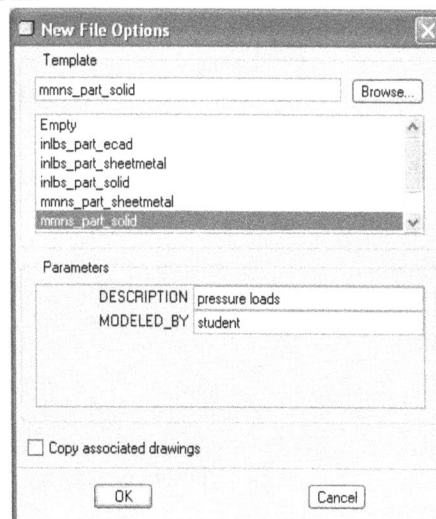

Click the icon of **Extrude.** Set the height value to 100. Activate **Placement** to define a sketch plane > **Define.**

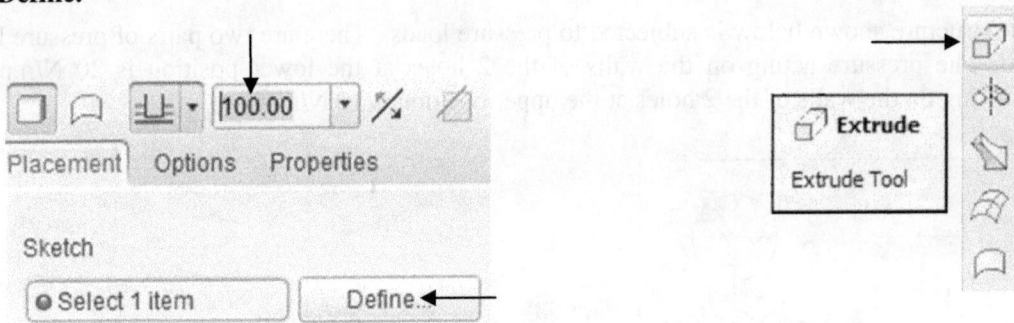

Pick the **FRONT** datum plane from the display, and click **Sketch** to accept the orientation of the sketch plane by positioning the **RIGHT** datum plane at the right side. Click the button of **Sketch** to complete the process of defining the sketch plane.

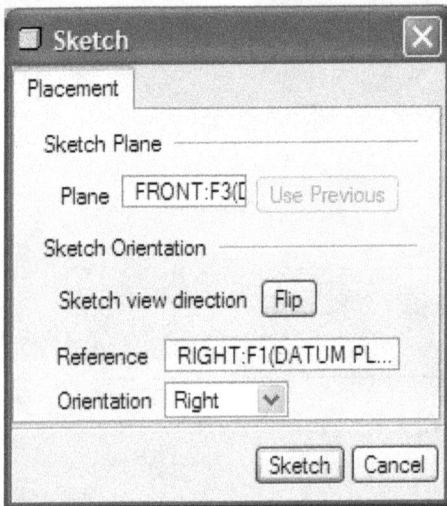

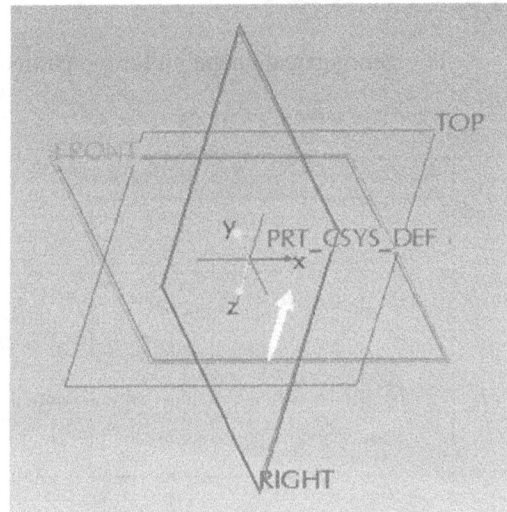

Click the icon of **Rectangle,** and sketch rectangle. The 2 dimensions are 200, and 150, respectively.

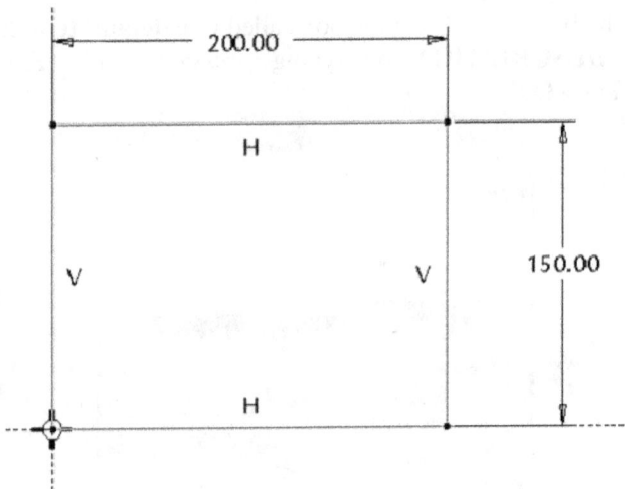

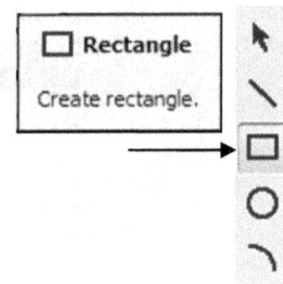

To create the 4 holes, click the icon of **Circle.** Specify 45 as the diameter dimension. Specify 40, 50, 110 and 150 as the position dimension, as shown.

○ **Center and Point**

Create circle by picking the center and a point on the circle.

200.00

150.00

50.00

45.00

150.00

110.00

40.00

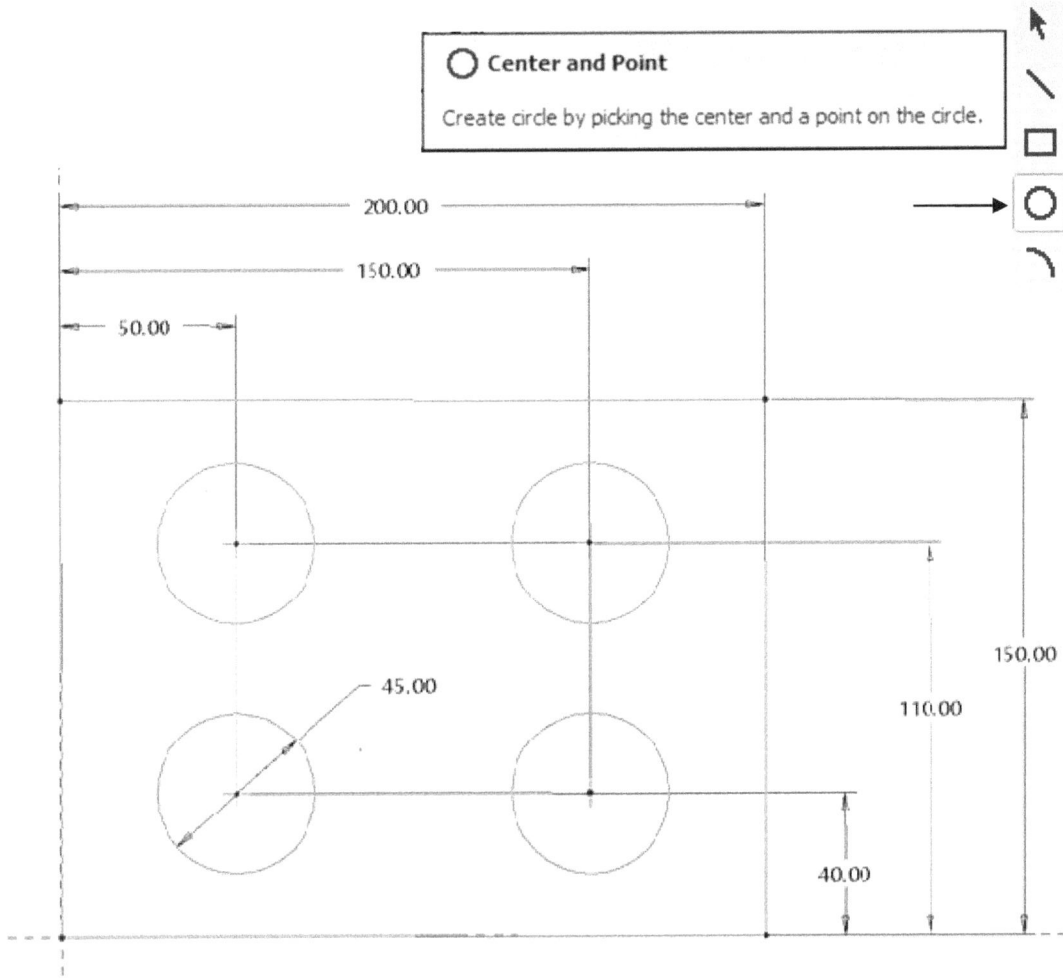

Upon completing the sketch, select the icon of **Done** and the icon of **Apply and Save**.

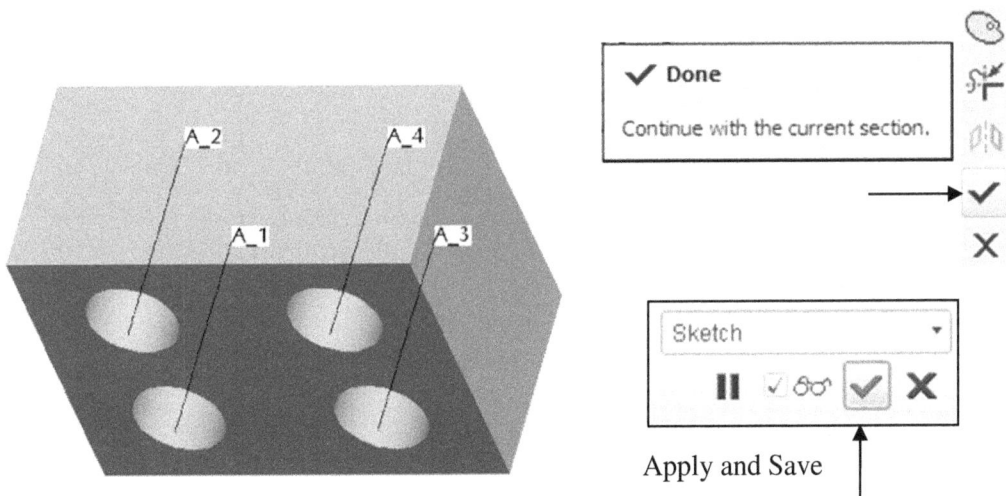

A_2

A_4

A_1

A_3

✔ **Done**

Continue with the current section.

Sketch

Apply and Save

Step 2: FEA under Pro/MECHANICA.

Applications > Mechanica > Continue > Structure > OK.

In the **Materials** window, select **STEEL** from the left side called Materials in Library > click the directional arrow so that the selected material type goes to the right side called Materials in Model.

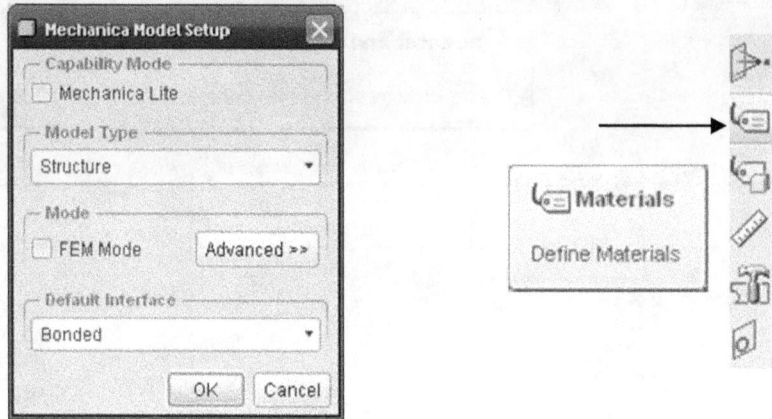

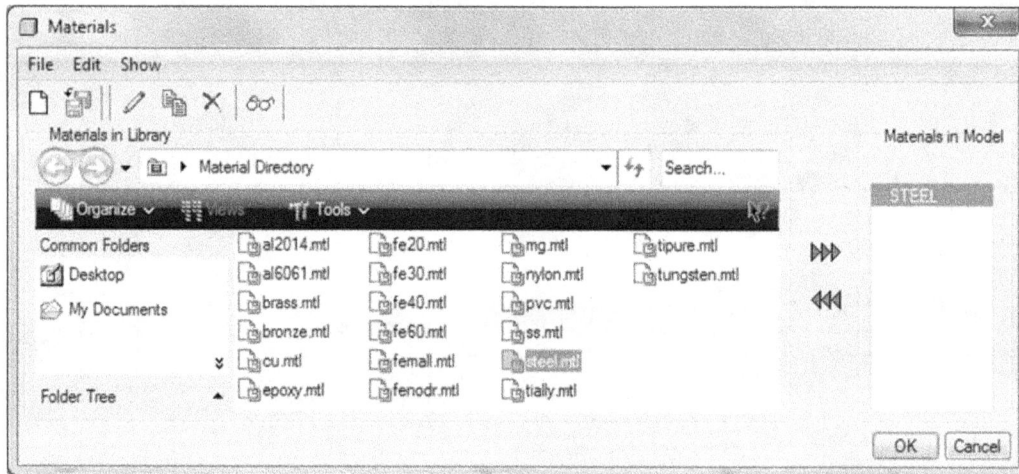

Click the icon of **Material Assignment**, the pop-up window indicates there is one component and the material type is STEEL. Just click **OK**.

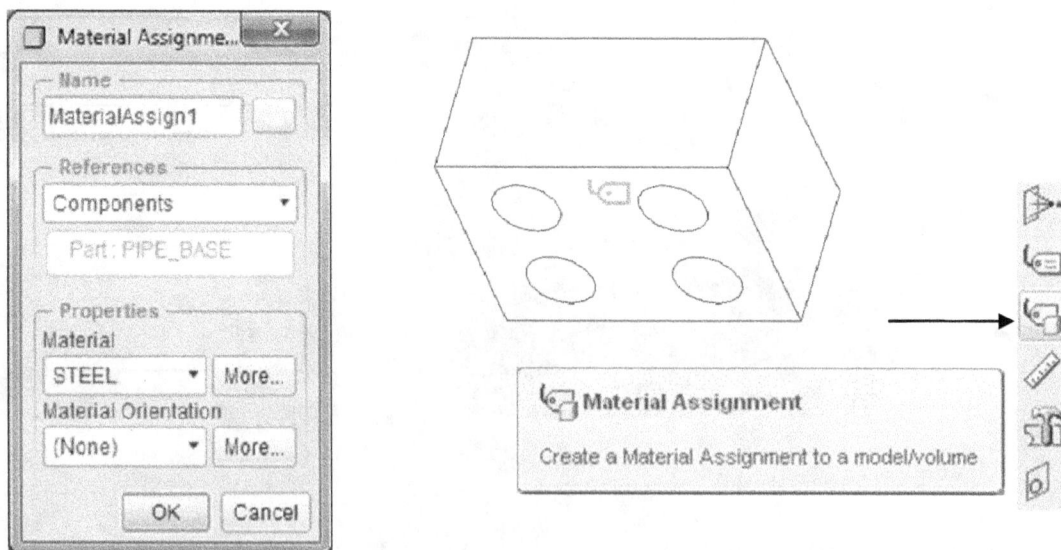

To define the fixed end constraint on the bottom of the pipe base component, click the icon of **Constraints >** select the surface at the bottom, and accept the default settings, namely, fixing all six degrees of freedom > **OK**.

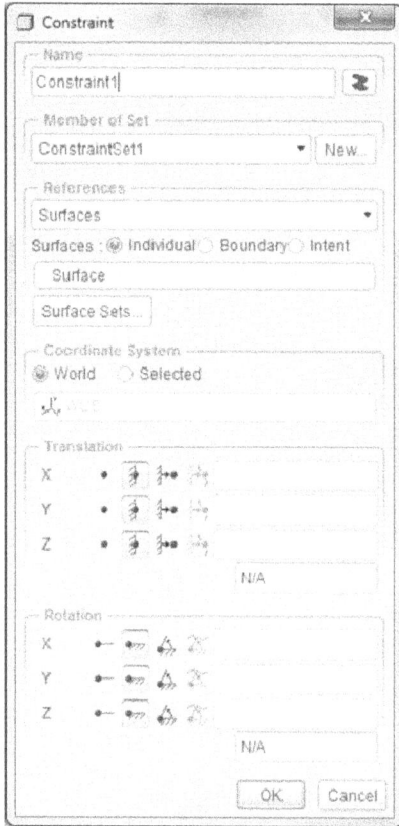

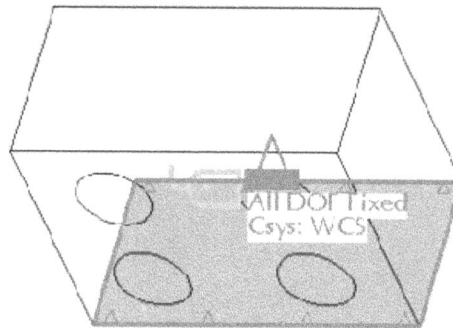

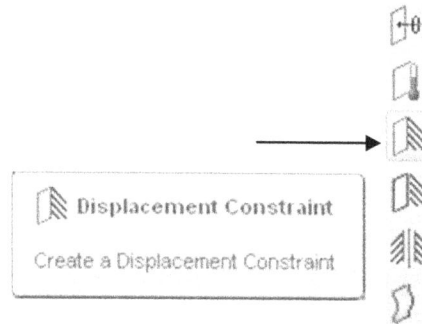

Step 3: Under the Pro/MECHANICA, define the load condition.

Click the icon of **Pressure Load** > holding down the **Ctrl** key, pick the 2 internal circular surfaces at the lower position, as shown > set the pressure value to 20 N/mm2 > **OK**.

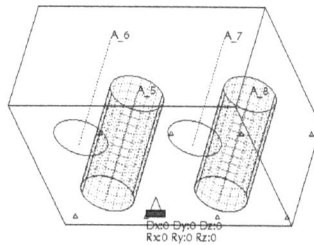

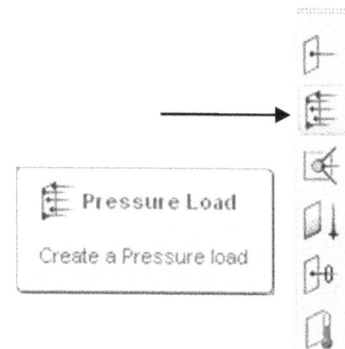

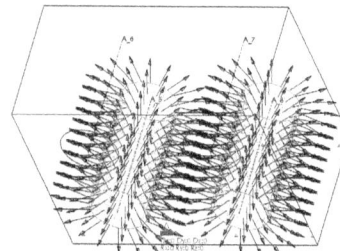

To define the second pressure load, click the icon of **New Pressure Load** > holding down the **Ctrl** key, pick the 2 internal circular surfaces at the upper position, as shown > set the pressure value to 10 N/mm2 > **OK**.

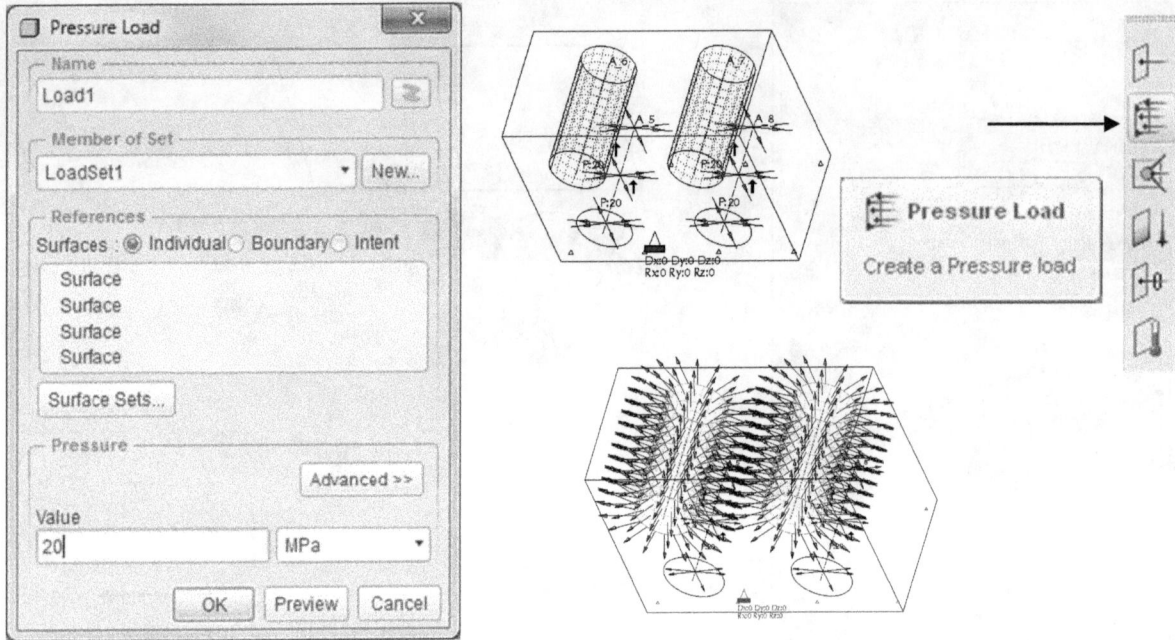

Step 4: Under the Pro/MECHANICA environment, set up **Analyses** and **Run** it

Click the icon of **Run an Analysis** from the main toolbar > **File** > **New Static** > type *pipe_base* as the name of the analysis folder > **OK.** Click the green flag of **Run.**

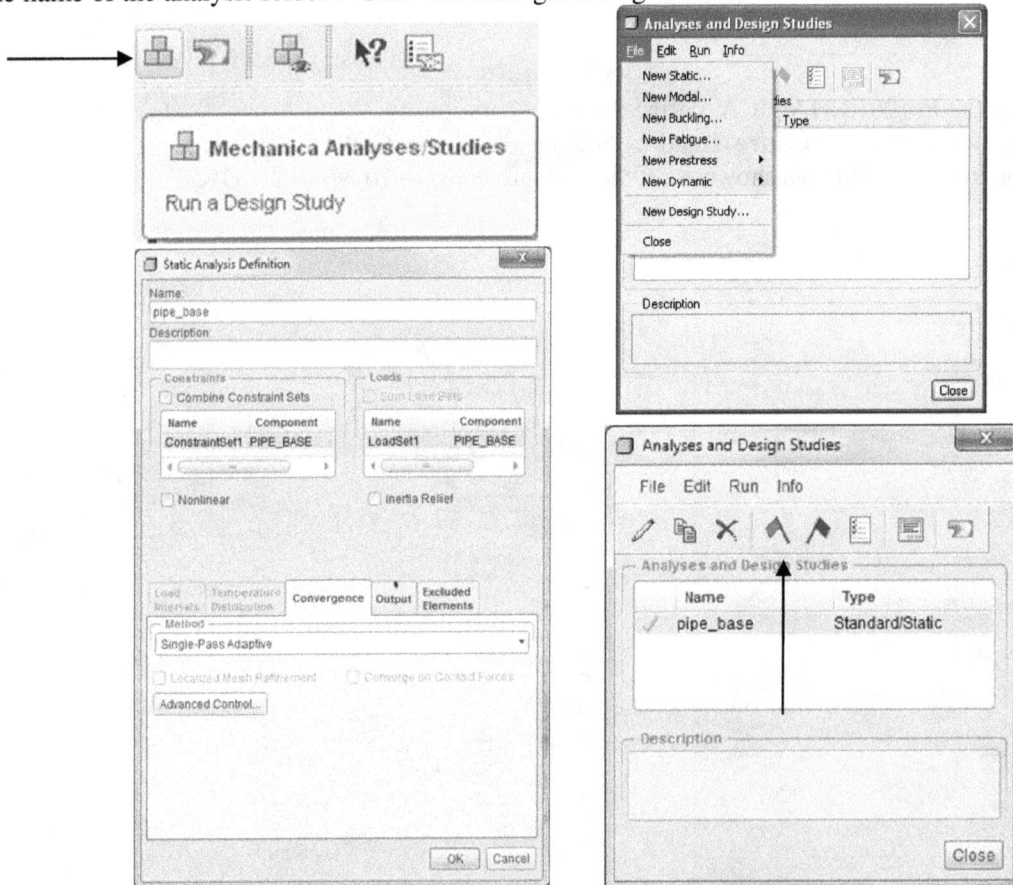

Step 5: Under the Pro/MECHANICA environment, study the **Results**

Select the icon of **Results.** Type *block deflection* as the title > locate the Design Study folder called pipe_base. Select **Fringe** > choose **Displacement** > **OK and Show.**

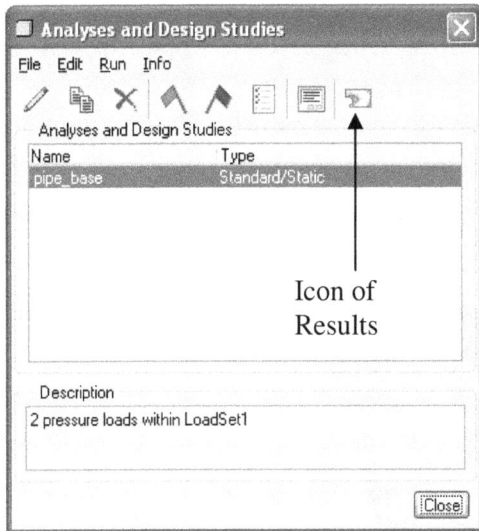

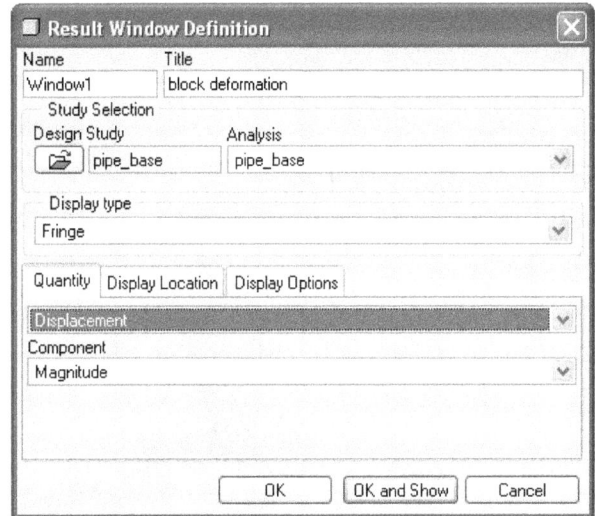

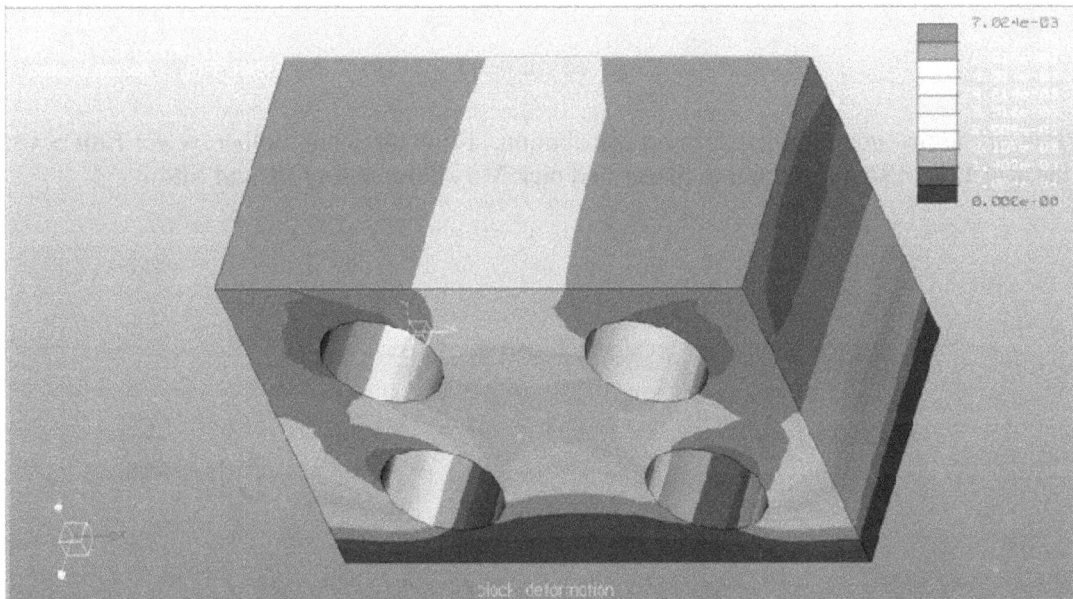

Plot of the displacement distribution with a capping surface crossing the 2 holes at the lower location. From the main toolbar, select **Insert > Cutting/Capping Surfs >** select **Capping Surface >** select **XZ >** the user may need to adjust the value of the Depth percentage to 30 or 40 > **Apply > OK**.

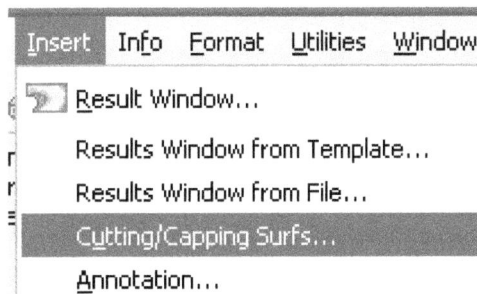

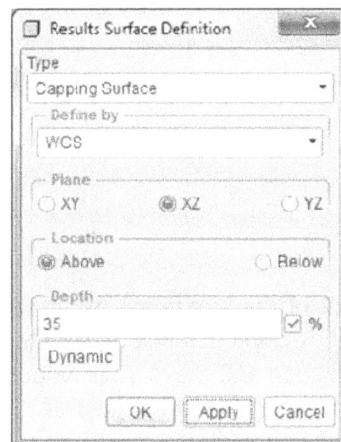

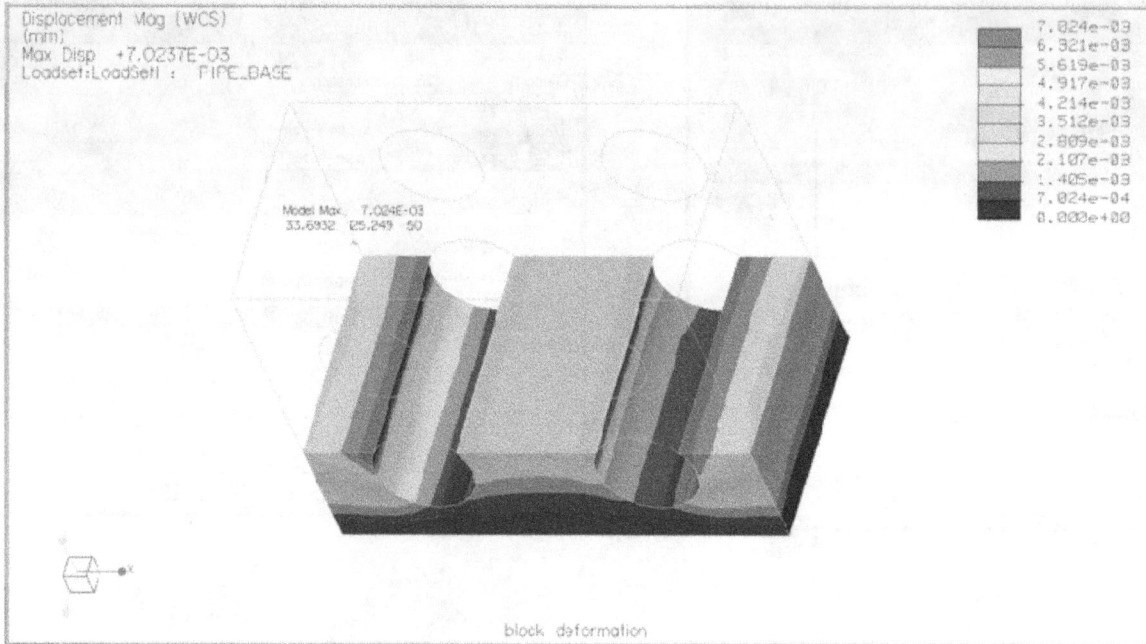

Plot of the maximum principal stress distribution. From the main toolbar, select **Edit > Copy >** change the selection of **Displacement** to **Stress** and pick **Max Principal> OK and Show**.

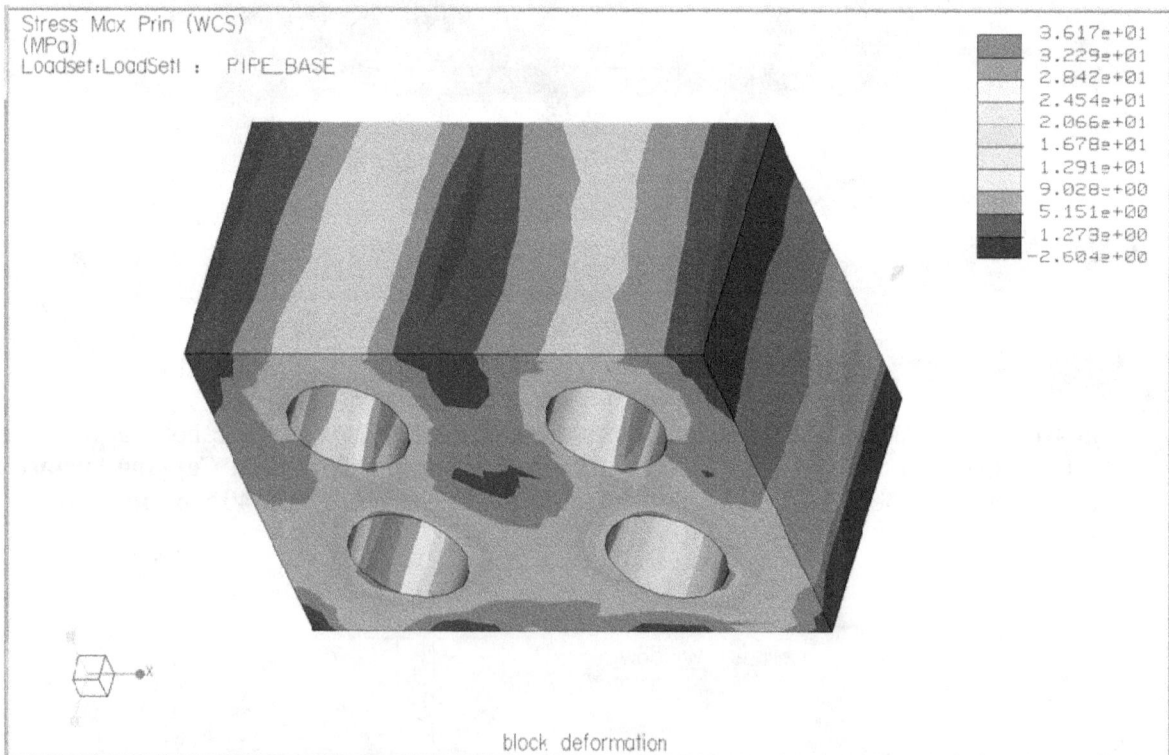

Plot of the maximum principal stress distribution with a horizontal capping surface. Note that, after **Apply** is clicked, a box called **Dynamic** is active. Click the box of **Dynamic** > locate your cursor to the cross-section and move your mouth. The position of capping surface moves following the movement of your mouth. This allows the user to dynamically view the stress distribution > click the middle button of mouth when it is done.

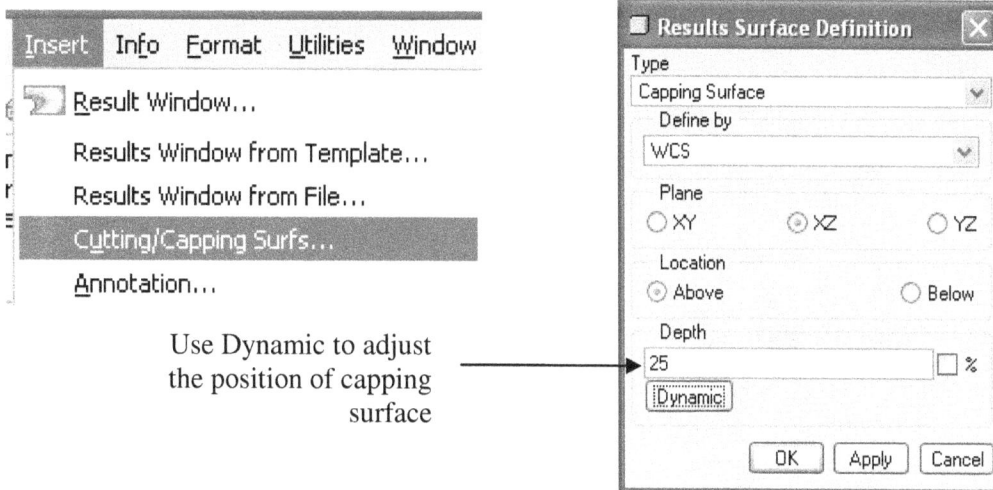

Use Dynamic to adjust the position of capping surface

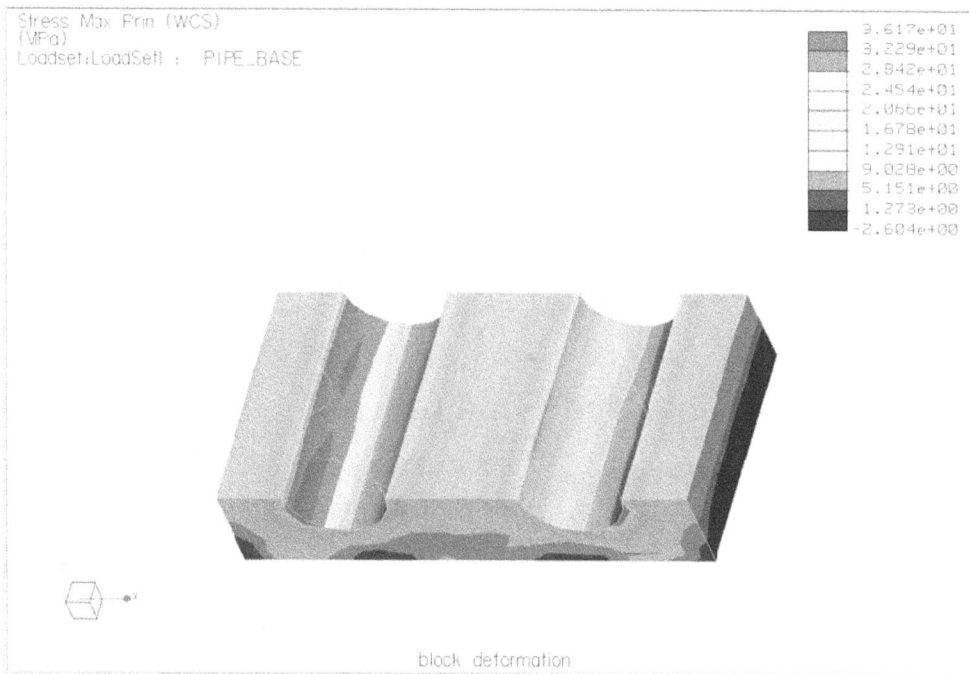

Plot of the maximum principal stress distribution with a vertical capping surface. In this case, select YZ, instead of selecting XZ. Using the function of **Dynamic** to relocate the capping surface.

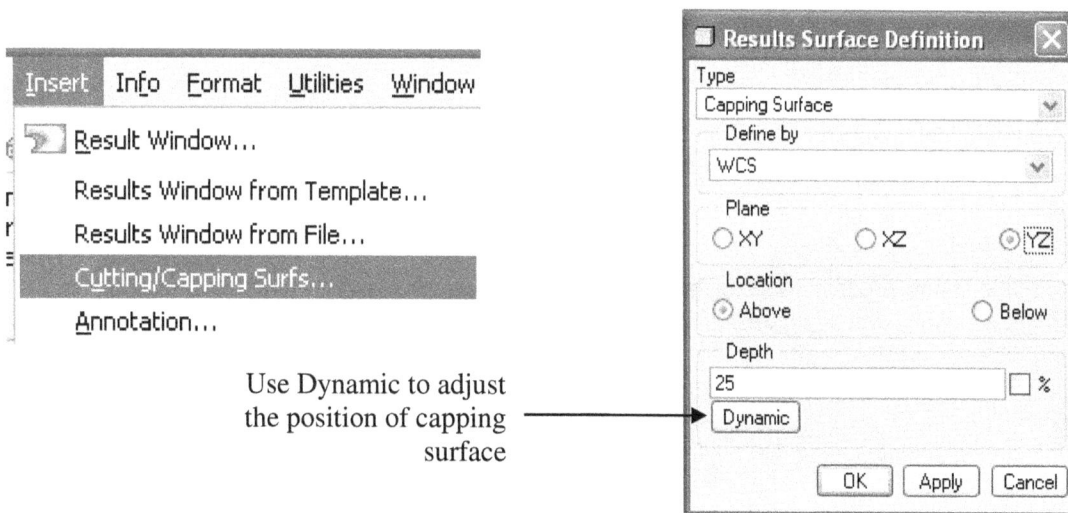

Use Dynamic to adjust the position of capping surface

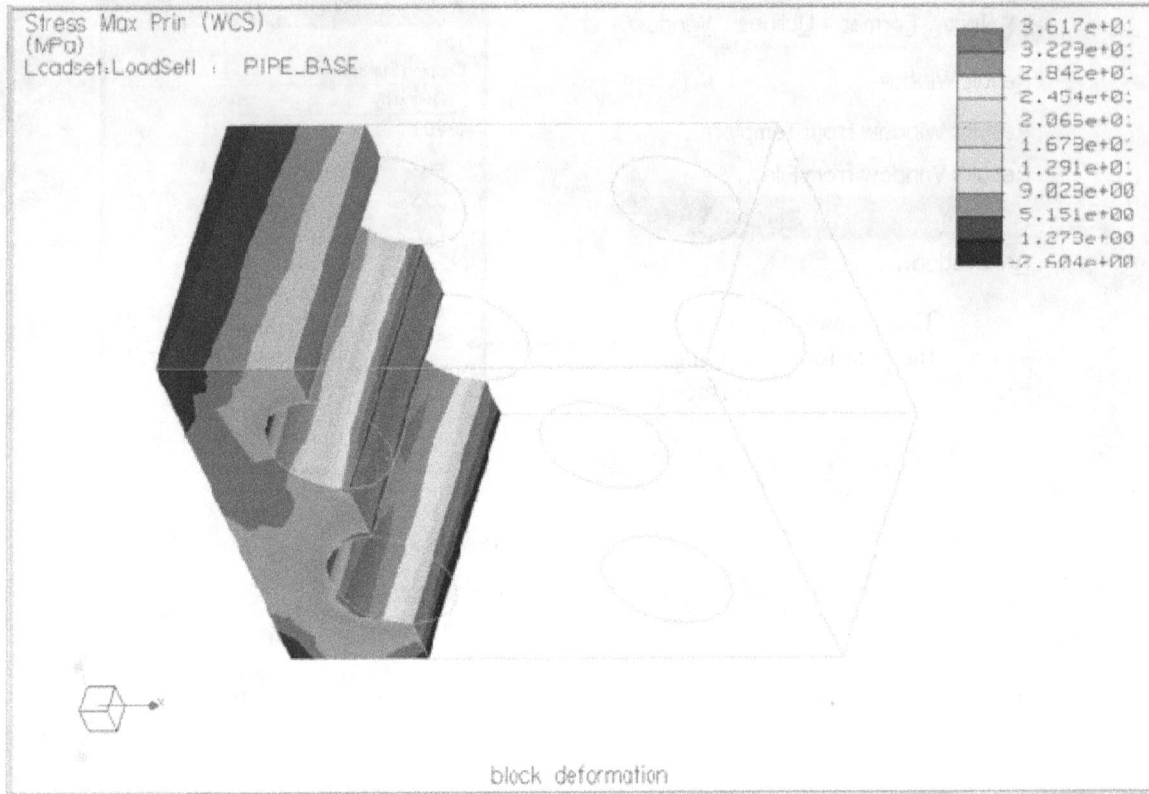

3.6 Modeling of a Point Load Condition and Gravity

When students take a course in mechanics of materials, they draw free-body diagrams when analyzing the forces acting on the object under study. On those free-body diagrams, some forces are assumed to act on points. The following figure presents a simple supported beam. The force acting on the beam is marked as P. Two forces marked as R1 and R2 are the reaction forces acting on the two supports. All these three forces are assumed to act on three point locations.

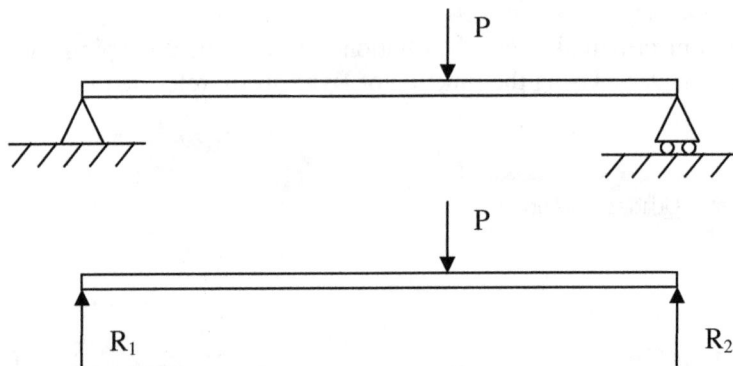

This modeling process used in mechanics of materials may pose difficulties in performing FEA where the stress calculation is related to an area. For example, the stress concentration occurred nearby the location where load P is acting can be viewed as

$$\sigma_{concentration} = \frac{P}{A}$$

In the above formula, a numerical value representing the size of an area occupied by a point is assumed to be zero. As a result, the FEA software system will send a warning message to the user, indicating that a calculation error has been occurred. To avoid such difficulties, it is a common practice in performing FEA that a surface region is defined at the point location so that the size of an area is specified before calculating the value of stress. In the following example, we demonstrate the procedure of modeling a point load.

The following figure illustrates a cantilever beam. The left end is a fixed end and the right end is a free end. A concentration load is acting on the free end. The magnitude is 200 Newton. The section area is a rectangle. The two dimensions are 50 and 10 mm, respectively. Assume that the material is steel and the Young's modulus is 2×10^{11} N/m^2.

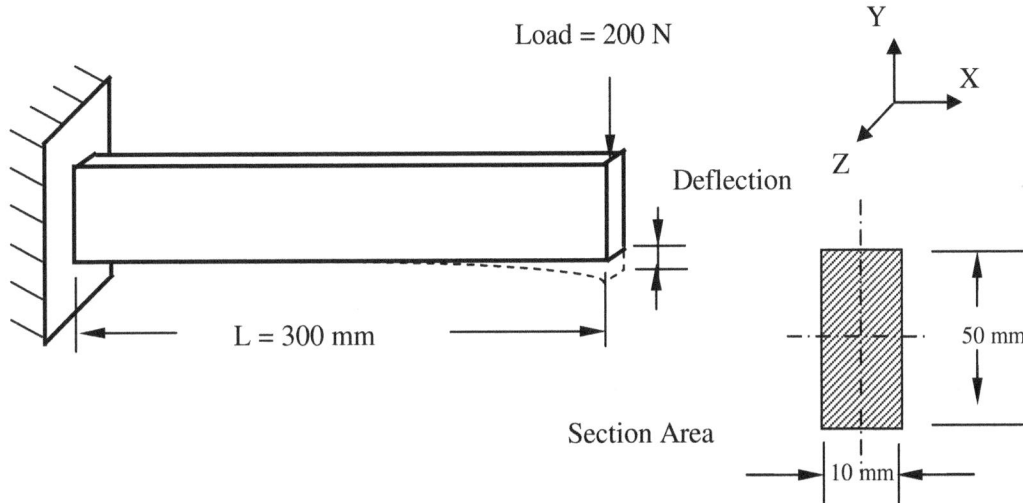

(1) Find the deflection at the free end based on an analytical formula.

The following formula was taken from a textbook titled Mechanics of Materials, authored by Geri and Thimoshenko (publisher: PWS-Kent, 1990). Using this formula to calculate the maximum deflection at the free end, the result is 0.0865 mm.

$$\delta_{free\_end} = \frac{PL^3}{3EI} = \frac{PL^3}{3E(\frac{bh^3}{12})} = \frac{(200)(0.3)^3}{3(2x10^{11})(\frac{(0.01)(.05^3)}{12})} = \frac{5.4}{(6*10^{11})(1.04x10^{-7})} = 0.0865x10^{-3}(meter)$$

(2) Find the deflection at the free end based on the result obtained from FEA analysis. Note that the loading area has to be defined as a surface region.

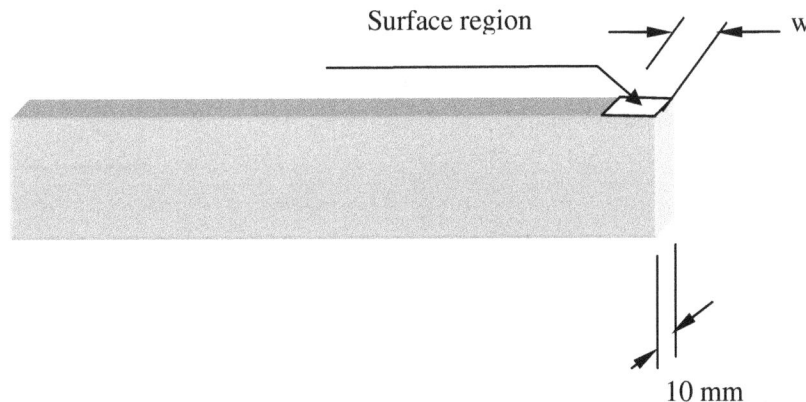

In this example, we will perform FEA through 3 runs. Each run will have a specific surface region. As shown in the following table, the surface region is set to 10 x 1 mm2 for Run 1. The surface region is set for 10 x 1 mm2 for Run 2. The surface region is set for 10 x 0.5 mm2 for Run 3. We will compare the results obtained from the 3 runs to gain a feeling on the sensitivity issue in modeling a point load.

	b	w	deflection (mm)
Run 1	10	1	
Run 2	10	2	
Run 3	10	0.5	

Step 1: Create a 3D solid model for the cantilever beam.

File > New > Part > type *cantilever_beam* as the file name. Clear the box called **Use default template > OK.** Select the **mmns_part_solid.** Fill in **DESCRIPTION** by typing *point load*. Fill in **MODELED_BY** by typing your name, say *student*.> **OK.**

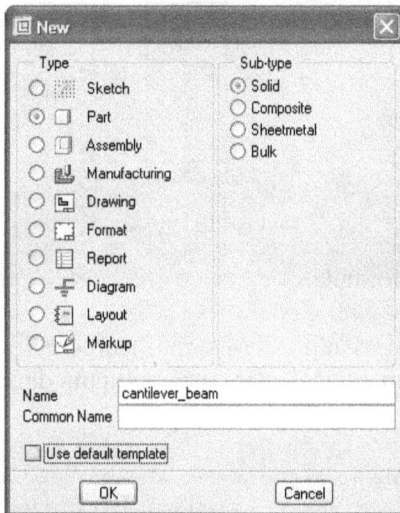

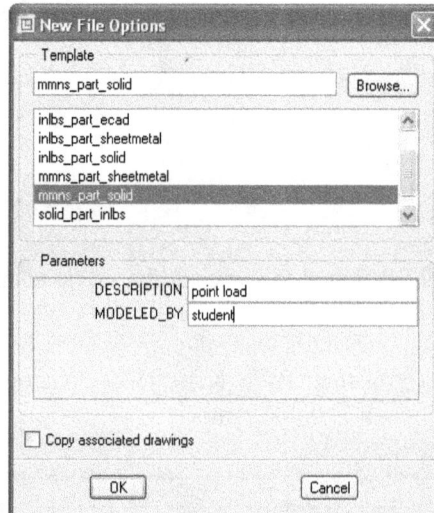

Click the icon of **Extrude**. Select Symmetry when specifying the depth value. Specify 10 as the depth value. Click the box of **Placement > Define**.

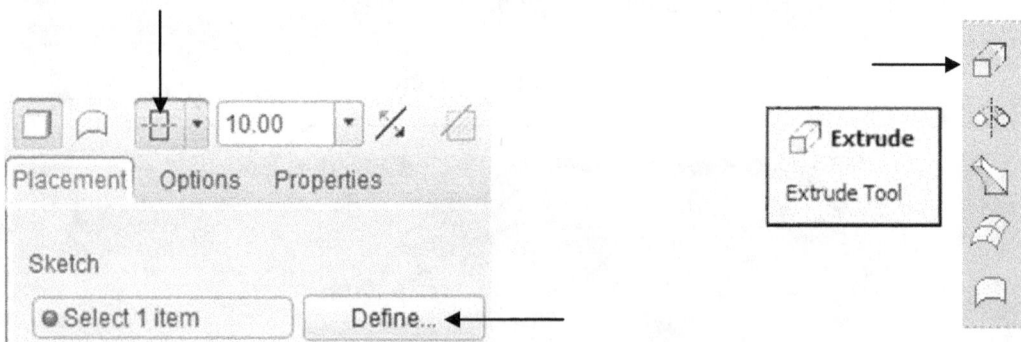

Select the **FRONT** datum plane displayed on screen as the sketch plane. Pick the box of **Sketch** to accept the **RIGHT** datum plane as the default reference to orient the sketch plane, as illustrated below:

Pick
FRONT as
the sketch
plane

Sketch

Placement

Sketch Plane

Plane | FRONT:F3(DA | Use Previous

Sketch Orientation

Sketch view direction | Flip

Reference | RIGHT:F1(DATUM PLA...

Orientation | Right |

| Sketch | | Cancel |

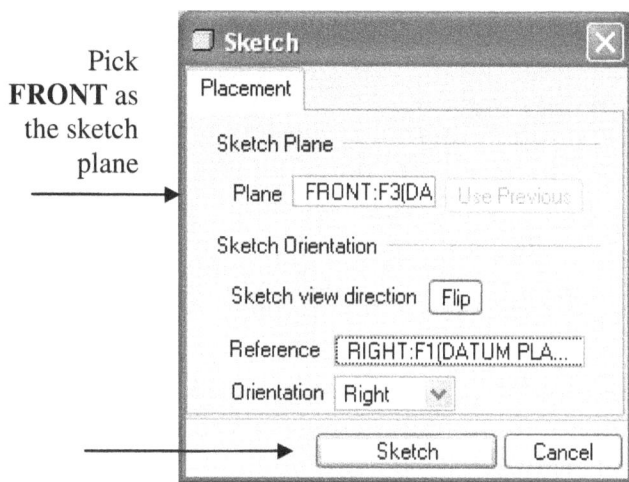

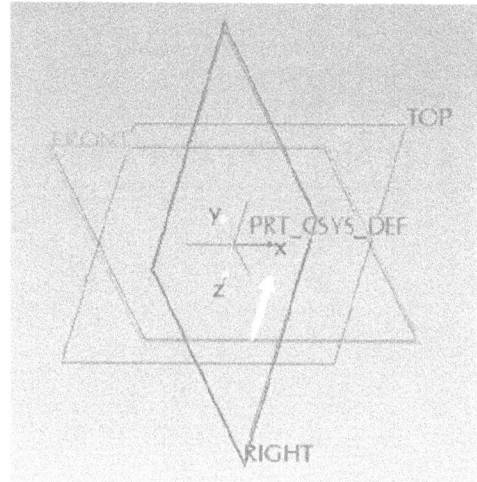

Let us first create 2 centerlines. Click the icon of **Centerlines**. Sketch a horizontal centerline and a vertical centerline, as shown.

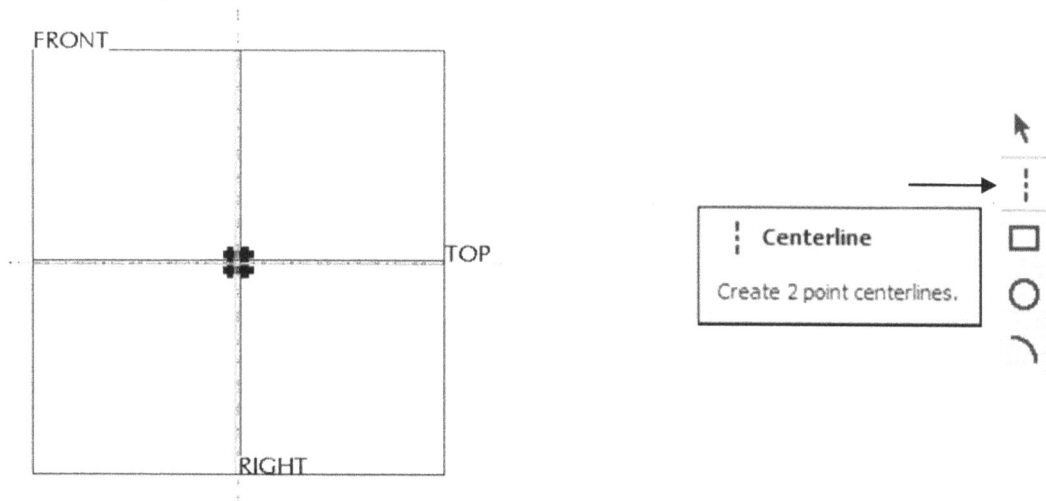

FRONT

TOP

RIGHT

| Centerline |

Create 2 point centerlines.

Click the icon of **Rectangle** and sketch a rectangle with the dimensions equal to 300 and 50, respectively. Keep the rectangle symmetric about the horizontal and the vertical centerlines.

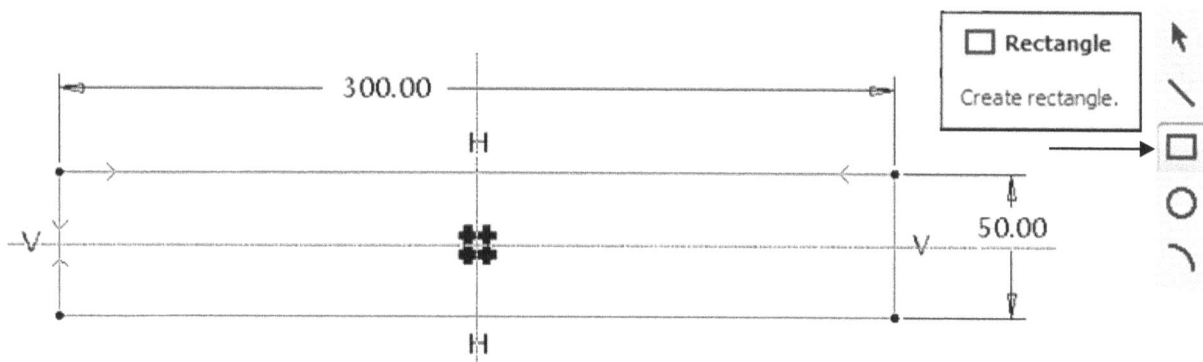

300.00

50.00

| Rectangle |

Create rectangle.

If your sketch is not symmetric about the horizontal centerline, pick the icon of **impose sketcher constraints** from the toolbar of sketcher, and select the icon of **vertices symmetric** from the pop up window called **Constraints** > click the horizontal centerline and click the 2 vertices so that the 2 vertices are symmetric about the horizontal centerline.

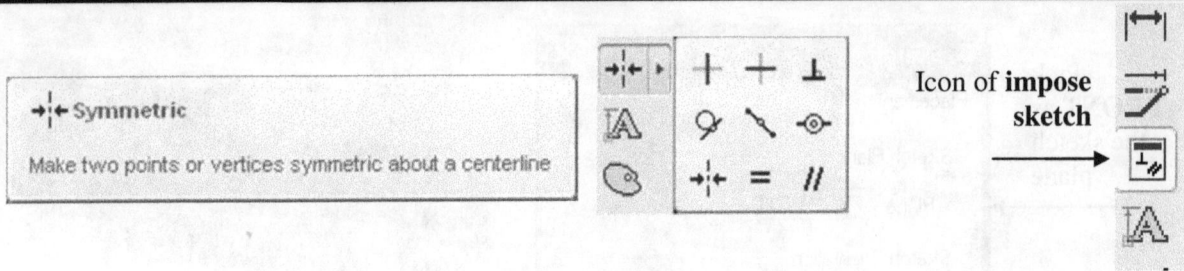

Icon of **impose sketch**

Upon completing the sketch, select the icon of **Done** and the icon of **Apply and Save**.

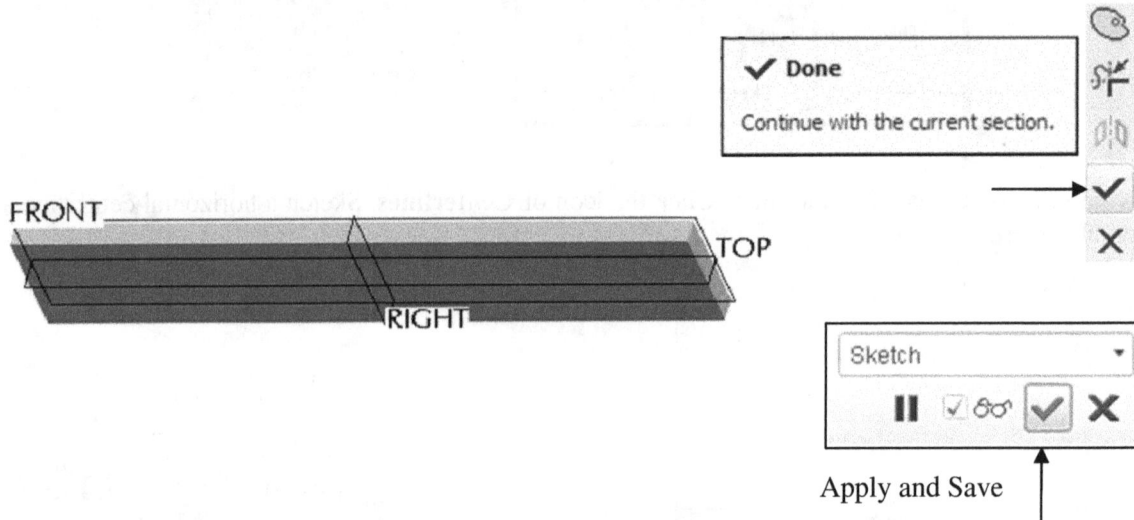

✔ **Done**

Continue with the current section.

Apply and Save

Step 2: Create a datum curve feature so that a surface regions can be define for modeling a point load.

To sketch a datum curve select the icon of **Sketch Tool** > Pick the top surface of the beam as the sketch plane > accept the default orientation of the sketch plan

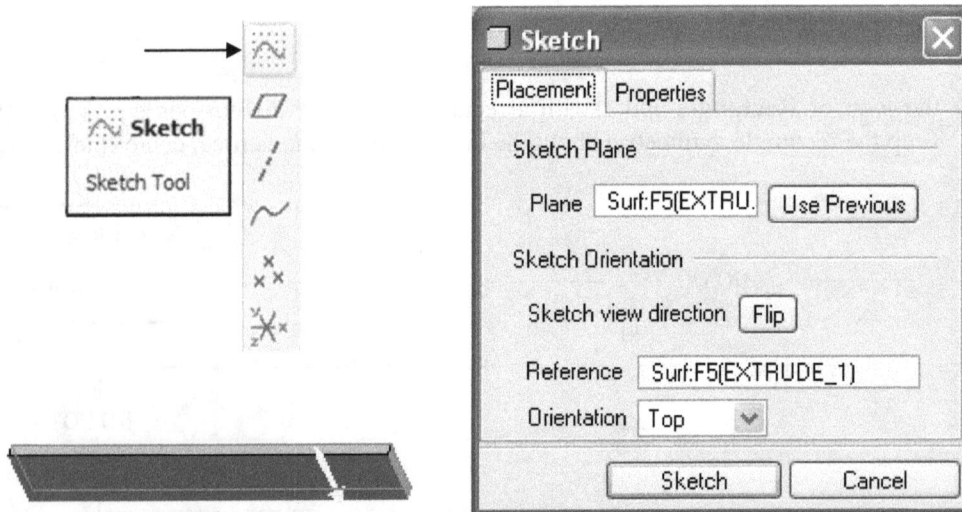

Sketch

Placement | Properties

Sketch Plane

Plane | Surf:F5(EXTRU. | Use Previous

Sketch Orientation

Sketch view direction | Flip

Reference | Surf:F5(EXTRUDE_1)

Orientation | Top

Sketch | Cancel

Before making a sketch, let us first add 2 new references. Click **Sketch**. These 2 new references are the surface at the right end and the surface in the front of the block, as shown. Afterwards, close the reference window.

Sketch   Analysis   Info   Applications   Tools   Window   Help

References...

References

Specify references which the section will be dimensioned and constrained to.

References

Surf:F5(EXTRUDE_1)
F1(RIGHT)
Surf:F5(EXTRUDE_1)
Surf:F5(EXTRUDE_1)

Select Use Edge/Offset   Delete

Reference status
Fully Placed

Close

Select this surface
as a new reference

Select this surface
as a new reference

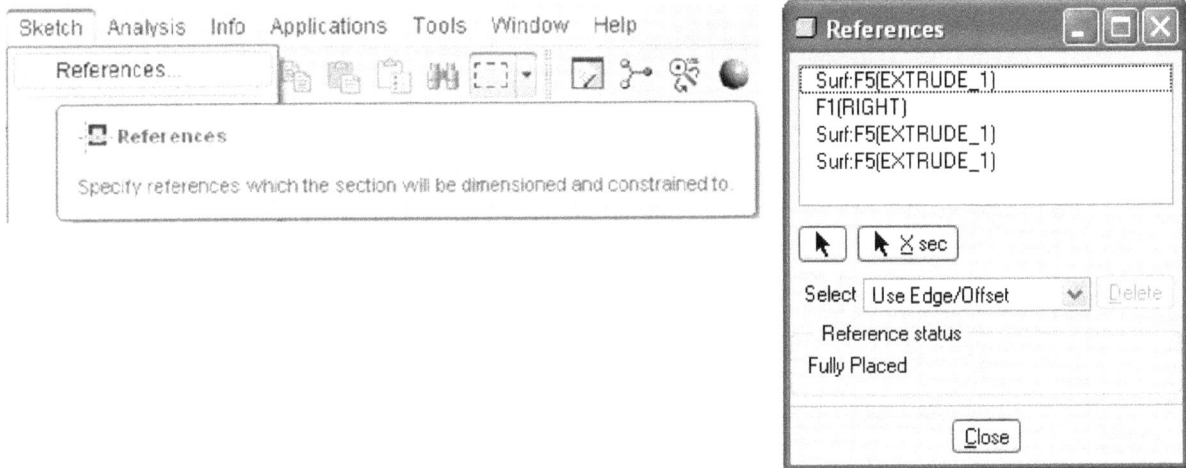

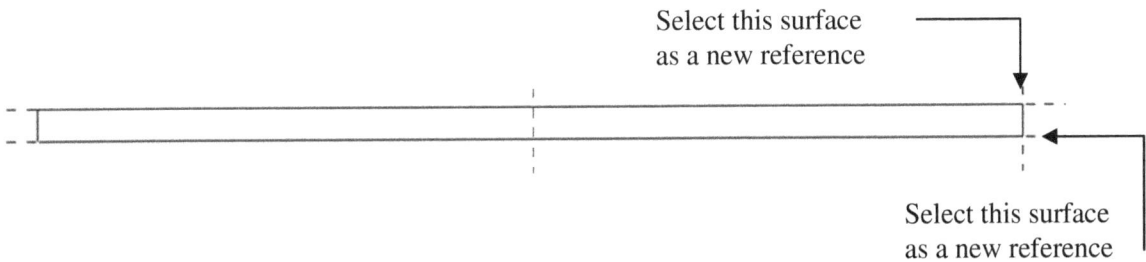

Click the icon of **Rectangle** and sketch a rectangle with the dimension of width equal to 1. Only one dimension is needed because the other sides of the sketched rectangle are aligned with the selected references. Pick the icon of **Done**.

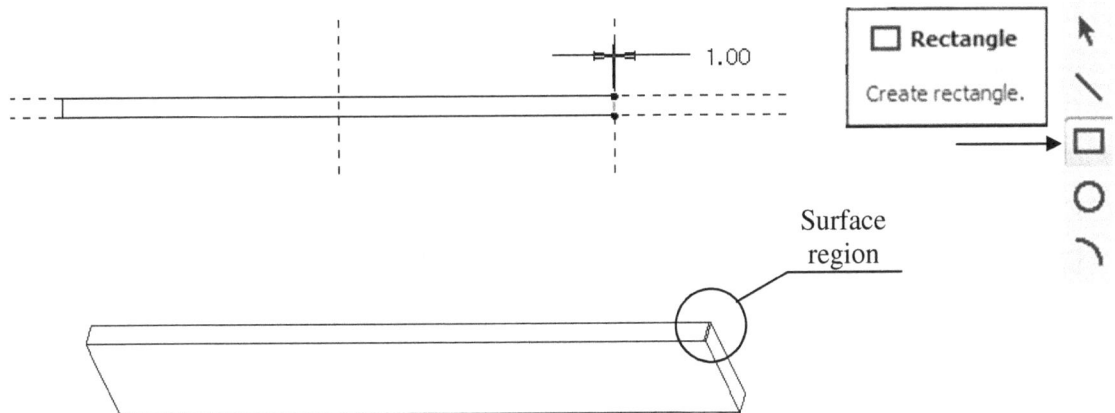

1.00

Rectangle

Create rectangle.

Surface
region

Step 3: Under the Pro/MECHANICA, perform FEA.
From the main toolbar or the menu toolbar, select **Applications > Mechanica.**
Continue after checking the unit system > **Structure > OK**.
In the **Materials** window, select **STEEL** from the left side called Materials in Library > click the directional arrow so that the selected material type goes to the right side called Materials in Model.

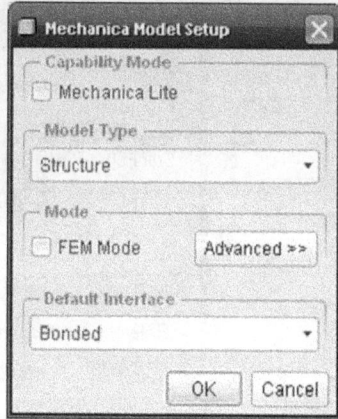

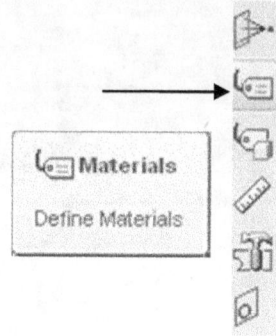

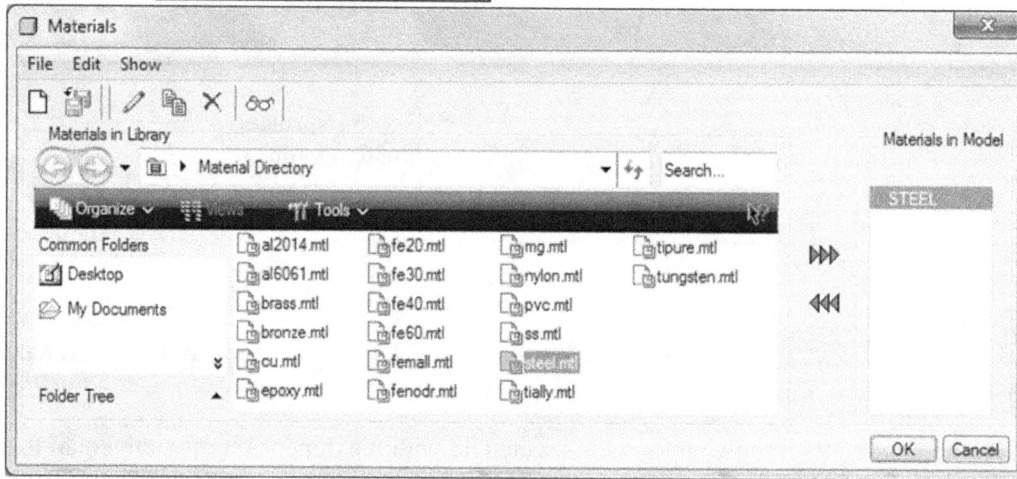

Click the icon of **Material Assignment**, the pop-up window indicates there is one component and the material type is STEEL. Just click **OK**.

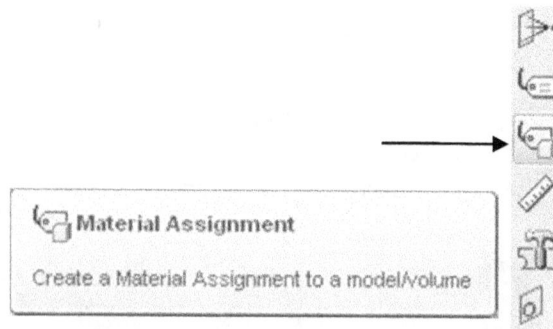

From the toolbar of functions, select the icon of **Materials** > select **STEEL** from the left side called Materials in Library > click the directional arrow so that the selected material type goes to the right side called Materials in Model.

To define the fixed end constraint on the left side of the beam, click the icon of **Constraints >** select the surface at the left end of the block as the fixed end, and accept the default settings, namely, fixing all six degrees of freedom > **OK**.

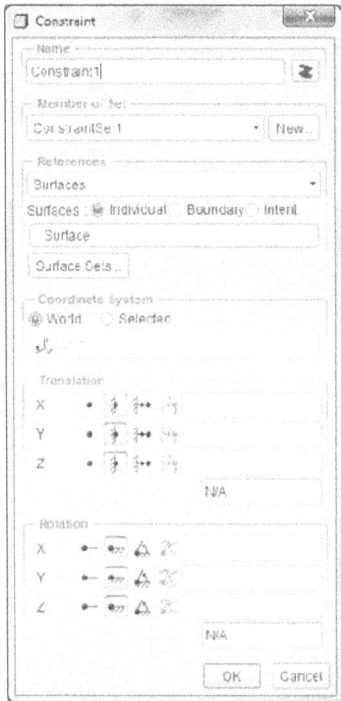

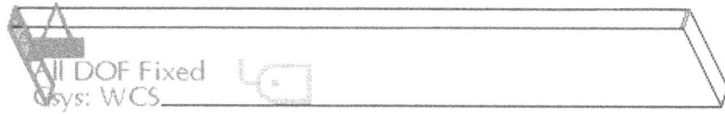

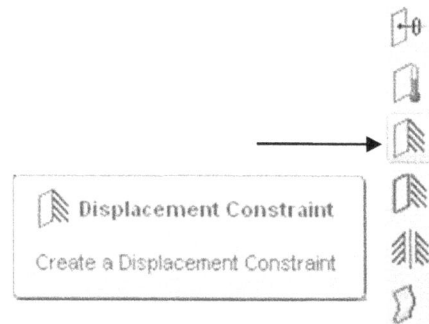

To define the load condition, We first define a surface region, on which the uniformly distributed load will be defined.

Click the icon of **Surface Region**.
Click the datum curve of rectangle > click the surface being split.
Click the icon of **Apply and Save**.

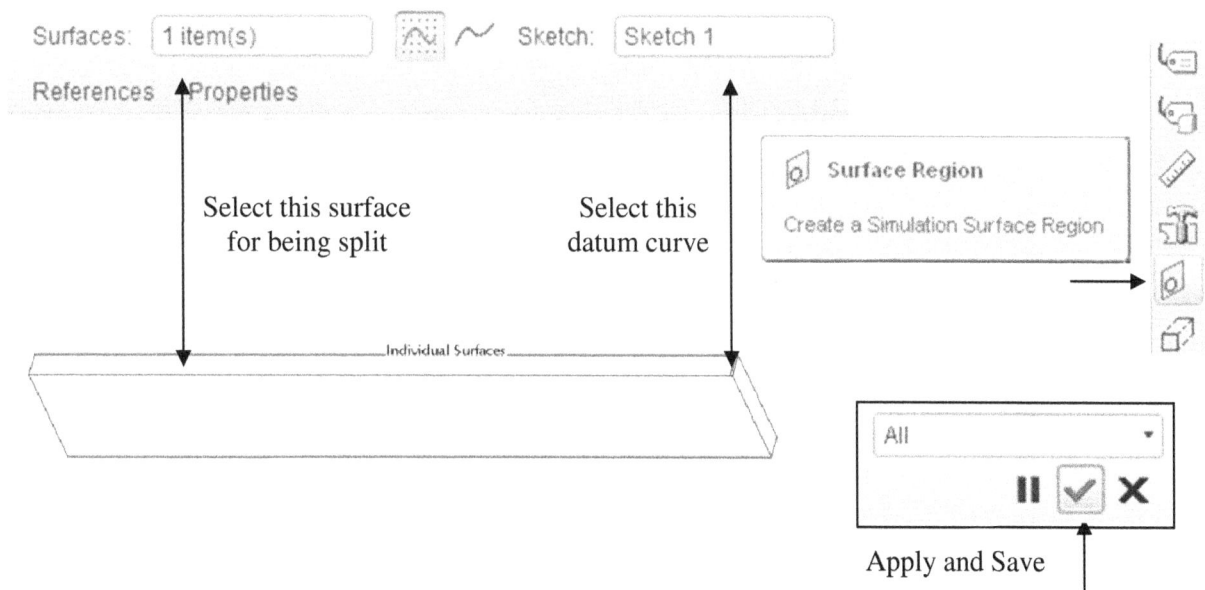

Select this surface for being split

Select this datum curve

Apply and Save

Click the icon of **Force/Moment Loads > Surface(s) >** select the surface region just defined > type -200 as the Y Force Component > **OK**.

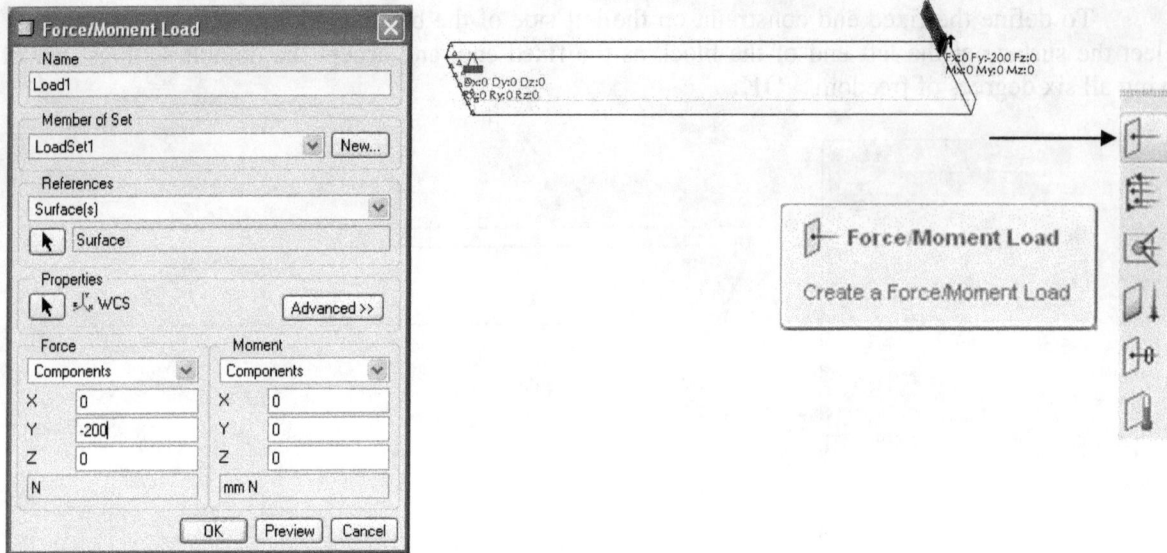

Step 4: Under the Pro/MECHANICA, set up **Analyses** and **Run.**
Select the icon of **Run an Analysis > File > New Static** > type in *cantilever_beam* > **OK**.

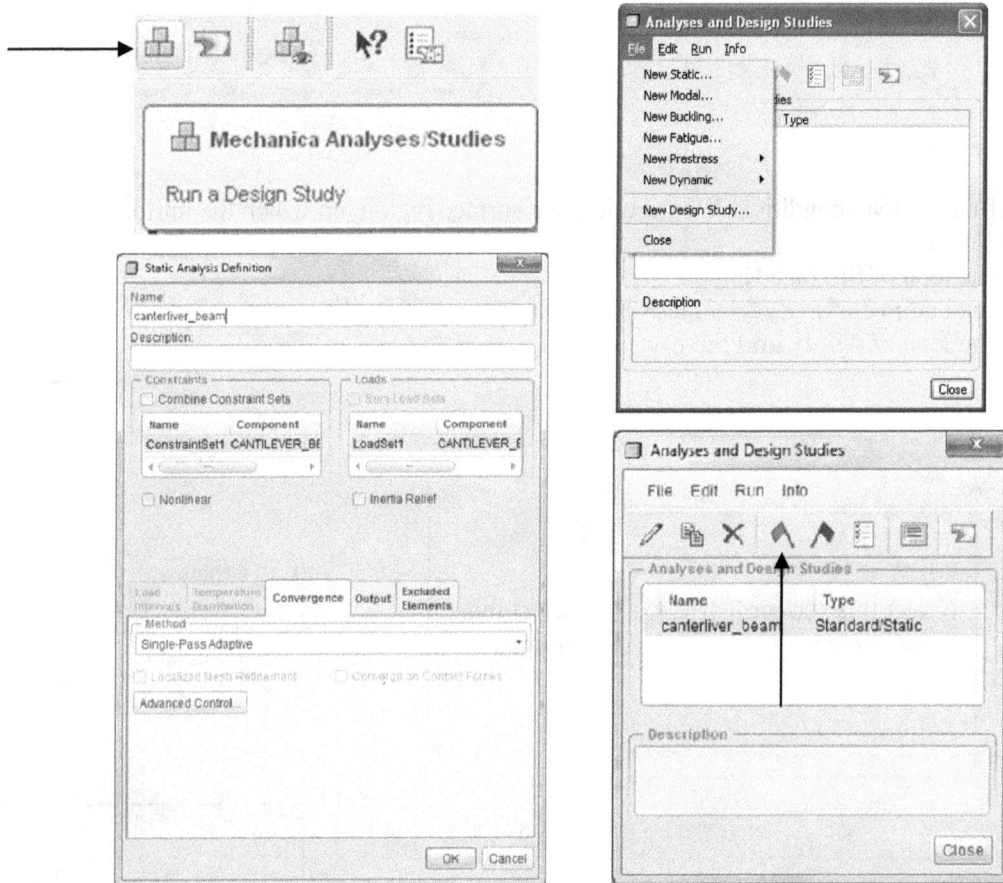

Step 5: Under the Pro/MECHANICA, study the **Results**
Select the icon of **Results.** Type *Beam Deflection* as the title > locate the Design Study folder called cantilever_beam.
Select **Fringe** > choose **Displacement** > **Magnitude** > **OK and Show**.

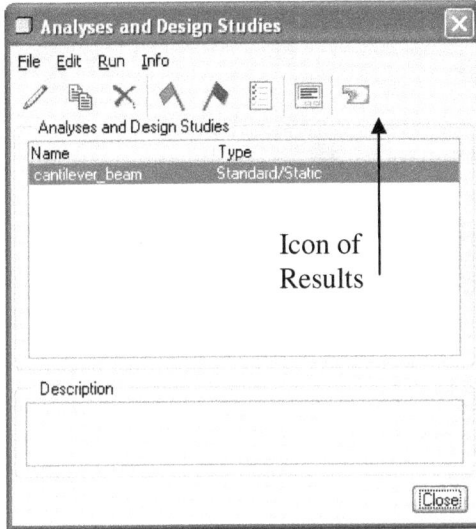

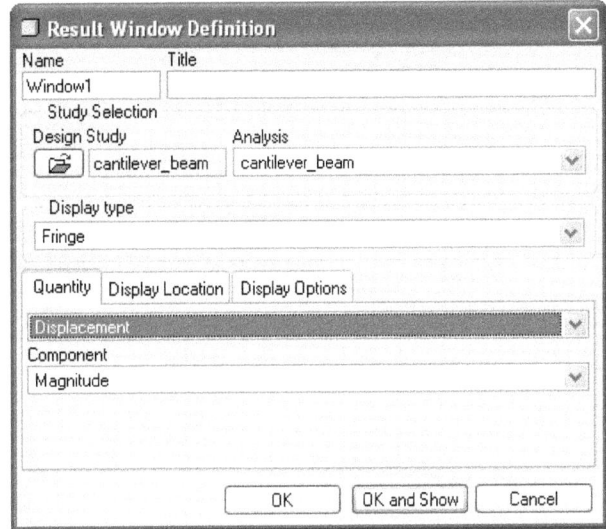

Examining the value of the maximum deflection shown at the free end, the numerical value is 0.08830 mm. As we recall, the numerical value obtained from the calculation using an analytical formula was 0.08650 mm. The difference is 0.00207 mm, or about 1.8 µm.

Note that the numerical value obtained from the calculation using an analytical formula was the deflection along the Y direction. Let us display the displacement in the Y direction. To do so, select the icon of **Edit** the selected definition from the main toolbar > change **Magnitude** to **Y** > **OK & Show**.

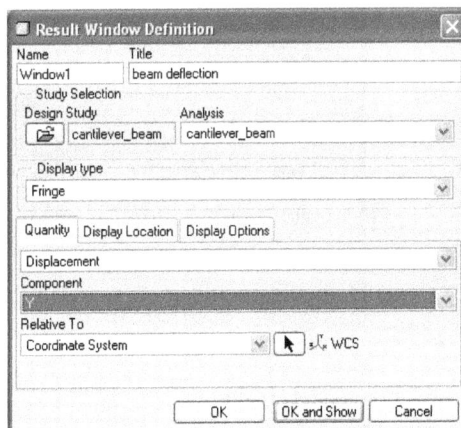

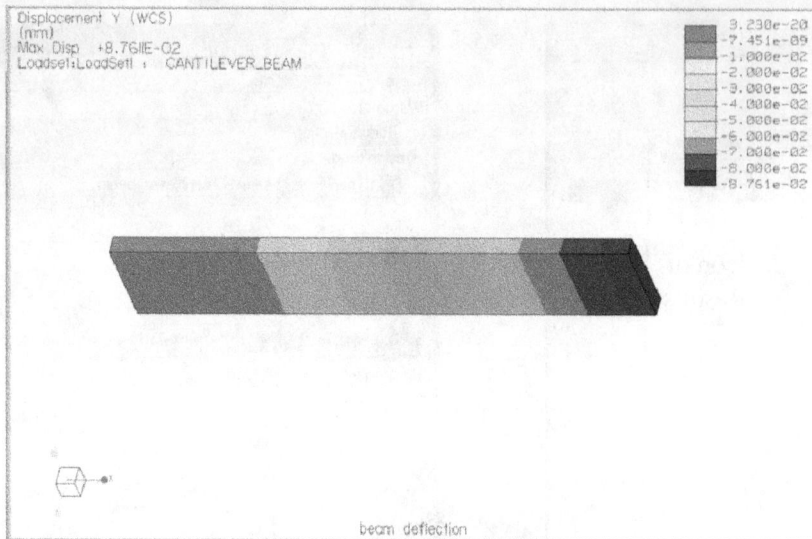

beam deflection

Examining the value of the maximum deflection shown at the free end, the numerical value is 0.08761 mm. As we recall, the numerical value obtained from the calculation using an analytical formula was 0.08650 mm. The difference is 0.00111 mm, or about 1 μm.

Now let us change the area of the surface region from 10 x 1 mm to 10 x 2 mm to perform Run 2. Let us go back to Pro/ENGINEER design system. From the menu bar, select **Applications > Standard**. From the model tree, select **Sketch1** > right click on the mouse to select **Edit Definition.**

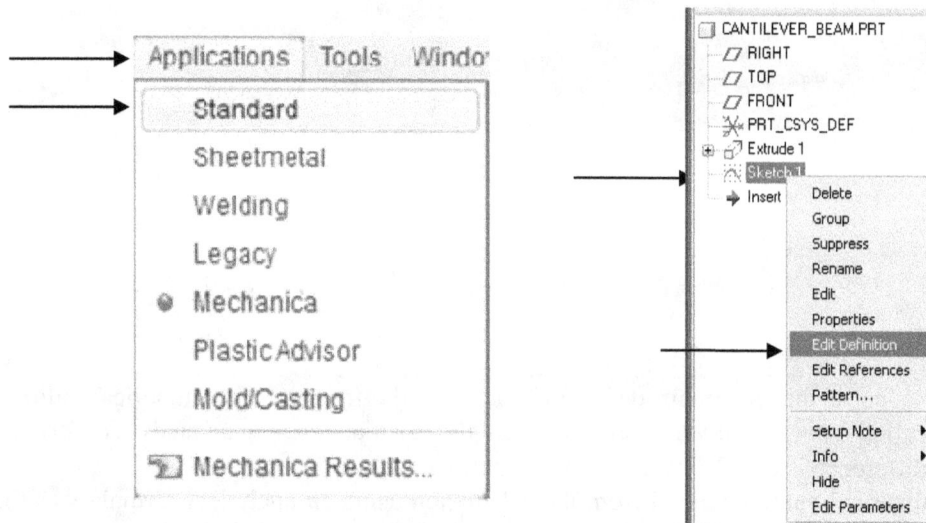

The Sketch window appears. Select **Use Previous** to go back to the previously selected sketch plane > **Sketch.**

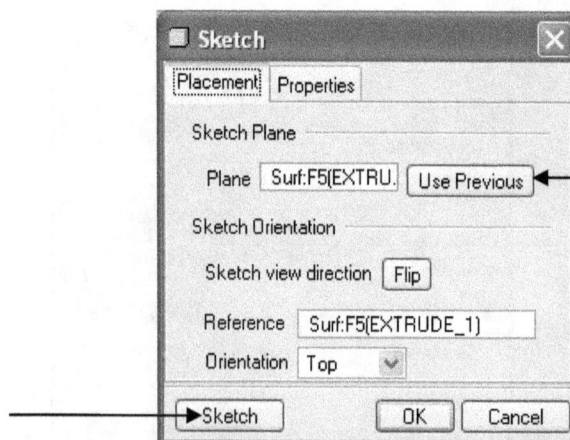

Modify the dimension from 1.00 to 2.00 by double clicking the dimension and typing 2 and press the **Enter** key > **click** the icon of **Done.**

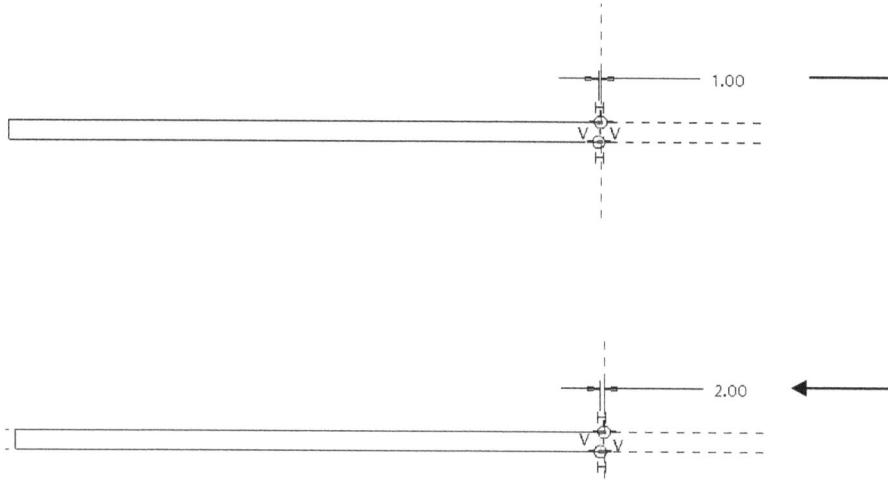

Step 6: Perform the FEA with the loading area equal to 10 x 2 mm^2.
From the menu bar, select **Applications** > **Mechanica**, going back to the Pro/Mechanica system.
Click **Continue** after checking the unit system, if asked > **Structure.**
Perform a new static analysis.
Select the icon of **Run an Analysis** > **File** > **New Static** > type *cantilever_beam_2* > **OK**.

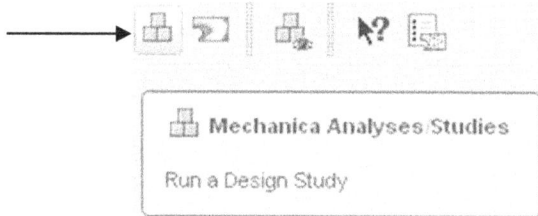

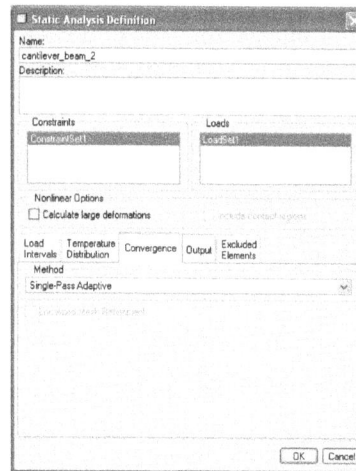

Click the icon of **Run**. When completed, select the icon of **Results.** Type *Beam Deflection* as the title > locate the Design Study folder cantilever_beam_2 > **Fringe** > **Displacement** > **Y** > **OK and Show**.

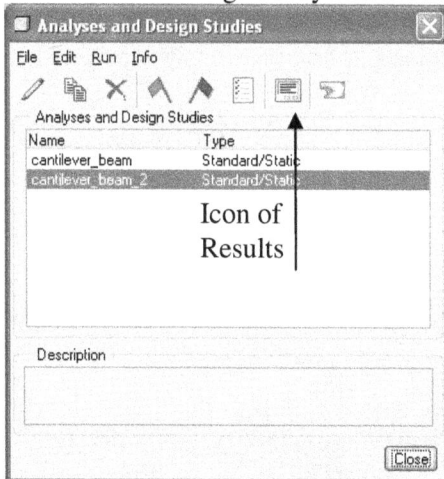

Y

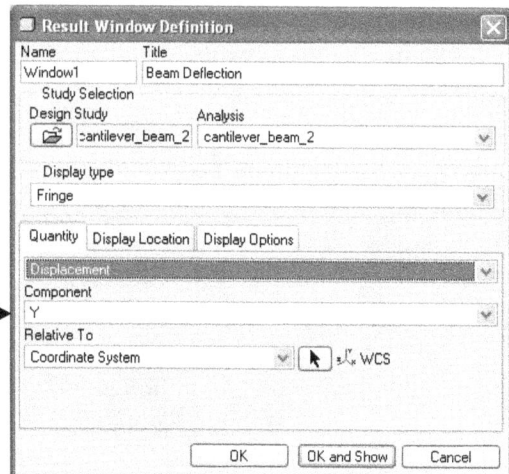

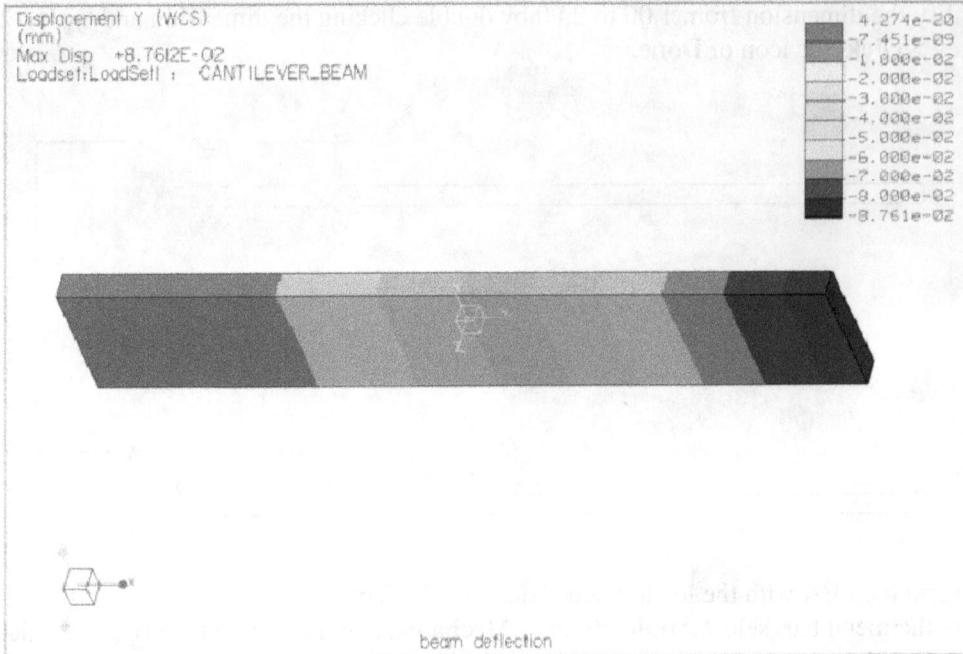

Displacement Y (WCS)
(mm)
Max Disp +8.76l2E-02
Loadset:LoadSetl : CANTILEVER_BEAM

```
 4.274e-20
-7.451e-09
-1.000e-02
-2.000e-02
-3.000e-02
-4.000e-02
-5.000e-02
-6.000e-02
-7.000e-02
-8.000e-02
-8.761e-02
```

beam deflection

Examining the value of the maximum deflection shown at the free end, the numerical value is 0.08761 mm. As we recall, the numerical value obtained from the calculation using an analytical formula was 0.08650 mm. The difference is 0.00106 mm, or about 1 μm.

Now let us change the area of the surface region from 10 x 2 mm to 10 x 0.5 mm to perform Run 3. Let us go back to Pro/ENGINEER design system. From the menu bar, select **Applications > Standard**. From the model tree, select **Sketch1** > right click on the mouse to select **Edit Definition.**

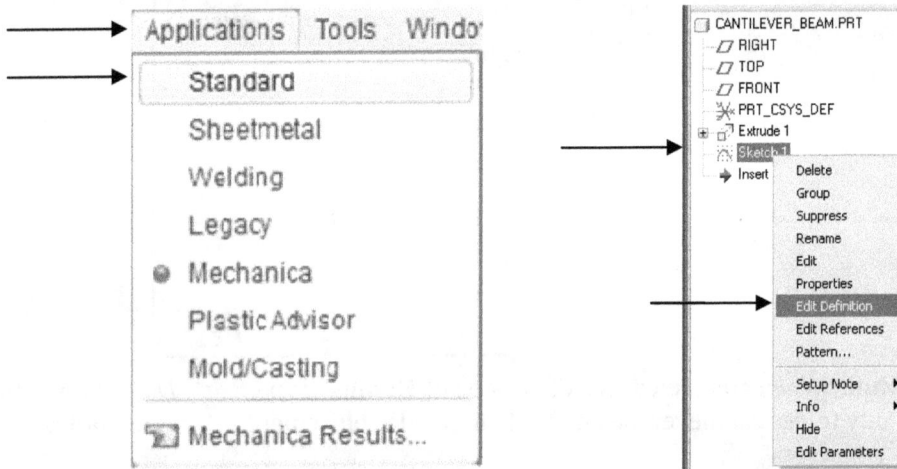

The Sketch window appears. Select **Use Previous** to go back to the previously selected sketch plane > **Sketch.**

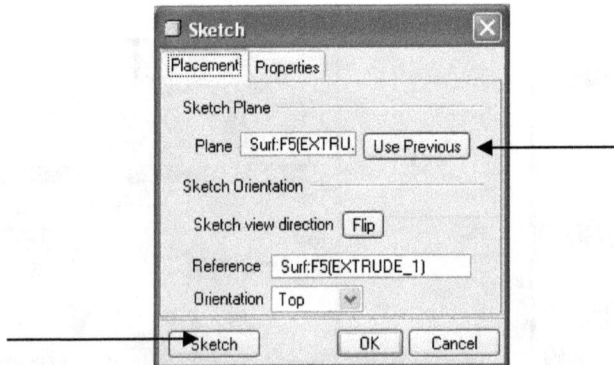

Modify the dimension from 2.00 to 0.5 by double clicking the dimension and typing 0.5 and press the **Enter** key > **click** the icon of **Done.**

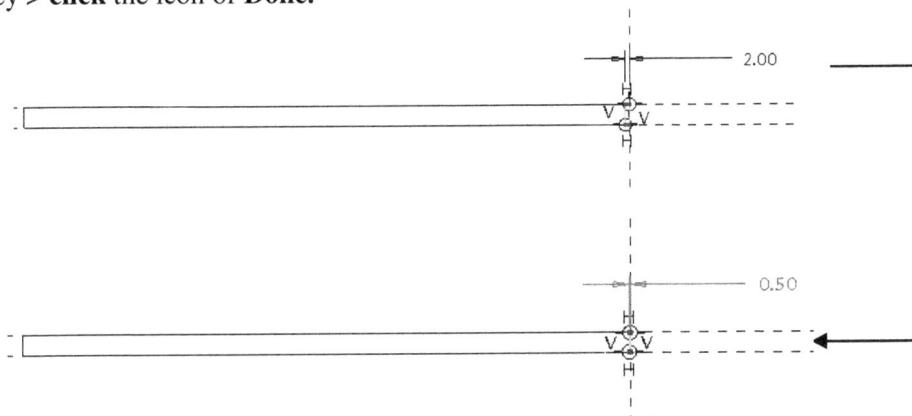

Step 6: Perform the FEA with the loading area equal to 10 x 0.5 mm^2.
From the menu bar, select **Applications > Mechanica**, going back to the Pro/Mechanica system.
Click **Continue** after checking the unit system, if asked > **Structure.**
Perform a new static analysis.
Select the icon of **Run an Analysis > File > New Static** > type *cantilever_beam_3* > **OK**.

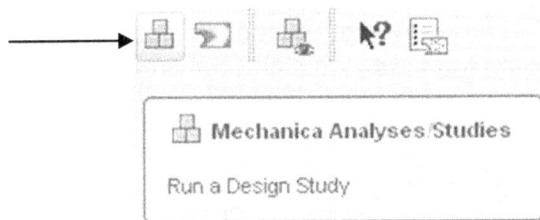

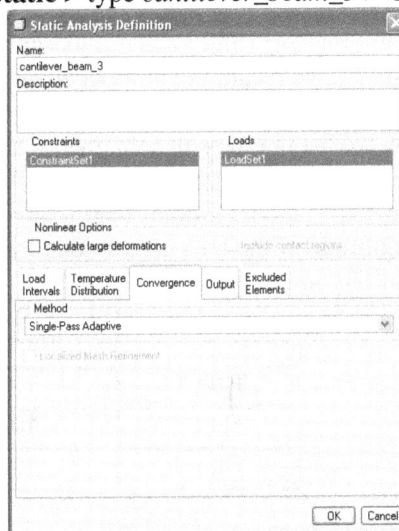

Click the icon of **Run**. When completed, select the icon of **Results.** Type *Beam Deflection* as the title > locate the Design Study folder cantilever_beam_3 > **Fringe > Displacement > Y > OK and Show**.

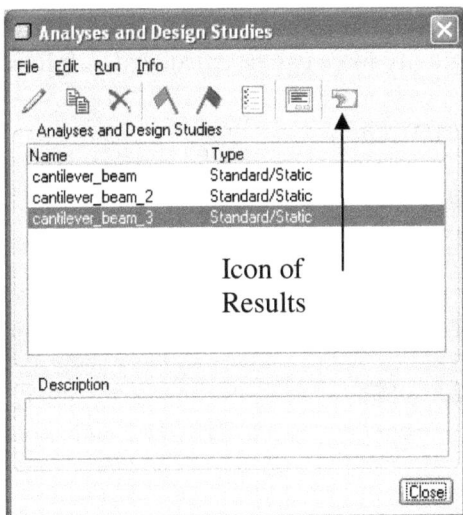

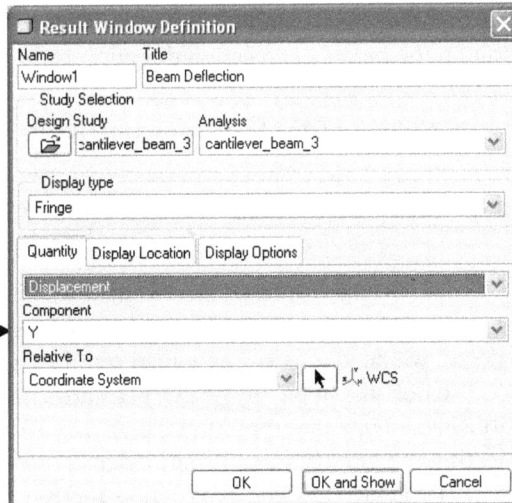

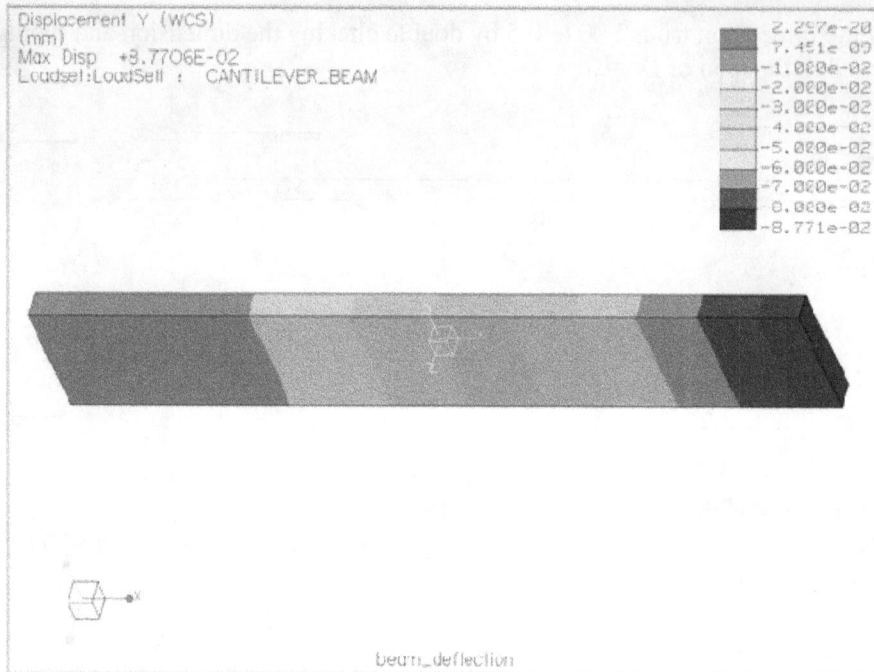

beam_deflection

Examining the value of the maximum deflection shown at the free end, the numerical value is 0.08771 mm. As we recall, the numerical value obtained from the calculation using an analytical formula was 0.08650 mm. The difference is 0.00121 mm, or about 1 μm.

Now we list the 3 results obtained from the 3 runs. There is no significant difference among the 3 results in terms of the numerical values of the maximum displacement. This observation indicates that the area selected to model a point load is not so sensitive to its width value. Certainly it is important to limit the magnitude of the area used so that it represents a point load.

	b (mm)	w (mm)	deflection (mm)
Run 1	10	1	0.08761
Run 2	10	2	0.08761
Run 3	10	0.5	0.08771

3.7 Consideration of the Gravity Force in Performing FEA

The cantilever beam studied in section 3.6 has its mass. As a result, there is a gravity force uniformly acting on the beam. The action direction is downward. Such a gravity force can be calculated as follows:

$$\text{Volume of the Beam} = (300)(50)(10) = 150000 \ (\text{mm}^3) = 150 \ (\text{cm}^3)$$

$$\text{Mass of the Beam} = (150)(7.85) = 1177.5 \ (\text{g}) = 1.1775 \ (\text{kg}) \quad (\rho = 7.85 \ \text{gram/cm}^3)$$

$$\text{Gravity Force} = (1.1775)(9.81) = 11.55 \ (\text{Newton}) \quad (g = 9.81 \ \text{m/sec}^2)$$

Now let us perform Run 4, which considers the effect of the gravity force on the beam deflection. To do so, we need to add a new load set. Let us call it LoadSet 2.

Click the icon of **Gravity Load.** In the Gravity Load window, click New and LoadSet2 is established. Type incorporating the gravity force > OK. and a new window called **Gravity Load** appears, indicating the gravity force is defined as LoadSet2 > **OK**.

Type -9810 in the Y direction > **OK**.

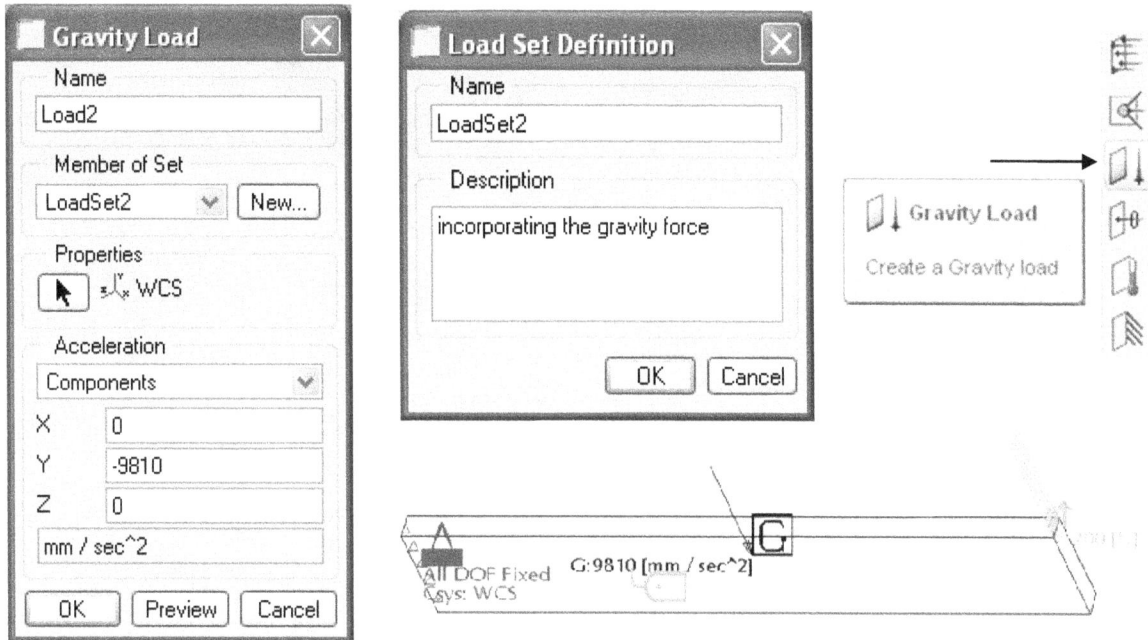

Perform a new static analysis

Select the icon of **Run an Analysis > File > New Static** > type *add_gravity_force* > make sure that both Loadset1 and Loadset 2 are selected while holding down the **Ctrl** key > check **Sum Load Sets** > **OK**.

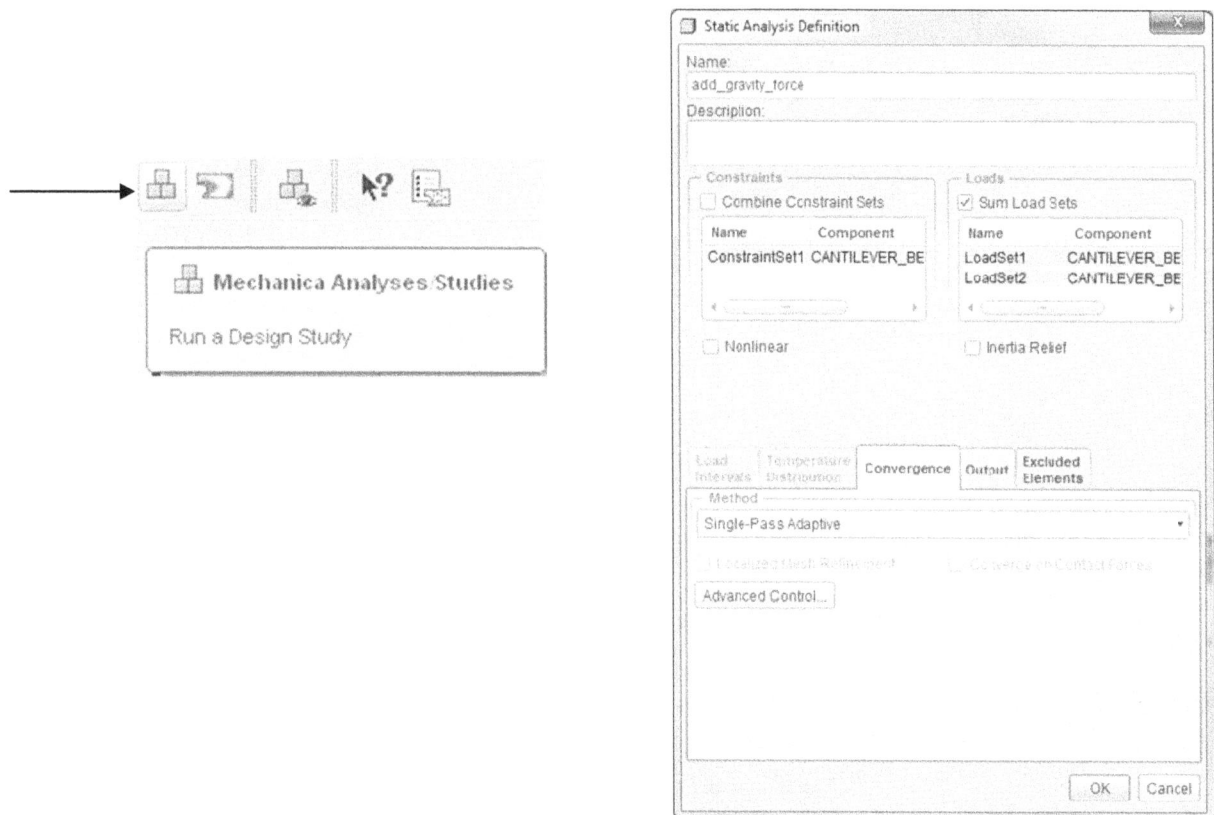

Click the icon of **Run**. When the run process is completed, select the icon of **Results.** Type *Beam Deflection* as the title > locate the Design Study folder called add_gravity_force. It is important to note that Loadset1 and Loadset2 are shown. Select both LoadSet1 and LoadSet2.

Select **Fringe** > choose **Displacement > Y > OK and Show.**

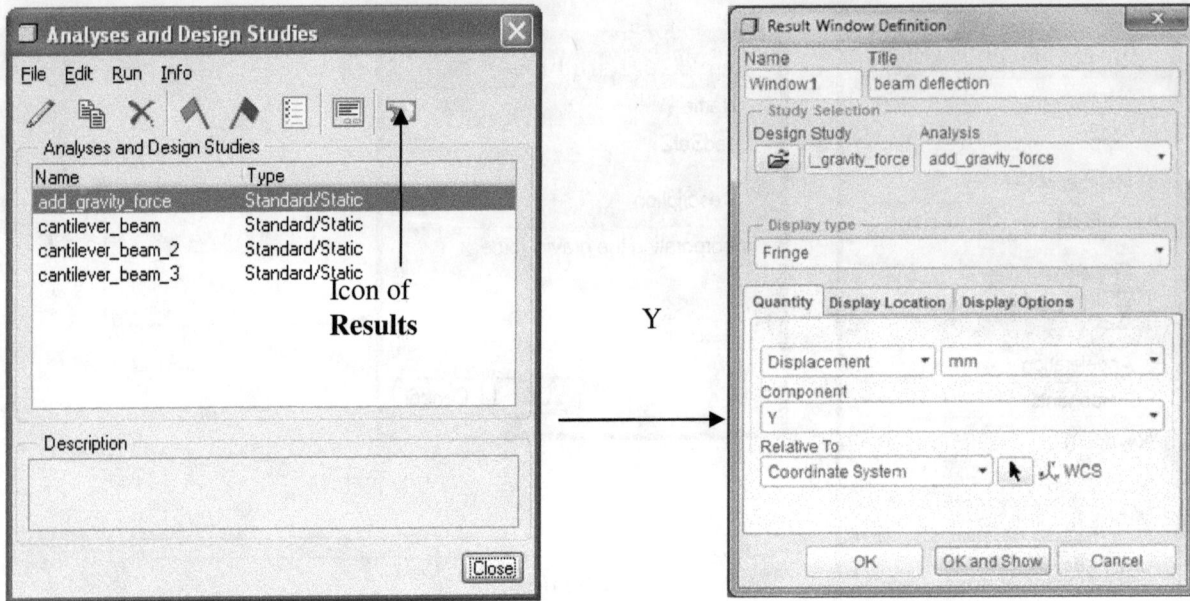

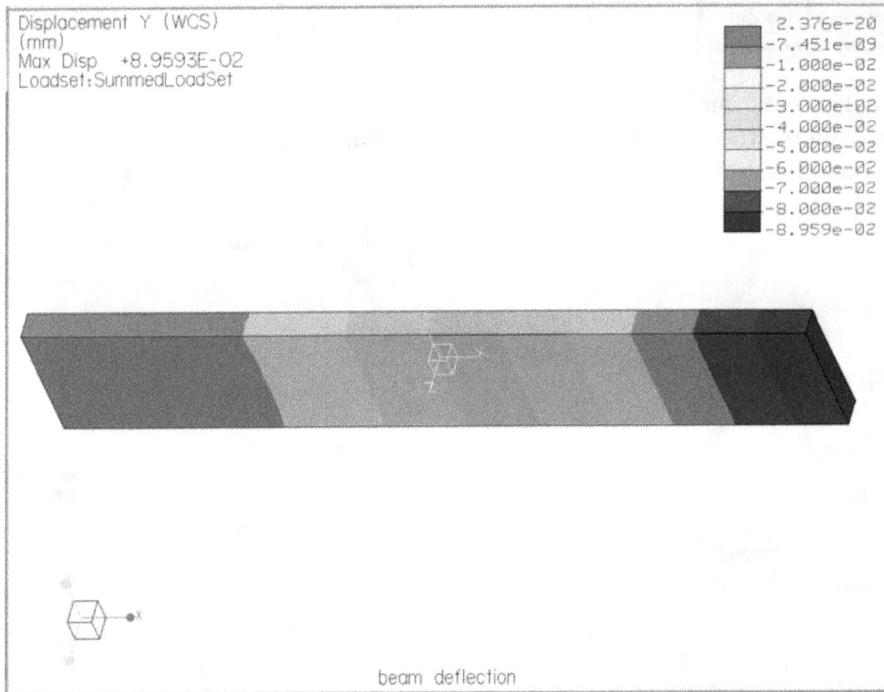

Examining the value of the maximum deflection shown at the free end, the numerical value is 0.08959 mm. As we recall, the numerical value obtained from FEA without considering the gravity force was 0.08761 mm for the surface region set at 10 x 0.5 mm². The difference is 0.00198 mm or 1.98 μm, which is rather small as compared with 0.08761 mm.

Now let us examine the contribution by the gravity force only. Click the icon of **Run an Analysis** from the main toolbar > **File** > **New Static** > type *gravity_only* as the name of the analysis folder > **OK.** Click the green flag of **Run.**

beam deflection

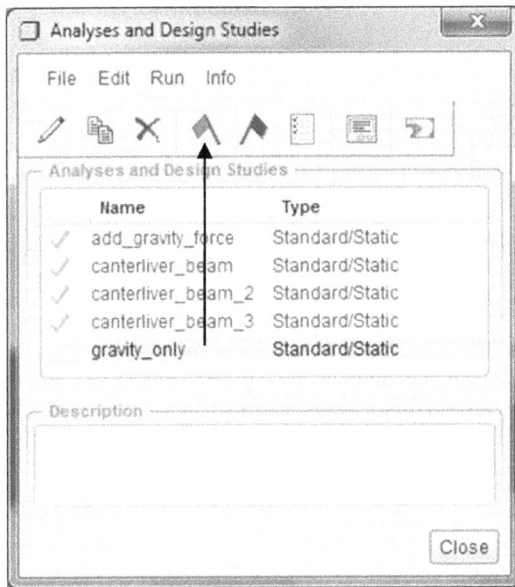

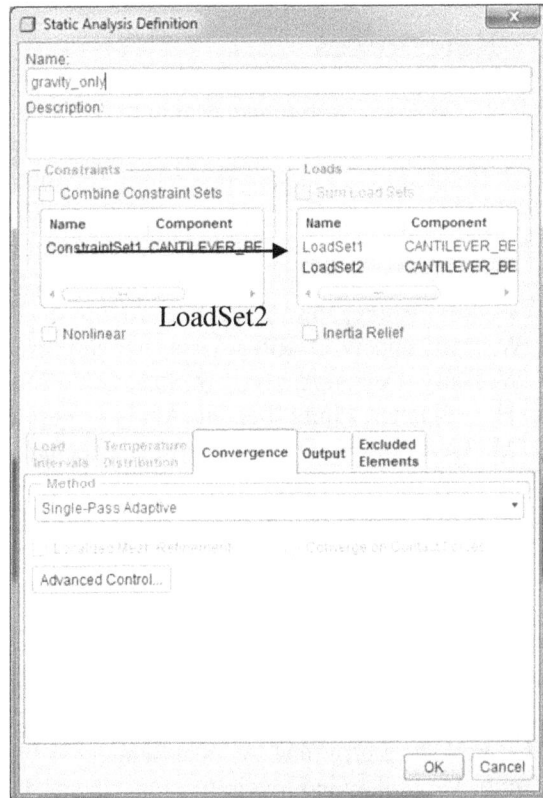

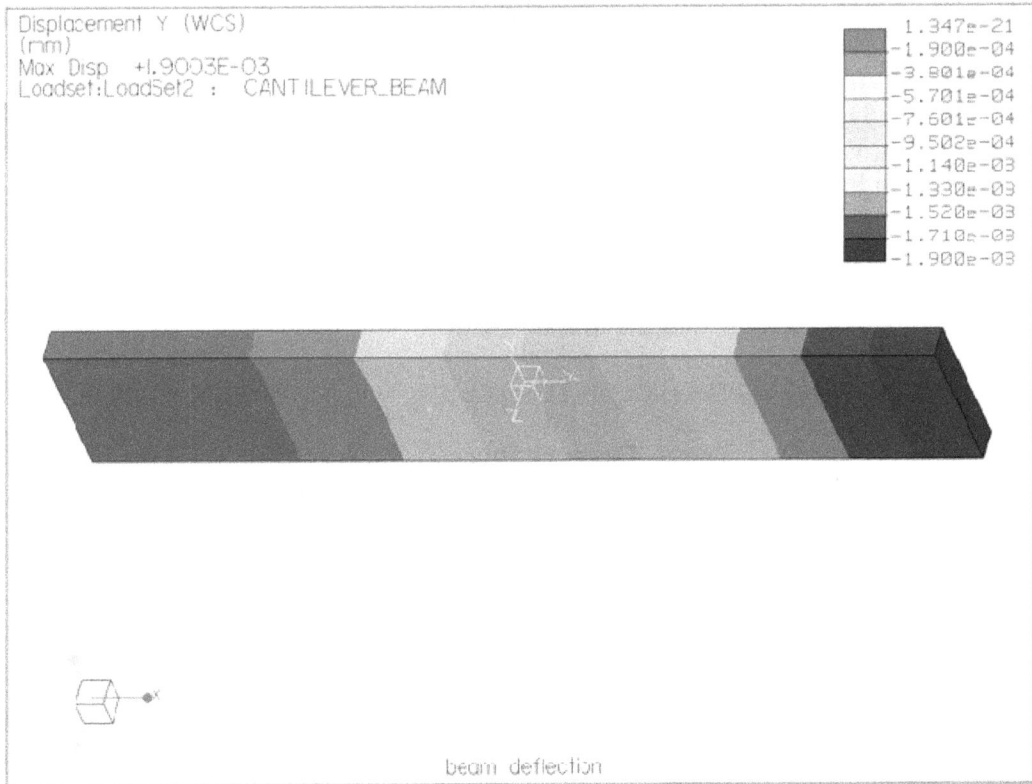

Examining the value of the maximum deflection shown at the free end, the numerical value is 0.0019 mm, which is a value very close to the difference of 0.00198 mm calculated above. In general, the gravity effect is ignored in performing FEA. However, under certain circumstances, we may still need to consider the effect due to the gravity force.

3.8   References

1. Algor Processor Reference Manual, Algor Interactive Systems, 150 Alpha Drive, Pittsburgh, PA 15238
2. K. J. Bathe, <u>Finite Element Procedure,</u> Prentice Hall, Upper Saddle River, New Jersey, 1996.
3. R. W. Clough, <u>The Finite Element in Plane Stress Analysis</u>, Proceedings of 2nd ASCE Conference on Electronic Computation, Pittsburgh, PA. September 1960.
4. R. D. Cook, D. S. Maikus and M. E. Plesha, <u>Concepts and Applications of Finite Element Analysis,</u> 3rd edition, Wiley, New York, NY, 1989.
5. J. W. Dally and W. F. Riley, Experimental Stress Analysis, McGraw Hill, 1965.
6. J. W. Dally, Design Analysis of Structural Elements, Goodington, New York, NY, 2003.
7. R. H. Gallagher, <u>Finite Element Analysis Fundamentals,</u> Prentice-Hall, Englewoood Cliffs, NJ, 1975.
8. J. M. Gere and S. P. Timoshenko, <u>Mechanics of Materials,</u> 3rd edition, PWS-KENT Publishing Company, Boston, 1990.
9. C.S. Krishnamoorthy, <u>Finite Element Analysis, Theory and Programming</u>, 2nd Ed., 1995.
10. D. L. Logan, A First Course in the Finite Element Method Using ALGOR, PWS Publishing, Boston, 1997.
11. H. C. Martin, <u>Introduction to Matrix Methods of Structural Analysis</u>, McGraw-Hill, New York, NY, 1966.
12. R. L. Norton, <u>Machine Design: An Integrated Approach</u>, Prentice Hall, Upper Saddle River, New Jersey, 1996.
13. M. Ortiz, Y. Leroy and A. Needleman, A <u>Finite Element Method for Localized Failure Analysis,</u> Comp. Meth. Applied Mech. Enging, 61, pp. 189-214, 1987.
14. M. Petyl, <u>Introduction to Finite Element Vibration Analysis,</u> Cambridge University Press, 1990.
15. T. H. H. Pian and P. Tong, <u>Basis of Finite Element Methods for Solid Continua,</u> Int. J. Numer. Methods Eng., Vol. 2, No. 7, April 1964, pp. 1333-1336.
16. P. G. Plockner, <u>Symmetry in Structural Mechanics, J. of the Structural Division,</u> American Society of Civil Engineers, Vol. 99, No. ST1, pp. 71-89, 1973.
17. J. N. Reddy, An Introduction to the Finite Element Method, McGraw-Hill Publishing Company, New York, NY, 1984.
18. T. Ross, <u>Finite Element Program in Structural Engineering and Continuum Mechanics,</u> Horwood Publishing Limited, Chichester, England, 1996.
19. T. Ross, <u>Advanced Applied Finite Element Methods,</u> Horwood Publishing Limited, Chichester, England, 1998.
20. L. Stasa, <u>Applied Finite Element Analysis for Engineers</u>, Holt, Rinehart and Winston, New York, NY, 1985, pp. 101-157.
21. S. P. Timoshenko and D. H. Young, <u>Theory of Structure</u>, McGraw-Hill, Kogaksha Limited, 1965.
22. M. J. Turner, R. W. Clough, H. C. Martin, and L. J. Topp, <u>Stiffness and Deflection Analysis of Complex Structures,</u> J. of the Aeronautical Sciences, Vol. 23, No. 9, pp. 805-824, Sept. 1956.
23. O. C. Zienkiewicz, <u>The Finite Element Method</u>, third edition, McGraw-Hill, New York, NY, 1977.
24. O. C. Zienkiewicz and R. L. Taylor, <u>The Finite Element Method</u>, 4th edition, Vol. 1, McGraw-Hill (UK),London, 1989.

3.9   Exercises

1. A cantilever beam is subjected to a non-uniformly distributed load. The total load is 500 Newton, acting downward. The following conditions are assumed:

    - The left side of the beam is fixed.
    - The right side of the beam is free.
    - The material of the cantilever beam is steel.

- The two weighting factors, which characterize the non-uniform distribution of the load condition, are 1.0 and 0.5 respectively.

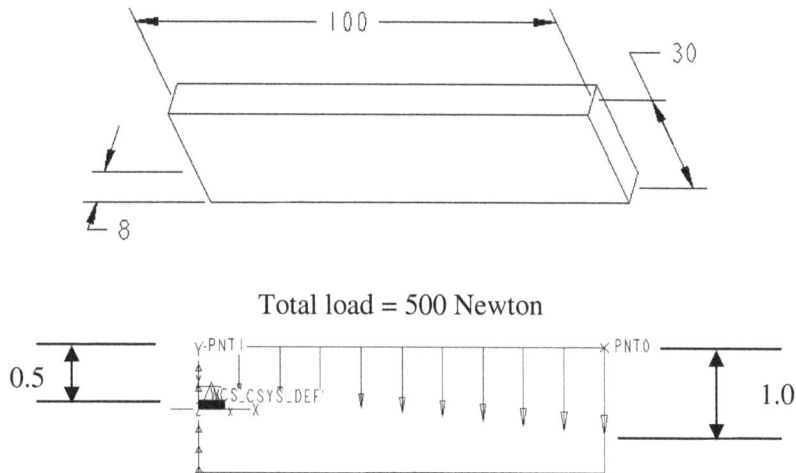

Total load = 500 Newton

- Plot the distribution of the deformation in the vertical direction.
- Plot the distribution of the maximum principal stress of the 3D solid model.

2. A cantilever beam is shown below. The left end is fixed and the right end is set to free. The structure is subjected to a uniform load, which acts on a surface region. The surface region is a rectangle. The two dimensions are 150 mm and 8 mm, respectively.

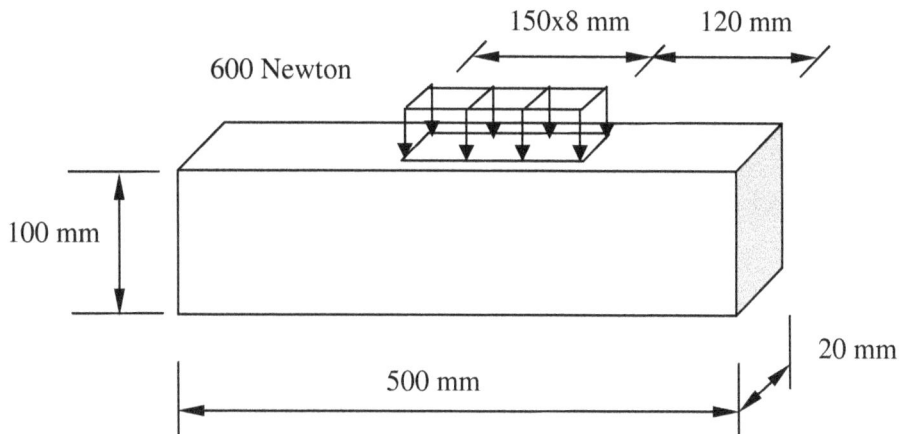

1) A 3D solid model is available and the datum curve for defining the load condition is also available. You are asked to create a surface region so that a uniform load condition can be defined.
2) Assume that the material is STEEL, as listed in the Pro/E material library. Define a constraint set for the fixed end at the left end of the beam, and define uniform load acting on the surface region as specified in (1).
3) Fill in the numerical values in the following table with their units.

	maximum displacement of the beam in Y	max principal stress of the beam	max Von mises stress of the beam
Location (x, y, z)			

3. Referring to the beam structure subjected to a uniformly distributed load in the text, do the following:

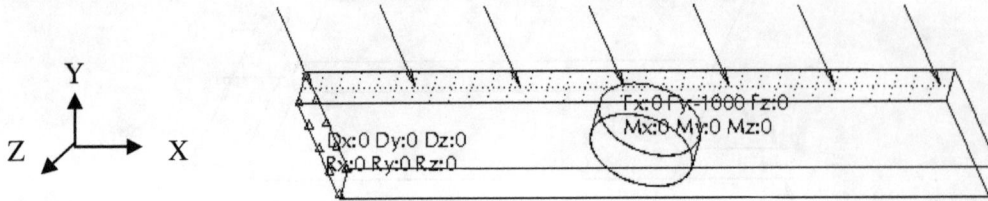

(1) Determine the maximum displacement on the beam when subjected to this uniformly distributed load and specify its location.
(2) Determine the maximum principal stress and specify its location.
(3) Determine the maximum von Mises stress and specify its location.

4. Referring to the shaft with step component in Chapter 4, use Design Study and Global Sensitivity to perform a single run of this program, based on the following conditions. We should be able to obtain the following plot, namely, the maximum principal stress vs dimension of the corner radius, which varies from 1 mm to 16 mm.

Calculation of Stress Concentration Factors (D/d=2 and D=100 mm)				
Ratio: r/d	Radius (mm)	σmax_principal (MPa)	σmax_reference (MPa)	SCF
0.05	2.50	1.5223	0.5098	2.99
0.10	5.00	1.1720	0.50958	2.30
0.15				
0.20				
0.25				
0.30				

Stress Concentration Factor vs Corner Radius

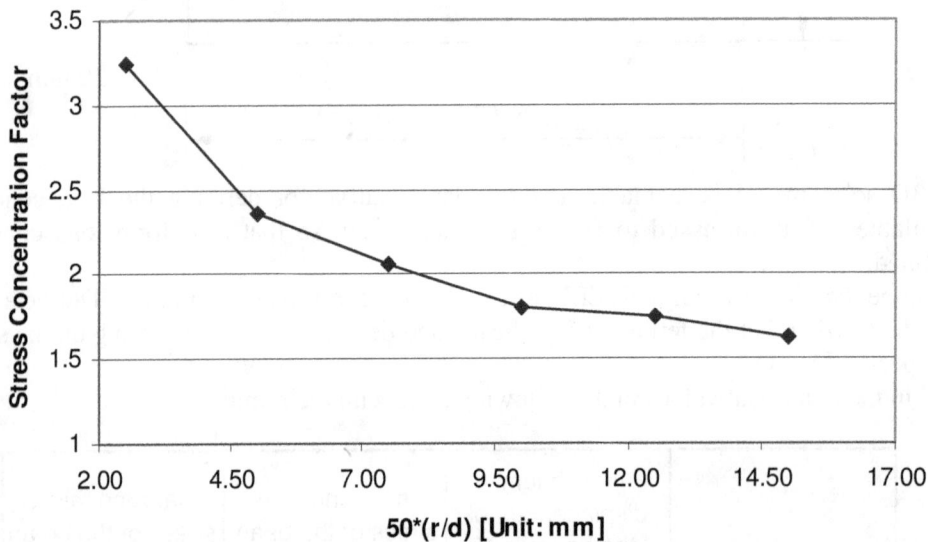

If time permits, change the dimensions of the 2 diameters while keeping the ratio of D to d at 2.0. For example, let D = 50 mm and d = 25 mm. Repeat the FEA runs with the radius value varies from 1.25 mm to 7.50 mm. Examine the results obtained from those runs.

CHAPTER 4

MODELING OF MEASURES, CONSTRAINTS AND ASSEMBLIES

4.1 Introduction

In Chapter 3, we discussed the process of modeling a variety of load conditions. There are three focuses in this chapter. The first focus is to use measures to quantify the performance of interest. The second focus is to model certain types of constraint conditions. The third focus is to perform FEA for assemblies, instead of components as we have discussed in chapters 2 and 3. In the chapter, we also continue our discussion on modeling the load conditions. The process of modeling load conditions with torque and centrifugal force are presented.

4.2 Definition of Measures for Stress Concentration Study

Stress concentration occurs when changes in the geometry of a structural material occur. The stress is literally concentrated at a point or over a very small area. Structural failures happen very often due to the presence of stress concentration. For safety consideration, locations of stress concentration under loading are often identified in mechanical structures. The following example illustrates the stress contraction occurs at the shoulder of a shaft component. As illustrated, the shaft component is under tension with its left end fixed to the wall.

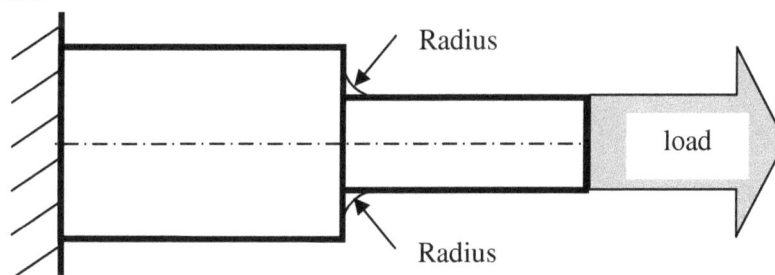

It is likely that the stress developed in the region of shoulder is not uniformly distributed. High stress is developed within the region of shoulder. The main reason for this stress irregularity is due to the presence of a shoulder fillet.

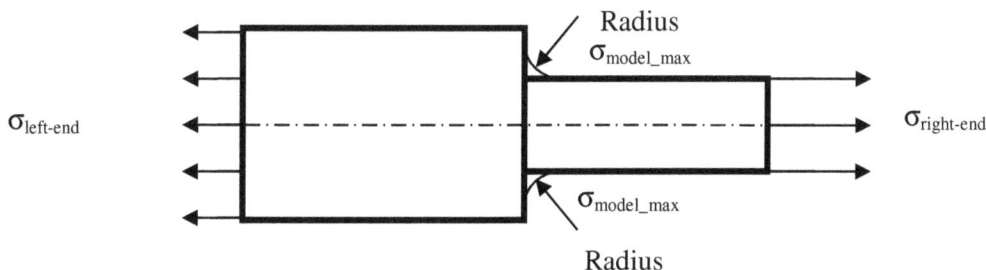

The stress concentration factor, or SCF, is defined as a ratio of $\sigma_{\text{max-model}}$ to $\sigma_{\text{right-end}}$ in the current case, or

$$Stress\_Concentration\_Factor = \frac{\sigma_{model\_max}}{\sigma_{right\_end}}$$

It is evident that the value of SCF is a function of the ratio of the radius to the diameter of the circular section at the right end.

In this example, let us create a shaft, which is shown below. The 2 diameters are 100 mm and 50 mm, respectively. Therefore, the sectional ratio in terms of diameter is 2:1. Assume that the material is steel and the load acting on the right side is 1000 Newton so that the shaft is under tension. The radius value shown in the figure is 15 mm. We will vary this value of radius from 2.5 mm to 15 mm.

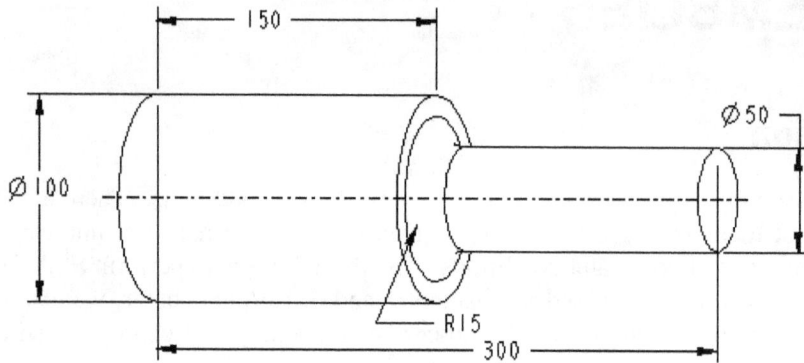

Step 1: Create a 3D solid model for the shaft component.
File > New > Part > type *bar_with_step* as the file name and clear the icon of Use default template. Select mmns_part_solid (units: Millimeter, Newton, Second) and type *stress concentration* in DESCRIPTION, and *student* in MODELED_BY, then OK.

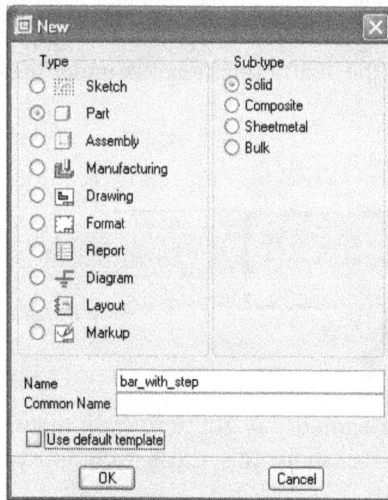

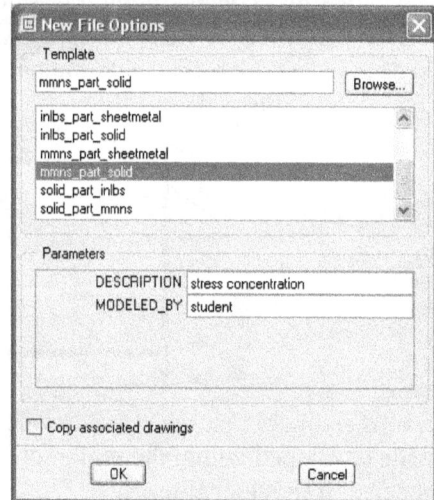

Step 2: Create a 3D solid model for the tube component.
Pick up the icon of Revolve. To initiate the process of sketching, activate Placement > Define.

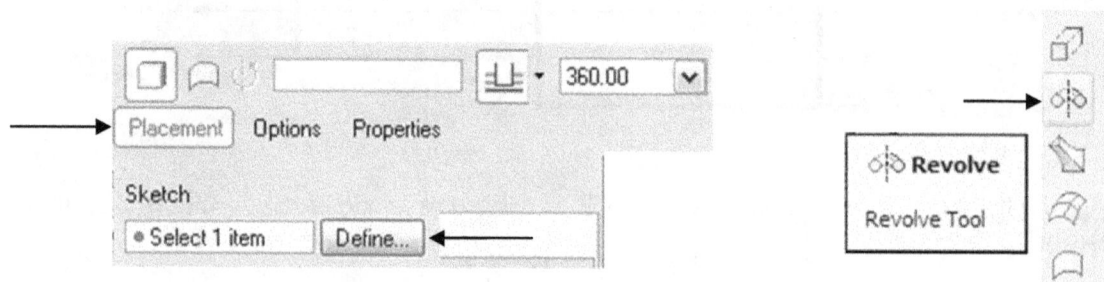

Select the **FRONT** datum plane displayed on screen as the sketch plane. Pick the box of **Sketch** to accept the **RIGHT** datum plane as the default reference to orient the sketch plane, as illustrated below:

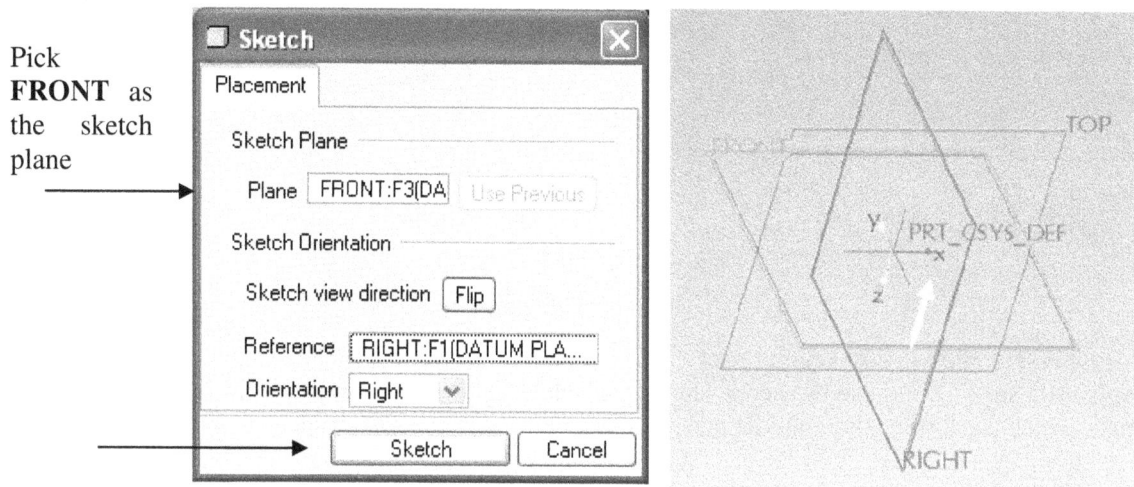

Pick **FRONT** as the sketch plane

Click the icon of **Centerline**. Sketch a horizontal centerline along the axis, as shown below.

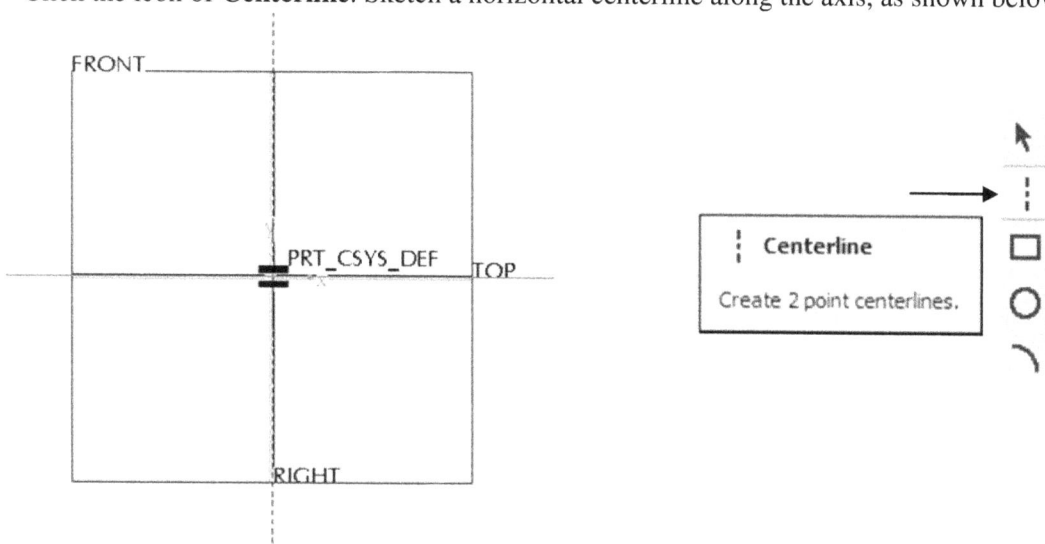

Select the icon of **Line** from the toolbar of sketcher and sketch the following shape. The 4 dimensions are 25, 50, 150 and 300, respectively.

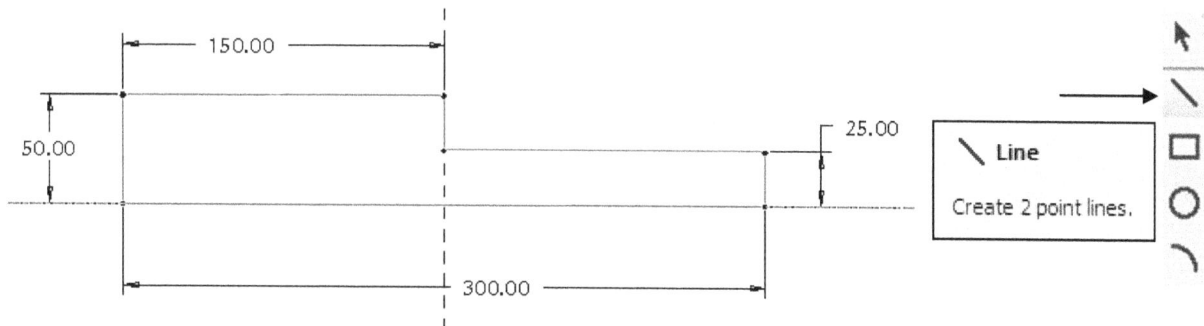

Pick the axis and right-click for selecting **Axis of Revolution** from the pop up window.

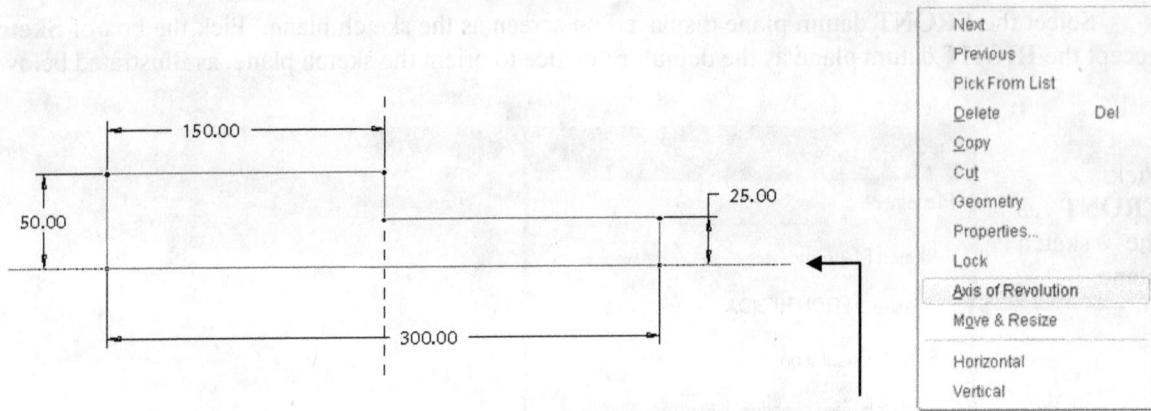

Next
Previous
Pick From List
Delete   Del
Copy
Cut
Geometry
Properties...
Lock
Axis of Revolution
Move & Resize

Horizontal
Vertical

Pick this axis and right-click

Upon completing the sketch, select the icon of **Done** from the toolbar of sketcher. Select the icon of **Apply and Save** from the feature control panel, completing the creation of a 3D solid model for the bar structure, which has a step in the middle position.

✓ **Done**
Continue with the current section.

Sketch

Apply and Save

Step 3: Create a round and the dimension is R2.5.
From the toolbar of feature creation, click the icon of **Round** > type 2.5 as the radius value > pick the corner, as shown. Upon completing, select the icon of Complete from the feature control panel.

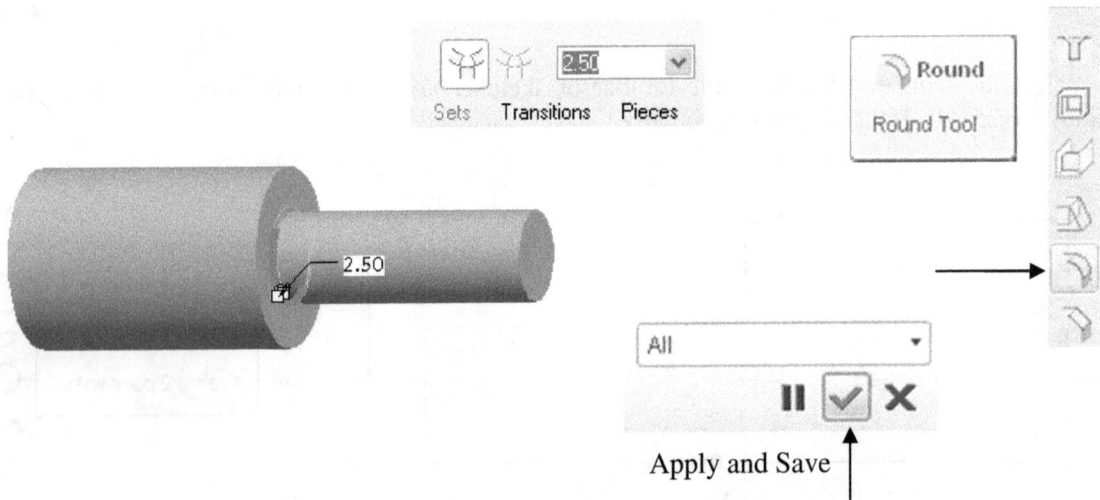

Sets   Transitions   Pieces

Round
Round Tool

All

Apply and Save

Step 4: Create a datum point so that a measure can be later defined.
To define a datum point, click the icon of **Datum Point Tool** from the toolbar of datum feature creation. While holding down the **Ctrl** key, pick the centerline of the bar and the surface on the right side > **OK**.

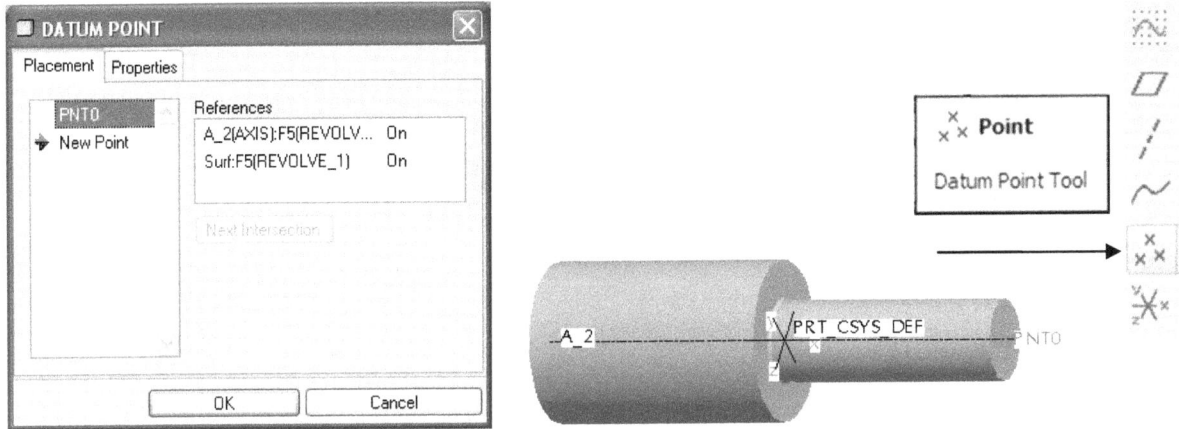

Step 5: To define a measure, we have to go to **Pro/Mechanica.**
From the top menu, click **Applications > Mechanica > Continue > Structure > OK.**

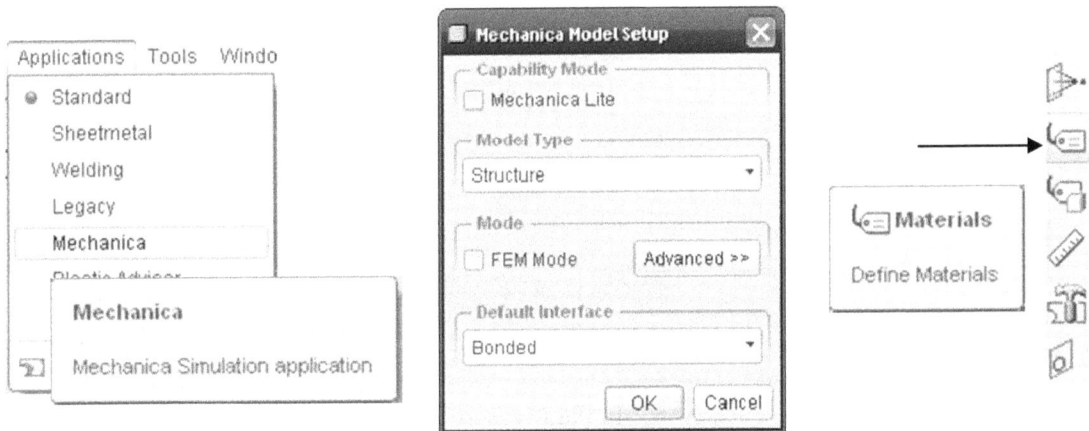

From the toolbar of functions, select the icon of **Materials >** select **STEEL** from the left side called Materials in Library > click the directional arrow so that the selected material type goes to the right side called Materials in Model.

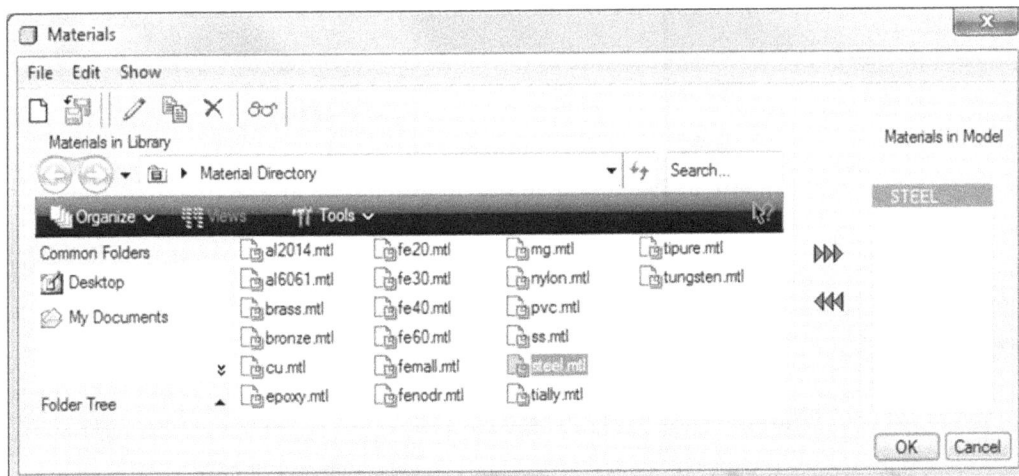

Click the icon of **Material Assignment**, the pop-up window indicates there is one component and the material type is STEEL. Just click **OK.**

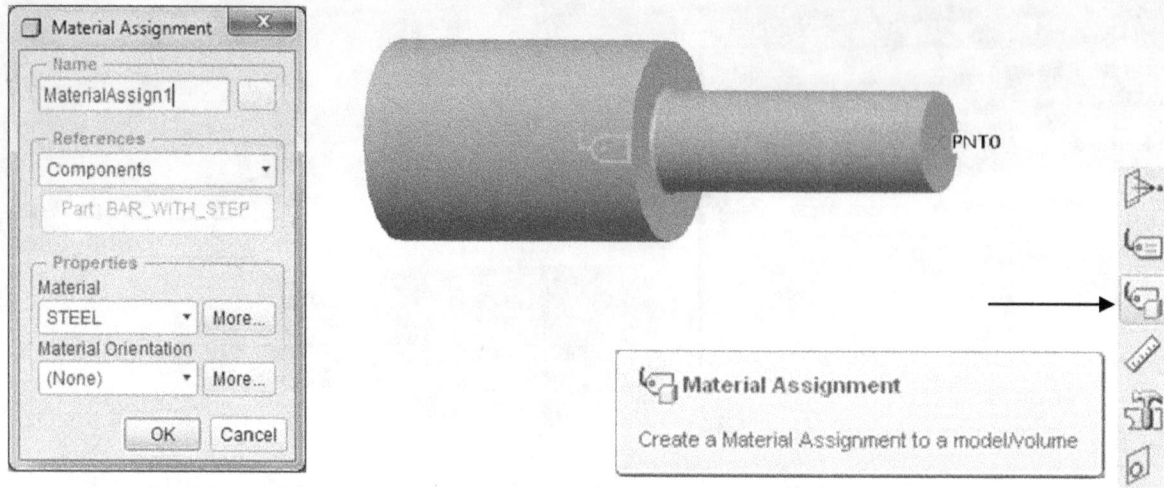

Step 6: Define a measure, which is associated with the created datum point.
Select the icon of **defining measures** from the toolbar of functions.

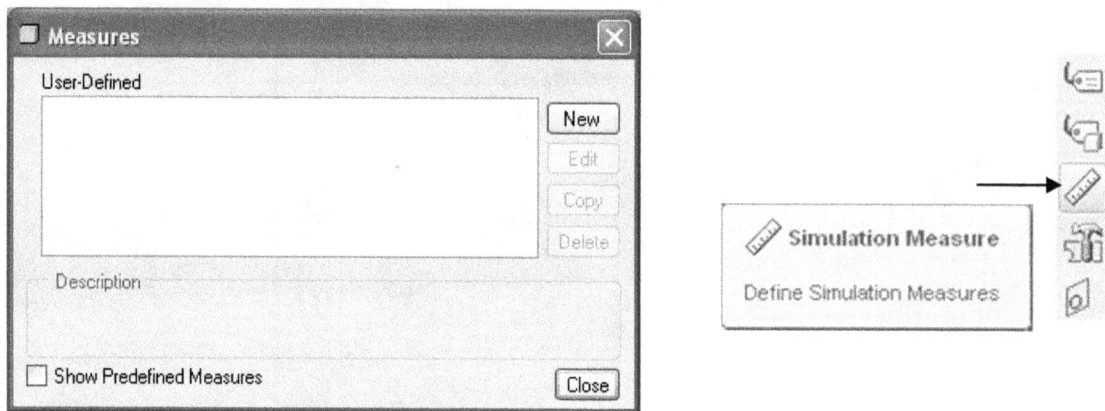

In the Measure window, click **New.** In the Measure Definition window, type *reference_stress* as the Name > **Stress** > select **Max Principal** > **At Point** > pick the data point **PNT0** > **OK**. In the Measure window, click **Close**.

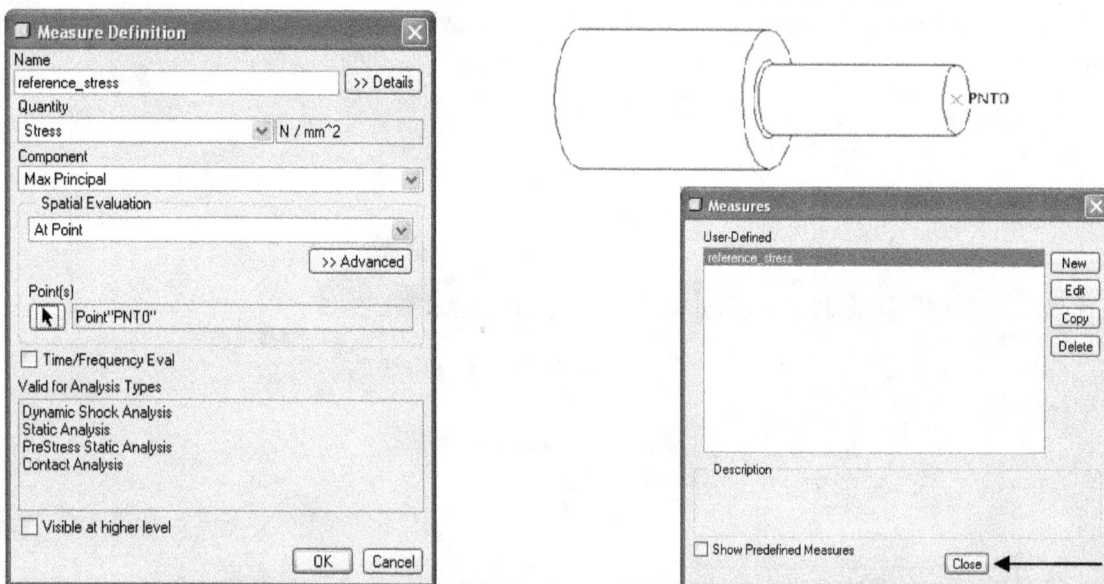

The symbol of a measure is on display.

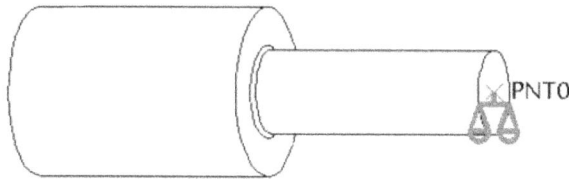

Step 7: Define the constraint condition.

To define the fixed end constraint on the left side of the bar structure, select the icon of **Displacement Constraint** from the toolbar of functions > **Surface(s)** > pick the surface at the left side of the tube.

Because the constraint condition is "a fixed end", we set all 6 degrees of freedom equal to zero, which is the default setting > **OK**.

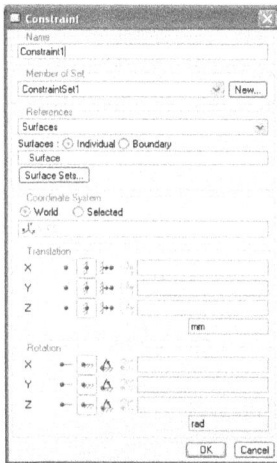

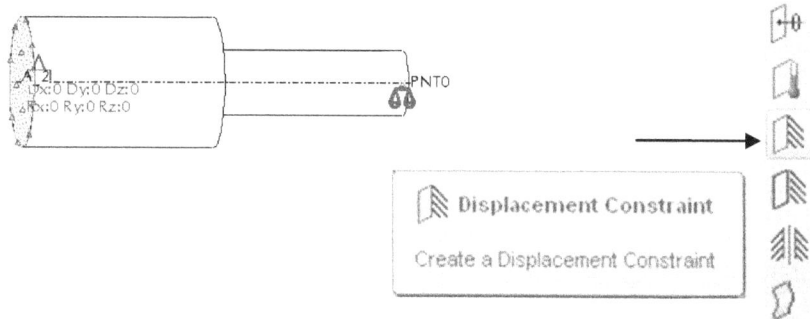

Step 8: Define the load condition.

Select the icon of **New Force/Moment Load** from the toolbar of functions > **Surface(s)** > pick the surface on the right side. Specify the force as 1000 N along the positive X direction > **OK**.

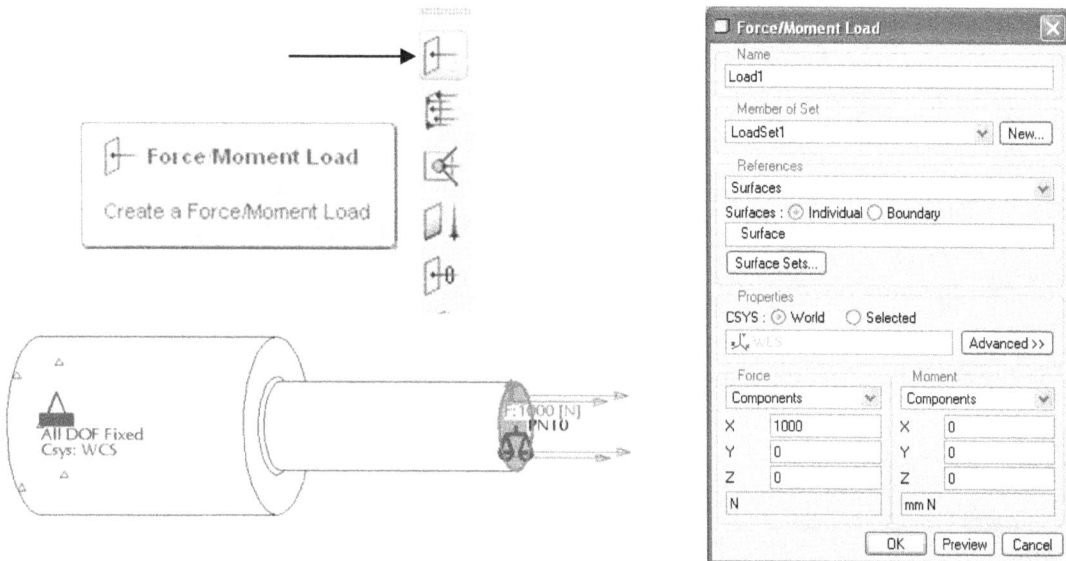

Step 9: Under the Pro/MECHANICA environment, set up **Analyses** and **Run** it

Select the icon of **Mechanica Analyses/Studies** from the main toolbar > **File** > **New Static** > type *bar_with_step* as the name of the analysis folder > **OK**. Click the box of **Run**.

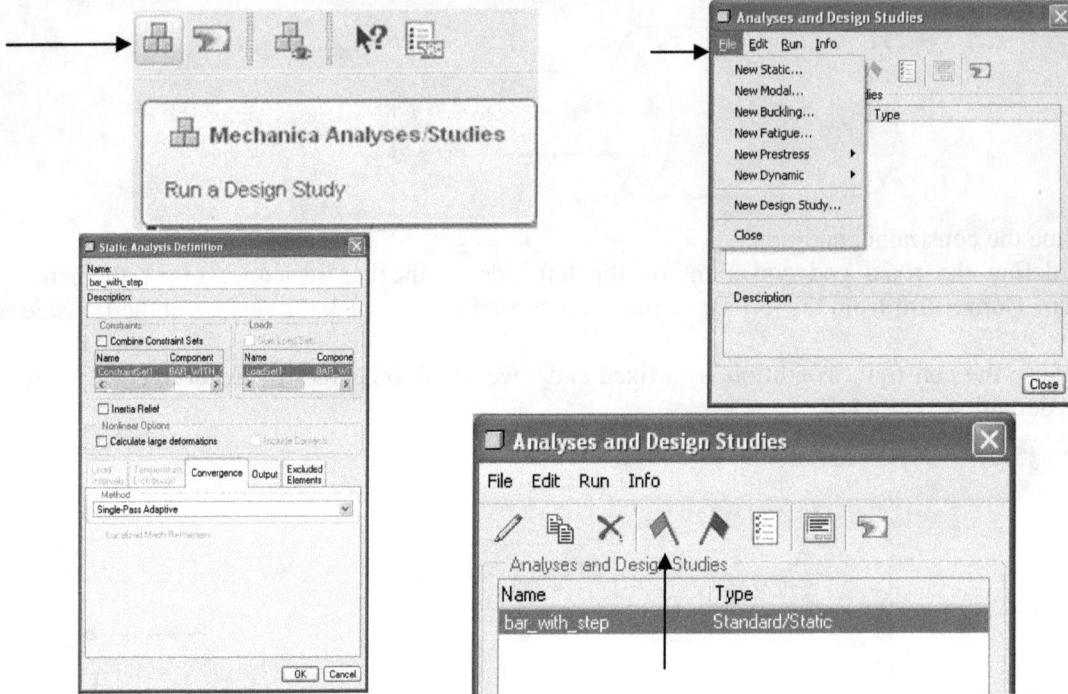

There is always a message displayed on screen. The message is asking the user "Do you want to run interactive diagnostics?" Click **Yes**. In this way a warning message will appear if an abnormal operation occurs.

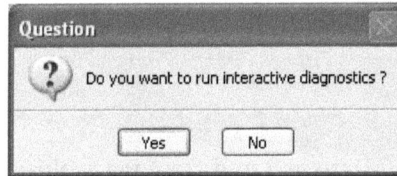

To monitor the computing process, users may click the icon of **Display study status**.

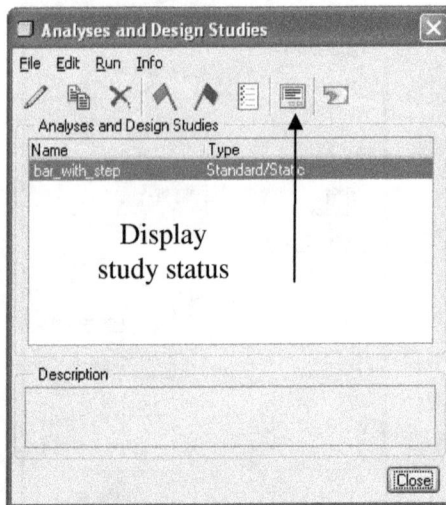

To study the obtained result, we write down the 2 values, namely, max_stress_prin and reference_stress. These 2 values are 1.506308 MPa and 0.5084501 MPa, respectively.

$$Stress\_Concentration\_Factor = \frac{\sigma_{model\_max}}{\sigma_{right\_end}} = \frac{1.505308}{0.5084501} = 2.96$$

To visualize the stress distribution, click the icon of **Review results > Stress > Max Principal >
OK and Show.**

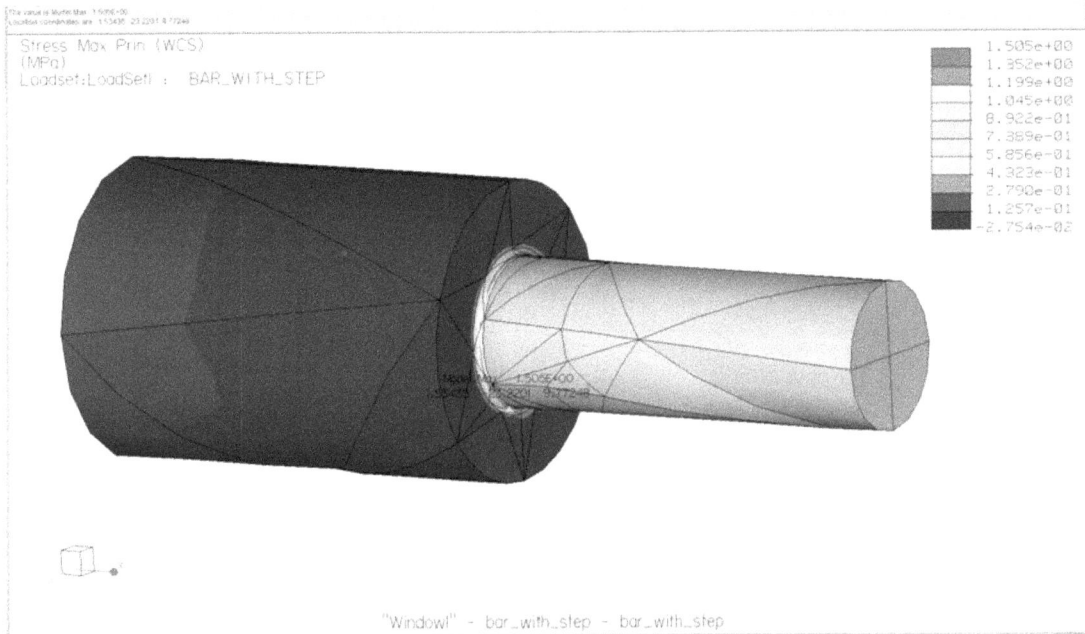

Step 10:   Let us change the radius value from 2.5 mm to 5.0, and run the program, again.
    In order to do so, we need to select the radius of the round feature as the design parameter, and set
its new value to 5.0 mm.
    How to define a design parameter? In the **Analyses and Design Studies window**, **File > New
Structural Design Study**. In the **Standard Study Definition** window, type radius_5 as the name of
design study > highlight bar_with_step (Static) > click the box called **Select dimension from model**.

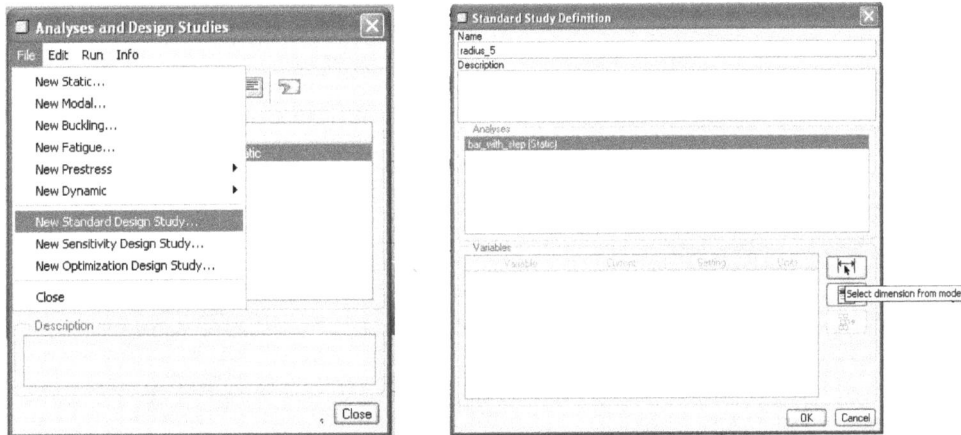

On the display, pick the dimension of corner radius, R2.5 and specify 5.0 as the new setting > **OK**.

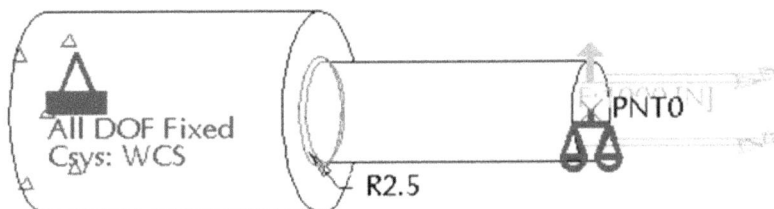

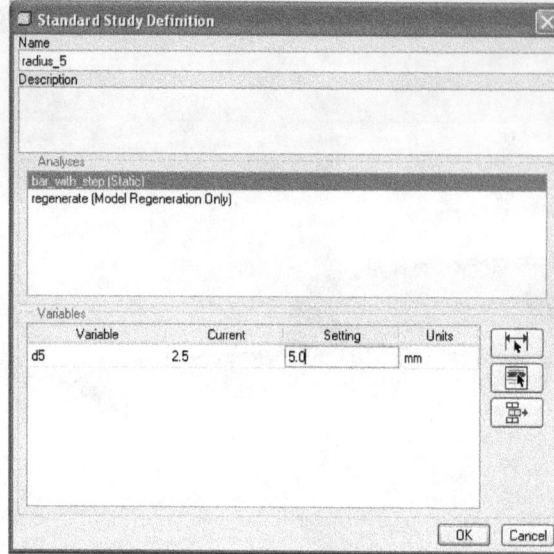

In the Analyses and Design Studies window, radius_5 is highlighted. Click the icon of **Run**. Click **Yes** to run interactive diagnostics for error checking.

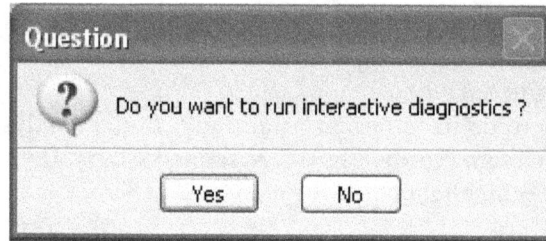

Click the icon of **Display study status** and check the values of maximum principal stress and the reference stresses. They are 1.178181 MPa, and 0.5084719 MPa, respectively.

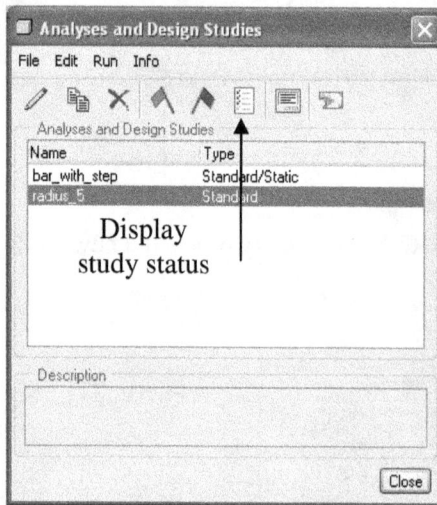

```
max_beam_bending:     0.000000e+00
max_beam_tensile:     0.000000e+00
max_beam_torsion:     0.000000e+00
max_beam_total:       0.000000e+00
max_disp_mag:         5.031412e-04
max_disp_x:           5.024705e-04
max_disp_y:           2.649079e-05
max_disp_z:          -1.777851e-05
max_prin_mag:         1.178181e+00
max_rot_mag:          0.000000e+00
max_rot_x:            0.000000e+00
max_rot_y:            0.000000e+00
max_rot_z:            0.000000e+00
max_stress_prin:      1.178181e+00
max_stress_vm:        1.043737e+00
max_stress_xx:        1.062412e+00
max_stress_xy:       -4.236308e-01
max_stress_xz:        3.786437e-01
max_stress_yy:        3.470053e-01
max_stress_yz:       -9.071569e-02
max_stress_zz:        3.550341e-01
min_stress_prin:     -1.416334e-01
strain_energy:        2.506120e-01
reference_stress:     5.084719e-01
```

Let us calculate the value of stress concentration factor. As shown below, the calculated value is 2.32.

$$Stress\_Concentration\_Factor = \frac{\sigma_{\text{mod}\,el\_max}}{\sigma_{right\_end}} = \frac{1.171961}{0.5095763} = 2.30$$

If we need to calculate a set of radius values, for example 6 radius values, should we repeat the above process six times? Instead of running the program 6 times for the 6 radius settings, let us use a new function called **New Sensitivity Study** so that a single run would be sufficient to accomplish this task.

In the **Analyses and Design Studies window, File > New Sensitivity Design Study**. In the **Sensitivity Study Definition** window, type *stress concentration* as the name of sensitivity design study > select **Global Sensitivity** > highlight bar_with_step (Static) > click the box called **Select dimension from model**.

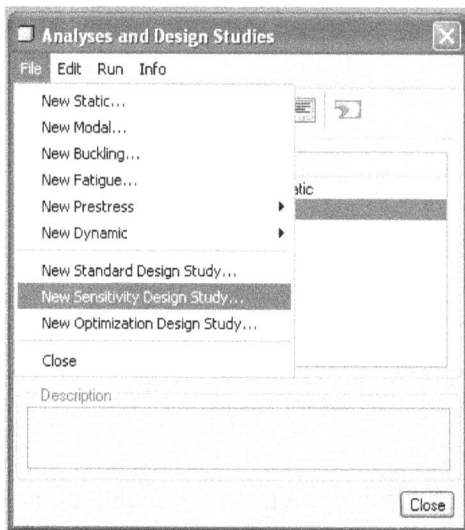

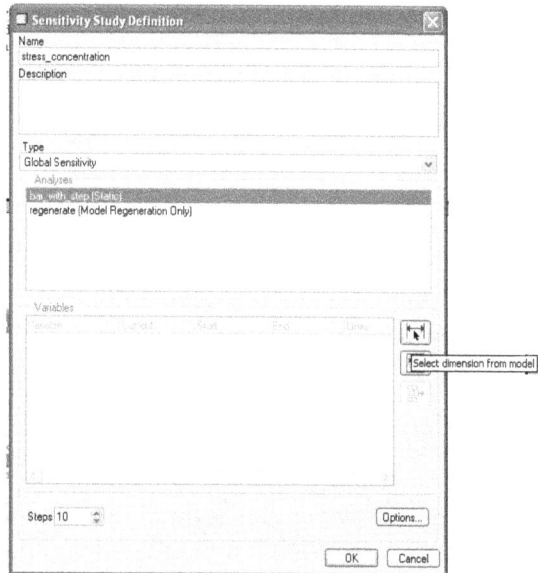

On the display, pick the dimension of corner radius, R2.5. Set the 2 limits to 1 and 16, respectively. Accept 10 as the number of steps > **OK**.

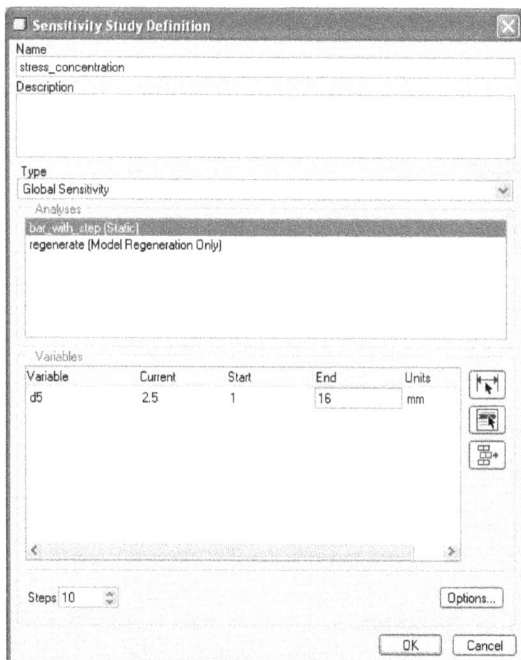

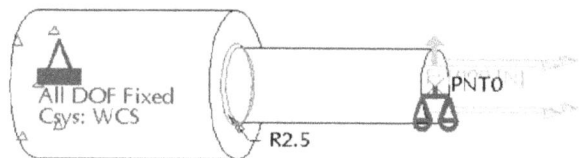

Now let us run the global sensitivity study called stress concentration. Click **Yes** to run interactive diagnostics for error checking.

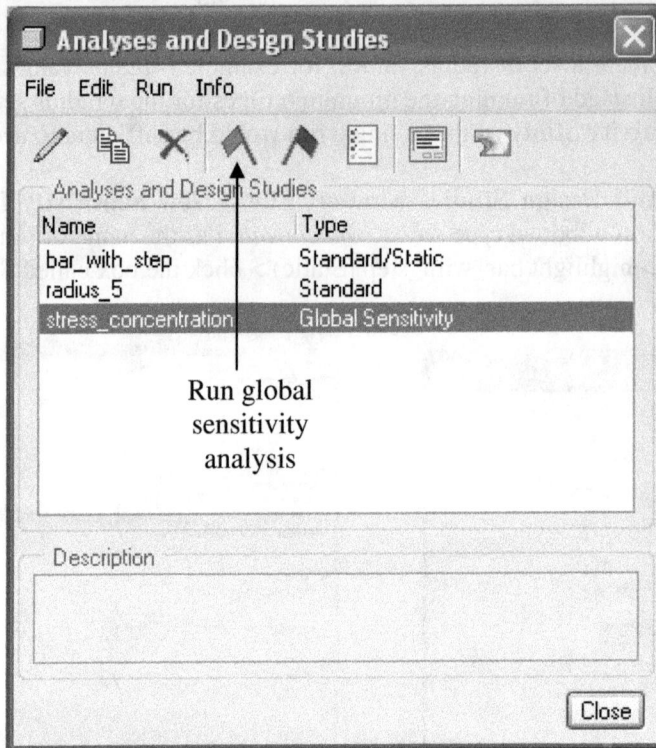

To show the plot of the maximum principal stress vs the design parameter (the radius of the round feature), click the icon of **Review result.**

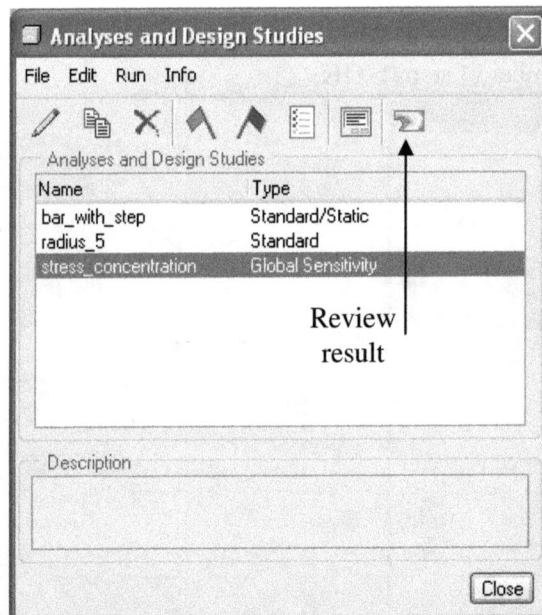

In the Result Window Definition window, type Max principal stress vs Corner Radius as the title **> Graph >** click the box under Measure to select the max_prin_mag **> OK > OK and Show**.

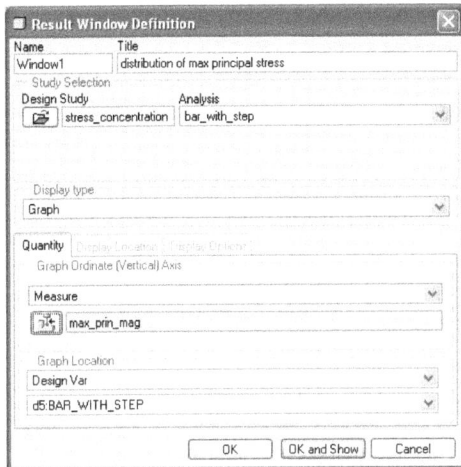

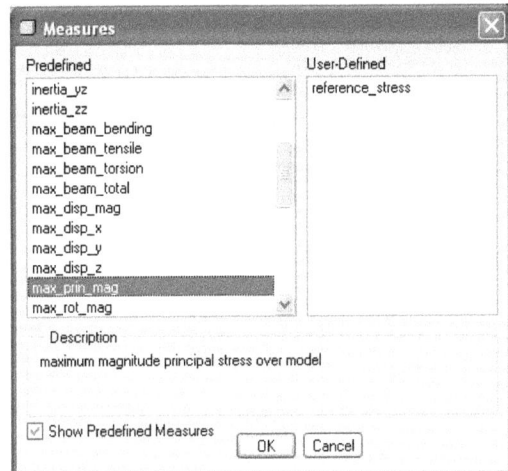

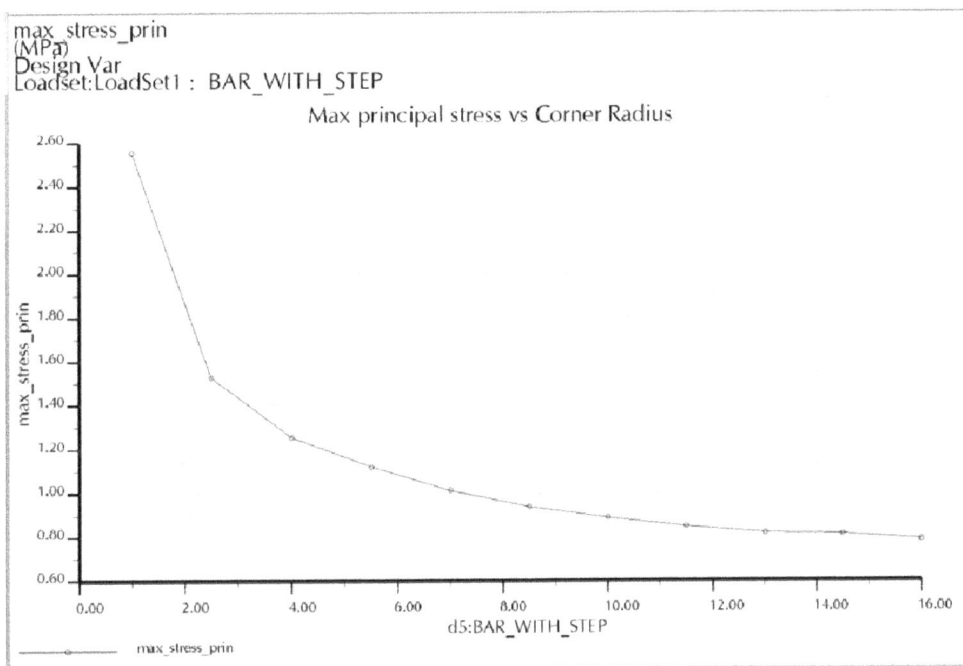

max_stress_prin
(MPa)
Design Var
Loadset:LoadSet1 : BAR_WITH_STEP

Max principal stress vs Corner Radius

To show the plot of the reference stress vs. the design parameter (corner radius), click the icon of **Review result > Graph** > type Reference Stress vs. Corner Radius as the title. Select the *reference_stress* **> OK > OK and Show**. Note the scale on the vertical axis, ranging from 0.5084 to 0.5076 MPa, which indicates an almost constant stress value, as compared with the magnitude of the variation of the maximum principal stress.

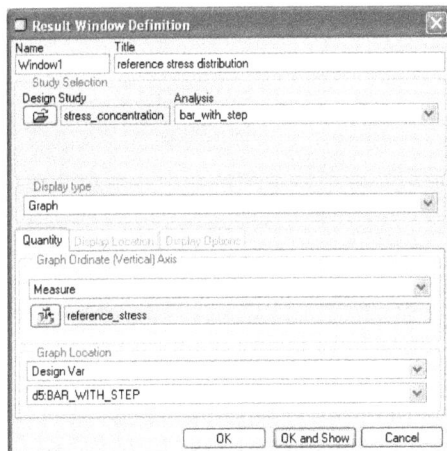

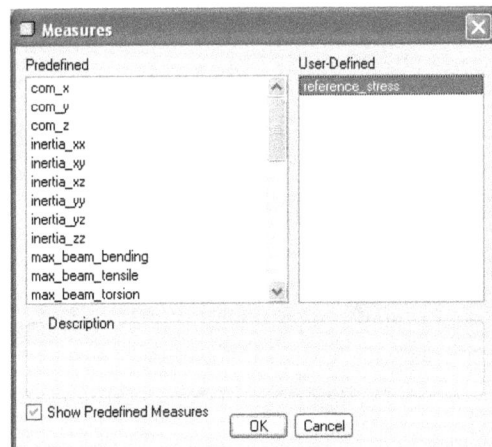

reference_stress
(MPa)
Design Var
Loadset:LoadSet1 : BAR_WITH_STEP

Reference Stress vs Corner Radius

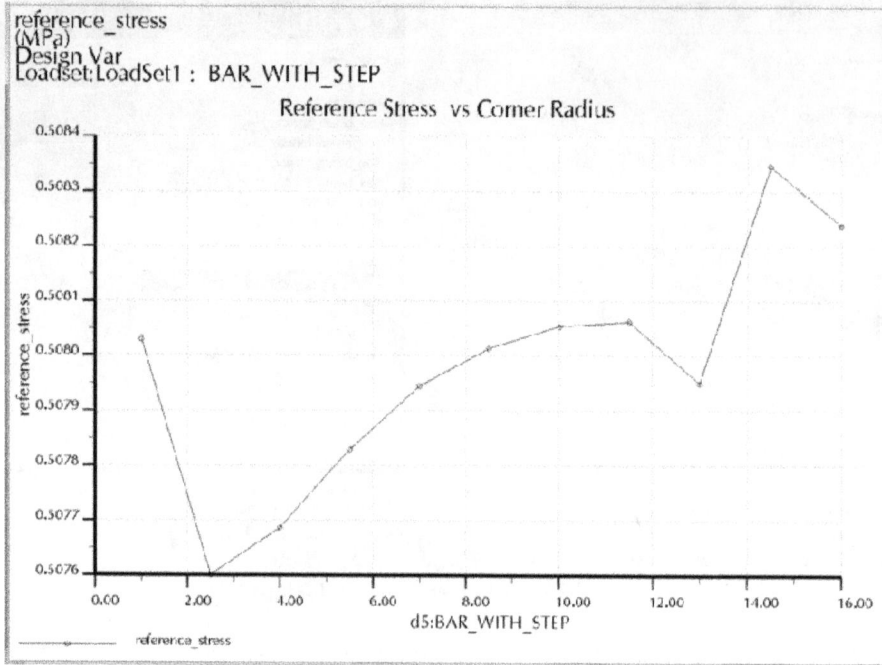

Users may modify the vertical scale of the graph. Click **Format > Graph**. In the Graph Window Options, click **Y Axis** and change the minimum value to 0.505 and Maximum value to 0.510 > **OK**. The plot of the reference stress turns to a "horizontal pattern", indicating the variation is legible vs. the change of corner radius value.

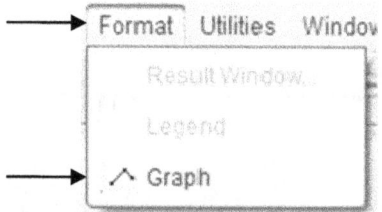

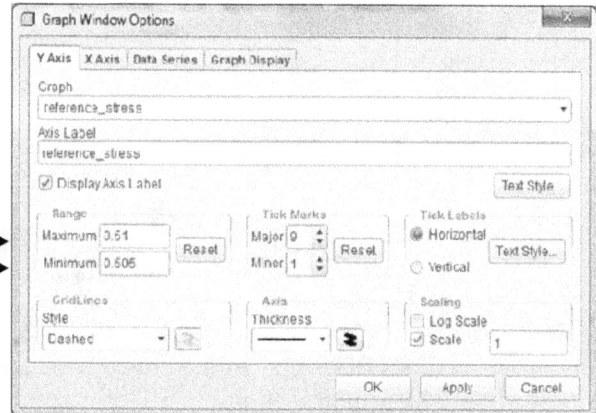

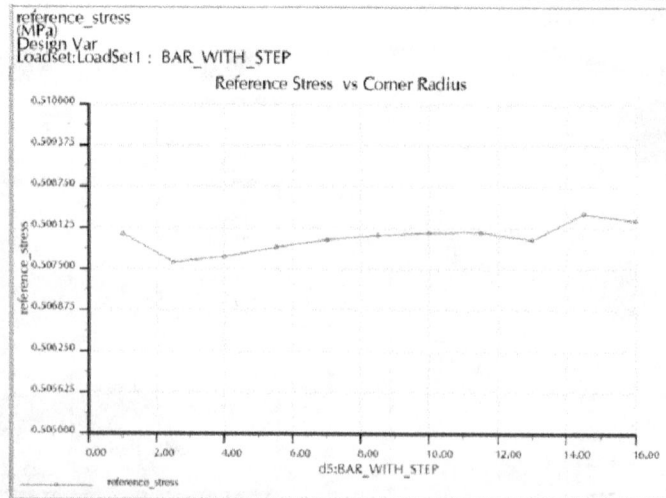

reference_stress
(MPa)
Design Var
Loadset:LoadSet1 : BAR_WITH_STEP

Reference Stress vs Corner Radius

4.3   Modeling of a Tube when Subjected to Torsion

Torsion is the twisting of an object due to an applied torque. The following figure illustrates the free end of the shaft tends to turn about its longitudinal axis while the other end is held fast.

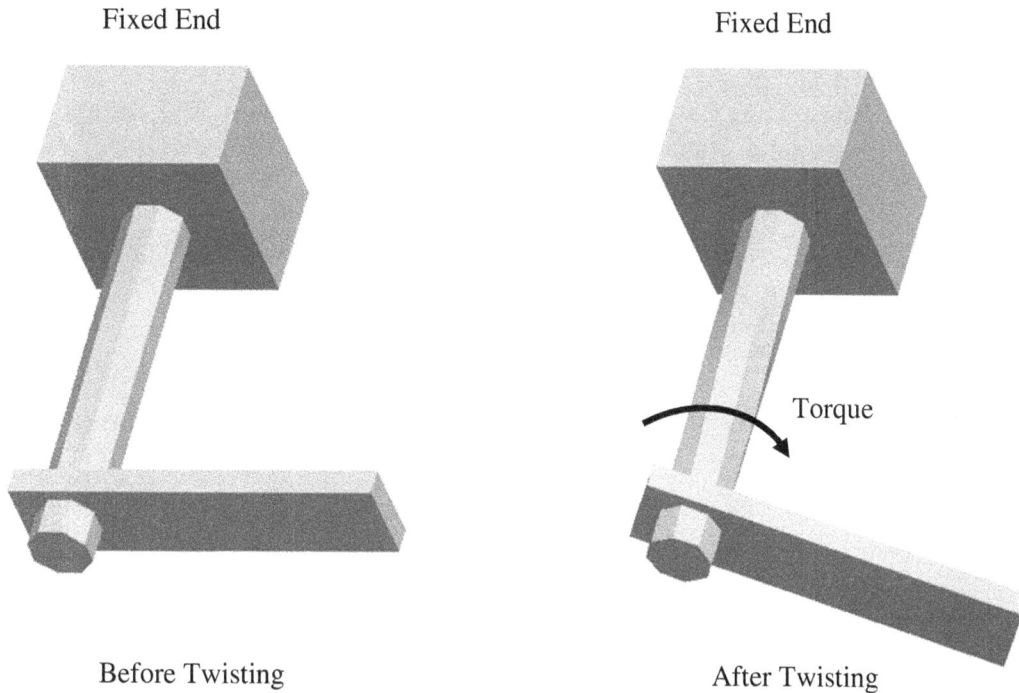

Fixed End   Fixed End

Before Twisting   After Twisting

A cylindrical tube is shown below. The left end is fixed and the right end is set free. A torque load is acting on the surface at the free end, as shown. A torque is acting on the free end. Its magnitude is 10 Newton-Meter.

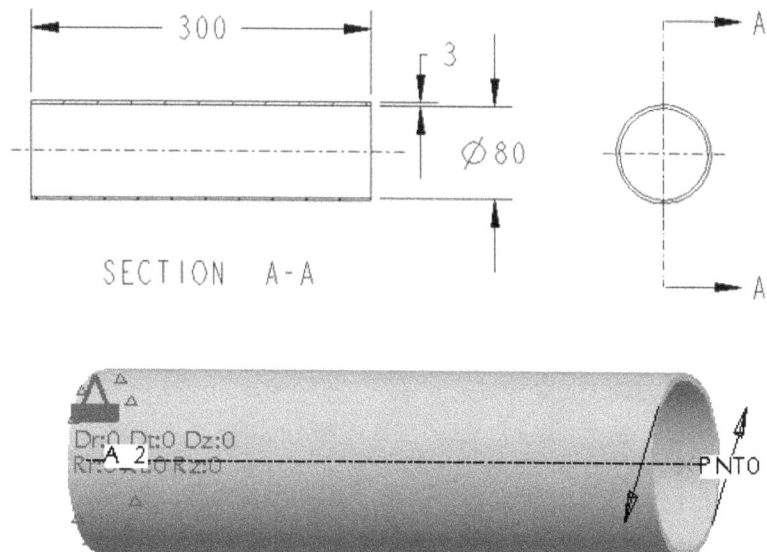

Let us use FEA to verify that the value of the maximum shear stress obtained from FEA is close to the value calculated by the analytical formula in the textbook of strength of materials, which is given by,

$$\tau_{max} = \frac{Torque * R_{outer}}{J} = \frac{10*(0.043)}{1.349x10^{-6}} = 0.3188x10^{6} \quad (pascal) = 0.3188 \ (MPa)$$

$$J = \frac{\pi}{2}(R_{outer}^4 - R_{inner}^4) = \frac{3.1416}{2}(0.043^4 - 0.04^4) = 1.349x10^{-6} \quad (M^4)$$

Step 1: Create a 3D solid model for the tube component.

File > New > Part > type *tube* as the file name and clear the icon of **Use default template**. Select **mmns_part_solid** (units: Millimeter, Newton, Second) and type *torsion* in **DESCRIPTION**, and *student* in **MODELED_BY**, then **OK**.

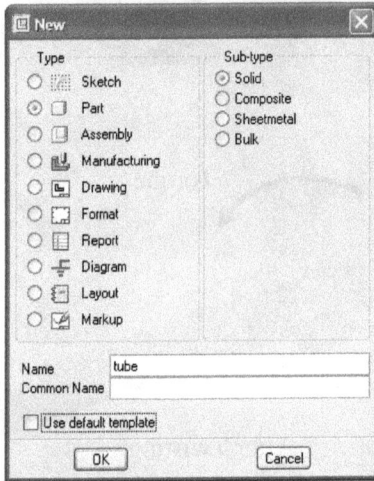

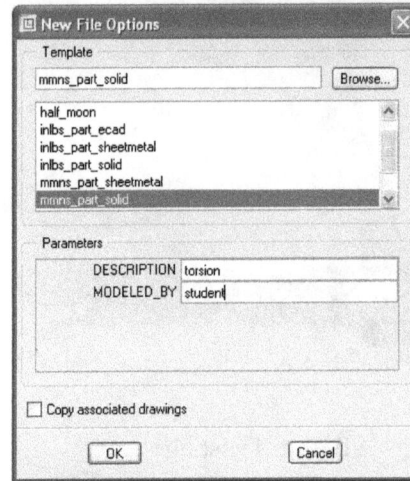

Pick up the icon of **Revolve**. To initiate the process of sketching, activate **Placement > Define**.

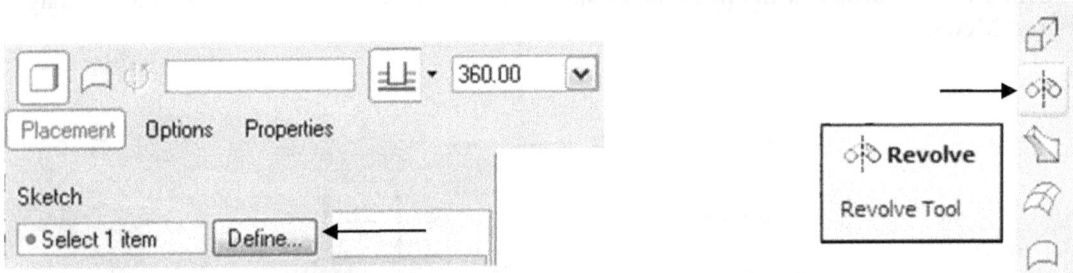

Select the **FRONT** datum plane displayed on screen as the sketch plane. Pick the box of **Sketch** to accept the **RIGHT** datum plane as the default reference to orient the sketch plane, as illustrated below:

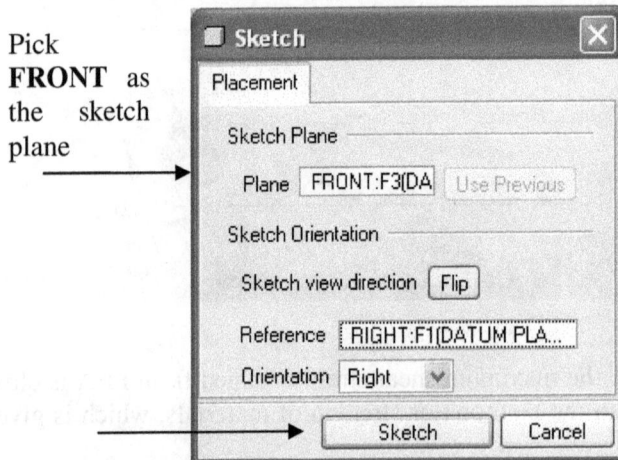

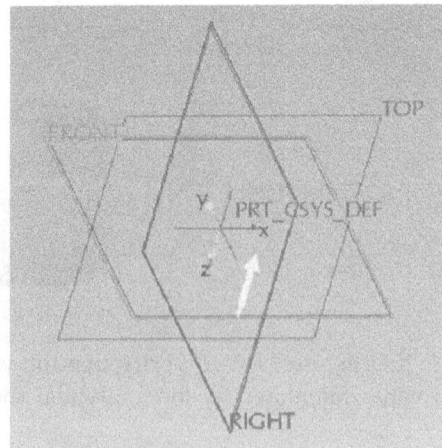

Pick **FRONT** as the sketch plane

Click the icon of **Centerline**. Sketch a horizontal centerline along the axis, as shown below.

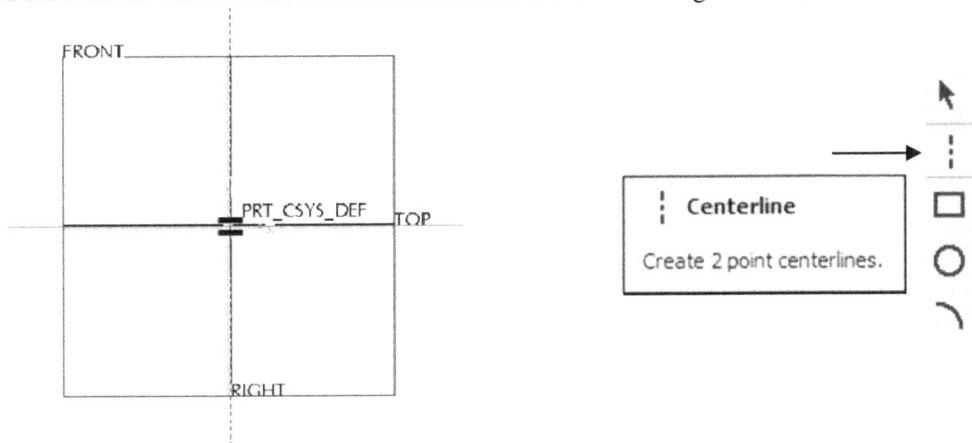

Select the icon of **Rectangle** and sketch a rectangle. The 3 dimensions are 40, 3 and 300, respectively.

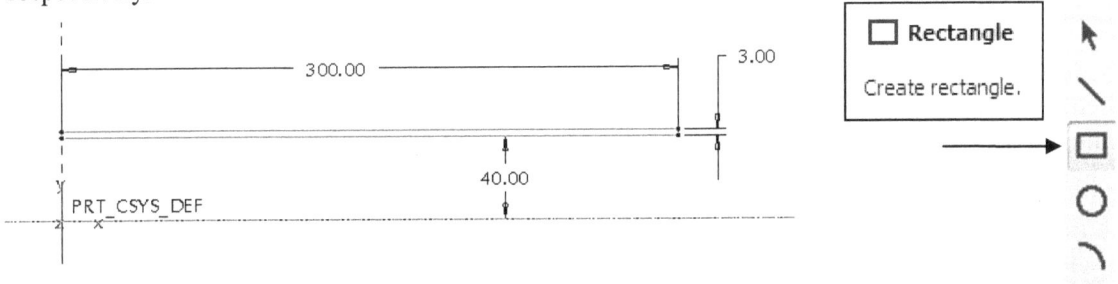

Pick the axis and right-click for selecting **Axis of Revolution** from the pop up window.

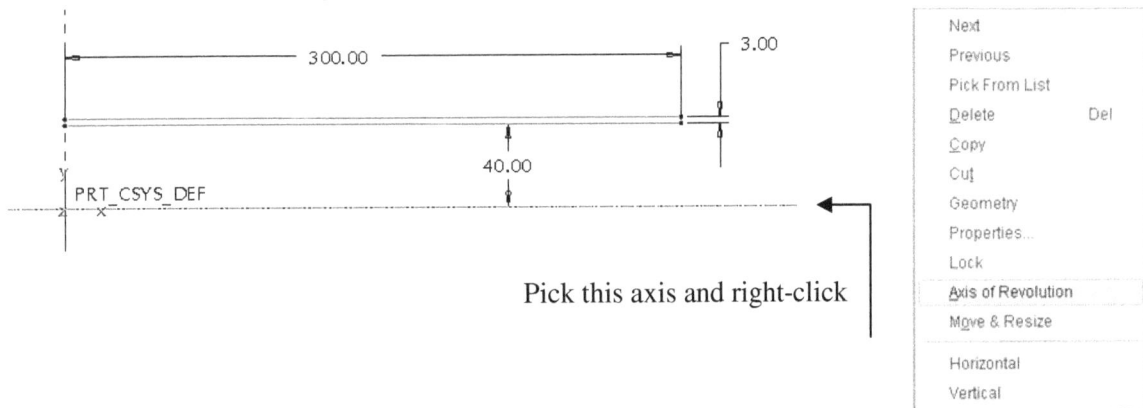

Pick this axis and right-click

Upon completing the sketch, select the icon of **Done** from the toolbar of sketcher. Select the icon of **Apply and Save** from the feature control panel, completing the creation of a 3D solid model for the tube structure.

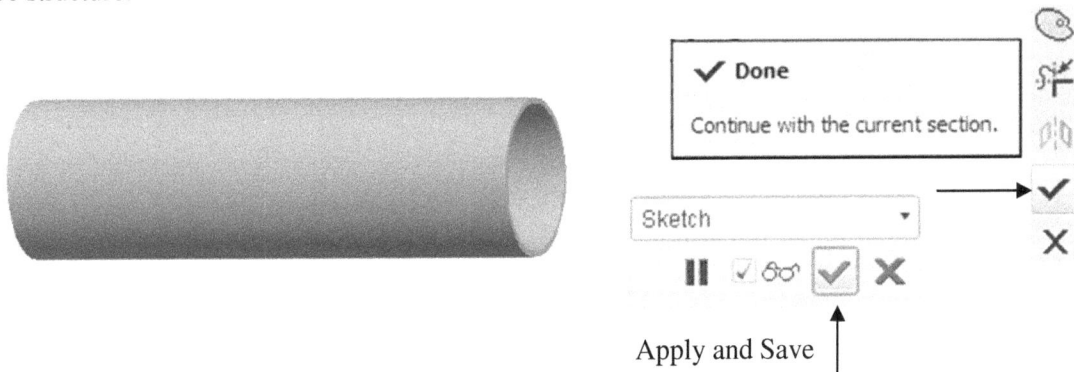

Apply and Save

Step 2: Create a datum point so that a torque load can be defined.

To define a datum point, select the icon of **Datum Point Tool.** Pick the axis of the tube and, while holding down the **Ctrl** key, pick the surface on the right side **> OK**.   PNT0 is the point defined at the central location of the circular surface on the right side, as shown.

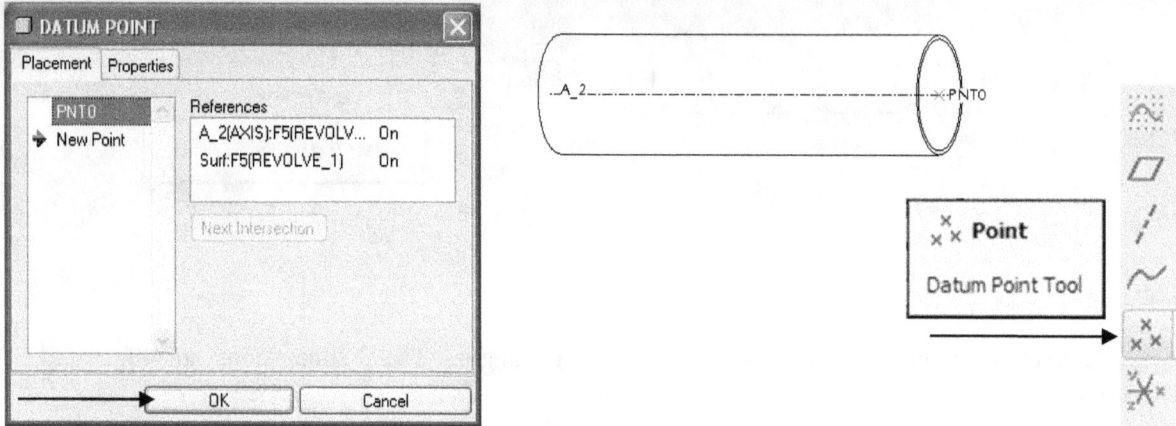

Step 3: Create 2 datum curves for measuring the displacement along the axial direction.

Highlight the **TOP** datum plane from the model tree. From the top menu, click **Edit > Intersect**. While holding down the **Ctrl** key, pick the half surface of the cylinder, as shown below. A datum curve is created. Click the check mark.

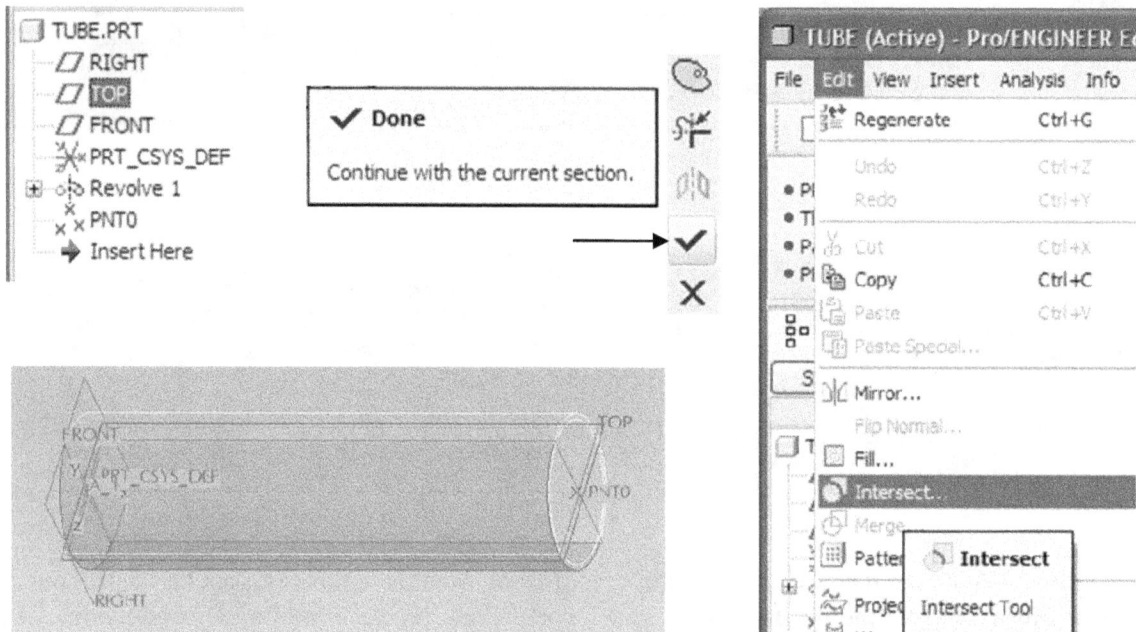

Repeat the above procedure to create the second datum curve on the opposite direction, as shown.

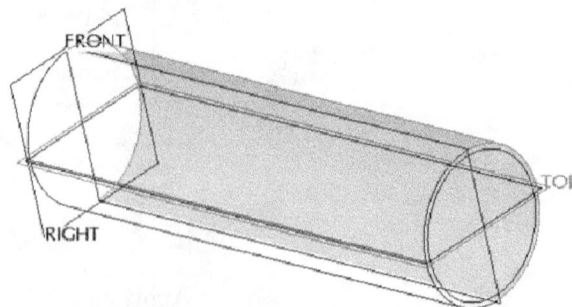

Step 4: Shift from the Pro/ENGINEER design system to Pro/Mechanica

From the main toolbar or the menu toolbar, select **Applications > Mechanica > Continue > Structure > OK.**

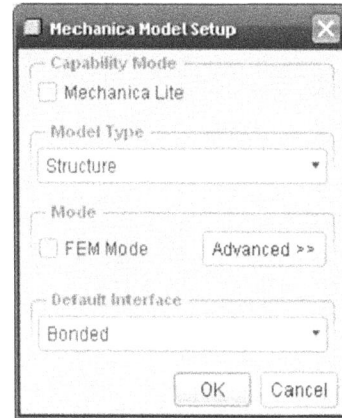

From the toolbar of functions, select the icon of **Materials >** select **STEEL** from the left side called Materials in Library > click the directional arrow so that the selected material type goes to the right side called Materials in Model.

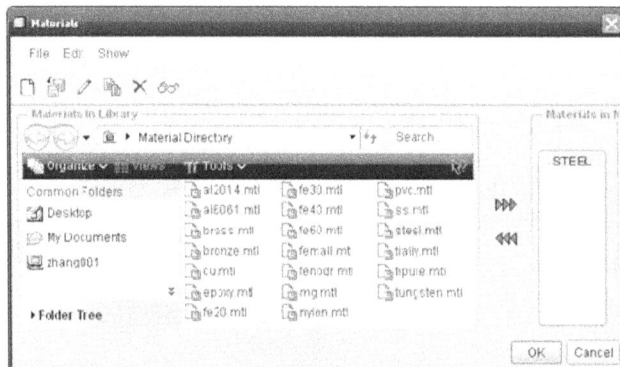

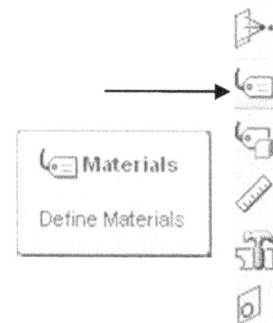

Click the icon of **Material Assignment.** In the Material Assignment window, the selected Steel is shown, and the component is also shown because there is only one volume or one component in the system. As a result, the software system automatically assigns Steel to the component. Just click **OK**.

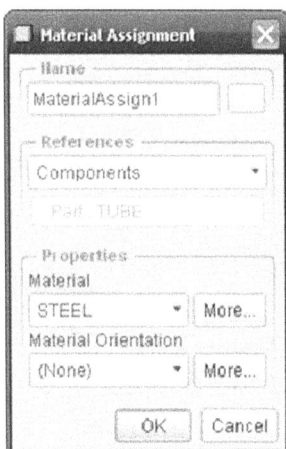

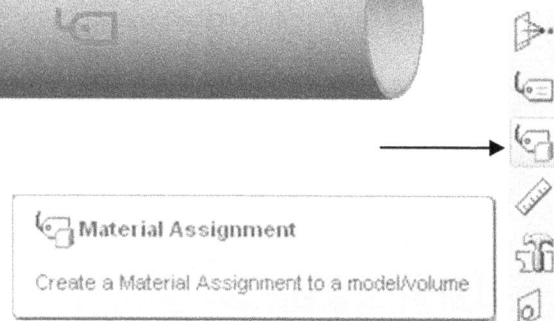

Step 5: Define the constraint condition.

To define the fixed end constraint on the left side of the tube, select the icon of **New Displacement Constraint > Surface(s) >** pick the surface at the left side of the tube. Because the constraint condition is "a fixed end", we set all 6 degrees of freedom equal to zero, which is the default setting > **OK.**

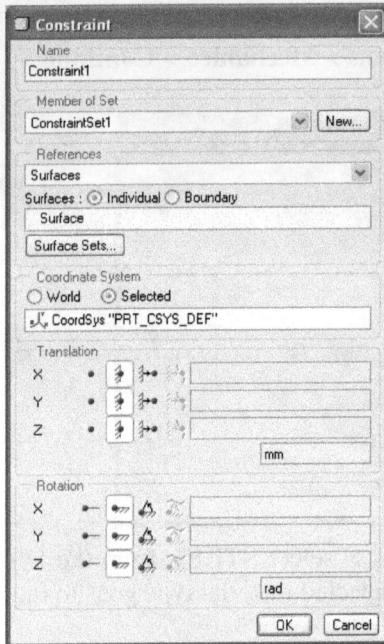

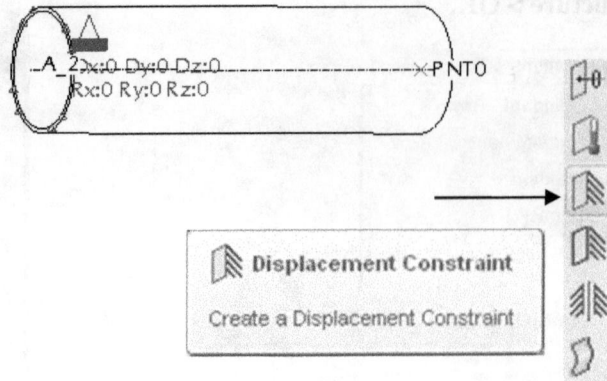

Step 6: Define the load condition.

Assume that the load condition is torque acting on the surface on the right side of the tube. Select the icon of **Force/Moment Load > Surface(s)** > pick the surface on the right side of the tube.

Advanced > Total Load at Point > click the box to activate the selection of the point > pick the defined datum point > type *10000* at the X-component under **Moment > Preview > OK**.

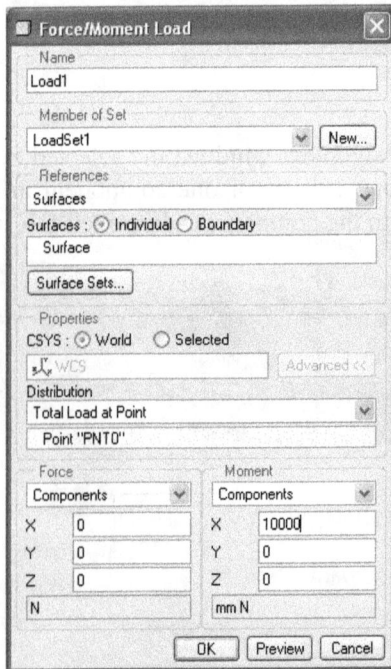

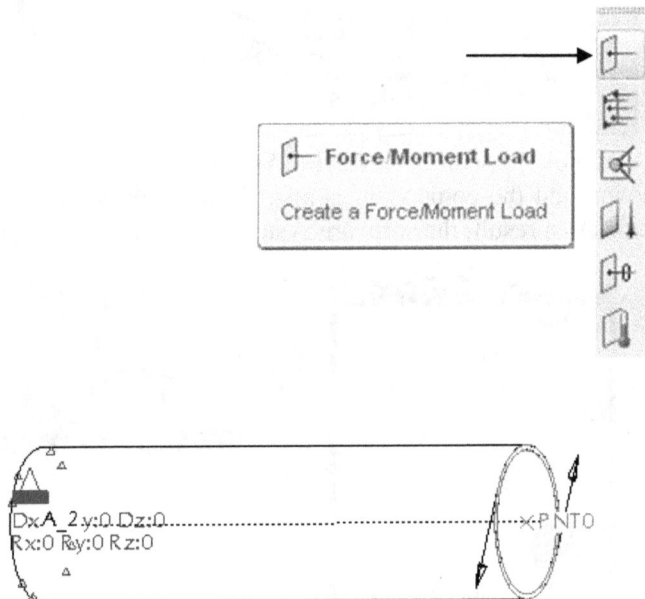

Step 7: Define surface regions using the created datum curves.

This step is critical for obtaining the data plots along the longitudinal direction.

Click the icon of **Surface Region**. Click **Reference** and select **Split by chain**.

Pick the sketched datum curve in the **Chain** box, and pick the half surface of the cylinder in the **Surfaces** box. Click the icon of **Apply and Save**.

Surfaces: 1 item(s)   Chain: One-by-One Chain

References Properties

Split by chain   ▼

Chain:
One-by-One Chain   Details...

Surfaces:
Individual Surfaces

Details...

All   ▼

Surface Region

Create a Simulation Surface Region

0.00

0.00

Apply and Save

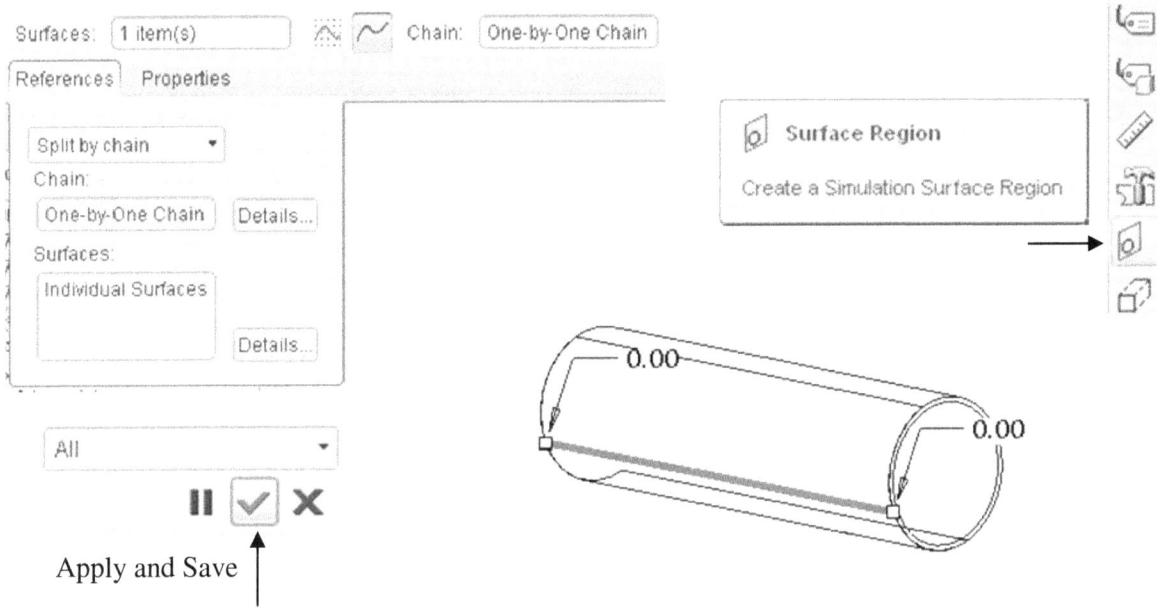

Repeat the above procedure to define the surface region on the other side of the cylinder.

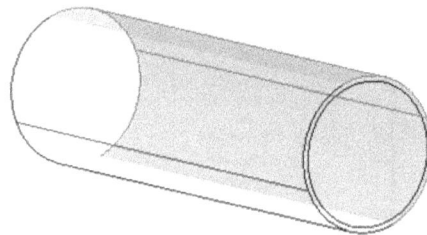

Step 8: Under the Pro/MECHANICA environment, set up **Analyses** and **Run** it
        Select the icon of **Run an Analysis** from the main toolbar > **File** > **New Static** > type *tube* as the name of the analysis folder > **OK.** Click the box of **Run.**

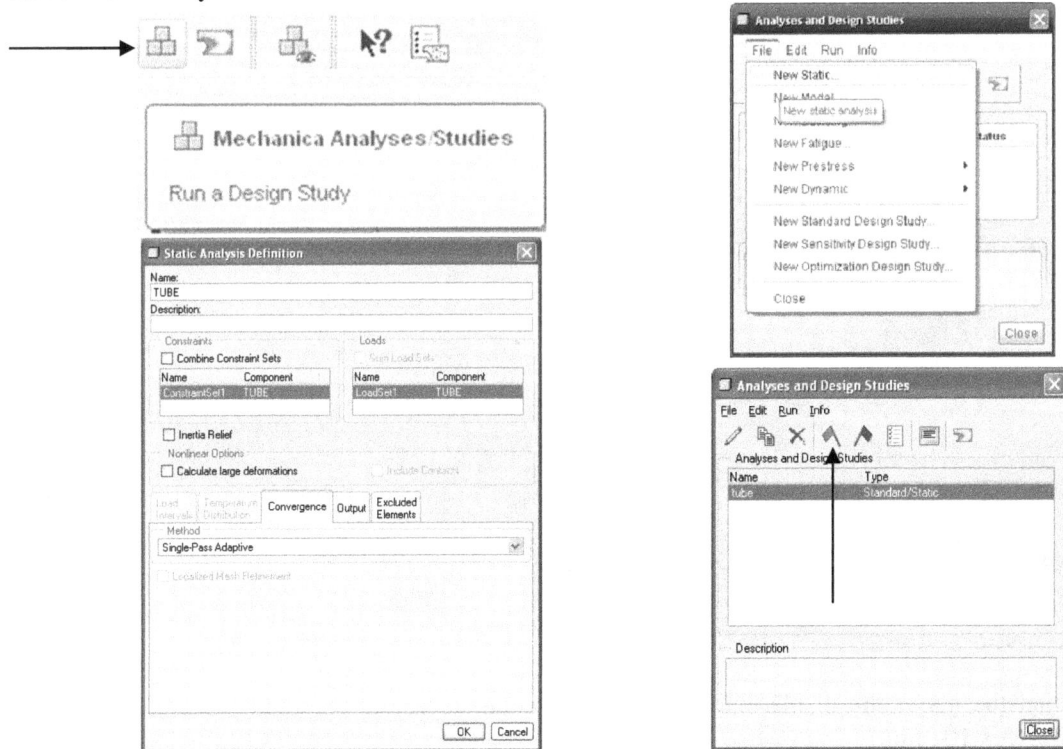

Mechanica Analyses/Studies

Run a Design Study

Static Analysis Definition

Name:
TUBE
Description:

Constraints   Loads
☐ Combine Constraint Sets   Sum Load Sets
Name   Component   Name   Component
ConstraintSet1 TUBE   LoadSet1   TUBE

☐ Inertia Relief
Nonlinear Options
☐ Calculate large deformations   ☐ Include Contacts

Load   Temperature   Excluded
Intervals Distribution Convergence Output Elements
Method
Single-Pass Adaptive   ▼

☐ Localized Mesh Refinement

OK   Cancel

Analyses and Design Studies

File Edit Run Info

New Static...
New Model...
    New static analysis
New Fatigue...
New Prestress   ▶
New Dynamic   ▶

New Standard Design Study...
New Sensitivity Design Study...
New Optimization Design Study...

Close

Close

Analyses and Design Studies

File Edit Run Info

Analyses and Design Studies
Name   Type
tube   Standard/Static

Description

Close

To monitor the computing process, users may click the icon of **Display study status**.

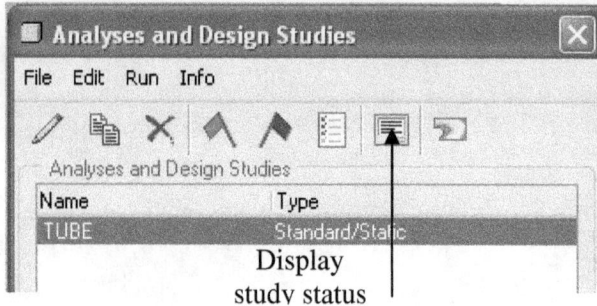

To study the obtained result, say the displacement of the tube under the static loading condition, click the icon of **Review results > Displacement > Magnitude > OK and Show.**

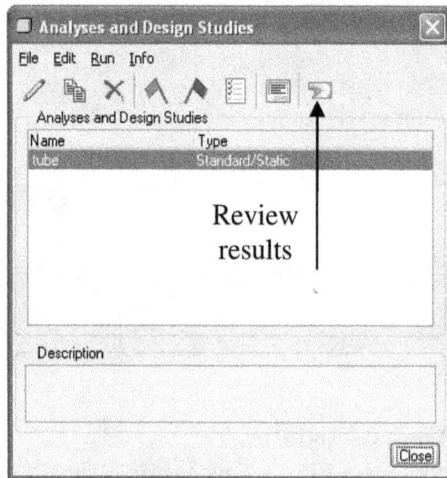

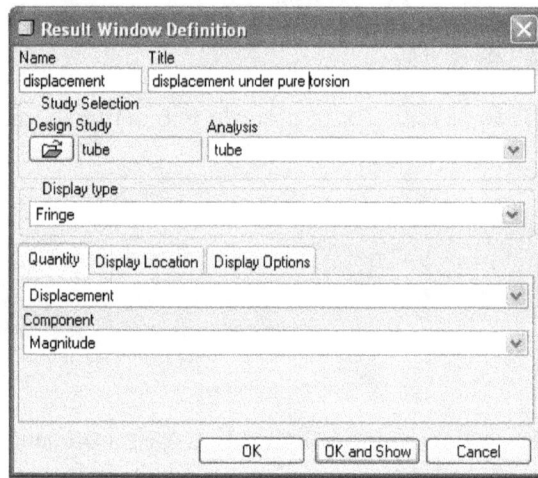

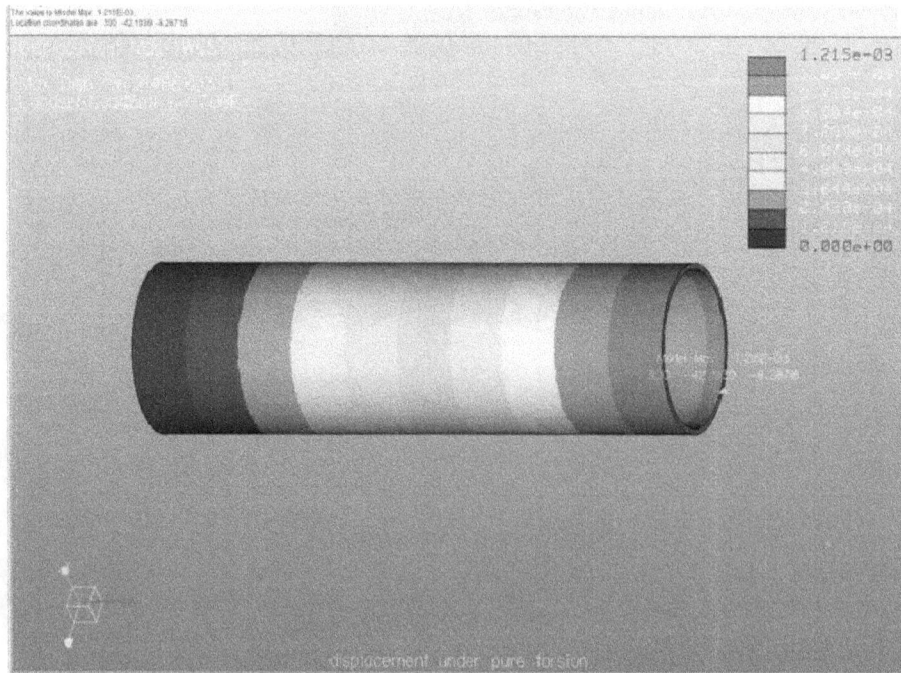

Note that the maximum displacement is located on the surface at the right side, as shown. Its magnitude is 1.215×10^{-3} mm.

To display the maximum shear stress, click the icon of **Copy the selected definition** > type *shear_stress* as the window name and type *maximum shear stress under pure torsion* as the title > **Stress > Maximum Shear Stress > OK and Show**.

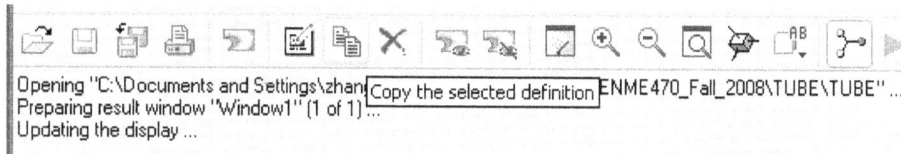

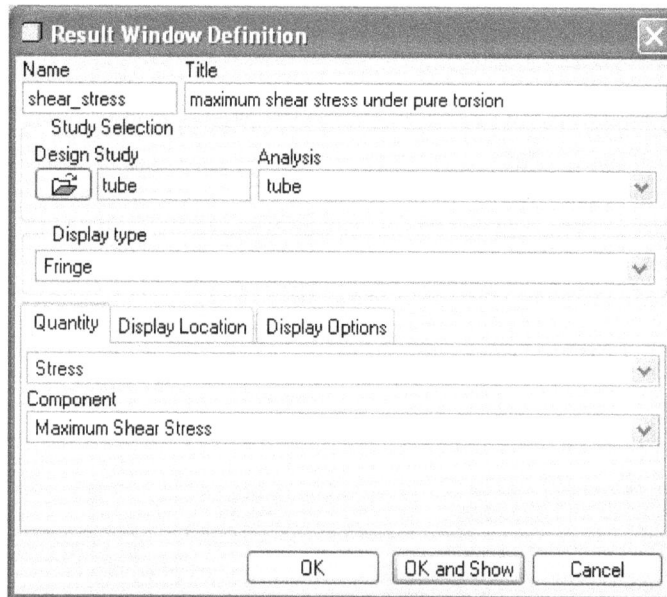

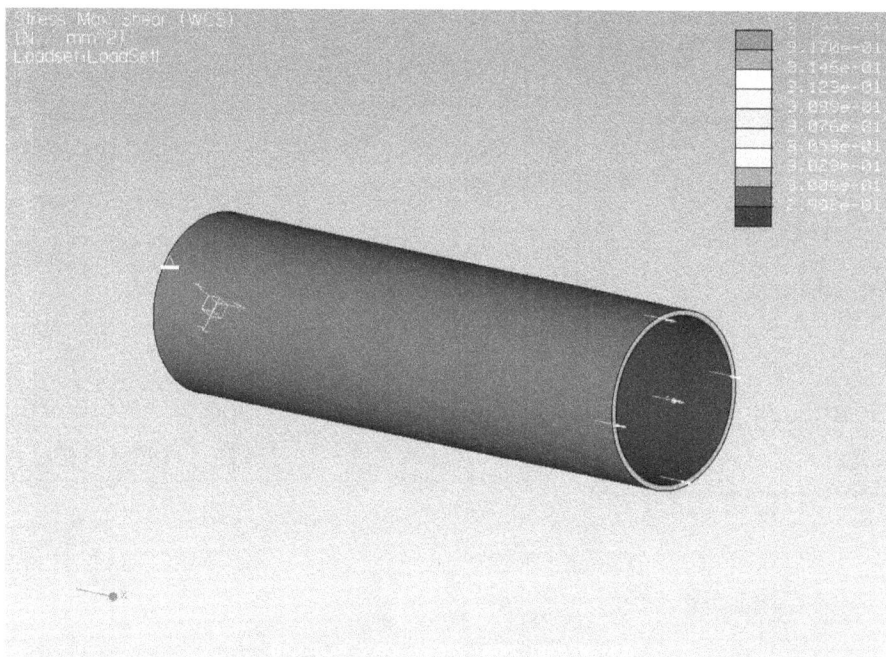

Note that the maximum shear stress is distributed on the entire outer surface of the tube. Its magnitude is 0.3193 MPa, which well matches the value calculated by the analytical formula of 0.3188 MPa.

Step 9: Verification of the results obtained from FEA as compared with the results calculated by the analytical formula in the textbook of strength of materials.

The calculation to obtain the angle of twist, φ, is shown below:

$$\varphi = \frac{Torque * Length}{J * G} = \frac{10*(0.3)}{(1.349x10^{-6})(78x10^{9})} = 0.028511x10^{-3}\,(radian)$$

where G = 78 GPa (modulus of elasticity in shear).

Note that the maximum deflection obtained from FEA is 1.215 x 10^{-3} mm. Let us convert the displacement to an equivalent angle of twist,

$$\varphi = \frac{1.215x10^{-3}}{43x10^{-3}} = 0.02825(radian)$$

The difference between the 2 values is 0.028251 – 0.028511 = -0.00026 in radian. These two values are close enough to each other.

Let us add one more plot, which is the displacement distribution along the longitudinal direction of the tube. To display this distribution, we use the graph function. Click the icon of **Copy the selected definition** > Select **Graph** as the Display type. Type *maximum displacement along the tube length* as the title. Select **Displacement > Magnitude >** click the box to activate the selection of the curve representing the length of the tube > pick the line as shown > **OK** to accept the default setting for the origin of the length > mid-button of your mouse to accept the selection > **OK and Show**. (If you are not able to pick the line, right click and select **Pick From List** and select **Edge**).

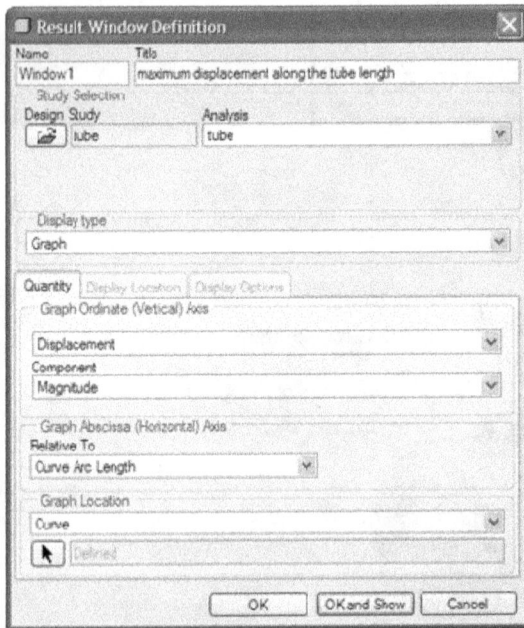

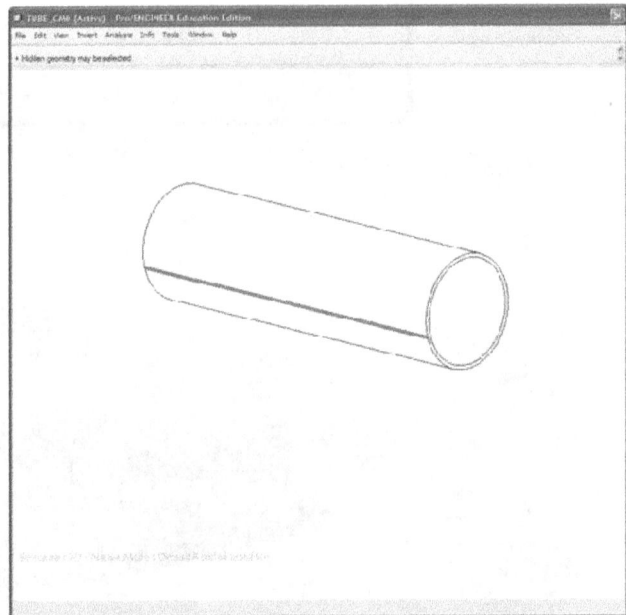

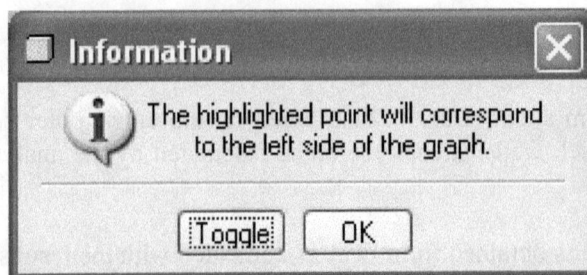

Displacement Mag (WCS)
(mm)
Curve
Max Disp +1.2148E-03
Loadset:LoadSet1

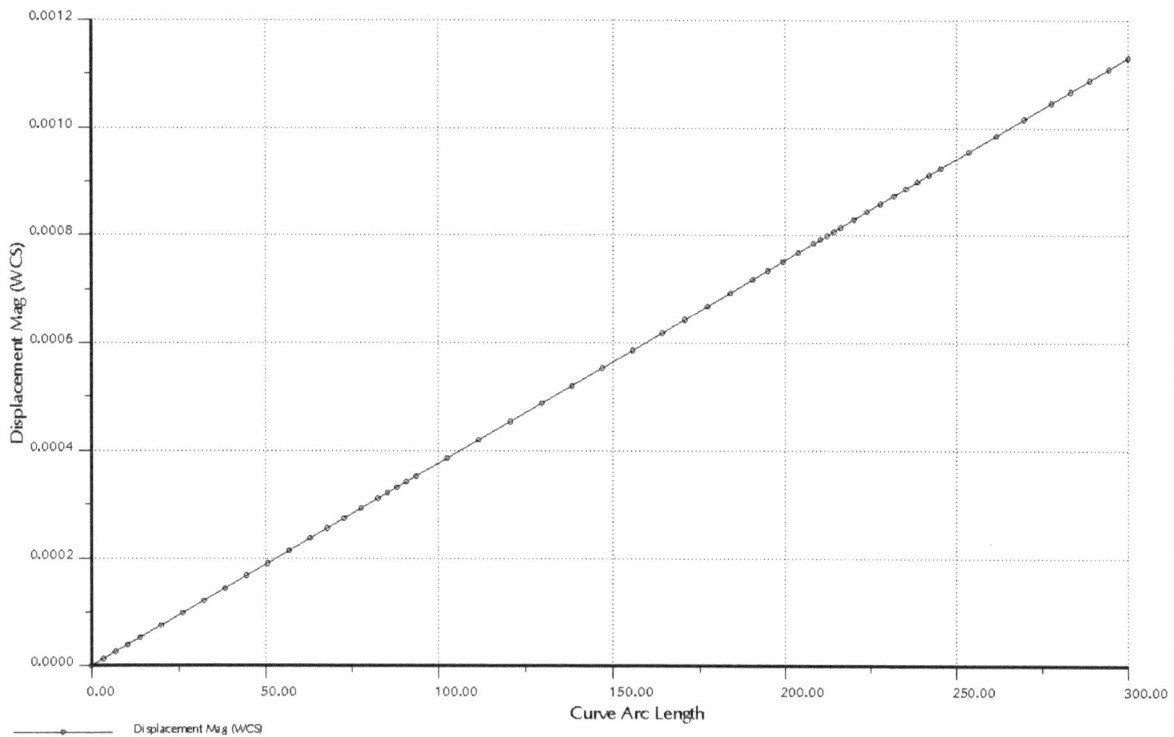

In the theory of mechanics of material, under the pure torsion condition and the total angle of twist of the circular bar is small, the cross sections along the longitudinal direction rotate as rigid bodies about the longitudinal axis, with radii remaining straight and the cross sections remaining plane and circular. During twisting, a rotation about the longitudinal axis of one end of the circular bar with respect to the other is the accumulation of an equal rotation between two cross sections. The angle of twist per unit length can be calculated as

$$\theta = \frac{\varphi}{L} = \frac{0.028511}{300} = 0.0095(radian\,/\,mm)$$

Therefore, the results obtained from FEA well match those values calculated by the analytical formulas.

4.4   Concept of Failure Index

As we know, most of the engineering structures function well in operation or in services. It is not uncommon to observe that some of the engineering structures do not work well in operation or in service. As a result, system failure of those engineering structures occurs. There are several types of mechanical failure, such as excessive deflection, ductile fracture, brittle fracture, corrosion, impact, etc. The concept of failure index has been widely used in the finite element implementation. For example, when the failure type of excessive deflection is the major concern, this is true for most of ductile materials such as metals; the distortion energy criterion is applied. A limiting value of the tensile yield stress is set for quantitatively evaluating the failure index, which is defined by

$$Failure\_Index = \frac{\sigma_{von\_Mises}}{\sigma_{defined\_tensile\_yield\_stress}}$$

In this section, we present a case study to evaluate the performance of a shaft under torsion in the mean time the values of failure index are evaluated.

Step 1: Create a 3D solid model for the shaft component.
File > New > Part > type *motor shaft* as the file name and clear the icon of **Use default template.** Select **mmns_part_solid** (units: Millimeter, Newton, Second) and type *torsion of the shaft* in **DESCRIPTION**, and *student* in **MODELED_BY**, then **OK.**

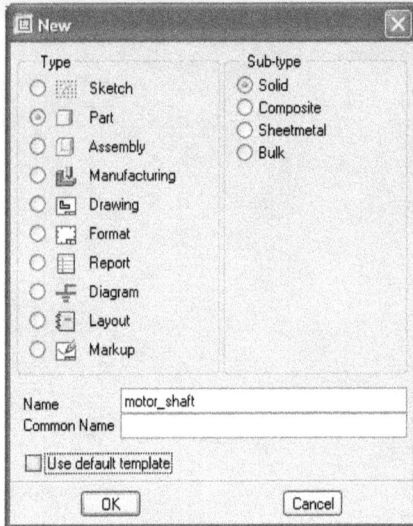

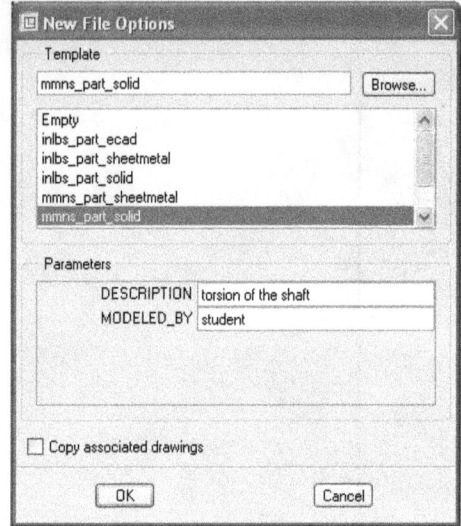

Step 2: Create a 3D solid model for the shaft component.
Click the icon of **Revolve.** To initiate the process of sketching, activate **Placement > Define.**

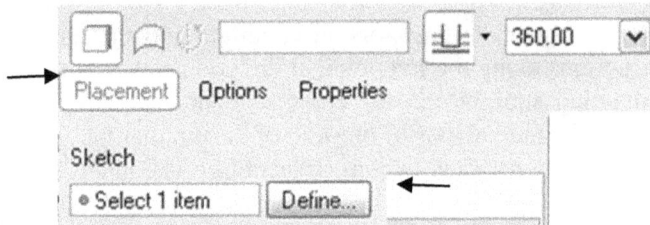

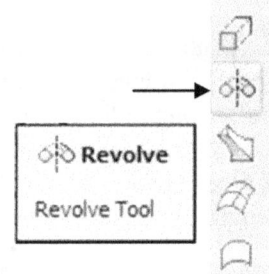

Select the **FRONT** datum plane displayed on screen as the sketch plane. Pick the box of **Sketch** to accept the **RIGHT** datum plane as the default reference to orient the sketch plane, as illustrated below:

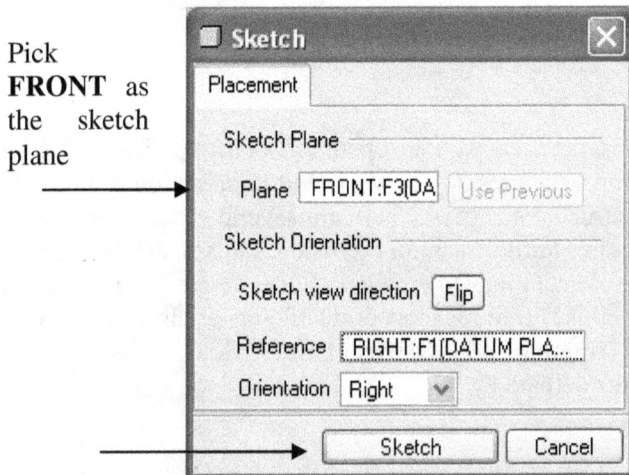

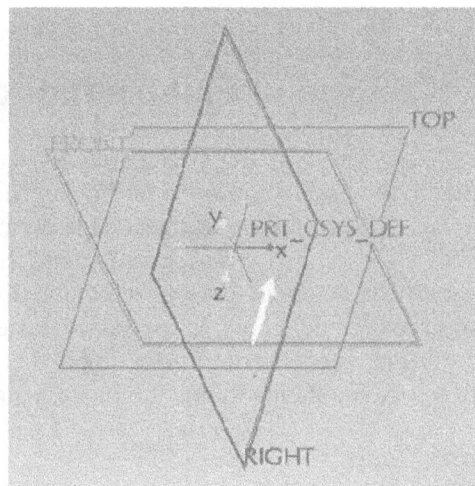

Pick **FRONT** as the sketch plane

Click the icon of **Centerline**. Sketch a horizontal centerline for the revolving operation. Afterwards, sketch a vertical centerline to be used for symmetry when sketching.

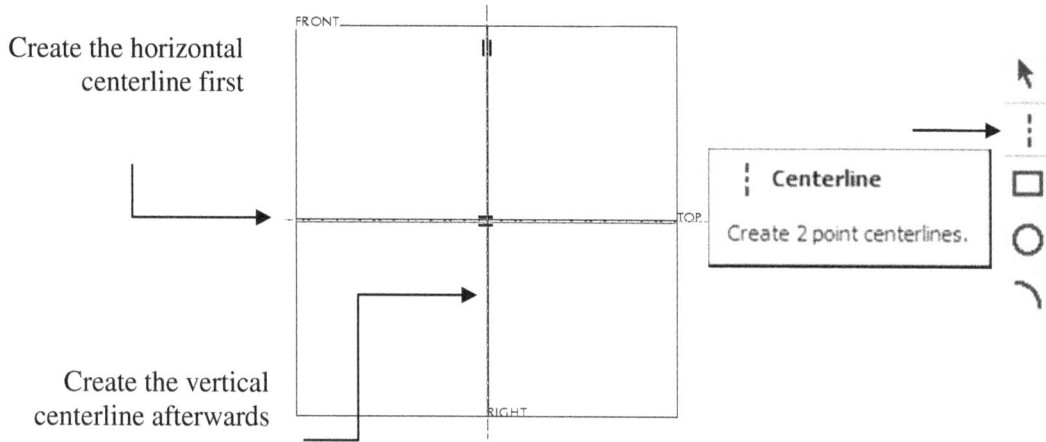

Create the horizontal
centerline first

Create the vertical
centerline afterwards

Select the icon of **Line** and sketch the revolving shape, as shown below. Make sure that the dimension of 500 is symmetric about the vertical centerline.

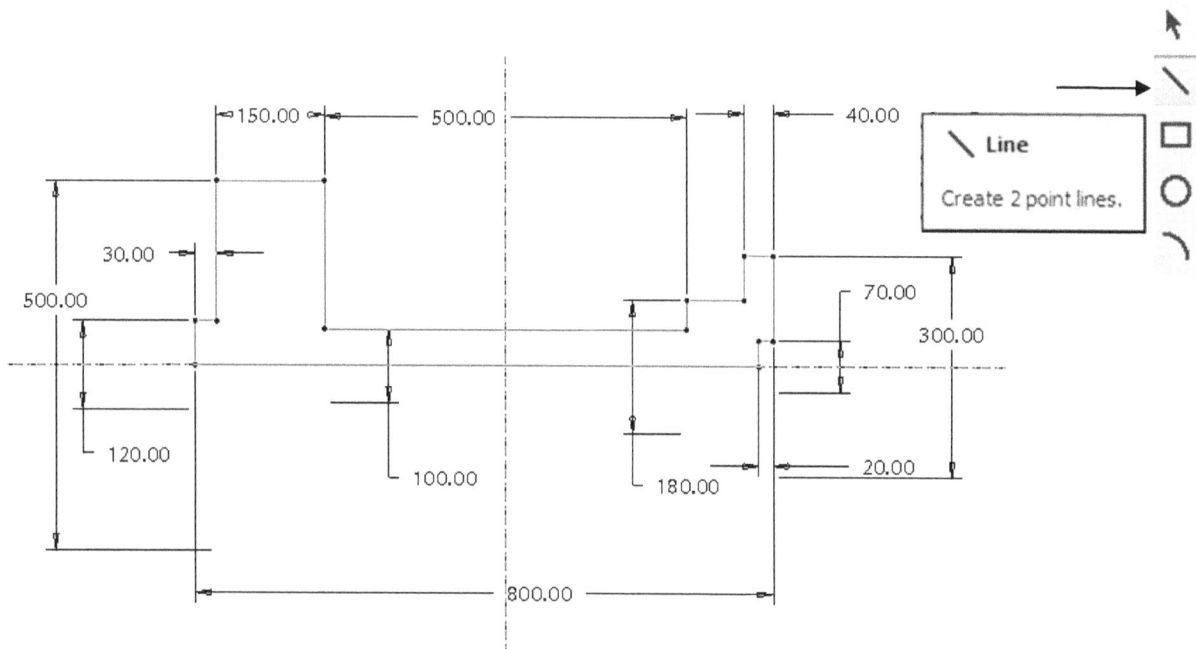

Pick the horizontal axis and right-click for selecting **Axis of Revolution** from the pop up window.

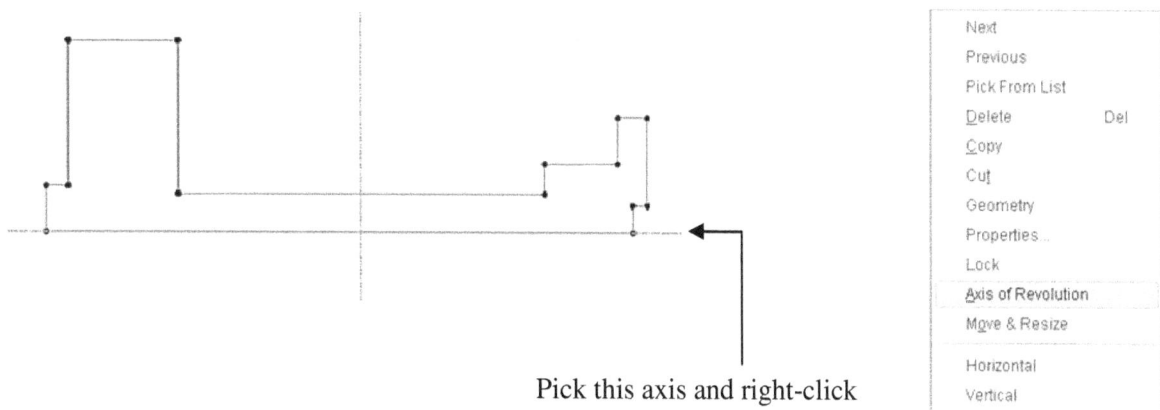

Pick this axis and right-click

Upon completing the sketch, select the icon of **Done** from the toolbar of sketcher. Select the icon of **Apply and Save** from the feature control panel, completing the creation of a 3D solid model for the tube structure.

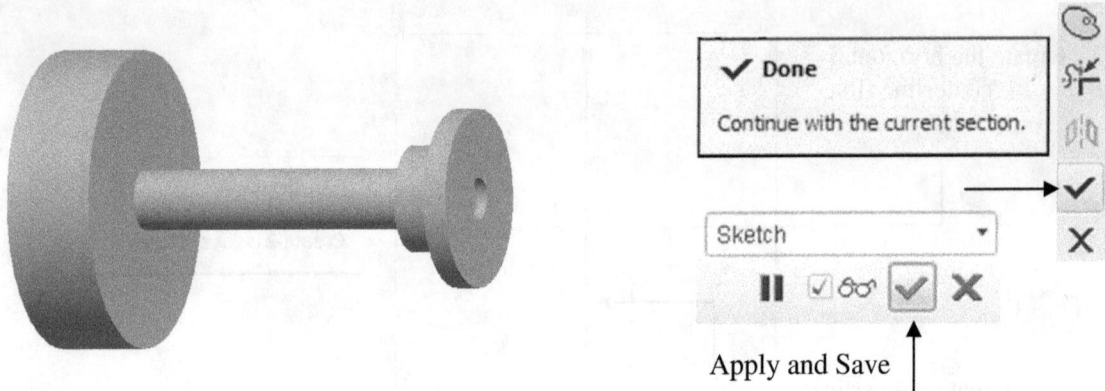

Apply and Save

Step 3: Create 2 datum planes, which will be used to create 2 datum curves. The 2 datum curves will be used to create a surface region to be used in the process of defining a constraint condition.

To define the first datum plane, select the icon of **Datum Plane Tool**. Pick the **RIGHT** datum plane and type 75 as the offset value > **OK**.

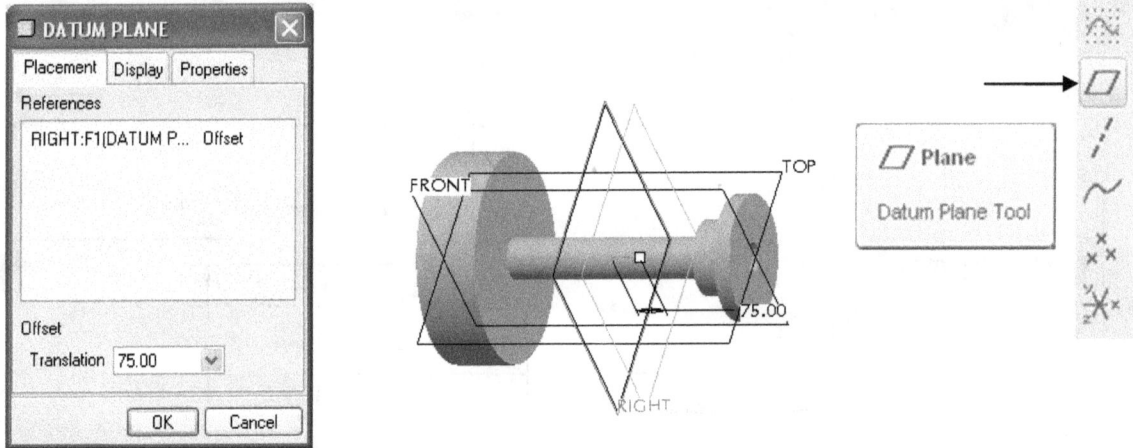

We use the created **DTM1** to create a datum curve through intersection. From the main toolbar, select **Edit > Intersect >** while holding down the **Ctrl** key, pick **DTM1** and the 2 halves of the cylindrical surface, as shown. The created datum curve is a circle because of the intersection of a plane and a cylinder.

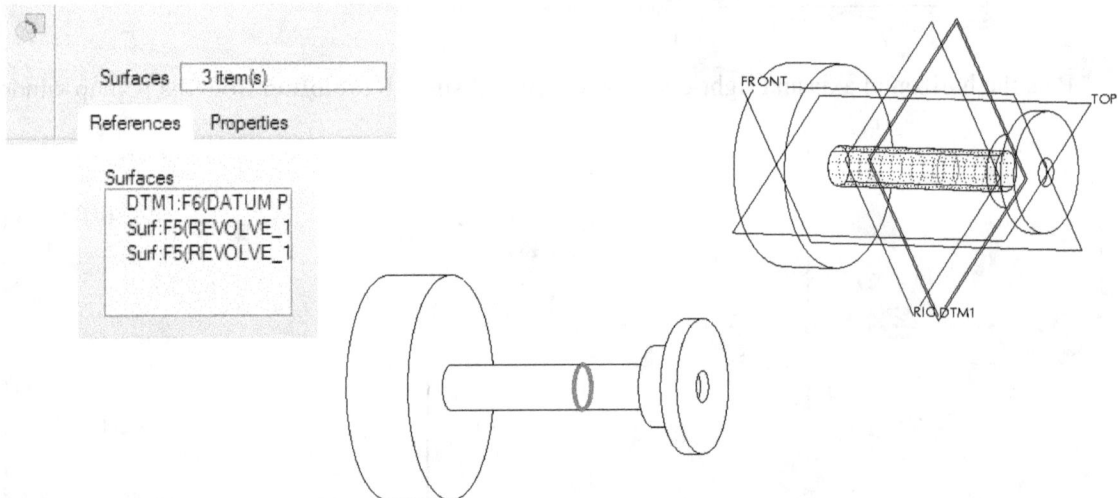

Following the same procedure, we create the second datum curve. Let us begin with the creation of a datum plane. From the toolbar of datum feature creation, select the icon of **Datum Plane Tool**. Pick the **RIGHT** datum plane and type -75 as the offset value > **OK.**

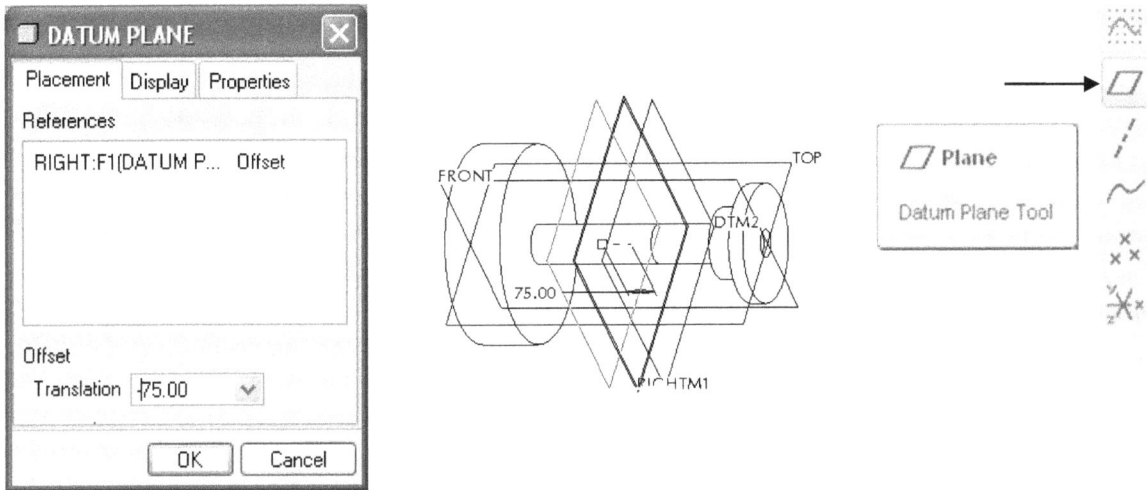

We use the created **DTM2** to create a datum curve through intersection. From the main toolbar, select **Edit > Intersect >** while holding down the **Ctrl** key, pick **DTM2** and the 2 halves of the cylindrical surface, as shown. The created datum curve is also a circle because of the intersection of a plane and a cylinder.

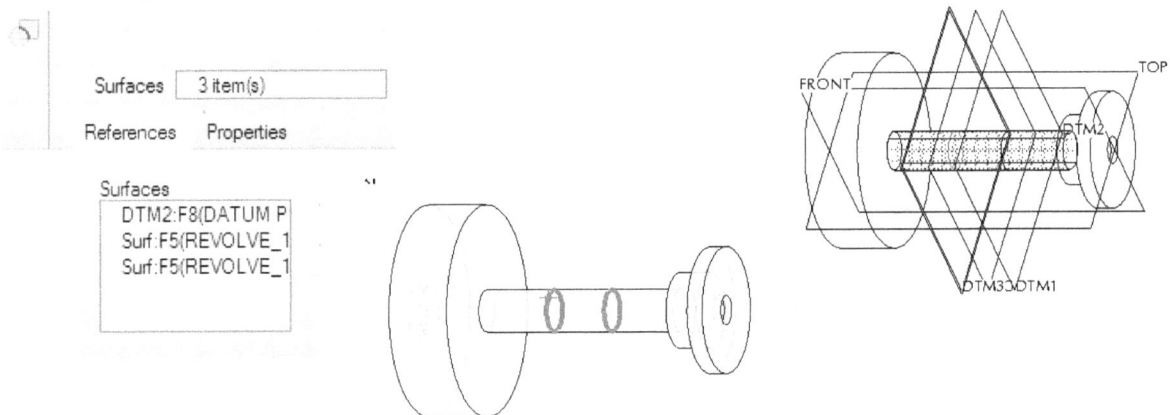

Step 4: Switch from the Pro/ENGINEER design system to Pro/Mechanica
From the main toolbar, select **Applications > Mechanica > Continue > Structure > OK.**

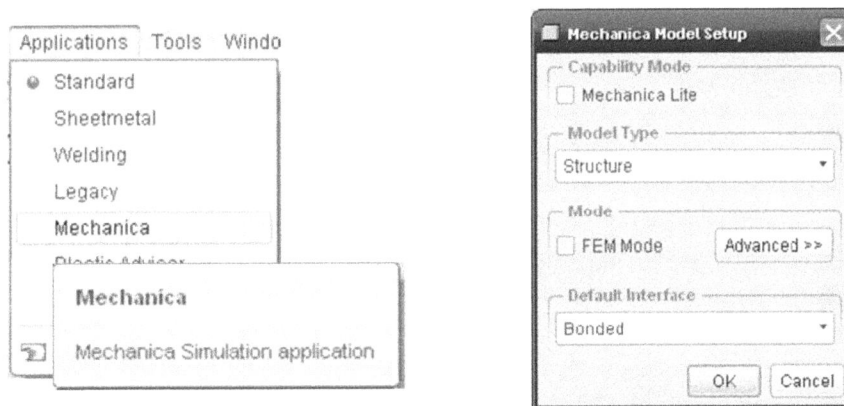

From the toolbar of functions, select the icon of **Materials >** select **STEEL** from the left side called Materials in Library > click the directional arrow so that the selected material type goes to the right side called Materials in Model.

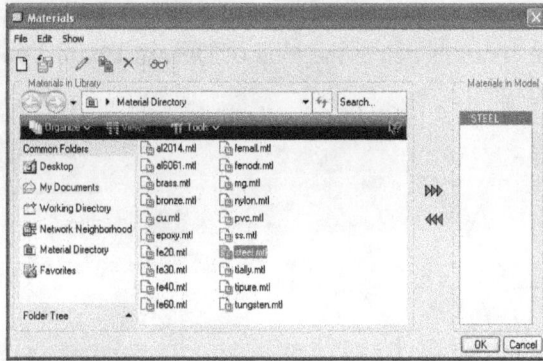

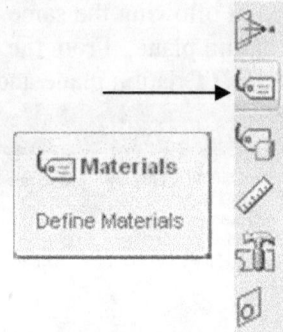

Click the icon of **Material Assignment.** In the Material Assignment window, the selected **Steel** is shown, and the component is also shown because there is only one volume or one component in the system. As a result, the software system automatically assigns **Steel** to the component. Just click **OK**.

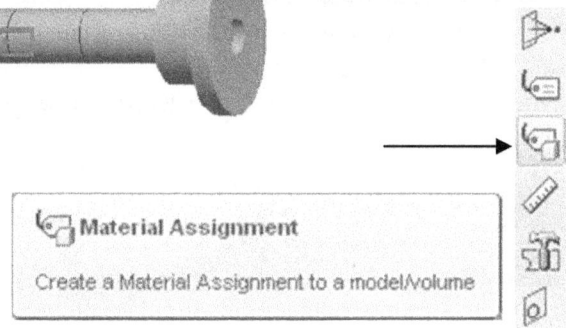

Step 5: Create a surface region so that the displacement constraint condition can be defined.

Click the icon of **Surface Region**. Click **Reference** and select **Split by chain**.

Pick the sketched datum curve on the left side in the **Chain** box, and pick the 2 half surfaces of the cylinder in the **Surfaces** box. Click the icon of **Apply and Save**.

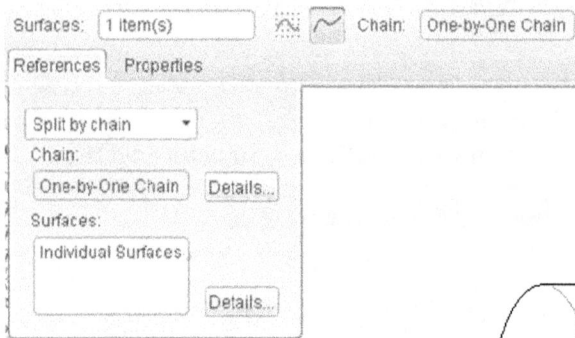

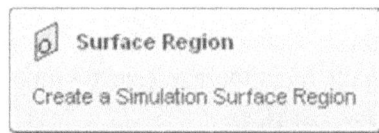

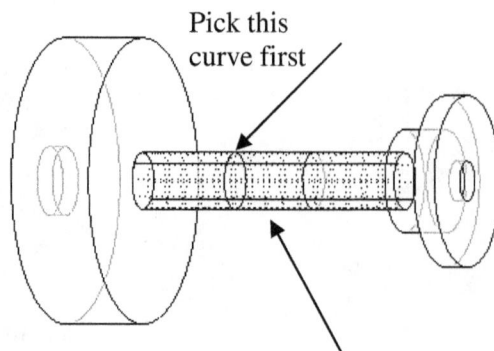

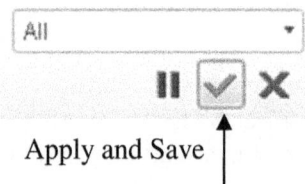

Apply and Save

Pick this curve first

Afterwards, pick both halves of the cylindrical surface while holding down the **Ctrl** key

distribution of the vm stress

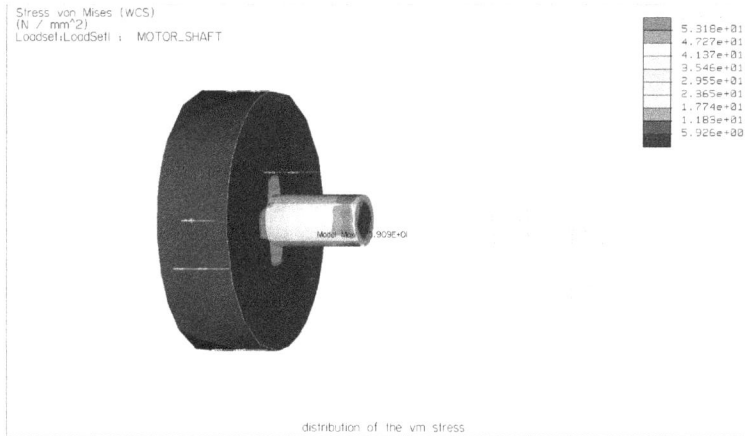

Step 8: Evaluate the safety factor using a parameter called Failure Index.

From the toolbar of functions, select the icon of **Materials. STEEL** is the selected material type. Click **Edit > Properties**.

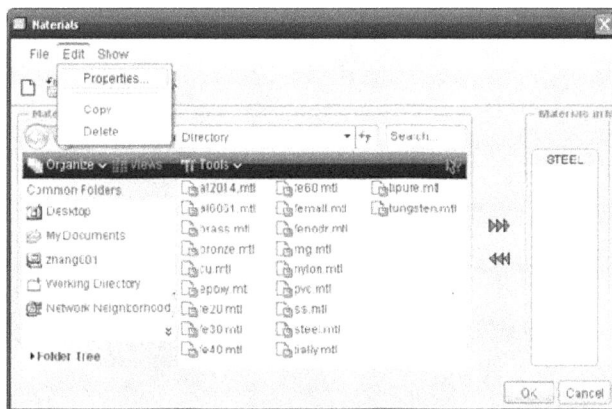

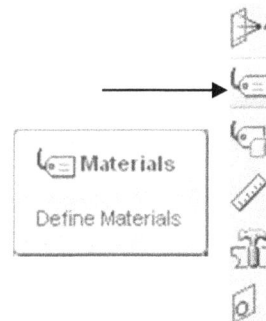

In the **Failure Criterion** box, select **Distortion Energy (von Mises),** specify 50 MPa as the limiting value, as shown. Click **OK > OK**. Now let us run a new design study under Static. Let us call it failure study > **OK**, and run the design study.

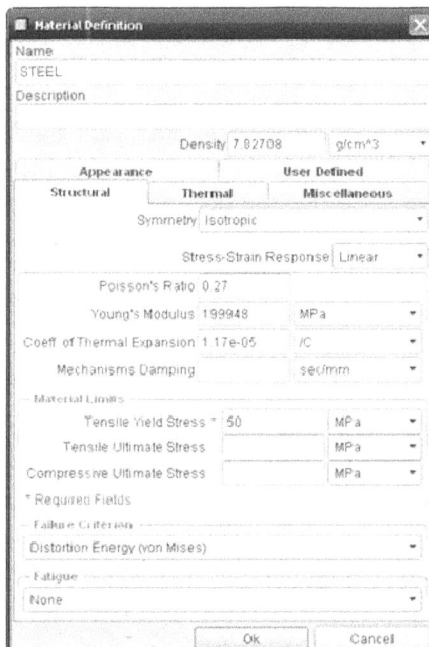

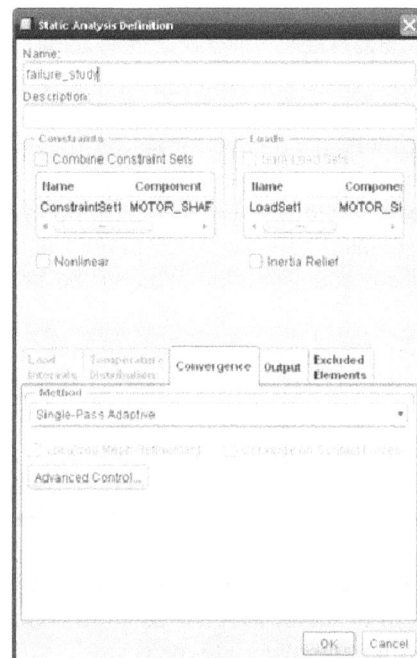

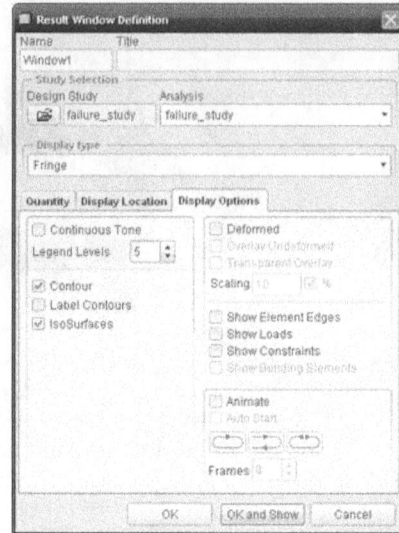

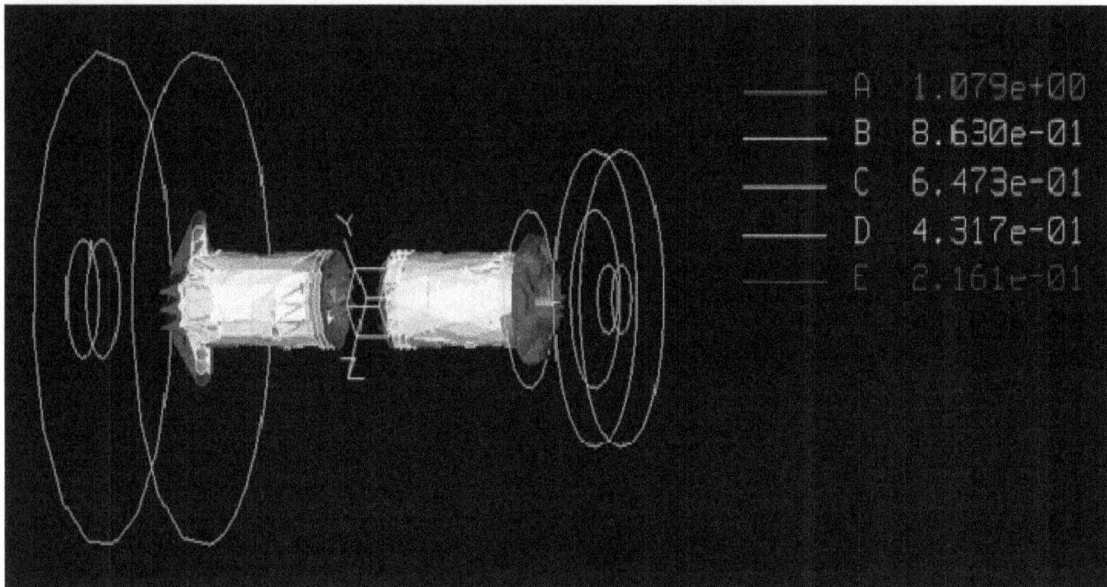

The importance of using the failure index approach together with the surface contour plot is to provide more detailed information on the stress distribution. Instead of notifying the location of the maximum value, whose information is related to a point, the surface contour plot provides a surface of a set of points, which have the same stress value. As we all know, the occurrence of failure mode is not only associated with the magnitude of the stress value at a single point, but also related to the location of the high stress distribution, which is best described by a contour surface.

4.5   Modeling of Centrifugal Load

When we are driving a car and the car is making a right or left turn, the force keeping the car moving on such a curved trajectory is termed as centrifugal force. An object is rotating about an axis, the inertia or the mass of the object will exert a force on the object, causing deformation and generation of stress within the object. In this section, we study the deformation and stress distributions of a spindle, which is rotating, as shown below. As illustrated, the spindle is supported through the use of 2 bearings, as shown below. The width of the bearings is equal to 80 mm. The spindle rotates at 1200 RPM (revolutions per minute). Assume that the material type is steel.

```
                                        200.00
              FRONT
Radius value                                           500.00
= 0.50 mm
                                               50.00
                                          TOP
                         800.00
```

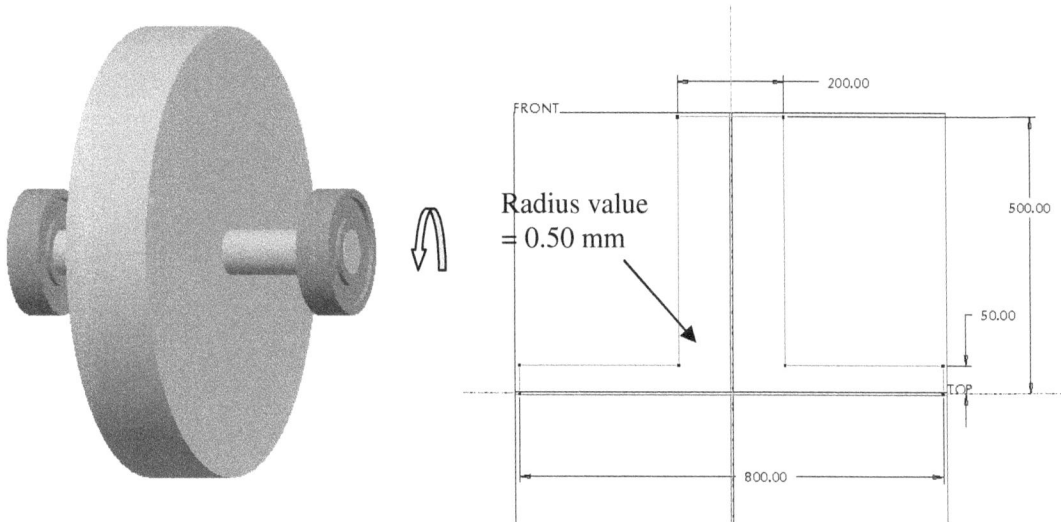

Step 1: Create a 3D solid model for the shaft component.

File > New > Part > type *spindle* as the file name, clear the box of **Use default template > OK**. Select the unit of **mmns_part_solid**, type *centrifugal force load* under the **description** of the model, and type *student* or *your name* under the **modeled_by > OK**.

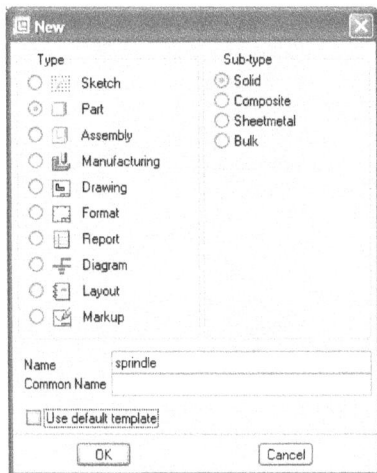

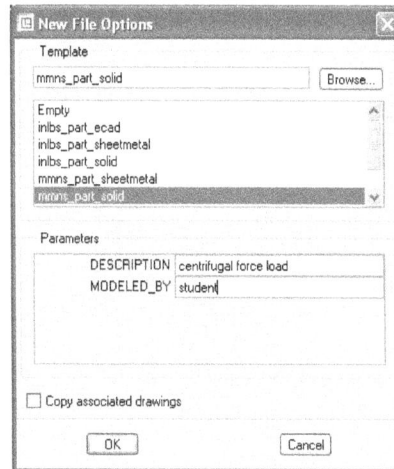

Step 2: Create a 3D solid model for the spindle component.

Pick up the icon of **Revolve** displayed on the toolbar of feature creation. To initiate the process of sketching, activate **Placement > Define.**

Select the **FRONT** datum plane displayed on screen as the sketch plane. Pick the box of **Sketch** to accept the **RIGHT** datum plane as the default reference to orient the sketch plane, as illustrated below:

Pick **FRONT** as the sketch plane

Sketch

Placement

Sketch Plane

Plane FRONT:F3[DA   Use Previous

Sketch Orientation

Sketch view direction Flip

Reference RIGHT:F1[DATUM PLA...

Orientation Right

Sketch   Cancel

TOP

FRONT

Y PRT_CSYS_DEF
   x
z

RIGHT

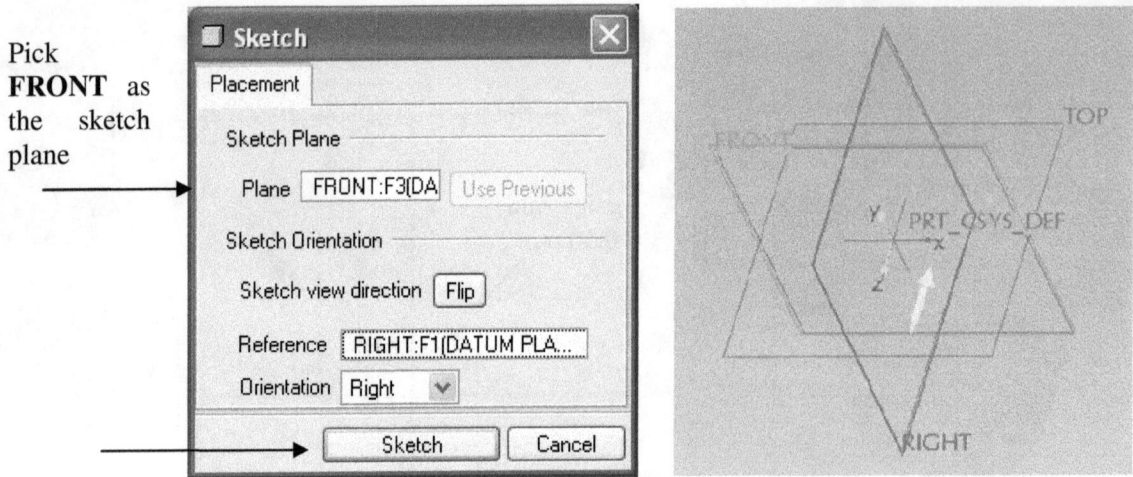

Click the icon of **Centerline**. Sketch a horizontal centerline for the revolving operation. Afterwards, sketch a vertical centerline to be used for symmetry when sketching.

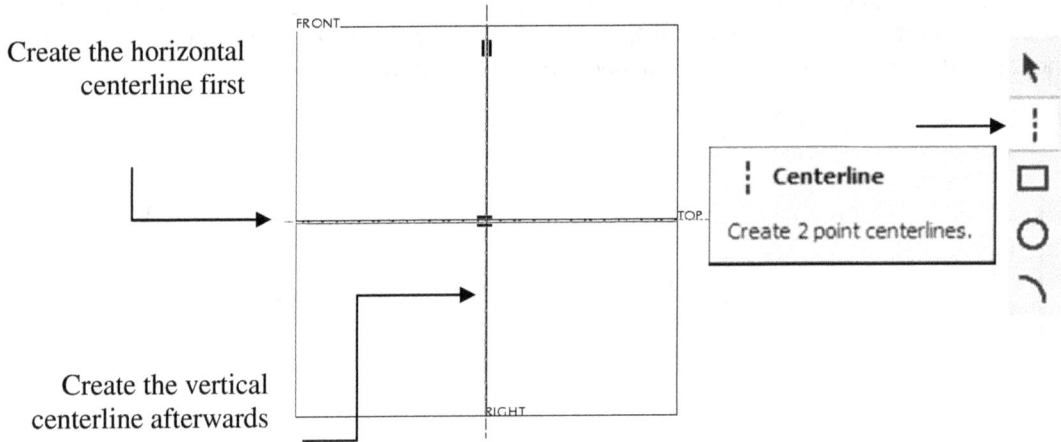

Create the horizontal centerline first

FRONT

TOP.

Centerline

Create 2 point centerlines.

Create the vertical centerline afterwards

RIGHT

Click the icon of **Line** to sketch the revolving shape, as shown below. Make sure that the entire sketch is symmetric about the vertical centerline. Pick the horizontal axis and right-click. Select **Axis of Rotation**.

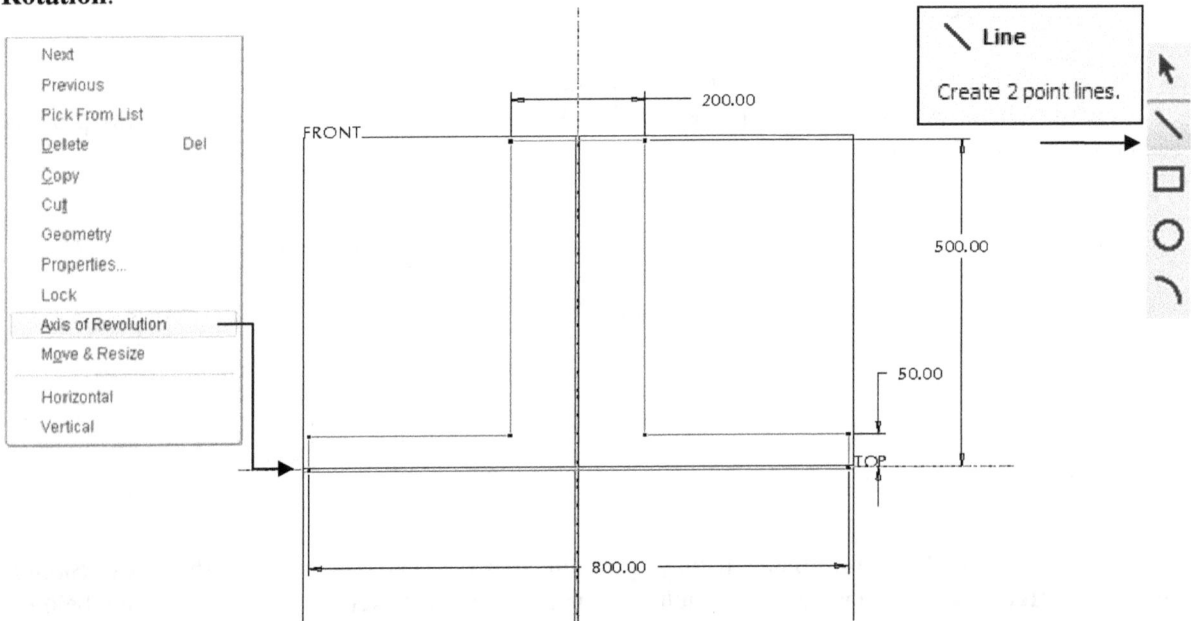

Next
Previous
Pick From List
Delete   Del
Copy
Cut
Geometry
Properties...
Lock
Axis of Revolution
Move & Resize
Horizontal
Vertical

FRONT

200.00

500.00

50.00

TOP

800.00

Line

Create 2 point lines.

Upon completing the sketch, click the icon of **Done**, and select the icon of **Apply and Save.**

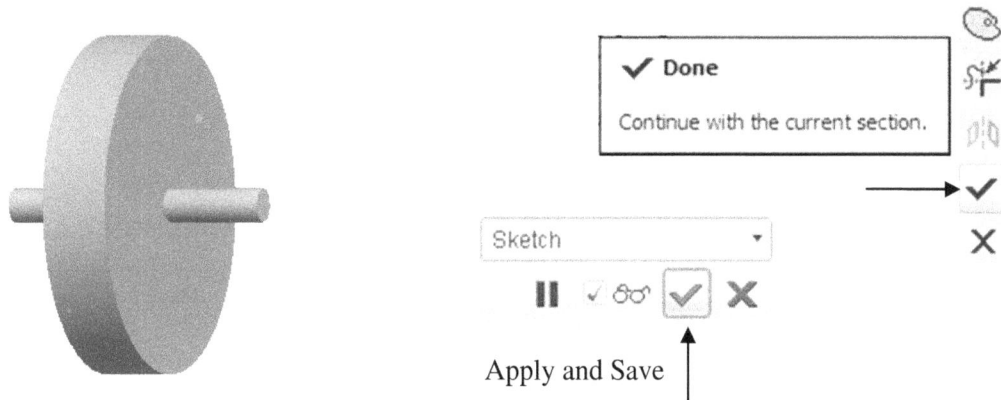

Apply and Save

Add a round at the sharp corner on each side. Select the icon of **Round** > type *0.5* as the value of radius > while holding down the **Ctrl** key, pick the 2 corners or 2 edges, as shown > **OK**.

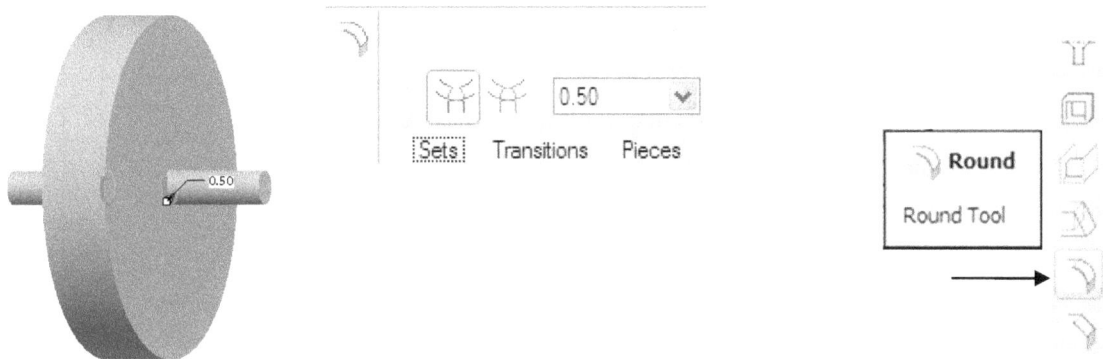

Step 3: Create 2 datum planes, which will be used to create 2 datum curves. The 2 datum curves will be used to create 2 surface regions to be used in the process of defining 2 constraints.

To define the first datum plane, click the icon of **Datum Plane Tool**. Pick the planar surface on the right side and type 80 as the offset value (you may need to type -80 if the direction is on the opposite side) > **OK. DTM1** is created.

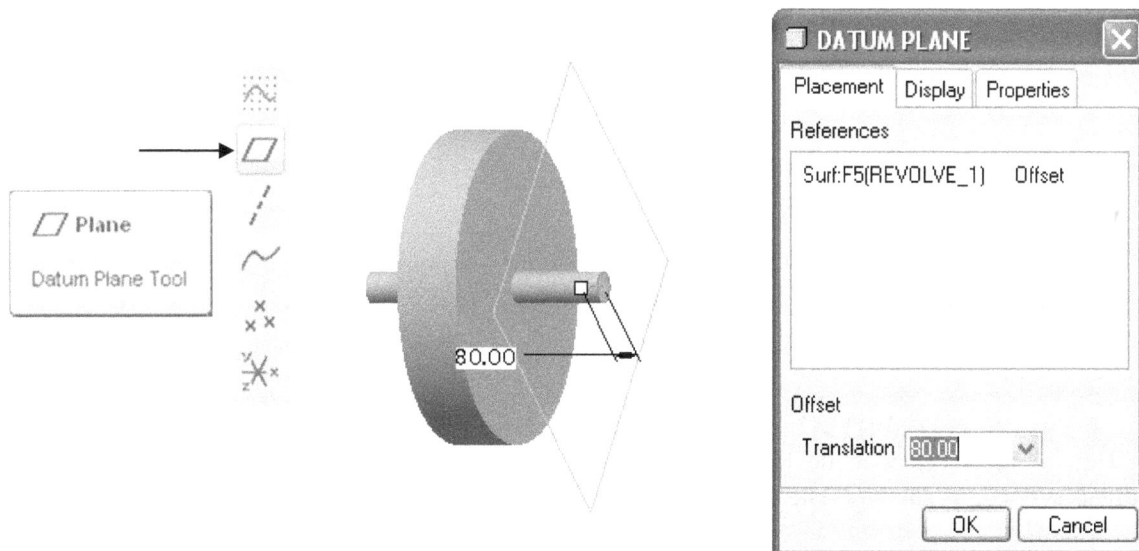

We now use the created **DTM1** to create a datum curve through intersection. From the main toolbar, select **Edit > Intersect >** while holding down the **Ctrl** key, pick **DTM1** and the 2 halves of the cylindrical surface on the right side, as shown. The created datum curve is a circle because of the intersection of a plane and a cylindrical surface.

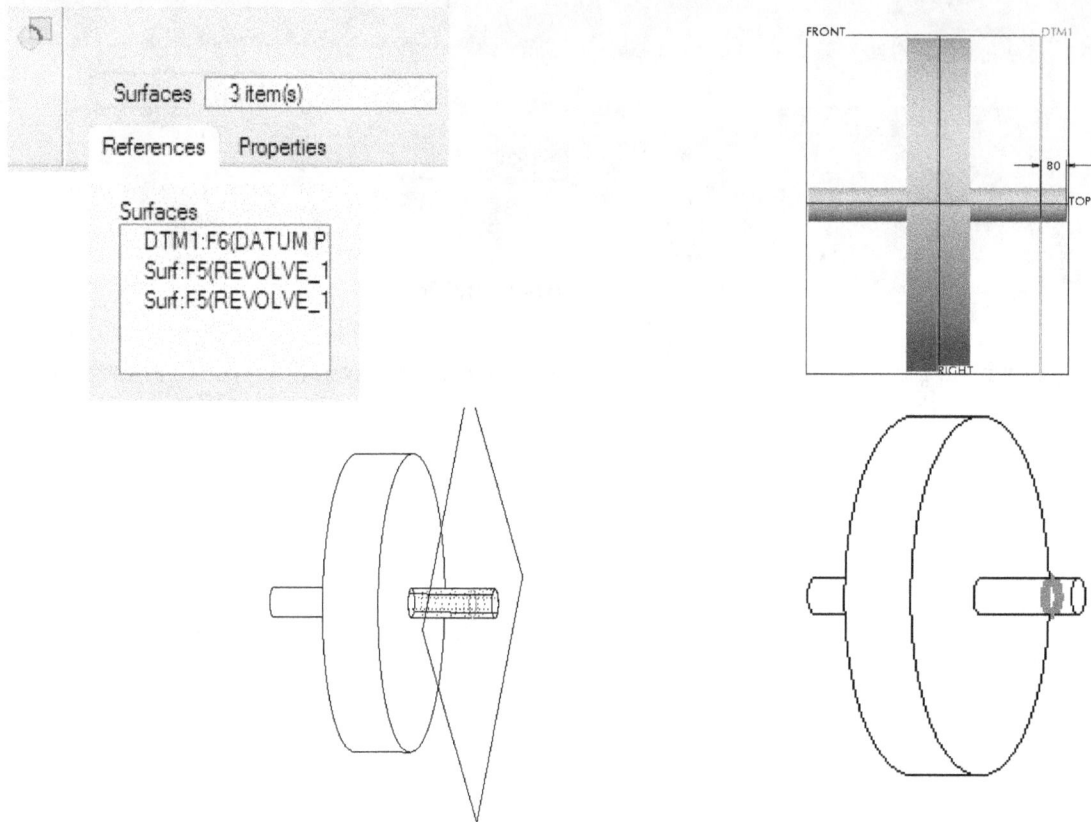

Following the same procedure, we create the second datum curve. Let us begin with the creation of a datum plane. Select the icon of **Datum Plane Tool**. Pick the planar surface on the left side and type 80 as the offset value > **OK.** DTM2 is created.

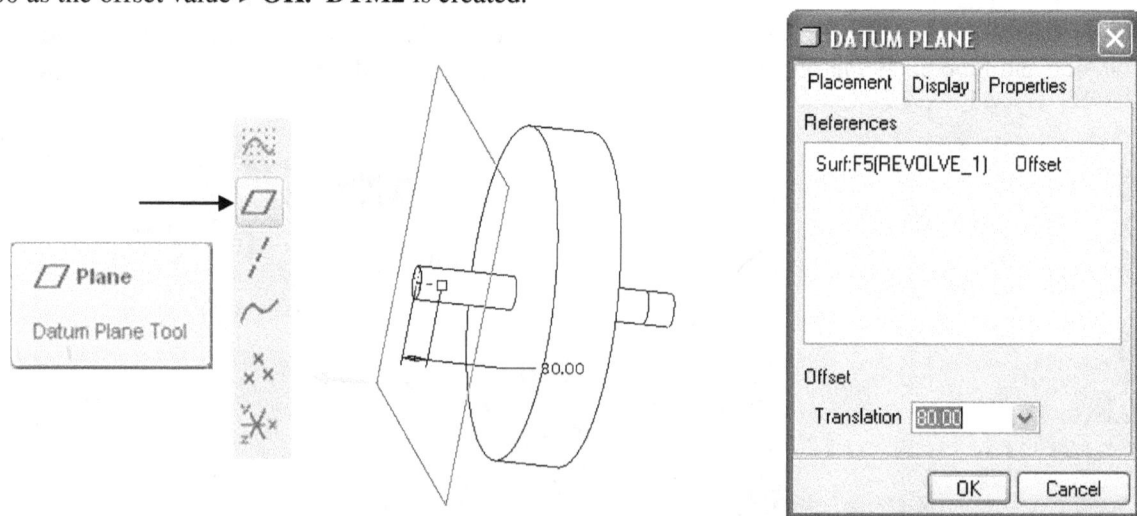

We use the created **DTM2** to create a datum curve through intersection. From the main toolbar, select **Edit > Intersect >** while holding down the **Ctrl** key, pick **DTM2** and the 2 halves of the cylindrical surface, as shown. The created datum curve is also a circle because of the intersection of a plane and a cylinder.

DTM1

Surfaces   3 item(s)

References   Properties

Surfaces
DTM2:F8(DATUM P
Surf:F5(REVOLVE_1
Surf:F5(REVOLVE_1

DTM2

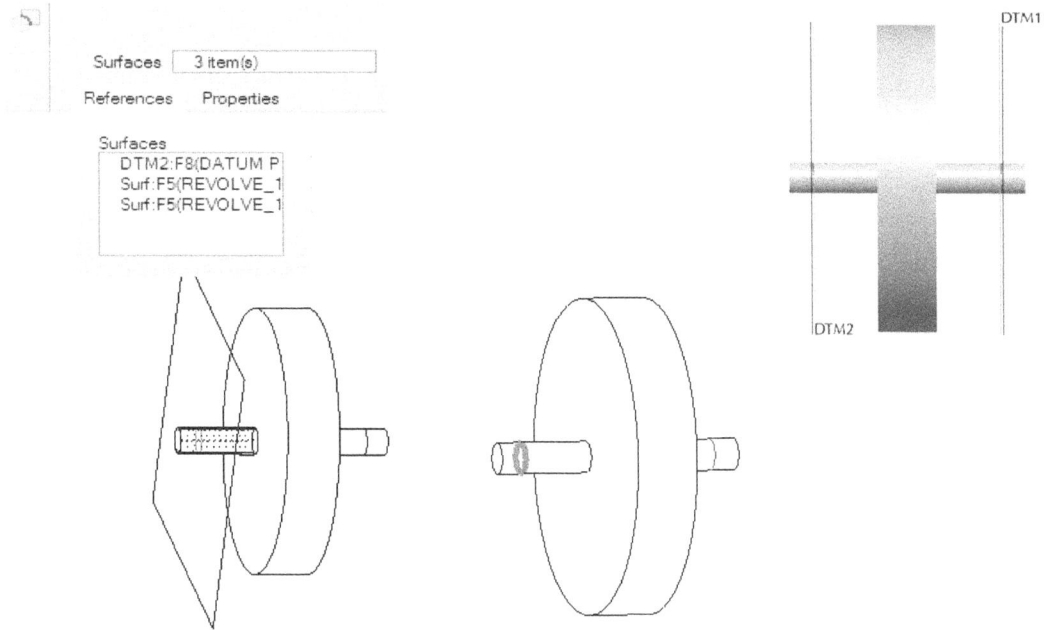

Step 4: Switch from the Pro/ENGINEER design system to Pro/Mechanica
From the main toolbar or the menu toolbar, select **Applications > Mechanica > Structure > OK.**

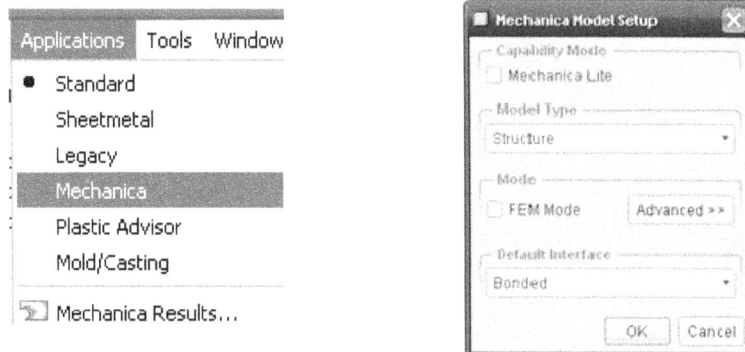

Applications   Tools   Window
• Standard
Sheetmetal
Legacy
Mechanica
Plastic Advisor
Mold/Casting
Mechanica Results...

Mechanica Model Setup
Capability Mode
Mechanica Lite
Model Type
Structure
Mode
FEM Mode   Advanced >>
Default Interface
Bonded
OK   Cancel

From the toolbar of functions, select the icon of **Materials >** select **STEEL** from the left side called Materials in Library > click the directional arrow so that the selected material type goes to the right side called Materials in Model.

Materials
File Edit Show
Materials in Library   Material Directory   Search...
Browse   Tools
Common Folders   al2014.mtl   femalt.mtl
Desktop   al6061.mtl   fenodr.mtl
My Documents   brass.mtl   mg.mtl
Working Directory   bronze.mtl   nylon.mtl
Network Neighborhood   cu.mtl   pvc.mtl
Material Directory   epoxy.mtl   ss.mtl
Favorites   fe20.mtl   steel.mtl
fe30.mtl   tially.mtl
fe40.mtl   tipure.mtl
Folder Tree   fe60.mtl   tungsten.mtl
Materials in Model   STEEL
OK   Cancel

Materials
Define Materials

Click the icon of **Material Assignment.** In the Material Assignment window, the selected Steel is shown, and the component is also shown because there is only one volume or one component in the system. As a result, the software system automatically assigns Steel to the component. Just click **OK**.

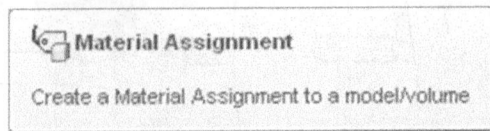

Step 5: Create 2 surface regions so that the 2 required constraints can be defined.

Click the icon of **Surface Region**. Expand the **References** box.

Select **Split by chain**. Pick the first created datum curve under Chain. Pick the 2 halves of the cylindrical surface under Surfaces. Click **Apply and Save**. As a result, the entire cylindrical surface is divided, and the required surface region to be used for defining a constraint is shown.

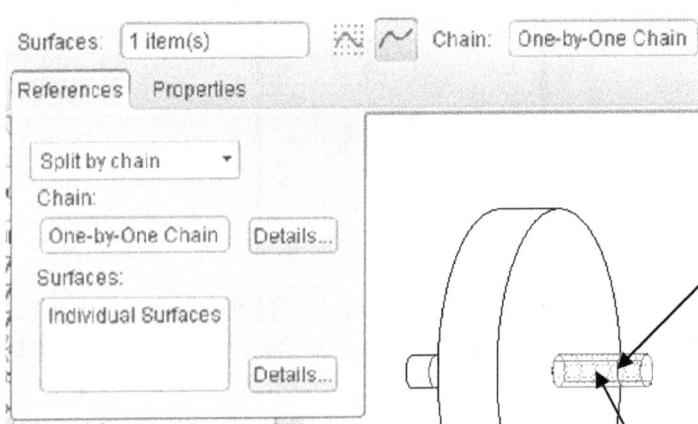

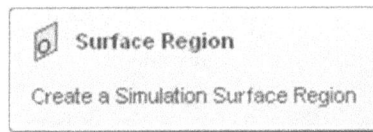

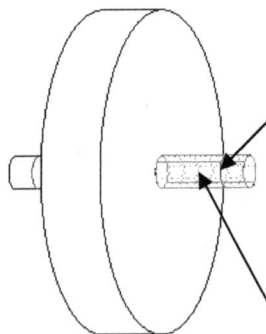

Pick this curve first

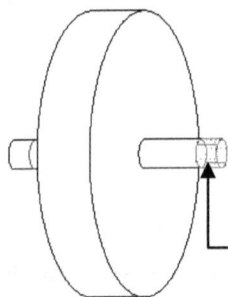

Afterwards, pick both halves of the cylindrical surface while holding down the **Ctrl** key

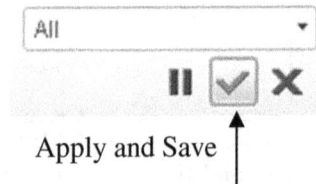

The required surface region

Apply and Save

Again, click the icon of **Surface Region.** Expand the **References** box.

Select **Split by chain.** Pick the secondly created datum curve under Chain. Select the 2 halves of the cylindrical surface under Surfaces. Click **Apply and Save**. As a result, the left portion of the cylindrical surface is divided, and the required surface region is shown.

Pick this curve first

Surface Region
Create a Simulation Surface Region

All

Apply and Save

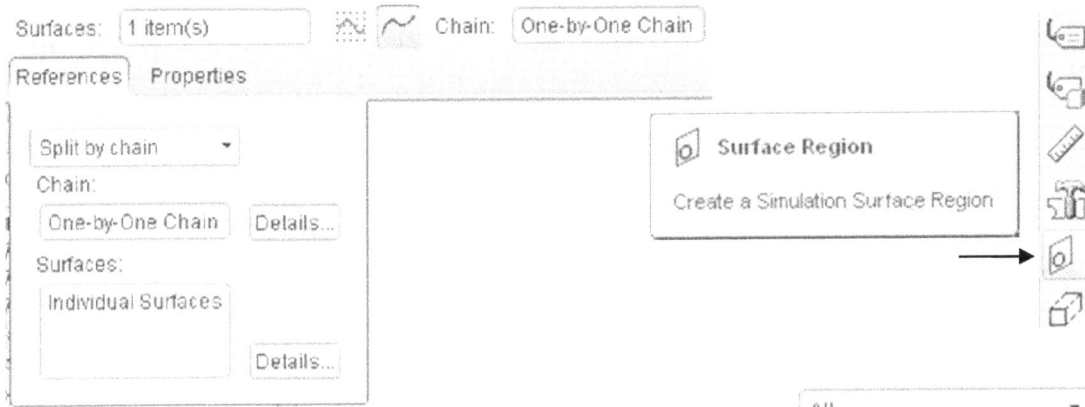

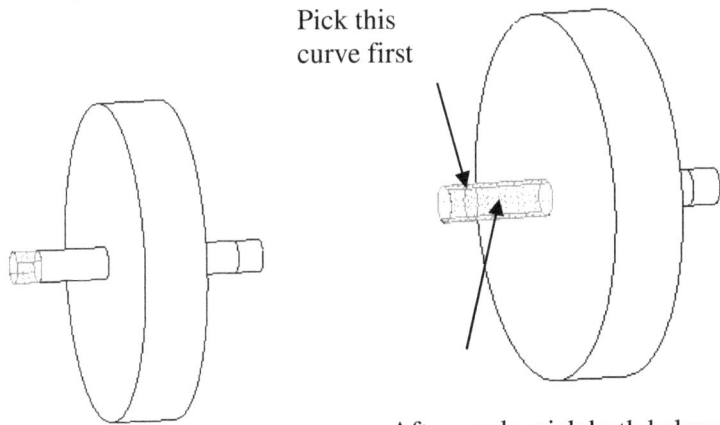

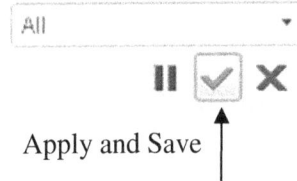

Afterwards, pick both halves of the cylindrical surface while holding down the **Ctrl** key

At this moment, we have created the 2 required surface regions. Let us define the 2 constraints now.

Assume that the constraint condition is fixed to the ground. Select the icon of **New Displacement Constraint** from the toolbar of Mechanica objects > **Surface(s)** > while holding down the **Ctrl** key, pick the 2 required surface regions. Set all 6 degrees of freedom equal to zero, which is the default setting > **OK**.

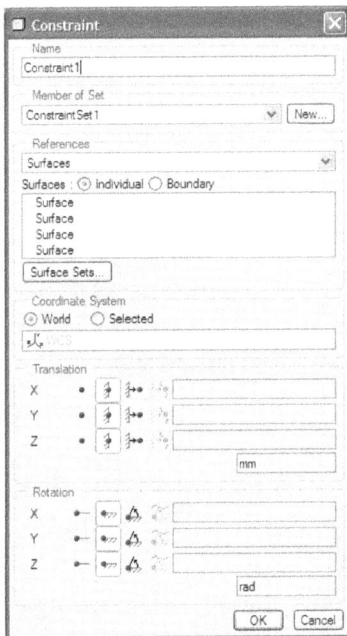

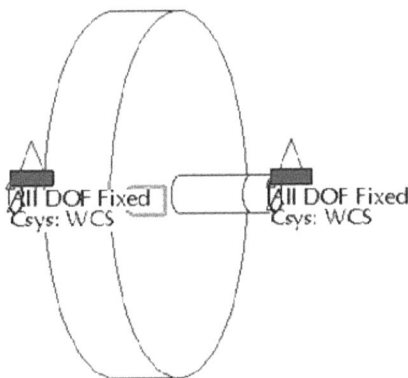

All DOF Fixed
Csys: WCS

All DOF Fixed
Csys: WCS

Displacement Constraint
Create a Displacement Constraint

Step 6: Define the centrifugal force load condition.

Select the icon of centrifugal load > in terms of the angular velocity, type *125* as the X component > **Preview > OK**.

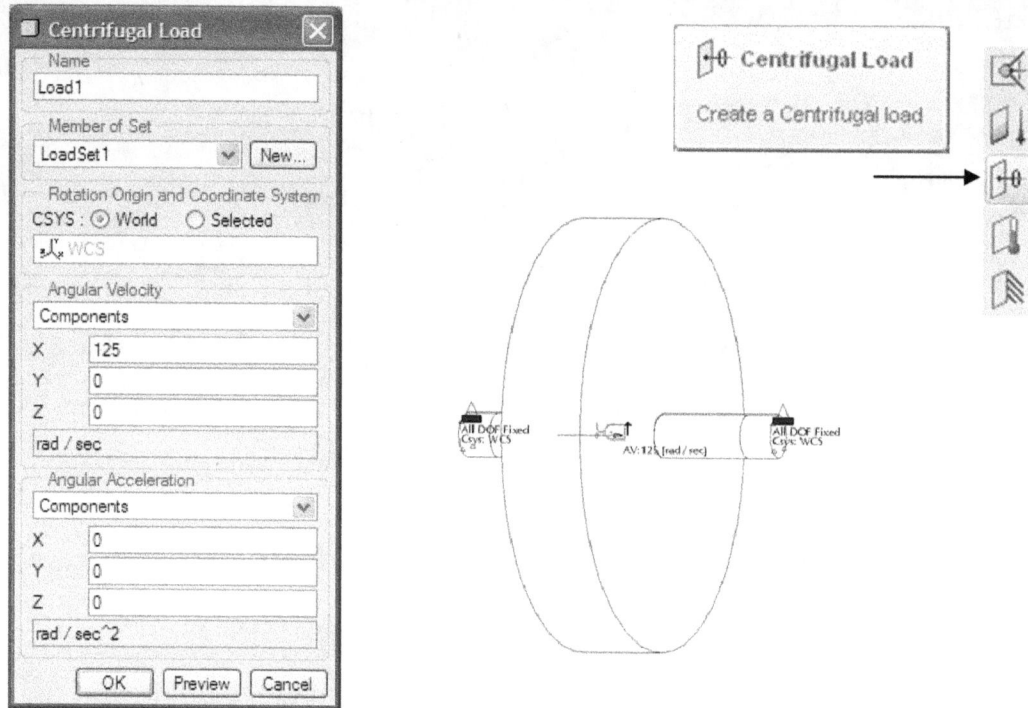

Step 7: Under the Pro/MECHANICA environment, set up **Analyses** and **Run** it

Select the icon of **Run an Analysis** from the main toolbar > **File** > **New Static** > type *centrifugal_load* as the name of the analysis folder > **OK**. Click the box of **Run**.

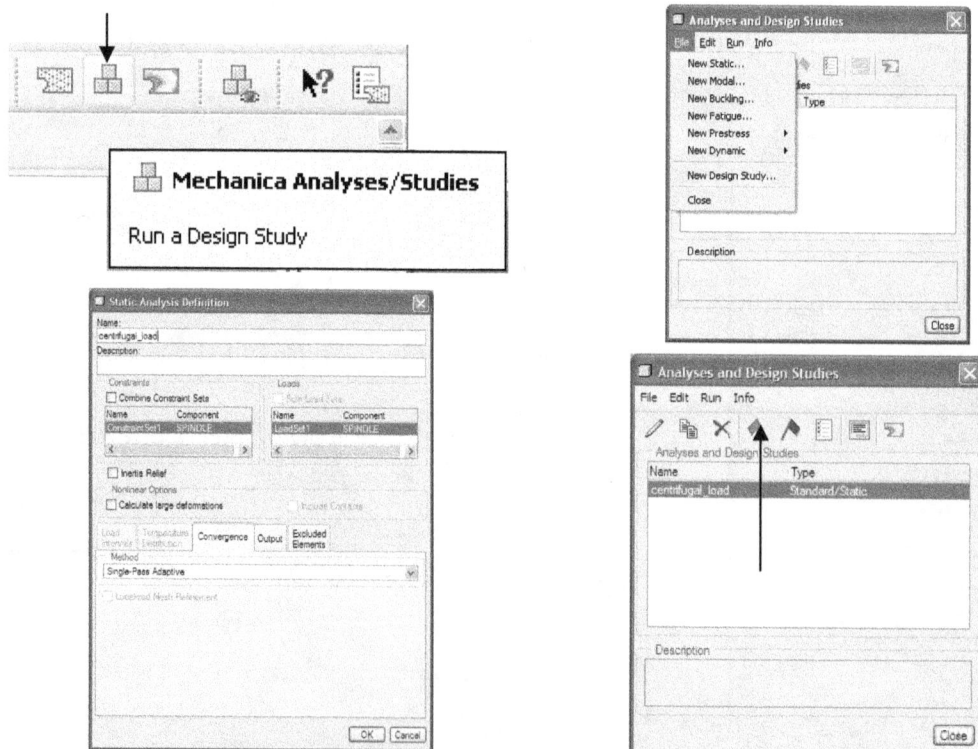

There is always a message displayed on screen. The message is asking the user "Do you want to run interactive diagnostics?" Click **Yes**. In this way a warning message will appear if an abnormal operation occurs.

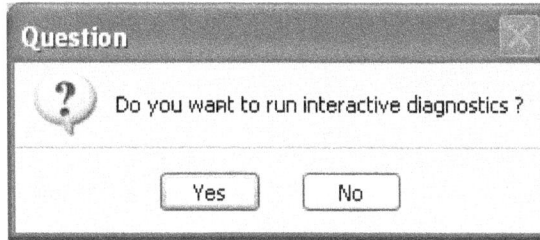

To monitor the computing process, users may click the icon of **Display study status**.

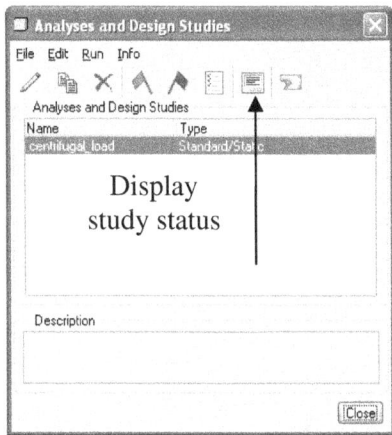

```
Memory and Disk Usage:

    Machine Type: Windows NT/x86
    RAM Allocation for Solver (megabytes): 128.0

    Total Elapsed Time (seconds): 35.95
    Total CPU Time     (seconds): 35.08
    Maximum Memory Usage (kilobytes): 221318
    Working Directory Disk Usage (kilobytes): 299008

    Results Directory Size (kilobytes):
    15328 .\centrifugal_load

    Maximum Data Base Working File Sizes (kilobytes):
    186368 .\centrifugal_load.tmp\kblk1.bas
    99328 .\centrifugal_load.tmp\kel1.bas
    13312 .\centrifugal_load.tmp\oel1.bas

---------------------------------------------
Run Completed
Wed Aug 11, 2010   13:52:18
---------------------------------------------
```

To study the obtained result, say the distribution of the max principal stress, click the icon of **Review results > Stress > Max Principal > OK and Show.**

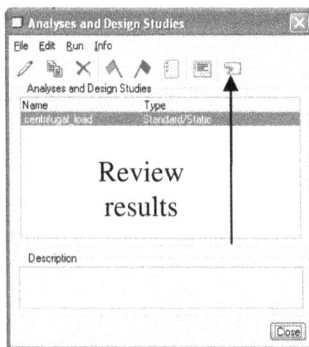

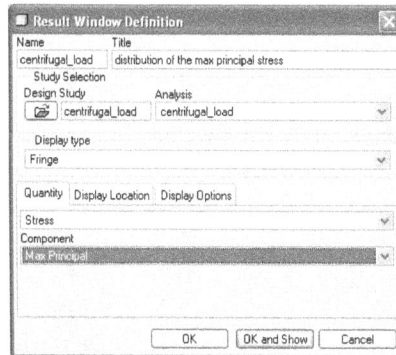

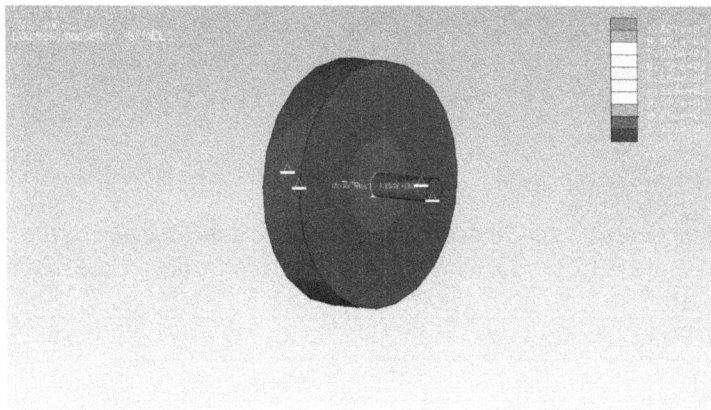

Note that the maximum magnitude of the max principal stress is located at the corner. For the radius value equal to 0.5 mm, the magnitude is 113.4 MPa.

Step 8: Go back to Pro/ENGINEER and change the radius value from 0.5 mm to 1.0 mm, and run the FEA program again and record the maximum value of the maximum principal stress.

Make sure you go back to the Pro/ENGINEER design environment, first. From the main toolbar, select Applications > Standard.

From the model tree, highlight **Round1** > right-click and hold, pick **Edit** from the pop up menu. Double click the displayed value 0.5, change it to 1.0 and press the **Enter** key. From the main toolbar, select the icon of **regenerate** to complete this modification.

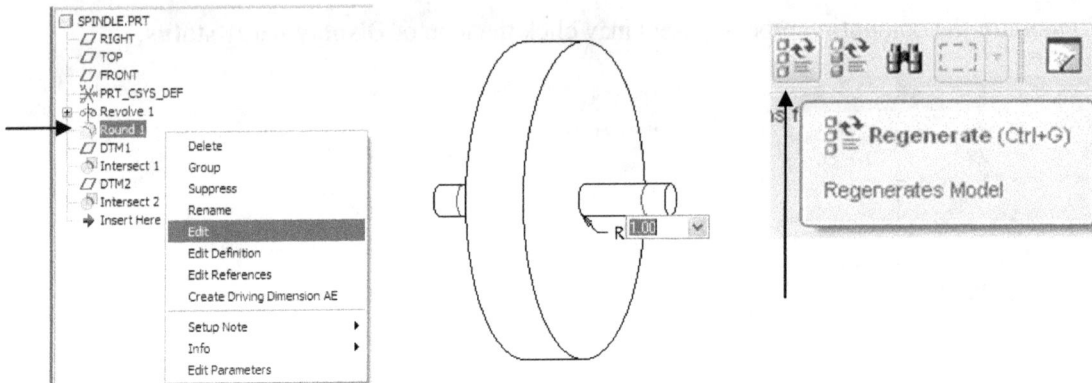

Make sure that you go back to Pro/Mechanica to run the program, again. From the main toolbar, select **Applications > Pro/Mechanica**. Select the icon of **Run an Analysis** from the main toolbar > the folder called *centrifugal_load* is on display > click the box of **Run** > plot the distribution of the maximum principal stress and record the maximum value of the maximum principal stress. As shown, the location of the maximum value of the maximum principal stress remains unchanged. Its magnitude is reduced from 113.4 MPa to 99.13 MPa.

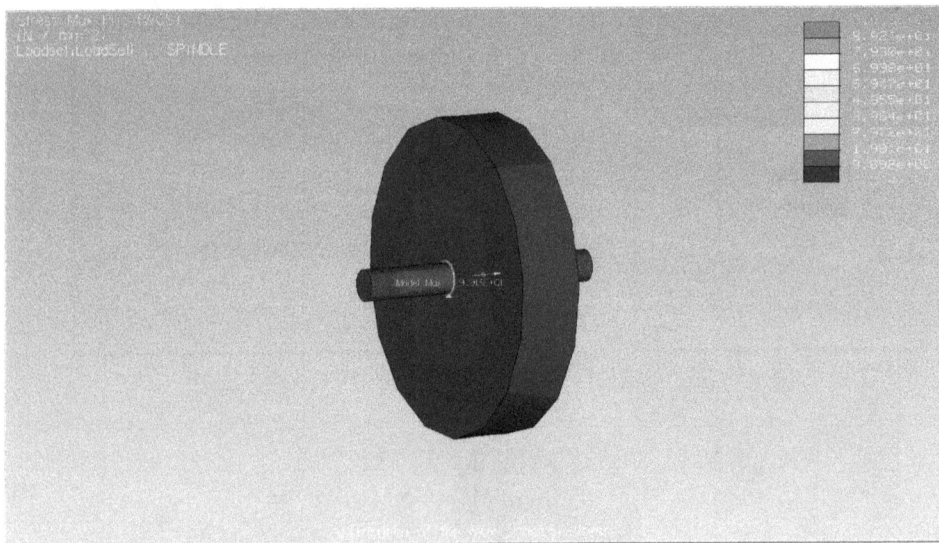

Step 9: Repeat the procedure described in Step 8 for the radius dimension equal to the following values: 2.0, 5.0, 10, 20, 30, and 40. Construct a curve with maximum value of the maximum principal stress vs. radius value.

Radius (mm)	0.5	1	2	5	10	20	30	40
$\sigma_{max\text{-}principal}$ (MPa)	113.2	99.13	74.58	46.5	34.54	25.51	21.61	20.29

Stress Concentration vs Radius Value

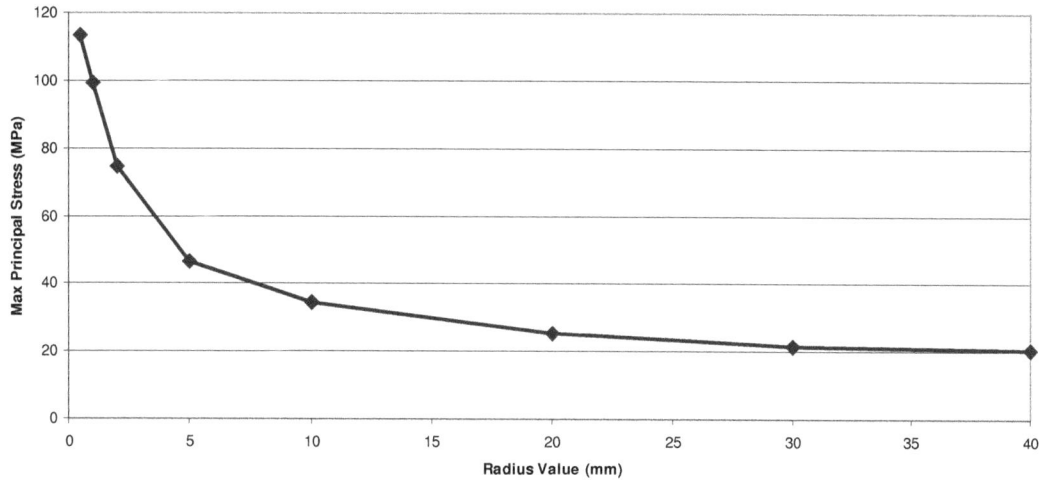

To facility the process of simulating a set of system responses, the students may use the method of constructing a design study and perform a global sensitivity analysis described in the previous studies. The software system will vary the radius value incrementally within a specified arrange.

Select the icon of **Run a Design Study** from the main toolbar > **File** > **New Sensitivity Design Study** > type *centrifugal_load* as the name of the analysis folder. Click the box of select the dimension from model > pick the radius > **OK.** Specify 0.5 and 40 as the start and end values. Click the box of **Run.**

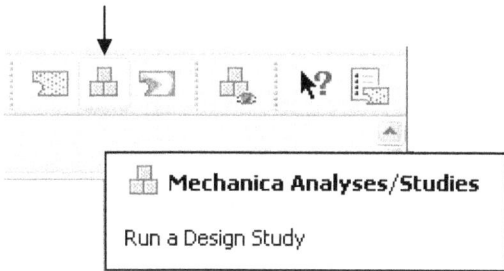

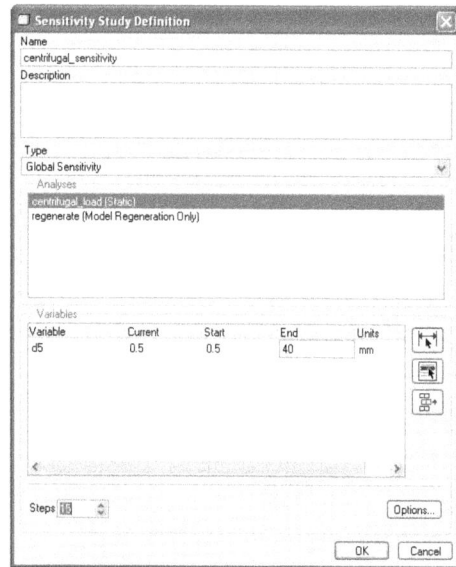

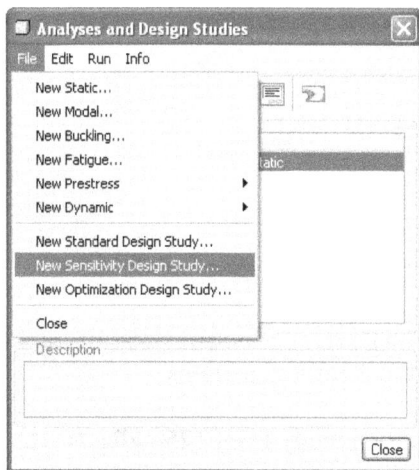

To monitor the computing process, users may click the icon of **Display study status**.

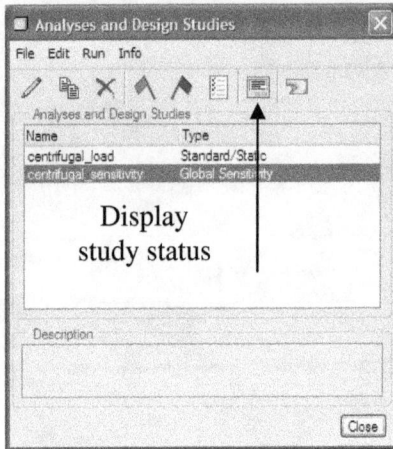

To study the obtained result, click the icon of **Review results > Graph >** Under Measure, click undefined to search for **max_stress_prin > OK.** In the Result Window Definition, click **OK and Show.**

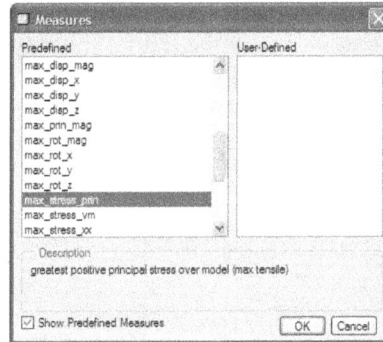

max_prin_mag
(N/mm^2)
Design Var
Loadset:LoadSet1 : SPINDLE

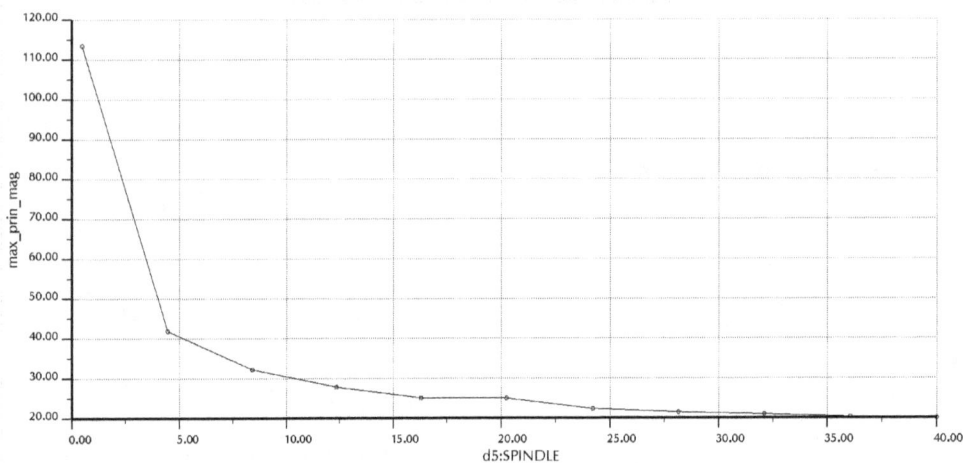

Examining the plot, it is evident that the magnitude of the maximum principal stress decreases as the radius value increases, easing the stress concentration effect at the 2 corner locations.

Step 10: To study the effect due to rotation speed, change the angular velocity from 125 radian/sec to 157 radian/sec, which is equivalent to change 1200 rpm to 1500 rpm in terms of the rotation speed.
    Click the icon of **Centrifugal Load >** click **New >** in the **Load Set Definition**, LoadSet2 appears, specify 157 rad/sec as the angular velocity, and **OK**.

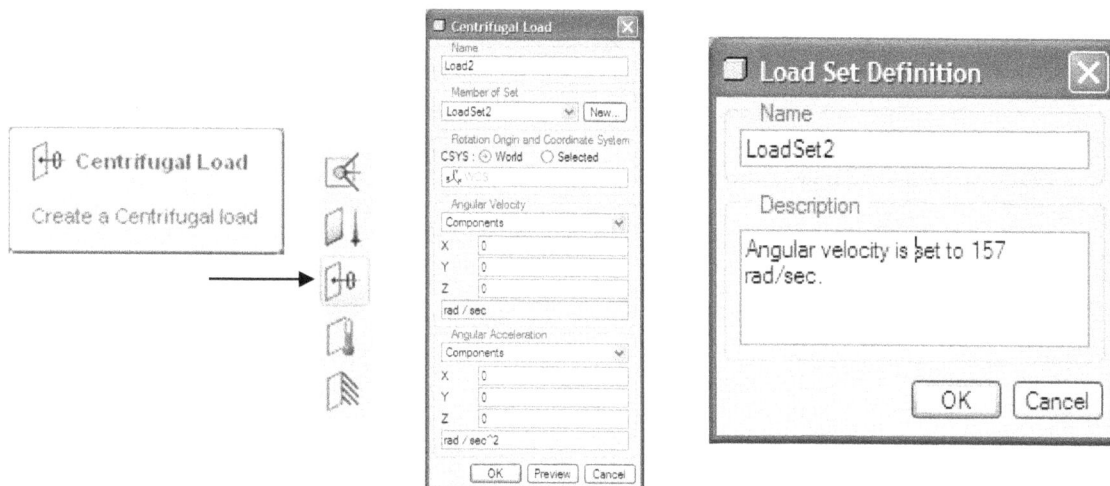

Specify 157 rad/sec as the magnitude of angular velocity about the X axis.

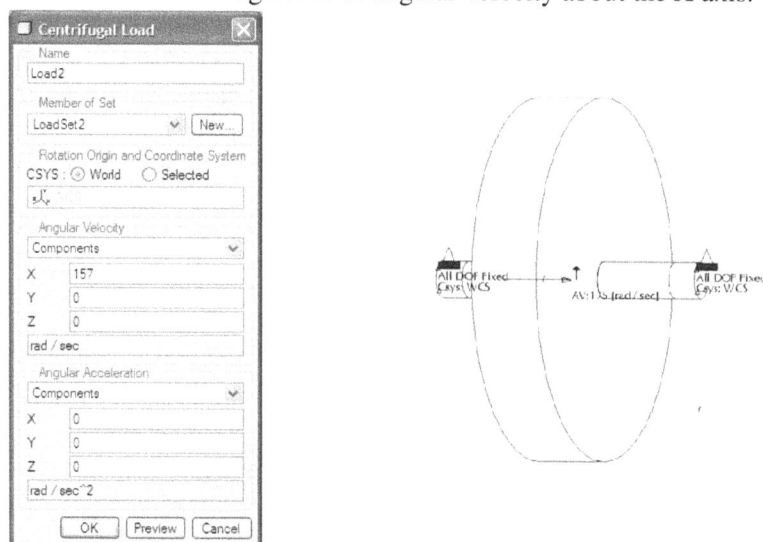

Now let us run the global sensitivity analysis for the second time. Should we do something else before the second run? Yes! You need to run the static analysis using the new load condition before running the global sensitivity analysis for the second time.

Select the icon of **Run an Analysis** from the main toolbar > highlight centrifugal_load and click **Edit** from the Analyses and Design Studies > only highlight LoadSet2 > **OK.** Click the box of **Run.**

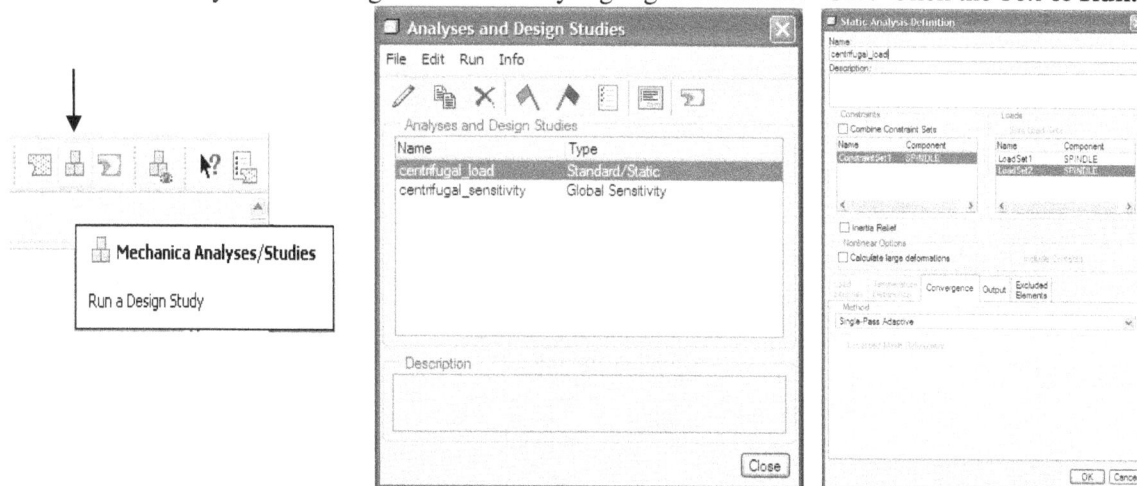

To monitor the computing process, users may click the icon of **Display study status**.

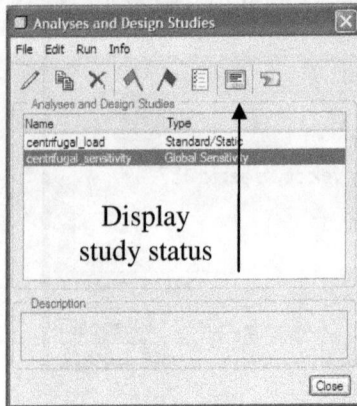

```
Analysis "centrifugal_load" Completed   (06:49:57)

Sensitivity Step 11 of 11

Parameters:
    d5                              40

Static Analysis "centrifugal_load":

    Convergence Method: Single-Pass Adaptive
    Plotting Grid:      4

    Convergence Loop Log:

    >> Pass  1 <<
        Calculating Element Equations
           Total Number of Equations:   23952
           Maximum Edge Order:              3
        Solving Equations
        Post-Processing Solution
        Checking Convergence
        Resource Check
```

To study the obtained result, click the icon of **Review results > Graph** > Under Measure, click undefined to search for **max_stress_prin > OK.** In the Result Window Definition, click **OK and Show.**

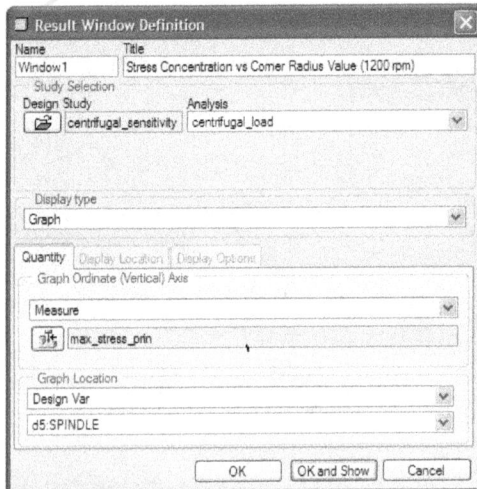

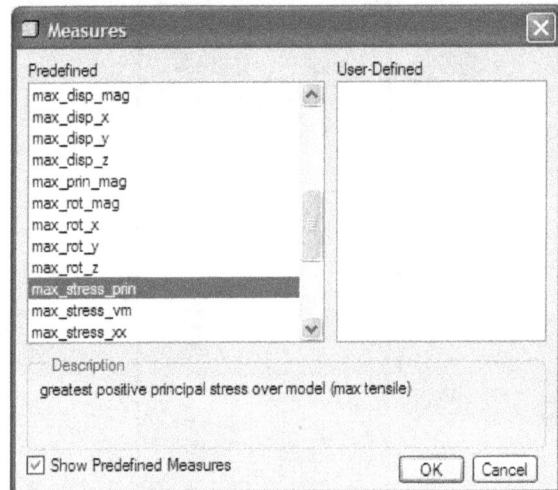

```
max_stress_prin
(N/mm^2)
Design Var
Loadset:LoadSet2 : SPINDLE
```

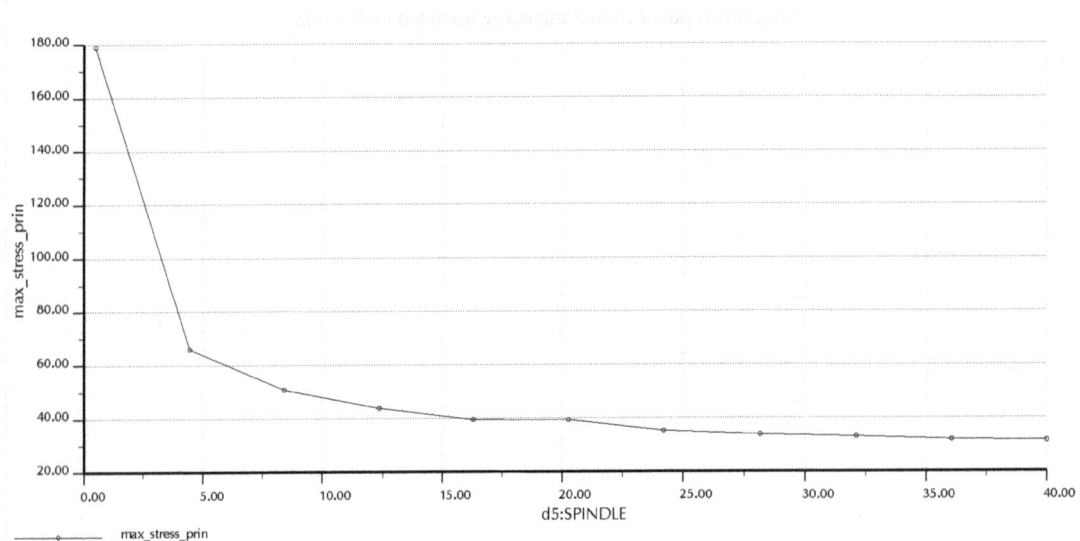

It is important that we compare the results obtained from the 2 global sensitivity studies. Users should come up with those observations: the higher the rotation speed, the severer the stress concentration becomes. Increasing the radius value has a significant effect on reducing the magnitude of the maximum principal stress. This effect is more evident when the rotation speed is set high. When the magnitude of the radius value increase to a certain value, say 30 mm in this case, the magnitude of the maximum principal stress tends to approach a constant value, say 20 MPa for the setting of low rotation speed and 35 MPa for the setting of high rotation speed.

4.6   Modeling of Fastener Connections

As illustrated, a support structure is held to the ground through 4 bolts. The load conditions include torque acting on the top surface, and a load uniformly acting on the top surface in the vertical direction.

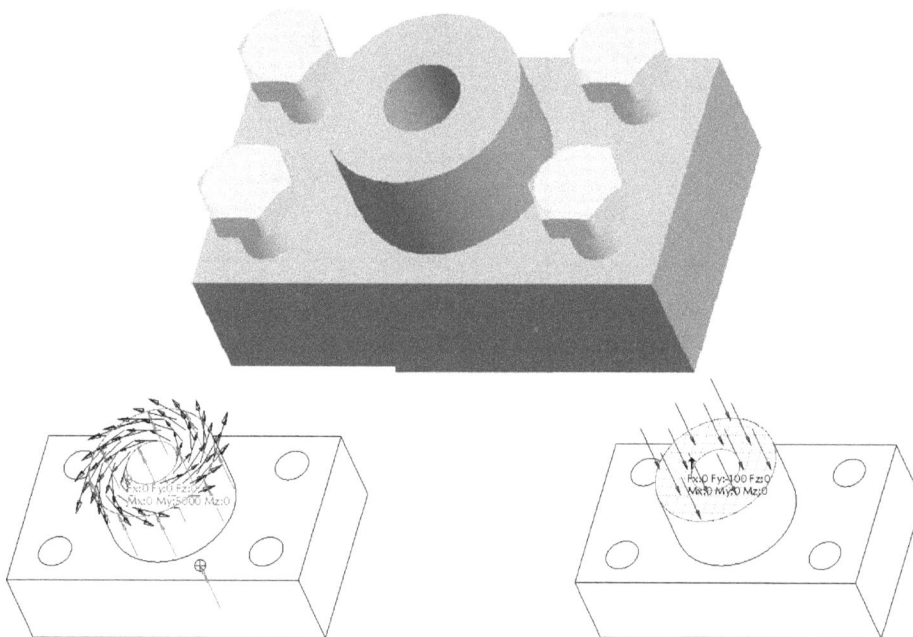

The geometry of the support structure and its dimensions are shown below. We will perform FEA to illustrate the deformation distribution and the distribution of the maximum principal stress. The constraint set consists of the 4 cylindrical surfaces of the 4 holes. In addition, we consider that the bottom surface is in contact with the ground, not fixed to the ground.

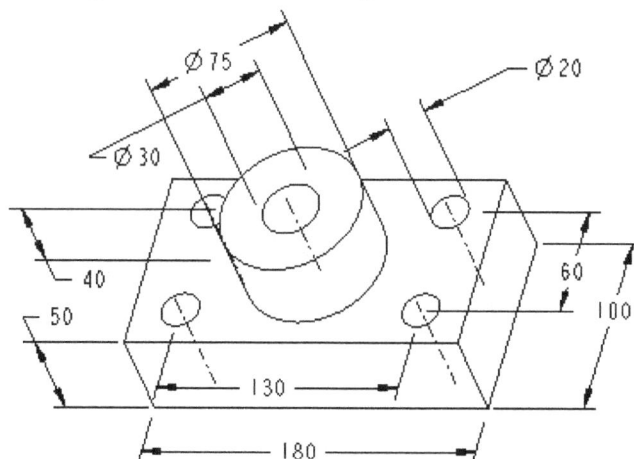

Step 1: Create a file for the 3D solid mode.

File > New > Part > type *support* as the file name and clear the icon of **Use default template > OK.** Select **mmns_part_solid** (units: Millimeter, Newton, Second) and type *torque and constraint* in **DESCRIPTION**, and *student* in **MODELED_BY**, then **OK.**

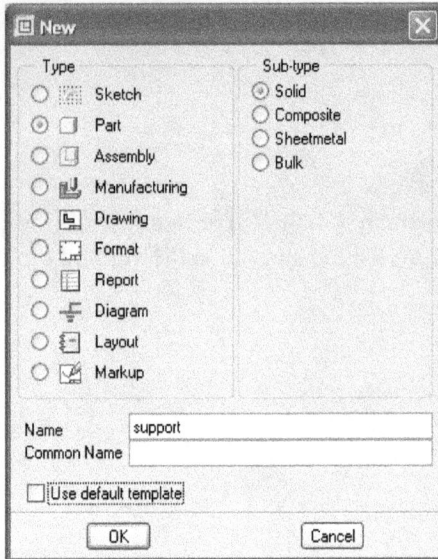

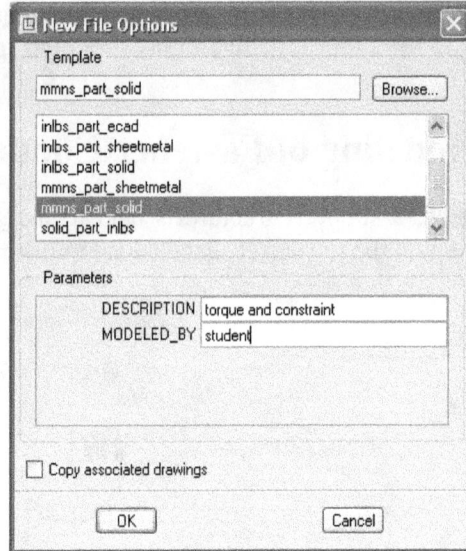

Create a 3D solid model and start with the block feature. The size is 180 x 100 x 50 mm.
Click the icon of **Insert > Extrude.** Activate **Placement** to define a sketch plane > **Define.**

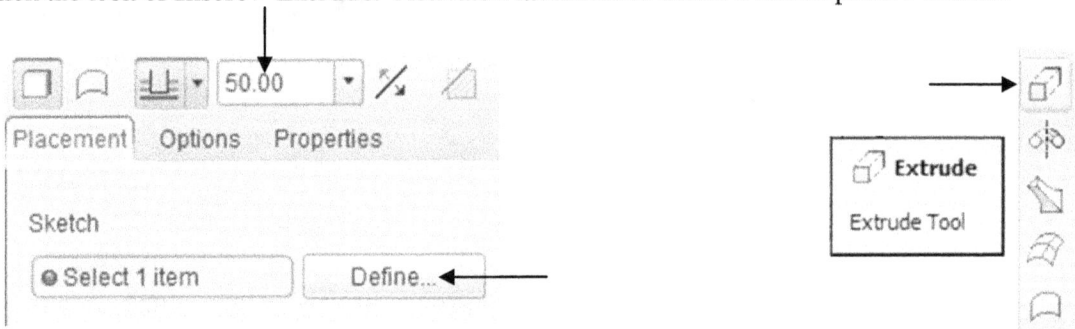

Select the **FRONT** datum plane displayed on screen as the sketch plane. Pick the box of **Sketch** to accept the **RIGHT** datum plane as the default reference to orient the sketch plane, as illustrated below:

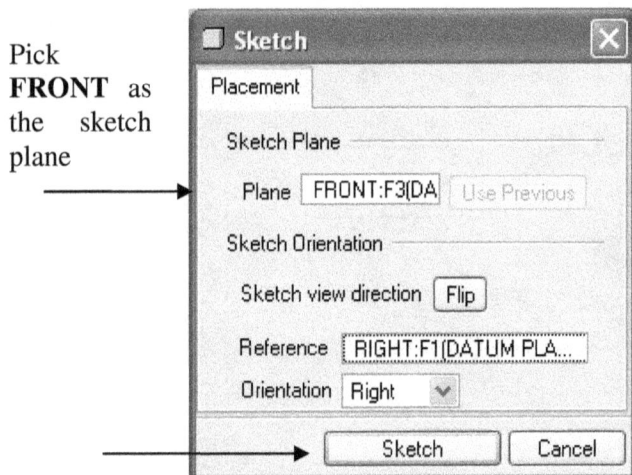

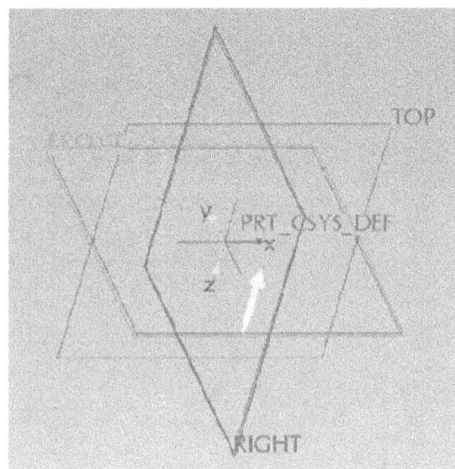

Pick **FRONT** as the sketch plane

Click the icon of **Centerline**. Sketch a horizontal centerline for the revolving operation. Afterwards, sketch a vertical centerline to be used for symmetry when sketching.

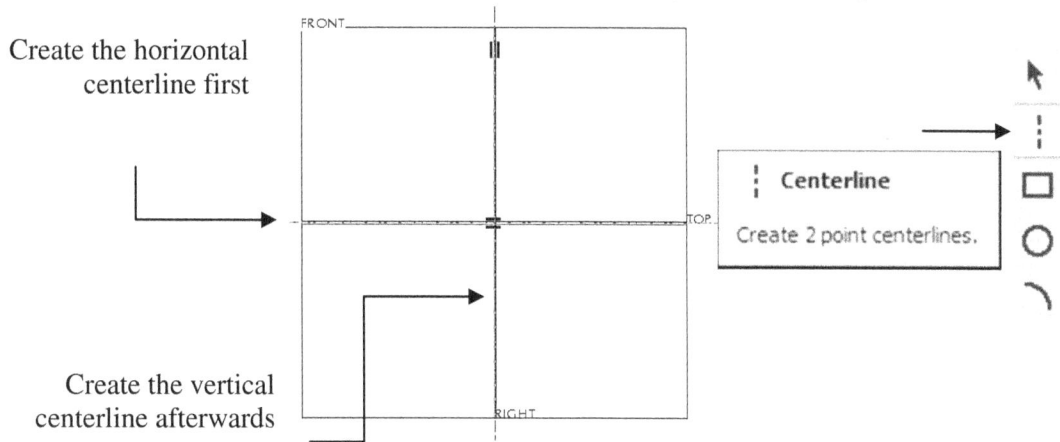

Create the horizontal
centerline first

Create the vertical
centerline afterwards

Select the icon of **Rectangle** to sketch a rectangle, which is symmetric with respect to the two centerlines, by allowing the rectangle to snap to symmetry.

Modify the two dimensions to 180 and 100, respectively. (if the sketched rectangle is not symmetric about the y axis or z axis or both axes, pick the icon of **Constraints** > select the icon of **Symmetry** > select a centerline, say y axis and, afterwards, pick the two corner points, which are constrained to meet the symmetrical requirement).

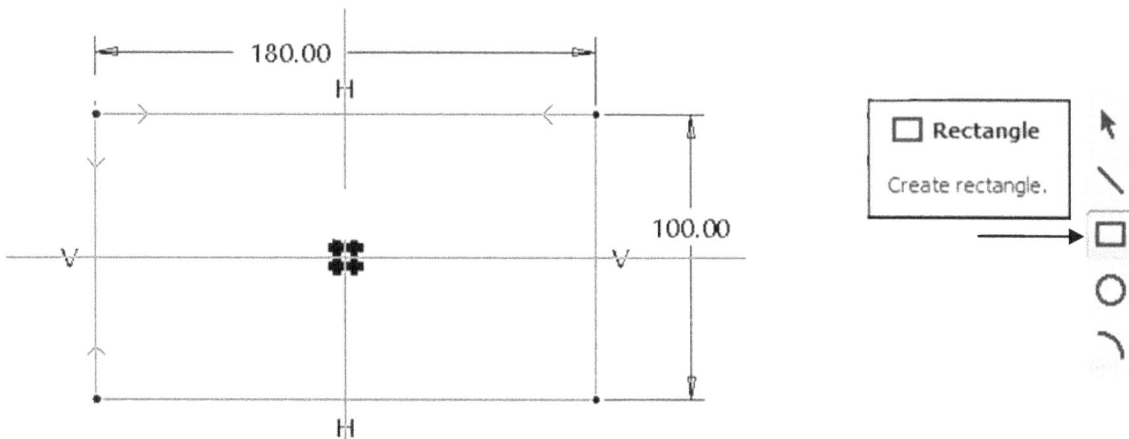

Click the icon of Circle and sketch the first circle with the 3 dimensions, as shown. The diameter dimension is 20. The 2 position dimensions are 65 and 30, respectively.

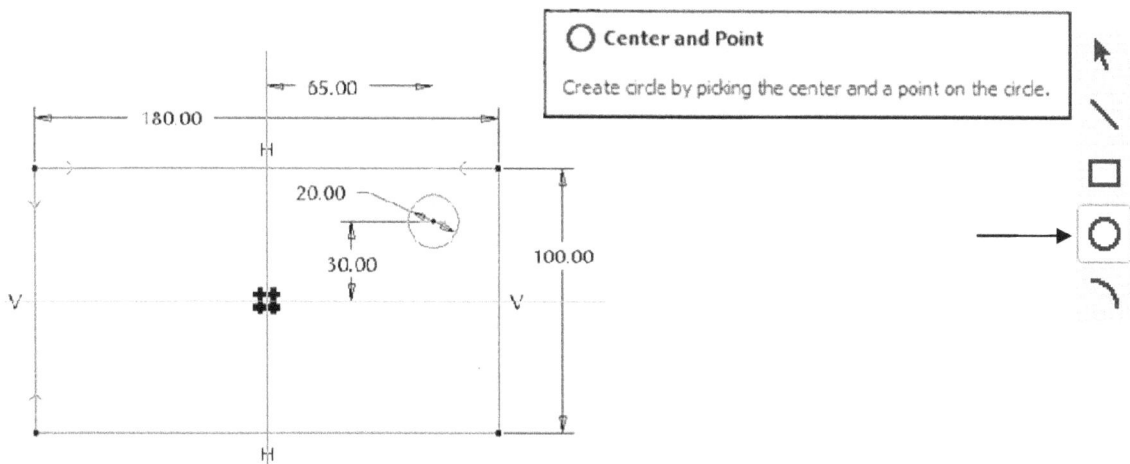

We need 3 more circles. Let us use the copy function. Click the icon of **Mirror** > pick the sketched circle first. Afterwards, pick the vertical centerline, completing the creation of the second circle.

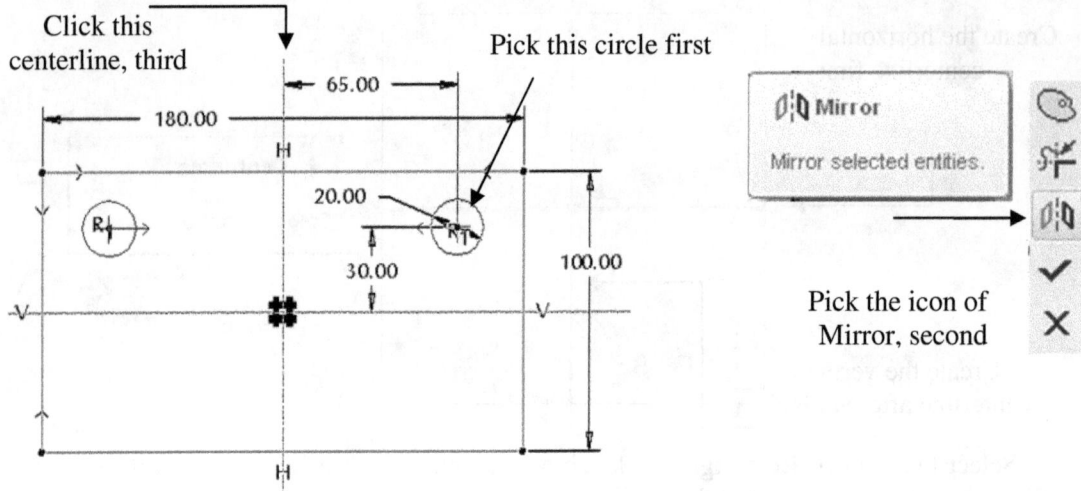

Click this
centerline, third

Pick this circle first

← 65.00 →

180.00

20.00

30.00

100.00

Ḋ Mirror

Mirror selected entities.

Pick the icon of
Mirror, second

While the copied circle is still actively selected, click the icon of **Mirror** > click the horizontal centerline, completing the creation of the third circle.

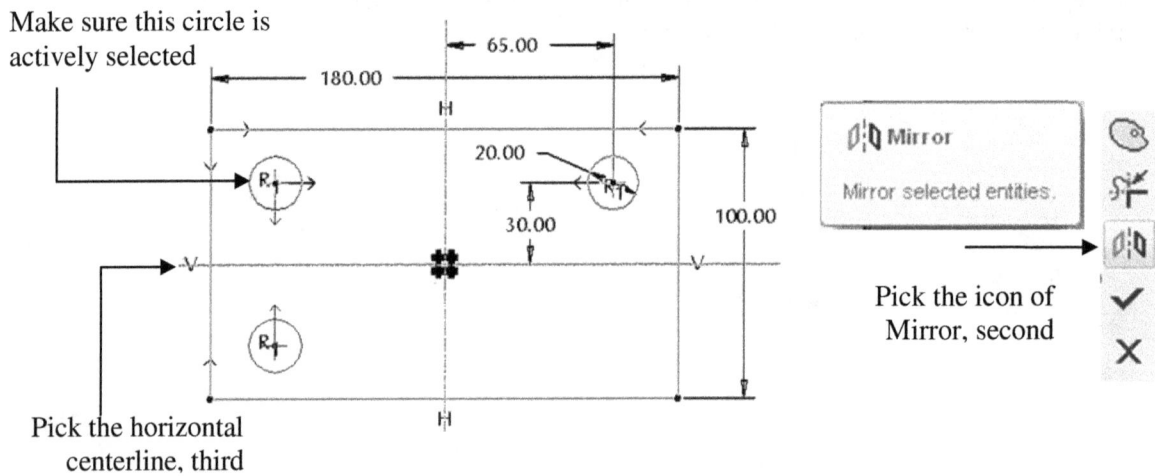

Make sure this circle is
actively selected

← 65.00 →

180.00

20.00

30.00

100.00

Ḋ Mirror

Mirror selected entities.

Pick the icon of
Mirror, second

Pick the horizontal
centerline, third

While the copied circle at the right and lower portion is still actively selected, click the icon of **Mirror** > click the vertical centerline, completing the creation of the 4th circle.

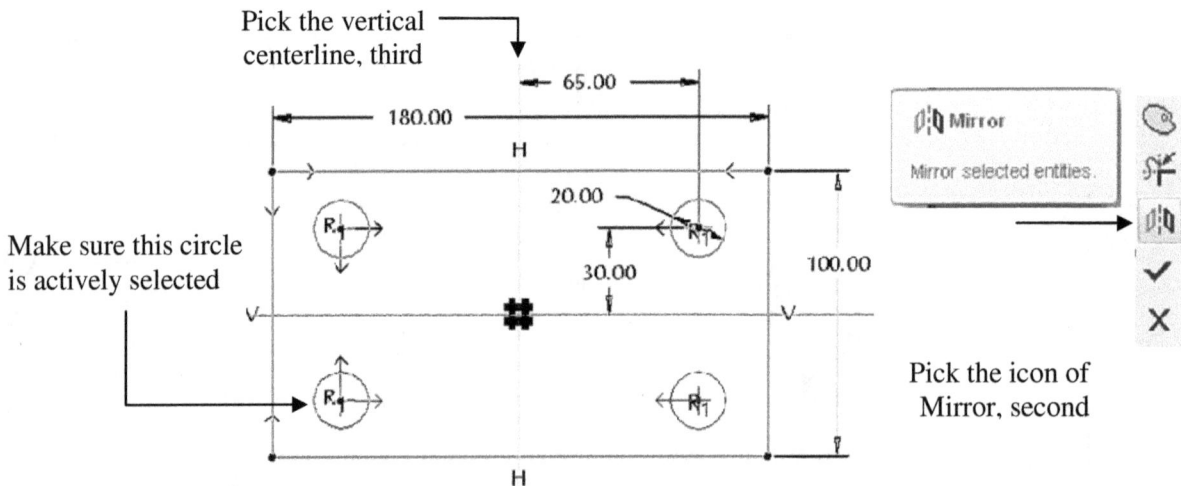

Pick the vertical
centerline, third

← 65.00 →

180.00

20.00

30.00

100.00

Make sure this circle
is actively selected

Ḋ Mirror

Mirror selected entities.

Pick the icon of
Mirror, second

Upon completing the sketch, click the icon of **Done**, and select the icon of **Apply and Save.**

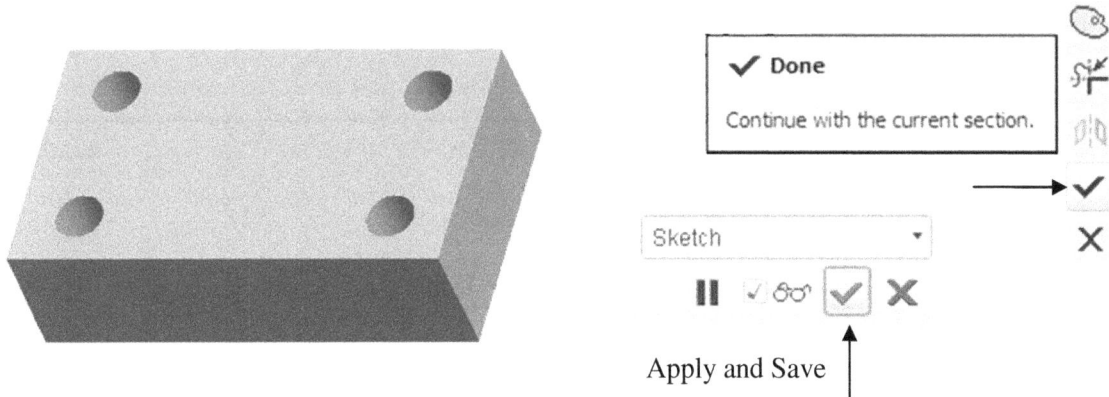

Apply and Save

After creating the block with holes feature, we create the cylinder feature. Its size is φ75 x 40 mm. Click the icon of **Extrude.** Specify 40 as the height value. Activate **Placement** to define a sketch plane > **Define**.

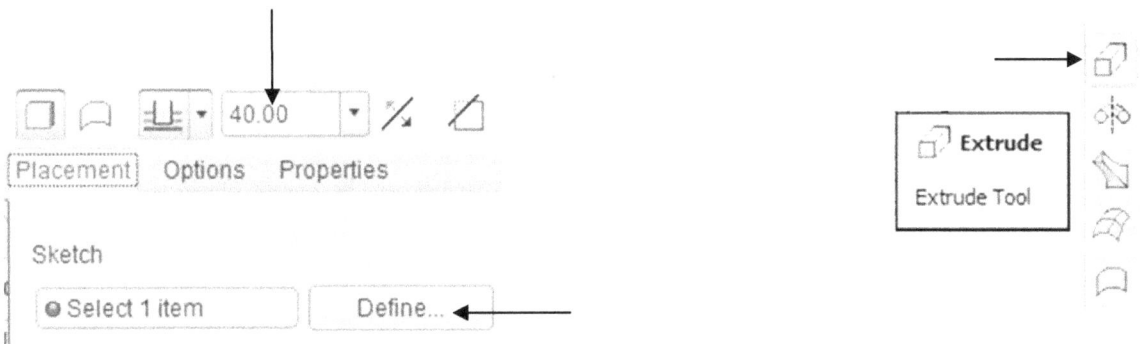

Select the top surface of the block (note: do not select the **TOP** datum plane) displayed on screen, and click **Sketch** to accept the **RIGHT** datum plane as the default reference to orient the sketch plane, as illustrated below:

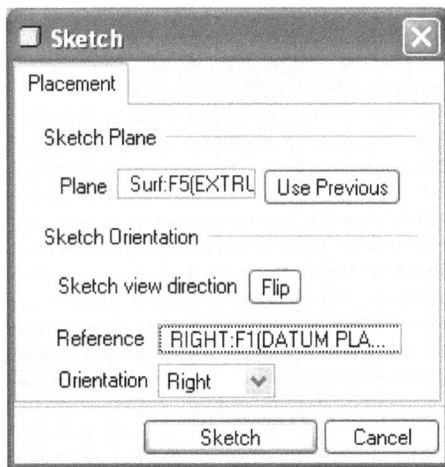

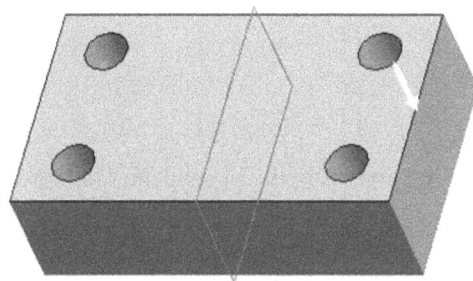

Click the icon of **Circle** to sketch a circle with diameter equal to 75.

75.00

○ **Center and Point**

Create circle by picking the center and a point on the circle.

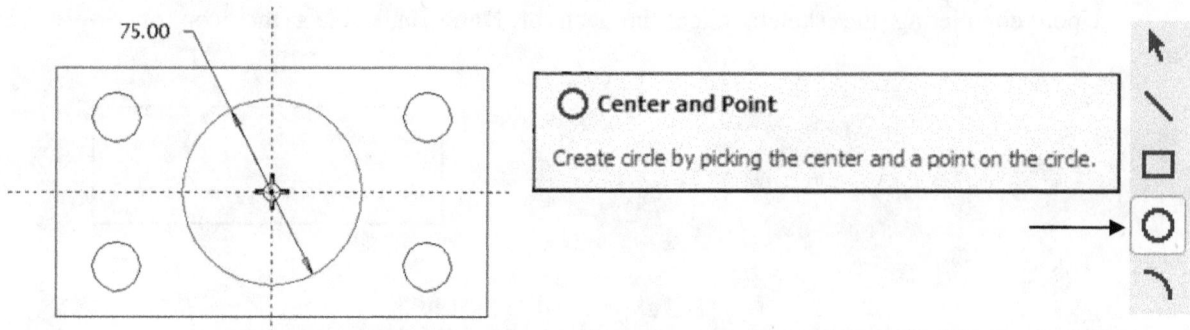

Upon completing the sketch, click the icon of **Done**, and select the icon of **Apply and Save.**

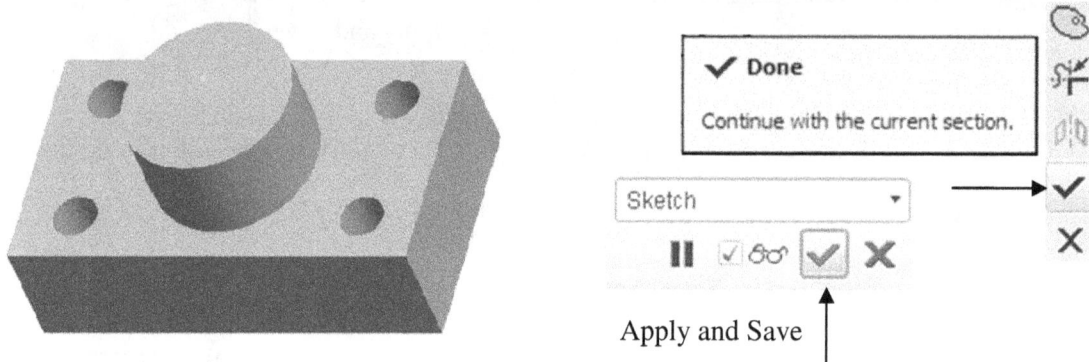

✓ **Done**

Continue with the current section.

Sketch

Apply and Save

To create a through hole feature of φ30, click the icon of **Hole**. Specify 30 as the diameter value and select **Through All** as the depth choice (click the black arrow on the Depth Options Flyout). Click the axis of the cylinder feature. While holding down the **Ctrl** key, pick the top surface of the cylinder.

Specify 30 as the diameter value

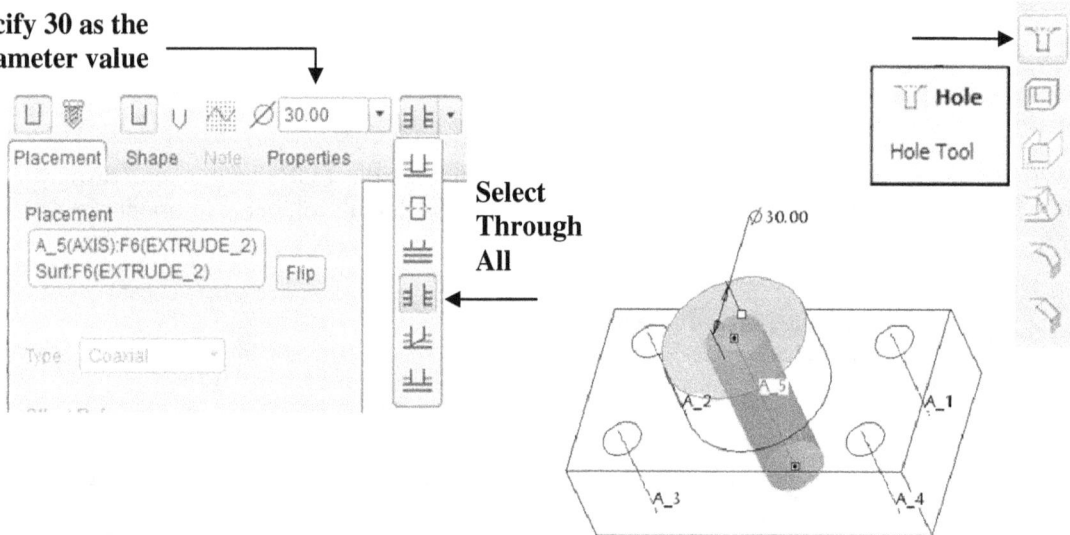

Ø 30.00

Placement | Shape | Note | Properties

Placement

A_5(AXIS):F6(EXTRUDE_2)
Surf:F6(EXTRUDE_2)   Flip

Type | Coaxial

Select Through All

Ⓣ **Hole**

Hole Tool

Ø 30.00

A_5

A_2

A_1

A_3

A_4

Upon completing the placement, click the icon of **Apply and Save.**

A_2   A_5   A_1

A_3   A_4

All

Apply and Save

Step 2: Perform FEA under **Pro/Mechanica (material, constraints and loads)**
From the menu bar, select **Applications > Mechanica > Continue** after checking the unit system > **Structure > OK**.

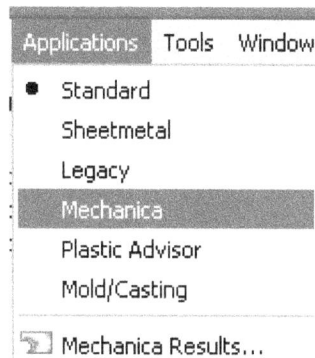

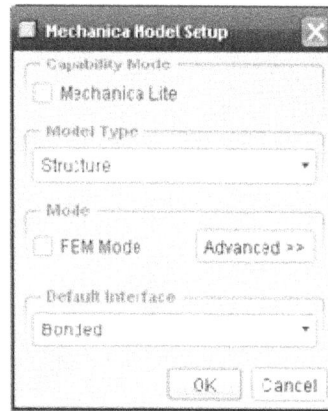

From the toolbar of functions, select the icon of **Materials >** select **STEEL** from the left side called Materials in Library > click the directional arrow so that the selected material type goes to the right side called Materials in Model.

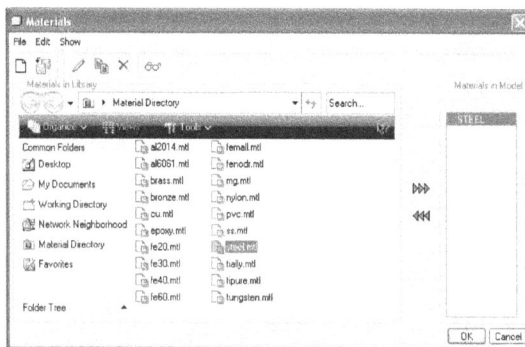

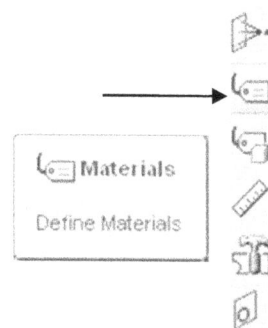

Click the icon of **Material Assignment.** In the Material Assignment window, the selected Steel is shown, and the component is also shown because there is only one volume or one component in the system. As a result, the software system automatically assigns Steel to the component. Just click **OK**.

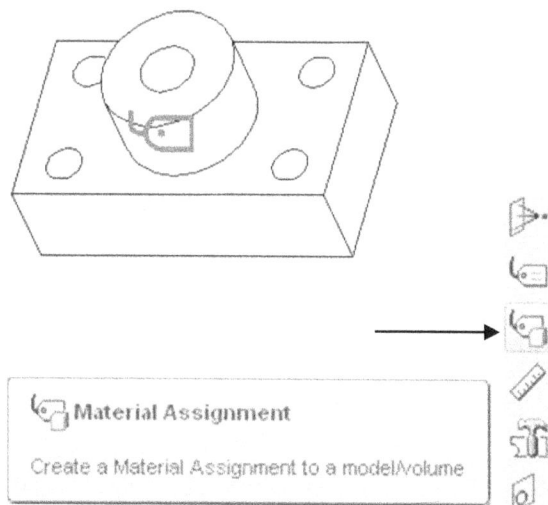

Now let us define Constraint1, simulating the bottom surface of the support structure is in contact with the ground.

Click the icon of **Displacement Constraint >** pick the bottom surface shown below. Set the translation in X, translation in Z and rotation about Y to free, and fully constrain the other 3 degrees of freedom > **OK.**

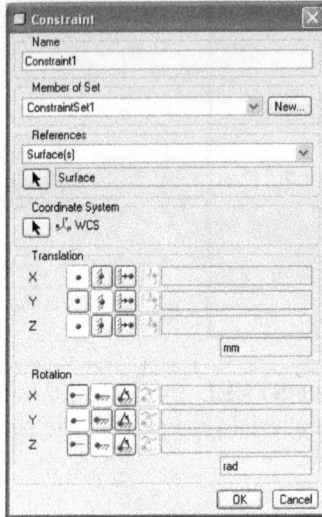

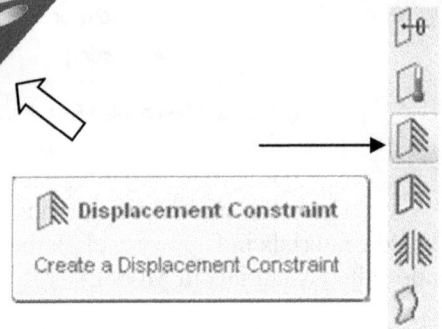

Now let us define Constraint2, simulating the support structure is firmly held to the ground through the use of 4 bolts.

In order to do so, we need to establish 4 local coordinate systems, which are cylindrical coordinate systems.

Let us create the first cylindrical coordinate system. We begin with creating a Cartesian coordinate system. Click the icon of **Datum Coordinate System Tool >** pick the axis of the hole and the edge shown below:

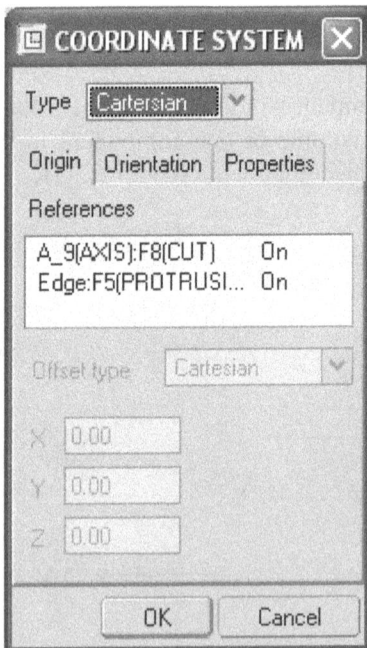

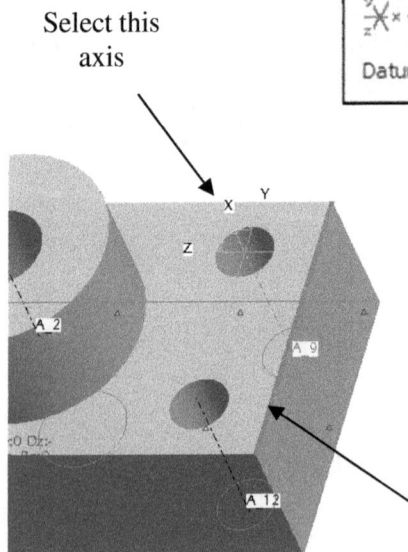

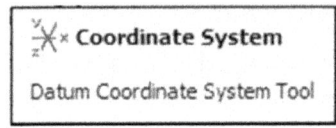

Change the vertical axis from X to Z before converting it to a cylindrical coordinate system.

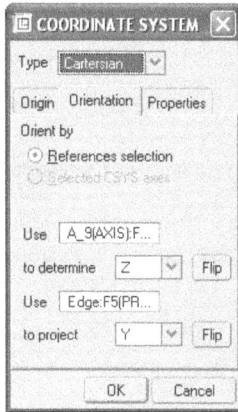

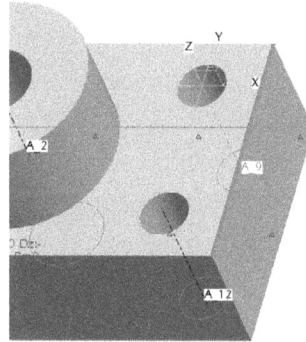

In the box marked as **Type**, change Cartesian to Cylindrical > **OK**.

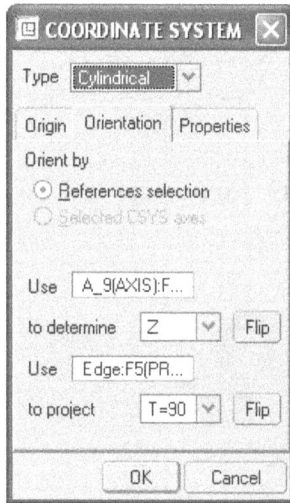

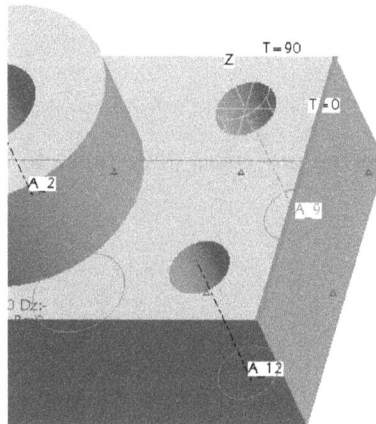

Repeat the procedure three times to create the other three cylindrical coordinate systems, as shown below:

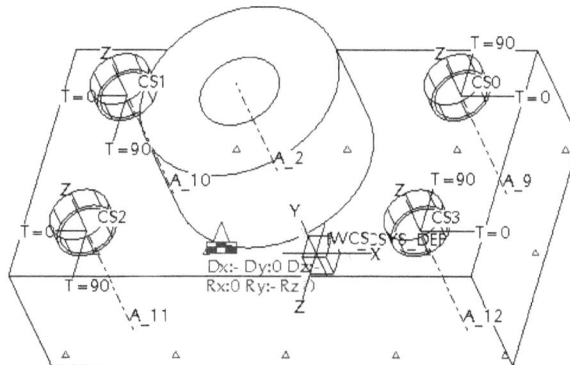

Click the icon of **Displacement Constraint** > Activate the surface selection > pick the cylindrical surface as shown below.

Activate the selection of Coordinate system > pick **CS0**.

Set the translation in R to free (indicating the presence of clearance between the hole and the bolt, and fully constrain the other 5 degrees of freedom > **OK**

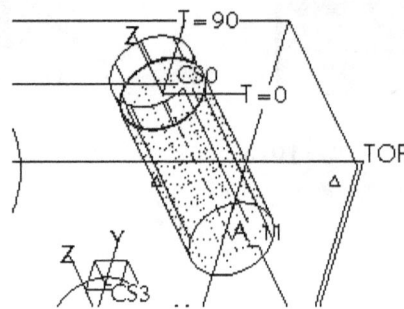

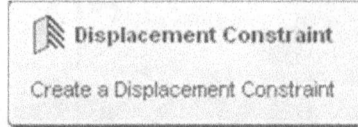

Repeat the procedure three times to define Constraint3, Constraint4 and Constraint5, as shown below:

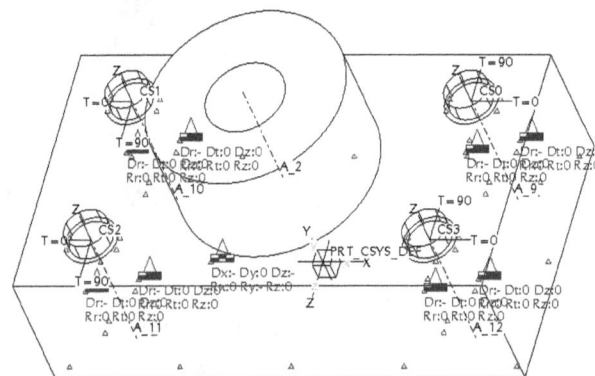

Now let us define the loads. There are two loads. The first load is torque acting on the top surface of the cylinder.

In order to define a torque, we need to define a point, about which the magnitude of the torque can be evaluated. To define a datum point, click the icon of Datum Point Tool. While holding down the **Ctrl** key, pick the axis of the cylinder and the top surface of the cylinder, PNT0 is defined > **OK**.

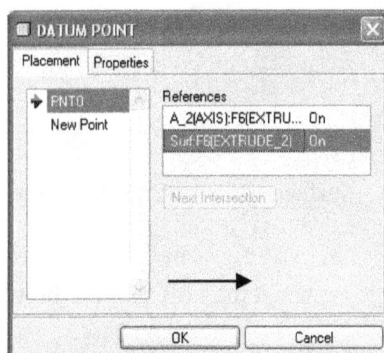

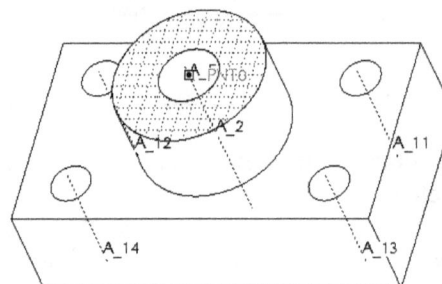

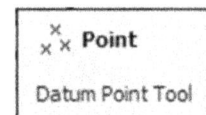

To define the torque load, click the icon of **Force/Moment Load** > **Surface**(s) > activate the selection of surface and pick the top surface of the cylinder.

Click **Advanced** > change the distribution from **Total Load** to **Total Load at Point** and select PNT0 > type in 5000 in Y of the Moment box > **OK**.

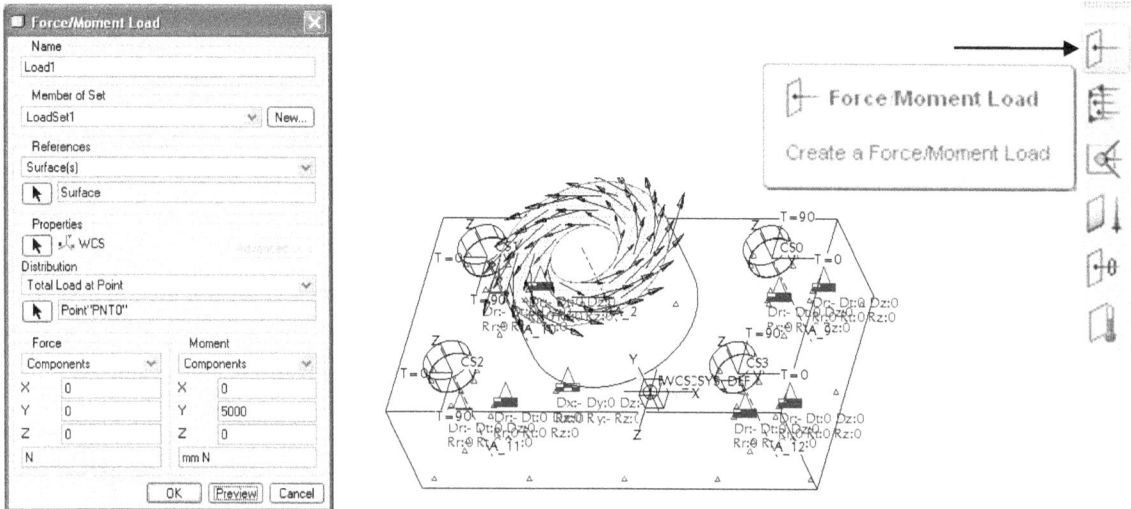

The second load is uniform load acting on the top surface of the cylinder in the Y direction (downward). To define it, click the icon of **Force/Moment Load** > **Surface**(s) > activate the selection of surface and pick the top surface of the cylinder.

Click **Advanced** > change the distribution back to **Total Load** > **Uniform** > type in -100 in Y of the Force box > **OK**.

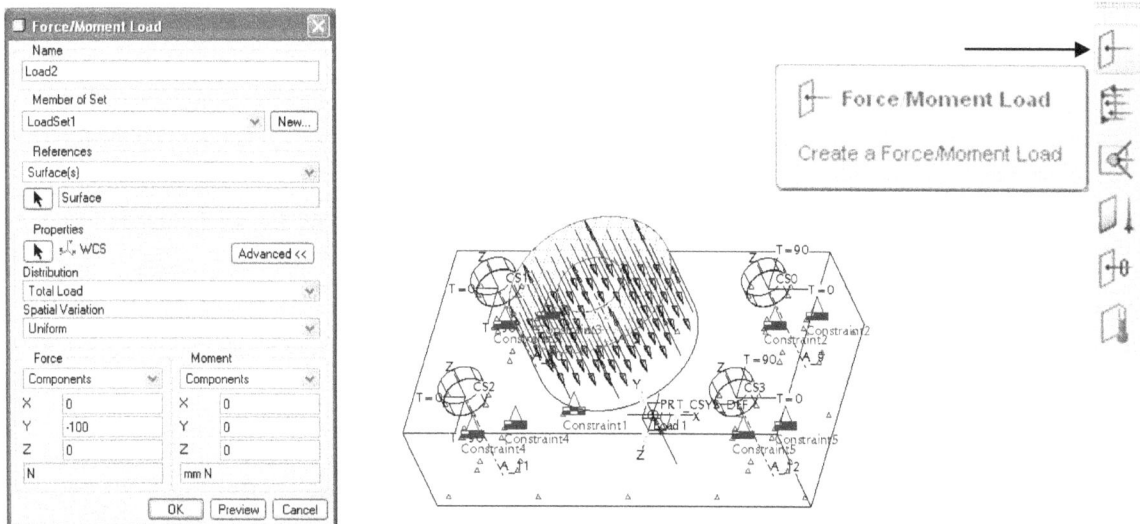

Select the icon of **Run an Analysis**> **File** > Select **New static analysis**.

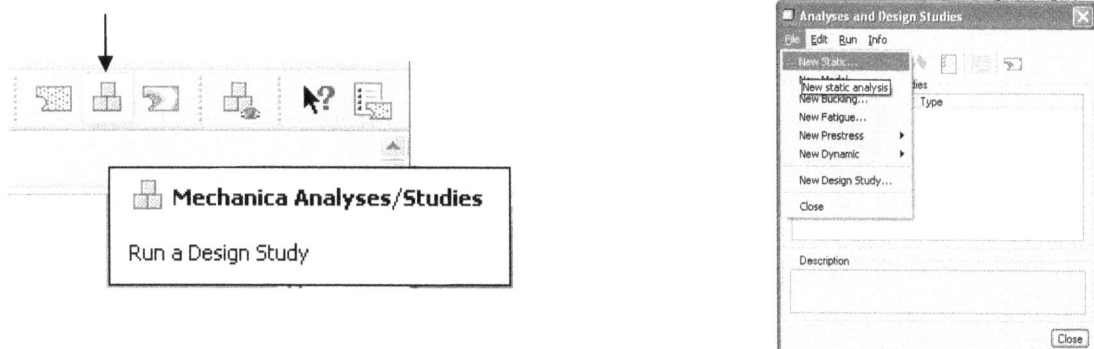

Specify support_structure_1 as the folder name > Make sure that ConstraintSet1 and LoadSet1 are highlighted > **OK**... Click the icon of Green Flag to run.

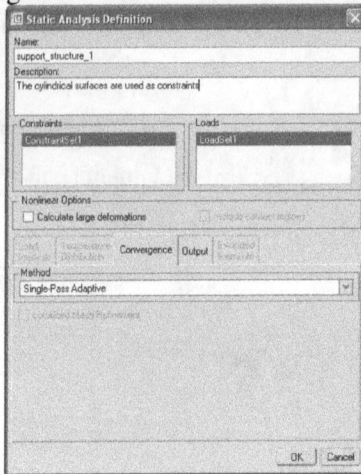

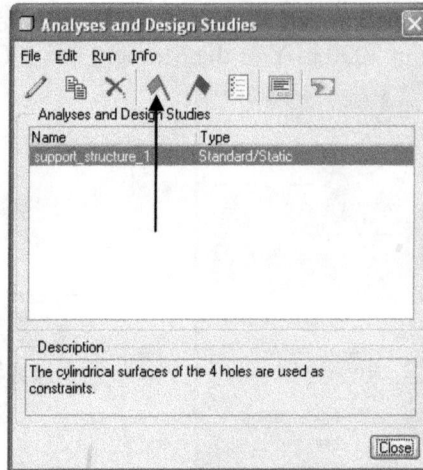

To review the results, click the icon of **Review Results** > type *deformation distribution (cylindrical surfaces)* as the title of the result window > select the folder called support_structure_1 > **Fringe** > **Displacement** > **Magnitude** > **OK and Show**.

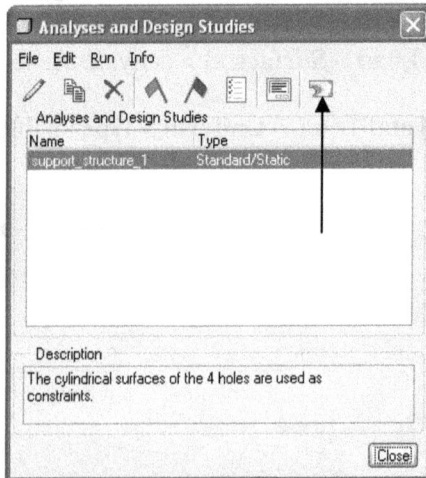

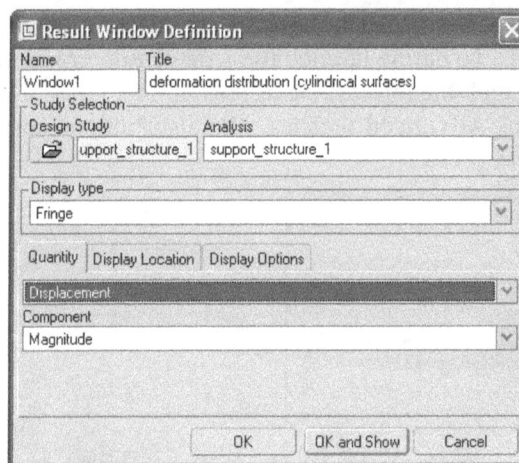

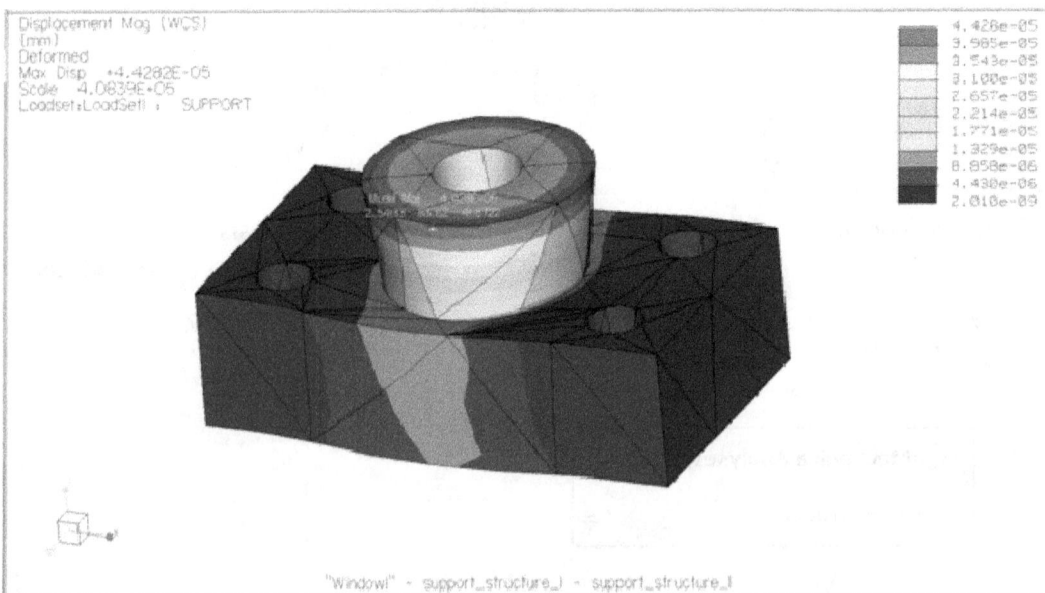

To display the distribution of the maximum principal stress, Edit > Copy > type in distribution of max prin stress (cylindrical surfaces) > change to **Stress** and **Max Principal > OK and Show.**

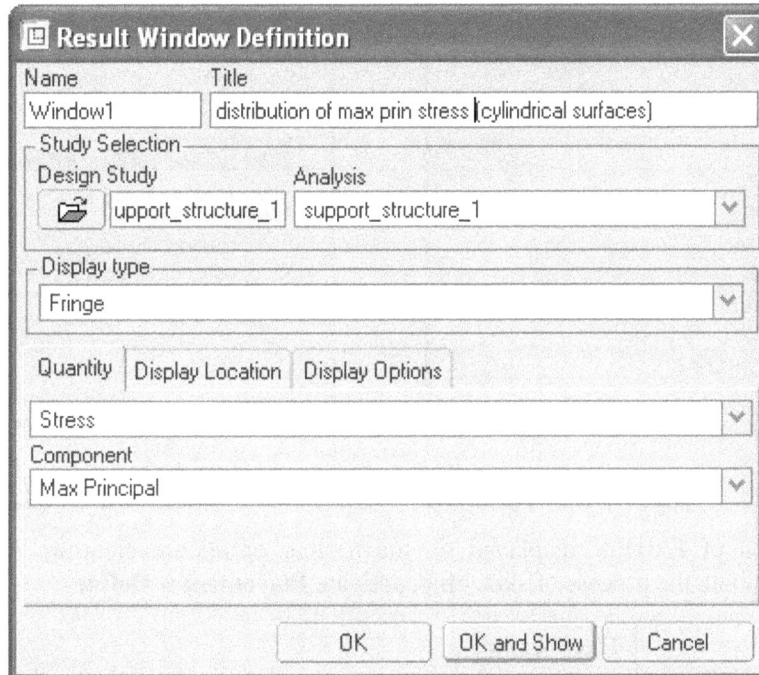

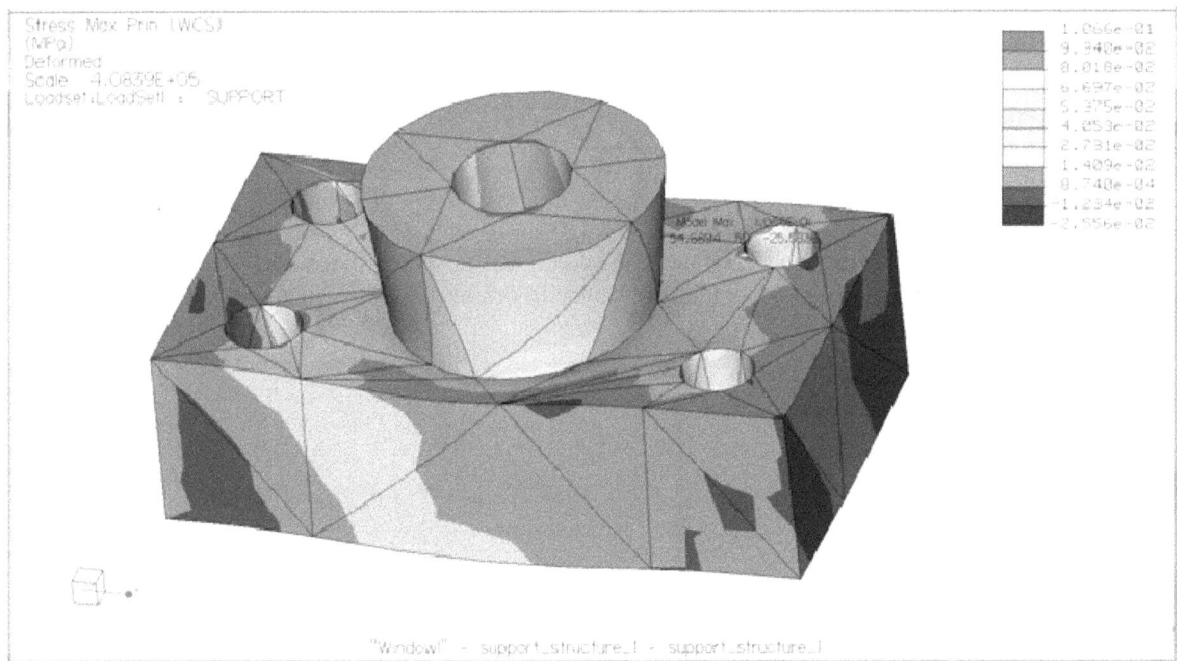

In Pro/Mechanica, there is a built-in function called fastener connection for modeling joints where bolts are used. However, the function of fastener connection is only available under an assembly environment. For this reason, in the following, we create the base plate first and assume the plate has 3 dimensions: 250x150x50 mm when considering the dimensions of the support component.

Step 1: Create a base plate.

File > New > Part > type *base_plate* as the file name and clear the icon of **Use default template > OK.** Select **mmns_part_solid** (units: Millimeter, Newton, Second) and type *base_plate* in **DESCRIPTION**, and *student* in **MODELED_BY**, then **OK.**

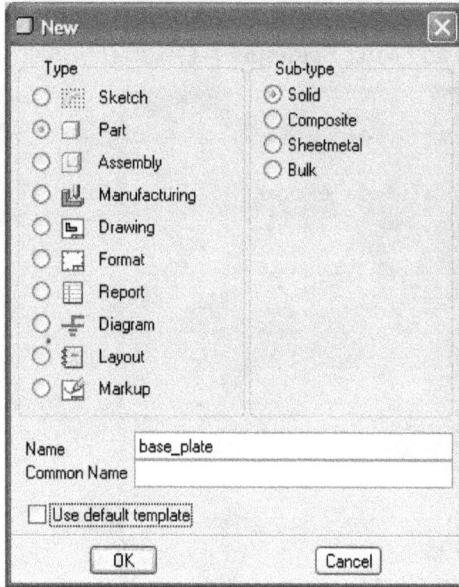

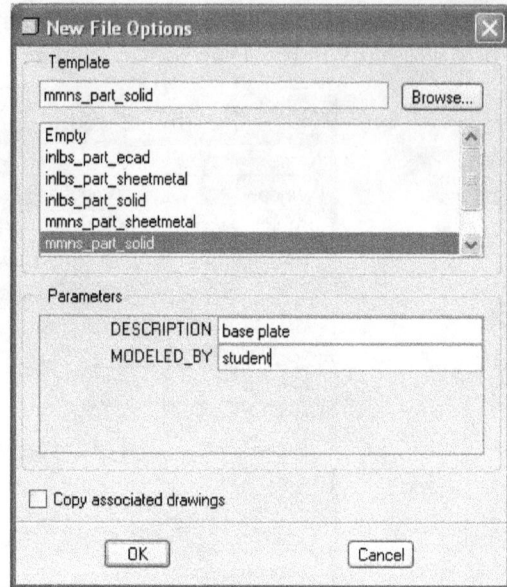

Click the icon of **Extrude** displayed on the toolbar of feature creation. Specify *50* as the thickness value. To initiate the process of sketching, activate **Placement > Define.**

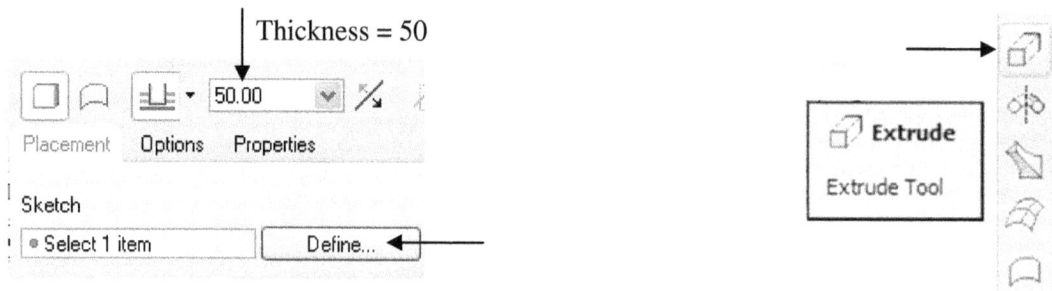

Select the **TOP** datum plane displayed on screen as the sketch plane. Pick the box of **Sketch** to accept the **RIGHT** datum plane as the default reference to orient the sketch plane, as illustrated below:

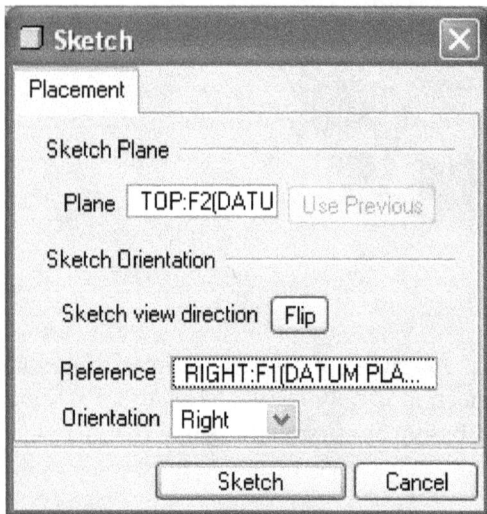

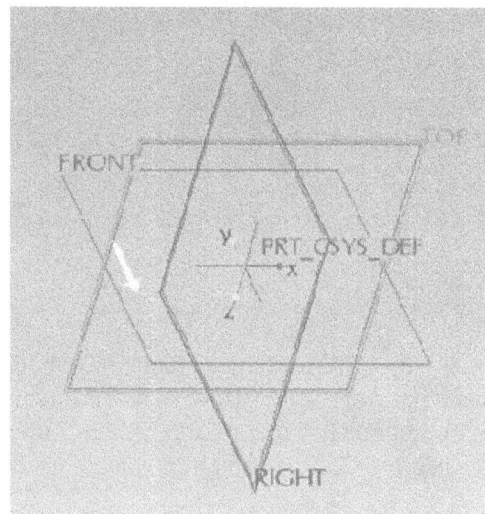

Let us sketch 2 centerlines. These 2 centerlines are used for symmetry when sketching a rectangle. Click the icon of **Centerlines**. Sketch a vertical and a horizontal centerline, as shown.

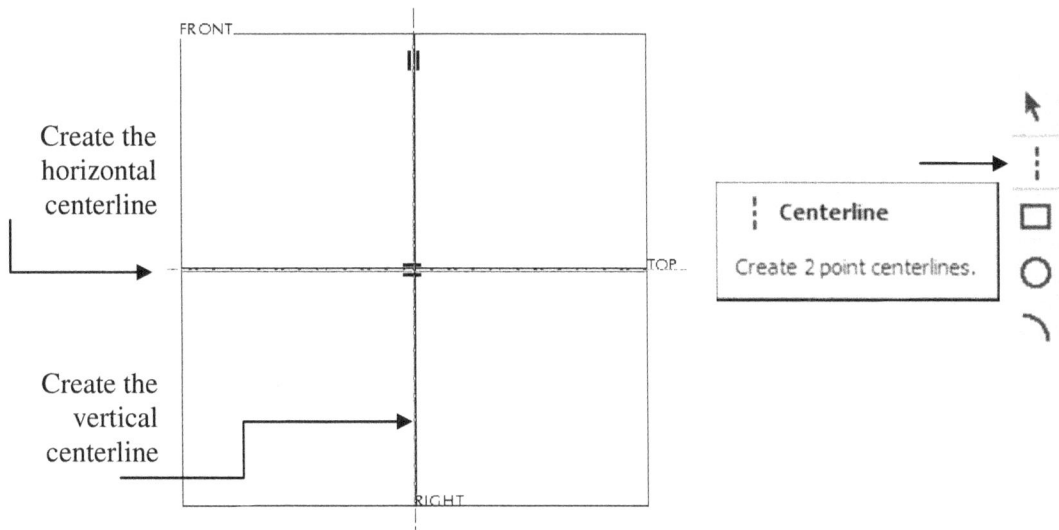

Create the horizontal centerline

Create the vertical centerline

Click the icon of **Rectangle** and sketch a rectangle, which is symmetric about the 2 centerlines. The 2 dimensions are 250 and 160, respectively.

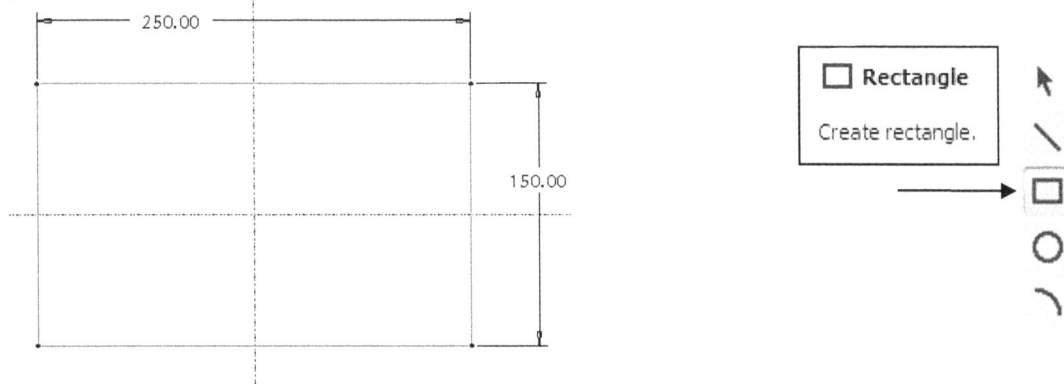

Upon completing the sketch, click the icon of **Done.**   Flip the arrow direction, or flip the extrusion direction so that the thickness is added downward, as shown. This arrangement will facilitate the assembly process greatly because the origin of the coordinate system of the base plate and the origin of the coordinate system of the support component will be at the same location with the identical orientation.

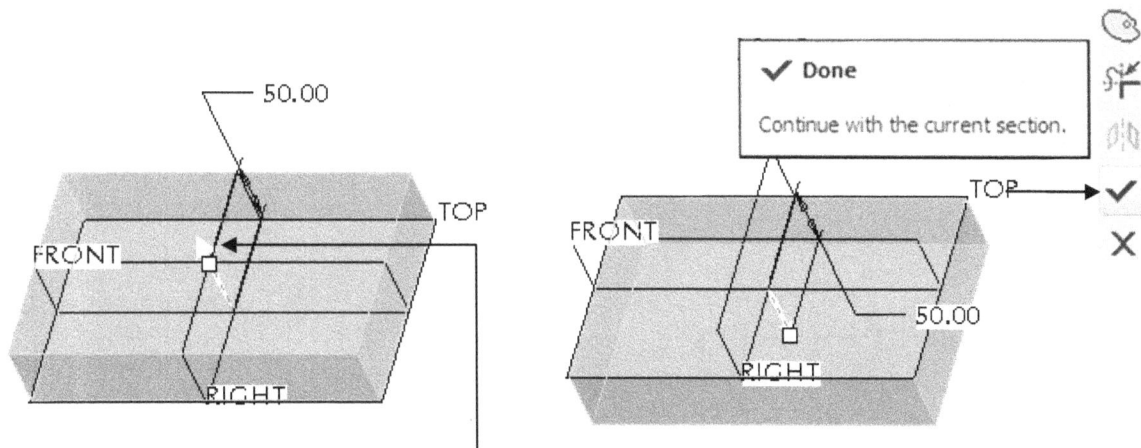

Flip this arrow direction so that the thickness is added downward.

Select the icon of **Apply and Save.**

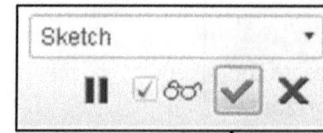

Apply and Save

Now let us create holes. The dimensions of these holes should be matched the dimensions of the 4 holes located in the support component.

Click the icon of **Extrude.** Make sure that the icon of **Cut** is selected. Specify **Through All** as the depth choice for the 4 holes. To initiate the process of sketching, activate **Placement > Define.**

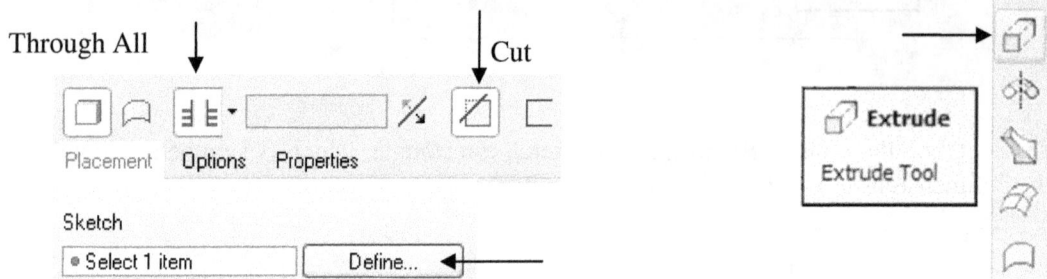

Through All

Cut

Placement   Options   Properties

Extrude

Extrude Tool

Sketch

Select 1 item   Define...

Select the top surface of the plate, not the **TOP** datum plane displayed, as the sketch plane. Pick the box of **Sketch** to accept the **RIGHT** datum plane as the default reference to orient the sketch plane, as illustrated below:

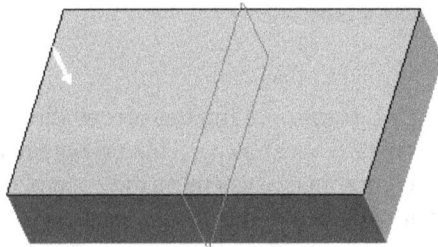

Sketch

Placement

Sketch Plane

Plane Surf:F5(EXTRL Use Previous

Sketch Orientation

Sketch view direction Flip

Reference RIGHT:F1(DATUM PLA...

Orientation Right

Sketch   Cancel

Let us sketch 2 centerlines. These 2 centerlines are used for symmetry when creating 4 circles through the mirror operation.

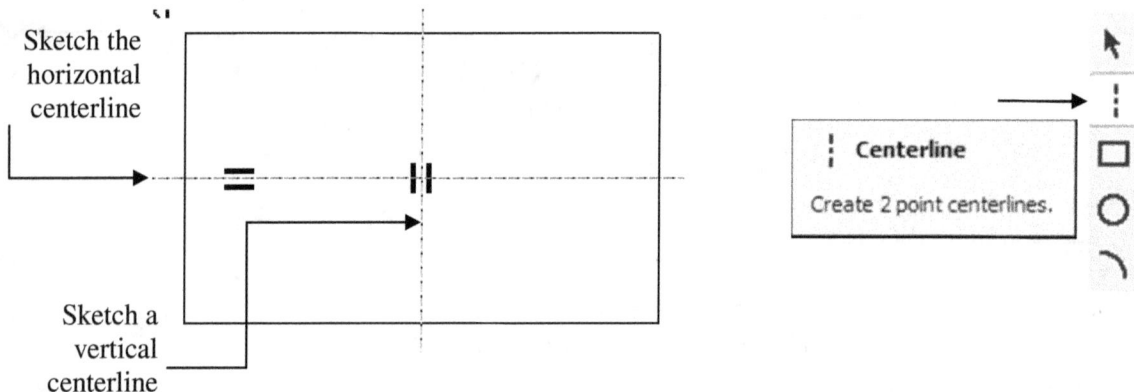

Sketch the horizontal centerline

Sketch a vertical centerline

Centerline

Create 2 point centerlines.

Click the icon of **Circle** and sketch a circle. The diameter dimension is 20. The 2 position dimensions are 65 and 30, respectively.

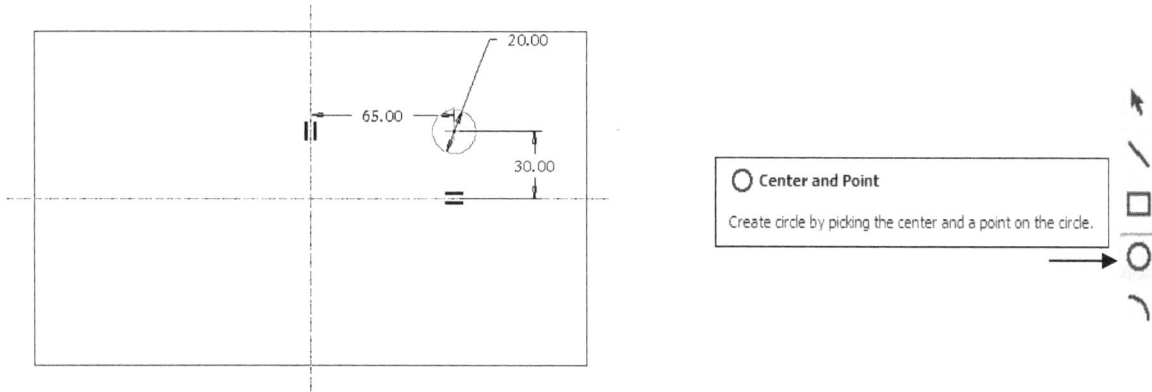

We need 3 more circles. Let us use the copy function. Click the icon of **Mirror** > pick the circle first. Afterwards, pick the vertical centerline, completing the creation of the second circle.

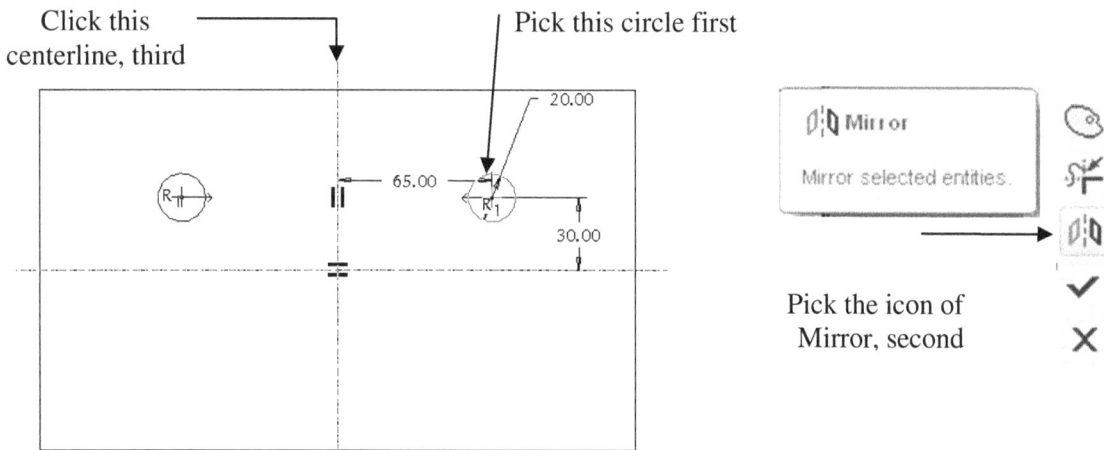

While the copied circle is still actively selected, pick the icon of **Mirror** > click the horizontal centerline, completing the creation of the third circle.

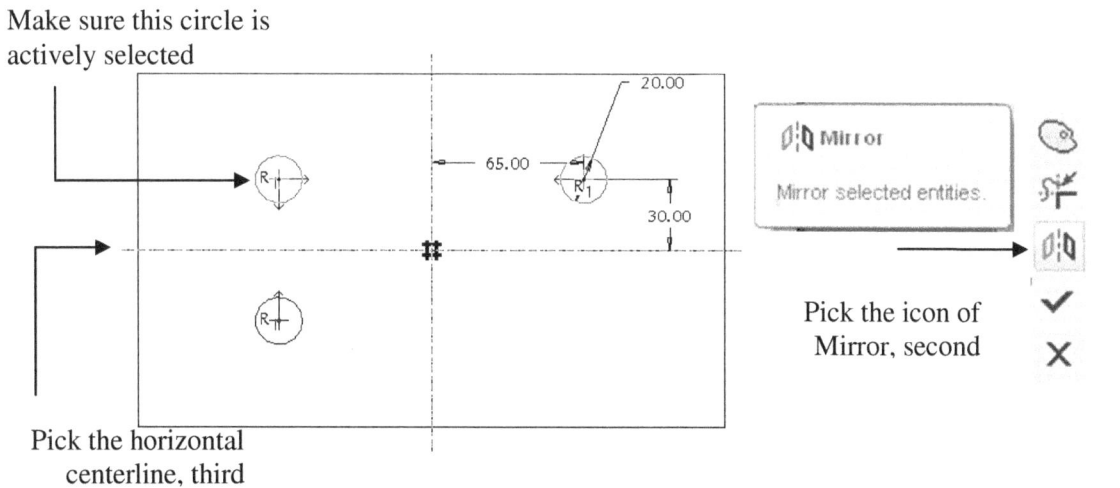

While the copied circle at the right and lower portion is still actively selected, pick the icon of **Mirror** > click the vertical centerline, completing the creation of the 4th circle.

Pick the vertical centerline, third

Make sure this circle is actively selected

20.00

65.00

30.00

Mirror

Mirror selected entities.

Pick the icon of Mirror, second

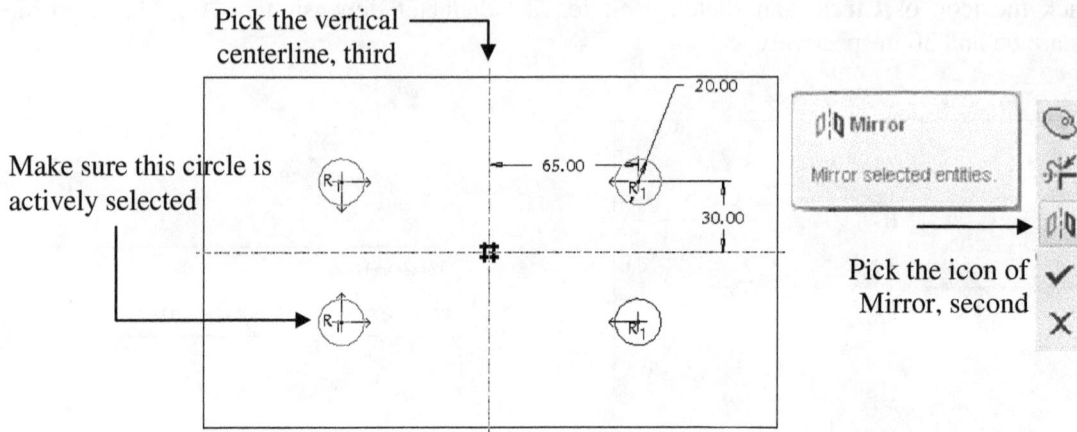

Upon completing the sketch, click the icon of **Done** and the icon of **Apply and Save.**

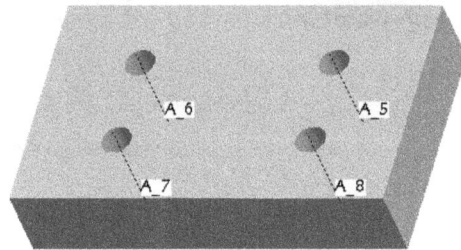

Sketch

Apply and Save

A_6   A_5

A_7   A_8

Step 2: Create an assembly using the support component and the base plate component.
Click the icon of **Create a new object.** Make sure Assembly is selected.

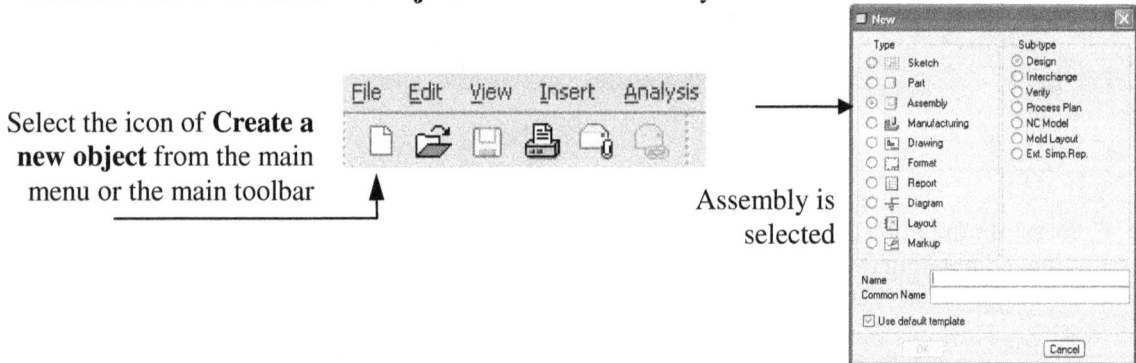

Select the icon of **Create a new object** from the main menu or the main toolbar

File Edit View Insert Analysis

Assembly is selected

New

Type
○ Sketch
○ Part
◉ Assembly
○ Manufacturing
○ Drawing
○ Format
○ Report
○ Diagram
○ Layout
○ Markup

Sub-type
◉ Design
○ Interchange
○ Verify
○ Process Plan
○ NC Model
○ Mold Layout
○ Ext. Simp.Rep.

Name
Common Name
☑ Use default template
OK   Cancel

Type *support_plate_assembly* as the file name, clear the box of **Use default template > OK**. Select the unit of **mmns_asm_design**, type *fastener connection* under the **description** of the model, and type *student* or *your name* under the **modeled_by > OK**.

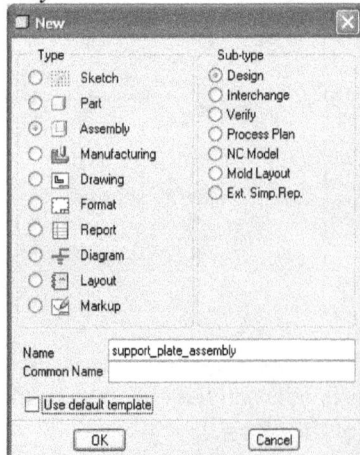

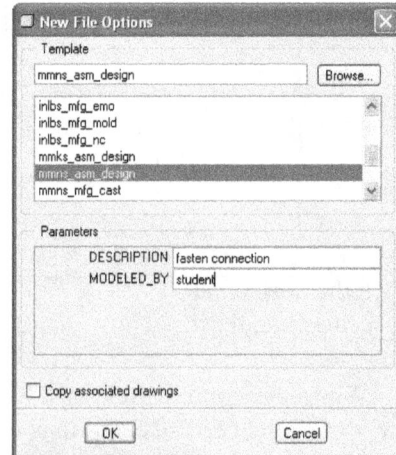

New

Type
○ Sketch
○ Part
◉ Assembly
○ Manufacturing
○ Drawing
○ Format
○ Report
○ Diagram
○ Layout
○ Markup

Sub-type
◉ Design
○ Interchange
○ Verify
○ Process Plan
○ NC Model
○ Mold Layout
○ Ext. Simp.Rep.

Name   support_plate_assembly
Common Name
☐ Use default template
OK   Cancel

New File Options

Template
mmns_asm_design   Browse...

inlbs_mfg_emo
inlbs_mfg_mold
inlbs_mfg_nc
mmks_asm_design
mmns_asm_design
mmns_mfg_cast

Parameters
DESCRIPTION fasten connection
MODELED_BY student

☐ Copy associated drawings
OK   Cancel

Click the icon of **Add component**.

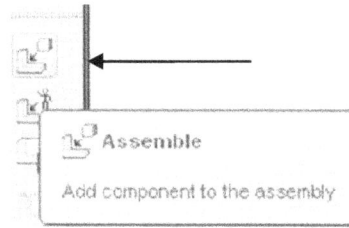

Make sure that you are at **In Session**. Highlight the component called base_plate > **Open**.

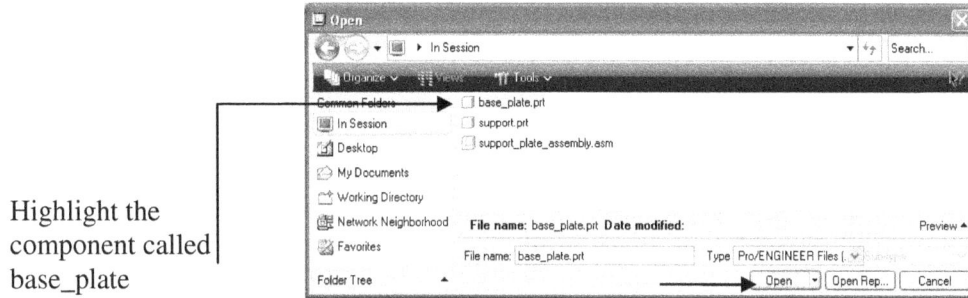

Highlight the
component called
base_plate

From the assembly window, select **Default**, meaning that the coordinate system of the base plate component is assembled at the location of the assembly coordinate system. Afterwards, click **OK** because the base plate component is fully constrained.

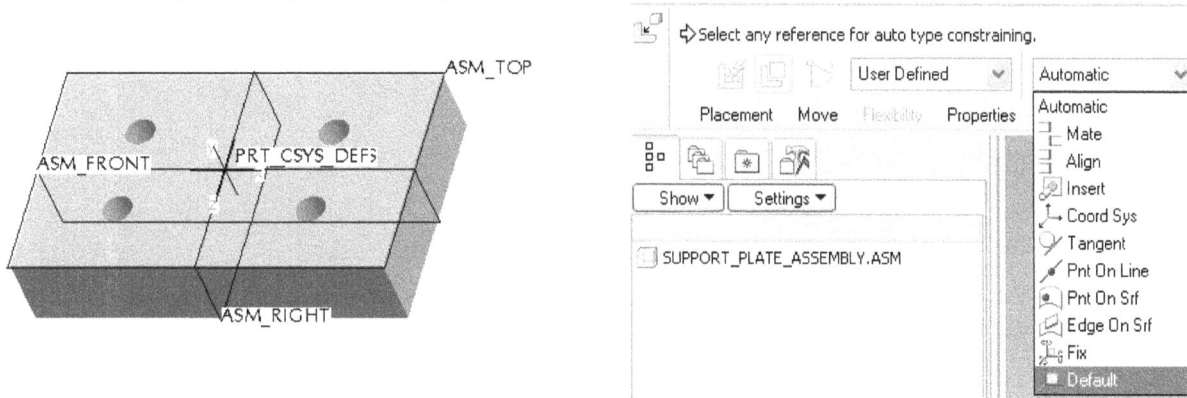

Click the icon of **Add component**. Make sure that you are at **In Session**. Highlight the component called support > **Open**.

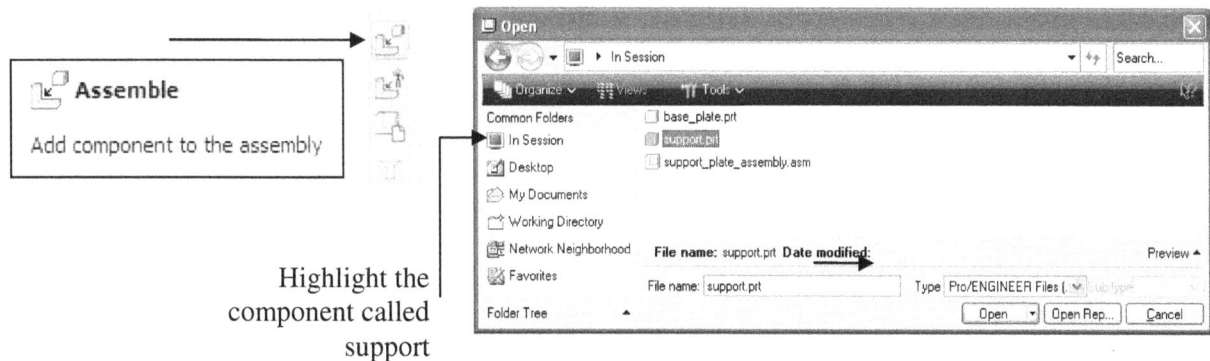

Highlight the
component called
support

From the assembly window, click **Default**, meaning that the coordinate system of the support component is assembled at the location of the assembly coordinate system. Afterwards, click **OK** because the support component is fully constrained.

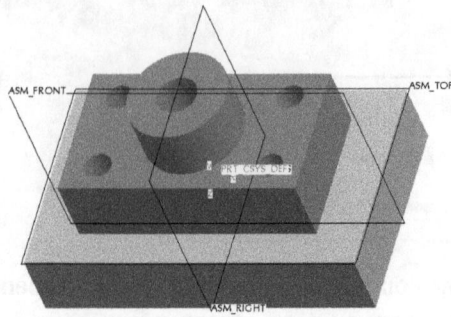

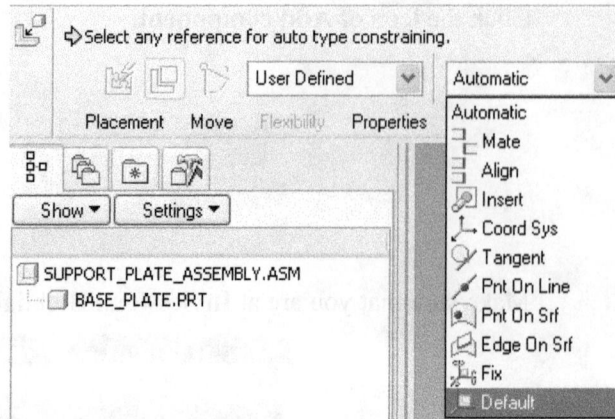

The assembly process is completed.

Step 3: Switch from the Pro/ENGINEER design system to Pro/Mechanica
From the main toolbar or the menu toolbar, select **Applications > Mechanica > Structure > OK.**

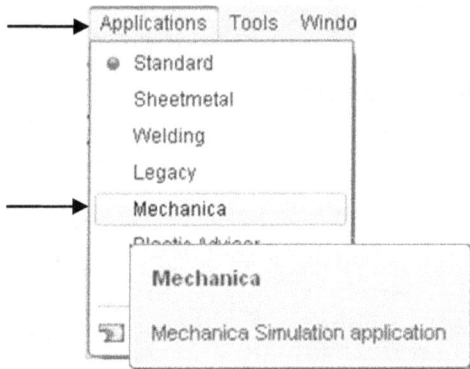

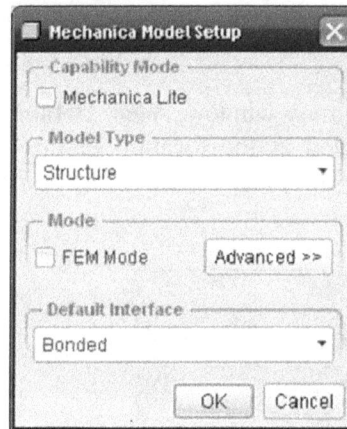

From the toolbar of functions, select the icon of **Materials >** select **STEEL** from the left side called **Materials in Library >** click the directional arrow so that the selected material type goes to the right side called **Materials in Model**.

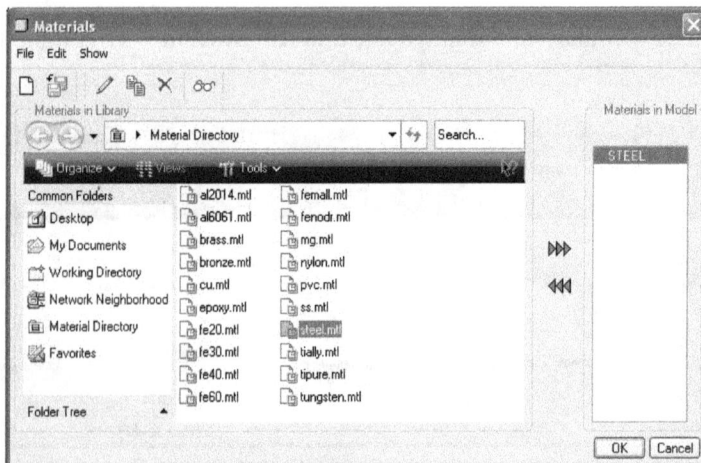

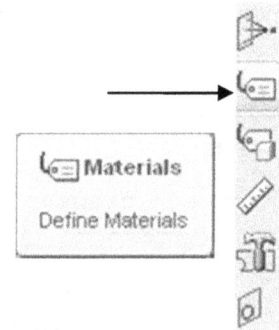

Click the icon of **Material Assignment.** In the Material Assignment window, the selected **Steel** is shown, and select both base plate and support components while holding down the **Ctrl** key > **OK.**

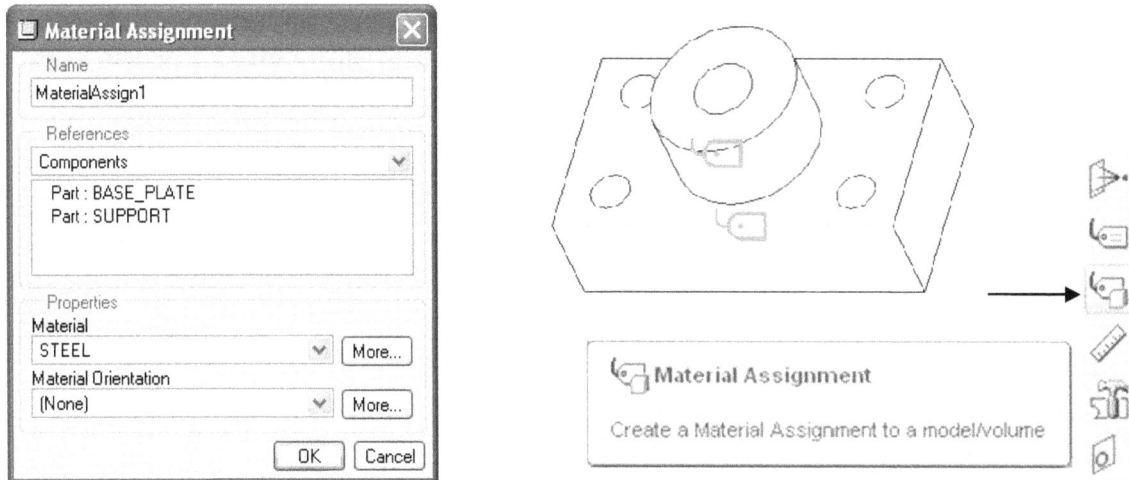

Material Assignment
Create a Material Assignment to a model/volume

Step 4: Define the constraint condition by fixing the bottom surface of the base plate to the ground.

From the toolbar of **Mechanica objects**, click the icon of **New Displacement Constraint >
Surface(s) >** pick the bottom surface of the base plate > **OK** to set all 6 degrees of freedom equal to zero,

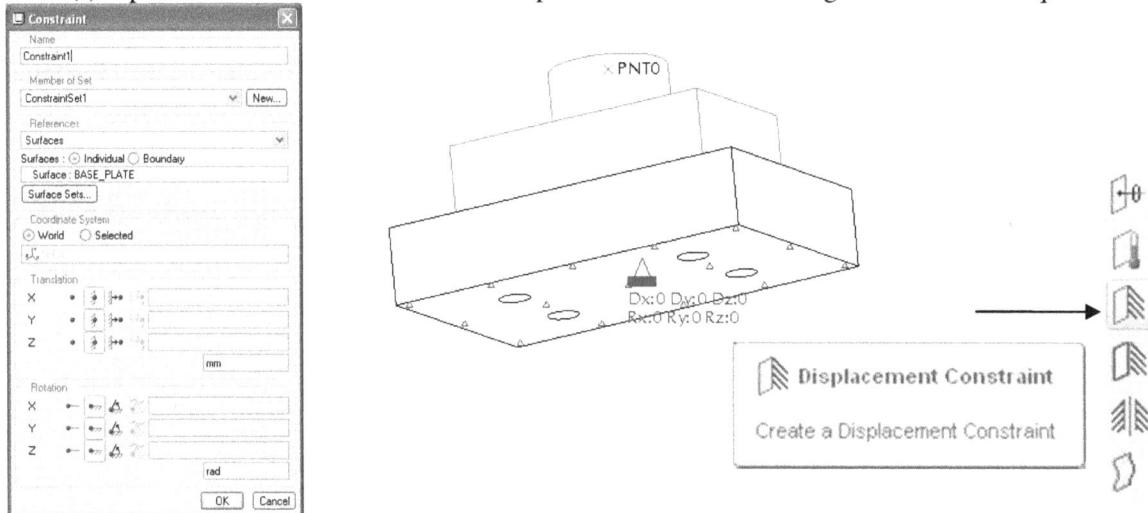

Displacement Constraint
Create a Displacement Constraint

Step 5: Define the load condition.

Select the icon of **Force and Moment** from the toolbar of Mechanica objects > **Surfaces >** pick
the top surface. Click **Advanced > Total Load at Point >** pick PNT0.

Type *5000* as the Y component under Moment > **Preview > OK**.

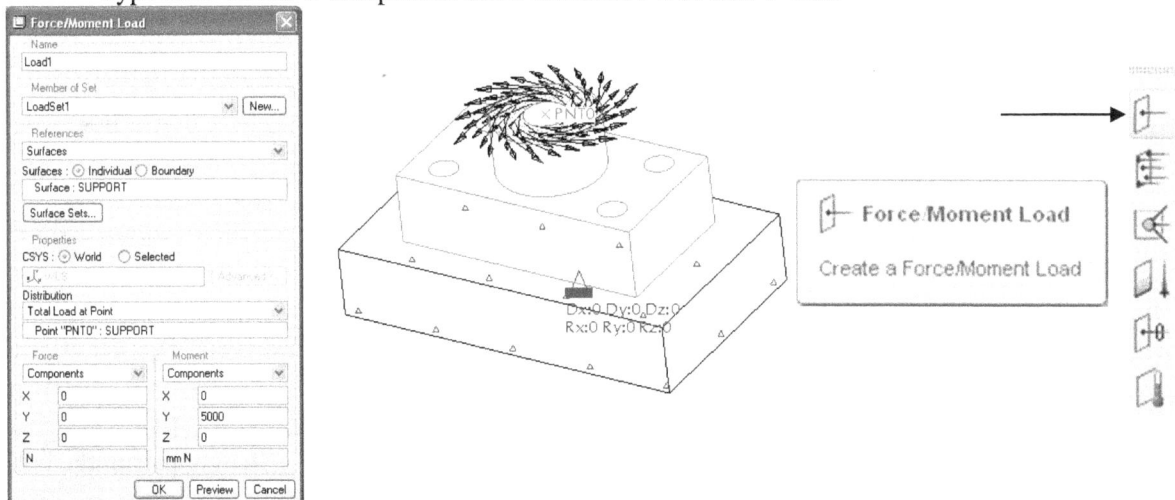

Force/Moment Load
Create a Force/Moment Load

Again, click the icon of **Force/Moment Load > Surfaces >** pick the top surface.
Click **Advanced > Total Load >** type *-100* as the Y component under **Force > Preview > OK.**

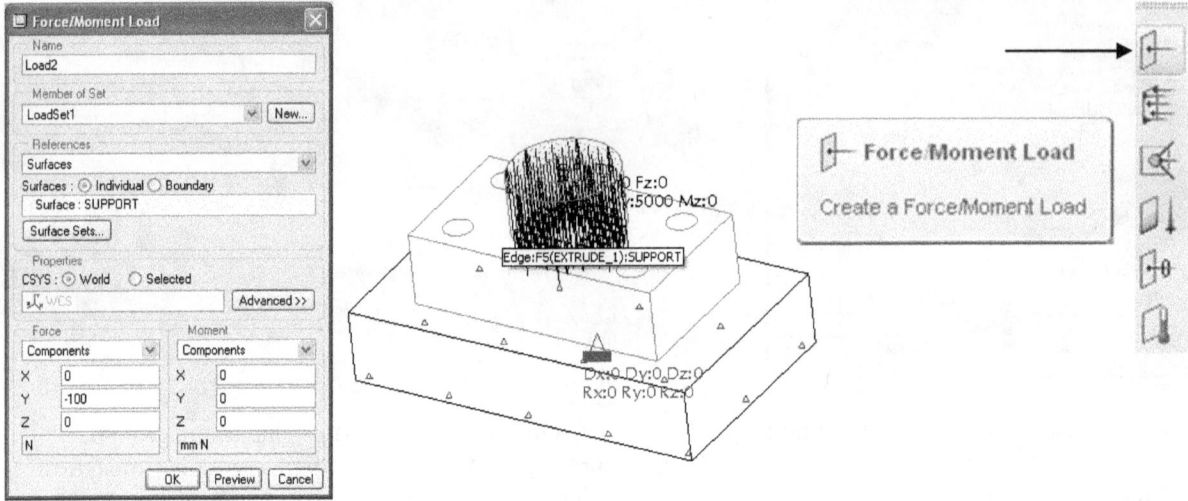

Step 6: Define the fastener connections representing the 4 bolt joints.
From the main toolbar, select **Insert > Connection > Fastener** and the window called **Fastener Definition** appears. On the screen, it asks the user to pick 2 edges.

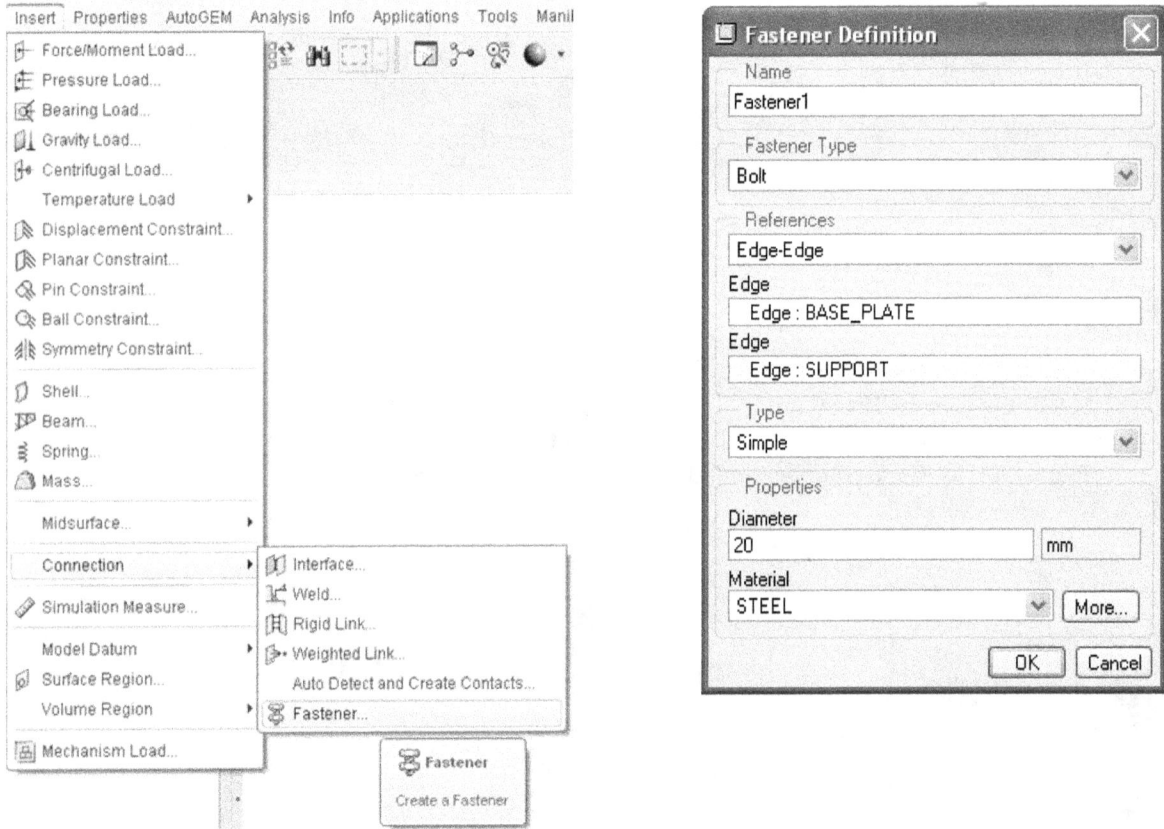

Pick the first edge from the base plate, as shown. While holding down the **Ctrl** key, pick the second edge from the support and make sure that both edges are taken from the matching pair in the bolt joint.

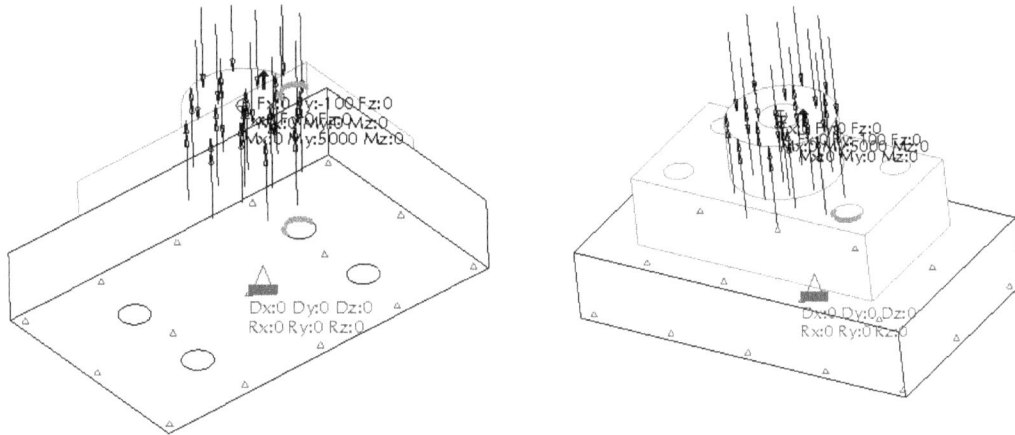

In the box called Material, click the box called **Material** and select **Steel** from the material library. Leave **Simple** unchanged. The software system has also detected that the hole size is 20 mm. Click *OK* to complete the fastener definition. On the screen, a warning window appears indicating the nature of the 2 contacting surfaces > **OK** to accept the separation condition.

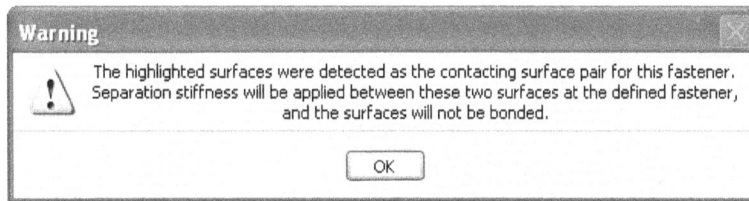

On the screen display, a bolt shaped sketch appears, indicating the fastener connection has been defined. Repeat the above procedure 3 more times to complete the other 3 bolt joints.

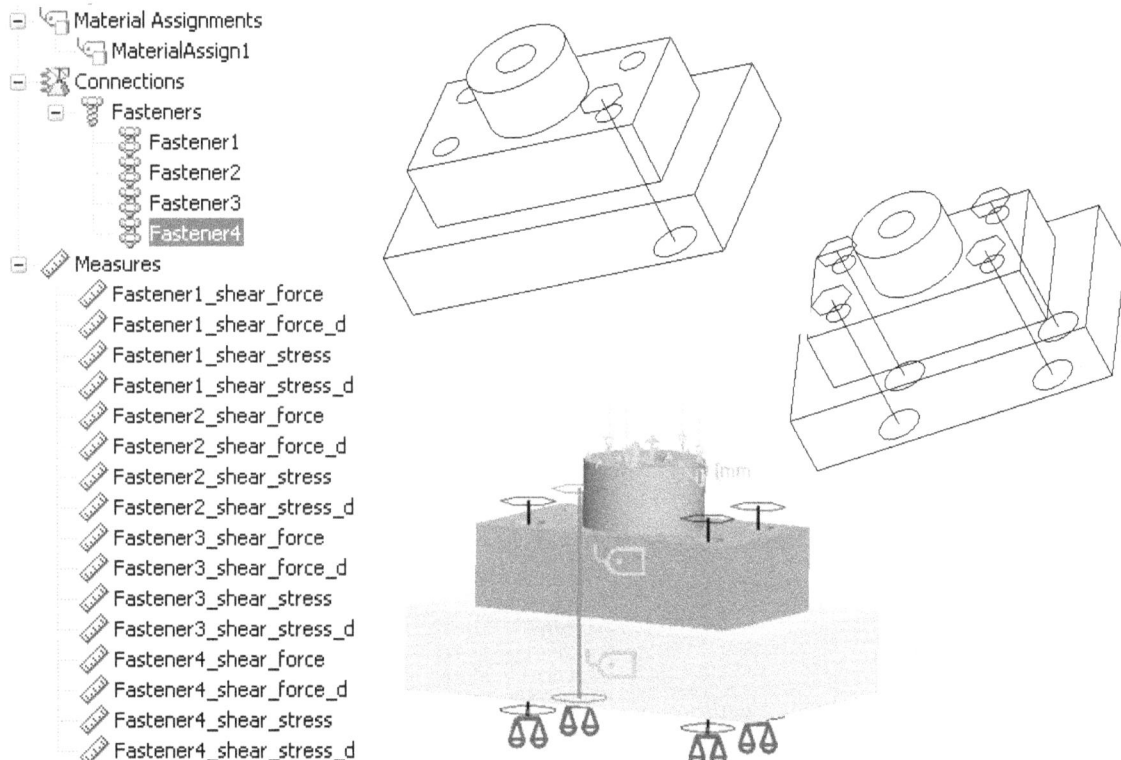

Step 7: Under the Pro/MECHANICA environment, set up **Analyses** and **Run** it
Select the icon of **Run an Analysis** > **File** > **New Static** > type *fastener_connection* as the name of the analysis folder > **OK.** Click the box of **Run.**

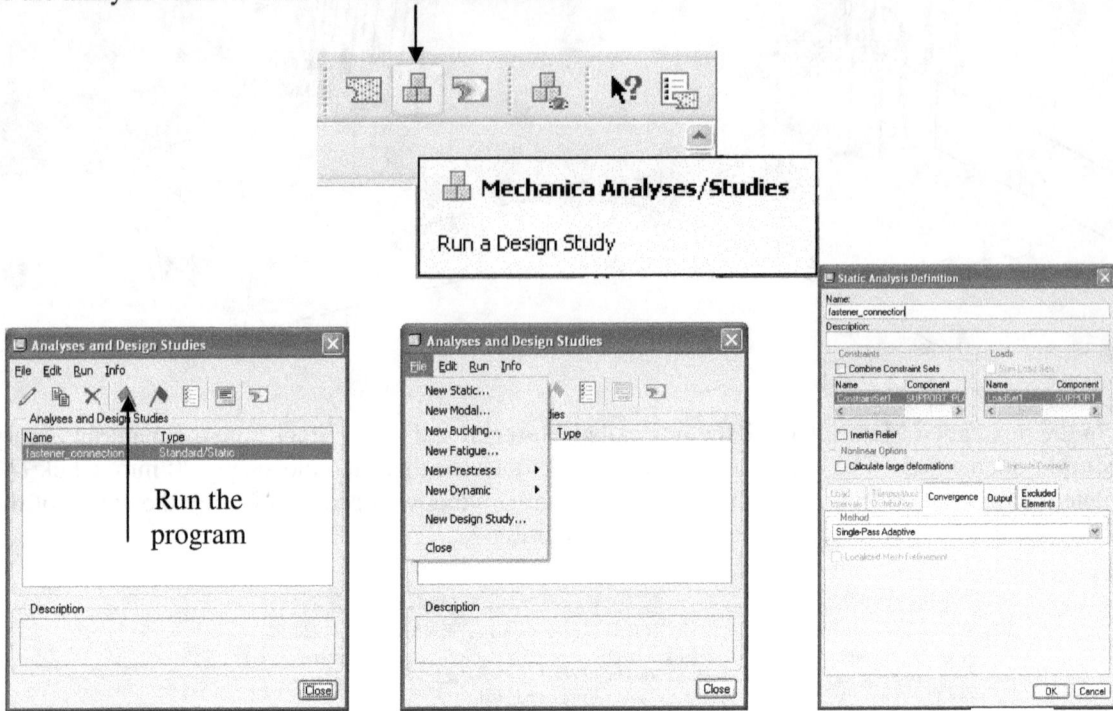

There is always a message displayed on screen. The message is asking the user "Do you want to run interactive diagnostics?" Click **Yes**. In this way a warning message will appear if an abnormal operation occurs.

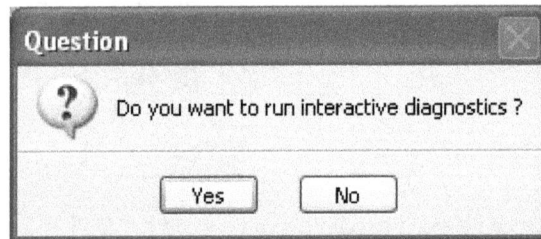

To monitor the computing process, users may click the icon of **Display study status**.

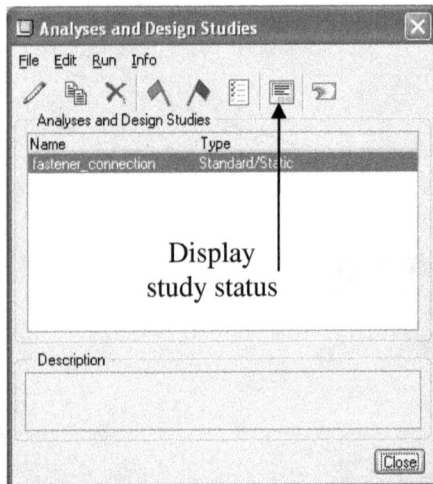

```
Constraint Set: ConstraintSet1

Load Set: LoadSet1

    Resultant Load on Model:
       in global X direction:  1.085873e-06
       in global Y direction: -1.000000e+02
       in global Z direction: -1.372149e-07

    Measures:

    Fastener1_shear_force:    1.837110e+01
    Fastener1_shear_stress:   5.847703e-02
    Fastener2_shear_force:    1.837590e+01
    Fastener2_shear_stress:   5.849230e-02
    Fastener3_shear_force:    1.681851e+01
    Fastener3_shear_stress:   5.353499e-02
    Fastener4_shear_force:    1.680750e+01
    Fastener4_shear_stress:   5.349995e-02
```

To study the obtained result, say the distribution of the max principal stress, click the icon of **Review results > Stress > Max Principal > OK and Show.**

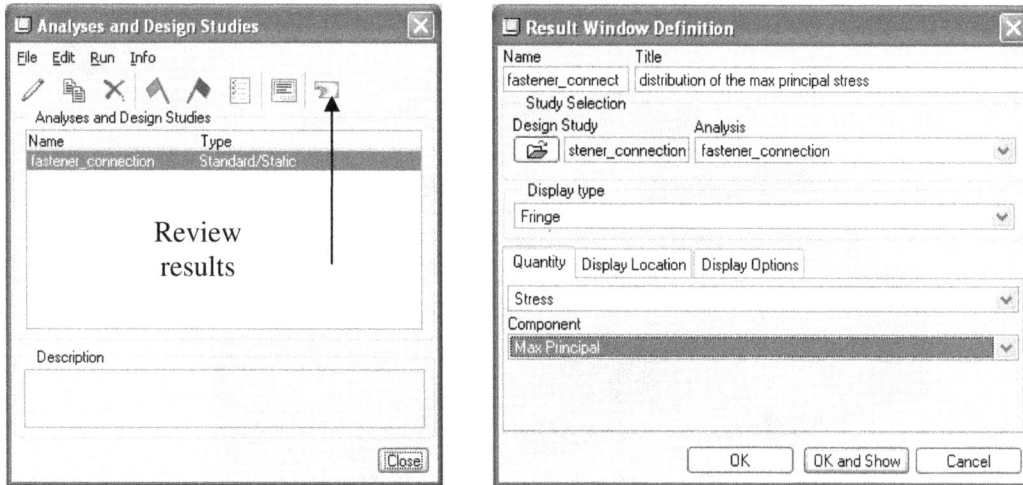

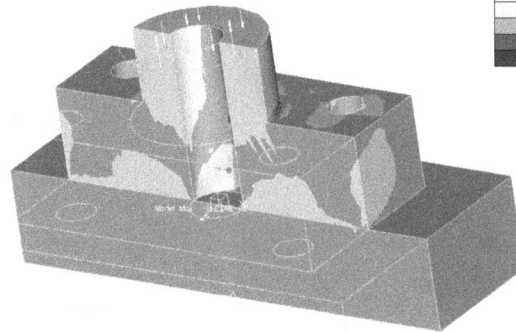

Note that the maximum magnitude of the max principal stress is located with the support. The magnitude is 0.0927 MPa.

To display the maximum von Mises stress, click the icon of **Copy the selected definition** > type *vm_stress* as the window name and type *distribution of the vm stress* as the title > **Stress > von Mises > OK and Show**. The maximum value is 0.2271 MPa.

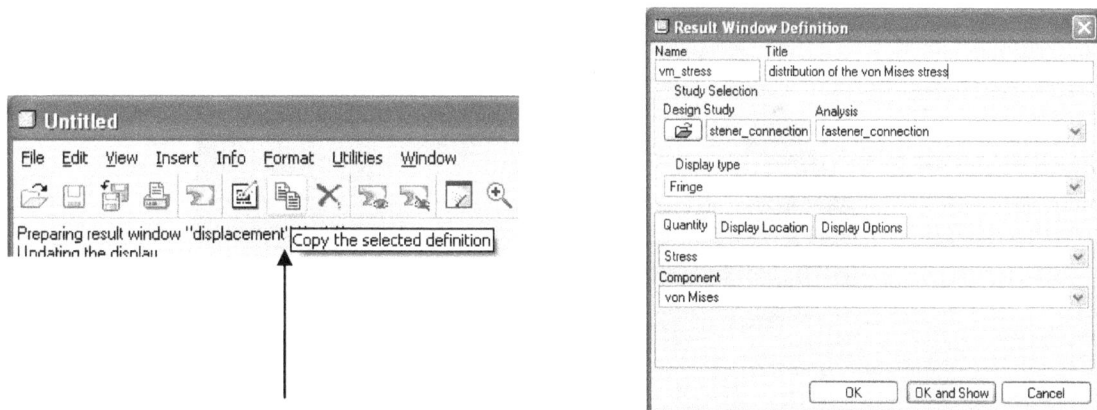

```
Stress von Mises (WCS)
(N / mm^2)
Deformed
Scale  8.5473E+04
Loadset:LoadSetl :  SUPPORT_PLATE_ASSEMBLY
```

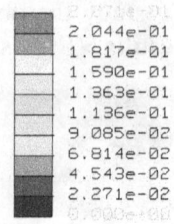

```
2.044e-01
1.817e-01
1.590e-01
1.363e-01
1.136e-01
9.085e-02
6.814e-02
4.543e-02
2.271e-02
```

"Windowl" - fastener_connection - fastener_connection

To display the maximum displacement, click the icon of **Copy the selected definition** > type *displacement* as the window name and type *distribution of displacement* as the title > **Displacement > Magnitude > OK and Show**. The maximum value is 0.000117 mm.

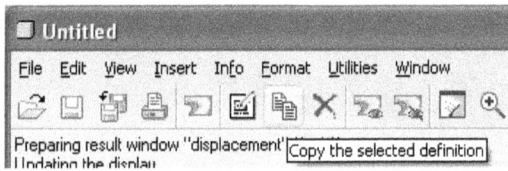

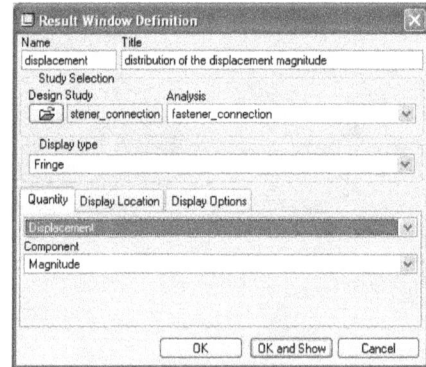

```
Displacement Mag (WCS)
(mm)
Deformed
Max Disp  +1.1703E-04
Scale  8.5473E+04
Loadset:LoadSetl :  SUPPORT_PLATE_ASSEMBLY
```

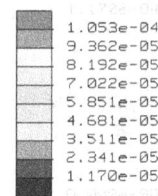

```
1.053e-04
9.362e-05
8.192e-05
7.022e-05
5.851e-05
4.681e-05
3.511e-05
2.341e-05
1.170e-05
```

"Windowl" - fastener_connection - fastener_connection

4.7 FEA with an Assembly of Gear and Shaft

The following figure illustrates a gear assembly. The assembly consists of a helical gear and a shaft. The fitting between the gear component and the shaft component is an interference fit so that the force acting on a tooth or teeth of the gear is capable of driving the shaft to rotate. The material of the gear and the material of the shaft are steel material. Let us assume that only one pair of teeth is in contact, and the force acting on the tooth of the gear is 150 Newton.

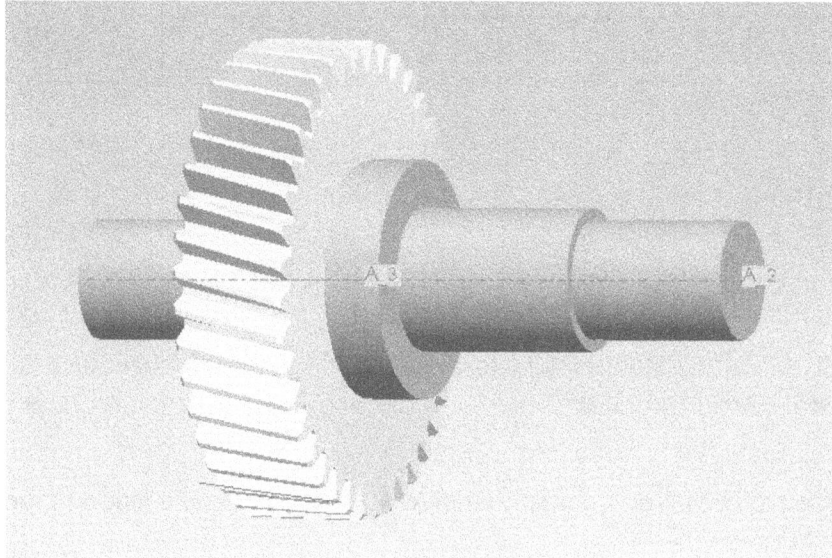

During operations of gear transmission, there are two failure modes, which are commonly observed. The first failure mode is the breakage of the tooth of the helical gear during operation. Such breakages occur when the force acting on the tooth induces excessively large stresses. The second failure mode is the twist of the shaft when subjected to torque due to the force acting on the tooth of the gear. In this example, we start with the study of analyzing the first failure mode. In this study, only the gear component is under consideration. To study the second failure mode, we take the entire assembly under consideration.

An engineering drawing of the gear component is shown below. We will use the shown dimensions to construct a 3D solid model of the gear component.

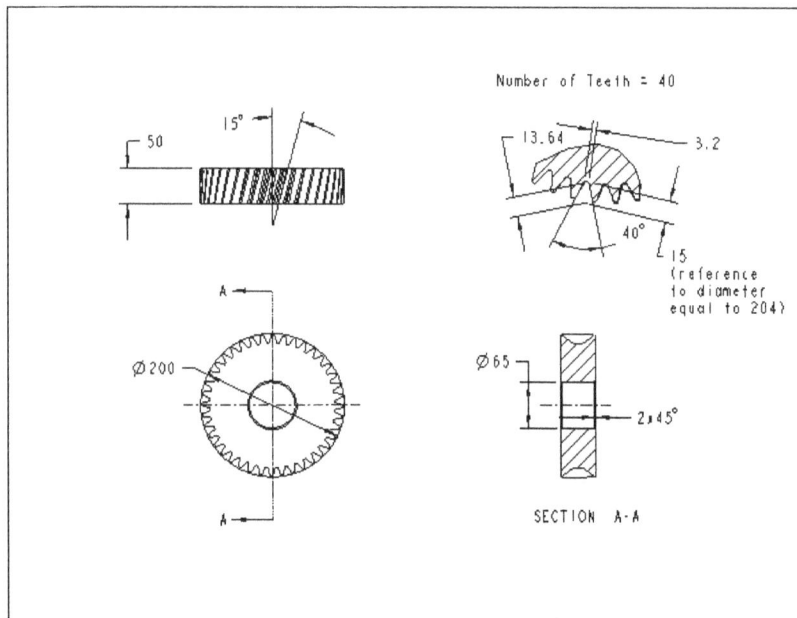

An engineering drawing of the shaft component is shown below. We will use the shown dimensions to construct a 3D solid model of the shaft component.

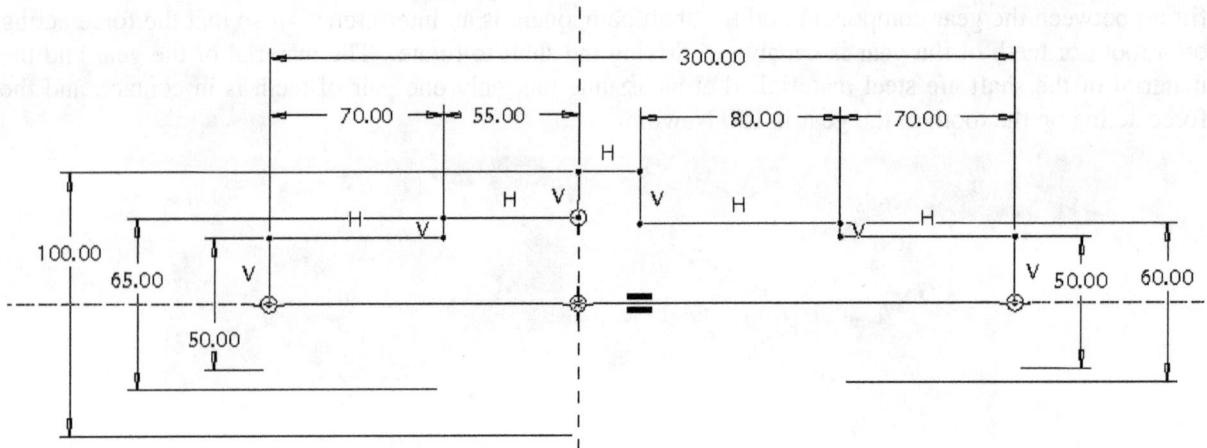

We will use an assemble process to assemble the gear component and the shaft component together before getting on performing FEA.

Step 1: Under the Pro/ENGINEER design environment, create a 3D solid model of the helical gear. The number of teeth is 40.

File > New > Part > type *helical_gear* as the file name and clear the icon of **Use default template > OK.** Select **mmns_part_solid** (units: Millimeter, Newton, Second) and type *helical gear* in **DESCRIPTION**, and student in **MODELED_BY**, then **OK.**

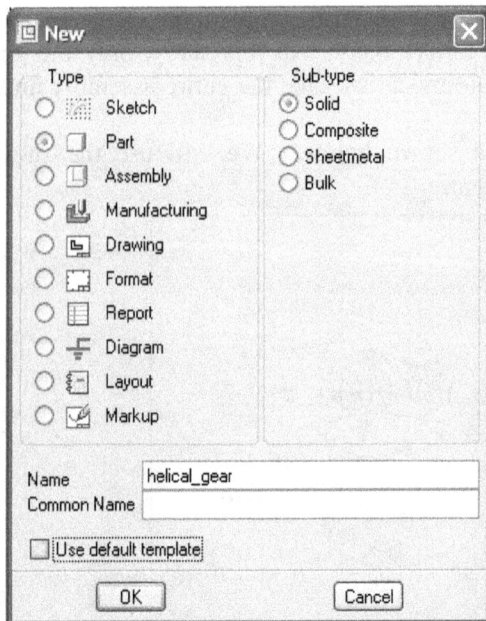

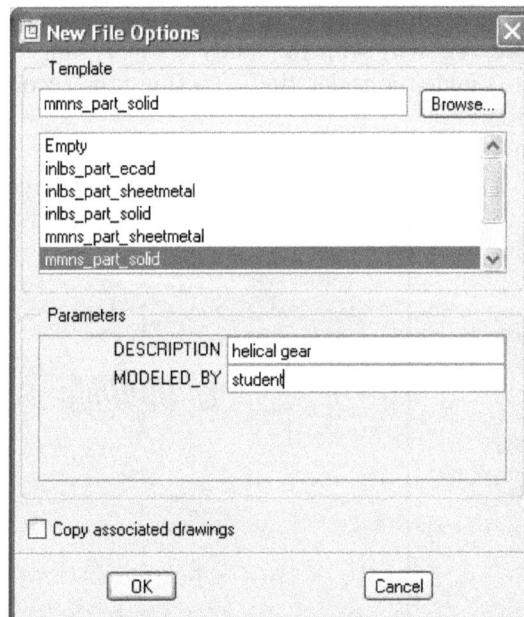

Let us create a cylinder feature: φ200x50 with a hole in the central location. The diameter of the hole is 65 mm. Directly select the icon of **Extrude** from the toolbar of feature creation. Make sure the icon of **Solid** is selected. Set the symmetric requirement and the depth value to 50. Click the box of **Placement > Define**.

Select the **FRONT** datum plane displayed on screen as the sketch plane. Pick the box of **Sketch** to accept the **RIGHT** datum plane as the default reference to orient the sketch plane, as illustrated below:

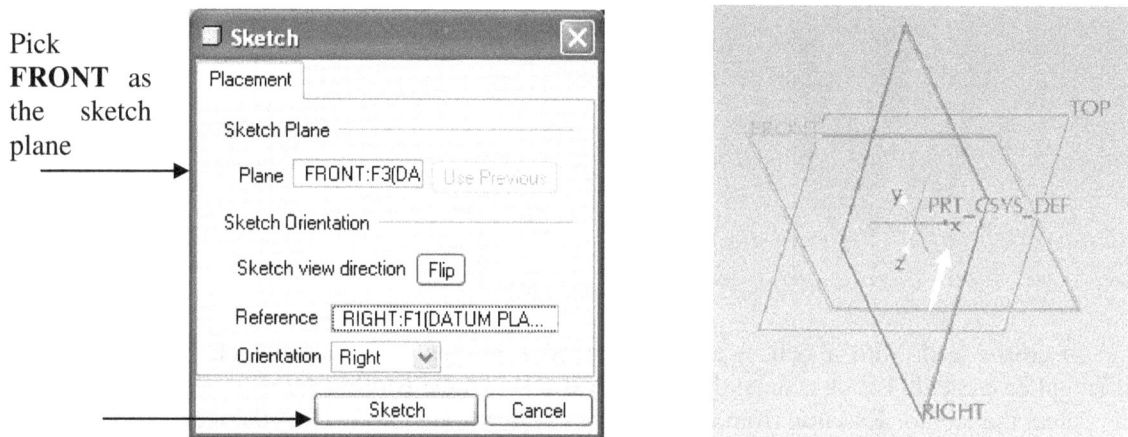

Pick **FRONT** as the sketch plane

Click the icon of **Circle** and sketch 2 circles. Their diameter values are 200 and 65, respectively.

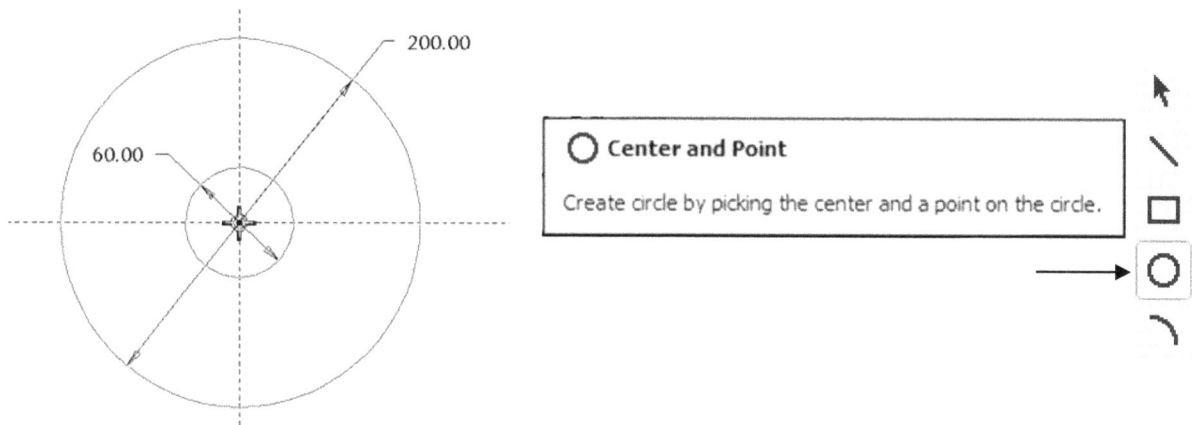

Upon completion, click the icon of **Done** and the icon of **Apply and Save**.

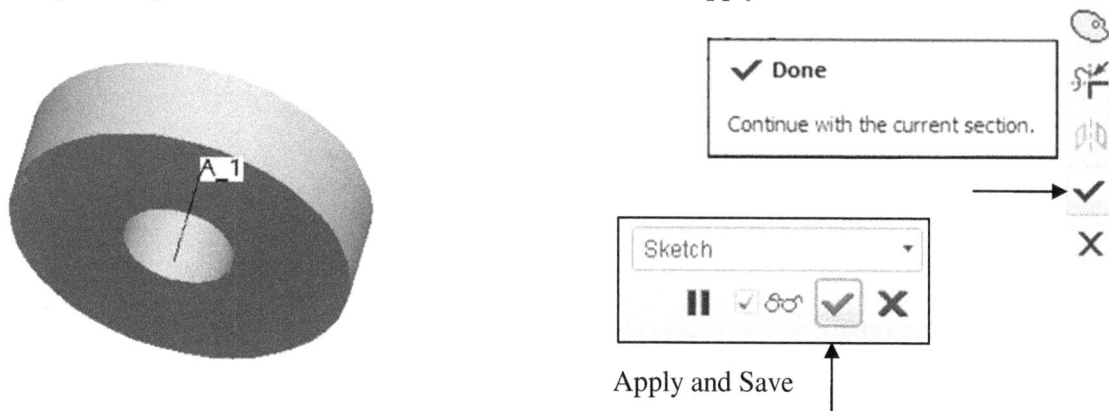

Apply and Save

Let us add chamfers at the sharp corner with dimension equal to 2x45°. Click the icon of **Chamfer** > select 45xD and type *2* as the dimension value > pick one of the 2 edges on the outer circumference first, afterwards, holding down the **Ctrl** key to pick the second edge > select the icon of **Apply and Save**.

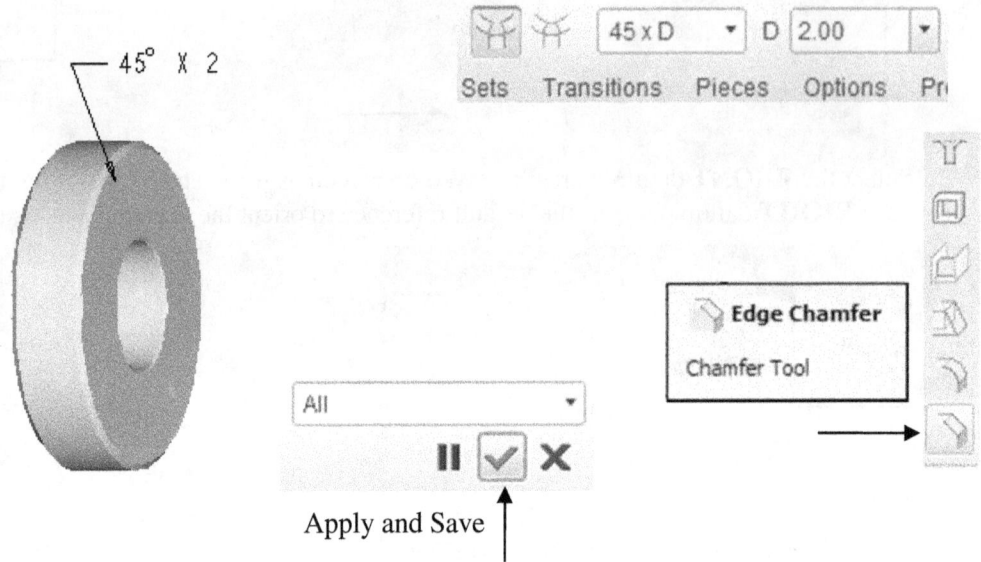

Apply and Save

To create teeth with a helical angle, we need a reference surface, which just goes over the cylindrical plate created. Let us assume the diameter is 204 and the width is 60.

Select the icon of **Extrude** from the toolbar of feature creation > select the icon of **Surface** > set the depth choice to Symmetry and specify 60 as the length of extrusion.

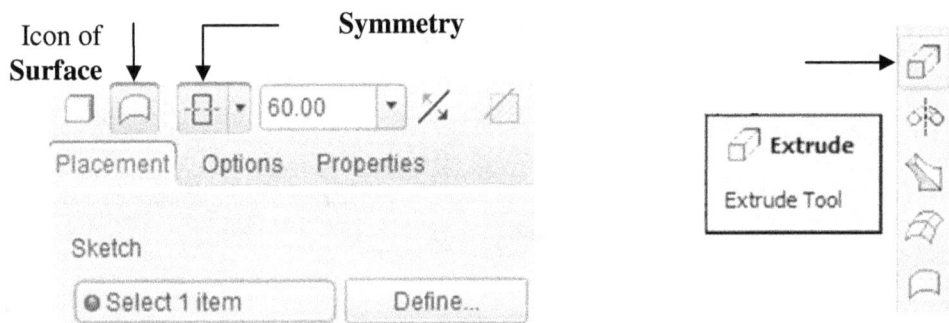

Select the **FRONT** datum plane as the sketch plane and accept the default setting to orient the sketch plane. Sketch a circle and the diameter dimension is 204.

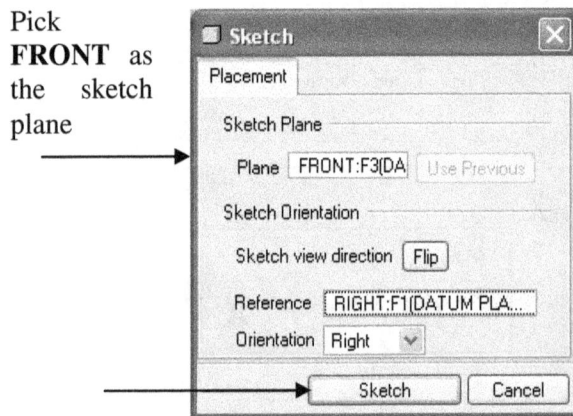

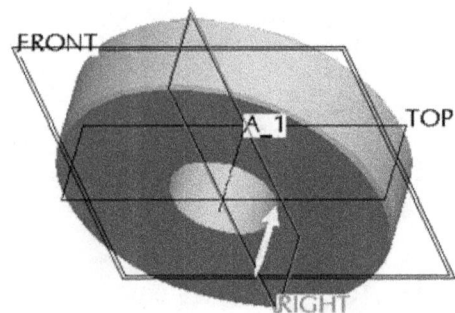

Click the icon of **Circle** and sketch a circle. The diameter value is 204.

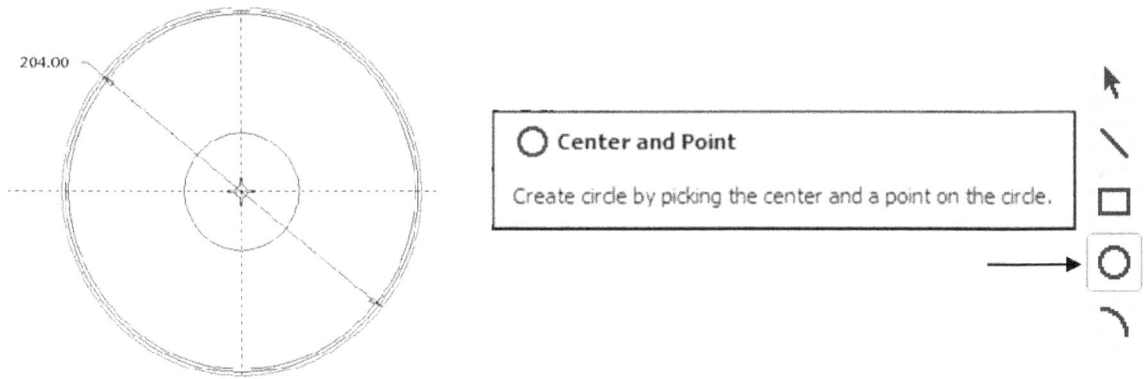

○ **Center and Point**

Create circle by picking the center and a point on the circle.

Upon completion, click the icon of **Done** and the icon of **Apply and Save**.

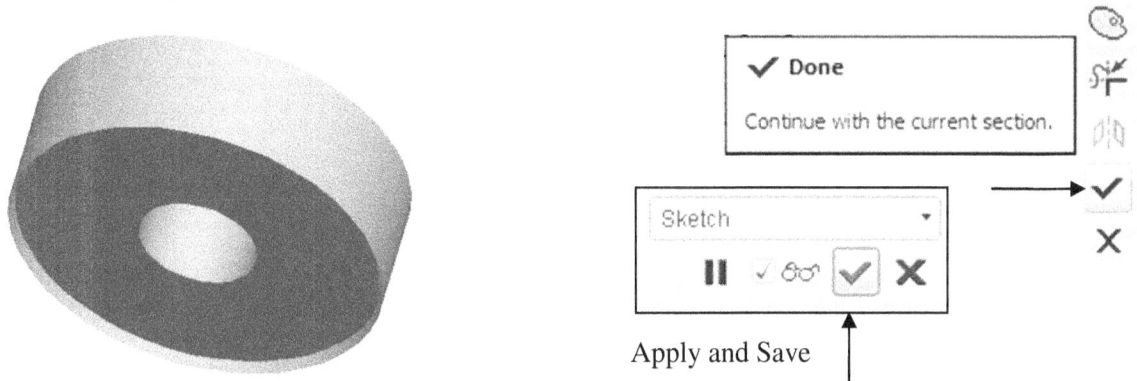

✓ **Done**

Continue with the current section.

Sketch

Apply and Save

Let us sketch a datum curve (a straight line) to define the helical angle. Click the icon of **Sketch Tool**. Select the **TOP** datum plane as the sketch plane. Select the **RIGHT** datum plane to orient the sketch plane.

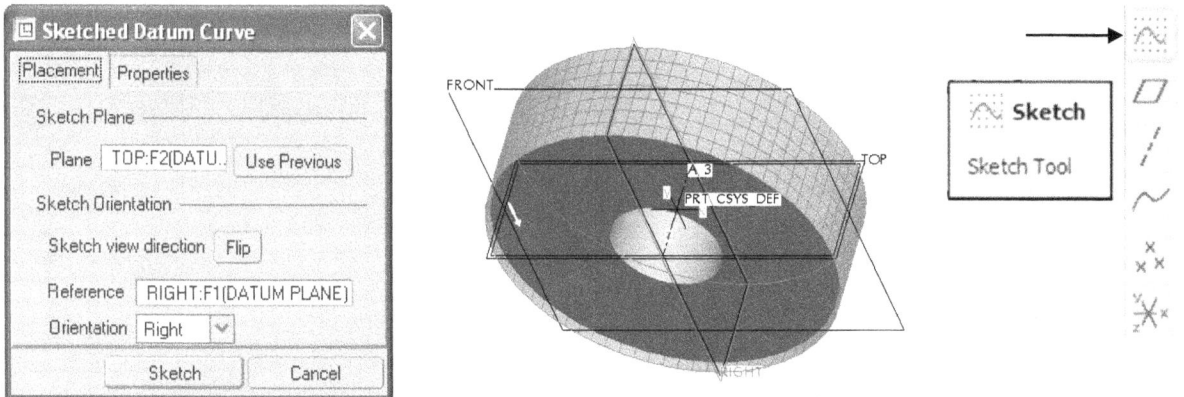

Sketched Datum Curve

Placement | Properties

Sketch Plane

Plane | TOP:F2(DATU.. | Use Previous

Sketch Orientation

Sketch view direction | Flip

Reference | RIGHT:F1(DATUM PLANE)

Orientation | Right

Sketch | Cancel

FRONT

TOP

PRT_CSYS_DEF

RIGHT

∿ **Sketch**

Sketch Tool

Before sketch a straight line, define a coordinate system. Click the icon of **Create coordinate system**. Click a location on the vertical axis; specify 60 as the distance, as shown.

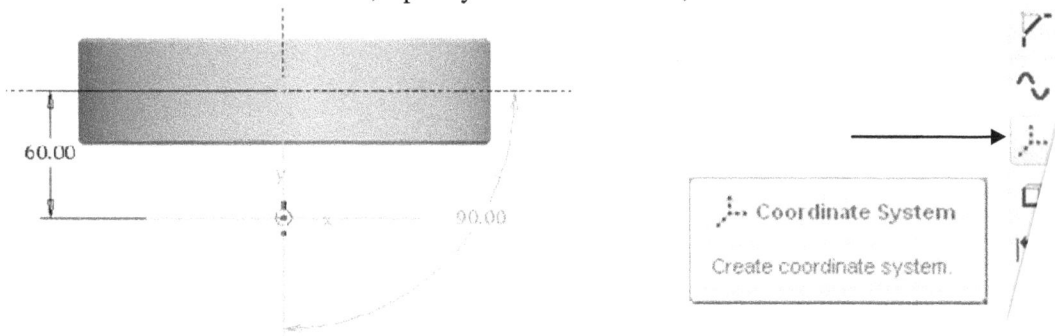

60.00

90.00

⋮ **Coordinate System**

Create coordinate system.

Click the icon of **Line** and sketch a straight line at an angle with respect to the vertical reference or the y-axis. The angle dimension of the straight line is 15 degrees, and the length is set by specifying a dimension of 30, as shown. Click the icon of **Done**.

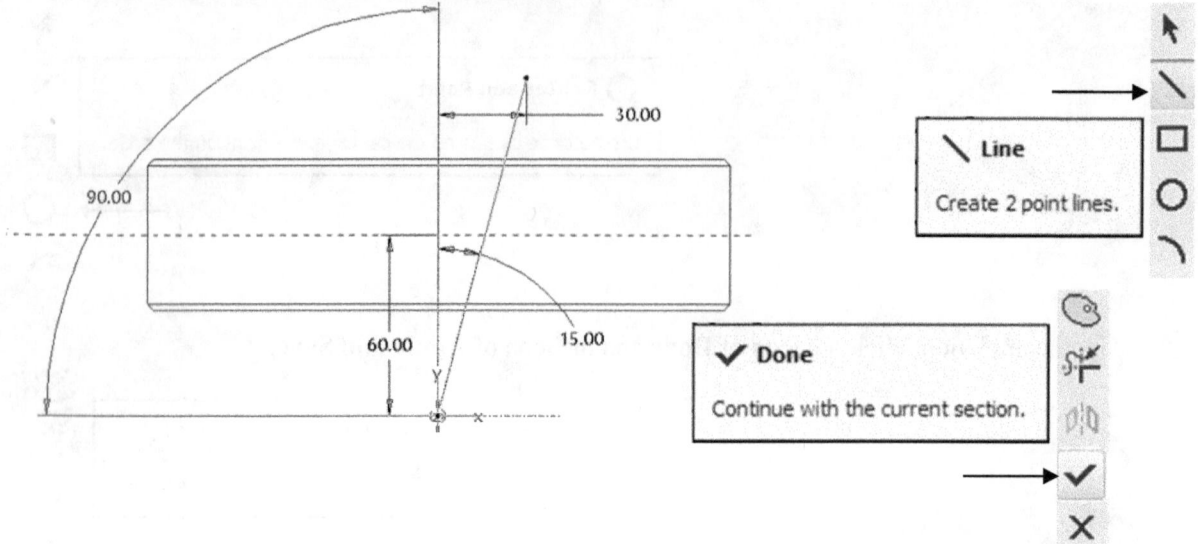

Now we need to project this straight line to the cylindrical surface. In the model tree, highlight the datum curve just created. From the main toolbar, select **Edit > Project** > pick the upper portion of the cylindrical surface (do not pick the upper surface of the solid cylinder) so that a projected datum curve shown on the upper portion of the cylindrical surface > select the icon of **Apply and Save**.

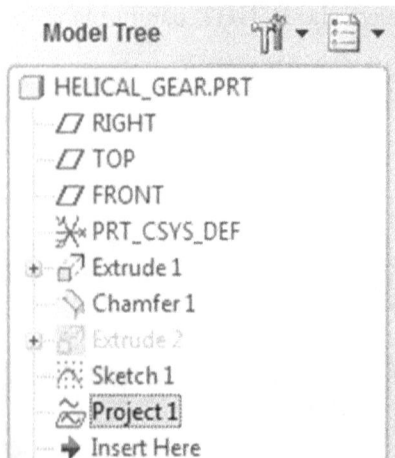

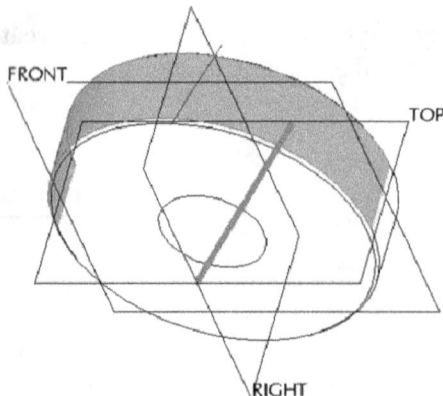

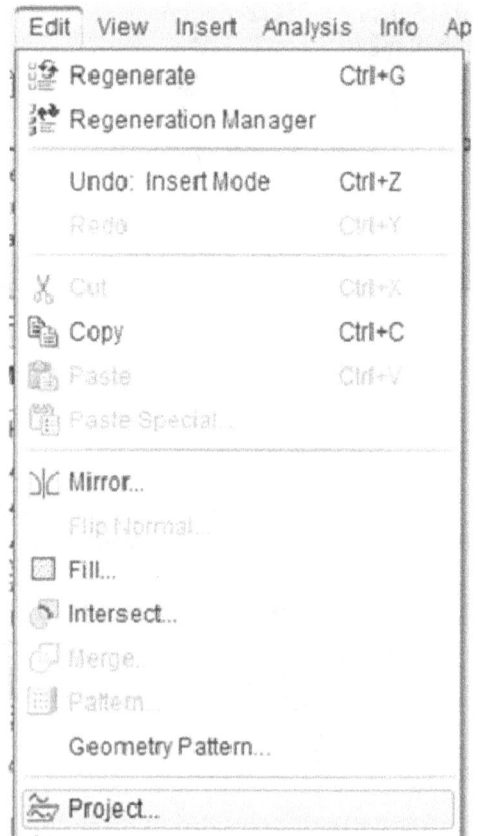

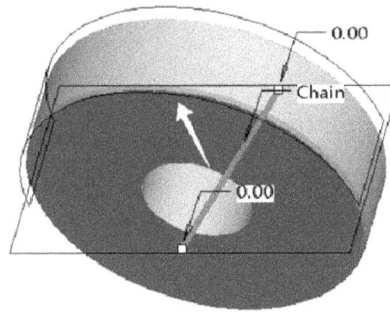

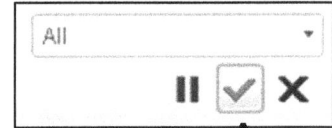

Apply and Save

Now let us use the function called **Sweep** to cut the first tooth. From the main toolbar, select **Insert > Sweep > Cut > Select Traj > One by One** > Select pick the projected datum curve > **Done.**
Click the icon of **Centerline**. Sketch a vertical centerline first.

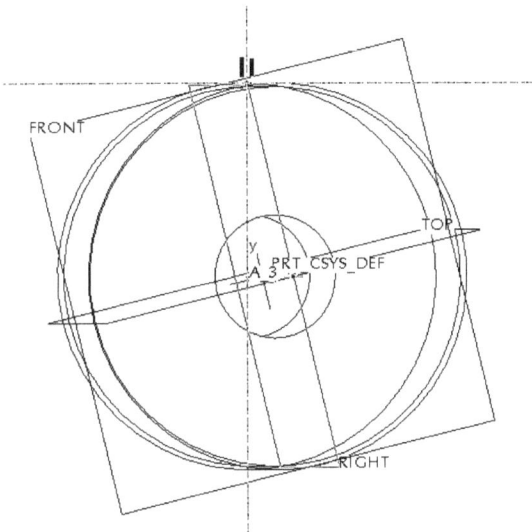

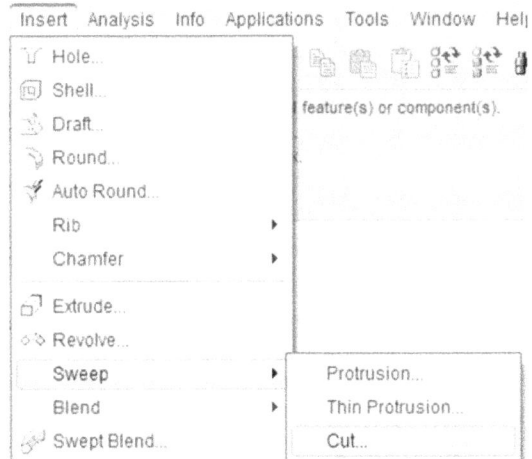

Click the icon of **Line** and sketch the profile of the tooth, paying attention to the symmetric issue so that the number of dimensions needed is minimized. Click the icon of **Done** and **Okay** to accept the cut direction for material side > **OK**.

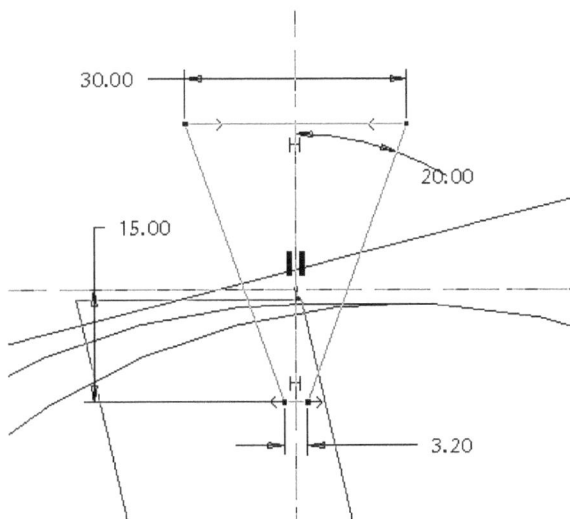

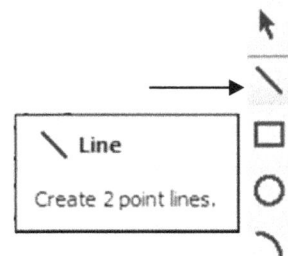

Upon completion, click the icon of **Done** and click the icon of **Apply and Save**.

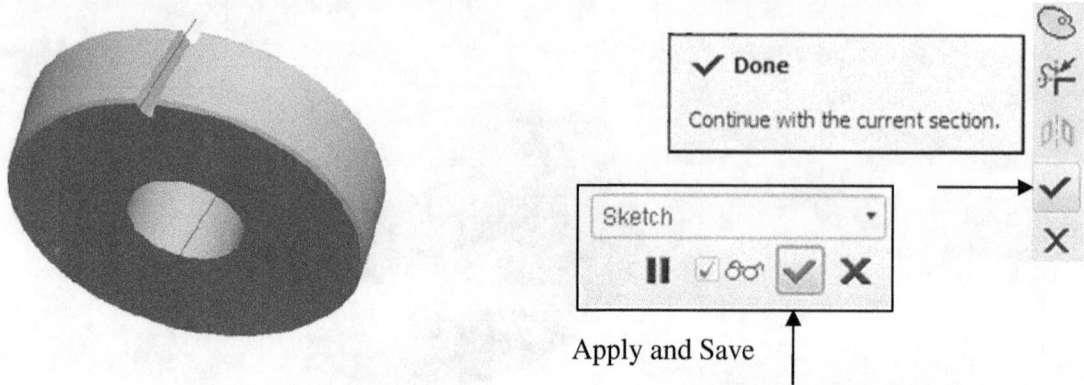

Apply and Save

In order to show the first cut tooth, we may need to hide the cylindrical surface feature. From the model tree, highlight the cylindrical surface feature > right-click and hold, select **Hide.** The first cut tooth is clearly depicted.

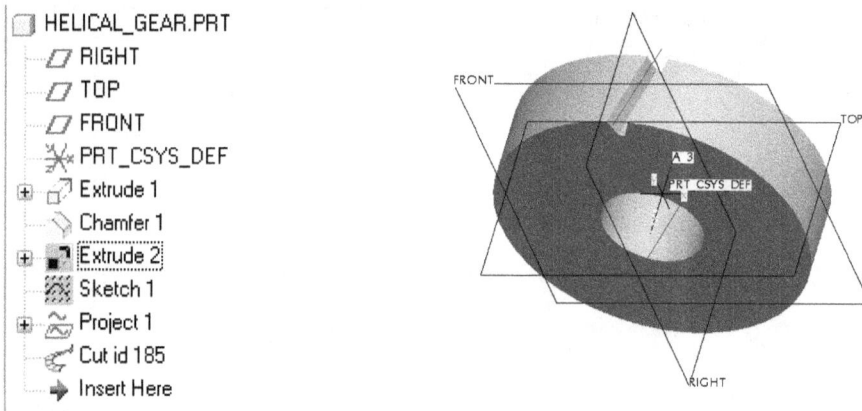

We use **Pattern** to create the other teeth. In the model tree, highlight the cut feature > right-click and hold, select **Pattern** > select Axis and pick the axis from the display, set the number of copies to 40 and the angle to 9 degrees > click the icon of **Apply and Save**.

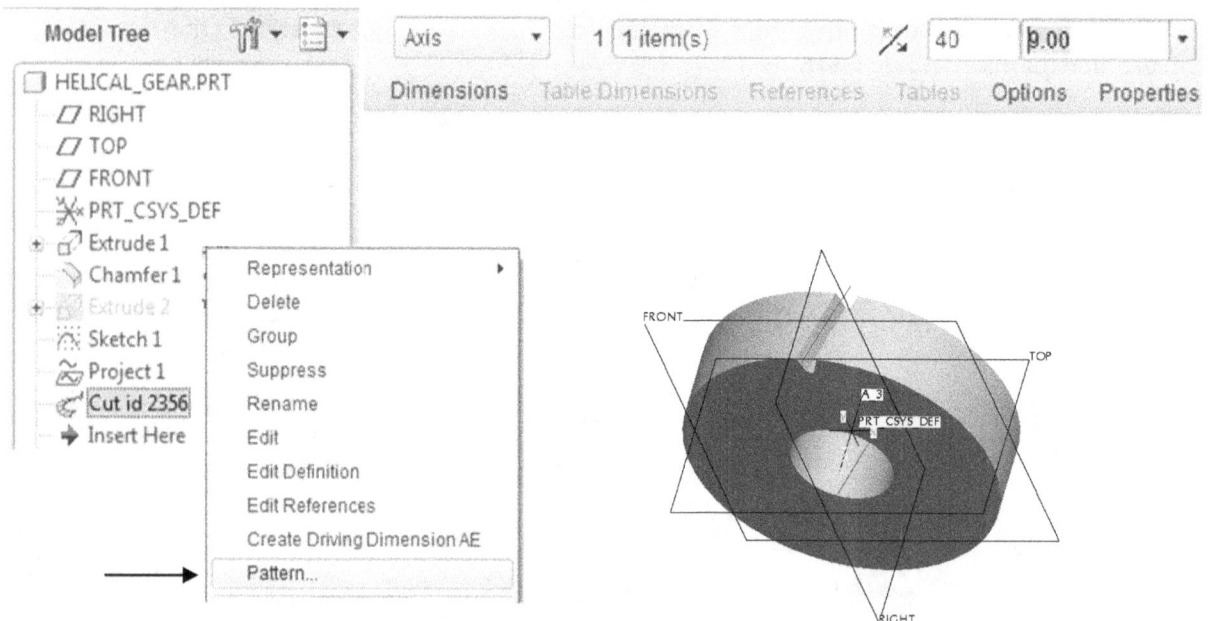

Apply and Save

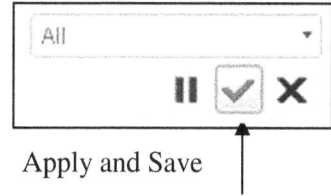

In order to define a load acting on the surface of a tooth, which has a helical angle, we need to define a local coordinate system so that the load condition can be added with ease.

First define a datum point serving as the origin of the local coordinate system. Click the icon of Datum Point Tool > select the projected curve and, while holding down the **Ctrl** key, pick the **FRONT** datum plane > **Close**. PNT0 is created, as shown.

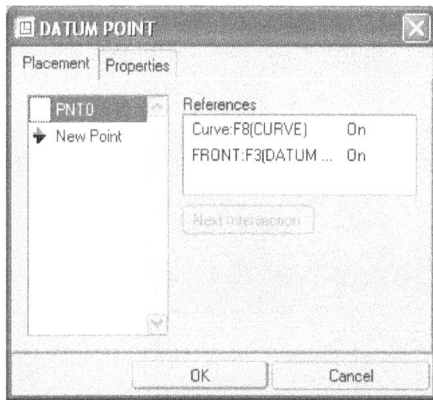

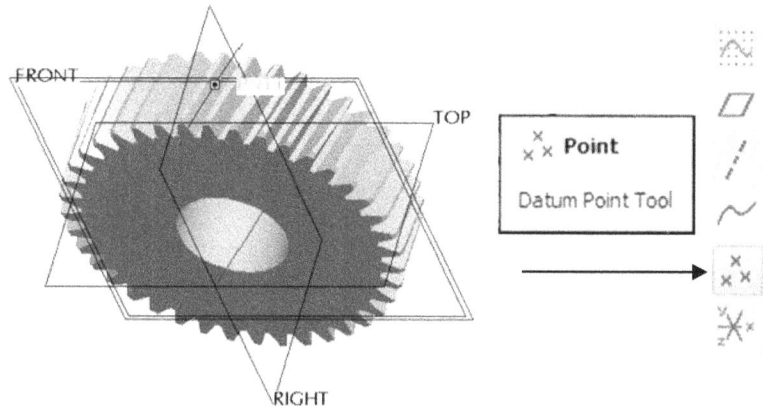

Let us create three datum planes. DTM1 is a plane normal to the projected curve through PNT0. Click the icon of **Datum Plane Tool**. Pick the projected curve and use Normal. While holding down the **Ctrl** key, pick PNT0 > **OK**. DTM1 is created, as shown.

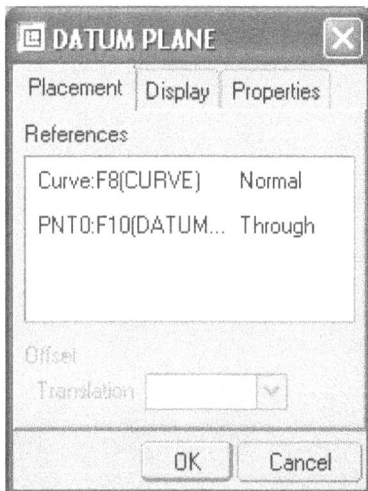

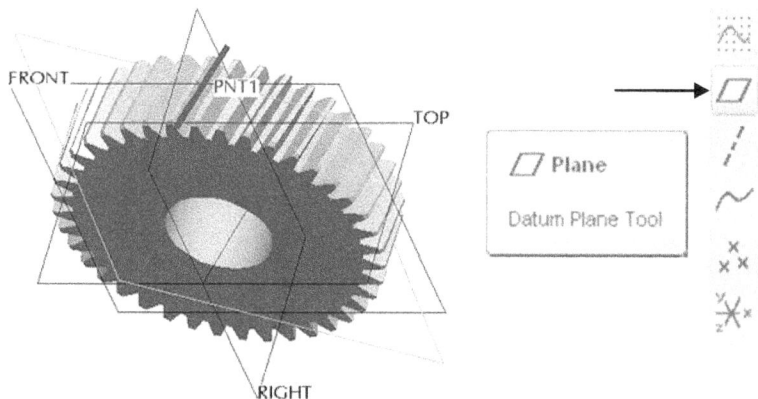

DTM2 is a datum plane normal to DTM1 and passing through the projected curve. Click the icon of **Datum Plane Tool**. Pick DTM1 and use **Normal**. While holding down the **Ctrl** key, pick the projected curve and **Through**. DTM2 is created, as shown.

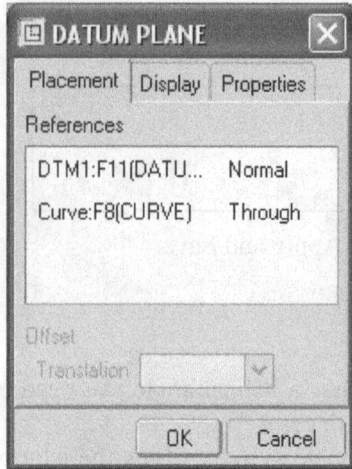

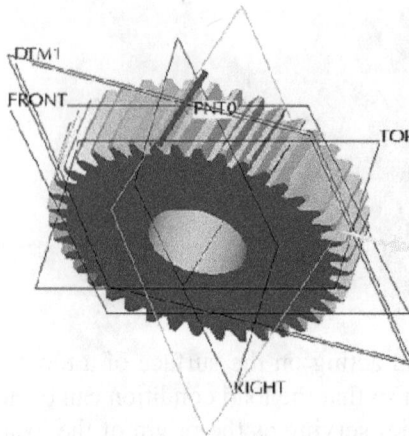

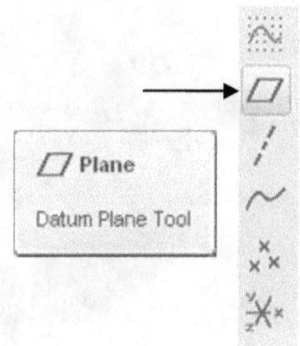

DTM3 is a plane normal to both DTM1 And DTM2. Click the icon of **Datum Plane Tool**. Pick DTM1 and use Normal. While holding down the **Ctrl** key, pick DTM2 and use Normal. While holding down the **Ctrl** key, pick PNT0. DTM3 is created, as shown.

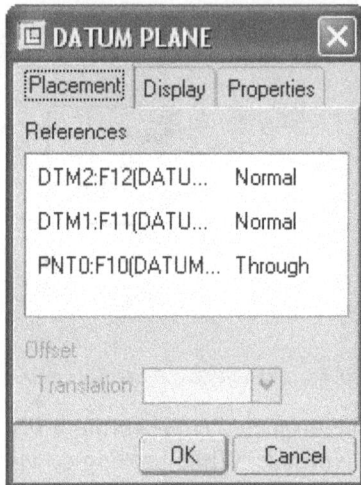

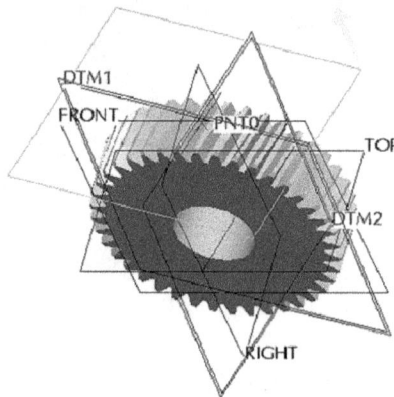

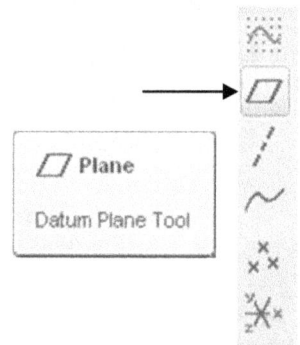

To create a local coordinate system using DTM1, DTM2 and DTM3, select the icon of Datum **Coordinate System Tool** > pick the three datum planes (holding down the **Ctrl** key) > orient the direction of the local coordinate system, as shown. Note that the direction of the x-axis will be the direction of the load to be applied.

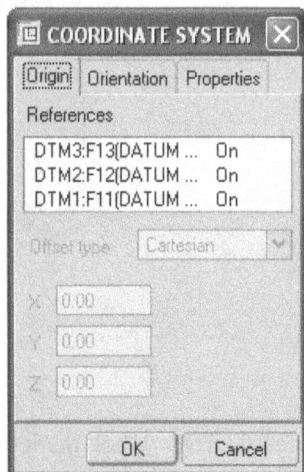

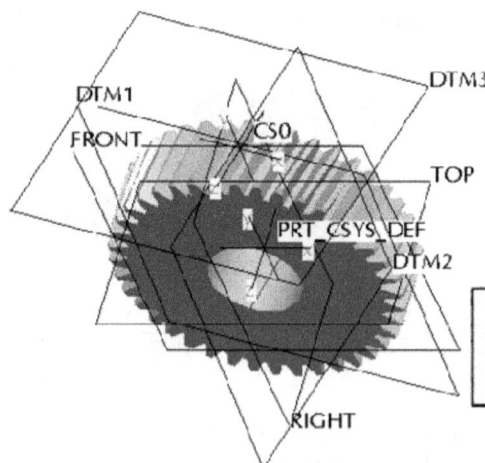

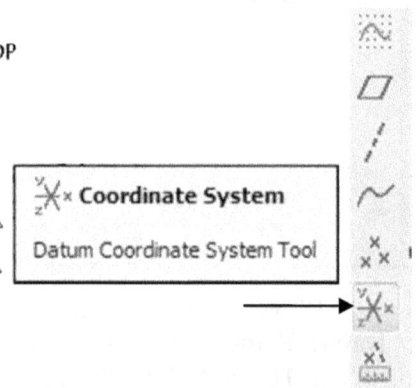

Step 2: Under the Pro/MECHANICA environment, define the material type.
        Applications > Mechanica > Continue > Structure > OK.

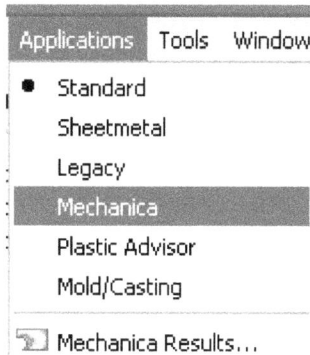

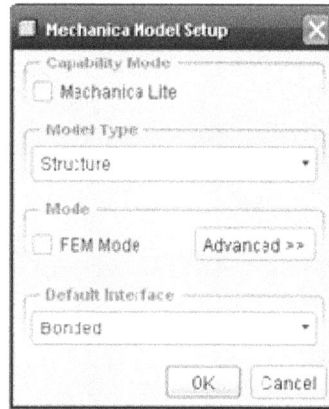

From the toolbar of functions, select the icon of **Materials >** select **STEEL** from the left side called Materials in Library > click the directional arrow so that the selected material type goes to the right side called Materials in Model.

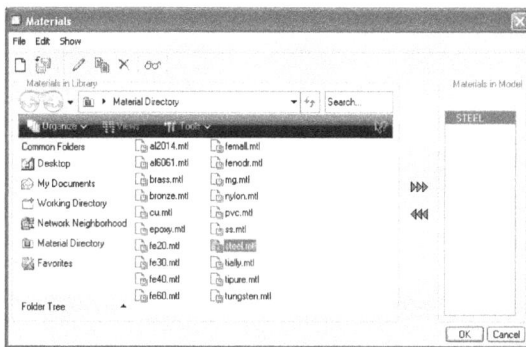

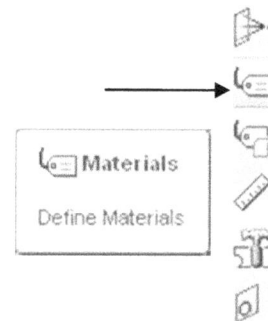

Click the icon of **Material Assignment.** In the Material Assignment window, the selected **Steel** is shown, and the component is also shown because there is only one volume or one component in the system. As a result, the software system automatically assigns Steel to the component. Just click **OK**.

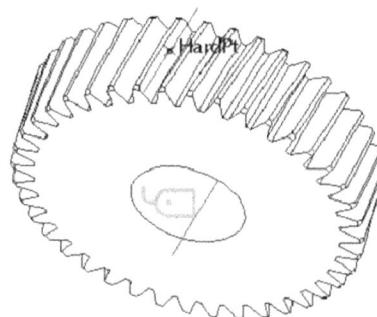

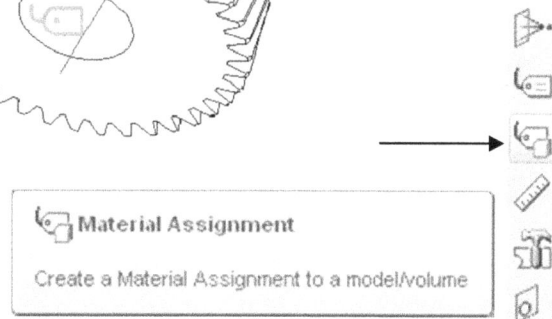

Step3:   Define the fixed end constraint condition and load condition.
   Click the icon of **Displacement Constraints >** pick the cylindrical surface of the hole. Set the 6 degrees of freedom to be fixed.

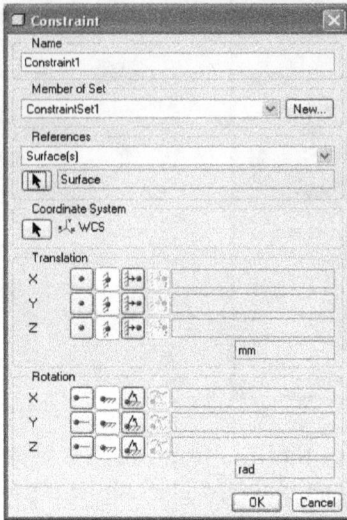

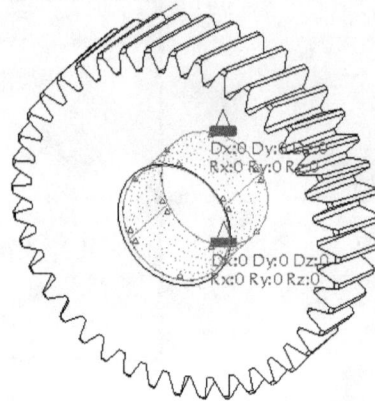

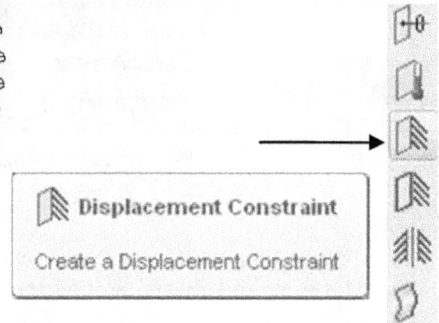

   To define the load condition, from the toolbar of functions, click the icon of **Force/Moment** Loads > Surface(s) > pick the surface of tooth > select **CSO** > specify 200 in the X Force Component > **OK**.

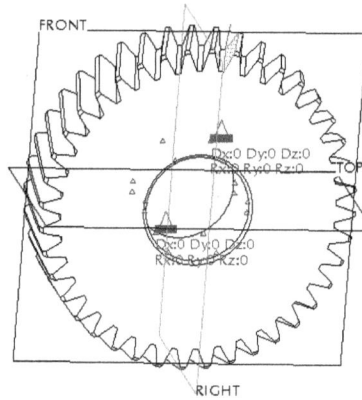

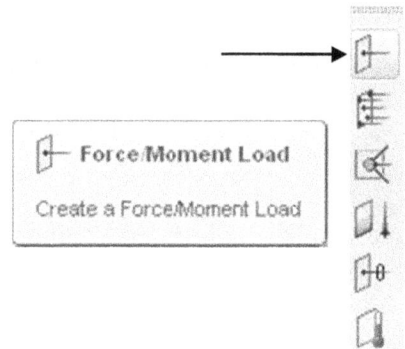

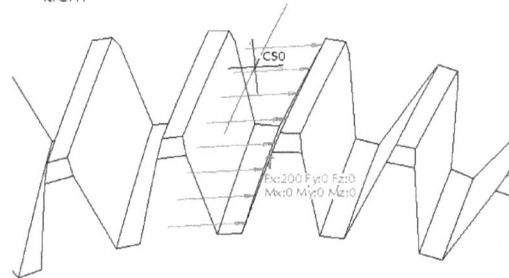

Step 4: Under the Pro/MECHANIC, set up **Analyses** and **Run.**
   Select the icon of **Run an Analysis > File > New Static >** type *helical_gear* > **OK > Run.**

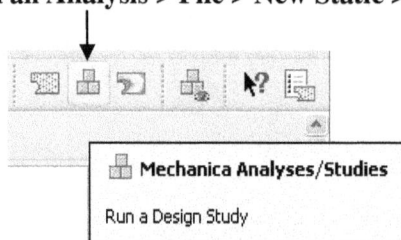

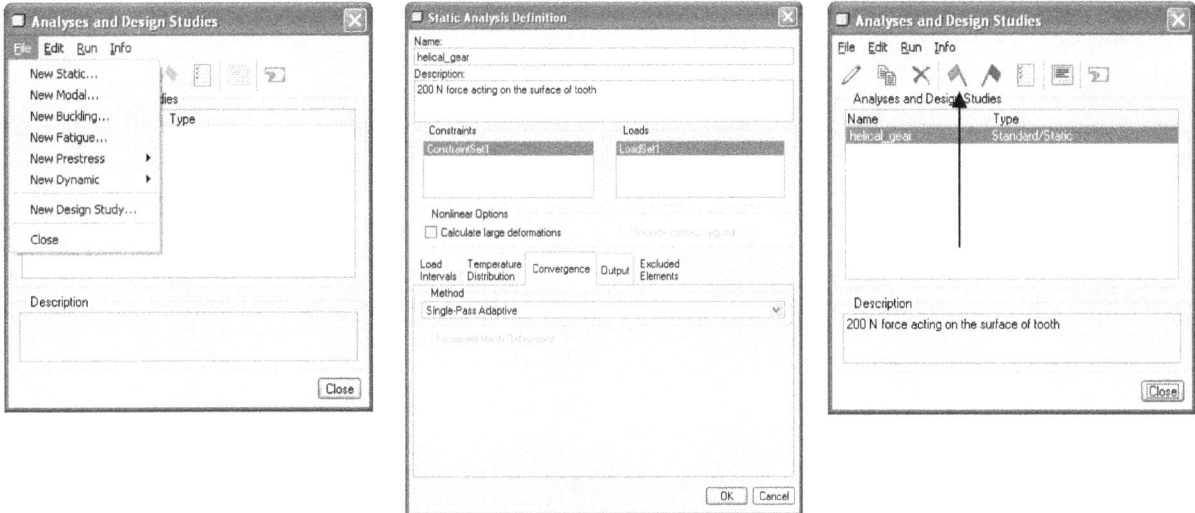

Step 5: Under the Pro/MECHANICA, study the **Results**

Select the icon of **Results.**

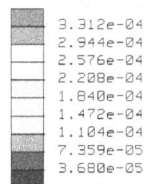

Type *Tooth Deformation* as the title > locate the Design Study folder called helical_gear.

Select **Fringe** > choose **Displacement > Magnitude > OK and Show**.

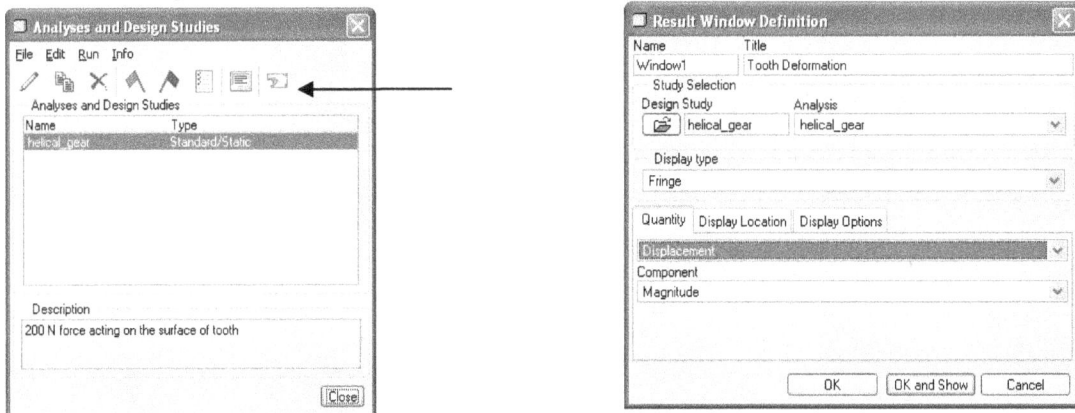

tooth deformation

To plot a distribution of stress, from the main toolbar, select **Edit > Copy**. Select **Stress > Max Principal > OK and Show**. Users should be able to obtain the plot of the maximum principal stress distribution.

Edit   View   Insert   Info   Format

Result Window...

Legend Value...

Capping Surf

Cutting Surf

Annotation

Copy...

Result Window Definition

Name   Title
Window2   distribution of max principal stress

Study Selection
Design Study   Analysis
helical_gear   helical_gear

Display type
Fringe

Quantity | Display Location | Display Options

Stress

Component
Max Principal

OK   OK and Show   Cancel

Stress Max Prin (WCS)
(MPa)
Deformed
Scale 1.6430E+04
Locdset:LoadSet1 : HELICAL_GEAR

4.241e+00
3.747e+00
3.253e+00
2.759e+00
2.264e+00
1.770e+00
1.276e+00
7.813e-01
2.870e-01
-2.073e-01
-7.015e-01

distribution of max principal stress

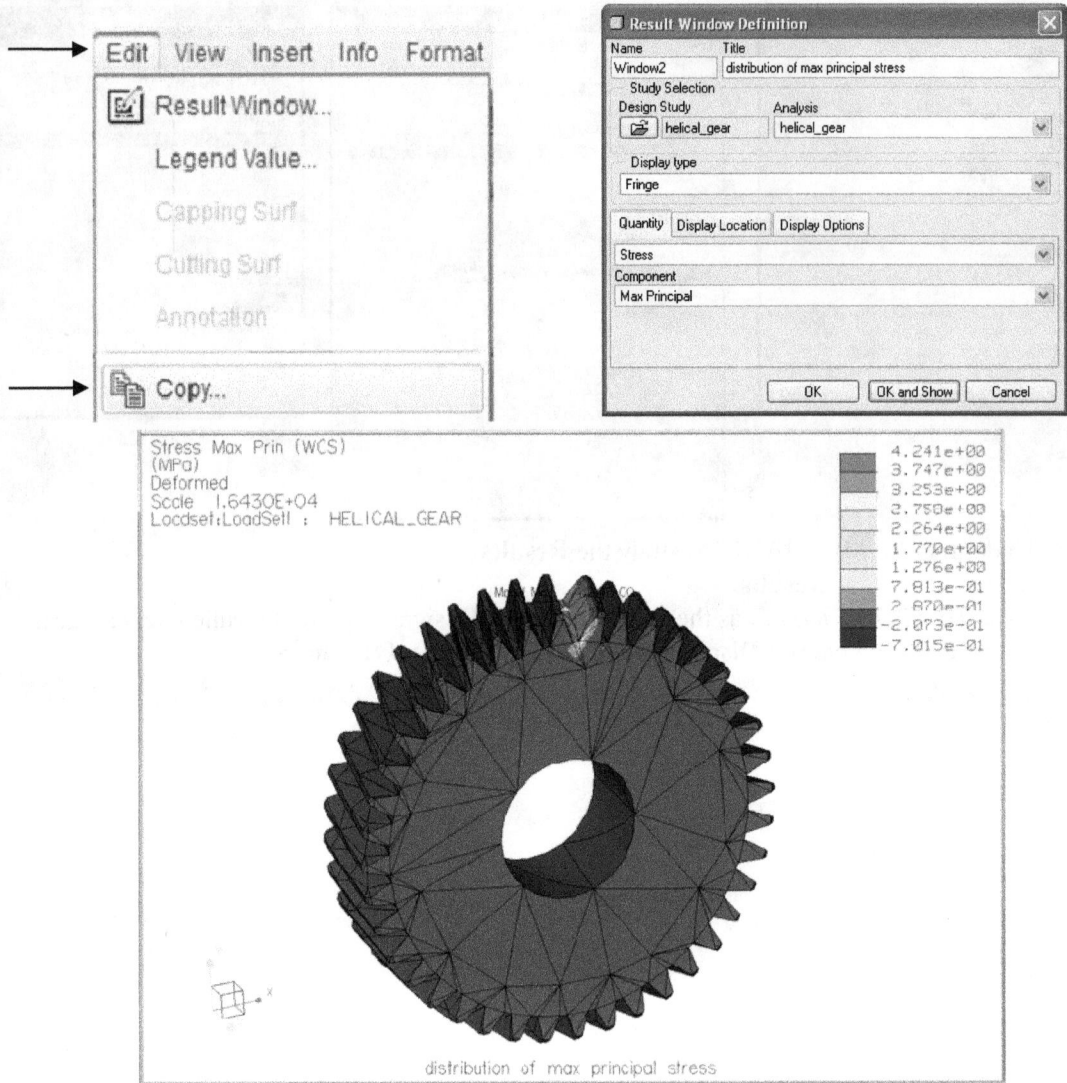

Step 6: Under the Pro/ENGINEER design environment, create a 3D solid model of the shaft component. **File > New > Part** > type *shaft_for_helical_gear* as the file name and clear the icon of **Use default template > OK.** Select **mmns_part_solid** (units: Millimeter, Newton, Second) and type *Shaft* in **DESCRIPTION**, and *student* in **MODELED_BY**, then **OK.**

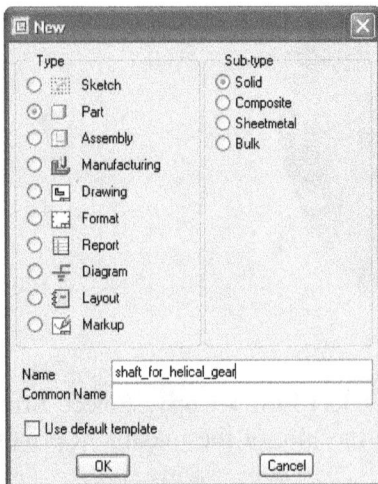

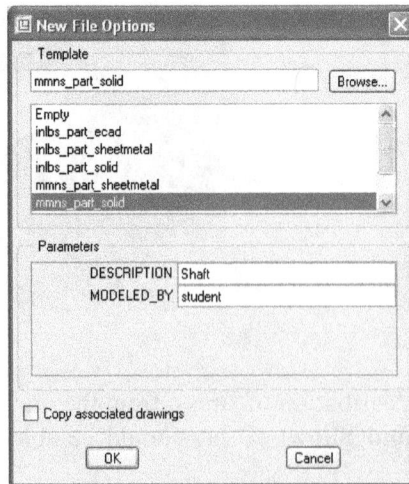

New

Type   Sub-type
○ Sketch   ⦿ Solid
⦿ Part   ○ Composite
○ Assembly   ○ Sheetmetal
○ Manufacturing   ○ Bulk
○ Drawing
○ Format
○ Report
○ Diagram
○ Layout
○ Markup

Name   shaft_for_helical_gear
Common Name

☐ Use default template

OK   Cancel

New File Options

Template
mmns_part_solid   Browse...

Empty
inlbs_part_ecad
inlbs_part_sheetmetal
inlbs_part_solid
mmns_part_sheetmetal
mmns_part_solid

Parameters

DESCRIPTION   Shaft
MODELED_BY   student

☐ Copy associated drawings

OK   Cancel

We use Revolve to create the solid model. Select the icon of **Revolve** from the toolbar of feature creation. Make sure the icon of **Solid** is selected. Click the box of **Placement > Define**, and select **FRONT** as the sketch plane, accept **RIHGT** to orient the sketch plane > click the box of **Sketch**.

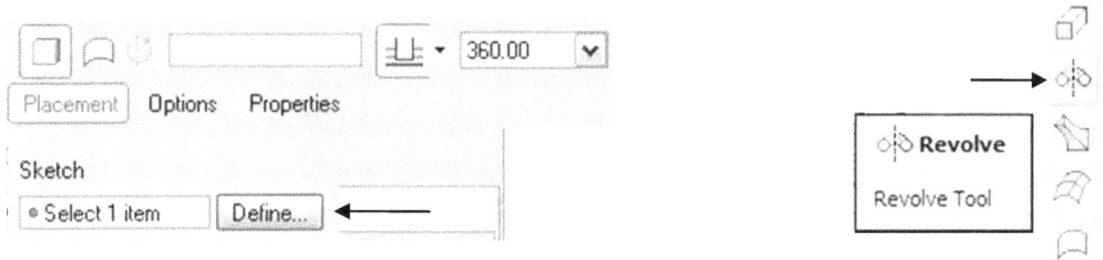

Click the icon of **Centerline**. Sketch a horizontal centerline for the revolving operation.

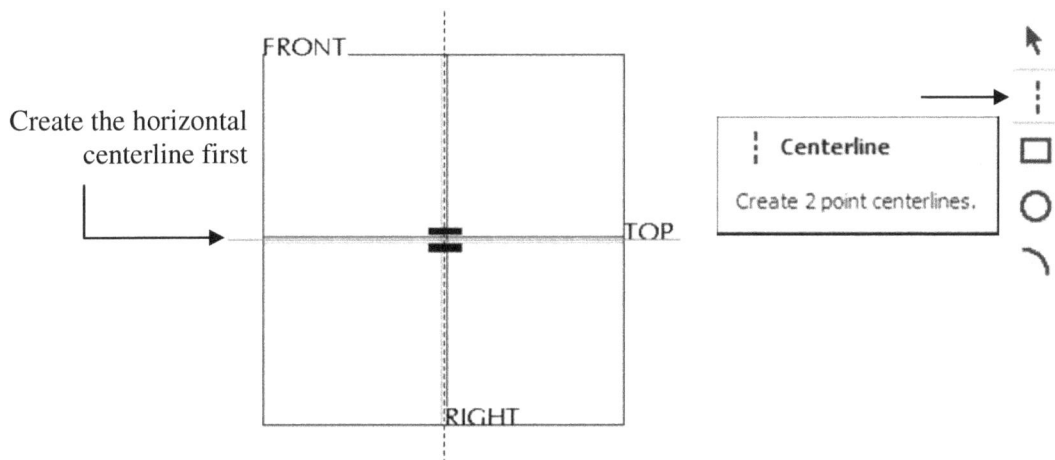

Click the icon of **Line** and sketch the following shape. Pay attention to dimensioning the diameters.

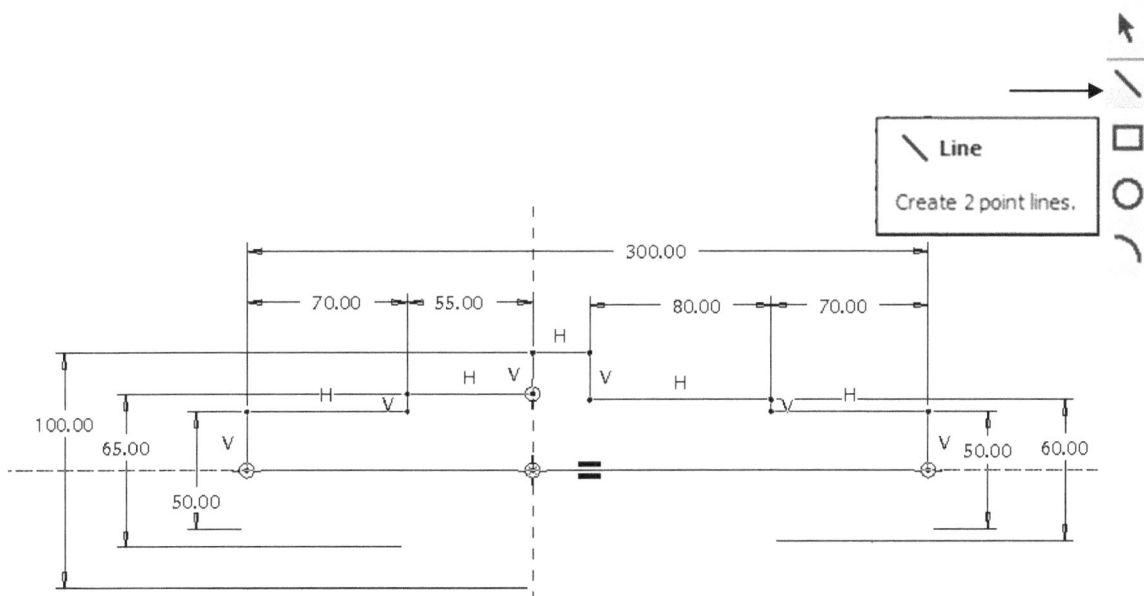

Upon completion, click the icon of **Done** and the icon of **Apply and Save**.

Apply and Save

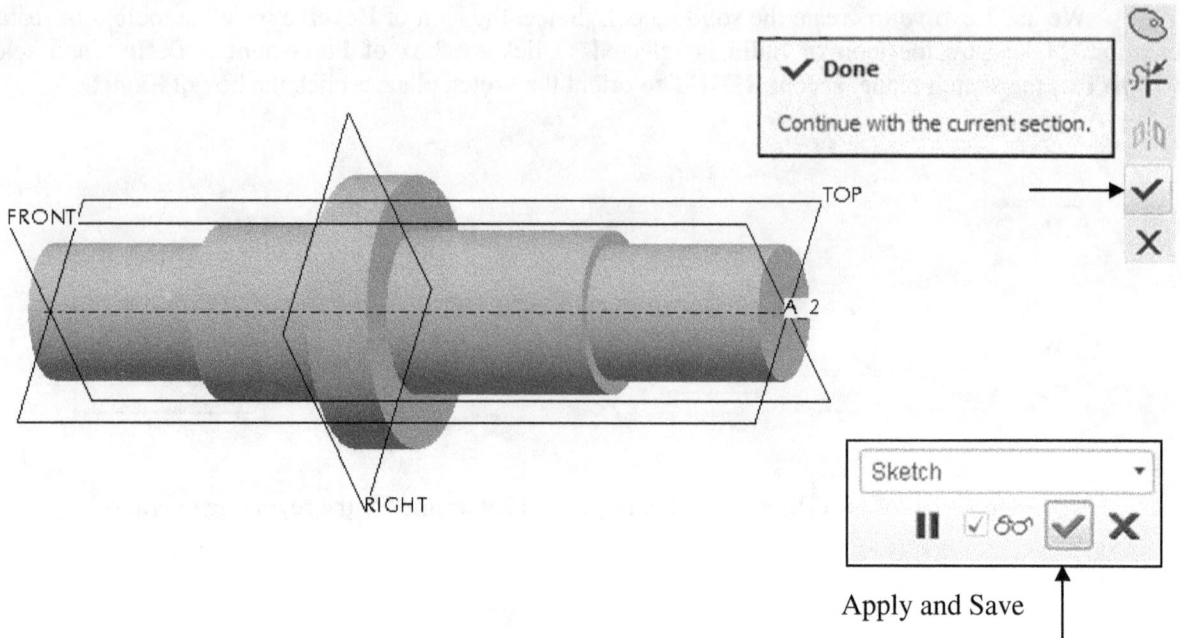

Step 7: Under the Pro/ENGINEER design environment, we assemble the gear and shaft components.

File > New > Assembly > type *gear_shaft_assembly* as the file name and clear the box of **Use default template > OK.**

Select **mmNs_asm_design** (units: mm, Newton, and second) and type in gear shaft assembly in **DESCRIPTION**, and student in **MODELED_BY**, then **OK.**

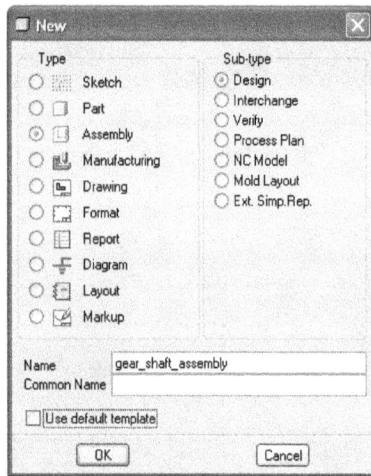

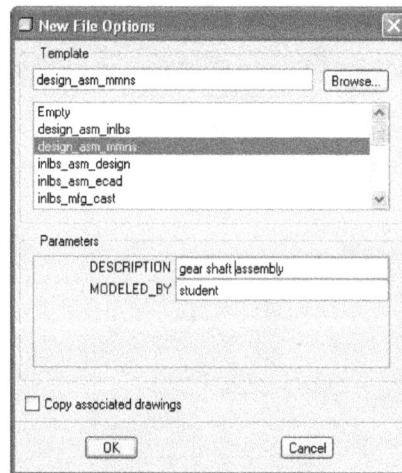

Click the icon of **Add component**.

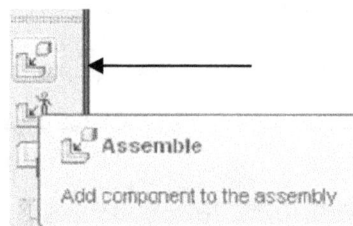

Make sure that **In Session** is activated. Click the icon of **Add component**. Highlight the component called Shaft_for_Helical_Gear > **Open**.

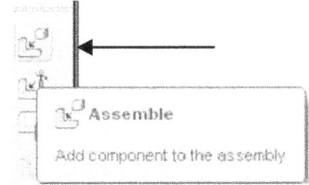

From the assembly window, select **Default**, meaning that the coordinate system of the shaft component is assembled at the location of the assembly coordinate system. Afterwards, click **OK** because the shaft component is fully constrained.

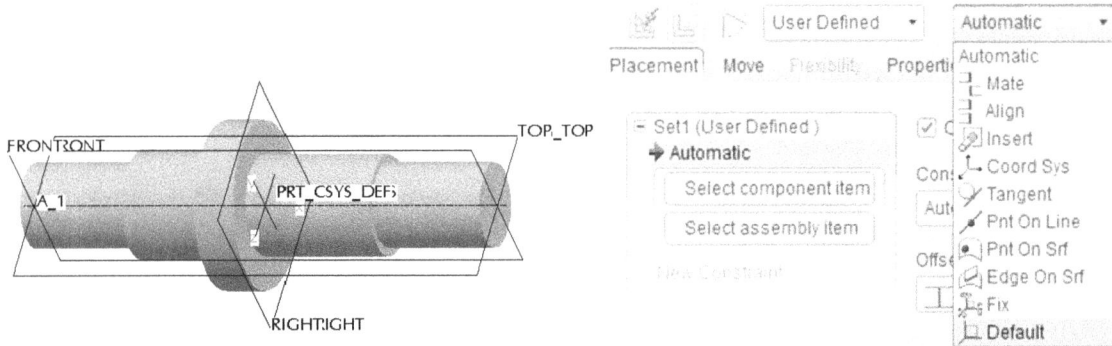

Click the icon of **Add component**. Make sure that you are at **In Session**. Highlight the component called helical gear > **Open**.

Click the axis from the helical gear component, and click the axis from the shaft component. An alignment constraint is defined, as shown below.

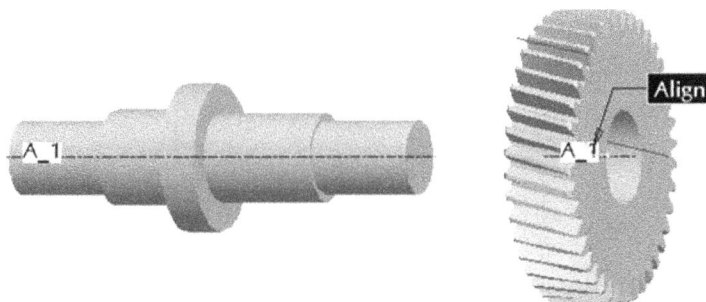

Click the surface from the helical gear component and click the surface from the shaft component. Afterwards, click the box of Coincident. A mate constraint is defined.

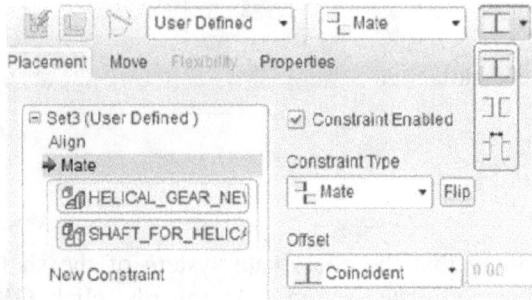

Step 8: Under the Pro/MECHANICA environment, define the 2 material types for the shaft and gear components. Both material types are steel.

Applications > Mechanica > Continue > Structure > OK.

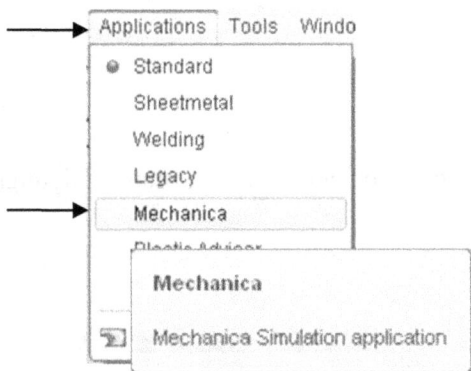

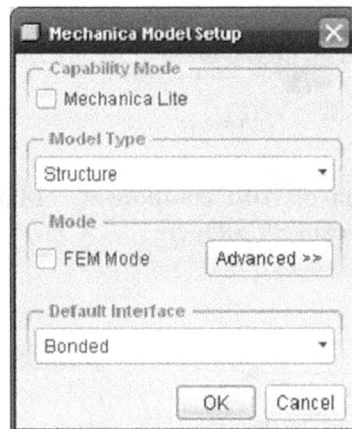

From the toolbar of functions, select the icon of **Materials >** select **STEEL** from the left side called **Materials in Library** > click the directional arrow so that the selected material type goes to the right side called **Materials in Model**.

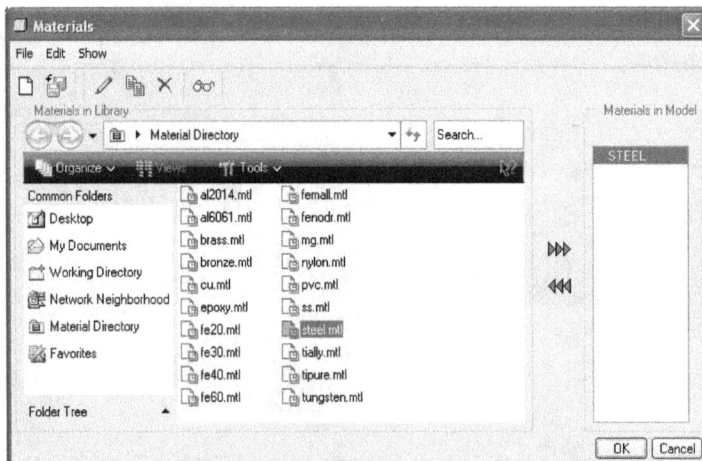

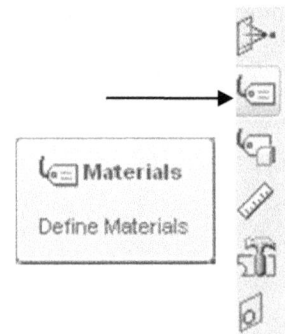

Click the icon of **Material Assignment.** In the Material Assignment window, the selected **Steel** is shown, and select both shaft and gear components while holding down the **Ctrl** key > **OK**.

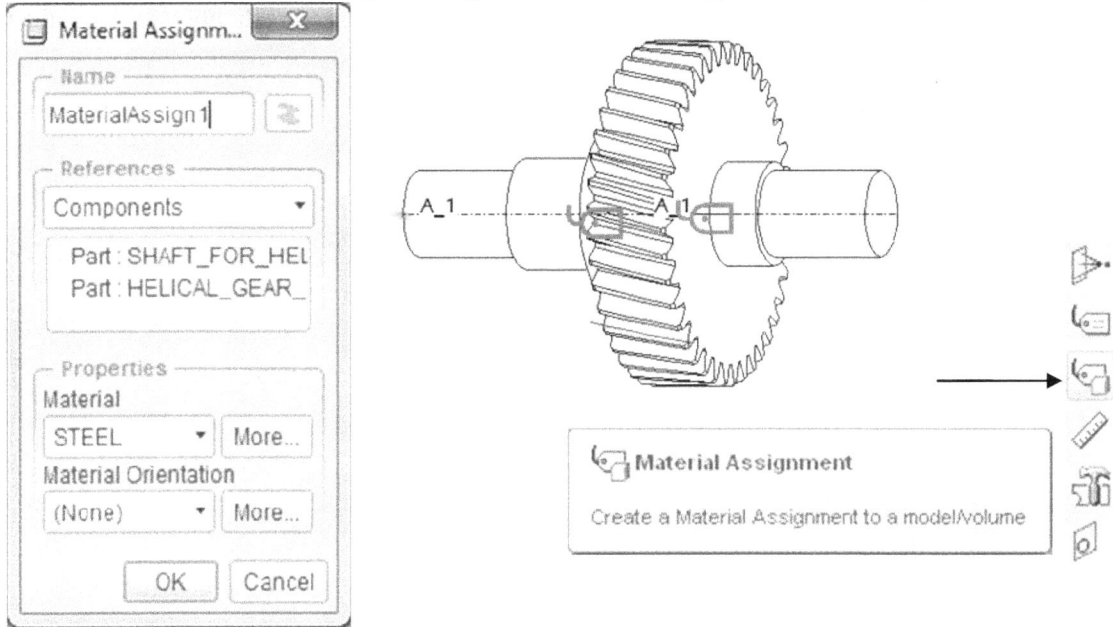

Step 9: Under the Pro/MECHANICA environment, define the fixed end constraint condition, (assume that both ends of the shaft are fixed)

From the toolbar of functions, select the icon of **Displacement Constraints >** select the two end surfaces of the shaft > **OK**. Set the 6 degrees of freedom to be fixed > **OK**.

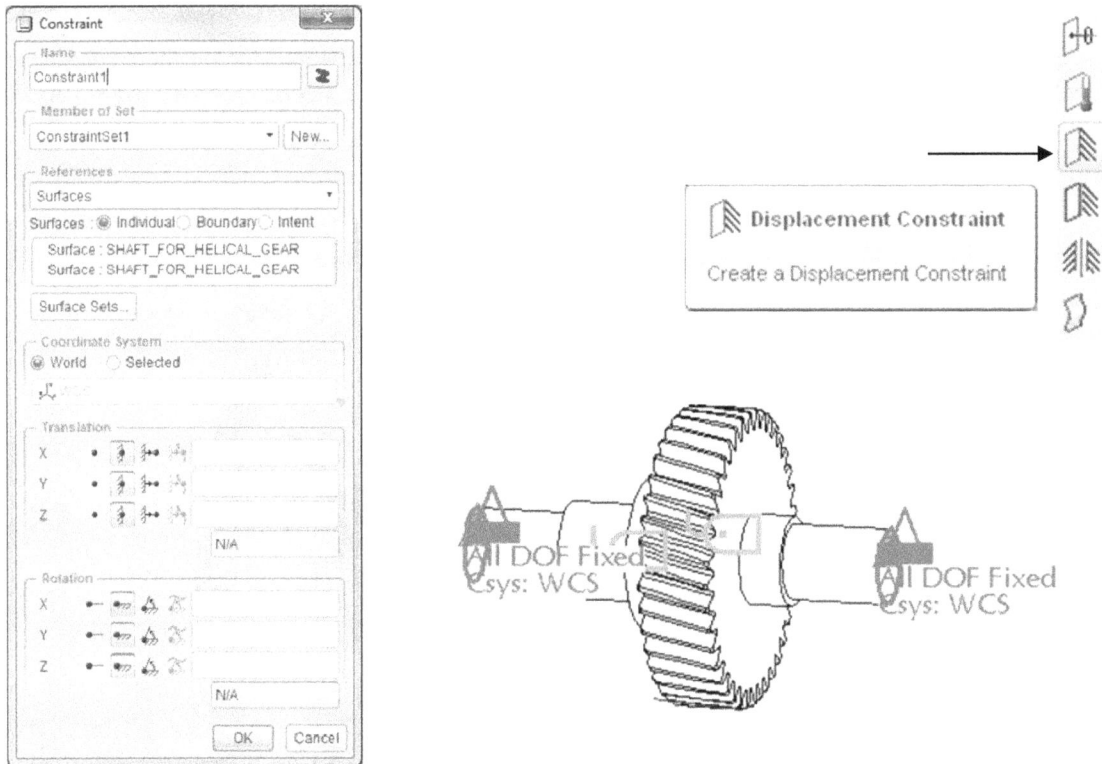

Step 10: Under the Pro/MECHANICA environment, define the load condition.

There are two loads. The first load is a torque about the rotation axis, and the torque is acting on the flat surface on the right side of the gear component.

Before defining the torque load, let us define a datum point. Click the icon of **Datum Point Tool > Select** > click Helical Gear component listed in the model tree. At this moment, the Datum Point window appears. Pick the surface on the right side of the gear component. While holding down the **Ctrl** key, pick the axis from the gear component > **OK**. PNT1 is created.

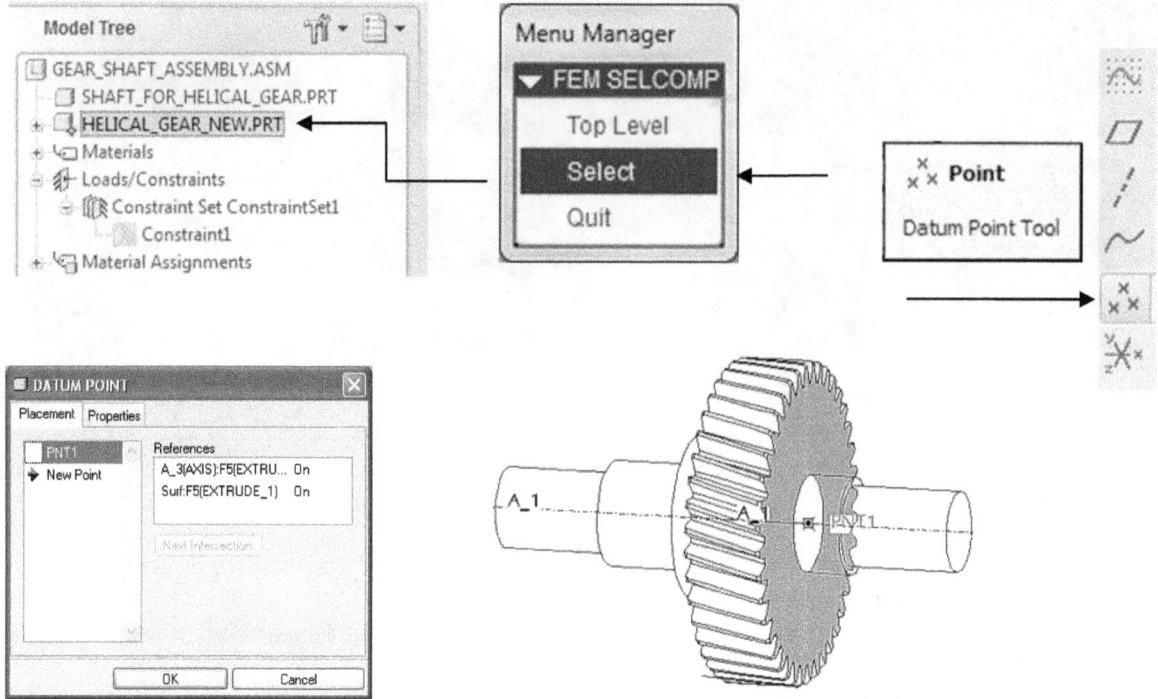

Click the icon of **Force/Moment Loads > Surface**(s) > pick the surface on the right side of the gear component. Click **Advanced > Total Load at Point** and pick **PNT1**. Specify 10000 in the X-component under Moment > **OK**.

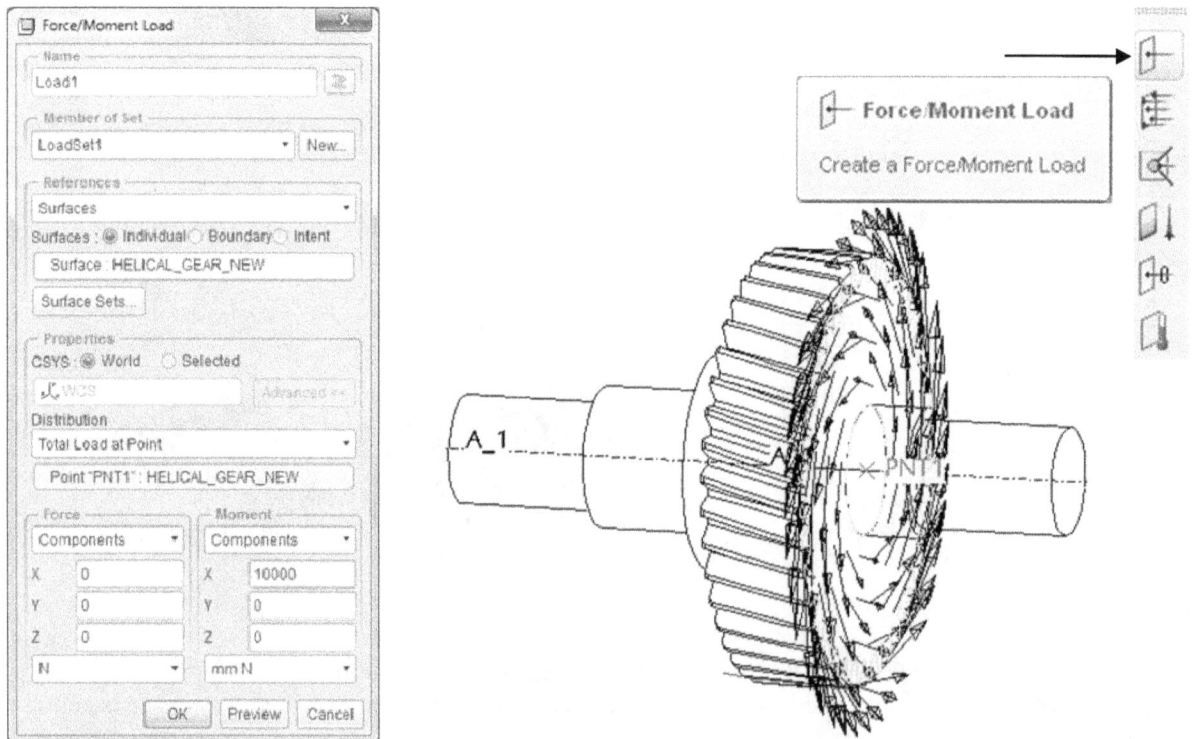

To define load 2, this is the force acting on the flat surface on the right side of the helical gear component, select the icon of **Force/Moment Load > Surface**(s) > pick the flat surface on the right side of the gear component > type 200 as the X-component under Force > **OK**.

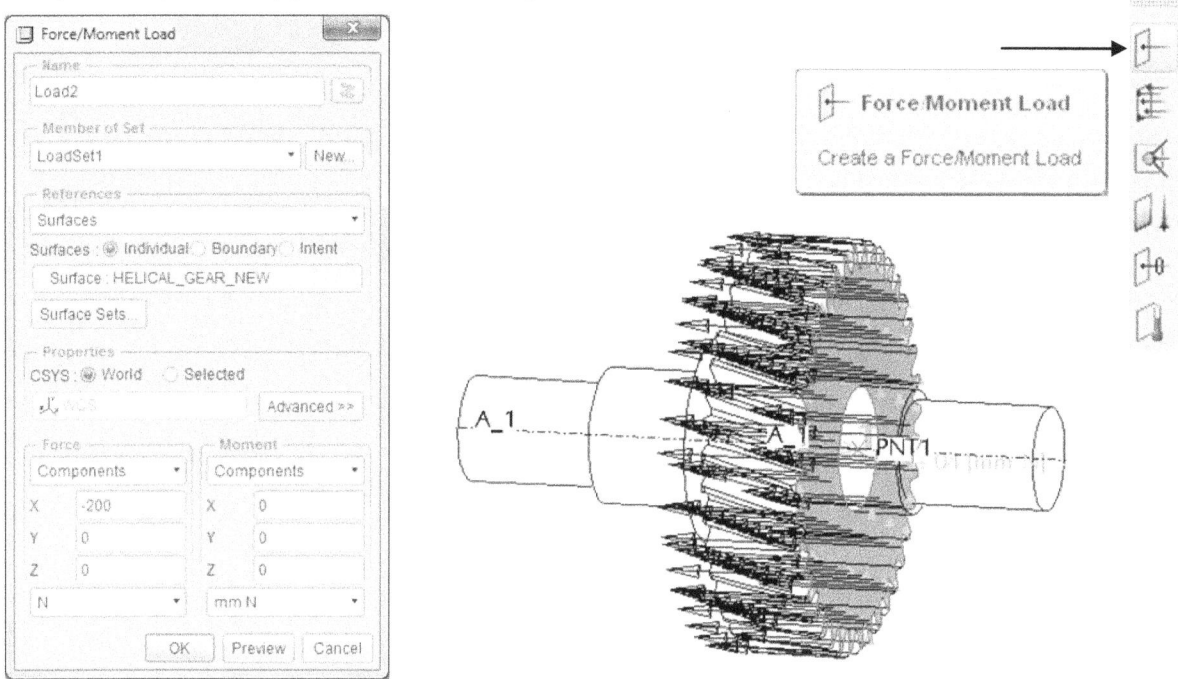

Step 11: Under Pro/MECHANICA, run FEA.
Click the icon of **Run an Analysis > File > New Static >** *type gear_shaft_assembly* as the name of the design study folder > **OK > Run.**

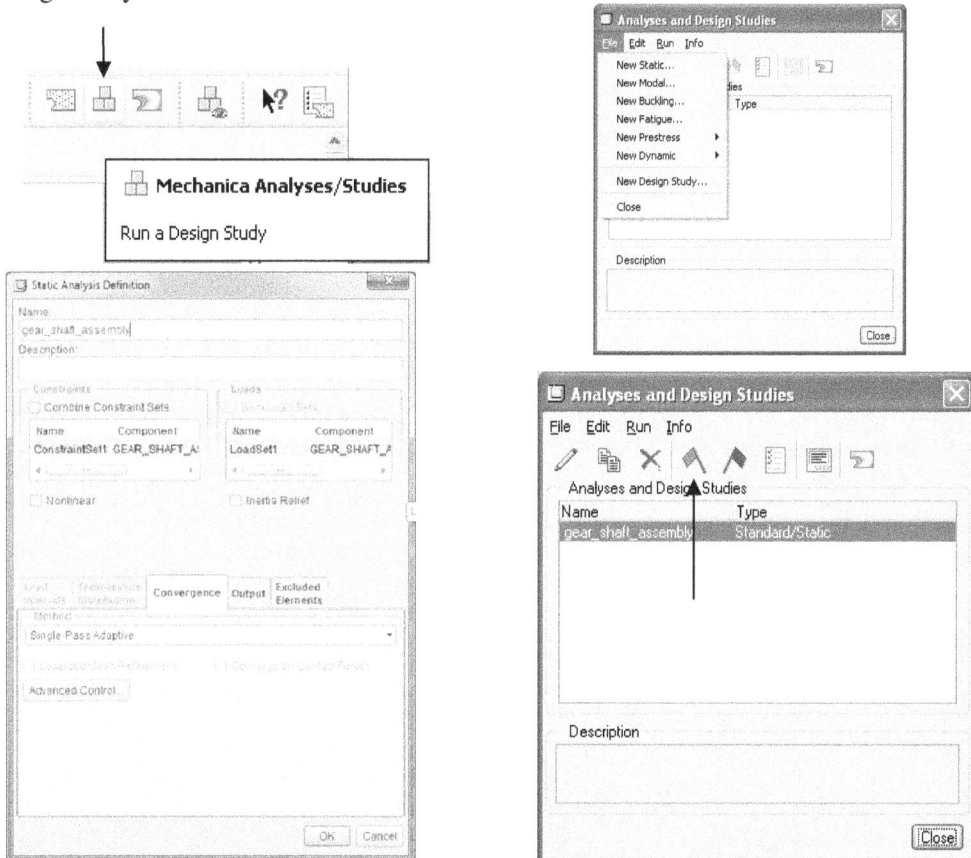

Click the icon of **Results.** Type *gear_shaft_deformation* as the title > locate the Design Study folder called *gear_shaft_assembly.* Select **Fringe** > choose **Displacement** > **OK and Show**.

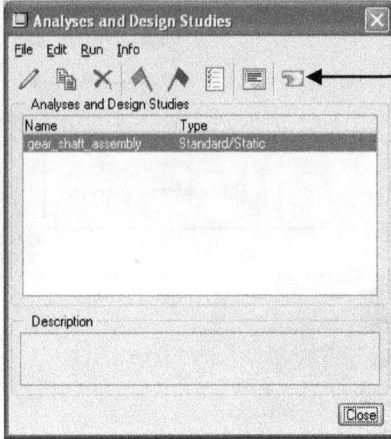

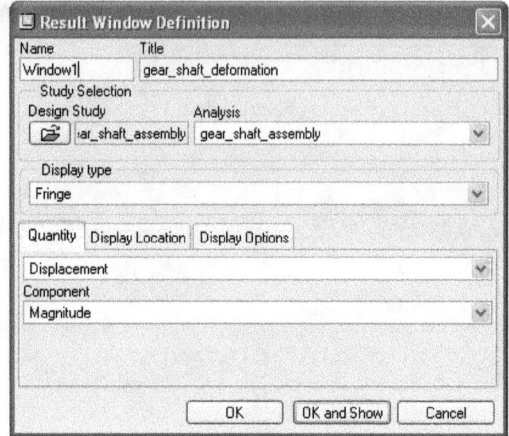

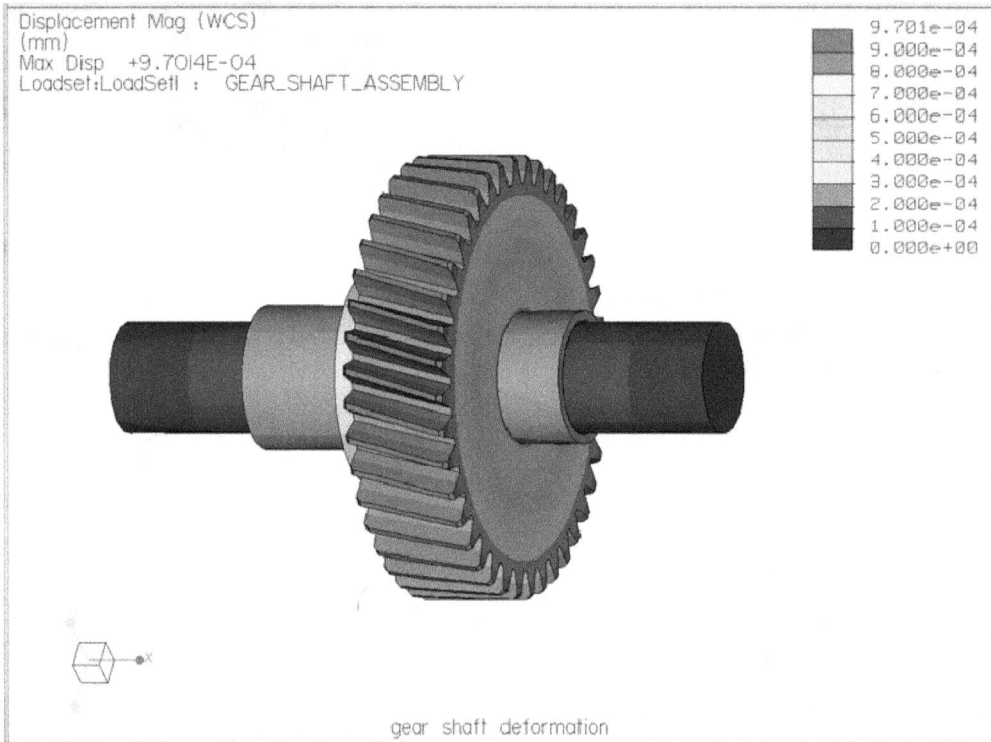

To plot a distribution of stress, from the main toolbar, select **Edit > Copy**. Select **Stress > von Mises > OK and Show.** Users should be able to obtain the plot of the von Mises stress distribution.

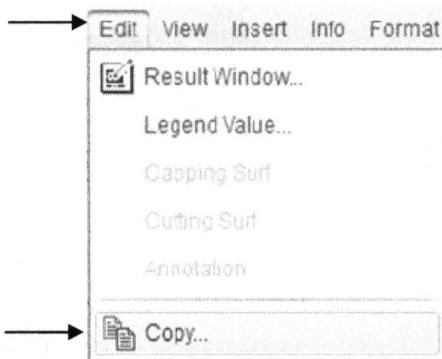

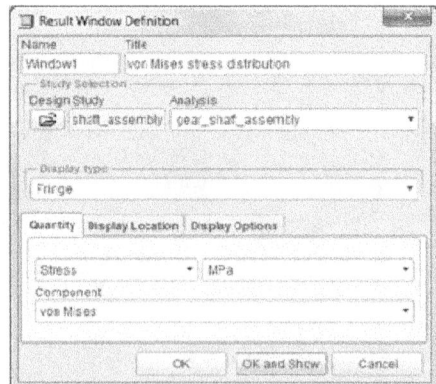

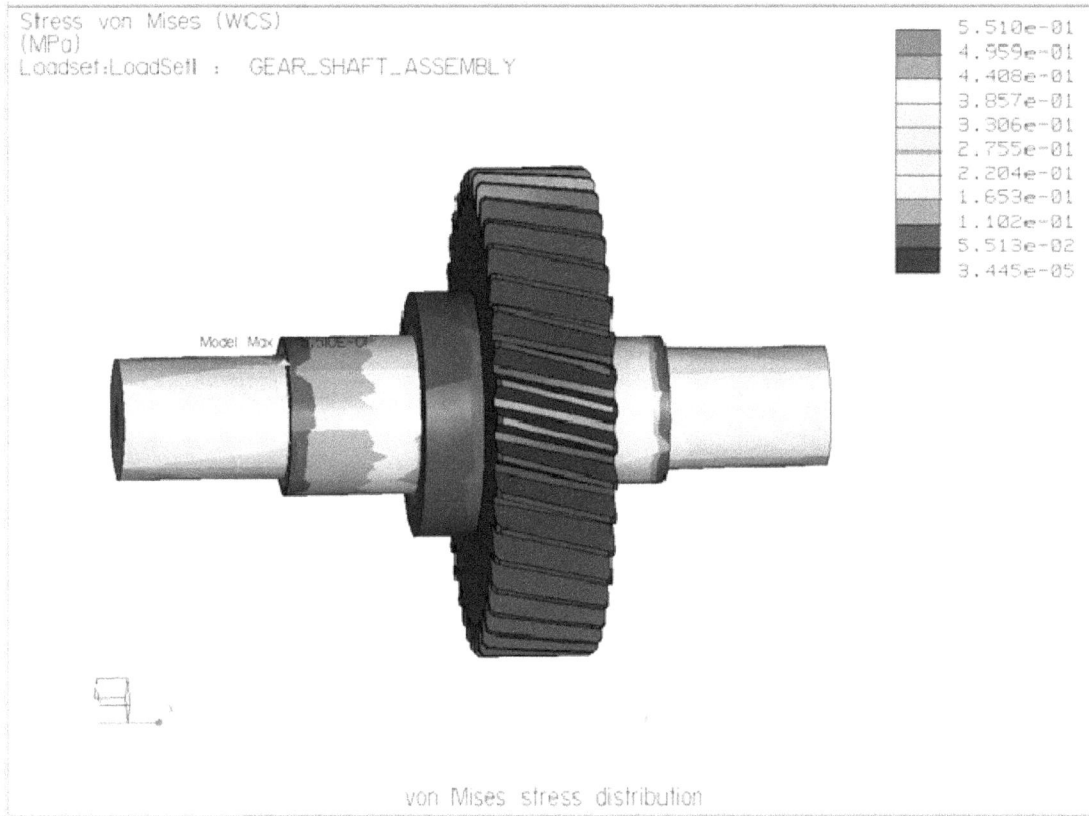

Stress von Mises (WCS)
(MPa)
Loadset:LoadSet1 : GEAR_SHAFT_ASSEMBLY

5.510e-01
4.959e-01
4.408e-01
3.857e-01
3.306e-01
2.755e-01
2.204e-01
1.653e-01
1.102e-01
5.513e-02
3.445e-05

von Mises stress distribution

Users are also able to plot the displacement and/or stress distribution at the component level after running the analysis with the assembly structure. Let us create a new window and have a new definition for the gear component. From the main toolbar, select **Edit > Copy >** type max principal stress of gear > click Display Location > change **All** to **Volume**.

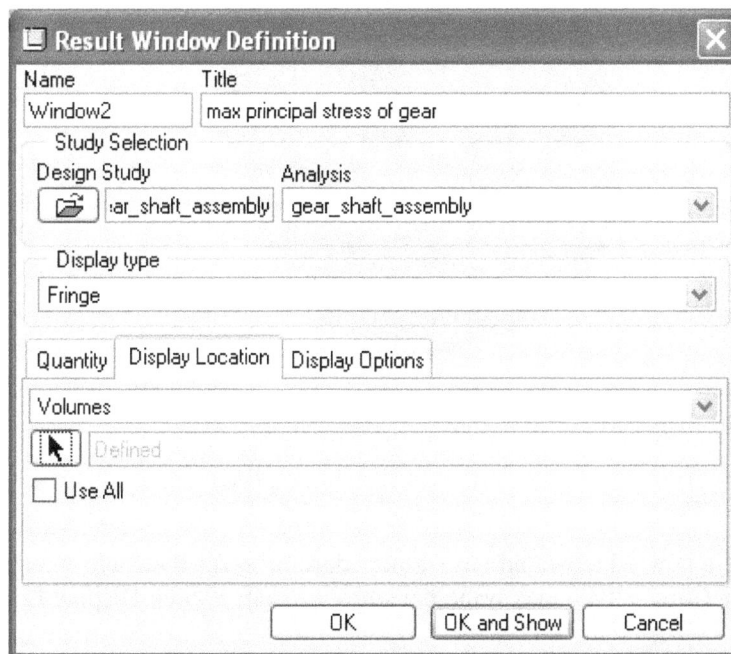

Click the icon of arrow to select the component from the assembly > pick the gear component, which is highlighted after picking > middle button of mouse to accept the selection > **OK and Show**.

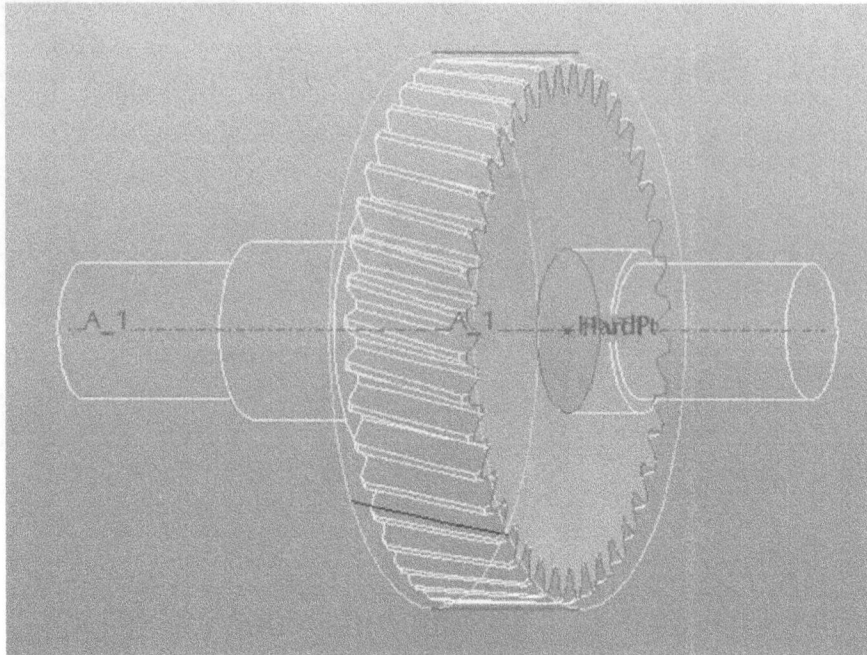

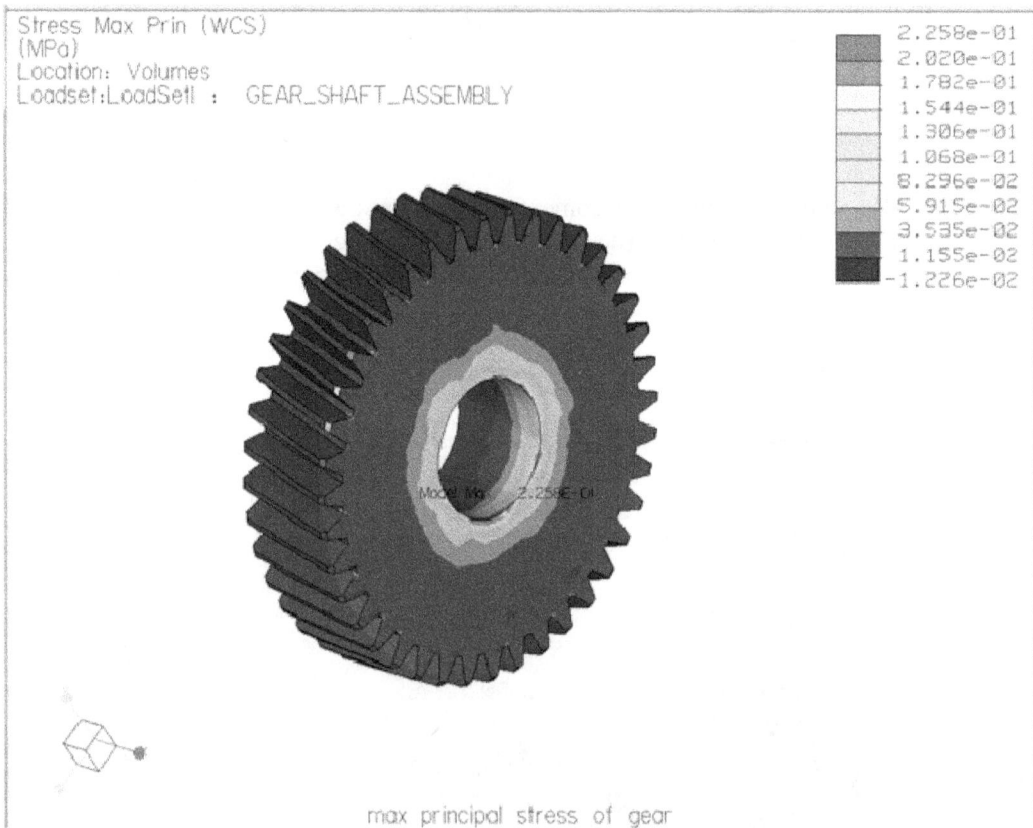

Stress Max Prin (WCS)
(MPa)
Location: Volumes
Loadset:LoadSetl :   GEAR_SHAFT_ASSEMBLY

	2.258e-01
	2.020e-01
	1.782e-01
	1.544e-01
	1.306e-01
	1.068e-01
	8.296e-02
	5.915e-02
	3.535e-02
	1.155e-02
	-1.226e-02

max principal stress of gear

Let us create a new window and have a new definition for the shaft component. From the main toolbar, select **Edit > Copy >** type max principal stress of shaft > click Display Location > change **All** to **Volume**.

Click the icon of arrow to select the component from the assembly > click the gear component first as the de-selection step > pick the shaft component, which is highlighted after picking > middle button of mouse to accept the selection > **OK and Show**.

max principal stress of shaft

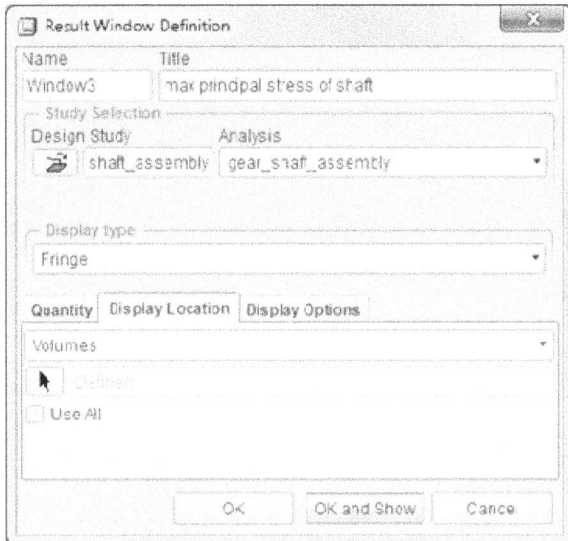

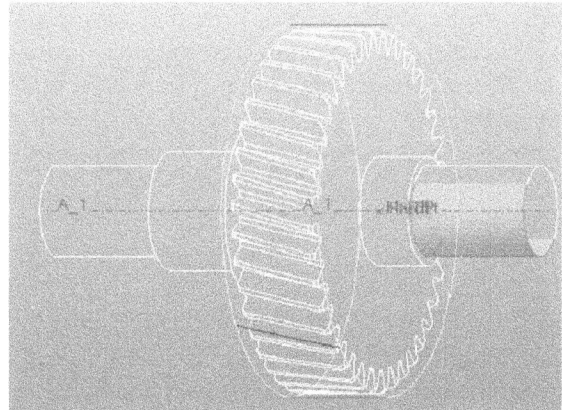

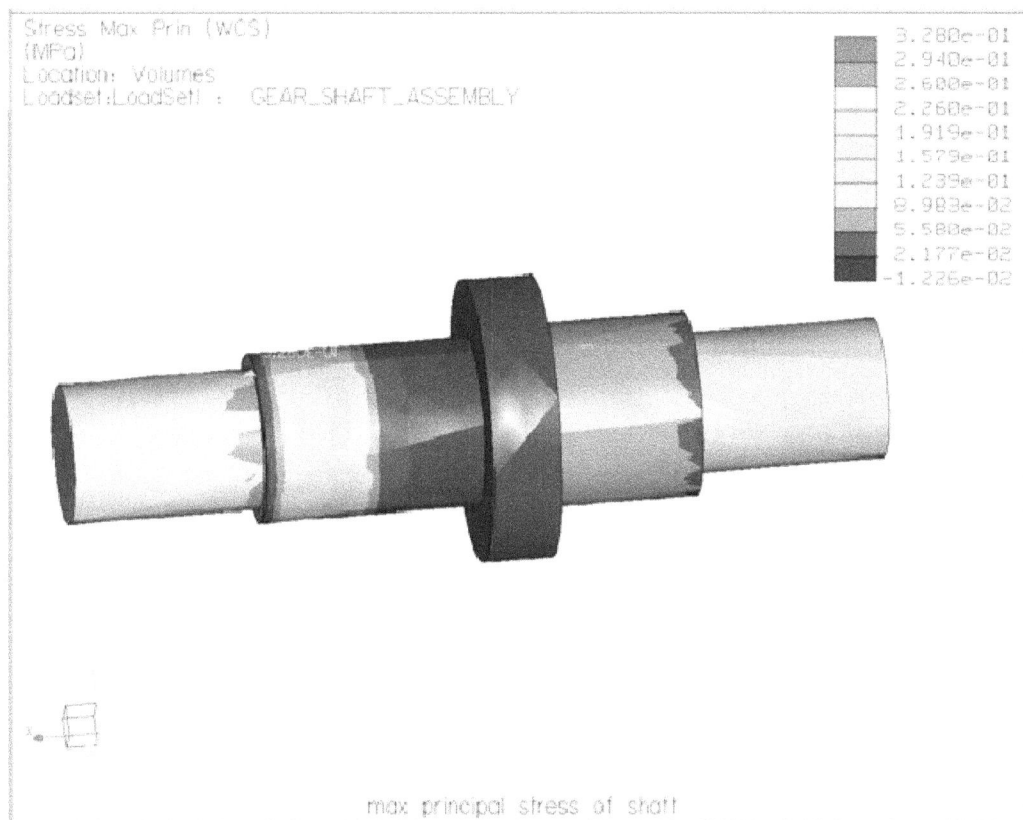

4.8 Modeling of Load Conditions due to Temperature Variation

As illustrated in the previous sections, an assembly consists of several components. The material type of each of the components in the assembly can vary. Some components may be made of steel. Some components may be made of copper. A few components could be made of plastic material. Different types of material demonstrate different characteristics in terms of material properties, such as Young's modulus, Poisson's ratio, coefficient of thermal expansion, etc. The mismatch of material properties in an assembly often leads to strain and stress developments.

For example, a change in the body temperature of a component leads to a change in its dimensions, such as the length and/or width. Such changes in dimension will induce strain and stress. To evaluate these effects quantitatively, the coefficient of thermal expansion plays a critical role. In the following, we present a case study, which deals with an assembly consisting of 2 layers and base substrate. The two layers are the cement layer and the glass layer. The base substrate represents a dentin structure, which is replaced by a type of composite material in dental research. Therefore, this layer structure resembles a crown-cement-dentin system used by the scientists and dentists in dentistry in the research of dental materials.

3	TOP_LAYER	glass
2	MIDDLE_LAYER	cement
1	BASE	composite
INDEX	PART NAME	MATERIAL

Let us examine the geometrical characteristics. The base structure is a rectangular block. Its size is 10 x 10 x 4 mm, as illustrated. The thickness of the cement layer is 0.1 mm. The thickness of the glass layer is 0.9 mm.

The properties of the three types of material are shown below:

	glass	cement	composite
mass density (g/cm^3)	5.0	1.3	1.7
coef. of thermal expansion (/C)	5.00E-06	6.00E-05	3.00E-05
Young's modulus (N/mm^2)	8000	6000	15000
Poisson's ratio	0.27	0.33	0.33

Investigate the stress development and the pattern of deformation when subjected temperature variation.

The temperature variation: reference temperature: 20°C

                                     model temperature: 40°C

For the boundary condition, we assume that the dentin structure is fixed to the ground through a local contact. The surface region is a circle. The diameter is 3 mm. Plot the deformation pattern and distribution of the maximum principal stress when subjected to the temperature variation.

Step 1: Under the Pro/ENGINEER design environment, create each of the three components, namely, cement_layer, glass_top and root.

Glass_top.prt

10 x 10

0.9

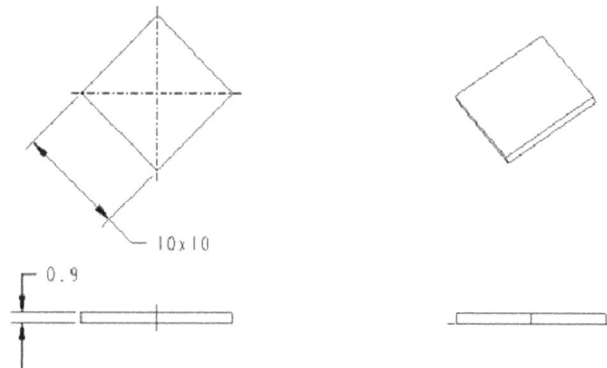

Cement_layer.prt

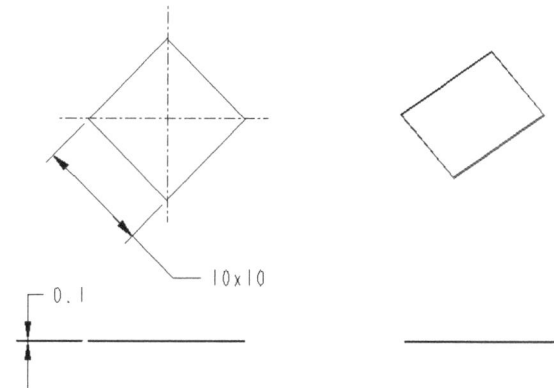

10 x 10

0.1

Root.prt

10 x 10

4

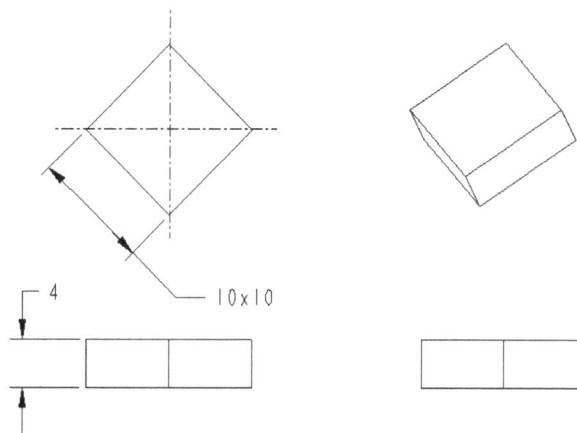

Let us create a 3D solid model for the root component first. From the menu tool bar, select **File > New > Part >** type in *root* as the file name and clear the icon of **Use default template > OK.**
Select **mmns_part_solid** (units: Millimeter, Newton, Second) and type *root* in **DESCRIPTION**, and *student* in **MODELED_BY**, then **OK.**

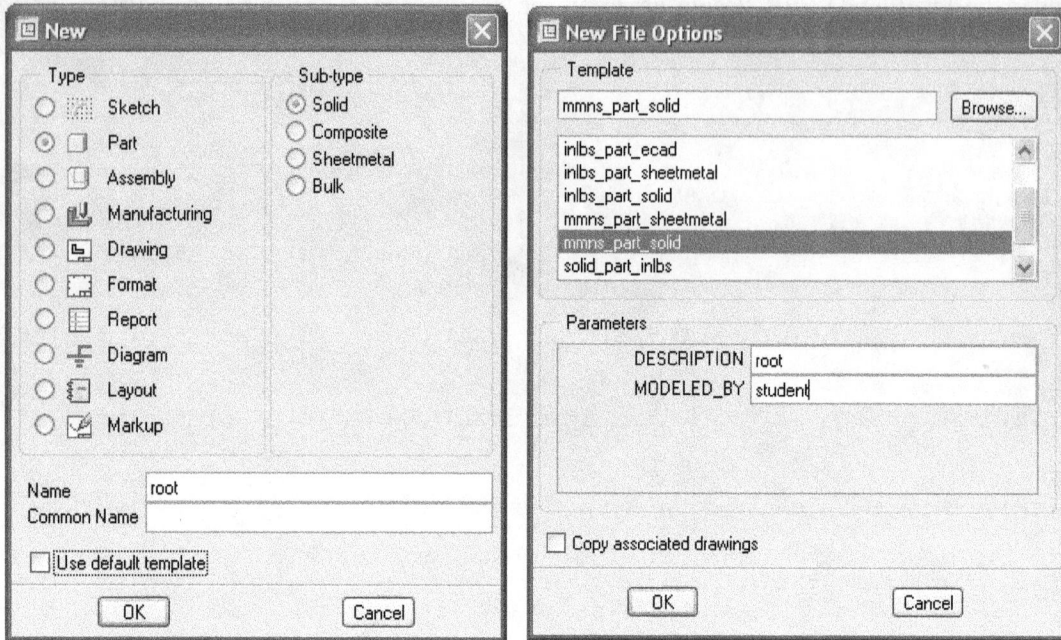

Click the icon of **Extrude.** Specify 4 as the value of height. Click **Placement** > Define to initiate the sketch process.

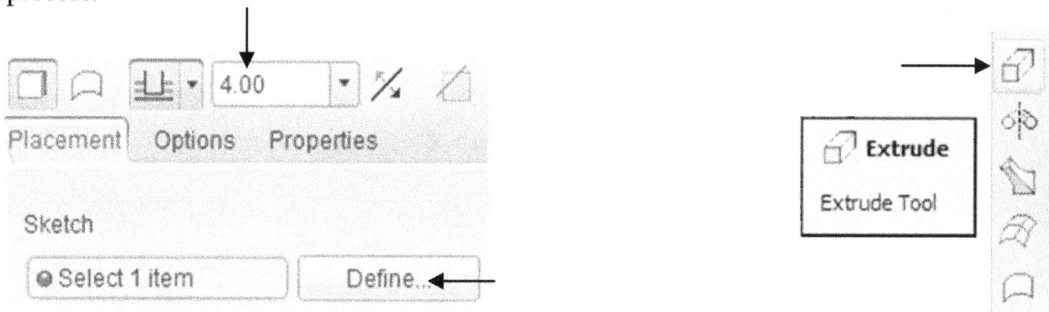

Pick the **TOP** datum plane as the sketch plane, and accept the default orientation setting

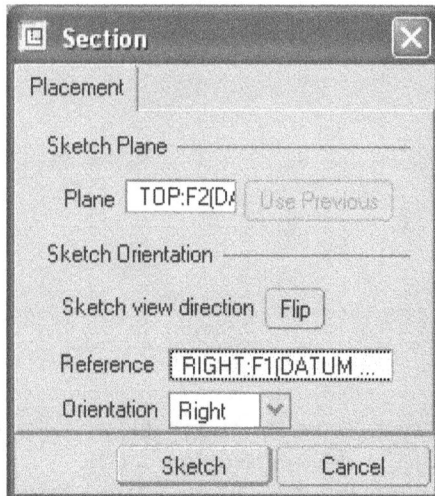

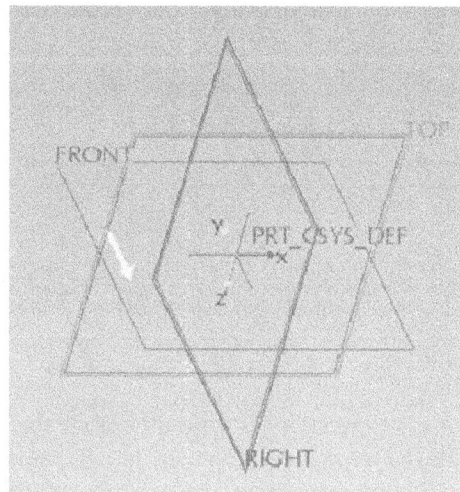

Click the icon of **Centerline**. Let us sketch 2 centerlines.

Create the
horizontal
centerline

Create the
vertical
centerline

Centerline

Create 2 point centerlines.

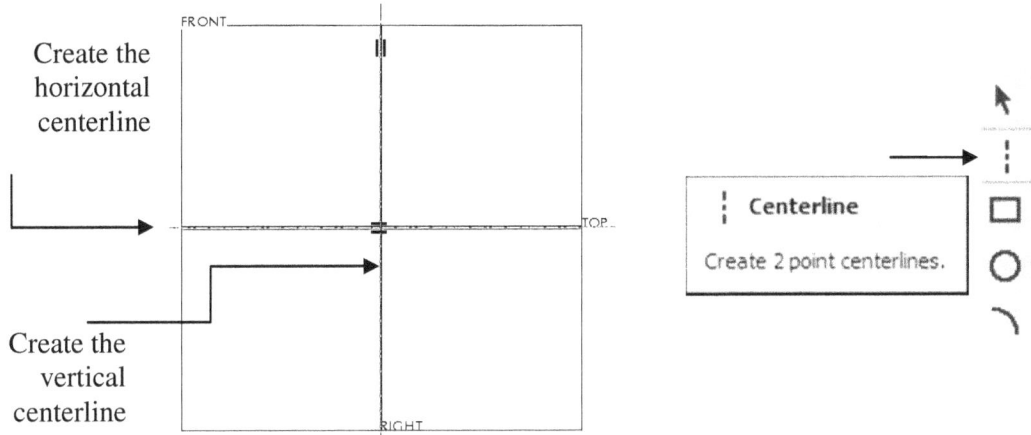

Click the icon of **Line** and sketch a square, as shown. The dimension of each side is 10.

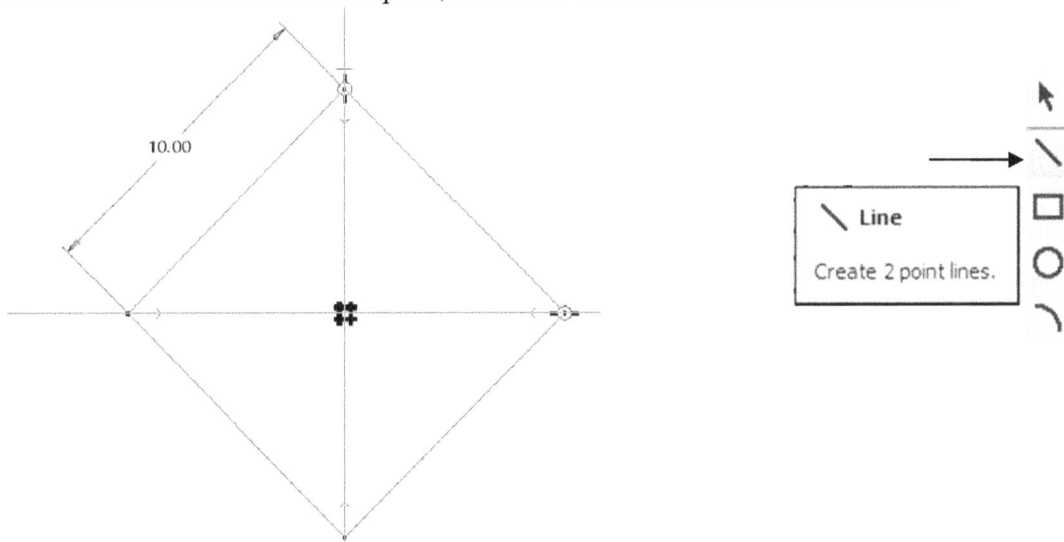

10.00

Line

Create 2 point lines.

Upon completing the sketch, select the icon of **Done** and click the icon of **Apply and Save**.

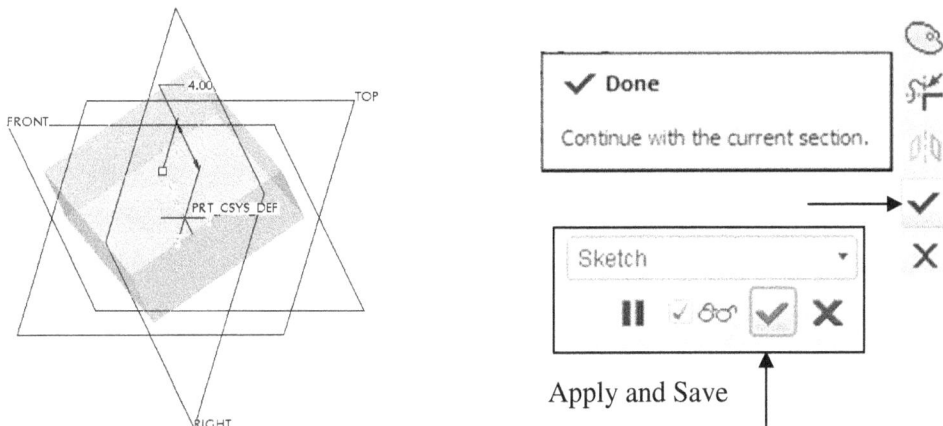

✓ **Done**

Continue with the current section.

Sketch

Apply and Save

Click the icon of **Save**.

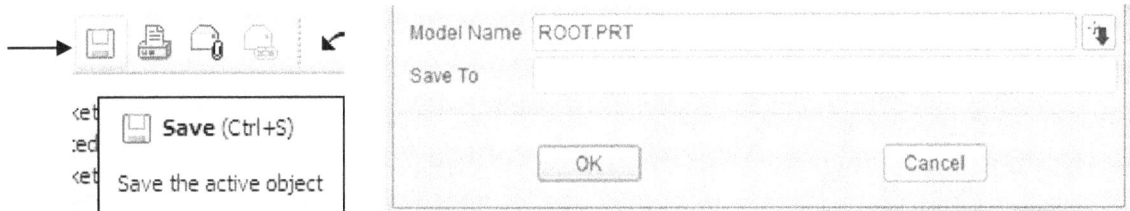

Save (Ctrl+S)

Save the active object

Model Name   ROOT PRT

Save To

OK   Cancel

To create a 3D solid model for the cement_layer component, we make use of the created 3D solid model for the root component. From the main toolbar, select **File > Save a Copy >** type *cement_layer >* **OK**.

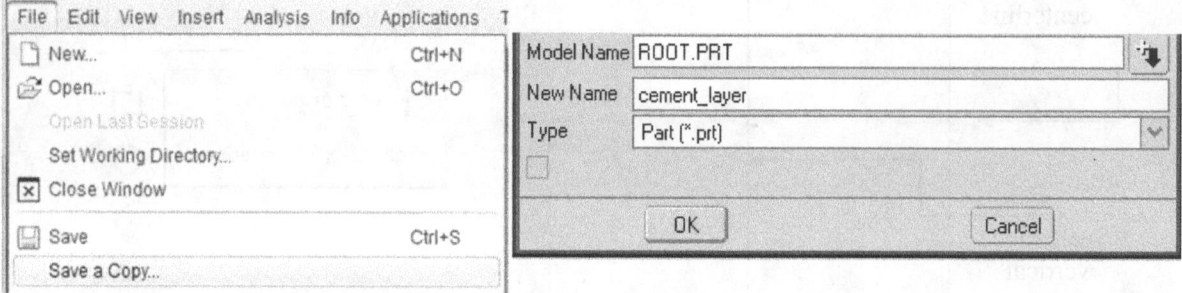

When this process is completed, we open the saved file, called cement_layer.prt. The file should be in the working directory and open it after locating it.

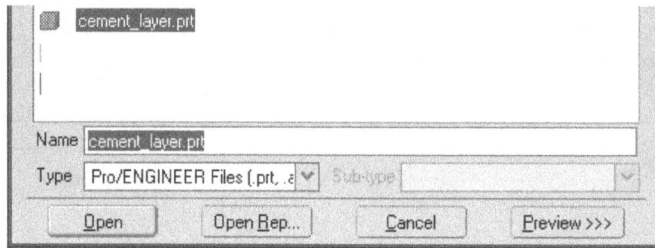

From the model tree, highlight the extrude feature – Extrude 1, right click and hold, pick **Edit Definition >** change the thickness value from 4 to 0.1. Click the icon of **Apply and Save**.

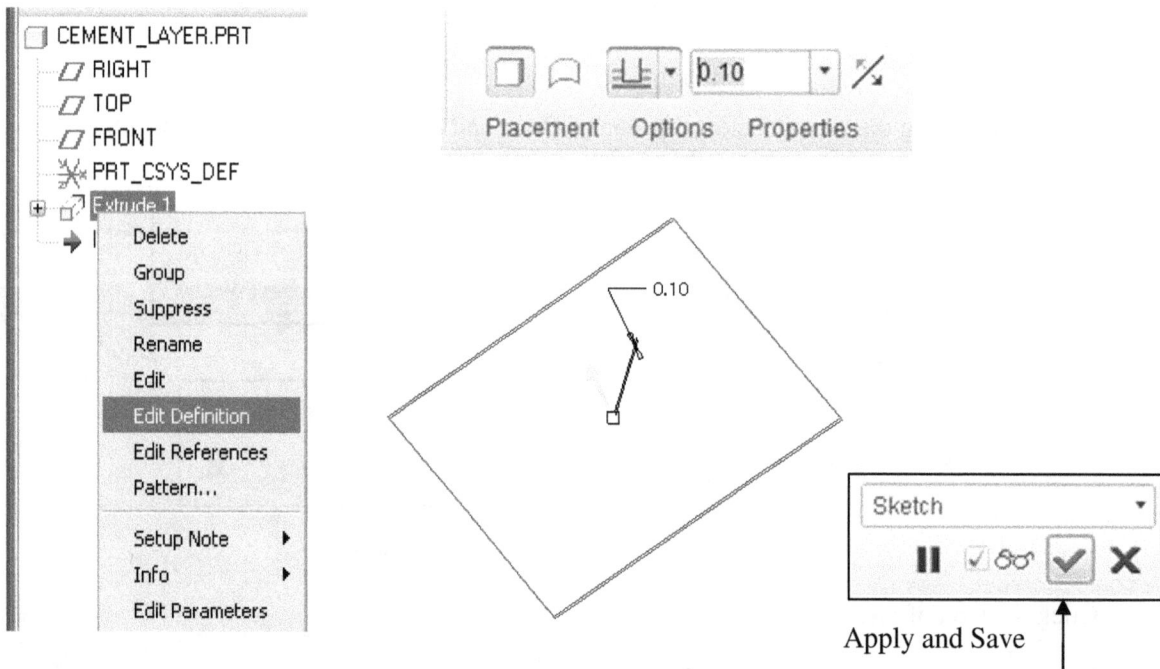

Apply and Save

Following the same procedure, we utilize the 3D solid model for the cement_layer component to create a 3D solid model for the glass_top component. From the menu toolbar, select **File > Save a Copy >** type **glass_top > OK**.

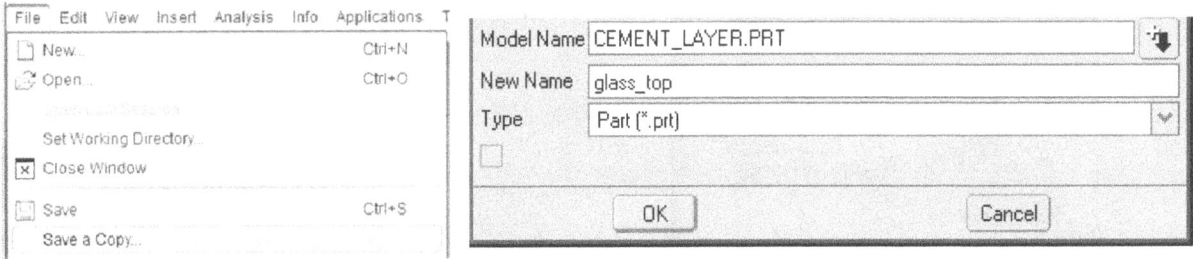

When this process is completed, we open the saved file, called glass_top.prt. The file should be in the working directory and open it after locating it.

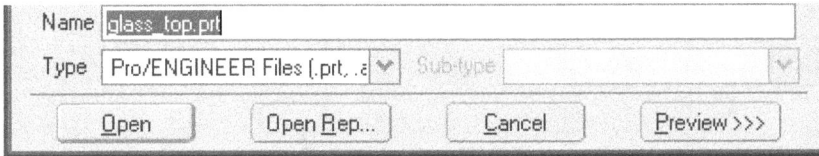

From the model tree, highlight the protrusion feature, right click and pick **Edit Definition** > change the thickness value from 0.1 to 0.9. Click the icon of **Apply and Save**.

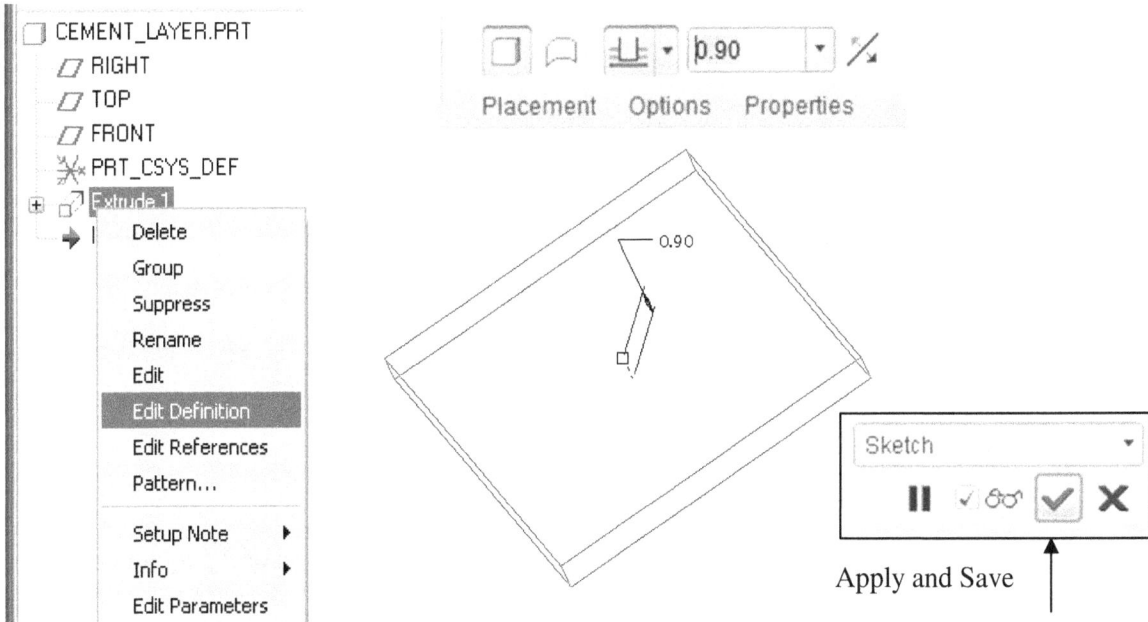

Apply and Save

Step 2: Under the Pro/ENGINEER design environment, open the root.prt and define a datum curve at the bottom surface. Afterwards, under the Pro/Mechanica, define a surface region using the defined datum curve.

Click the icon of sketch tool > pick the bottom surface as the sketch plane and accept the default setting to orient the sketch plane. The default setting is using the RIGHT datum plane and positioning it at the right side > **Sketch**.

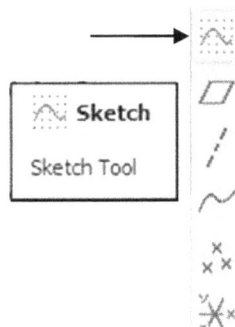

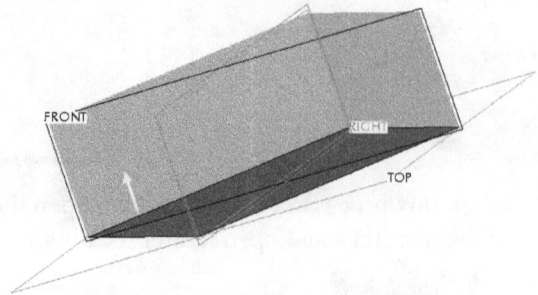

Click the icon of **Circle** and sketch a circle. The diameter dimension is 5, as shown.

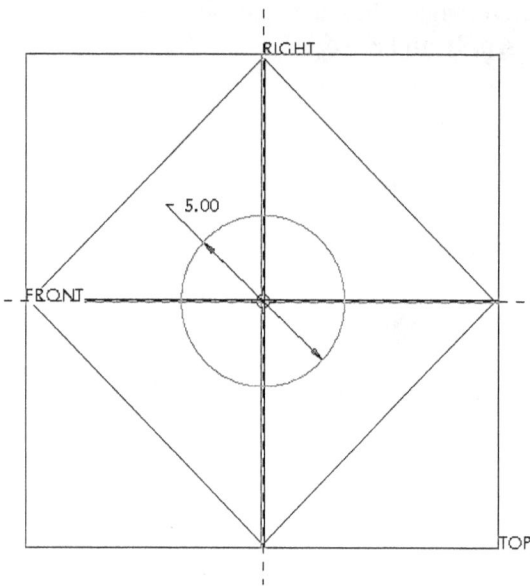

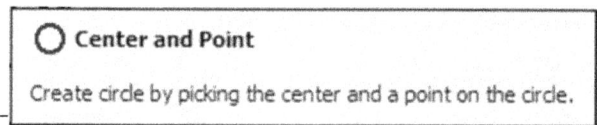

Upon completing the sketch, click the icon of **Done** and click the icon of **Apply and Save**.

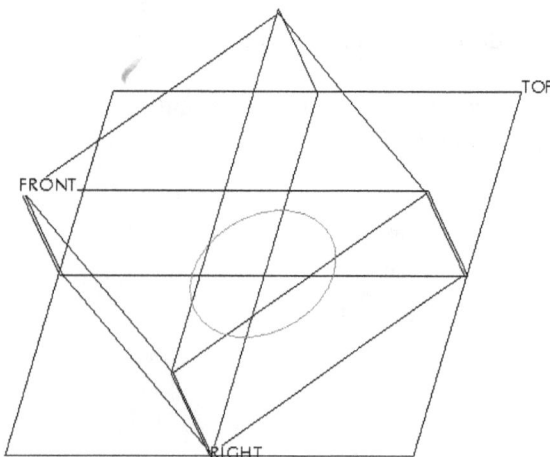

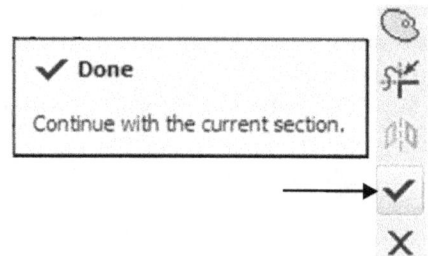

Apply and Save

Under the **Pro/MECHANICA** environment, define a surface region using the defined datum curve.

From the main toolbar, select **Applications > Mechanica**

Click **Continue** after checking the unit system > **Structure > OK**.

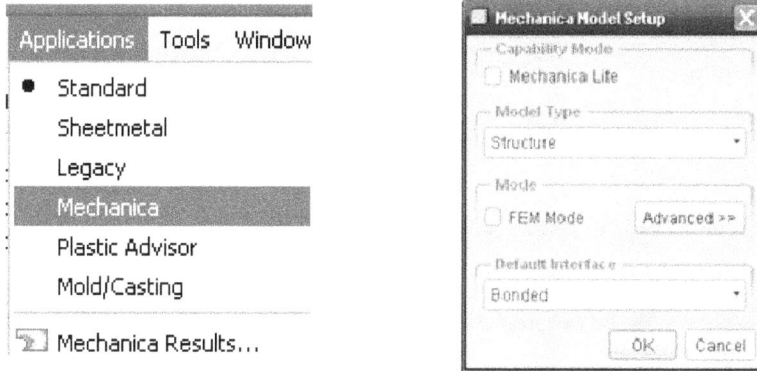

To define the surface region, click the icon of **Surface Region.**

Click the icon of **Surface Region**. Pick the sketched circle first. Afterwards, pick the flat surface for splitting. Click the icon of **Apply and Save**.

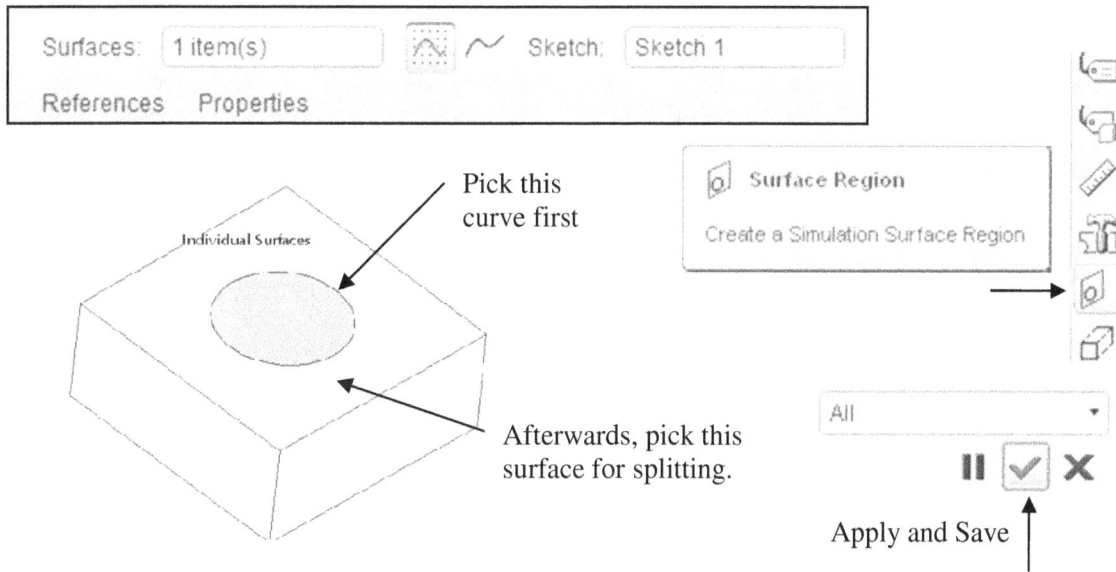

Step 3: Assemble the three components to define a three-layer structure.

From the menu tool bar, select **File > New > Assembly >** type *flat_layers* as the file name and clear the icon of **Use default template > OK.**

Select **mmns_asm_design** (units: Millimeter, Newton, Second) and type *flat layer structure* in **DESCRIPTION**, and *student* in **MODELED_BY**, then **OK.**

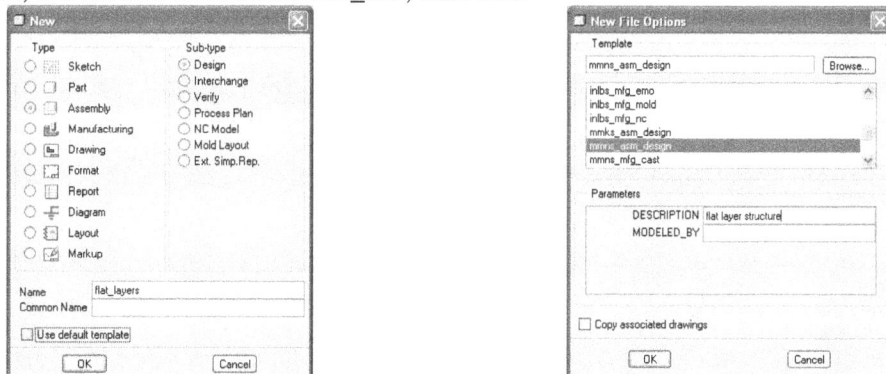

Click the icon of **Add component**.

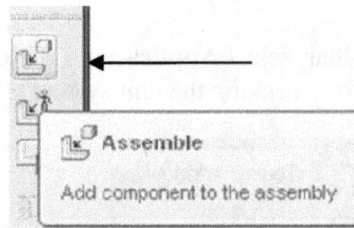

Make sure that **In Session** is activated. Click the icon of **Add component**. Highlight the component called Shaft_for_root.prt > **Open**.

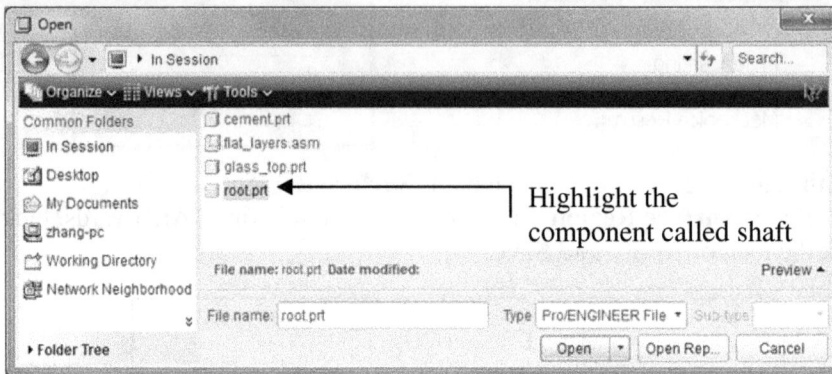

Highlight the component called shaft

From the assembly window, select **Default**, meaning that the coordinate system of the root component is assembled at the location of the assembly coordinate system. Afterwards, click **OK** because the root component is fully constrained.

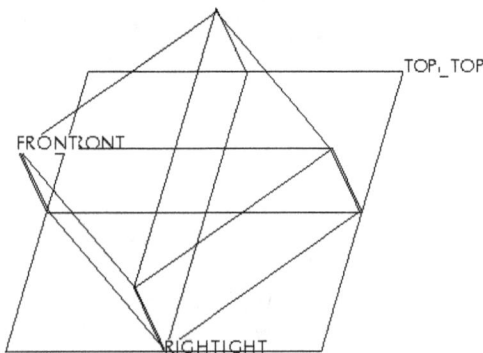

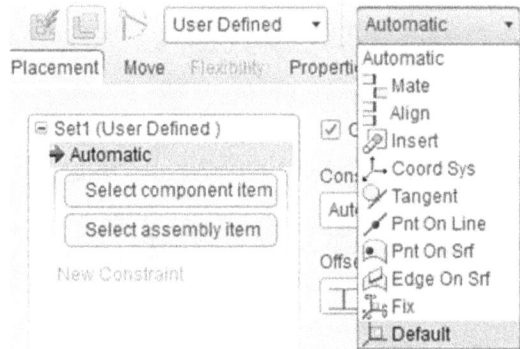

Click the icon of **Add component**. Make sure that you are at **In Session**. Highlight the component called cement.prt > **Open**.

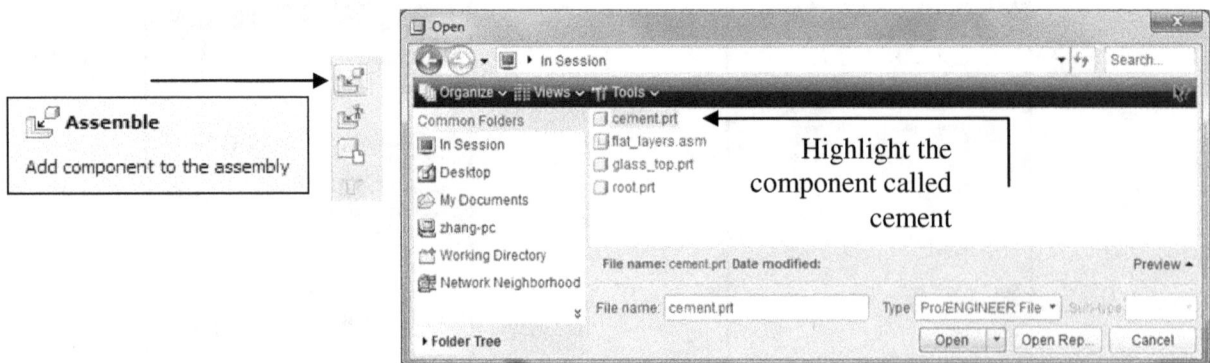

Highlight the component called cement

Pick the top surface of the root component to be in contact with the bottom surface of the cement component. A mate constraint is defined.

Pick this
surface

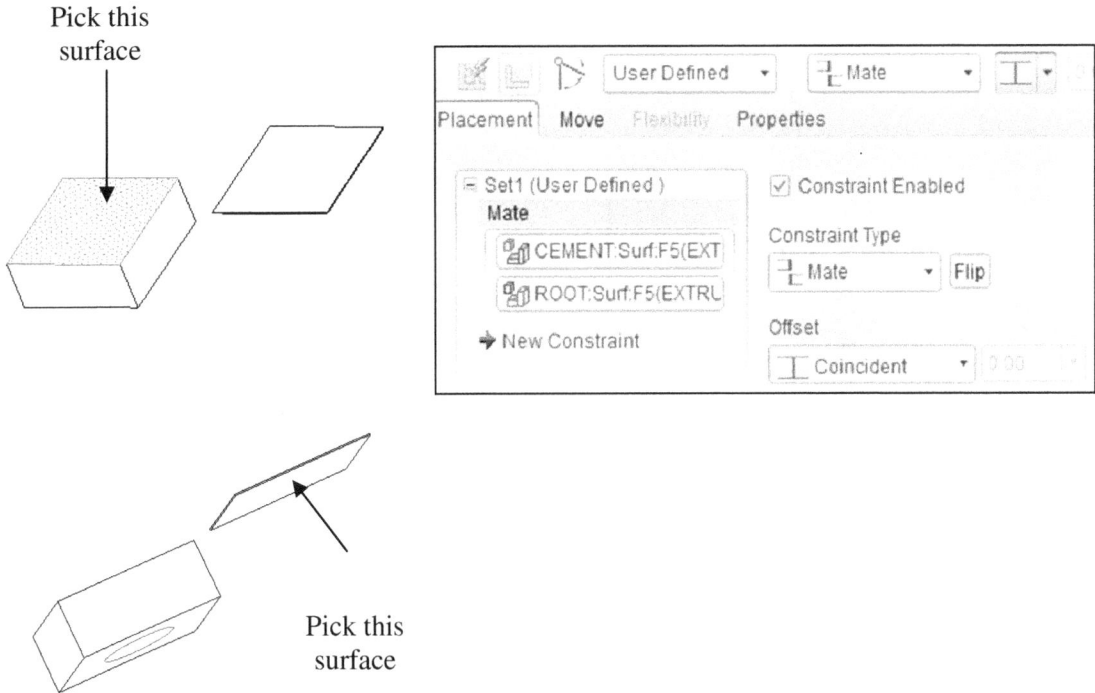

Pick this
surface

Pick the side surface from the root component and pick the side surface from the cement component. An alignment constraint is defined.

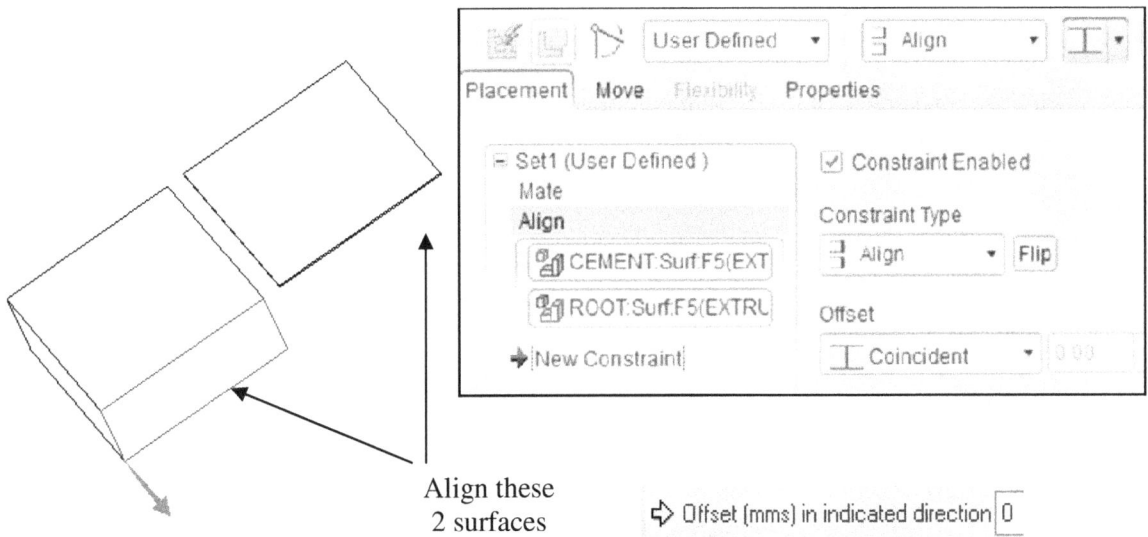

Align these
2 surfaces

⇨ Offset (mms) in indicated direction 0

Pick the other side surface from the root component and pick the other side surface of the cement component. An alignment constraint is defined. The cement component is fully constrained.

Align these
2 surfaces

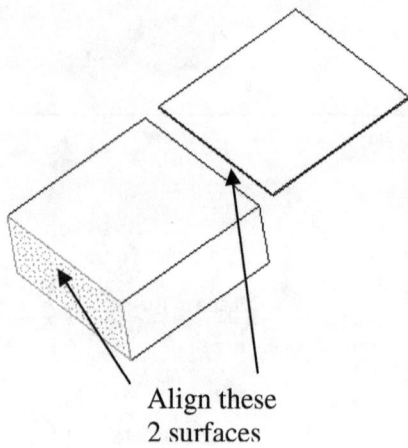

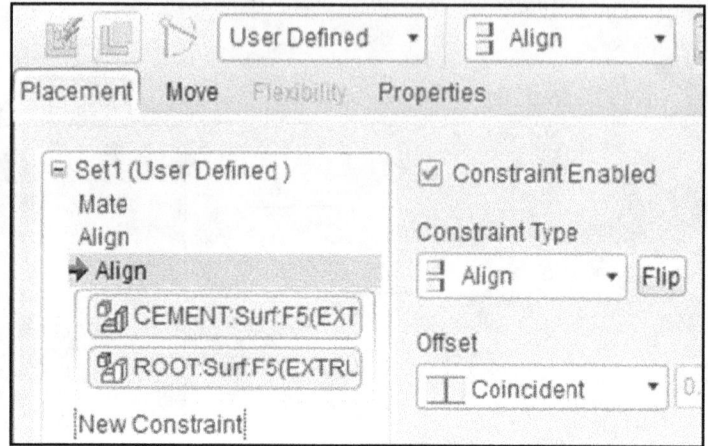

⇨ Offset (mms) in indicated direction [0]

To assemble the glass_top component, click the icon of Adding Component. From **In Session**, select *glass_top.prt* > **Open**.

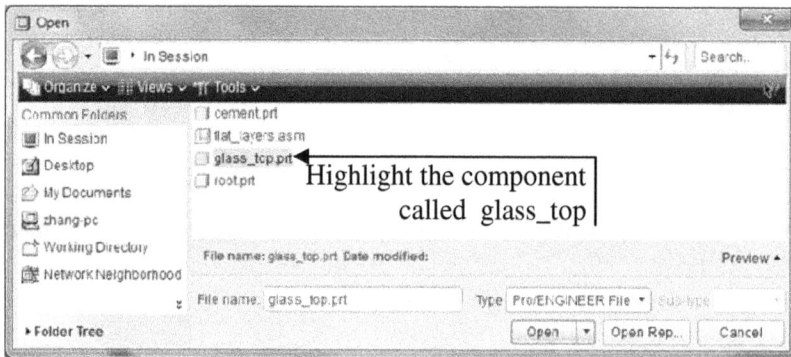

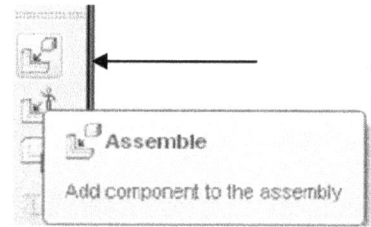

Highlight the component called glass_top

Pick the top surface of the cement component to be in contact with the bottom surface of the glass top component. A mate constraint is defined.

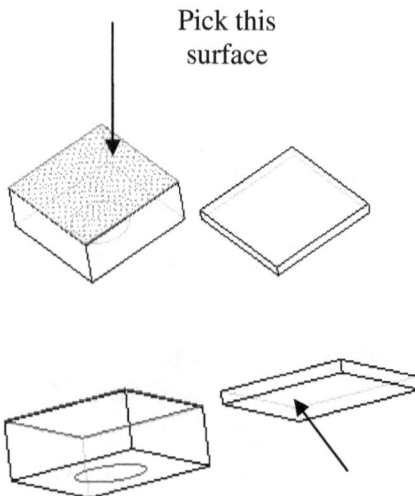

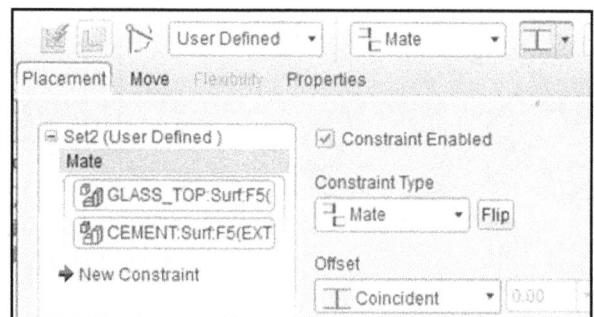

Pick this
surface

Pick this
surface

Pick the side surface from the cement component and pick the side surface from the glass top component. An alignment constraint is defined.

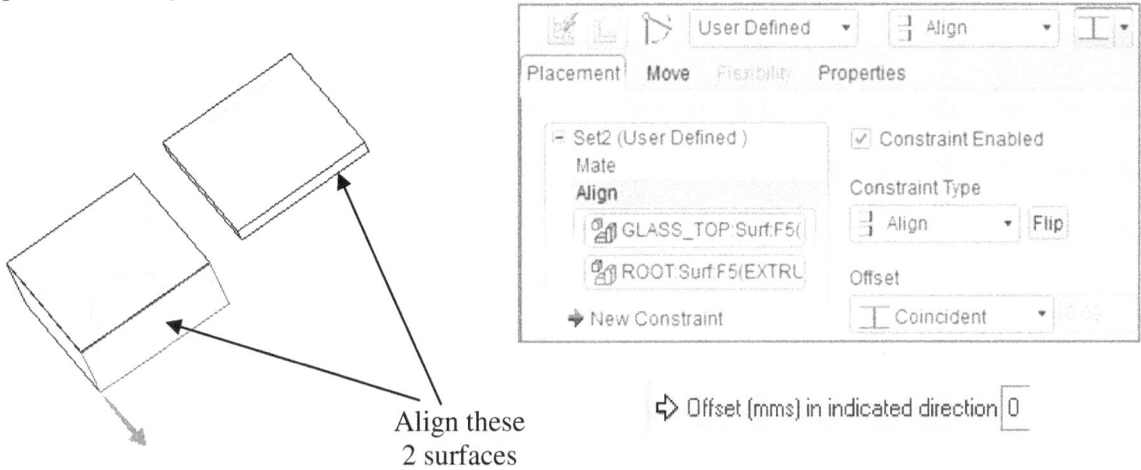

Align these
2 surfaces

Pick the other side surface from the root component and pick the other side surface of the cement component. An alignment constraint is defined. The cement component is fully constrained.

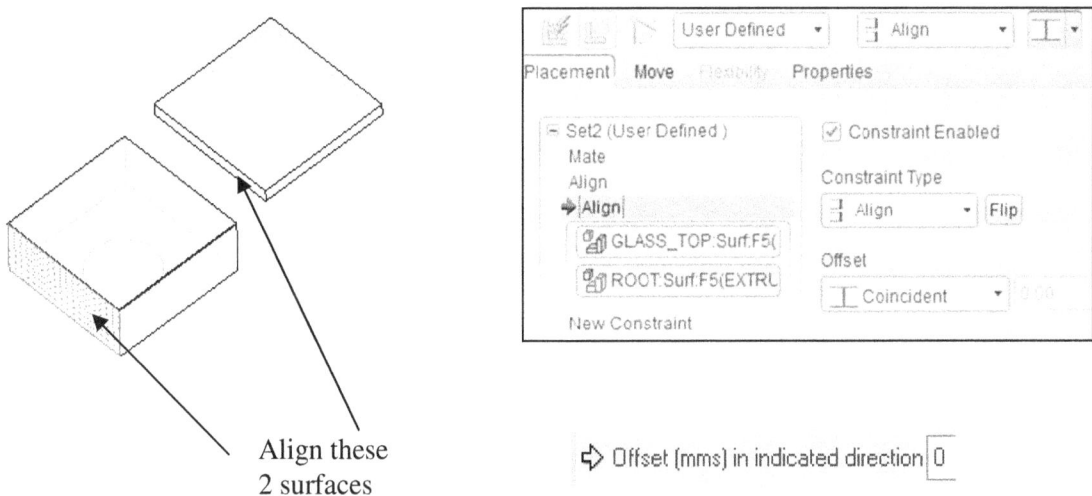

Align these
2 surfaces

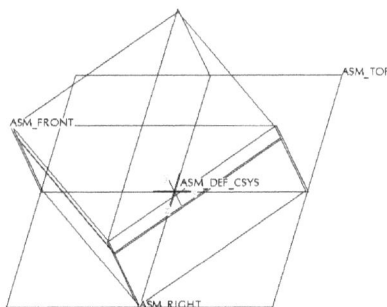

Step 4: Under the **Pro/MECHANICA** environment, specify the three different types of material, and then assign each of the material type to its corresponding component.
From the main toolbar, select **Applications > Mechanica**
Applications > Mechanica > Continue > Structure > OK.

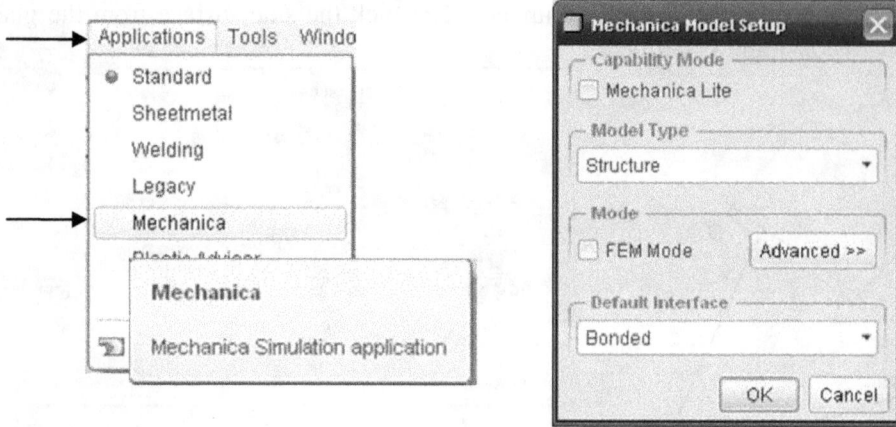

From the toolbar of functions, select the icon of **Materials > New.**

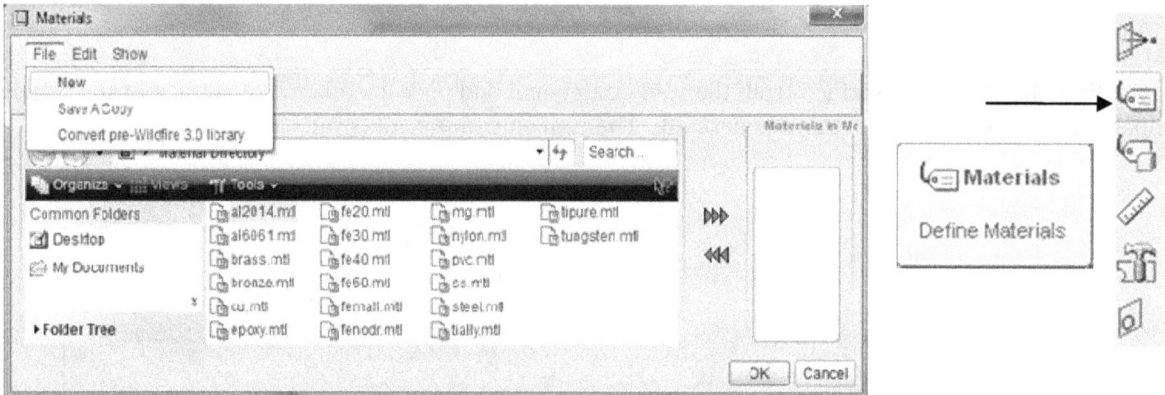

Specify glass as the name of new material. Change the unit of **Density** to g/cm3, and specify 5 as the value. Click **Structural**, and material properties as shown below > click Save to Model, thus completing the process of defining the material properties for GLASS.

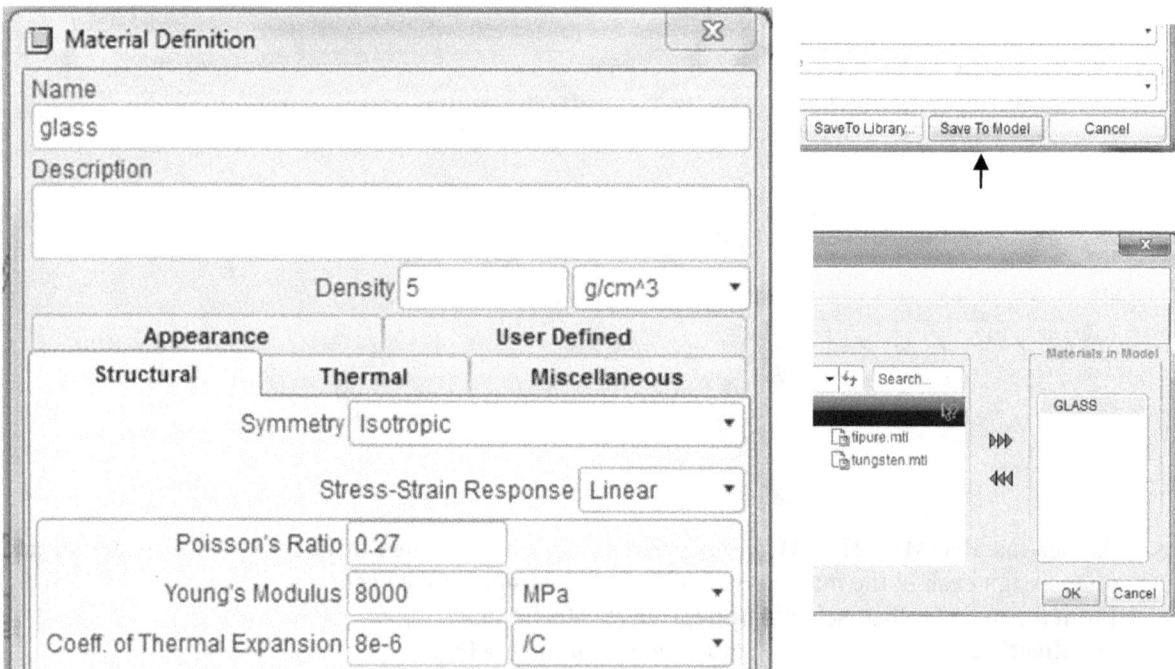

Repeat the above steps to define the material properties for the cement and the root components. Start from clicking **New**, and fill in the 2 tables of material properties as follows for the cement and root layers.

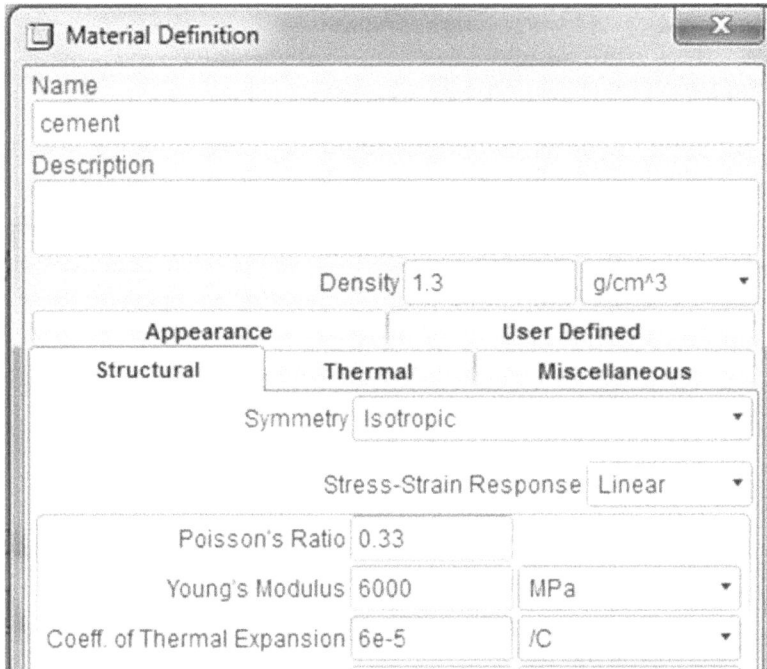

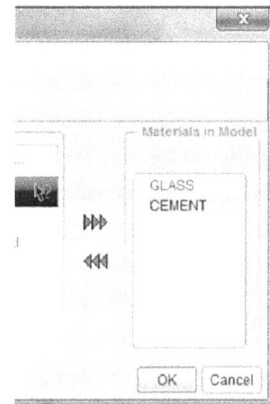

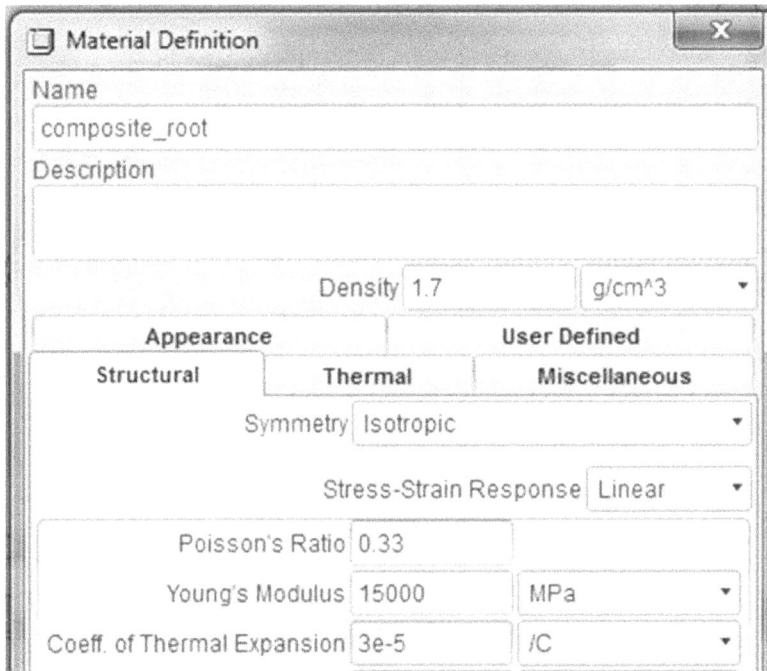

Click the icon of **Material Assignment.** In the Material Assignment window, select GLASS as **Material**, and select the glass_top component on display or from the model tree > **OK**.

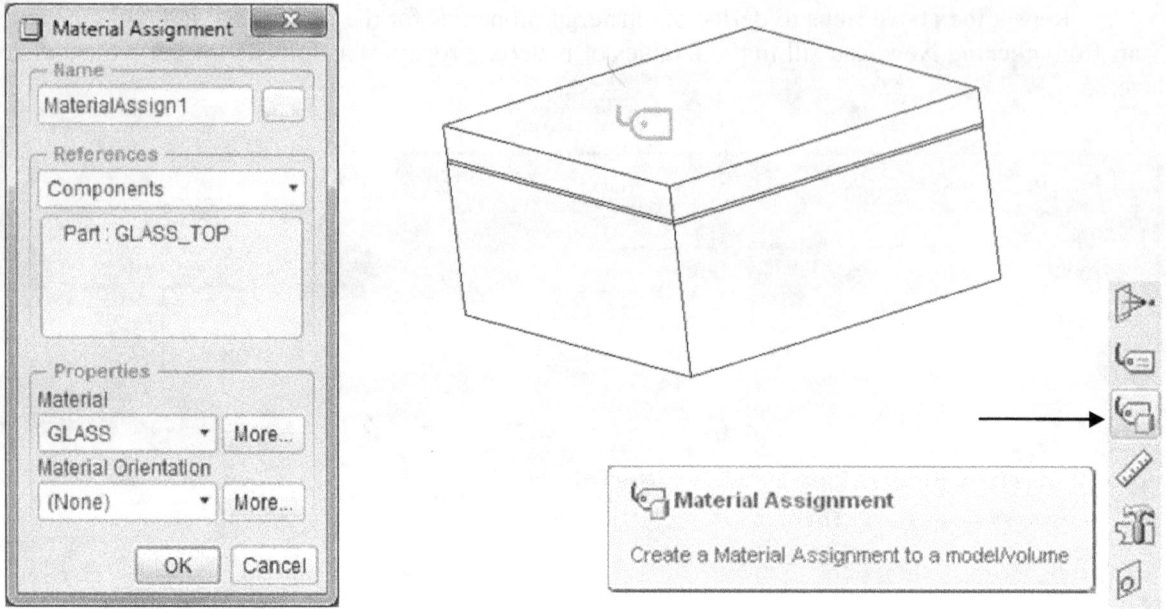

Repeat the above steps to assign the material properties for the cement and root components, as shown below.

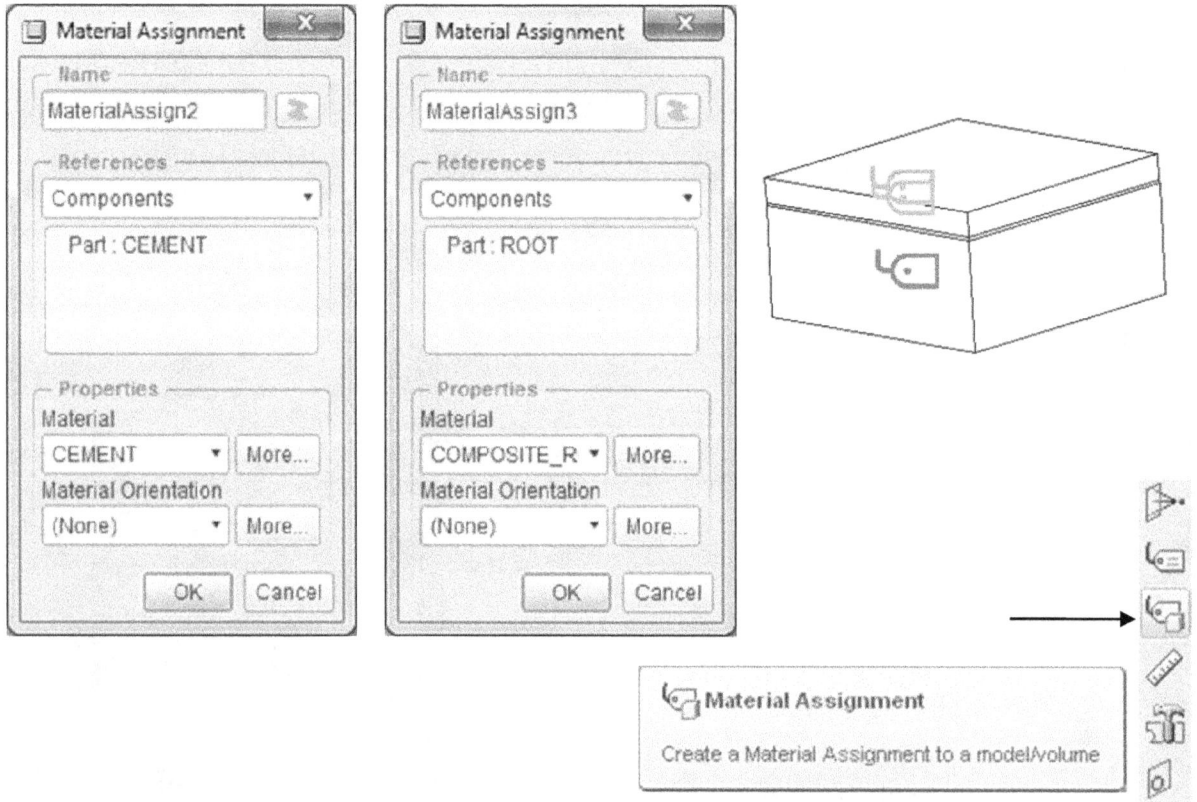

Step 5: Under the **Pro/MECHANICA** environment, define the constraint condition.

Click the icon of **Constraints > ** select the surface region defined at the bottom surface > **OK >** accept the default settings, namely, fixing all 6 degrees of freedom > **OK**.

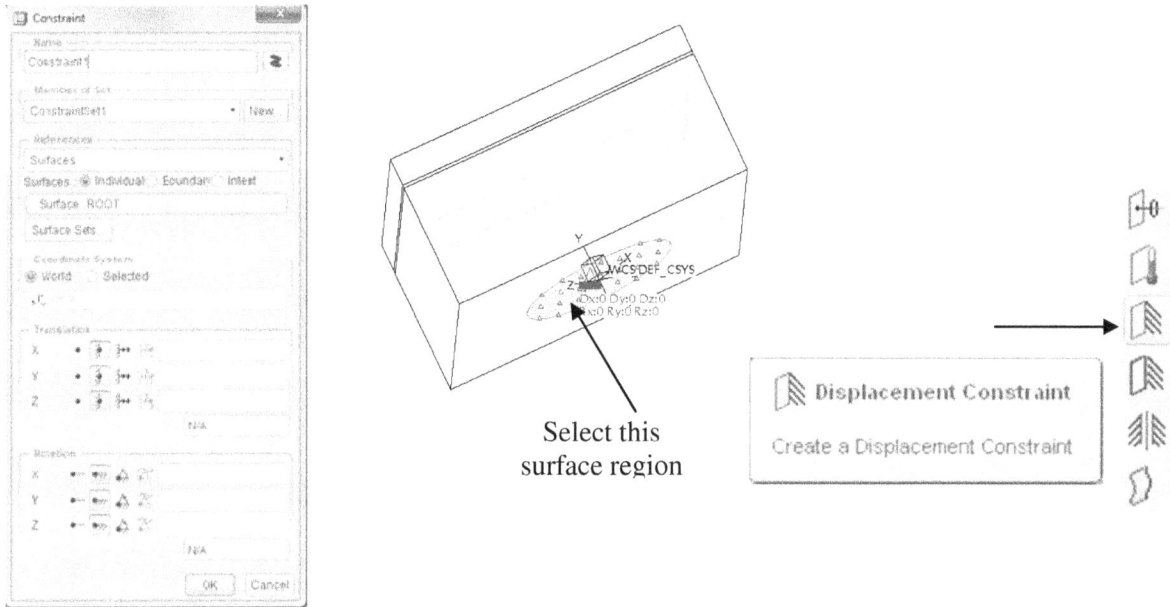

Select this
surface region

Step 6: Under the Pro/MECHANICA environment, define the temperature load condition.

The temperature variation:

reference temperature:	20°C
model temperature:	40°C

Click the icon of **Global Temp Load >** type 40 in the box of Model Temperature and 20 in the Reference Temperature > **OK**.

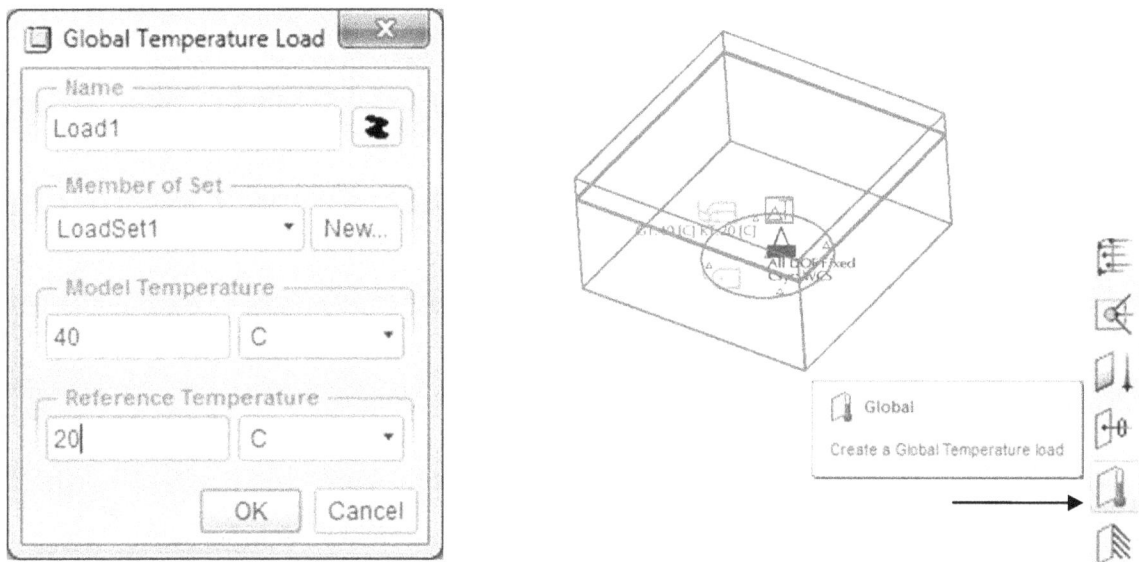

Step 7: Perform a new static analysis called flat_layers_1.
Click the icon of **Run an Analysis** > **File** > **New Static** > *type* type *flat_layers_1* as the name of the design study folder. Make sure that ConstraintSet1 and LoadSet1 are highlighted > **Single-Pass Adaptive** > **OK** > **Run.**

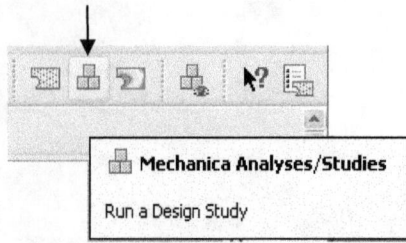

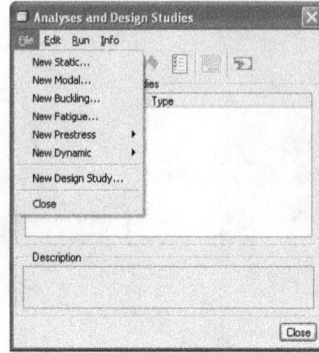

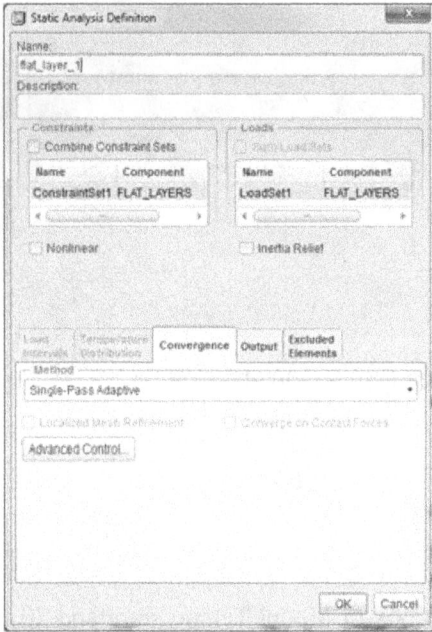

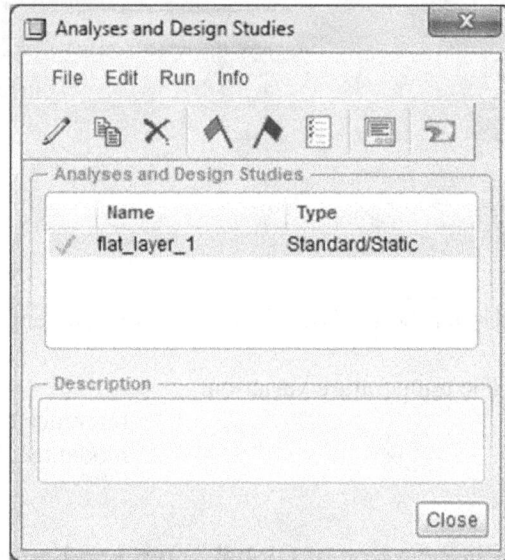

Click the icon of **Results.** Type *deflection distribution (assembly)* as the title and select the folder called flat_layers_1 > **Fringe > Displacement > Magnitude > OK and Show.**

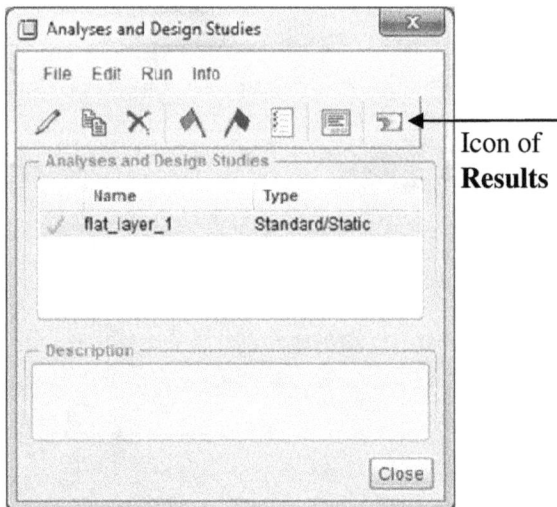

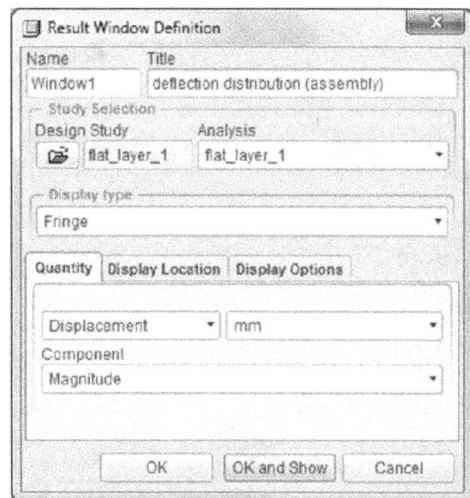

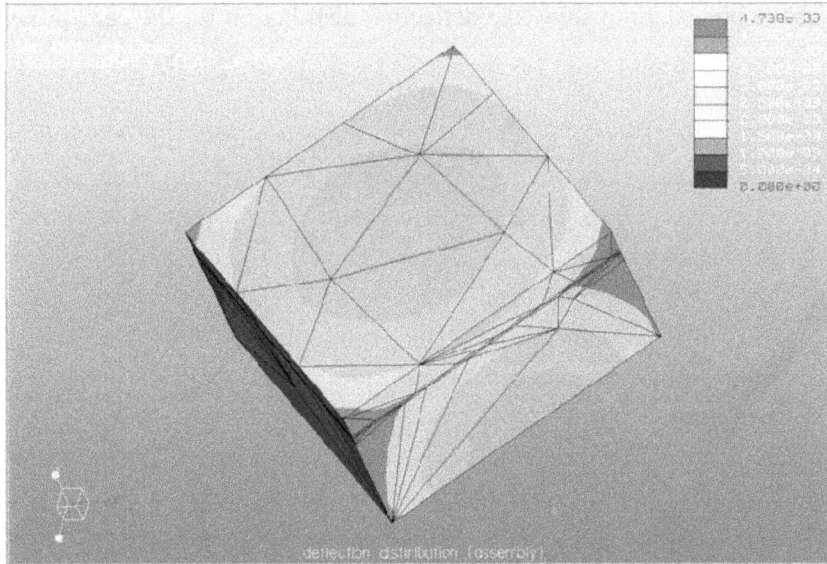

To show the deflection distribution for each of the three components, from the menu bar, select **Edit > Copy** > type *deflection distribution (glass)* > **Fringe > Displacement > Magnitude** > activate **Display Location** > select **Volume** > activate the selection and pick the glass_top component on display, and middle button > **OK and Show**.

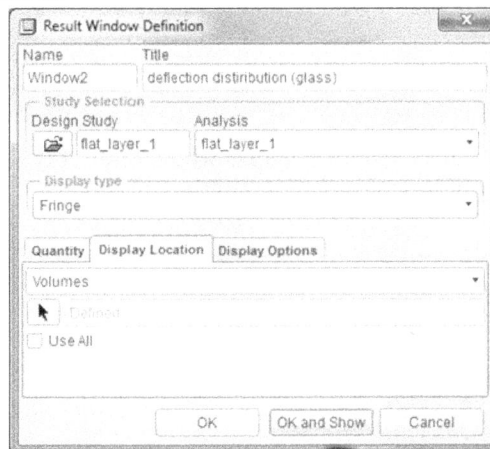

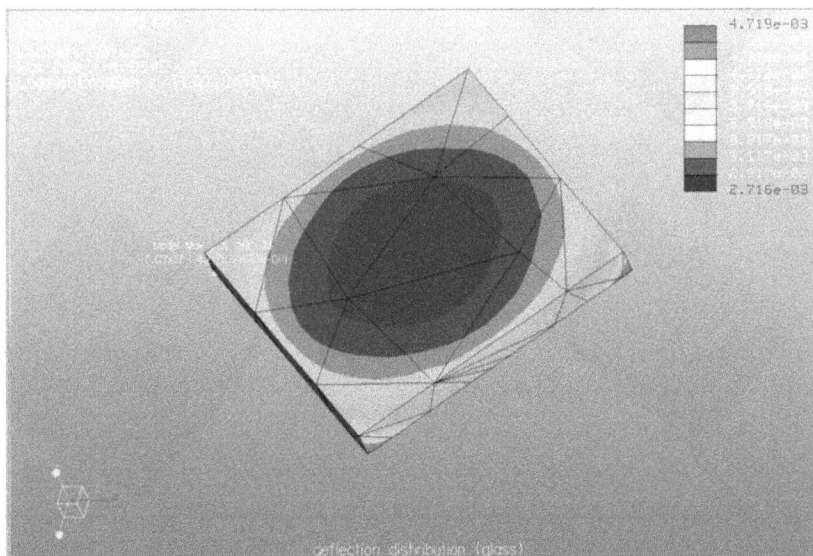

Repeat the above procedure to show the deflection distribution for the cement layer and the root component.

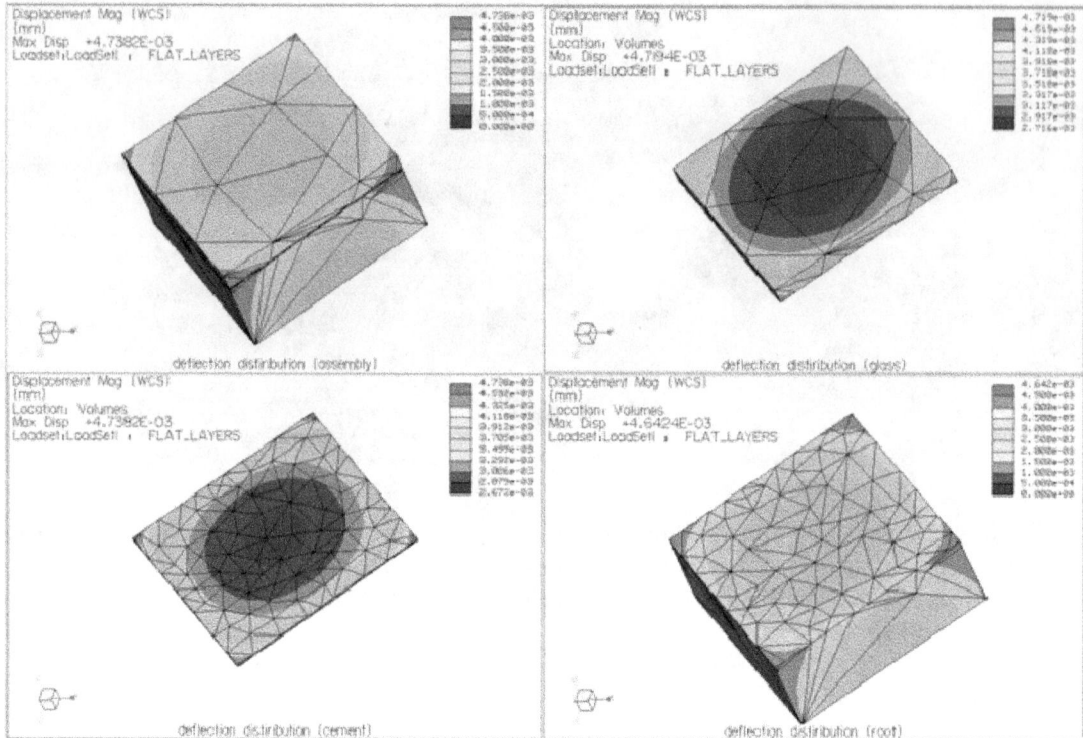

Repeat the above procedure to obtain 4 plots, which show the 4 distributions of the maximum principal stress.

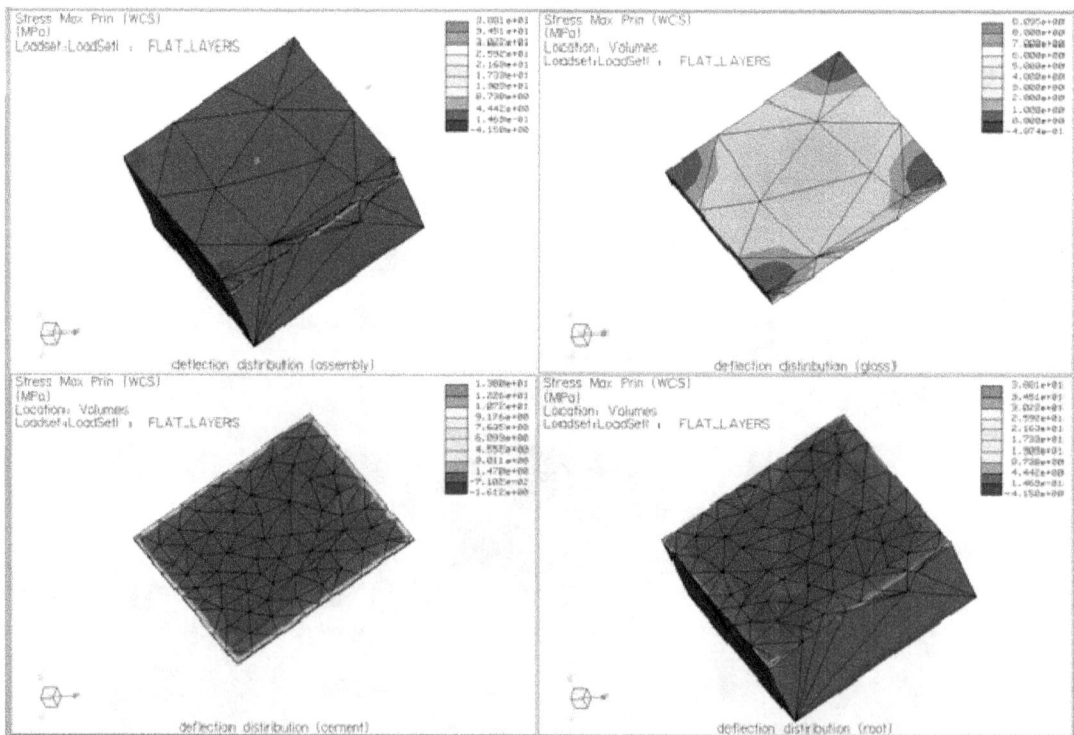

4.9   References

1. Algor Processor Reference Manual, Algor Interactive Systems, 150 Alpha Drive, Pittsburgh, PA 15238
2. O. Axelsson, Iterative Solution Methods, Cambridge University Press, New York, 1994.
3. K. J. Bathe, Finite Element Procedure, Prentice Hall, Upper Saddle River, New Jersey, 1996.
4. R. W. Clough, The Finite Element in Plane Stress Analysis, Proceedings of 2nd ASCE Conference on Electronic Computation, Pittsburgh, PA. September 1960.
5. J. W. Dally, Design Analysis of Structural Elements, Goodington, New York, NY, 2003.
6. C. S. Desal and J. F. Abel, Introduction to the Finite Element Method, Van Nostrand Reinhold, New York, NY, 1972.
7. J. H. Earle, Graphics for Engineers, AutoCAD Release 13, Addision-Wesley, Reading, Massachusetts, 1996.
8. R. H. Gallagher, Finite Element Analysis Fundamentals, Prentice-Hall, Englewoood Cliffs, NJ, 1975.
9. J. M. Gere and S. P. Timoshenko, Mechanics of Materials, 3rd edition, PWS-KENT Publishing Company, Boston, 1990.
10. W. W. Hager, Applied Numerical Linear Algebra, Prentice-Hall, Englewood Cliffs, NJ, 1988.
11. Holland, The Finite Element Method in Plane Stress Analysis, in the Finite Element Method in Stress Analysis, I. Holland and K. Bell (eds), Tapir Press, Trondheim, Norway, 1969, Chapter 2.
12. C.S. Krishnamoorthy, Finite Element Analysis, Theory and Programming, 2nd Ed., 1995.
13. D. L. Logan, A First Course in the Finite Element Method Using ALGOR, PWS Publishing, Boston, 1997.
14. H. C. Martin, Introduction to Matrix Methods of Structural Analysis, McGraw-Hill, New York, NY, 1966.
15. R. L. Norton, Machine Design: An Integrated Approach, Prentice Hall, Upper Saddle River, New Jersey, 1996.
16. M. Ortiz, Y. Leroy and A. Needleman, A Finite Element Method for Localized Failure Analysis, Comp. Meth. Applied Mech. Enging, 61, pp. 189-214, 1987.
17. M. Petyl, Introduction to Finite Element Vibration Analysis, Cambridge University Press, 1990.
18. T. H. H. Pian and P. Tong, Basis of Finite Element Methods for Solid Continua, Int. J. Numer. Methods Eng., Vol. 2, No. 7, April 1964, pp. 1333-1336.
19. P. G. Plockner, Symmetry in Structural Mechanics, J. of the Structural Division, American Society of Civil Engineers, Vol. 99, No. ST1, pp. 71-89, 1973.
20. J. N. Reddy, An Introduction to the Finite Element Method, McGraw-Hill Publishing Company, New York, NY, 1984.
21. L. Stasa, Applied Finite Element Analysis for Engineers, Holt, Rinehart and Winston, New York, NY, 1985, pp. 101-157.
22. Y. Saad, Iterative Methods for Sparse Linear Systems, PWS Publishing Company, Boston, 1996.
23. P. Silvester, Higher-Order Polynomial Triangular Finite Element for Potential Problems, Int. J. Eng. Soc., Vol. 7, No. 8, pp. 849-861.
24. Y. Tada and G. C. Lee, Finite Element Solution to an Elastic Problem of Beams, Int. J. Numer. Methods Eng., Vol. 2, No. 2, April 1970, pp. 229-241.
25. M. J. Turner, R. W. Clough, H. C. Martin, and L. J. Topp, Stiffness and Deflection Analysis of Complex Structures, J. of the Aeronautical Sciences, Vol. 23, No. 9, pp. 805-824, Sept. 1956.
26. O. C. Zienkiewicz, The Finite Element Method, third edition, McGraw-Hill, New York, NY, 1977.
27. O. C. Zienkiewicz and R. L. Taylor, The Finite Element Method, 4th edition, Vol. 1, McGraw-Hill (UK),London, 1989.

4.8   Exercises

1. The following engineering drawing illustrates an assembly. The assembly consists of two components. One is the support and the other is a shaft.

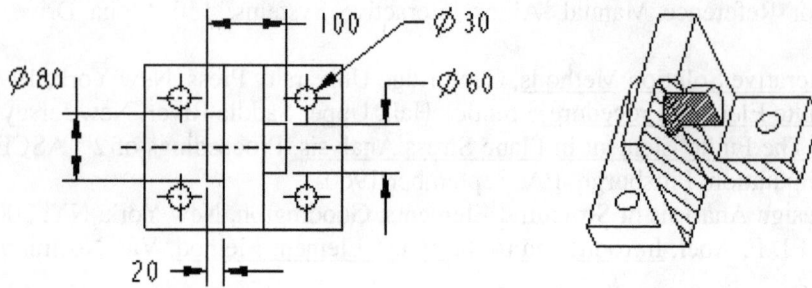

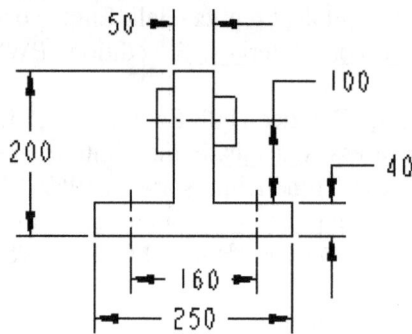

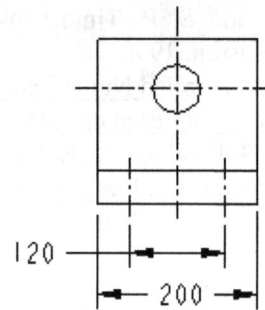

The dimensions of the support component are shown below. Pay attention to the four (4) holes, which are prepared for 4 bolts. The diameter is 30 mm. There are clearances between the bolts and the holes. The entire assembly is fixed to the ground through the 4 bolts, which are not shown here.

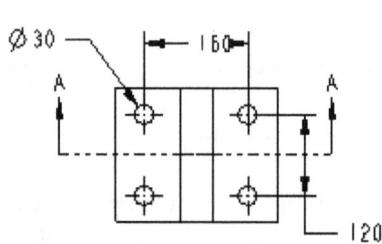

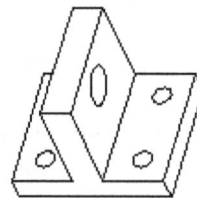

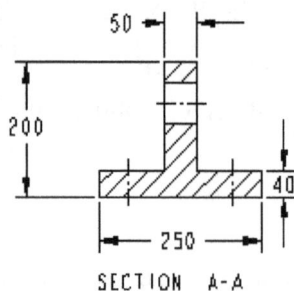

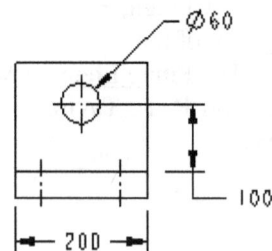

SECTION A-A

The dimensions of the shaft component are shown below. Pay attention to the end on the right side. A load is acting on the surface uniformly. The magnitude is 300 Newton.

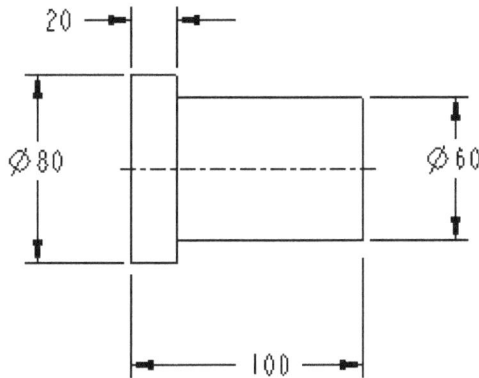

The 3D model of the assembly and the two components are available for performing an engineering analysis. The materials of both components are steel.

support structure

	maximum displacement in magnitude	max principal stress	max Von mises stress
Location (x, y, z)			

shaft structure

	maximum displacement in magnitude	max principal stress	max Von mises stress
Location (x, y, z)			

1) Run the system under the following conditions: the thickness of the middle part is 3 mm, the cement material is zinc-phosphate and the inclination angle is 15 degrees. Plot the distribution of the maximum principal stress, identify the location of the maximum principal stress, and identify the magnitude of the maximum principal stress. Compare the two locations obtained in (3) and (4). Is there a shift in the location of the maximum principal stress? Compare the 2 magnitudes of the maximum principal stress obtained in (3) and (4). Is the value increased or reduced as a result of the material type change?

2) Carry out a systematic study and obtain a quantitative relation which characterizes the value of the maximum principal stress as a function of the three design parameters associated with the middle part component.

2. Referring to the bearing load case in text, do the followings:

(1)   Plot the distribution of the deformation in the vertical direction.
(2)   Plot the distribution of the maximum principal stress of the 3D solid model.
(3)   What is the highest value of the maximum principal stress? Where does it occur?

3.   Referring to the motor shaft component in the text, do the followings:

(1) Define 2 load sets. One acts on the cylindrical surface on the left side. The other acts on the surface on the right side. These 2 torque actions keep the motor shaft in an equilibrium state. The torque magnitude is 5000 N-M.
(2) Assume that the material type is steel. Perform a static analysis and plot the distribution of the maximum principal stress and the distribution of the von Mises Stress.
(3) Assume that the failure mode is related to the degree of plastic deformation of the torque, and maximum von Mises stress allowable is 120 MPa. Calculate the safety factor of the current design.

4.   Referring to the gravity force case study, do the followings:

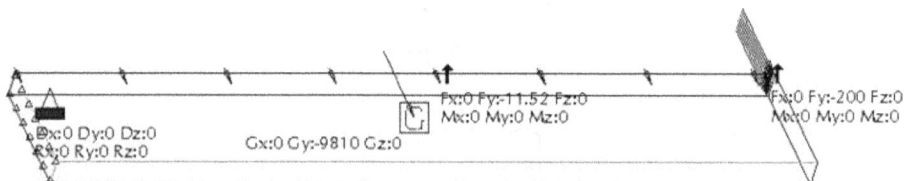

Run this model three times and record the values of the maximum principal stress using the following settings.

	200 N Load	Gravity Load (11.52 N)	Uniform Distributed (11.52N)	Max Principal Stress (MPa)
Run 1	X			
Run 2		X		
Run 3			X	
Run 4	X	X		
Run 5	X		X	

Analyze the results obtained in (4). What is the effect of the Gravity Force?

CHAPTER 5

THERMAL ANALYSIS

5.1 Introduction

In the previous chapters, we evaluate the deflection and stress distributions, using FEA, when a mechanical structure is subjected to a loading condition and/or loading conditions in terms of force, torque and displacements. In this chapter, we deal with temperature settings and variations. Thermal analysis is a branch of materials science where the properties of materials are studied as they change with temperature. The objective of a thermal study is basically to understand response and performance of the structure in response to temperature settings or variations so as to understand the behavior of materials/structure in terms of stress, deformation, etc.

5.2 Material Properties Used in Thermal Analysis

When performing an analysis in FEA, we need to specify the material properties. The basic requirements set for performing the structural analysis of a mechanical system are mass density, Young's modulus and Poisson's ratio. When considering the gravity effect or the gravitational force, we also need to specify the gravitational constant, which is equal to 9.81 m/s^2.

The basic requirements set for performing the thermal analysis of a mechanical system are mass density, specific heat capability, the coefficient of thermal expansion and the coefficient of thermal conductivity. In the following, we give a brief review on these concepts related to the thermal analysis.

Specific heat capability

The thermal energy transferred between a system and its surroundings as a result of temperature differences. When heating an object, say a piece of metal, the amount of heat, or the amount of thermal energy (ΔQ), required to raise its temperature is governed by

$$\Delta Q = cm\Delta T$$

where m is the mass, ΔT is the temperature difference before and after being heated and c is a constant of proportionality, which characterizes the material being heated. This constant is called the specific heat capacity. Values of the specific heat capacity for some materials are listed below:

	Aluminum	Steel	Glass
kcal/(kgC°)	0.231	0.113	0.120
mm²/(s²C°)	9.665x10⁸	9.665x10⁸	5.021x10⁸

In the above table, we have listed 2 values of the specific heat capability for each material type because of the use of 2 different unit systems. From the historic viewpoint, the fields of mechanics and thermal science developed independently. As a result, the work done by force was measured in mechanical units (joules). On the other hand, heat was measured in thermal units (calories or kilocalories).Thanks to the hard work completed by the scientist called Sir James Joule, the quantitative relationship between the unit of joule and the unit of kilocalories was established in the 19th century, and now this quantitative relationship is given by

$$1 \text{ kilocalories} = 4.184 \times 10^3 \text{ joule}$$

where 4.184 is named as Joule's equivalent.

For example, the specific heat capability for steel is 0.113 kcal/(kgC°) listed in the first row of the above table. Using the above equation, or considering the Joule's equivalent, the specific heat capability for steel is given by, in terms of the unit of joule,

$$\text{Specific heat capability of steel} = (0.113)(4.184)10^3 \text{ joule/(kgC°)} = 0.473 \times 10^3 \text{ joule/(kgC°)}$$

Note that the unit of joule is given by

$$1 joule = (1 Newton)(1 meter) = [1\frac{(kg)(meter)}{s^2}](1 meter)$$

Therefore, the specific heat capability of steel can also be given by, as listed in the second row of the above table.

$$specific\_heat\_capability\_of\_steel = 0.473x10^3 \frac{m^2}{s^2 C^o}$$

In the Pro/Mechanica library system, the specific heat capability is given by 4.7334×10^8 (mm^2/(s^2C°) because of the use of mm, instead of meter.

Coefficient of Thermal Expansion or Coefficient of Linear Expansion

A change in the temperature of an object tends to produce a change in its dimensions. A simple illustration of this effect is shown in the following figure where a block is being heated and is free to expand in all directions, assuming that the block is heated uniformly. As a result, the temperature of the block will be raised while it is being heated. To measure the degree of expansion in relation to the strain development due to the increase of body temperature, the coefficient of thermal expansion, α, is introduced. In SI units, α has the dimension of 1/K, or the reciprocal of kelvins. If the temperature is measured in degrees Celsius, α has the dimension of 1/C, or the reciprocal of degrees Celsius. The relation between a temperature in Kelvin and degrees Celsius is K = 273.15 + C°.

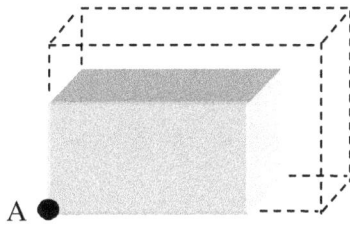

$$\varepsilon_t = \alpha(\Delta T)$$

A ●

Values of the specific heat capacity for some materials are listed below:

	Aluminum	Steel	Glass
$1/C^o$	23×10^{-6}	$(10 \sim 18) \times 10^{-6}$	$(3 \sim 11) \times 10^{-6}$
$1/F^o$	13×10^{-6}	$(5.5 \sim 9.9) \times 10^{-6}$	$(1.7 \sim 6) \times 10^{-6}$

In the above table, we also listed the values of the coefficient of thermal expansion in the unit of $1/F^o$. The relation between a temperature in degrees Celsius and a temperature in degrees Fahrenheit is $F^o = 1.8*C^o + 32$. As a result, the conversion factor between the two scales is 1.785 for values converted from $1/F^o$ to $1/C^o$, and 0.560 from $1/C^o$ to $1/F^o$.

In the Pro/Mechanica library system, the unit of the coefficient of thermal expansion is given by $1/C^o$.

Coefficient of Thermal Conductivity or Coefficient of Heat Conductivity

In physics, we have learned that in heat conduction, molecule, while remaining at relatively fixed positions in space, absorb and pass on heat in the form of random mechanical motion. In this process, a certain amount of heat ΔQ is conducted through a substance from a higher temperature T_2 to a lower temperature T_1. The amount of heat conducted per second depends on both the number and the kind of molecules involved.

A simple illustration of this phenomenon is shown in the following figure. We consider the conduction of heat through a solid metal bar of length L. The two ends of the bar are kept at temperatures T_2 and T_1. Assume that the sides of this bar are perfectly insulated so that no heat can escape in this conduction process.

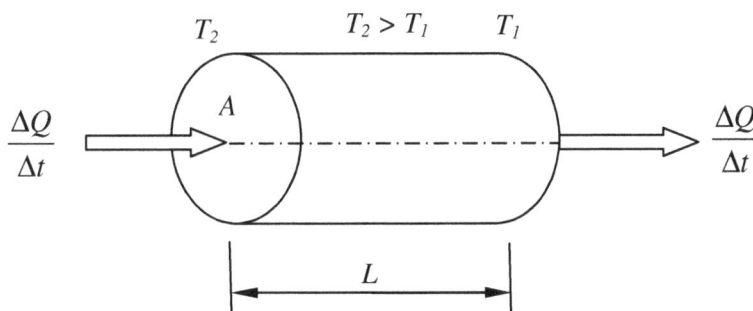

We naturally believe that the amount of heat transferred to be directly proportional to the cross-section area A, the bar length L, and the time Δt the heat flows. However, there is another parameter, called the coefficient of thermal conductivity k, playing an equal effect on the amount of heat transferred.

$$\frac{\Delta Q}{\Delta t} = k \frac{A(T_2 - T_1)}{L}$$

The coefficient of thermal conductivity is one of the material properties characterizing the material behavior in the process of heat transfer. Its unit is given by $W/(mC^o)$.

Note that 1 Watt is equal to a ratio of a joule to second. As a result, we may have the unit of a coefficient of thermal conductivity, which is given by:

$$\frac{Watt}{mC^o} = \frac{\dfrac{joule}{s}}{mC^o} = \frac{joule}{msC^o} = \frac{(Newton)(m)}{msC^o} = \frac{Newton}{sC^o}$$

Values of the coefficient of thermal conductivity for some materials are listed below:

	Aluminum	Steel	Glass
W/(mC°)	180~250	40~60	70~90
N/(sC°)	180~251	40~61	70~91

In the Pro/Mechanica library system, the unit of the coefficient of thermal conductivity is given by N/(sC°).

5.3   Conduction and Convection Modes

Heat transfer can be defined as the transmission of energy from one region to another as a result of a temperature difference between them. Since differences in temperature exist all over the universe, the phenomena of heat flow are as universal as those associated with gravitational attractions. Unlike gravity, however, heat flow is governed not by a unique relationship but rather by a combination of various independent laws of physics. In general, there are 3 distinct modes of heat transmission: conduction, convection and radiation.

It is important to note that, when applying FEA under the Pro/Mechanica environment, we only deal with 2 modes of hear transmission: conduction and convection. Furthermore, from an engineering viewpoint, the key problem is the determination of the rate of heat transfer at a specified temperature difference. When applying FEA under the Pro/Mechanica environment, we mainly deal with 2 types of thermal analysis. They are:

Type1: Thermodynamics dealing with systems in equilibrium. When applying this type of thermal analysis, our interest lies in with a steady-state of the temperature distribution, not to quantify the rate of change. Many engineering problems are concerned only with the steady-state heat transfer of a part or assembly. It is important to note that a steady state of the temperature distribution has nothing to do with time.

Type 2: Transient heat transfer. This type of thermal analysis deals with heating and cooling. Under those circumstances, the transitional period of time is of great interest, but not the steady-state. The transient heat transfer process takes into account the change in internal heat energy of the body with time.

As we have emphasized, no matter what type of a specific thermal analysis we deal with, we only consider the 2 modes: conduction and convection, not radiation. In the following, we briefly present the basic concepts related to the conduction mode and the convection mode.

Thermal Conduction Mode

Conduction is a heat transfer mode where the transfer of heat through materials without net mass motion of the material.

The general equation for heat transfer by conduction is

$$q_k = -kA\frac{dT(x)}{dx}$$

In this relation, T(x) is the local temperature and x is the distance in the direction of the heat flow. A is the area normal to x direction through which heat flows. Parameter k is the thermal conductivity, a physical property of the material. The minus sign indicates heat must flow in the direction from higher to lower temperature.

Thermal Convection Mode

Convection is a heat transfer mode where heat transfer occurs between a surface and moving fluid when they are at different temperatures. It is the process, in which thermal energy is transferred between a solid and a fluid flowing past it.

The general equation for heat transfer by convection, or the rate of heat transfer q_c by convection, is given by

$$q_c = hA\Delta T$$

where A is the heat transfer area, ΔT is the difference between the surface temperature T_s and a temperature of the fluid T_∞ at some specified location (usually far away from the surface, and h is the heat transfer coefficient over the area A. Parameter h is also called as the surface coefficient of heat transfer or the convection heat transfer coefficient.

The unit of the convection heat transfer coefficient is $W/(m^2 C^o)$. It can also be expressed as

$$\frac{Watt}{m^2 C^o} = \frac{\dfrac{joule}{s}}{m^2 C^o} = \frac{joule}{m^2 s C^o} = \frac{(Newton)(m)}{m^2 s C^o} = \frac{Newton}{ms C^o}$$

Evaluation of the convection heat transfer coefficient for a given circumstance is not an easy task. The following table lists some ranges of the value of the convection heat transfer coefficient for 3 circumstances so that readers have a general feeling about the magnitude range when dealing with the analysis of convection heat transfer.

	air, free convection	air, forced convection	water, forced convection
$W/(m^2 C^o)$	0.02 ~ 0.12	0.12 ~ 1.20	1.20 ~ 72.0
$N/(ms C^o)$	0.02 ~ 0.12	0.12 ~ 1.20	1.20 ~ 72.0

Conduction and Convection Modes in Thermal Analysis

Whenever a temperature gradient exists within a system, or whatever 2 or more systems at different temperatures are brought into contact, energy is transfer. The scope of heat transfer is so wide because the process of energy transfer and conversion is inevitable. Numerical methods, such as FEA, have been widely employed in the study of heat transfer, and more important is the fact that many applications are with great success in terms of the results, as compared with the results obtained from traditional analytical methods.

When using the FEA to solve problems related to heat transfer, it is important to pay attention to the dimensions and units. Dimensions are the basic concepts of measurements, such as length, time and temperature. Units are the means of expressing dimensions numerically. Therefore, dimensions are quantified by units. Before any numerical calculations can be made, applying proper units or sustaining the consistency in units is critical. Readers have to pay great attention to this issue. Otherwise, the results obtained from FEA with improper units may differ from the values, which should be obtained, making the effort of performing FEA meaningless, even harmful to the working process, which seeks an accurate and reliable solution.

5.4 Thermal Analysis of a Firewall

In this case study, we use a firewall as an example. First, we create a 3D solid model of a firewall, which is a solid block with the 3 dimensions equal to 1000 x 500 x 2000 mm. To control the temperature distribution, there are 3 water channels designed to disputed excessive heat confined within the firewall. Note the 3D model is part of the entire firewall. The material type is brick. We need to define the properties of brick because the library system in ProE does not have the material type of brick. The material properties are listed below:

Properties	Values
Density	2.50 (g/cm^3)
Poisson's ratio	0.15
Young's modulus	40000 (MPa)
Coefficient of thermal expansion	1.20 E -05 (1/C^o)
Specific heat capacity	5.0 E +08 (mm^2/sec^2/C^o)
Thermal conductivity	0.008 (N/sec/C^o)

Assume that a fire has already started. We define such a fire as a heat load of 0.5 W. The surrounding temperature is at 20C^o. Let us first construct a solid block with the 3 dimensions equal to 1000 x 500 x 2000 mm. The diameter of the 3 water channels is 180 mm. The objects of the thermal analysis are to obtain temperature distributions of the firewall with and without the cooling condition.

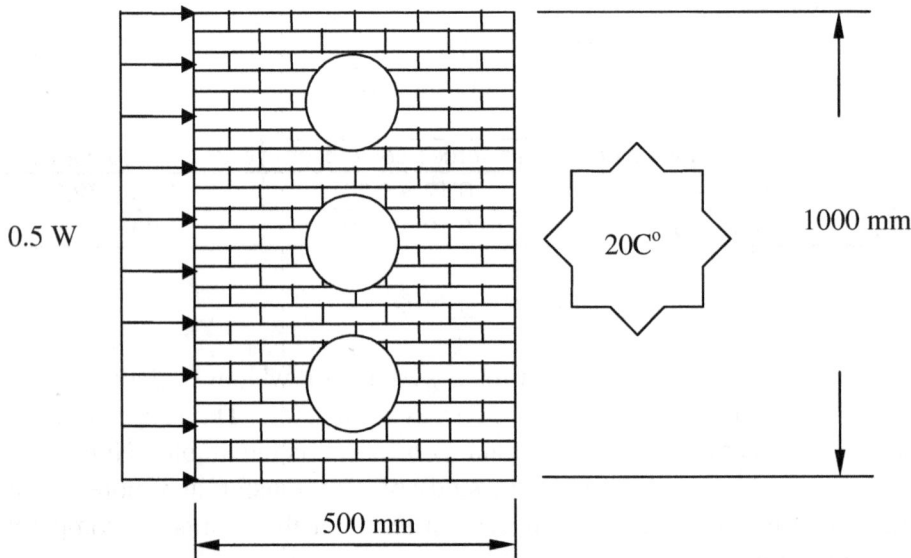

Step 1: Create a 3D solid model for the firewall structure.
Click the icon of **Create a new object** from the menu toolbar to initiate the creation of a 3D model. Type *firewall* as the file name, clear the box of **Use default template > OK**. Select the unit of **mmns_part_solid**, type *firewall* under the **description** of the model, and type *student* or *your name* under the **modeled_by > OK**.

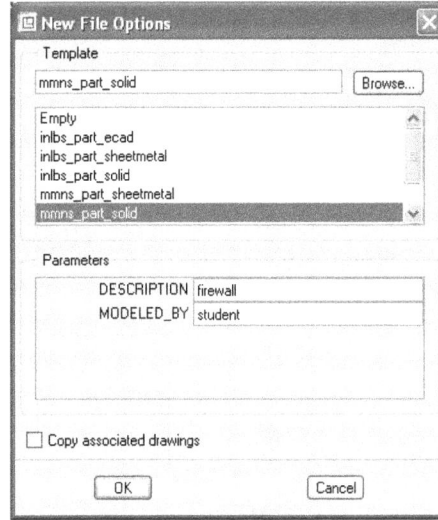

Pick up the icon of **Extrude** displayed on the toolbar of feature creation. Specify the depth value equal to 2000. To initiate the process of sketching, activate **Placement > Define.**

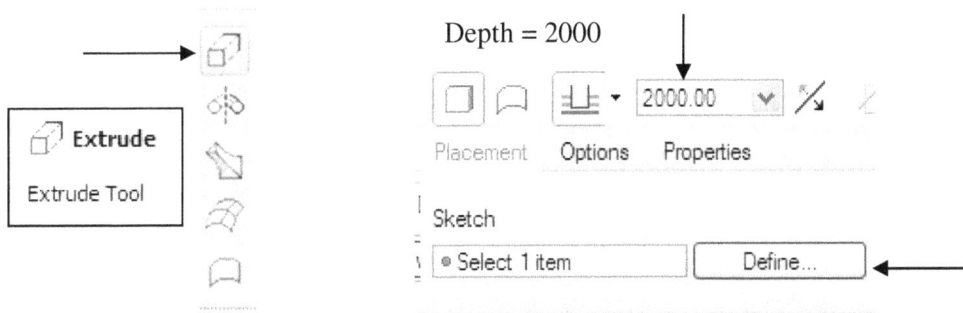

Depth = 2000

Extrude

Extrude Tool

Placement   Options   Properties

Sketch
Select 1 item   Define...

Select the **FRONT** datum plane displayed on screen as the sketch plane. Pick the box of **Sketch** to accept the **RIGHT** datum plane as the default reference to orient the sketch plane, as illustrated below:

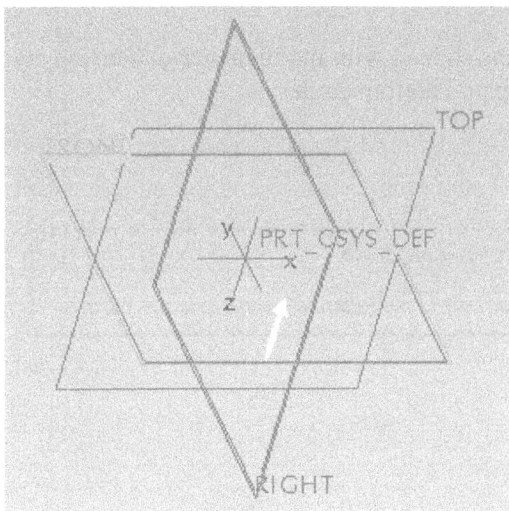

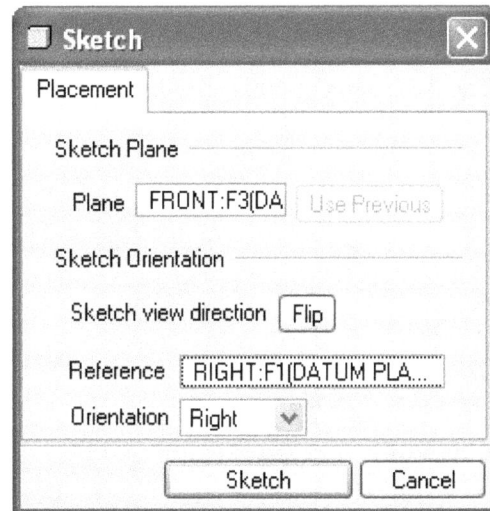

Click the icon of **Centerline**. Sketch a horizontal centerline and a vertical centerline to be used for symmetry when sketching.

Create the horizontal
centerline first

Create the vertical
centerline afterwards

Centerline

Create 2 point centerlines.

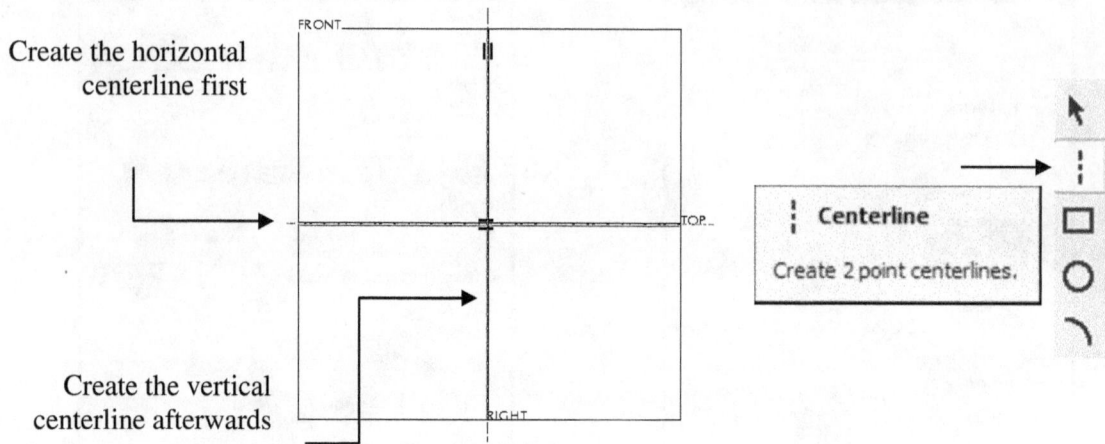

Click the icon of **Rectangle** and sketch a rectangle, which is symmetric about the 2 sketched centerlines. The 2 dimensions of the rectangle are 500 and 1000, respectively.

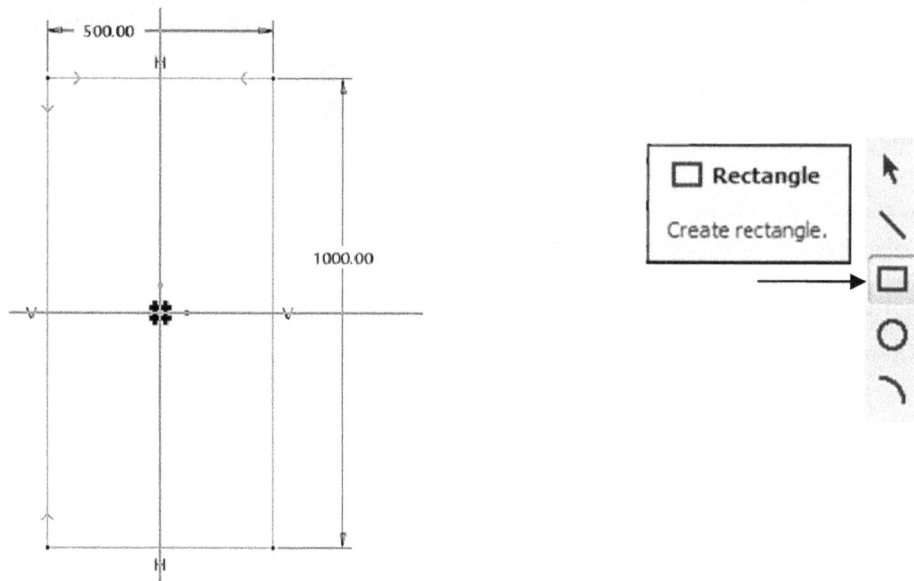

Rectangle

Create rectangle.

Click the icon of **Circle** and sketch the 3 circles, as shown, with the diameter dimension equal to 180. The dimension of 600 is the distance between the top and bottom circles.

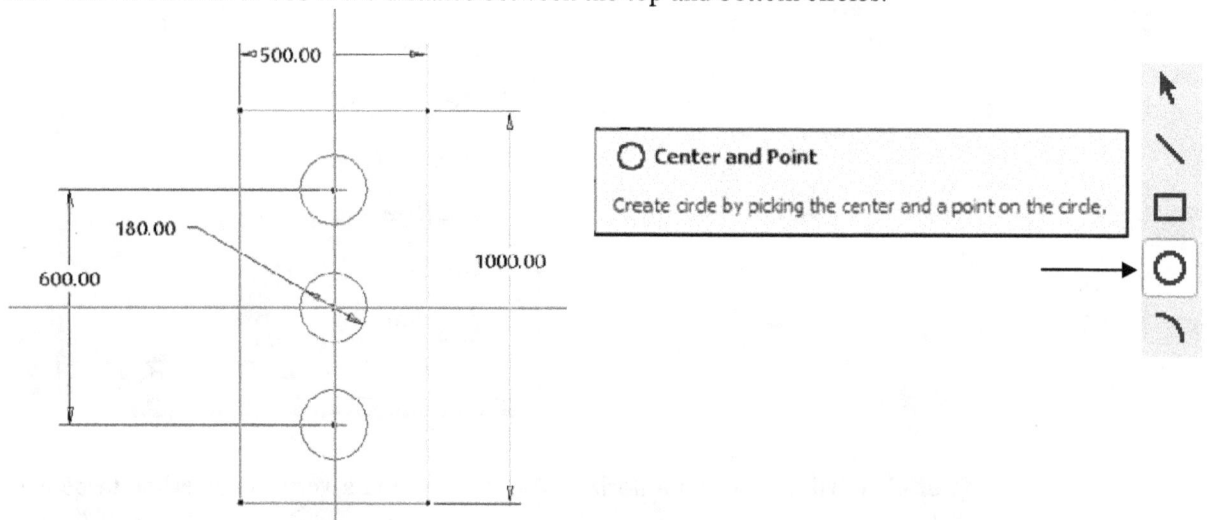

Center and Point

Create circle by picking the center and a point on the circle.

Upon completing the sketch, click the icon of **Done** and click the icon of **Apply and Save.**

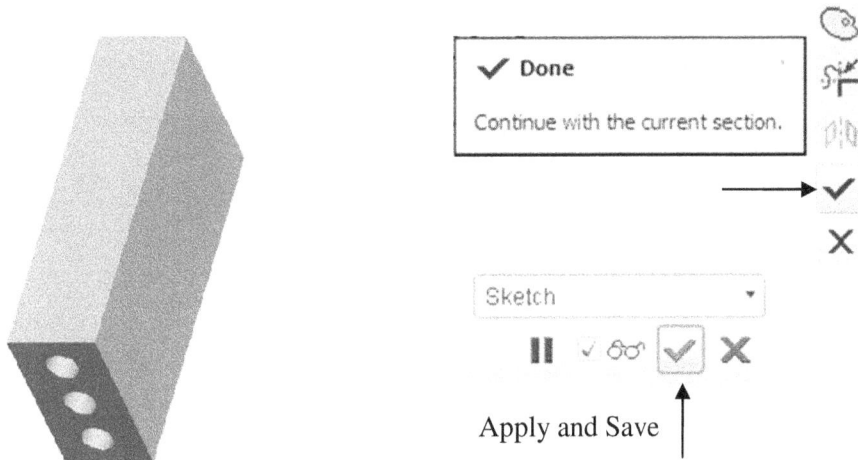

Apply and Save

Step 2: Switch from the Pro/ENGINEER design system to Pro/Mechanica

From the main toolbar, select **Applications > Mechanica > Continue > Thermal > Bonded > OK.**

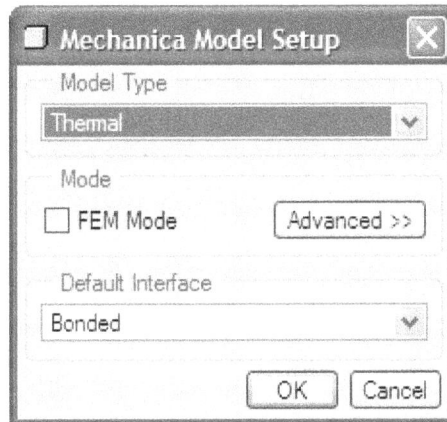

From the toolbar of Mechanica objects, select the icon of **Materials >** select **New** from the left side called **Materials in Model >** because the material type of brick does not exist in the Pro/Mechanica Library. We need to input and install this information related to the material properties based on the given information as listed.

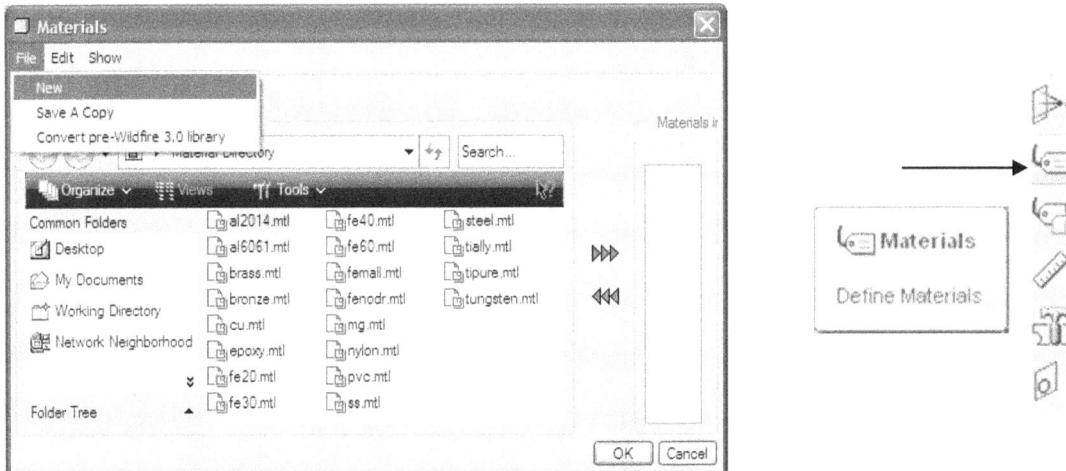

Specify the material properties, both Structural and Thermal, as shown below:

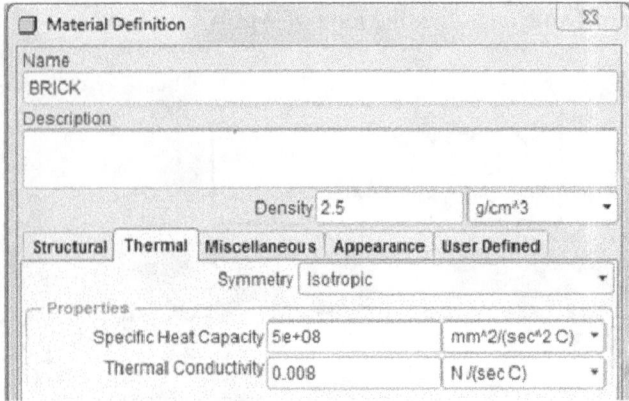

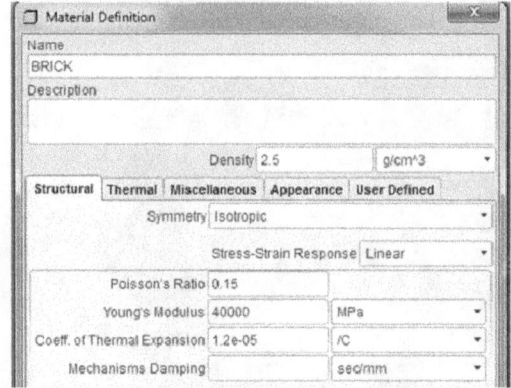

Upon completing the information > **Save to Model > OK.** We need to assign this set of material properties to the firewall component. Click the icon of **Assign.** The software system has automatically completed the assignment because there is only one volume in the system > **OK.**

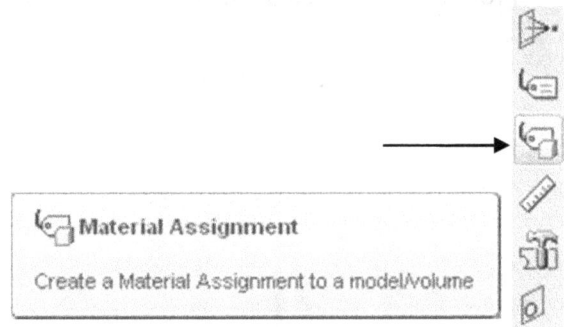

Step 3: Define the heat load and convection boundary condition.

Click the icon of **Heat Load > Surfaces** > pick the surface on the left side of the firewall > . Enter 500 for the load value > **OK**.

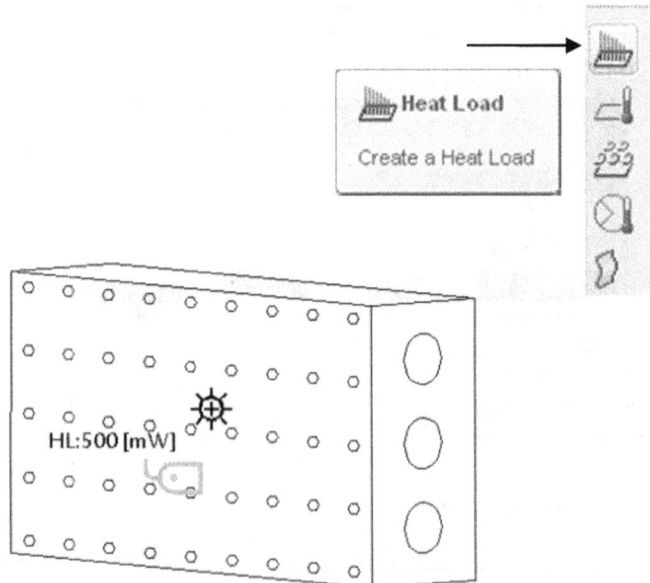

Click the icon of **Convection** > **Surfaces** > pick the surface on the right side of the firewall > enter 0.001 as the coefficient value and 20 as the temperature value > **OK**.

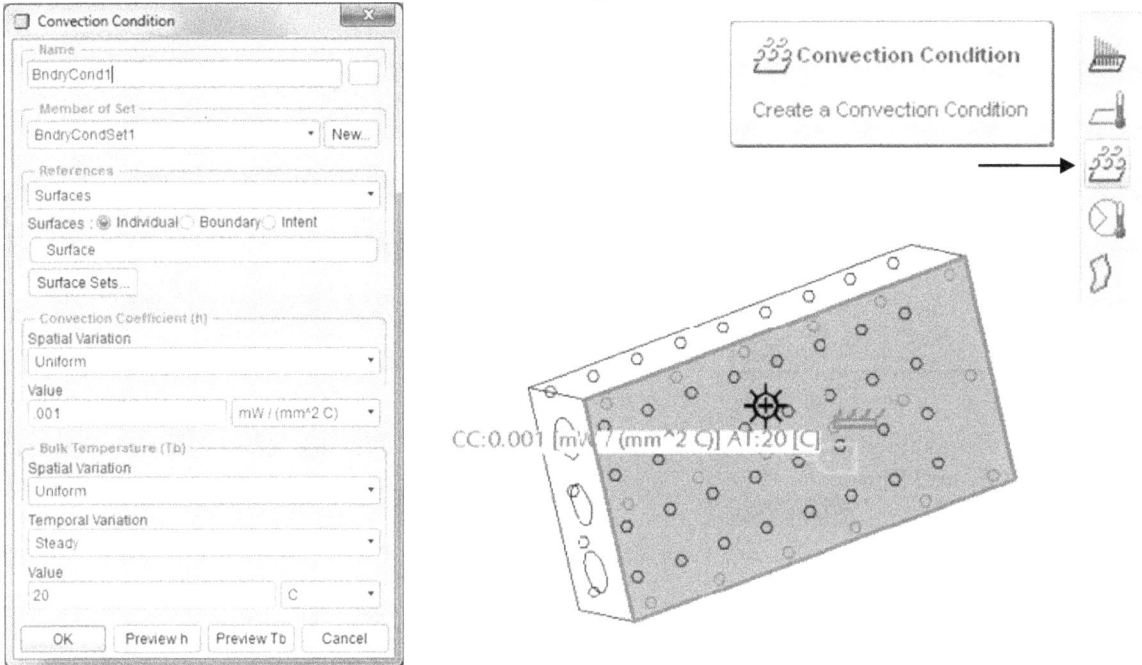

Step 4: Under the Pro/MECHANICA environment, set up **Analyses** and **Run** it

Select the icon of **Run an Analysis** > **File** > **New steady state thermal** > type *firewall* as the name of the analysis folder > **OK**. Click the **Run Flag**.

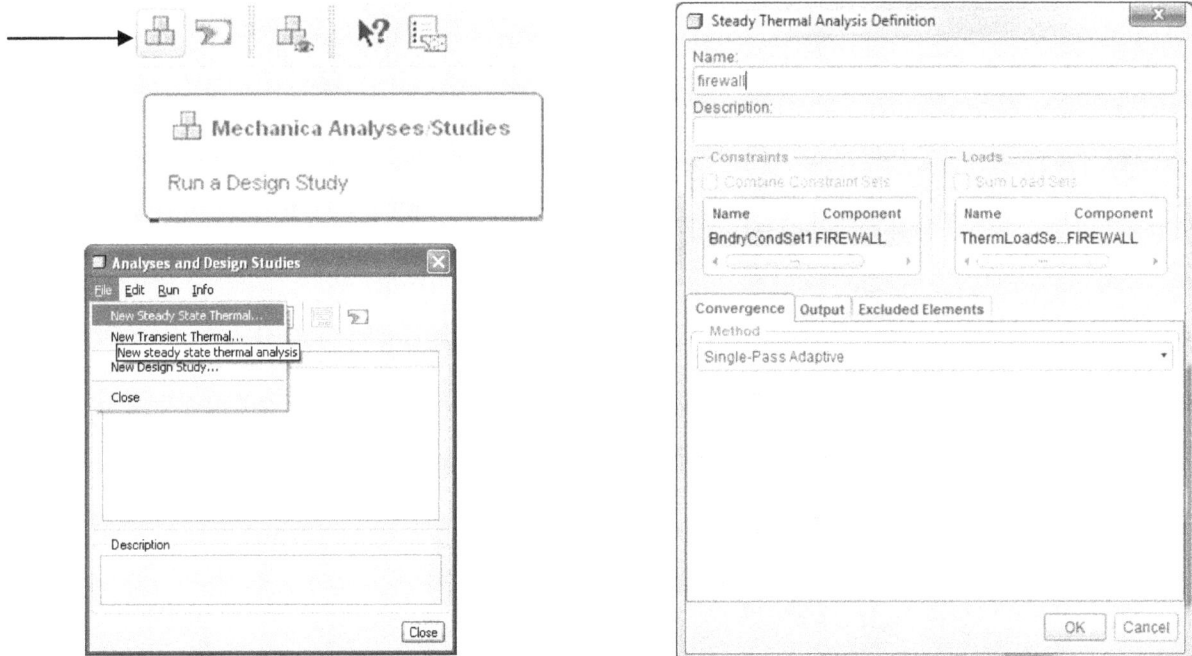

There is a message displayed on screen. The message is asking the user "Do you want error detection?" Click **Yes**. In this way a warning message will appear if an abnormal operation occurs.

To monitor the computing process, users may click the icon of **Display study status**.

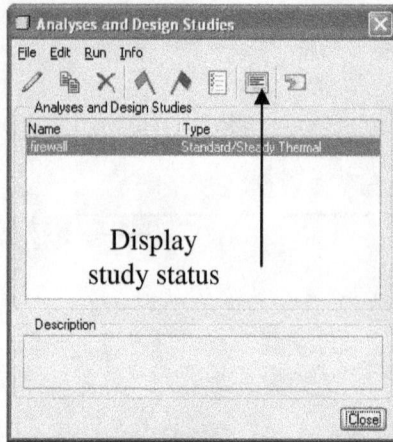

```
Analysis "firewall" Completed  (23:08:16)

---------------------------------------------------

Memory and Disk Usage:

    Machine Type: Windows NT/x86
    RAM Allocation For Solver (megabytes): 128.0

    Total Elapsed Time (seconds): 12.26
    Total CPU Time    (seconds): 11.53
    Maximum Memory Usage (kilobytes): 207093
    Working Directory Disk Usage (kilobytes): 6144

    Results Directory Size (kilobytes):
    4441 .\firewall

    Maximum Data Base Working File Sizes (kilobytes):
    6144 .\firewall.tmp\kel1.bas

---------------------------------------------------
Run Completed
Fri Jul 01, 2011   23:08:16
```

To study the obtained result, say the temperature distribution, click the icon of **Review results > Temperature > OK and Show.**

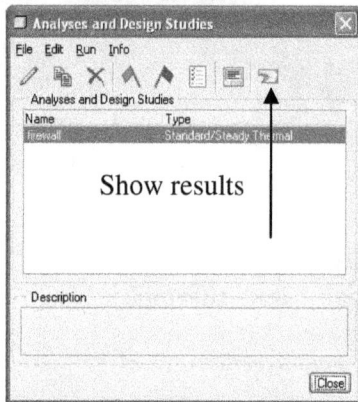

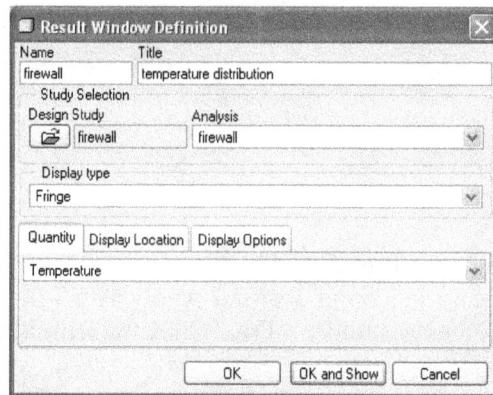

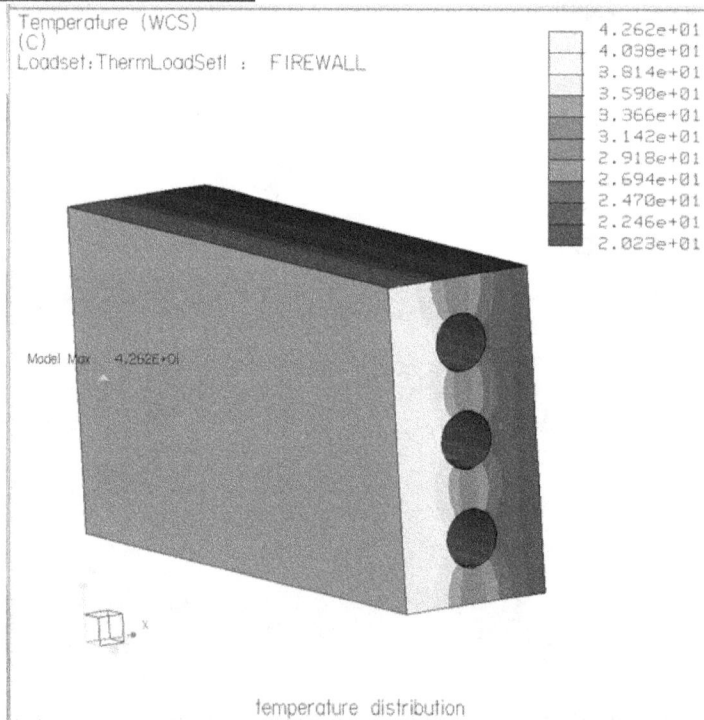

To show heat flux motion, click the icon of **Edit**. Change **Fringe** to **Vector**. Change **Temperature** to **Flux** and click **OK and Show**.

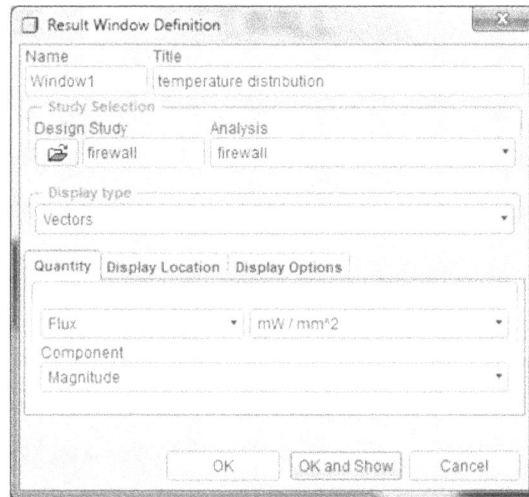

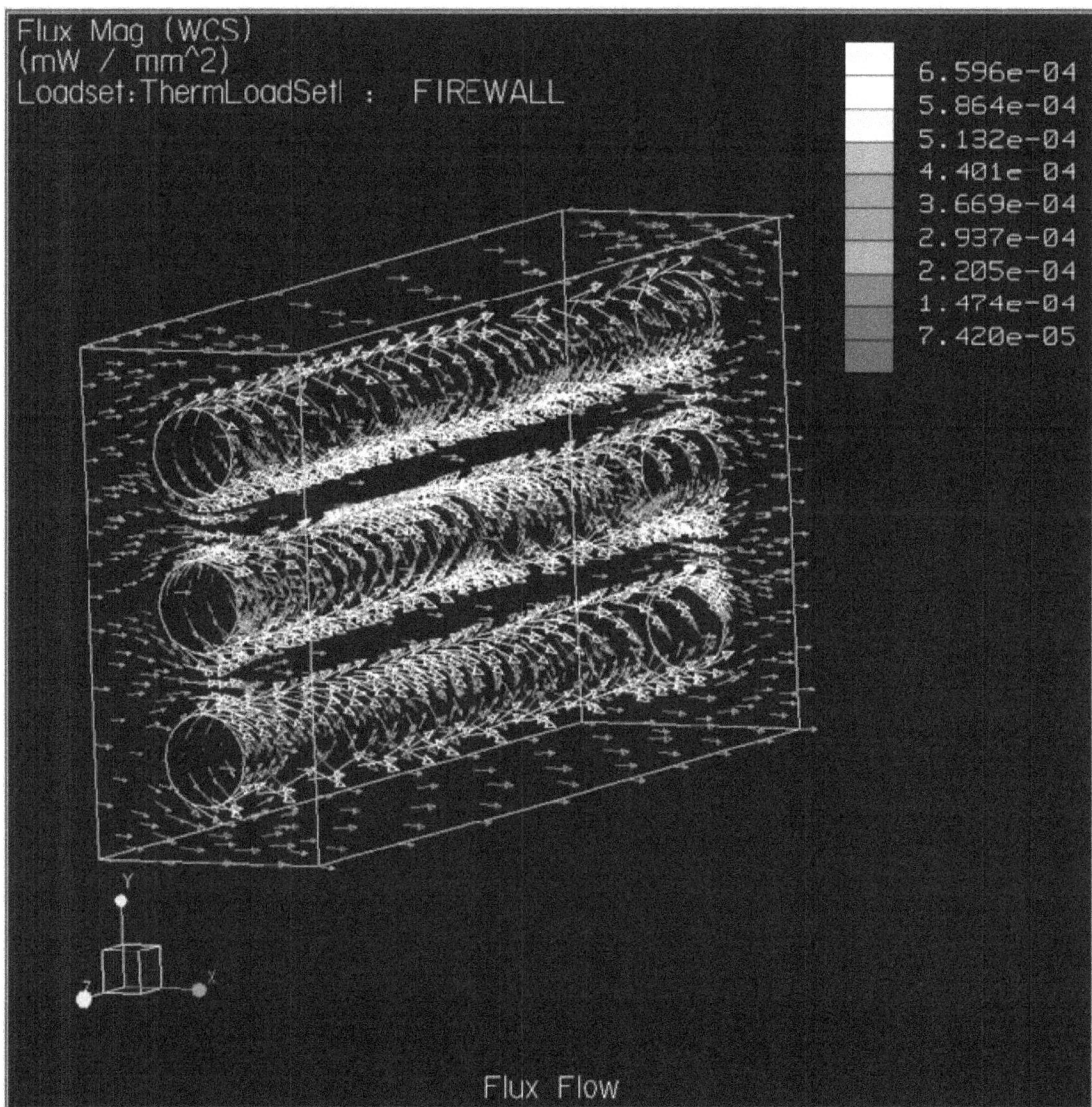

Step 5: Add the convection of cooling and run the program again.

Click the icon of **Convection** > **Surfaces** > pick the 3 cylindrical surfaces > enter 0.02 as the coefficient value and 5 as the temperature value > **OK**.

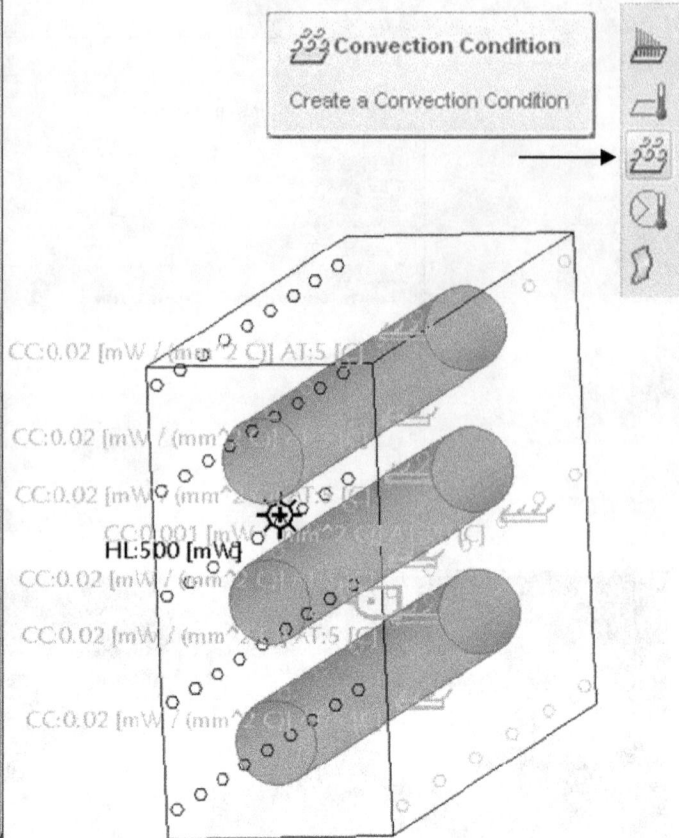

Select the icon of **Run an Analysis > File > New steady state thermal >** type *cooling* as the name of the analysis folder > **OK.** Click the **Run Flag.**

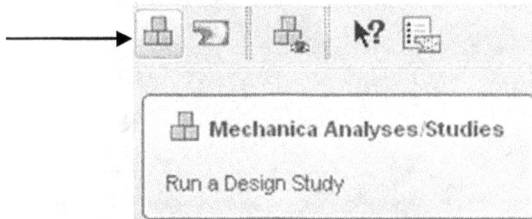

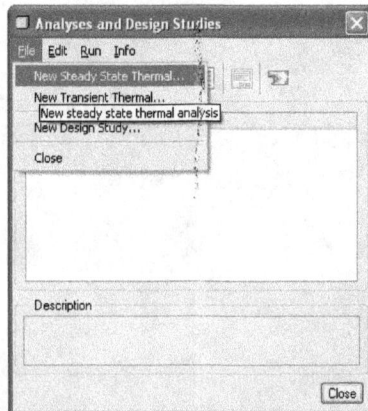

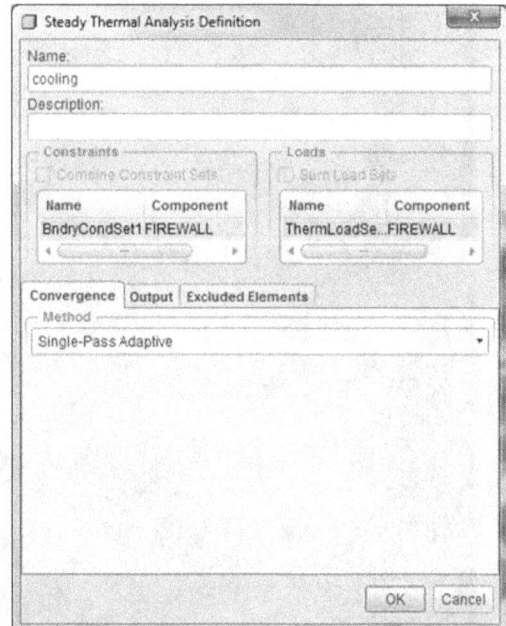

To study the obtained result, say the temperature distribution, click the icon of **Review results > Temperature > OK and Show.**

Temperature (WCS)
(C)
Loadset:ThermLoadSetl :   FIREWALL

```
1.956e+01
1.811e+01
1.617e+01
1.422e+01
1.228e+01
1.034e+01
8.400e+00
6.459e+00
5.003e+00
```

Model Max   1.956E+01

temperature distribution

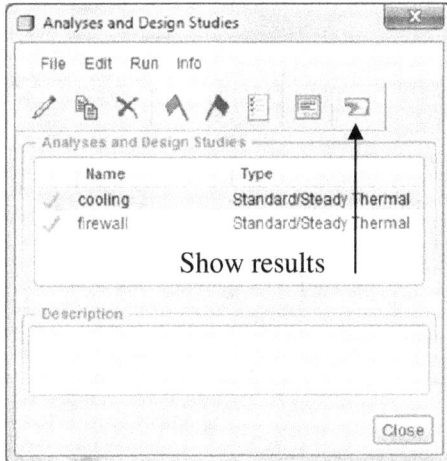

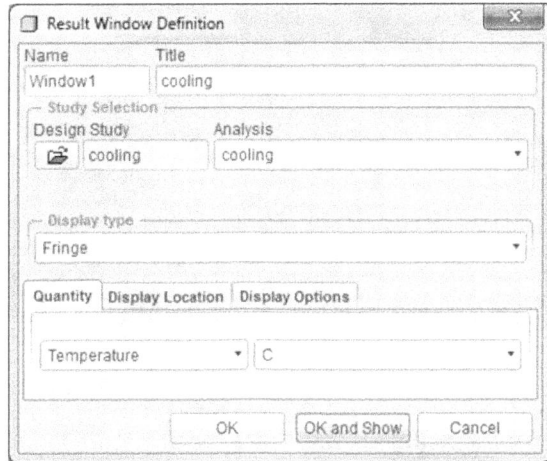

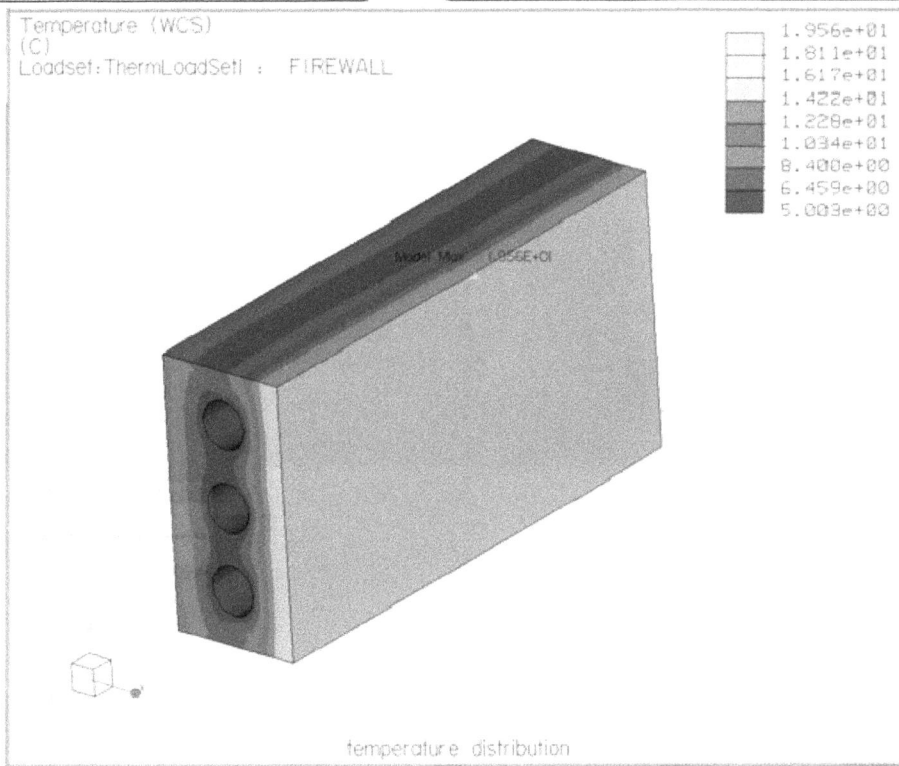

To show heat flux motion, click the icon of **Edit**.   Change **Fringe** to **Vector**.   Change **Temperature** to **Flux** and click **OK and Show**.

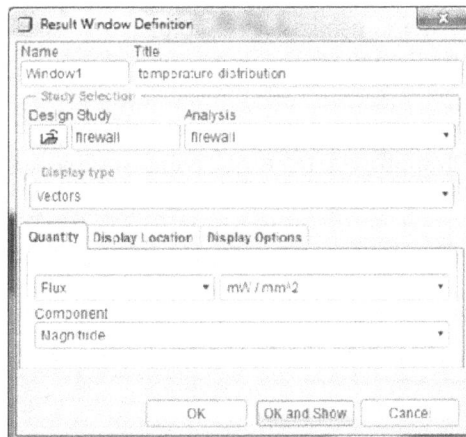

Flux Mag (WCS)
(mW / mm^2)
Loadset:ThermLoadSet1 : FIREWALL

	1.172e-03
	1.005e-03
	8.377e-04
	6.705e-04
	5.033e-04
	3.360e-04
	1.688e-04

flux flow

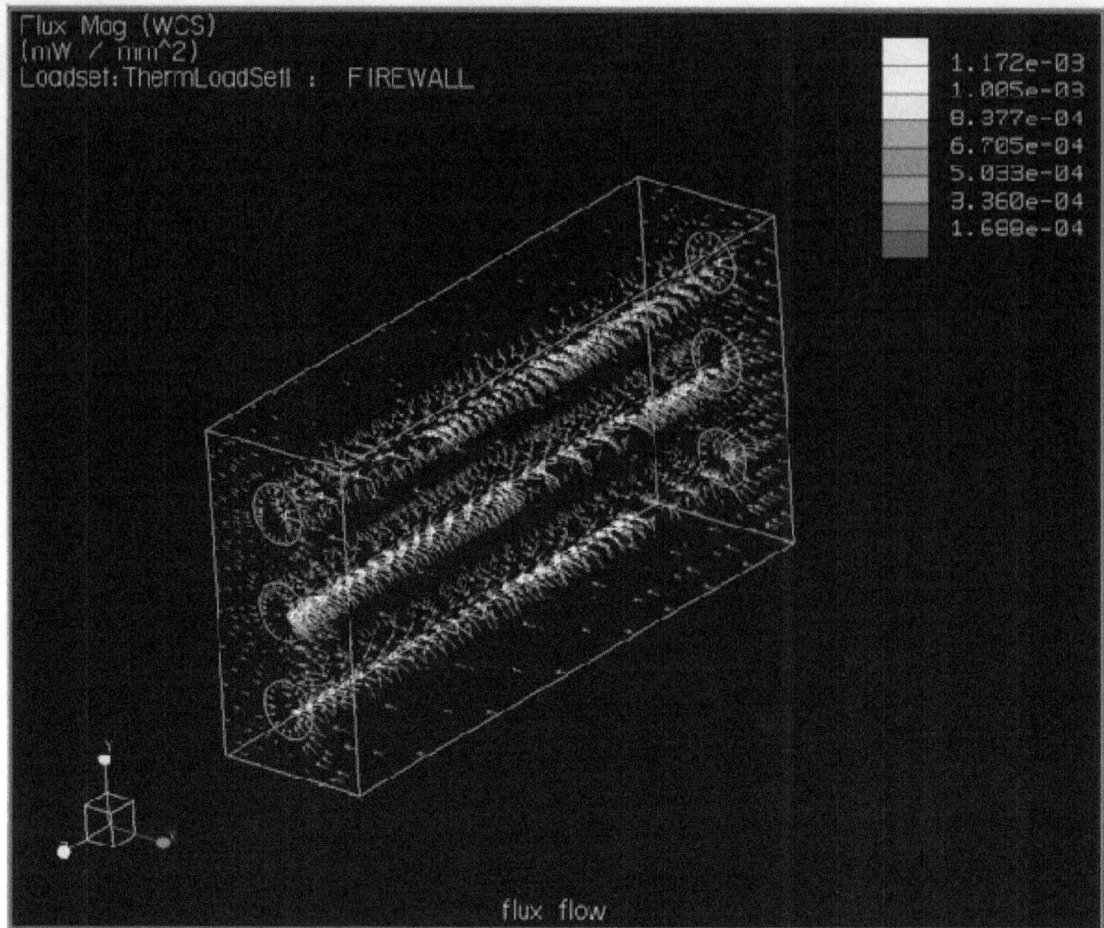

5.5   Thermal Analysis of a Heat Sink

Computers are used in office and at home. Computers are electronic devices and consume electrical power. As a result, electric currents cause power dissipation in the electronic components. The most significant effect is the increase of the temperature of the CPU device. The reliability and life expectancy of a computer is closely related to its operating temperature. To protect the UPC from heat related failure, a heat sink is usually designed so that the operating temperature of the CPU can be controlled.

In this section, a heat sink design is used as an example to show the process of heat dissipation during operation through a steady thermal analysis

Step 1: Create a 3D solid model for heat-sink component.

Click the icon of **Create a new object** from the menu toolbar to initiate the creation of a 3D model.

Select the icon of **Create a new object** from the main menu or the main toolbar

File   Edit   View   Insert   Analysis

Type *heat_sink* as the file name, clear the box of **Use default template > OK**. Select the unit of **mmns_part_solid**, type *heat sink for CPU* under the **description** of the model, and type *student* or *your name* under the **modeled_by > OK**.

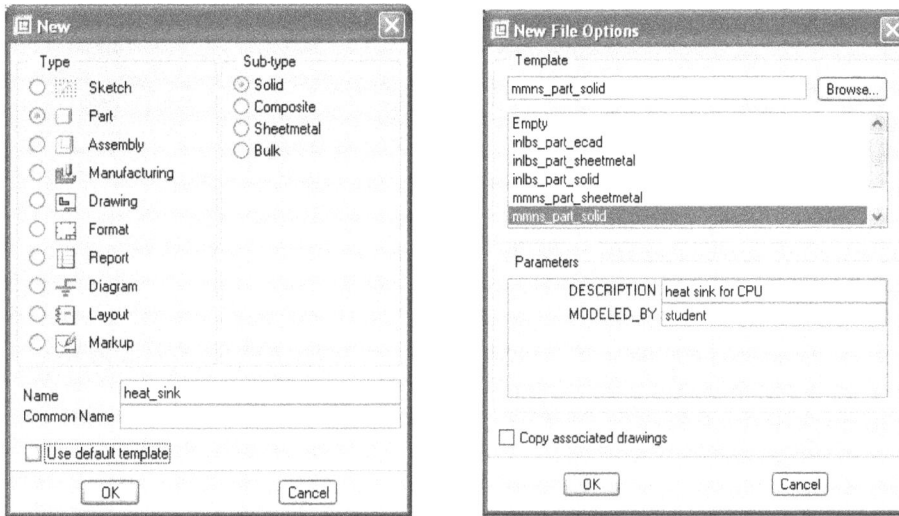

Click the icon of **Extrude** displayed on the toolbar of feature creation. Specify the depth value equal to 45. To initiate the process of sketching, activate **Placement > Define.**

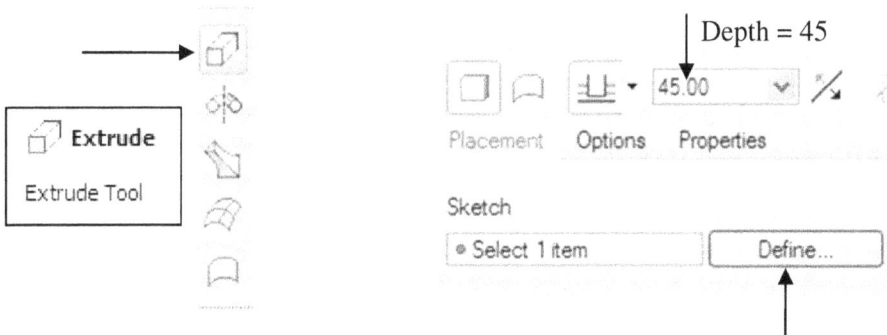

Select the **FRONT** datum plane displayed on screen as the sketch plane. Pick the box of **Sketch** to accept the **RIGHT** datum plane as the default reference to orient the sketch plane, as illustrated below:

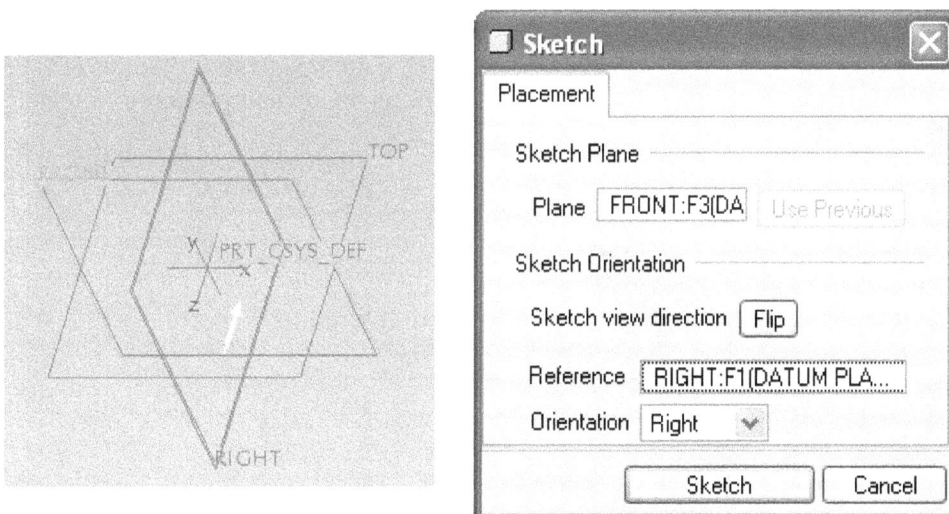

Click the icon of **Rectangle** from the toolbar of sketcher and sketch a rectangle. The left side and the bottom side of the rectangle are on the 2 references selected by the default. The 2 dimensions of the rectangle are 45 and 22, respectively.

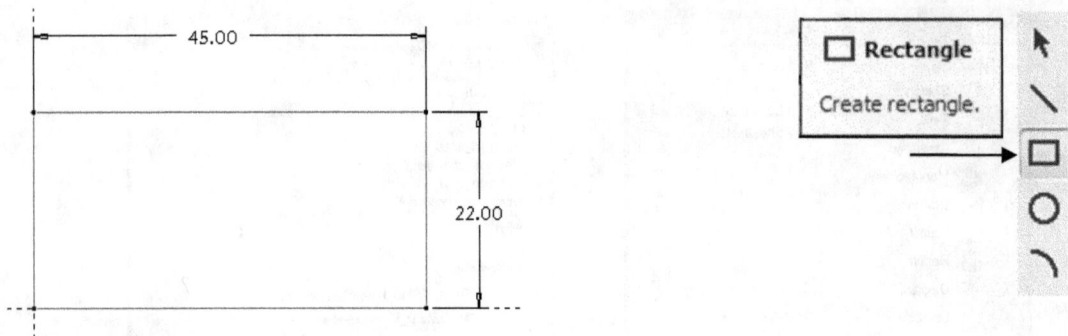

Upon completing the sketch, click the icon of **Done** and click the icon of **Apply and Save.**

Apply and Save

Now let us create parallel slots designed for air flow and heat dissipation. Click the icon of **Extrude**. Make sure that the icon of **Cut** is selected. Specify **Through All** as the depth for the slots. To initiate the process of sketching, activate **Placement > Define.**

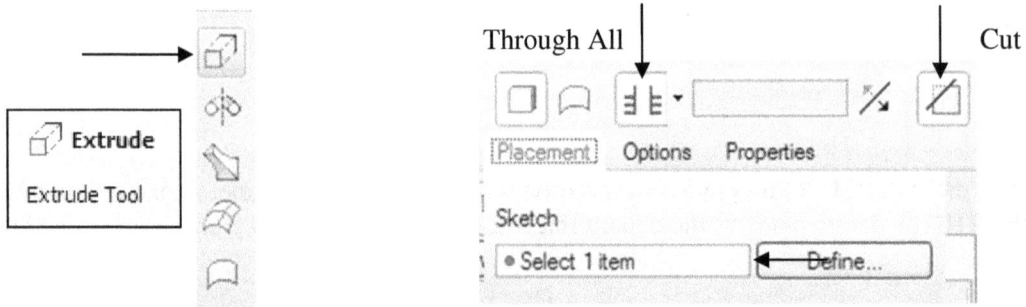

Select the front surface of the plate, not the **FRONT** datum plane displayed, as the sketch plane. Click the box of **Sketch** to accept the **RIGHT** datum plane as the default reference to orient the sketch plane, as illustrated below:

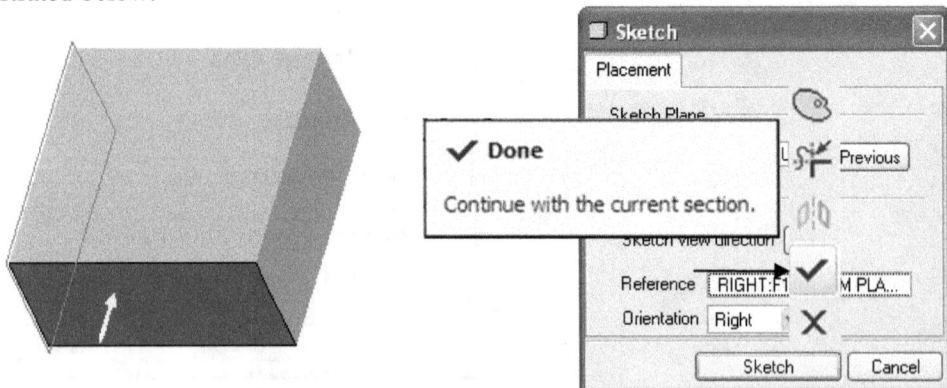

Click Sketch and select References. The references window appears. Let us add a new reference, which is the top surface of the block, as shown.

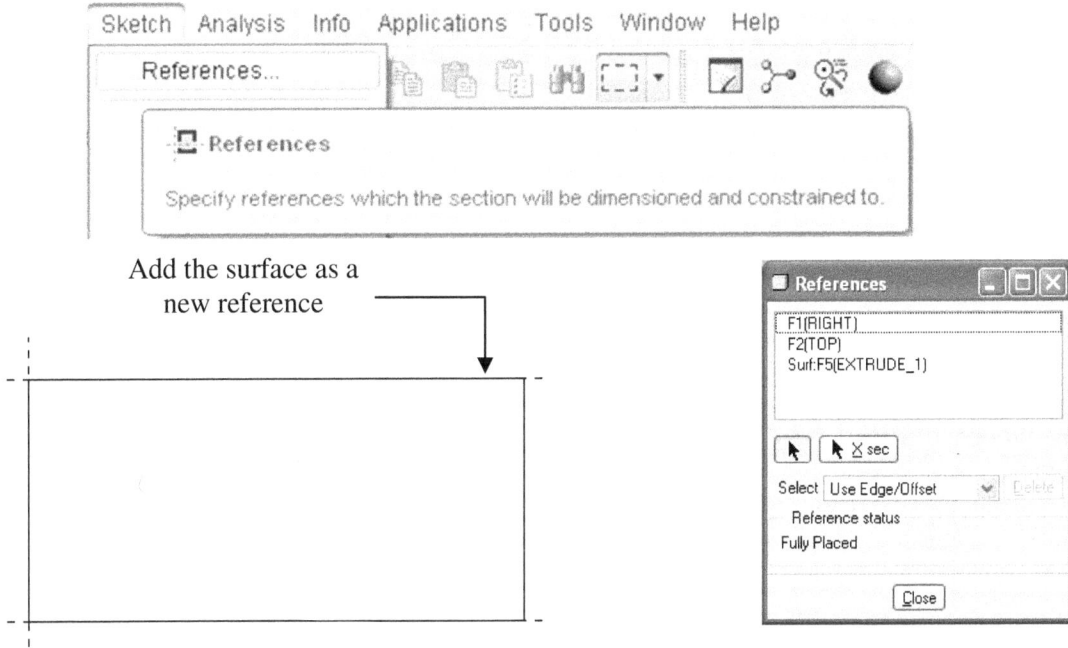

Add the surface as a new reference

Click the icon of **Rectangle** from the toolbar of sketcher and sketch the first slot, as shown below. The width of the slot is 3 mm. There are 2 position dimensions, which are 3 and 2, respectively.

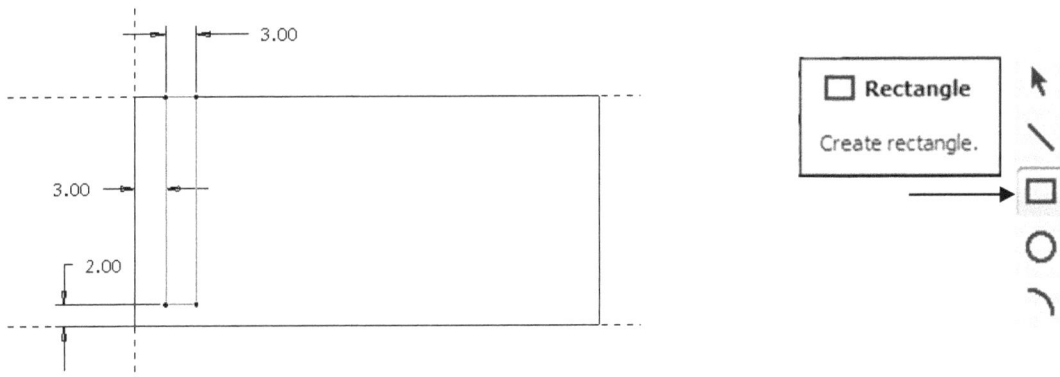

Afterwards, sketch the other 6 equal sized slots. The distance between 2 neighboring slots is 3 mm, as illustrated below.

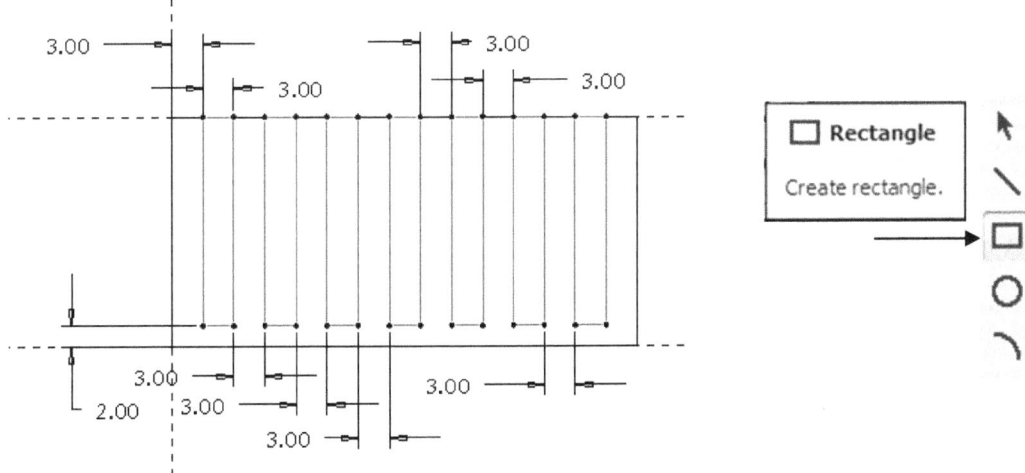

Upon completing the sketch, click the icon of **Done** and click the icon of **Apply and Save.**

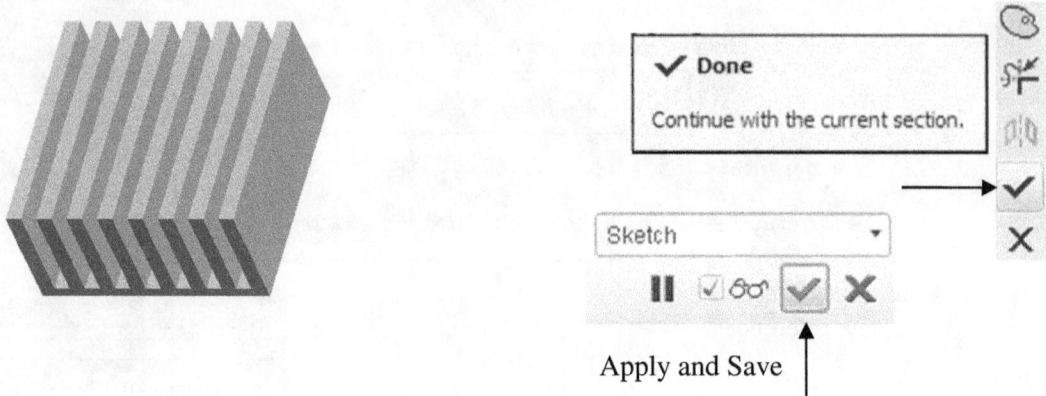

Apply and Save

Step 2: Create 2 datum curves. The 2 datum curves will be used to create 2 surface regions to be used in the process of defining an area residing the head load and an area for a prescribed temperature region.

To create the first datum curve, click the icon of **Sketch**. Pick the planar surface on the bottom of the heat sink component as the sketch plane > **Sketch** to accept the default setting for orientation.

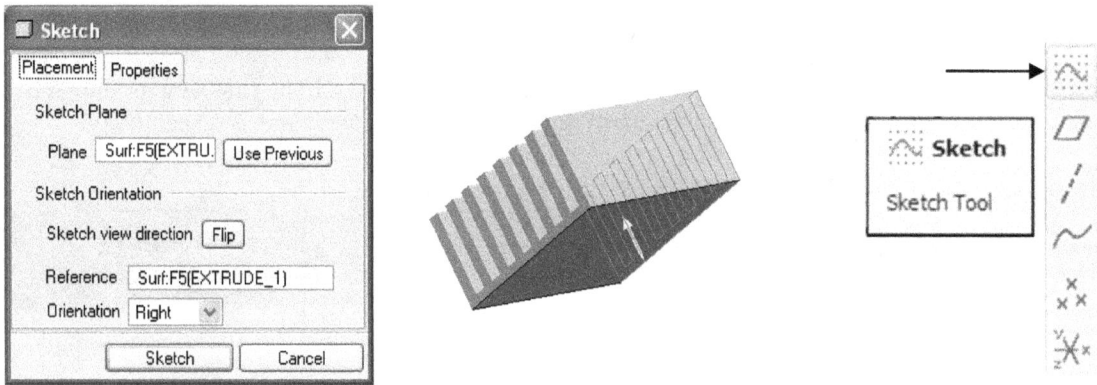

Click the icon of **Rectangle**.and sketch a rectangle, as shown below. The 2 sides of the rectangle are 20 mm. The 2 position dimensions are 12.5 mm. Upon completing the sketch, click the icon of **Done**.

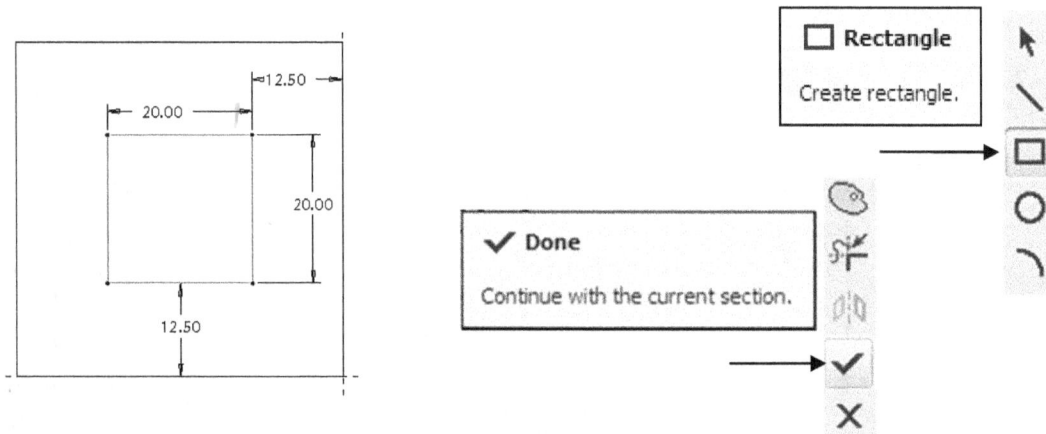

To create the second datum curve, we follow the same procedure stated above. Click the icon of **Sketch**. Pick the planar surface on the bottom of the heat sink component as the sketch plane > click **Sketch** to accept the default setting for orientation.

Click the icon of **Rectangle** from the toolbar of sketcher and sketch a rectangle located in the central position, as shown below. The distance between each side of the rectangle and the wall of the block is 5 mm, as shown.

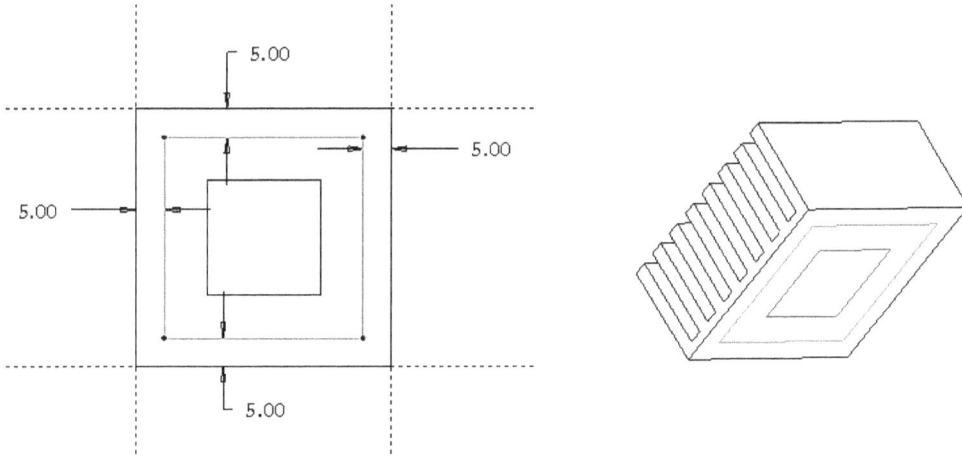

Step 3: Switch from the Pro/ENGINEER design system to Pro/Mechanica
From the main toolbar, **Applications > Mechanica > Continue > Thermal > Bonded > OK.**

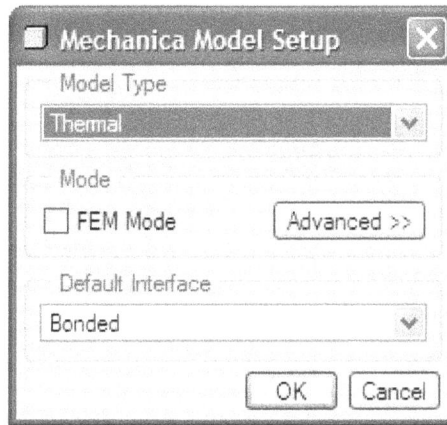

From the toolbar of Mechanica objects, select the icon of **Materials >** select **New** from the left side called **Materials in Model >** because the material type of brick does not exist in the Pro/Mechanica Library. We need to input and install this information related to the material properties based on the given information as listed.

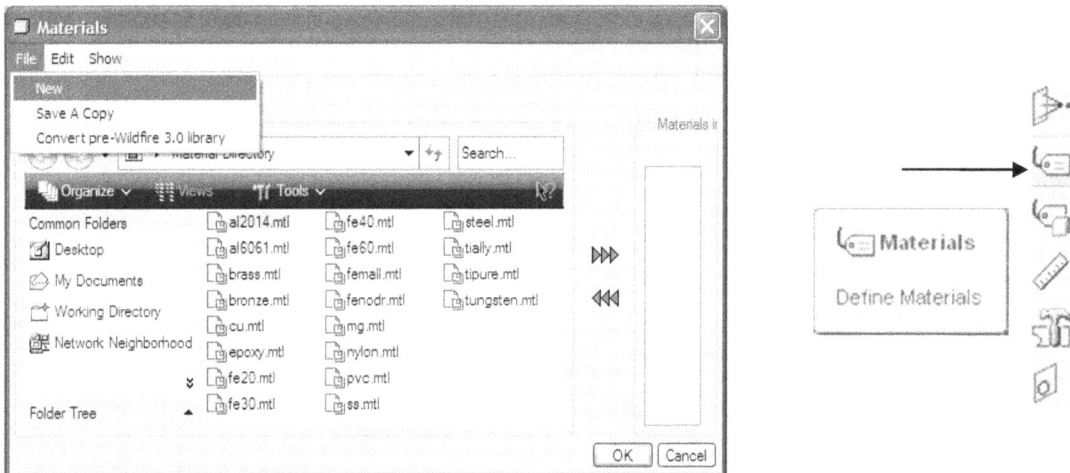

Specify the material properties, both Structural and Thermal, as shown below:

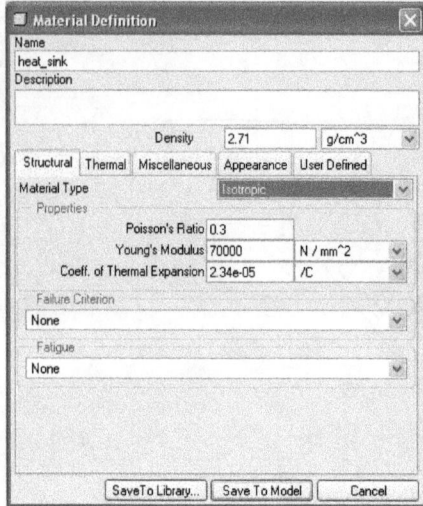

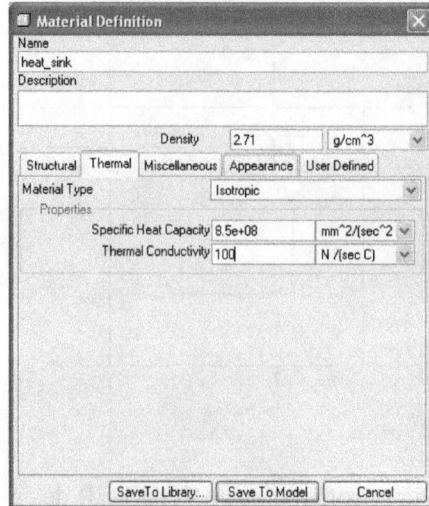

Upon completing the information > **Save To Model** > **OK**.

To assign the selected material to the heat sink component, click the icon of Material Assignment > click **OK** because there is one volume in the model and, by default, the selected material type is assigned to the heat sink component.

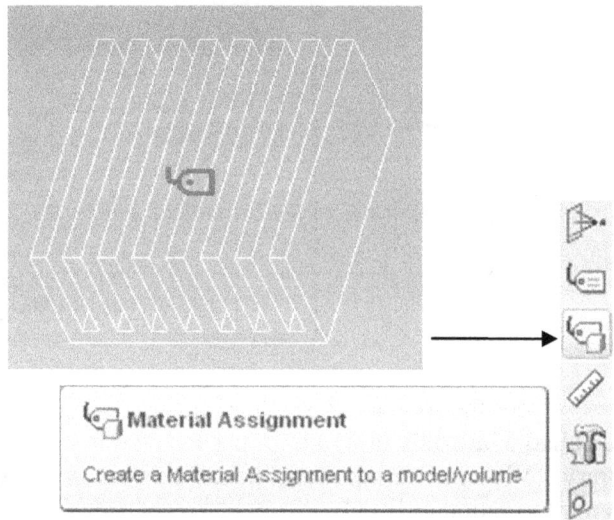

Step 4: Create the first surface regions to be used for defining the heat load.

Click the icon of **Surface Region**. Pick the sketched datum curve first, and pick the surface nearby afterwards. Click the icon of **Apply and Save**.

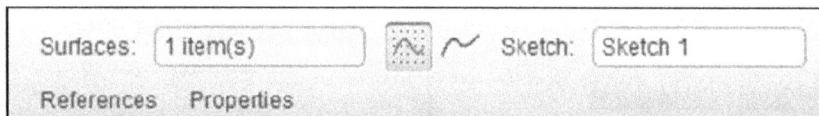

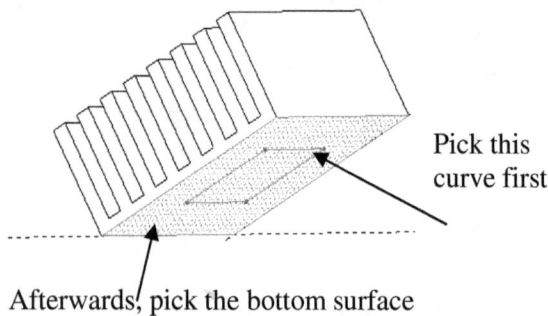

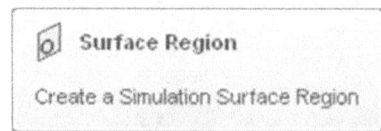

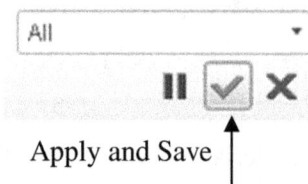

To define the head load representing the CPU as a heat source during the operation, click the icon of **Heat Load** > **Surfaces** > pick the surface region just created. Enter *1000* as the Q value, assuming that the CPU dissipates 1 Watt of heat during the normal operation > **OK**.

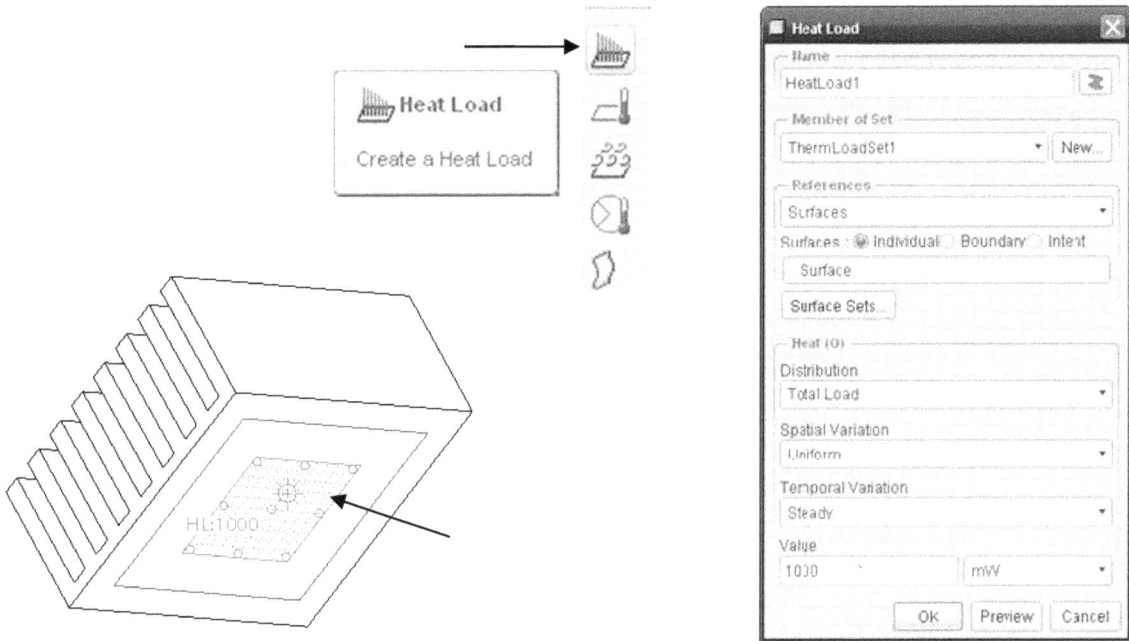

Step 5: Create the second surface regions to be used for defining a prescribed temperature area.

Click the icon of **Surface Region.** Pick the sketched datum curve first, and pick the surface nearby afterwards. Click the icon of **Apply and Save**.

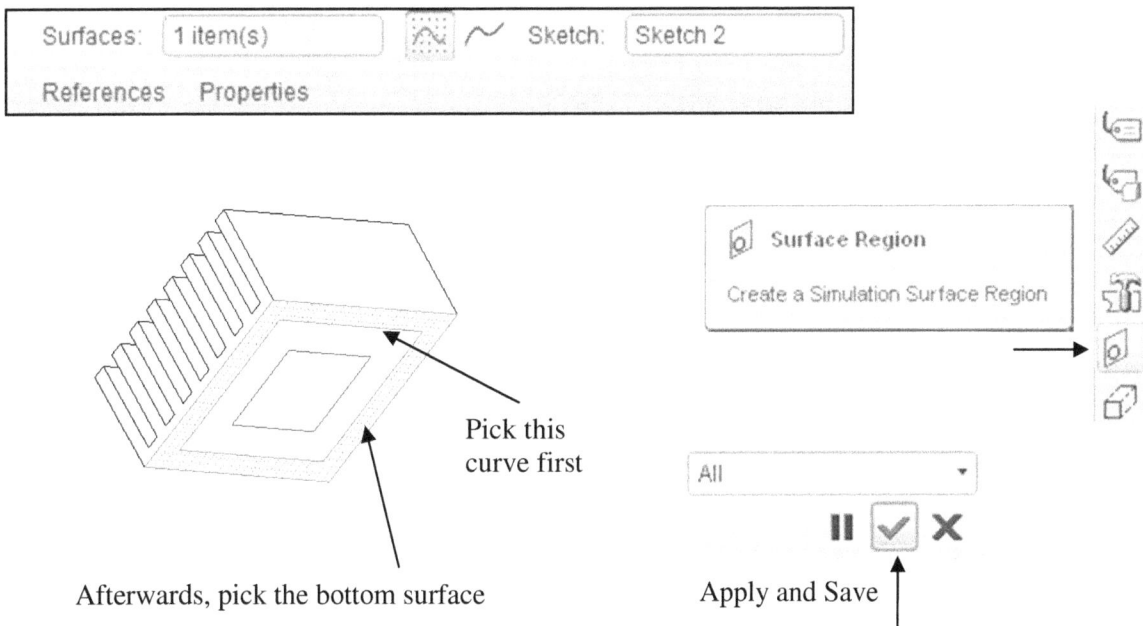

Click the icon of **Prescribed Temperature** > **Surfaces** > pick the surface region just created. Enter 18 for the temperature value > **OK**.

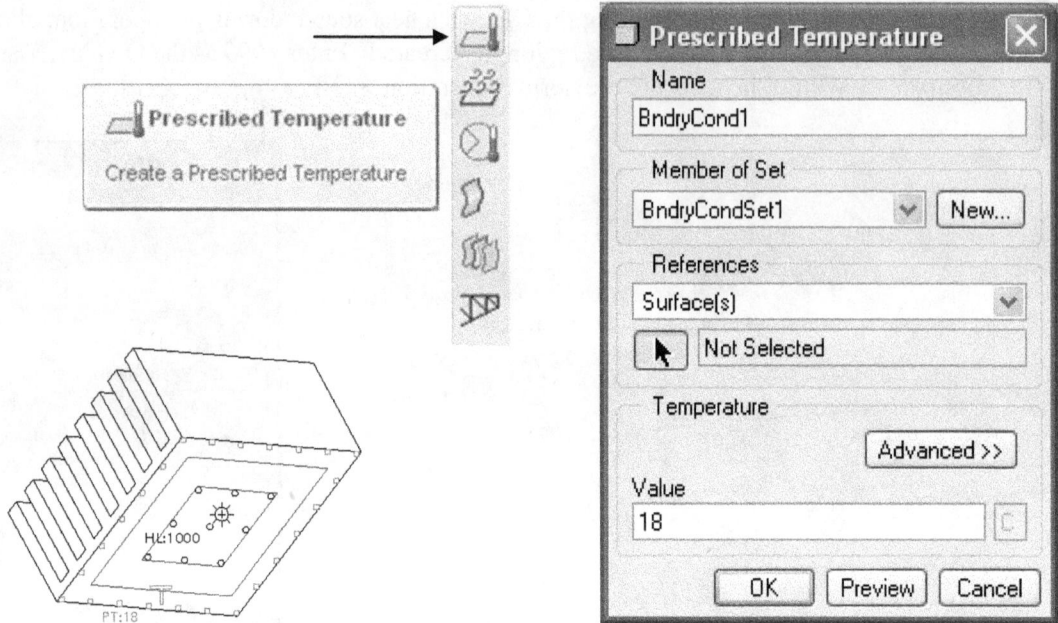

Step 6: Create the convection boundary condition to simulate the natural convection over the entire heat-sink component. Note the bottom surface has to be excluded as the heat load and the prescribed temperature areas are already defined on the bottom surface.

Click the icon of **Convection > Surfaces >** while holding down the **Ctrl** key, pick the 5 surfaces, as shown.

Enter *0.01* as the value of the Convection Coefficient and Enter *25* as the value of the Bulk temperature > **OK**.

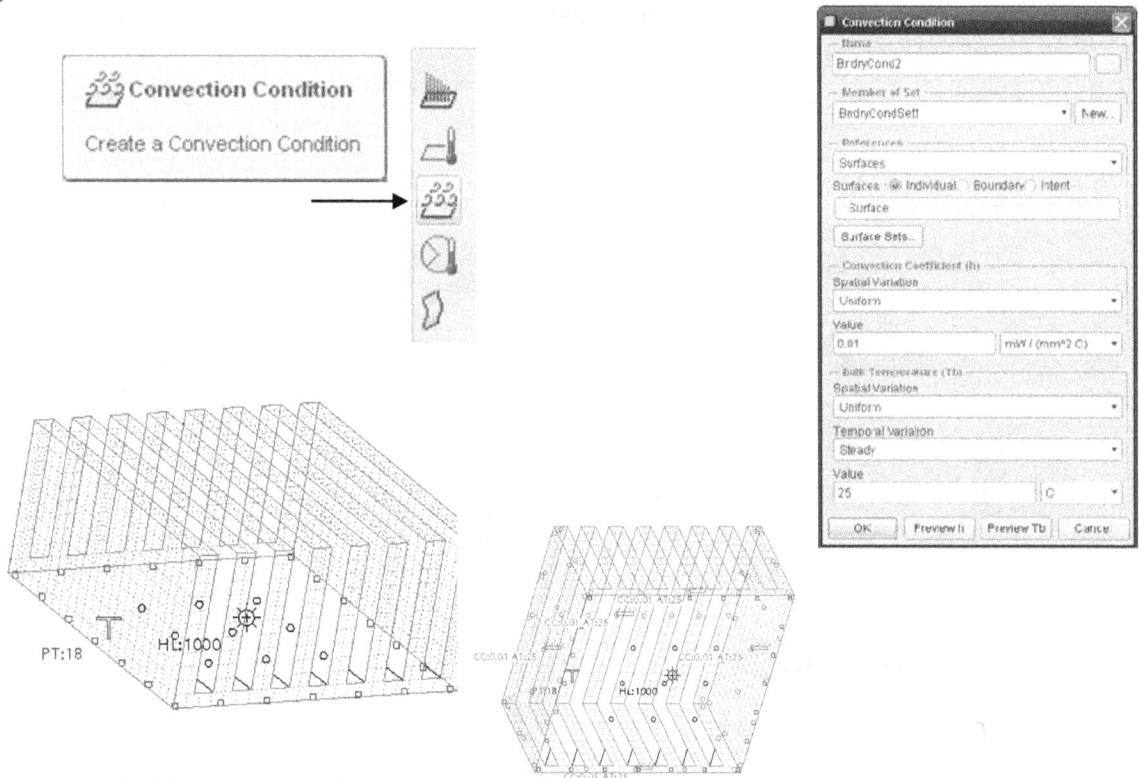

Step 7: Under the Pro/MECHANICA environment, set up **Analyses** and **Run** it

Select the icon of **Mechanica Analyses/Studies > File > New steady state thermal >** type *heat_sink* as the name of the analysis folder > **OK**. Click the box of **Run.**

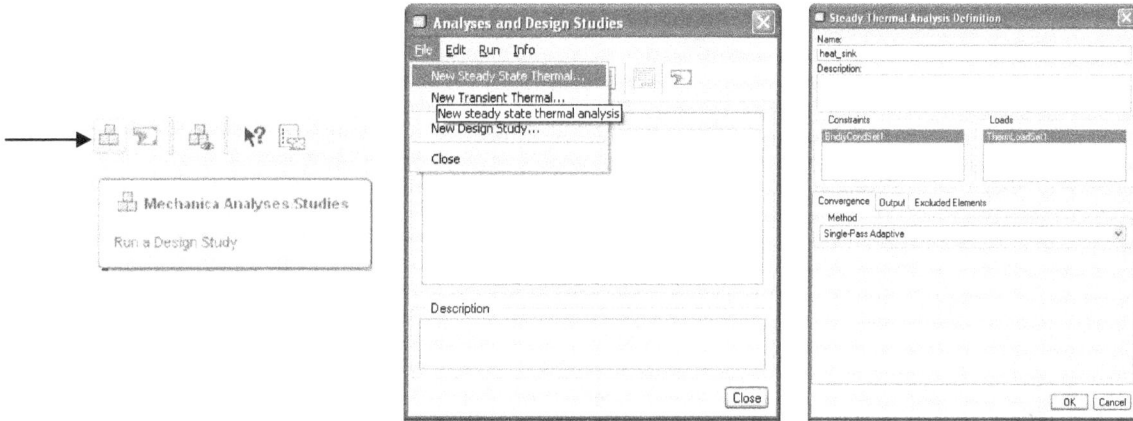

There is a message displayed on screen. The message is asking the user "Do you want error detection?" Click **Yes**. In this way a warning message will appear if an abnormal operation occurs.

To study the obtained result, say the temperature distribution, click the icon of **Review results > Temperature > OK and Show.**

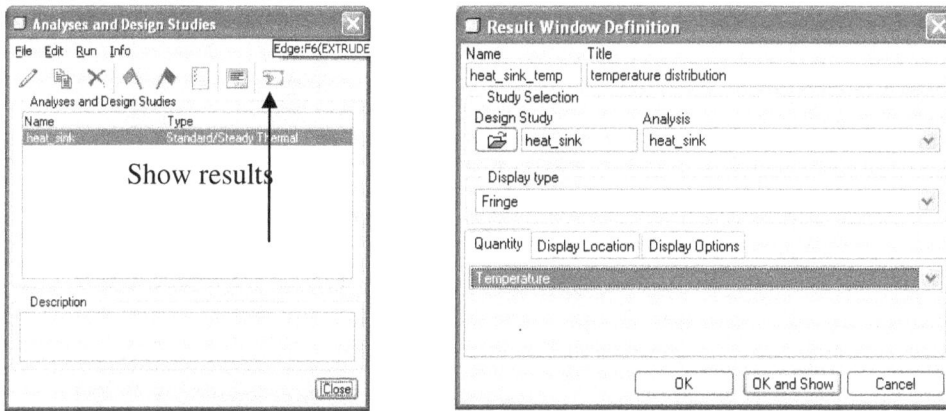

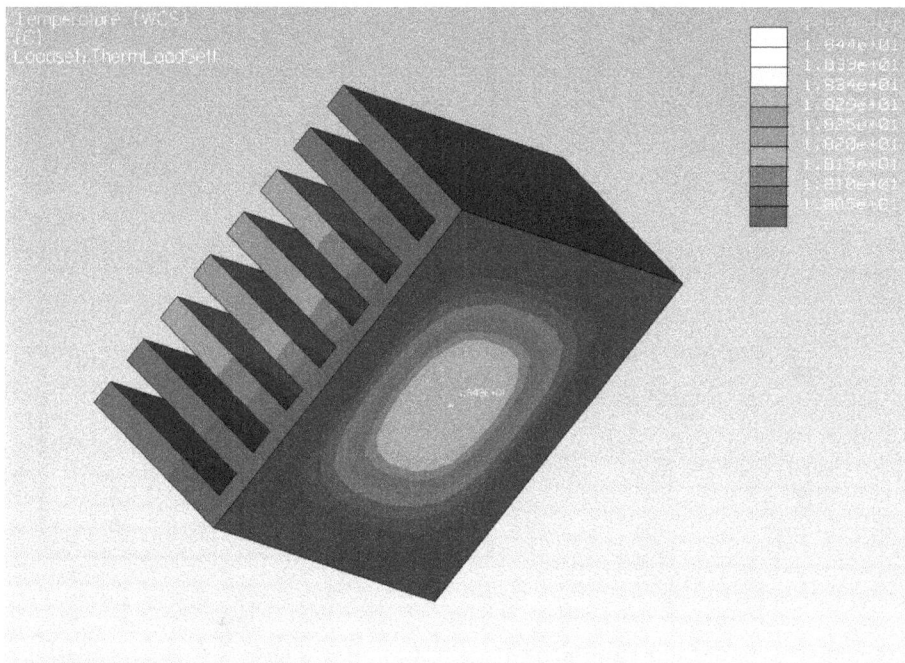

Note that the maximum temperature is located at the center of the bottom surface. The magnitude is 18.49 C°.

Step 8: Let us modify the heat load and the boundary conditions to simulate an over loaded case.
From the model tree, highlight HeadLoad1, right-click and hold > **Edit Definition**. Change the Q value from 1000 to 2000, indicating that the CPU dissipates 2 Watt of heat > **OK**.

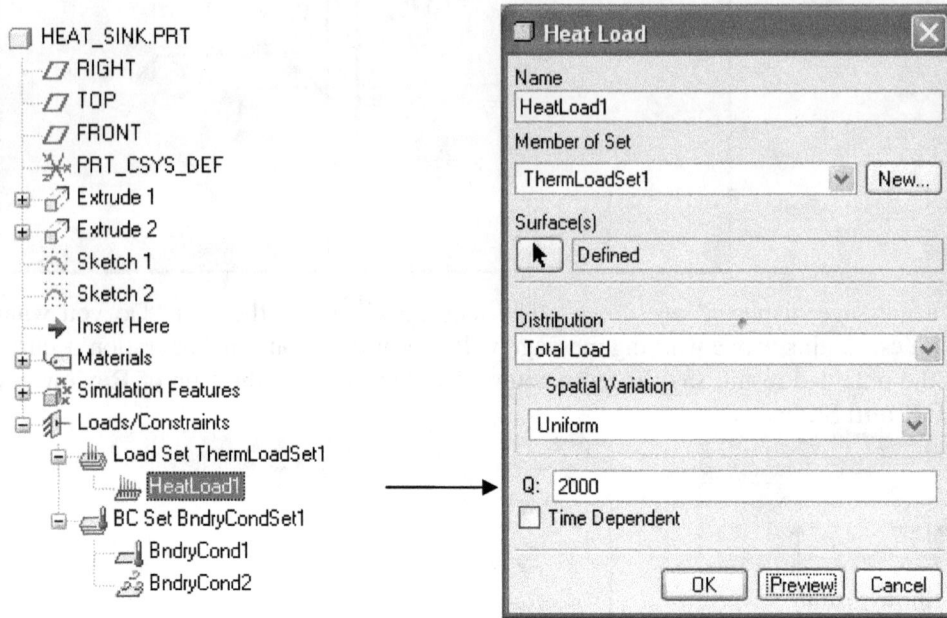

From the model tree, highlight **BndryCond1**, right-click and hold > **Edit Definition**. Change the temperature value from 18 to 25, indicating that the temperature of the support base is increased > **OK**.

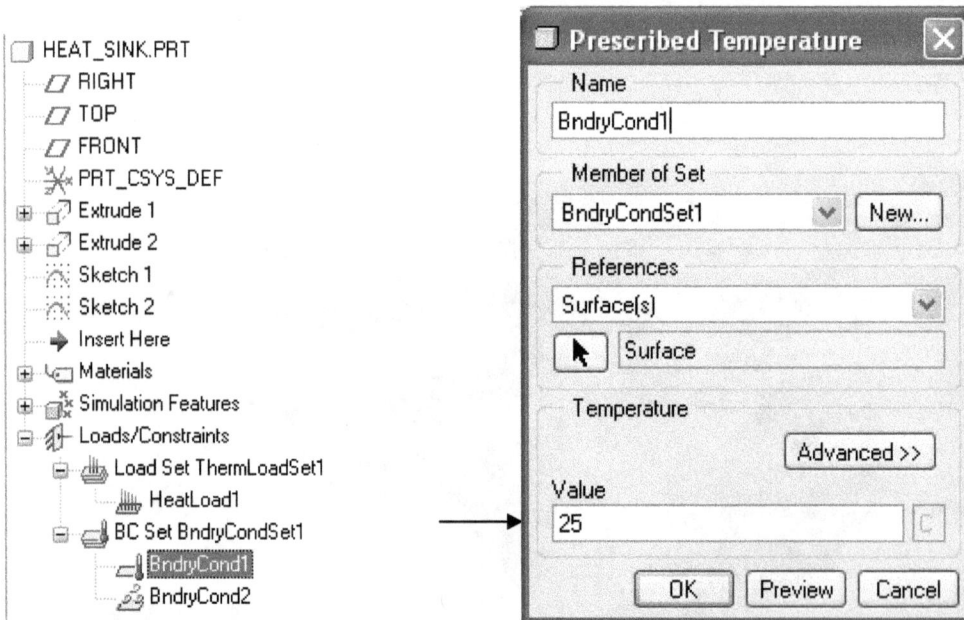

From the model tree, highlight **BndryCond2**, right-click and hold > **Edit Definition**. Change the Bulk Temperature value from 25 to 30, indicating that the temperature of the surroundings has been changed > **OK**.

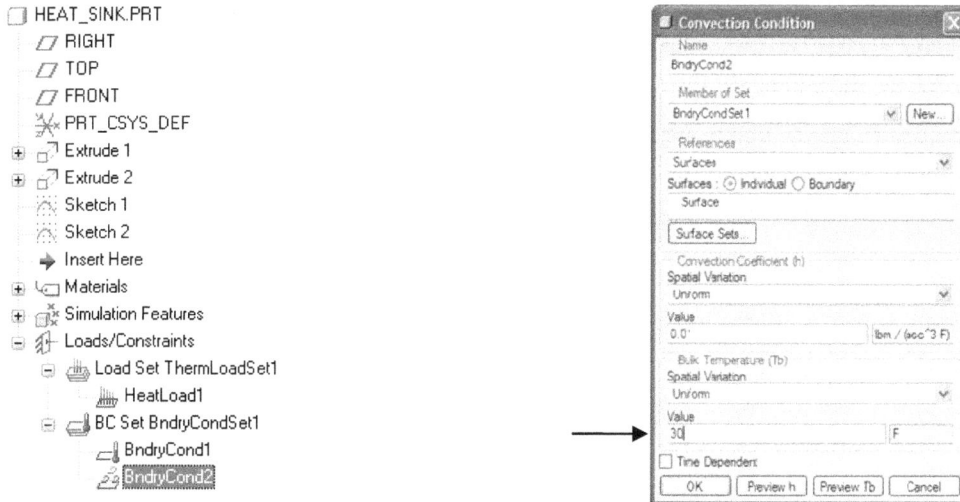

Click the Run Flag to run the program with these new settings. When the run is completed, display the results.

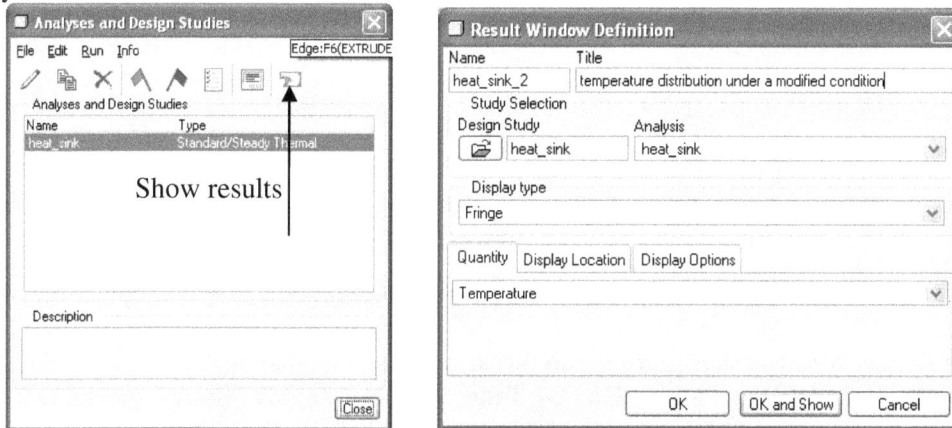

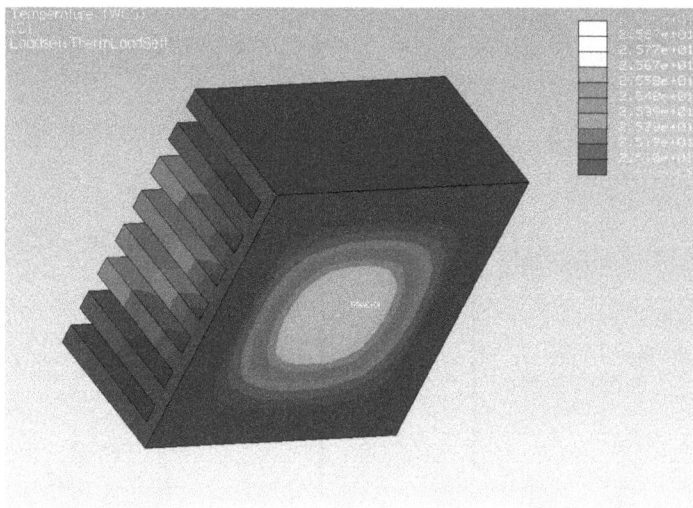

Note that the maximum temperature is still located at the center of the bottom surface. The magnitude is 25.96 C°.

Step 9: Let us add a forced convection boundary condition to simulate that a cooling fan is used to drive a steam of cooling air through the 7 slots. The convection heat transfer coefficient is 0.10, which is 10 times as high as the value for the free convection. Also assume that the temperature of the cooling air is 10C°.

From the toolbar of Mechanica objects, select the icon of surface convection > **Surfaces** > while holding down the **Ctrl** key, pick the surfaces associated with the 7 slots, as shown > **OK**.

Enter *0.1* as the value of the Convection Coefficient and Enter *10* as the value of the Bulk temperature > **OK**.

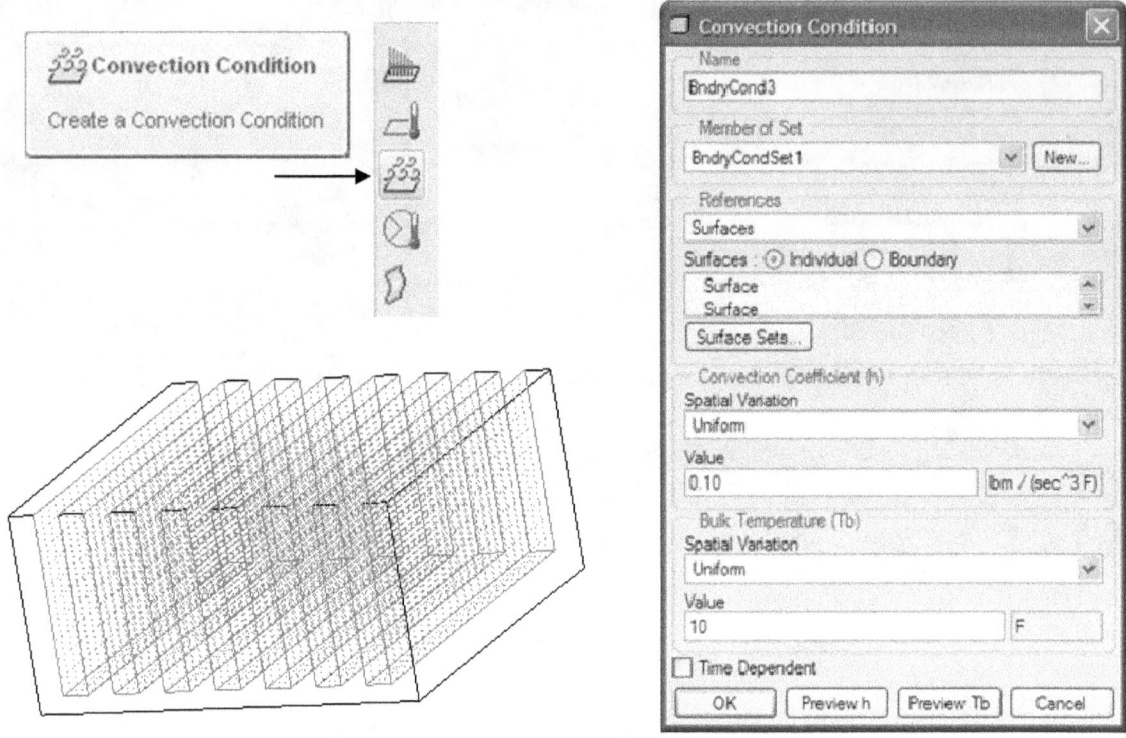

Now let us create a new design study to run this steady state thermal analysis. Select the icon of **Run an Analysis** from the main toolbar > **File** > **New steady state thermal** > type *add_forced_convection* as the name of the analysis folder > **OK**. Click the **Run Flag**.

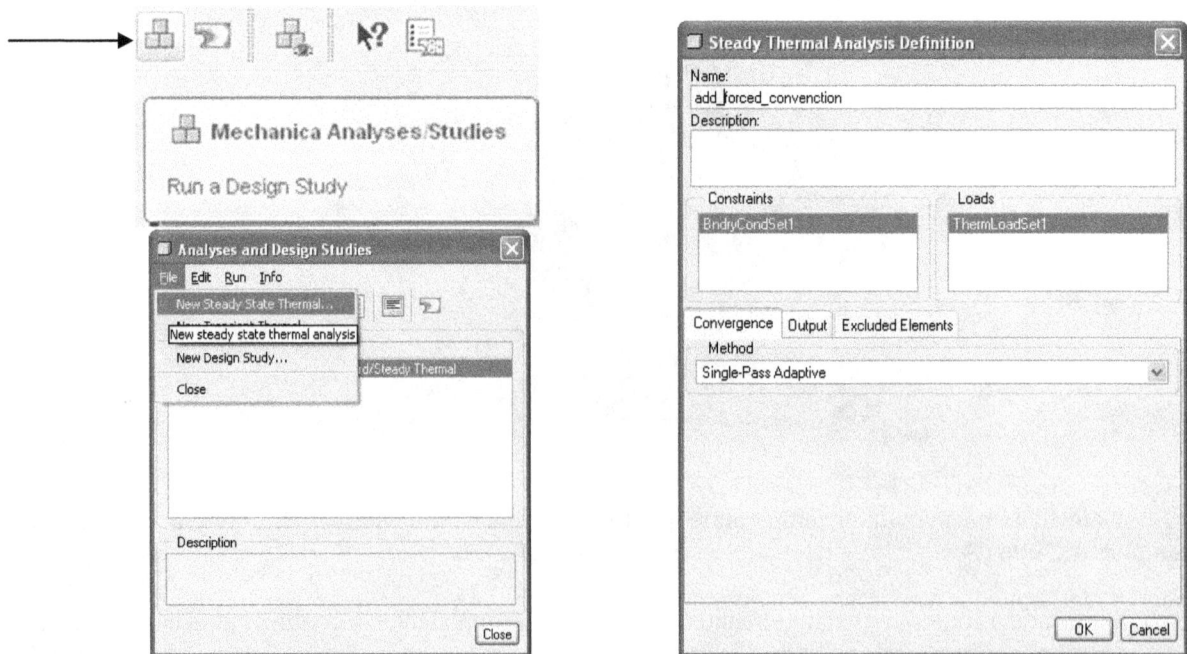

There is a message displayed on screen. The message is asking the user "Do you want error detection?" Click **Yes**. In this way a warning message will appear if an abnormal operation occurs.

When the program run is completed, click the icon of show results.

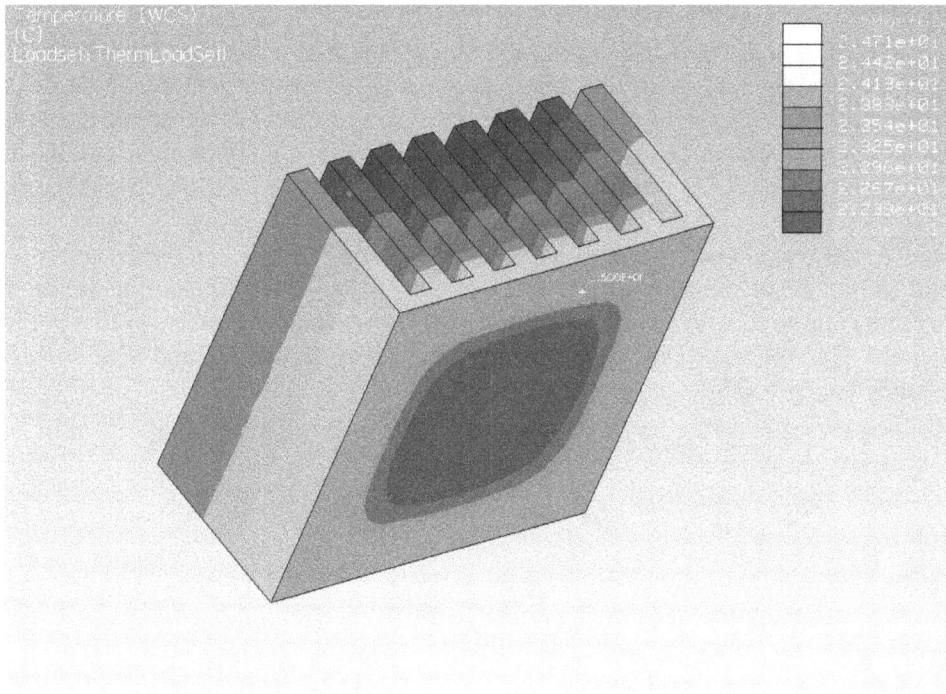

Note that the maximum temperature still remains at 25C°. However, its location is not at the center of the bottom surface where the CPU resides. The temperature at the central location has been dramatically reduced to about 23 C°.

Let us use the method of creating a "Capping Surface" to closely view the temperature distribution inside. From the main toolbar, select **Insert > Cutting/Capping Surfs** > select **Capping Surface** > select **XY > OK** to accept the half section view.

The displayed plot clearly depicts that the temperature gradient distribution, which is inside of the heat sink component. The added cooling fan has brought a stream of cooling air to the system, which has effectively taken the heat away in those areas closed to the head load, or those areas closed to the CPU installation site.

5.6 Thermal Analysis Combined with Structural Analysis

Results obtained from thermal analyses usually indicate the variation of temperature distribution. It is not uncommon to observe, in the real time, that internal stress may develop under circumstances where structural constraints present. In this section, we present 2 examples to demonstrate the procedure to perform a combined thermal-structural analysis. The first example is a thermal plate. In this case study, we first create the 3D solid model. Afterwards, we perform a combined thermal-structural analysis.

Step 1: Create a 3D solid model for the plate structure.
Click the icon of **Create a new object** from the menu toolbar to initiate the creation of a 3D model. Type *thermal_plate* as the file name, clear the box of **Use default template > OK**. Select the unit of **mmns_part_solid**, type *thermal plate* under the **description** of the model, and type *student* or *your name* under the **modeled_by > OK**.

Pick up the icon of **Extrude** displayed on the toolbar of feature creation. Specify the depth value equal to 20. To initiate the process of sketching, activate **Placement > Define.**

Select the **FRONT** datum plane displayed on screen as the sketch plane. Pick the box of **Sketch** to accept the **RIGHT** datum plane as the default reference to orient the sketch plane, as illustrated below:

Before making a 2D sketch, let us first create 2 centerlines, a horizontal centerline and a vertical centerline passing through the origin of the coordinate system, as shown. To do so, right-click and hold, select **Centerline**. Users may select the icon of **Centerline**. Sketch a vertical centerline and a horizontal centerline.

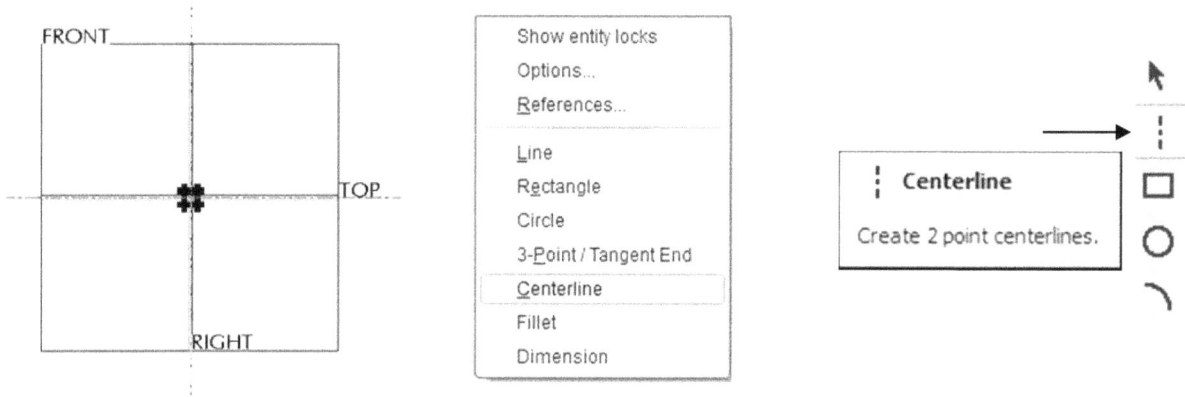

Select the icon of **Rectangle** and sketch a rectangle, which is symmetric about the 2 sketched horizontal and vertical centerlines, as shown. The 2 dimensions are 500 and 400, respectively.

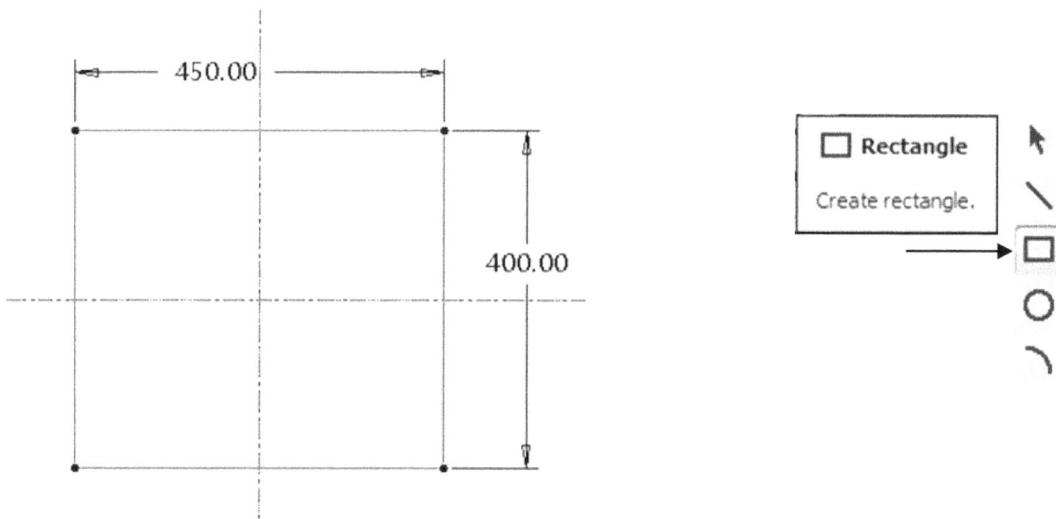

Click the icon of **Circle** and sketch a circle, as shown. The diameter dimension is 150.

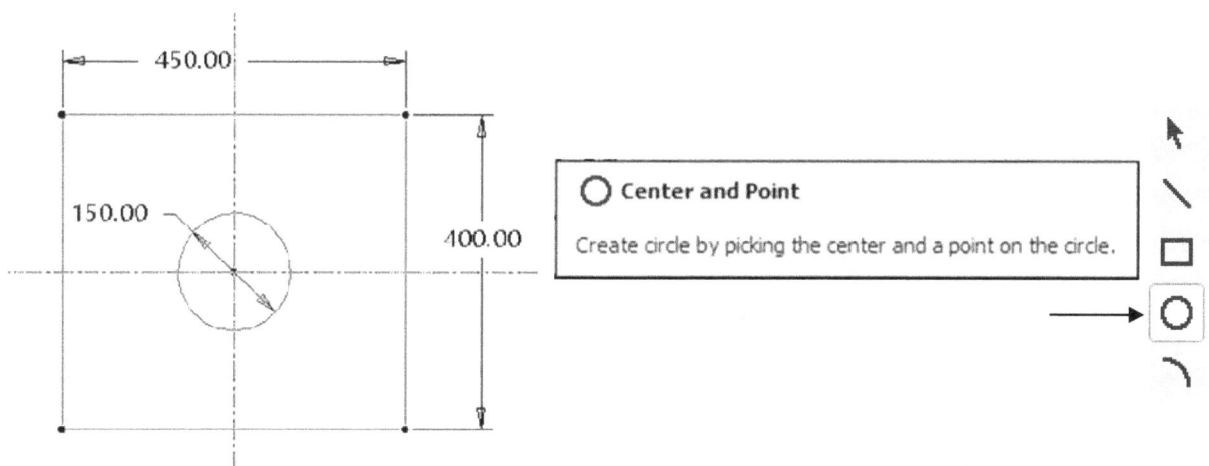

Click the icon of **Rectangle** and sketch a rectangle, as shown. The 2 dimensions are 100 and 60, respectively.

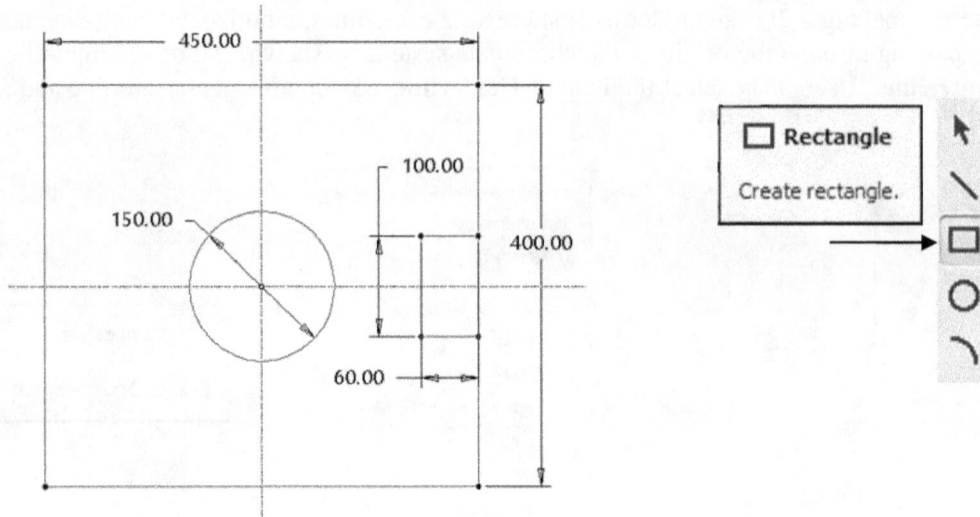

Click the icon of **Delete** and remove those parts from the 2 sketched rectangles.

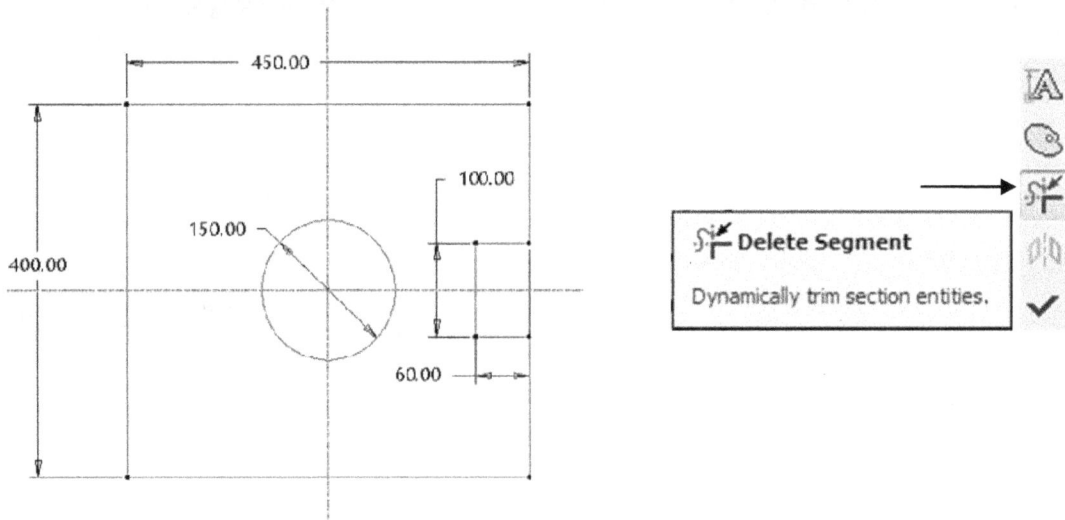

Upon completing the sketch, select the icon of **Done** from the toolbar of sketcher. Select the icon of **Apply and Save.**

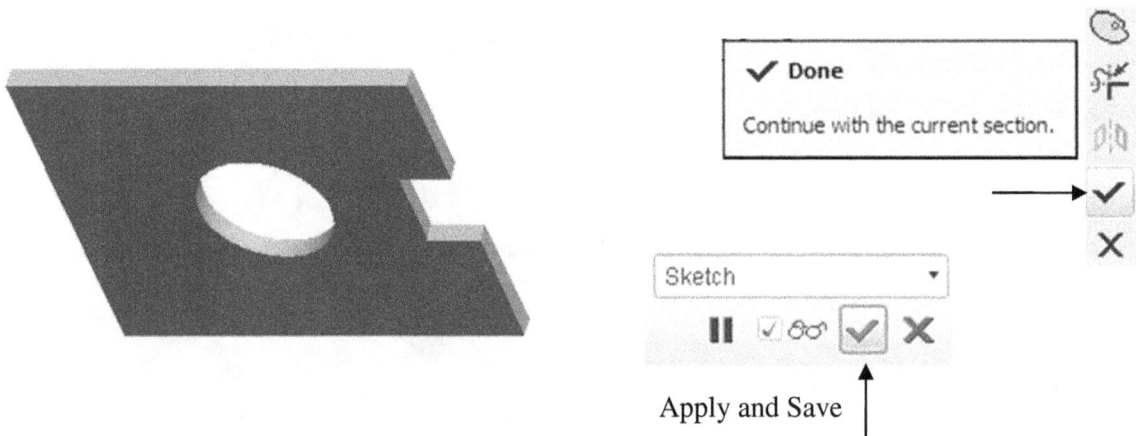

Apply and Save

To control the amount of heat energy going through the thermal plate, we need to reduce the contact area. Click the icon of **Extrude**. Select **Cut** and **Through All**. Activate **Placement > Define**.

Through All   Cut

Extrude

Extrude Tool

Sketch

• Select 1 item   Define...

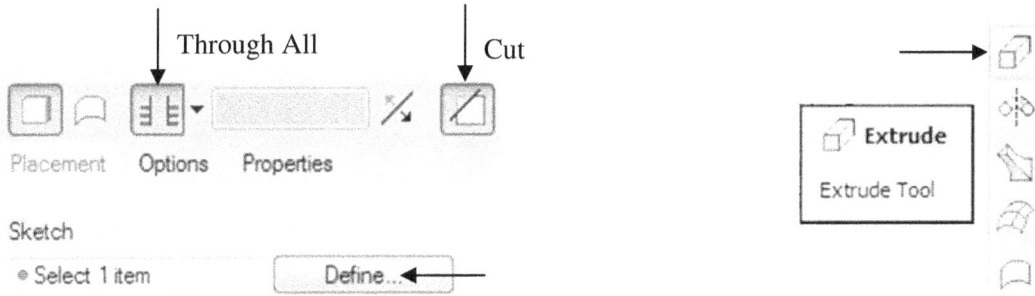

Select the surface from the thermal plate as the sketch plane and click Sketch to accept the default setting for the orientation.

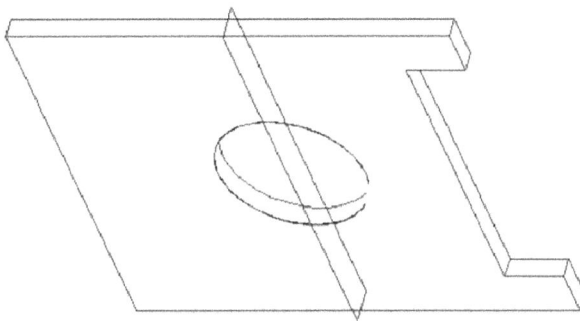

Sketch

Placement

Sketch Plane

Plane Surf:F5(EXT Use Previous

Sketch Orientation

Sketch view direction Flip

Reference RIGHT:F1(DATUM PL...

Orientation Right

Sketch Cancel

Click **Sketch > References**. Add 3 new references, as shown below.

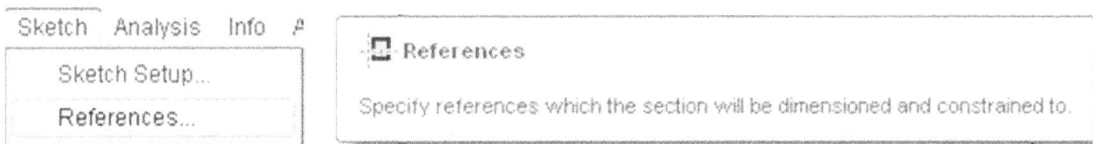

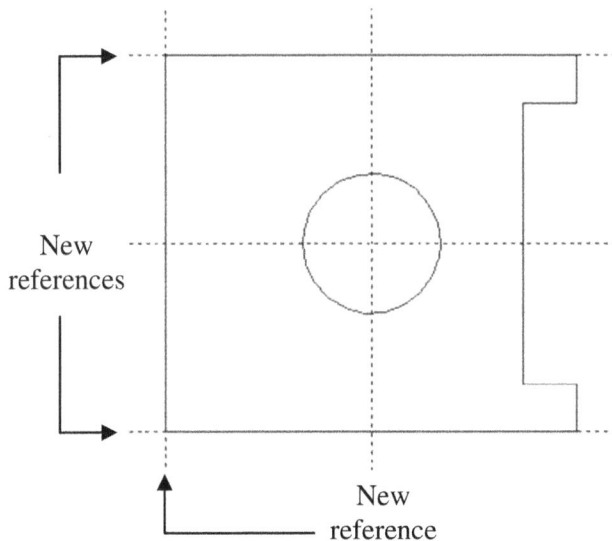

Sketch Analysis Info A

Sketch Setup...

References...

References

Specify references which the section will be dimensioned and constrained to.

References

F1(RIGHT)
F2(TOP)
Surf.F5(EXTRUDE_1)
Surf.F5(EXTRUDE_1)
Surf.F5(EXTRUDE_1)

X sec Select Use Edge/Offset

Replace Delete Solve

Reference status

Unsolved sketch

Close

New references

New reference

Select the icon of **Rectangle** and sketch 2 rectangles, which are symmetric about the sketched horizontal centerline, as shown. The 2 dimensions are 500 and 400, respectively.

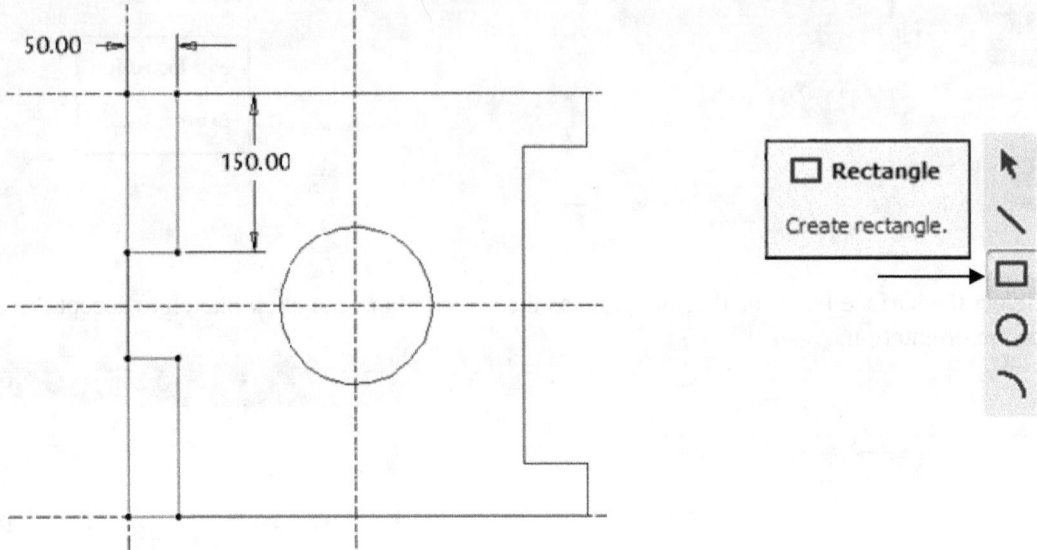

Upon completing the sketch, select the icon of **Done** from the toolbar of sketcher. Select the icon of **Apply and Save** from the feature control panel, completing the creation of a 3D solid model for the plate structure.

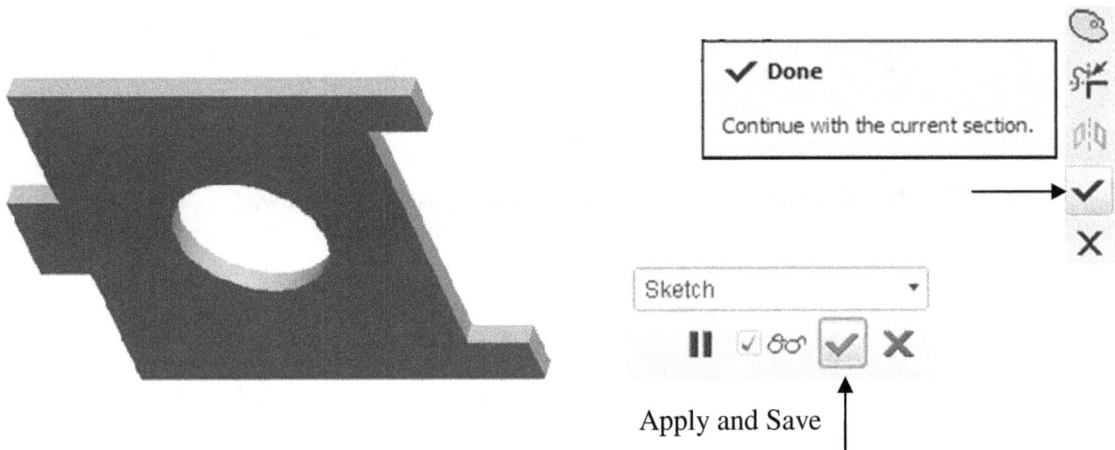

Apply and Save

Step 2: Switch from the Pro/ENGINEER design system to Pro/Mechanica

To enable Pro/Mechanica, from the main toolbar, select **Applications > Mechanica > Continue > Thermal > Bonded > OK.**

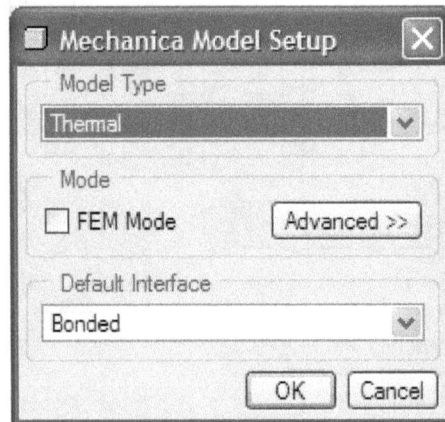

To specify the material type, select the icon of **Define Materials >** select **STEEL** from the list of Materials in Library > click the directional arrow to switch the selection to the list of **Materials in Model > OK**.

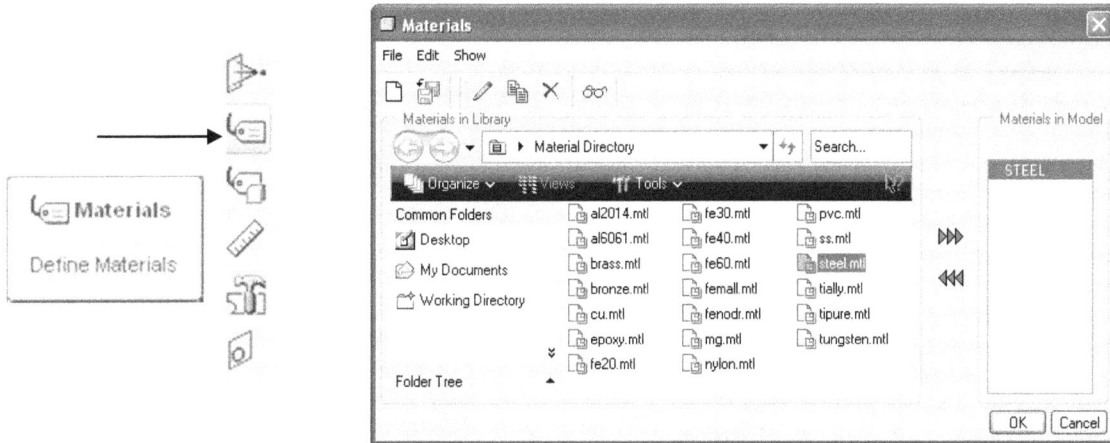

Select the icon of **Material Assignment >** in the Material Assignment window, the material type **STEEL** is already shown and the assignment has already been done because there is only one volume or one component in this case > click **OK** to accept the material assignment. In the model tree, the material type and the material assignment are also listed.

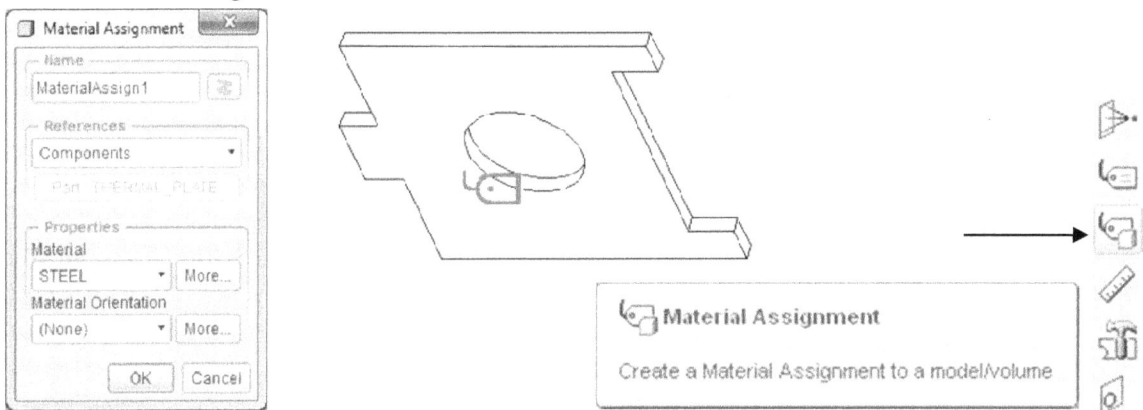

Step 3: Define the prescribed temperature boundary condition.

From the toolbar of Mechanica objects, select the icon of **prescribed temperature > Surfaces >** pick the surface on the left side of the plate. Enter 20 for the temperature value > **OK**.

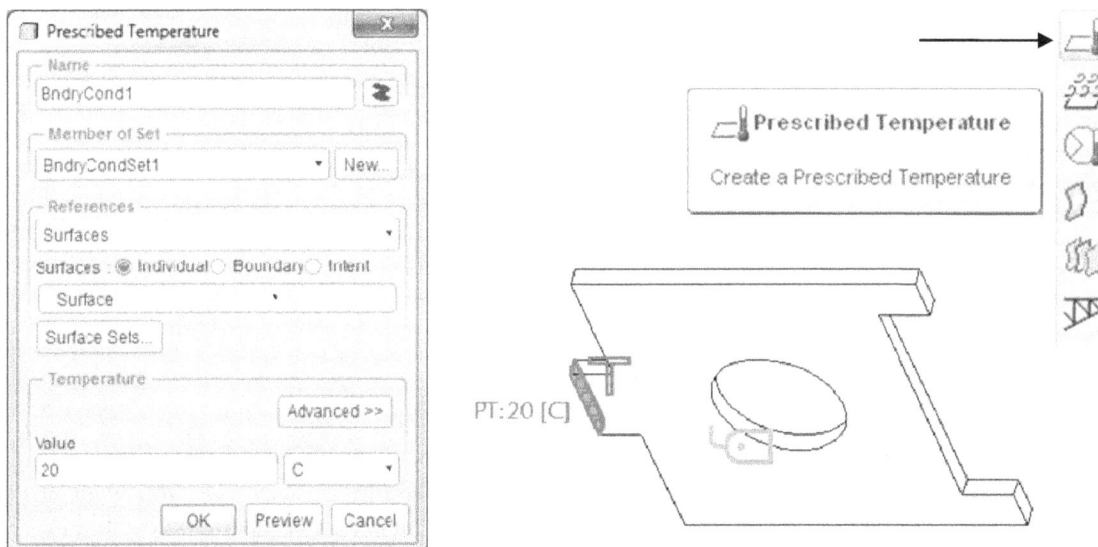

Step 4: Define the convection boundary condition.
     Select the icon of **Create a convection condition** > **Surfaces** > pick the surface on the right side of the plate. Enter 4 for the convection coefficient and enter 10 for the temperature value > **OK**.

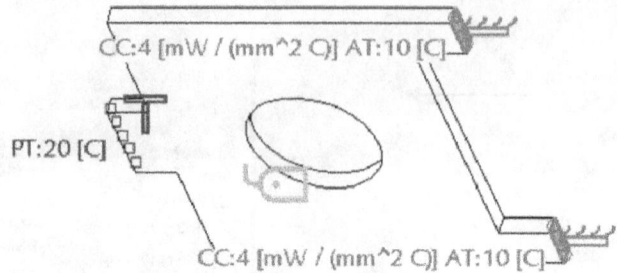

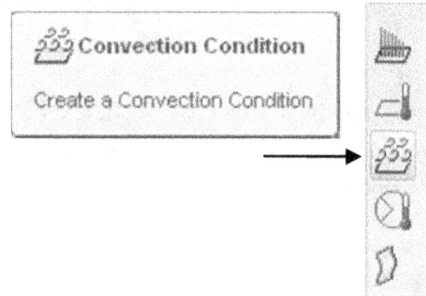

Step 5: Under the Pro/MECHANICA environment, set up **Analyses** and **Run** it
     Select the icon of **Mechanica Analyses/Studies** from the main toolbar > **File** > **New steady state thermal** > type *thermal_plate* as the name of the analysis folder > **OK.** Click the **Run Flag**. It is important to note that there is no load in this case. The heat transfer process is due to the prescribed temperature and convection condition on the left and right sides of the plate.

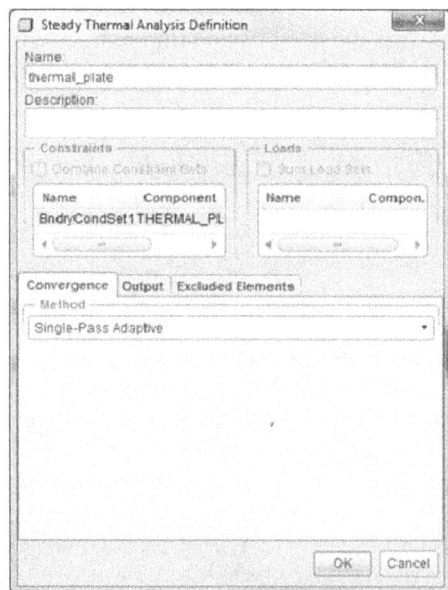

     There is a message displayed on screen. The message is asking the user "Do you want error detection?" Click **Yes**. In this way a warning message will appear if an abnormal operation occurs.
     To monitor the computing process, users may click the icon of **Display study status**.

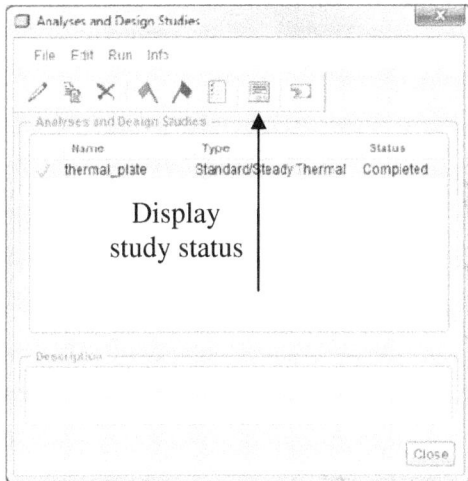

Measures:

```
energy_norm:        1.454209e+04
max_flux_mag*:      6.512570e+00
max_flux_x*:        4.034123e+00
max_flux_y*:        4.934951e+00
max_flux_z*:       -3.107411e+00
max_grad_mag*:      1.514110e-01
max_grad_x*:       -9.378950e-02
max_grad_y*:       -1.147329e-01
max_grad_z*:        7.224433e-02
max_temperature:    2.000000e+01
min_temperature:    1.033400e+01
```

To study the obtained result, say the temperature distribution, click the icon of **Review results > Temperature > OK and Show.**

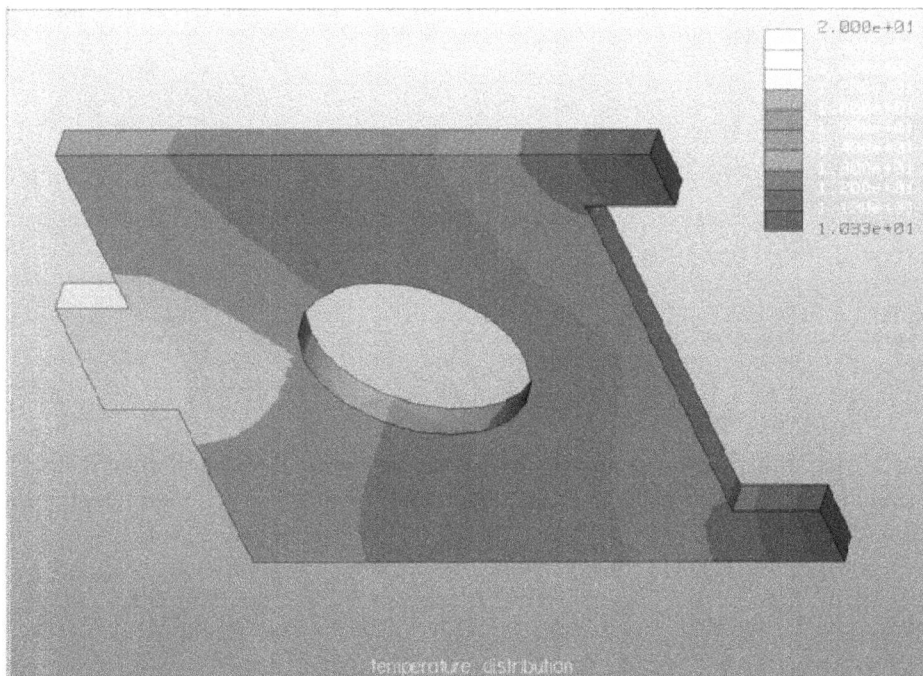

Let us edit the temperature fringe plot. Go to Display Option and select Contour, and Isosurfaces > **OK and Show**.

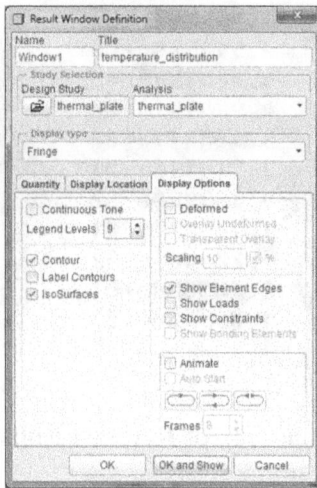

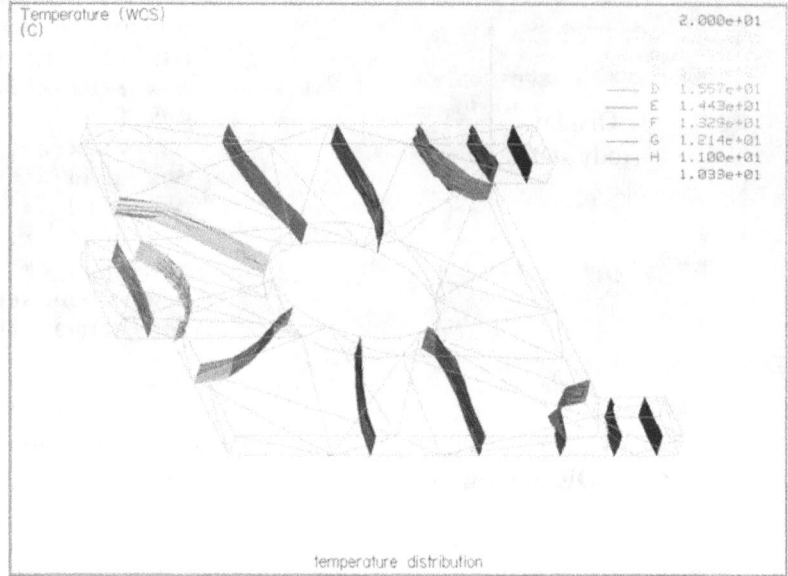

Let us create a new window called *Heat_Flux*. Select **Vectors** under Display Type. Select **Flux** under Quantity. Select **Magnitude** under Component. In the Display Options tab, select the **Animation** option > **OK and Show**. An animated representation of the heat flux through the part is displayed. Notice the flux vectors are scaled and colored according to the magnitude of the local heat flux. Also, the vector direction is perpendicular to the isotherm.

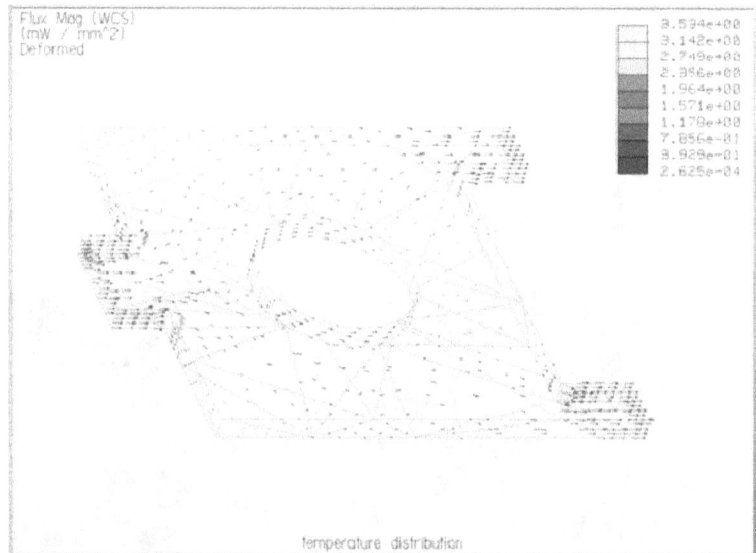

Step 6: Transfer the temperature distribution to a thermal load and apply it to the thermal plate.

First, we need to go to the Structural Module. Click **Edit** from the top menu > **Mechanica Model Setup** > **Structure** > **OK**.

To apply the thermal load, click **Insert** from the top menu > **Temperature Load** > **MEC/T**. From the pop up window, type temperature_load as the name. Make sure Previous Design Study is checked. Set the Reference Temperature to be 10°C to match the convection condition used > **OK**.

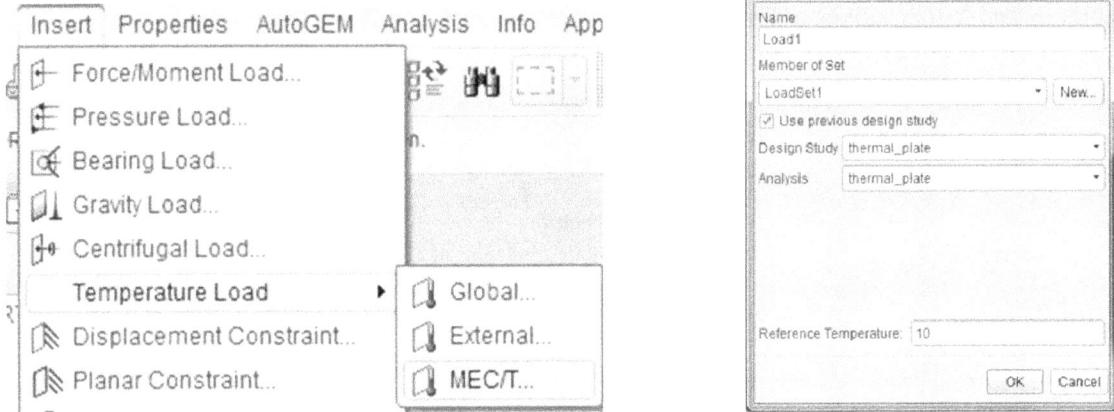

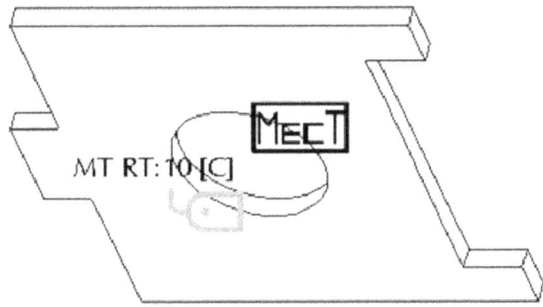

Step 7: Define the displacement constraint conditions.

Click the icon of **Constraints >** select the surface at the left side of the thermal plate > **OK** to accept the default settings, namely, fixing all 6 degrees of freedom > **OK**.

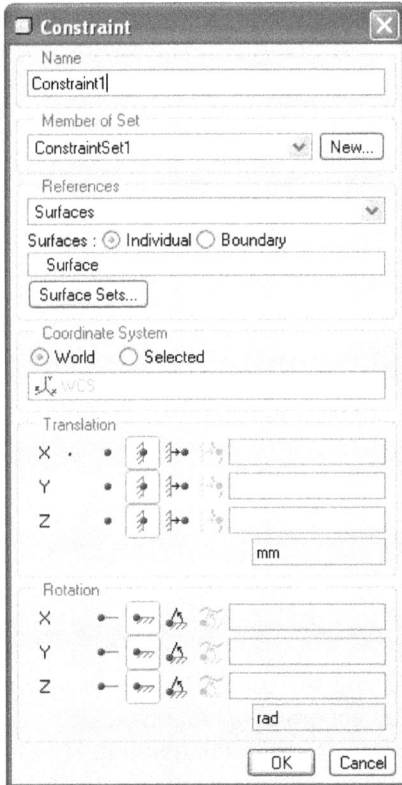

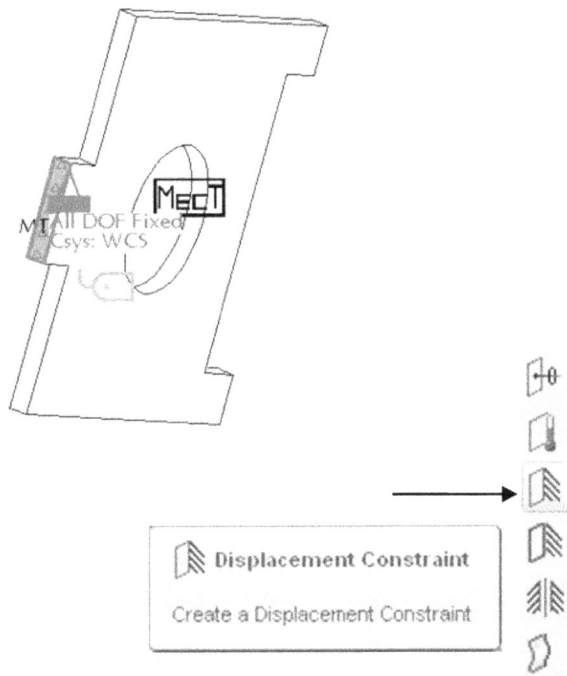

Click the icon of **Constraints,** again **>** select the surface at the front end of the ring support **>** **OK >** set the X and Y translations to Free and fix the other 4 degrees of freedom **> OK**.

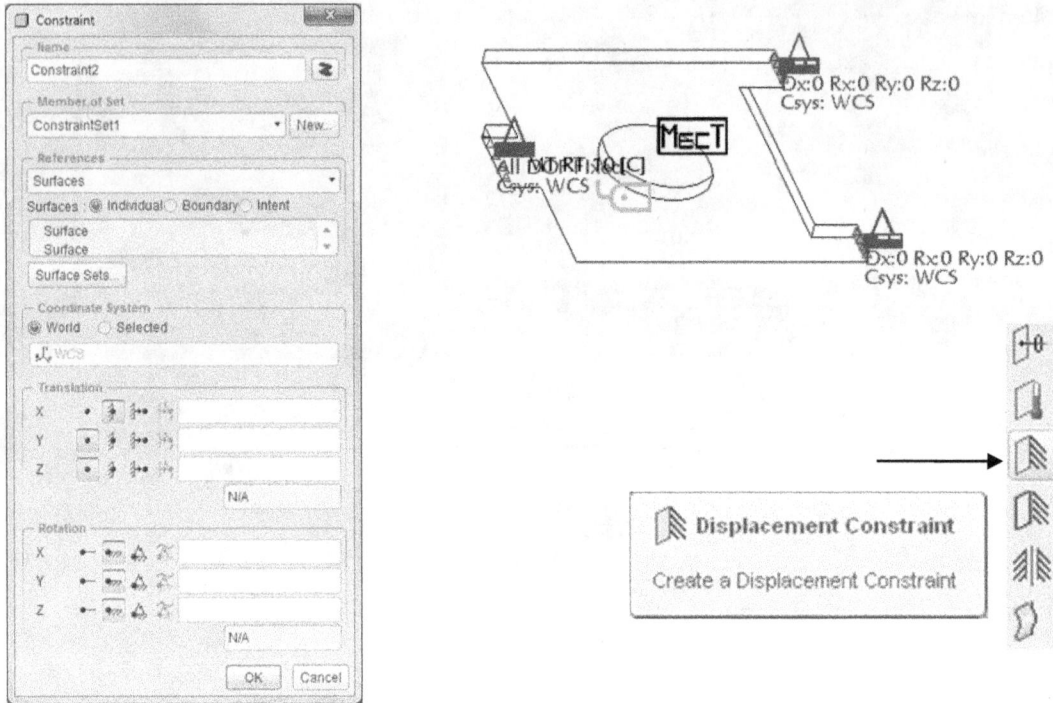

Step 7: Under the Pro/MECHANICA environment, set up **Analyses** and **Run** it

Select the icon of **Mechanica Analyses/Studies > File > New Static >** type *temperature load* the name of the Design Study Folder **> OK**.

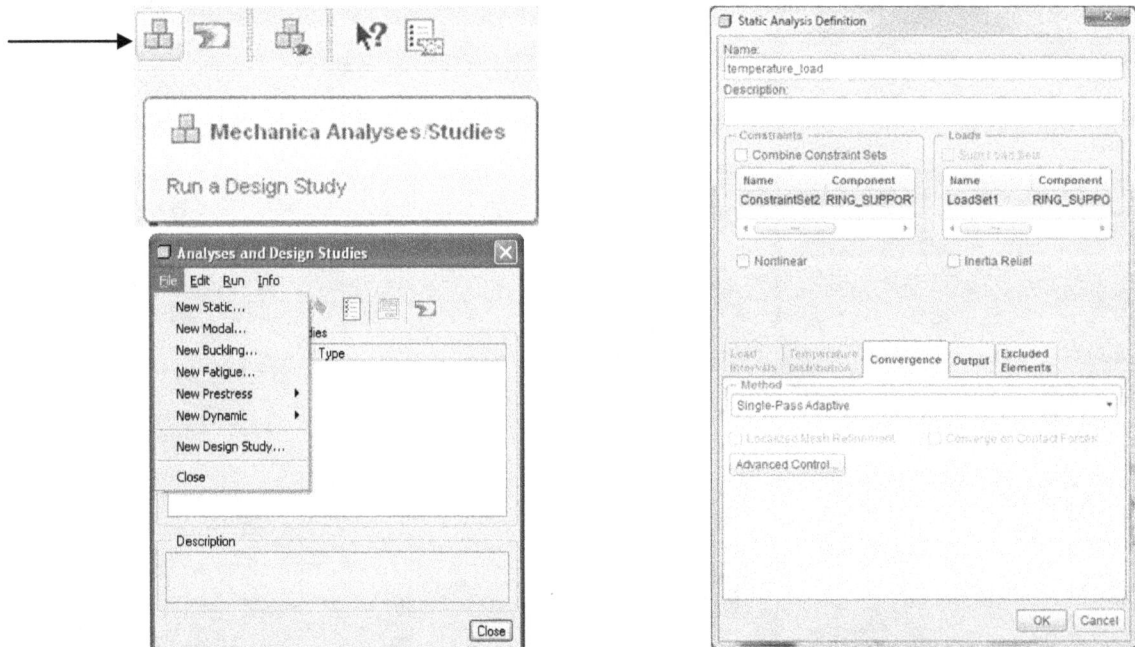

There is a message displayed on screen. The message is asking the user "Do you want error detection?" Click **Yes**. In this way a warning message will appear if an abnormal operation occurs.

To study the obtained result, say the distribution of the von Mises stress distribution, click the icon of **Review results > Stress > von Mises Stress > OK and Show**. The maximum value is 111.1 MPa.

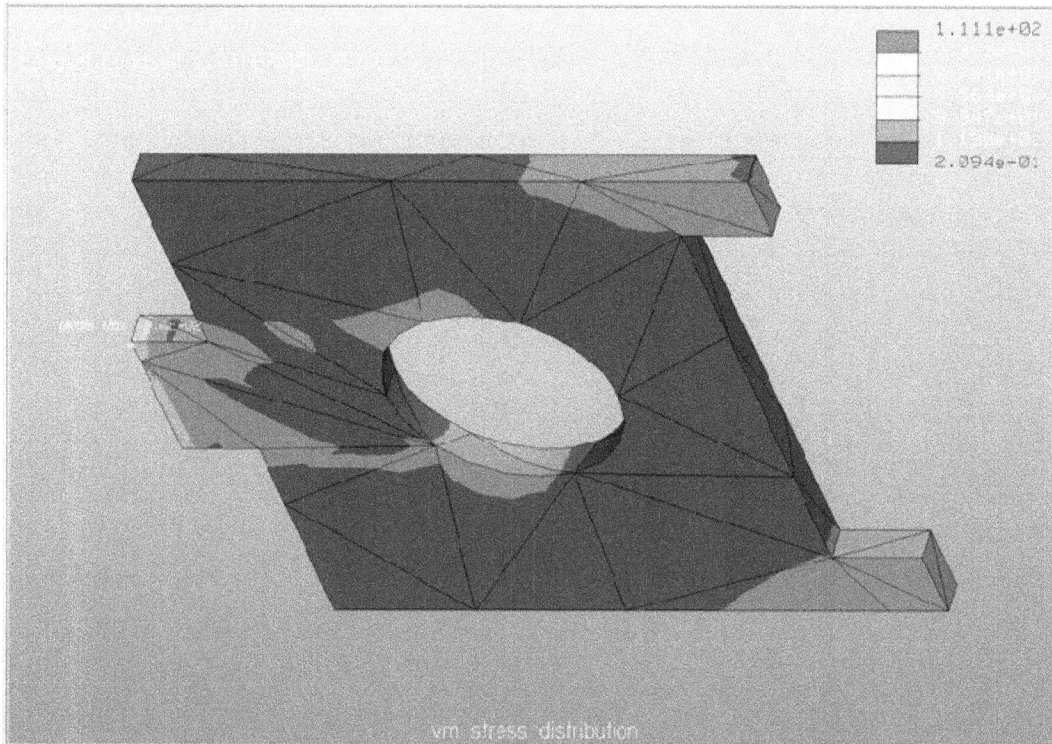

vm stress distribution

To plot the distribution of the displacement due to the temperature load applied to the ring support component, select **Displacement > Magnitude > OK and Show**. The maximum value is 0.01723 mm.

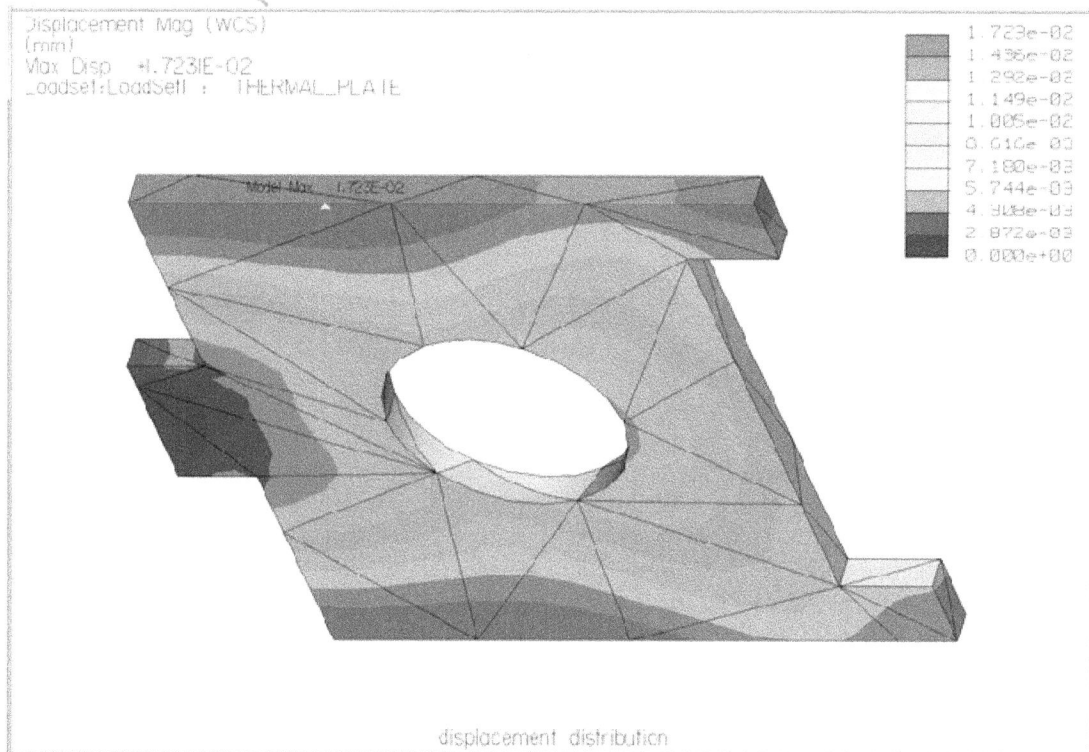

displacement distribution

Let us plot the distribution of the displacement using Model, Change **Fringe** to **Model**, and select **Displacement > Magnitude > OK and Show**. The maximum value is 0.125 mm.

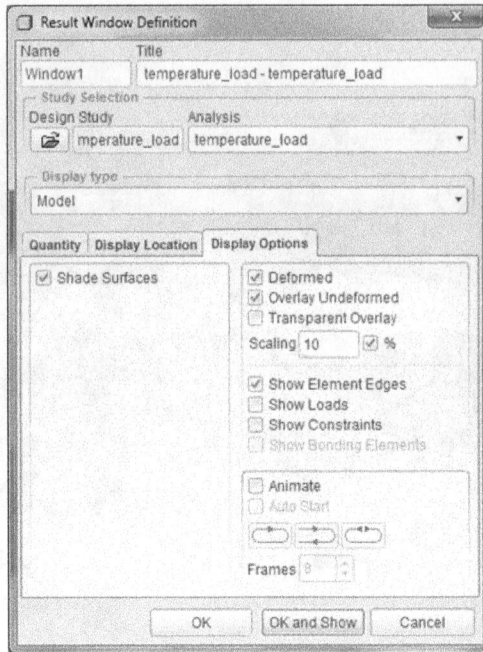

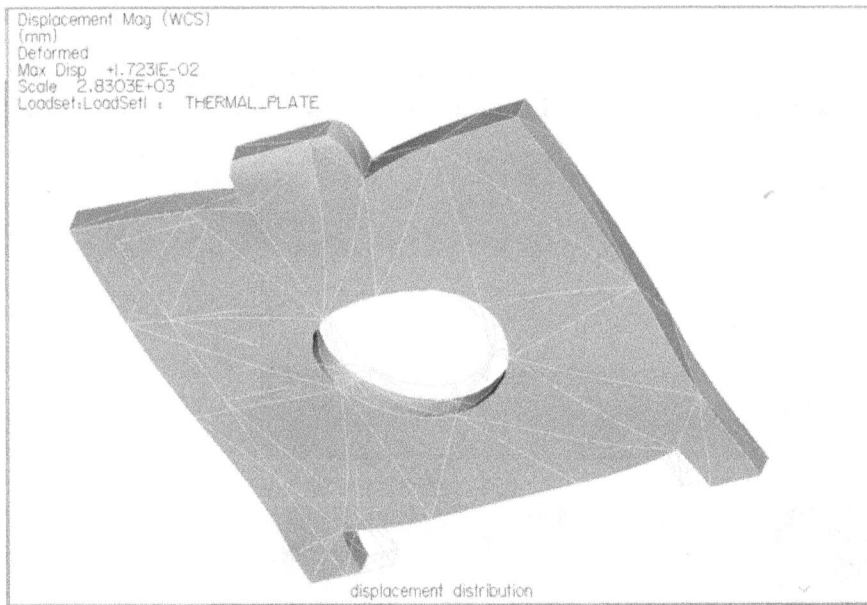

displacement_distribution

In the following example, the procedure to perform a combined thermal-structural PEA is demonstrated. The 3D solid model is a ring support component. There is a pressure load acting on the ring support component. We first perform a structural analysis to obtain the maximum value of the vm stress. Secondly, we evaluate the temperature distribution due to the heat load acting on the hole (simulating the effect of friction force during rotation). Afterwards, we evaluate the stress distribution due to the temperature variation. Finally, we combine both the pressure load and the temperature load to perform FEA to study the vm stress pattern, and compare it with the results previously obtained.

Step 1: Create a 3D solid model for the ring support structure.

Click the icon of **Create a new object** from the menu toolbar to initiate the creation of a 3D model. Type *ring_support_thermal* as the file name, clear the box of **Use default template > OK**. Select the unit of **mmns_part_solid**, type *ring support thermal* under the **description** of the model, and type *student* or *your name* under the **modeled_by > OK**.

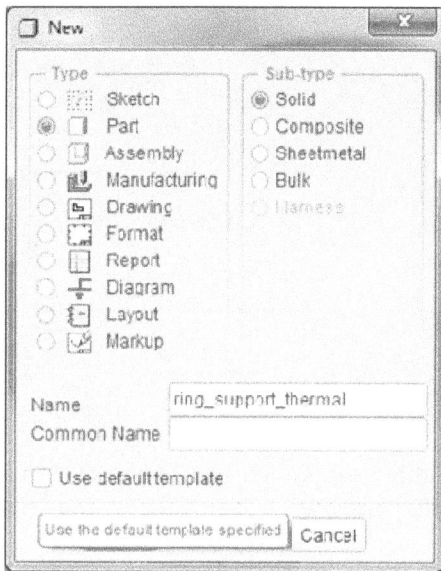

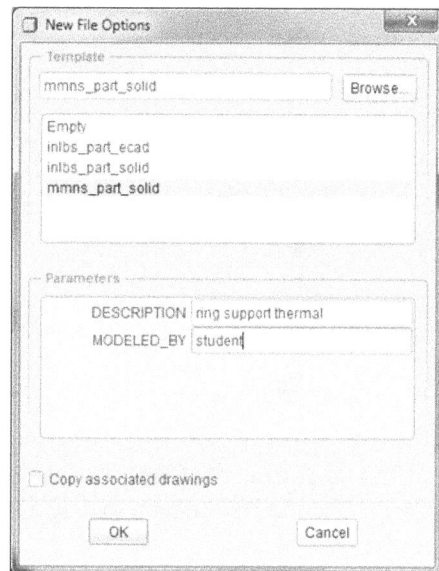

Pick up the icon of **Extrude**. Specify 30 as the height value. To initiate the process of sketching, activate **Placement > Define.**

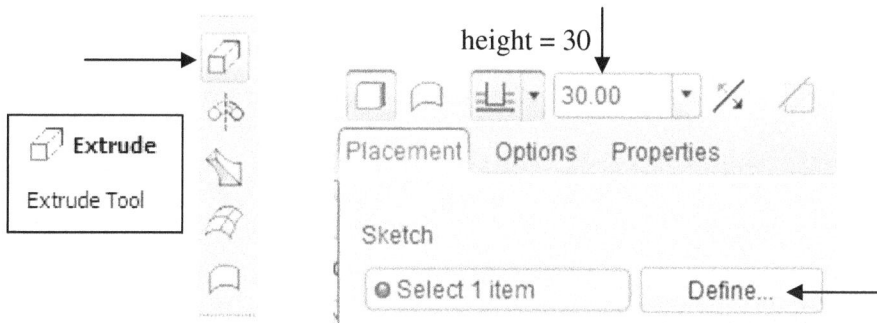

Select the **TOP** datum plane displayed on screen as the sketch plane. Pick the box of **Sketch** to accept the **RIGHT** datum plane as the default reference to orient the sketch plane, as illustrated below:

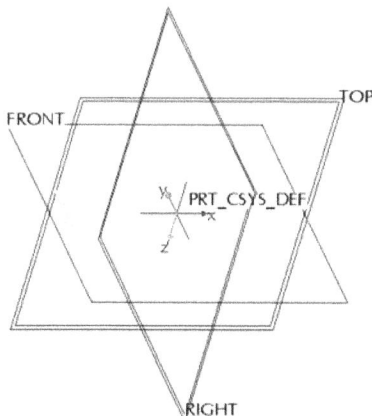

Before making a 2D sketch, let us first create 2 centerlines, a horizontal centerline and a vertical centerline passing through the origin of the coordinate system, as shown. To do so, right-click and hold, select **Centerline**. Users may select the icon of **Centerline**. Sketch a vertical centerline and a horizontal centerline.

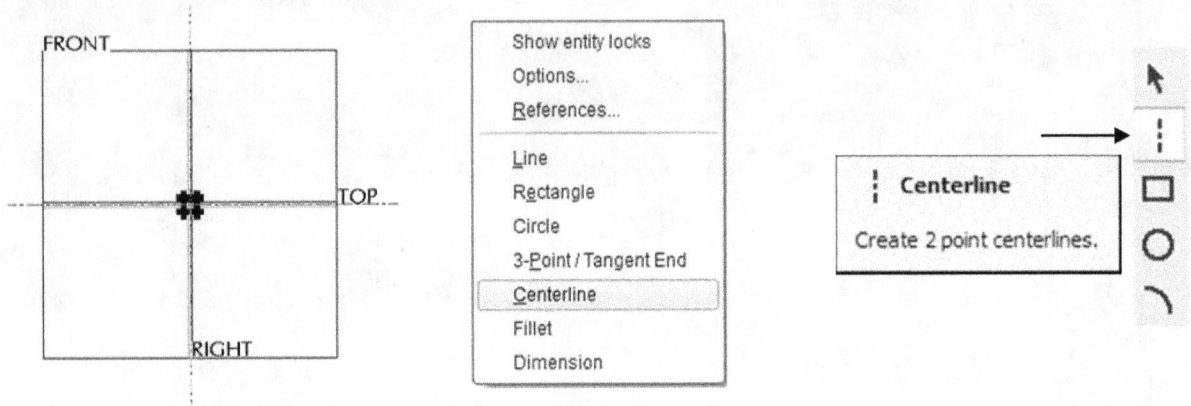

Click the icon of **Circle** to sketch 2 circles. Their diameter dimensions are 300 and 400 respectively. Use the **Delete** icon to delete the extra segments in the sketch.

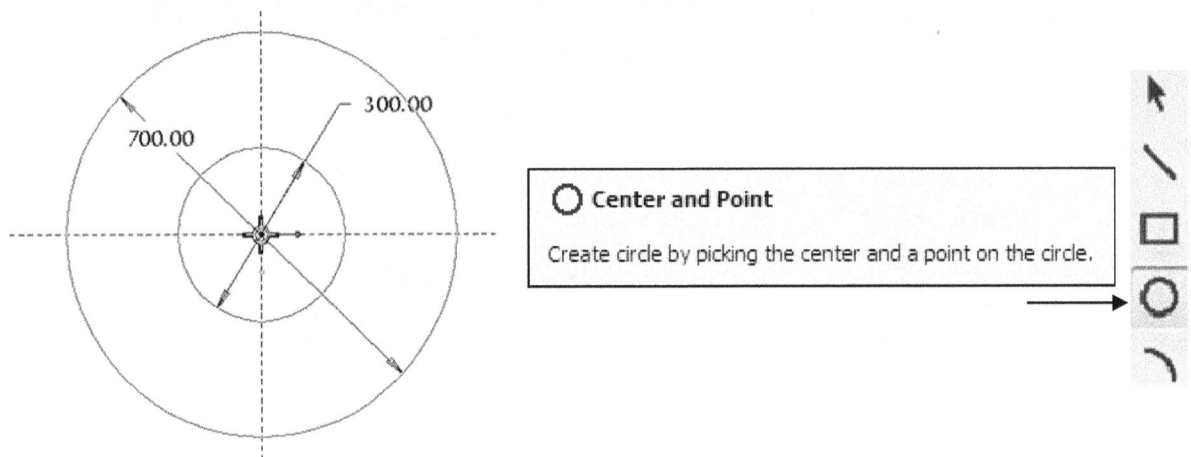

Click the icon of **Rectangle** to sketch a rectangle, which is symmetric about the 2 sketched centerlines. The 2 dimensions are 2200 and 560, respectively.

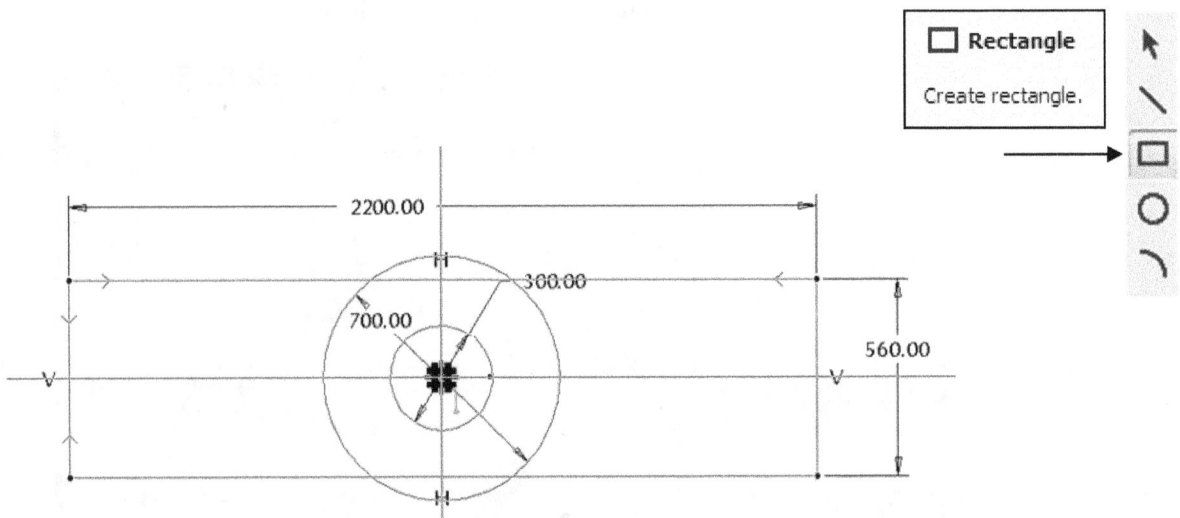

Click the icon of **Delete** to remove a set of arc and line segments from the sketch, as shown.

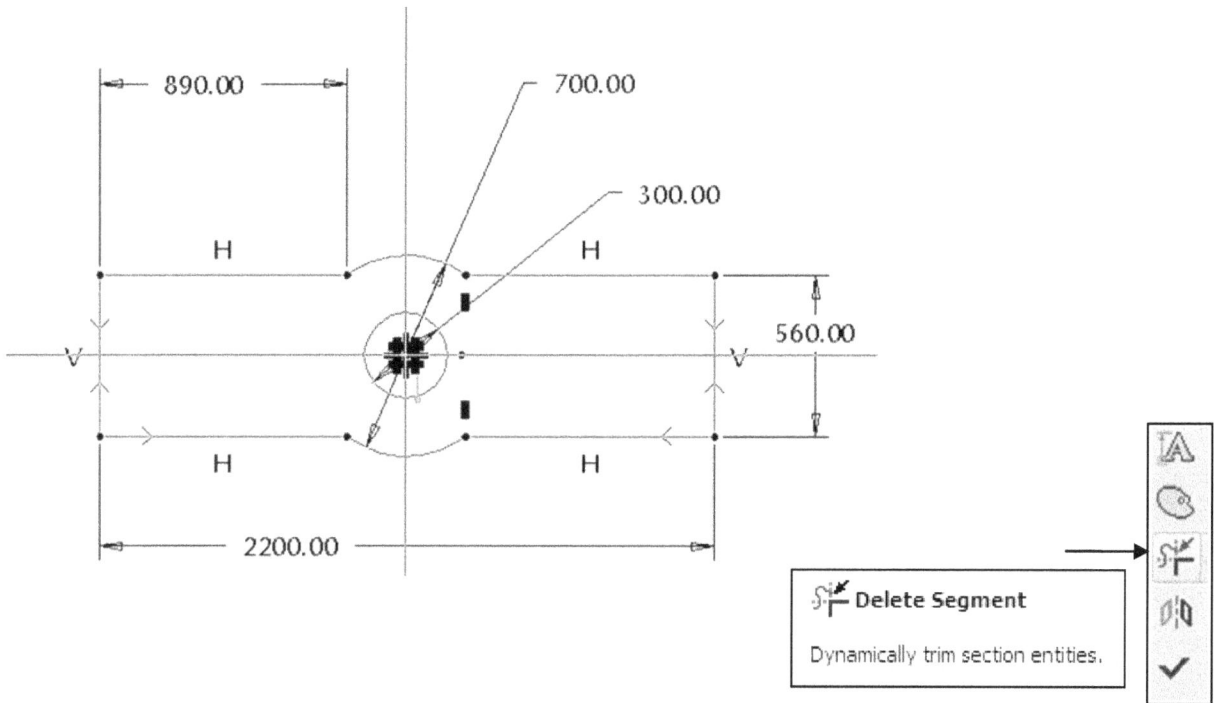

890.00 700.00 300.00 H H 560.00 H H 2200.00

Delete Segment
Dynamically trim section entities.

Click the icon of **Circle** to sketch 2 circles. Their diameter dimensions are 200. The position dimensions are 500.

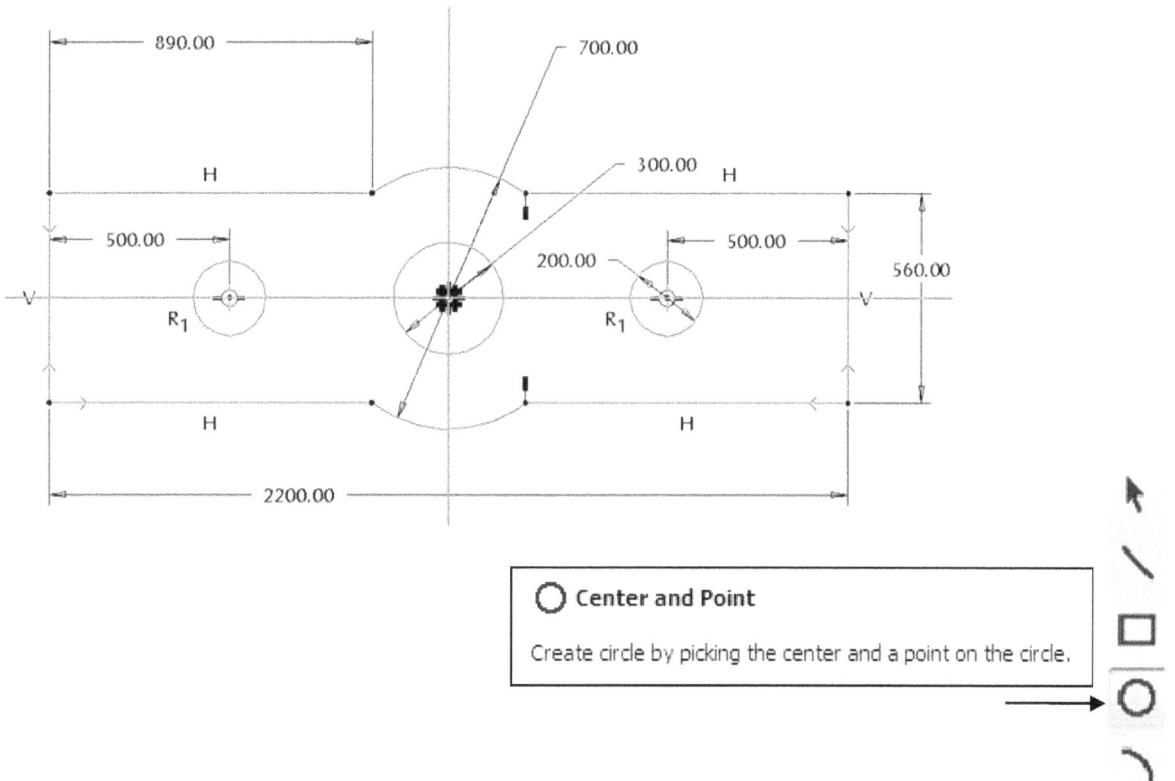

890.00 700.00 300.00 H H 500.00 500.00 560.00 R₁ 200.00 R₁ H H 2200.00

Center and Point
Create circle by picking the center and a point on the circle.

Click the icon of **Line** and sketch 2 lines on each of the two sides, as shown.

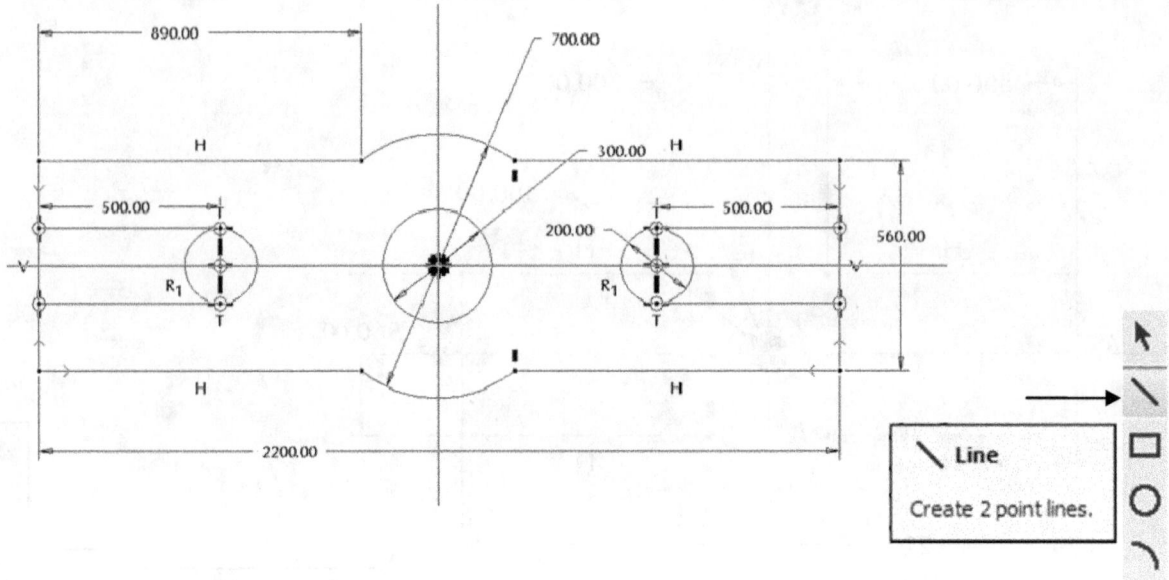

Click the icon of **Delete** to remove a set of arc and line segments from the sketch, as shown.

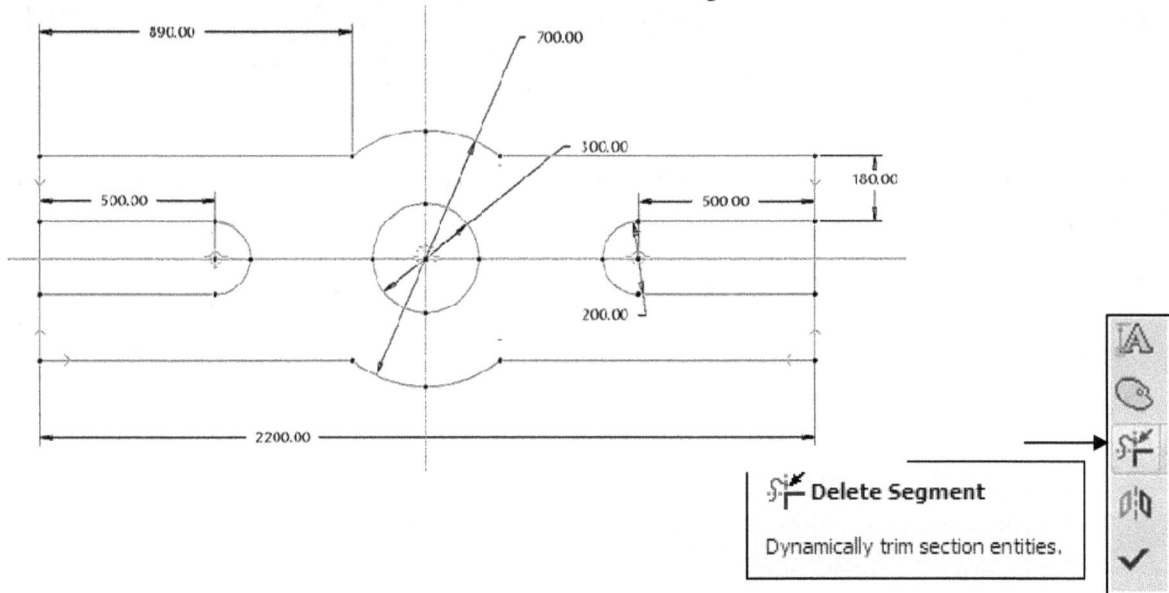

\Upon completing the sketch, select the icon of **Done** and click the icon of **Apply and Save,** completing a 3D solid model for the ring support component.

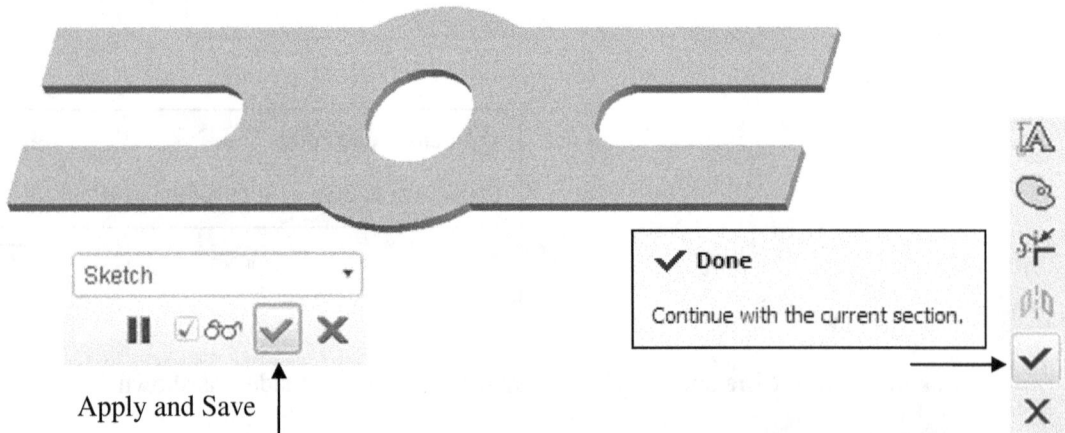

Apply and Save

Now let us add rounds at the 4 corners. Click the icon of Round. Specify 50 as the radius value. Pick the 4 edges while holding down the **Ctrl** key, as shown. Click the icon of **Apply and Save** from the feature control panel, completing the creation of 4 rounds.

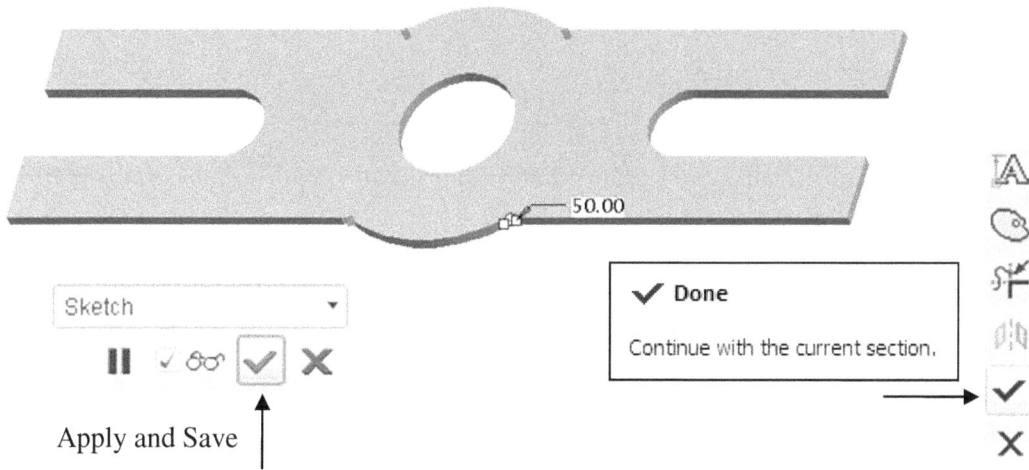

Step 2: Switch from the Pro/ENGINEER design system to Pro/Mechanica
Click **Applications > Mechanica > Thermal > OK.**

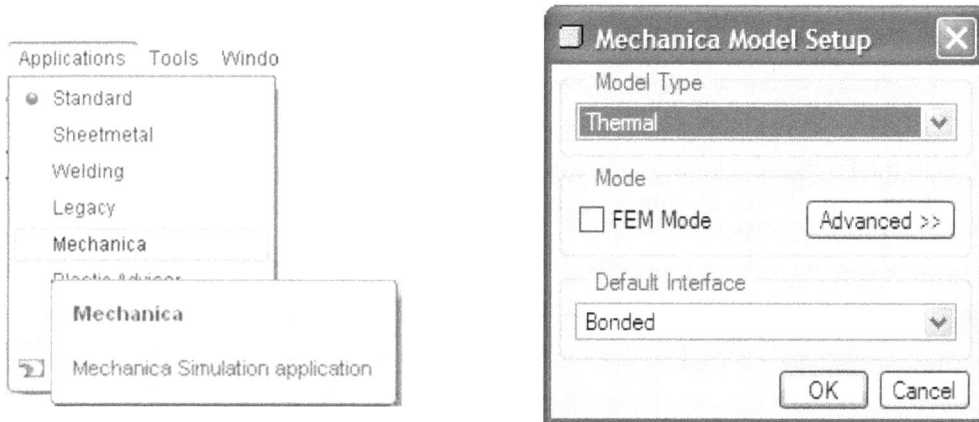

To specify the material type, select the icon of **Define Materials >** select **STEEL** from the list of Materials in Library > click the directional arrow to switch the selection to the list of **Materials in Model > OK**.

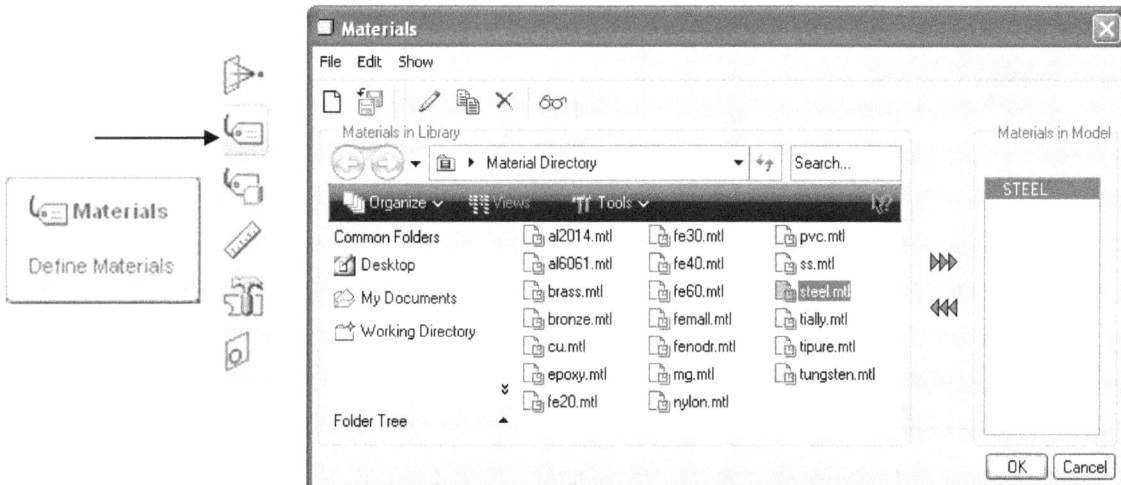

Select the icon of **Material Assignment** > in the Material Assignment window, the material type **STEEL** is already shown and the assignment has already been done because there is only one volume or one component in this case > click **OK** to accept the material assignment. In the model tree, the material type and the material assignment are also listed.

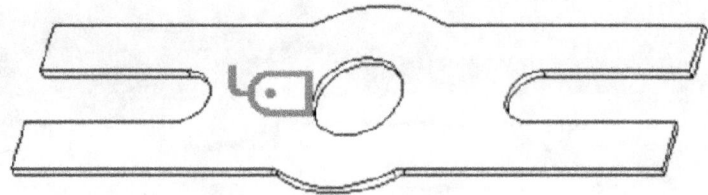

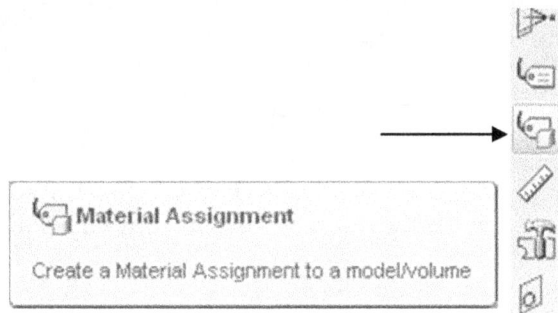

Step 3: Under the Pro/MECHANICA environment, define the constraint condition: Fixed end.

Click the icon of **Constraints** > select **Surface(s)** > pick the surface on the right end > **OK** to fix all 6 degrees of freedom > **OK**.

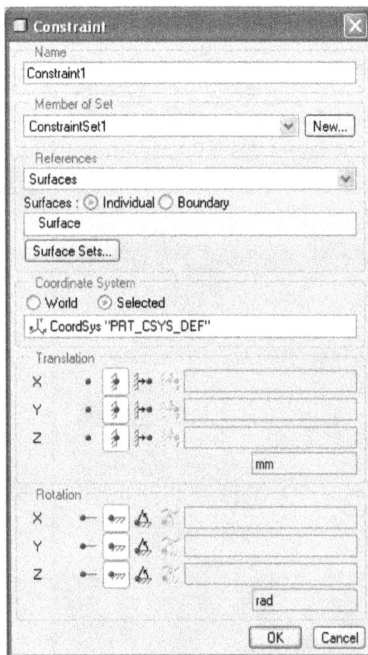

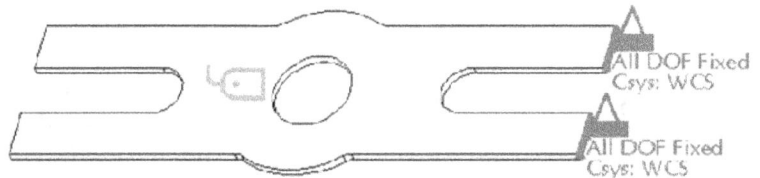

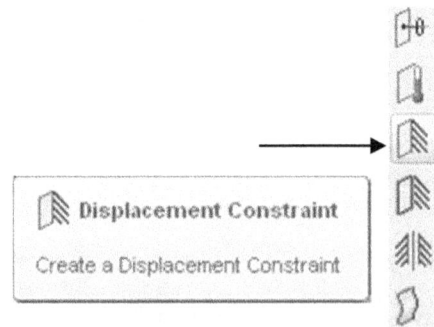

Click the icon of **Constraints** > select **Surface(s)** > pick the surface on the left end > **OK** to fix 4 degrees of freedom and leave the translation in the Y and Z direction free > **OK**.

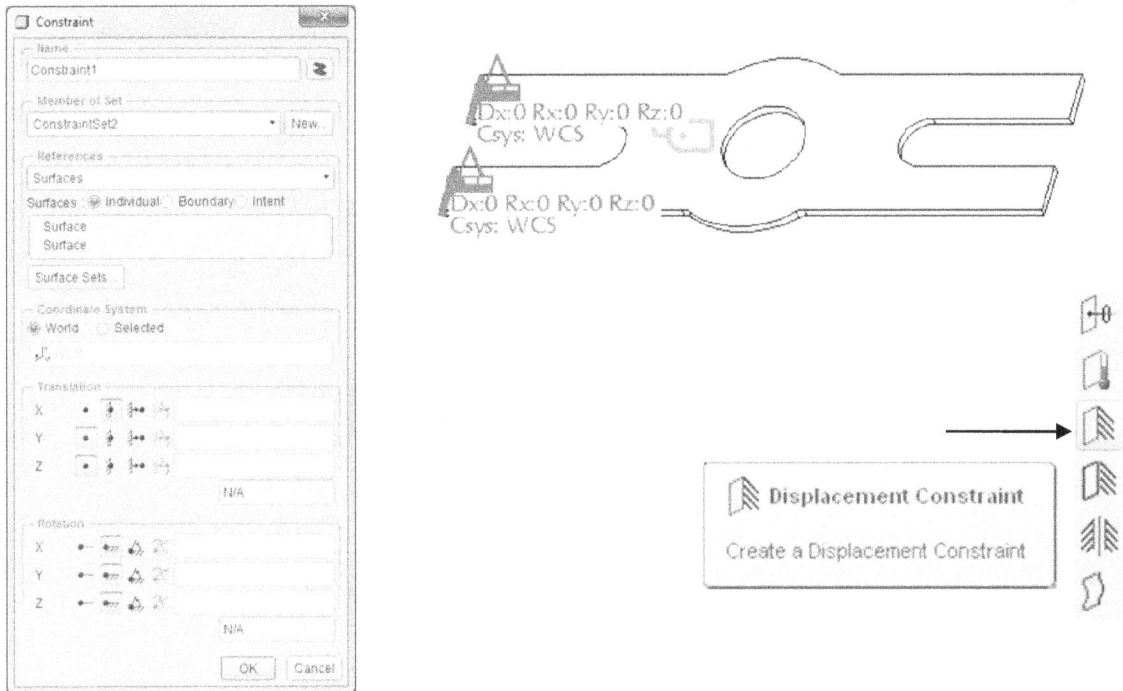

Step 4: **Under the Pro/MECHANICA environment, define the load condition.**

To define a pressure load acting on the 2 half cylindrical surfaces, click the icon of Pressure > **Surface(s) >** pick the 2 half cylindrical surfaces >. Specify 10 as the magnitude > **OK**.

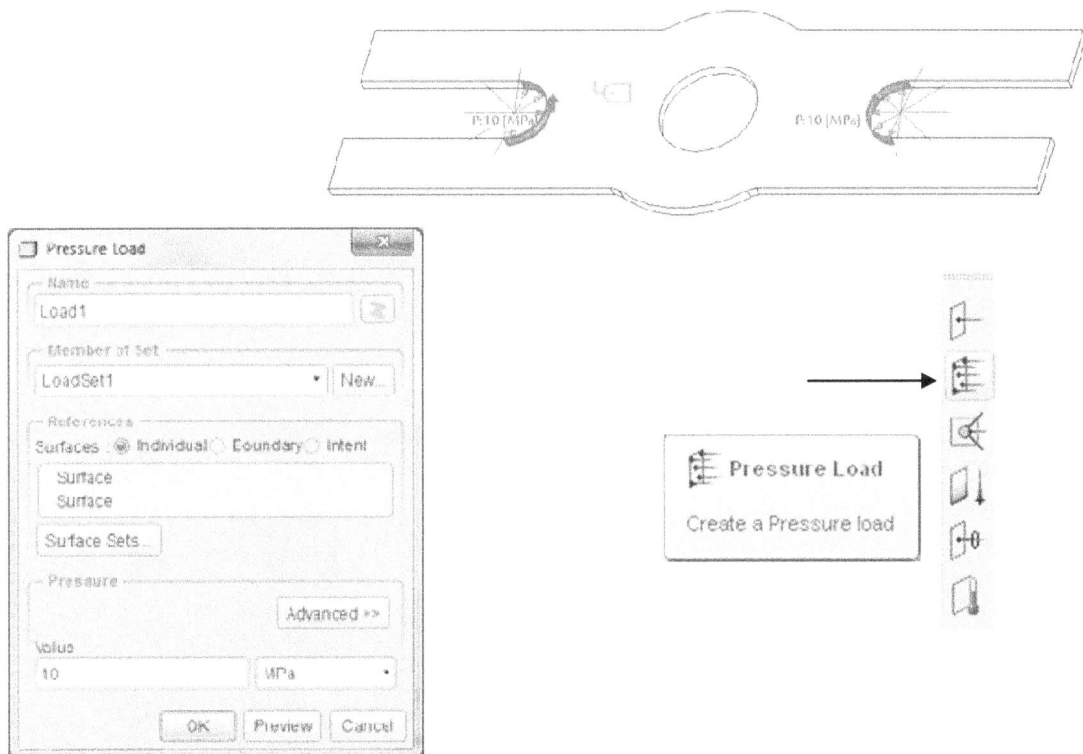

Step 5: Under the Pro/MECHANICA environment, set up **Analyses** and **Run** it

Select the icon of **Mechanical Analyses/Studies > File > New Static >** type *pressure_load_only* as the name of the analysis folder > **OK.** Click the box of **Run**.

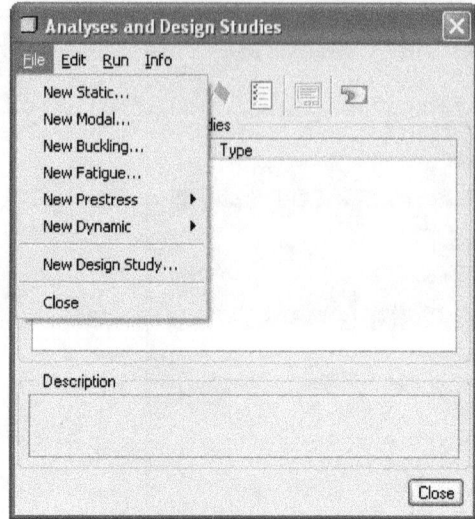

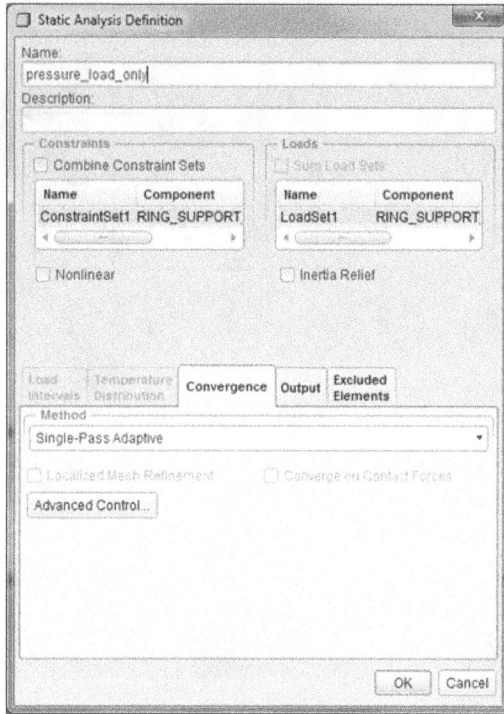

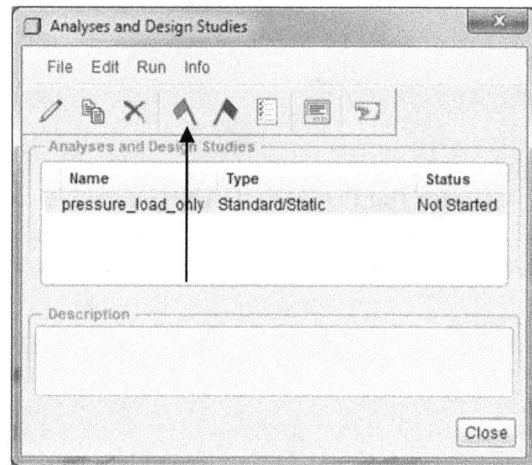

The vm stress distribution is shown below. The maximum value is 21.13 MPa.

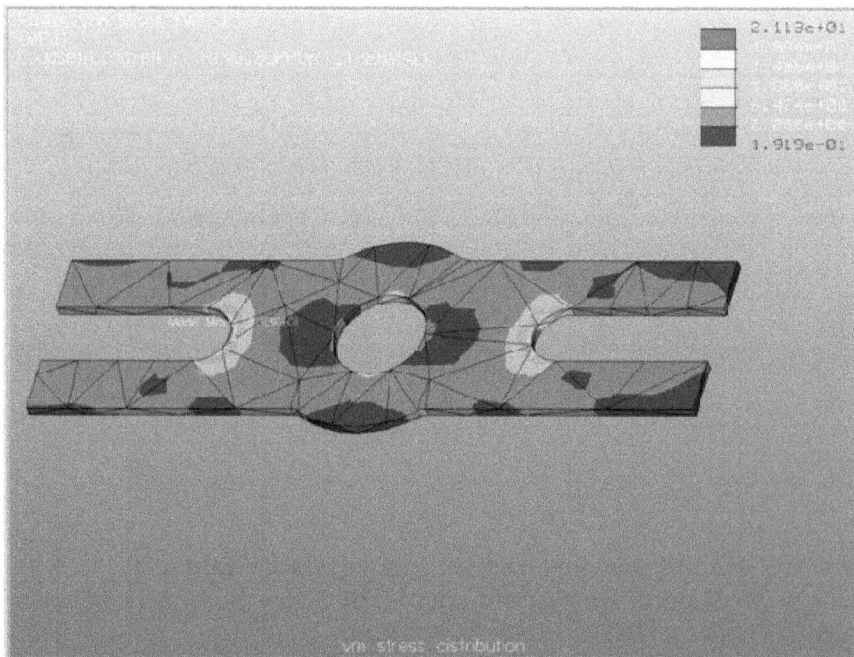

Step 6: Switch to thermal module and perform a thermal analysis
Click **Edit** from the top menu > **Mechanica Model Setup** > **Thermal** > **OK**.
Select the icon of **Heat Load** > **Surfaces** > pick the inner cylindrical surface > enter 4000 > **OK**.

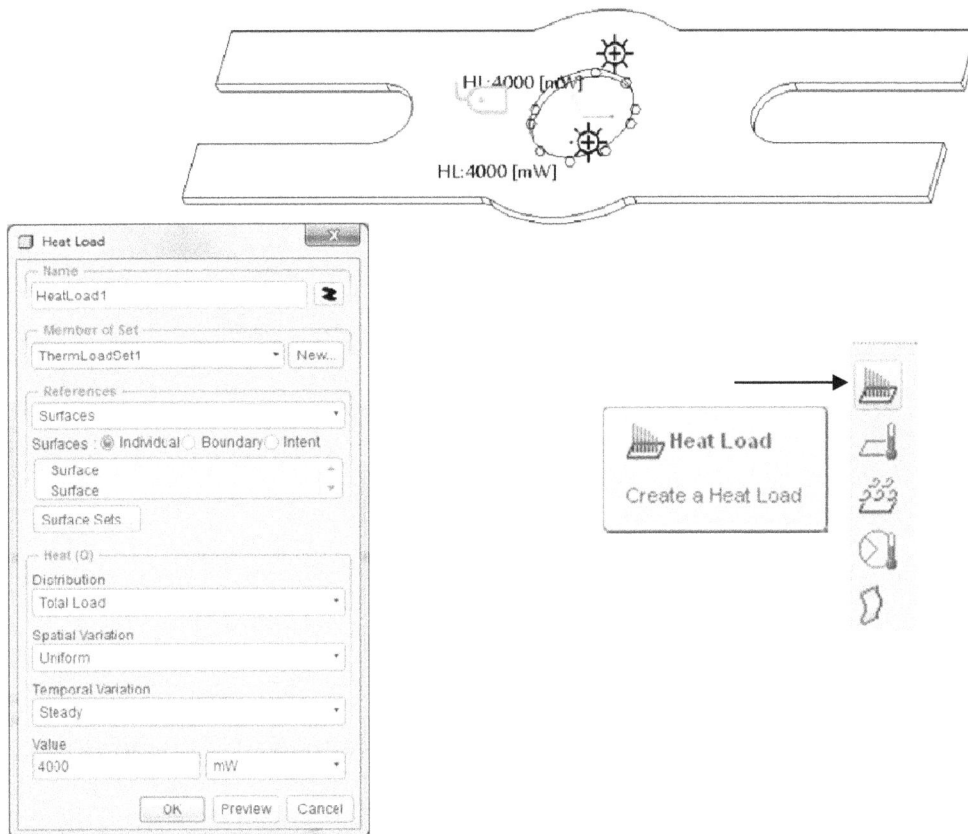

Select the icon of **Create a convection condition** > **Surfaces** > pick the 2 surfaces on both ends > enter 0.1 as the convection coefficient and enter1 0 as the environmental temperature > **OK**.

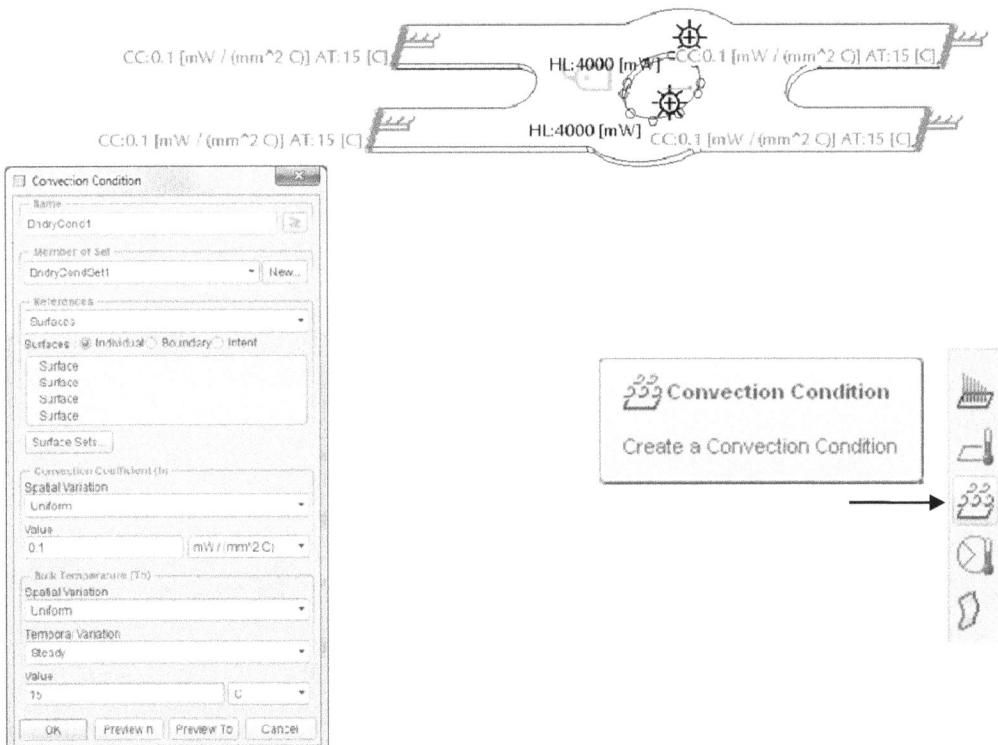

Step 7: Under the Pro/MECHANICA environment, set up **Analyses** and **Run** it
    Select the icon of **Mechanica Analyses/Studies > File > New Steady State Thermal** > type *ring_support_thermal* as the name of the Design Study Folder > **OK**.

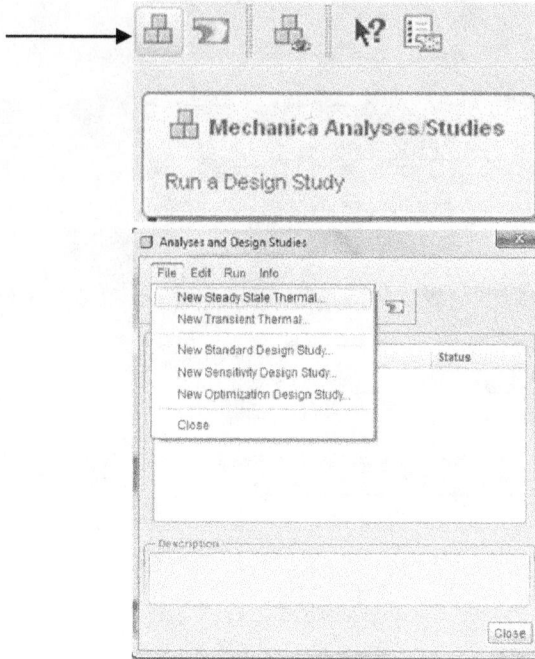

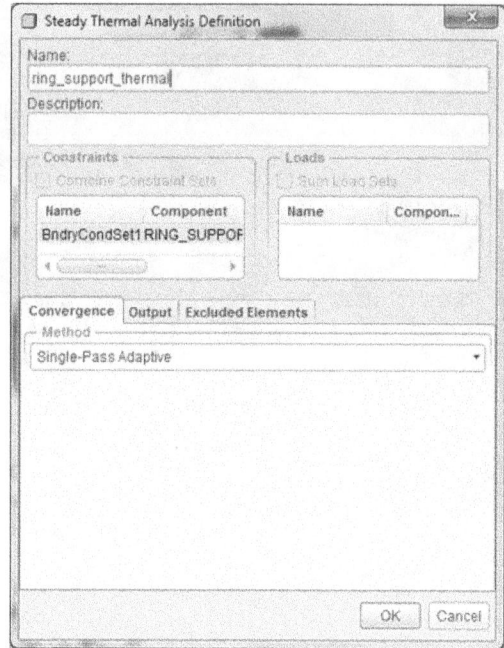

The obtained temperature distribution is shown below. The maximum value is 20.7°C.

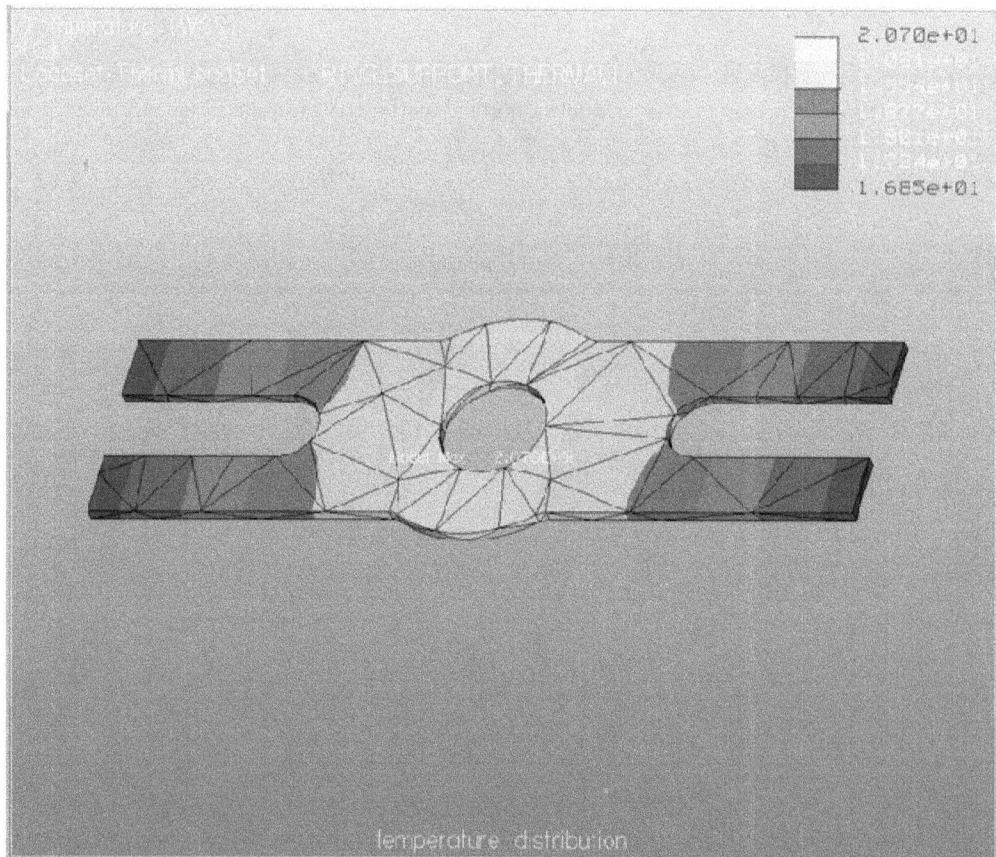

Let us edit the temperature fringe plot. Go to Display Option and select **Contour**, and **Isosurfaces > OK and Show**.

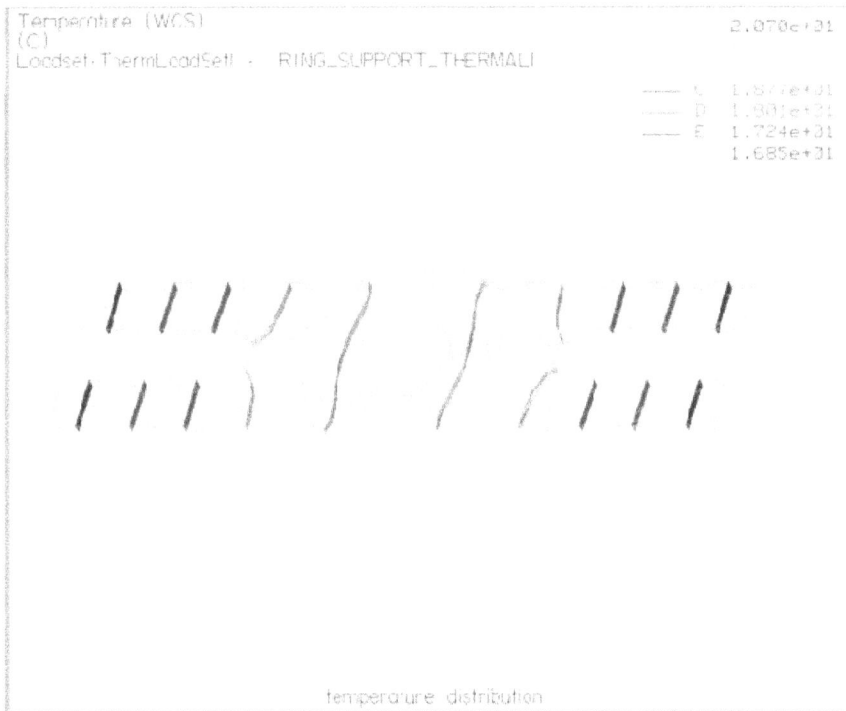

temperature distribution

Let us create a new window called *Heat_Flux*. Select **Vectors** under Display Type. Select **Flux** under Quantity. Select **Magnitude** under Component. In the Display Options tab, select the **Animation** option > **OK and Show**. An animated representation of the heat flux through the part is displayed. Notice the flux vectors are scaled and colored according to the magnitude of the local heat flux. Also, the vector direction is perpendicular to the isotherm.

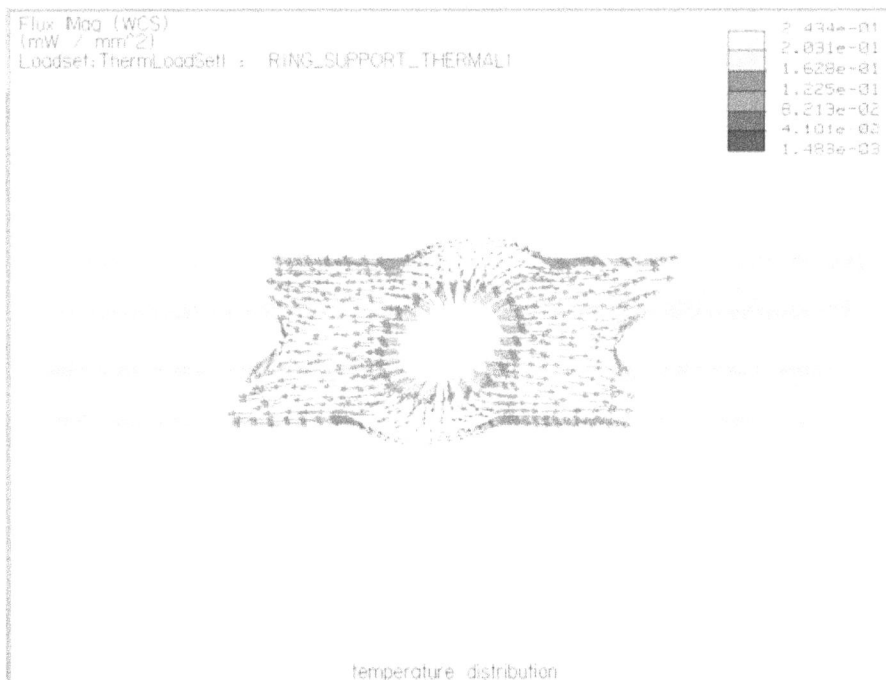

temperature distribution

Step 8: Transfer the temperature distribution to a thermal load and apply it to the ring support component.

First, we need to go to the Structural Module. Click **Edit** from the top menu > **Mechanica Model Setup > Structure > OK**.

To apply the thermal load, click **Insert** from the top menu > **Temperature Load > MEC**/T. Make sure "Click New so that the temperature load is separate from LoadSet 1". From the pop up window, type temperature_load as the name. Make sure Previous Design Study is checked. Set the Reference Temperature to be 15°C > **OK**.

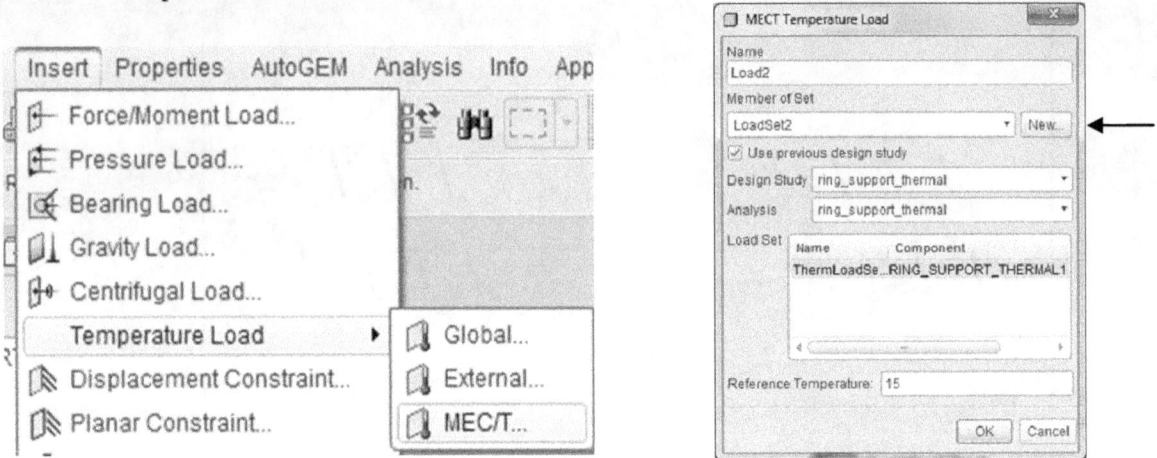

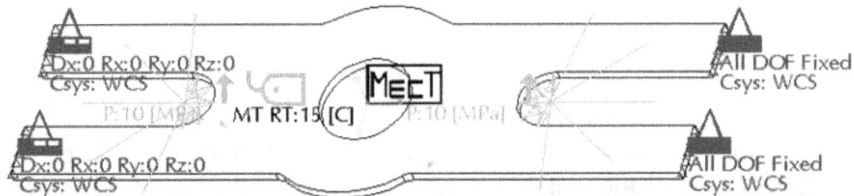

Step 9: Perform a static analysis considering the temperature only.

Select the icon of **Mechanica Analyses/Studies > File > New Static** > type *temperature_load* the name of the Design Study Folder > **OK**. Make sure that LoadSet2 is selected.

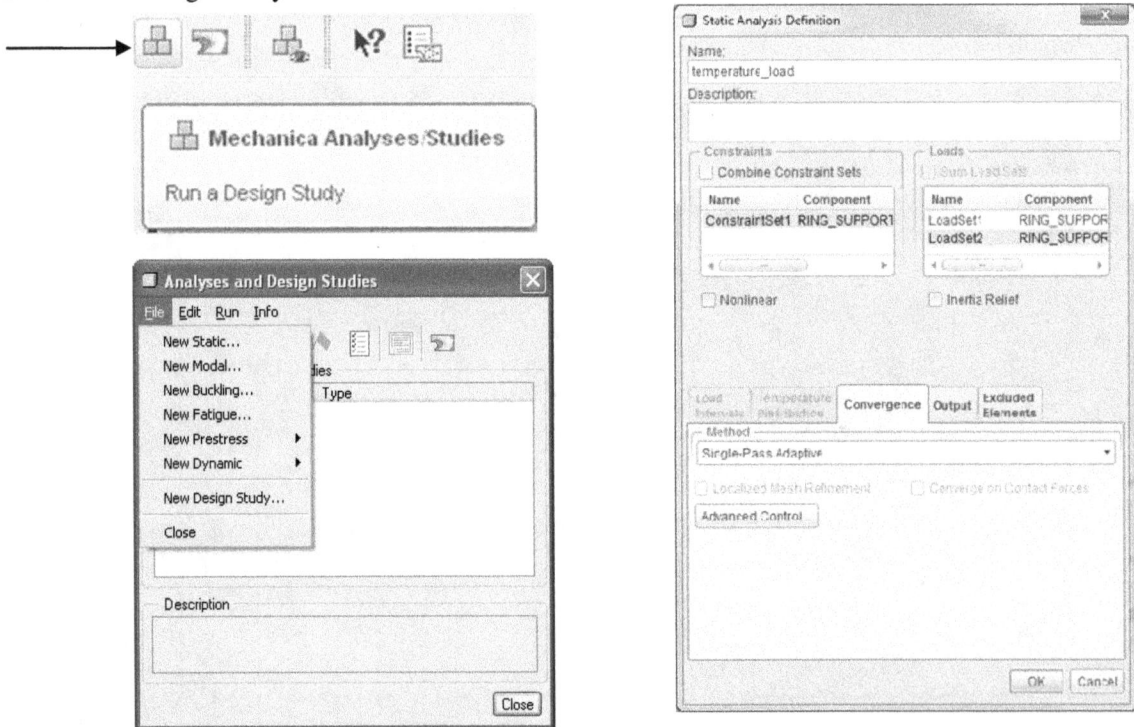

To study the obtained result, say the distribution of the von Mises stress distribution, click the icon of **Review results > Stress > von Mises Stress > OK and Show.** The maximum value is 26.65 MPa.

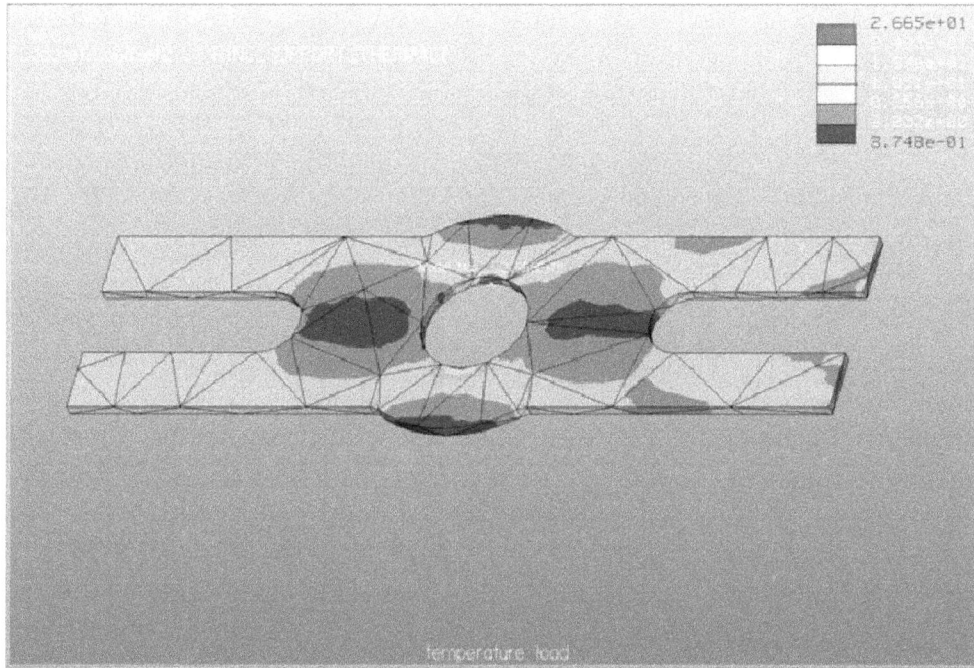

Step 10: Under the Pro/MECHANICA, perform a combined structural and thermal analyses.

Select the icon of **Mechanica > File > New Static >** type *thermal_mech_static* as the name of the analysis folder > make sure that both Loadset1 and LoadSet2 (MechTloadset) are selected > **OK.**

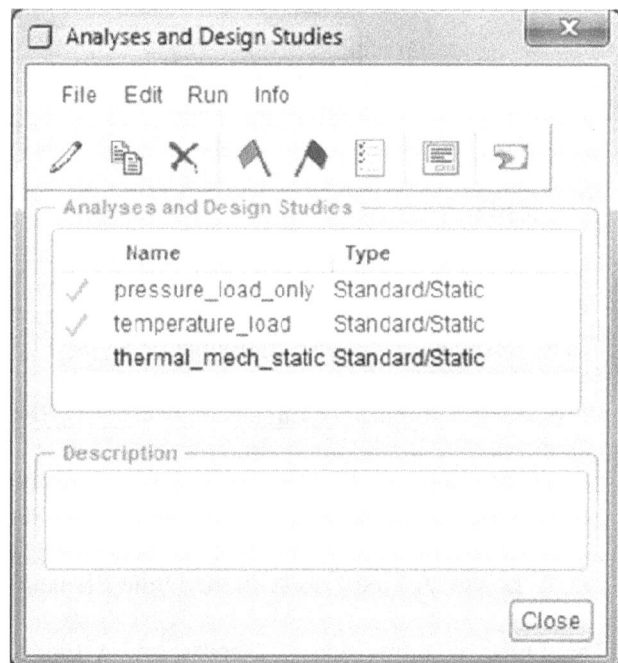

To study the obtained result, say the distribution of the von Mises stress distribution, click the icon of **Review results > Stress > von Mises Stress > OK and Show.** The maximum value is 35.23 MPa.

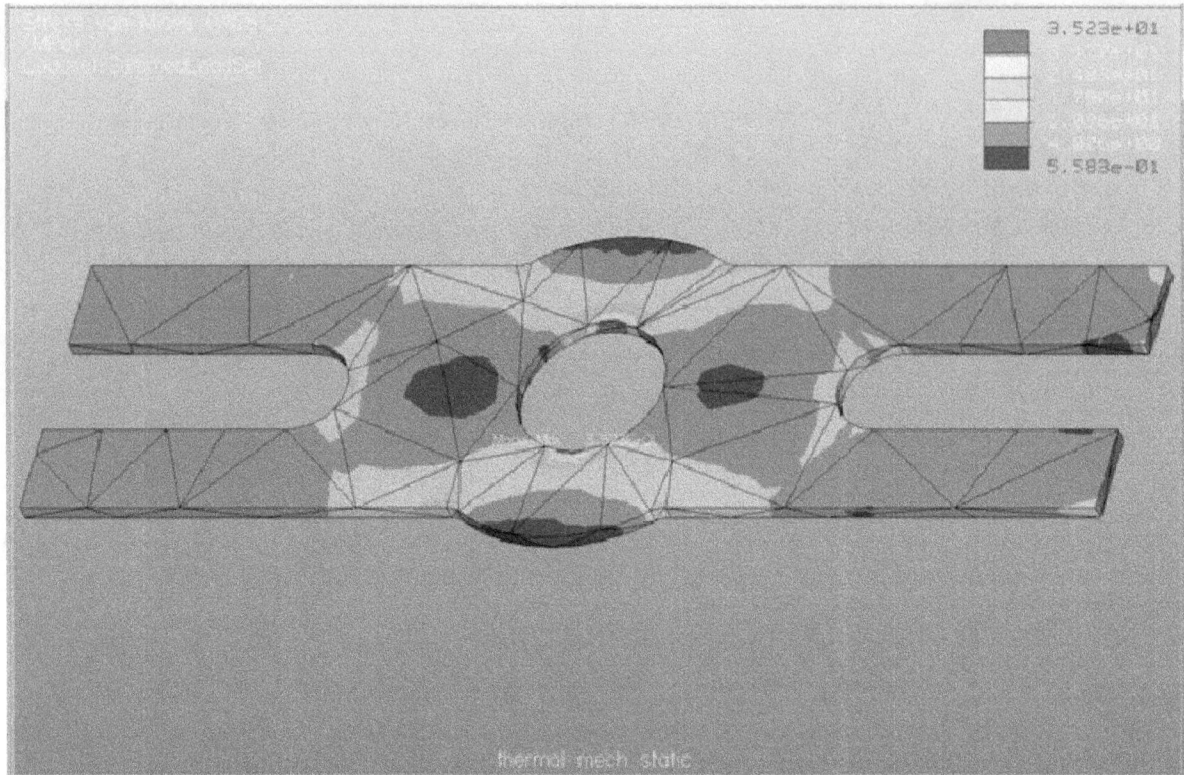

Let us summarize the results obtained from the combined structural and thermal analysis. We should notice that a simple arithmetic addition and/or subtraction do not apply to the data obtained from FEA. Why?

load condition	max [σmax_vm]
pressure load	21.13 MPa
temperature load	26.65 MPa
combined loads	35.23 MPa

5.7   References

1. J. W. Dally and W. F. Riley, Experimental Stress Analysis, McGraw Hill, 1965.
2. J. W. Dally, Design Analysis of Structural Elements, Goodington, New York, NY, 2003.
3. C. S. Desal and J. F. Abel, Introduction to the Finite Element Method, Van Nostrand Reinhold, New York, NY, 1972.
4. J. H. Earle, Graphics for Engineers, AutoCAD Release 13, Addision-Wesley, Reading, Massachusetts, 1996.
5. K. H. Huebner, D. L. Dewhirst, D. E. Smith, and T. G. Byrom, The Finite Element Method for Engineers, 4th edition, John Wiley & Sons, Inc., 2001.
6. C.S. Krishnamoorthy, Finite Element Analysis, Theory and Programming, 2nd Ed., 1995.
7. D. L. Logan, A First Course in the Finite Element Method Using ALGOR, PWS Publishing, Boston, 1997.
8. M. Ortiz, Y. Leroy and A. Needleman, A Finite Element Method for Localized Failure Analysis, Comp. Meth. Applied Mech. Enging, 61, pp. 189-214, 1987.
9. M. Petyl, Introduction to Finite Element Vibration Analysis, Cambridge University Press, 1990.

10. T. H. H. Pian and P. Tong, Basis of Finite Element Methods for Solid Continua, Int. J. Numer. Methods Eng., Vol. 2, No. 7, April 1964, pp. 1333-1336.
11. P. G. Plockner, Symmetry in Structural Mechanics, J. of the Structural Division, American Society of Civil Engineers, Vol. 99, No. ST1, pp. 71-89, 1973.
12. S. P. Timoshenko and D. H. Young, Theory of Structure, McGraw-Hill, Kogaksha Limited, 1965.
13. M. J. Turner, R. W. Clough, H. C. Martin, and L. J. Topp, Stiffness and Deflection Analysis of Complex Structures, J. of the Aeronautical Sciences, Vol. 23, No. 9, pp. 805-824, Sept. 1956.
14. O. C. Zienkiewicz, The Finite Element Method, third edition, McGraw-Hill, New York, NY, 1977.
15. O. C. Zienkiewicz and R. L. Taylor, The Finite Element Method, 4th edition, Vol. 1, McGraw-Hill (UK),London, 1989.

5.8 Exercises

1. A ring component is shown below. The material type is steel. The right is subjected to a temperature load, as shown. Obtain the temperature distribution. Assume both ends are constrained, as shown. Convert the temperature distribution to a mechanical load and obtain the value of the maximum von Mises stress and its location.

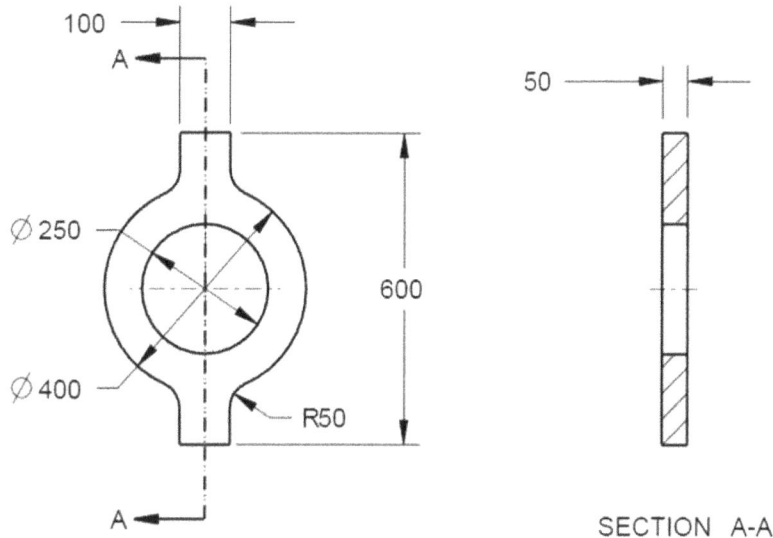

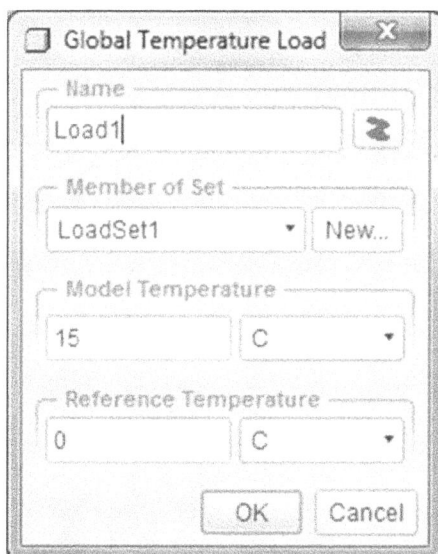

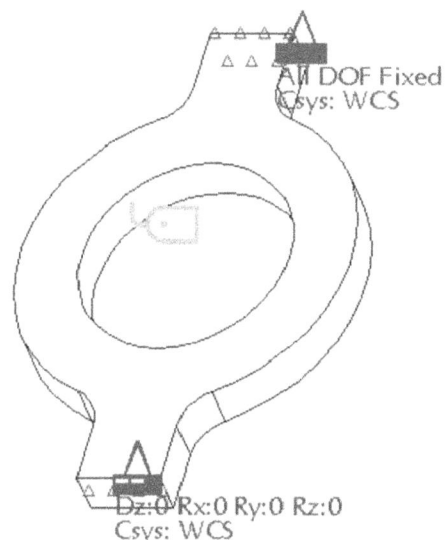

SECTION A-A

2. Use the ring component in previous question. Both ends are subjected to the boundary conditions of pre-scribed temperature. The cylindrical surface is subjected to the convection boundary conditions. Determine the temperature distribution and heat flux flow.

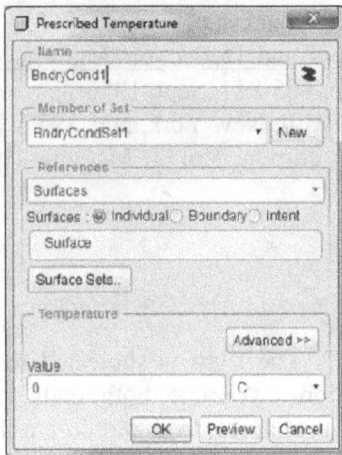

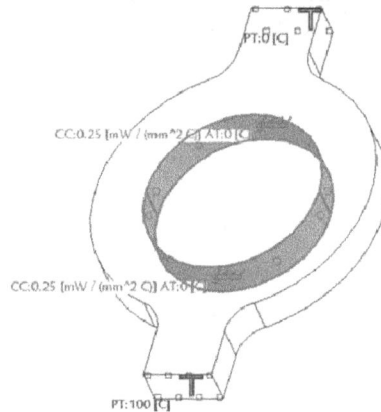

3. Perform a thermal analysis for a plate component. Its shape and dimensions are shown below. The boundary conditions are shown below. The left end is subjected to a pre-scribed temperature of 500°C. The right end is subjected to a convection boundary where the convection coefficient is 5 mW/mm²C and the bulk temperature is set at 0°C.

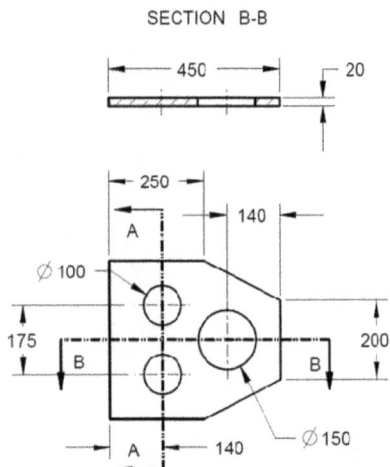

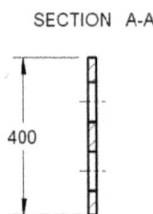

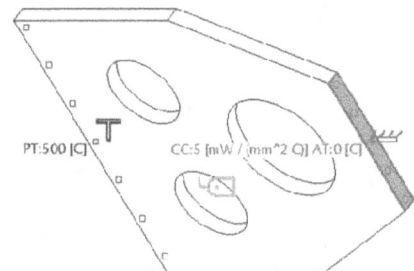

CHAPTER 6

IDEALIZATION: MODELING WITH SHELLS

6.1   Introduction

In this chapter, modeling with shells will be covered. In solid mechanics, many problems can be treated satisfactorily or approximately by a two-dimensional approach. For example, plane stress is one of them. The normal stress in the third direction or σ_z and the shear τ_{xz} and τ_{yz}, directed perpendicular to the x-y plane are assumed to be zero. As a result, a 2D model can be used to replace a 3D solid model in the process of evaluating the displacement, stress and strain. FEA dealing with a 2D model demands much less computational time while the accuracy is still acceptable.

6.2   Concept of Surface Pairs

Before performing an analysis, we need a 3D solid model in general. Figure 6-1 illustrates a plate used to support three plants, which are on display. The three dimensions of the plate and the three surface regions where the three plants are placed are shown below:

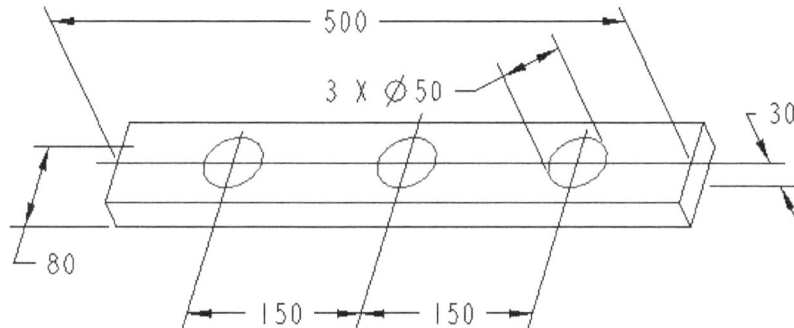

Figure 6-1 Geometrical Shape and Dimensions of the Plate

Assume that the weight of each plant is 200 N. The plate material is Nylon. Use a shell modeling approach to evaluate the deflection distribution and the distribution of the von Mises stress.

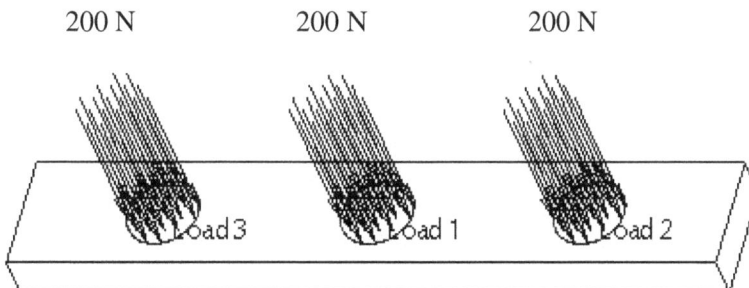

To model the constraint condition, assume the plate is fixed to the ground through two supports. The dimensions of the two supports are shown below.

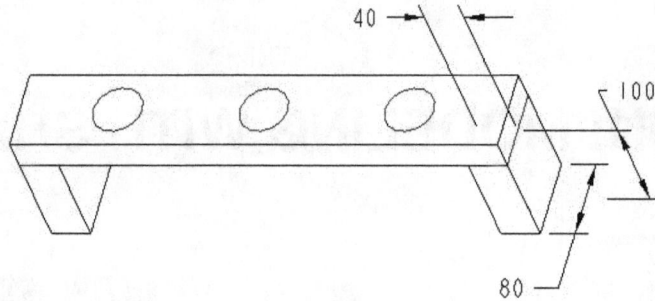

Step 1: Create a 3D solid model

File > New > Part > type *plate_for_display* as the file name and clear the icon of **Use default template.** Select **mmns_part_solid** (units: Millimeter, Newton, Second) and type *plate_shell* in **DESCRIPTION**, and *student* in **MODELED_BY**, then **OK.**

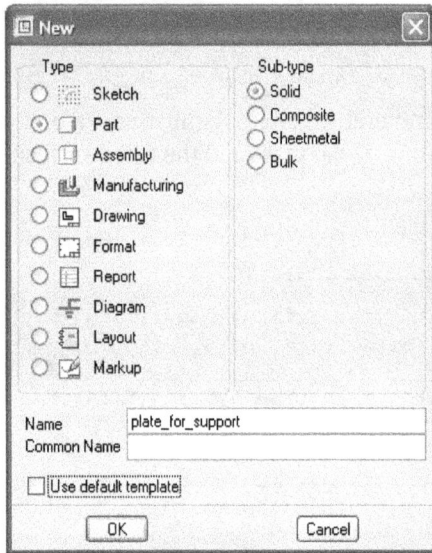

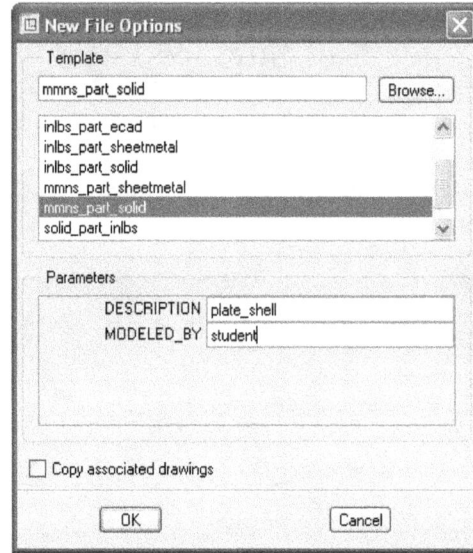

Pick up the icon of **Extrude** displayed on the toolbar of feature creation. Make sure that the icon of **Solid** is selected. Set the thickness value to 30. To initiate the process of sketching, activate **Placement > Define.**

Select the **TOP** datum plane displayed on screen as the sketch plane. Pick the box of **Sketch** to accept the **RIGHT** datum plane as the default reference to orient the sketch plane, as illustrated below:

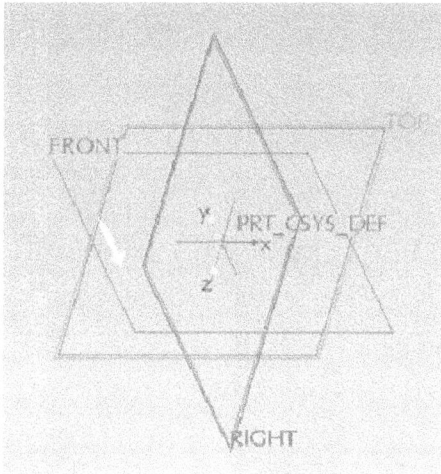

Click the icon of **Centerline**. Sketch a horizontal centerline and a vertical centerline to be used for symmetry when sketching.

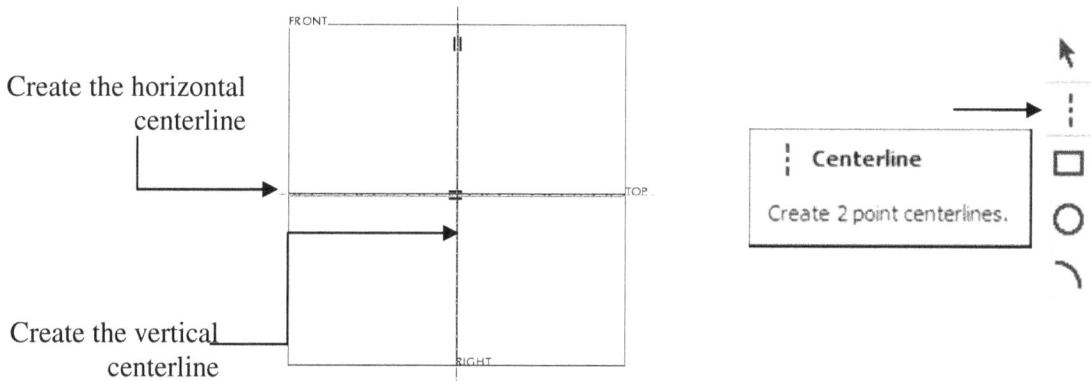

Create the horizontal centerline

Create the vertical centerline

Click the icon of **Rectangle** and sketch a rectangle, which is symmetric about the 2 sketched centerlines. The 2 dimensions of the rectangle are 500 and 80, respectively.

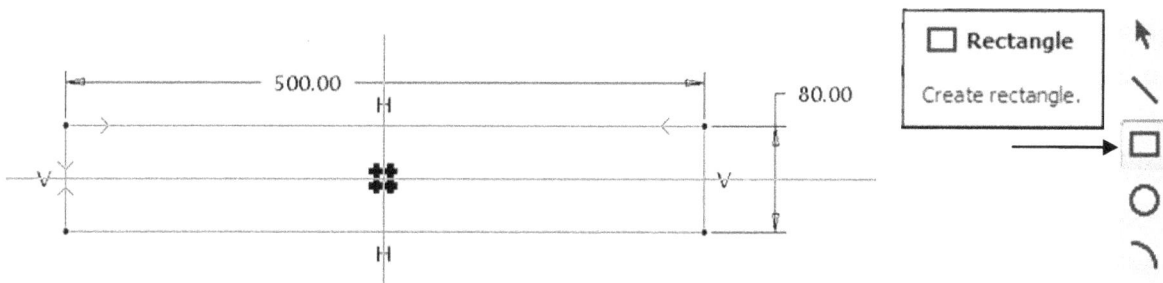

Upon completing the sketch, select the icon of **Done** and the icon of **Apply and Save**.

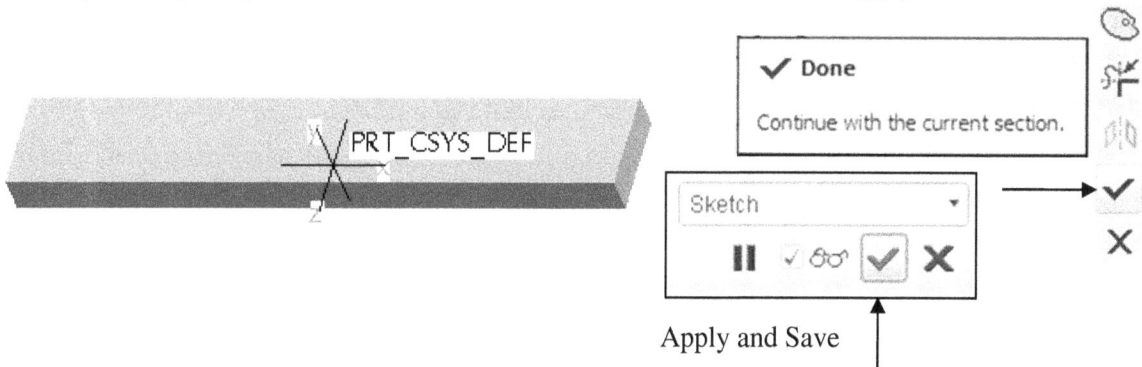

Apply and Save

Step 2: Create five (5) datum curves.

Select the icon of **Sketch Tool.** Pick the top surface of the plate, and click Sketch to accept the default setting for orientation.

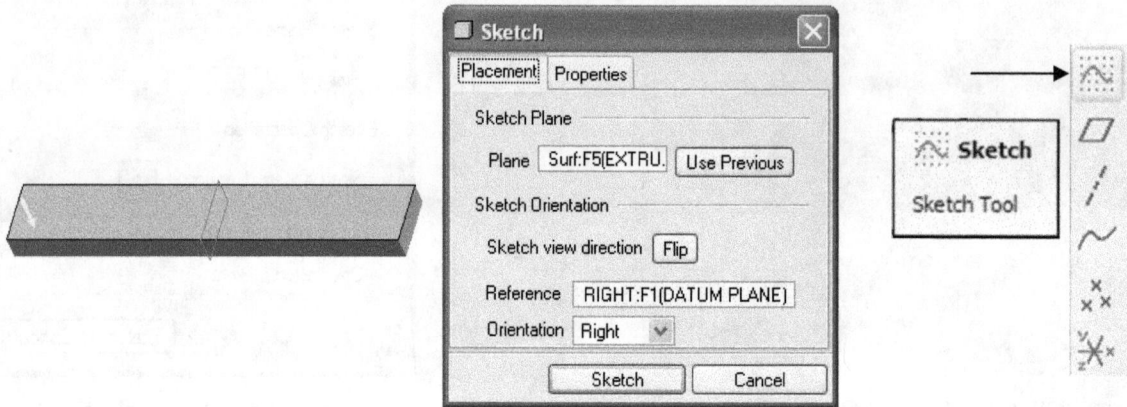

Pick the icon of circle to sketch a circle. The diameter dimension is 50. Upon completion, click the icon of **Done.**

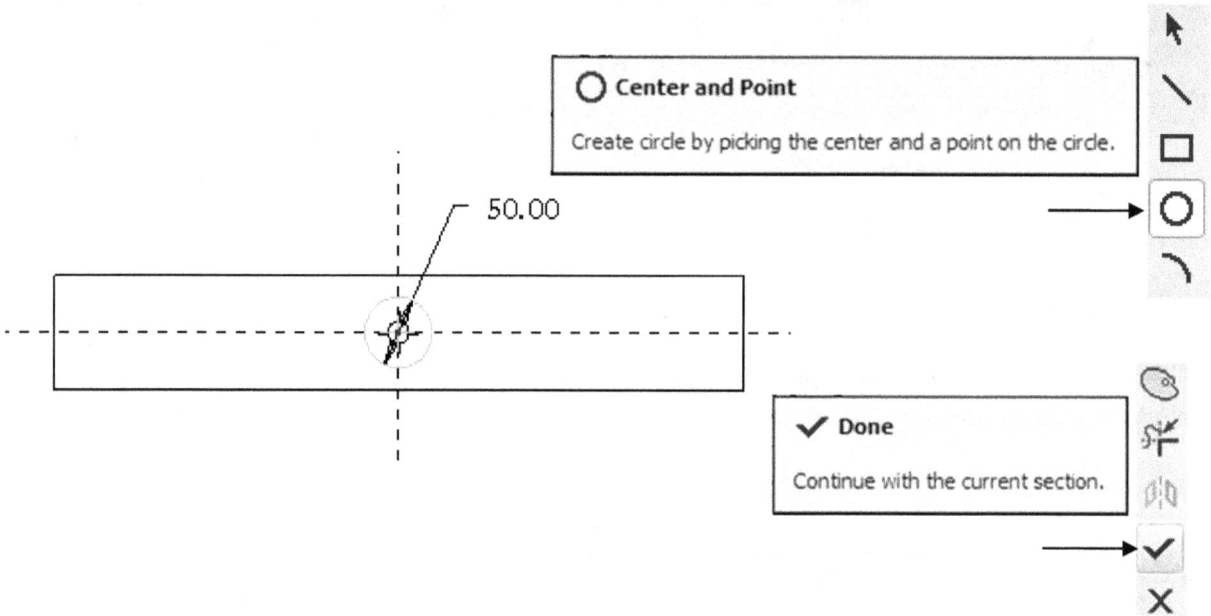

Repeat the above procedure to create the second datum curve, as shown.

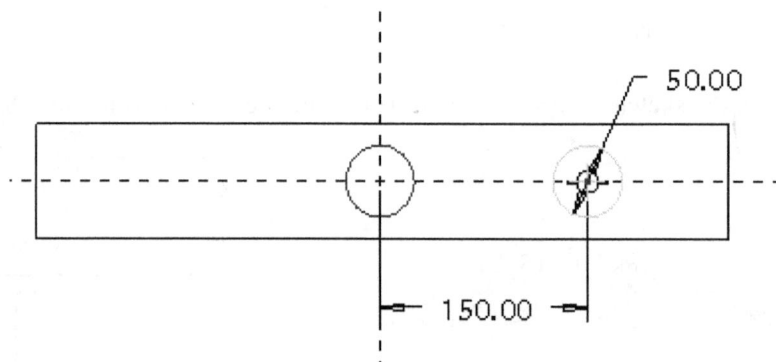

Repeat the above procedure to create the third datum curve, as shown.

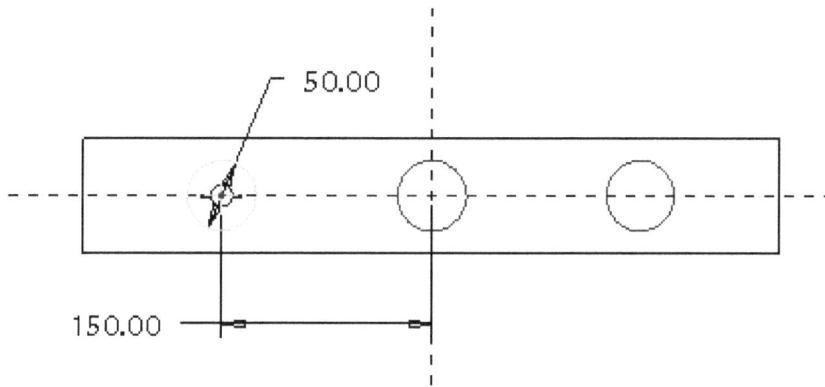

To create the fourth datum curve, we need to select the bottom surface of the plate as the sketch plane, as shown.

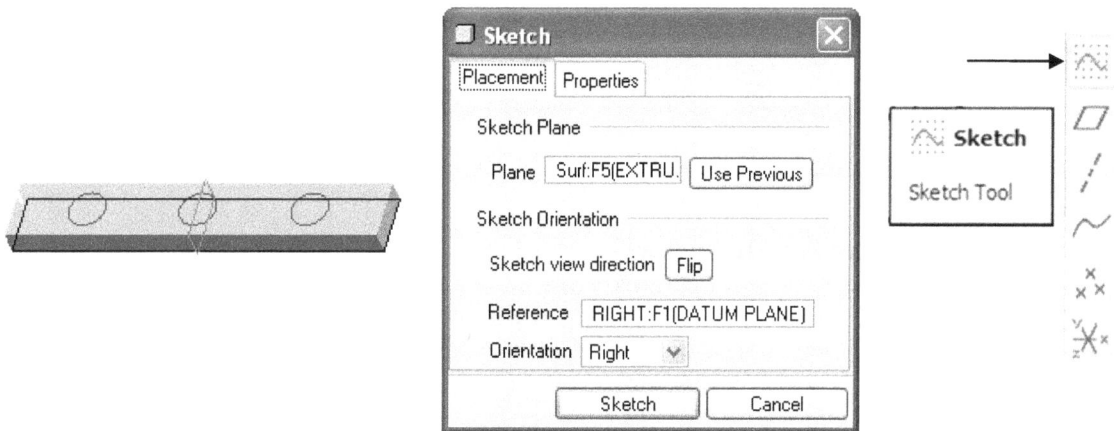

Click **Sketch > References** and add 3 new references, which are shown below:

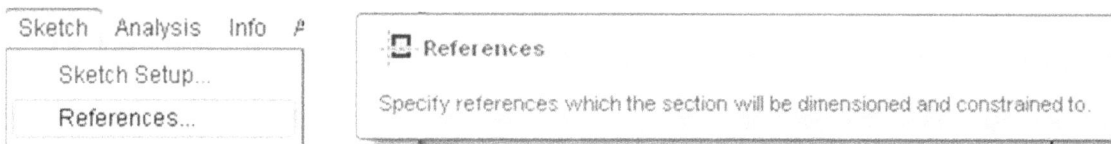

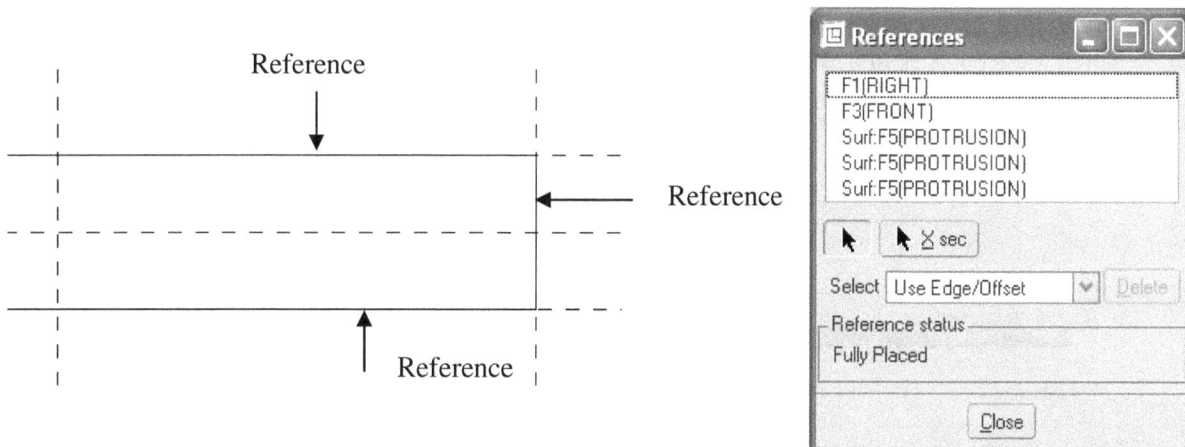

Pick the icon of **Rectangle** to sketch a rectangle. Only one dimension is needed because there are 3 sides already selected as the references. The needed dimension is the width of the rectangle, which is equal to 40. Upon completion, select the icon of **Done**.

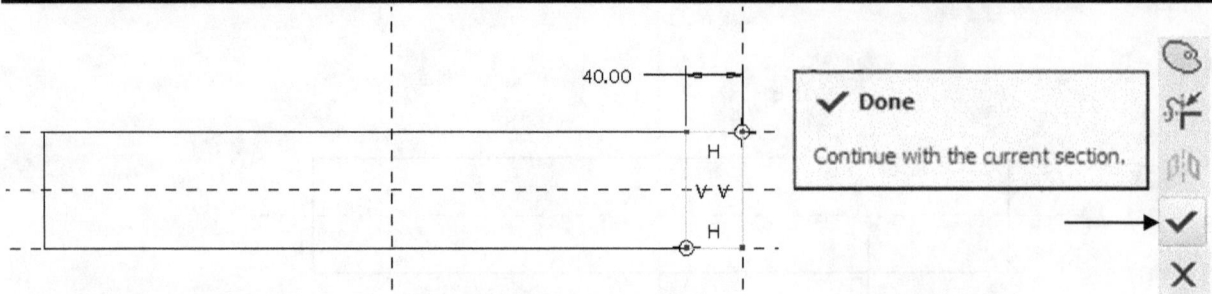

✓ Done

Continue with the current section.

Repeat the above procedure to create the 5th datum curve, as shown.

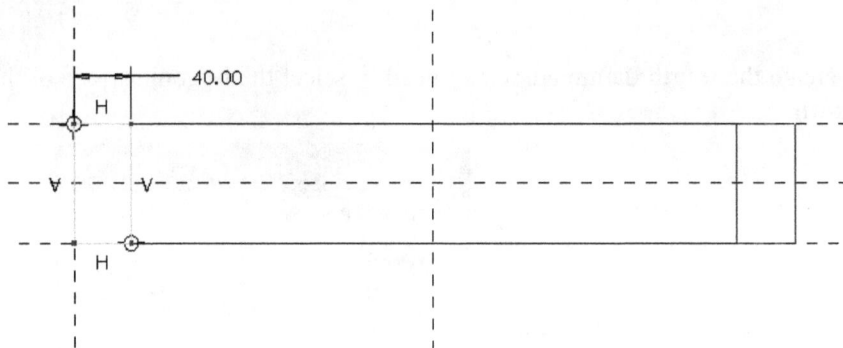

Step 3: Perform FEA under **Pro/Mechanica**

From the top menu, click **Applications > Mechanica > Continue > Structure > OK.**

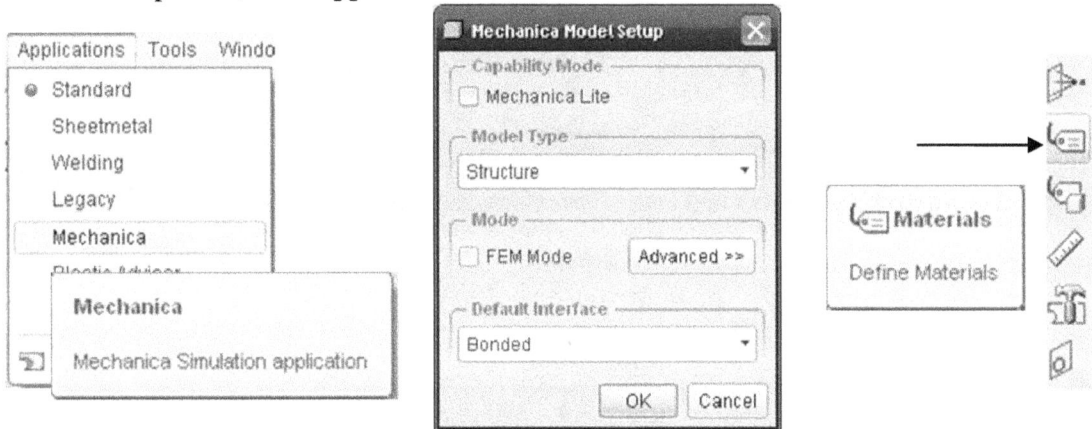

Click the icon of **Materials >** select **Nylon** from the left side called Materials in Library > click the directional arrow so that the selected material type goes to the right side called Materials in Model.

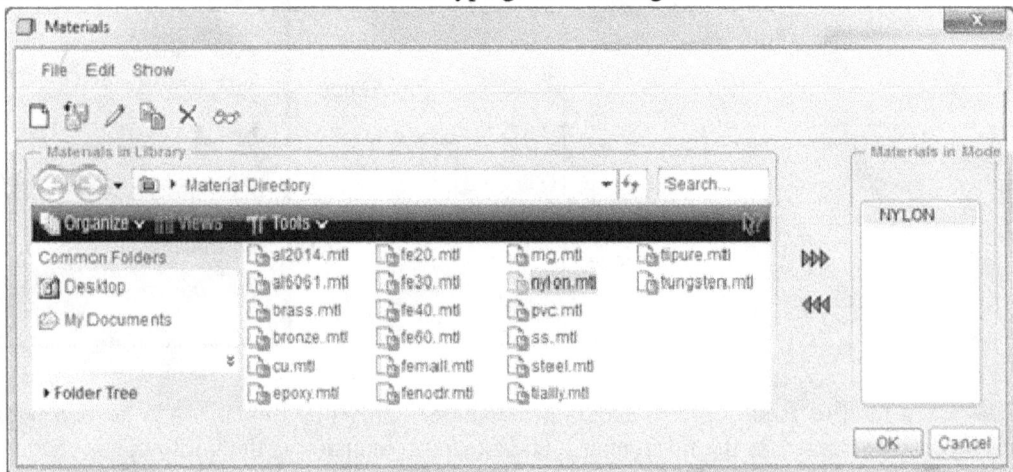

Click the icon of **Material Assignment**, the pop-up window indicates there is one component and the material type is Nylon. Just click **OK**.

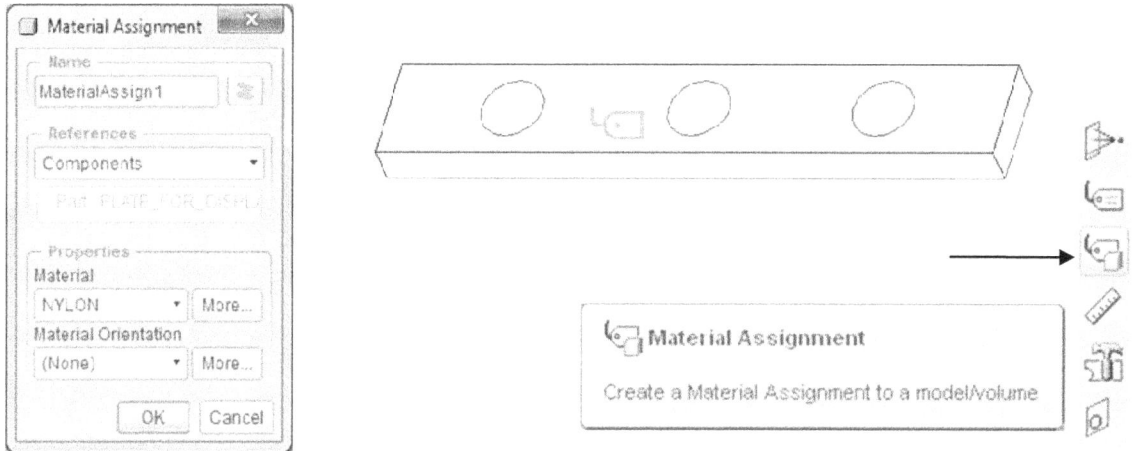

To define the surface region, click the icon of **Surface Region.** Pick the sketched rectangle first. Afterwards, pick the flat surface for splitting. Click the icon of **Apply and Save**.

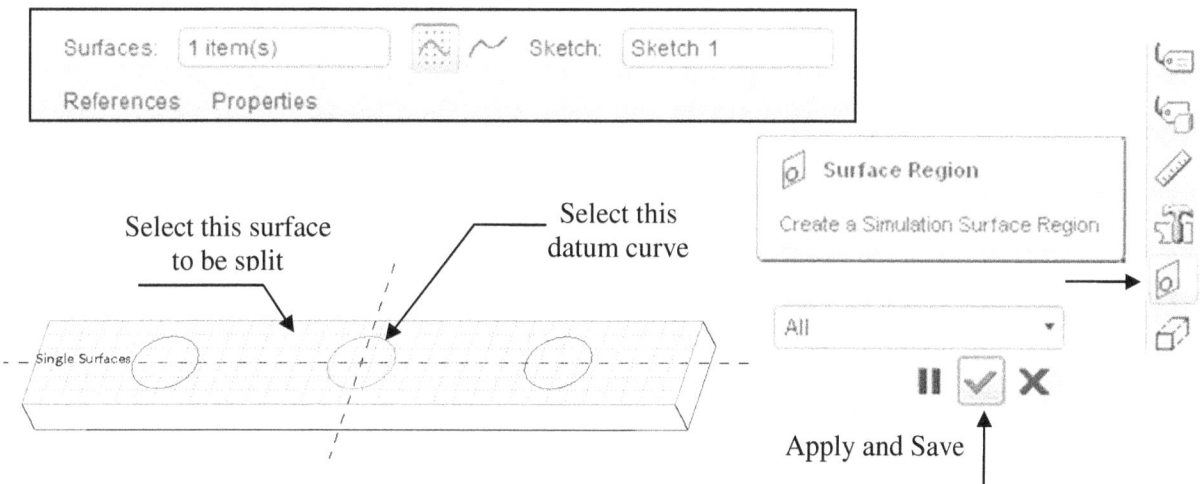

Repeat the above process 2 more times so that the 3 surface regions on the top of the plate are defined. To define the two surface regions on the bottom of the plate, we follow the above procedure as well.

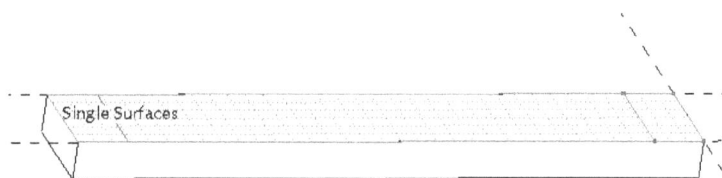

From the model tree, the 5 created surface regions are listed under Simulation Feature.

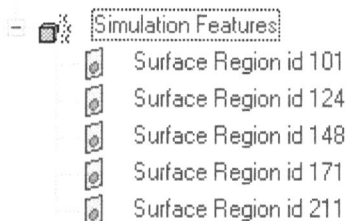

To define the shell mode, click the icon of **Shell Pair**.

Pick the top surface of the plate first, while holding down the **Ctrl** key, pick the three surface regions located on the top surface. The software system picks up the opposing surfaces automatically. Click the check mark.

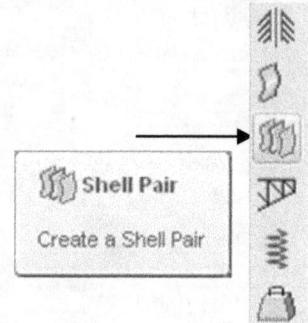

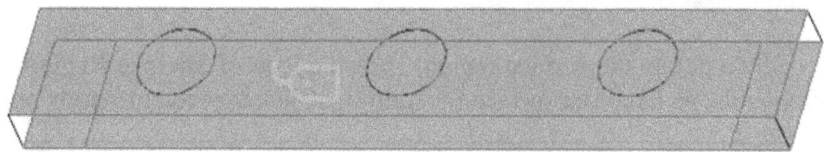

To define the constraint condition, select the icon o **Displacement Constraint > Surface(s) >** pick the surface region at the right side and the surface region on the left side on the bottom surface of the plate. Because the constraint condition is "fixed to the ground", we set all 6 degrees of freedom equal to zero, which is the default setting **> OK**.

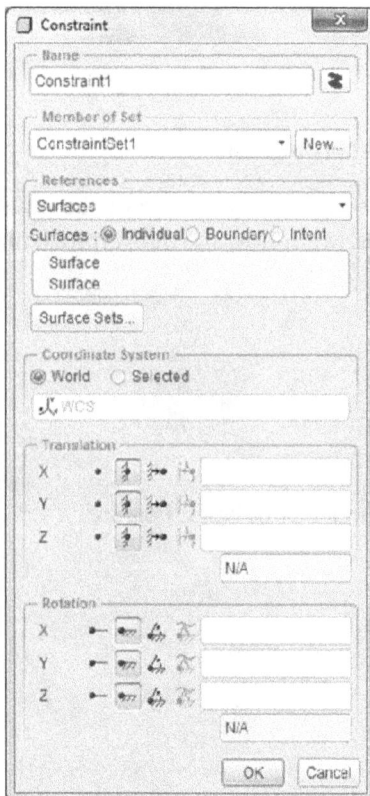

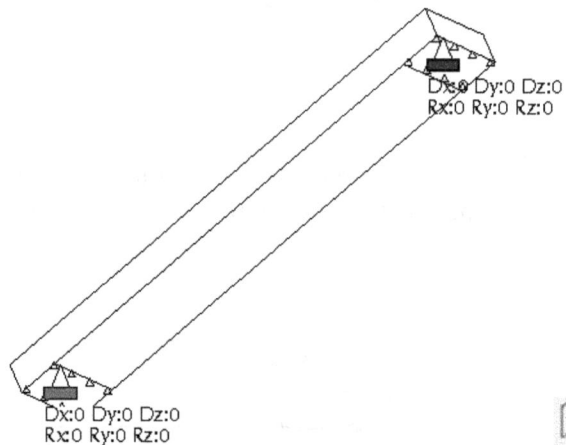

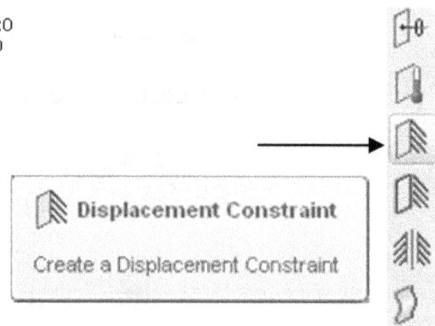

To define the load condition on the top surface of the plate, select the icon of **Force/Moment Load** from the toolbar of functions > **Surface(s)** > pick the surface region at the right side > **OK** to accept the surface selection.

Because the load is acting along the negative Y direction, we type -200 in the Y force component box > **OK**.

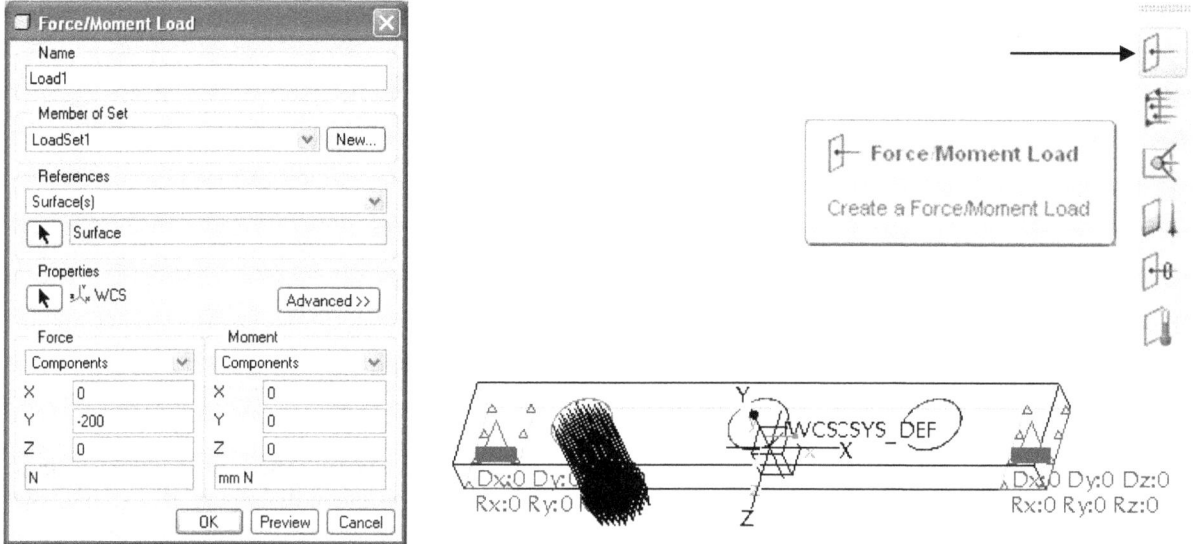

Repeat this procedure, to define Load2 and Load3 at the other 2 surface regions with the magnitude equal to -200 N.

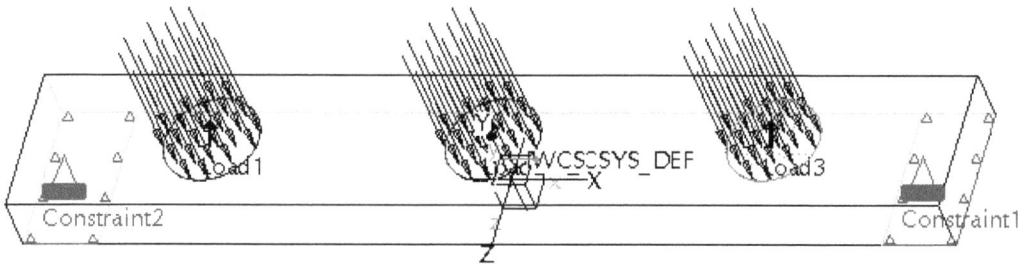

Step 4: Run the FEA program

Select the icon of **Mechanica Analyses/Studies** > **File** > **New Static** > type *plate* as the name of the analysis folder > **OK.** Click the box of **Run.**

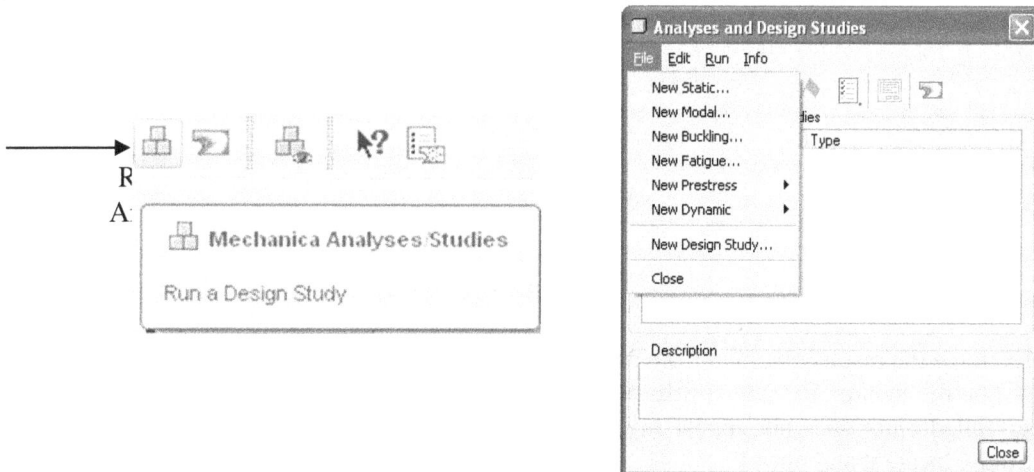

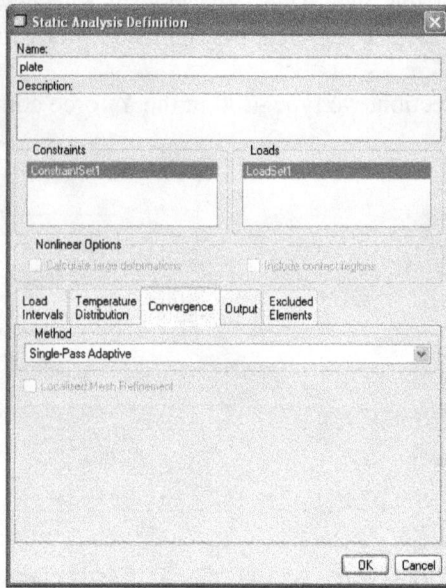

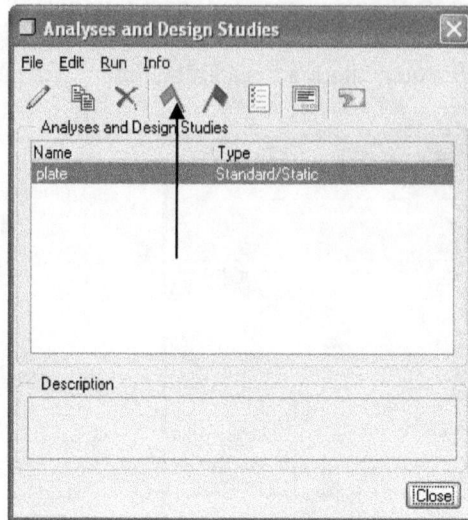

There is always a message displayed on screen. The message is asking the user "Do you want to run interactive diagnostics?" Click **Yes**. In this way a warning message will appear if an abnormal operation occurs.

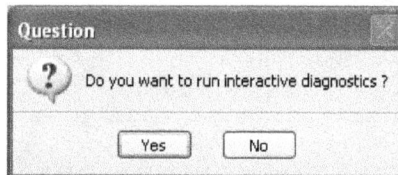

To monitor the computing process, users may click the icon of **Display study status**. The computation time is very short because of using a 2D model.

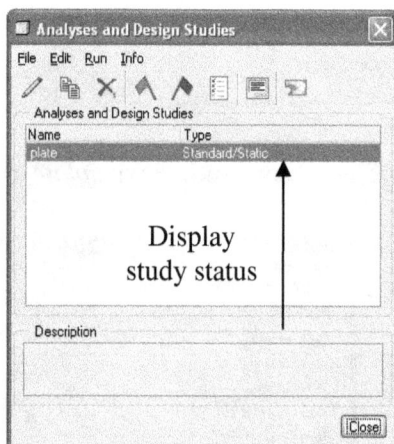

```
Mechanica Structure Version L-03-40:spg
Summary for Design Study "plate"
Tue Jul 05, 2011   20:03:35
```

```
Run Completed
Tue Jul 05, 2011   20:03:36
```

Step 5: Examine the results obtained from running the FEA program
        Click the icon of **Review results > Displacement > Magnitude > OK and Show.** Note that if the folder called *plate* does not appear in the Result Window Definition, click the icon of folder under Design Study, locate it and open it.

plate deformation

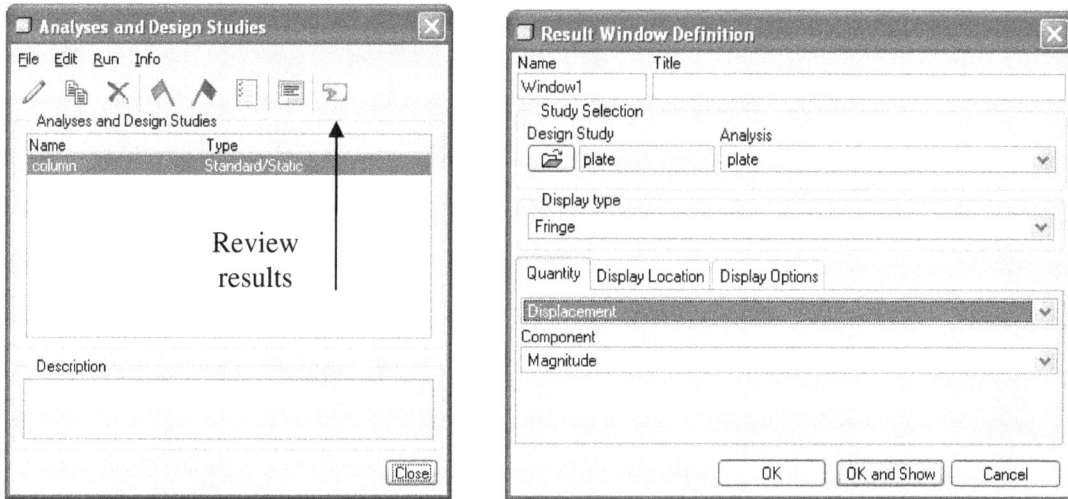

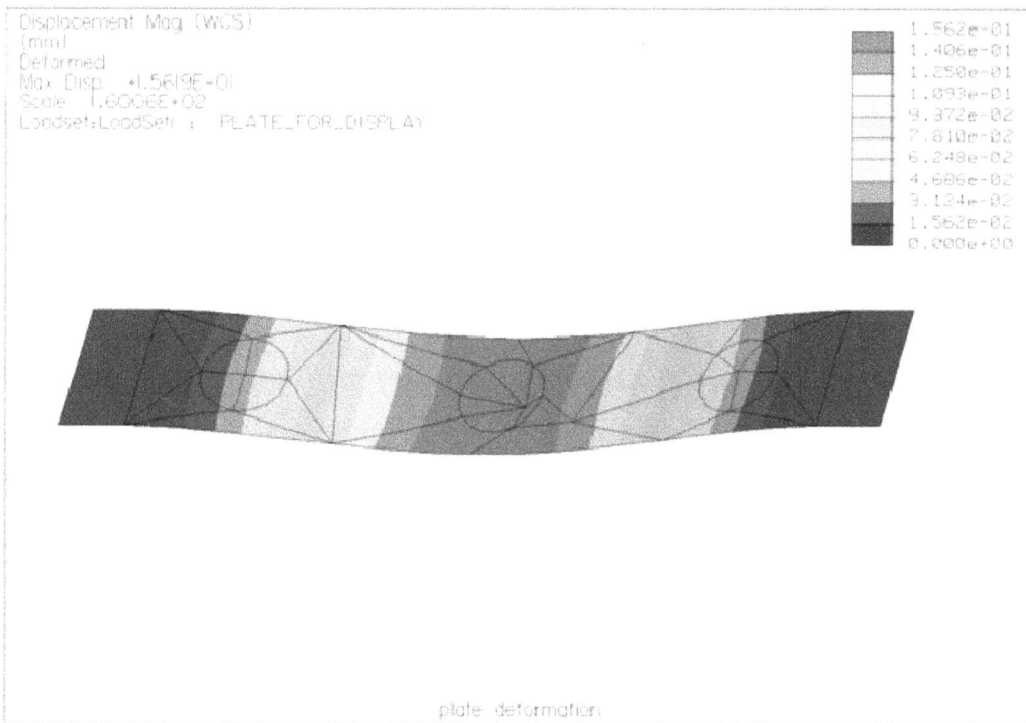

To study the distribution of von Mises stress, select **Edit** from the main toolbar **> Copy** and a new Result Window appears **> Stress > von Mises >** type Distribution of von Mises as the title **> OK and Show**.

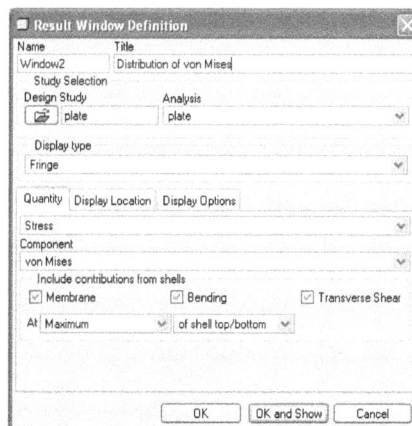

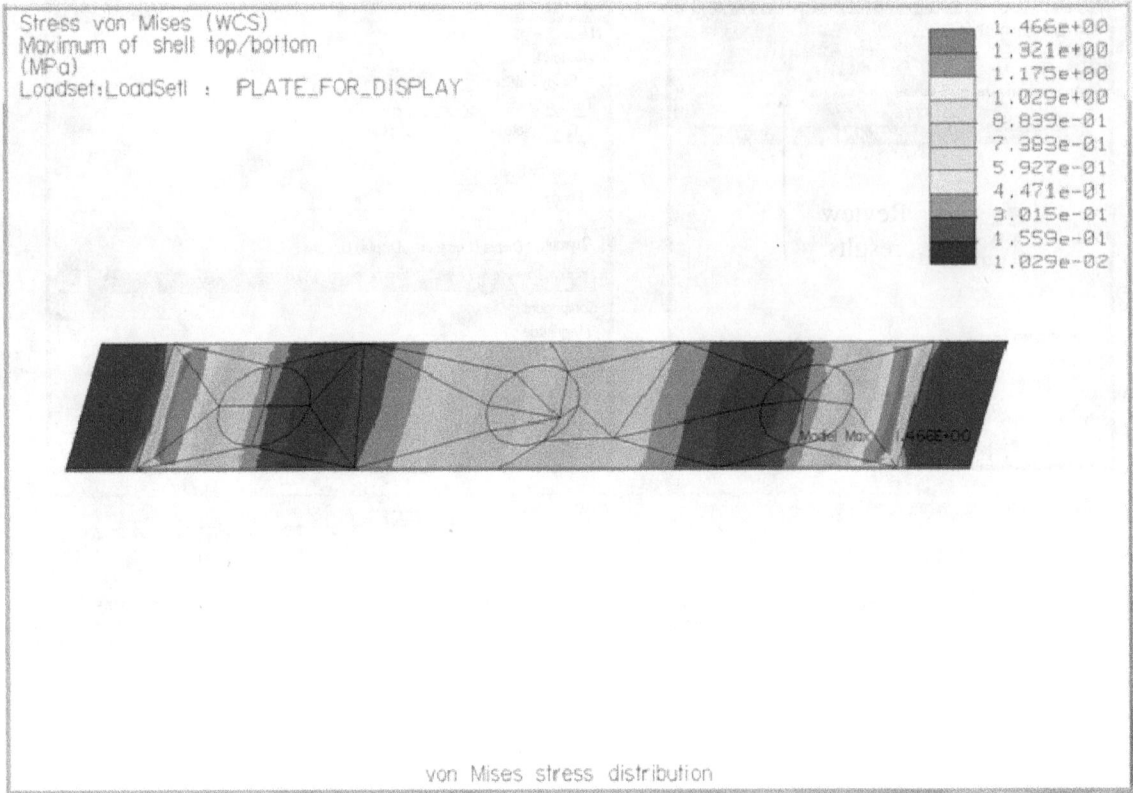

```
Stress von Mises (WCS)
Maximum of shell top/bottom
(MPa)
Loadset:LoadSetl :  PLATE_FOR_DISPLAY

1.466e+00
1.321e+00
1.175e+00
1.029e+00
8.839e-01
7.383e-01
5.927e-01
4.471e-01
3.015e-01
1.559e-01
1.029e-02
```

von Mises stress distribution

6.3   Shell Modeling Using Auto Detect

In section 6.2, we discussed the creation of a shell model, or the creation of a middle surface, by picking a pair of surfaces, which are parallel to each other. Under certain circumstances, if the geometrical model is created such a way that the software system may be capable of automatically detecting pairs of surfaces and generate the middle surfaces.   The following example is used to demonstrate this unique characteristic in shell modeling.

A t-shaped bracket is shown below. This object consists of two plates. The thickness of both plates is 25 mm. The dimensions of the plate in the horizontal position are 250 x 250 mm. The dimensions of the plate in the vertical position are 250 x 250 mm.

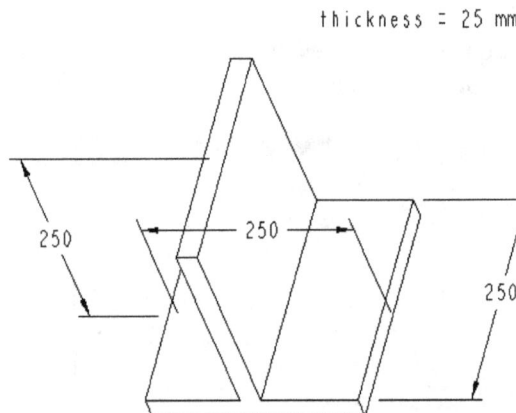

thickness = 25 mm

250   250   250

Assume that the load acts on the top surface of the support structure. The load is a uniformly distributed load with a total value equal to 500 Newton. The entire support structure is held to the ground by the two lateral surfaces, as shown. Assume that the material type of the two plates is steel.

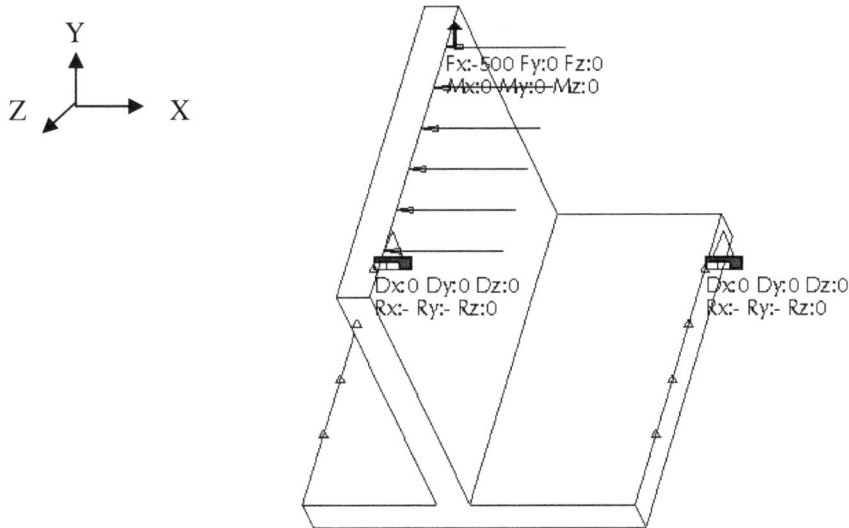

Y

Z ← → X

Fx:-500 Fy:0 Fz:0
Mx:0 My:0 Mz:0

Dx:0 Dy:0 Dz:0
Rx:- Ry:- Rz:0

Dx:0 Dy:0 Dz:0
Rx:- Ry:- Rz:0

There are 2 objectives for performing FEA:

(1) Find the location of the maximum displacement on the beam when subjected to this uniformly distributed load and specify its magnitude.
(2) Find the location of the maximum principal stress and specify this magnitude.

Step 1: Create a 3D solid model

File > New > Part > type in *support* as the file name and clear the icon of **Use default template > OK.** Select **mmns_part_solid** (units: Millimeter, Newton, Second) and type in *shell model* in **DESCRIPTION**, and *student* in **MODELED_BY**, then **OK**

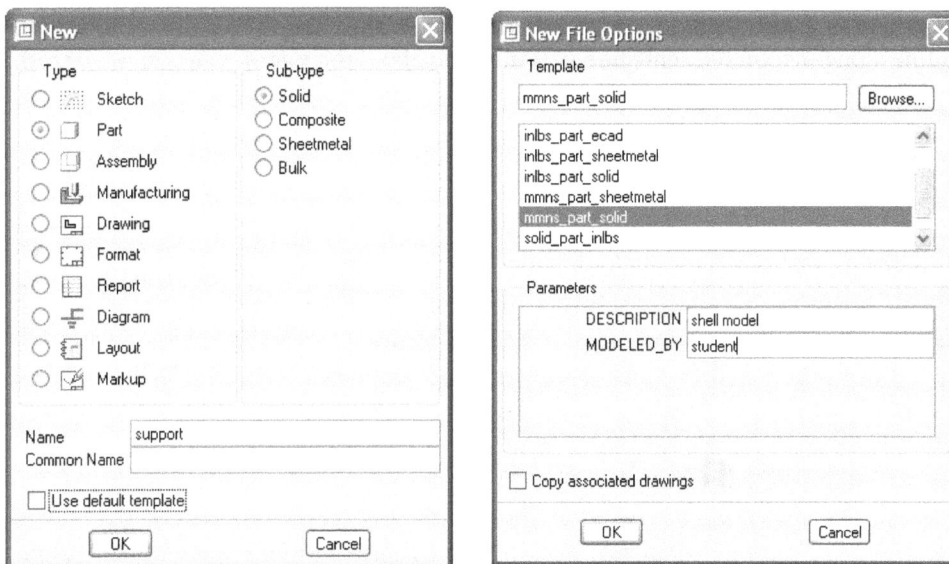

To create the plate in the horizontal position, from the main toolbar, select **Insert > Extrude.** From the dashboard, select **Thin Protrusion** > set the thickness value to 25 > select the symmetric depth setting > the depth value equal 250. Activate **Placement** to define a sketch plane > **Define.**

Symmetric
depth setting

Thin
Protrusion

Define a
sketch plane

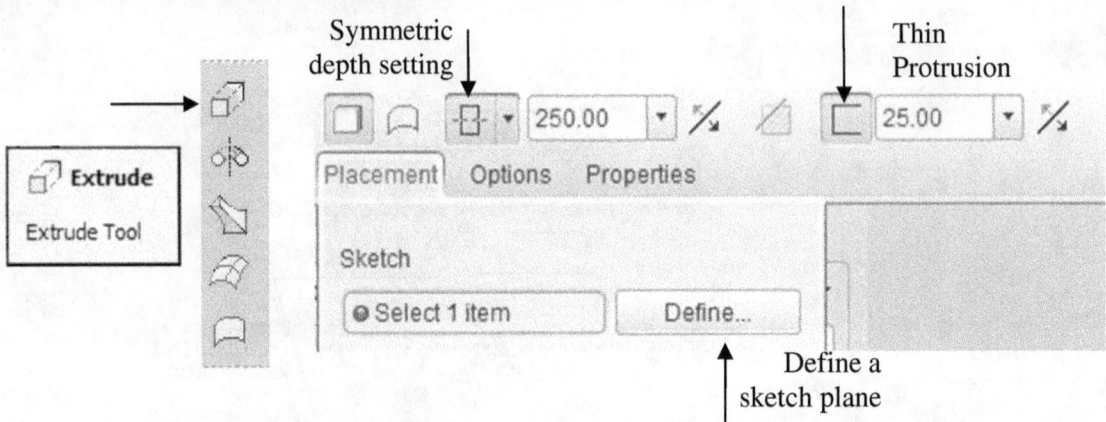

Select the **FRONT** datum plane as the sketch plane, and click the box of **Sketch** to accept the **RIGHT** datum plane as the default reference to orient the sketch plane, as illustrated below.

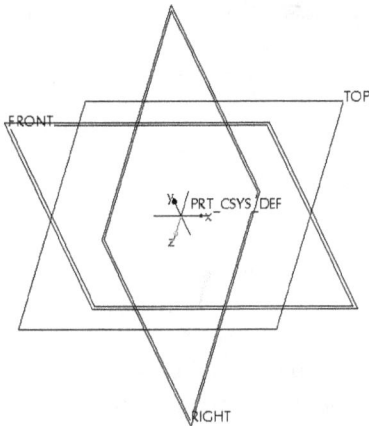

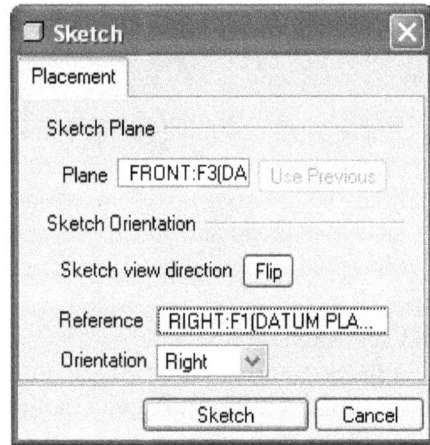

Select the icon of **Centerline** to sketch a vertical centerline long the y-axis. Select the icon of **Line** to sketch a line along the x-axis. This line is symmetric about the y-axis. The length is 250.

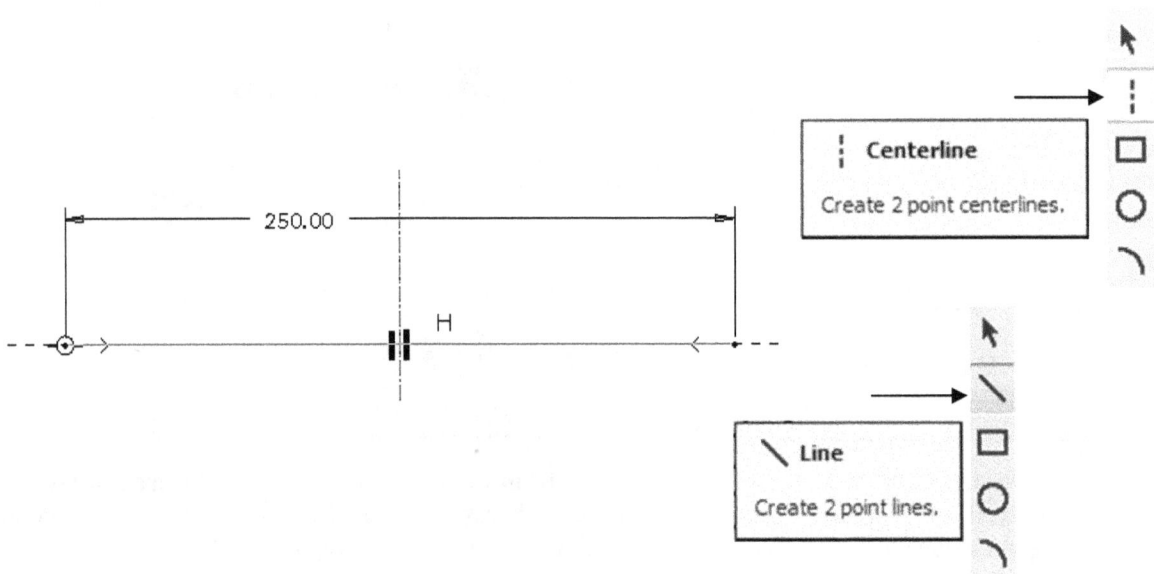

Upon completion, pick the icon of **Done** and click the arrow, as shown, to set the thickness for both sides.

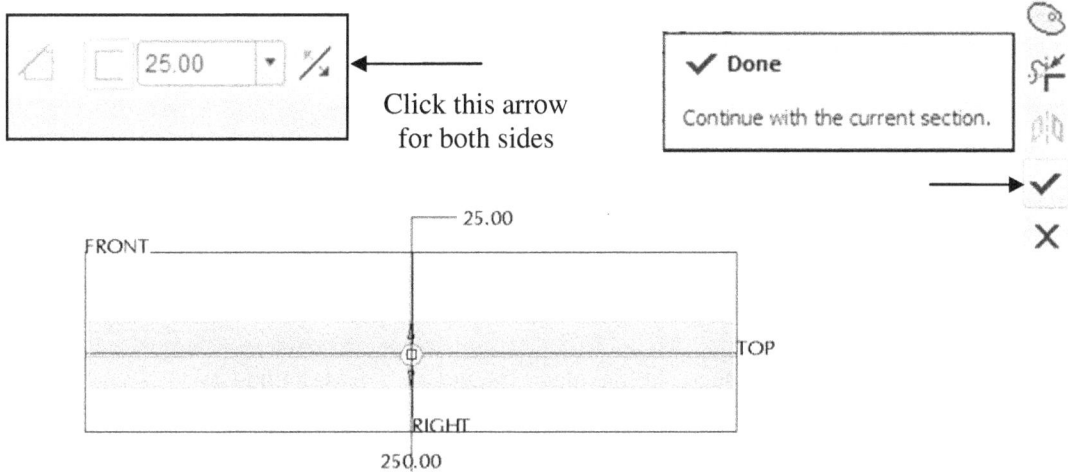

Click this arrow
for both sides

✓ Done
Continue with the current section.

Click the icon of **Apply and Save**.

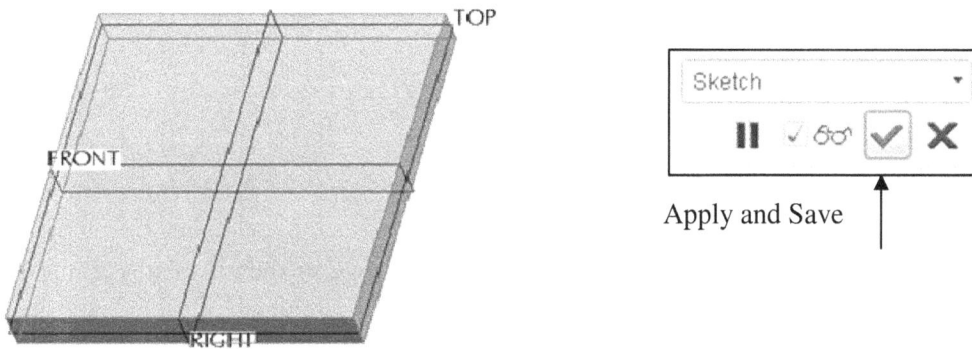

Apply and Save

To create the vertical plate, select the icon of **Extrude.** Select **Thin Protrusion** > set the thickness value to 25 > select the symmetric depth setting > the depth value equal 250. Activate **Placement** to define a sketch plane > **Define.**

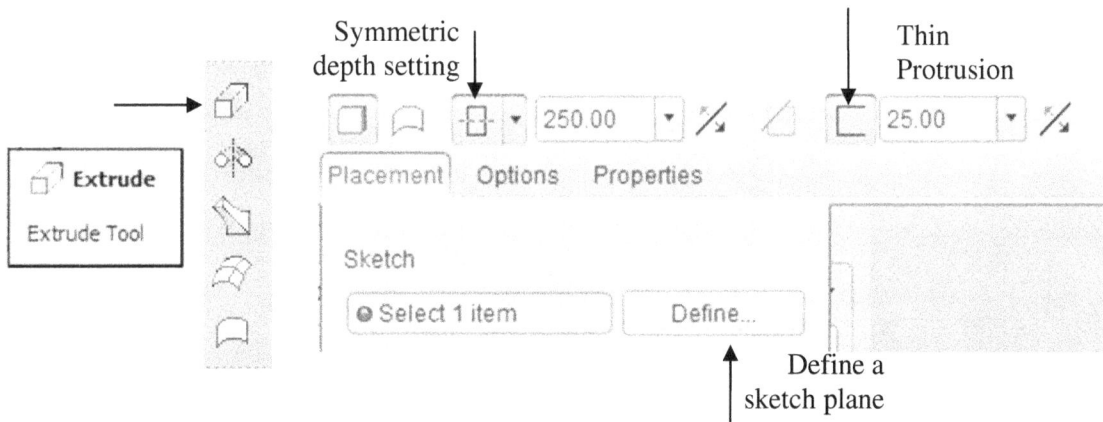

Symmetric
depth setting

Thin
Protrusion

Extrude
Extrude Tool

Placement   Options   Properties

Sketch
Select 1 item   Define...

Define a
sketch plane

Select the **FRONT** datum plane as the sketch plane, and click the box of **Sketch** to accept the **RIGHT** datum plane as the default reference to orient the sketch plane, as illustrated below.

Sketch

Placement

Sketch Plane

Plane   FRONT:F3[DA   Use Previous

Sketch Orientation

Sketch view direction   Flip

Reference   RIGHT:F1[DATUM PLA...

Orientation   Right

Sketch   Cancel

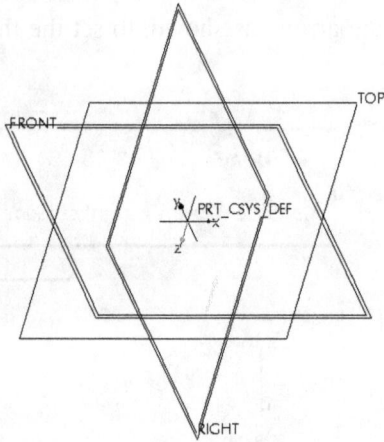

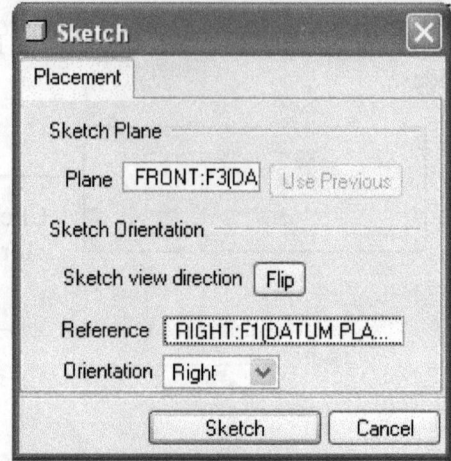

Select the icon of **Line** to sketch a line along the y-axis. The length is 250.

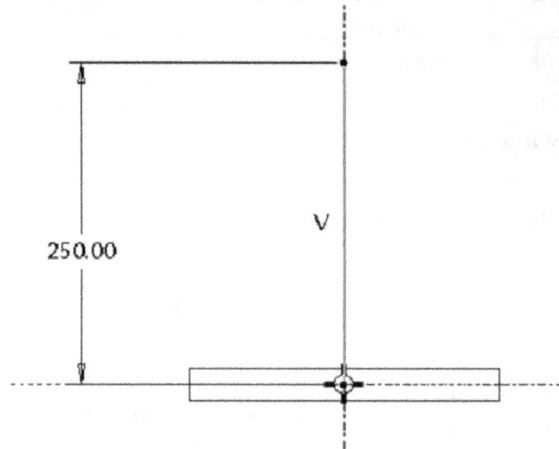

250.00

V

Upon completion, pick the icon of **Done** and click the arrow, as shown, to set the thickness for both sides.

25.00

Click this arrow
for both sides

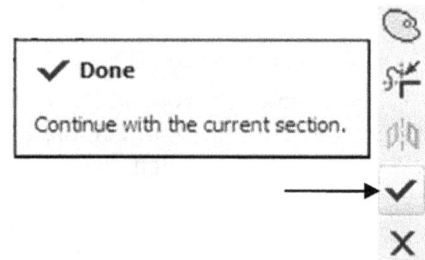

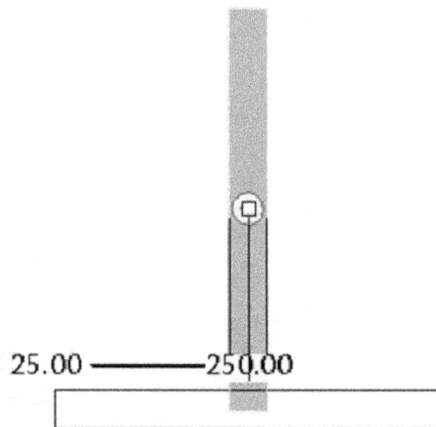

✔ **Done**

Continue with the current section.

25.00 ———— 250.00

Click the icon of **Apply and Save**.

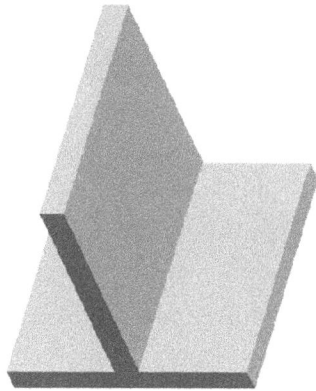

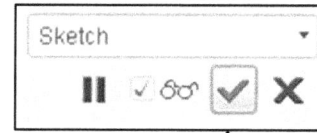

Apply and Save

Step 2: FEA under Pro/MECHANICA.

From the top menu, click **Applications > Mechanica > Continue > Structure > OK.**

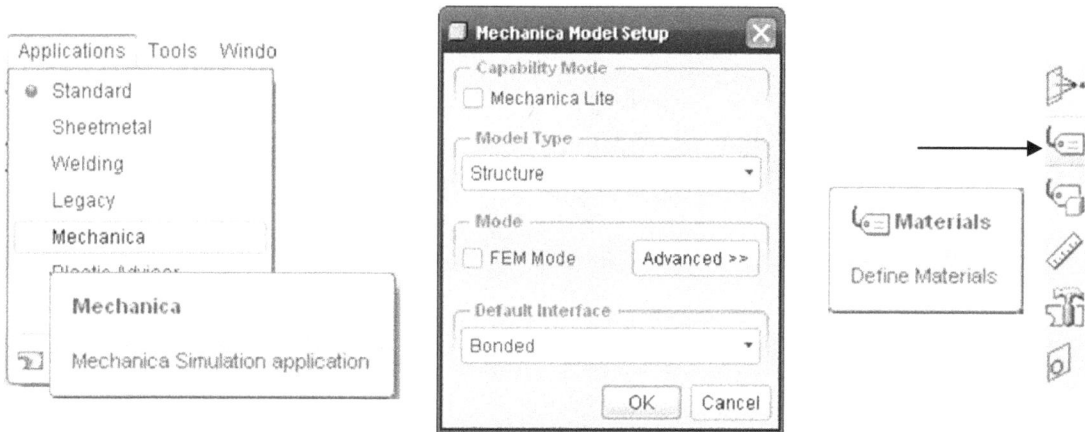

Click the icon of **Materials >** select **STEEL** from the left side called Materials in Library > click the directional arrow so that the selected material type goes to the right side called Materials in Model.

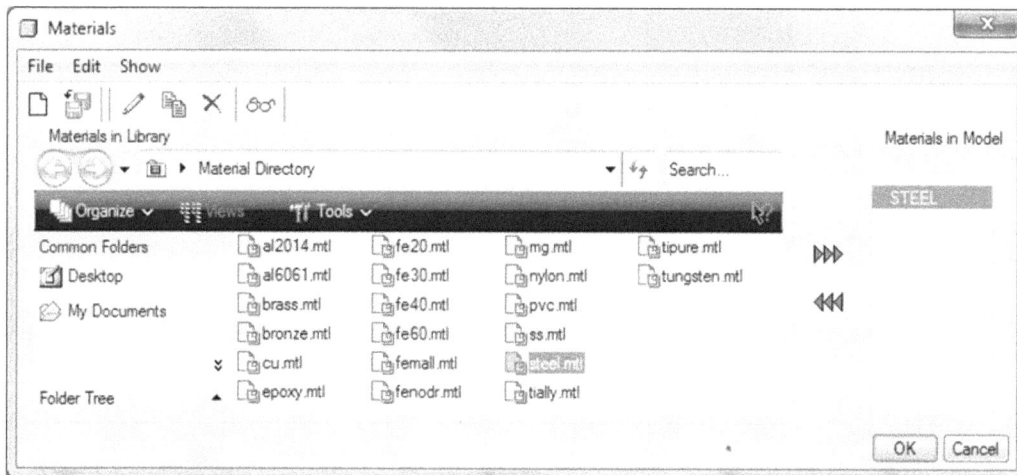

Click the icon of **Material Assignment**, the pop-up window indicates there is one component and the material type is STEEL. Just click **OK.**

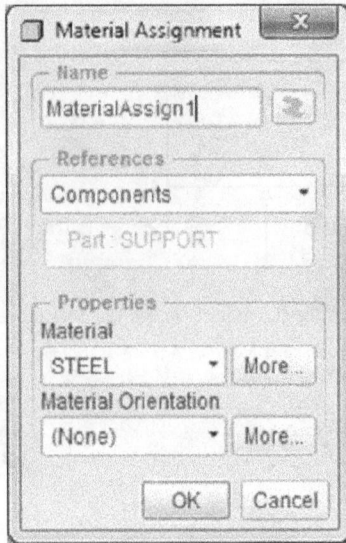

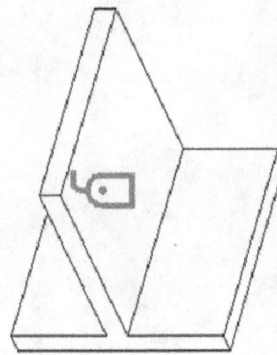

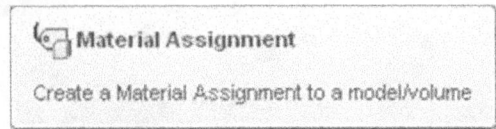

From the toolbar of functions, select the icon of **Insert > Midsurface > Auto Detect Shell Pairs.** Specify 25 as the thickness value > **Start**. In the model tree, 2 shell pairs are listed.

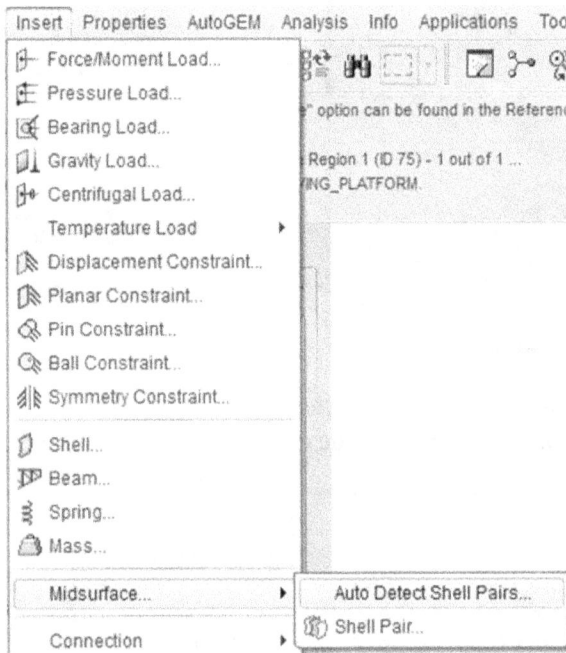

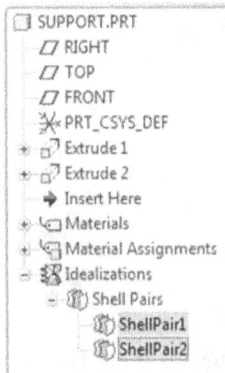

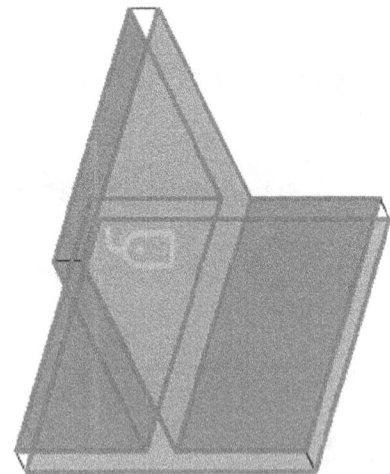

From the toolbar of functions, select the icon of **Constraints** > select **Edge(s)/Curve(s)** > pick the edge on the left side and the edge on the right side while holding down the **Ctrl** key > **OK**.

Note the constraint conditions for fixing a straight line are, in this case, fixing Translation X, Translation Y, Translation Z and Rotation Z.

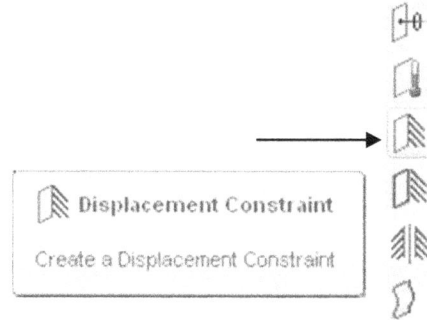

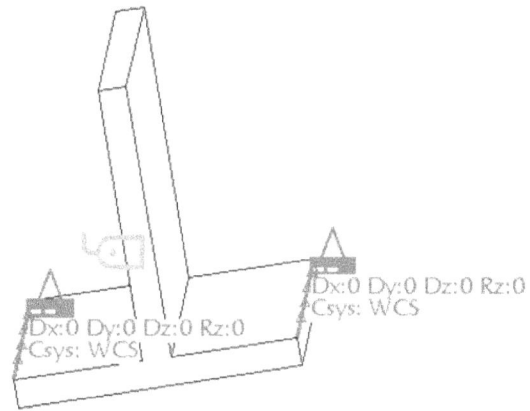

To define a uniform load acting on the top edge in the negative x direction, select the icon of **Force and Moment** > **Edge(s)/Curve(s)** > pick the edge on the top surface > type -*500* in X of the Force box > **OK**.

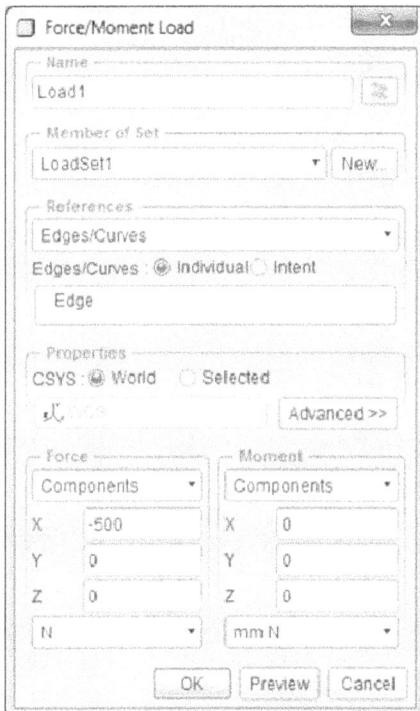

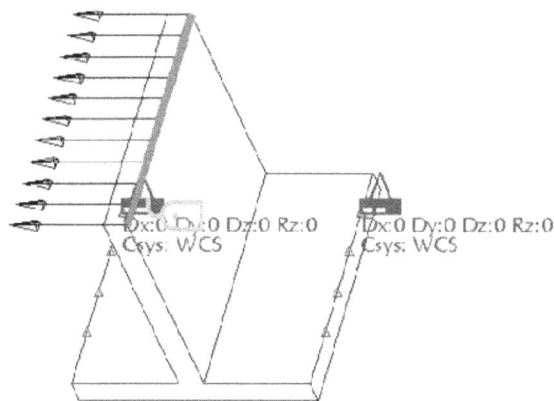

Perform a static analysis

Click the icon of Mechanica Analyses/Studies > **File** > Select **New static analysis**. Type t_bracket as the name of design study and **Run**.

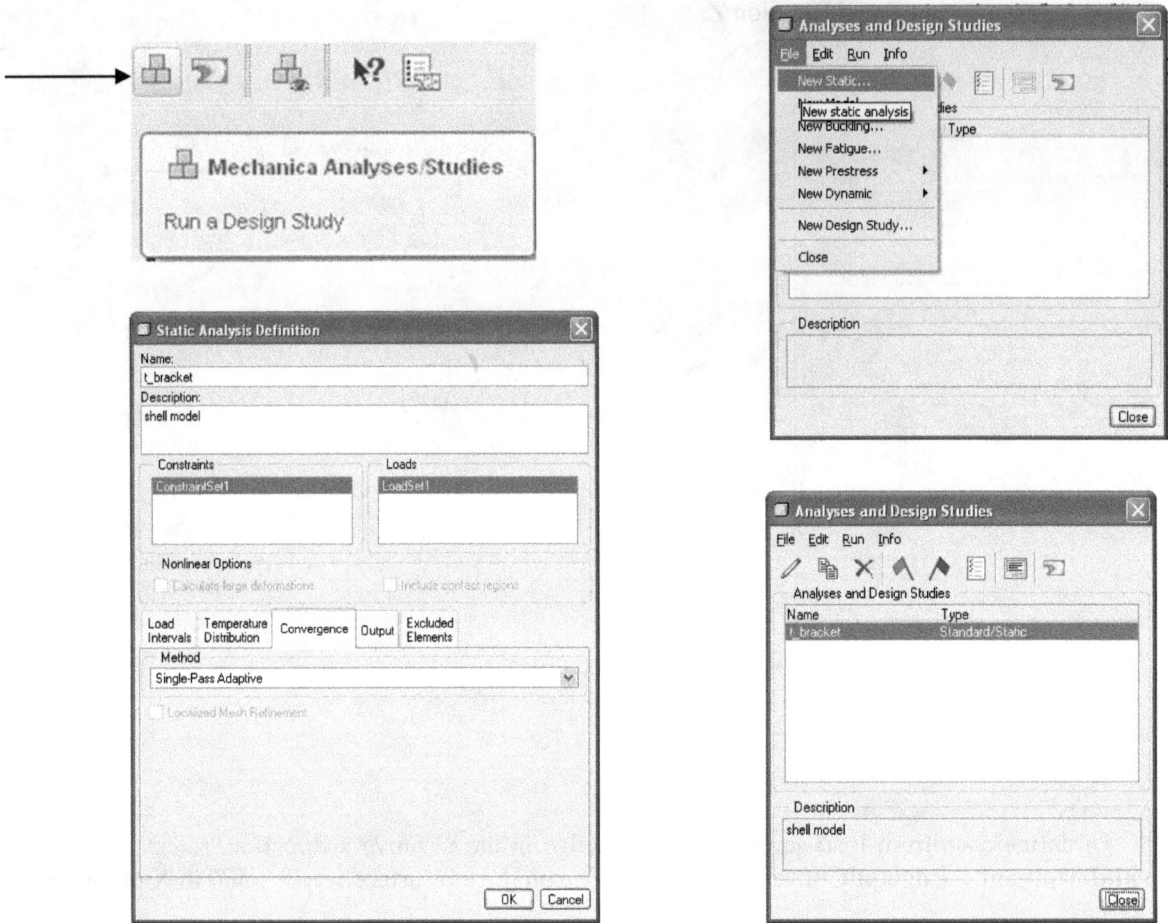

To review the results, click the icon of review results > type deformation distribution as the title of the result window > select the folder called t_bracket > **Fringe > Displacement > Magnitude > OK and Show**.

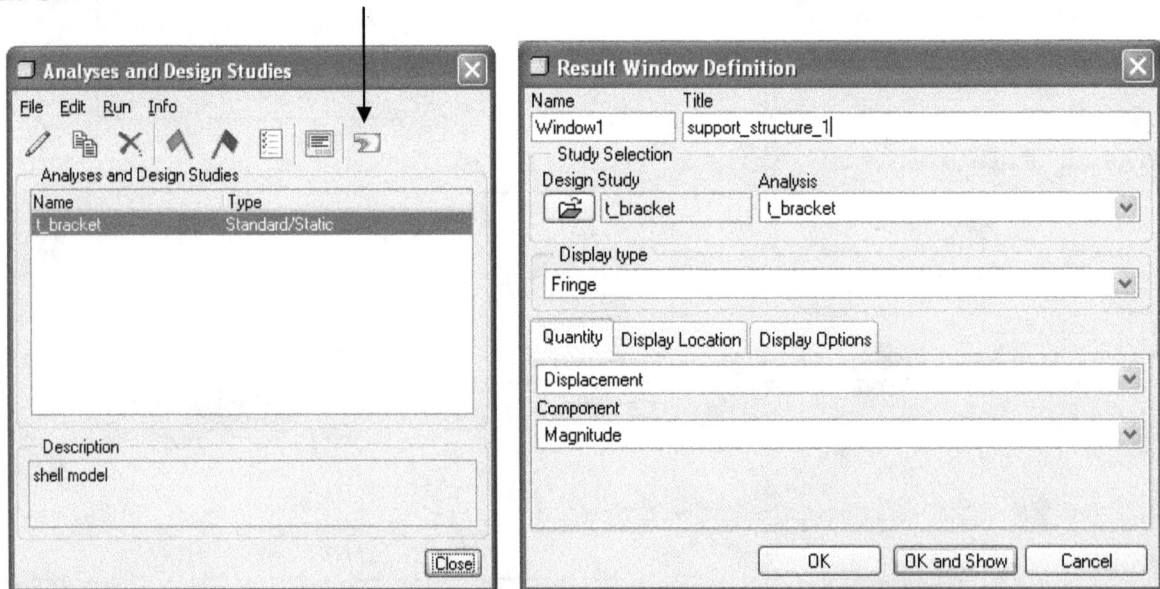

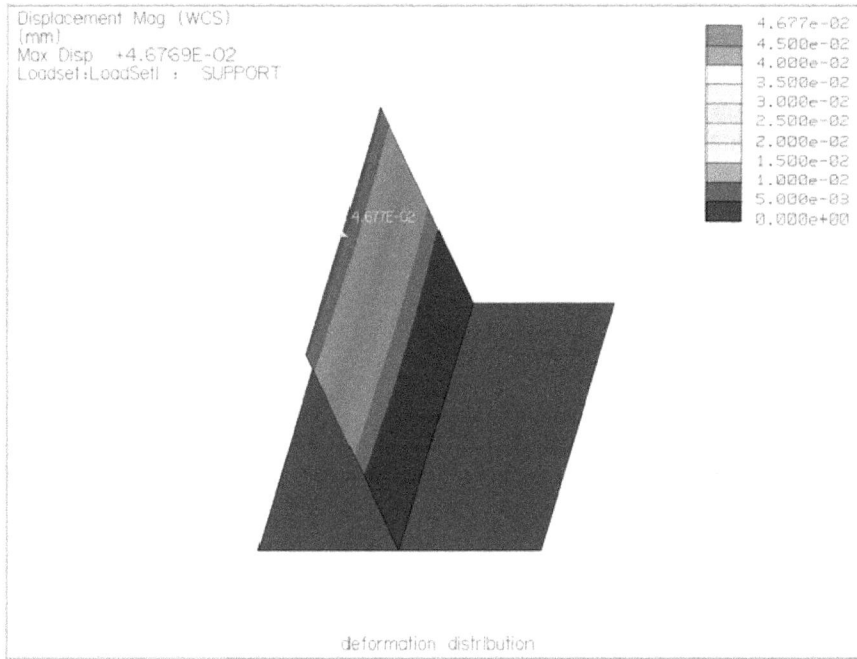

deformation distribution

To study the distribution of the von Mises stress distribution, select **Edit** from the main toolbar > **Copy** and a new Result Window appears > **Stress** > **von Mises** > type von Mises stress distribution > **OK and Show**.

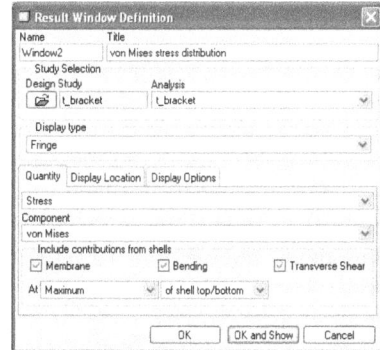

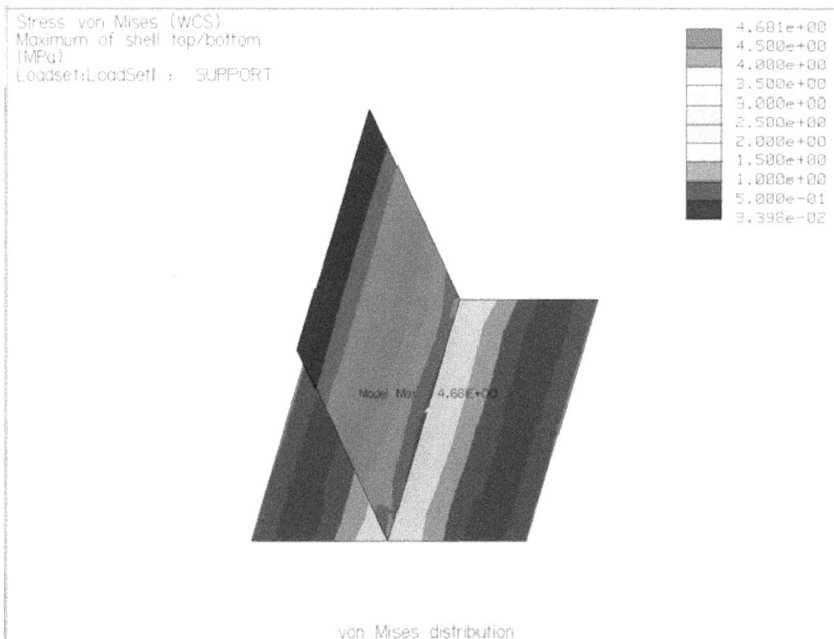

von Mises distribution

6.4   Case Study: Shell Modeling for a Welding Structure

In this section, we will work with an assembly. The assembly is a welding structure. As shown below, there are two components. One is called Base, and the other Cover. The base component is fixed to the ground. The thickness of the two components is 8 mm. There will be 8 spot welds to connect the two components. The size of each spot weld is 20 mm. The material for the two components and the welds is assumed to be STEEL.

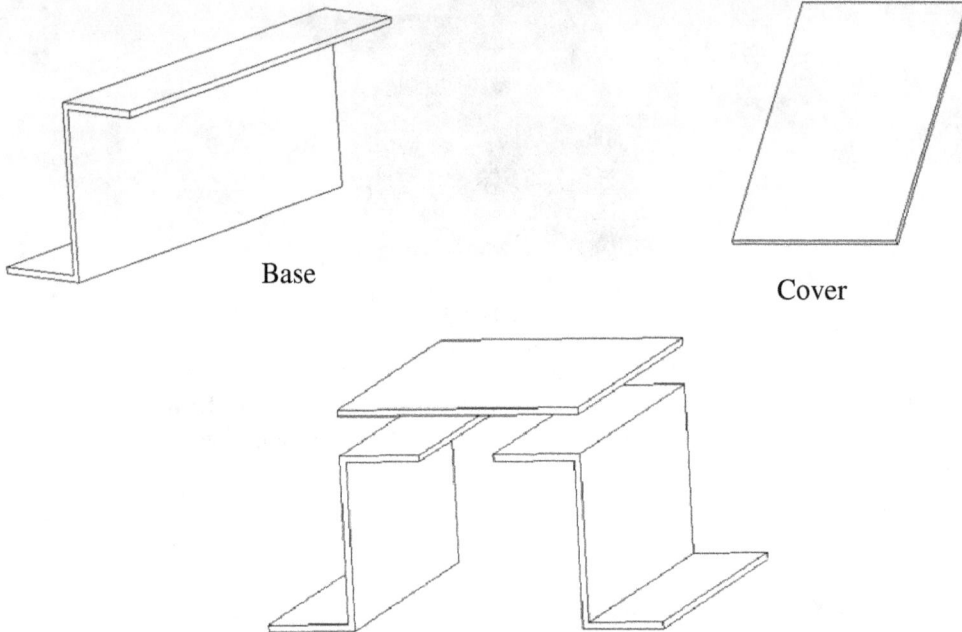

Base

Cover

The geometrical shape and key dimensions of the welding structure are shown below.   To illustrate the dimensions involved in creating two (2) 3D solid models for the 2 components and a 3D solid model of the welding assembly, an engineering drawing is shown below.

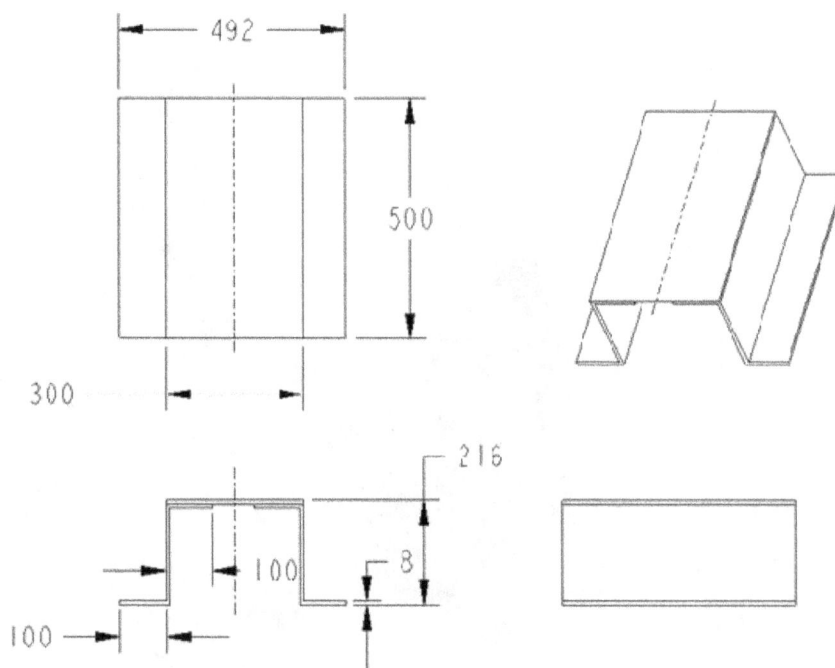

Step 1: Create a 3D solid model for the base component

File > New > Part > type *cover* as the file name and clear the icon of **Use default template > OK.** Select mmns_part_solid (units: Millimeter, Newton, Second) and type *cover* in DESCRIPTION, and *student* in MODELED_BY, then **OK.**

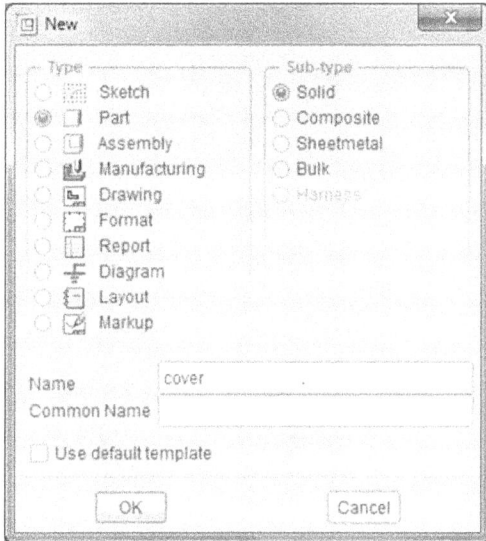

Feature 1: a plate with the dimensions: 300 x 500 x 8 mm.

Click the icon of **Extrude. S**elect **Thin Protrusion** > set the thickness value to *8* > select the symmetric depth setting > set the depth value equal *500*. Activate **Placement > Define.**

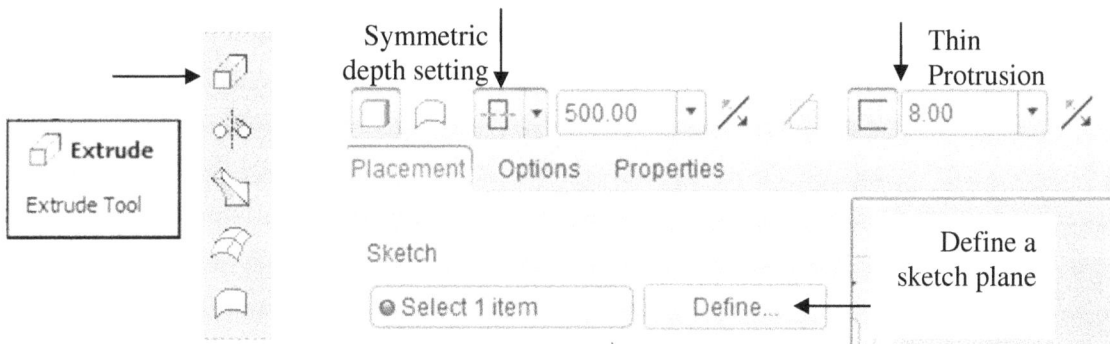

Select the **FRONT** datum plane as the sketch plane, and click the box of **Sketch** to accept the **RIGHT** datum plane as the default reference to orient the sketch plane, as illustrated below.

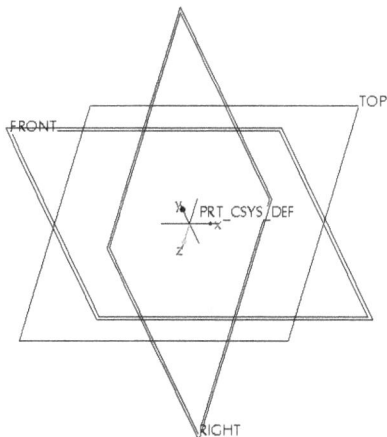

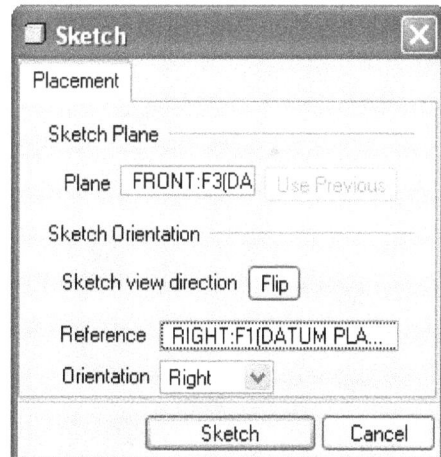

Select the icon of **Centerline** to sketch a vertical centerline long the y-axis. Select the icon of **Line** to sketch a line along the x-axis. This line is symmetric about the y-axis. The length is 300.

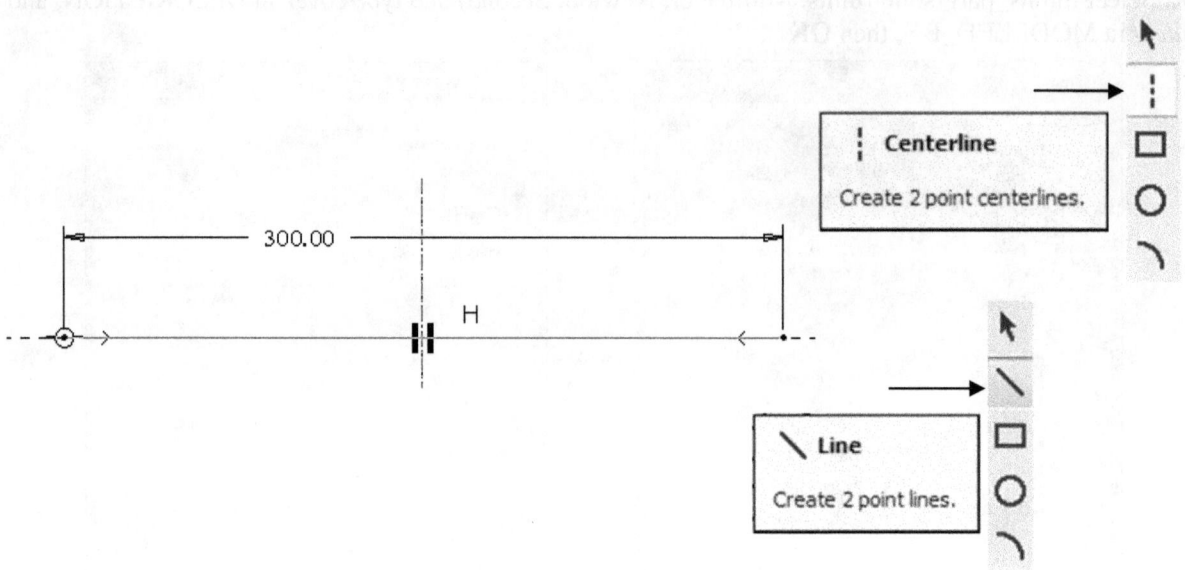

300.00

H

Centerline
Create 2 point centerlines.

Line
Create 2 point lines.

Upon completion, pick the icon of **Done** and click the arrow, as shown, to set the thickness for one side (UPPER).

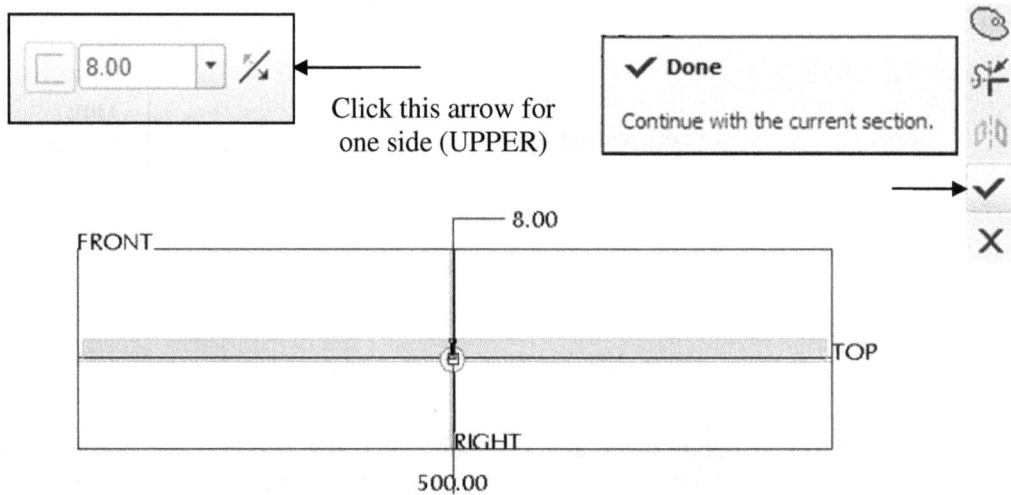

8.00

Click this arrow for one side (UPPER)

✔ Done
Continue with the current section.

8.00

FRONT

TOP

RIGHT

500.00

Click the icon of **Apply and Save**.

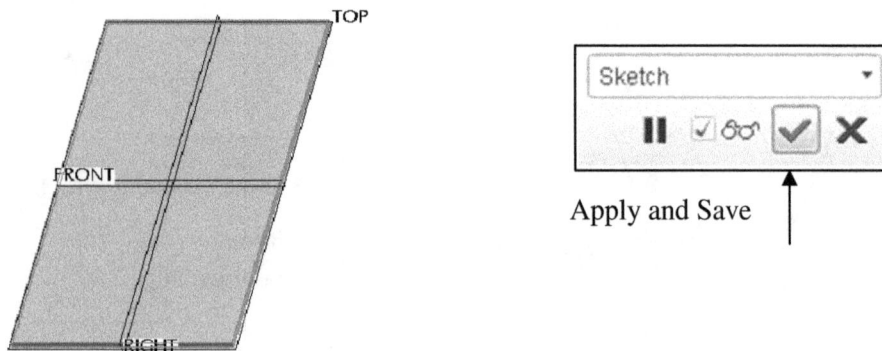

TOP

FRONT

RIGHT

Sketch

Apply and Save

Click the icon of **Datum Point Tool** and expand it, and select the default coordinate system. Type the following coordinates for the 8 datum points.

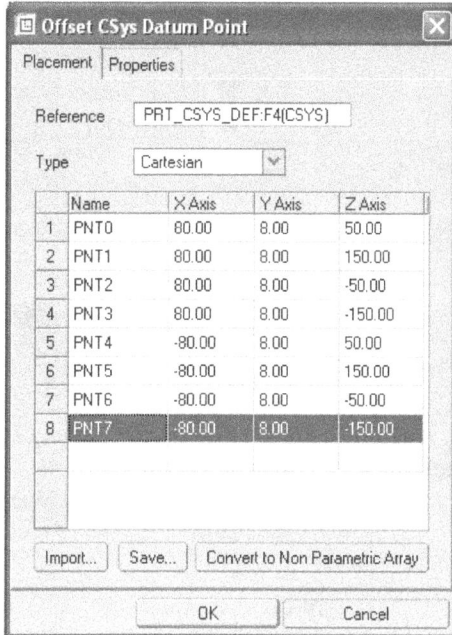

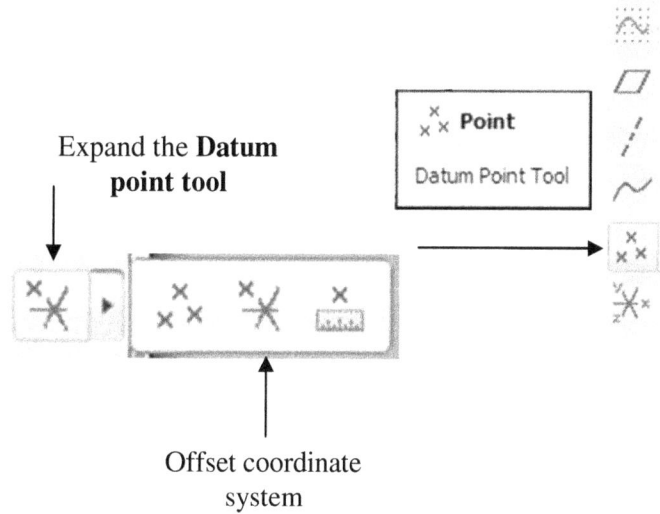

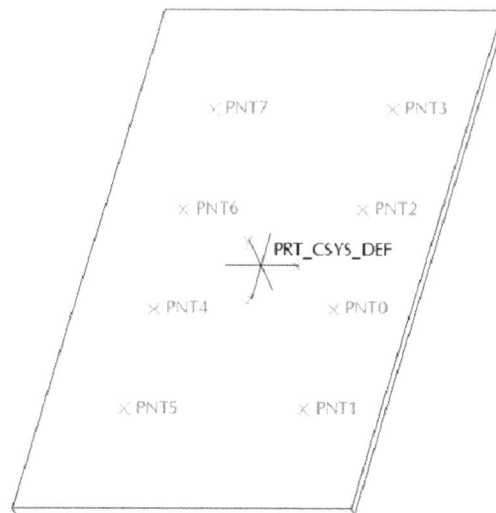

Expand the **Datum point tool**

Offset coordinate system

Step 2: Create a 3D solid model for the U-shape component

File > New > Part > type *base* as the file name and clear the icon of **Use default template > OK.** Select **mmns_part_solid** (units: Millimeter, Newton, Second) and type *base* in **DESCRIPTION**, and *student* in **MODELED_BY**, then **OK.**

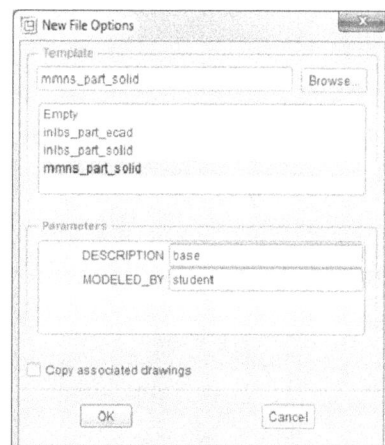

Feature 1: a bent base plate.

Feature 1: a bent plate with the dimensions: 300 x 500 x 8 mm.

Click the icon of **Extrude.** Select **Thin Protrusion** > set the thickness value to *8* > select the symmetric depth setting > set the depth value equal *500*. Activate **Placement > Define.**

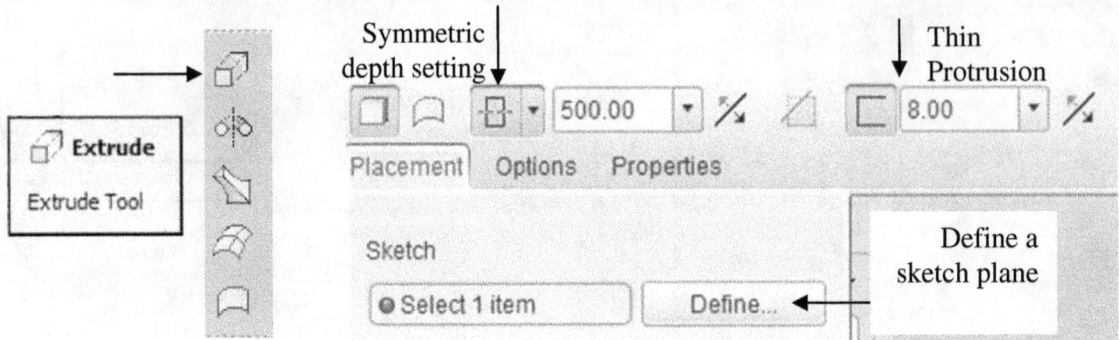

Symmetric depth setting

Thin Protrusion

Extrude

Extrude Tool

Placement   Options   Properties

Sketch

● Select 1 item   Define...

Define a sketch plane

Select the **FRONT** datum plane as the sketch plane, and click the box of **Sketch** to accept the **RIGHT** datum plane as the default reference to orient the sketch plane, as illustrated below.

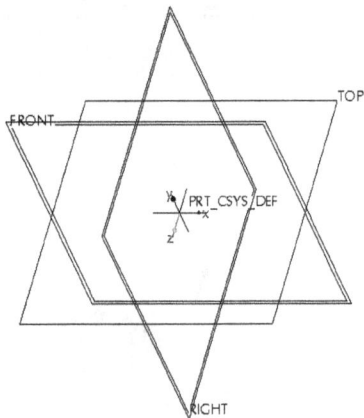

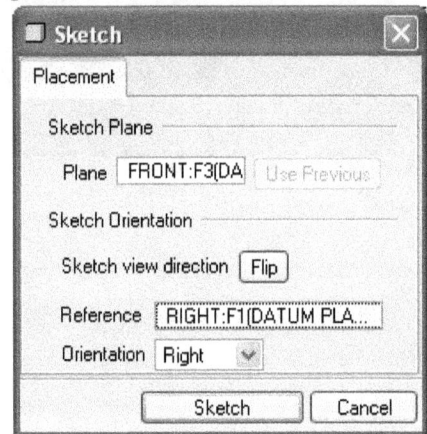

Sketch

Placement

Sketch Plane

Plane FRONT:F3(DA   Use Previous

Sketch Orientation

Sketch view direction Flip

Reference RIGHT:F1(DATUM PLA...

Orientation Right

Sketch   Cancel

Select the icon of **Line** to sketch a bent shape. The dimensions are shown below.

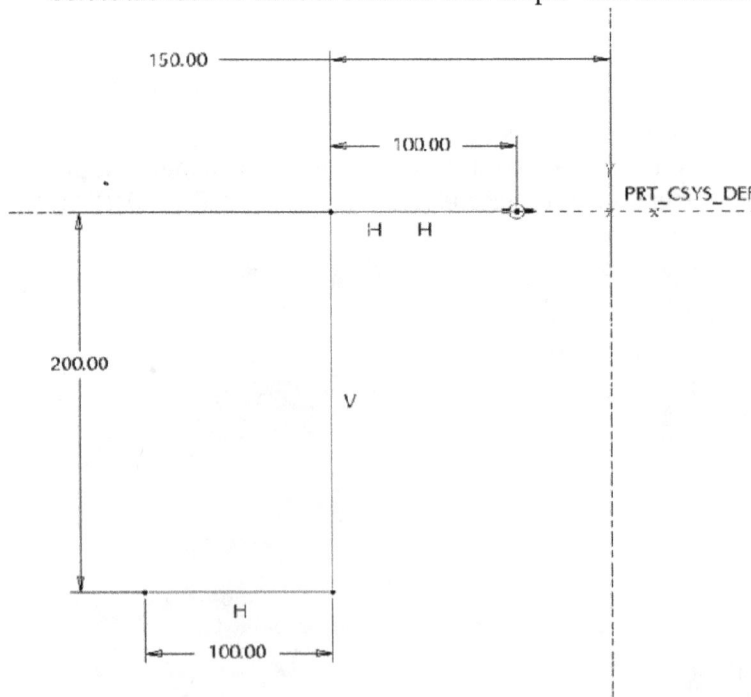

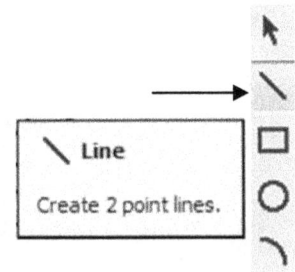

150.00

100.00

PRT_CSYS_DEF

H   H

200.00

V

H

100.00

Line

Create 2 point lines.

Upon completion, pick the icon of **Done** and click the arrow, as shown, to set the thickness for one side (UPPER).

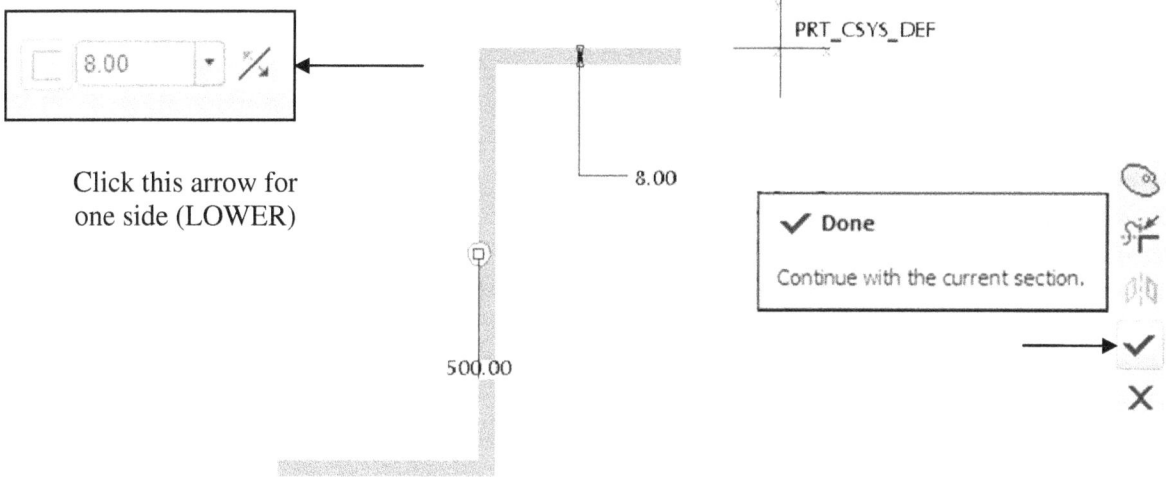

PRT_CSYS_DEF

8.00

Click this arrow for one side (LOWER)

8.00

✓ Done

Continue with the current section.

✓

✕

500.00

Click the icon of **Apply and Save**.

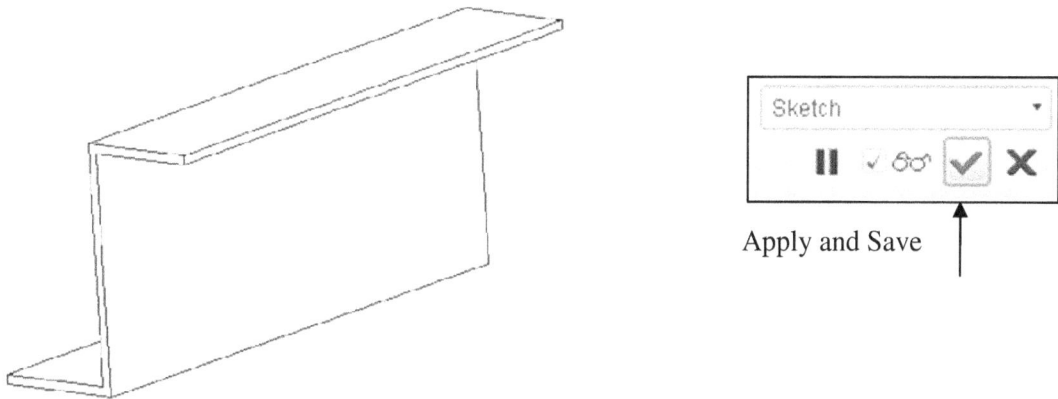

Sketch

❚❚ ✓ 👓 ✓ ✕

Apply and Save

Step 3: Create a 3D solid model for the assembly, which consists of the cover and 2 base components.

File > New > Part > type *spot_welding* as the file name and clear the icon of **Use default template > OK.** Select **mmns_part_solid** (units: Millimeter, Newton, Second) and type in *spot welding* in **DESCRIPTION**, and *student* in **MODELED_BY**, then **OK.**

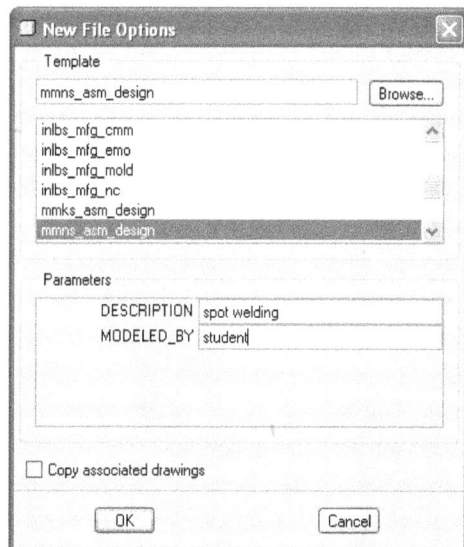

New

Type	Sub-type
○ Sketch	⊙ Design
○ Part	○ Interchange
⊙ Assembly	○ Verify
○ Manufacturing	○ Process Plan
○ Drawing	○ NC Model
○ Format	○ Mold Layout
○ Report	○ Ext. Simp.Rep.
○ Diagram	
○ Layout	
○ Markup	

Name   spot_welding
Common Name

☐ Use default template

OK   Cancel

New File Options

Template

mmns_asm_design   Browse...

inlbs_mfg_cmm
inlbs_mfg_emo
inlbs_mfg_mold
inlbs_mfg_nc
mmks_asm_design
mmns_asm_design

Parameters

| DESCRIPTION | spot welding |
| MODELED_BY | student |

☐ Copy associated drawings

OK   Cancel

Click the icon of **Assembly**. Select the file called base.prt from the **In Session** window > **Open**.

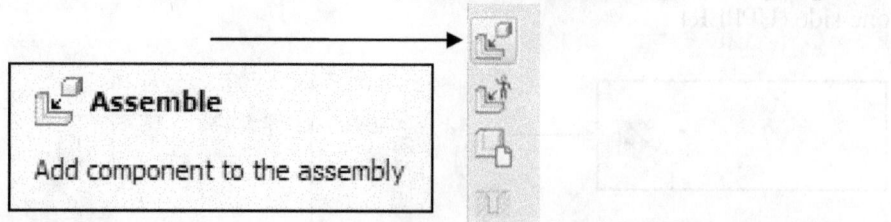

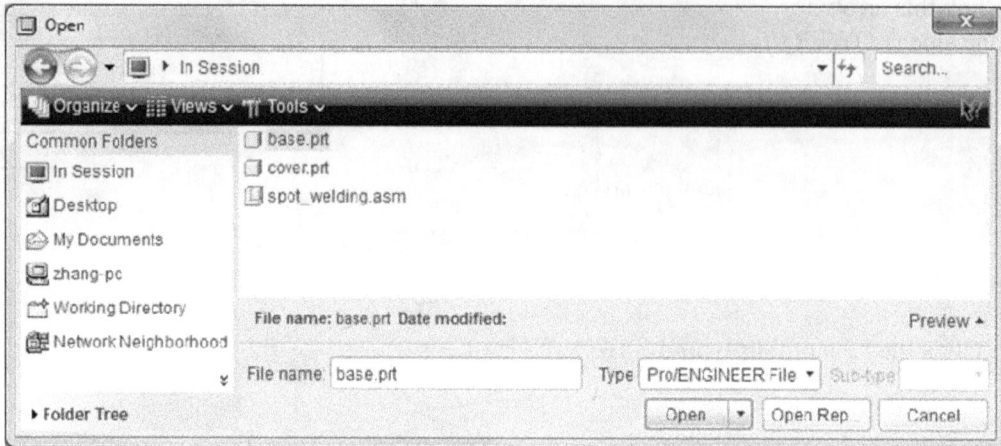

Select the icon of **Default** > **OK** to assemble the base component. Click the icon of **Apply and Save**.

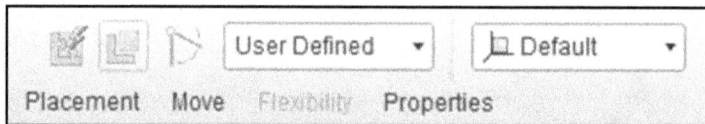

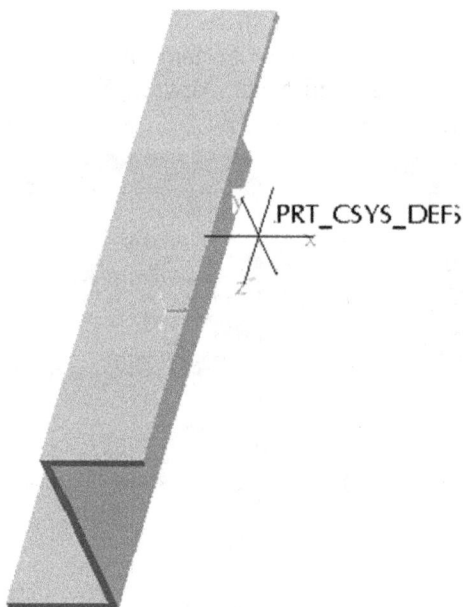

Click the icon of **Assembly**. Select the file called cover.prt from the **In Session** window > **Open**.

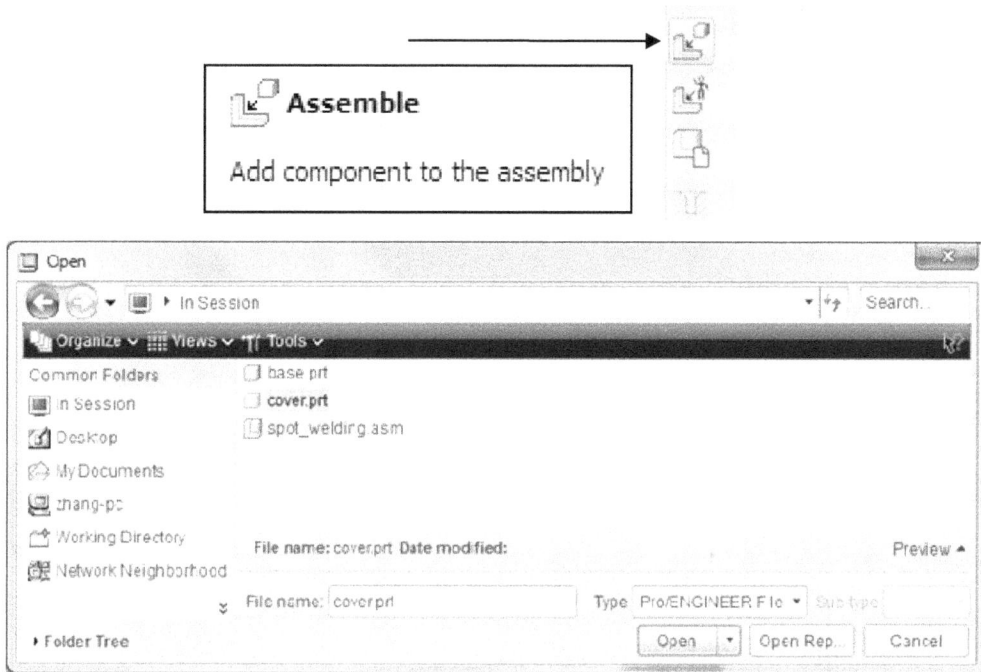

Select the icon of default > **OK** to complete assembling the cover component. Click the icon of **Apply and Save**.

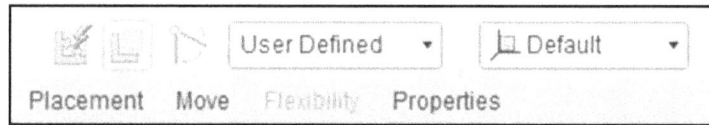

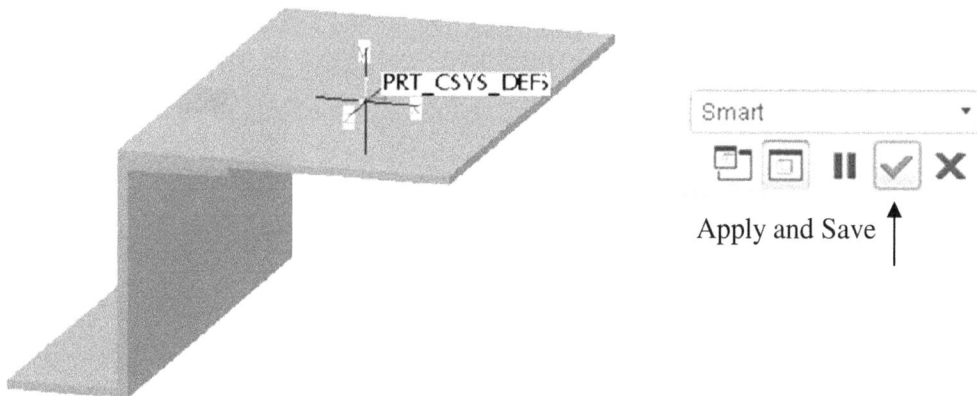

Click the icon of **Assembly**. Select the file called base.prt from the **In Session** window > **Open**.

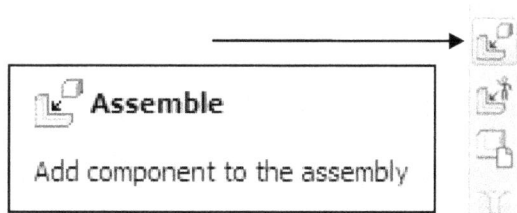

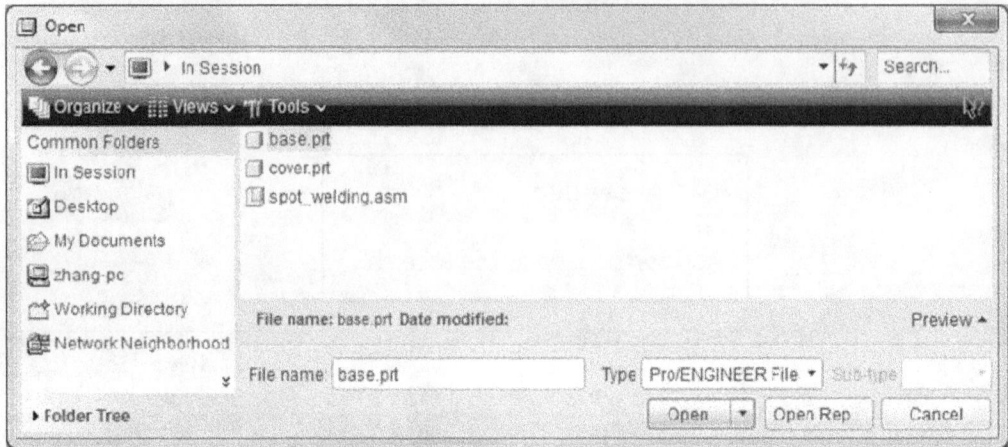

Select surfaces, as shown. Select Coincident.

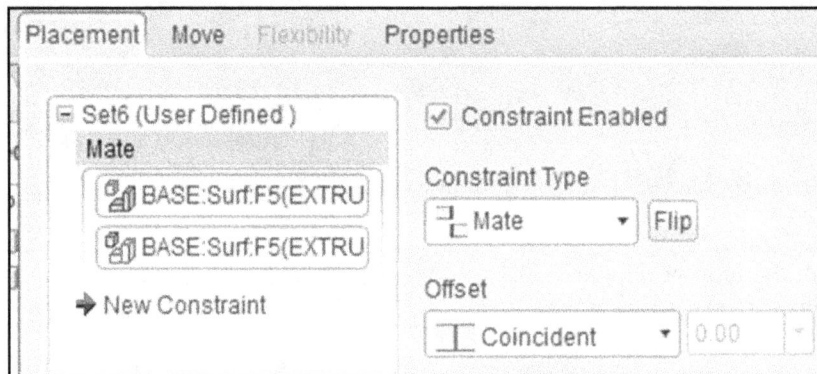

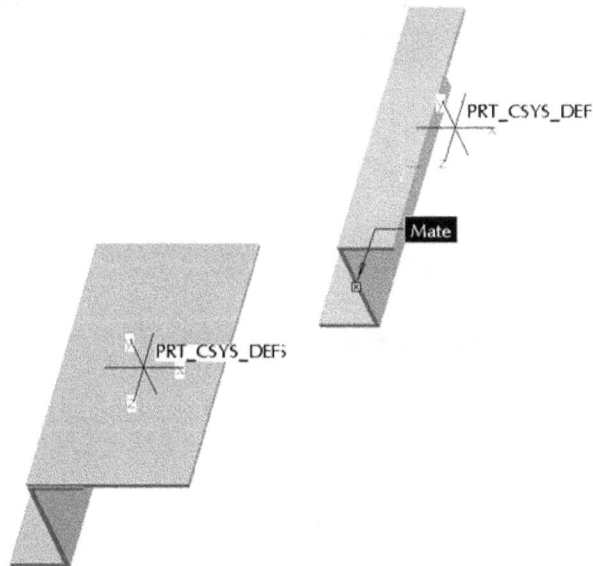

Clip **Flip** to reverse the orientation, as shown.

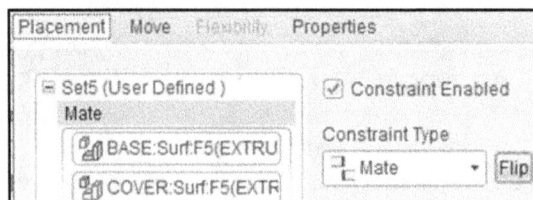

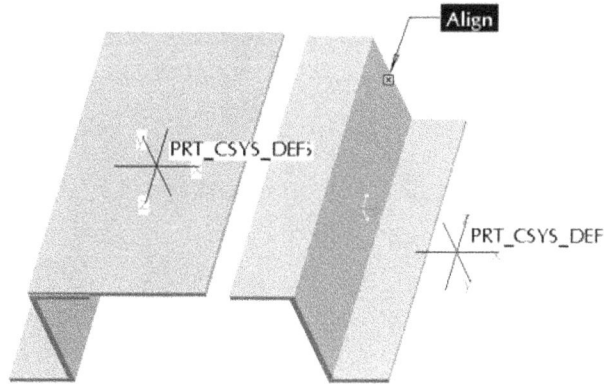

To define the second constraint, select the bottom surfaces from the 2 base components.

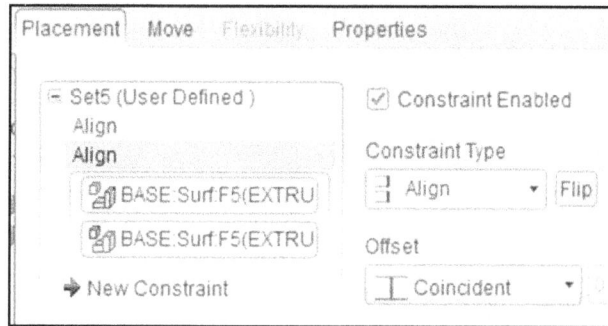

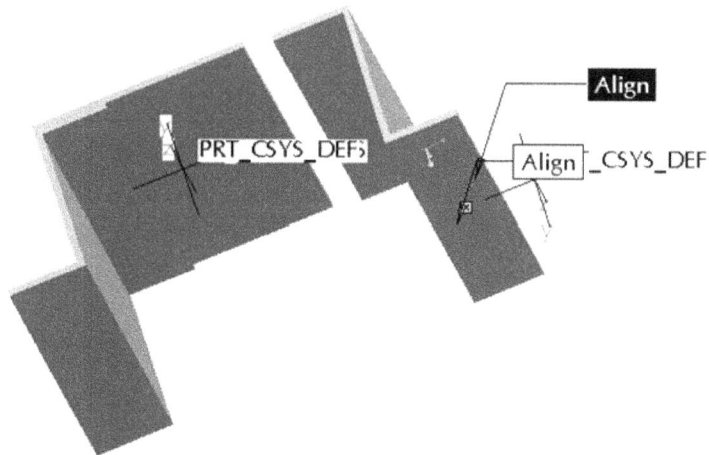

To define the third constraints, pick the 2 surfaces as shown.

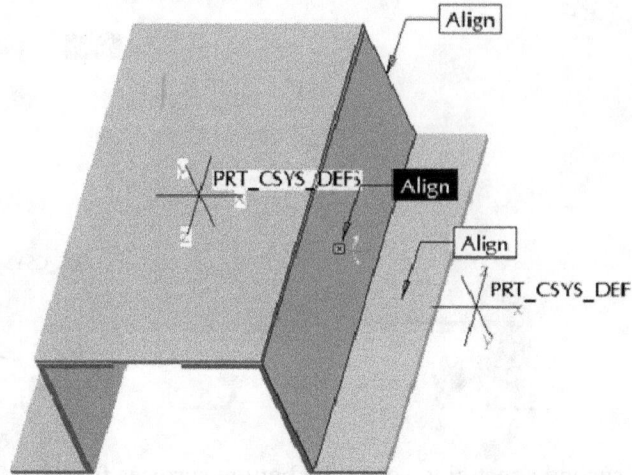

Step 5: Define the shell model.

From the top menu, click **Applications > Mechanica > Continue > Structure > OK.**

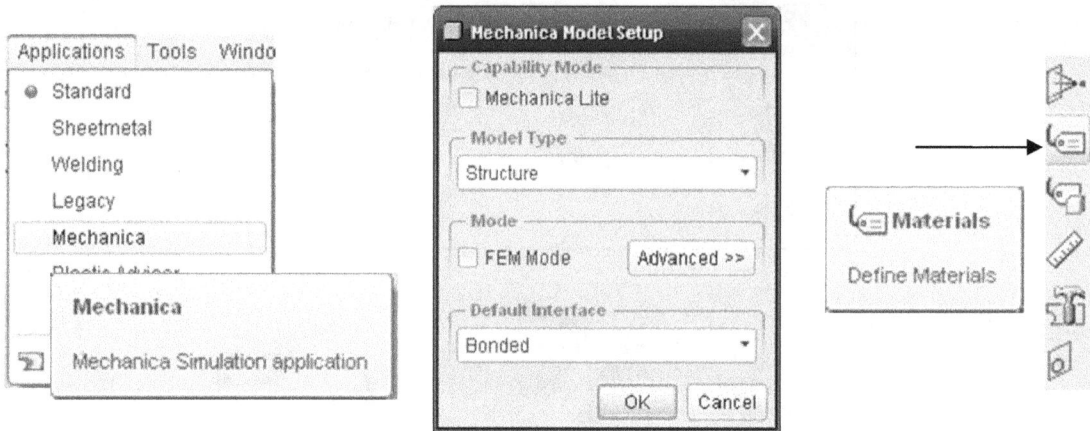

Click the icon of **Materials >** select **STEEL** from the left side called Materials in Library > click the directional arrow so that the selected material type goes to the right side called Materials in Model.

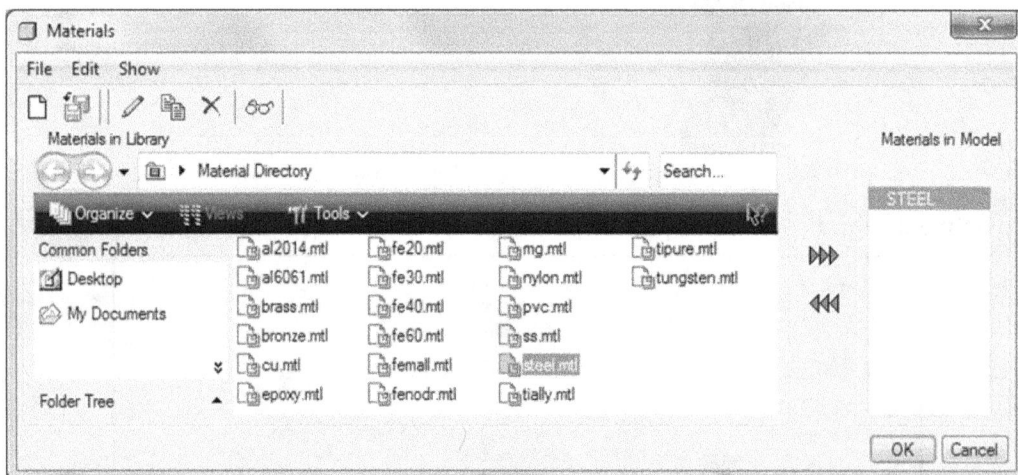

Click the icon of **Material Assignment**; select all components from the model tree while holding down the **Ctrl** key. Click **OK**.

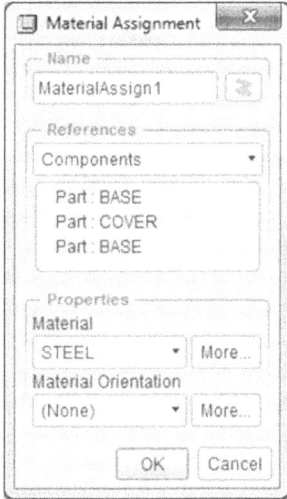

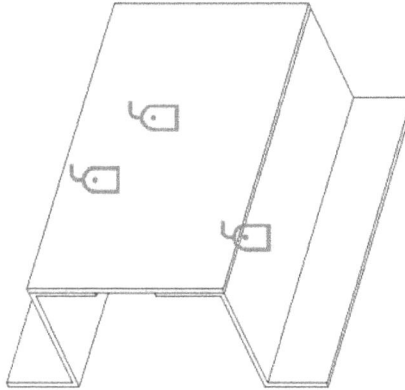

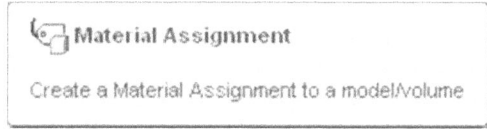

From the toolbar of functions, select the icon of **Insert > Midsurface > Auto Detect Shell Pairs.**
Select the 3 components from the model tree while holding down the **Ctrl** key. And specify 8 as
the thickness value **> Start**. In the model tree, 7 shell pairs are listed.

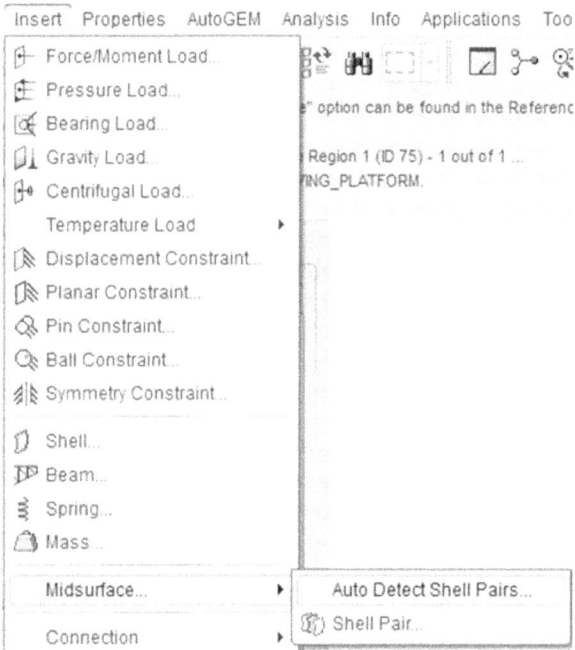

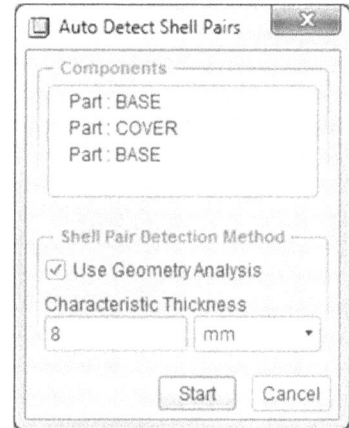

Highlight ShellPair3 and right-click to select **Edit Definition**. In the Shell Pair Definition window, Use **Selected Surface** and pick the bottom surface from the base as shown > click the check mark.

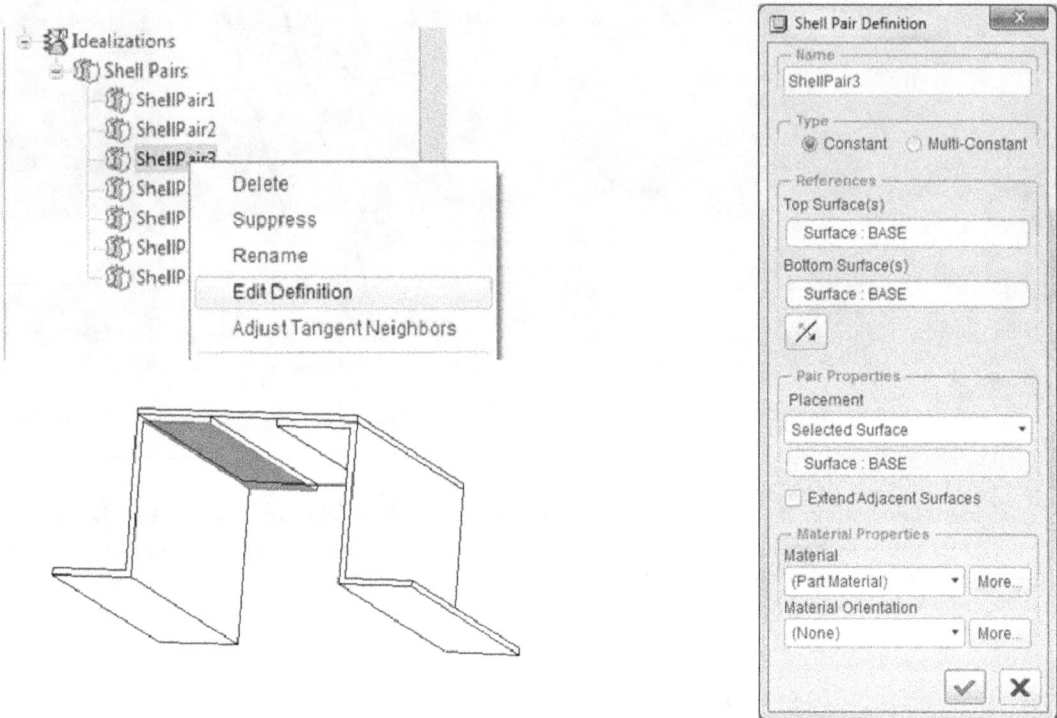

Highlight ShellPair4 and right-click to select **Edit Definition**. In the Shell Pair Definition window, Use **Selected Surface** and pick the top surface from the cover as shown > click the check mark.

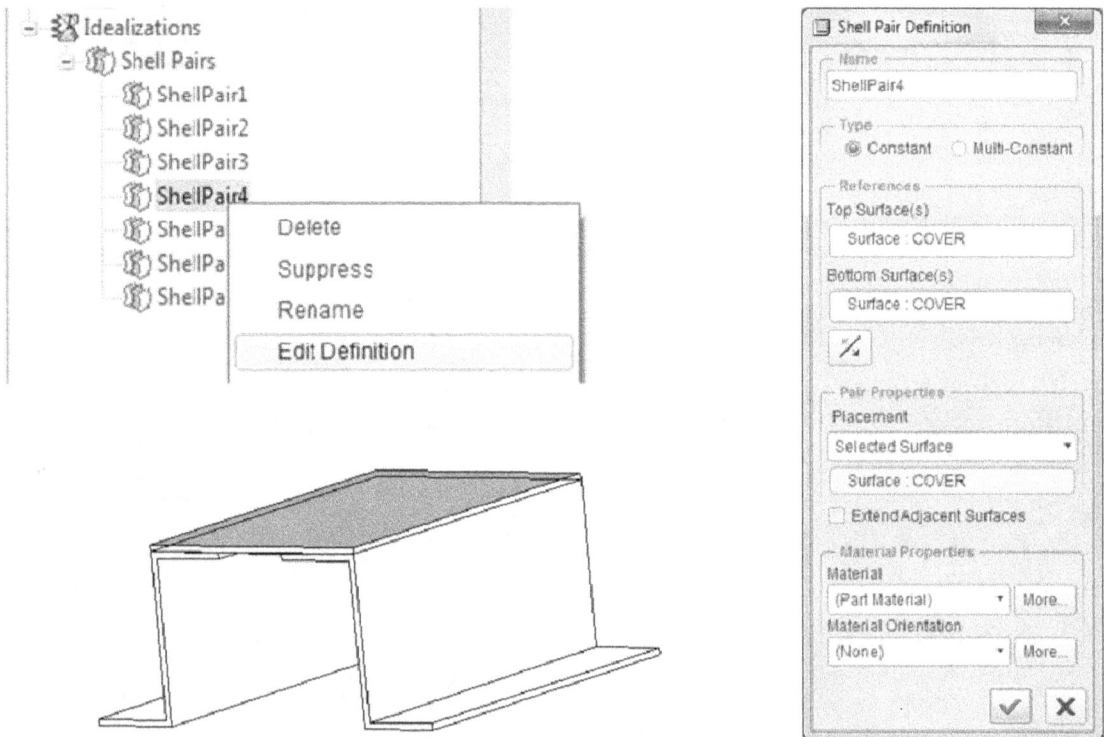

Highlight ShellPair5 and right-click to select **Edit Definition**. In the Shell Pair Definition window, Use **Selected Surface** and pick the bottom surface from the base as shown > click the check mark.

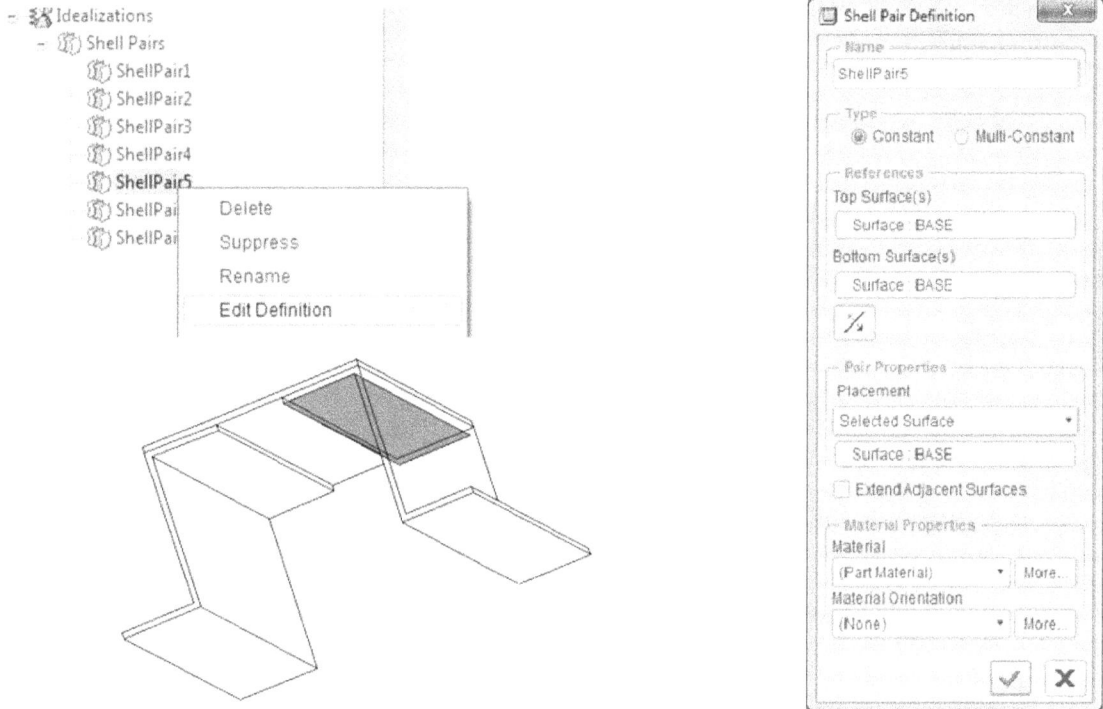

Step 5: Define the constraint condition and the load condition.

Define the bolt constraint conditions.

Select the icon of **Constraint** > > **Surfaces** > select the bottom surfaces from the 2 base components > **OK** to fix the 6 degrees of freedom.

Define the 8 spot welds.

A spot weld is considered as a beam element. This beam element, when it is defined, is fused to the surface over an area defined by diameter of the weld and material type specified. The length of this beam element is determined by the location of the fusion points on each surface.

Click the icon of Weld > Spot Weld > select the top surface of the cover component as the first surface of the contact, and select the bottom surface of the base component on the left side as the second surface of the contact.

Select > pick PNT0 first, holding down the Ctrl key, pick PNT7, PNT6, PNT4 and PNT5.

Specify the diameter of the spot weld: 20 mm.

Select STEEL as the material type > OK.

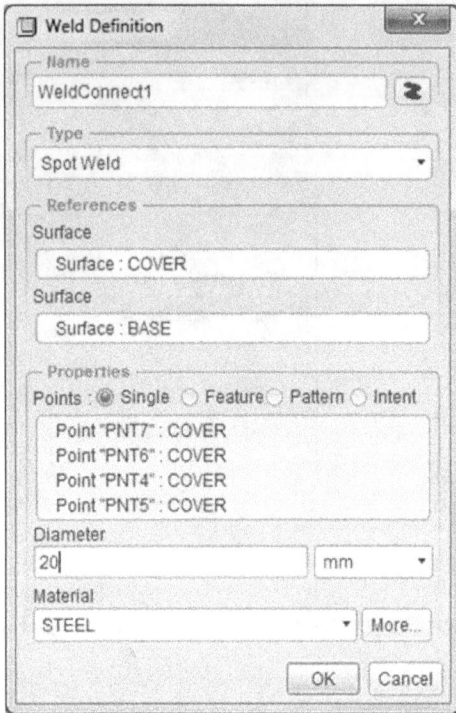

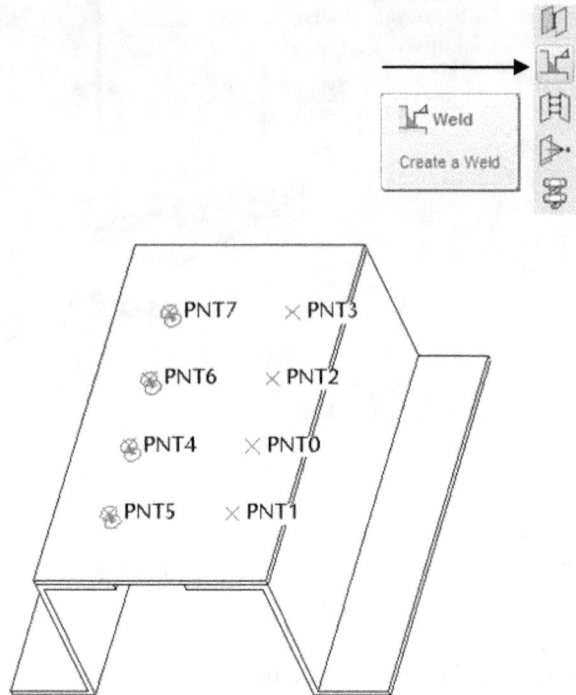

Repeat the procedure for the other side of the model in terms of selecting the two contact surfaces, picking up PNT0, PNT1, PNT2 and PNT3. Specify the diameter of the spot weld and the material type > OK.

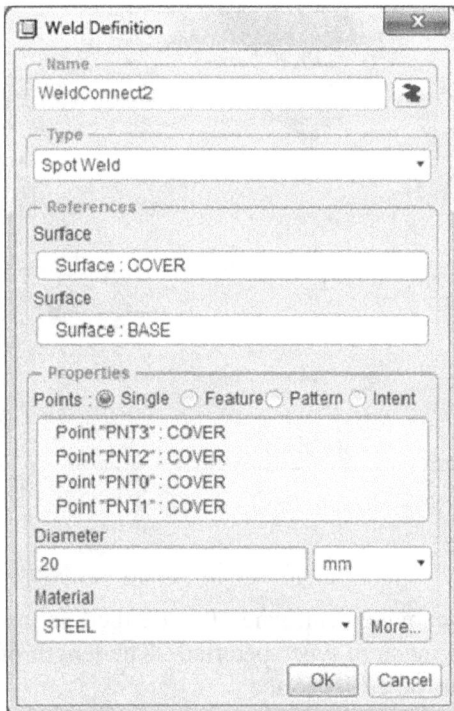

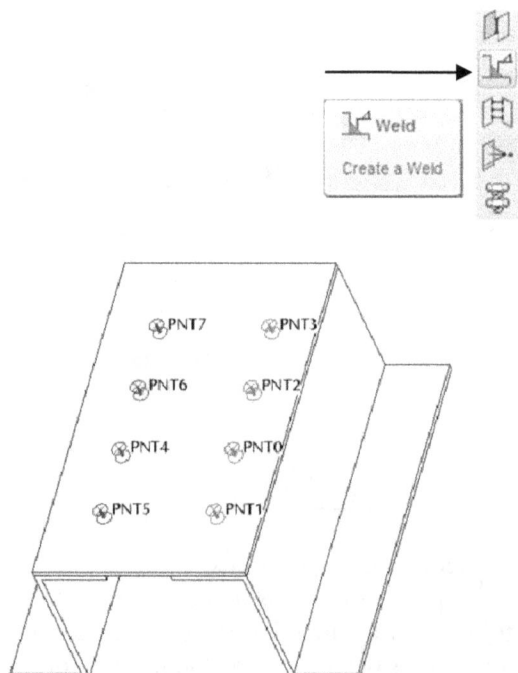

Define the load condition. Click the icon of **Force and Moment > Surface(s)** > pick the top surface from the cover component > type *-1000* in Y of the Force box > **OK**.

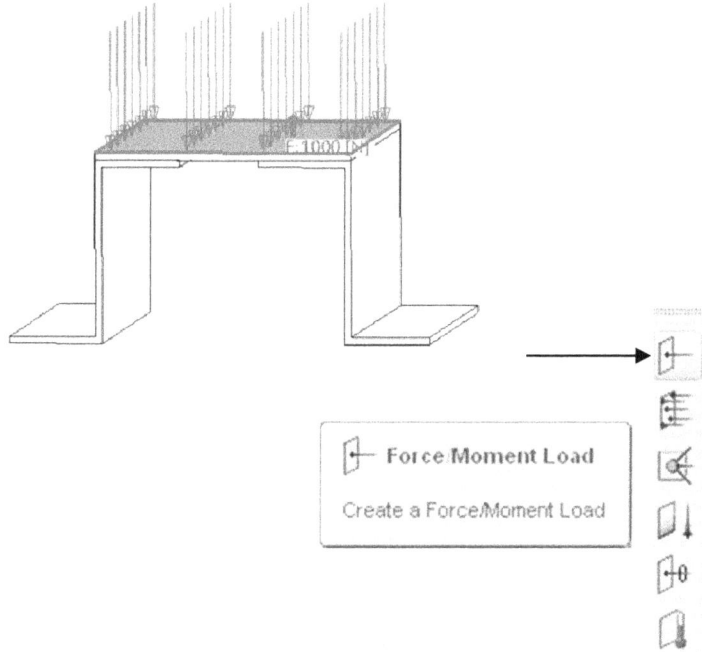

Perform a static analysis

Click the icon of **Mechanica Analyses/Studies** > **File** > Select **New static analysis**. Type *spot_weld* as the name of design study. Run the program

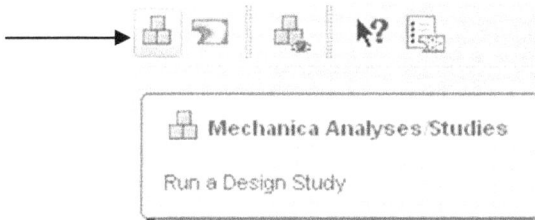

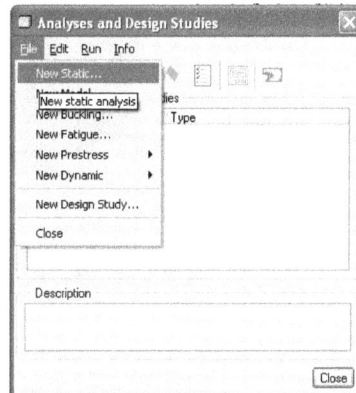

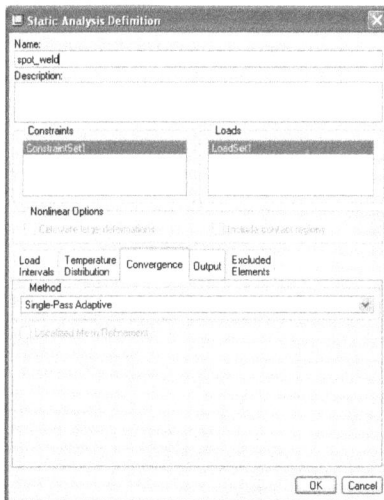

To review the results, click the icon of review results > type *deflection distribution* as the title of the result window > select the folder called bathtub_1 > **Fringe** > **Displacement** > **Magnitude** > **OK and Show**.

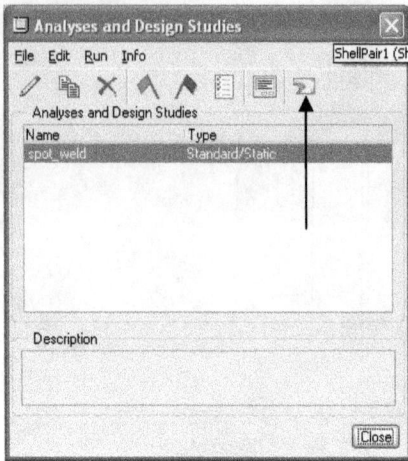

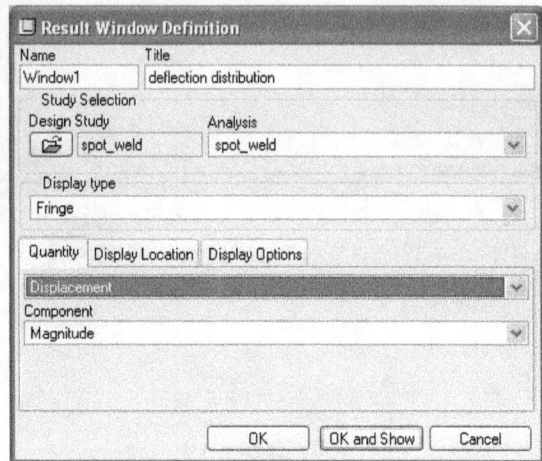

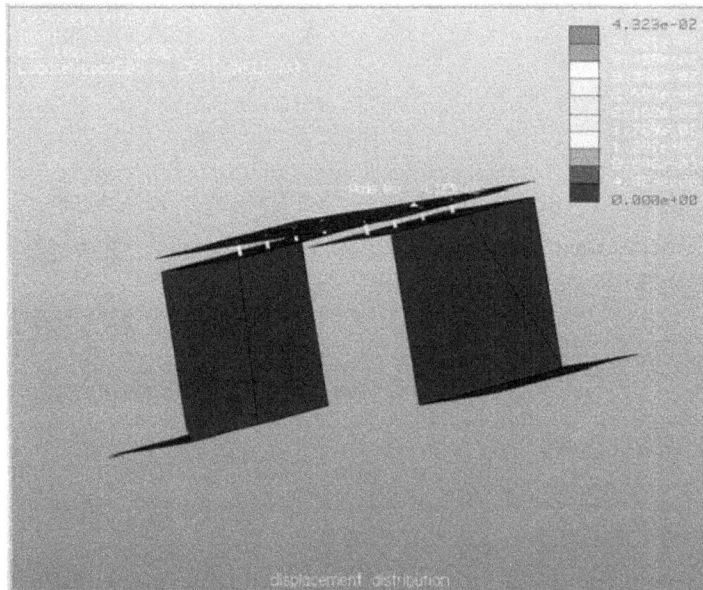

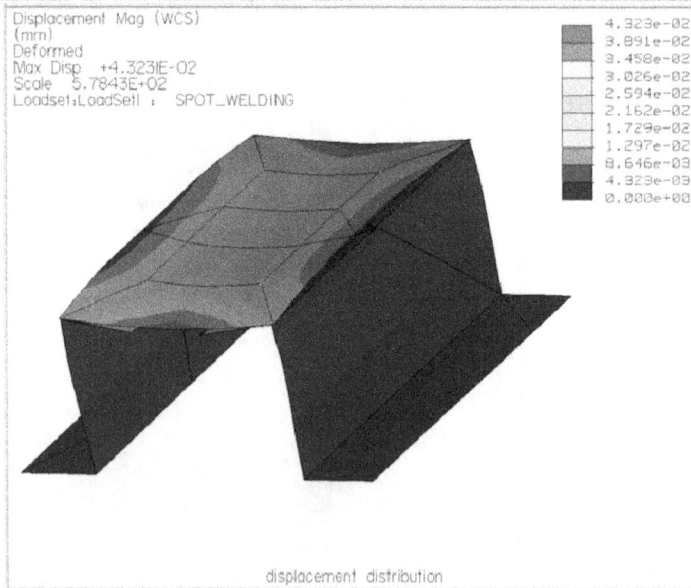

displacement distribution

The following plots are the distribution of von Mises stress.

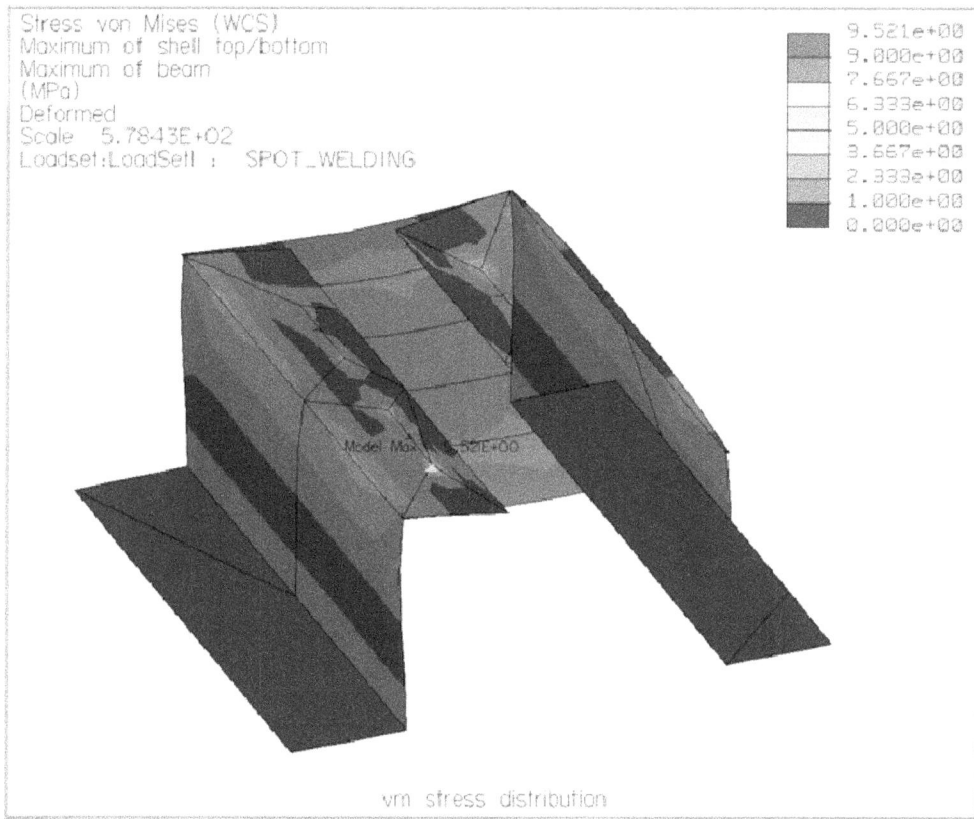

vm stress distribution

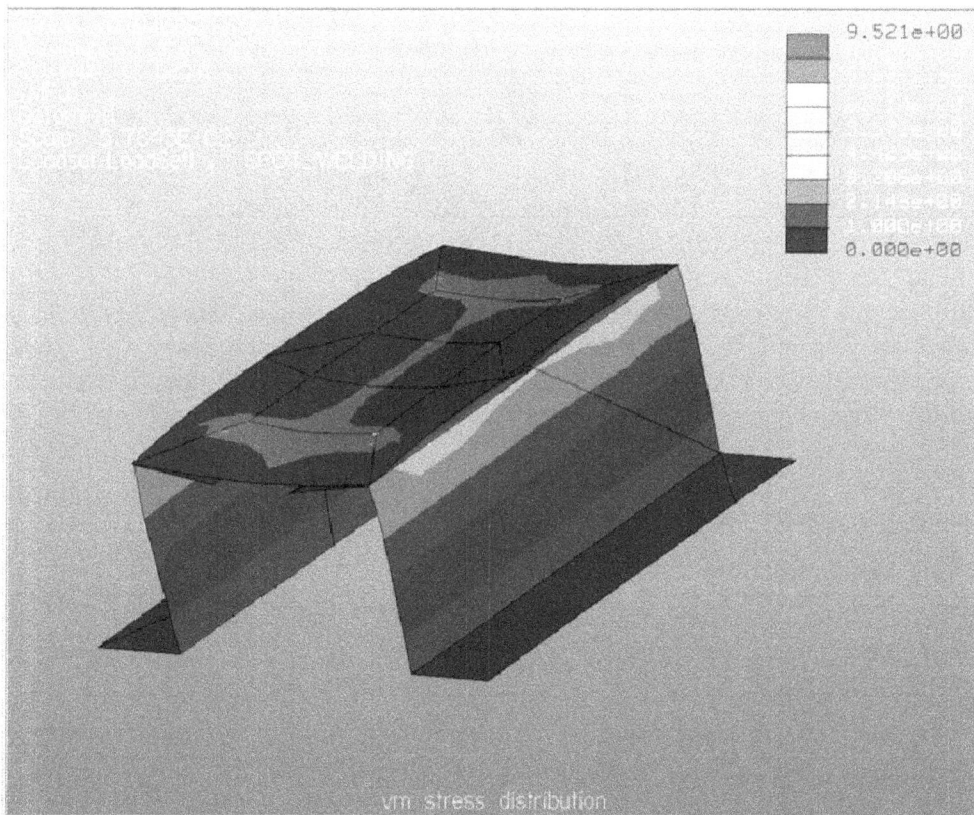

vm stress distribution

6.5 Evaluation of the 3 Principal Stresses and von Mises Stress

As illustrated below, a beam has a rectangular cross section with its left end fixed and the right end set to free. The two dimensions of the rectangular section are 30 mm and 10 mm. The overhanging length of the beam is 180 mm.

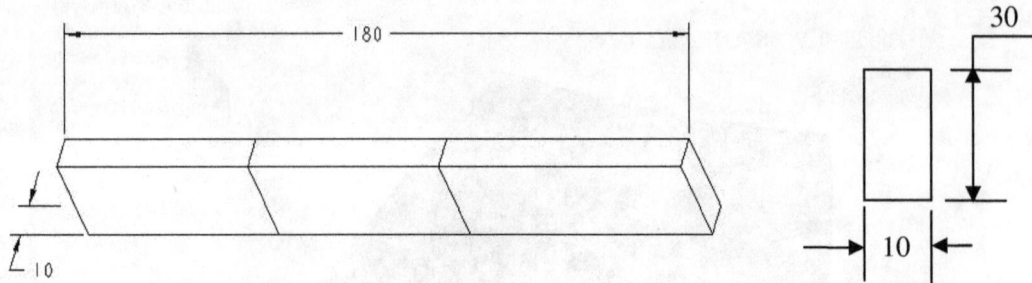

In the process of carrying out a structural analysis, such as the study of a stress distribution or the study of a strain distribution, people may be interested in knowing the value of stress or the value of strain at a given location or at a set of locations. In order to obtain the numerical values, those locations must be the locations of element nodes in performing FEA so that performance indices, such as maximum principal stress, von Mises stress, etc., can be evaluated. To do so, datum points must be defined at those locations before carrying out FEA.

Generally speaking, the process of creating datum points that are on surfaces is manageable as we experienced when working with the examples in the previous homework assignments. However, for those locations which are not on surfaces, instead of, which are inside the solid model, special caution has to be taken to make sure that node assignments are assured at those locations.

As illustrate in the following, there are 6 locations where the evaluation of stress and/or strain is required. Four (4) of those 6 locations are inside the solid model. They are PNT0, PNT1, PNT2 and PNT3. Both PNT4 and PNT5 are located on the boundary surface (or at the free end of the beam).

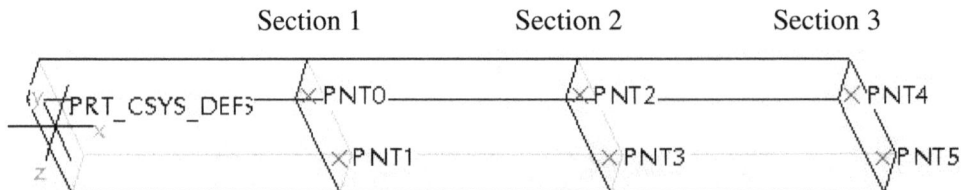

In this example, we first divide the entire beam into three portions. These 3 portions are treated as 3 independent components. We call them beam_left, beam_middle and beam_right. Their material assignments are identical among the three components. We assemble them together for the evaluation. In this way, the node assignment to each location can be assured.

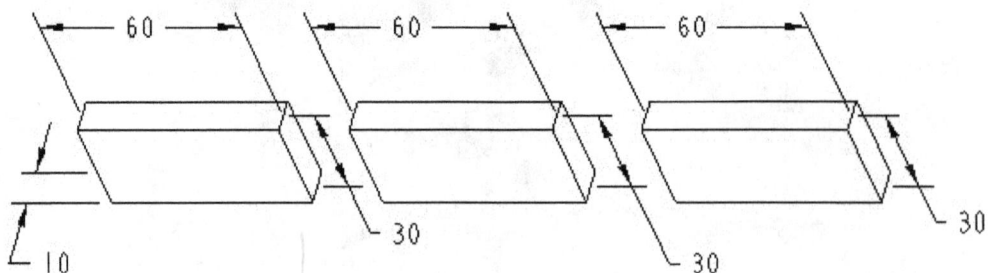

Step 1: Create a 3D solid model for beam_left.
From the main toolbar, select the icon of **Create a new object**.

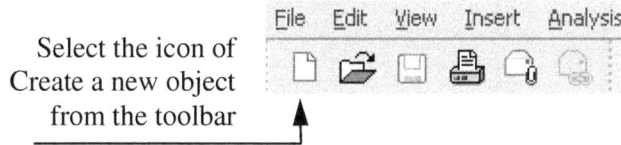

Select the icon of
Create a new object
from the toolbar

File Edit View Insert Analysis

Make sure that the Part mode is selected, and type *beam _left* as the file name. Clear the box called **Use default template > OK**, and select the **mmns_part_solid**. Fill in **DESCRIPTION** by typing *fixed end part*. Fill in **MODELED_BY** by typing *student > OK.**

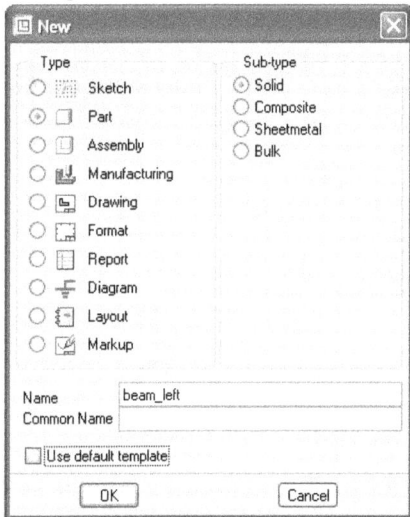

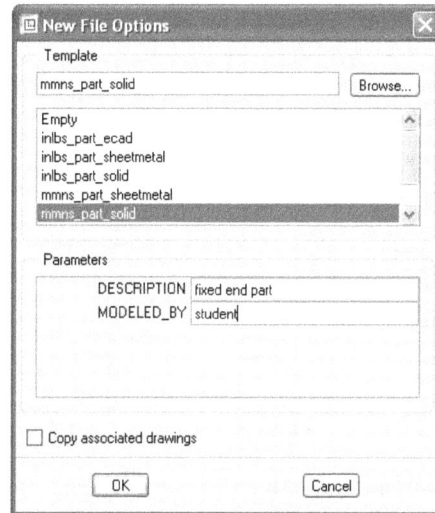

Create a 3D solid model and start with a block. The 3 dimensions are 60 x 30 x 10 mm. Click the icon of Extrude. Select Symmetry and specify 10 as the thickness value. **Placement > Define**.

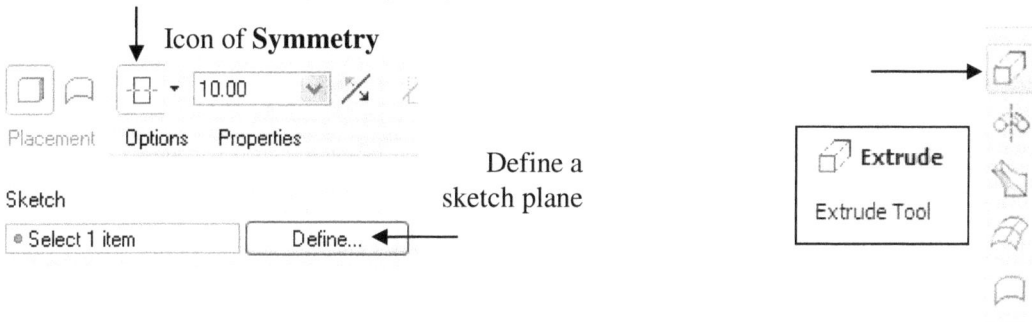

Icon of **Symmetry**

Define a
sketch plane

Placement Options Properties

Sketch

Select 1 item   Define...

Extrude
Extrude Tool

Select the **FRONT** datum plane as the sketch plane, and click the box of **Sketch** to accept the **RIGHT** datum plane as the default reference to orient the sketch plane, as illustrated below.

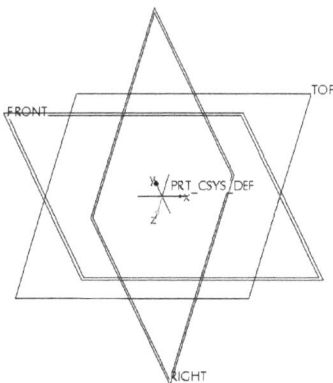

Select the icon of **Centerline** to draw a horizontal centerline or a centerline along the x-axis.

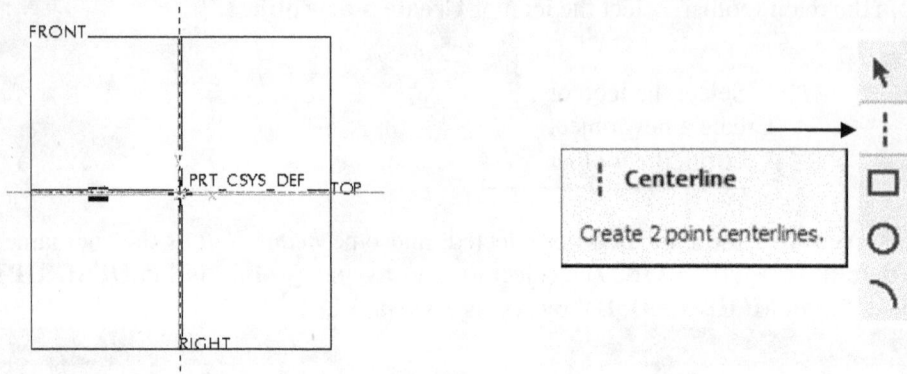

Select the icon of **Rectangle** to make a 2D sketch as shown below. The 2 dimensions are 60 and 30, respectively. Pay attention to the symmetric requirement about the horizontal centerline.

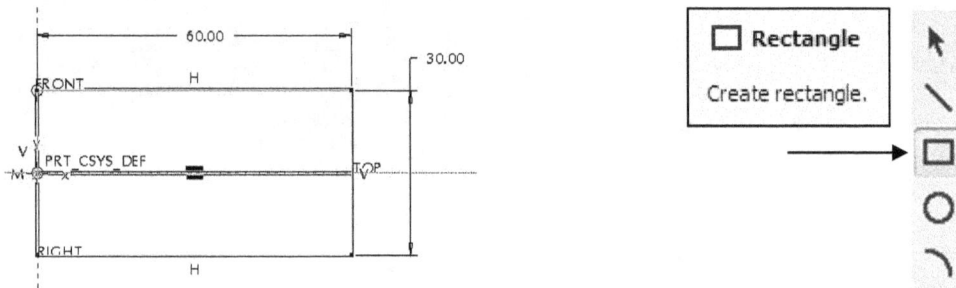

Upon completing the sketch, select the icon of **Done** from the toolbar of sketcher. On the Feature Control Panel, pick the icon of **Apply and Save** to complete the creation of the cylindrical feature, as shown below:

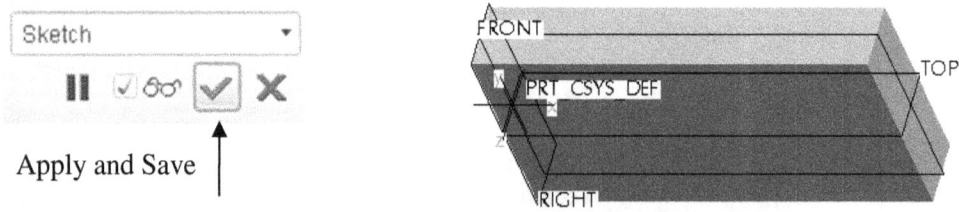

Apply and Save

Click the icon of **Save**. In the **Save Object** window, click **OK**.

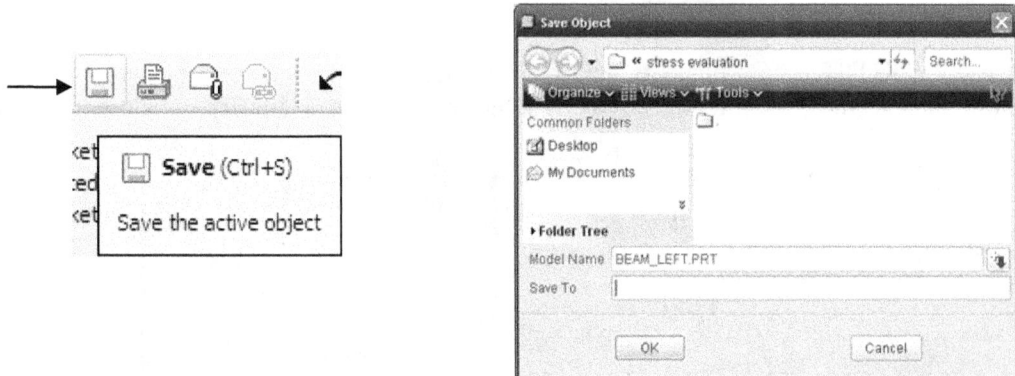

Step 2: Create an assembly

Make sure that the **Assembly** is selected, and type *beam_3_parts* as the file name. Clear the box called **Use default template**, and select the **mmns_asm_design**. Fill in **DESCRIPTION** by typing in *beam_assembly*. Fill in **MODELED_BY** by typing *student* > **OK.**

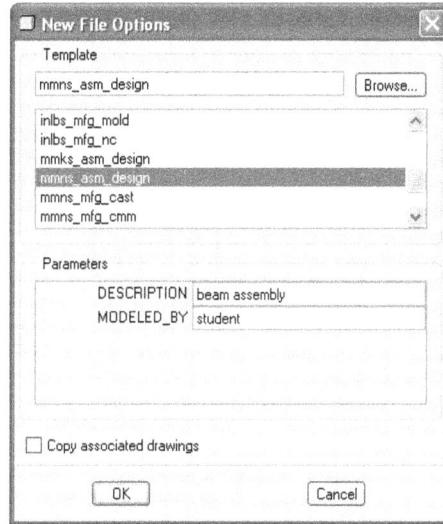

Select the icon of **Add** component to the assembly, and select the file of beam_left.prt > **Open**.

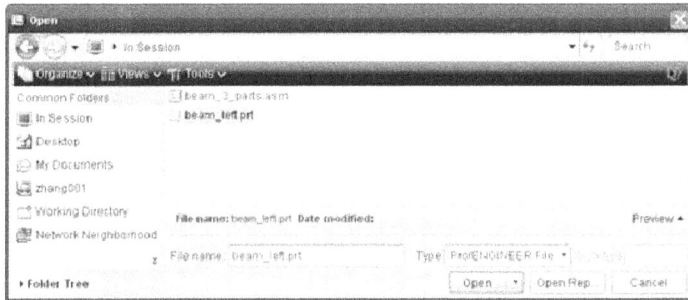

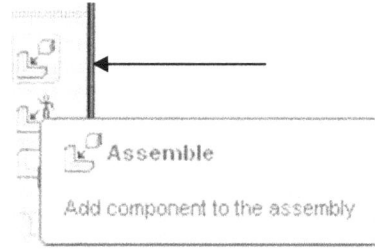

In the **Component Placement** window, select the icon of **Default** location > **OK** to complete the assembly process. This is equivalent to assemble the part_default_coordinate system at the location of the assembly_default_coordinate system. Click the icon of **Apply and Save** to complete the assembly of beam left.

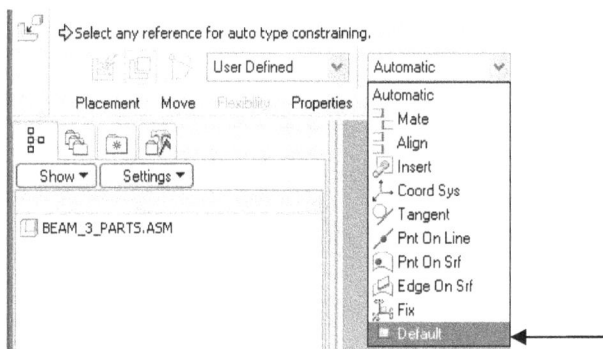

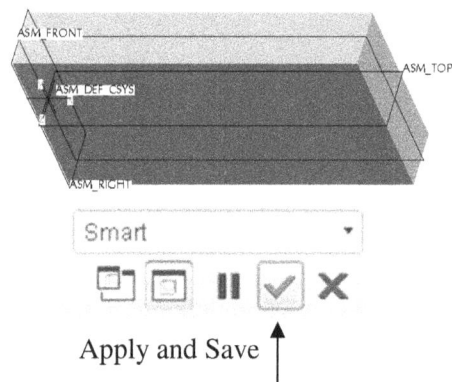

Apply and Save

Click the icon of **Save**. In the **Save Object** window, click **OK**.

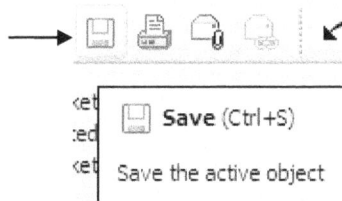

Select the icon of **Create** a component in the assembly mode from the toolbar of assembly.

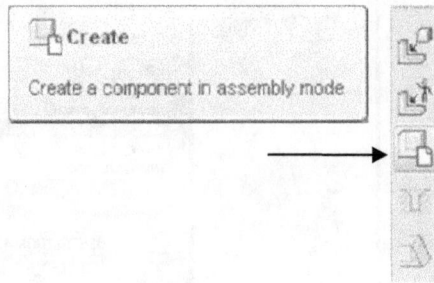

Type *beam_middle* as the name of the new component, and select the mmns_part_solid.prt as the startup file for the new component > **OK**.

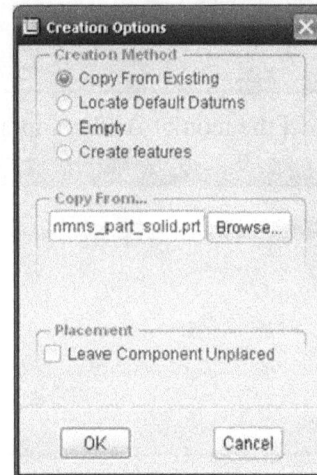

Select the icon of **Default** > click the check mark to complete the assembly process.

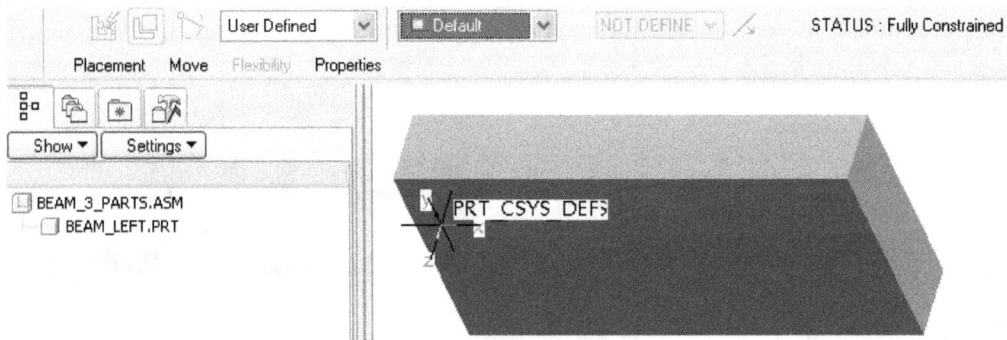

To activate beam_middle.prt in the assembly mode, pick beam_middle from the model tree, right-click and hold, select **Activate**.

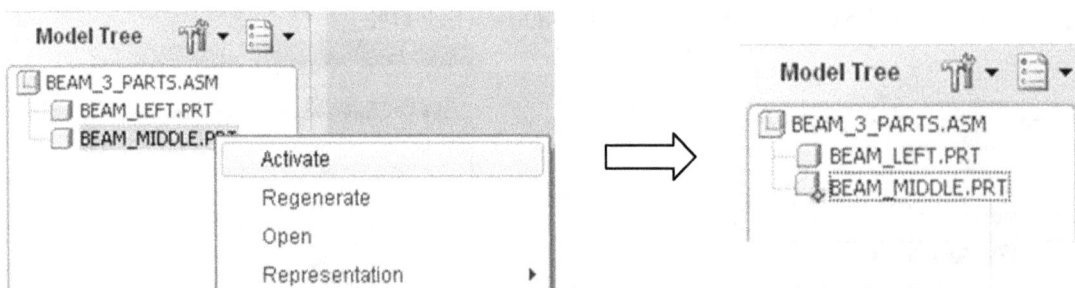

From the toolbar of datum feature, select the icon of datum plane, select the surface on the right side of beam_left and specify the offset value equal to zero > **OK**. This means that a datum plane called **DTM1** has been created within beam_middle.prt.

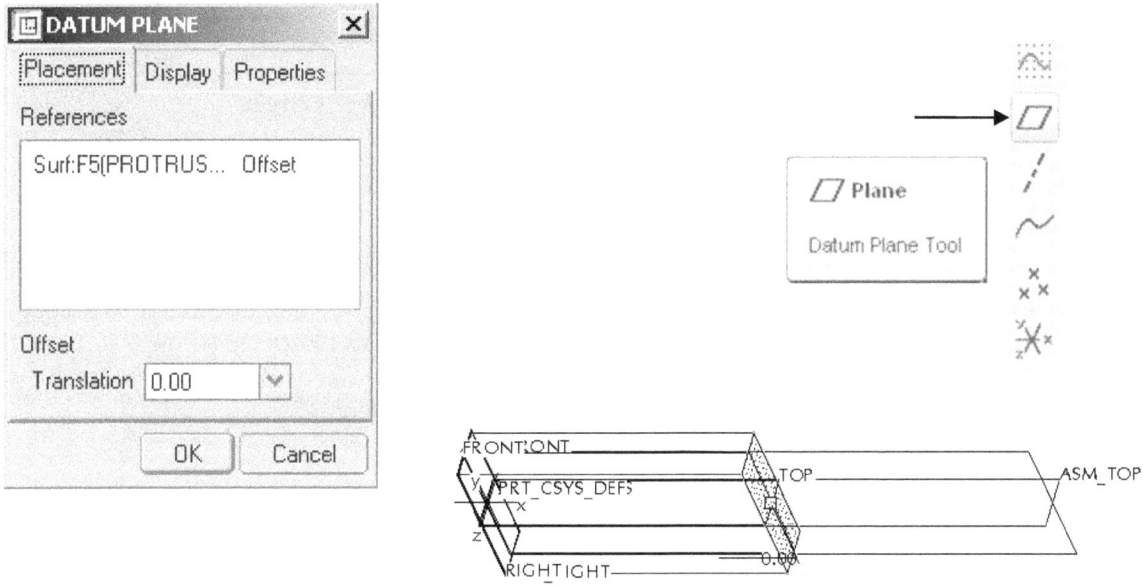

Now let us activate the assembly by right-click on beam_3_parts.

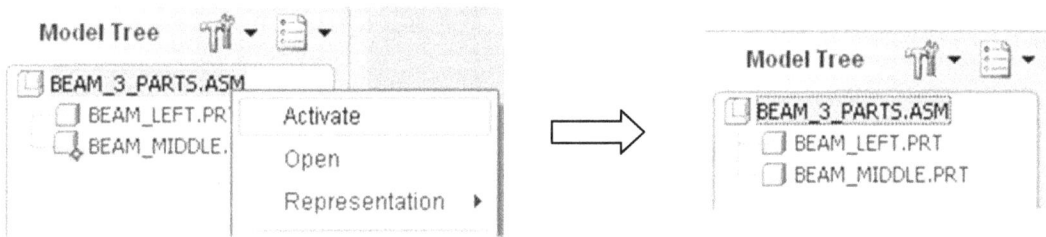

Afterwards, pick beam_middle.prt > right-click and hold > select **Open**.

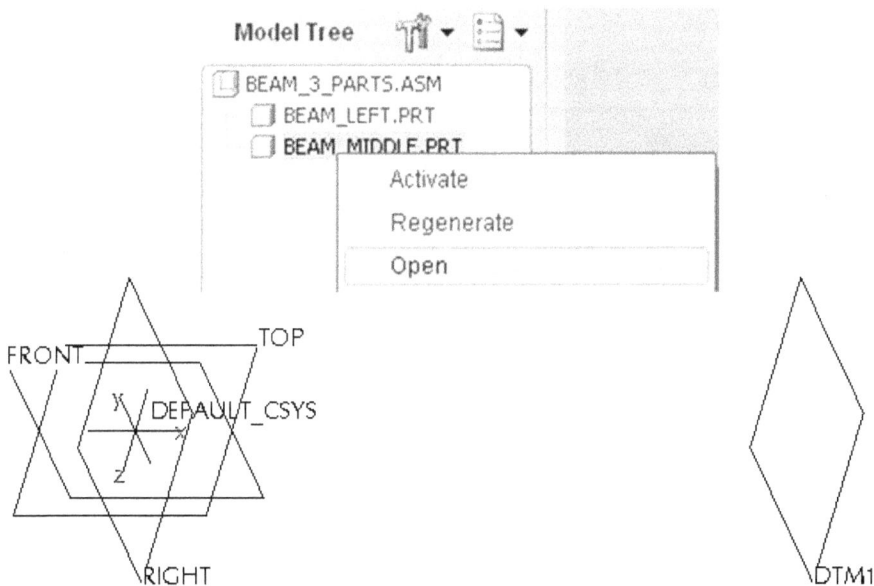

To create a 3D solid model, select the icon of **Extrude** from the toolbar of feature creation. The dashboard appears. Set the depth choice to **Symmetry** and type 10 as the thickness value. Activate **Placement** to define a sketch plane > **Define.**

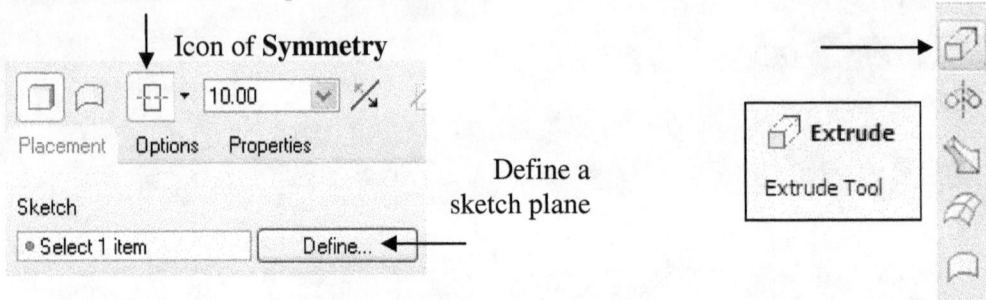

Icon of **Symmetry**

Placement   Options   Properties

Define a
sketch plane

Sketch

● Select 1 item   Define...

Extrude

Extrude Tool

Select the **FRONT** datum plane as the sketch plane, and click the box of **Sketch** to accept the **RIGHT** datum plane as the default reference to orient the sketch plane, as illustrated below.

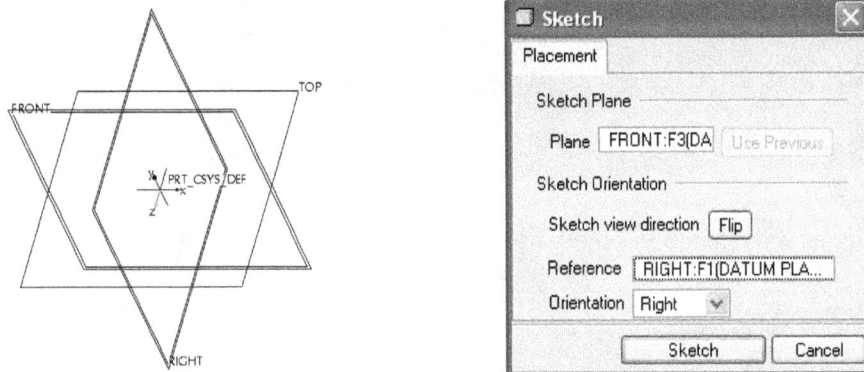

Sketch

Placement

Sketch Plane

Plane   FRONT:F3(DA   Use Previous

Sketch Orientation

Sketch view direction   Flip

Reference   RIGHT:F1(DATUM PLA...

Orientation   Right

Sketch   Cancel

From the top menu, select **Sketch > References**. Select **DTM1** as a new reference > **Close.**

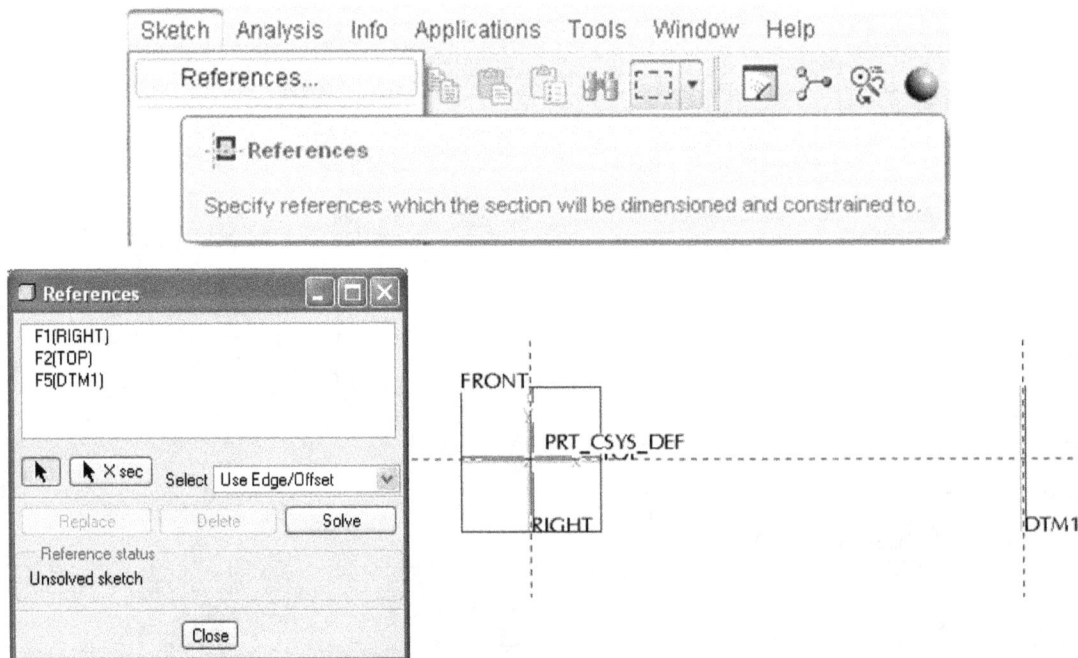

Sketch   Analysis   Info   Applications   Tools   Window   Help

References...

References

Specify references which the section will be dimensioned and constrained to.

References

F1(RIGHT)
F2(TOP)
F5(DTM1)

Select   Use Edge/Offset

Replace   Delete   Solve

Reference status
Unsolved sketch

Close

FRONT

PRT_CSYS_DEF

RIGHT

DTM1

Click the icon of **Centerline**. Sketch a horizontal centerline.

FRONT TOP RIGHT DTM1

Centerline

Create 2 point centerlines.

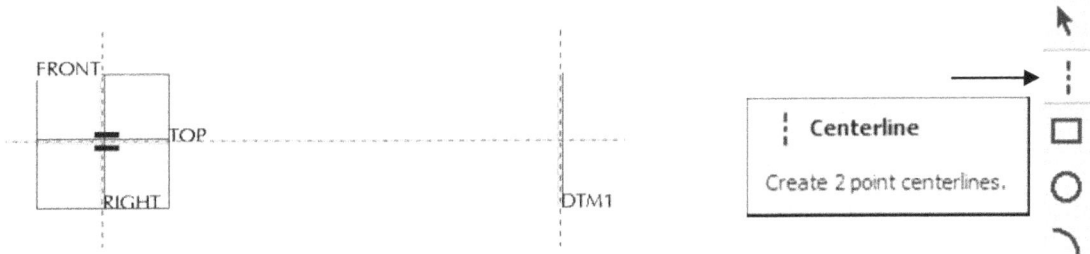

Click the icon of **Rectangle**, sketch a rectangle that starts from **DTM1** and is symmetric about the horizontal centerline. The 2 dimensions are 60 and 30, as shown below:

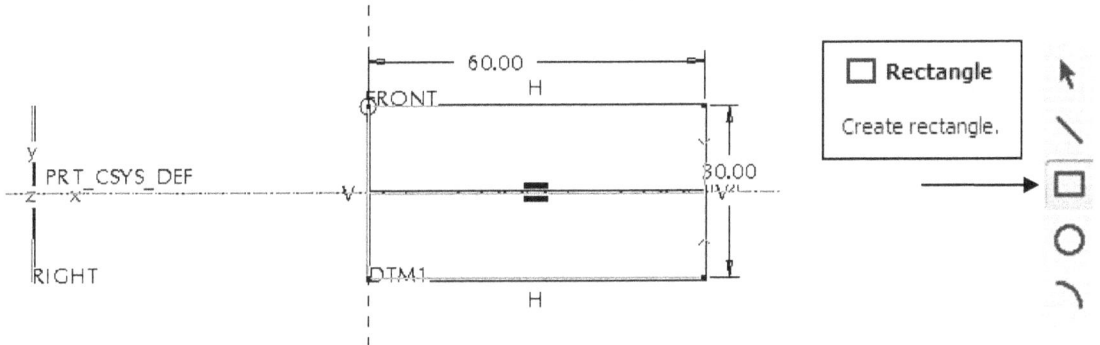

60.00 H FRONT PRT_CSYS_DEF RIGHT 30.00 DTM1 H

Rectangle

Create rectangle.

Upon completing the sketch, select the icon of **Done** from the toolbar of sketcher. On the Feature Control Panel, pick the icon of **Apply and Save** to complete the creation of the cylindrical feature, as shown below:

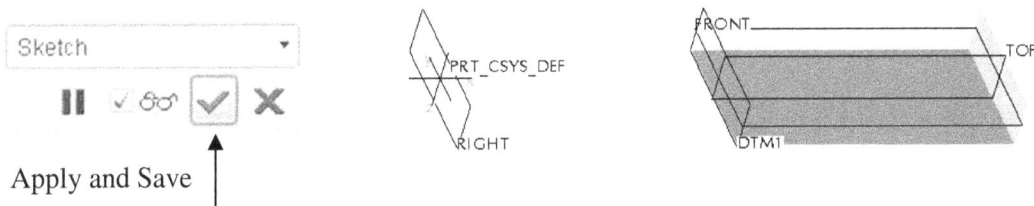

Sketch

Apply and Save

PRT_CSYS_DEF RIGHT

FRONT TOP DTM1

Click the icon of **Save**. In the **Save Object** window, click **OK**.

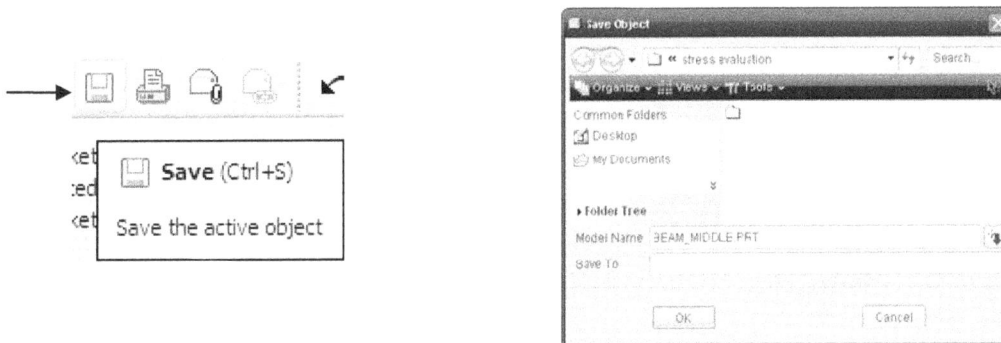

Save (Ctrl+S)

Save the active object

Save Object

Organize Views Tools

Common Folders
Desktop
My Documents

Folder Tree

Model Name BEAM_MIDDLE.PRT

Save To

OK   Cancel

Let us now go back to the assembly. The newly created block feature is already shown in the assembly through data sharing.

PRT_CSYS_DEF

Now let us create another new component, namely, the portion on the right side of middle part. Select the icon of **Create** a component in the assembly mode from the toolbar of assembly.

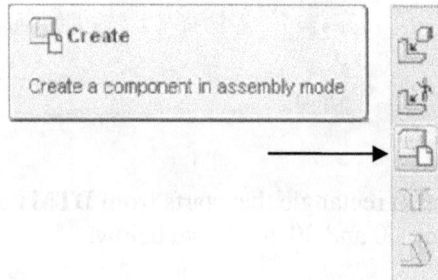

Type *beam_right* as the name of the new component, and select the mmns_part_solid.prt as the startup file for the new component > **OK**.

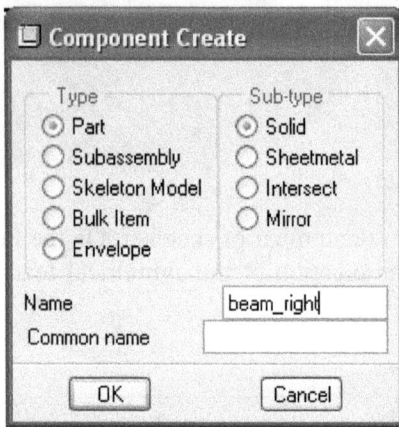

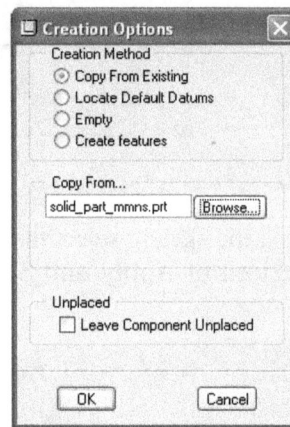

Select the icon of **Default** > click the check mark to complete the assembly process.

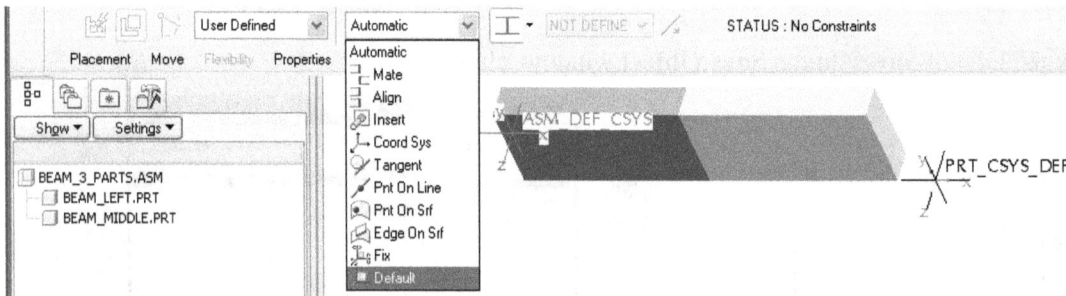

To activate beam_right.prt in the assembly mode, pick beam_right from the model tree, right-click and hold, select **Activate**.

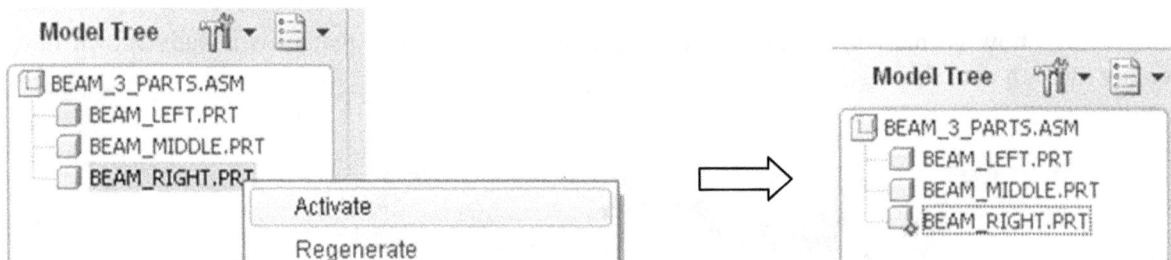

From the toolbar of datum feature, select the icon of **Datum Plane Tool**, select the surface on the right side of beam_middle and specify the offset value equal to zero > **OK**. This means that a datum plane called **DTM1** has been created within beam_right.prt.

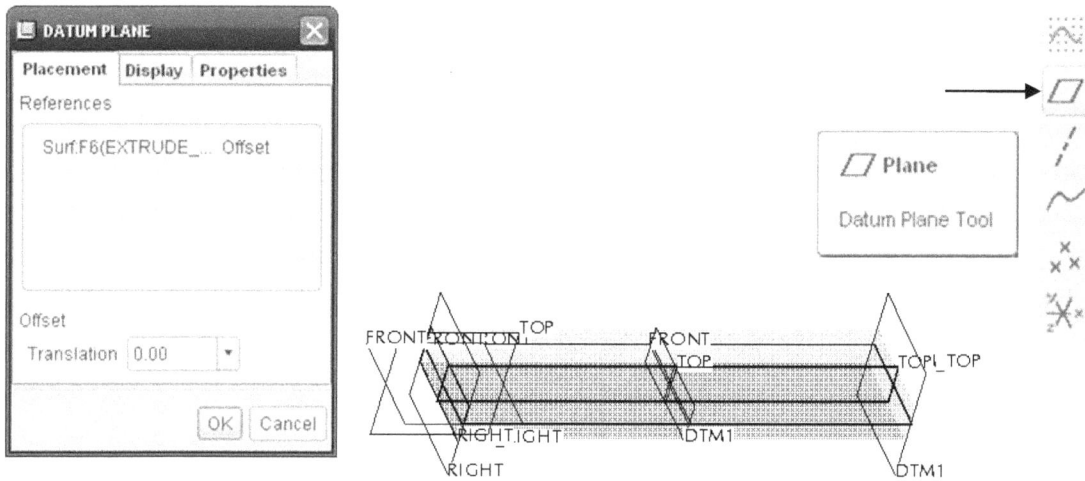

Now let us activate the assembly by right-click on beam_3_parts.

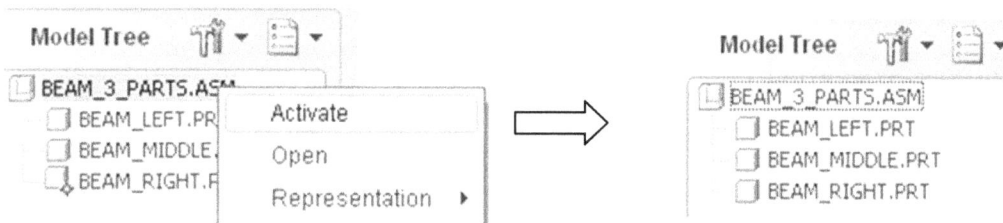

Afterwards, pick beam_right.prt > right-click and hold > select **Open**.

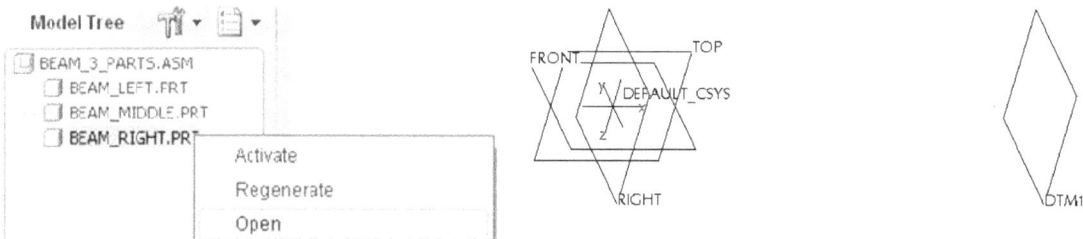

To create a 3D solid model, select the icon of **Extrude**. Set the depth choice to **Symmetry** and type 10 as the thickness value. Activate **Placement** to define a sketch plane > **Define.**

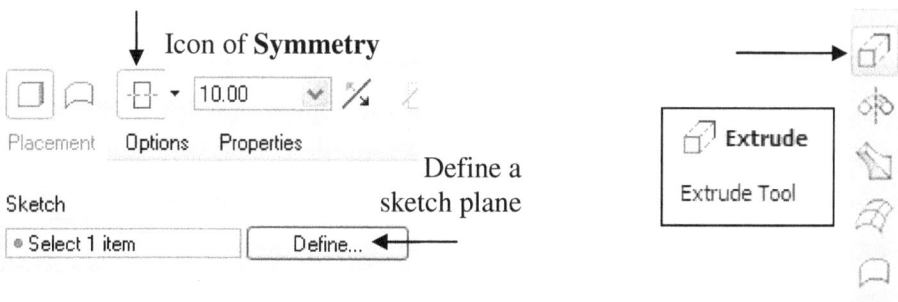

Select the **FRONT** datum plane as the sketch plane, and click the box of **Sketch** to accept the **RIGHT** datum plane as the default reference to orient the sketch plane, as illustrated below.

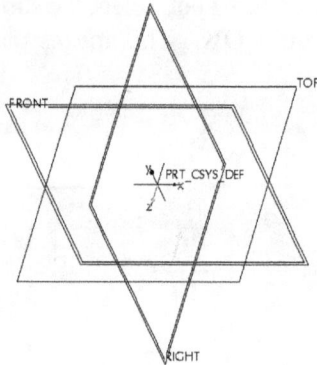

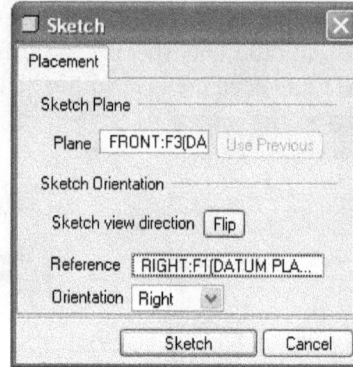

From the top menu, select **Sketch > References**. Select **DTM1** as a new reference > **Close**.

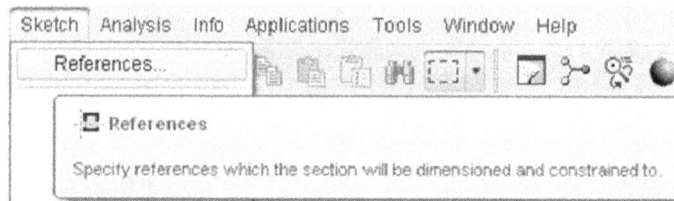

Click the icon of **Centerline**. Sketch a horizontal centerline.

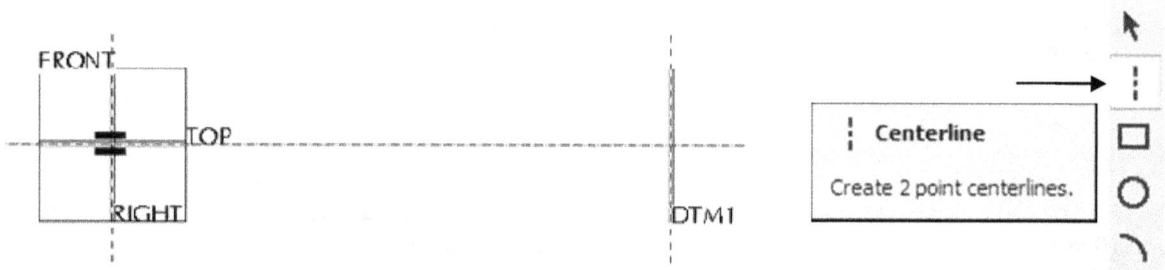

Click the icon of **Rectangle**, sketch a rectangle that starts from **DTM1** and is symmetric about the horizontal centerline. The 2 dimensions are 60 and 30, as shown below:

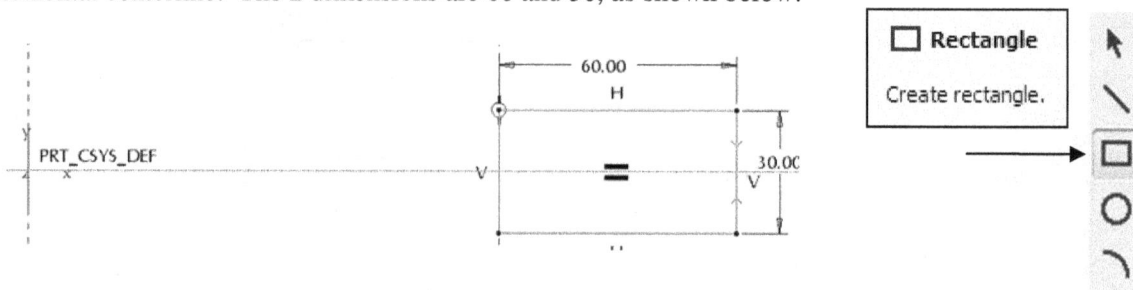

Upon completing the sketch, select the icon of **Done** from the toolbar of sketcher. Click the icon of **Apply and Save** to complete the creation of the cylindrical feature, as shown below:

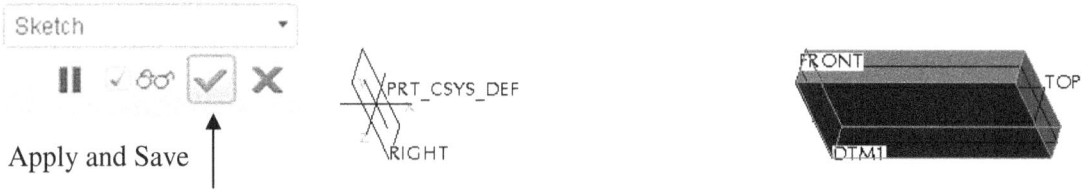

Click the icon of **Save**. In the **Save Object** window, click **OK**.

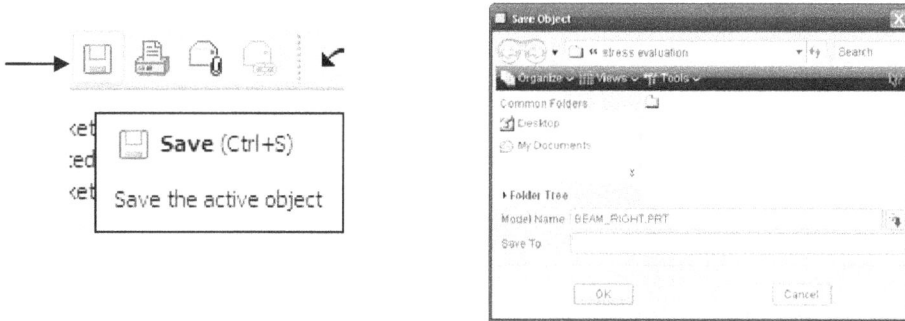

Let us now go back to the assembly. The newly created block feature is already shown.

Step3: Create Datum Points on the right side of the beam left component.

Open the beam_left.prt. From the toolbar of datum feature, select the icon of **Datum Plane Tool**, select the **TOP** datum plane and specify 10 as the offset value > **OK,** thus a datum plane called **DTM1** has been created.

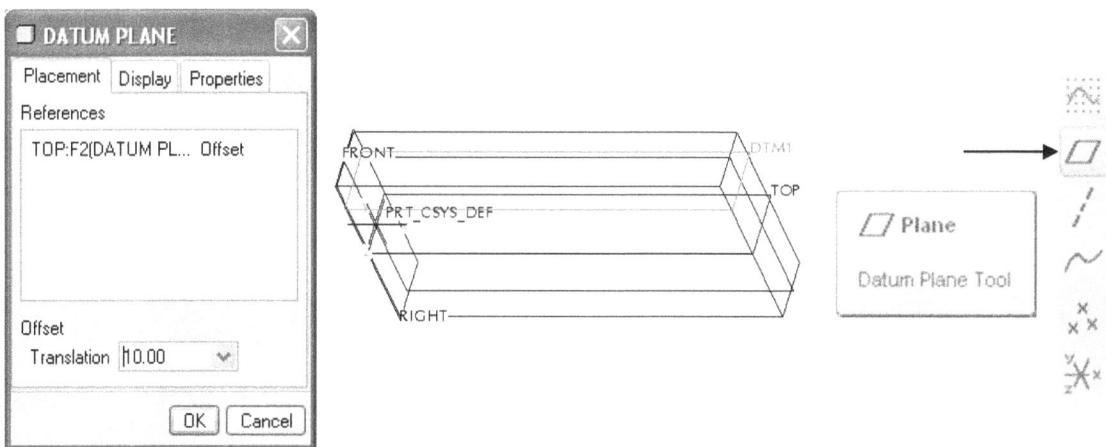

From the top menu, click **Edit > Intersect** and pick the surface on the right side, as shown. A datum curve or a straight line is created.

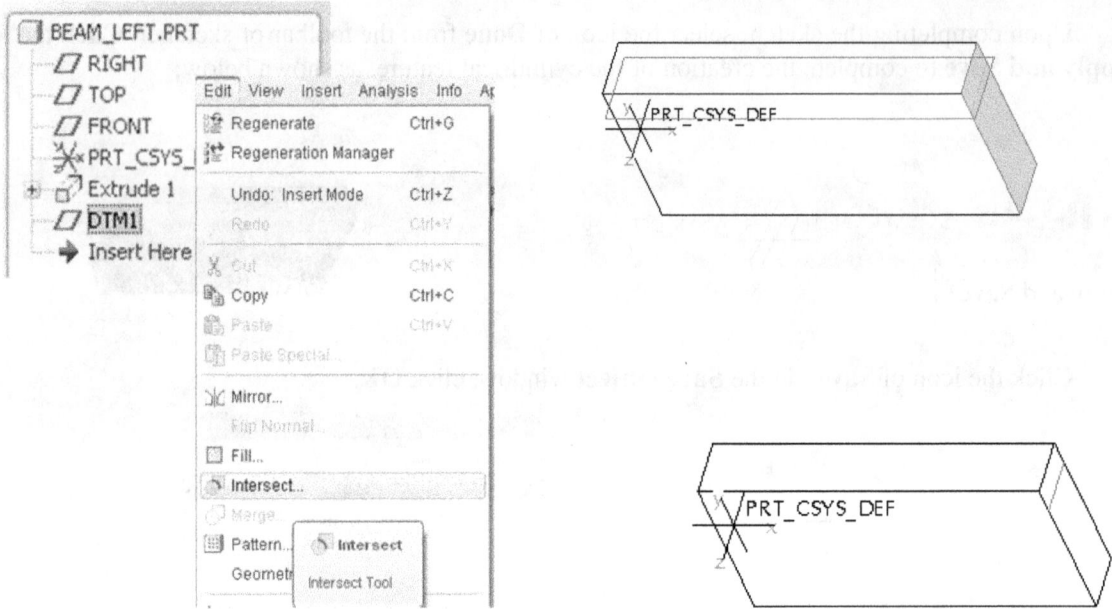

Click the icon of **Datum Point Tool**, and make a click on the datum curve and specify 0.50 as the offset value for **Ratio**, as shown, thus **PNT0** is created.

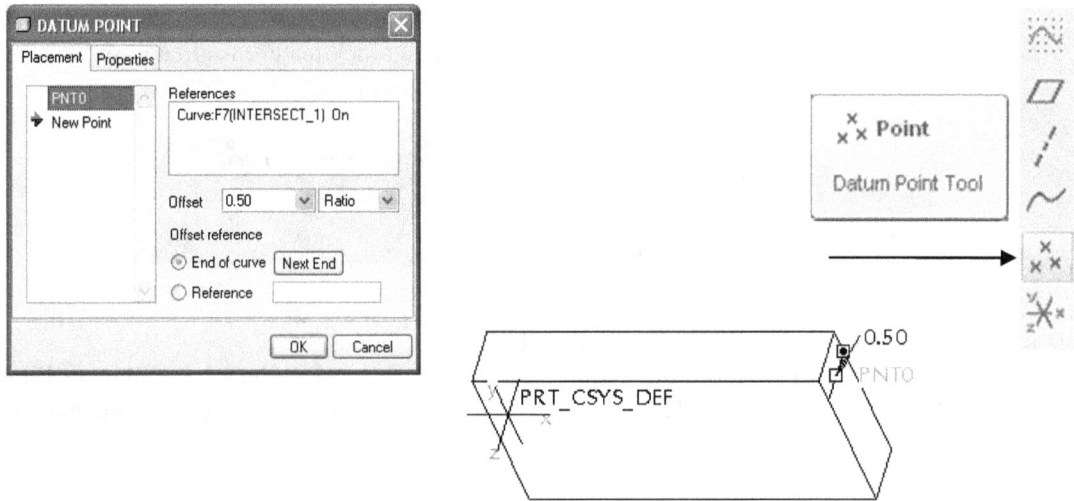

Step 4: Go to Pro/Mechanica to perform FEA.
From the main toolbar or the menu toolbar, select **Applications > Mechanica > Structure > OK.**

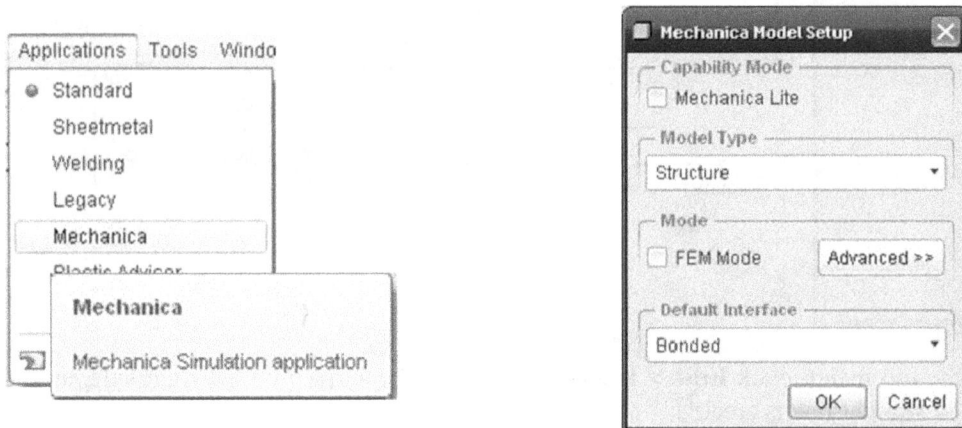

From the toolbar of functions, select the icon of **Materials >** select **STEEL** from the left side called **Materials in Library >** click the directional arrow so that the selected material type goes to the right side called **Materials in Model**.

Click the icon of **Material Assignment.** In the Material Assignment window, the selected **Steel** is shown, and select the 3 components while holding down the **Ctrl** key > **OK**.

To define the constraint condition, click the icon of **New Displacement Constraint** from the toolbar of Mechanica objects > **Surface(s) >** pick the left surface of the beam assembly and click **OK** to set all 6 degrees of freedom equal to zero, which is the default setting.

Now let us define Constraint1, simulating the assembly is fixed to the platform or the ground on the left side. Click the icon of **Displacement Constraints > Surface >** pick the left surface from beam left > set the 6 degrees of freedom to be fixed to the ground > **OK**.

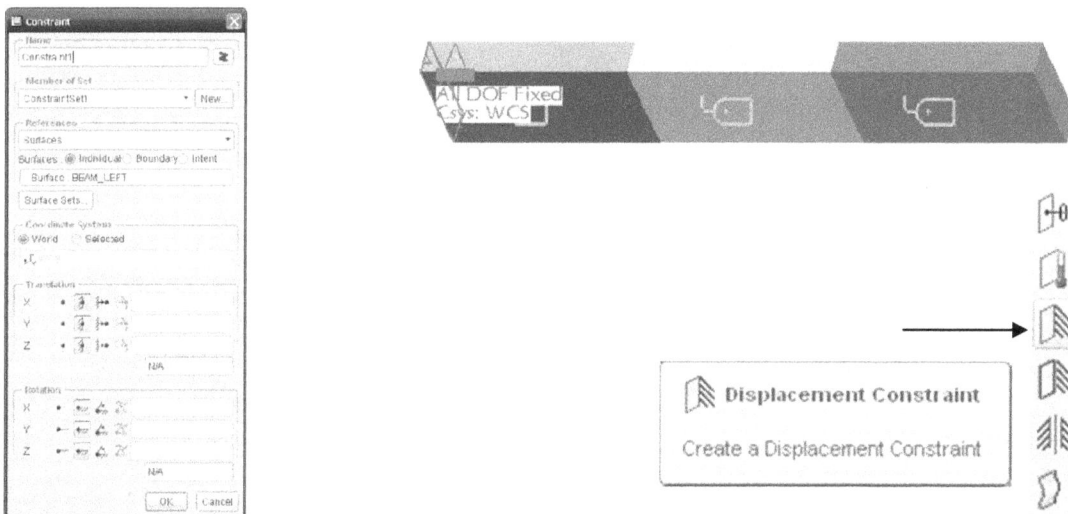

To define a load condition, we need to create a surface region on the beam right component. In the model tree, highlight BEAM_RIGHT and right click to open the beam_right.prt.

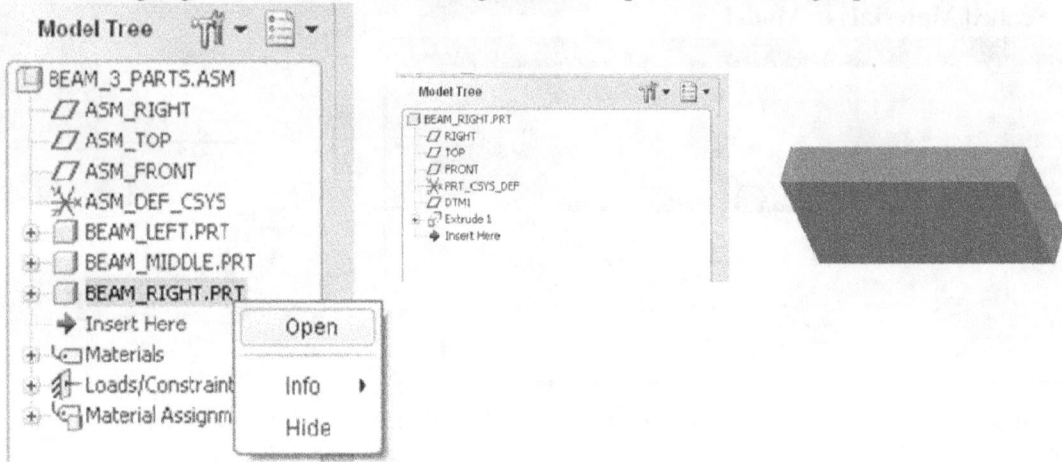

Click the icon of **Sketch** and pick the top surface of the beam as the sketch plane and click **Sketch**.

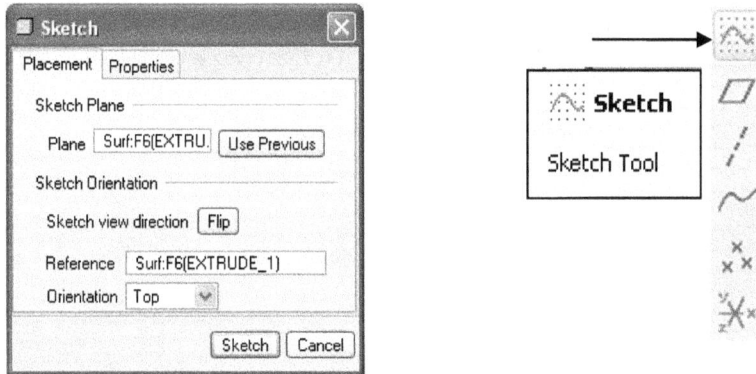

From the top menu, select **Sketch > References**. Select the 2 surfaces, as shown, as new references > **Close**.

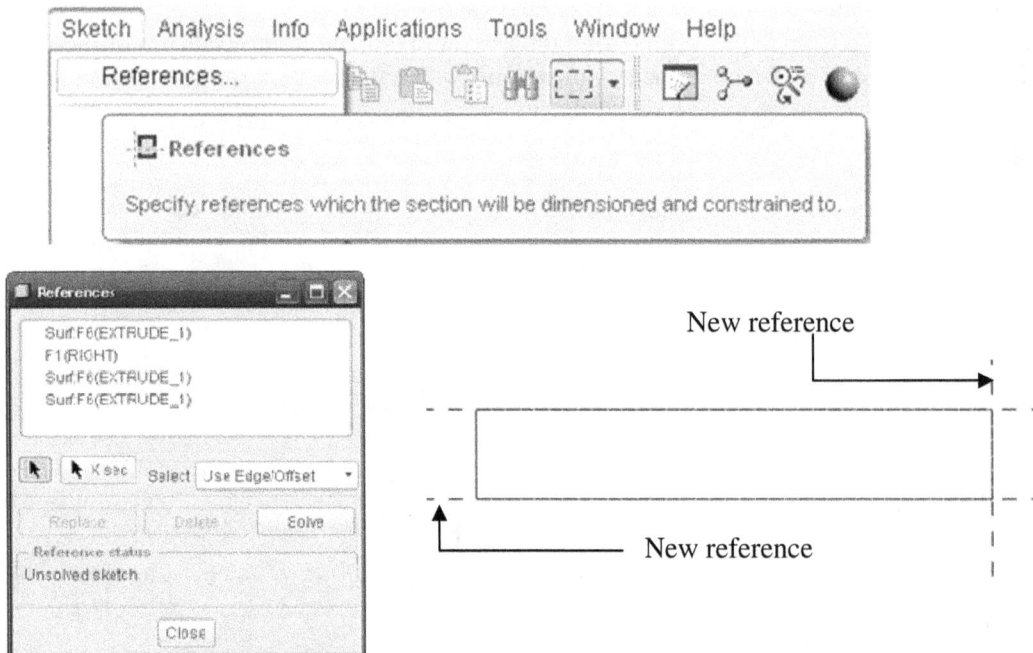

Click the icon of **Rectangle** and sketch a rectangle. The needed dimension is 30.

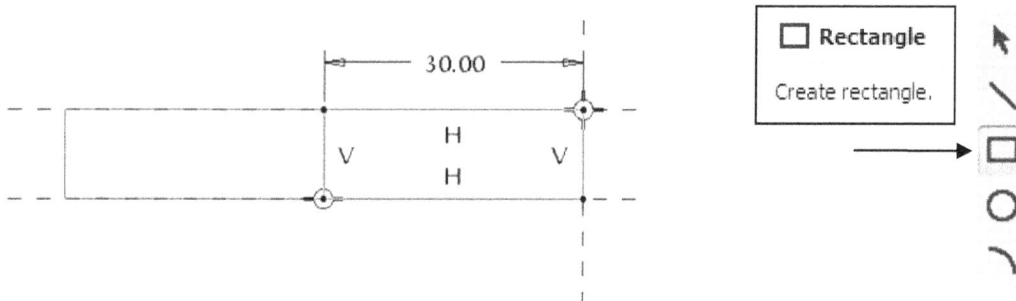

Go to Pro/Mechanica and click the icon of **Surface Region**. Expand **References**. Select **Split by chain** and pick the sketched curve in rectangle. Activate **Surfaces** and pick the surface nearby as shown. Click **Apply and Save**.

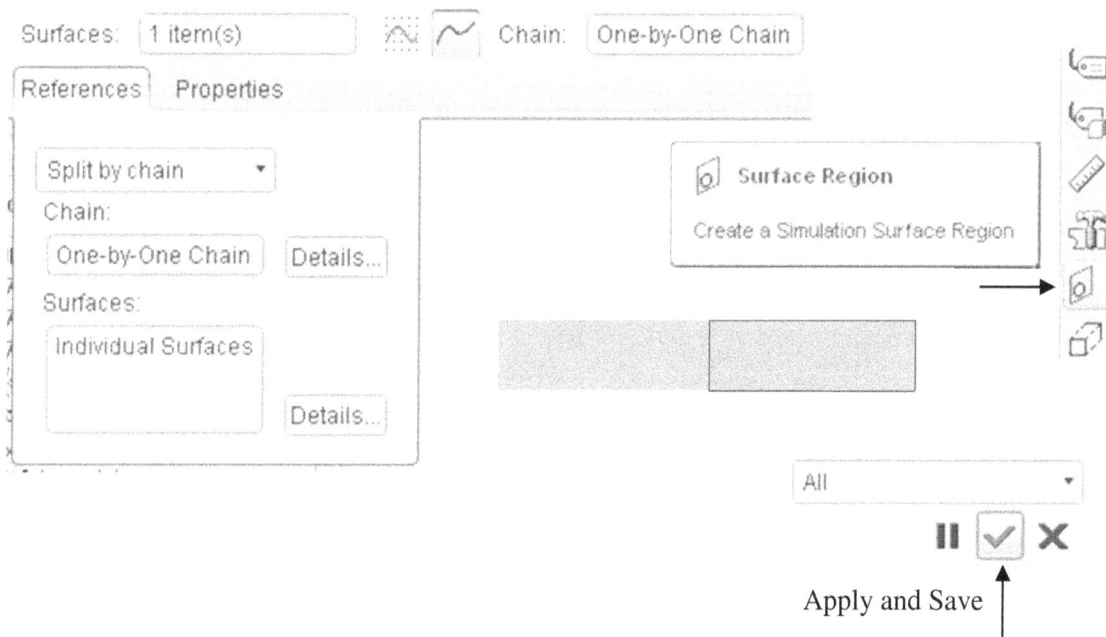

Apply and Save

Now let us go back to the assembly system. Under Pro/Mechanica, a surface region is shown on the beam right component.

Surface Region

Select the icon of **Force/Moment** from the toolbar of Mechanica objects > **Surfaces** > pick the surface region. Type *-500* as the Y component under Force > **Preview > OK**.

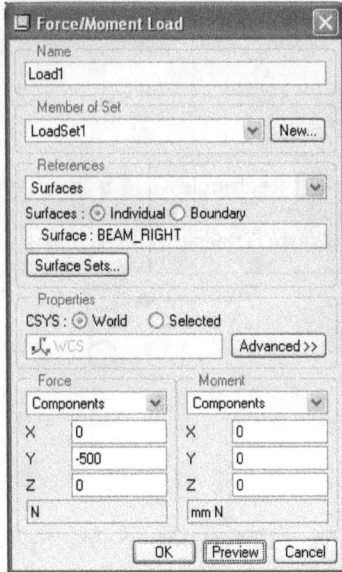

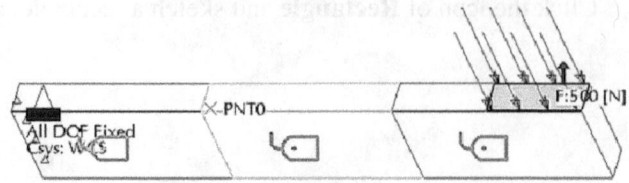

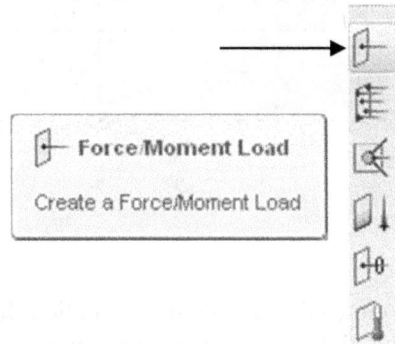

Step 5: Create Measures to calculate the principal stresses and von Mises stress

Select a datum point and define 6 measures associated with the selected datum point. The 6 measures are

$$\sigma_x, \sigma_y, \sigma_z, \tau_{xy}, \tau_{xz}, \tau_{yz}$$

To define σ_x at PNT0, select the icon of **Simulation Measure** > **New** > type *sigma_xx_* as the Name > select **Stress** and select **XX**, as shown > select **At Point** and pick the datum point **PNT0** > **OK**.

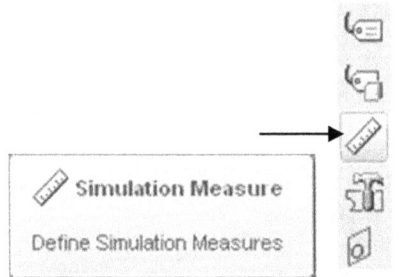

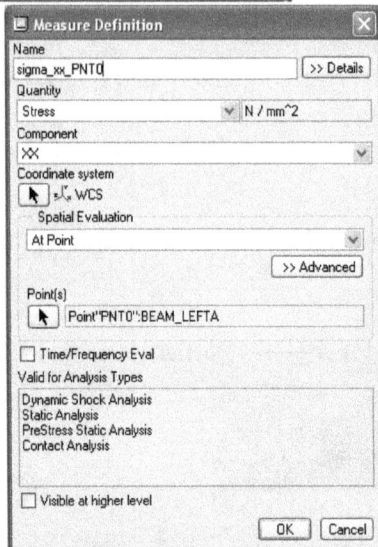

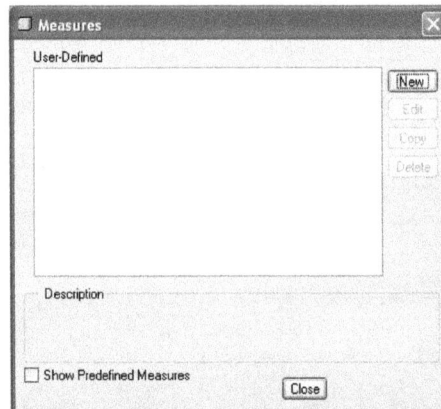

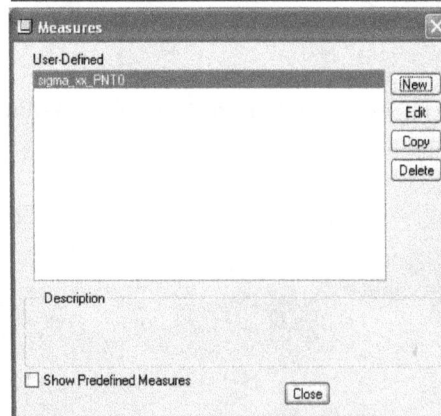

To define σ_y at PNT0, following the process of defining σ_x, select **New** > type *sigma_yy_* as the Name > select **Stress** and select **YY**, as shown > select **At Point** and pick the datum point **PNT0** > **OK**.

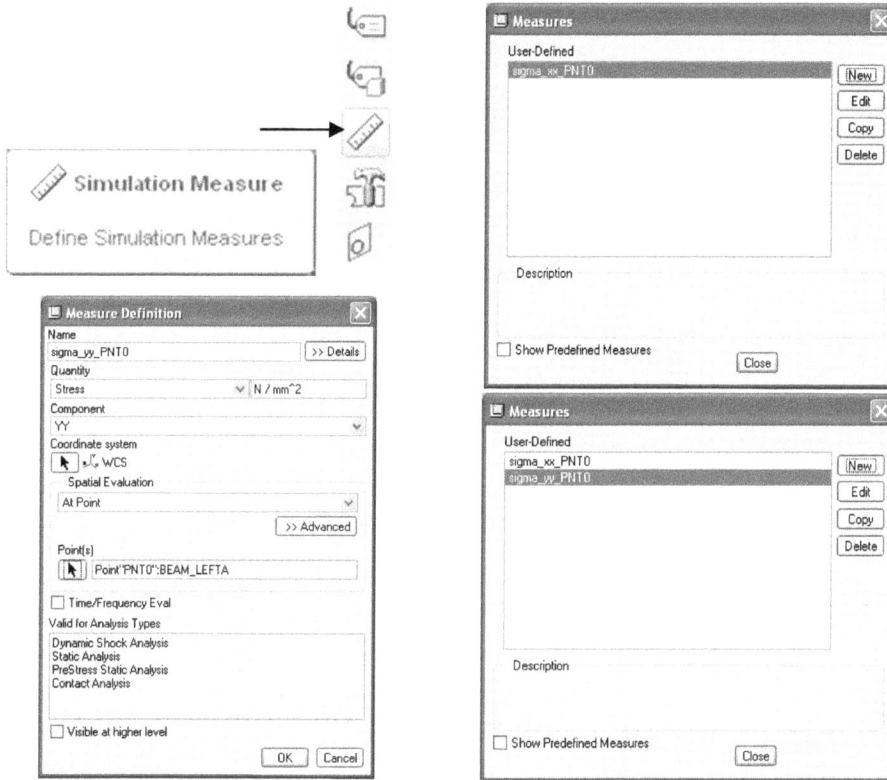

To define σ_z at PNT0, following the process of defining σ_y, select **New** > type *sigma_zz_* as the Name > select **Stress** and select **ZZ**, as shown > select **At Point** and pick the datum point **PNT0 > OK.**

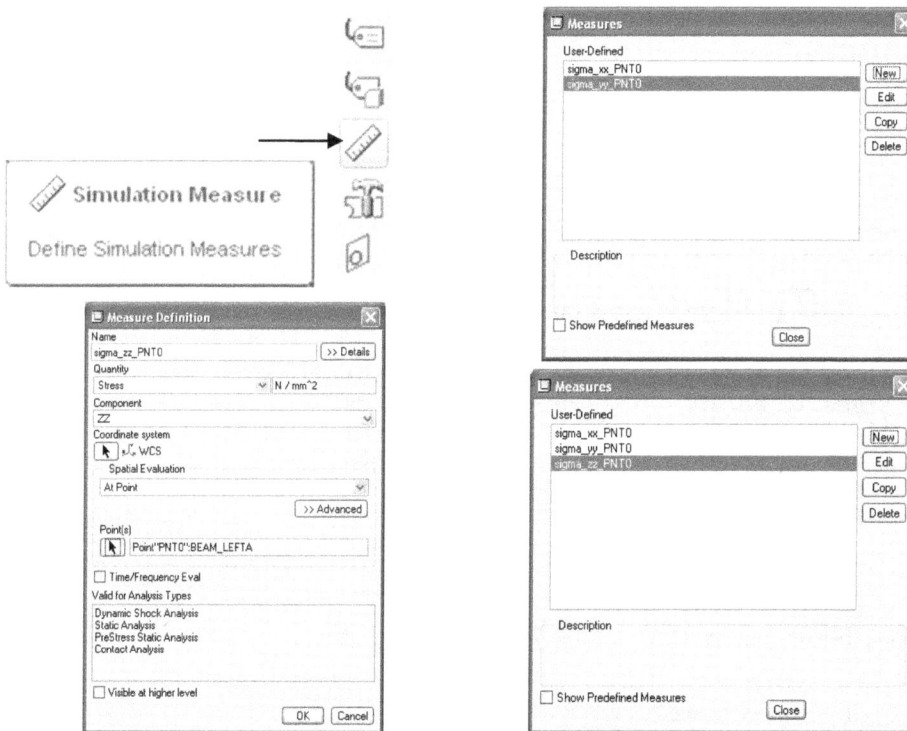

Follow the above procedure to define τ_{xy}, τ_{xz}, τ_{yz}, σ_{vm}, $\sigma_{max\text{-}principal}$ and τ_{max},

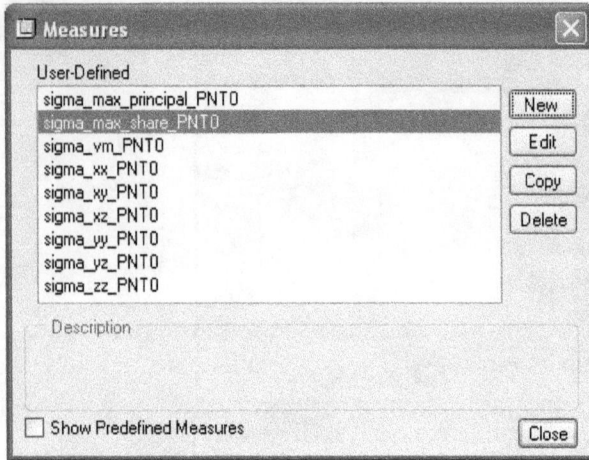

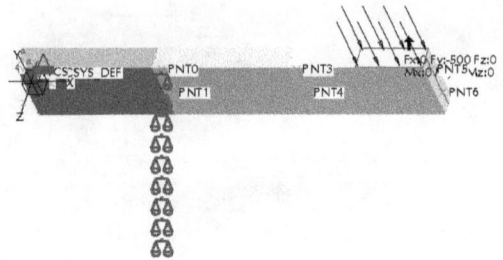

Step 6: After specifying the material type, and defining the constraint condition and the load condition, run the program.

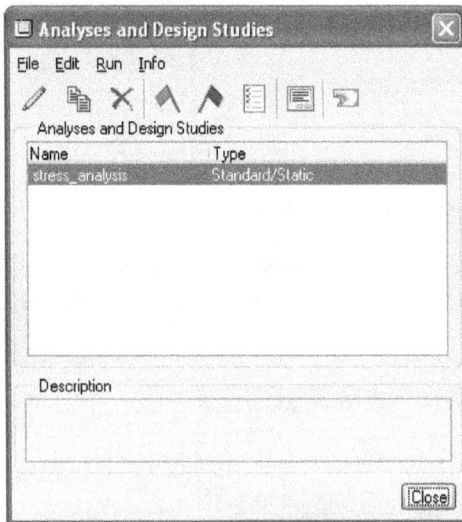

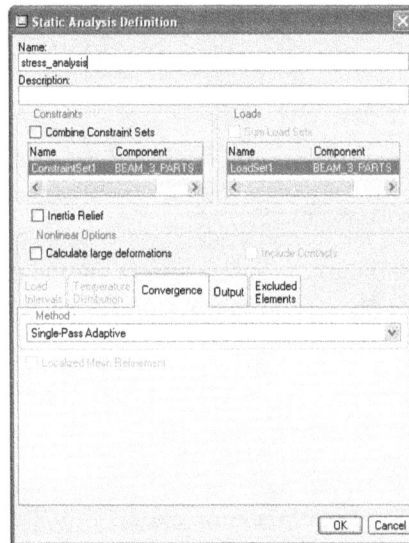

Step 7: Use the numerical values obtained from the FEA run to calculate the max principal stress and von Mises stress and max shear stress at PNT0.

$$\sigma_x, \sigma_y, \sigma_z, \tau_{xy}, \tau_{xz}, \tau_{yz}$$

sigma_xx_PNT0:	2.319494e+01
sigma_xy_PNT0:	-1.371437e+00
sigma_xz_PNT0:	-1.440686e-01
sigma_yy_PNT0:	9.666577e-02
sigma_yz_PNT0:	-2.587550e-01
sigma_zz_PNT0:	-1.869021e-01

Use these values to calculate A, B, C, which are defined as follows:

$$A = \sigma_x + \sigma_y + \sigma_z$$
$$B = \sigma_x\sigma_y + \sigma_x\sigma_z + \sigma_y\sigma_z - \tau_{xy}^2 - \tau_{xz}^2 - \tau_{yz}^2$$
$$C = \sigma_x\sigma_y\sigma_z + 2\tau_{xy}\tau_{xz}\tau_{yz} - \sigma_x\tau_{yz}^2 - \sigma_y\tau_{xz}^2 - \sigma_z\tau_{xy}^2$$

Users may use Matlab to solve the following 3rd order equation to obtain $\sigma_{max\_principal}$, $\sigma_{mid\_principal}$, and $\sigma_{min\_principal}$.

$$\sigma^3 - A\sigma^2 + B\sigma - C = 0$$

After obtaining the numerical values of $\sigma_{max\_principal}$, $\sigma_{mid\_principal}$, and $\sigma_{min\_principal}$, use the following equation to calculate τ_{max}:

$$\tau_{max} = \frac{\sigma_{max-principal} - \sigma_{min-principal}}{2}$$

Use the following equation to calculate σ_{vm}:

$$von\_Mises\_stress = \frac{1}{\sqrt{2}}\sqrt{(\sigma_1 - \sigma_2)^2 + (\sigma_2 - \sigma_3)^2 + (\sigma_1 - \sigma_3)^2}$$

Step 8:   Verify your calculations through comparing the calculated $\sigma_{max-principal}$, τ_{max}, σ_{vm} with their corresponding values calculated by the FEA program.

```
sigma_max_principal_PNT0:    2.327679e+01
sigma_max_share_PNT0:        1.182416e+01
sigma_vm_PNT0:               2.336806e+01
```

Step 9:   Calculate the stress values at PNT2.

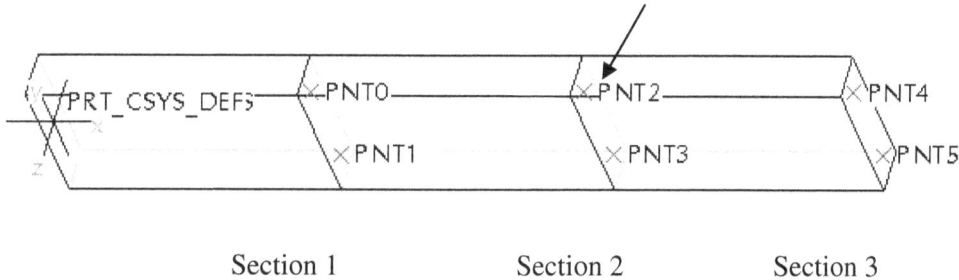

Section 1   Section 2   Section 3

We repeat the same process as we did in the process of calculating the stress values at PNT0. We open BEAM_MIDDLE, and create a datum point. Name the date point as PNT2.

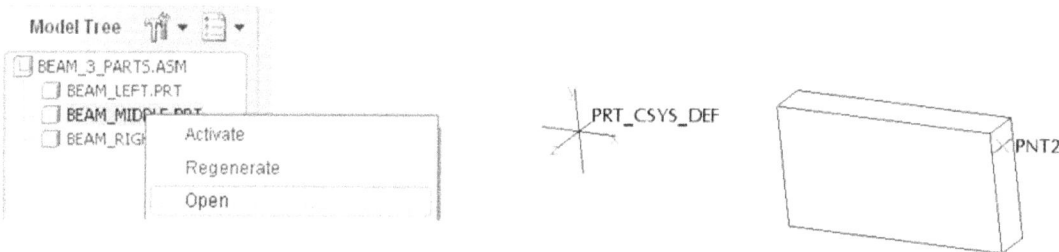

From the main toolbar or the menu toolbar, select **Applications > Mechanica > Structure > OK.**

To define the following 6 measures at PNT2, users may use an EDIT process.

$$\sigma_x, \sigma_y, \sigma_z, \tau_{xy}, \tau_{xz}, \tau_{yz}$$

Select the icon of **Simulation Measure > Edit.**

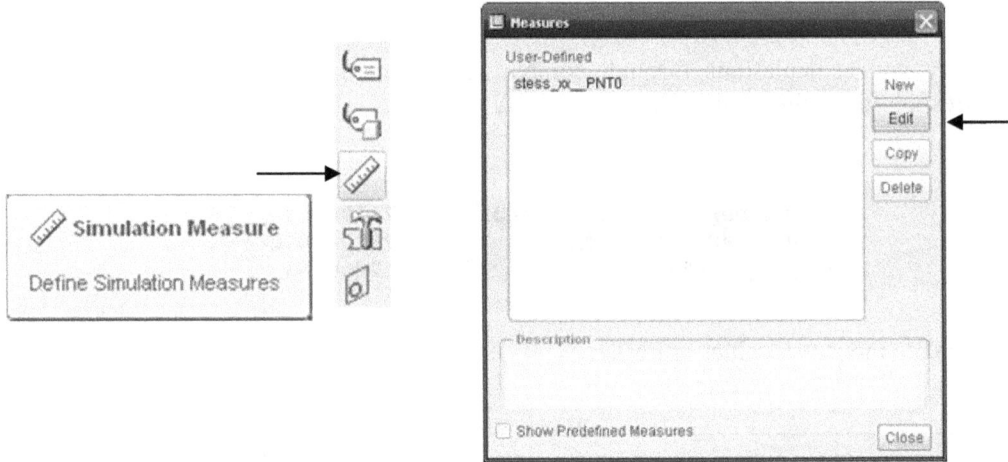

In the **Measure Definition** window, click the box called **Point**(s). While holding down the **Ctrl** key, pick PNT2 > Click **OK** to accept PNT2 > Click **OK** to complete the process of adding the new measure: stress_xx_PNT2, as shown.

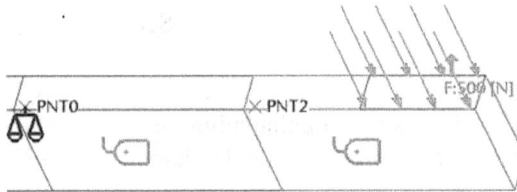

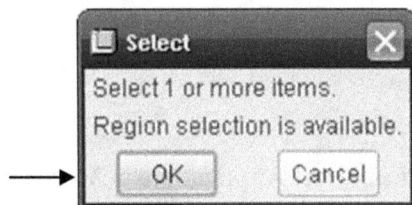

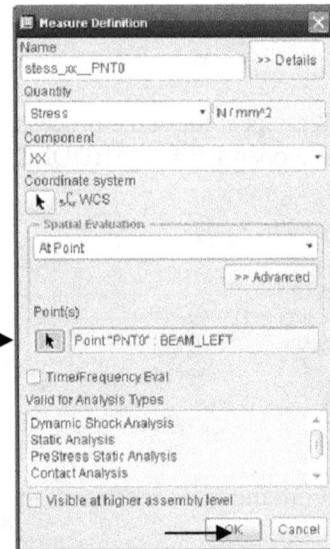

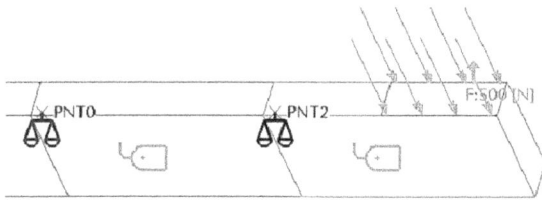

Users may repeat the above procedure to define those new measures. Afterwards, run the FEA program again.

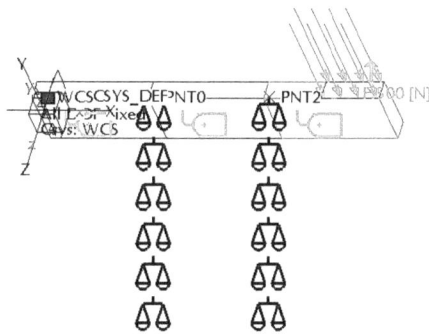

As a summary, we need to know the values of the three normal stresses and three shear stresses. We will be able to calculate the 3 principal stress values.

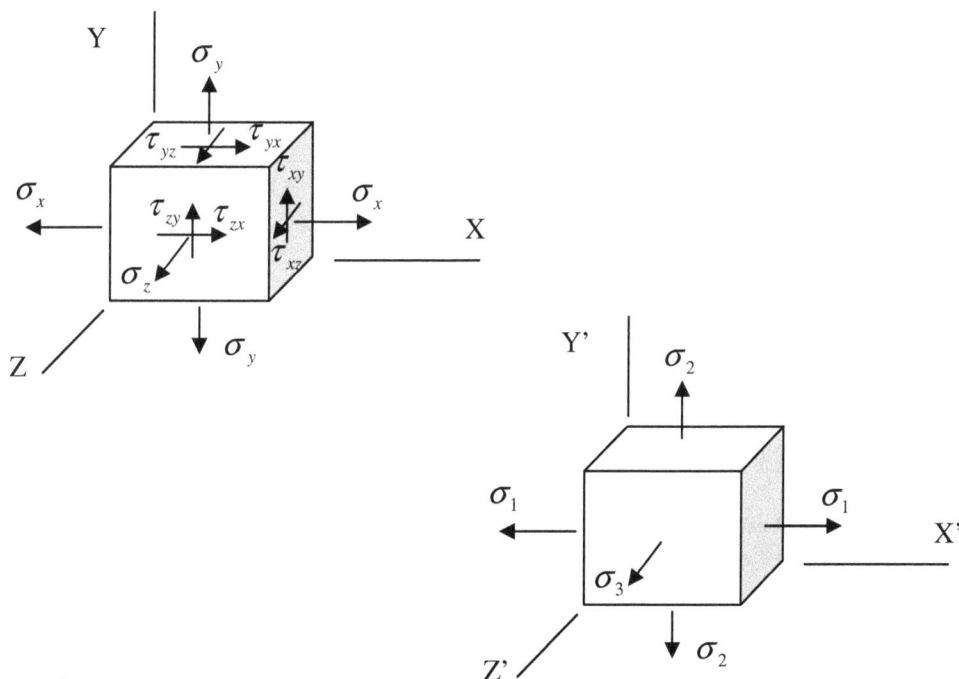

The derivation of the maximum shear stress is based on Mohr's circle. For plane stress, we have the relations between the set of principal stresses and the set of three maximum shear stresses as follows:

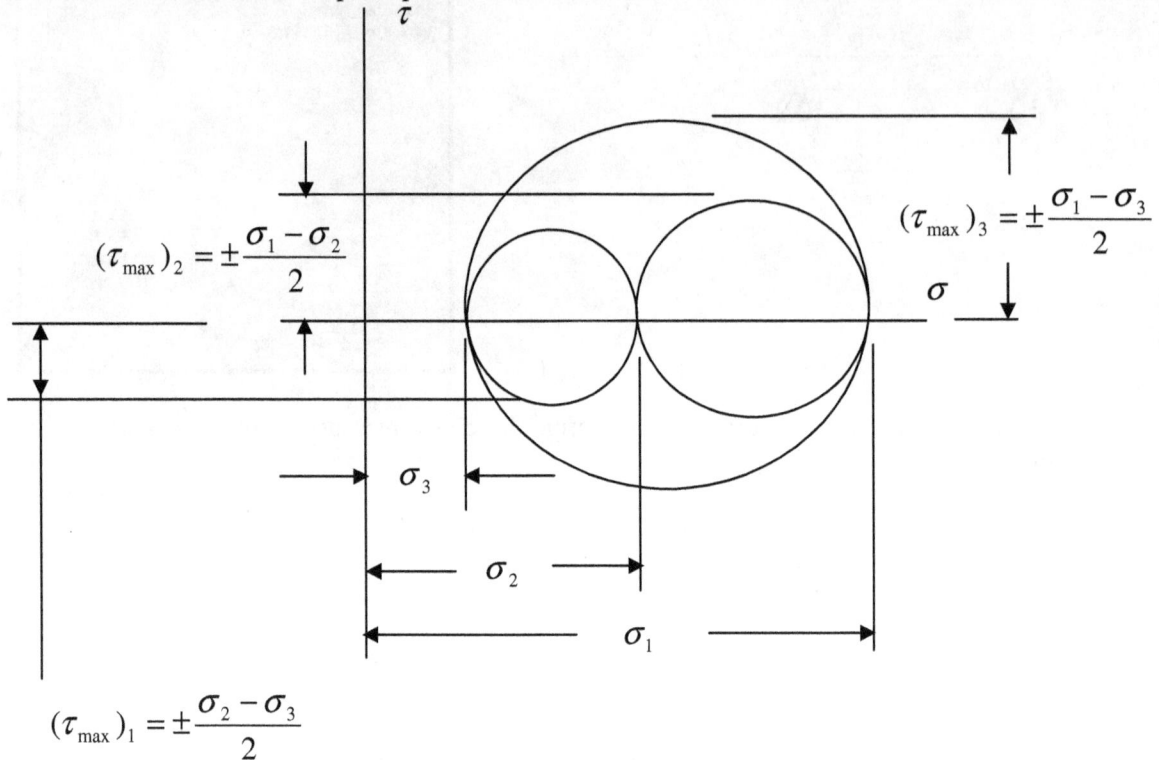

$$(\tau_{max})_2 = \pm \frac{\sigma_1 - \sigma_2}{2}$$

$$(\tau_{max})_3 = \pm \frac{\sigma_1 - \sigma_3}{2}$$

$$(\tau_{max})_1 = \pm \frac{\sigma_2 - \sigma_3}{2}$$

It is important to note that the absolute maximum shear stress is the numerically largest of the stresses determined from these three equations shown in the above figure.

There are several criteria to determine the onset of inelastic behavior or yield for ductile materials under the combined stresses described above. In Pro/MECHANICA, von Mises yield condition is implemented as default. Note that the von Mises yield condition is based on the principle that the strain energy of stressed component can be divided into volumetric strain energy and shear strain energy components. The volumetric strain energy is related to the mean "hydrostatic stress", leading to volume change and no distortion. The shear strain energy is related to the deviatory or distortional stress, leading to no volumetric change, but instead distorting or deviating the element from its initial cubic shape. The definition of the von Mises stress is given by

$$von\_Mises\_stress = \frac{1}{\sqrt{2}} \sqrt{(\sigma_1 - \sigma_2)^2 + (\sigma_2 - \sigma_3)^2 + (\sigma_1 - \sigma_3)^2}$$

6.6   References

1. Algor Processor Reference Manual, Algor Interactive Systems, 150 Alpha Drive, Pittsburgh, PA 15238
2. R. D. Cook, D. S. Maikus and M. E. Plesha, Concepts and Applications of Finite Element Analysis, 3rd edition, Wiley, New York, NY, 1989.
3. J. W. Dally and W. F. Riley, Experimental Stress Analysis, McGraw Hill, 1965.
4. J. W. Dally, Design Analysis of Structural Elements, Goodington, New York, NY, 2003.
5. C. S. Desal and J. F. Abel, Introduction to the Finite Element Method, Van Nostrand Reinhold, New York, NY, 1972.
6. J. H. Earle, Graphics for Engineers, AutoCAD Release 13, Addision-Wesley, Reading, Massachusetts, 1996.
7. R. H. Gallagher, Finite Element Analysis Fundamentals, Prentice-Hall, Englewoood Cliffs, NJ, 1975.

8. J. M. Gere and S. P. Timoshenko, Mechanics of Materials, 3rd edition, PWS-KENT Publishing Company, Boston, 1990.
9. W. W. Hager, Applied Numerical Linear Algebra, Prentice-Hall, Englewood Cliffs, NJ, 1988.
10. Holland, The Finite Element Method in Plane Stress Analysis, in the Finite Element Method in Stress Analysis, I. Holland and K. Bell (eds), Tapir Press, Trondheim, Norway, 1969, Chapter 2.
11. P. G. Plockner, Symmetry in Structural Mechanics, J. of the Structural Division, American Society of Civil Engineers, Vol. 99, No. ST1, pp. 71-89, 1973.
12. J. N. Reddy, An Introduction to the Finite Element Method, McGraw-Hill Publishing Company, New York, NY, 1984.
13. T. Ross, T. Johns and Emile, Dynamic Buckling of Thin Walled Domes under External Water Pressure, Applied Solid Mechanics 2nd Conference, Ed. Tooth & Spence, Elsevier, pp. 211-224, 1987.
14. O. C. Zienkiewicz, The Finite Element Method, third edition, McGraw-Hill, New York, NY, 1977.
15. O. C. Zienkiewicz and R. L. Taylor, The Finite Element Method, 4th edition, Vol. 1, McGraw-Hill (UK),London, 1989.

6.7 Exercises

1. A plastic plate is used to support three plants, which are on display. The three dimensions of the plate and the three surface regions where the three plants are placed are shown below.

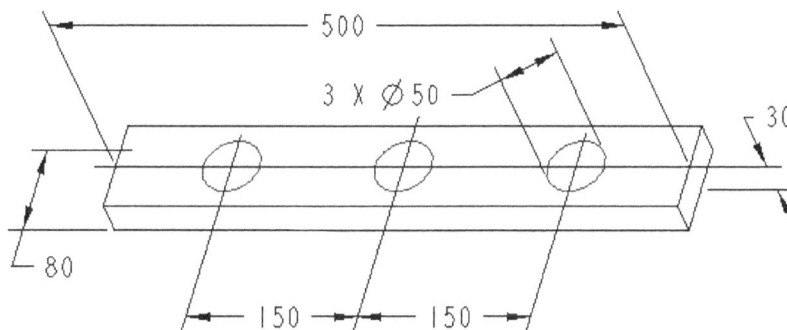

Assume that the weight of each plant is 200 N. The plate material is Nylon. Use a shell modeling approach to evaluate the deflection distribution and the distribution of the von Mises stress.

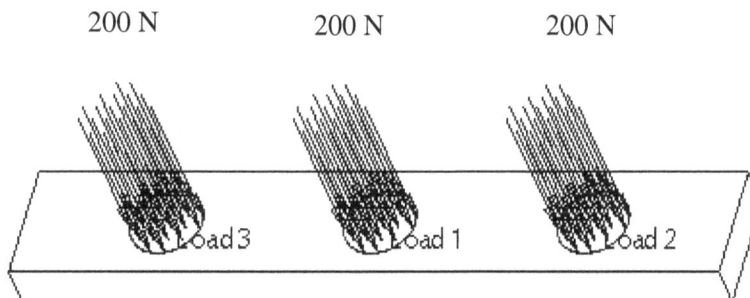

Assume that the plate is fixed to the 2 supports through the use of 4 bolts and 4 washers.

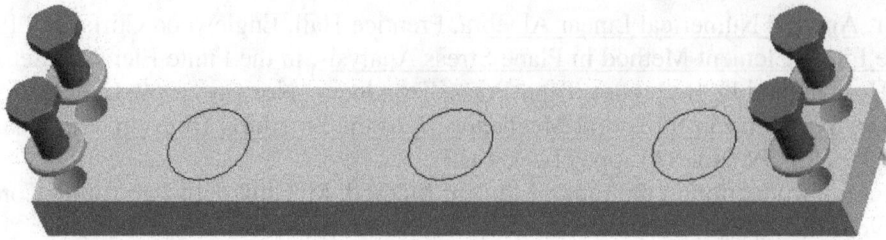

2.   Use a shell modeling approach. A welding structure is shown below. It consists of 2 components, namely, a base plate and a top plate, as shown below:

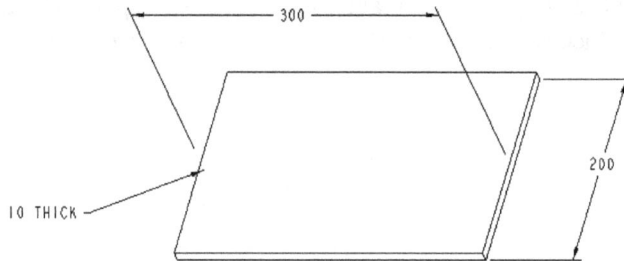

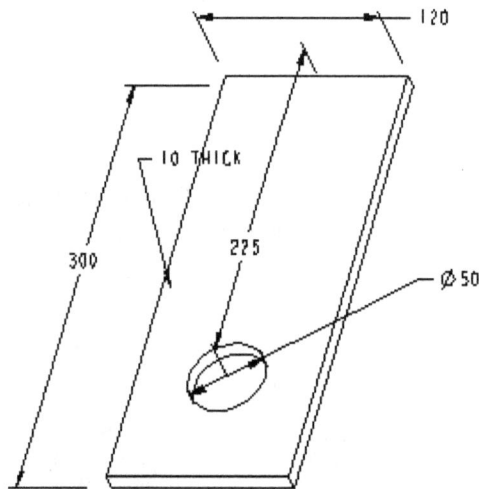

    In your previous homework, you dealt with a spot welding structure. This structure is not a spot welding structure. The welding part is a rectangular shaped string. As a result, you need to assemble it with the base plate and the top plate. The welding part is shown below:

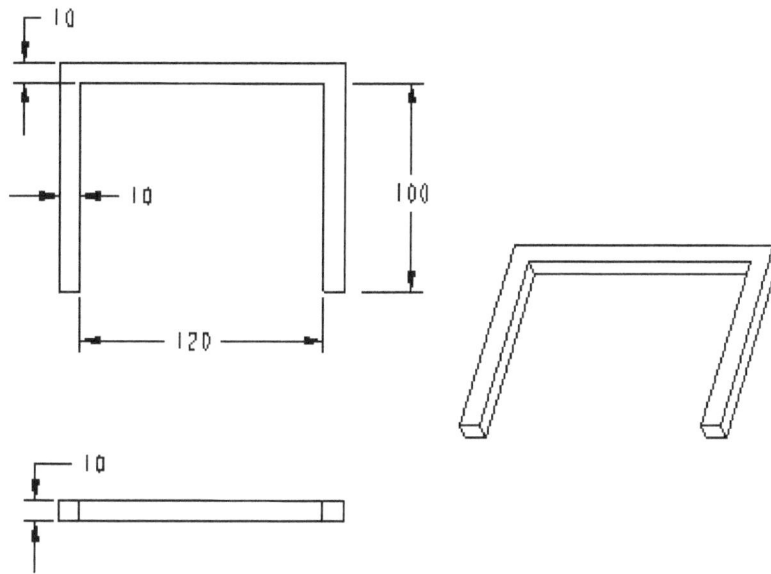

The following figure shows the entire welding structure after the assembly.

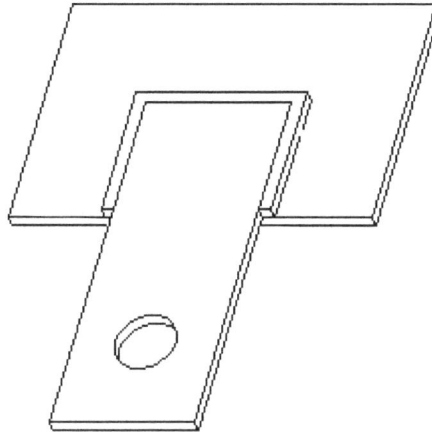

(1)   You are asked to create a 3D solid model for the base plate component.
(2)   You are asked to create a 3D solid model for the top plate component.
(3)   You are asked to create a 3D solid model for the welding part.
(4)   You are asked to create an assembly, which consists of the base plate, the top plate and the welding part.
(5)   You are asked to create a shell model for each of the three components.
(6)   You are asked to verify your shell model at the assembly level.
(7)   You are asked to use Edit > Pair Place to adjust the locations of the compressed planes so that the entire assembly is in the following setting.

Top View

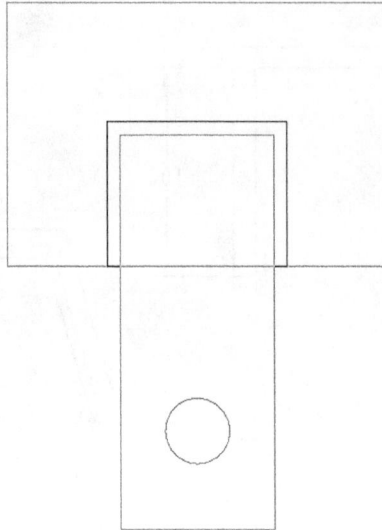

Right Side View

(8)   Assume that the material for the base plate and the top plate components is STEEL, as listed in the Pro/E material library. Assume that the material for the weld part is TUNGSTEN, as listed in the Pro/E material library.

(9)   Perform two (2) design studies, assuming that the bottom surface of the base plate is fixed to the ground:

Design study 1: Bearing load with magnitude equal to 1000 N.
Design study 2: Uniform load with magnitude equal to 1000 N.

In each design study, prepare 2 plots of the displacement distribution and the max principal stress distribution. Identify the locations of the max displacements and the max principal stress of the entire assembly structure.

CHAPTER 7

MODAL, BUCKLING, PRESTRESS AND VIBRATION ANALYSES

7.1 Introduction

Finite element methods have been successfully used in structural analyses in engineering. They have also been widely used in other fields in engineering. In the chapter, we present several applications, covering modal analysis, buckling analysis, pre-stress analysis and vibration analysis. The procedure to perform each of these applications will be demonstrated through examples.

7.2 Applications on Modal Analyses

As we discussed in Chapter 1, finite element methods treat a continuous system as an assembly of elements through discretization. The global stiffness matrix obtained through an assembly of individual stiffness matrices characterizes the system response when subjected to a load condition or load conditions. The global stiffness matrix has complex eigenvalues. Transformation of this global stiffness matrix into a non-diagonal matrix which is called the modal form is viewed as modal analysis, which focuses on the mode shapes of the vibrating system.

     The following figure illustrates a driving board used in sport competition. As illustrated, a person is jumping from the board and diving to the swimming pool.

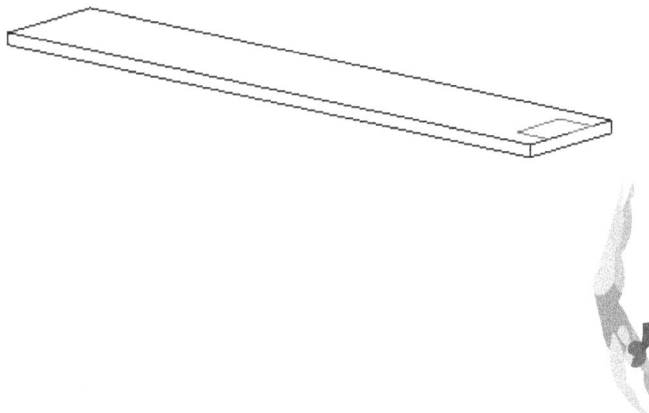

Assume the load condition is 600 Newton and loading area is a rectangle. The two dimensions of the rectangular area are 400 x 250 mm. The geometrical shape of the diving board is a rectangular plate. The three dimensions are 4000 x 750 x 75 mm. We assume that the back end of the diving board is a fixed end.

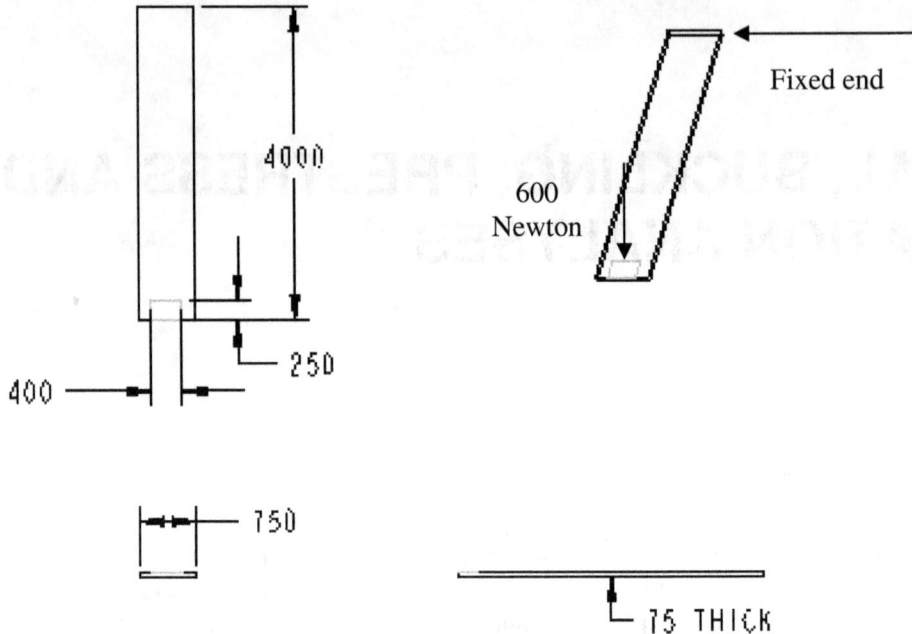

Step 1: Under the Pro/ENGINEER design environment, create a 3D solid model of the diving platform.

From the main toolbar, select **File > New > Part >** type diving_*platform* as the name of the file > clear the icon of **Use default template > OK**.

Select **mmns_part_solid** (unit system: millimeter, Newton and Second) and type *diving platform* in **DESCRIPTION** and *student* in **MODELED_BY > OK**.

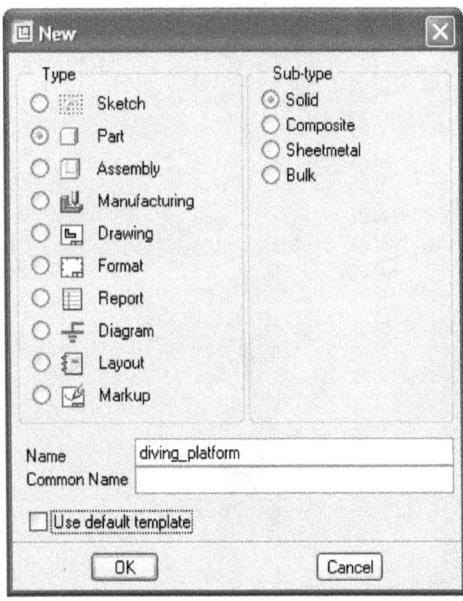

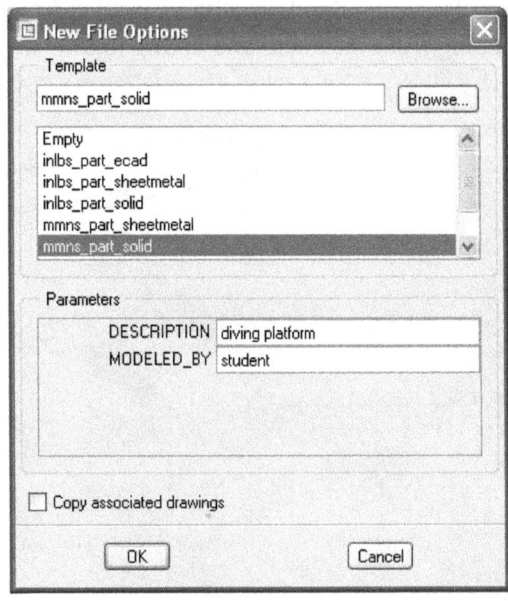

Let us use **Thin Protrusion** to create the 3D solid model so that **Auto Detect** can be used to generate a shell model later.

From the toolbar of feature creation, select the icon of **Extrude.** The dashboard appears. Select the icon of **Thicken sketch**, set the thickness value to *75* and set the depth value to *4000*. Click the box of **Placement > Define**.

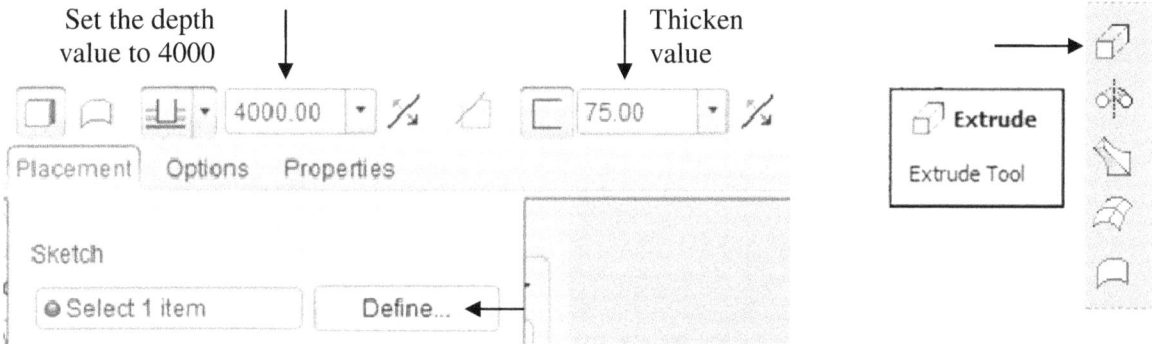

Set the depth value to 4000

Thicken value

Extrude
Extrude Tool

Placement   Options   Properties

Sketch

Select 1 item   Define...

Select the **FRONT** datum plane as the sketch plane, and click the box of **Sketch** to accept the **RIGHT** datum plane as the default reference to orient the sketch plane, as illustrated below.

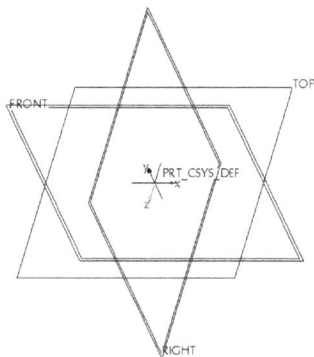

Sketch

Placement

Sketch Plane

Plane   FRONT:F3(DA   Use Previous

Sketch Orientation

Sketch view direction   Flip

Reference   RIGHT:F1(DATUM PLA...
Orientation   Right

Sketch   Cancel

Select the icon of **Centerline** and draw a horizontal centerline along the y-axis. Select the icon of **Line** to sketch a line along the x-axis. The line is symmetric about the y-axis. The length is 750.

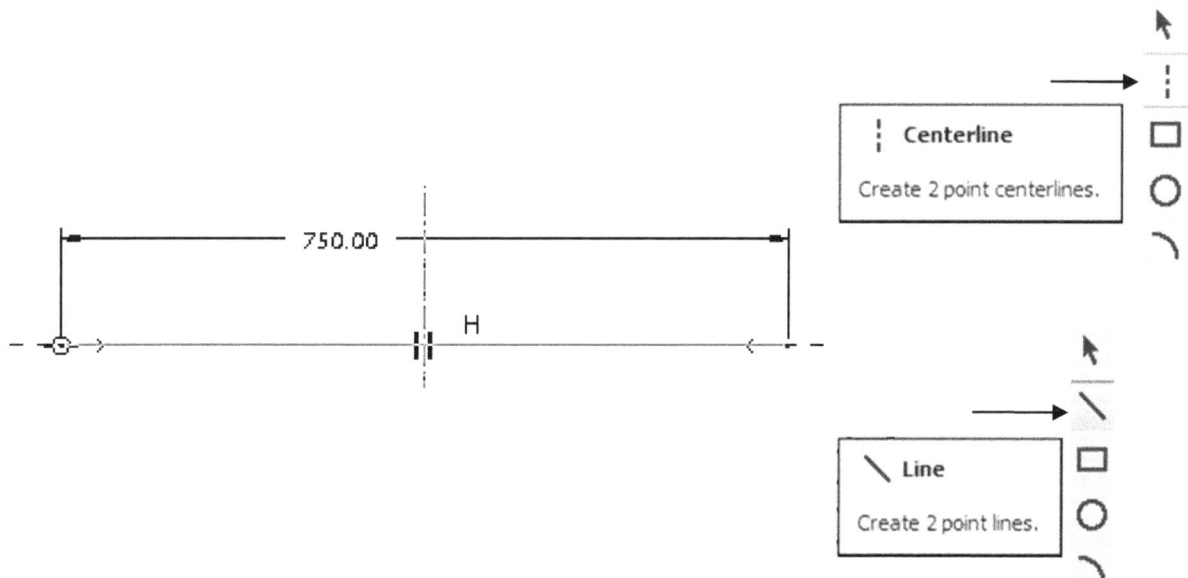

Centerline
Create 2 point centerlines.

750.00

H

Line
Create 2 point lines.

Upon completion, click the icon of **Done** and click the icon of **Apply and Save**.

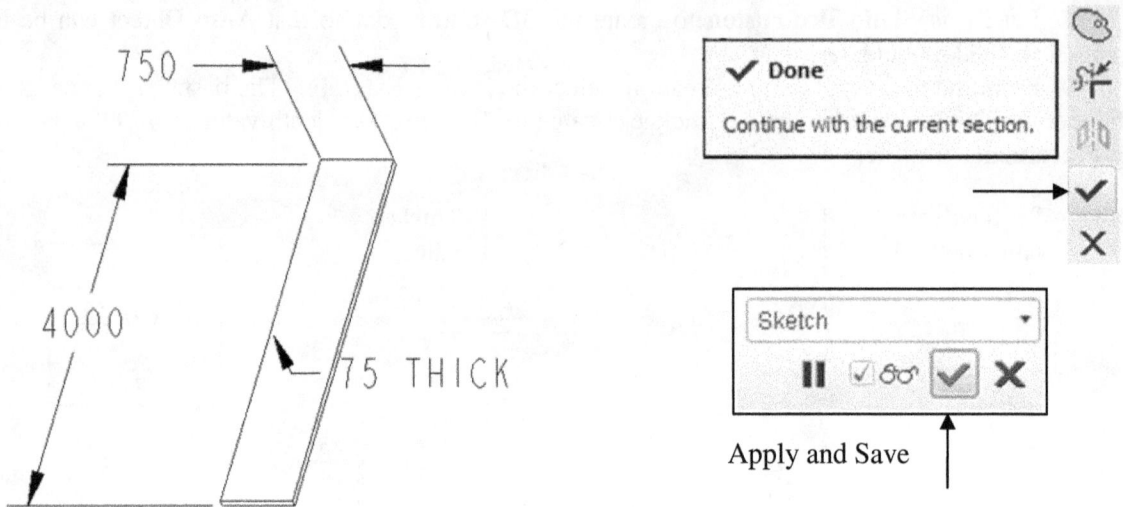

750 —

4000

75 THICK

✓ **Done**

Continue with the current section.

✓

✕

Sketch ▾

❚❚ ☑ 👓 ✓ ✕

Apply and Save

Note that a datum curve is needed for defining a surface region where a load can be applied. To sketch the datum curve, click the icon of **Sketch** Tool > pick the top surface of the platform as the sketch plane and accept the default setting to orient the sketch plane.

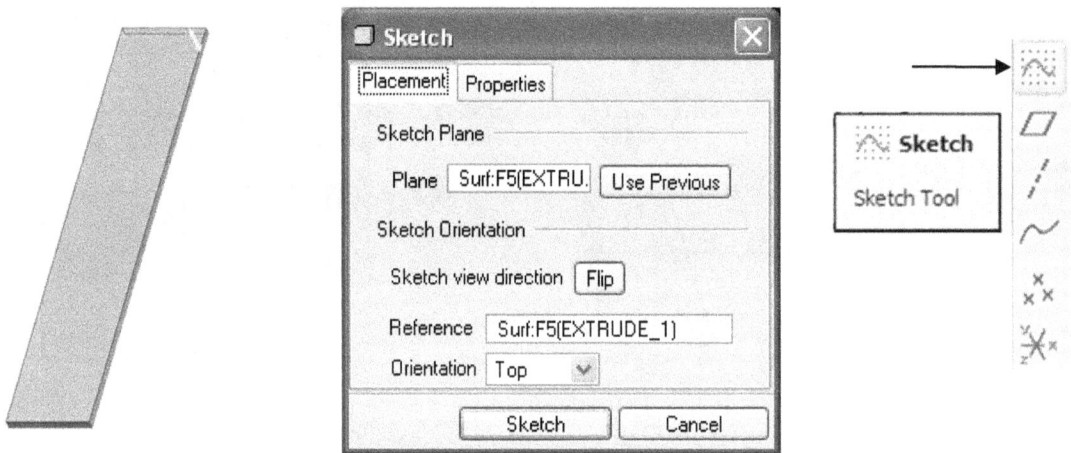

■ **Sketch**   ✕

[Placement] Properties

Sketch Plane

Plane [Surf:F5(EXTRU.] [Use Previous]

Sketch Orientation

Sketch view direction [Flip]

Reference [Surf:F5(EXTRUDE_1)]

Orientation [Top   ∨]

[Sketch]   [Cancel]

〰 **Sketch**

Sketch Tool

Click the icon of Centerline and sketch a vertical centerline, as shown.

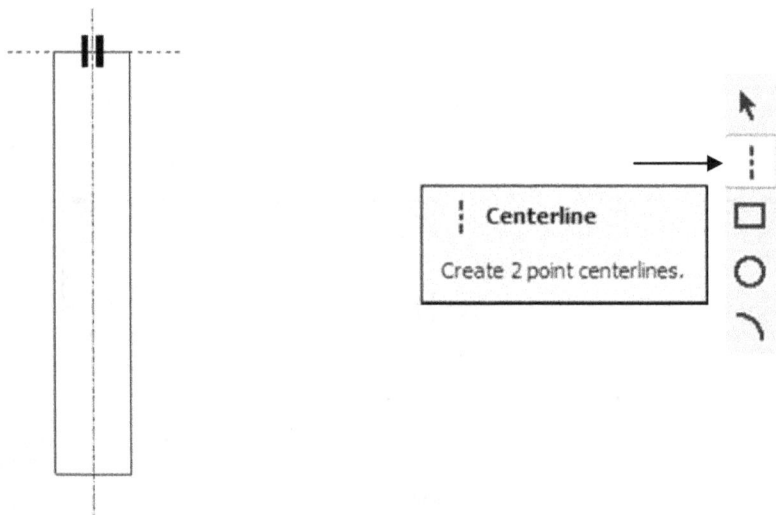

┆ **Centerline**

Create 2 point centerlines.

To add a new reference, click **Sketch > References**. Pick the bottom line (surface) of the rectangle as a new reference, as shown.

Sketch   Analysis   Info   A

Sketch Setup...

References...

References

Specify references which the section will be dimensioned and constrained to.

References

Surf:F5(EXTRUDE_1)
F1(RIGHT)
Surf:F5(EXTRUDE_1)

X sec   Select   Use Edge/(

Replace   Delete   Solve

Reference status
Unsolved sketch

Close

New Reference

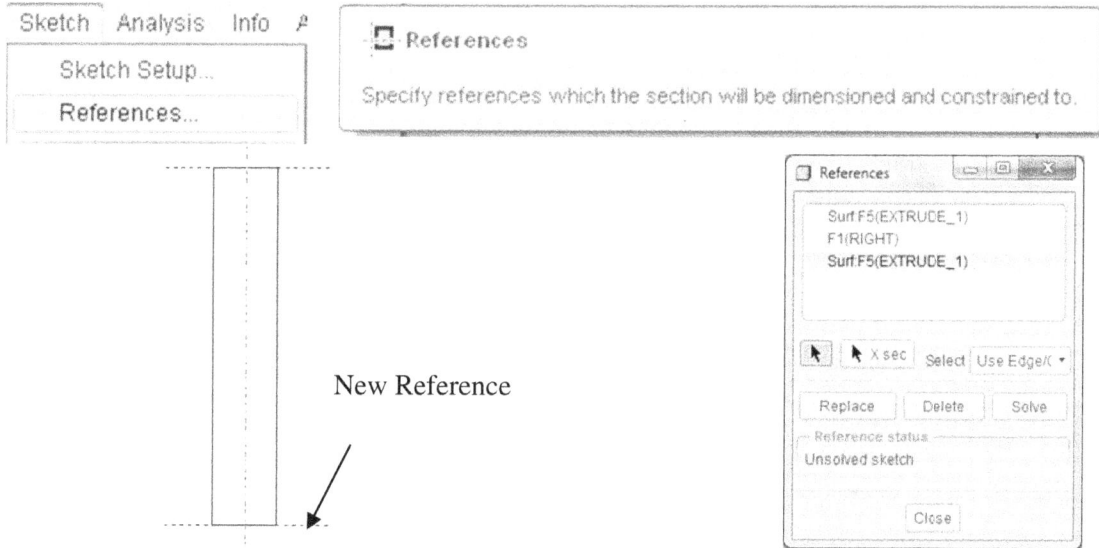

Click the icon of **Rectangle** and sketch a rectangle, which is symmetric about the sketched vertical centerline. The 2 side dimensions are 400 and 250, respectively.

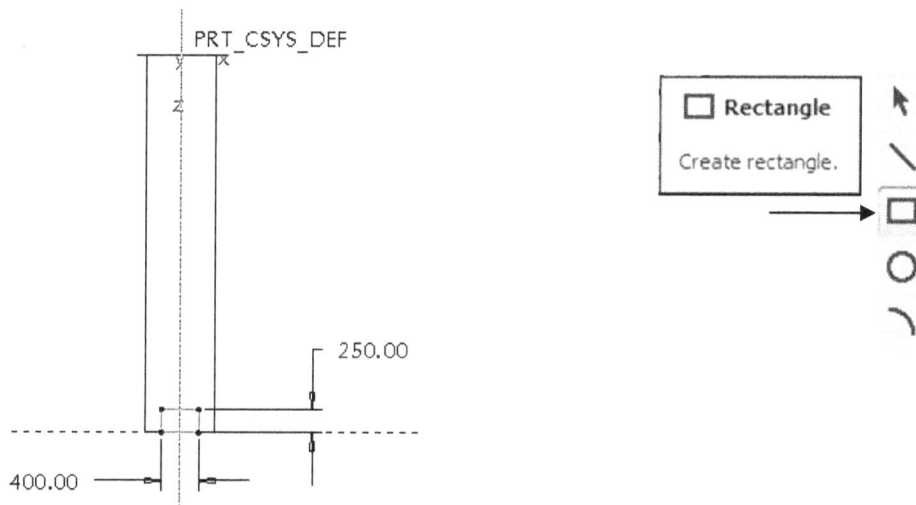

PRT_CSYS_DEF

250.00

400.00

Rectangle

Create rectangle.

Step 2: Under the Pro/MECHANICA environment, specify the material type. Select **Nylon**.
From the top menu, click **Applications > Mechanica > Continue > Structure > OK.**

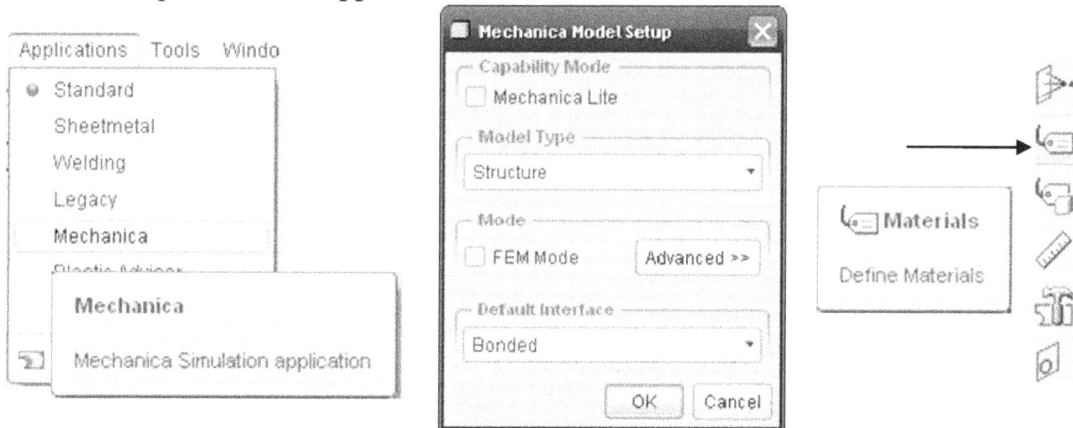

Applications   Tools   Windo

● Standard
Sheetmetal
Welding
Legacy
Mechanica

Mechanica

Mechanica Simulation application

Mechanica Model Setup

Capability Mode
☐ Mechanica Lite

Model Type
Structure ▾

Mode
☐ FEM Mode   Advanced >>

Default Interface
Bonded ▾

OK   Cancel

Materials

Define Materials

From the toolbar of functions, select the icon of **Materials >** select **Nylon** from the left side called Materials in Library > click the directional arrow so that the selected material type goes to the right side called Materials in Model.

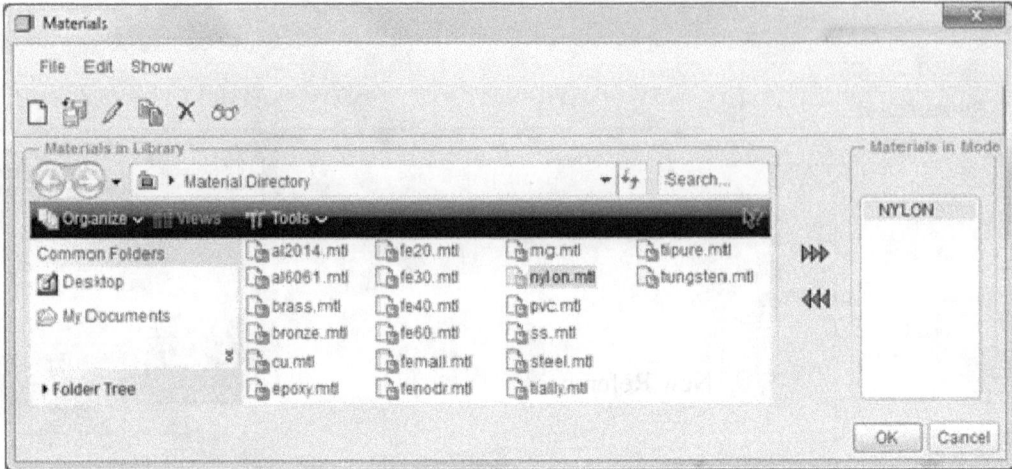

Click the icon of **Material Assignment**, the pop-up window indicates there is one component and the material type is Nylon. Just click **OK**.

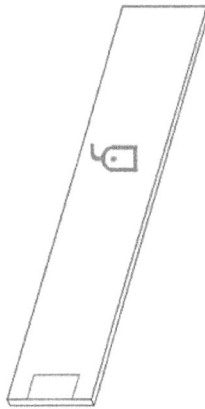

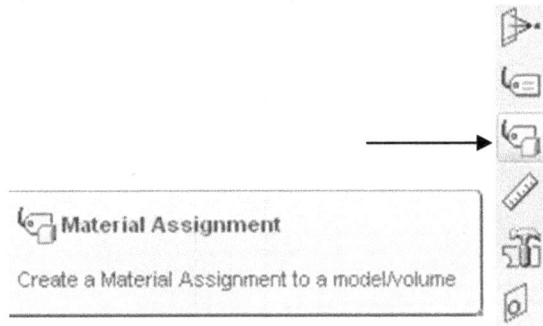

Step 3: Under the Pro/MECHANICA environment, define the surface region.

To define the surface region, click the icon of **Surface Region.** Pick the sketched rectangle first. Afterwards, pick the flat surface for splitting. Click the icon of **Apply and Save**.

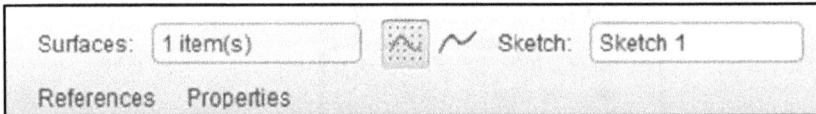

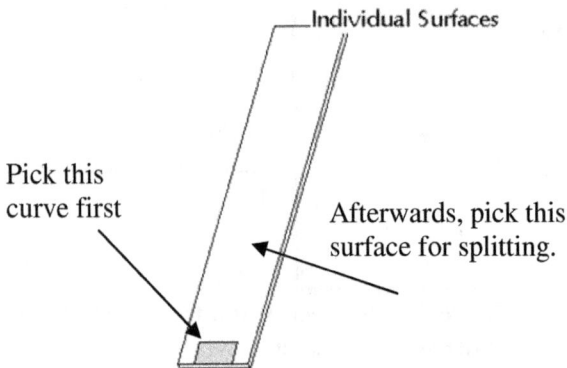

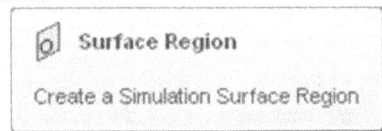

Step 4: Under the Pro/MECHANICA, define a shell model.
From the toolbar of functions, select the icon of **Insert > Midsurface > Auto Detect Shell Pairs.** Specify 75 as the thickness value > **Start**.

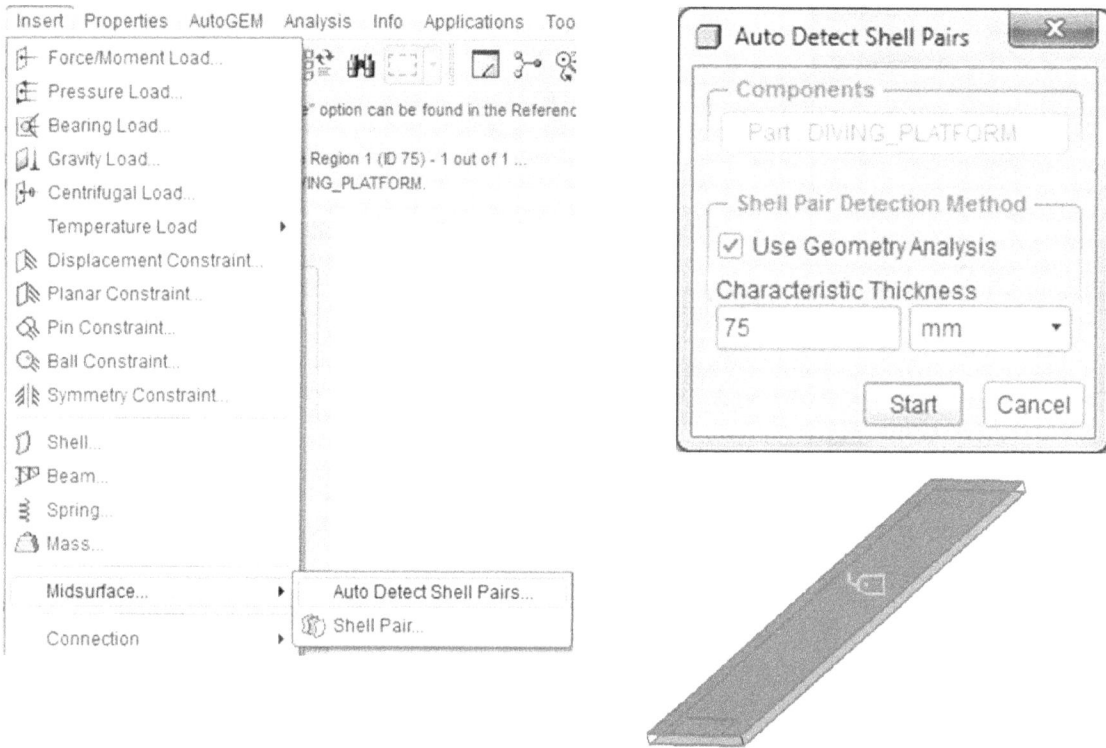

Step 5: Under the Pro/MECHANICA environment, define the constraint condition: Fixed end.
Note that we need to use **Edge/Curve**, and cannot use Surface.
Click the icon of **Constraints** > select **Edge(s)/Curve(s)** > pick the edge on the back side > **OK** to fix all 6 degrees of freedom > **OK**.

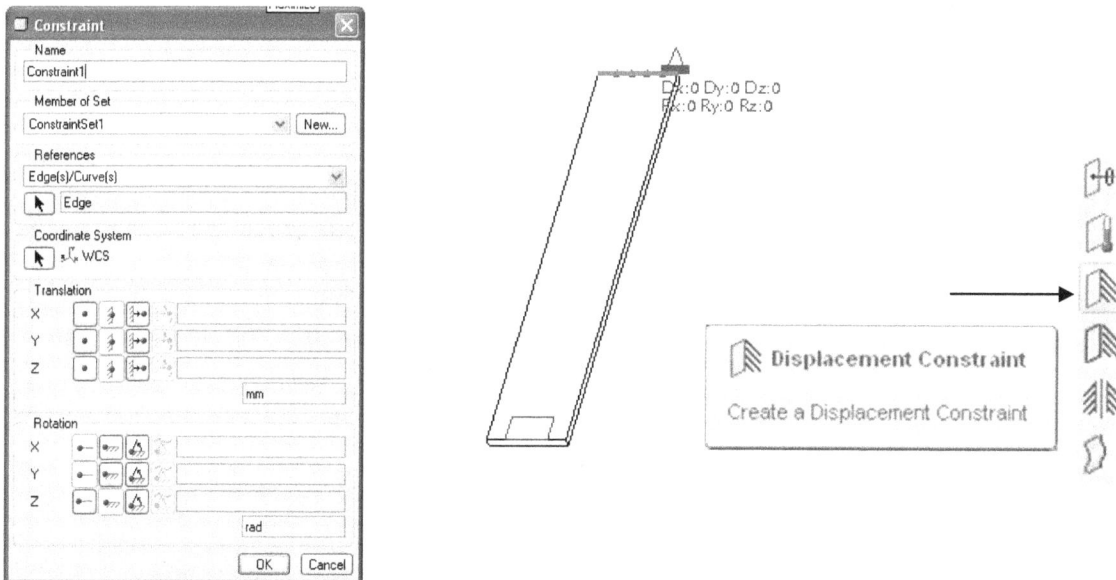

Step 6**: Under the Pro/MECHANICA environment, define the load condition.**
To define a uniform load acting on the surface region in the negative y direction, select the icon of **Force/Moment > Surface** > pick the surface region. Specify -600 in Y of the Force box > **OK**.

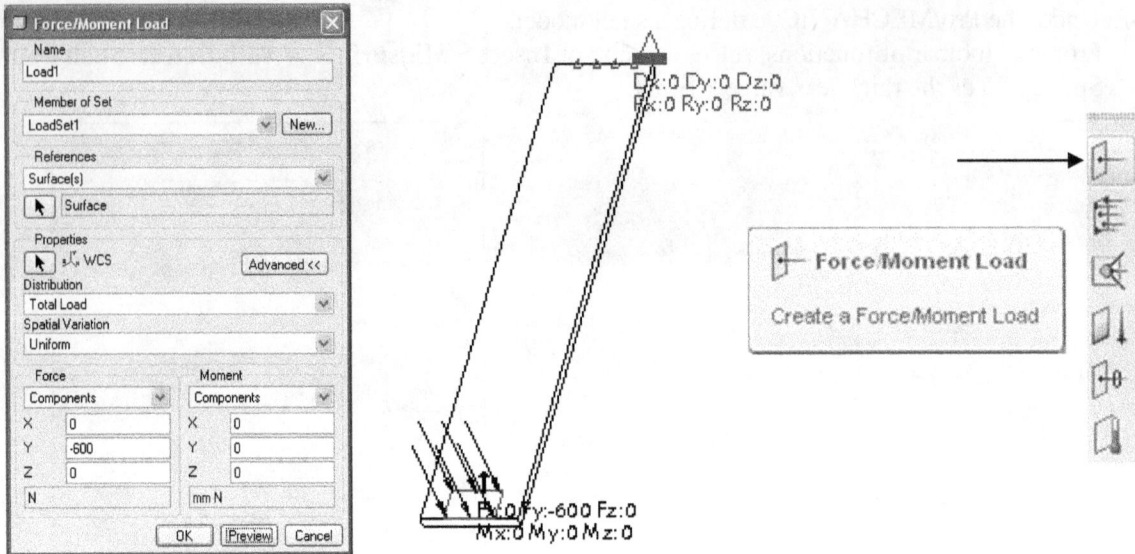

Step 7: Perform a modal analysis and run it.
Click the icon of **Mechanica Analyses/Studies > File >** Select New Modal.

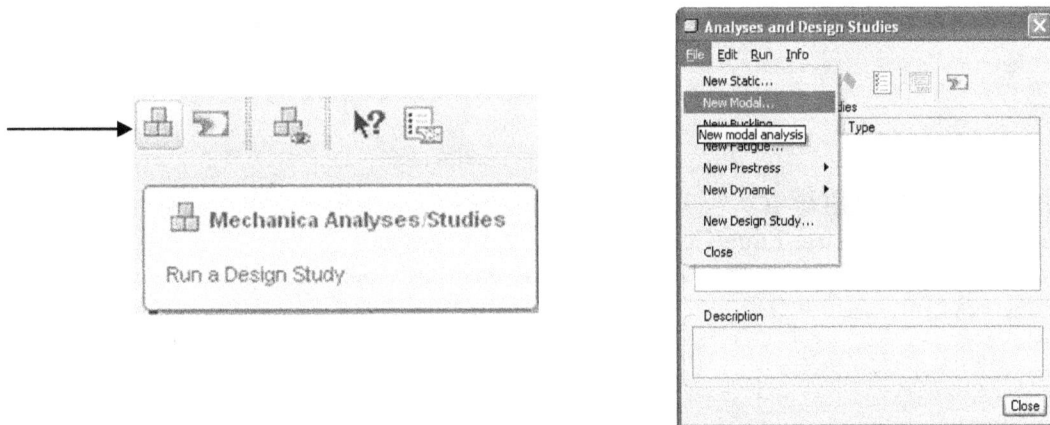

Type the file name: *modal_beam* and run the program.

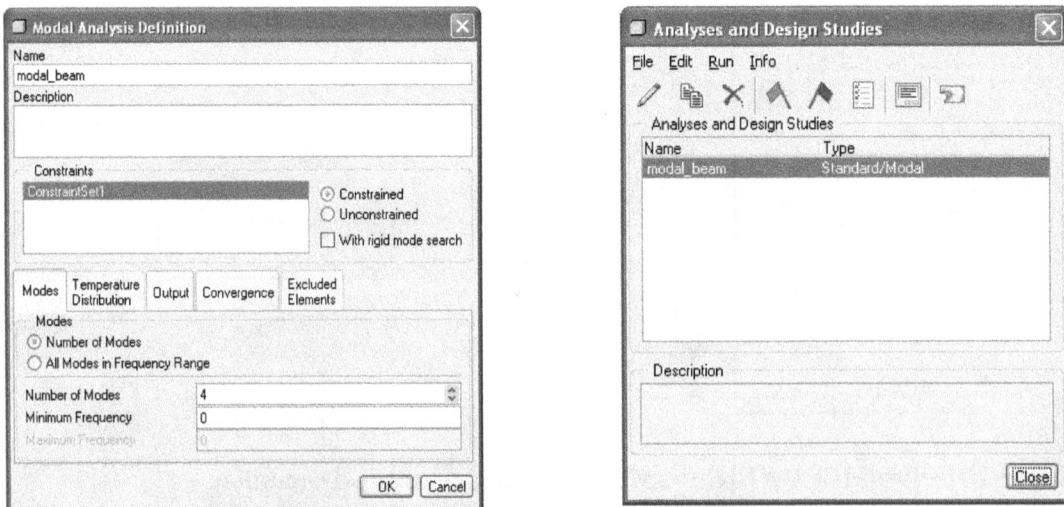

Note that the load condition defined is not shown. Let us run the program.

Step 8: Display the results by showing the shapes of the first 4 modes.

```
Number of Modes: 4

  Mode   Frequency (Hz)
  ----   --------------
    1    1.412859e+00
    2    8.819251e+00
    3    1.352389e+01
    4    1.386584e+01
```

Frequency of Mode 1:	1.41 Hz
Frequency of Mode 2:	8.82 Hz
Frequency of Mode 3:	13.5 Hz
Frequency of Mode 4:	13.9 Hz

The 4 mode shapes are shown below:

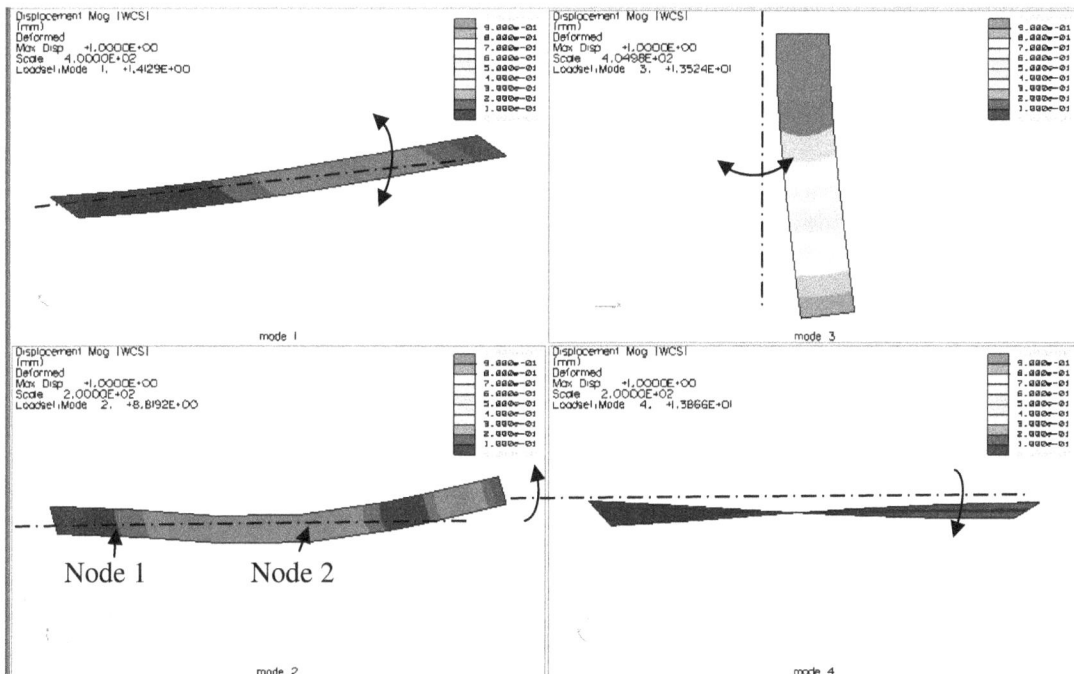

If readers are asked a question: If the load condition is changed from the current 600 Newton to 750 Newton while maintaining its direction downwards, will the first 4 mode shapes change accordingly? To verify your judgment, run the program again with the load condition equal to 750 Newton, instead of

600 Newton. Readers should recognize that the magnitude of the load has no effect on the 4 mode shapes. From the analytical viewpoint, mode shapes are independent of the input format.

If the material type is being changed, will the 4 mode shapes change accordingly? How about the numerical values of the frequencies of the 4 modes? We suggest that readers also try a new run with a material type different from Nylon as we selected. Readers should observe that the 4 mode shapes will change in terms of the frequencies of the 4 new modes. The reason is changing of the material type leads to the change of the system characteristics.

7.3   Applications on Designing a Music Tuning Fork

The following figure illustrates a tuning fork used to tune a piano. A tuning fork is a simple two-pronged metal instrument with a handle and tines that form a U shape. Tuning forks usually made of steel, aluminum, or magnesium-alloy, will vibrate at a set frequency to produce a musical tone when struck and therefore is used for many applications such as piano tuning as well as assessing a person's ability to hear. A tuning fork acts as an example of a simple mechanical system with normal modes of resonance. A tuning fork can also be modeled as two particles of equal mass m moving in the x-y plane under the influence of restoring forces. Tuning forks work by holding the handle and striking the tines with a force great enough for the tines to vibrate at a frequency, resulting in a sound.

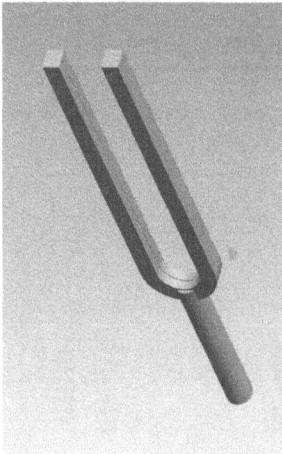

The geometry of a music fork is shown below. Assume that the material is steel. Determine the frequencies of the first 4 modes and the mode shapes of the first 4 modes.

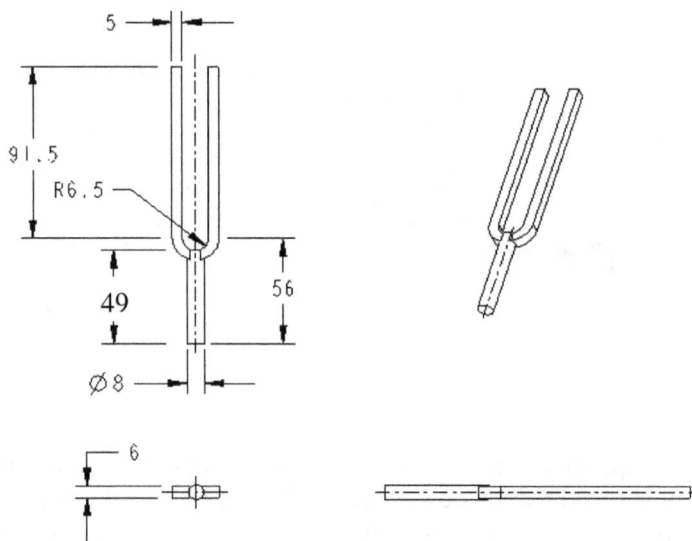

Step 1: Under Pro/ENGINEER, create a 3D solid model of the tuning fork.

   Use **Sweep** to create the U-shape first (create the U-shape datum curve before sweeping).

   From the main toolbar, select **File > New > Part >** type *fork_g* as the name of the file > clear the icon of **Use default template > OK**. Select **mmns_part_solid** (unit system: millimeter, Newton and Second) and type *fork G tune* in **DESCRIPTION** and student in **MODELED_BY > OK**.

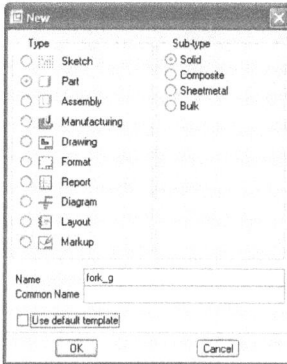

From the main toolbar, select **Insert > Sweep > Protrusion > Sketch Traj**.

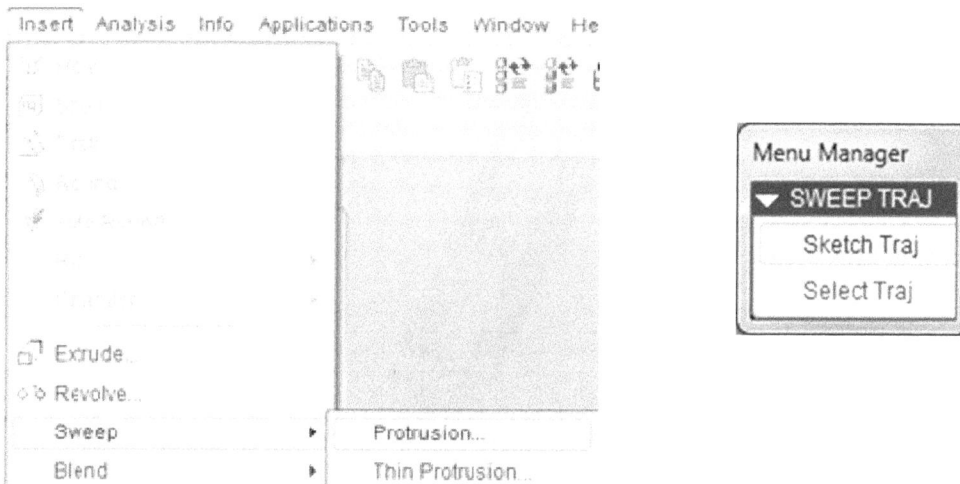

Setup New > Plane > Pick the **FRONT** datum plane as the sketch plane > **Okay** and **Default** for orientation setting.

   Sketch a circle and the diameter is 13, first.

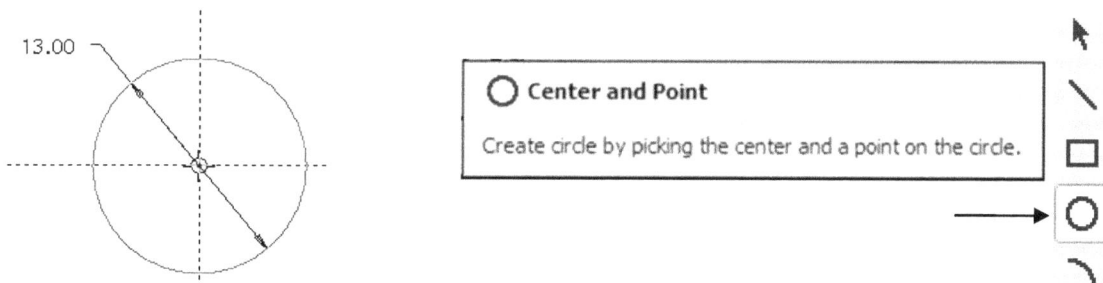

   Sketch 2 lines tangential to the sketched circle. The dimension of length is 91.5, as shown. Afterwards, delete the extra 2 segments, as shown > select the icon of Done from the toolbar of sketcher.

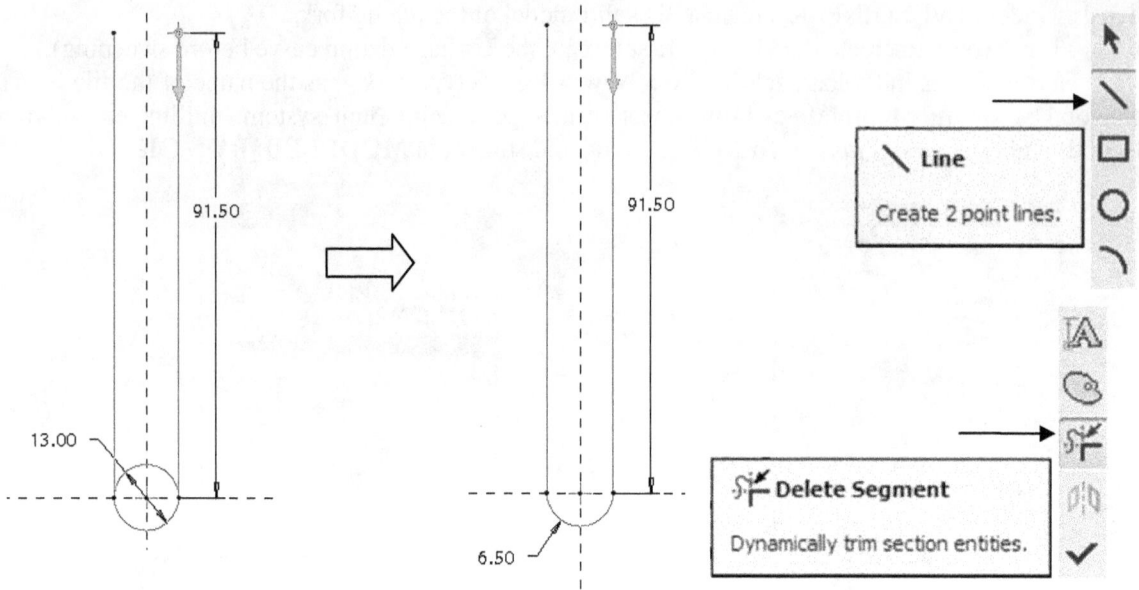

Click the icon of **Rectangle** and sketch a rectangle, and the 2 dimensions are 5 and 6, respectively > select the icon of **Done** > **OK**.

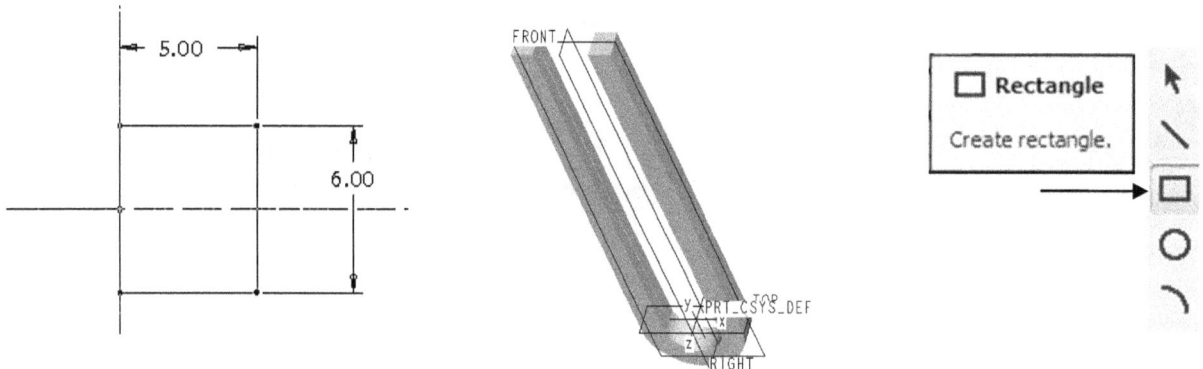

Use **Extrude** to add a cylinder of φ8 to the U-shape. The extrusion starts at a plane which is 56 away from the **TOP** datum plane. The extruding distance is set to 49.

To add this feature, the first step is to create a datum plane. Click the icon of **Datum Plane Tool** > pick the **TOP** datum plane > type *56* as the distance of translation (you may need to type -56 if you need to reverse the direction) > **OK**.

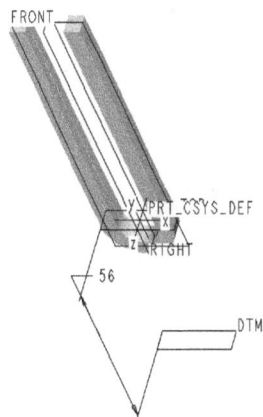

The second step is to use Extrude to create the cylindrical feature. Select the icon of **Extrude** from the toolbar of feature creation > Set the depth value to 49 > **Placement** > **Define**.

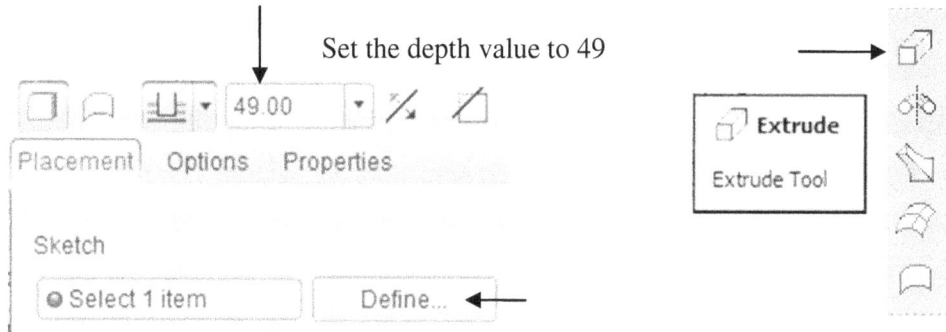

Pick **DTM1** as the sketch plane and accept the default setting to orient the sketch plane. Sketch a circle and the dimension of the diameter is 8 > click the icon of **Done** and click the icon of **Apply and Save**.

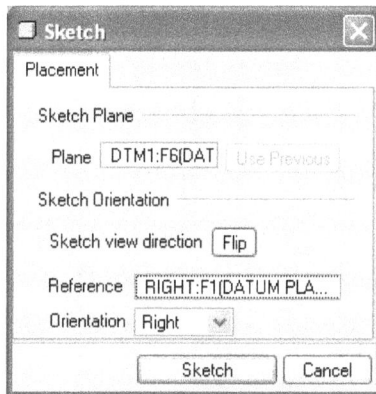

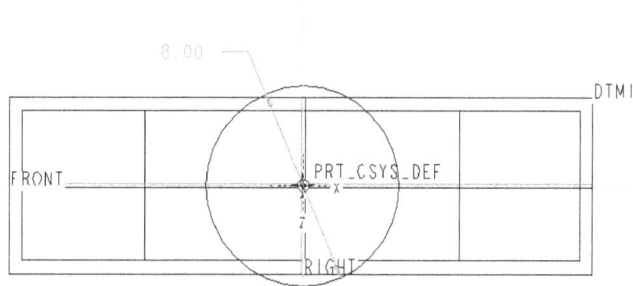

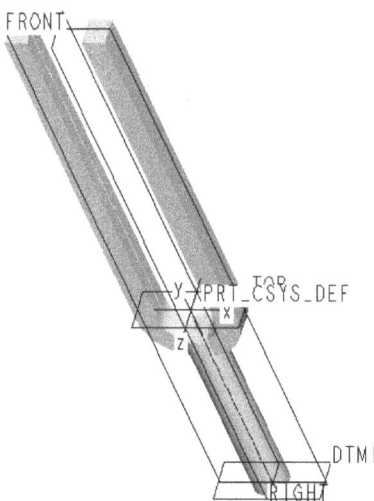

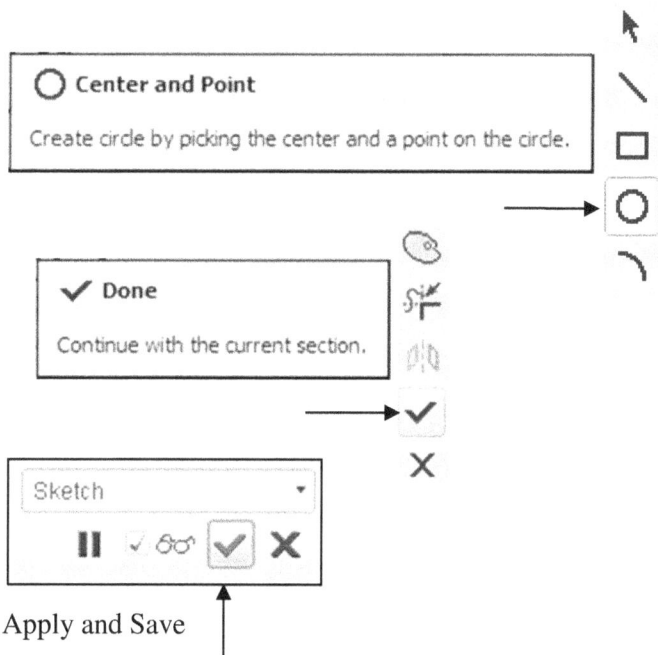

Step 2: Under the Pro/MECHANICA environment, specify the material type. Select STEEL. From the top menu, click **Applications** > **Mechanica** > **Continue** > **Structure** > **OK**.

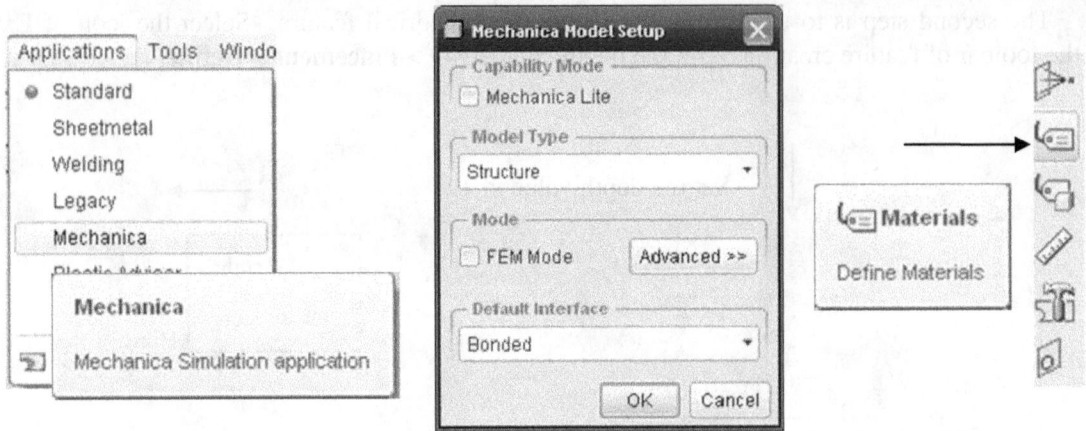

From the toolbar of functions, select the icon of **Materials >** select **Steel** from the left side called Materials in Library > click the directional arrow so that the selected material type goes to the right side called Materials in Model.

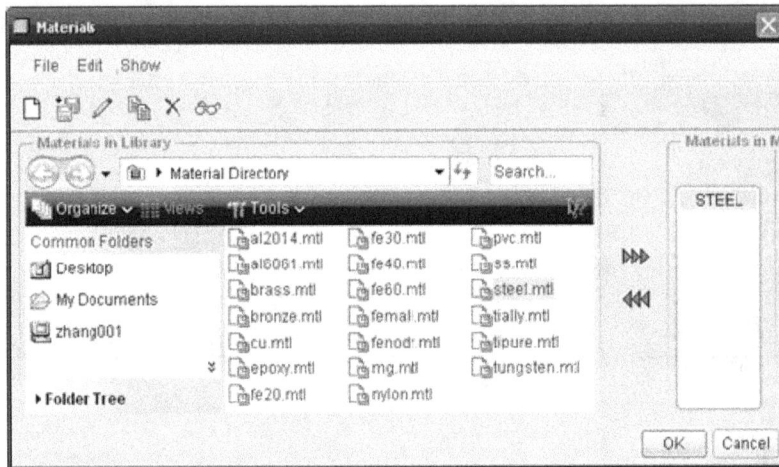

Click the icon of **Material Assignment**, the pop-up window indicates there is one component and the material type is Steel. Just click **OK**.

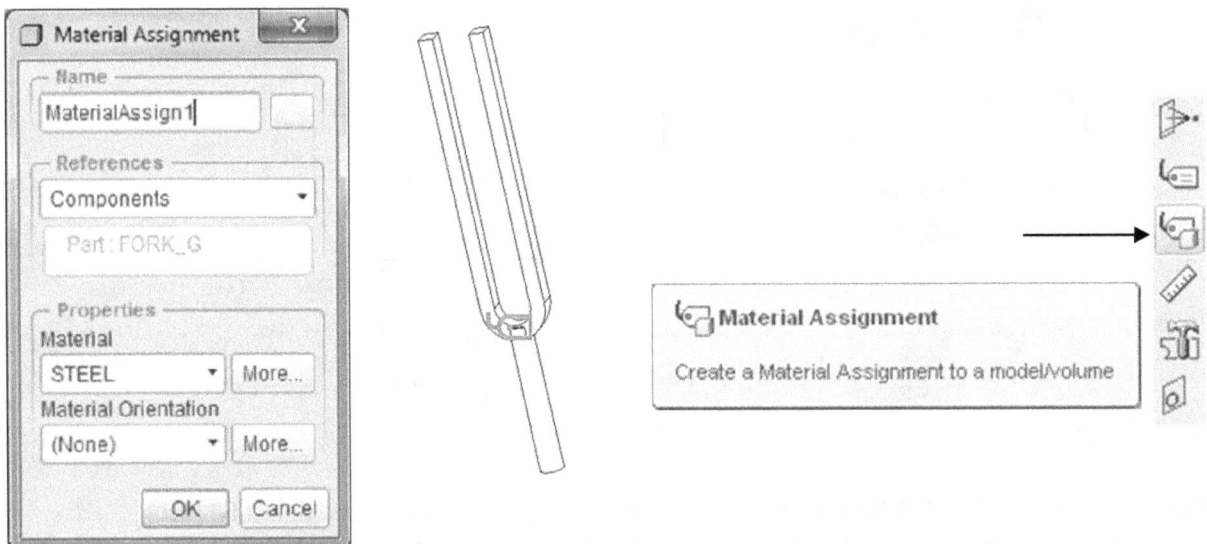

Step 3: Under the Pro/MECHANICA environment, define the constraint condition. The condition is to fully constrain the handle surface of the fork (the cylindrical protrusion).

Click the icon of **Displacement Constraints > Surface >** select the cylindrical surface. Click **OK** to fix the six degrees of freedom.

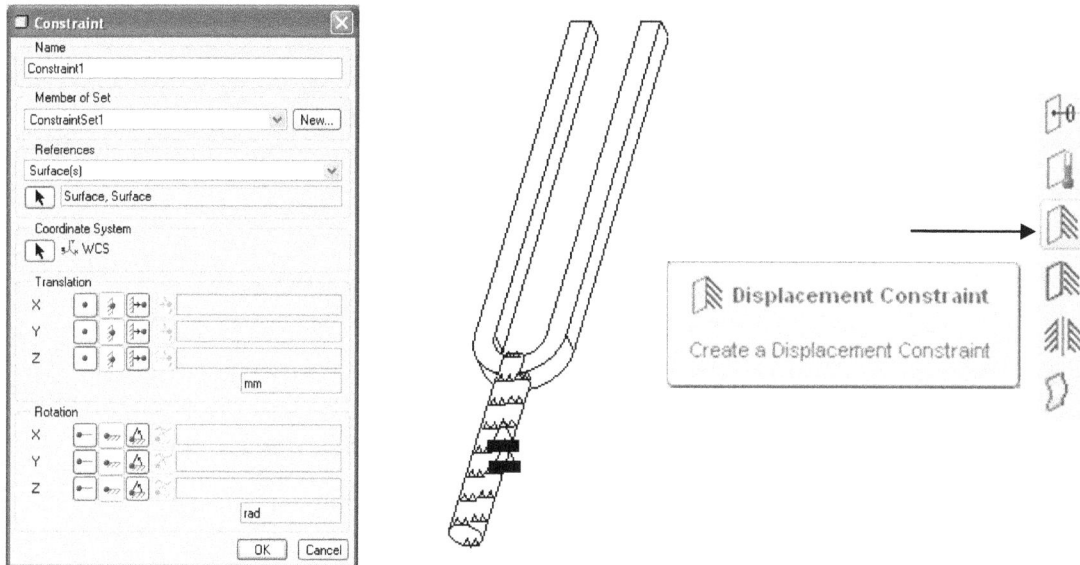

Step 4: Create an analysis and run it.

Note that there is no load defined. It is important to remember that the shapes of individual modes are independent of the direction and magnitude of the exciting force.

Click the icon of **Mechanica Analyses Studies > File >** Select **New Modal**.

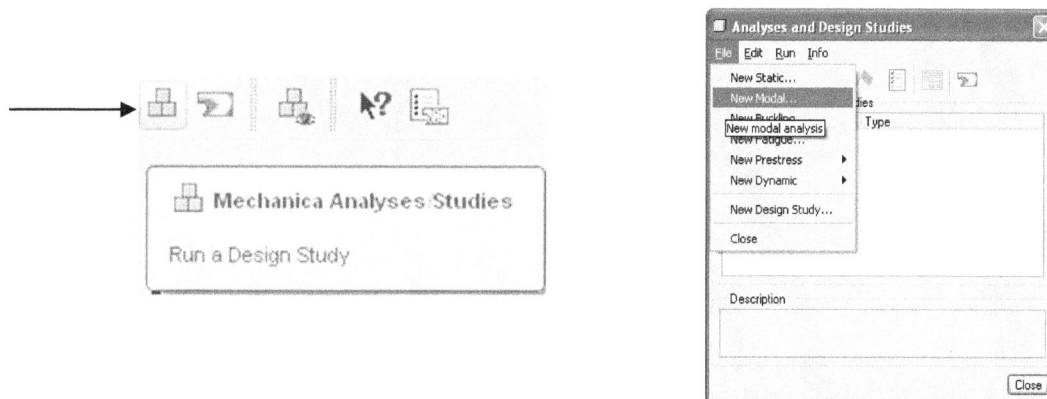

Type the file name: *modal_fork* and run the program.

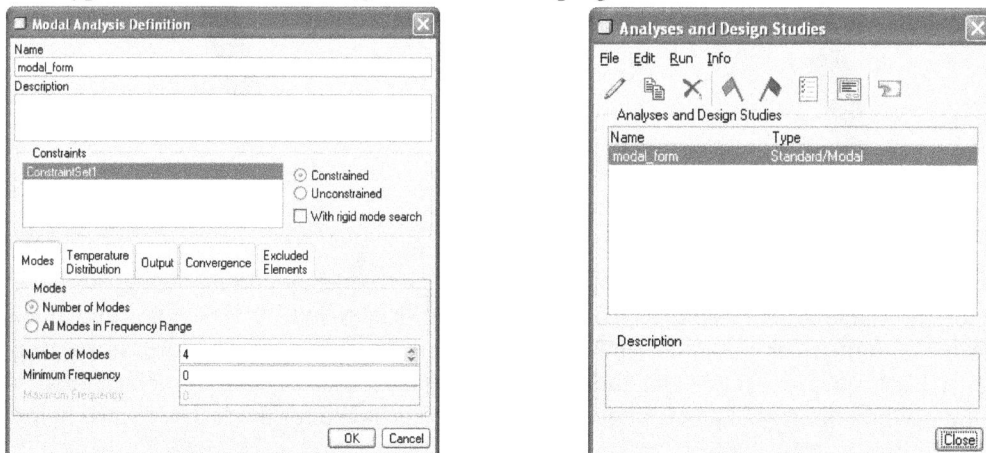

Step 5: Display the results by showing the shape of the first mode through animation.

Note that the frequency of the first mode is 381Hz, which is very close to 384 Hz, the frequency of G-sharp.

.

```
Number of Modes: 4

Mode    Frequency (Hz)
----    --------------
   1    3.810660e+02
   2    3.826385e+02
   3    4.405395e+02
   4    4.418484e+02
```

Readers may search this information on the web. Two (2) of the web addresses are

http://www.indigo.com/tuning/tuning.html.
http://mathforum.org/library/drmath/view/52470.html.

Let us examine the first mode shape for the G note fork, for a frequency of 381 Hz is illustrated below.

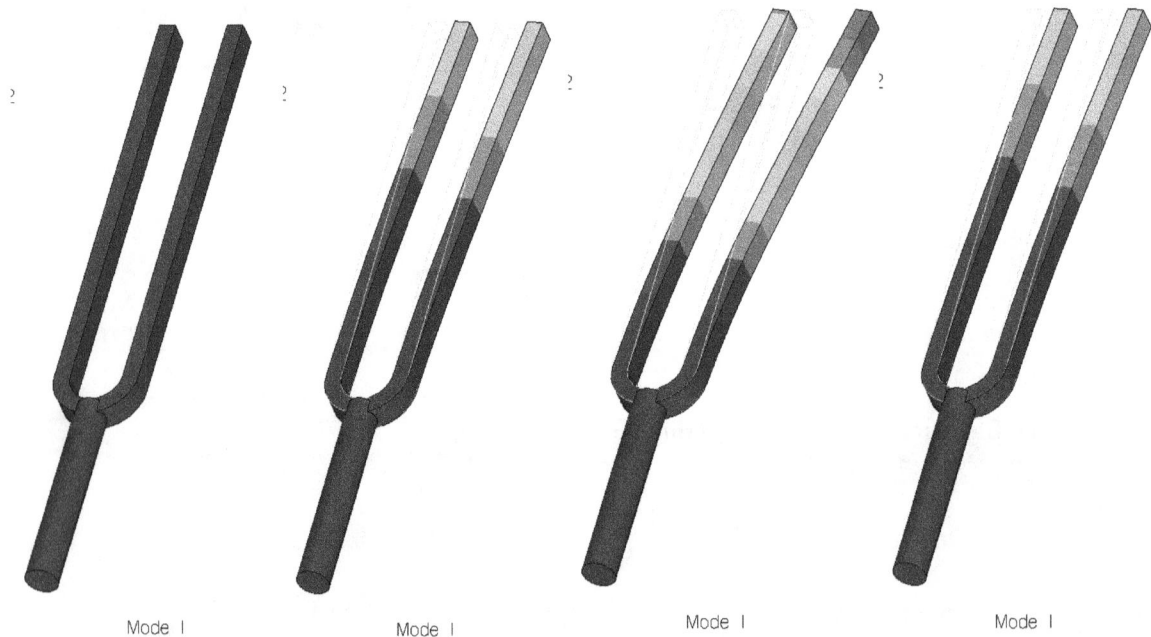

Mode 1   Mode 1   Mode 1   Mode 1

The first mode is the dominant mode of fork vibration. The frequency corresponding to this mode is the frequency of the musical note denoted G. Tuning a piano is based on this note, and thus this frequency. The way the tuning fork is excited directly corresponds to the harmonics heard. Impacting one of the tines with a hard surface excites the tuning fork. The direction of impact determines which harmonics, or modes, will be present. The excitation force and mode shape for mode 1 is shown below:

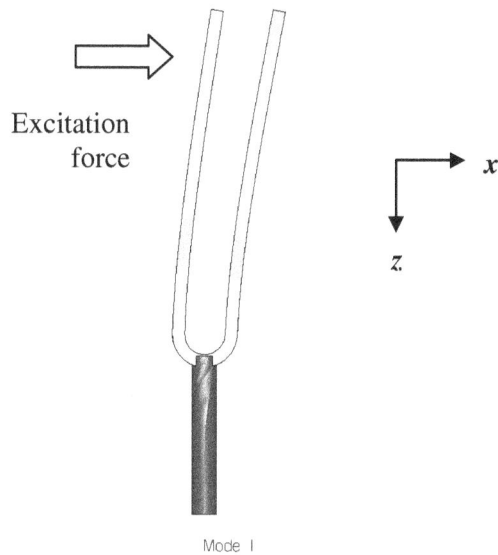

Mode 1

This figure shows that the excitation force must be along the same direction as the mode motion. For mode 1, the excitation force should therefore be along the x-direction.

For mode 2, the tines move in the x-z plane in directions either towards, or away from each other. Two opposite forces oriented along the x-direction excite mode 2. This phenomenon is illustrated below. So, by striking only one tine of the tuning fork along the x-direction, the first mode will be excited without presence of the second mode.

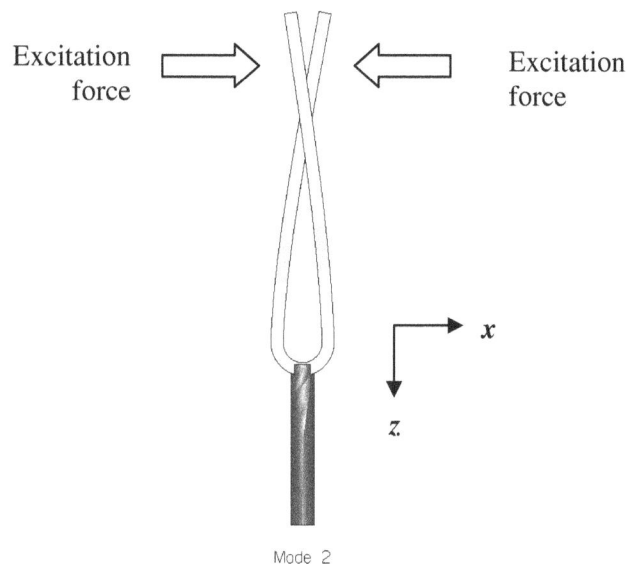

Mode 2

Mode 3 consists of motion in the y-z plane along the y-direction. To excite this mode, an excitation force must strike one of the tines along the y-direction, as shown. Again, by striking a tine in the x-direction, only mode 1 will be excited. So in summary, by striking one of the tuning fork tines along the x-direction, mode 1 will be the dominant mode of oscillation, which causes the mode 1 frequency to be the dominant tone. While the higher harmonics of modes 2 and 3 will still be present, by striking the fork in the above manner, their roles in the tone will be reduced.

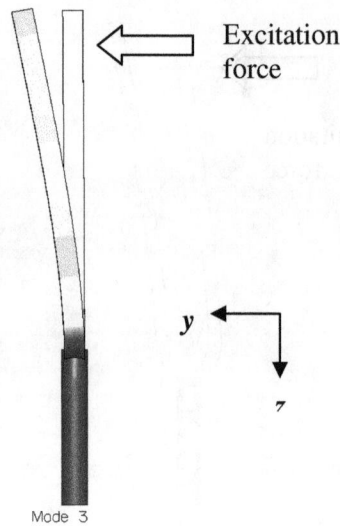

Mode 3

In the following, we use the Design Study to search for the length to optimize the frequency setting. First, we search the length value, which is corresponding to G-sharp (384 Hz). Afterwards, we search for the length value set for middle-C (256 Hz), and the length value set for the note, E (320 Hz), assuming that the 3D solid model is available.

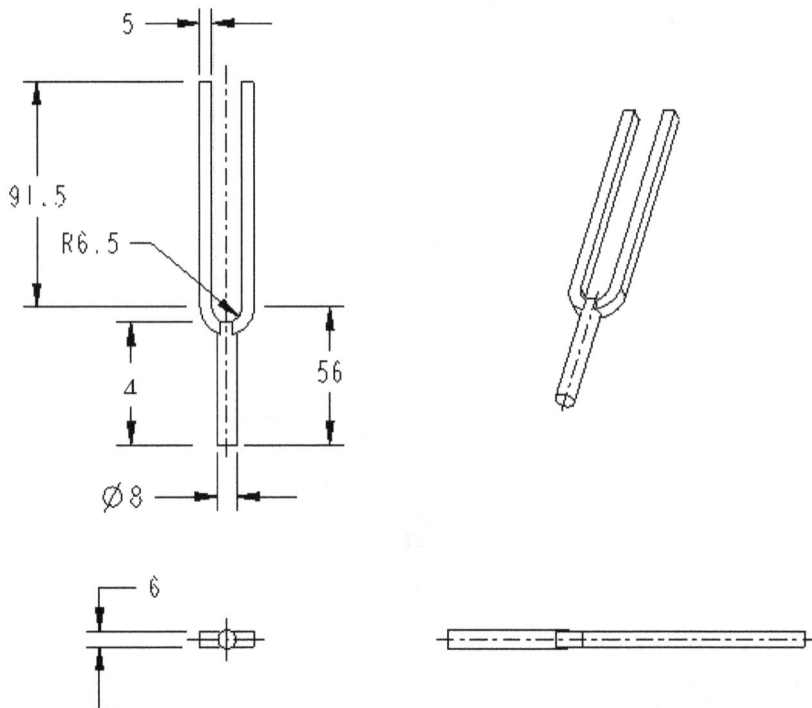

Step 1: Define a design parameter and set an arrange of variation

Click the icon of **Mechanica Analyses/Studies > File > New Optimization Design Study** > type *opt_search_for_G_sharp* as the name of the analysis folder.

Make sure that **Optimization** is selected under **Type**. Accept the default setting for the goal function or do not make changes to the Goal section.

Under **Design Limits**, set the model frequency equal to 384 (1/sec or Hz).

Under **Measures**, click the selection box and select modal_frequency > **OK**.

Click the box of select the dimension from model > pick the length dimension or 91.5, as shown > **OK.** Accept the default setting with the current value as 91.5, minimal value as 90 and maximum value as 100 mm.

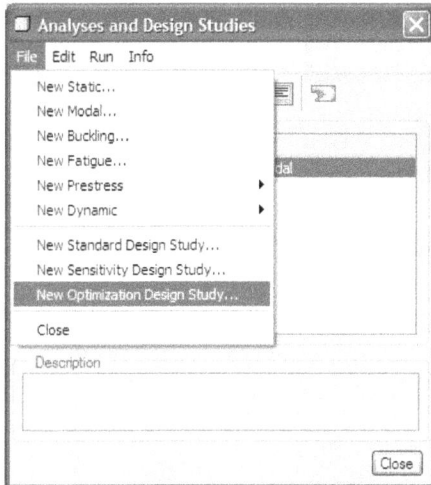

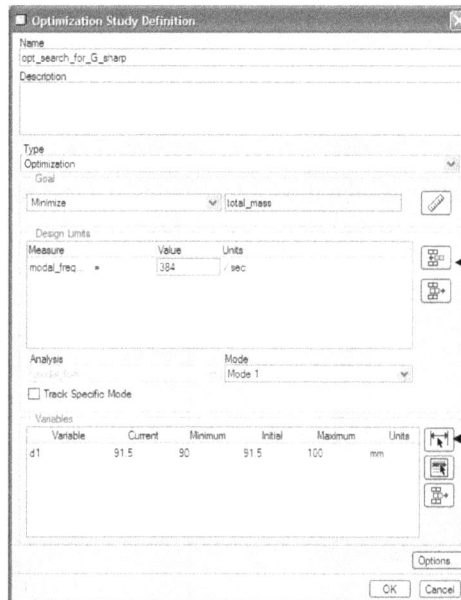

Pick the dimension equal to 91.5

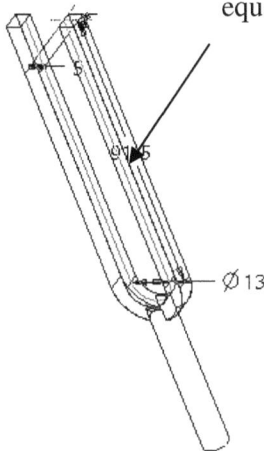

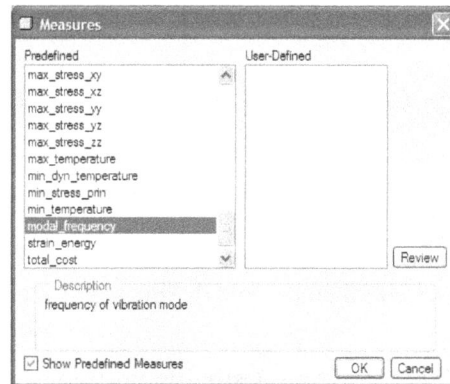

Click the box called **Options**. In the Design Study Option window, change the Optimization Convergence value from 1% to 0.1%. Check the box called Repeat P-Loop Convergence and check the box called Remesh after each shape update > **Close**. A question appears on screen to ask the user if the convergence value set to 0.1% is necessary. Click **Yes**.

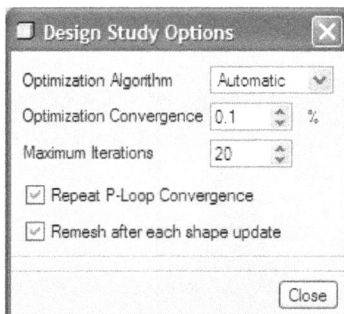

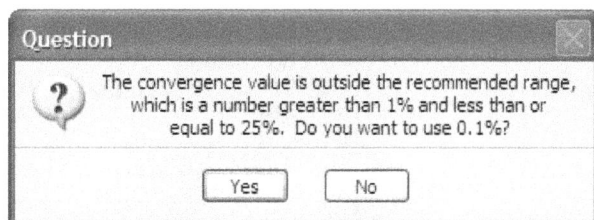

In the Optimization Study Definition window, click **OK** and click the box of **Run.**
There is always a message displayed on screen. The message is asking the user "Do you want to run interactive diagnostics?" Click **Yes.** In this way a warning message will appear if an abnormal operation occurs.

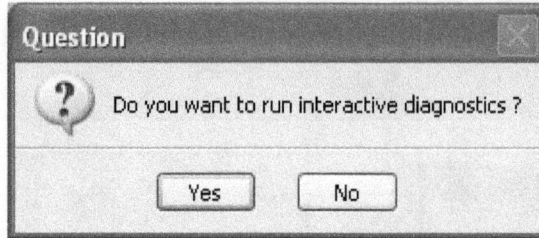

To monitor the computing process, users may click the icon of **Display study status**. A violated constraint condition on the modal frequency is indicated before starting the search.

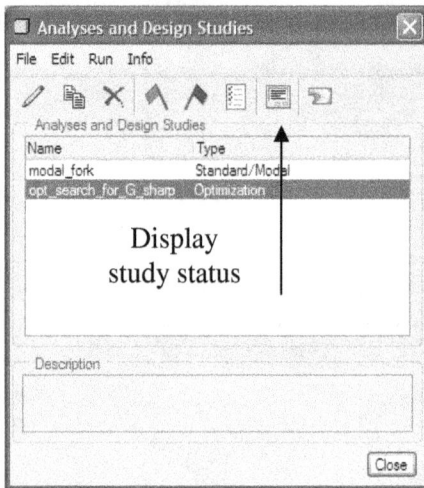

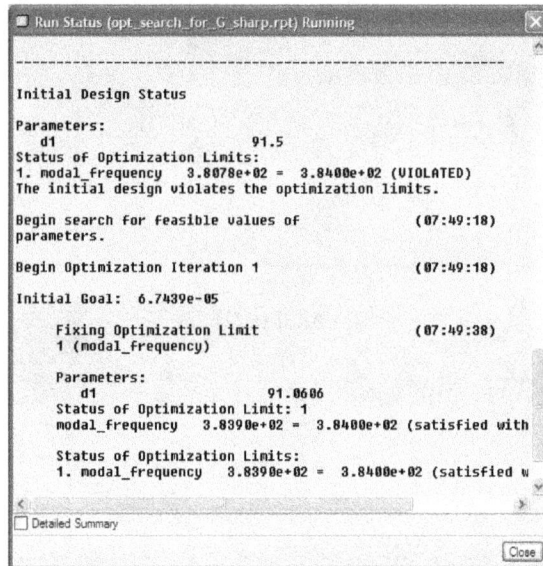

The search is very effective because the result obtained from the first iteration indicates the modal frequency is 383.9 Hz when the length dimension is changed from 91.5 mm to 91.06 mm.

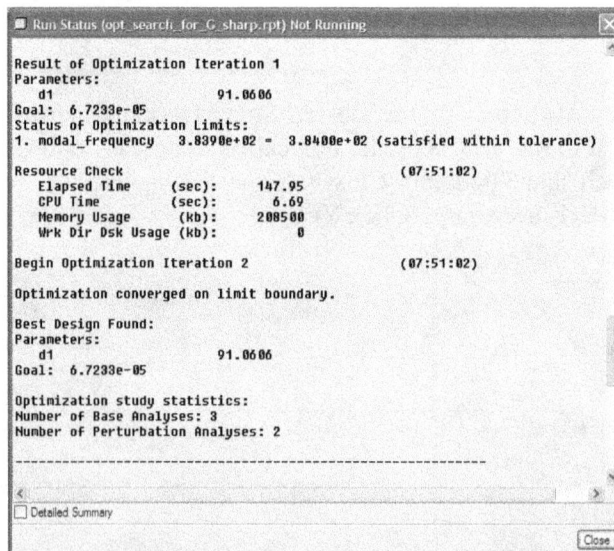

The search is also completed because the convergence criterion is set at 0.1% level.

Users are also able to review the search process step by step to better understand the search process to set the length value to 91.16 mm, which is so-called optimal value. To do so, from the window of Analyses and Design Studies, click **Info > Optimize Hist.**

When being asked to review the next shape for the first time, the length dimension is set to 91.5, as shown. Click **Yes**.

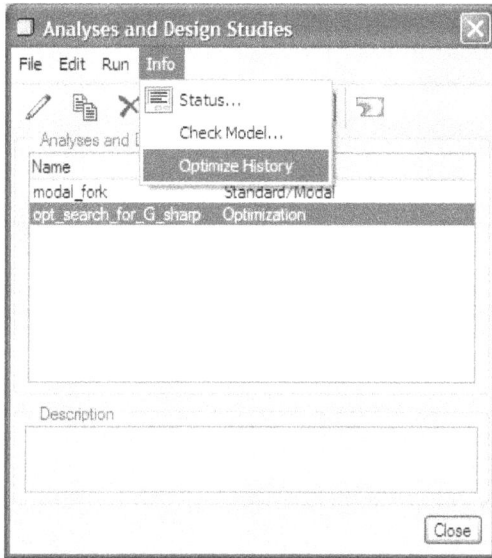

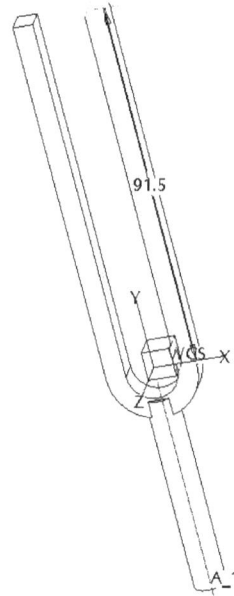

Do you want to review the next shape? ▢   Yes

When being asked the same question for the second time, click **No** (only one iteration to reach the optimal in this case). When being asked to leave the model at the optimized shape, click **Yes**. The dimension 91.06 is displayed on screen.

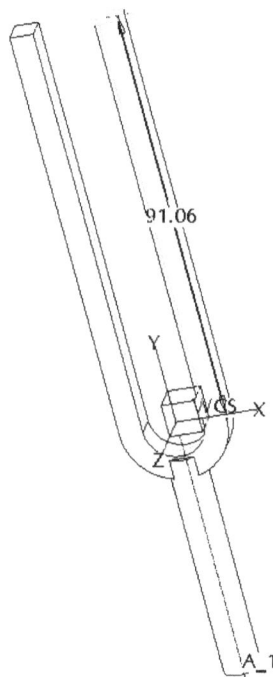

Do you want to review the next shape? ▢

No

Leave the model at the optimized shape? ▢

Yes

Step 3: We repeat the procedure described in Step 2 to search the length value set for middle C (256 Hz). In the window of Analyses and Design Studies, highlight opt-search_for_G_sharp, and right click > **Copy**.

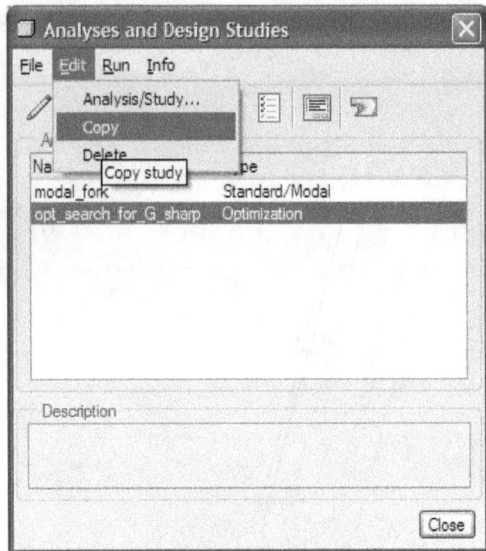

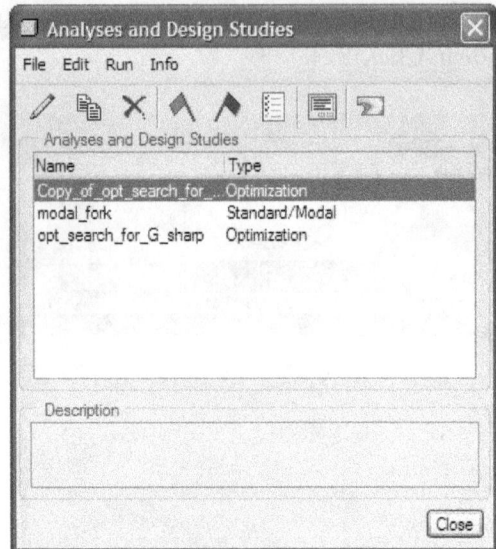

While the Copy of opt_search_for_G_sharp is highlighted, right click and select **Edit**. In the window of **Optimization Study Definition**, change the name to *opt_search_for_middle_C*. Change the modal frequency value to *256*. Reset the minimum limit and maximum limit to *80* and *120*, respectively > **OK**.

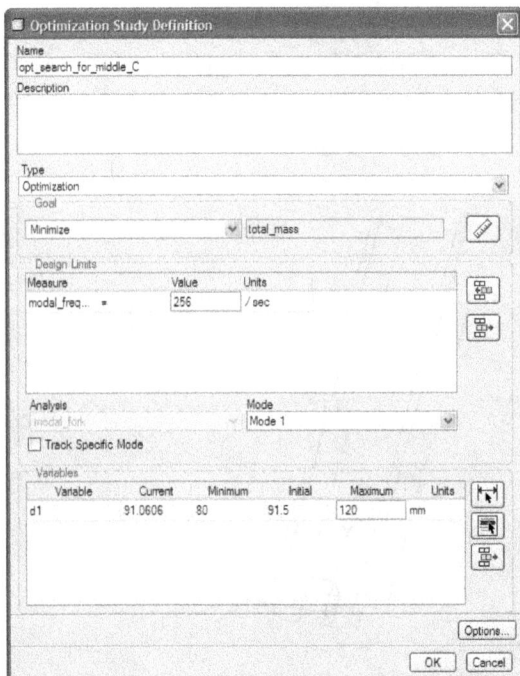

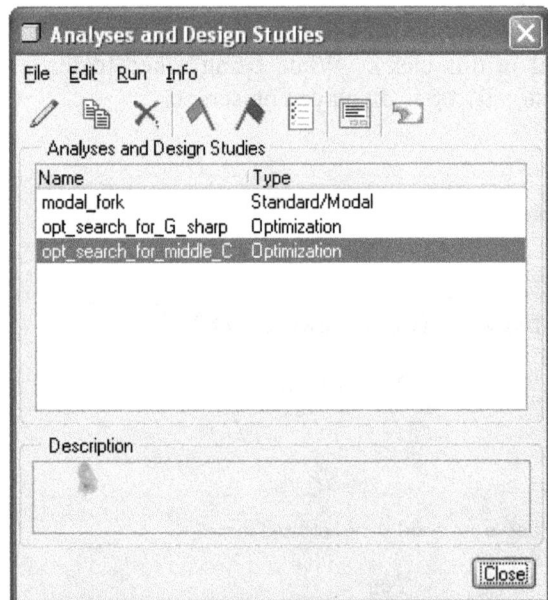

Now we start to run the design study called opt_search_for_middle_C. Click the Green Flag or the **Run** icon.

```
Parameters:
    d1                    91.5
Status of Optimization Limits:
1. modal_frequency   3.8078e+02 =  2.5600e+02 (VIOLATED)
The initial design violates the optimization limits.

Begin search for feasible values of          (10:26:38)
parameters.

Begin Optimization Iteration 1               (10:26:38)

Initial Goal:  6.7439e-05

    Fixing Optimization Limit                (10:26:58)
    1 (modal_frequency)

    Parameters:
        d1                108.591
    Status of Optimization Limit: 1
    modal_frequency   2.8081e+02 =  2.5600e+02 (VIOLATED)

    Parameters:
        d1                113.952
    Status of Optimization Limit: 1
    modal_frequency   2.5734e+02 =  2.5600e+02 (VIOLATED)

    Parameters:
        d1                114.282
```

```
    d1                114.282
Status of Optimization Limit: 1
modal_frequency   2.5640e+02 =  2.5600e+02 (VIOLATED)

Parameters:
    d1                114.382
Status of Optimization Limit: 1
modal_frequency   2.5539e+02 =  2.5600e+02 (VIOLATED)

Moving to limit boundary.

Parameters:
    d1                114.334
Status of Optimization Limit: 1
modal_frequency   2.5563e+02 =  2.5600e+02 (VIOLATED)

Parameters:
    d1                114.318
Status of Optimization Limit: 1
modal_frequency   2.5631e+02 =  2.5600e+02 (VIOLATED)

Status of Optimization Limits:
1. modal_frequency   2.5631e+02 =  2.5600e+02 (VIOLATED)

Fixing Optimization Limit                (10:30:30)
1 (modal_frequency)

Parameters:
```

```
Parameters:
    d1                114.396
Status of Optimization Limit: 1
modal_frequency   2.5518e+02 =  2.5600e+02 (VIOLATED)

Moving to limit boundary.

Parameters:
    d1                114.348
Status of Optimization Limit: 1
modal_frequency   2.5543e+02 =  2.5600e+02 (VIOLATED)

Status of Optimization Limits:
1. modal_frequency   2.5543e+02 =  2.5600e+02 (VIOLATED)

Fixing Optimization Limit                (10:31:35)
1 (modal_frequency)

Parameters:
    d1                114.205
Status of Optimization Limit: 1
modal_frequency   2.5599e+02 =  2.5600e+02 (satisfied with

Status of Optimization Limits:
1. modal_frequency   2.5599e+02 =  2.5600e+02 (satisfied w

The parameter values are now feasible.
```

```
        The parameter values are now feasible.

Result of Optimization Iteration 1
Parameters:
    d1                114.205
Goal: 7.8102e-05
Status of Optimization Limits:
1. modal_frequency   2.5599e+02 =  2.5600e+02 (satisfied within

Resource Check                          (10:32:59)
    Elapsed Time    (sec):    425.18
    CPU Time        (sec):     25.91
    Memory Usage    (kb):    220020
    Wrk Dir Dsk Usage (kb):       0

Begin Optimization Iteration 2          (10:32:59)

Optimization converged on limit boundary.

Best Design Found:
Parameters:
    d1                114.205
Goal: 7.8102e-05

Optimization study statistics:
Number of Base Analyses: 11
Number of Perturbation Analyses: 7
```

The report indicates that the length value should be changed from 91.16 mm to 114.205 mm. Accordingly, the frequency of the first mode is changed from 384 Hz to 256 Hz.

To review the optimal history, from the window of Analyses and Design Studies, click **Info > Optimize Hist.**

When being asked to review the next shape for the first time, click **Yes.** When being asked the same question for the second time, click No (only one iteration to reach the optimal in this case). When being asked to leave the model at the optimized shape, click **Yes**. The dimension 114.21 is displayed on screen.

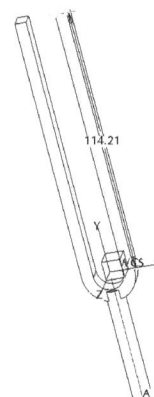

Step 4: We repeat the procedure described in Step 3 to search the length value set for the note, E (300 Hz).

In the window of Analyses and Design Studies, highlight opt-search_for_middle_C, and right click > **Copy**.

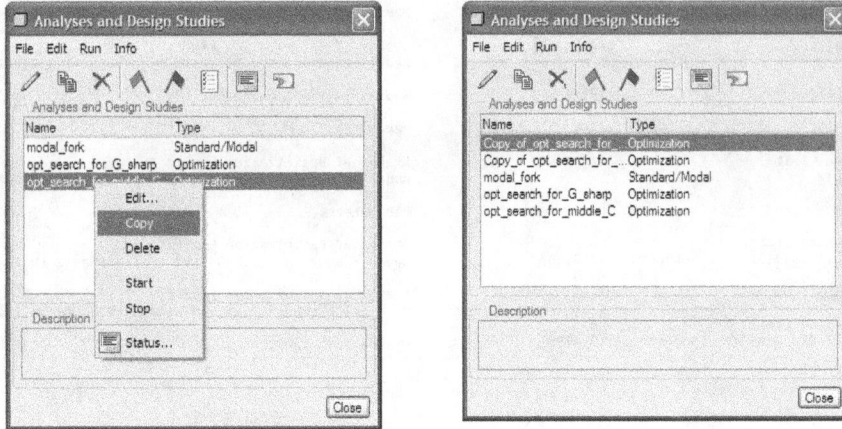

While the Copy of opt_search_for_middle_C is highlighted, right click and select **Edit**. In the window of **Optimization Study Definition**, change the name to *opt_search_for_E*. Change the modal frequency value to *300*. There is no need to reset the minimum limit and maximum limit, which are *80* and *120*, respectively, because we know that the variation range (80 ~ 120 mm) is sufficiently to cover the search space needed, we start to run a new design study.

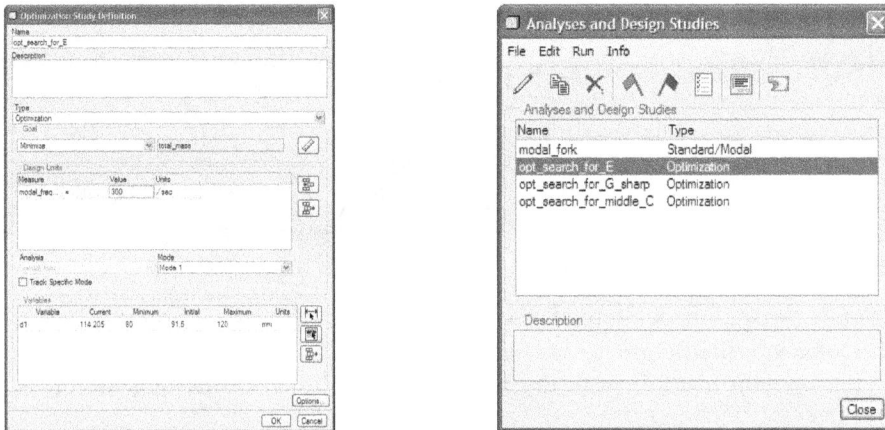

Now we start to run the design study called opt_search_for_E.

The report indicates that the length value should be changed from 91.5 mm to 104.607 mm. Accordingly, the frequency of the first mode is changed from 384 Hz to 300 Hz.

To review the optimal history, from the window of Analyses and Design Studies, click **Info > Optimize Hist.** When being asked to leave the model at the optimized shape, click **Yes**. The dimension 104.61 is displayed on screen.

Do you want to review the next shape?

Yes

Do you want to review the next shape?

No

Leave the model at the optimized shape?

Yes

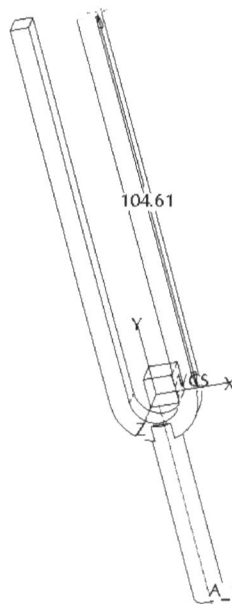

7.4   Applications on Buckling Analyses

Structures fail in a variety of ways, depending on their geometrical shapes, types of material, kinds of loads, and boundary conditions. Structures made of ductile materials, such as metals, may stretch and/or bend excessively when subjected to severe and unexpected large loads. The locations of failure could be those sections having small areas or with sharp corners where the maximum stress developed exceeds the strength limit. For brittle materials, cracking is a major concern causing structural failures. When cycling or repeated loads are present, fatigue is the form of failure.

The following figure shows a column. The lower end is a fixed end and the upper end is a free end. The dimension of its length is L, and the dimension of its section is $a \times a$. When the ratio of L to a is larger than 10, the column is viewed as a slender column. When such a slender column, or such a slender bar, is subjected to a compressive load, such as P, the column could fail if it bends and deflects laterally.

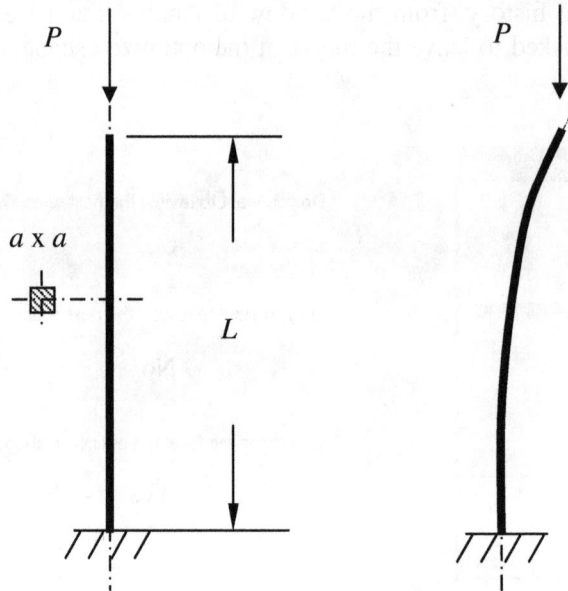

This type of failure is called buckling. Under an increasing axial load, the lateral deflection increases too, and eventually the column collapses completely. Have you seen this kind of failure in your daily life? For example, a tornado storm hits an area, hundreds of houses collapse under the heavy blow exerted by the tornado. Buckling is usually one of the major failure modes.

Evaluation of the load, which leads to a buckling failure mode, is a technical challenge because there is a significant amount of factors related to such an evaluation. For example, variation of the boundary condition, or the forms of the supports at both ends, can lead to a different mode shape (the first mode shape) when buckling occurs.

The following figure illustrates three types of the support form, resulting three different mode shapes at the onset of buckling. As a result, the critical load leading to buckling could be significantly different from one another.

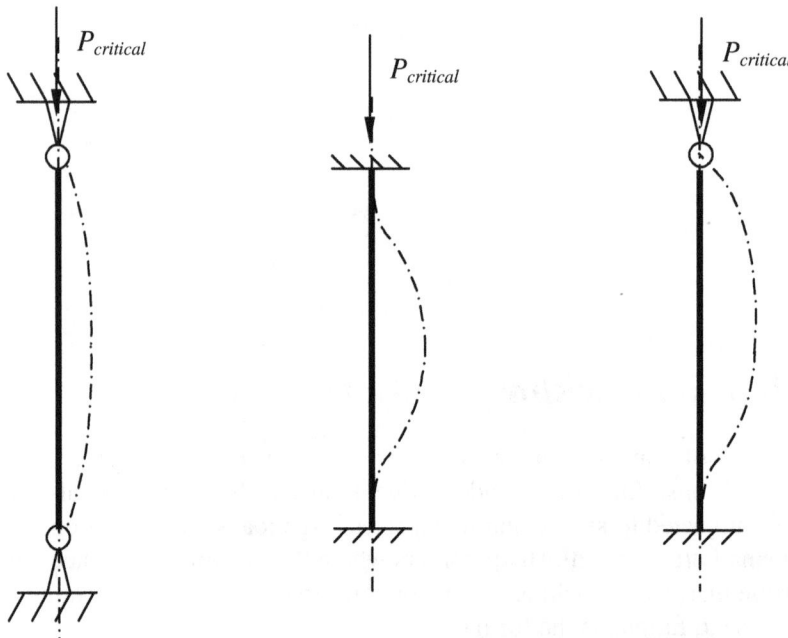

To understand the fundamentals behind buckling, we consider an idealized structure shown in the following figure. The slender bar is a rigid body that is pinned at the base and supported by an elastic spring at the top. The stiffness of the spring, or the spring constant is K.

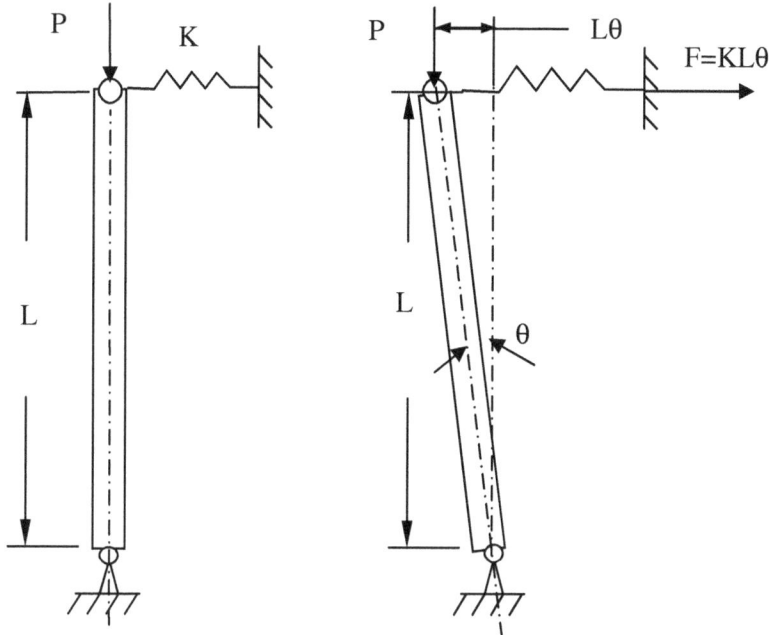

Assume that the slender bar is subjected to a load, P, which is perfectly aligned with its axis in the vertical direction. Assume that there is no stretch of the spring at its initial stage so that the slender bar is at its equilibrium position. Note that this equilibrium is unstable. This means that a small disturbance from the lateral side could lead the slender bar move away from its equilibrium position.

The analysis of this slender bar – spring system starts from giving a small rotation of the slender bar with respect to its axis, simulating a circumstance where an external force of disturbance causes such a small rotation of the slender bar. Because the small rotation gives rise to the presence of a distance, which is equal to $L*\theta$, it induces two moments acting on the slender bar:

The first moment is due to the acting compression force, P. The moment is called overturning moment, and it is given by

$$Overturning\ Moment = P*(L*\theta)$$

The second moment is due to the restoring force exerted by the spring. The moment is called restoring moment, and is given by

$$Restoring\ Moment = K*(L*\theta)*L$$

If Overturning Moment > Restoring Moment, the slender bar – spring system is unstable.
If Overturning Moment < Restoring Moment, the slender bar – spring system is stable.
If Overturning Moment = Restoring Moment (it is called a bifurcation point), the slender bar – spring system is in its neutral equilibrium, or at its bifurcation point. Under such a circumstance, the critical load for the buckling analysis can be determined by equating the two moment equations:

$$P_{critical}*(L*\theta) = K*(L*\theta)*L$$
$$P_{critical} = K*L$$

Therefore, $P < P_{critical} = K*L$, the system is stable, and $P > P_{critical} = K*L$, the system is unstable. To quantify the stability condition, a new parameter called Buckling Load Factor (BLF) is introduced. It is defined by a ratio of the applied force to the critical load:

$$BLF = \frac{P_{critical}}{P_{applied}}$$

Under the Pro/MECHANICA environment, a buckling load factor is calculated for each of the buckling mode shapes. Therefore, the buckling analysis is a linear eigenvalue bifurcation instability analysis. It is important to note that before defining a buckling analysis, a static analysis must be defined before hand. In the static analysis, Pro/MECHANICA calculates the stress stiffening of the 3D solid model due to the applied forces. In the buckling analysis, Pro/MECHANICA calculates the model's elastic stiffness due to geometry and material properties. Afterwards, Pro/MECHANICA uses these two solutions to calculate the BLF.

In the following figure, a slender column is shown. The length of the column is 4000 mm. The cross-section is a rectangle. The two dimensions of the rectangle are 75 and 50, respectively. Assume that one end of the column is fixed to the ground, as shown. The loading condition is uniform load acting at the other end. The magnitude of the load is 5000000 Newton.

Now let us determine the numerical value of the critical load, *Pcritical*, under which buckling failure occurs for the slender bar shown below. The left end of the slender bar is the fixed end. The right end is free to move. The compression load is acting on the center of the hole with the diameter equal to 500000 mm. The material of the slender column is steel.

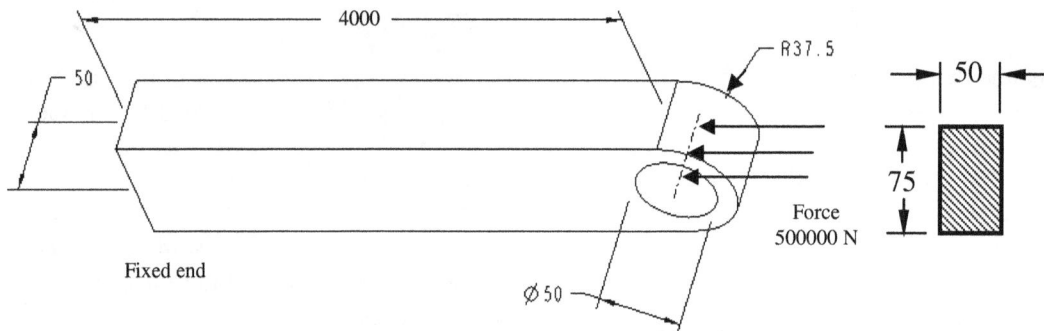

Step 1: create a 3D solid model
   File > New > type in *slender_column* as the file name and select the mmNs as the unit system.

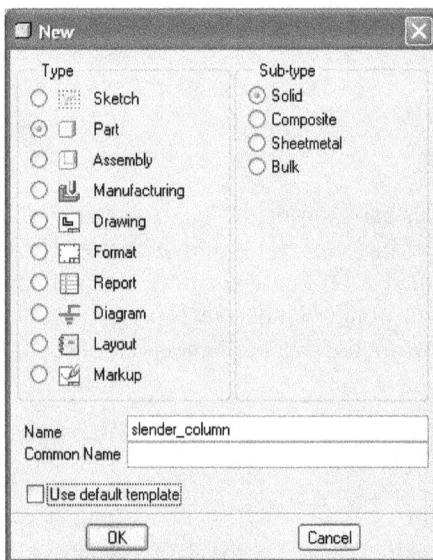

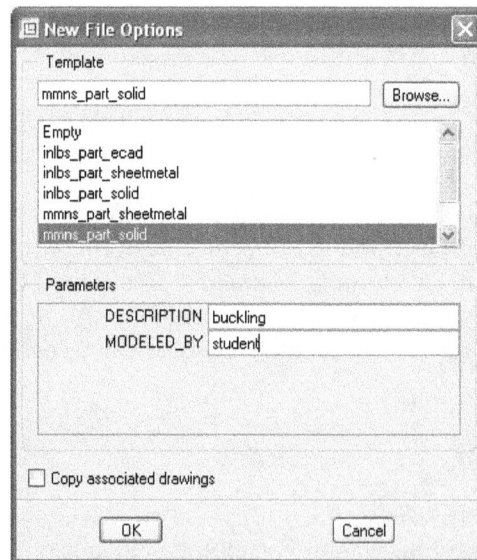

From the toolbar of feature creation, select the icon of **Extrude.**
The dashboard appears. Select the icon of symmetry, set the depth value to *50*. Click the box of **Placement > Define**.

Symmetry   Depth value = 50

Placement   Options   Properties

Sketch

⊙ Select 1 Item   Define...

Extrude
Extrude Tool

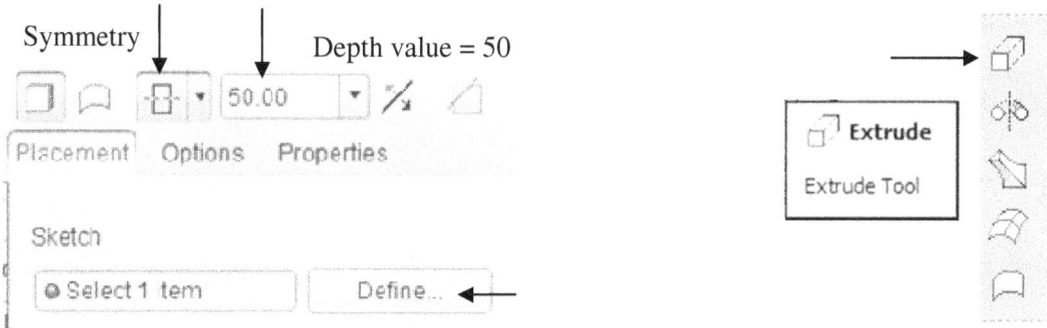

Select the **FRONT** datum plane as the sketch plane, and click the box of Sketch to accept the **RIGHT** datum plane as the default reference to orient the sketch plane > **Sketch**.

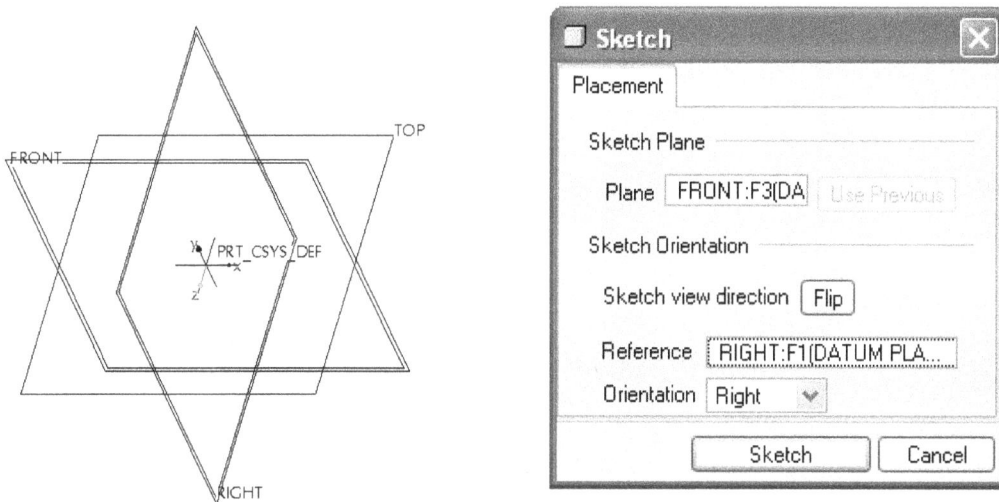

Sketch

Placement

Sketch Plane

Plane   FRONT:F3[DA   Use Previous

Sketch Orientation

Sketch view direction   Flip

Reference   RIGHT:F1[DATUM PLA...
Orientation   Right

Sketch   Cancel

Click the icon of **Circle** and sketch a circle. Click the icon of **Delete** and delete the half circle on the left side. Specify 37.5 as the radius value.

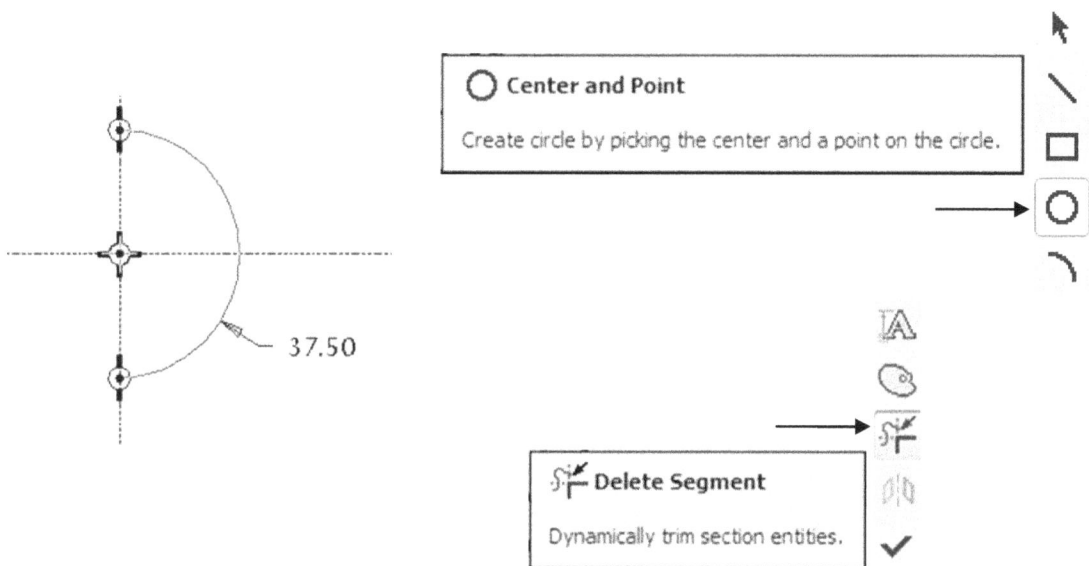

37.50

◯ Center and Point

Create circle by picking the center and a point on the circle.

⌇ **Delete Segment**

Dynamically trim section entities.

Click the icon of **Line** and sketch 3 lines, as shown. Specify 4000 as the dimension of length, as shown.

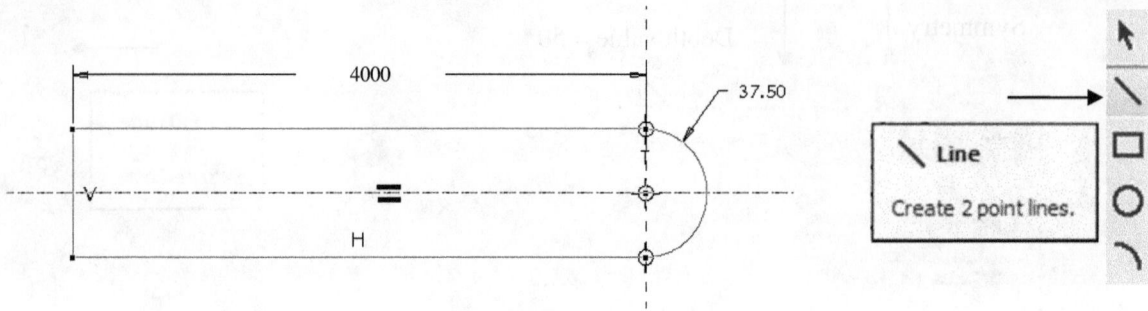

Click the icon of **Circle** and sketch a circle. The dimension of diameter is 50.

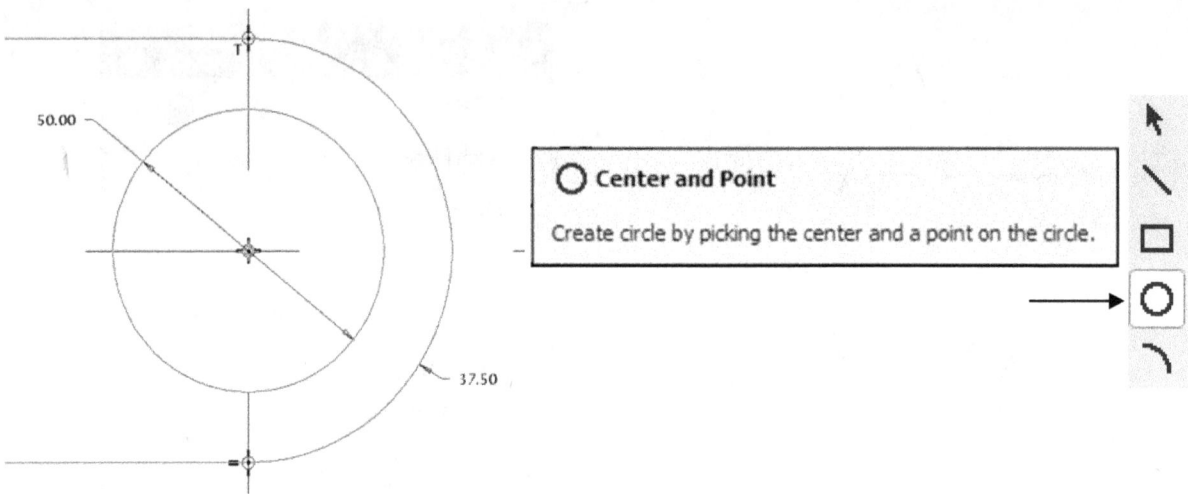

Upon completion, click the icon of **Done** and click the icon of **Apply and Save**.

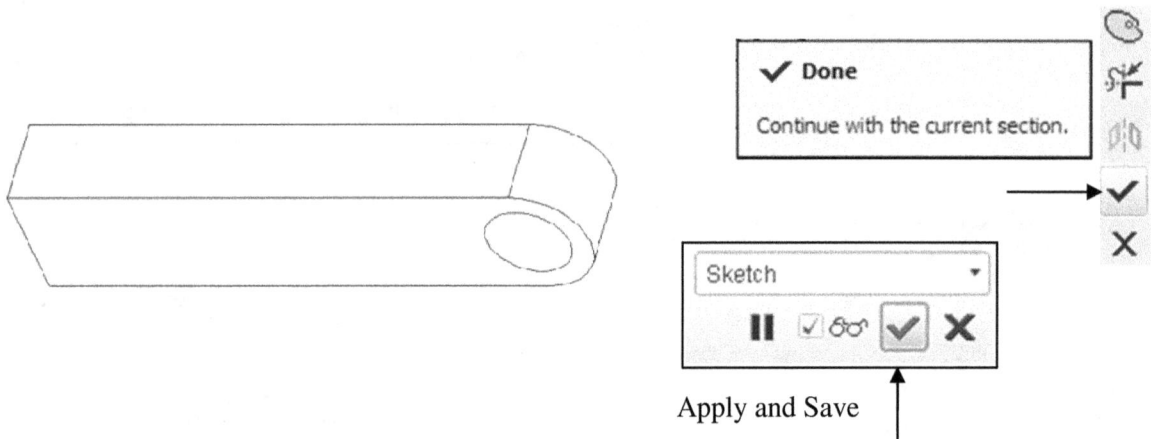

Apply and Save

Upon completion, select the icon of **Done** from the toolbar of sketcher. Select the icon of **Complete** from the feature control panel.

Let us define a datum point, which will be used when specifying the load condition. Click the icon of **Datum Point Tool**. While holding down the **Ctrl** key, pick the **FRONT** datum plane and the axis of the hole, as shown. Click **OK** when completing this process. **PNT0** is created.

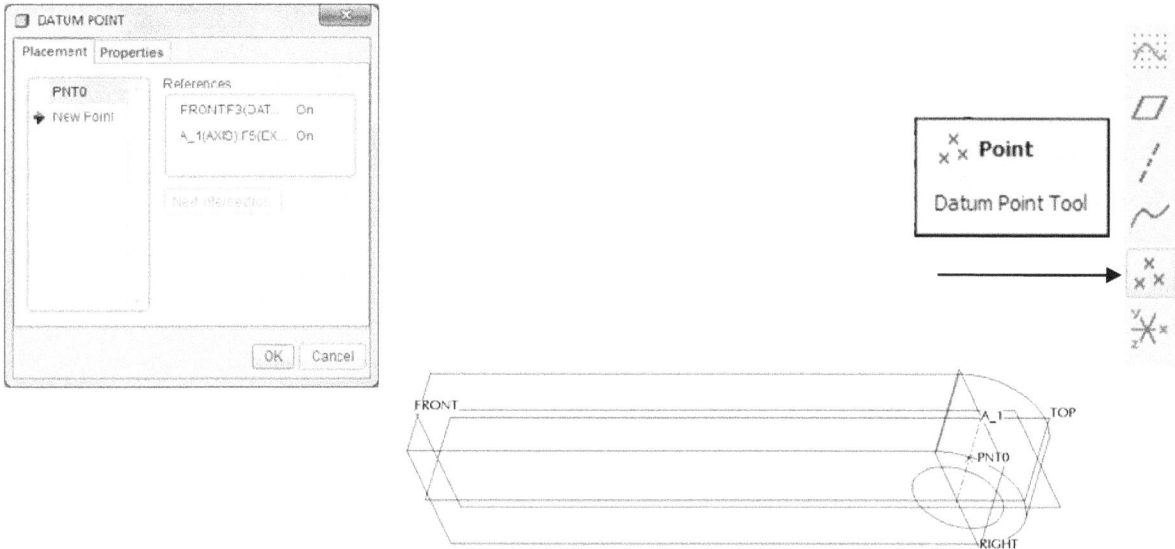

Step 2: under the Pro/Mechanica environment, specify the material type, the constraint condition and the load condition.

From the top menu, click **Applications > Mechanica > Continue > Structure > OK.**

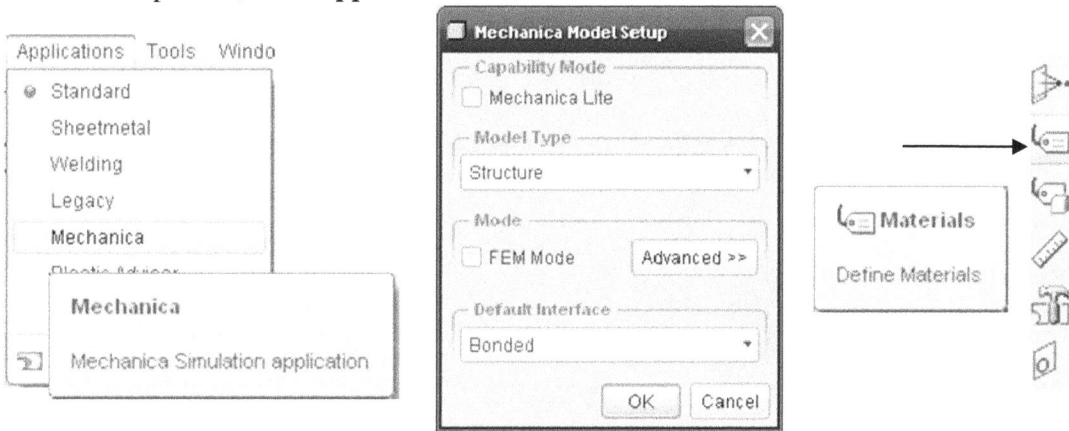

From the toolbar of functions, select the icon of **Materials >** select **Steel** from the left side called Materials in Library > click the directional arrow so that the selected material type goes to the right side called Materials in Model.

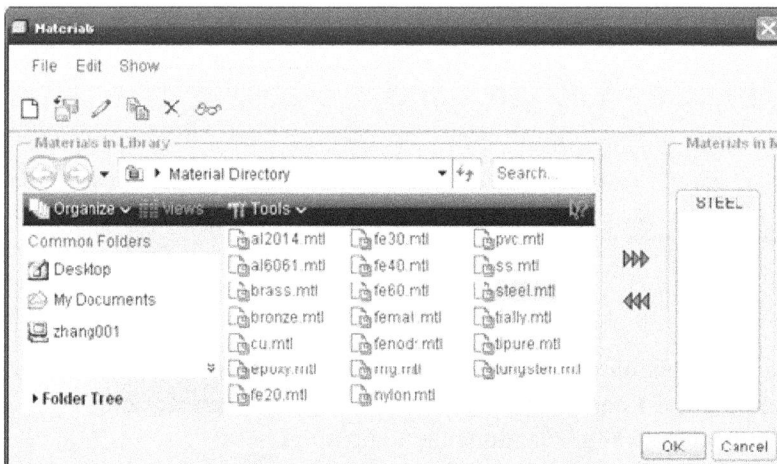

Click the icon of **Material Assignment**, the pop-up window indicates there is one component and the material type is Steel. Just click **OK**.

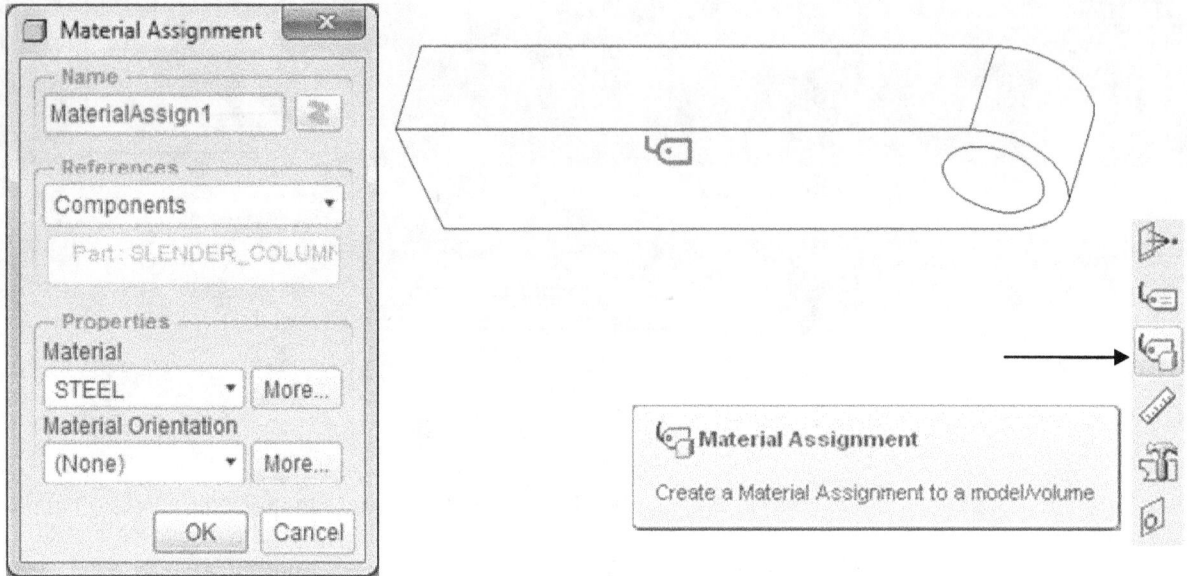

To define the constraint condition, select the icon of constraint > select the surface on the left end and fix all six degrees of freedom > **OK**.

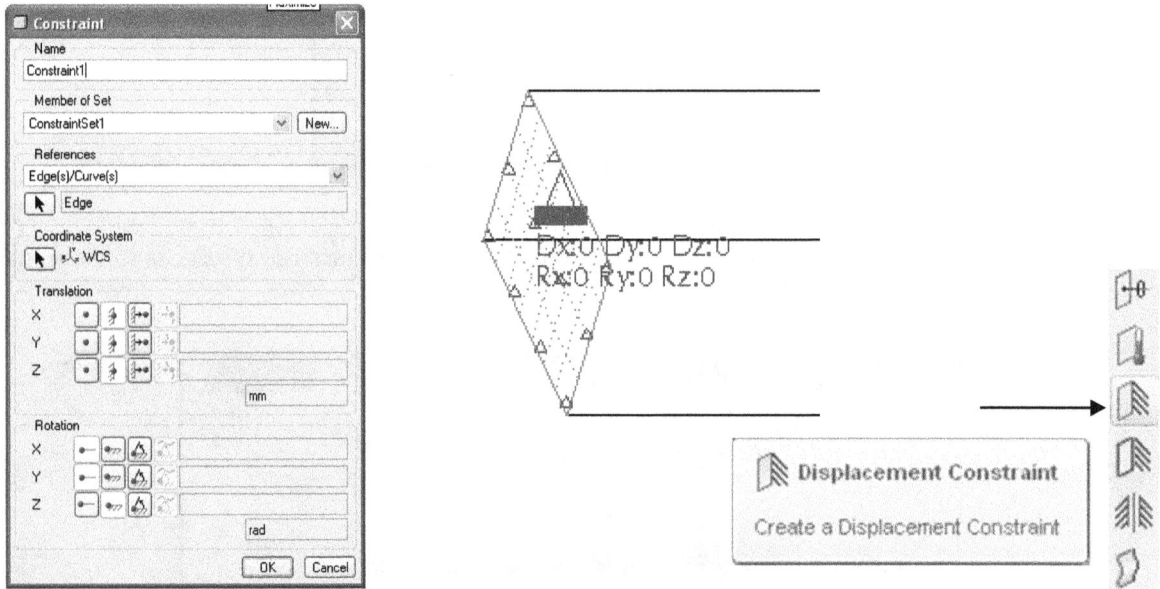

To define the load condition, select the icon of **Force/Moment** > pick the cylindrical surface. Click **Advanced** > select **Total Load at Point** and select **PNT0**, which was created in the part modeling. Specify the magnitude equal to -500000 N, along the x-direction.

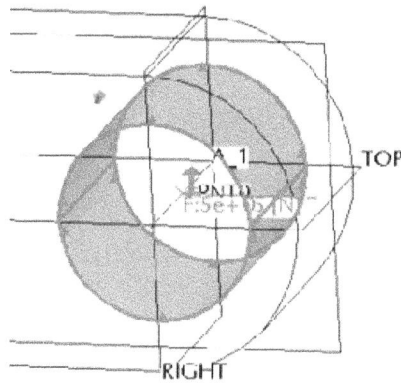

Step 3: perform buckling analysis

Click the icon of **Mechanica Analyses/Studies** > **File** > Select **New Static**. Type *static_for_buckling* as the file name > **OK**. Run the static analysis.

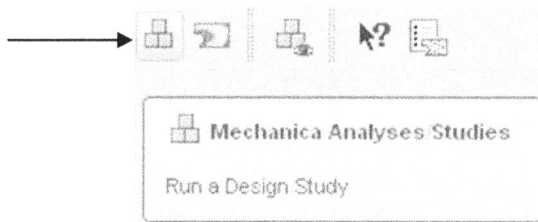

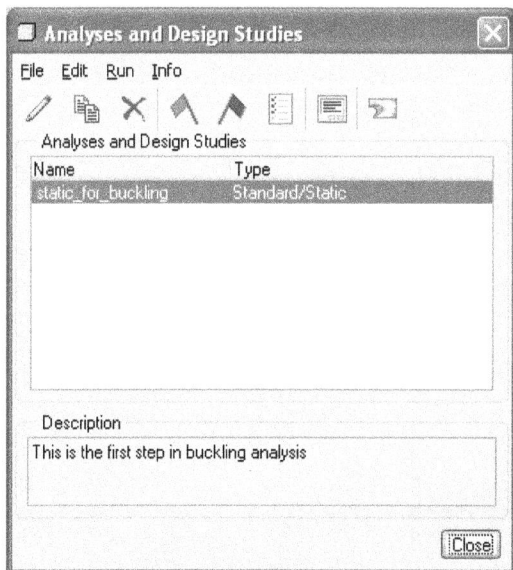

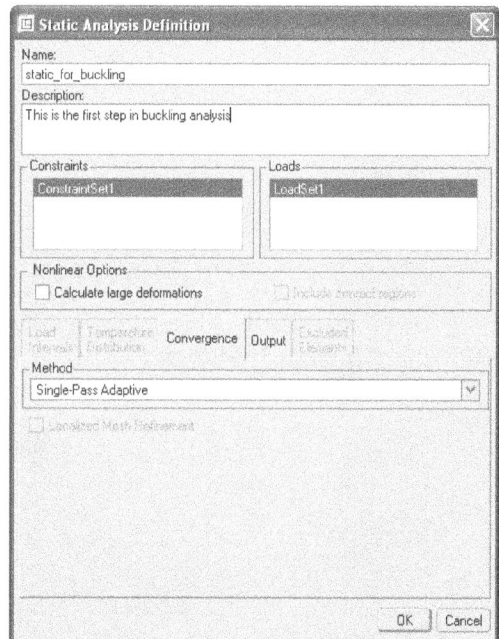

File > New Buckling > type *buckling_1* as the name. Specify the number of buckling modes to be 3. Therefore, we expect to have 3 **BLF** values. Each of the 3 **BLF** values is associated with its own buckling mode. Check the box called Use static analysis results from previous design study > **OK**.

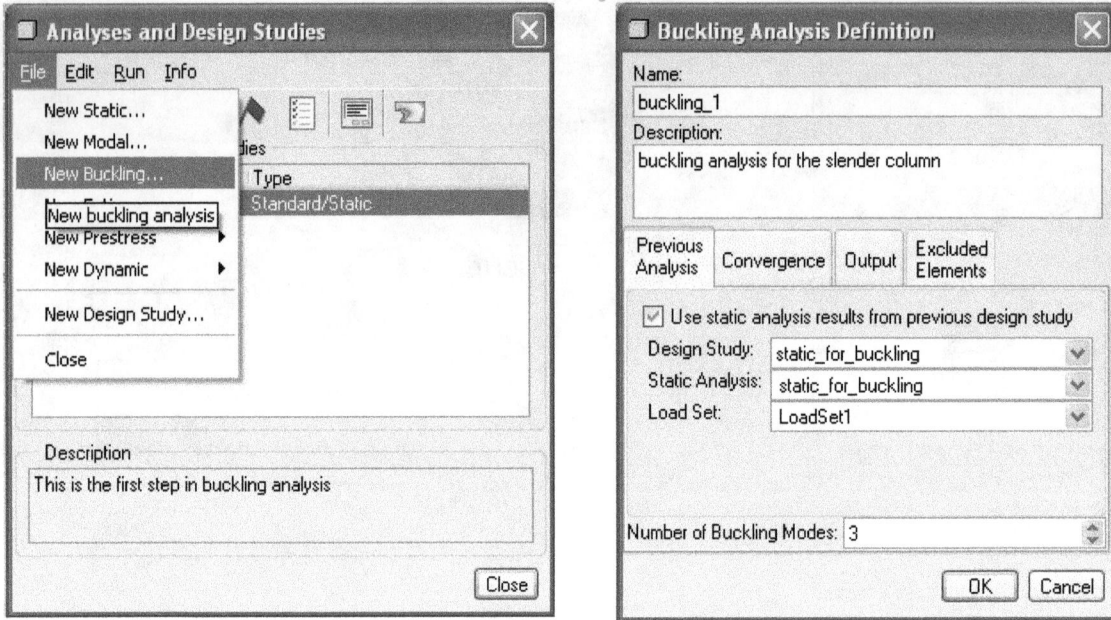

Run the buckling analysis.

The obtained BLF values are shown below. They are 0.0483, 0.1086 and 0.4358 for the 3 modes, respectively. All these 3 values are less than 1, indicating that they are all candidates for buckling failure. Note that these BLF values are associated with the magnitude of the applied load, which is 500000 N.

```
Number of Modes: 3

Mode    B. L. F.
----    ------------
   1    4.830472e-02
   2    1.086358e-01
   3    4.358160e-01
```

To plot the three mode shapes, select Results > Insert > Result Window > type in The first mode shape as the title > select the folder called buckling_1 > select Mode 1 > select Displacement > from Display Option, select Deformed > OK and Show. Follow the same procedure to plot Mod 2 and Mod 3.

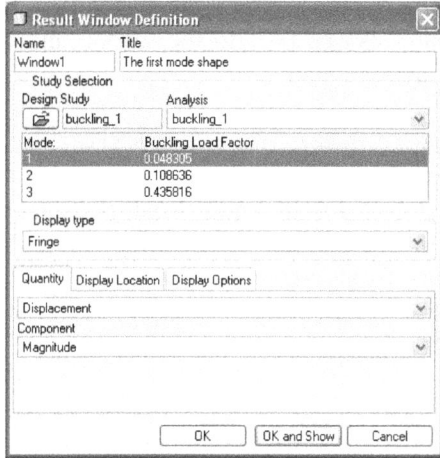

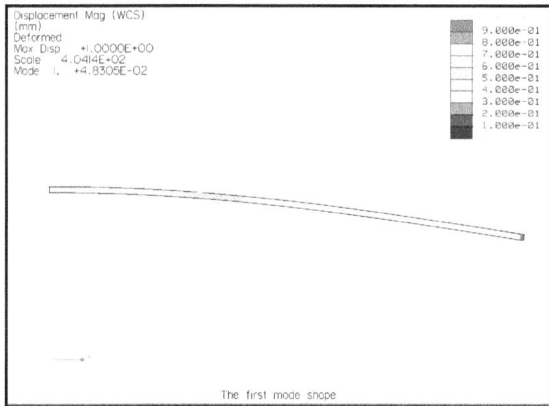

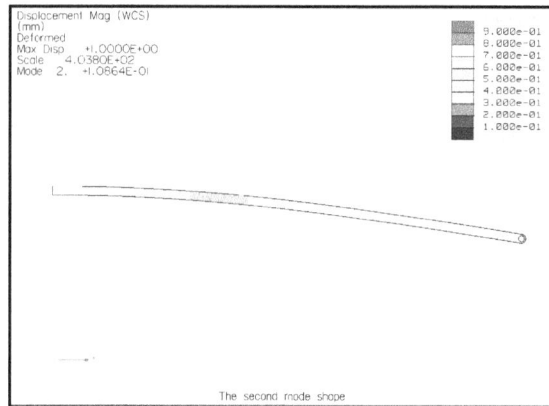

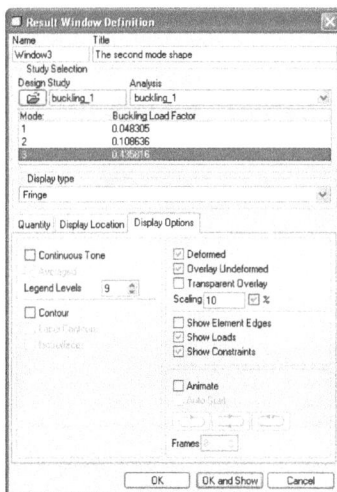

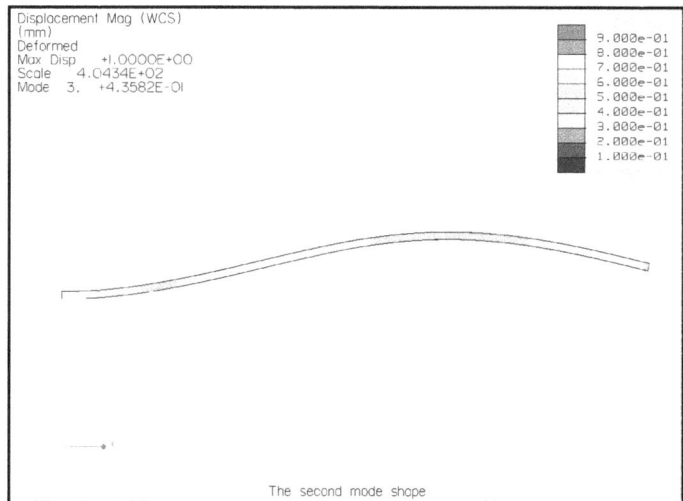

The first mode shape

The second mode shape

The second mode shape

Calculations of the critical loads for the three buckling modes are shown below:

$$P_{critical\_mode\_1} = 0.048305 * 500000 = 24153 \text{ N}$$
$$P_{critical\_mode\_2} = 0.108636 * 500000 = 54318 \text{ N}$$
$$P_{critical\_mode\_3} = 0.435816 * 500000 = 217908 \text{ N}$$

Because $P_{critical\_mode\_1} = 24153$ N is the smallest one among the three critical loads, the first mode is most likely to be the failure mode. However, attention should also be paid to the second and third modes because their associated critical loads are smaller than the applied load.

7.5   Pre-Stress Analysis of a Guitar String

Prestressed structures are widely used in engineering design to provide additional resistance to deformation, fatigue and certain failure modes in engineering. In this section, guitar strings are used to demonstrate the procedure required to carry out a prestress analysis in FEA. Guitar strings are metal cylinders placed under tension. The amount of tension T present in the string is directly related to the modal frequency at which the string resonates. The frequency of a string is also related to its length L and mass per unit length μ by the following equation:

$$f = \frac{1}{2L}\sqrt{\frac{T}{\mu}}$$

Step 1: Prepare a 3D solid model of a string with a circular section. The diameter is 0.049 inch and the length is 25.5 inches. This length is the standard length used in an electric guitar.

Click the icon of **Create a new object** from the menu toolbar to initiate the creation of a 3D model.

Select the icon of **Create a new object** from the main menu or the main toolbar

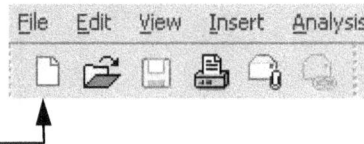

Type *guitar_string* as the file name, clear the box of **Use default template > OK**. Select the unit of **inlbs_part_solid**, type *guitar_string* under the description of the model, and type *student* or *your name* under the **modeled_by > OK**.

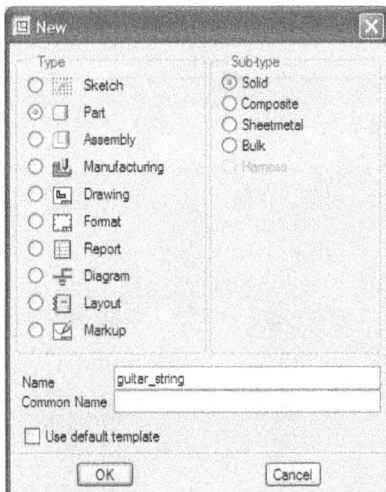

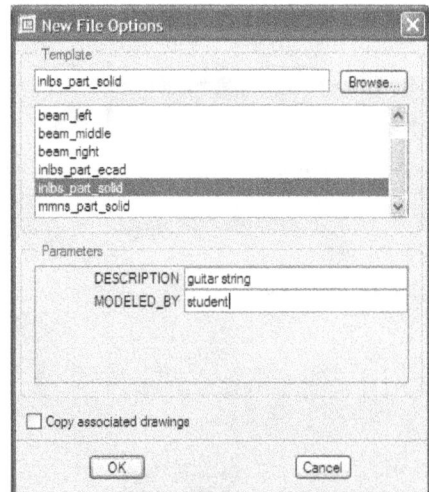

Click the icon of **Extrude**. Select **Symmetry** and set the string length value equal to 25.5. Click **Placement > Define**.

Thickness value

□ ◱ | ⊟ ▾ | 25.50 | ∨ | ⅍ | ⅃

Placement   Options   Properties

Extrude

Extrude Tool

Sketch

● Select 1 item   Define...

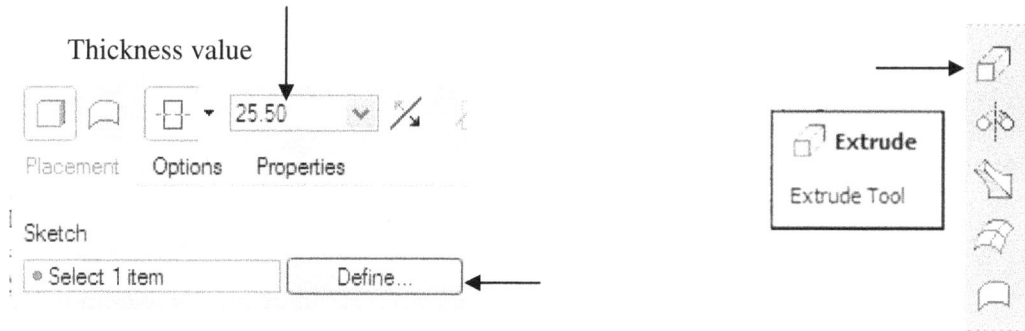

Select the **FRONT** datum plane displayed on screen, and click the button of **Sketch** to accept the **RIGHT** datum plane as the default reference to orient the sketch plane, as illustrated below > **Sketch**.

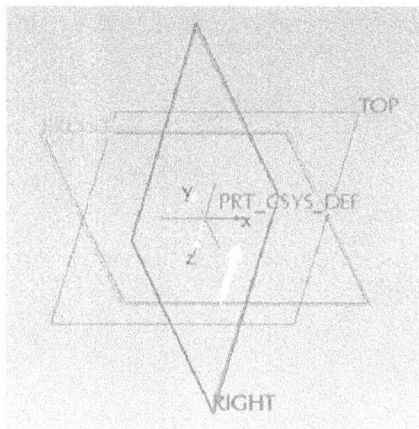

Sketch

Placement

Sketch Plane

Plane FRONT:F3(DA   Use Previous

Sketch Orientation

Sketch view direction Flip

Reference RIGHT:F1(DATUM PLA...

Orientation Right ∨

Sketch   Cancel

Click the icon of **Circle** and sketch a circle. The diameter value is 0.049. Upon completing the sketch, click the icon of **Done**. Click the icon of **Apply and Save**.

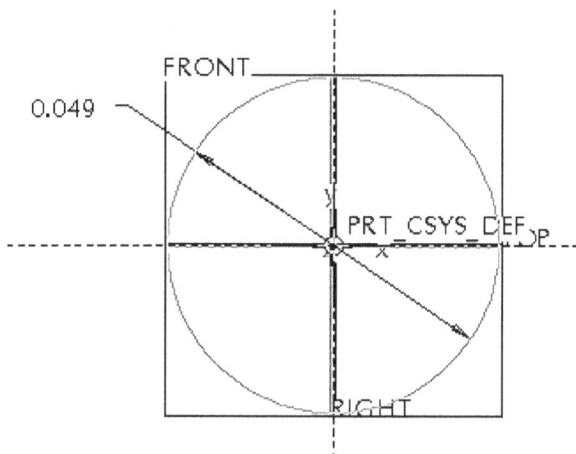

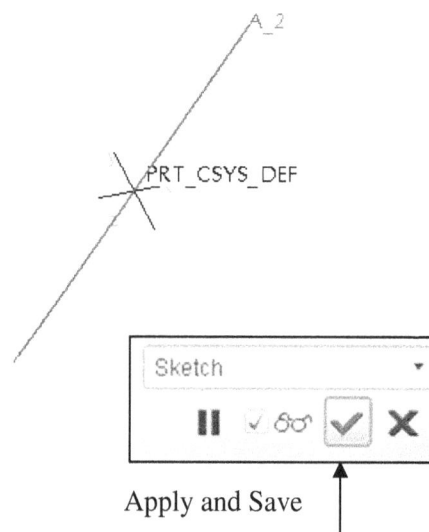

FRONT
0.049

PRT_CSYS_DEF
RIGHT

A_2

PRT_CSYS_DEF

Sketch

‖ ✓ 𝟞𝟞 ✓ ✗

Apply and Save

Step 2: Switch from the Pro/ENGINEER design system to Pro/Mechanica
From the main toolbar or the menu toolbar, select **Applications > Mechanica > Continue > Structure > OK.**

406 Engineering Analysis with Pro/Mechanica and ANSYS

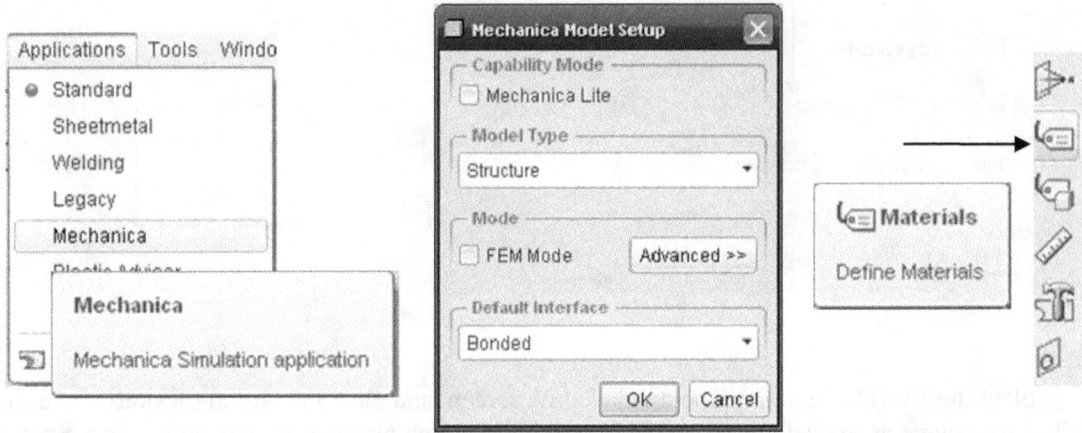

From the toolbar of functions, select the icon of **Materials** > select **Steel** from the left side called Materials in Library > click the directional arrow so that the selected material type goes to the right side called Materials in Model.

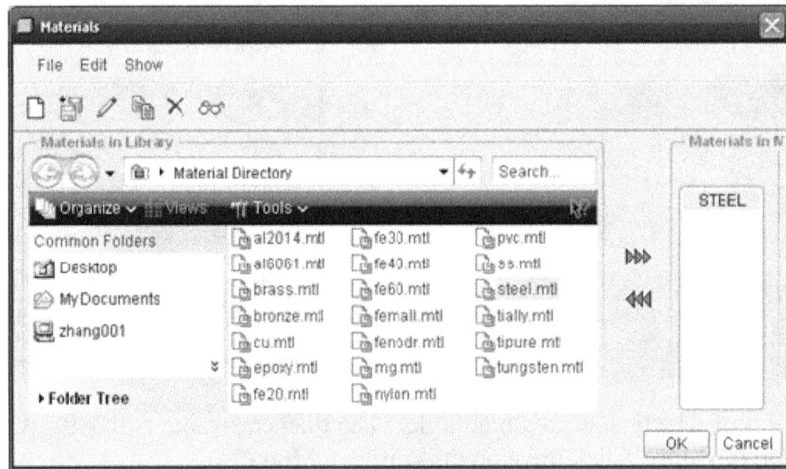

Click the icon of **Material Assignment**, the pop-up window indicates there is one component and the material type is Steel. Just click **OK**.

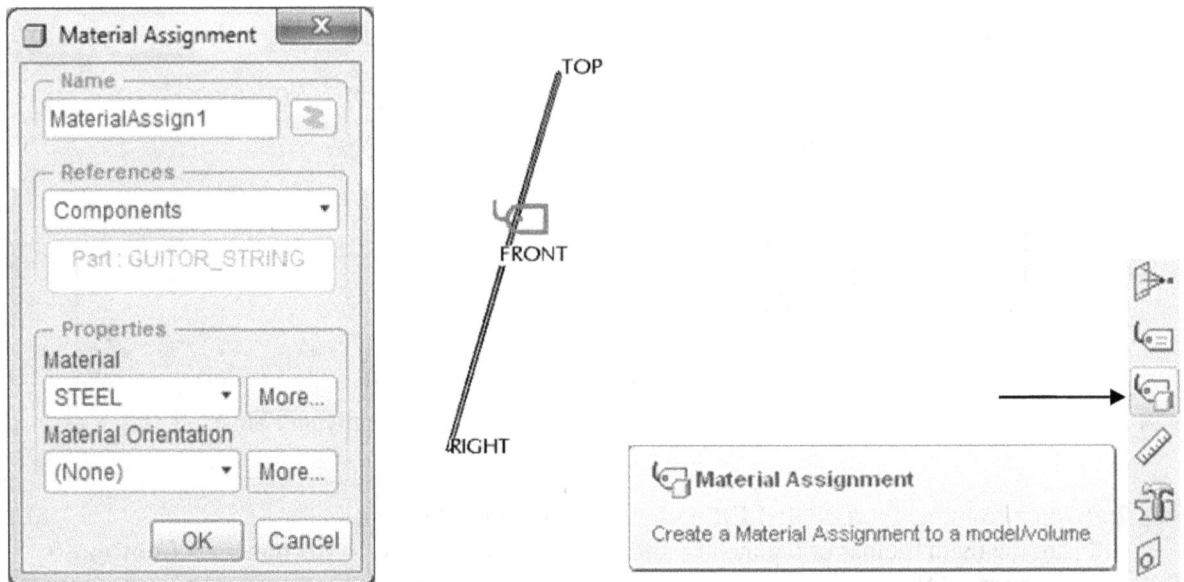

Step 3: Define the constraint condition.

For the constraint condition, one end of the string is fixed to the ground in all six degrees of freedom. The other end of the string is constrained in all but the z direction, or setting the translation along the Z axis FREE.

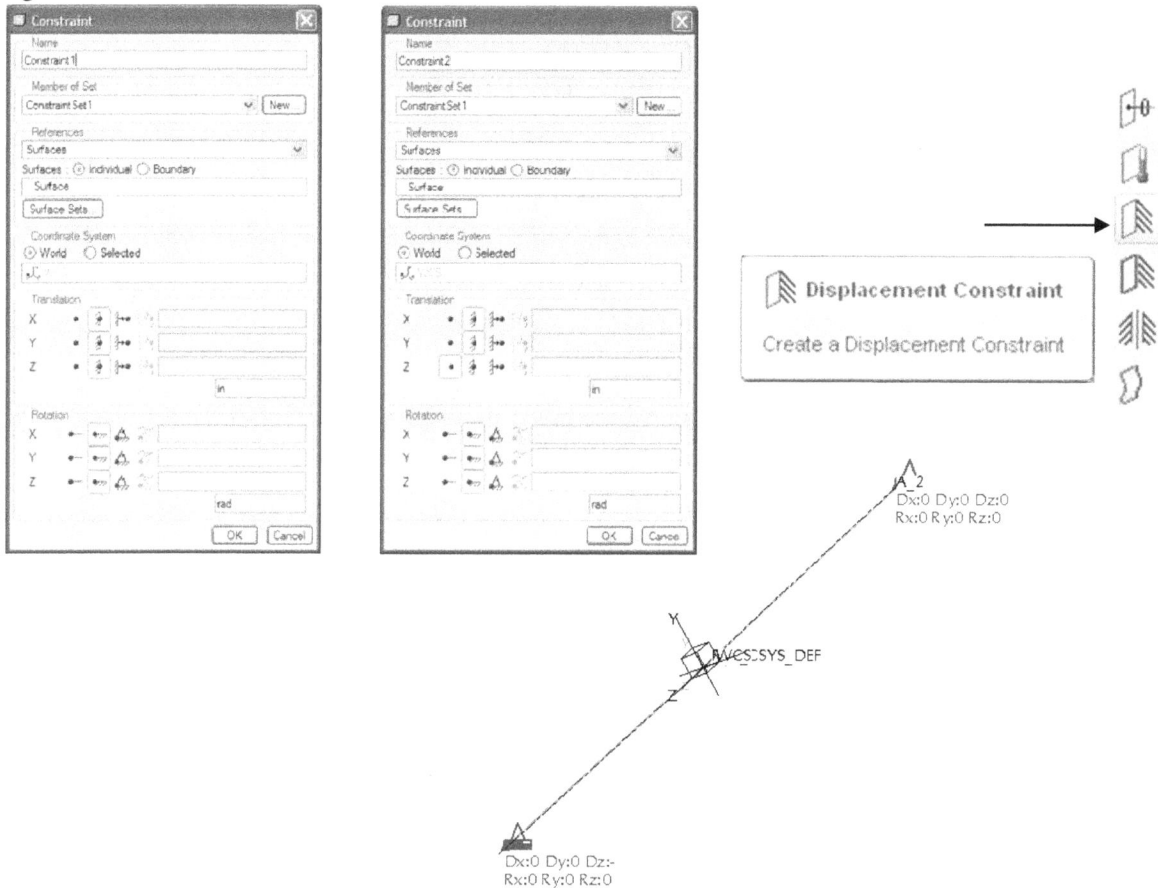

Step 4: Define the load condition.

A load is applied to the end with the degree of freedom in z set to free. Note that the force magnitude is set to 7573.44.

Select the icon of **New Force/Moment Load** from the toolbar of functions > **Surface(s)** > pick the surface at the end. Specify 7573.44 in the positive Z direction.

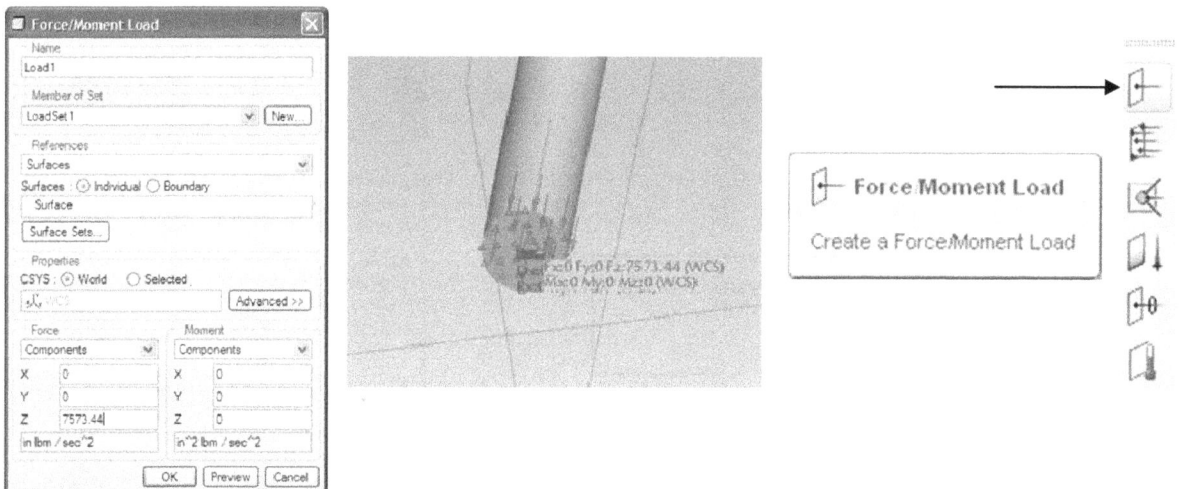

Where is the number of 7573.44 from? What is the unit?

The tension loads required to tune each string can be found using D'addario **strings, as shown below:**

EXL115 - Blues/Jazz Rock

D'Addario XL Electric Guitar strings are world-renowned as "The Player's Choice" amongst guitar players of all genres and styles. XL strings are wound with nickelplated steel and are known for their distinctive bright tone and excellent intonation.

Item#	Note	Diameter inches	mm	Tension lbs	kg
PL011	E	0.011	0.28	19.6	8.89
PL014	B	0.014	0.36	17.8	8.07
PL018	G	0.018	0.46	18.6	8.44
NW028	D	0.028	0.71	21.3	9.66
NW038	A	0.038	0.97	21.6	9.80
NW049	E	0.049	1.24	19.7	8.93

Note that the tensions in the table above are given in pounds. However, the units used by Pro/E are in inlbm/s^2.

$$1lb = 32.2 \frac{ft \cdot lbm}{s^2}$$

The tension in the low E string, as listed from D'addario, is 19.6 lb. The conversion comes up with the value of 7573.44.

$$19.6lb = 19.6 \cdot 32.2 \frac{ft \cdot lbm}{s^2} \left(\frac{12in}{ft} \right) = 7573.44 \frac{in \cdot lbm}{s^2}$$

Step 5: Under the Pro/MECHANICA environment, set up **Analyses** and **Run** it

As we know, the property of natural frequency of a metal string is subjected to the degree of tension. When the degree of tension is high, the natural frequency shifts to a high frequency region, and vice versa.

Because of this reason, unlike the tuning fork assignment listed in the previous homework, which required only a modal analysis to analyze natural frequencies, the guitar strings require a static analysis before running a pre-stressed modal analysis.

Select the icon of **Mechanical Analyses/Studies** > **File** > **New Static** > type *static_for_prestress_eval* as the name of the analysis folder > **OK.** Click the box of **Run.**

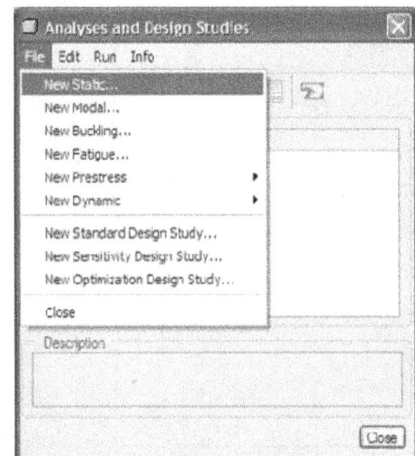

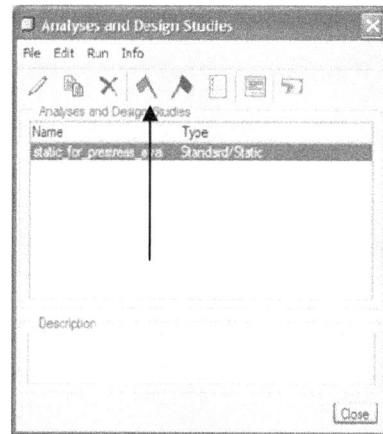

To monitor the computing process, users may click the icon of **Display study status**.

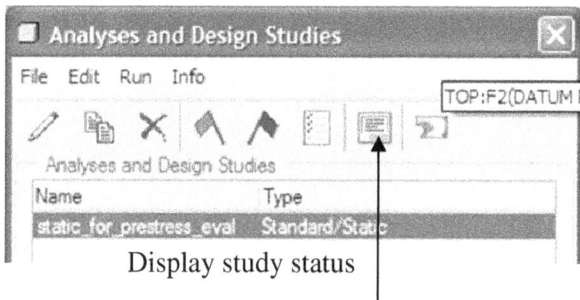

```
max_stress_prin:      6.331864e+06
max_stress_vm:        4.441219e+06
max_stress_xx:        2.468707e+06
max_stress_xy:       -2.396084e+05
max_stress_xz:       -1.579037e+06
max_stress_yy:        2.216517e+06
max_stress_yz:        1.356581e+06
max_stress_zz:        5.707924e+06
min_stress_prin:     -4.352295e+05
strain_energy:        3.475249e+01
```

Afterwards, run the pre-stress modal analysis. Specify guitar_string as the name. Make sure that users activate, or check the box called **Use static analysis**, as shown below > **OK** > **Run**.

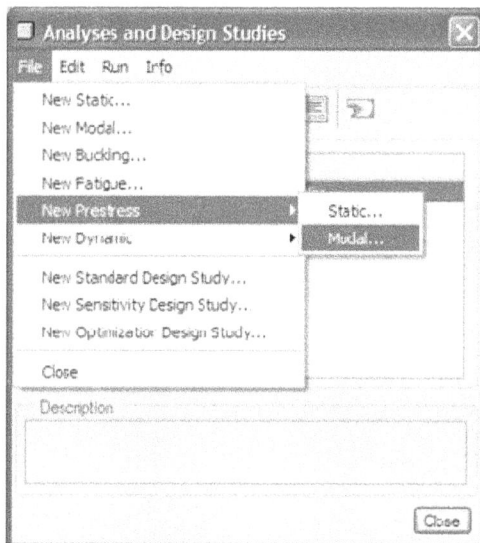

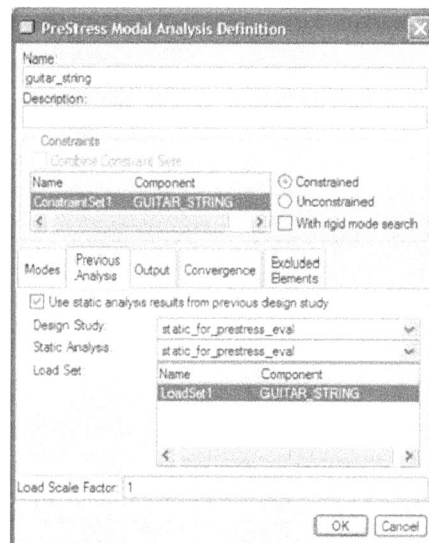

If we modeled the string correctly, the natural frequency of the first mode should be around 82.4 Hz. However, the obtained value is 76.21 Hz, as shown below:

```
Constraint Set: ConstraintSet1: GUITAR_STRING

Number of Modes: 4

    Mode  Frequency (Hz)
    ----  --------------
       1  7.620618e+01
       2  7.621091e+01
       3  1.538550e+02
       4  1.538657e+02
```

Now let us adjust the tension load applied from 19.6 lb to 23 lb. Based on the conversion, we have the magnitude equal to 8887.20.

$$T = 23 *32.2 *12 = 8887.20 \text{ inlbm/s}^2$$

To modify the magnitude of the load, highlight **Load1** listed in the model tree. Right-click and pick **Edit Definition**. Change the value to 8887.20 > **OK**.

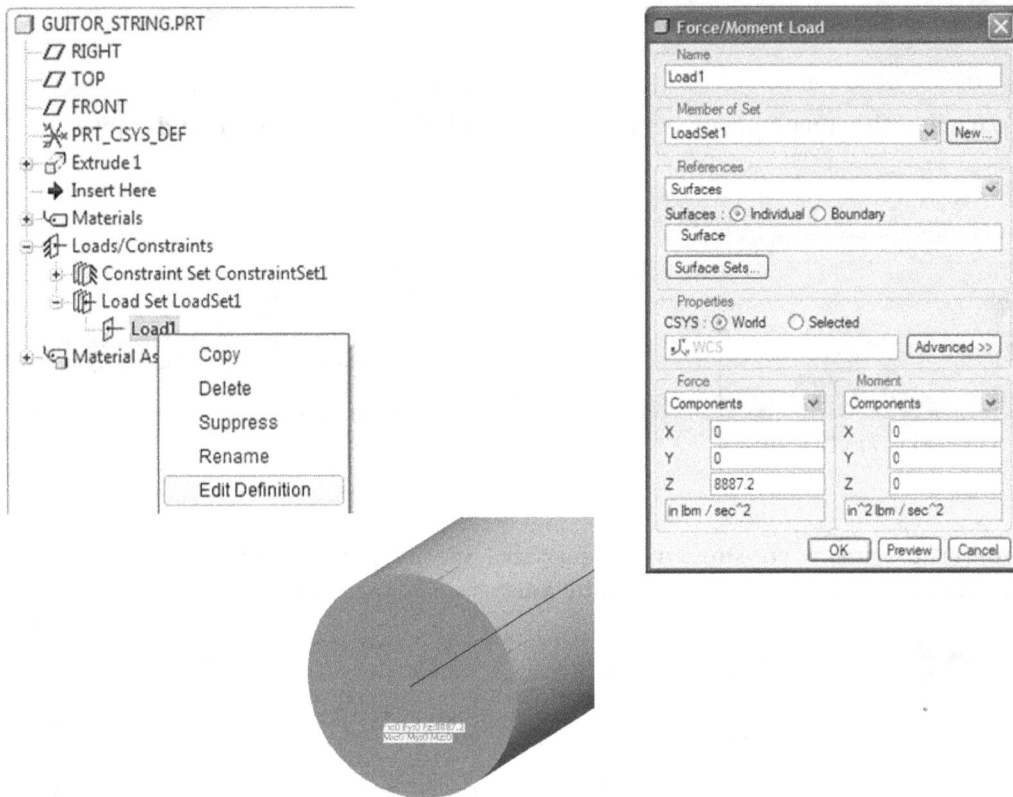

After the adjustment, run the static analysis again, and afterwards, run the pre-stress modal analysis for the second time. The desired natural frequency of 82.4 Hz is obtained, as shown below.

```
Constraint Set: ConstraintSet1: GUITAR_STRING

Number of Modes: 4

    Mode  Frequency (Hz)
    ----  --------------
       1  8.234769e+01
       2  8.235223e+01
       3  1.660246e+02
       4  1.660349e+02
```

The mode shape of the first mode is shown below:

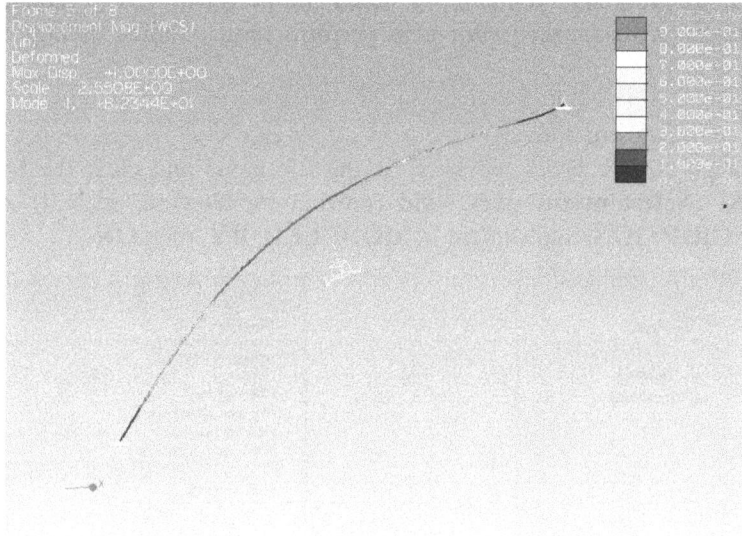

For the reference purpose, a guitar fingerboard layout published on the web is copied below:

http://www.daddariostrings.com/Resources/JDCDAD/images/tension_chart.pdf

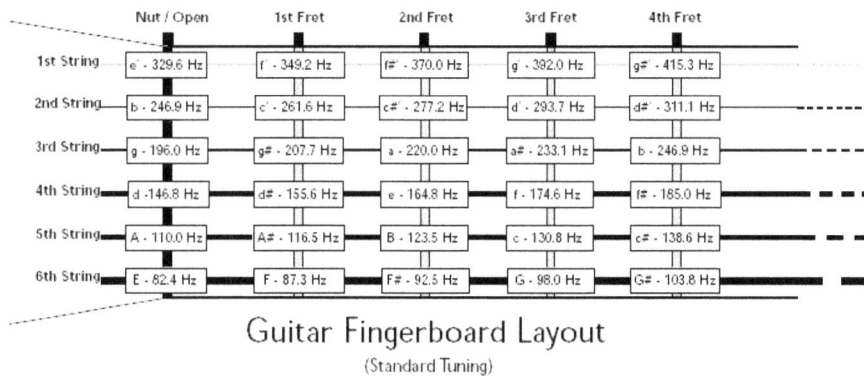

Guitar Fingerboard Layout
(Standard Tuning)

7.6   Vibration Analysis of a Cantilever Beam

Vibration refers to mechanical oscillation about an equilibrium point. Academically speaking, a vibration system consists of an object to vibrate, a spring and a dash-pot. As illustrated below, M stands for mass of the object. K stands for the stiffness of the spring. C represents the damping coefficient of the dash-pot. There are 2 types of vibration: free vibration and forced vibration. Free vibration is defined as the vibration is caused by its initial displacement or initial velocity or both with no external excitation. Forced vibration is viewed as the system response to an external excitation, which also varies as a function of time.

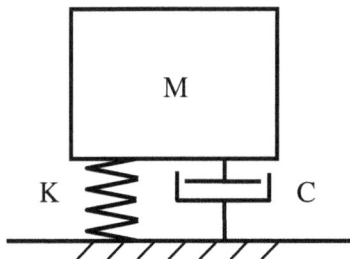

$$M\frac{d^2x}{dt^2}+C\frac{dx}{dt}+Kx(t)=0 \;\; \text{given} \;\; x(t=0) \text{ and } \frac{dx}{dt}|_{t=0}$$

$$M\frac{d^2x}{dt^2}+C\frac{dx}{dt}+Kx(t)=Af(t) \;\; \text{given} \; x(t=0) \text{ and } \frac{dx}{dt}|_{t=0}$$

In the following, we work with a cantilever beam and perform FEA to the free vibration, force vibration and the vibration at resonance. We also perform these studies in both time domain and frequency domain.

Step 1: Create a cantilever beam structure.
File > New > Part > type *beam_vibration* as the file name and clear the box of **Use default template > OK**. Select **mmns_part_solid** (units: mm, Newton, second) and type *vibration analysis* in **DESCRIPTION**, and *student* in **MODELED_BY**, then **OK.**

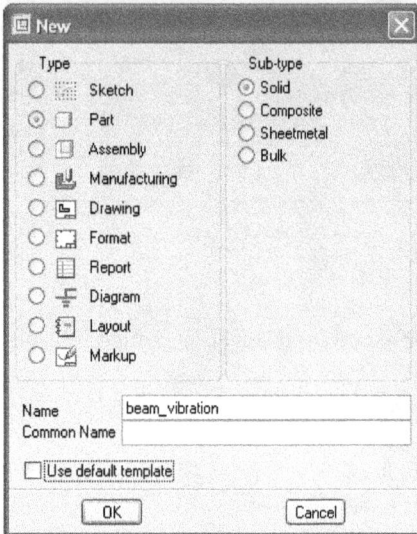

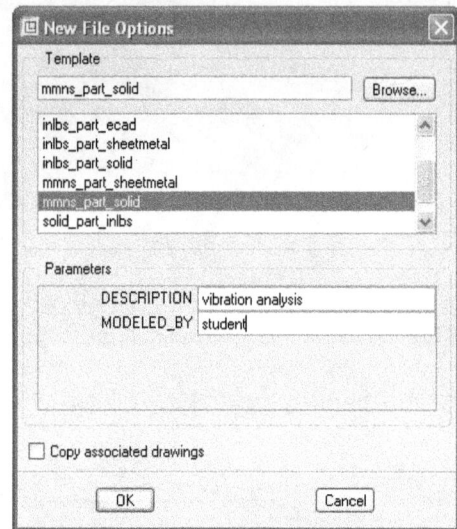

Click the icon of **Extrude**. Select **Symmetry** and set the thickness value equal to 60. **Placement > Define.**

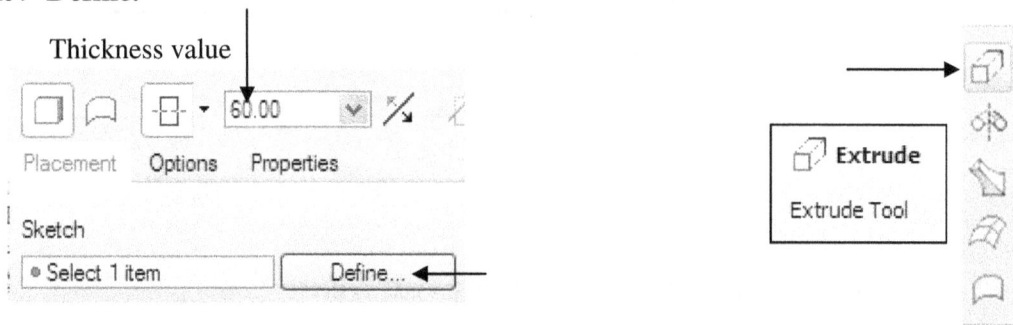

Select the **FRONT** datum plane displayed on screen, and click the button of **Sketch** to accept the **RIGHT** datum plane as the default reference to orient the sketch plane, as illustrated below > **Sketch**.

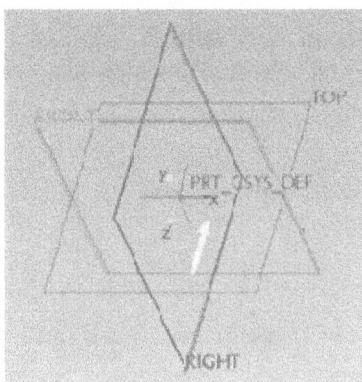

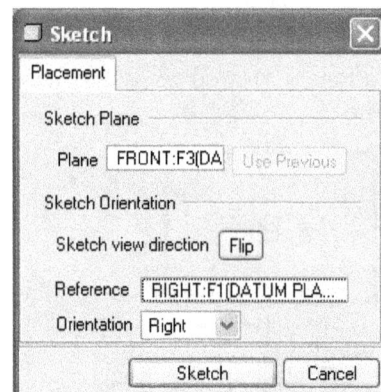

Create a horizon centerline, as shown below, and sketch a rectangle with the 2 dimensions equal to 300 and 80, respectively. Make sure this rectangle is symmetric about the created horizontal centerline.

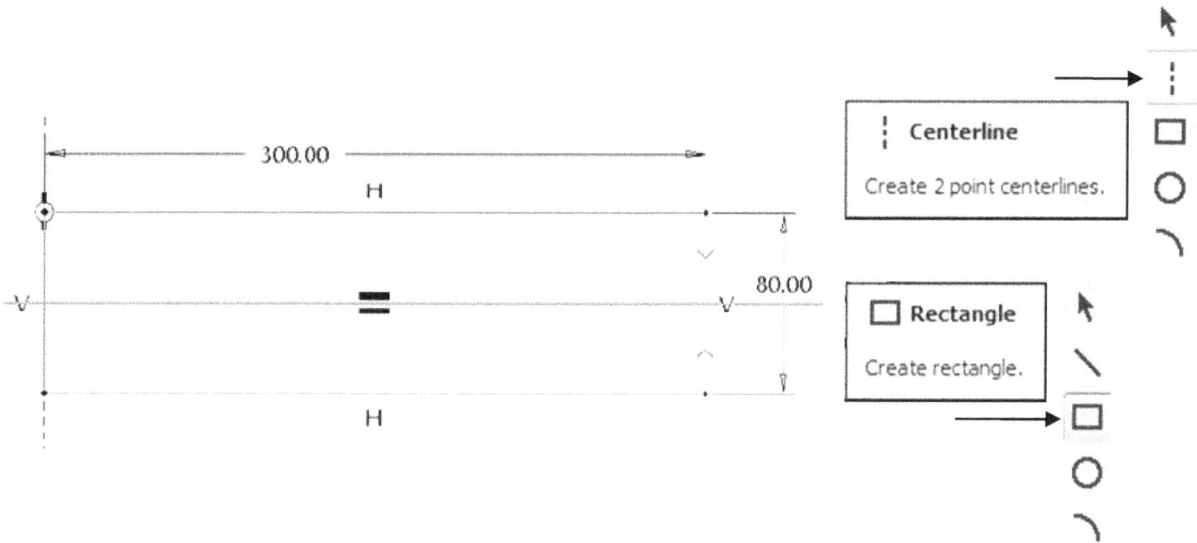

Upon completing the sketch, click the icon of **Done**. Click the icon of **Apply and Save.**

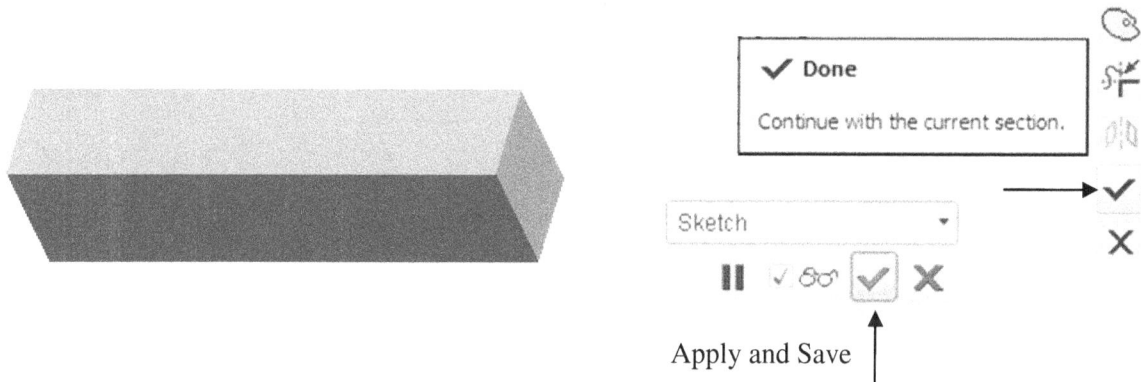

Apply and Save

Step 2: Create a rectangular datum curve for load application.
Select the icon of **Sketch Tool** > pick the top surface of the beam as the sketch plane > pick the **RIGHT** datum plane at the right for orientation > click **Sketch**.

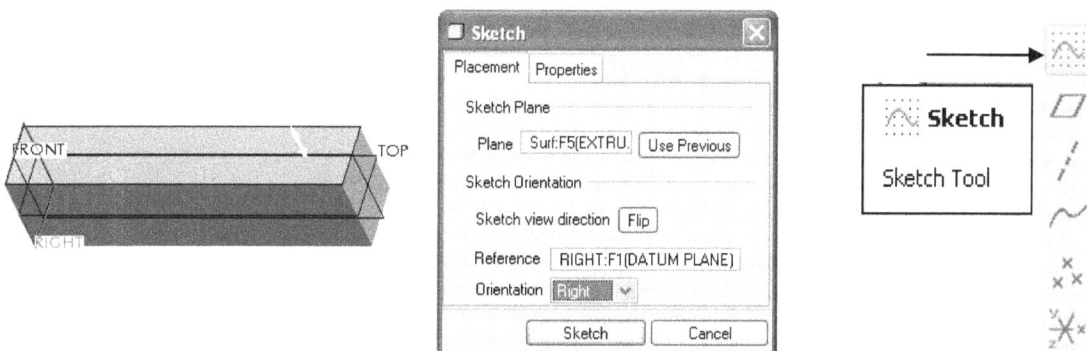

From the top menu, click **Sketch > References**. Add 2 new references to facilitate the process of sketching a rectangle.

Sketch   Analysis   Info   A

Sketch Setup...
References...

References
Specify references which the section will be dimensioned and constrained to.

References
F1(RIGHT)
Surf:F5(EXTRUDE_1)
Surf:F5(EXTRUDE_1)
Surf:F5(EXTRUDE_1)

Select Use Edge/Offset   Delete
Reference status
Fully Placed

Close

New reference

TOP
FRONT
RIGHT

New reference →

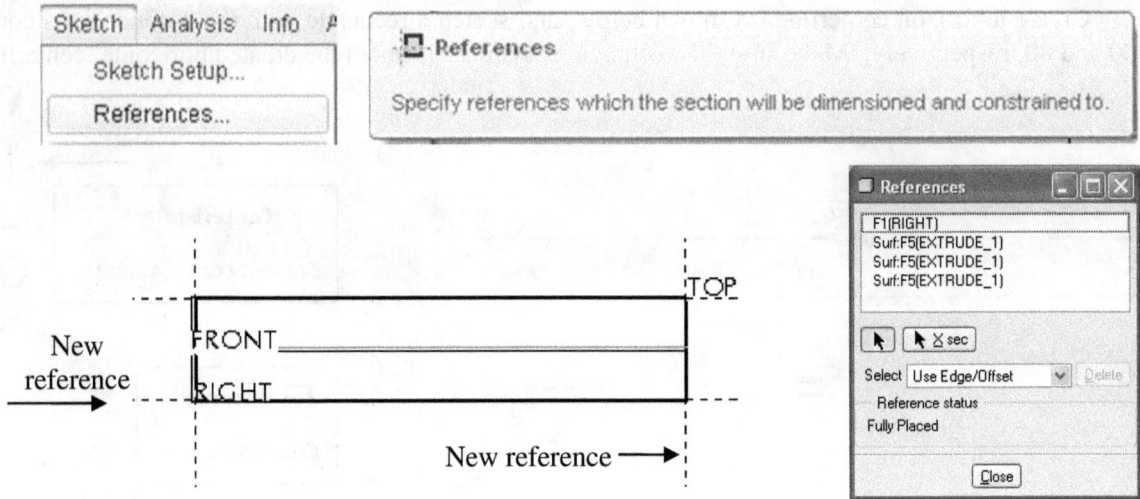

Pick the icon of **Rectangle** and sketch a rectangle. The only dimension needed is its width. Specify it equal to 10 > click the icon of **Done**.

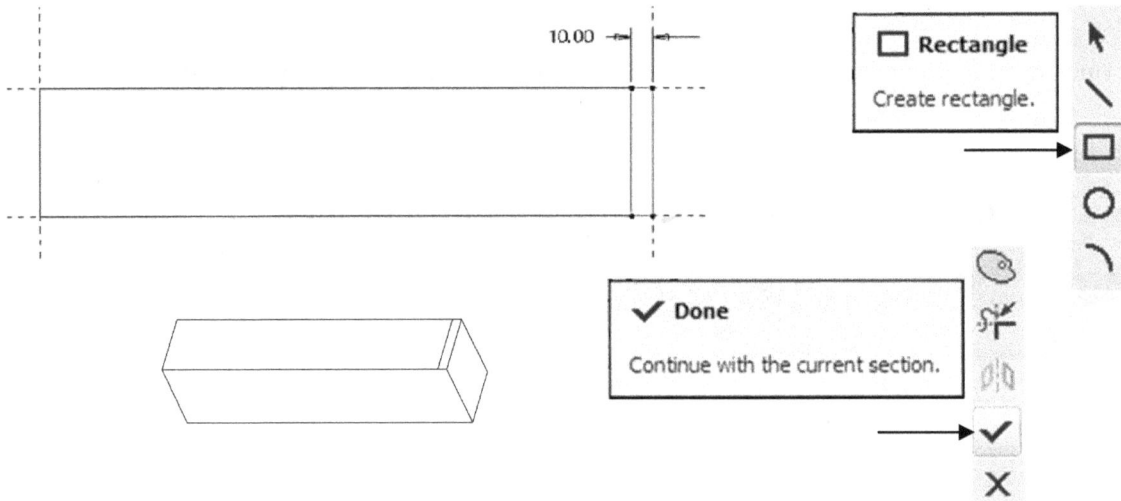

10.00

☐ Rectangle
Create rectangle.

✔ Done
Continue with the current section.

Step 3: Define a datum curve for defining a datum point to be used for measuring displacement at the free-end. Define a datum point at the fixed end for measuring stress.
        Pick or highlight the **TOP** datum plane from the model tree. From the top menu, click **Edit > Intersect** > pick the right side surface of the beam while holding down **Ctrl > Done**.

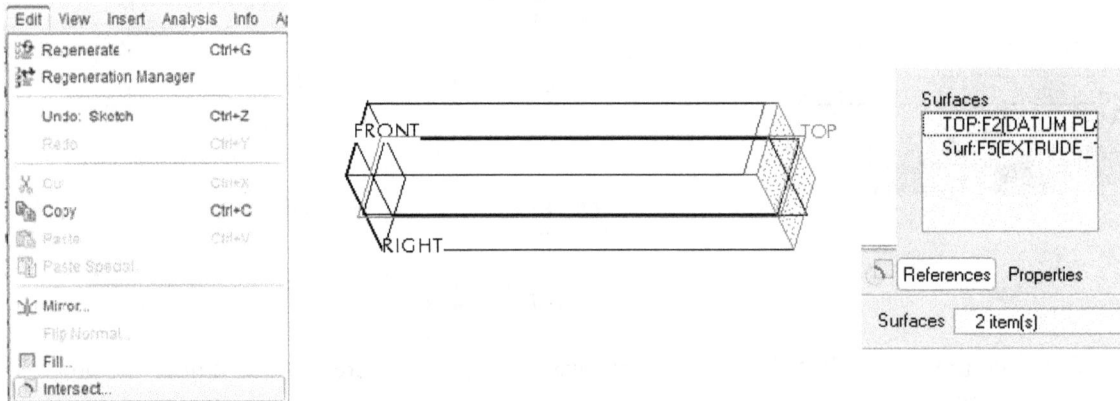

Edit   View   Insert   Analysis   Info   A
Regenerate   Ctrl+G
Regeneration Manager

Undo: Sketch   Ctrl+Z
Redo   Ctrl+Y

Cut   Ctrl+X
Copy   Ctrl+C
Paste   Ctrl+V
Paste Special...

Mirror...
Flip Normal...
Fill...
Intersect...

FRONT
TOP
RIGHT

Surfaces
TOP:F2(DATUM PLA
Surf:F5(EXTRUDE_1

References Properties

Surfaces   2 item(s)

Pick the icon of **Datum Point Tool**> select a location on the created datum curve > specify the value of length ratio equal to 0.5 > **OK**.

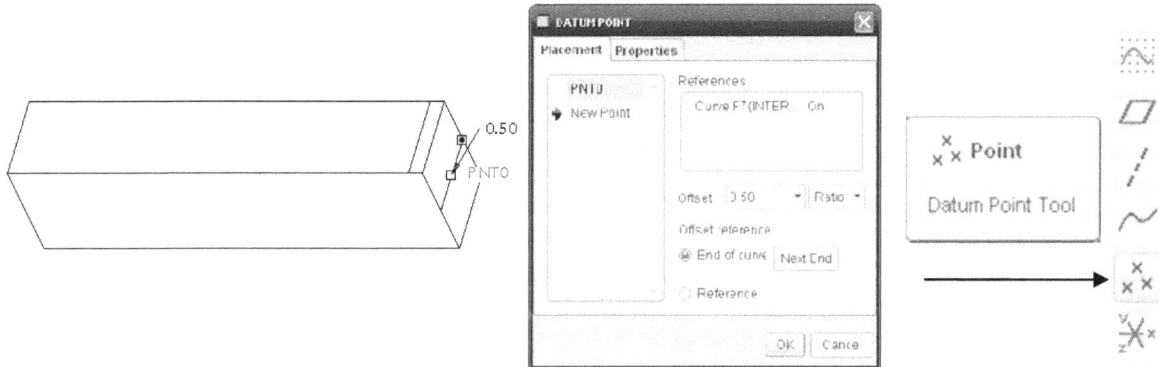

To define the second datum point, pick the icon of **Datum Point Tool**> select a location on the edge on the left side, as shown > specify the value of length ratio equal to 0.5 > **OK**.

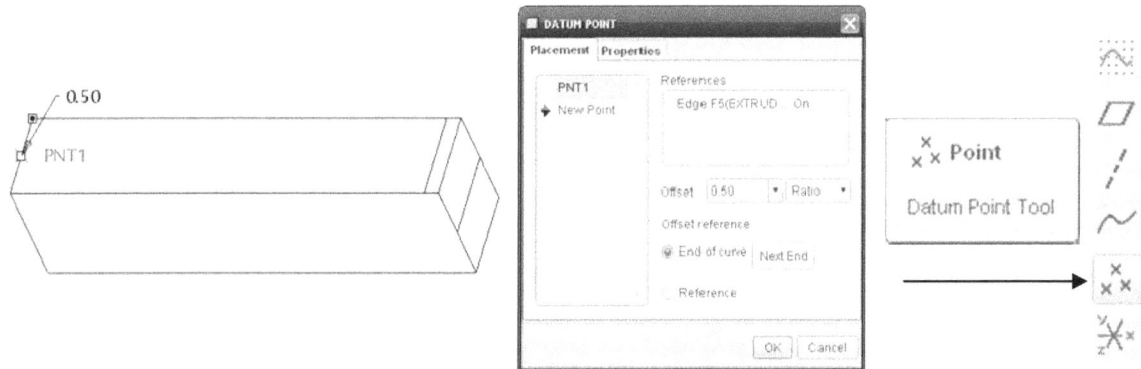

Step 4: Switch from the Pro/ENGINEER design system to Pro/Mechanica
From the main toolbar or the menu toolbar, select **Applications > Mechanica > Structure > OK.**

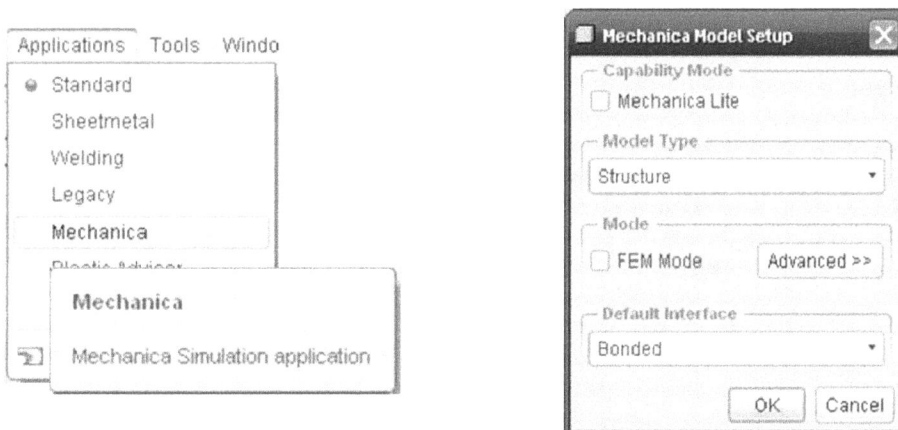

From the toolbar of functions, select the icon of **Materials >** select **STEEL** from the left side called Materials in Library > click the directional arrow so that the selected material type goes to the right side called **Materials in Model**.

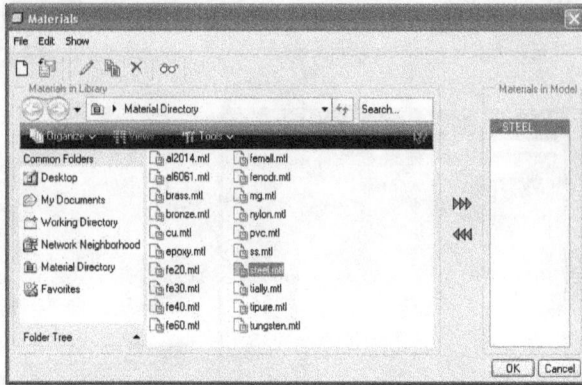

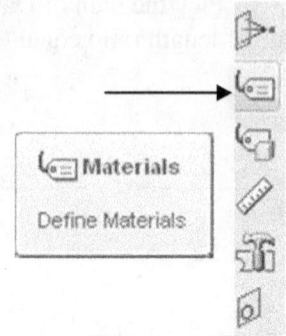

Click the icon of **Material Assignment.** In the Material Assignment window, the selected Steel is shown, and the component is also shown because there is only one volume or one component in the system. As a result, the software system automatically assigns Steel to the component. Just click **OK**.

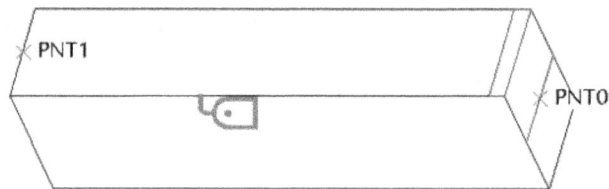

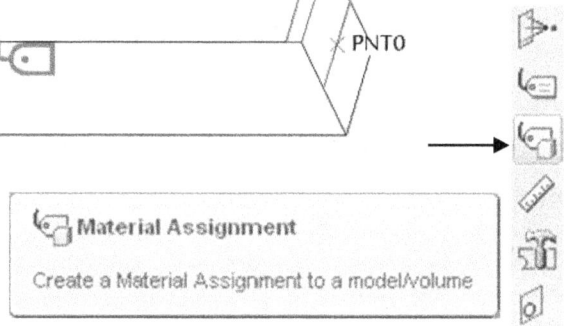

Step 5: Define 2 measures so that the displacement in the y direction and stress can be recorded.
Click the icon of **Creating a measure > New**.
In the **Measure Definition** window, type *displacement_y* as the name > select **Displacement** and **Y** > change **Maximum** to **At Point** > activate the point selection and pick **PNT0** > **OK**.
Activate **Time/Frequency Eval** > change **Maximum** to **At Each Step > OK > Close**.

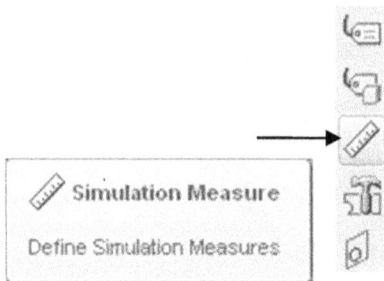

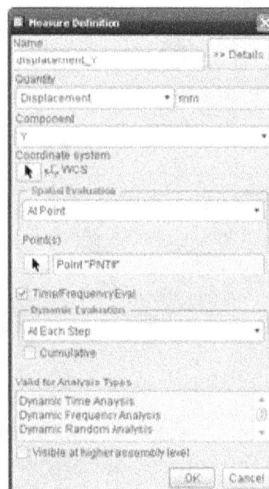

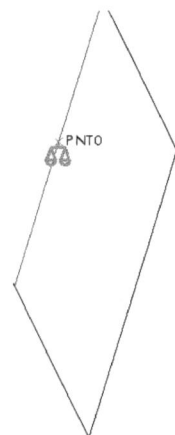

Click the icon of **Creating a measure > New**.

In the **Measure Definition** window, type *stress_vm* as the name > select Stress and von Mises > change **Maximum** to **At Point** > activate the point selection and pick **PNT1** > **OK**.

Activate **Time/Frequency Eval** > change **Maximum** to **At Each Step** > **OK** > **Close**.

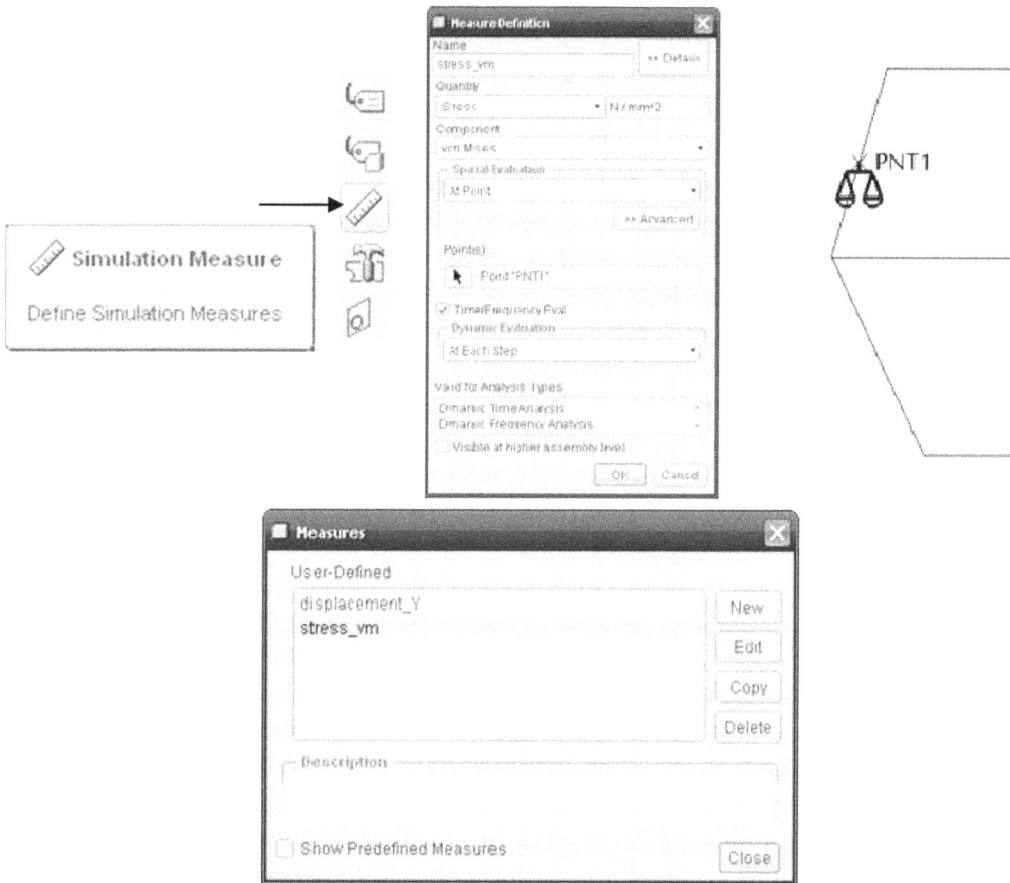

Step 6: Define the constraint condition.

To define the fixed end constraint on the left side of the beam, select the icon of **New Displacement Constraint** > **Surface(s)** > pick the surface at the left side of the beam.

Because the constraint condition is "a fixed end", we set all 6 degrees of freedom equal to zero, which is the default setting > **OK**.

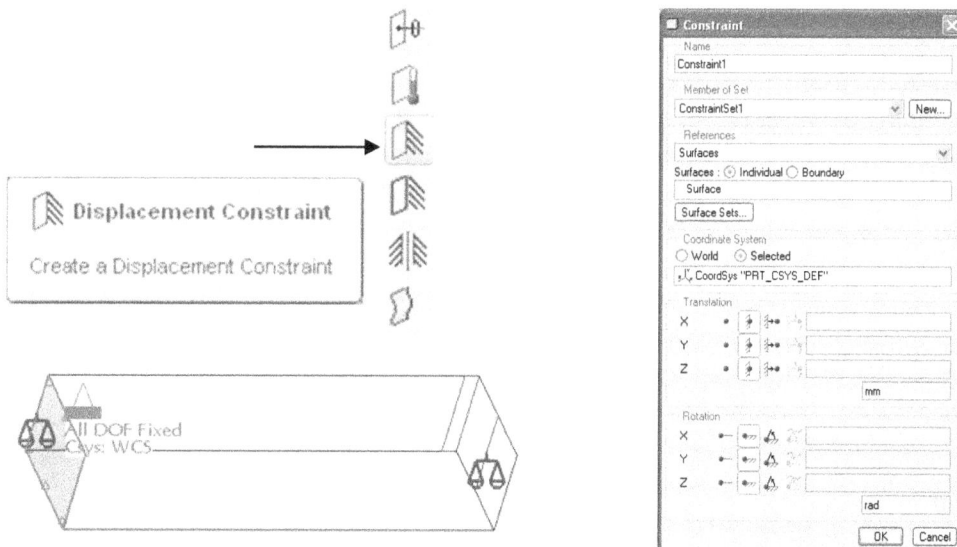

Step 7: Define the surface region.

Select the icon of **Surface Region. P**ick the datum curve > pick the top surface for splitting. Click **Apply and Save**.

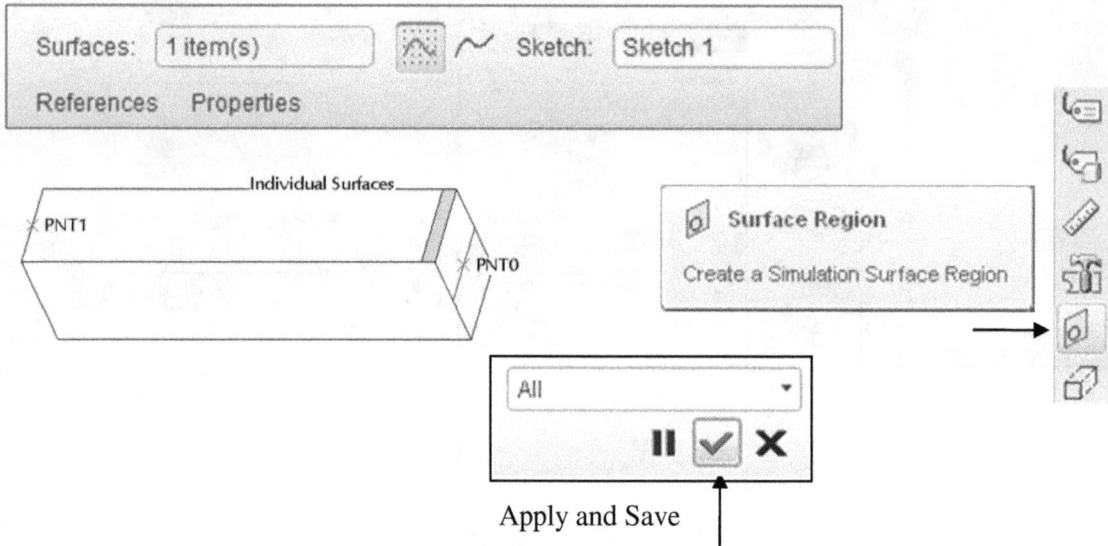

Apply and Save

Step 8: Define the load condition.

After defining the surface region, select the icon of **New Force/Moment Load** from the toolbar of functions > **Surface(s)** > pick the surface region just defined.

Specify -1000 N as the magnitude and the direction of downward along the y-axis.

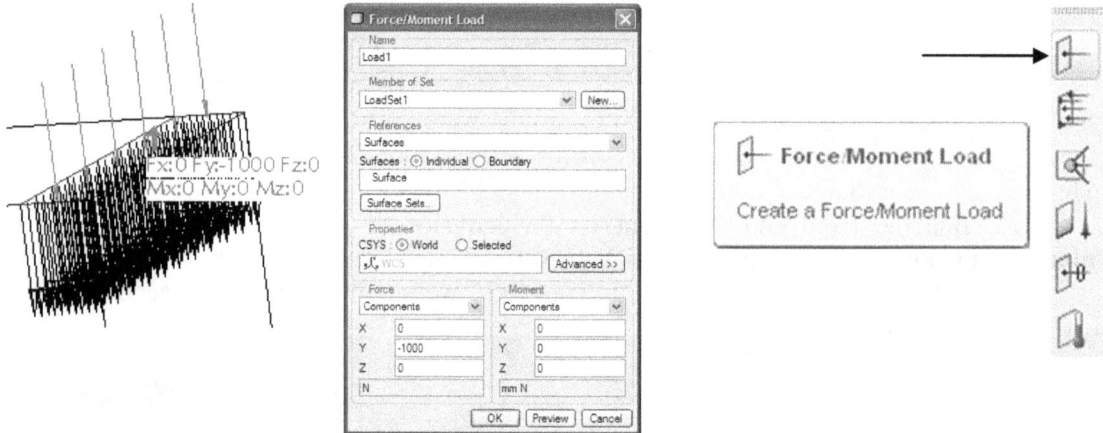

Step 9: Under the Pro/MECHANICA environment, set up **Analyses** and **Run** it

Select the icon of **Mechanical Analyses/Studies** > **File** > **New Static** > type *static_vibration* as the name of the analysis folder > **OK.** Click the box of **Run.**

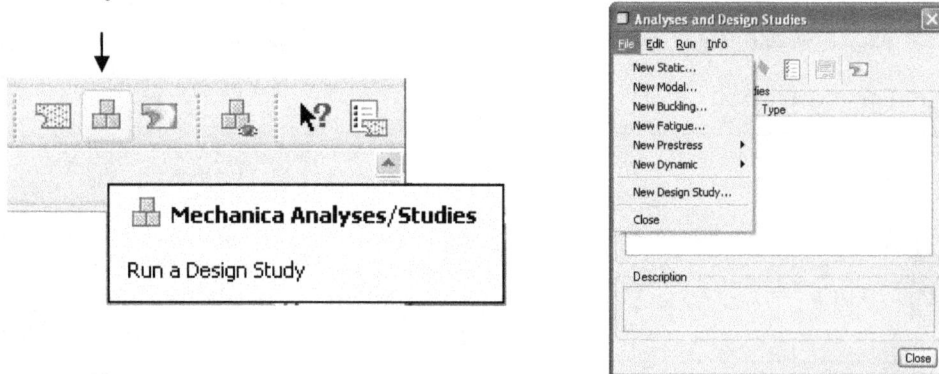

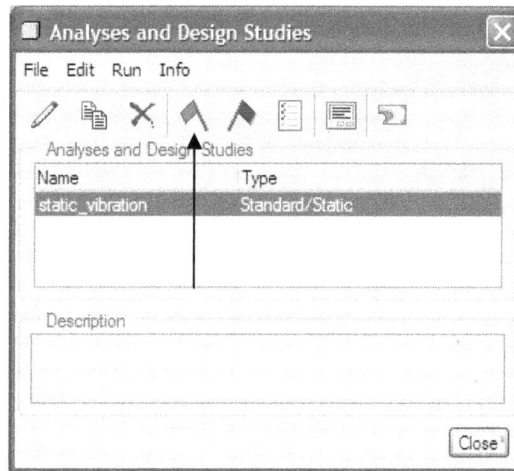

To monitor the computing process, users may click the icon of **Display study status**.

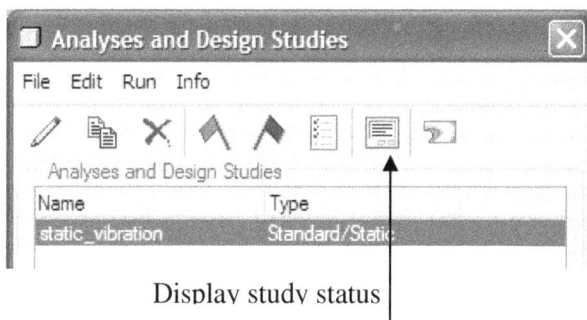

Measures:

max_beam_bending:	0.000000e+00
max_beam_tensile:	0.000000e+00
max_beam_torsion:	0.000000e+00
max_beam_total:	0.000000e+00
max_disp_mag:	1.814973e-02
max_disp_x:	3.448600e-03
max_disp_y:	-1.782576e-02
max_disp_z:	1.662817e-04

The following figure is the displacement plot.

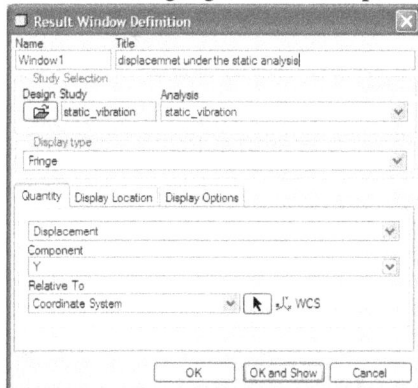

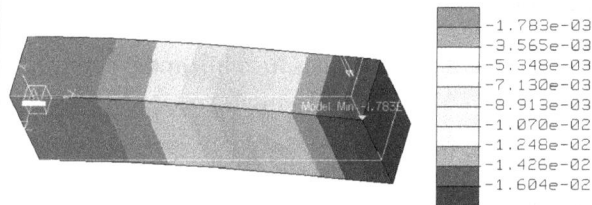

Step 10: Under the Pro/MECHANICA environment, perform a modal analysis to determine the natural frequencies of the first 4 modes.

Select the icon of **Mechanica Analyses/Studies > File > New Modal** > type *mode_frequencies* as the name of the analysis folder > **OK.** Click the green flag of **Run.**

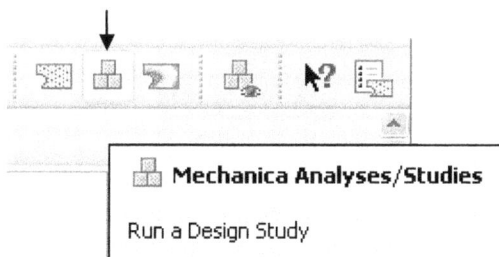

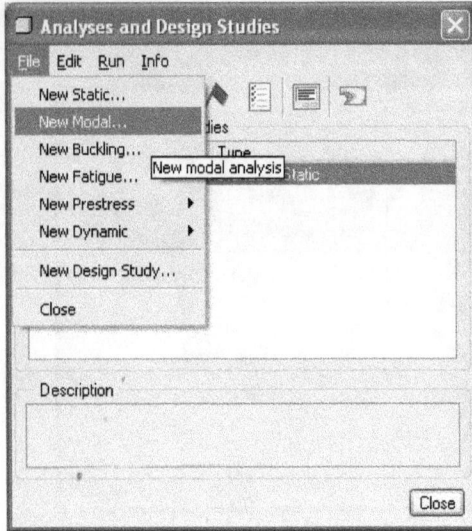

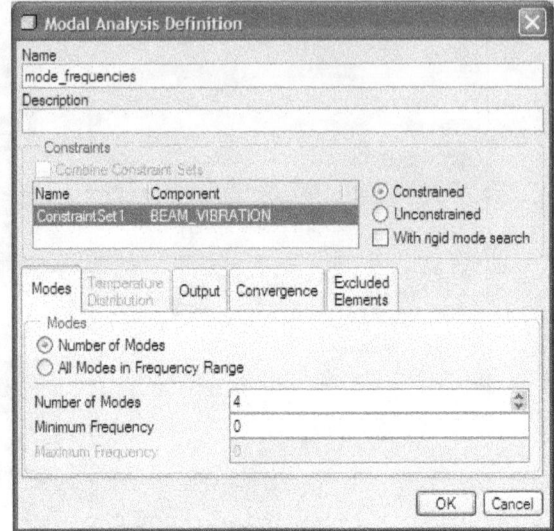

When the run is completed, review the report, which provides the information of the natural frequencies of the first 4 modes of this beam. These values are 537, 699, 2370 and 2900 Hz, respectively.

```
Number of Modes: 4

  Mode    Frequency (Hz)
  ----    --------------
     1    5.367029e+02
     2    6.994058e+02
     3    2.369843e+03
     4    2.899950e+03
```

Step 11: Perform a dynamic analysis to confirm the natural frequencies identified by the modal analysis.
File > New Dynamic > Frequency.

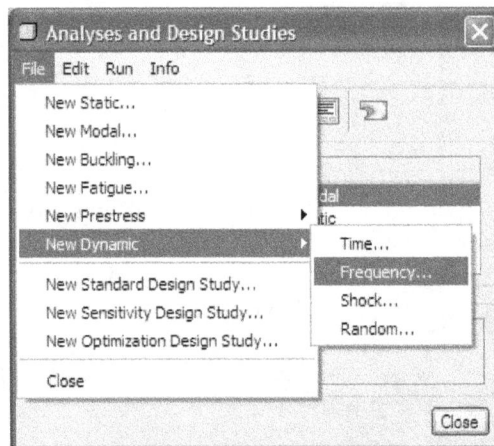

In the **Dynamic Frequency Analysis Definition** window, type *dynamic_frequency* as the name > accept the default load setting > under **Modes**, type *3* as the damping coefficient ($\zeta=3\%$). Under **Previous Analysis**, check the **Use Modes has previous design study** and select **mode_frequencies > OK.**

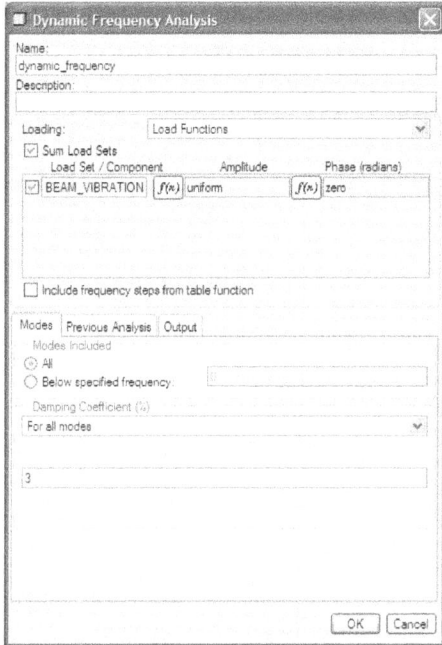

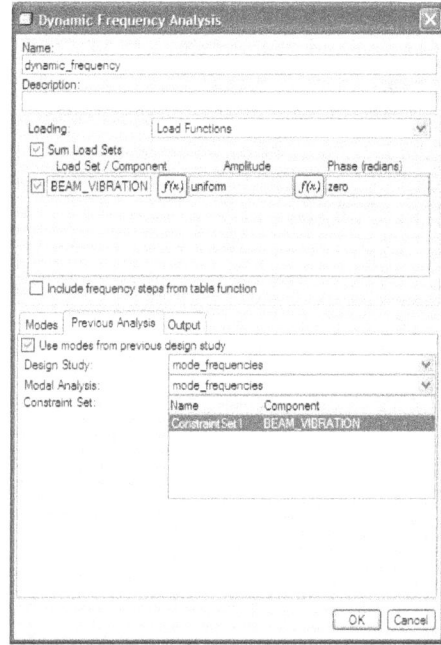

Click the green flag of **Run,** and upon completing, show the result. Type *spectrum_plot* as the name > **Graph > Measure >** select the measure defined as **displacement_y > OK > Frequency > OK and Show**.

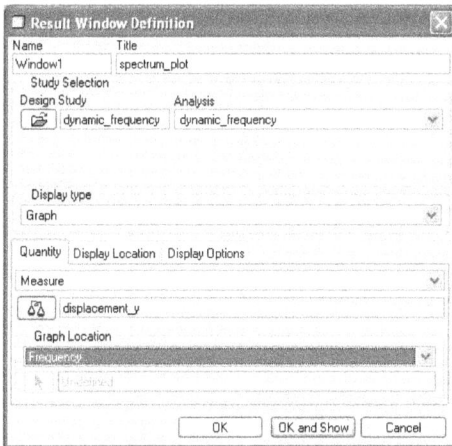

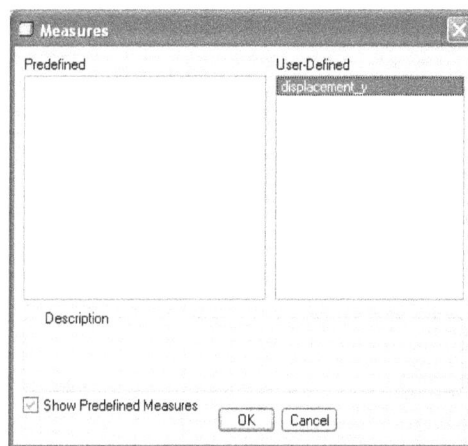

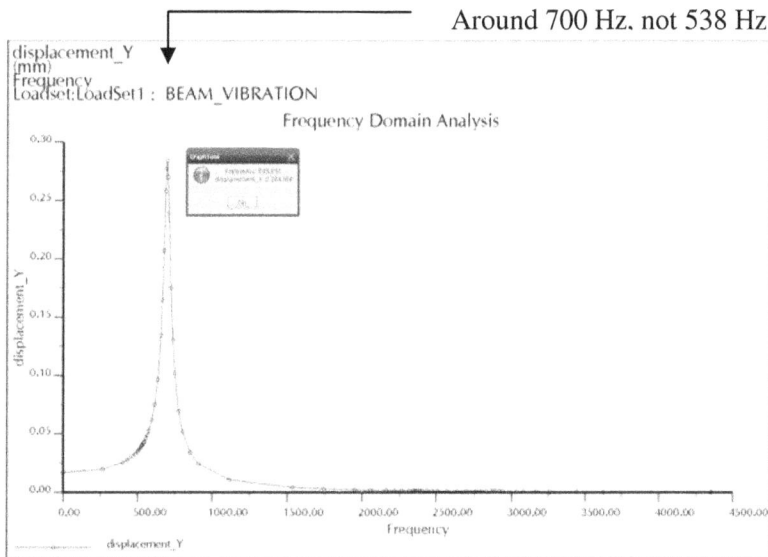

Around 700 Hz. not 538 Hz

As illustrated above, the first mode with its frequency, which is approximately located around 700 Hz. Note that the natural frequency of the first mode obtained from the model analysis was 538 Hz. The natural frequency of the second mode was 700 Hz. A question is now presented: "Which study gives an accurate solution in terms of the natural frequency of the first mode?" We will provide an answer to this question in the later part of this document.

Step 12: Perform a dynamic analysis in the time domain with the exciting force given by a unit impulse function (transient response).

File > New Dynamic > Time.

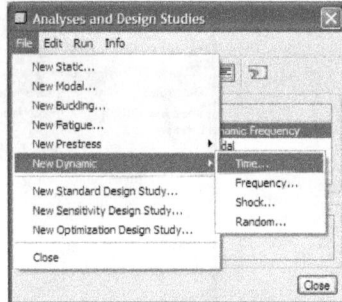

In the **Dynamic Time Analysis Definition** window, type *dynamic_time* as the name > accept the default load setting, which is impulse > under **Modes**, type *3* as the damping coefficient. Under **Previous Analysis,** check the **Use Modes has previous design study** and select **mode_frequencies**.

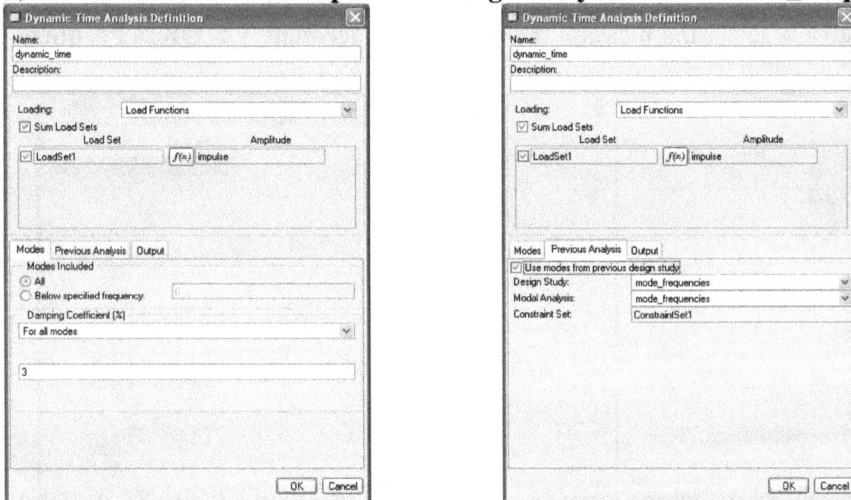

Under **Output**, set the **Maximum Time** to **Automatic**, and check the box for calculating stresses > **OK**.

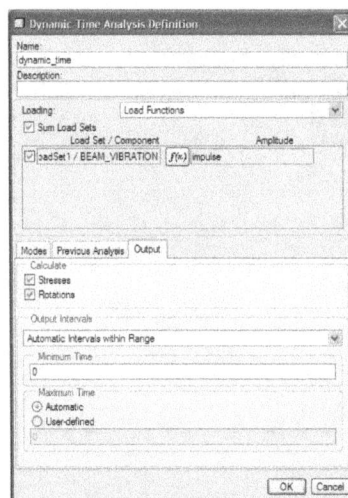

Click the green flag of **Run,** and upon completing, show the result. Type *impulse_response* as the name > **Graph > Measure >** select the measure defined as **displacement_y > OK > Time > OK and Show**.

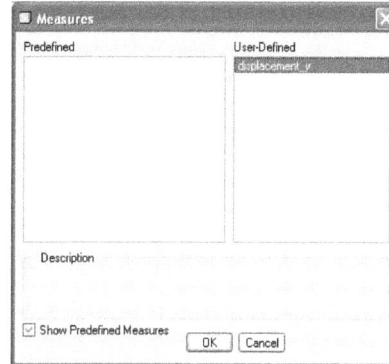

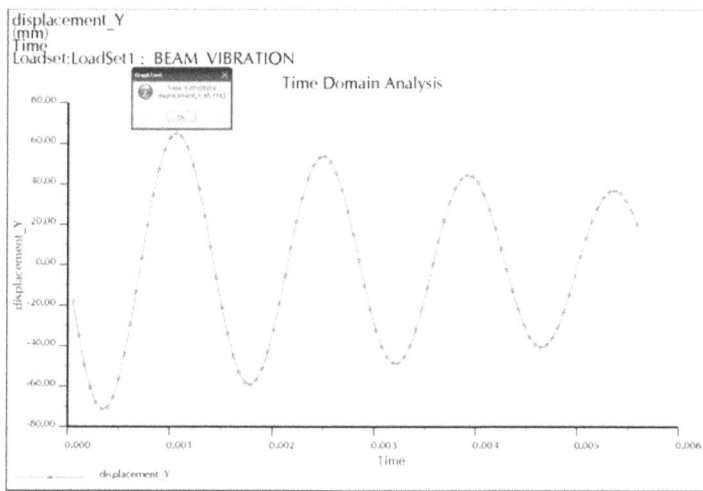

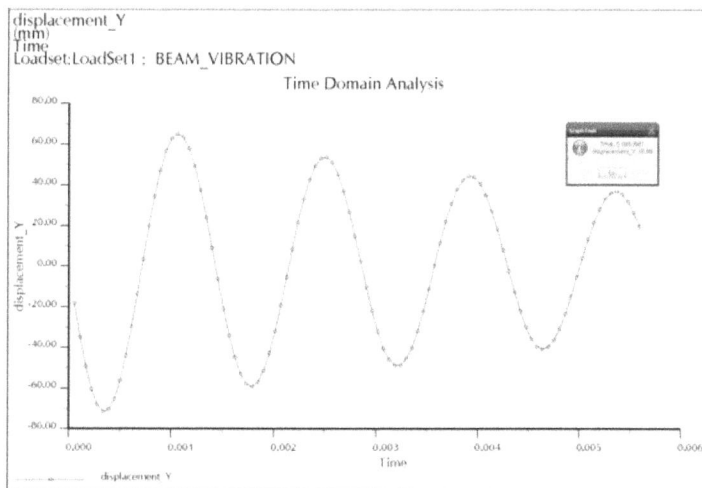

From the above plot, users should have the following data if collected:

	First Peak	Second Peak	Third Peak	Fourth Peak
Time (second)	0.00106204	0.00251536	0.00391278	0.0053661
Magnitude Y (mm)	65.12	53.75	44.58	36.96

Users may use the following formulas to estimate the damping coefficient of the beam system.

Logarithmic decrement, δ, is used to find the damping ratio of a damped system in the time domain. The logarithmic decrement is the natural log of the amplitudes of any two successive peaks:

$$\delta = \frac{1}{n} \ln \frac{x_0}{x_n},$$

where x_0 is the greater of the two amplitudes and x_n is the amplitude of a peak n periods away. The damping ratio is then found from the logarithmic decrement:

$$\zeta = \frac{1}{\sqrt{1 + \left(\frac{2\pi}{\delta}\right)^2}}.$$

The **damping ratio** can then be used to find the undamped natural frequency ω_n of vibration of the system from the damped natural frequency ω_d:

$$\omega_d = \frac{2\pi}{T},$$

$$\omega_n = \frac{\omega_d}{\sqrt{1 - \zeta^2}},$$

where T, the period of the waveform, is the time between two successive amplitude peaks.

Let us the data associated with the First Peak and the Fourth Peak to do the calculations.

$$\delta = \frac{1}{4} \ln \frac{65.12}{36.96} = \frac{1}{4} \ln(1.762) = \frac{1}{4}(0.566) = 0.1416$$

$$\zeta = \frac{1}{\sqrt{1 + \left(\frac{2\pi}{\delta}\right)^2}} = \frac{1}{\sqrt{1 + \left(\frac{4\pi^2}{\delta^2}\right)}} = \frac{1}{\sqrt{1 + \left(\frac{39.48}{0.1416^2}\right)}} = \frac{1}{\sqrt{1 + 1968.975}} = \frac{1}{44.38} = 0.023$$

$$T = \frac{t_4 - t_1}{3} = \frac{0.0053661 - 0.00106204}{3} = \frac{0.0043}{3} = 0.001435$$

$$\omega_d = \frac{2\pi}{T} = \frac{6.2832}{0.001435} = 4379.49 \text{ (rad/sec)} \quad \text{and} \quad f_d = \frac{1}{T} = \frac{1}{0.001435} = 697 \text{ (1/sec or Hz)}$$

$$\omega_n = \frac{\omega_d}{\sqrt{1 - \zeta^2}} = \frac{4379.49}{\sqrt{1 - (0.023)^2}} = \frac{4379.49}{0.9997} = 4380.6 \text{ (rad/sec)}$$

$$f = \frac{\omega_n}{2\pi} = \frac{4380.6}{6.2832} = 697.2 \text{ (1/sec or Hz)}$$

Step 13: Perform a dynamic analysis in the time domain with the exciting force given by a periodic function, say a sine wave function (forced vibration in its steady-state).

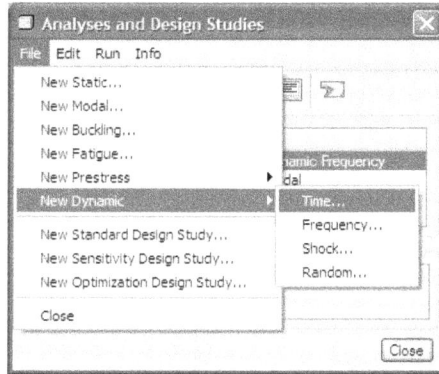

In the **Dynamic Time Analysis Definition** window, type *forced_vibration* as the name > to define a new time function, click *f(x)* > **New** > type sin(2*pi*10*time) > **Review** > change the Upper Limit to 1 > **Graph** > after reviewing the force function, **File > Exit > Done > OK > OK**.

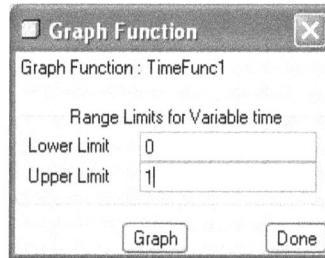

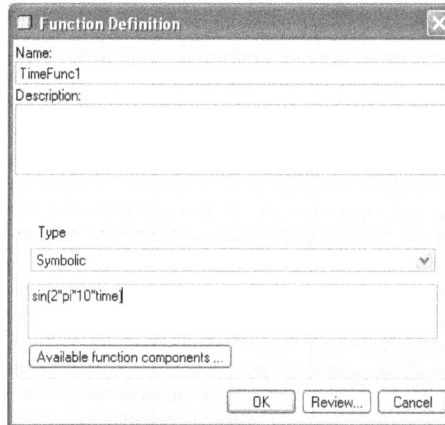

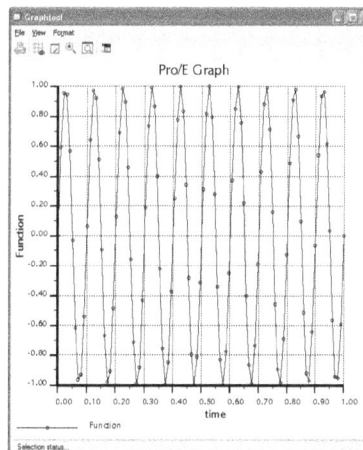

Under **Modes**, type *3* as the damping coefficient. Under **Previous Analysis**, check the **Use Modes has previous design study** and select **mode_frequencies**.

Under **Output**, select **Stresses** > select **User-defined** > specify *1* as the **Maximum Time** > **OK**.

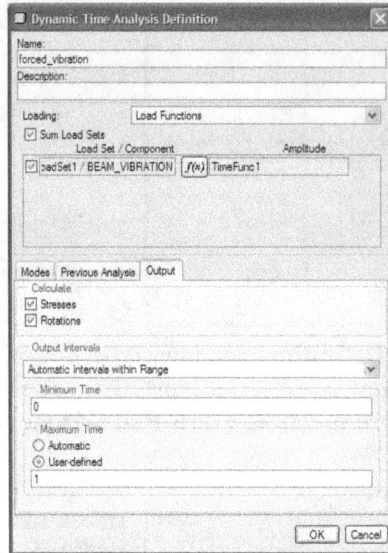

Click the green flag of **Run,** and upon completing, show the result. Type *forced_vibration* as the name > **Graph** > **Measure** > select the measure defined as **displacement_y** > **OK** > **Time** > **OK and Show**.

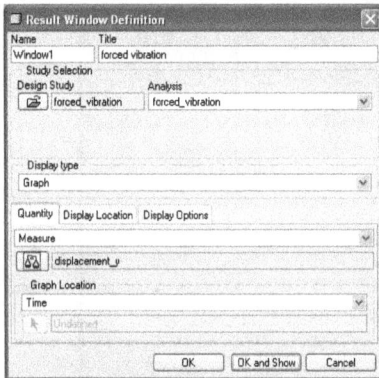

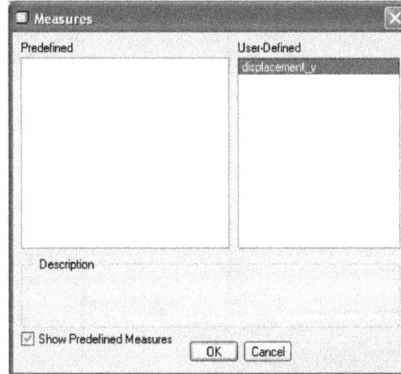

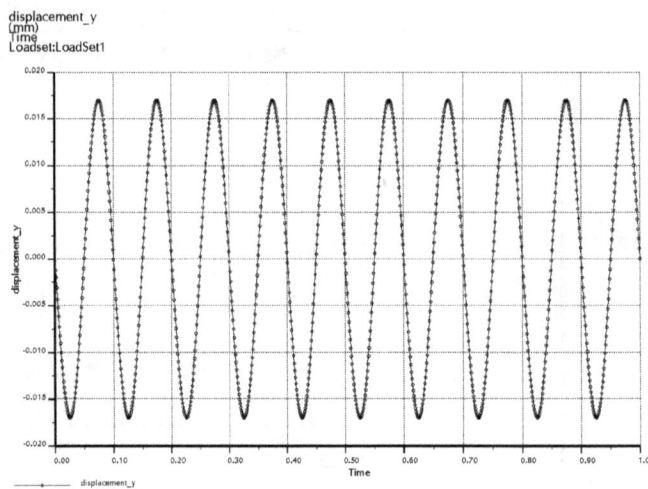

Step 14: Perform a dynamic analysis in the time domain with the exciting force given by a periodic function, say a sine wave function with the frequency equal to *538* Hz (forced vibration at resonance).

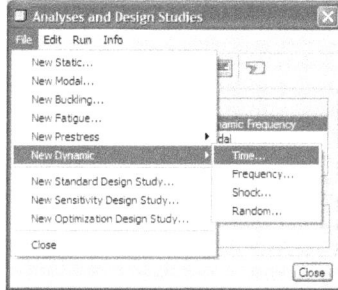

In the **Dynamic Time Analysis Definition** window, type *vibration_at_resonance* as the name > to define a new time function, click *f(x)* > **New** > type *sin(2*pi*538*time)* > **Review** > change the **Upper Limit** to *0.02* > **Graph** > after reviewing the force function, **File > Exit > Done > OK > OK**.

Under **Modes**, type *3* as the damping coefficient. Under **Previous Analysis**, check the **Use Modes has previous design study** and select **mode_frequencies**. Under **Output**, select **Stresses** > select **User-defined** > specify *0.02* as the **Maximum Time** > **OK**.

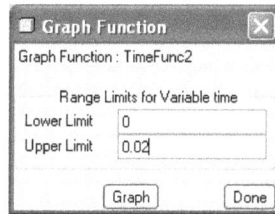

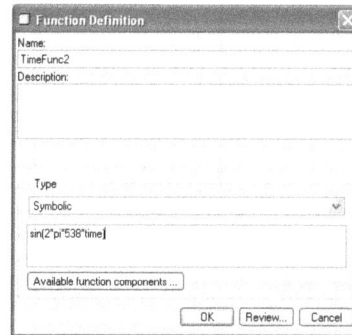

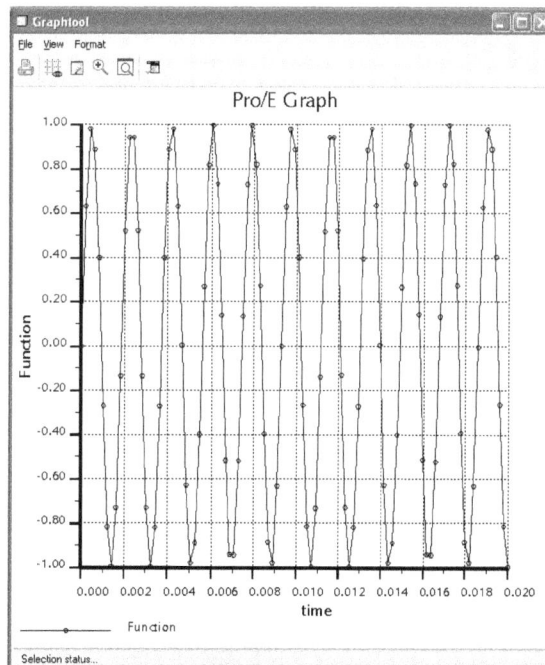

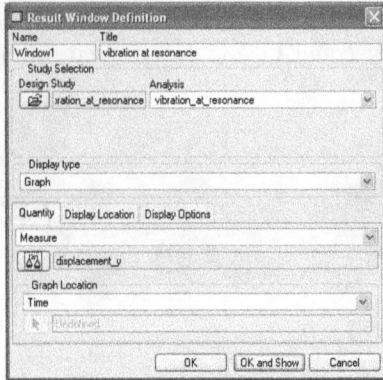

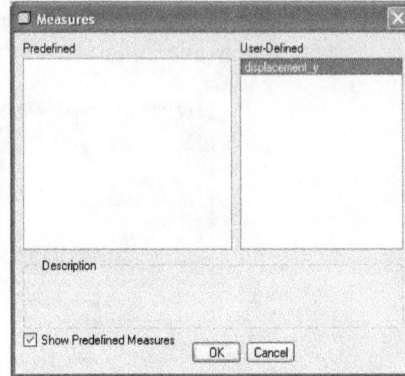

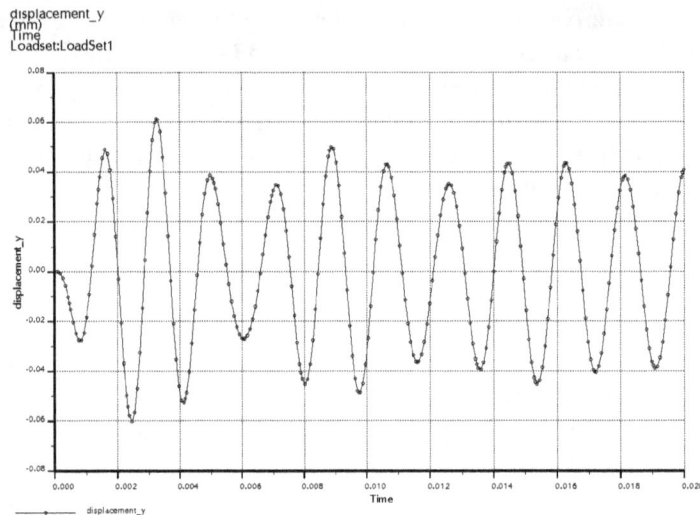

displacement_y
(mm)
Time
Loadset:LoadSet1

After examining the displacement plot, there is no sign indicating the presence of resonance at 538 Hz.

Now let us change the frequency of the excitation force from 538 Hz to 540Hz and run the program. There is no sign indicating the presence of resonance at 540 Hz. Let us repeat the run at 550 Hz, 560Hz, 570 Hz, 580 Hz, 590Hz, 600 Hz, 620 Hz, 640 Hz, and 660 Hz. There is no sign indicating the presence of resonance.

Now let us run at 680 Hz. First highlight the design study called *vibration_at_resonance* > Edit > Analysis/Study > to modify the frequency value, click *f(x)* > highlight TimeFunc2 > **Edit** > Change 660 to 680, or type *sin(2*pi*680*time)* > **Review** > change the **Upper Limit** to *0.02* > **Graph** > after reviewing the force function, **File > Exit > Done > OK > OK**. Run the program and show the result.

Edit

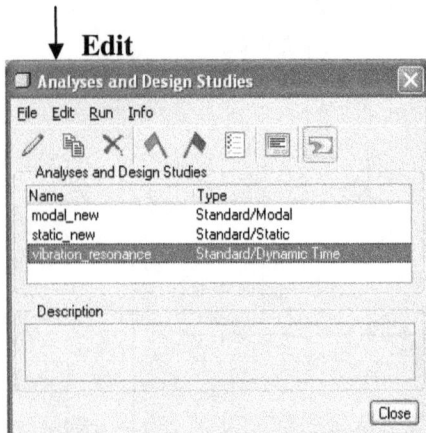

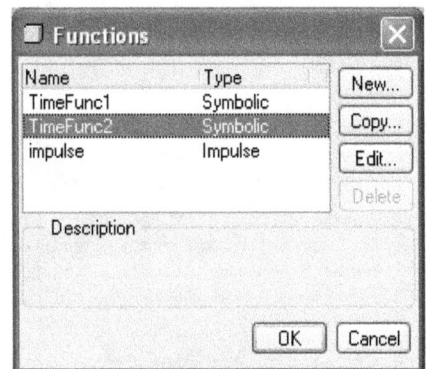

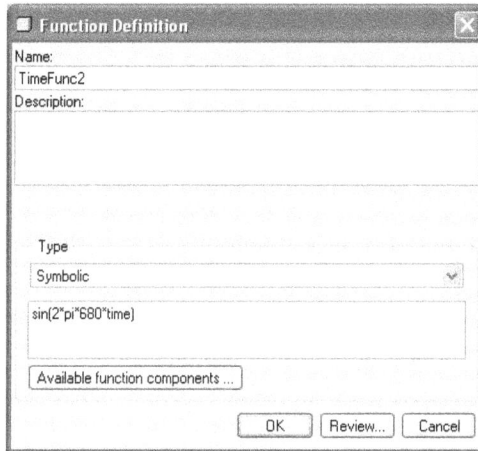

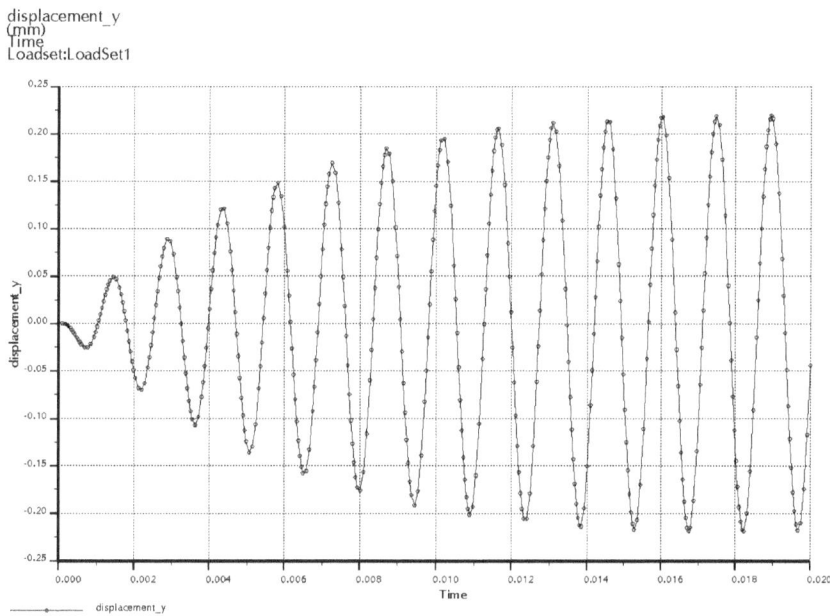

After examining the displacement plot, there is sign indicating the presence of resonance at 680 Hz.

Now let us run the program at 690 Hz. After examining the displacement plot, there is sign indicating the presence of resonance at 690 Hz.

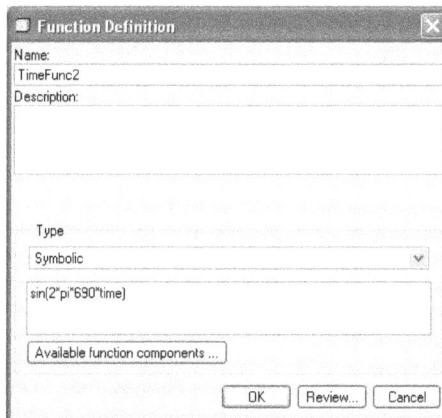

displacement_y
(mm)
Time
Loadset:LoadSet1

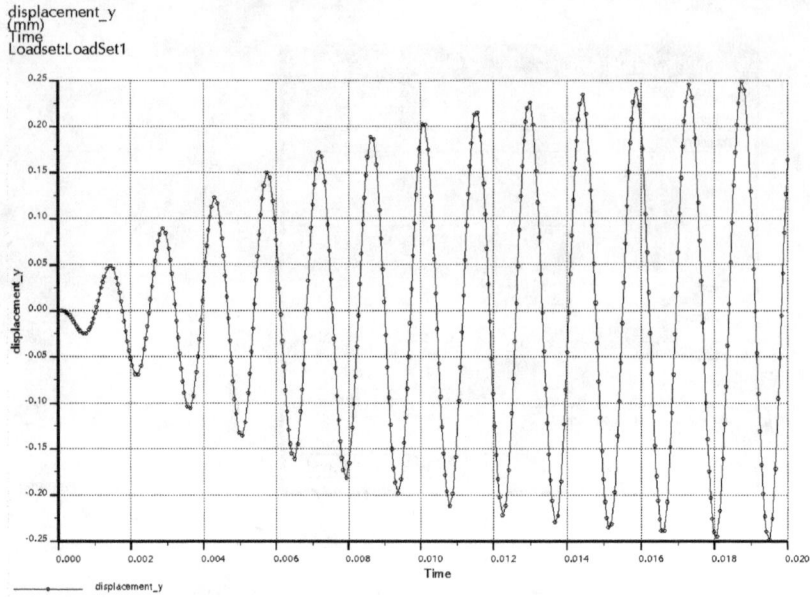

Now let us run the program at 700 Hz. After examining the displacement plot, there is sign indicating the presence of resonance at 700 Hz.

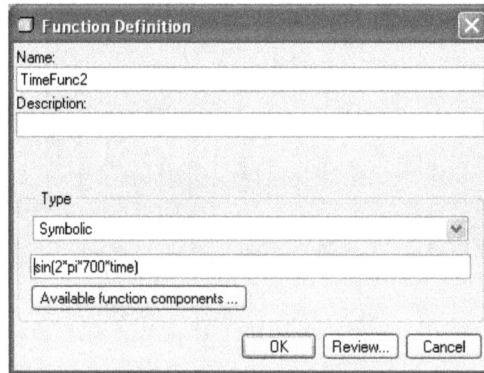

displacement_y
(mm)
Time
Loadset:LoadSet1

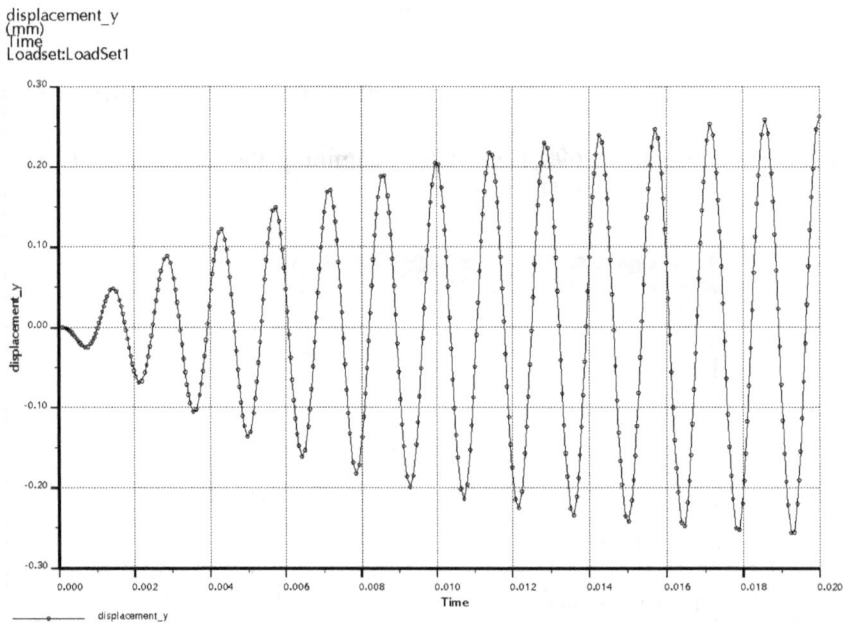

Now let us run the program at 730 Hz. After examining the displacement plot, there is sign indicating the presence of resonance at 730 Hz. However, the strength of resonance has been reduced significantly.

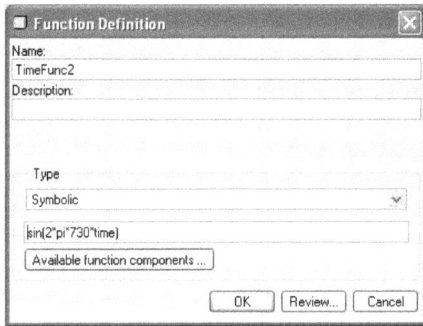

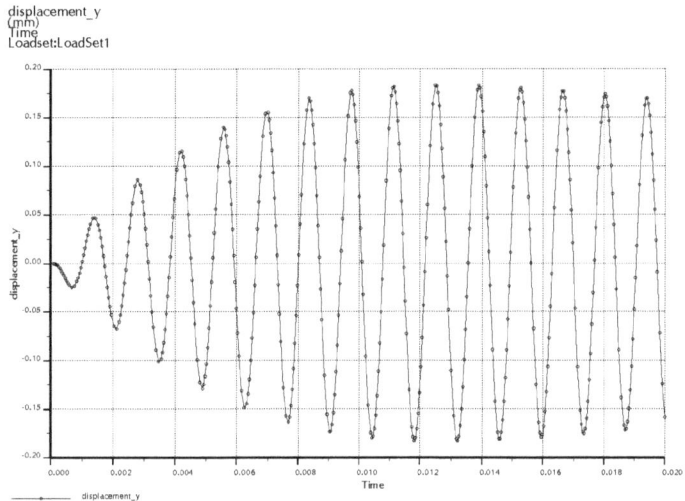

It is very interesting to observe that the resonance frequency is in about 700 Hz. Following the vibration theory, the resonance frequency is given by

$$\omega = \omega_n \sqrt{(1 - \zeta^2)}$$

where ω_n stands for the natural frequency of the first mode and ζ is the damping coefficient (set to 3% in the current case study).

Step 15: Investigate the VM stress distribution at resonance.
    Click the icon of Review Results. Select **Graph > Measure** > select stress_vm as the measure to be plotted > **OK > Time > OK and Show.**

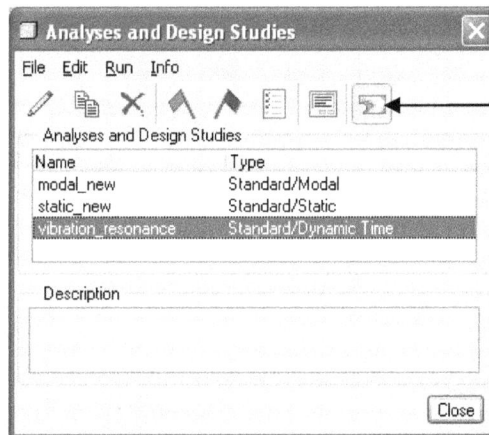

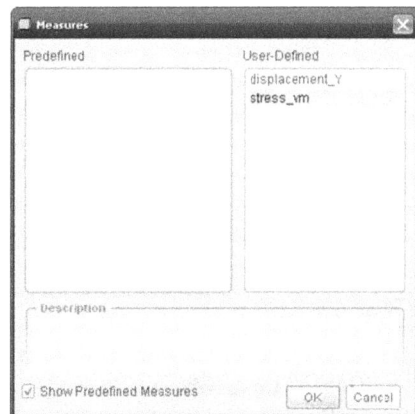

The following plot will be on display. By examining the pattern of magnitude increase, it is not difficult to image the damage possibly to be generated at the resonance. This fact would be evident if data are collected from the plot, as shown in the following table as well.

Time (second)	0.0014	0.0093	0.02
Magnitude σ_{vm}(MPa)	10.4	43.3	57.0

t = 0.0014 sec
σ_{vm} = 10.4 MPa

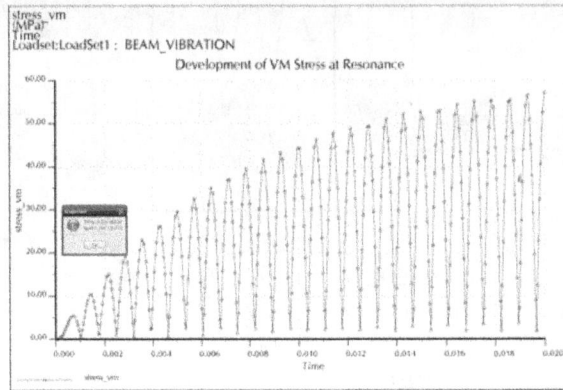

t = 0.0093 sec
σ_{vm} = 43.3 MPa

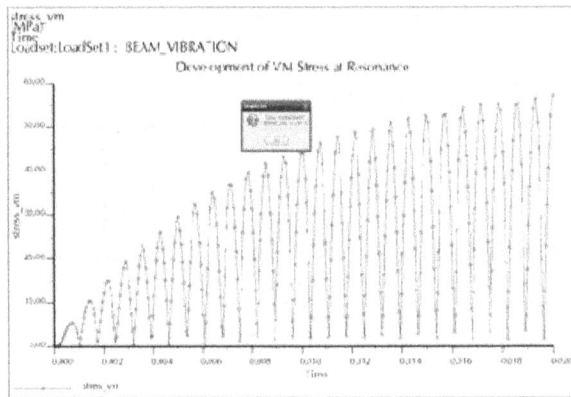

t = 0.02 sec
σ_{vm} = 57 MPa

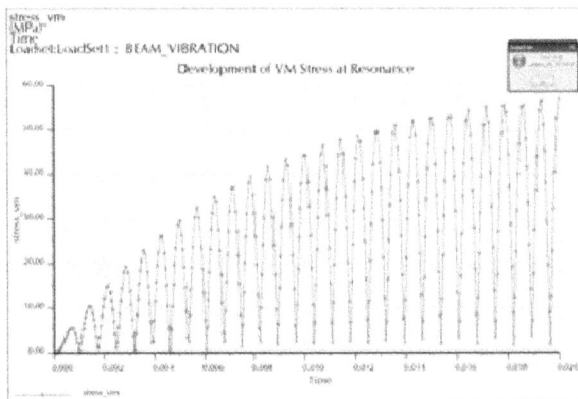

7.7   References

1. Algor Processor Reference Manual, Algor Interactive Systems, 150 Alpha Drive, Pittsburgh, PA 15238
2. R. D. Cook, D. S. Maikus and M. E. Plesha, Concepts and Applications of Finite Element Analysis, 3rd edition, Wiley, New York, NY, 1989.
3. J. W. Dally, Design Analysis of Structural Elements, Goodington, New York, NY, 2003.

4. 1996.
5. R. H. Gallagher, Finite Element Analysis Fundamentals, Prentice-Hall, Englewoood Cliffs, NJ, 1975.
6. J. M. Gere and S. P. Timoshenko, Mechanics of Materials, 3rd edition, PWS-KENT Publishing Company, Boston, 1990.
7. W. W. Hager, Applied Numerical Linear Algebra, Prentice-Hall, Englewood Cliffs, NJ, 1988.
8. C.S. Krishnamoorthy, Finite Element Analysis, Theory and Programming, 2nd Ed., 1995.
9. T. Ross, T. Johns and Emile, Dynamic Buckling of Thin Walled Domes under External Water Pressure, Applied Solid Mechanics 2nd Conference, Ed. Tooth & Spence, Elsevier, pp. 211-224, 1987.
10. P. Silvester, Higher-Order Polynomial Triangular Finite Element for Potential Problems, Int. J. Eng. Soc., Vol. 7, No. 8, pp. 849-861.
11. O. C. Zienkiewicz, The Finite Element Method, third edition, McGraw-Hill, New York, NY, 1977.
12. O. C. Zienkiewicz and R. L. Taylor, The Finite Element Method, 4th edition, Vol. 1, McGraw-Hill (UK),London, 1989.

7.8   Exercises

1.   Design a turning fork for the C note. The frequency of the first mode for the C note is 260.97 Hz. The geometrical shape shown below is for a turning force for the G-sharp fork. There are 2 dimensions are allowed to adjust so that a turning fork for the C note can be designed. The 2 dimensions are 5 mm and 6 mm, respectively. Plot the modal shapes of Mode 1, Mode 2, Mode 3 and Mode 4. What are the frequencies associated with these 4 modes?

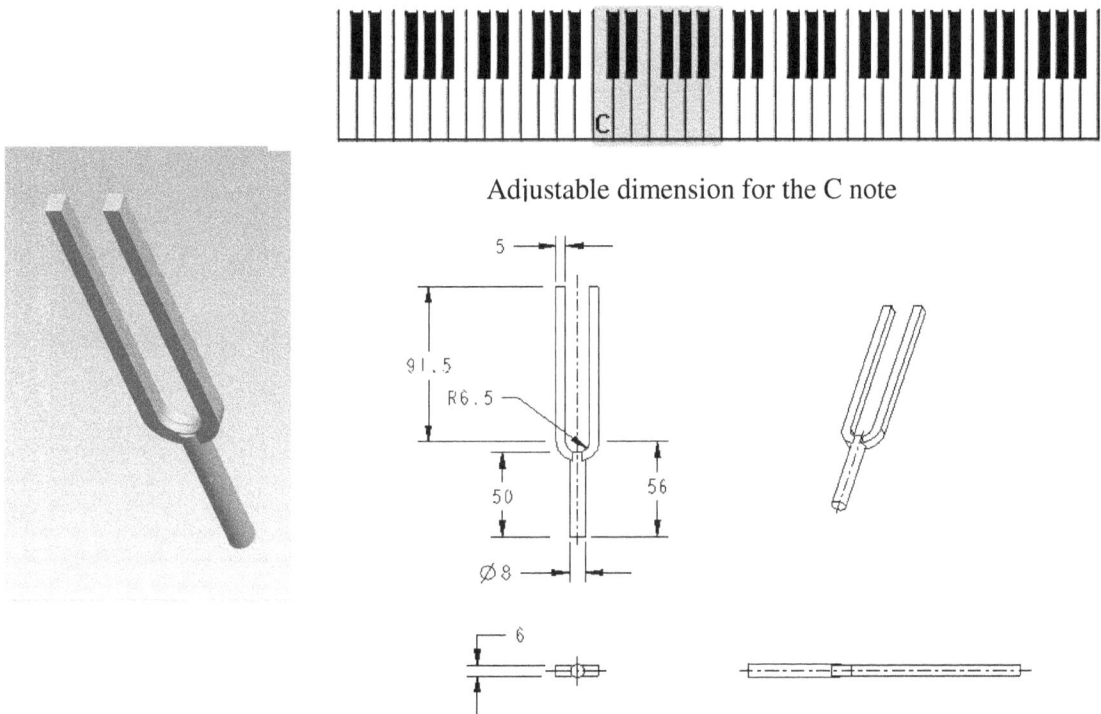

Adjustable dimension for the C note

2. Determine the first 4 modal shapes and their associated frequencies for the beam structure shown below. There are 2 cases. In case 1, the exciting force acts in the Y direction and downward. In case 2, the exciting force acts in the Z direction and towards the inside of the paper. Plot the 4 modal shapes for each of the 2 cases.

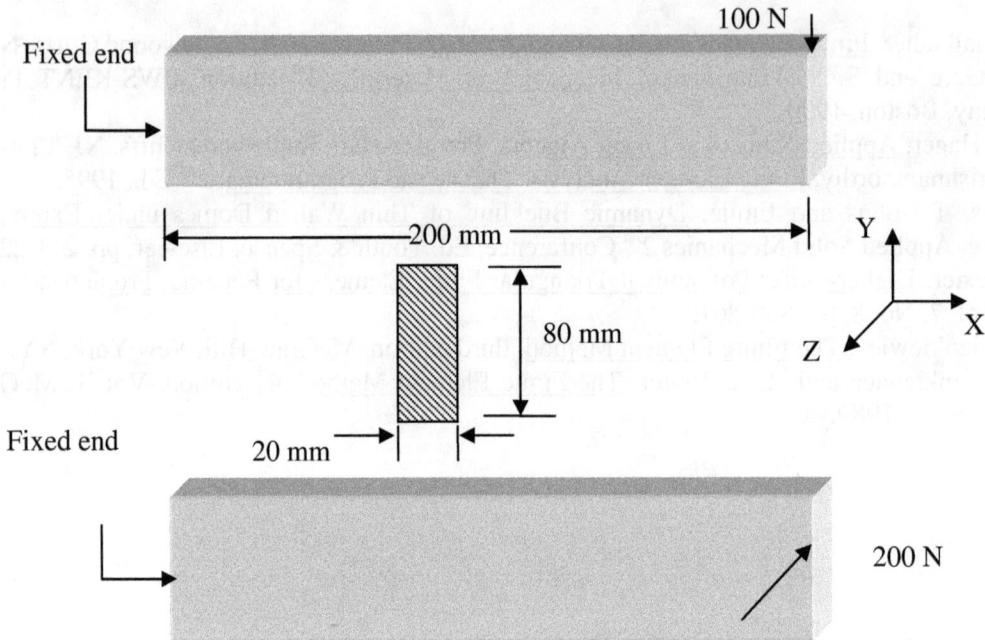

100 N

Fixed end

200 mm

Y

80 mm

Z

X

Fixed end

20 mm

200 N

3.   In section 7.4, a column structure is under investigation for buckling stability. At that time, we assumed that one end is firmly fixed to the ground. Now let us investigate the effects due to the boundary condition on the mode shape and BLF magnitude. In this exercise, let us change the constraint condition on the left side of the slender column from a fixing end to a pin joint with a small surface region to restrict the movement in the x-direction. Therefore, first cut a through hole at the left side, as shown. Afterwards, define the constraint condition which is closely related to a pin joint.

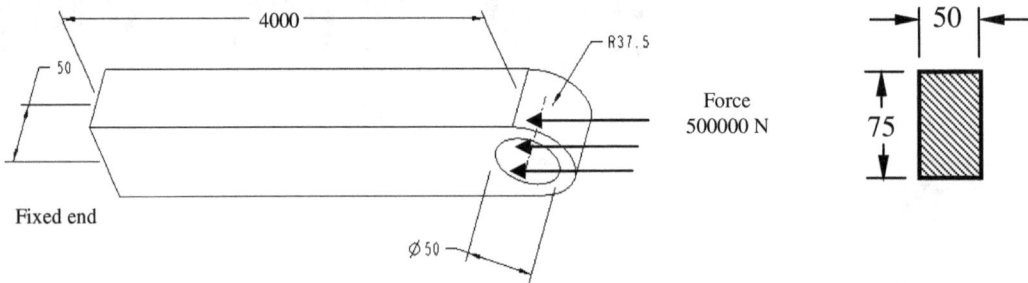

4000

50

R37.5

Force
500000 N

50

75

Fixed end

Ø 50

Construct a surface region at the bottom surface

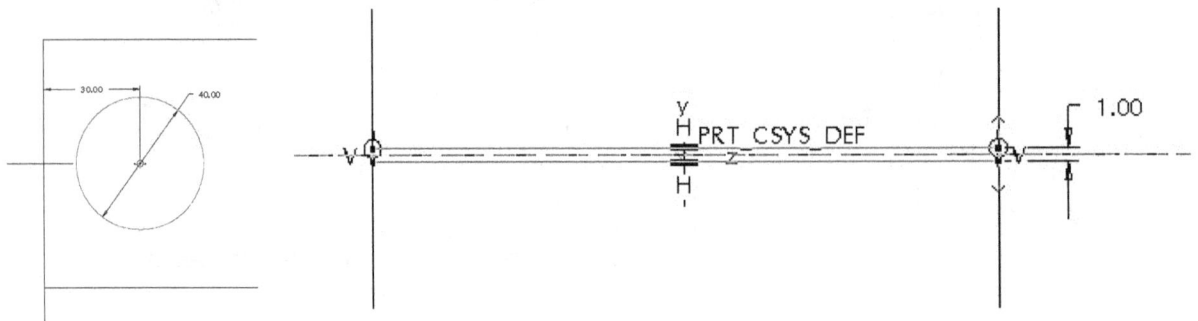

30.00   40.00

y
H PRT_CSYS_DEF
z
H

1.00

Determine the first 3 modal shapes and their associated BLF values.

CHAPTER 8

EVALUATION OF ENGINEERING DESIGN AND DESIGN OPTIMIZATION

8.1 Introduction

Design engineers design components and assemble them in their offices or under a virtual environment. A question, which is naturally to be asked, would be "Do their designed components or products meet the required specifications?" More important is such a question "What is the safety factor of the designed component or the product when considering the possible failure mode(s)?" Making design changes is a common practice for the purpose of improving the performance of the component and/or product or reducing the cost in the realization process. In this chapter, we present a case study. The component in this study is a bracket. The entire evaluation process represents an initial evaluation, a re-design process and re-evaluation so that numerical values of certain parameters of the designed bracket have been modified so that the performance of the re-design bracket meets the design specifications

8.2 Fundamentals of a Decision Making Process

The process of evaluation and redesign is a decision process. Before getting on the case study, let us go over the general guidelines and mathematical formulas used in a decision making process. For example, a company manufactures furniture. The company produces two types of furniture based on the demand in market. They are chairs and tables. The raw material used to manufacture the chairs and the tables is wood. Certainly, workers are needed to shape and assemble the components of the chairs and the tables. We will use normalized numbers to quantitatively describe this production process.

We assume that the objective of making a decision is to maximize the profit from selling the chairs and tables in market. Assume that the profit gained from selling a chair is \$4 and the profit for selling a table is \$5. We may mathematically represent this objective function as

Maximize the profit: profit is determined by $4*x_1 + 5*x_2$
where x_1 and x_2 are the number of the chairs and the number of the tables sold in market.

How many chairs and how many tables can be manufactured? There are certain constraint conditions, which limit the number of chairs and the number of table to be manufactured. Assume that total wood volume available at this moment for making furniture is 10 units. The number of workers available is 36. In order to make a chair, the consumption of wood volume is 1 unit. To make a table, the consumption of wood volume is 2 units. Therefore, the first constraint condition is

$$x_1 + 2x_2 \leq 10 \quad \text{(wood)}$$

To make a chair, 6 workers are needed. To make a table, 6 workers are needed. Therefore, the second constraint condition is

$$6x_1 + 6x_2 \leq 36 \quad \text{(workers)}$$

For example, we may add a new constraint condition, indicating that the maximum number of chairs needed in market is 4. Therefore, the third constraint condition is

$$x_1 \leq 4 \quad \text{(limit)}$$

When combining the objective function and the constraint conditions together, we characterize such a decision process mathematically using the following format:

$$\text{Maximize} \quad Z = 4x_1 + 5x_2$$

Subject to

$$x_1 + 2x_2 \leq 10 \quad \text{(wood)}$$

$$6x_1 + 6x_2 \leq 36 \quad \text{(workers)}$$

$$x_1 \leq 4 \quad \text{(limit)}$$

$$x_1 \geq 0 \quad \text{(non - negative)}$$

$$x_2 \geq 0 \quad \text{(non - negative)}$$

The solution to this so-called optimization process is the maximum profit achievable is $28 when $x_1 = 2$ and $x_2 = 4$. Such a combination makes the fullest use of the availability of wood and the number of workers available while it meets the limit on the number of tables set by the market demand.

In the following we use a graphical method to demonstrate the procedure of obtaining such an optimal solution. In this example, we have 2 design variables. As a result, we start the search for an optimal solution in a 2 dimensional space, as shown below. We only consider those regions where both x_1 and x_2 are positive.

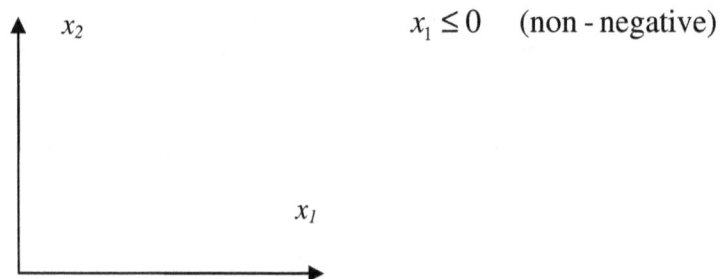

x_2

$$x_1 \leq 0 \quad \text{(non - negative)}$$

x_1

The second step in the search process is to identify a feasible region to search. In this example, there are three constraint conditions. Mathematically speaking, each constraint condition can be represented by a straight line.

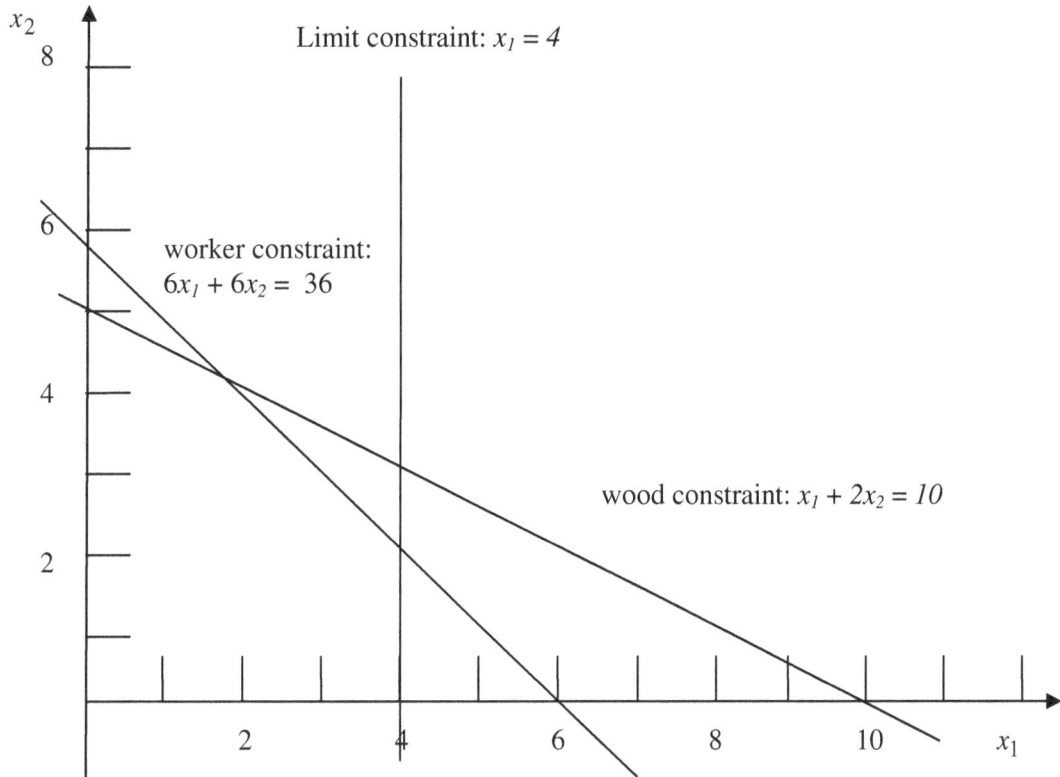

As illustrated, the three constructed lines intersect with each other. They also intersect with the x_1-axis and x_2-axis. When connecting all these segments, a closed area is formed. The shadow area shown in the following figure represents the feasible region to search for an optimal solution.

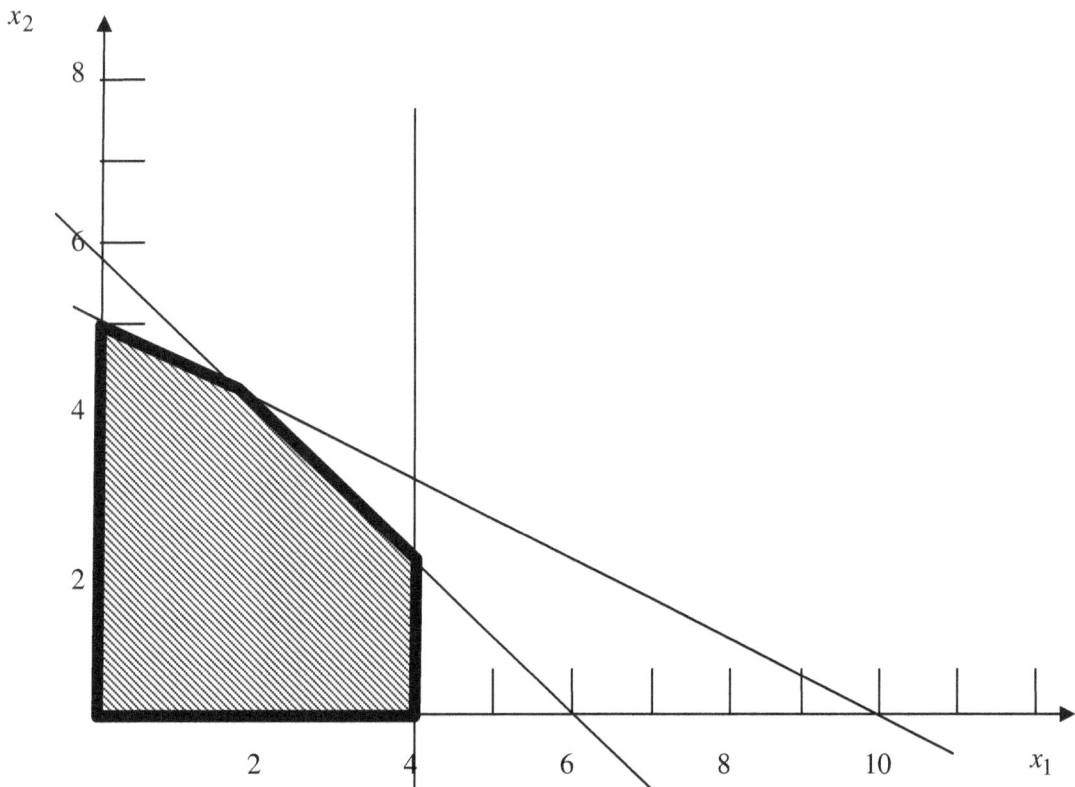

How to derive at an optimal solution? Readers may imagine moving an inclined line starting from the origin (x1 = x2 = 0) in a parallel fashion. The slope of this inclined line is determined by the 2 coefficients in the objective function. When this inclined line reaches its limiting position, the location (x_1 = 2 and x_2 = 4) is the optimal solution.

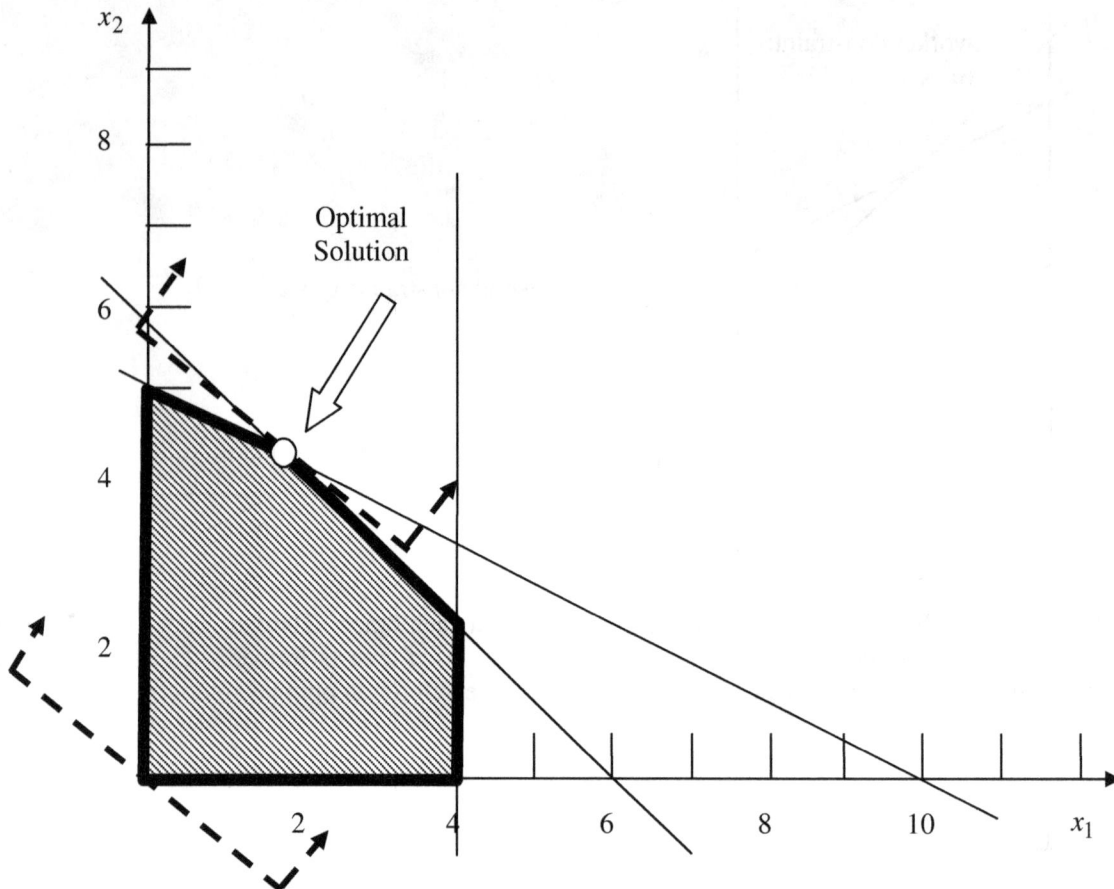

8.3   Case Study: Redesign a Bracket Component

The geometric shape of a bracket is shown below with its dimensions. The material is steel. The objective of this re-design process is to reduce the weight of the bracket mass so that the cost of the bracket can be reduced accordingly in terms of raw material utilization, packaging, and transportation.

In the re-design process, there are certain constraints. The first constraint is that the maximum von Mises stress should be kept below 85 MPa. The reason is that the yield stress of the steel material is about 225 MPa. For a safety factor of 2.5, the limit for the maximum von Mises stress should be set at 85 MPa. However, we are informed that the maximum von Mises stress at this design stage when it is subjected to loading is 90 MPa, a violation to the safe constraint condition.

The second constraint is the limit on the maximum displacement magnitude. The current value of this performance index is 0.028 mm. It has to be kept below 0.02 mm.

The third constraint is related to the product process. The bracket components will be fabricated using a punch process where a mold is used. We are limited to change a few dimensions among the many shown on the following engineering drawing. Therefore, we have to select the design parameters with care in the redesign process so that the changes of their dimensions can be accommodated in the process of modifying the mold design accompanied.

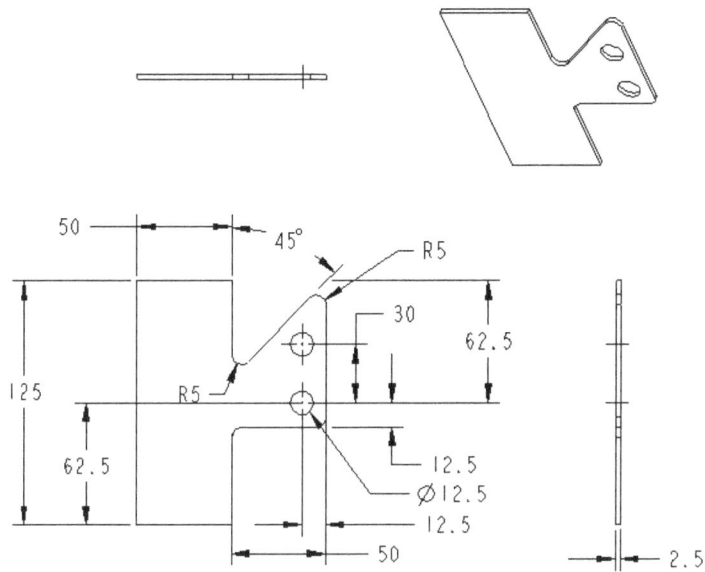

The boundary condition is that the top surface and the bottom surface on the right side of the bracket are fixed.

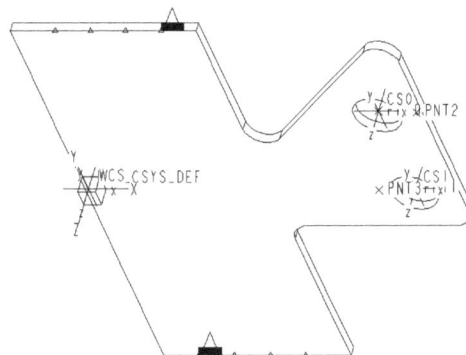

The loading condition consists of 2 bearing loads. One is acting on the upper hole and the other on the lower hole. The magnitudes of these 2 bearing load are 1000 Newton.

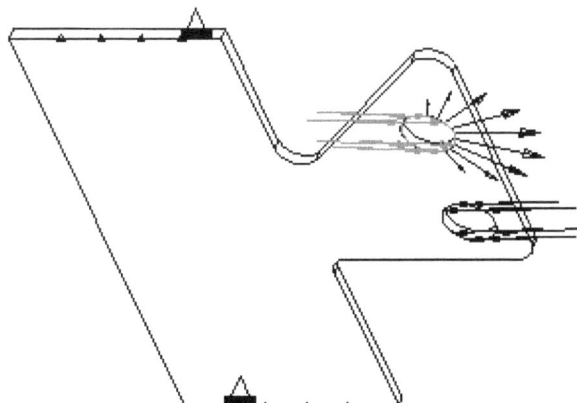

Step 1: Under the **Pro/ENGINEER** design environment, create a 3D solid model of the bracket. Assume that the file name is *bracket.prt*.

In the process of creating this bracket, pay special attention to the two inner fillets. They are created as equally to each other. The same is applied to the two outer tips. Their dimensions will be used as two (2) design parameters in the optimization process.

Step 2: Define design parameters.

Assume that the following 4 dimensions will be used as the design parameters.

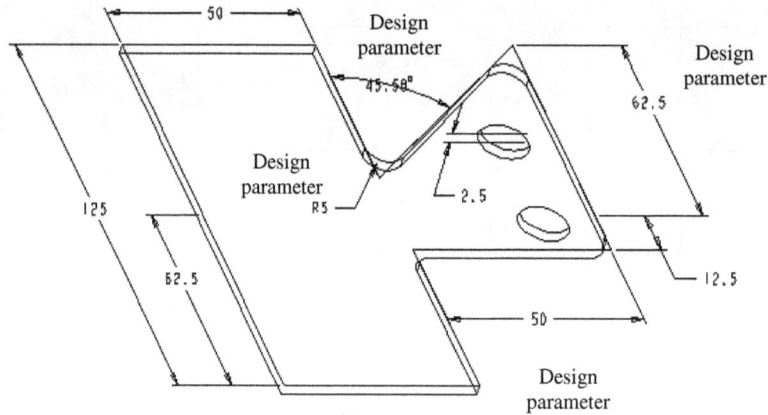

- The angle dimension: 45°
- The distance between TOP to the tip of the nose: 62.5
- The radius of the inner fillets: R5
- The over width: 50

Let us first use symbolic names for the 4 design parameters selected. From the model tree, highlight the protrusion feature > right click and pick **Edit**.

From the geometry on display, pick the dimension of 45 degrees, right click and pick **Properties**. Click the box called **Dimension Text**. In the box called **Name**, type in *ang*.

Repeat the above procedure by picking the dimension of 62.5 and type in *top*, picking the dimension of R5 and type in *inner_fillets* and picking the dimension of 50 and type over_width.

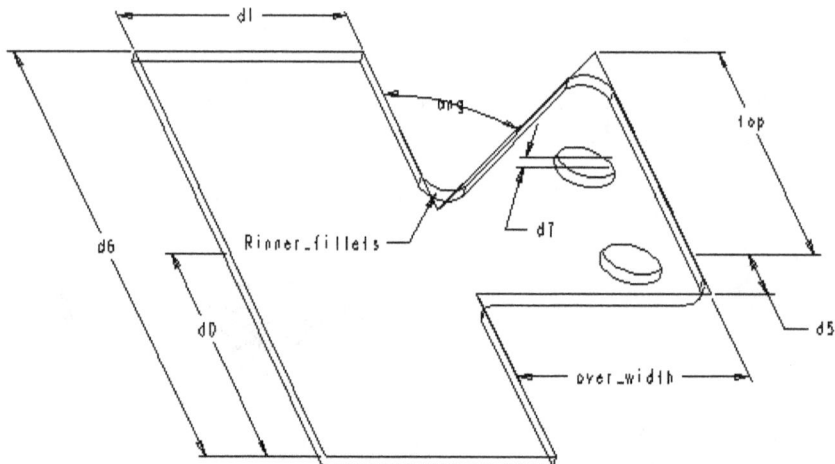

To switch between the dimensions in numerical values and the dimensions in symbolic names, from the main toolbar: **Info > Switch Dimensions**

Step 3: Use Relations to Assure the Compatibility of Geometric Shape in the Process of Optimization.

Note that the tip has to rotate as the angle dimension varies to maintain the desired part shape. The following relation is established.

From the main toolbar, **Tools > Relations >** in the window of **Relations**, type in the following relation, using three lines:

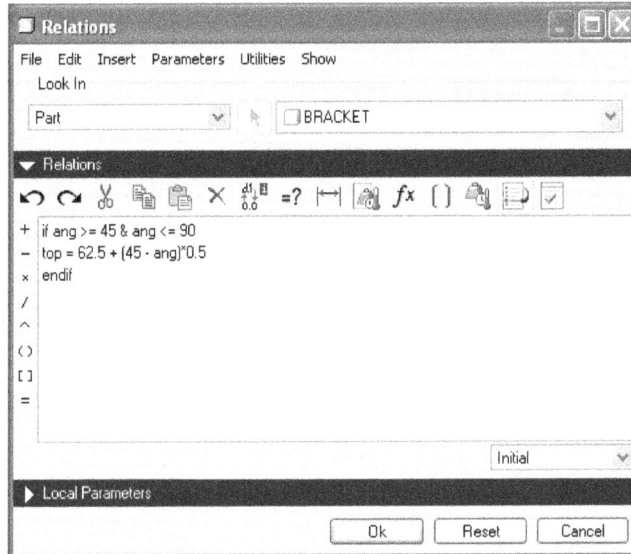

if ang >= 45 & ang <= 90
top = 62.5 + (45-ang)*0.5
endif

Upon completion, click **OK**.

To verify the compatibility and if the defined relation works, highlight the protrusion feature displayed on the model tree, right click and select Edit.

From the geometry on display, select the angle dimension and change it from 45 to 60 > click the icon of **Regenerate,** which is on the menu bar.

Select the angle dimension and change it from 60 to 90 > click the icon of **Regenerate,** which is on the main toolbar.

Select the angle dimension and change it back from 90 to 45 > **Regenerate.**

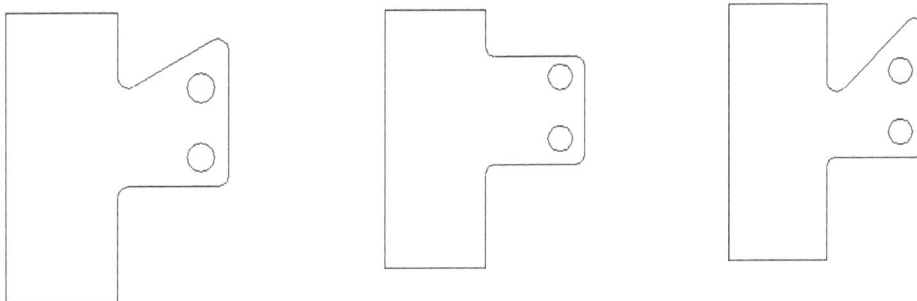

Step 4: Under the Pro/ENGINEER design environment, define 4 datum points for specifying the 2 bearing loads.

Define PNT0 and PNT1

Select the icon of datum point from the toolbar of creating datum features.
Select the **FRONT** datum plane and, while holding down the Ctrl key, select the axis of the upper hole **> OK. PNT0** is defined.
Repeat the above procedure to define **PNT1** by selecting the axis of the lower hole.

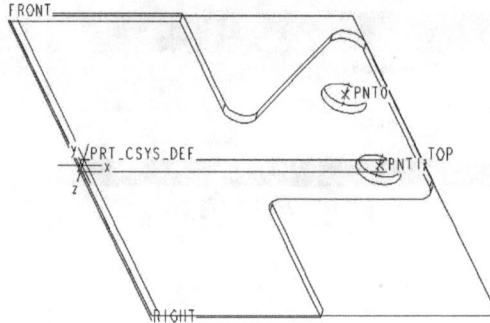

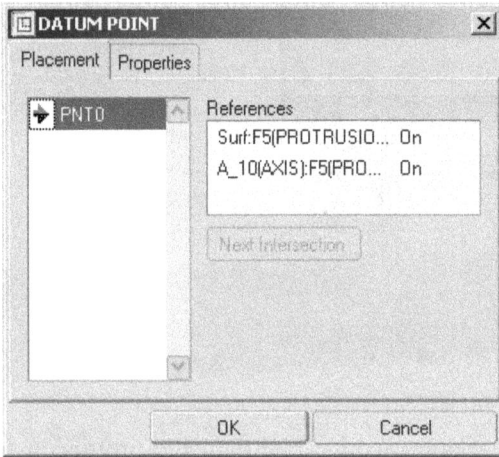

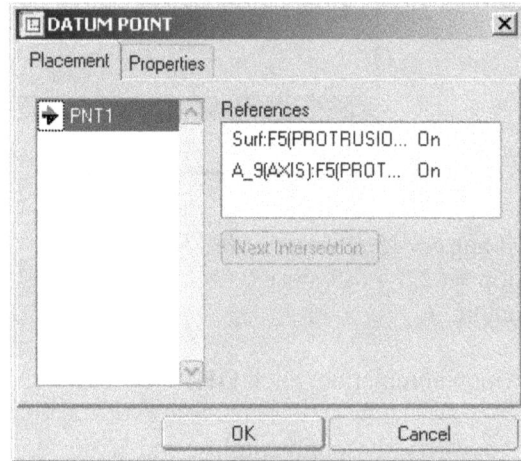

Define PNT2 and PNT3

To define **PNT2**, select the icon of datum point > click the middle location of the arc on the right side, as shown > OK. **PNT2** is defined.

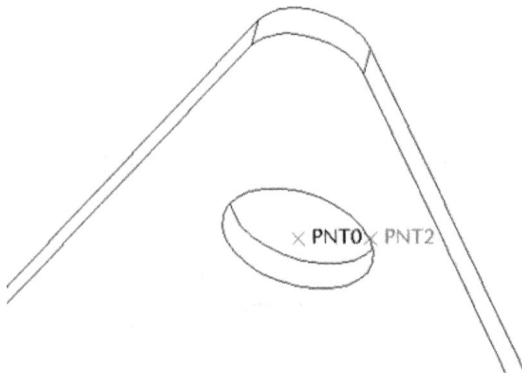

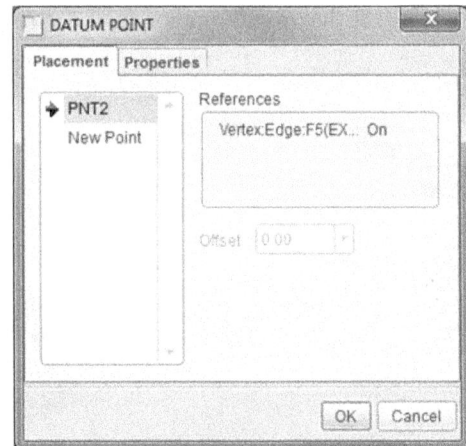

Repeat the above procedure to **PNT3**, as shown.

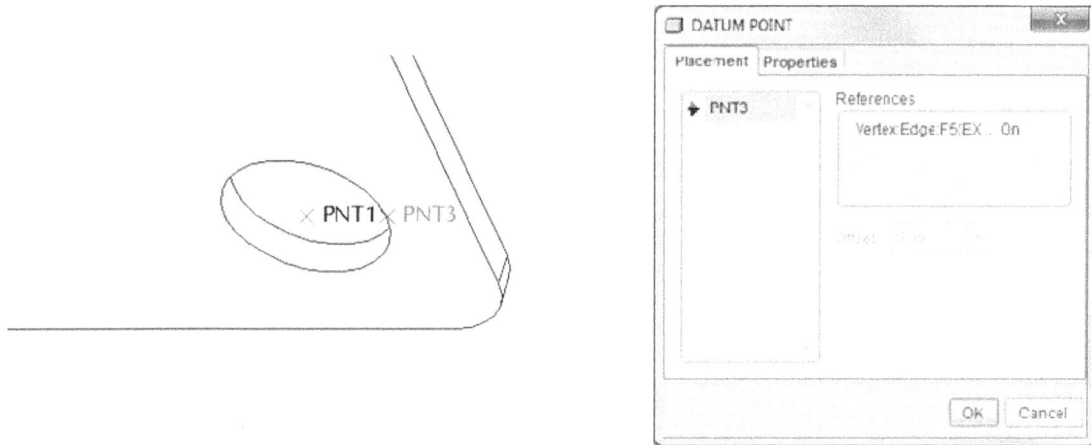

Step 5: Under the Pro/MECHANICA environment, develop a shell model.
From the top menu, click **Applications > Mechanica > Continue > Structure > OK.**

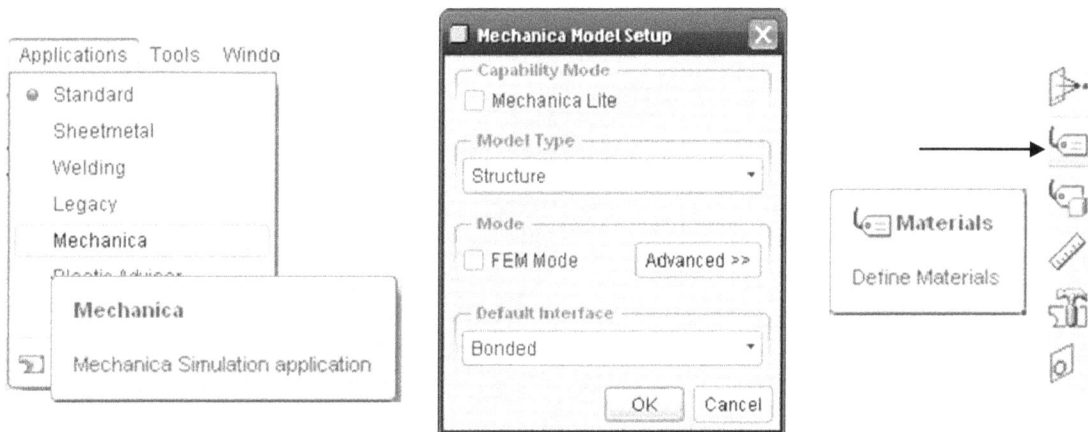

Click the icon of **Materials >** select **Steel** from the left side called Materials in Library > click the directional arrow so that the selected material type goes to the right side called Materials in Model.

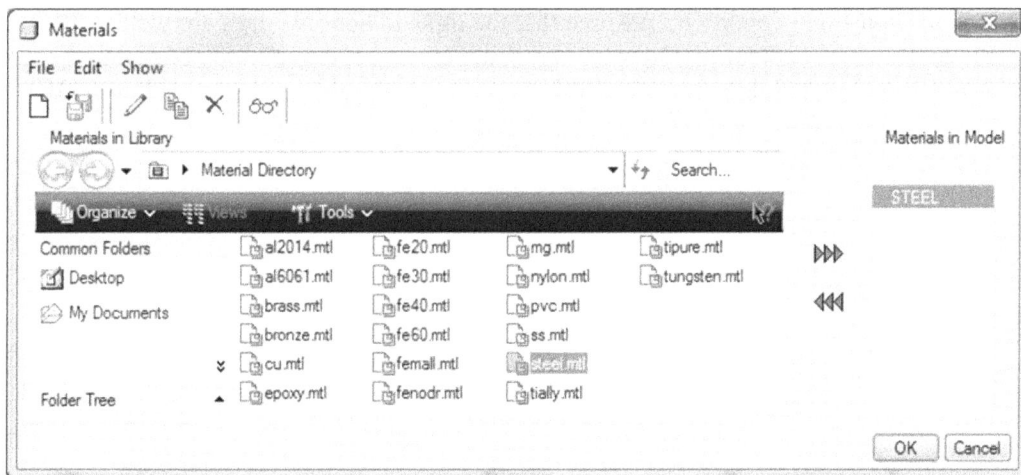

Click the icon of **Material Assignment**, the pop-up window indicates there is one component and the material type is Steel. Just click **OK**.

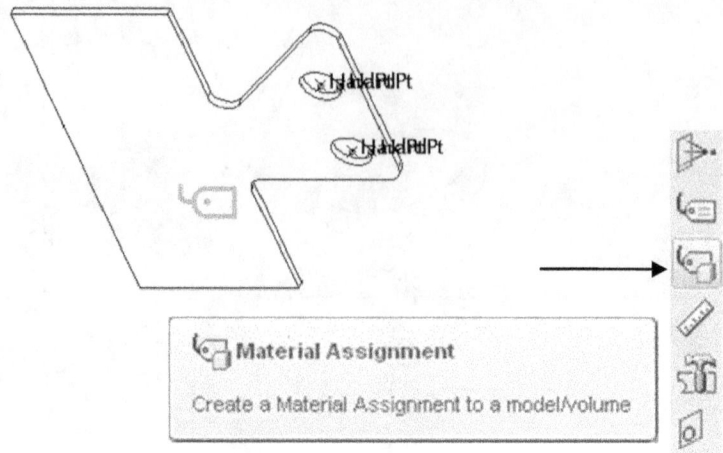

To define the shell mode, click the icon of **Shell Pair**.

Pick the front surface of the plate first. The software system picks up the opposing surfaces automatically. Click the check mark.

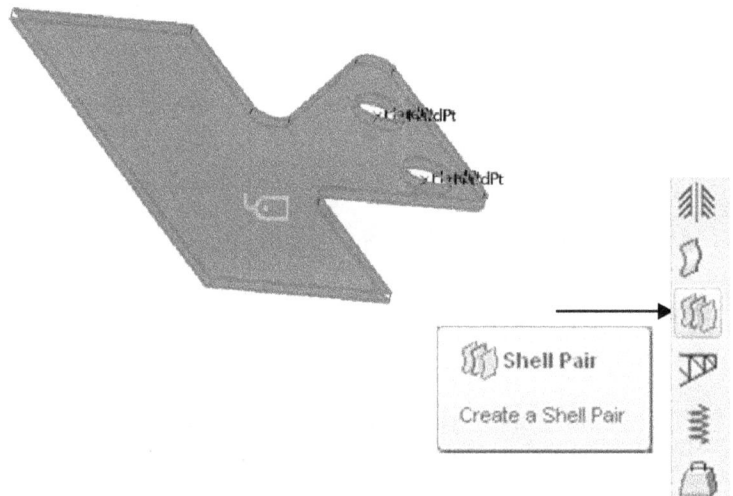

Step 6: Define the constraint condition. Assume that the top and bottom edges are welded.
Click the icon of **Constraint > New > Edge/Curve >** select the front top edge and the front bottom edge while holding down the **Ctrl** key, and fix all six degrees of freedom.> **OK**.
Fix all 6 degrees of freedom > **OK**.

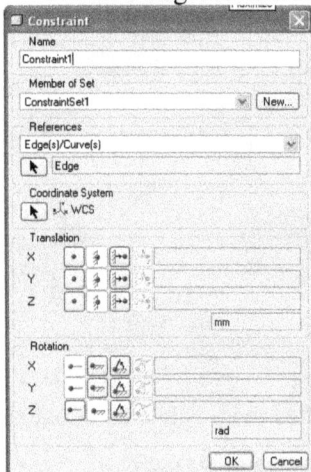

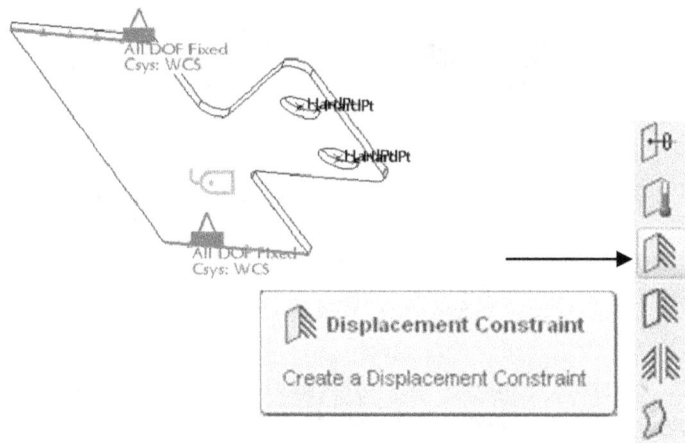

Step 7: Define the load condition. There are two bearing loads. One is acting on the upper hole and the other on the lower hole.

We have already defined four (4) datum points. To define the bearing load acting on the upper hole, select the icon of **Bearing Load > Edges/Curve >** pick the edge the upper hole > change to the **Dir Points & Mag**, select **PNT0** as the starting point and **PNT2** as the ending point. Specify 1000.

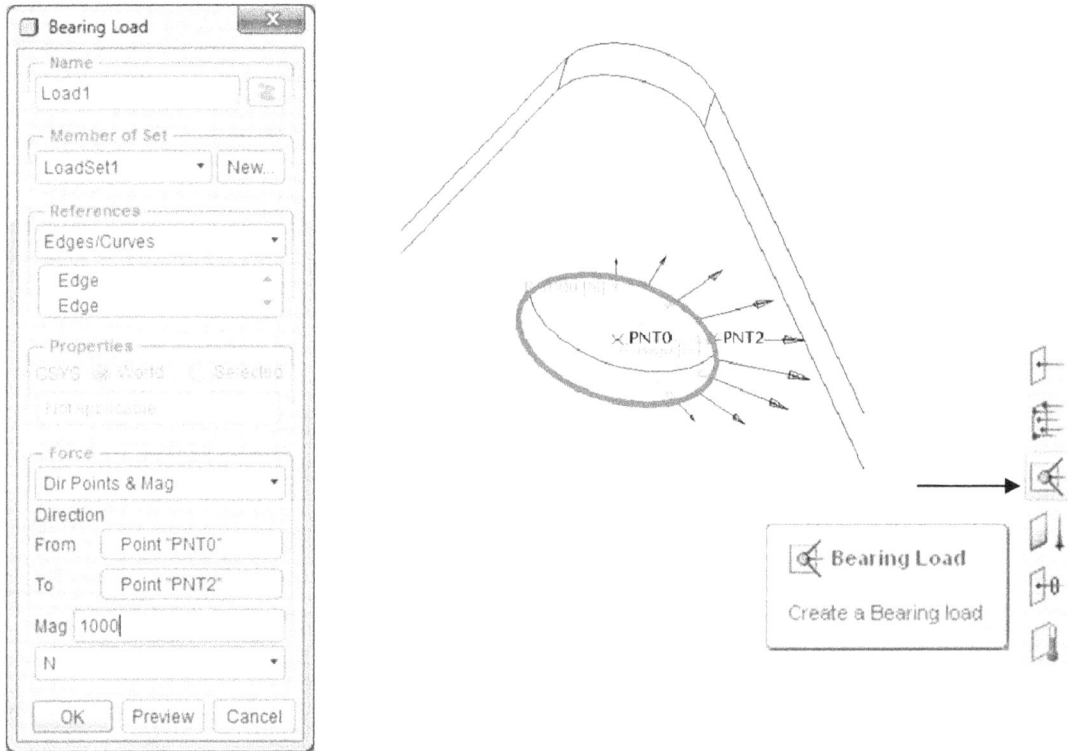

Repeat the above procedure to define the bearing load on the lower hole. Pay attention to the order of picking the 2 datum points. Pick PNT3 first and PNT1 afterwards.

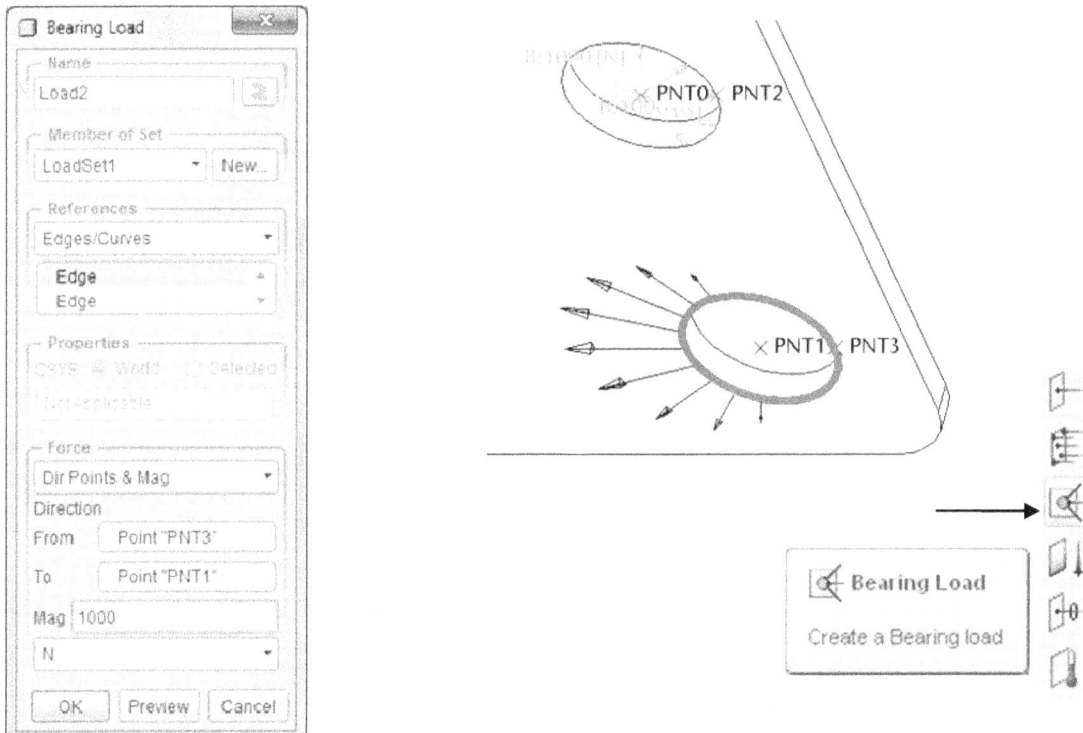

Step 8: Perform a static analysis and run it.

Click the icon of performing a design study from the main toolbar > File > Select New Static. Type bracket_base_run as the folder name and run the program.

Icon of performing a design study

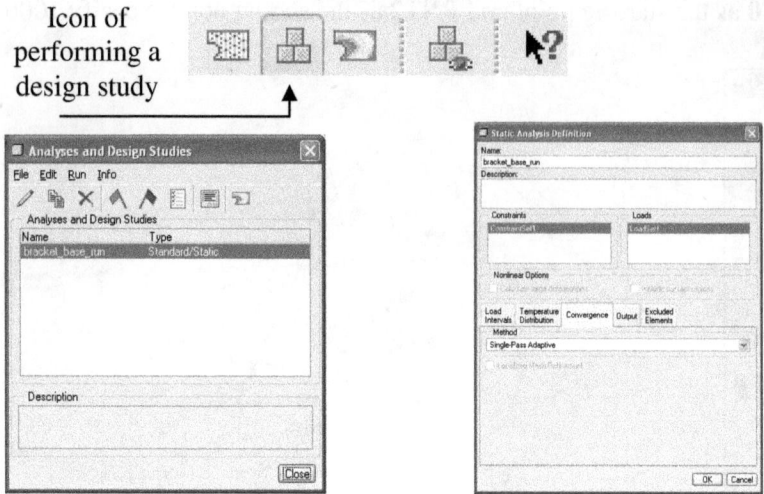

Step 9: Study the evaluation results.

The von Mises stress distribution is shown below. The maximum von Mises stress is 142 MPa at the location of the upper inner fillet.

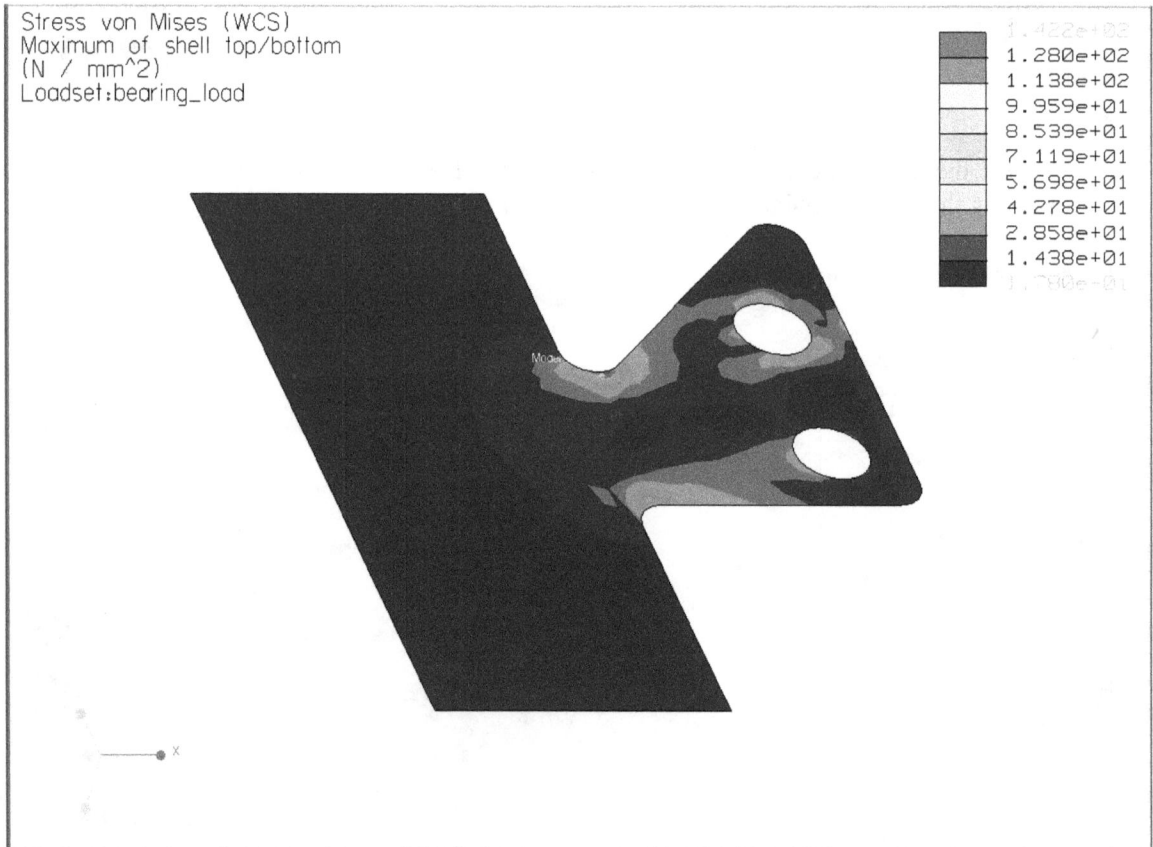

The displacement distribution is shown below. The maximum displacement is 0.05279 mm at the location of the upper tip.

```
Displacement Mag (WCS)
(mm)
Deformed
Max Disp     +5.2792E-02
Scale     2.3678E+02
Loadset:bearing_load
```

	4.751e-02
	4.223e-02
	3.695e-02
	3.167e-02
	2.640e-02
	2.112e-02
	1.584e-02
	1.056e-02
	5.279e-03

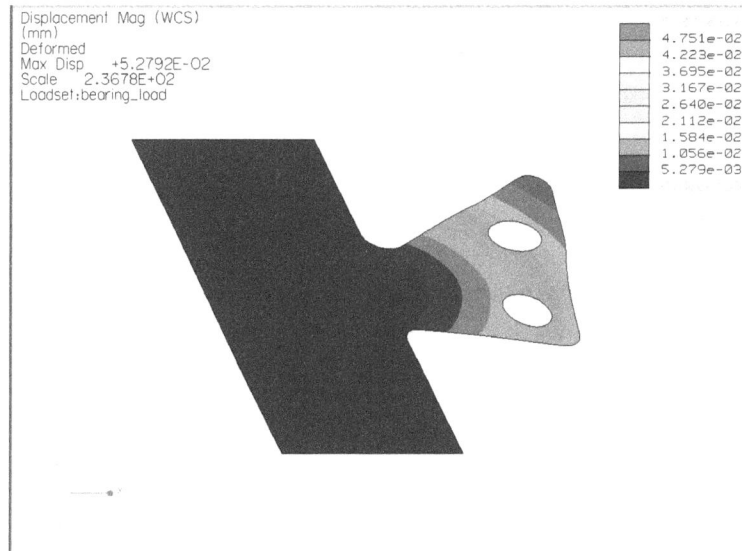

Step 10: Define the design parameters for the optimization.

Because the maximum Von Mises stress is 142 MPa at the current design, which is larger than 90 MPa (we do not use 80 at this moment), a redesign process is needed. The following flow chart outlines the basic steps involved in a re-design process.

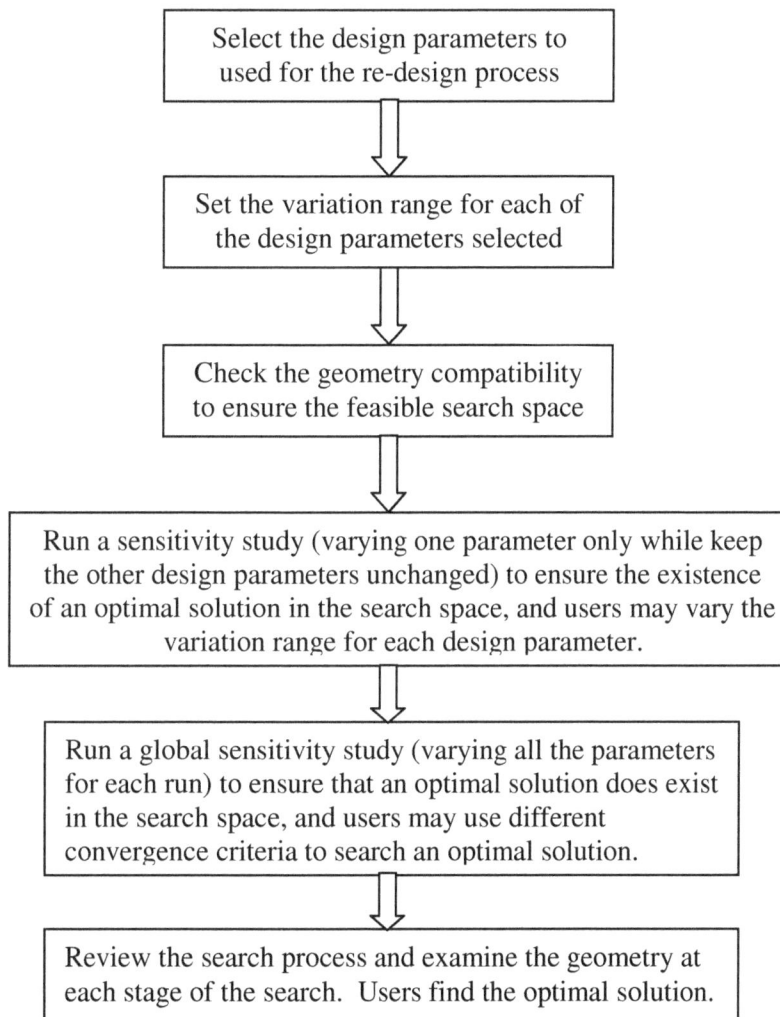

```
┌────────────────────────────────────┐
│   Select the design parameters to   │
│   used for the re-design process     │
└────────────────────────────────────┘
                  ↓
┌────────────────────────────────────┐
│   Set the variation range for each of │
│   the design parameters selected     │
└────────────────────────────────────┘
                  ↓
┌────────────────────────────────────┐
│   Check the geometry compatibility   │
│   to ensure the feasible search space │
└────────────────────────────────────┘
                  ↓
┌──────────────────────────────────────────────┐
│   Run a sensitivity study (varying one parameter only while keep │
│   the other design parameters unchanged) to ensure the existence │
│   of an optimal solution in the search space, and users may vary the │
│   variation range for each design parameter.    │
└──────────────────────────────────────────────┘
                  ↓
┌──────────────────────────────────────────────┐
│   Run a global sensitivity study (varying all the parameters │
│   for each run) to ensure that an optimal solution does exist │
│   in the search space, and users may use different │
│   convergence criteria to search an optimal solution. │
└──────────────────────────────────────────────┘
                  ↓
┌──────────────────────────────────────────────┐
│   Review the search process and examine the geometry at │
│   each stage of the search.  Users find the optimal solution. │
└──────────────────────────────────────────────┘
```

At the end of search, users should be able to obtain a set of solutions, which represent the redesigned bracket components. Users should be able to examine each of these redesigned bracket components through the use of **Optimal History**, and find the optimal design based on the design specifications.

Step 11: Local Sensitivity Study with the Defined Design Parameters

The purpose of carrying out a sensitivity study is to establish a search space, in which an optimal solution resides.

Click **File > New Sensitivity Study.** Type local_sen as the name. Select Local Sensitivity. Highlight bracket_base_run. Click the icon of **select dimensions from model**. On the display, pick angle, corner fillets and over_width > click OK and run the local sensitivity study.

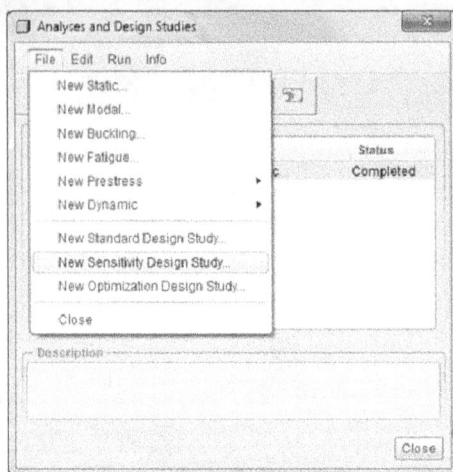

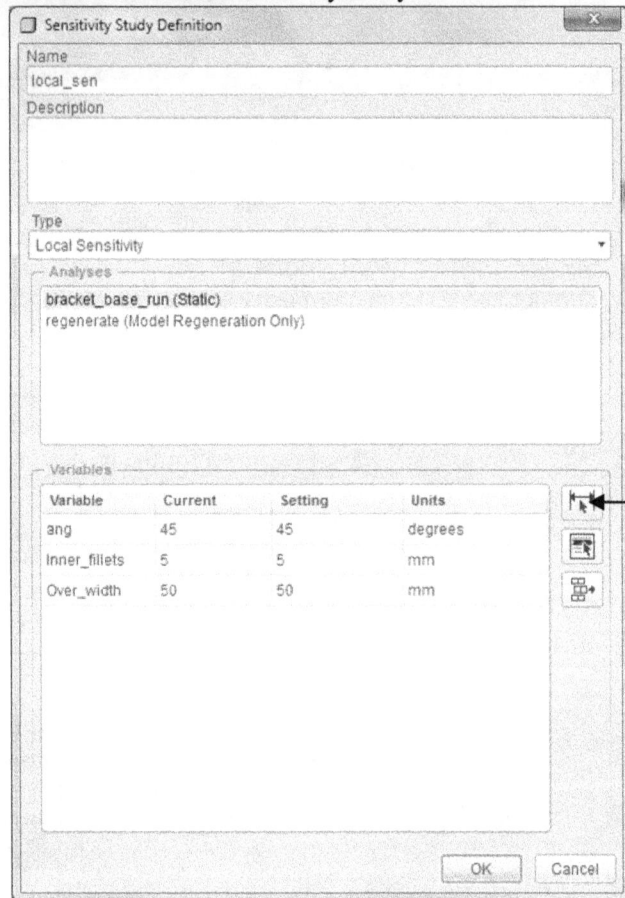

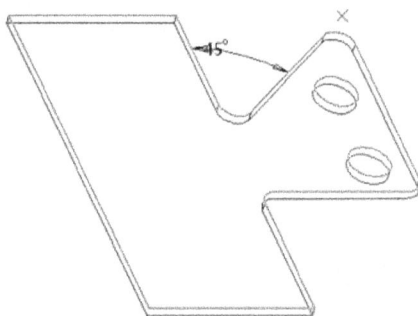

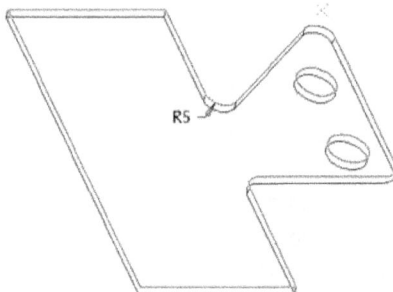

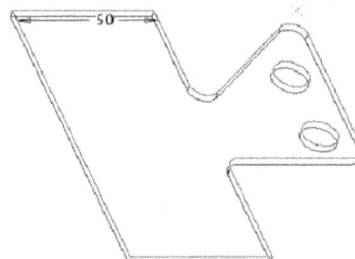

After running the local sensitivity study, plot the maximum von Mises stress vs a design parameter for ang, inner_fillets, and over_width, respectively. Click the icon of review results > select max_stress_vm > OK. Pick each of the 3 design parameters and plot the 3 graphs, as shown.

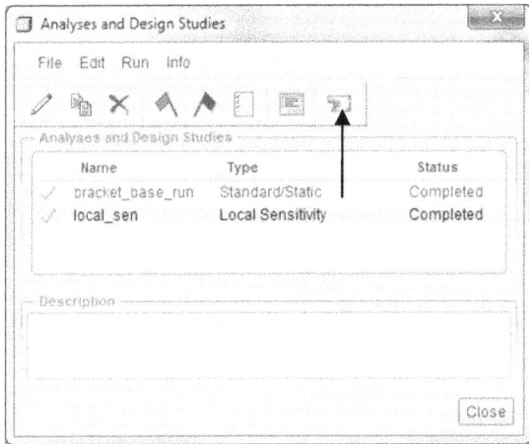

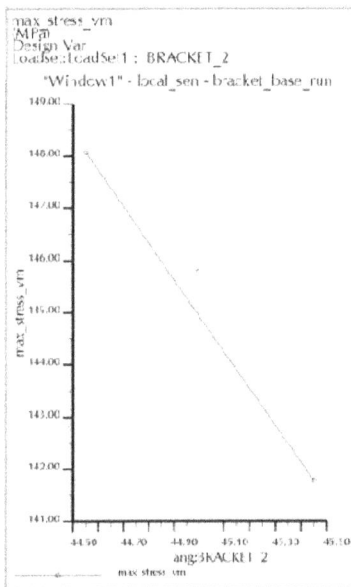

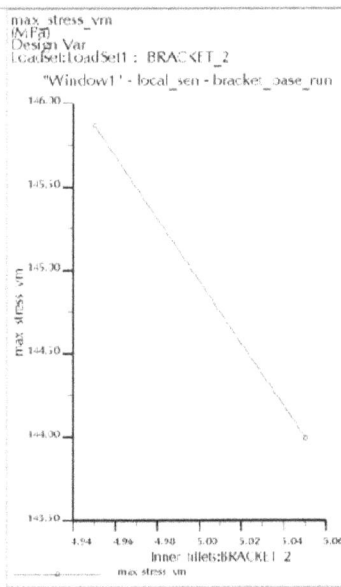

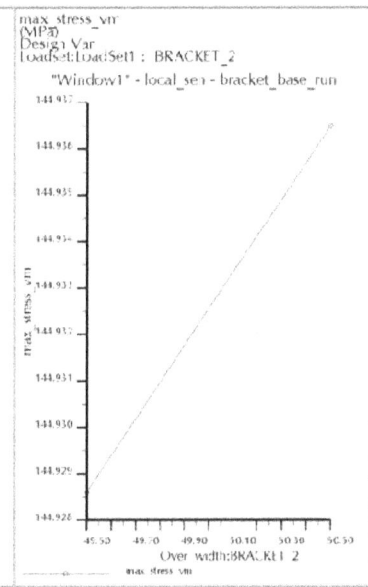

Parameters	Parameter variation ranges	Max VM variation ranges
Ang	44.5-45.5	148-141 MPa
Inner_fillets	4.94-5.06	146-144 MPa
Over_width	49.5-50.5	144.93-144.94 MPa

Examining those graphs, the maximum value of von Mises stress would increase as the values of angle and corner radius increase. The maximum value of von Mises stress would decrease as the values of over width increase. This information is very, very important for setting the global sensitivity study.

Next, review the results for the displacement requirement. The following three plots illustrate the variation of the maximum displacement when each of the three parameters varies about the setting values.

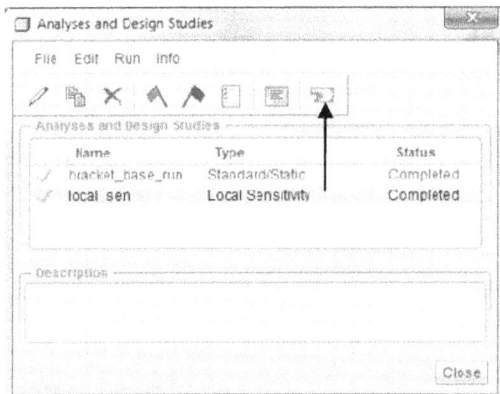

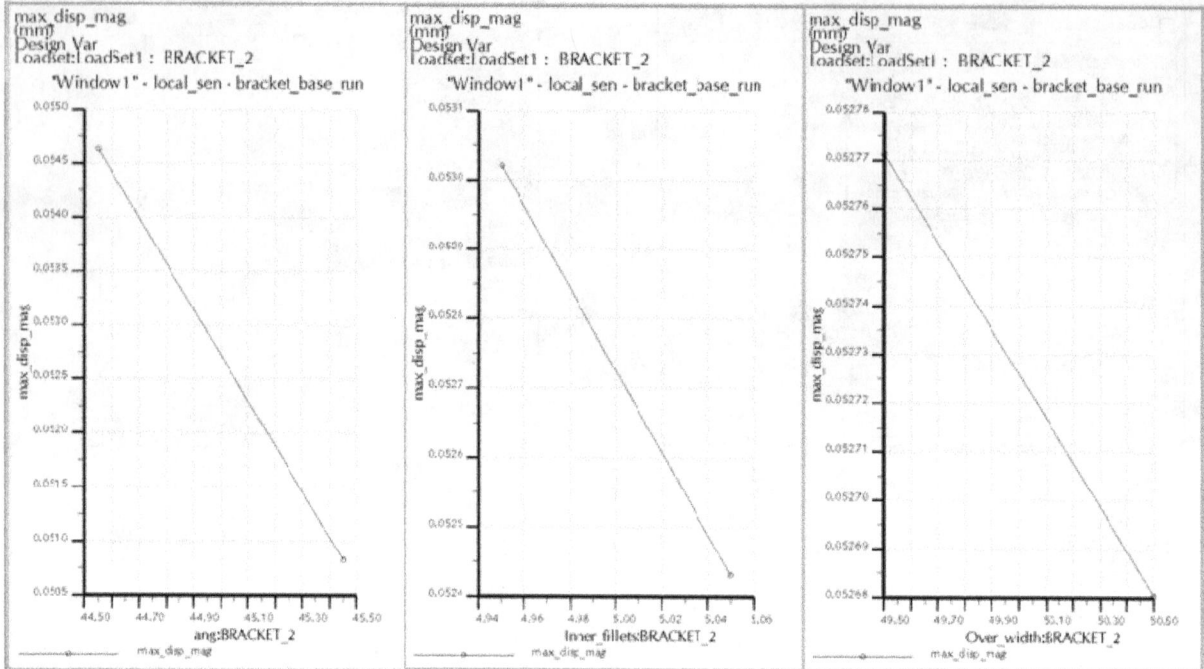

Parameters	Parameter variation ranges	Max Displacement ranges
Ang	44.5-45.5	0.055 – 0.050 mm
Inner_fillets	4.94-5,06	0.053 – 0.0524 mm
Over_width	49.5-50.5	0.05280 – 0.05270 mm

Examining those graphs, the information is similar to what has been obtained from examining the 3 graphs of von Mises stress.

Step 12: Global Sensitivity Study with the Adjusted Variation Ranges of the 3 Design Parameters

Click **File > New Sensitivity Study.** Type global_sen as the name**.** Select Global Sensitivity. Highlight bracket_base_run. Click the icon of **select dimensions from model**. On the display, pick angle, corner fillets and over_width > click OK and run the global sensitivity study.

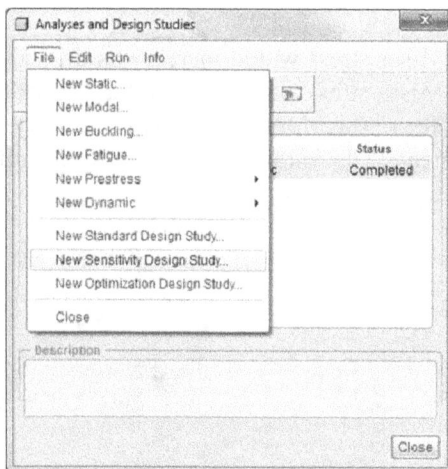

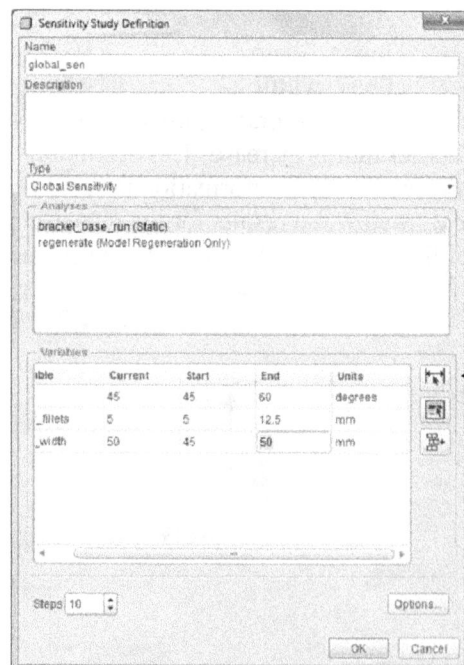

Set the ranges for the 3 design parameters as follows and run the program.

       ang: 45 – 60

       inner_fillets: 5 - 12.5

       over_width: 45 - 50

Two graphs are shown below to illustrate the variation of the maximum von Mises stress and the variation of the maximum displacement magnitude as the settings of the three design parameters vary at the same time. Note that the fundamentals behind the global sensitivity study are to evaluate the system responses to a set of design parameter settings. Those settings are so selected that the obtained results present a general picture of how the system measure or measures respond to the changing of the settings of the design parameters within the search space.

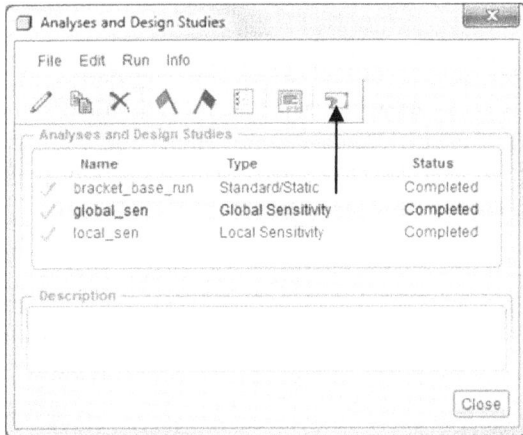

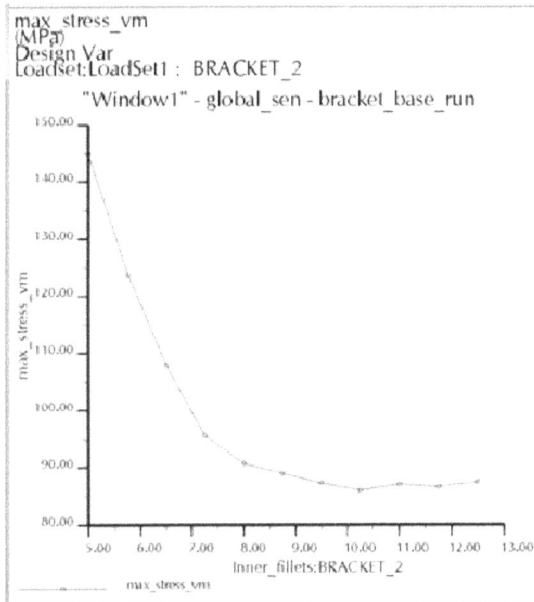

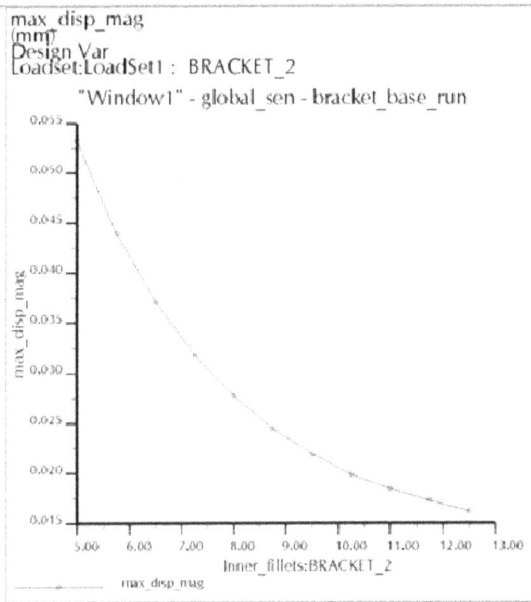

Examining these two curves, the maximum von Mises stress approaches to 80 MPa, and the maximum displacement magnitude approaches to 0.015 mm. Upon completing the global study, we feel comfortable to establish the two constraint conditions. They are

Maximum Von Mises stress <= 85 MPa

Maximum Displacement Magnitude <= 0.020 mm

Step 13: Perform a design study to search for an optimal solution

The objective function is to minimize the total mass of the bracket design. The two constraint conditions are the two limits on the maximum von Mises stress and the maximum displacement.

Minimize: Total Mass of the Bracket Design
Subjected to:

The Maximum Von Mises Stress <= 85 MPa
The Maximum Displacement <= 0.020 mm

Where

45 <= Ang <= 60
5 <= Inner_fillets <= 12.5
45 <= Over_width <= 50

Click **File > New Optimization Design Study.** Type optimal_bracket as the name. Click the icon of **select dimensions from model**. Set up 2 constraints: max von Mises less than 95 MPa and max displacement magnitude less than 0.020 mm. On the display, pick angle, corner fillets and over_width > set the parameter values listed above > click **OK.** Click Options set the convergence criterion to 1%, and the maximum number of runs is set at 20. Run the optimal study.

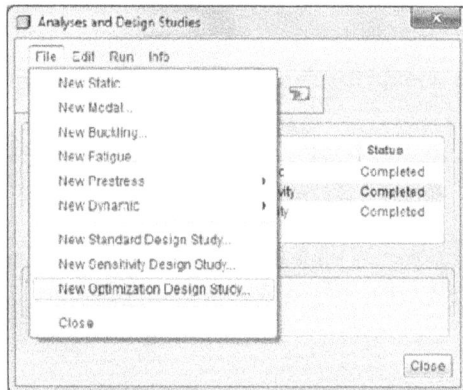

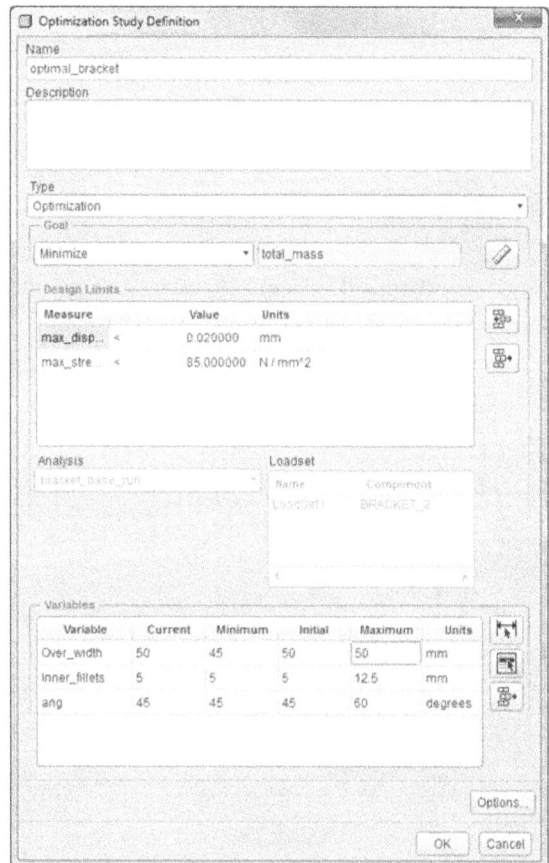

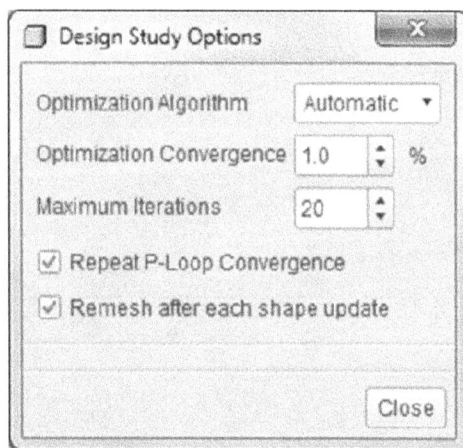

Pick the icon of **performing a design study > New Design Study > select Optimization** and type in optimal_bracket as the study name. Set up the two constraint conditions and specify the variation ranges of the three design parameters. The convergence criterion is 1%, and the maximum number of runs is set at 20. Run the optimal study.

It may take sometime to complete the search for an optimal solution. When the optimization process is completed, plot the graph showing the objective function (total mass in this study) values during the iterations.

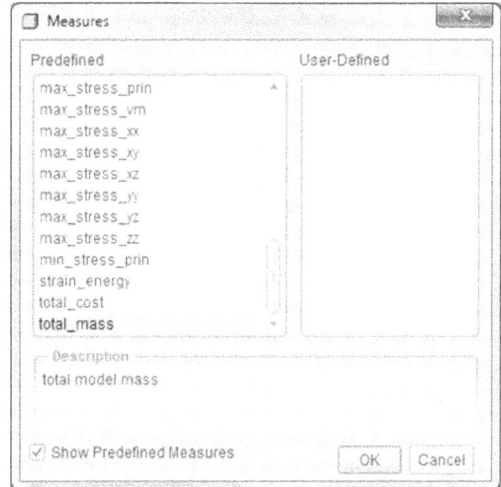

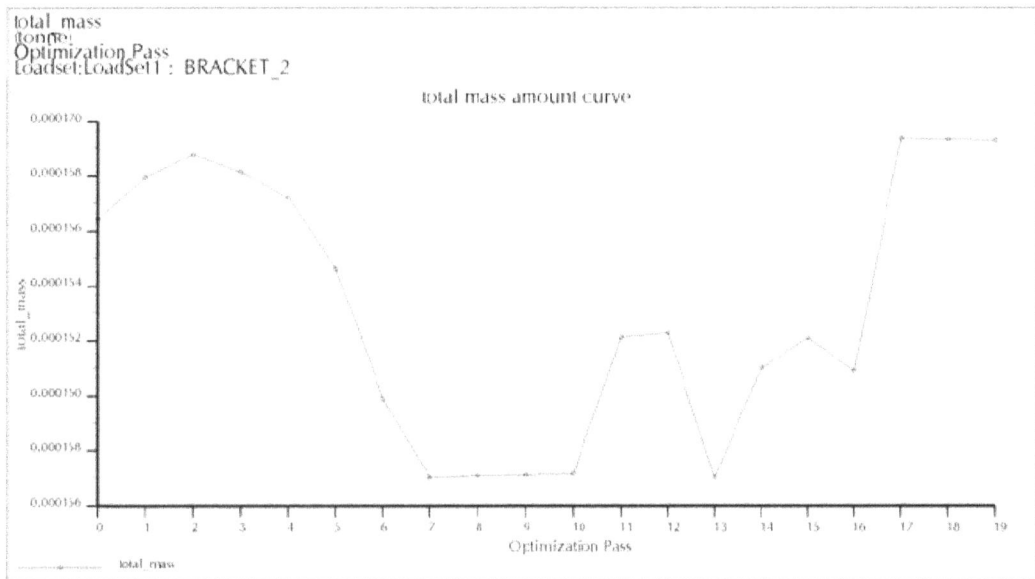

The final result listed in the summary report is shown below, indicating that the search for the optimal solution was terminated because the number of iterations has reached 20, which was the limit set when starting the optimal search. The report also indicates, among the 20 solutions, the results obtained from iteration 13 were the best. The values of the 3 design parameters were angle equal to 58.029 degrees, radius value equal to 9.21123 and the width value equal to 45. The maximum von Mises stress value was 85.075 MPa and maximum deformation value was 0.01998 mm. The total mass was 157.01 grams. Readers may continue the search process for a better solution.

```
Result of Optimization Iteration 19
Parameters:
    ang                  59.0751
    Inner_fillets         8.38474
    Over_width               50
Goal:  1.6934e-04
Status of Optimization Limits:
1. max_stress_vm     8.5066e+01 <  8.5000e+01 (satisfied within tolerance)
2. max_disp_mag      2.0008e-02 <  2.0000e-02 (satisfied within tolerance)

Resource Check                            (13:48:05)
    Elapsed Time    (sec):    123.42
    CPU Time        (sec):     48.72
    Memory Usage     (kb):    747453
    Wrk Dir Dsk Usage (kb):       86

Begin Optimization Iteration 20          (13:48:05)

Optimization stopped; iteration limit exceeded.

Best Design Found:

Best Design Found:
Parameters:
    ang                  58.0291
    Inner_fillets         9.21123
    Over_width               45
Goal:  1.5700e-04
```

```
Begin Optimization Iteration 13                    (13:47:26)

Result of Optimization Iteration 13
Parameters:
    ang                    58.0887
    Inner_fillets          9.13253
    Over_width                  45
Goal: 1.5701e-04
Status of Optimization Limits:
1. max_stress_vm    8.5075e+01 <  8.5000e+01 (satisfied within tolerance)
2. max_disp_mag     1.9988e-02 <  2.0000e-02 (satisfied within tolerance)
```

The two plots shown below are the plot of the maximum VM stress and the plot of the maximum displacement. Examining them carefully, the maximum VM stress converged to 85 MPa, and the maximum displacement value converges to 0.020 mm. The optimal search was terminated because the convergence criterion set for the optimality, which was at the level of 1.0%, was met.

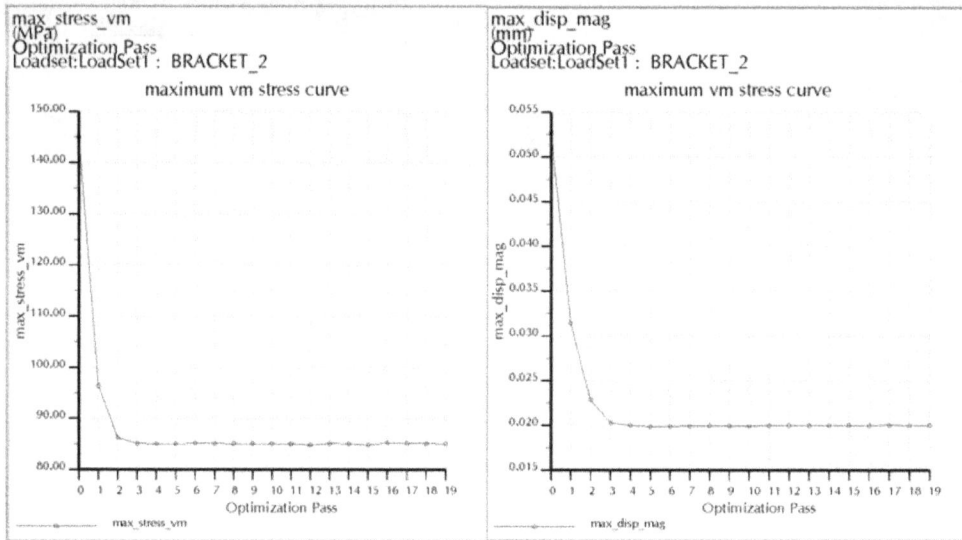

Step 15: plot the geometrical shape of the bracket design at its optimality and compare it with its original design.

Click **Info > Optimize History>** stop at the optimized shape and save.

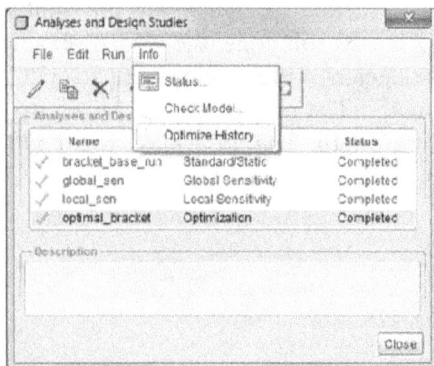

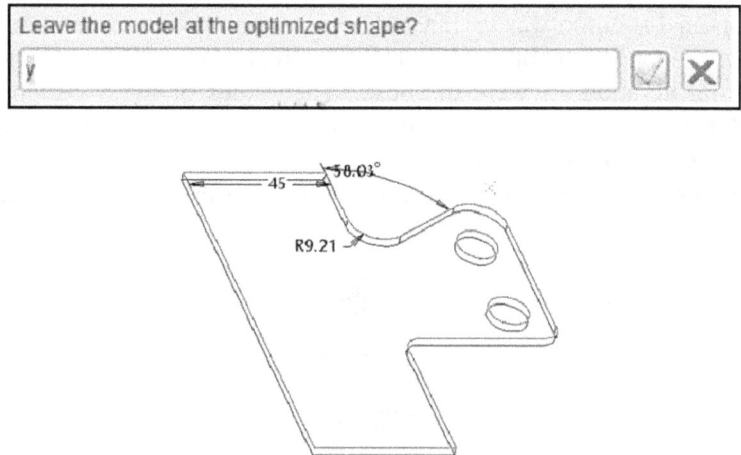

Let us perform FEA at this optimal setting. The von Mises distribution and displacement distribution are shown below. When compared those new results with the results before optimization, the improvement made through the optimization process is evident because the reduction of maximum value of von Mises stress is from 142 MPa to 85MPa and the reduction of maximum displacement is from 0.053 mm to 0.020 mm.

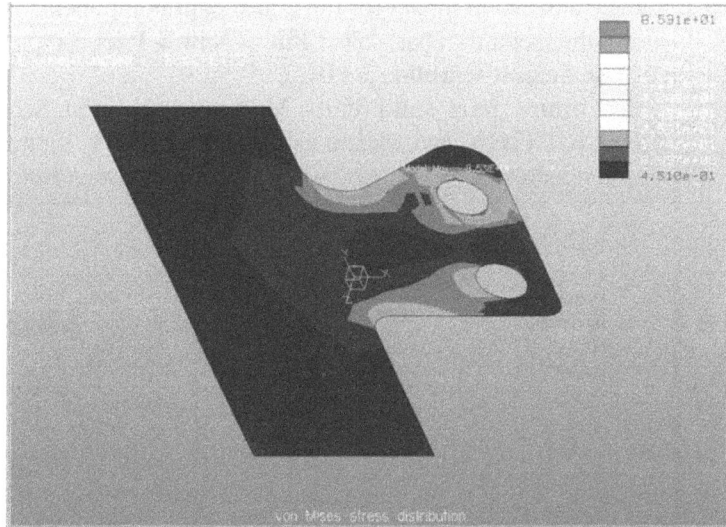

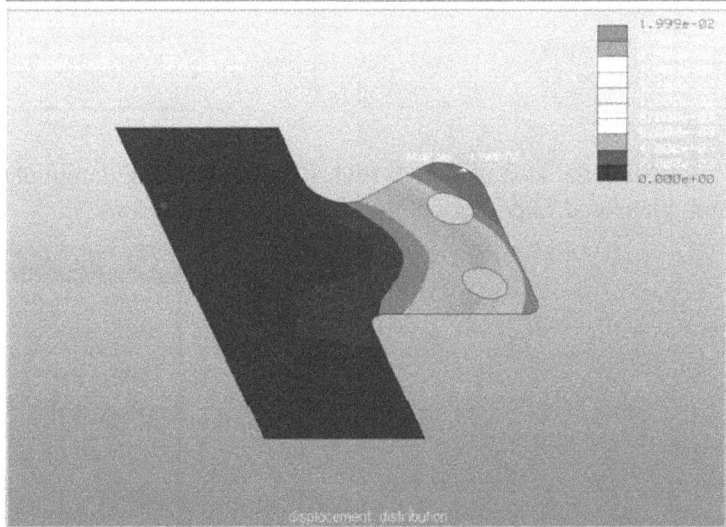

8.4   Case Study: Optimization of Crankshaft Design

The geometric shape of a crankshaft is shown below with its dimensions. The material is steel. The objective of this re-design process is to reduce the weight of the crankshaft so that the maximum value of the von Mises stress can be controlled below 50 MPa.

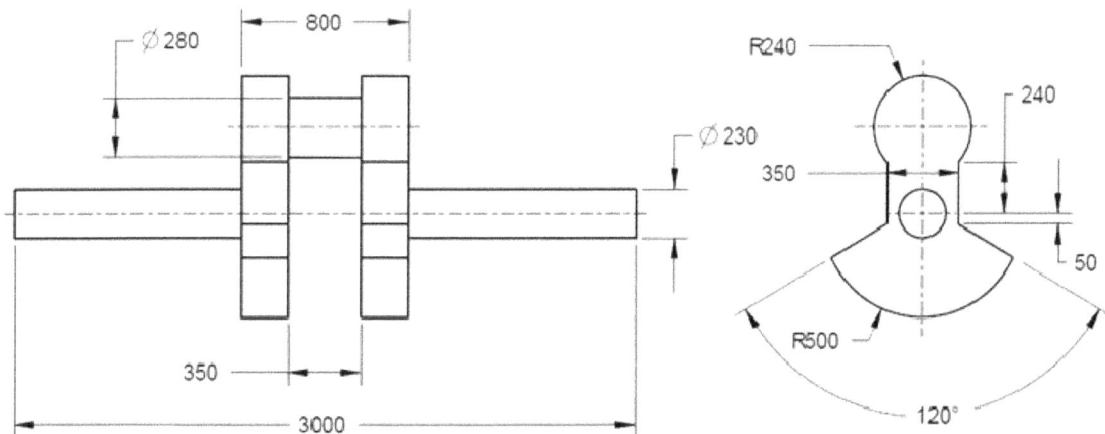

Step 1: Create a 3D solid model for the crank shaft component.
    From the menu tool bar, select **File > New > Part >** type in *crank_shaft* as the file name and clear the icon of **Use default template > OK.**
    Select **mmns_part_solid** (units: Millimeter, Newton, Second) and type crank shaft as **DESCRIPTION**, and *student* as **MODELED_BY**, then **OK.**

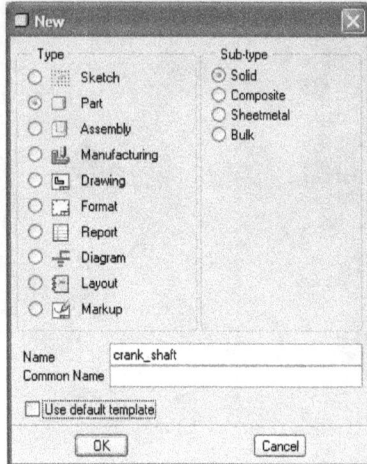

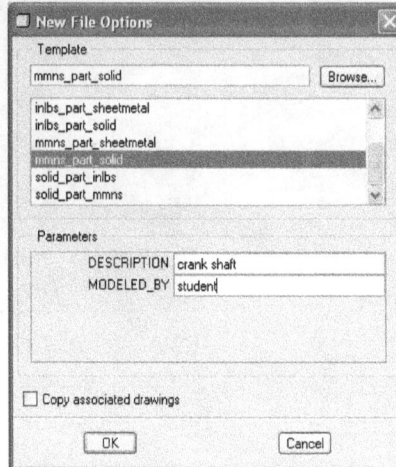

    Click the icon of **sketch tool.** Pick the **RIGHT** datum plane as the sketch plane, and use the **TOP** datum plane and **Top** to orient the sketch plane, as shown.

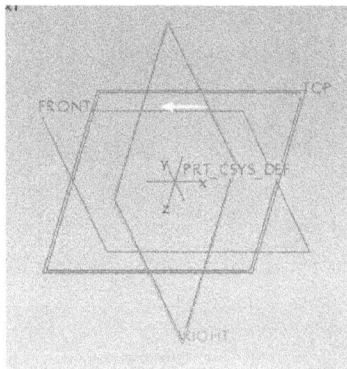

    Before making a 2D sketch, let us first create a vertical centerline passing through the origin of the coordinate system, as shown. To do so, expand the icon of **Line**, and pick the icon of **Centerline**. Draw the centerline, as shown. The user may just make a right click and pick **Centerline**.

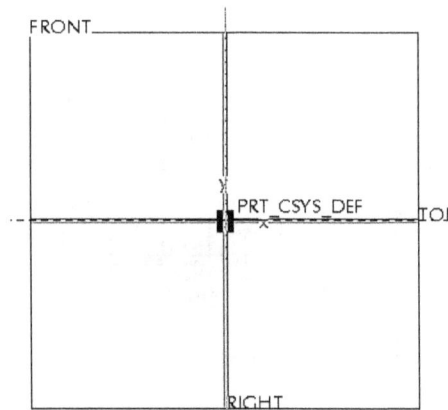

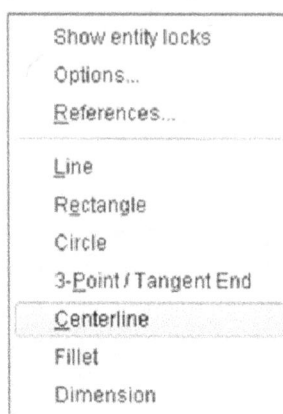

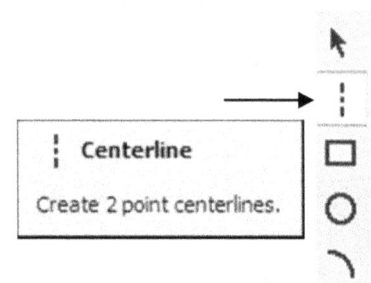

    Click the icon of **Line** to sketch 1 vertical line with 3 dimensions, as shown. The 3 dimensions are 175, 50 and 240.

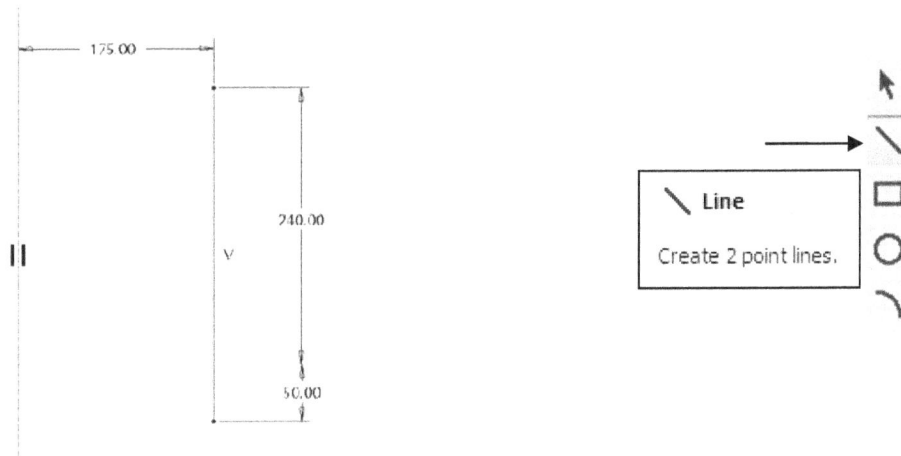

Pick the sketched line and click the icon of **Mirror**. Click the vertical centerline to obtain another line on the left side of the vertical centerline.

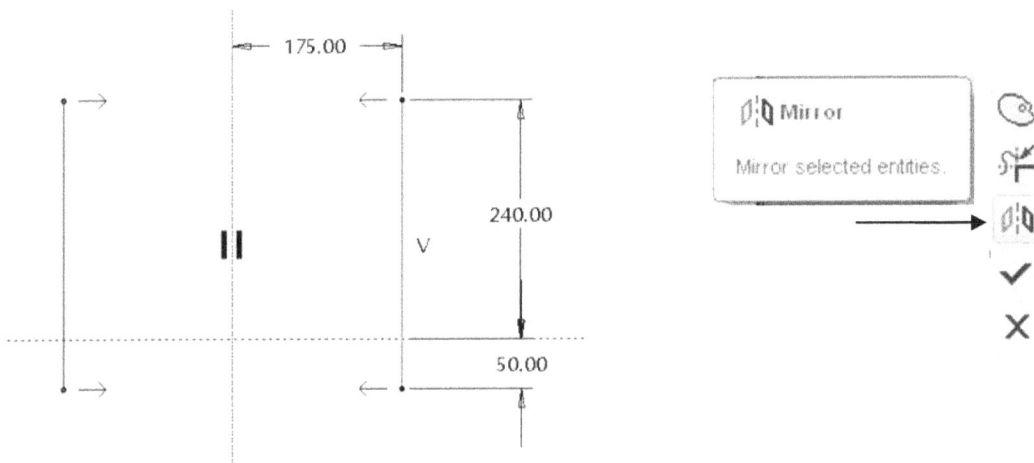

Sketch 2 more lines, as shown. These lines are also symmetric about the vertical centerline. Note the angle of 120 degrees.

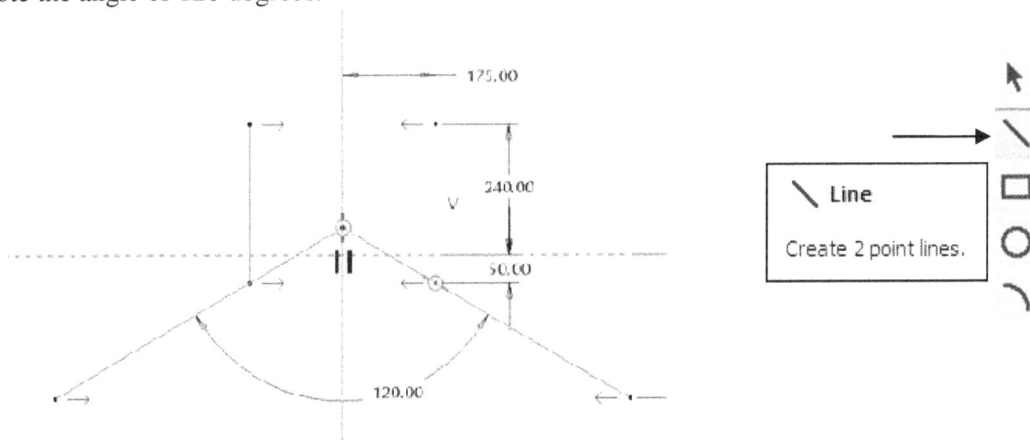

Click the icon of **Circular** to create a circular fillet. The center location is at the intersection of the 2 sketched lines, as shown. The radius value is 500.

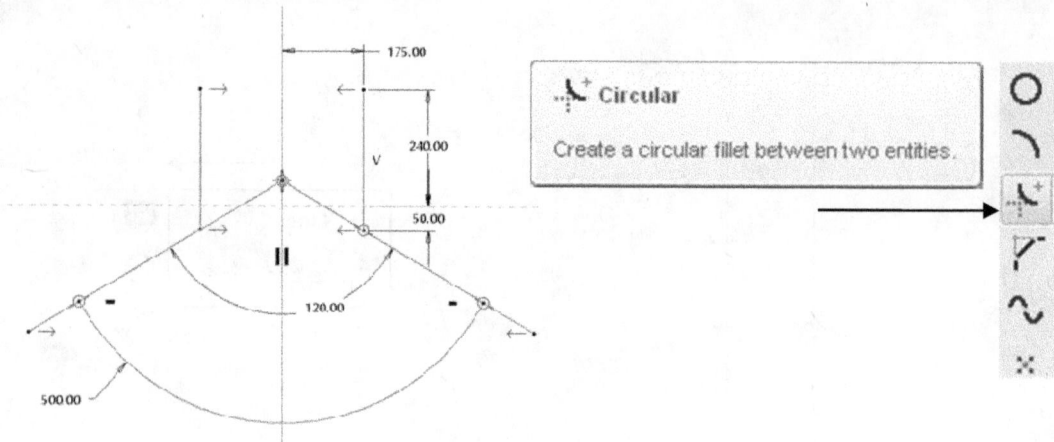

Circular

Create a circular fillet between two entities.

Click the icon of **Delete** to delete extra line segments.

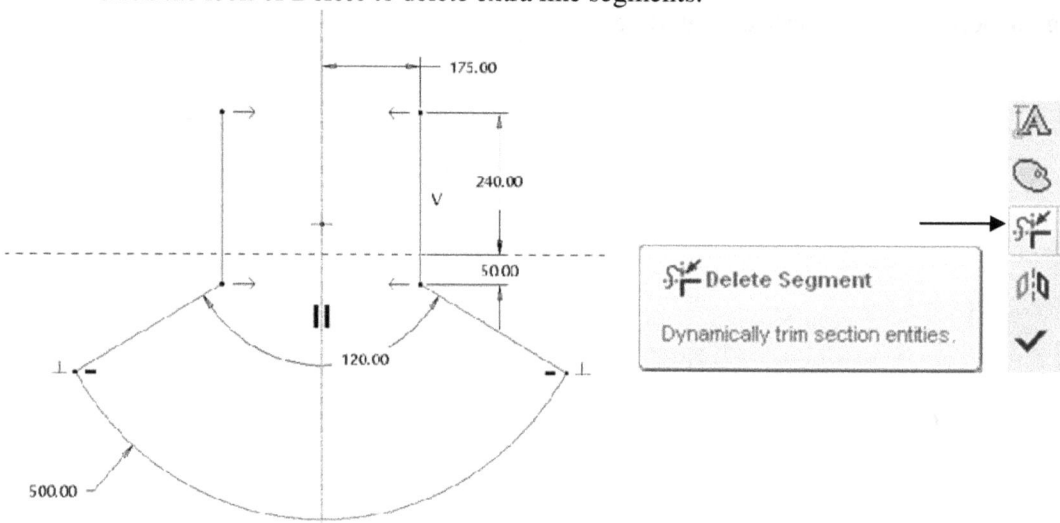

Delete Segment

Dynamically trim section entities.

Click the icon of **Circle** and sketch a circular fillet, as shown. The dimension of diameter is 240.

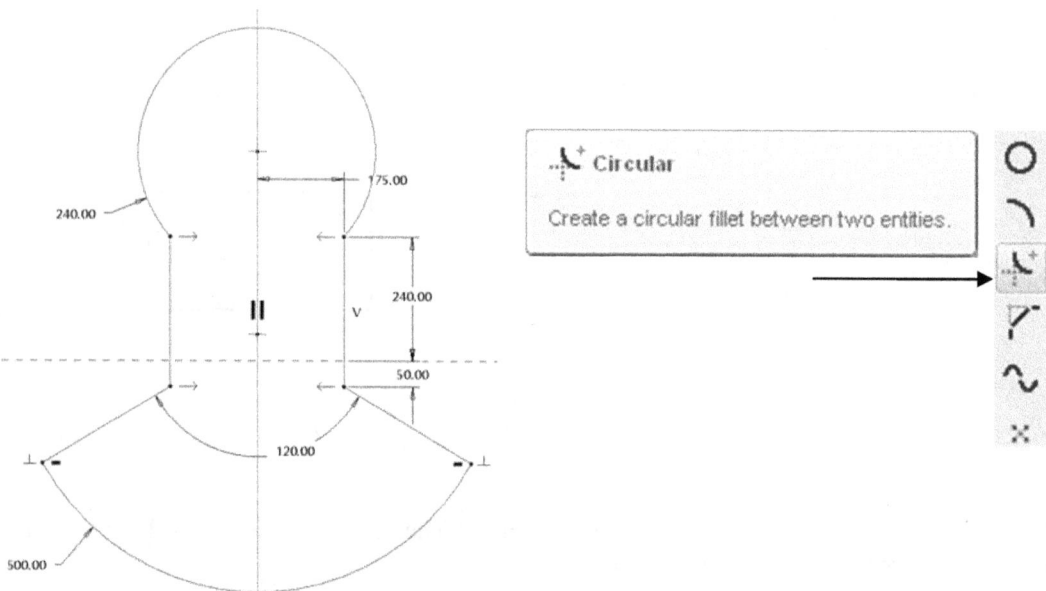

Circular

Create a circular fillet between two entities.

Select the icon of **Done** to complete the 2D sketch.

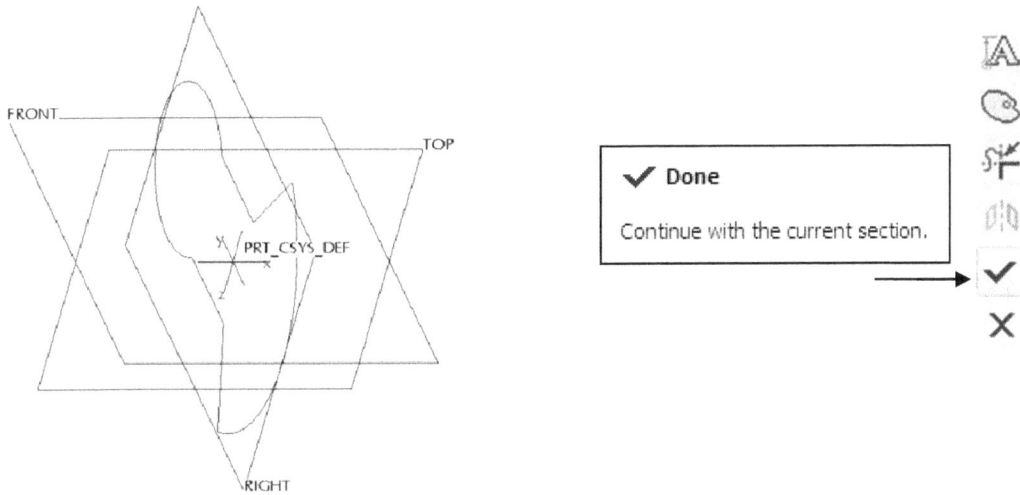

From the toolbar of feature creation, select the icon of **Extrude**.

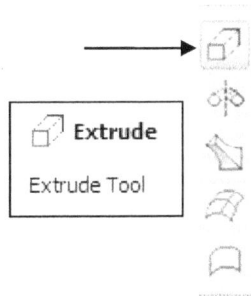

Because the 2D sketch is still active, just select the symmetric choice and set the extrusion value to 800 > click the icon of **Apply and Save**, thus completing this feature creation.

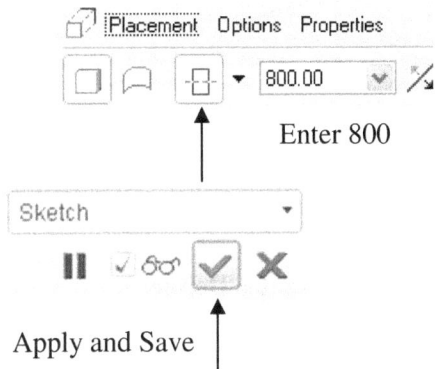

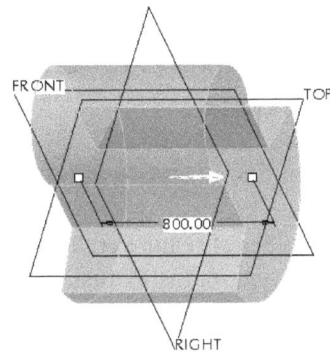

Click the icon of **Extrude**. Select the symmetric choice and set the extrusion value to 3000 > Click **Placement > Define.**

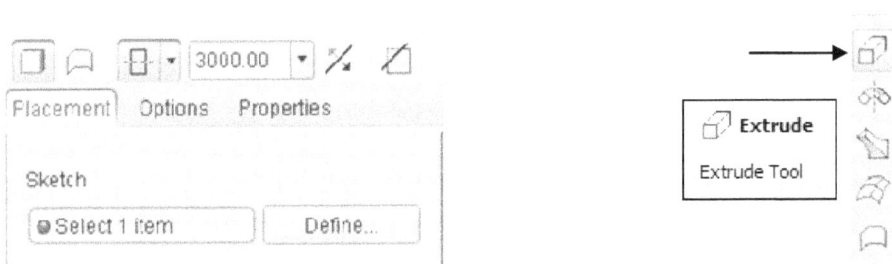

Pick the **RIGHT** datum plane as the sketch plane, or just click the box called **Use Previous**.

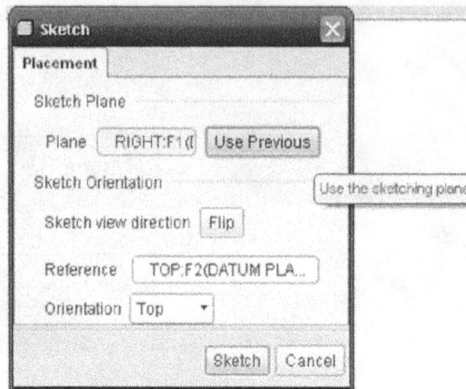

Click the icon of **Circle** and sketch a circle, as shown. The diameter dimension is 230.

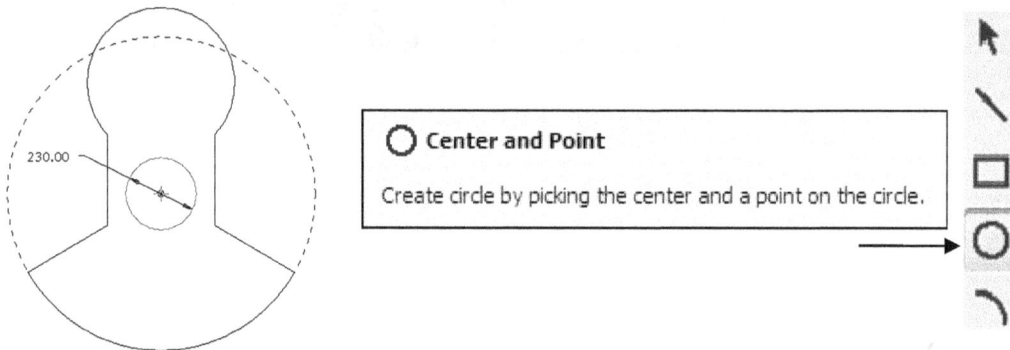

When completing the sketch, select the icon of **Done** and the icon of **Apply and Save**.

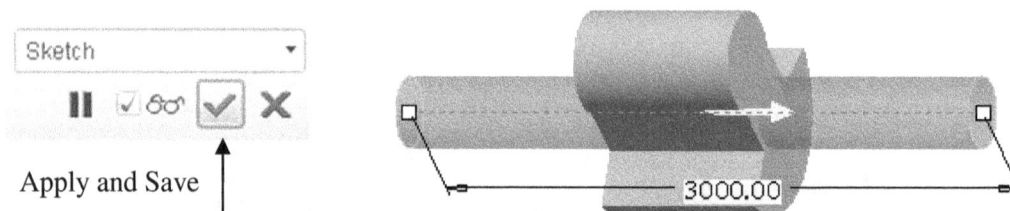

From the toolbar of feature creation, select the icon of **Extrude**. Select the icon of **Cut**. Select the symmetric choice and set the extrusion value to 350 > Click **Placement > Define.**

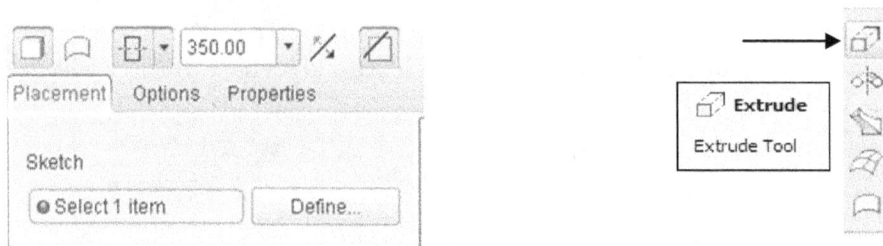

Pick the icon of **Use Previous.** Click **Sketch > References** from the top menu. Pick the upper arc as a new reference so that the center to be used for sketch is on display, as shown.

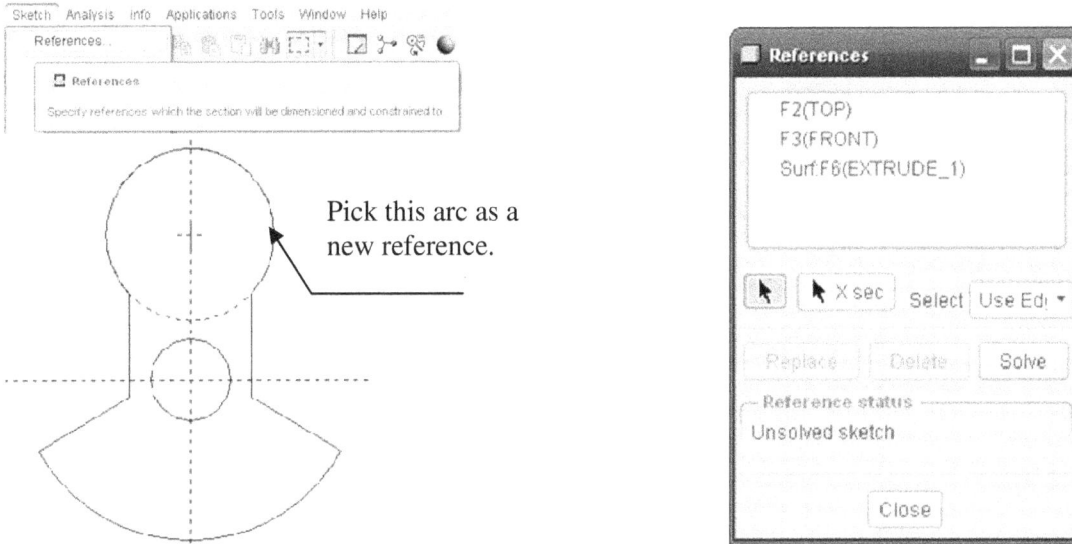

Click the icon of **Circle** and sketch a circle, as shown. The diameter dimension is 280.

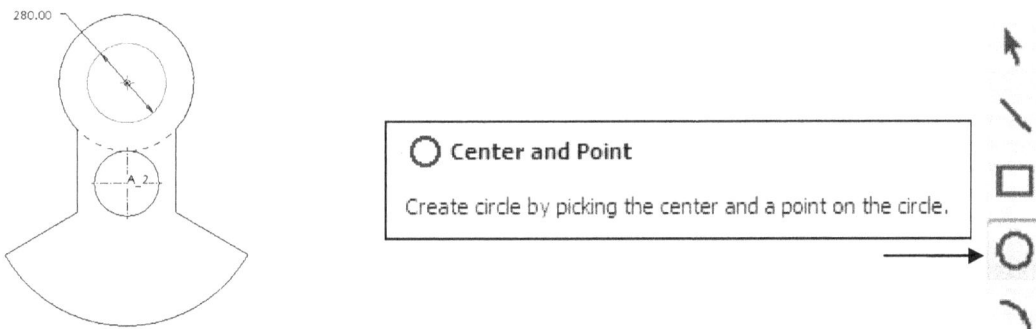

When completing the sketch, select the icon of **Done**. Pay attention to the cutting direction, and select the icon of **Apply and Save** from the feature control panel.

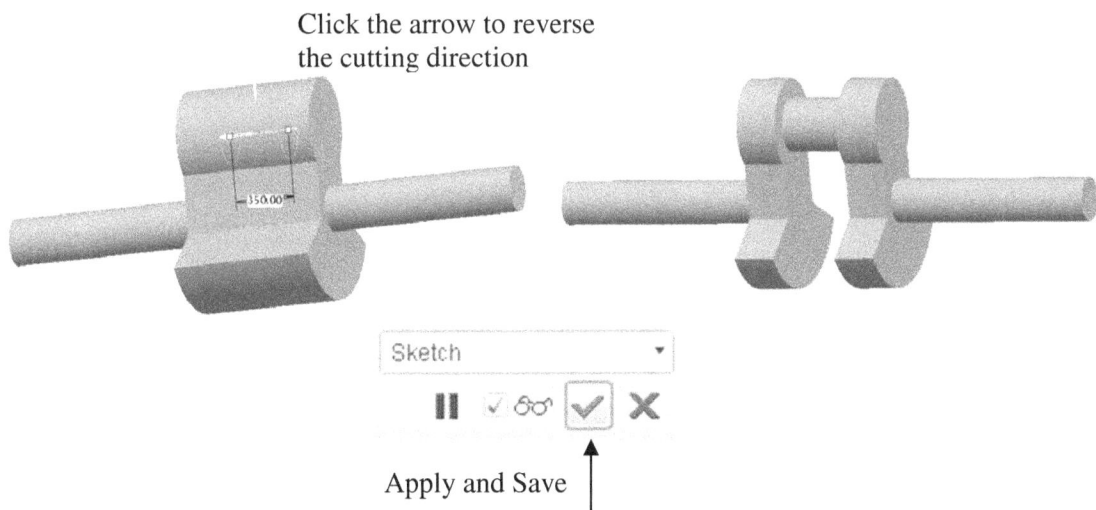

Step 2: Specify the mass density value. Assume the material type is steel. From the main menu, select **File > Mass Properties >** click **Change** > type 7.8e-9 (tonne/mm^3) > **OK > Close**.

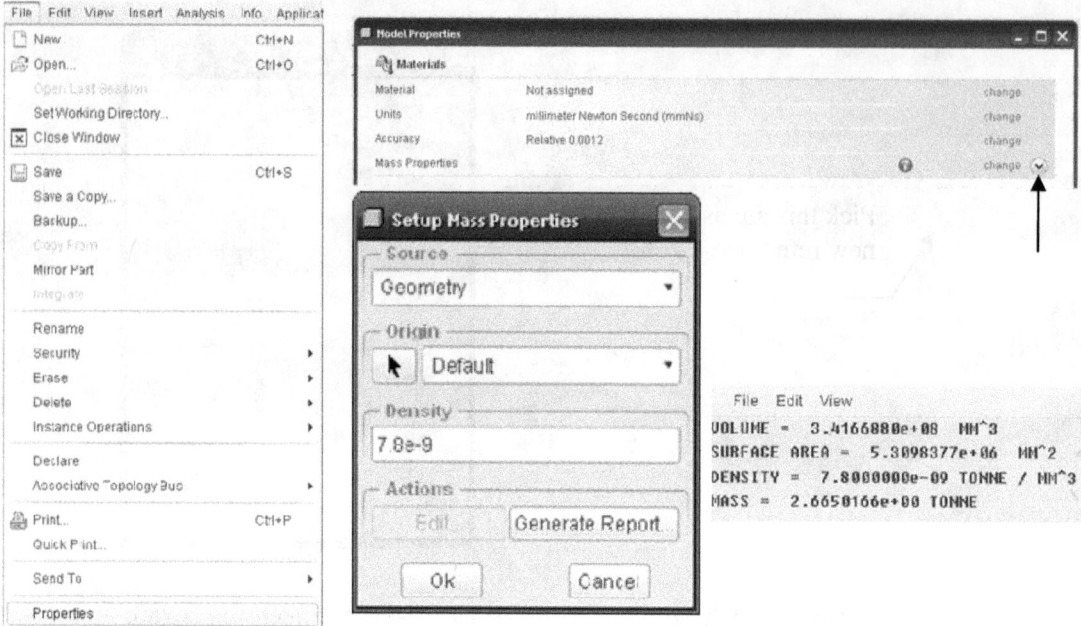

Shift from the Pro/ENGINEER design system to Pro/Mechanica From the main toolbar or the menu toolbar, select **Applications > Mechanica > Structure > OK.**

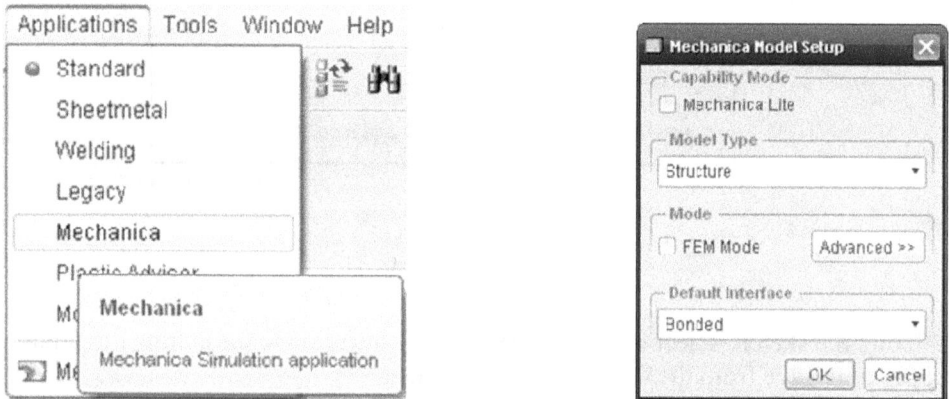

Select the icon of **Materials >** select **STEEL** from the left side called Materials in Library > click the directional arrow so that the selected material type goes to the right side called Materials in Model > **OK**.

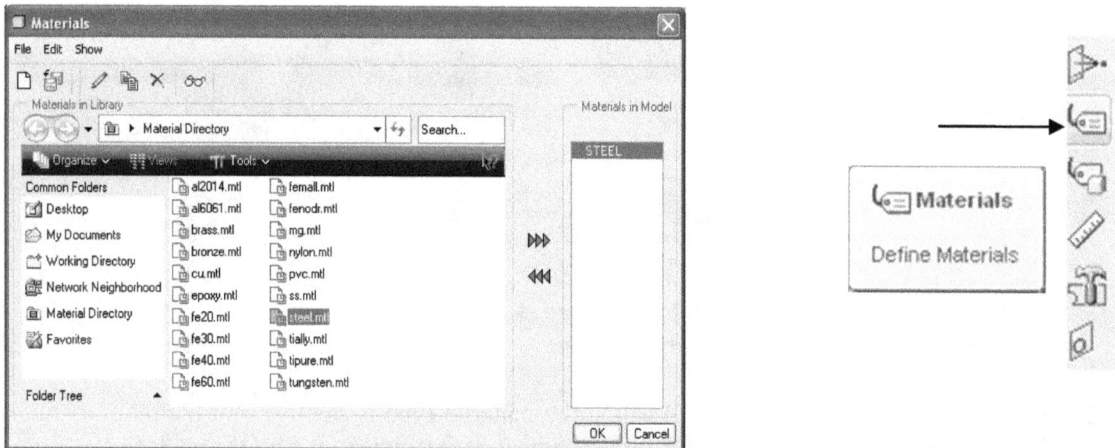

Click the icon of **Material Assignment.** In the Material Assignment window, the selected Steel is shown, and the component is also shown because there is only one volume or one component in the system. As a result, the software system automatically assigns **Steel** to the component. Just click **OK**.

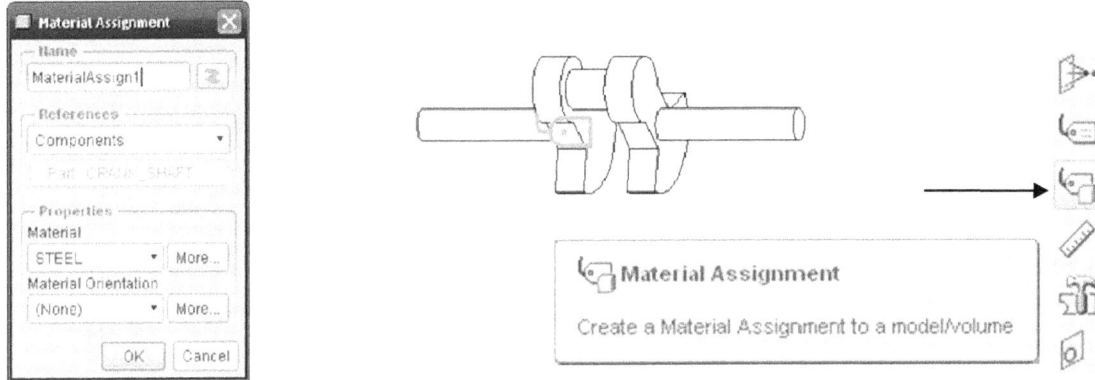

To define the fixed end constraint on the right and left sides of the crankshaft, select the icon of **New Displacement Constraint > Surface(s)** > pick the surfaces at the right and left side of the crankshaft. Because the constraint condition is "fixed ends", we set all 6 degrees of freedom equal to zero, which is the default setting > **OK**.

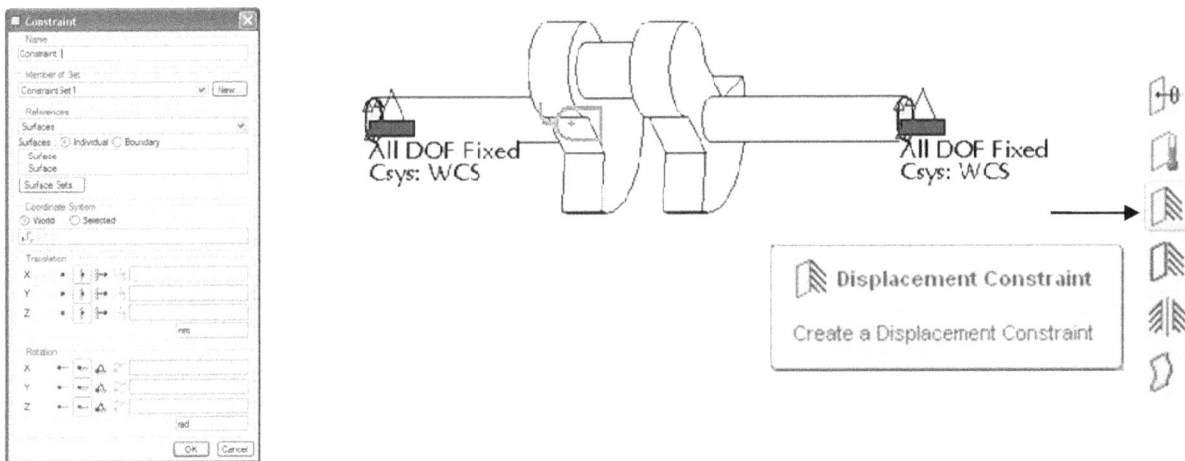

Assume the load is due to centrifugal force. Select the icon of **Centrifugal Load** > in terms of the angular velocity, type *125* as the X component > **Preview > OK**.

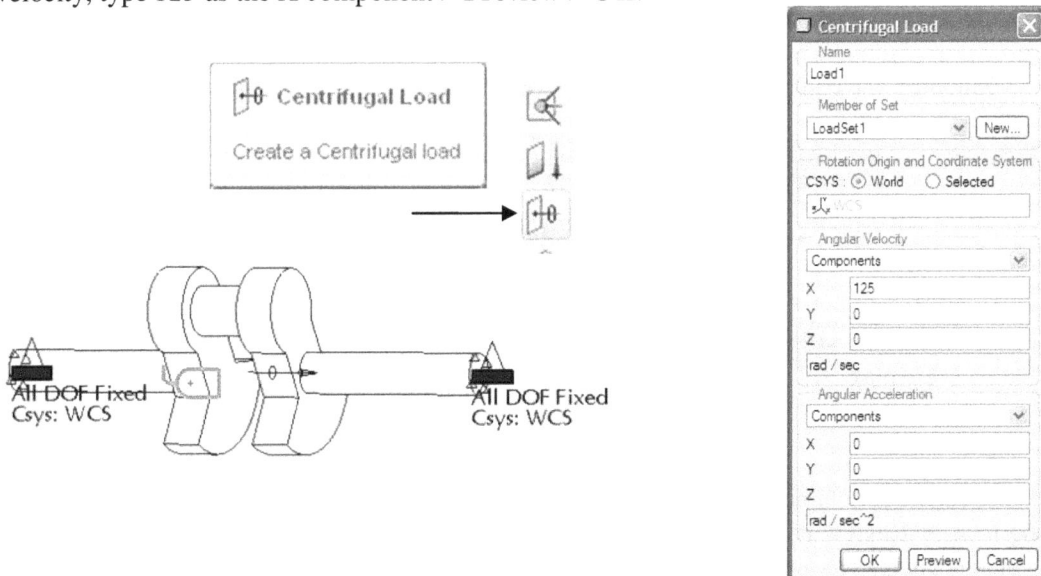

Set up **Analyses** and **Run** it. Select the icon of **Run an Analysis** from the main toolbar > **File** > **New Static** > type *original_design* as the name of the analysis folder > **OK.** Click the box of **Run.**

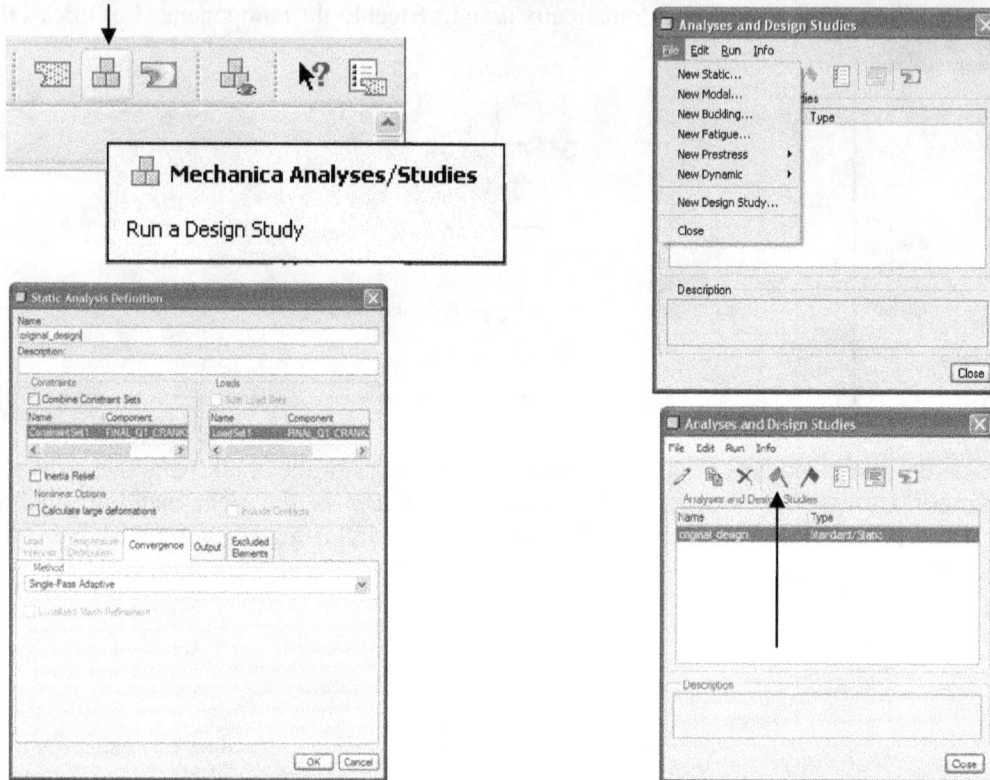

There is always a message displayed on screen. The message is asking the user "Do you want to run interactive diagnostics?" Click **Yes**. In this way a warning message will appear if an abnormal operation occurs.

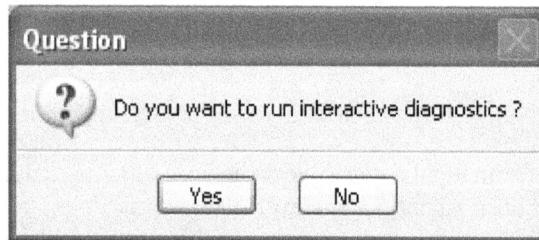

To monitor the computing process, users may click the icon of **Display study status**.

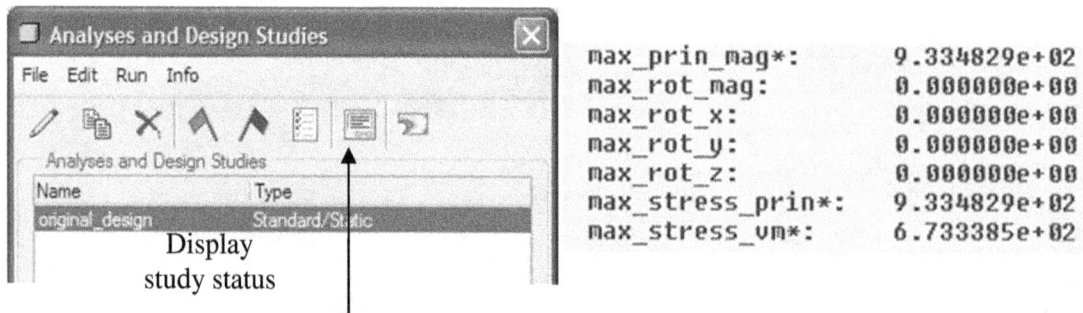

max_prin_mag*:	9.334829e+02
max_rot_mag:	0.000000e+00
max_rot_x:	0.000000e+00
max_rot_y:	0.000000e+00
max_rot_z:	0.000000e+00
max_stress_prin*:	9.334829e+02
max_stress_vm*:	6.733385e+02

To study the obtained result, say the displacement of the tube under the static loading condition, click the icon of **Review results > Stress > von Mises > OK and Show.** The maximum value is 673 MPa, which is completely unacceptable.

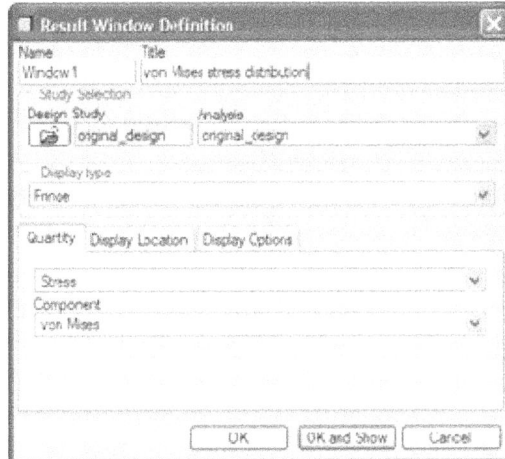

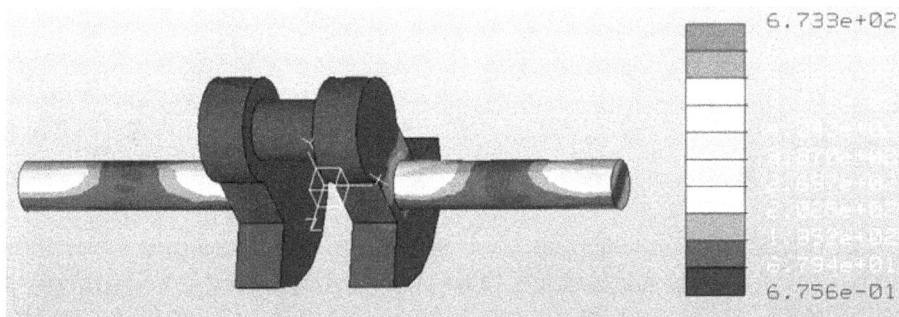

Step 3: For design optimization, we need to investigate the mass properties and the location of the center of gravity under the current design. Select **Analyses** from the top menu > **Model** > **Mass Properties**.

From the display, pick the default coordinate system so that the location of the center of mass can be specified. Select Feature and keep **Pass_Prop_1** as the name for this analysis

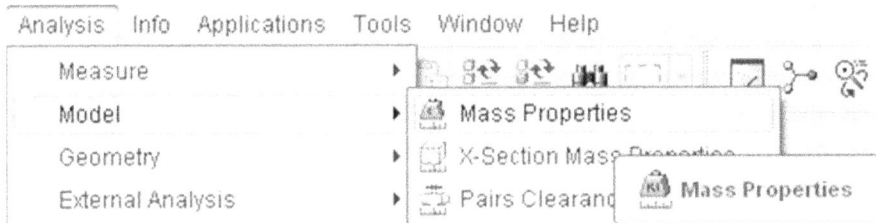

Select **Feature** from the top menu. Check 4 boxes in the Parameters field, namely, MASS, XCOG, YCOG and ZCOG. In the Datum's field, check CSYS_COG > click the check mark to complete

the process of defining the analysis feature. On the display, CSYS_COG is shown, and the distance between the two coordinate systems is 52.4 mm, as characterized by the value of YCOG.

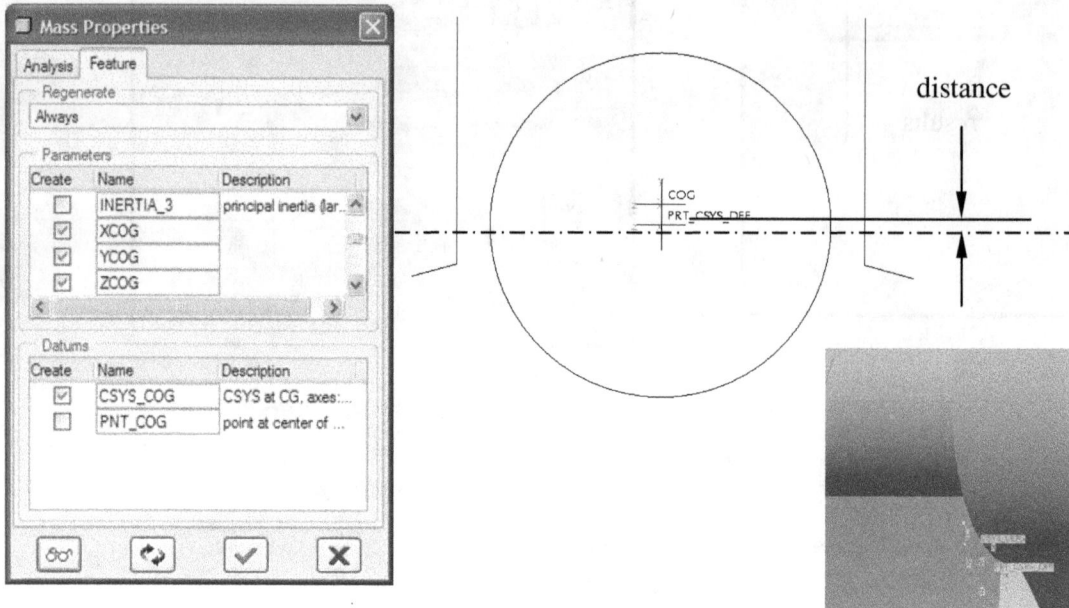

To display the **YCOG** in the model tree, click **Settings > Tree Columns**. From the window of **Model Tree Column**, select **Feat Params** > type **YCOG** and press the **Enter** key > **Apply > OK**. There is a new column called **YCOG** shown and the measured value of 52.4.

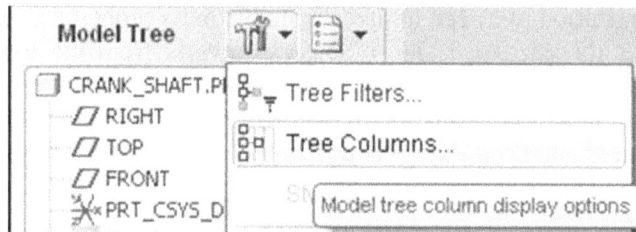

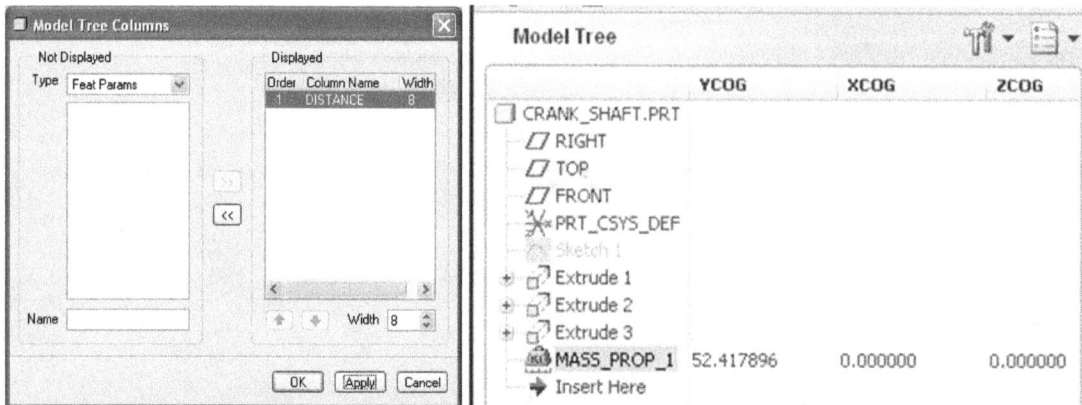

Step 4: Perform sensitivity study. We first need to select a set of design parameters, which should be varied in the search for an optimal solution.
Dimension d2 controls the distance between the two vertical lines.
Dimension rd5, controls the diameter of the top circle.
Dimension rd3, controls the diameter of the lobe curve.
Dimension d0 controls the starting position of the lobe curve.

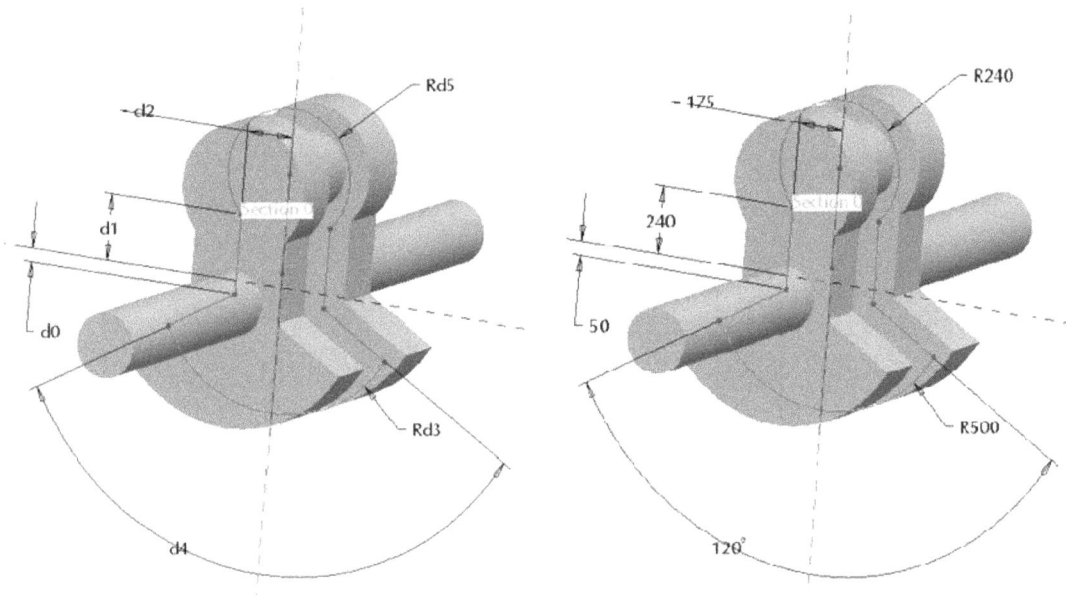

From the main menu, select **Analysis > Sensitivity Analysis.**

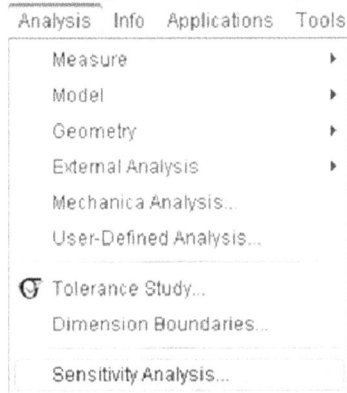

Type SENS_d0 as the name for the sensitivity analysis. Click Dimension and select d0. The current value is 50. Accept the default setting, which varies from 45 to 55. Select YCOG as the parameter to plot.

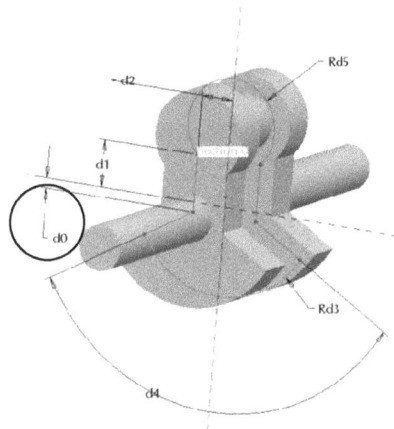

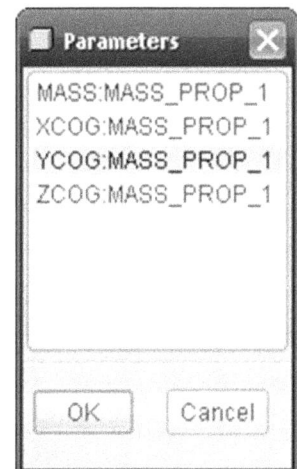

Click **Compute** and we have a graph indicating that the value of **YCOG** decreases when increasing the value of d0. The variation range is from 54 to 50.5. Therefore, the effectiveness on YCOG reduction is very limited. Click **File > Exit > Close.**

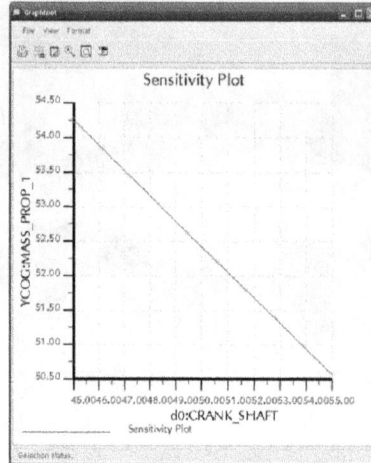

From the main menu, select **Analysis > Sensitivity Analysis.**

Type SENS_rd3 as the name for the sensitivity analysis. Click Dimension and select rd3. The current value is 500. Accept the default setting, which varies from 450 to 550. Select YCOG as the parameter to plot.

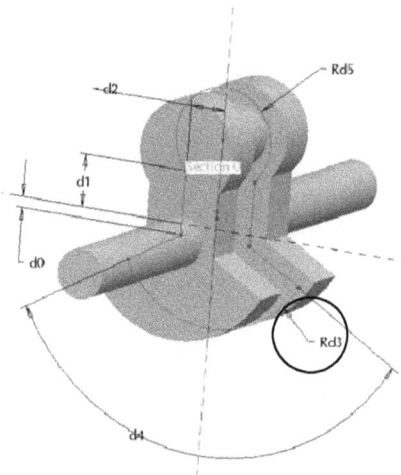

Click **Compute** and a graph indicates that the value of **YCOG** decreases when increasing the value of rd3. The effectiveness on YCOG reduction is very significant. When rd3 = 550, the YCOG value is about 20 mm between the center of gravity and the rotation axis. Click **File > Exit > Close.**

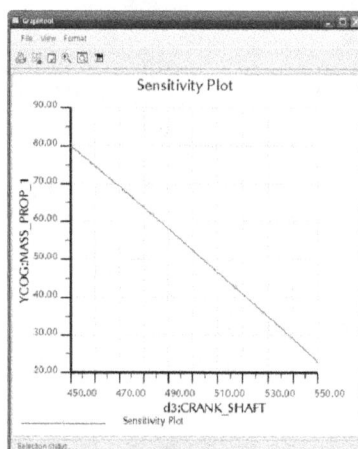

From the main menu, select **Analysis > Sensitivity Analysis.**

Type SENS_rd13 as the name for the sensitivity analysis. Click Dimension and select rd13. The current value is 240. Accept the default setting, which varies from 216 to 264. Select YCOG as the parameter to plot.

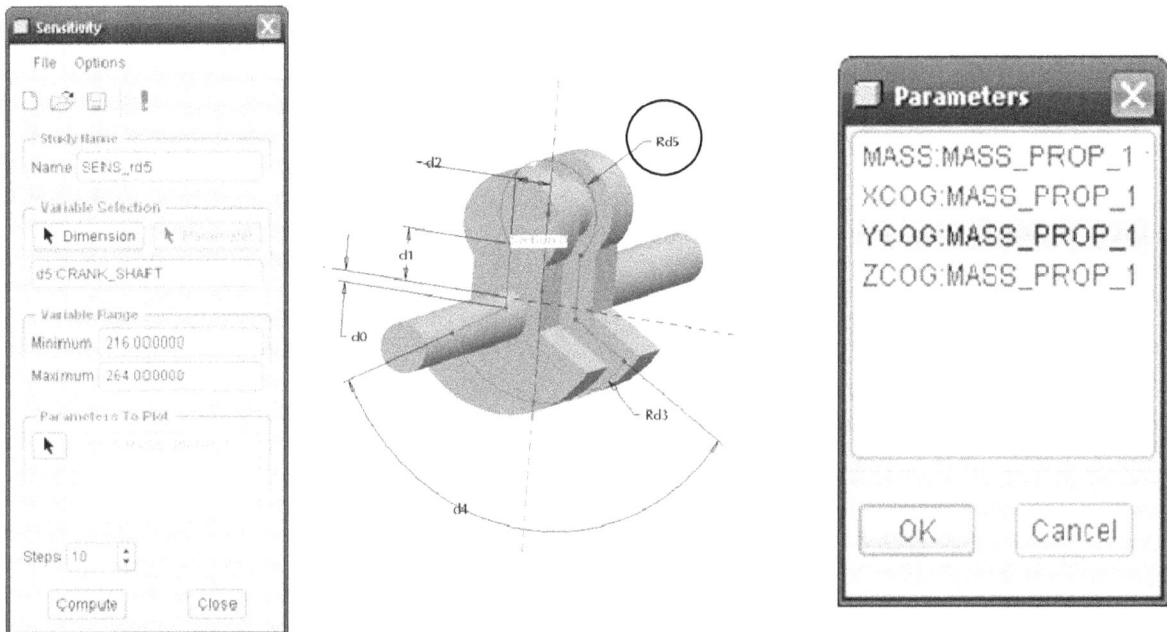

Click **Compute** and we have a graph indicating that the value of **YCOG** increases as the increase of rd5. The effectiveness on YCOG reduction is very significant. When rd5 = 216, the YCOG value is about 24 mm between the center of gravity and the rotation axis. Click **File > Exit > Close.**

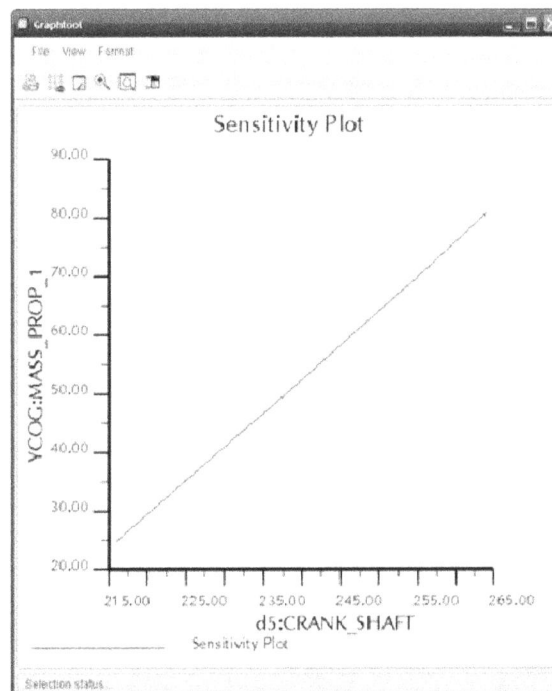

From the main menu, select **Analysis > Sensitivity Analysis.**

Type SENS_d2 as the name for the sensitivity analysis. Click Dimension and select d2. The current value is 175. Accept the default setting, which varies from 157.5 to 192.5. Select **YCOG** as the parameter to plot.

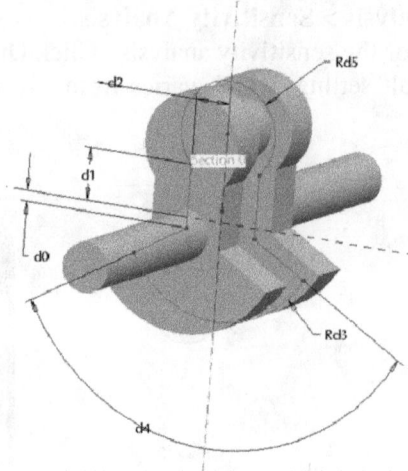

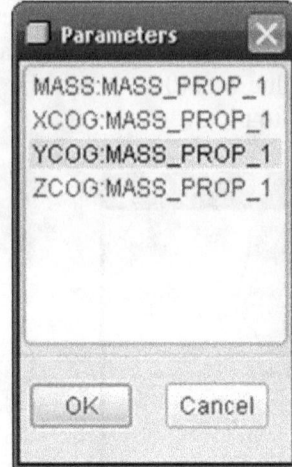

Click **Compute** and a graph shows a convex shape indicating that the YCOG value decreases as the value of d2 increases. However, such decrease is very limited. Click **File > Exit > Close.**

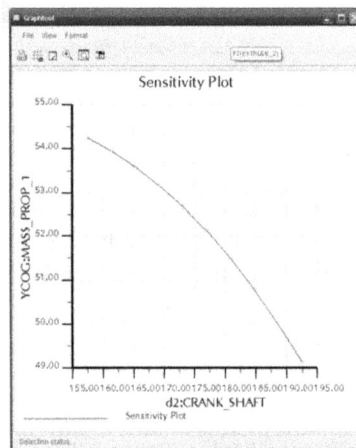

Summary of the local sensitivity study is shown below.

d_0	Rd_3	Rd_5	d_2

Step 5: From the main menu, select Analysis > Feasibility/Optimization > select Feasibility to identify, at least, one feasible solution within a given search space.

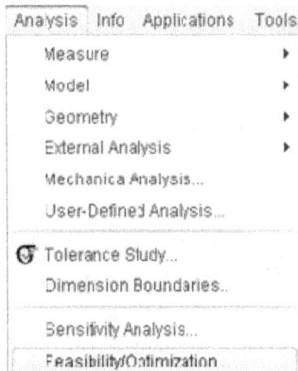

Set the design constraint to solve for YCOG = 0. The search space is defined by setting

d0:	50 - 55
Rd3:	500 - 550
Rd5:	210 – 240
d2:	175 – 195

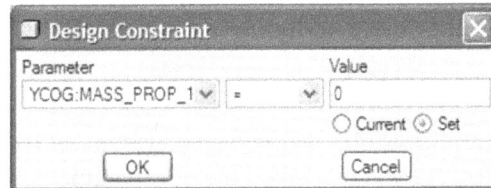

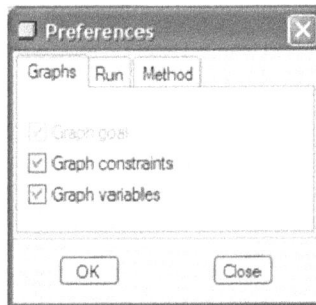

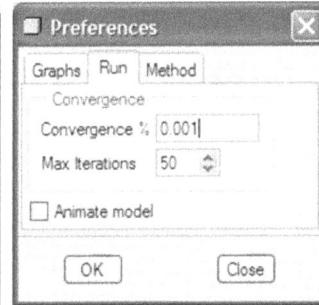

Click **Options > Preferences** and select the **Run** tab. Enter 0.001% as the convergence.

Select the **Graphs** tab and enable **Graph Variables** checkbox and **Graph** constraints box. Click **OK > Compute**.

A feasible solution at its best is found. The smallest value of YCOG is 0 mm.

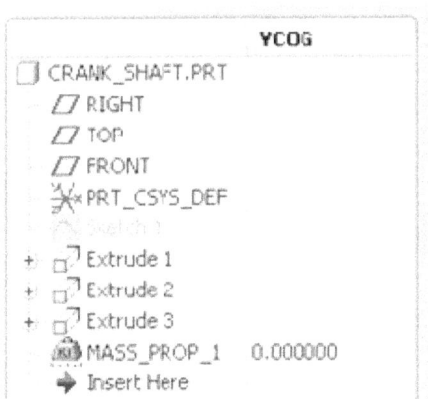

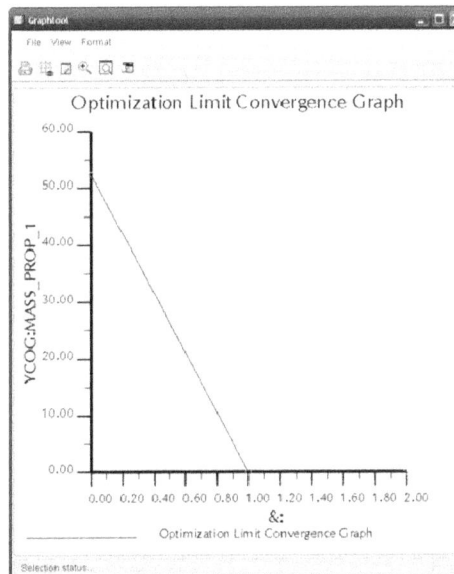

The software system is asking the user to confirm the new settings of all design parameters for the best possible feasible solution obtained > click **Confirm**.

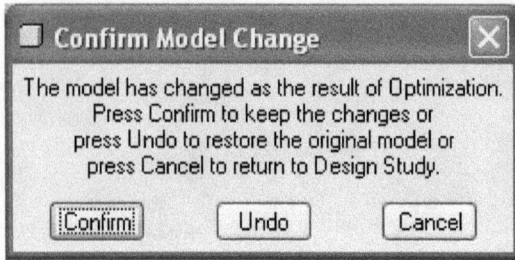

d0 = 50.24

d2 = 176.47

rd3 = 536.95

Rd5 = 213.06

Confirm Model Change

The model has changed as the result of Optimization. Press Confirm to keep the changes or press Undo to restore the original model or press Cancel to return to Design Study.

[Confirm] [Undo] [Cancel]

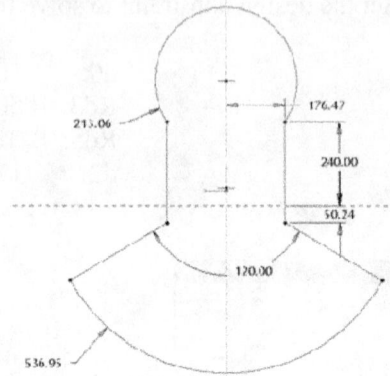

The variations of the 4 design parameters in the process of seeking the best possible feasible solution are shown below, confirming the new parameter settings.

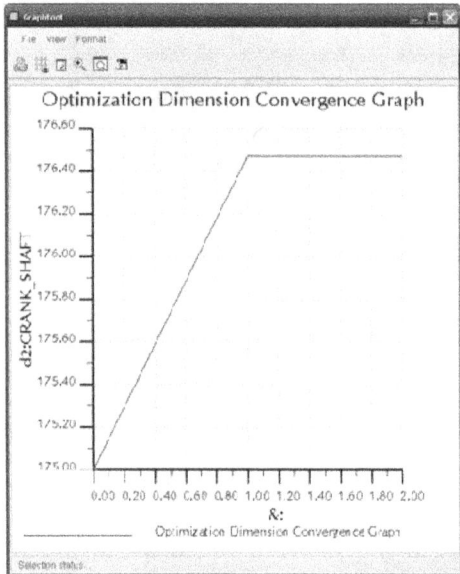

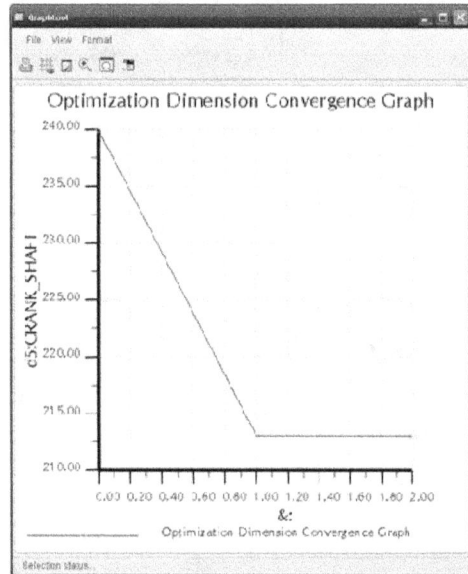

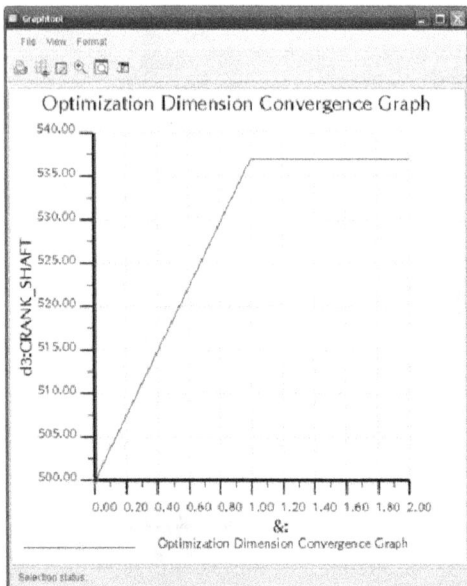

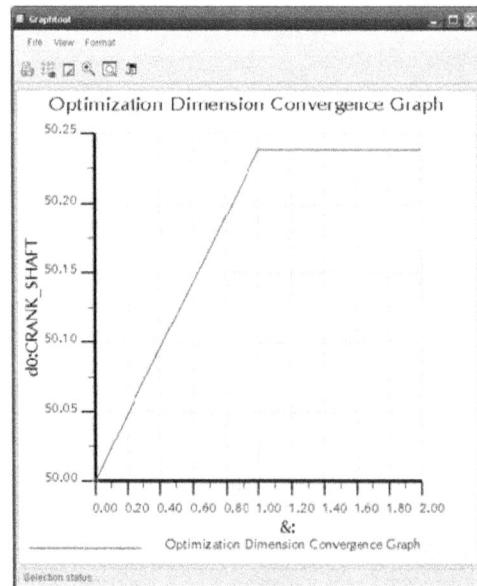

Step 6: From the main menu, select Analysis > Feasibility/Optimization > select Optimization.
The objective function or the goal of search is to minimize the mass amount of the crank shaft.

The design constraint is to keep the distance equal to zero between the location of the center of gravity and the rotation axis.

The variation ranges of the four (4) design variables remain the same as before.

d0:   50 - 55
Rd3:   500 - 550
Rd5:   210 – 240
d2:   175 - 195

Make sure that the plot of total mass, the plot of YCOG, and the 4 plots of the 4 design variables are all available during the computing.

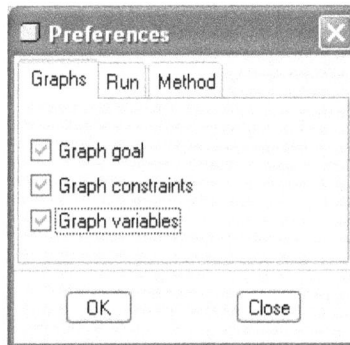

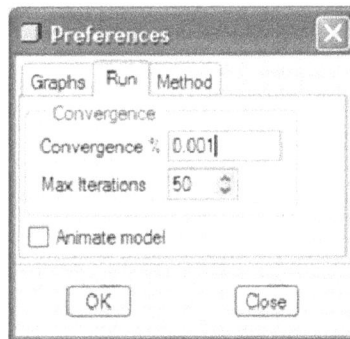

The plot of the distance between the location of the center of gravity and the rotation axis is shown below. Therefore, the objective to achieve YCOG = 0 is realized.

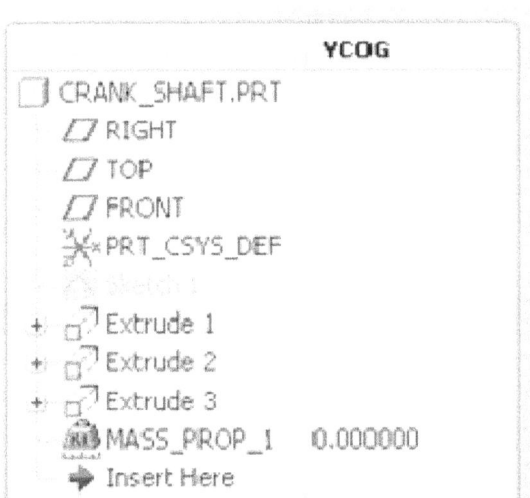

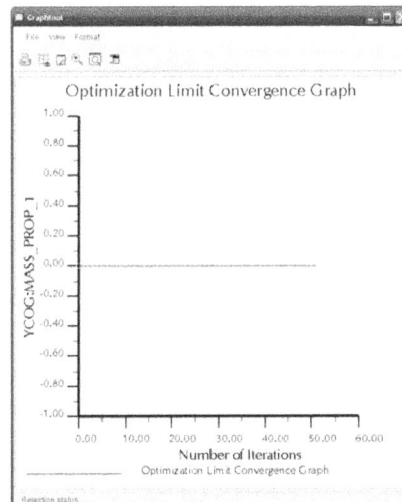

The plot of the total mass of the crank shaft during the search process is shown below.

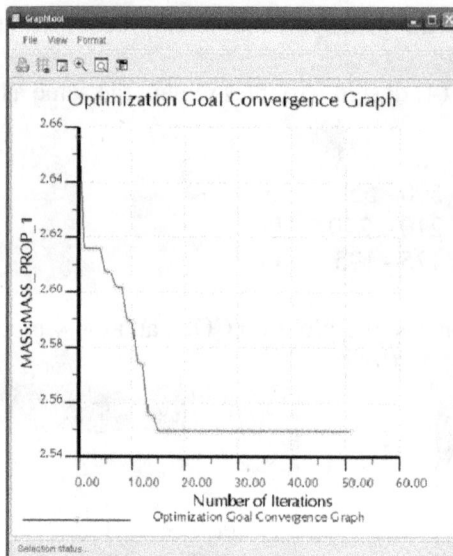

The plot of dimension d0 during optimization

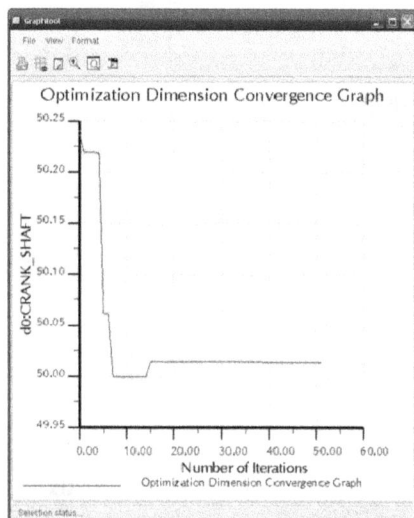

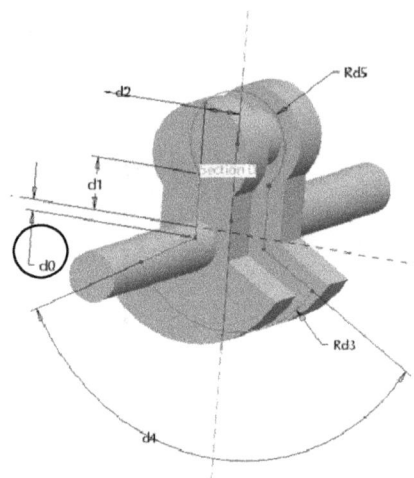

The plot of dimension rd26 during optimization

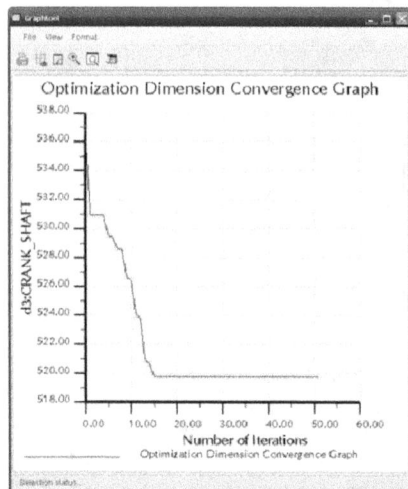

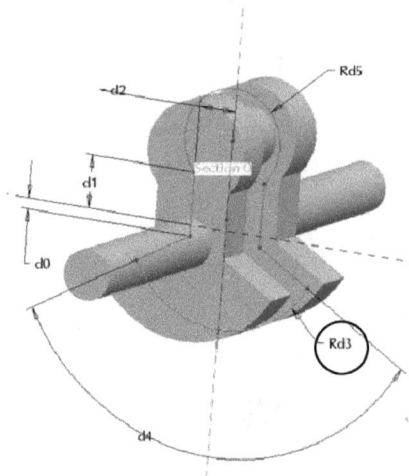

The plot of dimension Rd5 during optimization

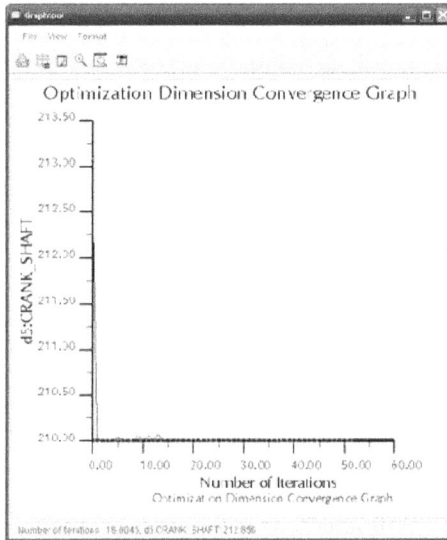

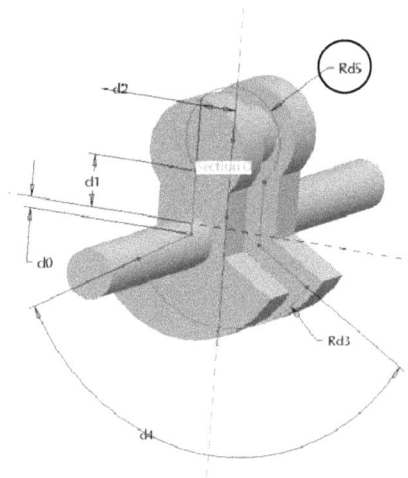

The plot of dimension d2 during optimization

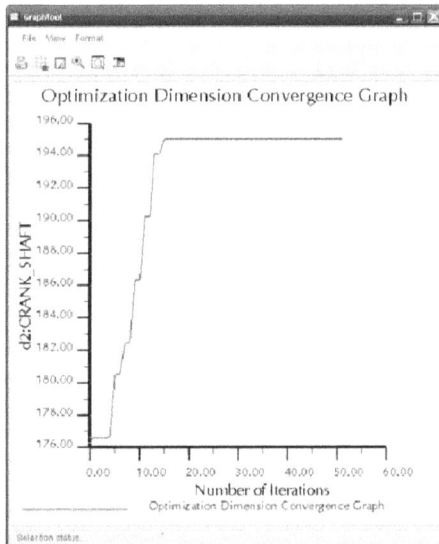

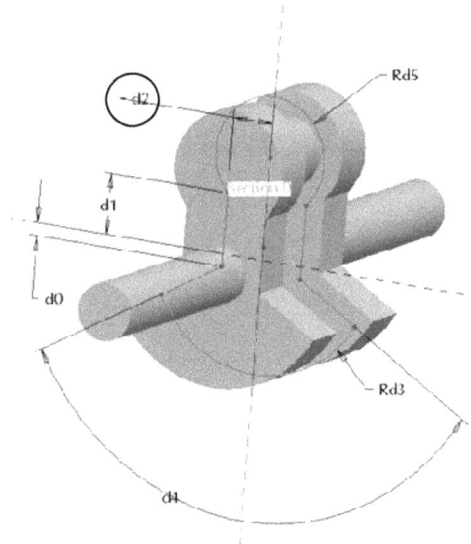

The software system is asking the user to confirm the new settings of all design parameters for the best possible feasible solution obtained > click **Confirm**.

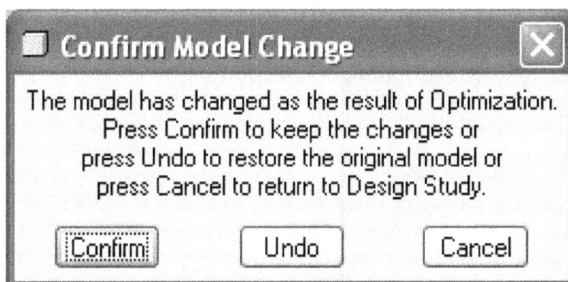

d9 = 55

d14 = 314.76

rd13 = 211.33

rd26 = 550

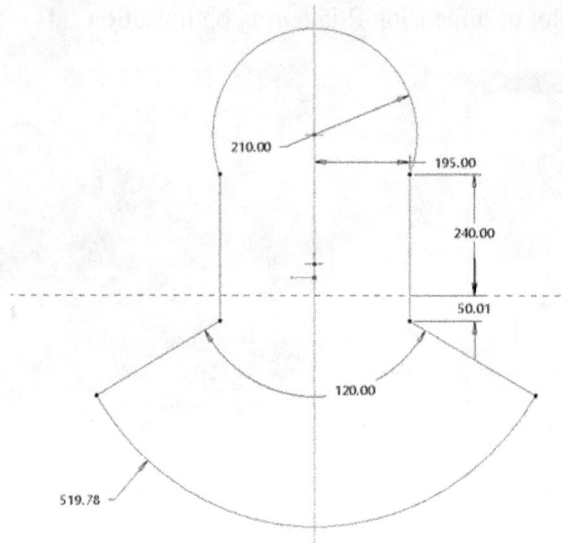

Set up **Analyses** and **Run** it. again. Select the icon of **Run an Analysis** from the main toolbar > **File** > **New Static** > type *optimal_design* as the name of the analysis folder > **OK.** Click the box of **Run.**

There is always a message displayed on screen. The message is asking the user "Do you want to run interactive diagnostics?" Click **Yes**. In this way a warning message will appear if an abnormal operation occurs.

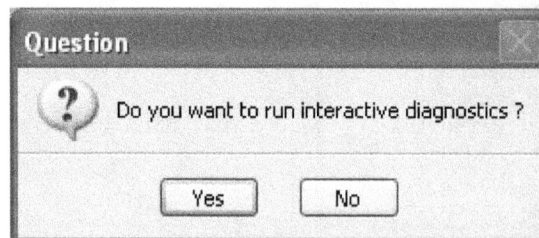

To monitor the computing process, users may click the icon of **Display study status**.

max_prin_mag*:	5.512750e+01
max_rot_mag:	0.000000e+00
max_rot_x:	0.000000e+00
max_rot_y:	0.000000e+00
max_rot_z:	0.000000e+00
max_stress_prin*:	5.512750e+01
max_stress_vm*:	4.757625e+01

Display
study status

To study the obtained result, say the von Mises stress distribution under the centrifugal load condition, click the icon of **Review results > Stress > von Mises > OK and Show.** The maximum value is 47.57 MPa, which is acceptable. When comparing to the previous value, which was 673 MPa, the reduction is very significant.

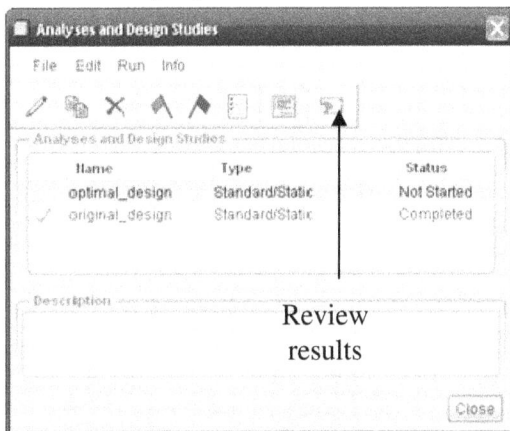

Review
results

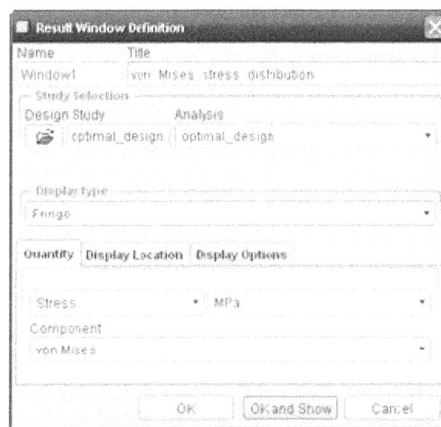

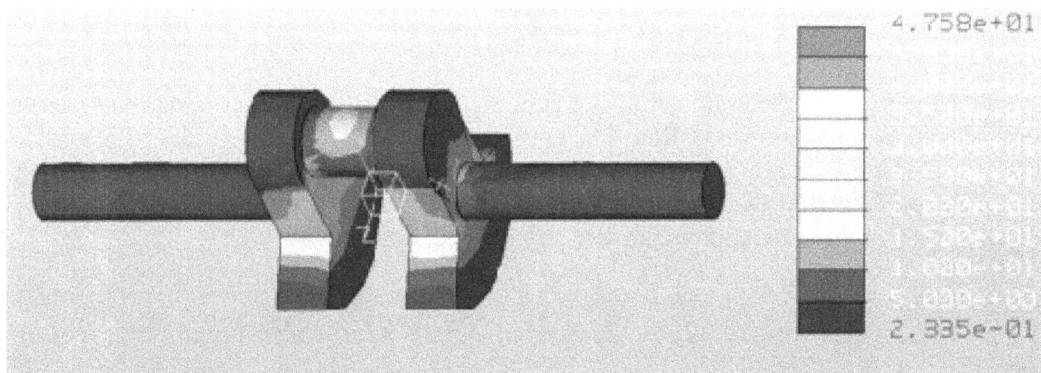

1. For the purpose of practice, the students are asked to go through the design optimization on his/her own. Assume that the rotation speed is 100 rpm, evaluate the maximum principal stress due to the presence of centrifugal force. The students should calculate the 2 values under the two circumstances: before optimization and after optimization. Which one is larger? What is the magnitude of reduction after optimization?

2. An engineer is now making a suggestion to drill holes on the lobe portion of the crank shaft. The dimensions of hole size and hole position are given below. Cut those holes from the 3D solid model. Evaluate the maximum principal stress due to the presence of centrifugal force. The students should calculate the 2 values under the two circumstances: without the hole feature, and

with the hole feature. Which one is larger? What is the magnitude of reduction after drilling the hole feature?

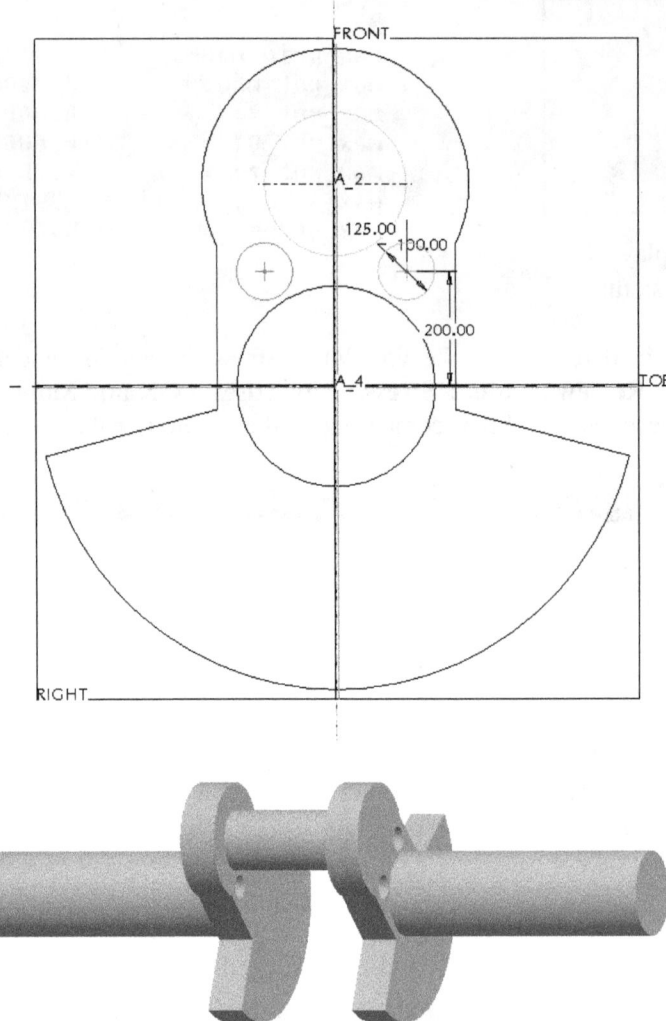

In the re-design process, there are certain constraints. The first constraint is that the maximum von Mises stress should be kept below 85 MPa. The reason is that the yield stress of the steel material is about 225 MPa. For a safety factor of 2.5, the limit for the maximum von Mises stress should be set at 85 MPa. However, we are informed that the maximum von Mises stress at this design stage when it is subjected to loading is 90 MPa, a violation to the safe constraint condition.

8.5   References

1. Algor Processor Reference Manual, Algor Interactive Systems, 150 Alpha Drive, Pittsburgh, PA 15238
2. O. Axelsson, Iterative Solution Methods, Cambridge University Press, New York, 1994.
3. K. J. Bathe, Finite Element Procedure, Prentice Hall, Upper Saddle River, New Jersey, 1996.
4. K. Bell, A Refined Triangular Plate Bending Finite Element, Int. J. Numer. Methods Eng., Vol. 1, No. 1 January 1969, pp. 101-122.
5. B. Carnahan, H. A. Luther, and J. O. Wikes, Applied Numerical Methods, Wiley New York, NY, 1969.

6. R. W. Clough, <u>The Finite Element in Plane Stress Analysis</u>, Proceedings of 2nd ASCE Conference on Electronic Computation, Pittsburgh, PA. September 1960.

7. R. D. Cook, D. S. Maikus and M. E. Plesha, <u>Concepts and Applications of Finite Element Analysis</u>, 3rd edition, Wiley, New York, NY, 1989.

8. J. W. Dally and W. F. Riley, Experimental Stress Analysis, McGraw Hill, 1965.

9. J. W. Dally, Design Analysis of Structural Elements, Goodington, New York, NY, 2003.

10. M. A. N. Hendriks, H Jongedijk, J. G. Rots and W. J. E. van Spange (eds), <u>Finite Elements in Engineering and Science</u>, DIANA Computational Mechanics, 1997.

11. K. H. Huebner, D. L. Dewhirst, D. E. Smith, and T. G. Byrom, <u>The Finite Element Method for Engineers</u>, 4th edition, John Wiley & Sons, Inc., 2001.

12. C.S. Krishnamoorthy, <u>Finite Element Analysis, Theory and Programming</u>, 2nd Ed., 1995.

13. D. L. Logan, A First Course in the Finite Element Method Using ALGOR, PWS Publishing, Boston, 1997.

14. H. C. Martin, <u>Introduction to Matrix Methods of Structural Analysis</u>, McGraw-Hill, New York, NY, 1966.

15. R. L. Norton, <u>Machine Design: An Integrated Approach</u>, Prentice Hall, Upper Saddle River, New Jersey, 1996.

16. P. G. Plockner, <u>Symmetry in Structural Mechanics, J. of the Structural Division,</u> American Society of Civil Engineers, Vol. 99, No. ST1, pp. 71-89, 1973.

17. Y. Saad, Iterative Methods for Sparse Linear Systems, PWS Publishing Company, Boston, 1996.

18. Soc., Vol. 7, No. 8, pp. 849-861.

19. O. C. Zienkiewicz, <u>The Finite Element Method</u>, third edition, McGraw-Hill, New York, NY, 1977.

20. O. C. Zienkiewicz and R. L. Taylor, <u>The Finite Element Method</u>, 4th edition, Vol. 1, McGraw-Hill (UK),London, 1989.

8.6 Exercises

1. The following drawing shows the design of a notch plate. We have interests on 4 design parameters, which are listed below

Design Parameters	Symbolic Names
Radius of the notch: R2.5	notch_radius
Thickness of the plate: 2.5	thickness
Cut distance: 60	cut_distance
Total height: 30	total_height

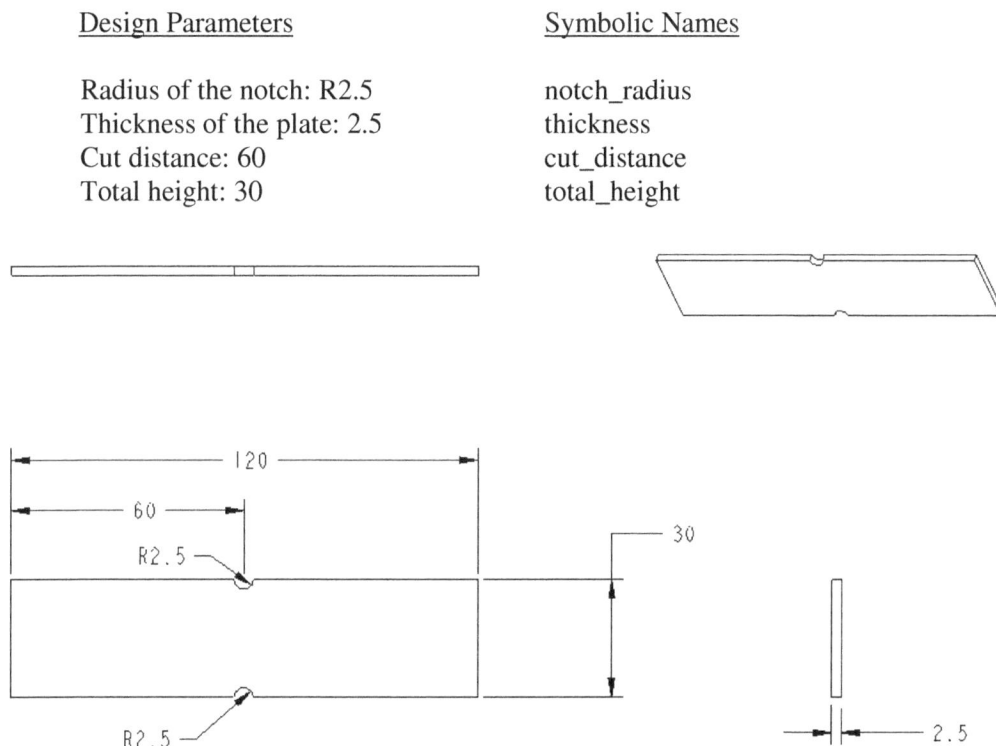

The working condition is that the left end of the notch plate is fixed. The right end of the notch plate is the free end where a load is acting on it. The two force components are 80 N along the x-direction and –200 N along the y-direction.

$F_x = 80$ Newton $F_y = -200$ Newton

The material of the notch plate is steel. Its yield strength is assumed to be 240 MPa. The safety factor is 3. Therefore, the maximum allowable stress is 80 MPa. The objective of this study is to search for a new design, if necessary, so that the mass of the notch plate can be reduced. In the meantime, the maximum von Mises stress is limited to 80 MPa, and the maximum displacement is limited to 0.1 mm at the upper right corner.

Minimize the total mass of the notch component

Subject to

$$\sigma_{max\_vm} \leq 80$$

$$\delta_{max\_deflection} \leq 0.10 \quad (mm)$$

and $2.5 \leq notch\_radius \leq 3.5$

$2.5 \leq thickness \leq 3.0$

Units: mm

$60 \leq cut\_distance \leq 70$

$2.5 \leq total\_height \leq 3.5$

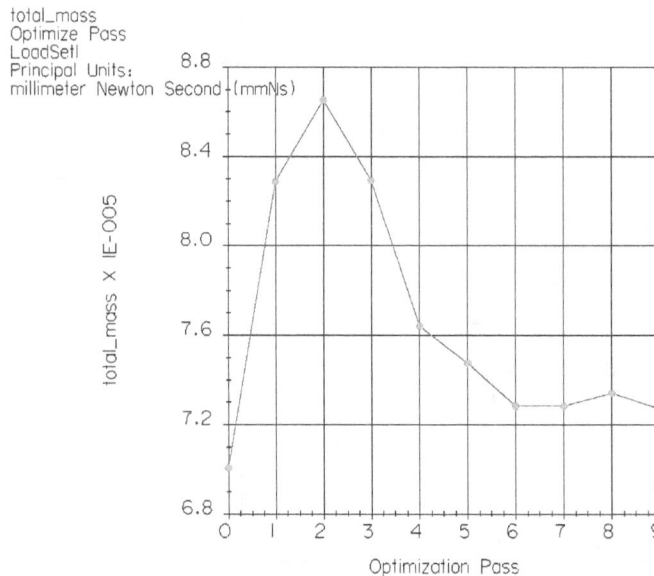

Minimized Total Mass vs Iterations

max_disp_mag
Optimize Pass
LoadSet1
Principal Units:
millimeter Newton Second (mmNs)

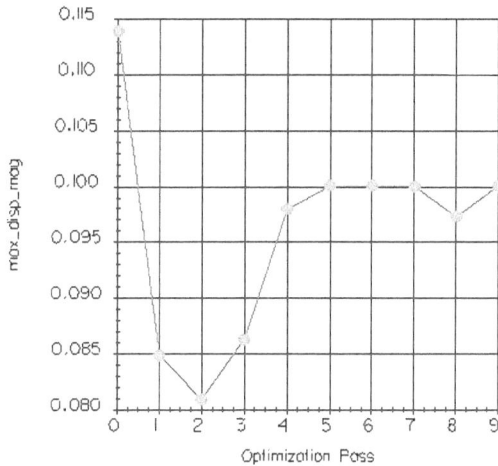

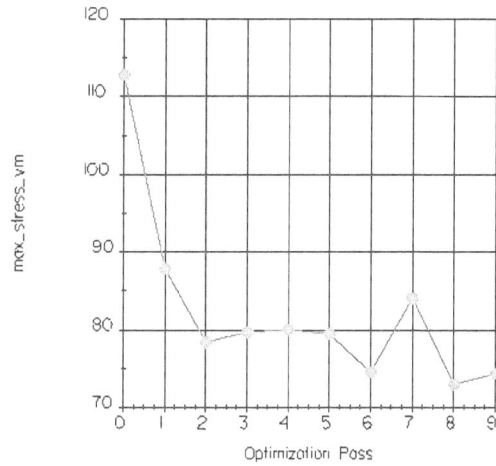

max_stress_vm
Optimize Pass
LoadSet1
Principal Units:
millimeter Newton Second (mmNs)

2.   In section 8.3, we performed a redesign process for the bracket component. In that study, three design parameters were under consideration. In this exercise, let us perform the redesign process for the bracket component again. Let us add a new design parameter to the set of the design parameters previously selected. The newly added design parameter is the total height of the bracket component. As illustrated in the following figure, the current value of this new design parameter is 125 mm. Let us allow it to vary between 120 mm and 130 mm.

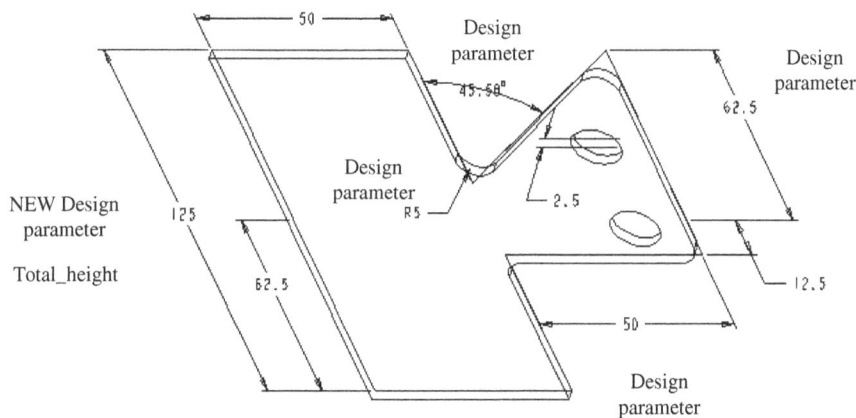

Let us keep the constraint condition, the load condition and the material type unchanged. The mathematical format for this resdesign process is listed below:

Minimize the total mass of the bracket component

Subject to

$$\sigma_{max\_vm} \le 90 \quad (\text{MPa})$$

$$\delta_{max} \le 0.20 \quad (\text{mm})$$

and $\qquad 45^o \le angle \le 60^o$

$\qquad 120 \le Total\_height \le 130 \qquad$ Units: mm

$\qquad 5 \le inner\_fillets \le 15$

$\qquad 45 \le Over\_width \le 55$

CHAPTER 9

MODELING WITH BEAM AND TRUSS STRUCTURES

9.1 Introduction

In Chapter 6, we discussed the idealization of mechanical structures with shells. When using the shell model, we need to detect the middle surface from a pair of surfaces, which are parallel to each other. A shell model is a combination of middle surfaces, namely, an idealization of a 3D mechanical structure. In this chapter, we deal with truss and beam structures through idealization. Instead of using middle surfaces, we will use datum curves to represent the truss and beam structures. The geometrical shape and dimensions are characterized through a process of defining the section area(s).   In section 9.2, we use Pro/Mechanica to perform FEA for a simple truss structure.

9.2 Application on a Simple Truss Structure

The following figure illustrates a truss structure. It consists of 4 bars. The lengths of these 4 bars are shown in the figure.   The three dimensions characterizing the truss structure are 1000, 350 and 250 mm. Regarding the geometrical shape, each bar is a pipe. Its cross section is a hollow circle. The outer diameter is 30 mm and the inner diameter is 26 mm. The load is acting on the free end (PNT2). PNT0 and PNT1 are fixed to the ground.

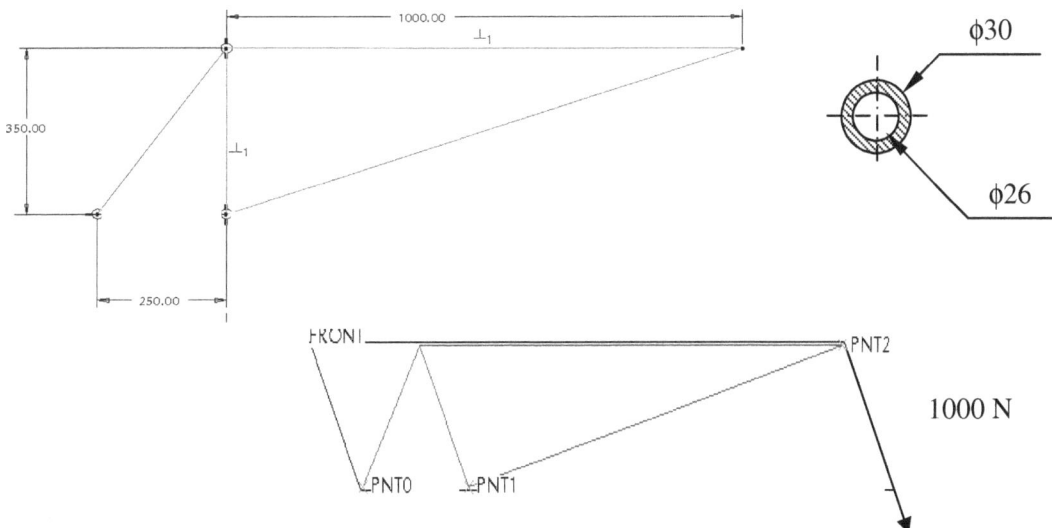

There are 2 objectives of this study. The first objective is to plot the deflection pattern of this truss structure. The second objective is to plot the internal forces induced to each bar when subjected to the load condition.

Step 1: Model the truss structure

From the mail toolbar, select the icon of **Create** a new object. Make sure the **Part** mode is selected. Type *truss_structure* and clear the box of **use default template > OK**

Select the unit system to be **mmns_part_solid** > type *truss structure* in **DESCRIPTION** and *student* in **MODELED_BY > OK**.

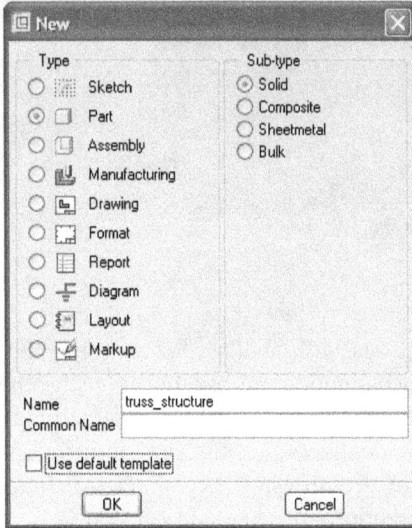

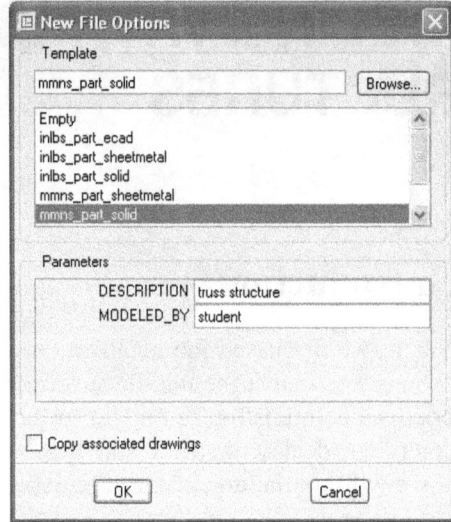

Select the icon of **Sketch Tool.** Pick the **FRONT** datum plane as the sketch plane, and accept the default orientation setting, which uses the **RIGHT** datum plane and position it at the right side > **Sketch.**

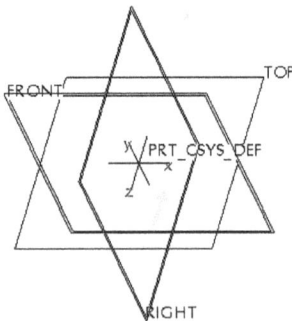

Pick the icon of **Line**, and sketch the first line with 2 dimensions (250 and 350), as shown below.

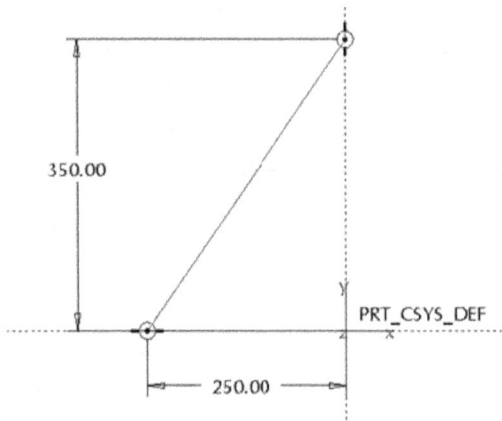

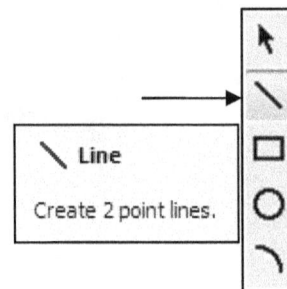

Sketch a horizontal line with the dimension equal to 1000, as shown below:

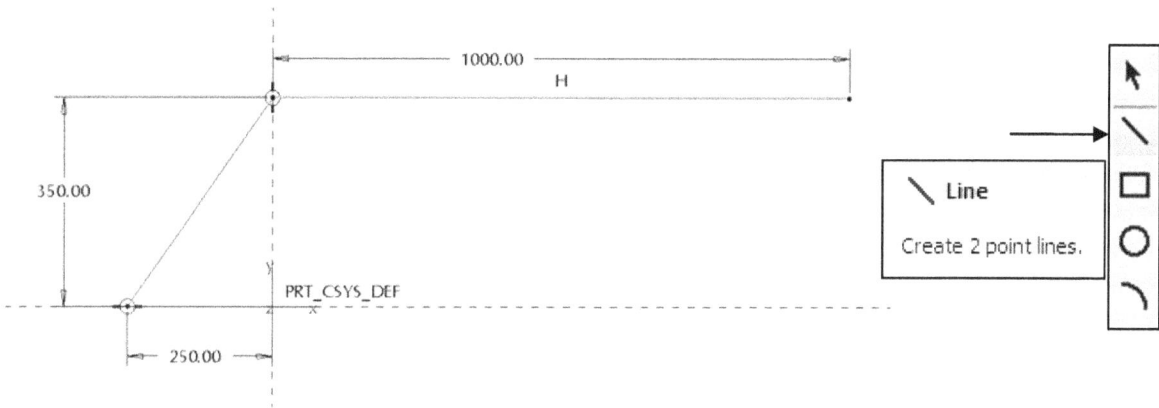

Sketch an inclined line by connecting 2 points, as shown.

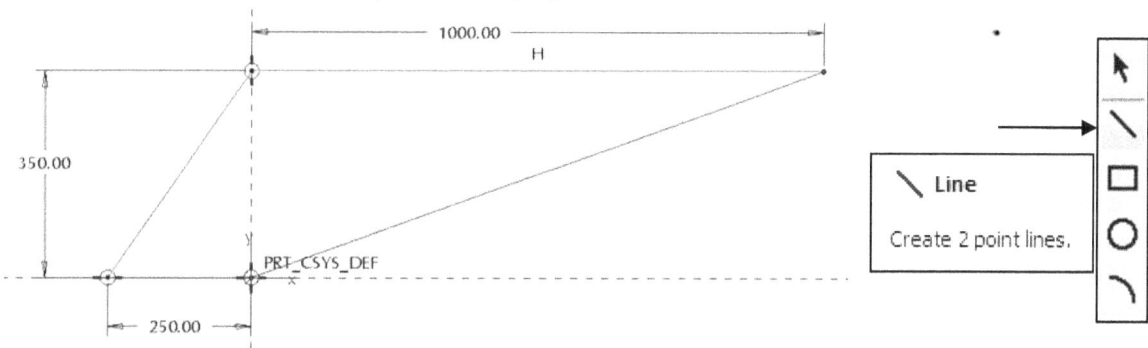

Sketch a vertical line by connecting two points, as shown below:

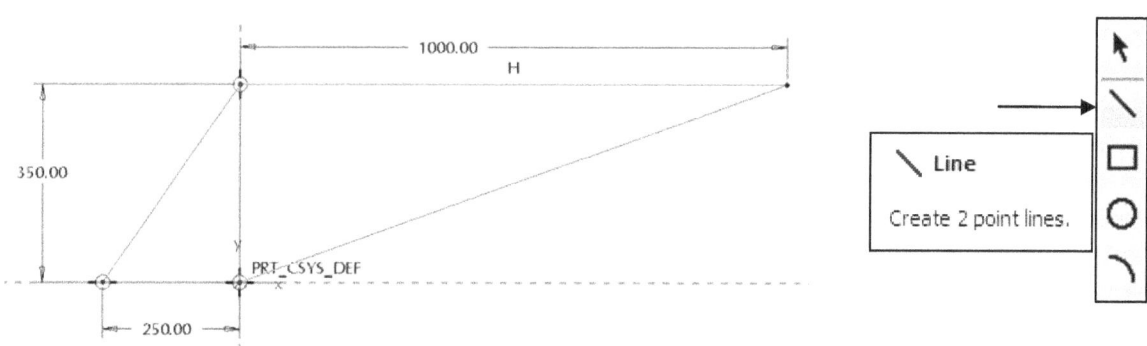

Upon completion, select the icon of **Done** from the toolbar of sketcher. The sketched 4 datum curves (actually, they are straight lines) represent the 4 bars of the truss structure.

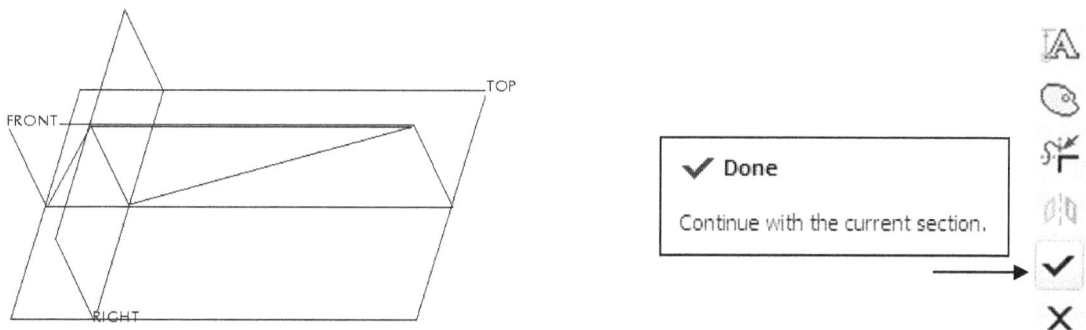

Now let us create four datum points. The first datum point is at the end of the datum curve located at the very left side. Select the icon of **Datum Point Tool**. Make sure the offset value is set at zero > **OK**. PNT0 is on display.

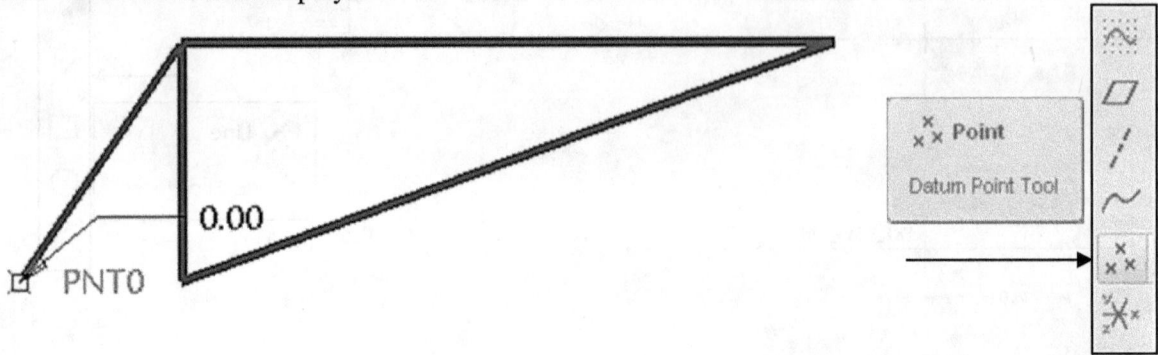

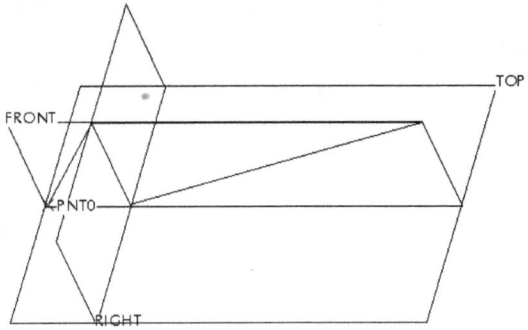

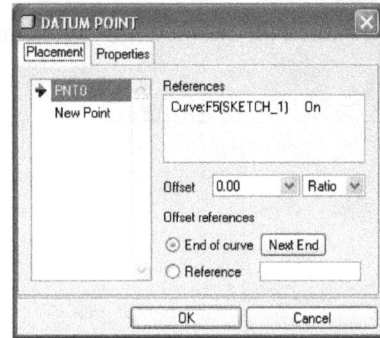

To define the second datum point, select the icon of **Datum Point Tool**. Pick the point at the corner of the datum curve shown below. Make sure the offset value is set at 0.00.

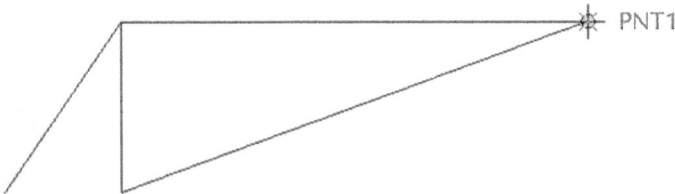

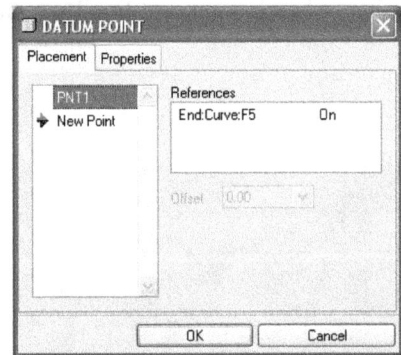

To define the third datum point, select the icon of **Datum Point Tool**. Pick the point at the right-upper corner of the datum curve on display. Make sure the offset value is set at 0.00.

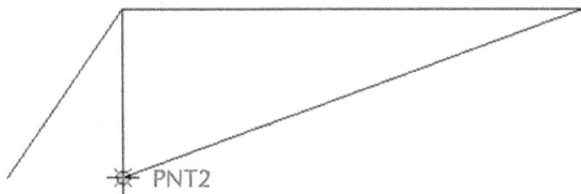

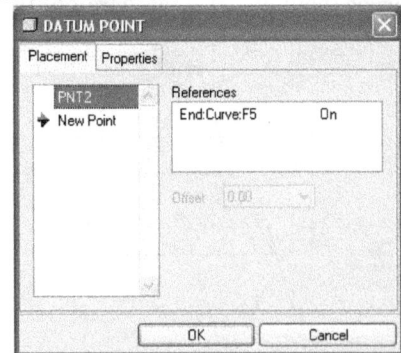

To define the 4th datum point, select the icon of Datum Point Tool. Pick a point on the inclined line and set the offset value is set at 0.25, as shown below.

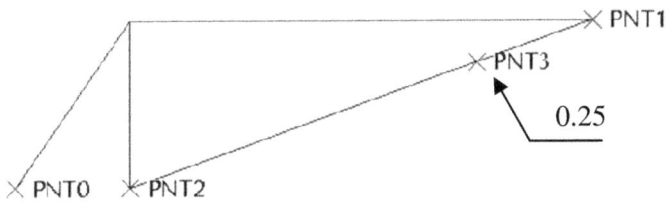

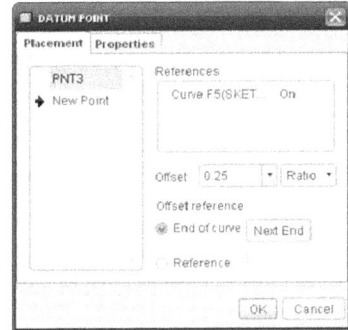

Step 2: under the Pro/Mechanica environment, perform FEA.
From the main toolbar or the menu toolbar, select **Applications > Mechanica > Structure > OK.**
Pick the icon of **Beam** from the toolbar of functions.

Accept the name: Beam1

Select Edge/Curve

Select Steel

Y vector

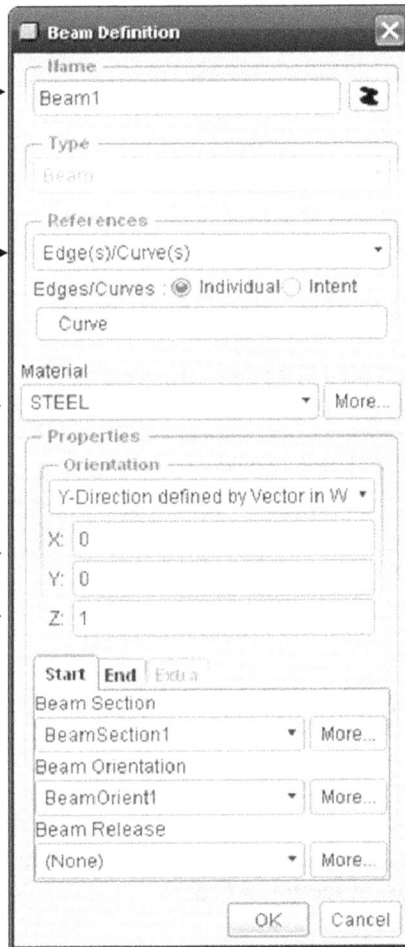

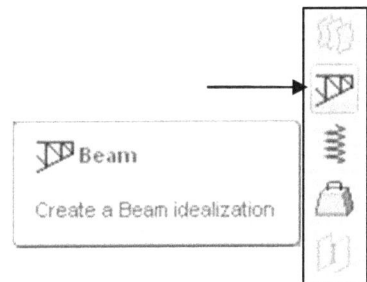

On the appeared **Beam Definition** window, do the followings:
Select **Edge/Curve** > pick the datum curve.

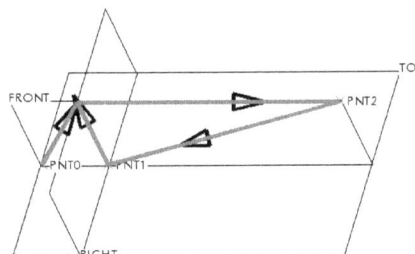

Select More from **Material** > Select **STEEL** > **Close**

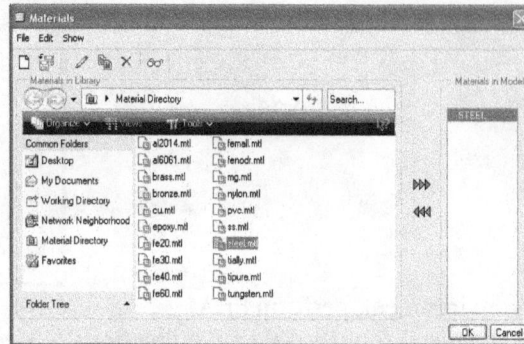

In the Properties field, set x = 0, y = 0 and z = 1 as the normal vector for orientation.

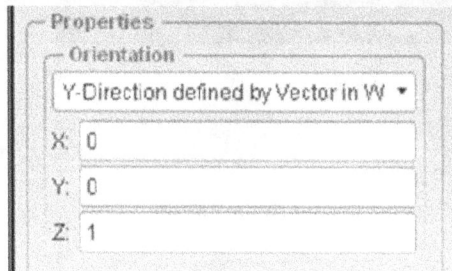

Select More from **Section** > New > select **Hollow circle** > type *30* and *26* as R and Ri, respectively > **OK**.

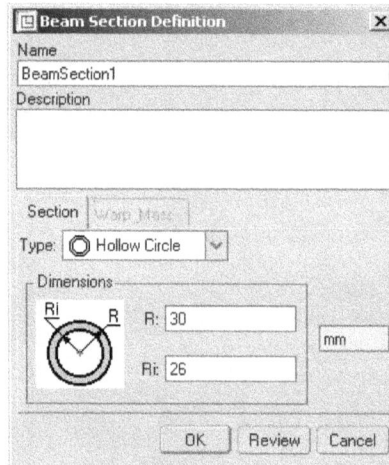

Define the Beam Orientation: click **More** > **New**, and make sure that DY=0 and DZ=0 to complete the process of defining the beam section > **OK**.

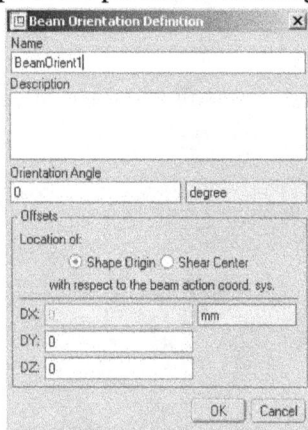

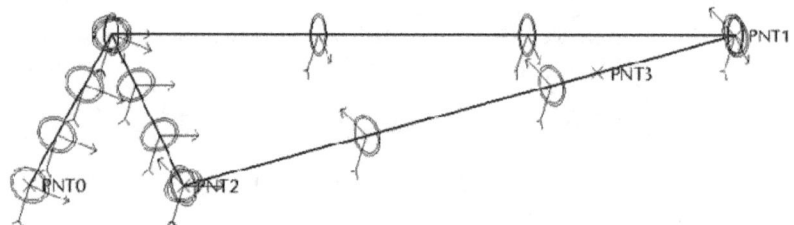

Define the constraint set to present the 2 fixed ends at PNT0 and PNT1. Select the icon of **Displacement Constraint**. Select **Point(s)** > pick **PNT0** > set all 6 degrees of from to fixed > OK.

Repeat the above steps to define the second constraint. Select **Point(s)** > pick **PNT1** > set all 6 degrees of from to fixed > **OK**.

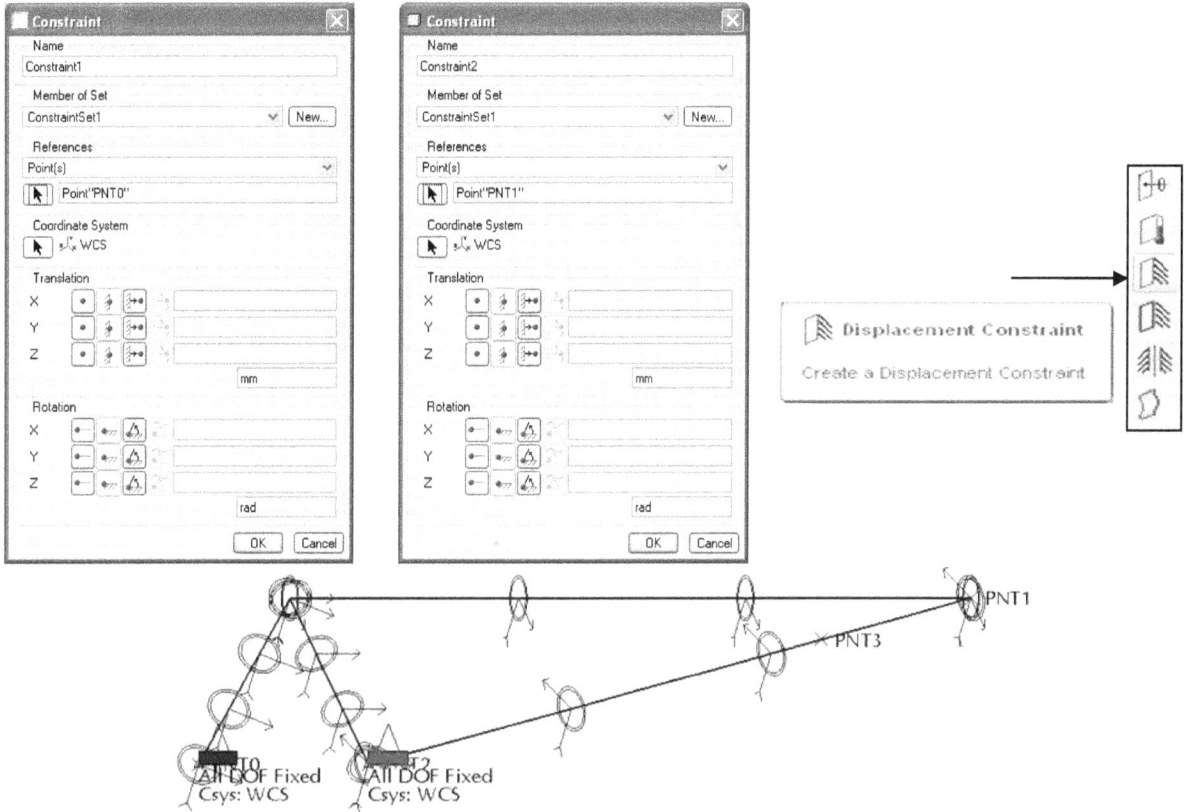

Now let us define the load condition. Select the icon of **Force/Moment**. Select **Point**(s) > Pick PNT2 > type *–1000* as the magnitude in the Y direction.

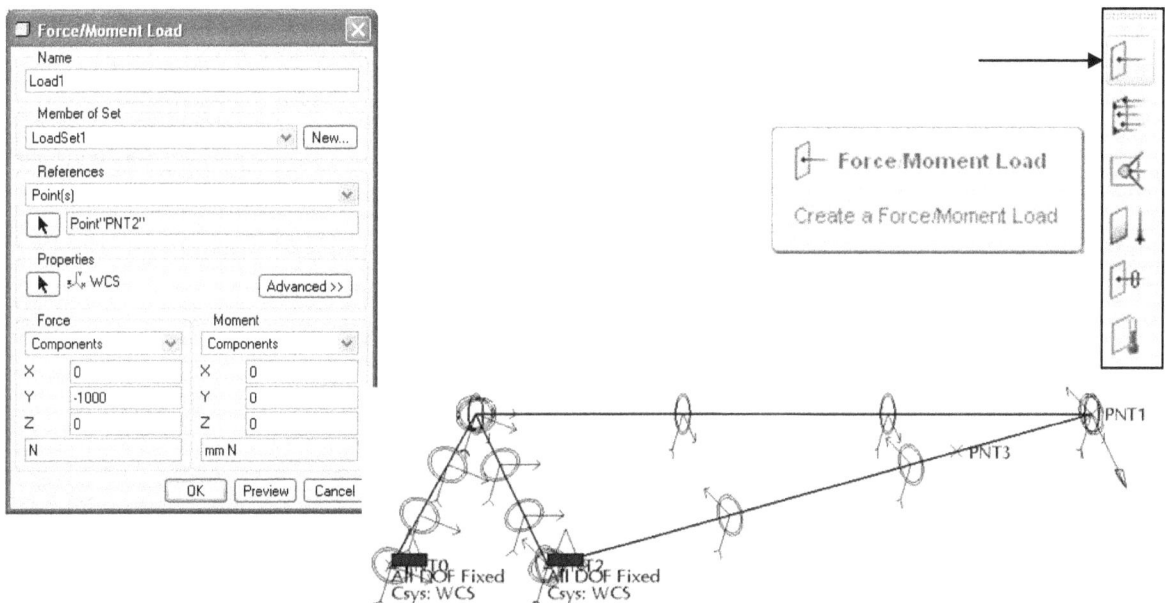

Step 3: Perform a static analysis

From the top menu, click the icon of Run a Design Study > **File > New Static.**

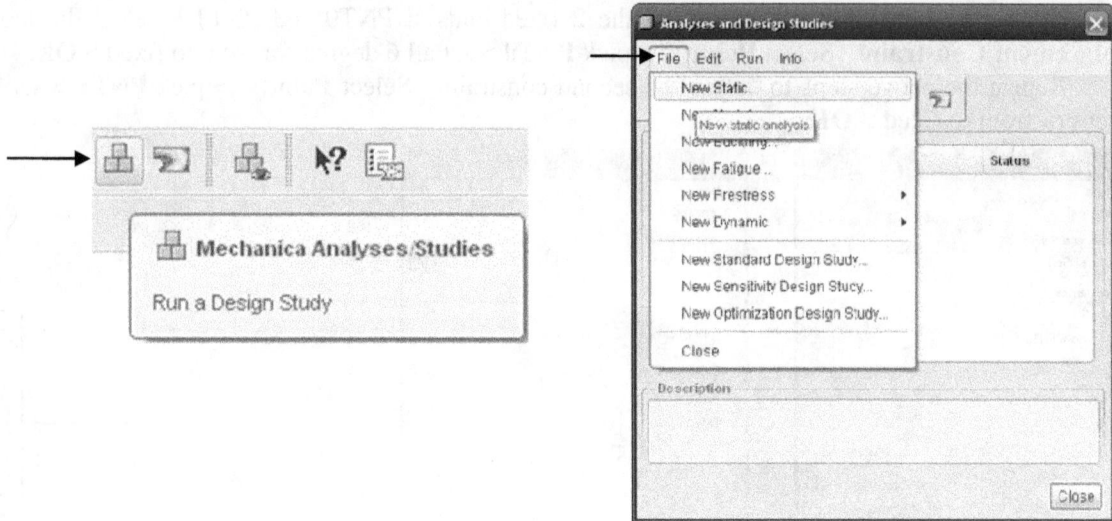

Type *truss_structure* as the name of analysis, and **OK** to accept the default setting, which is **Mult-Pass Adaptive > OK**.

Run the program > **Close** upon completion.

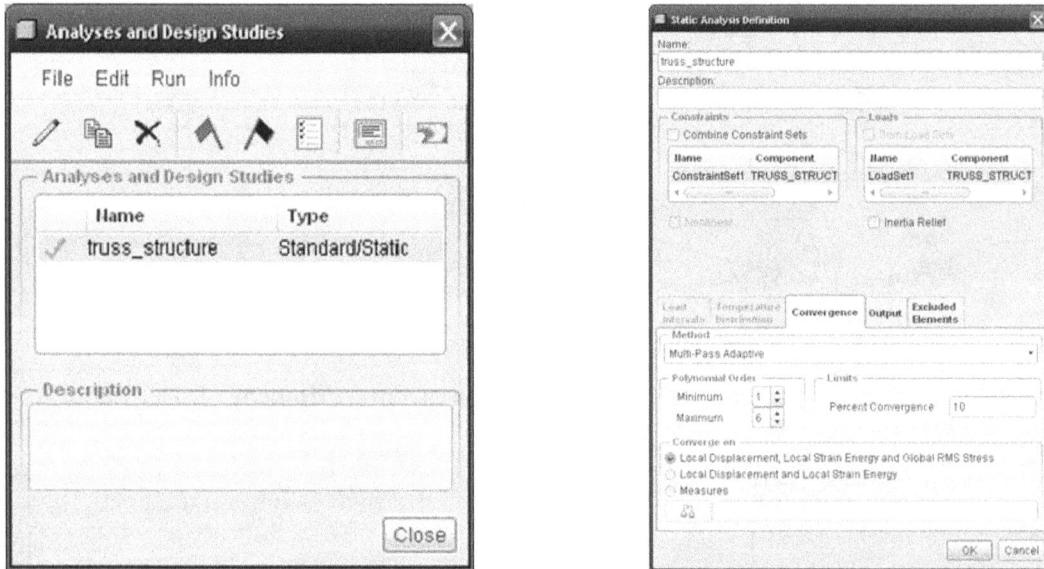

There is always a message displayed on screen. The message is asking the user "Do you want to run interactive diagnostics?" Click **Yes**. In this way a warning message will appear if an abnormal operation occurs.

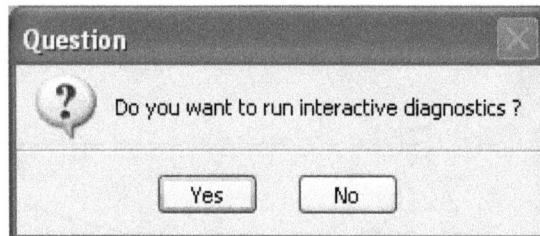

Step 4: Show the results obtained from FEA by plotting the distribution of deflection.

To study the obtained result, say the distribution of the max principal stress, click the icon of **Review results > Displacement > Magnitude > OK and Show.** The numerical value of the max deflection is 0.2297 mm at the tip of the truss structure.

Distribution of the deflection in the XY plane

When using **Graph**, users are able to plot the axial force for individual bars. Click the icon of **Review results** > **Graph** > **Shear & Moment** > select **P**. Click **Relative To** and pick the bar on the left side > press the middle button of mouse to accept the selection > click **OK** to accept the starting point set by default > **OK and Show.**

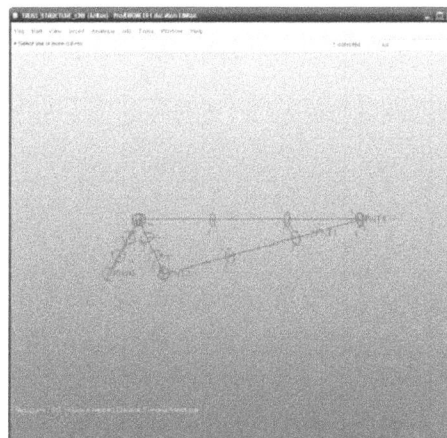

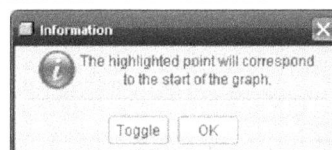

Distribution of the axial forces in the left bars, using <u>GRAPH</u>

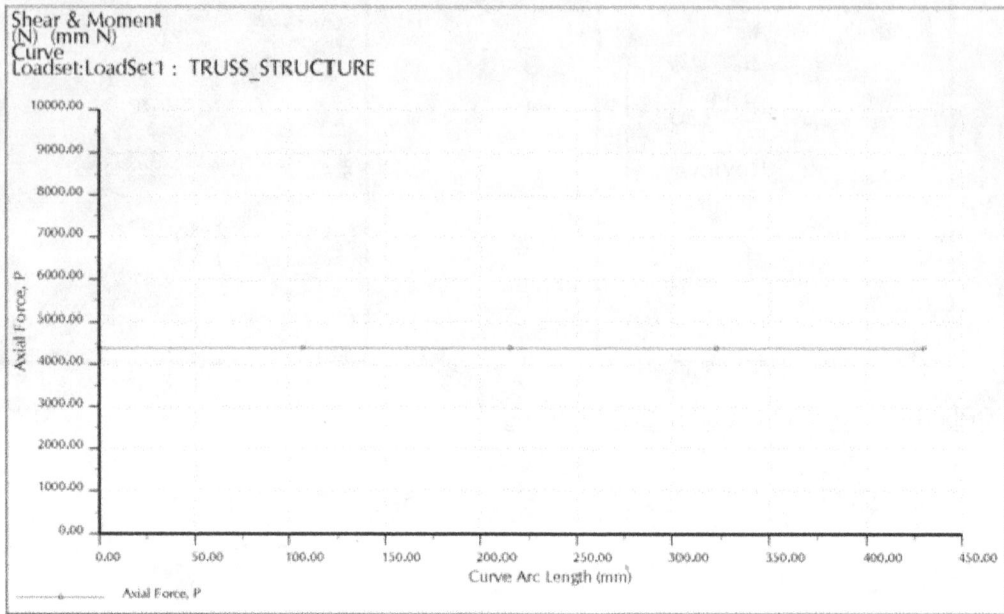

Users may follow the above procedure to plot the axial force for other individual bars.

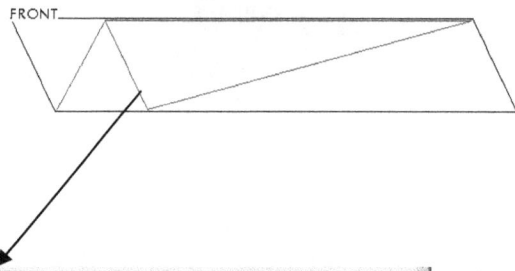

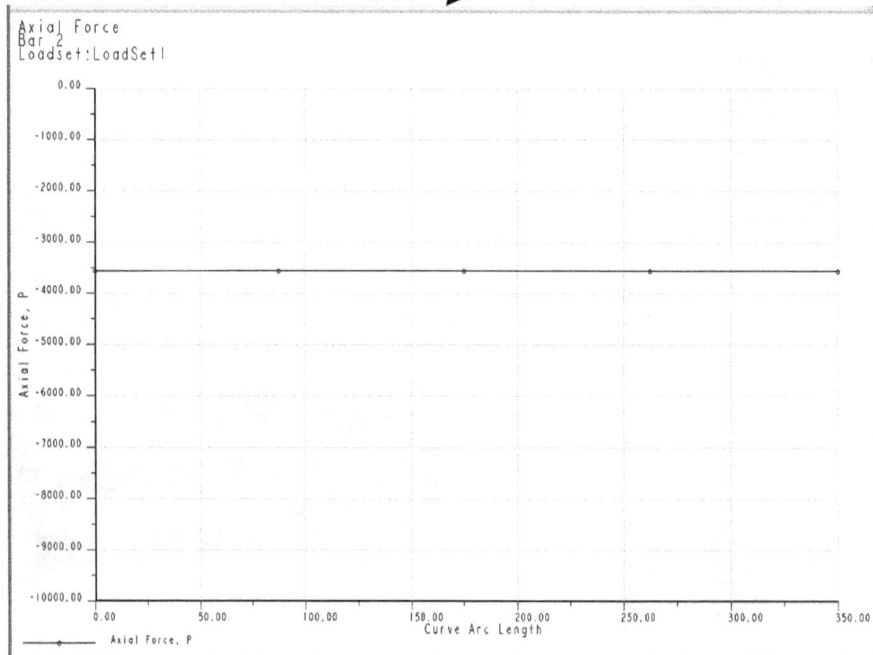

Axial Force
Bar 3 (horizontal)
Loadset:LoadSet1

Axial Force
Bar 4 (inclined)
Loadset:LoadSet1

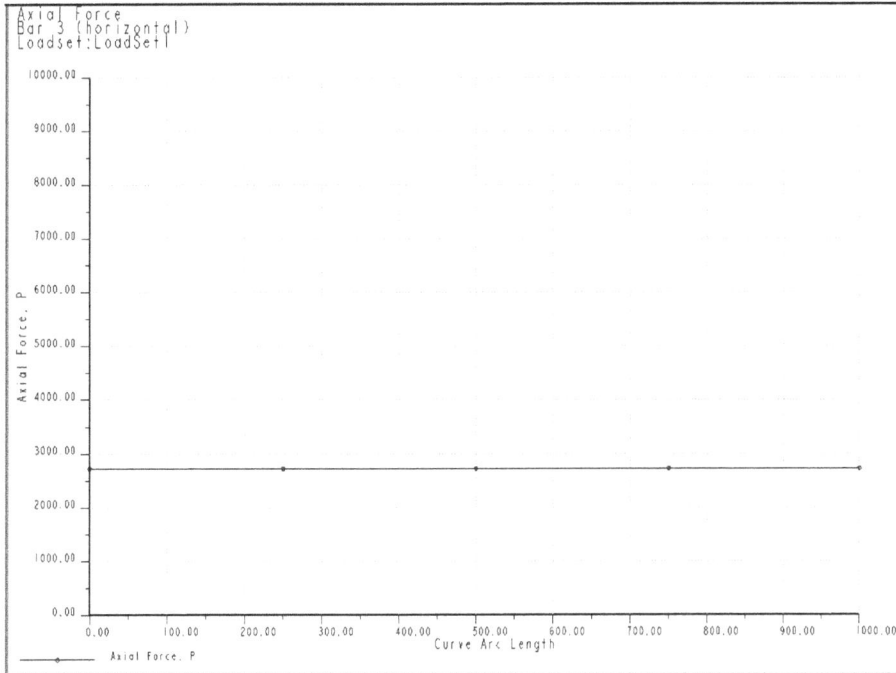

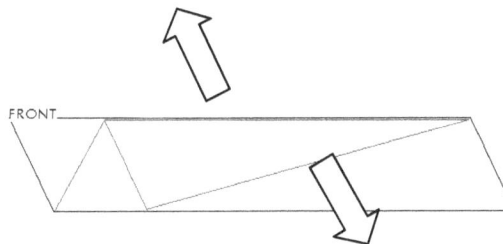

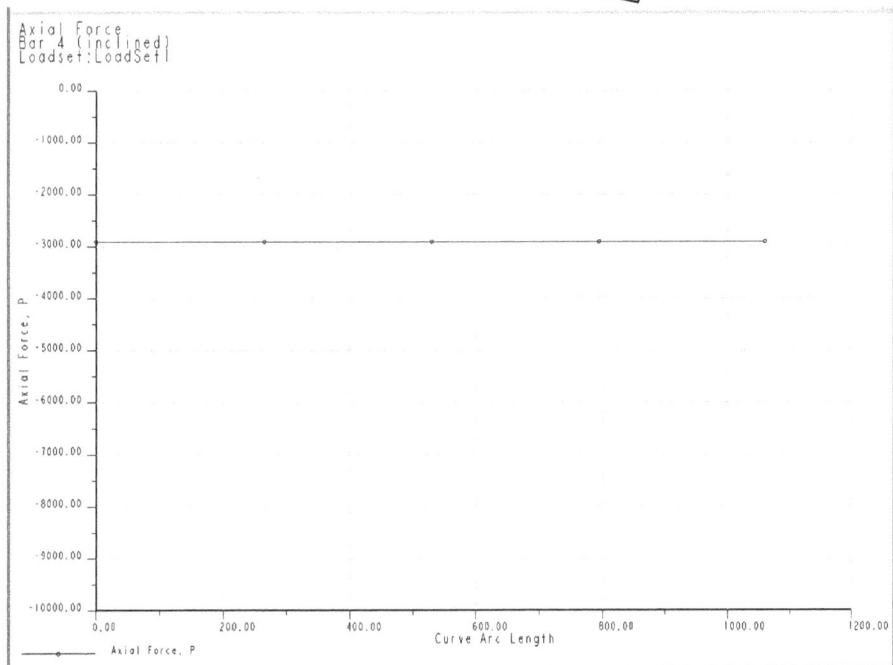

Step 5: Add a spring component to enhance the rigidity of the truss structure so that the numerical value of the max displacement at the tip is controlled below 0.2 mm and the current value is 0.2297 mm at the tip of the truss structure.

As shown below, a spring is created. The two ends are connected to the inclined beam and the ground respectively. In the following we provide the procedure to create such a spring component.

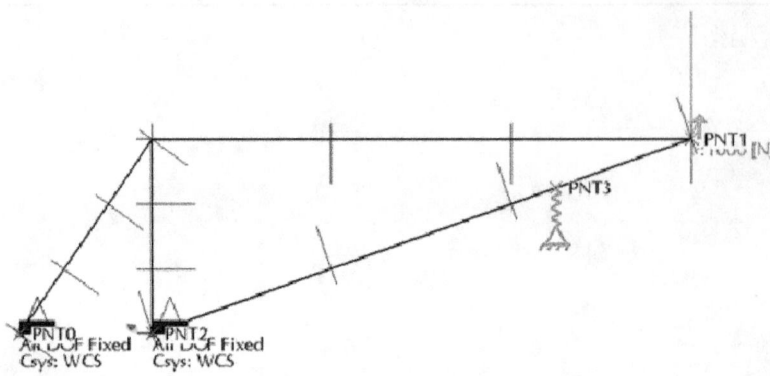

Click the icon of **Spring**. Accept the default name: Spring1. In the **Type** box, select **To Ground**. In the References box, select PNT3 from the inclined beam. In the **Properties** box, click **More > New >** set 2000 in Kyy and click **OK**. Click **OK** to complete the spring definition.

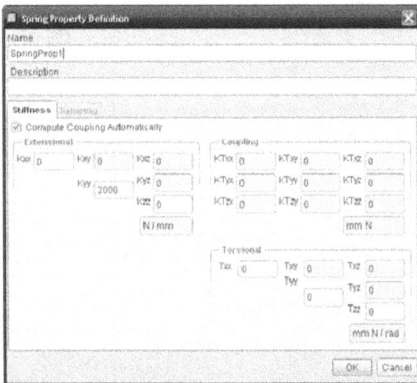

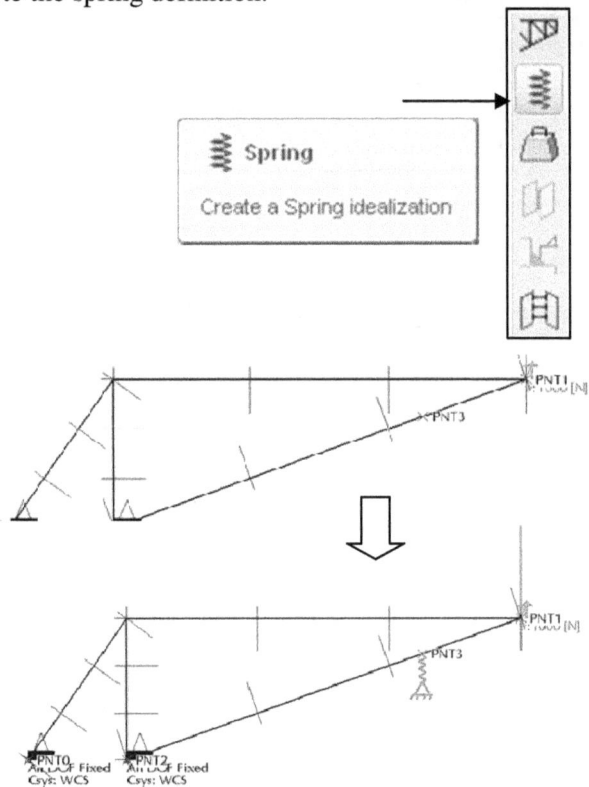

From the top menu, click the icon of Run a Design Study **> File > New Static.**

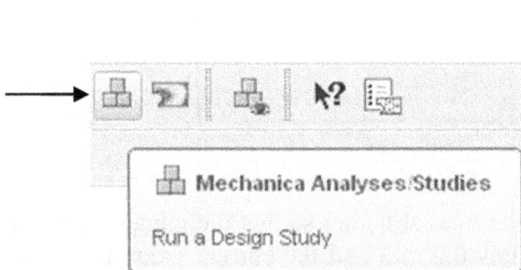

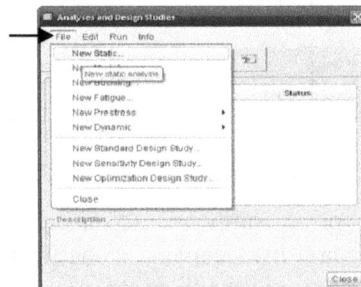

Type *truss_spring* as the name of analysis, and **OK** to accept the default setting, which is **Mult-Pass Adaptive > OK**. Run the program > **Close** upon completion.

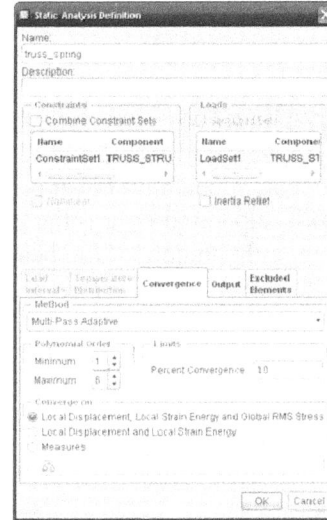

To study the obtained result or check the numerical value of the max displacement at the tip, click the icon of **Review results > Displacement > Magnitude > OK and Show.** The numerical value of the max deflection is 0.1975 mm at the tip of the truss structure.

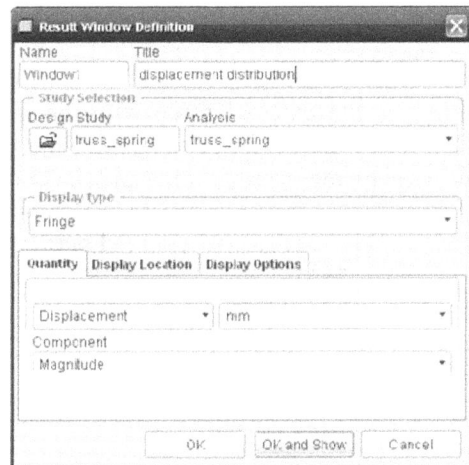

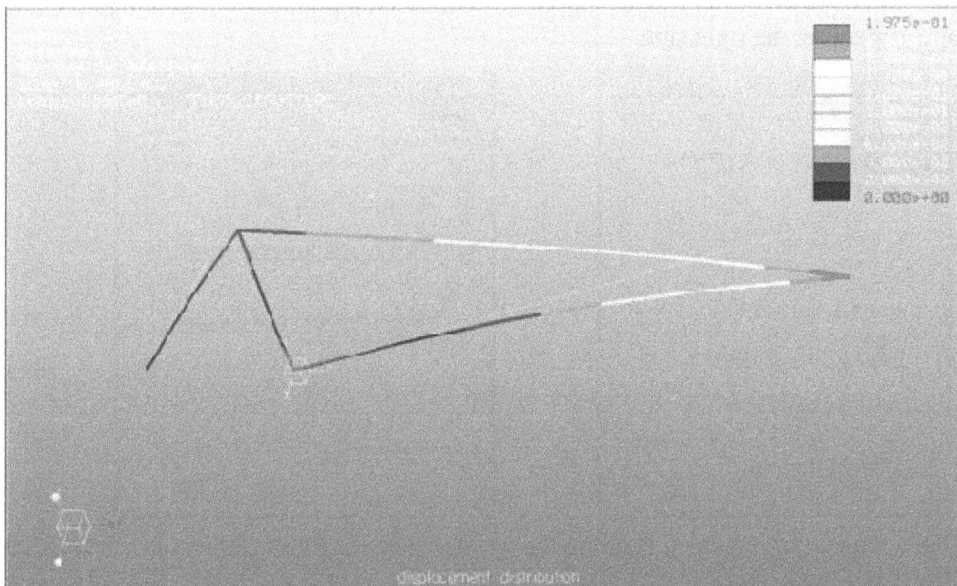

The conclusion obtained from this study is "an additional beam is needed to enhance the rigidity of truss structure." If this additional beam is to locate at the PNT3 position and is connected to the ground, the stiffness of this additional beam should be 2000 N/mm so that the max displacement occurred at the tip could be limited to 0.2 mm.

9.3 Application on Simple Supported Beam Structures

A simple supported beam is shown below. The left end of the beam is a pin connection and the right end of the beam is supported by a roller connection. The material type of the beam structure is steel. The load is acting at the middle of the beam length. The magnitude of the load is 100 N and the direction of the load is downward. The cross section of the beam is a rectangle and the two dimensions are 25 mm and 50 mm, respectively.

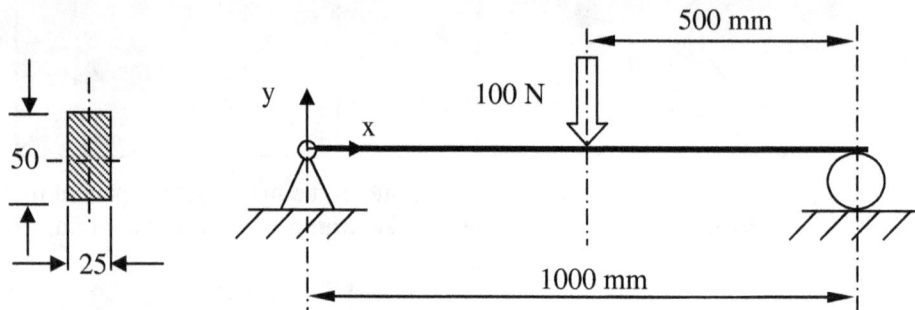

In this example, we will have 3 case studies for three load conditions. In case study 1, the load condition is the external force with the magnitude equal to 100 Newton. In case study 2, we only consider the effect on the deflection of the beam structure due to the gravity force. In case study 3, both loads, namely, the external force and the gravity force, are considered when performing FEA.

	Load 100 N	Gravity Force	Deflection (mm)
Case 1	X		
Case 2		X	
Case 3	X	X	

Step 1: Create a 2D model for the simple supported beam with a pin connection at the left side and with a simple support at the right side.

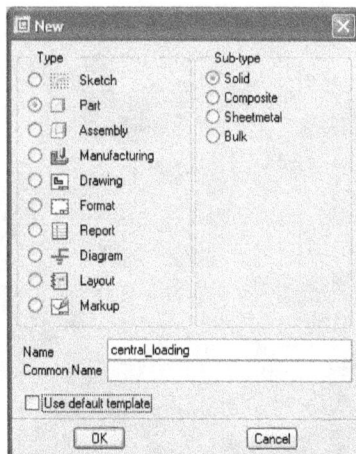

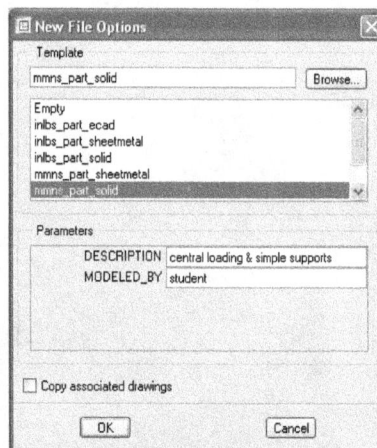

Define a datum curve, which is along the x-axis. The length of the datum curve is 1000 mm and it is symmetric about the y-axis.

Select the icon of **Sketch Tool.** Pick the **FRONT** datum plane as the sketch plane, and accept the default orientation setting, which uses the **RIGHT** datum plane and position it at the right side > **Sketch**.

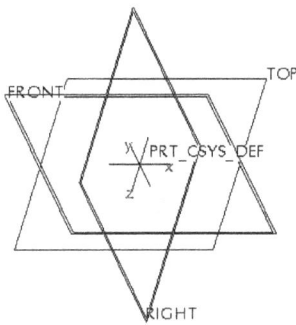

Click the icon of **Centerline** and sketch a centerline along the y-axis. This vertical centerline is used for symmetry.

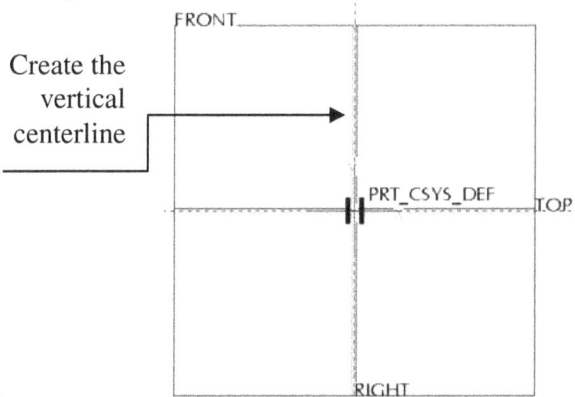

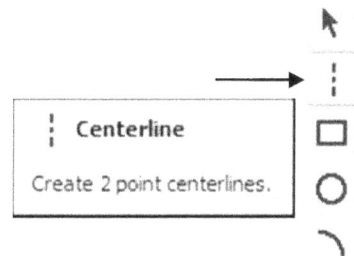

Create the
vertical
centerline

Click the icon of **Line** and sketch a line along the x-axis. The sketched line is symmetric about the vertical centerline. The dimension is 1000. Click the icon of **Done**.

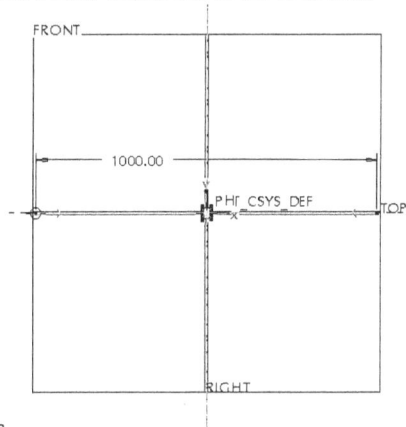

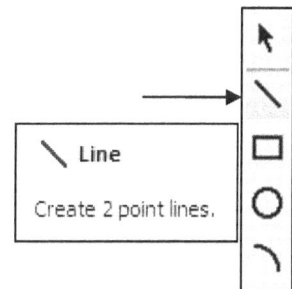

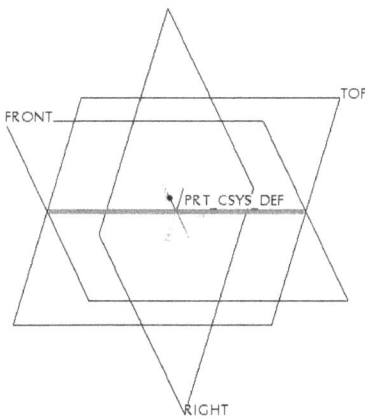

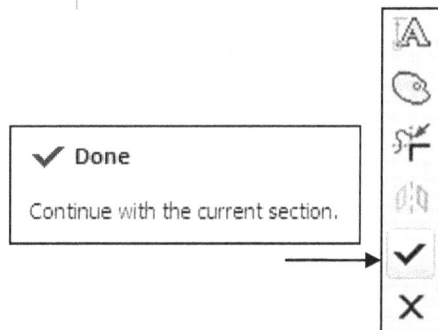

Create three datum points. The first datum point is at the central location of the datum curve. Select the icon of **Datum Point Tool**. Pick the point at the central location of the datum curve on display. Make sure the offset value is set at 0.50.

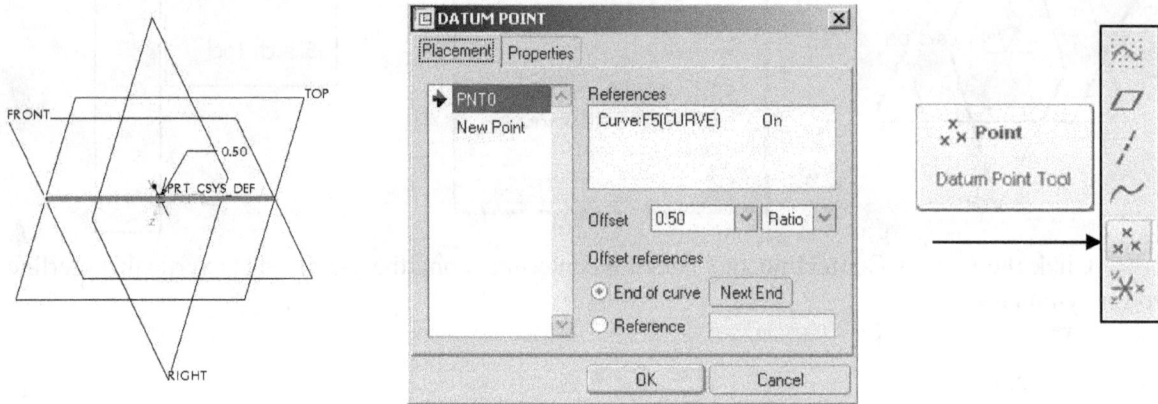

To define the second datum point, select the icon of Datum Point Tool. Pick the point at the left end of the datum curve on display. Make sure the offset value is set at 0.00.

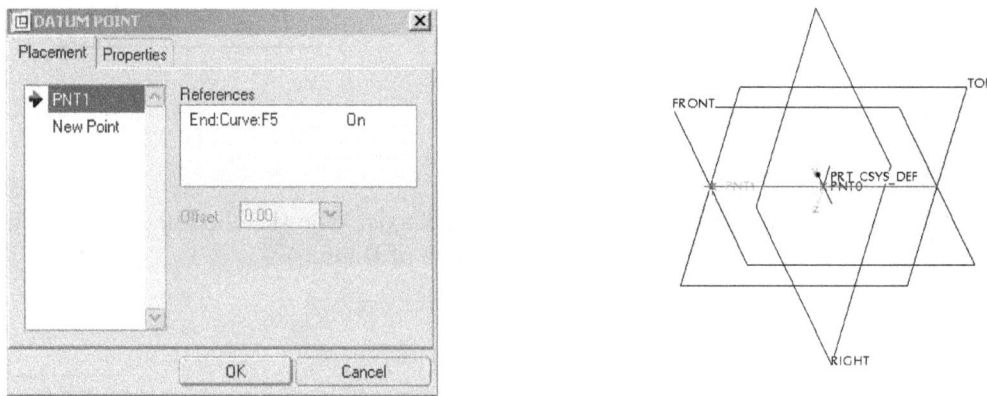

To define the third datum point, select the icon of Datum Point Tool. Pick the point at the left end of the datum curve. Make sure the offset value is set at 0.00.

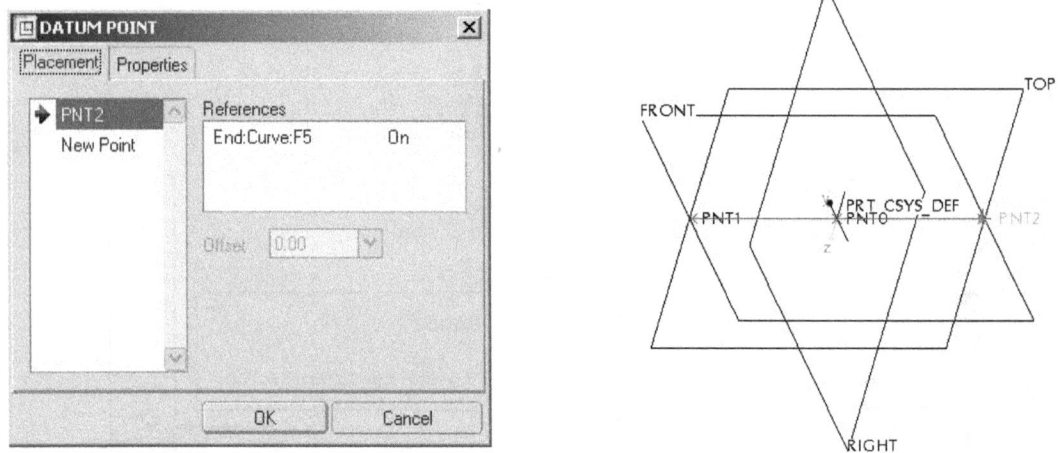

Step 2: Perform FEA under **Pro/Mechanica**
From the main toolbar or the menu toolbar, select **Applications > Mechanica > Structure > OK**. Pick the icon of **Beam**. Accept Beam1 as the name.

Accept the name: Beam1

Select Edge/Curve

Select Steel

Y vector

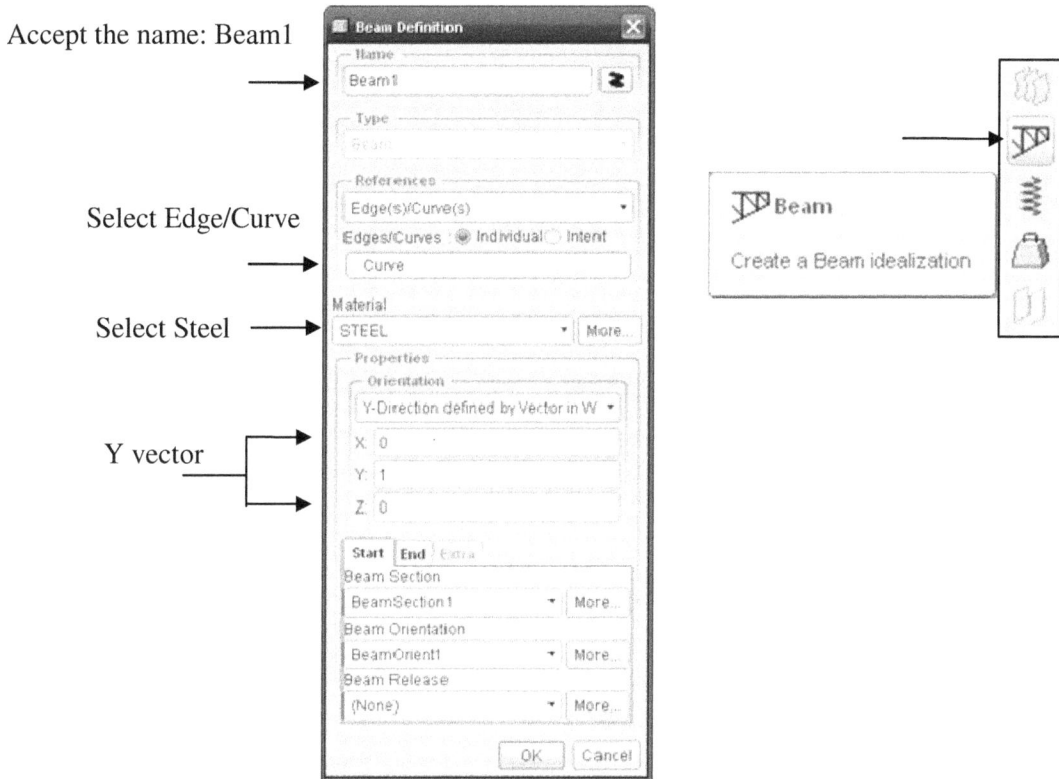

On the appeared **Beam Definition** window, do the followings:
Select **Edge/Curve** > pick the datum curve.

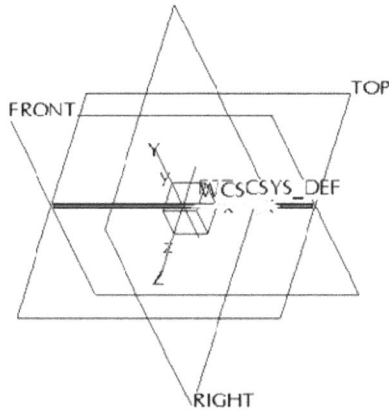

Select More from **Material** > Select **STEEL** > **Close**

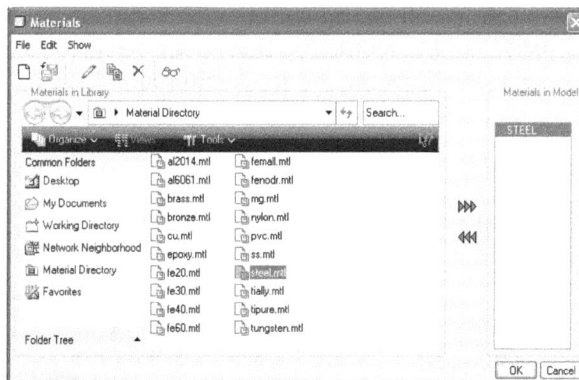

In the Properties field, set x = 0, y = 1 and z = 0 as the normal vector for orientation.

Properties

Orientation

Y-Direction defined by Vector in W ▾

X: 0

Y: 1

Z: 0

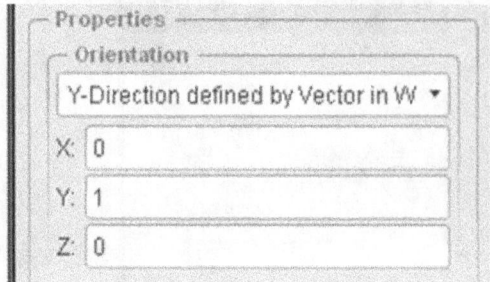

Select More from **Section** > New > select **Rectangle** > type *25* and *50* as b and d, respectively > **OK**.

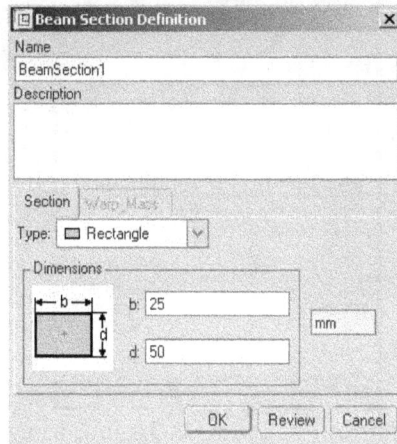

Beam Section Definition

Name
BeamSection1

Description

Section | Warp Mass

Type: ▢ Rectangle ▾

Dimensions

b: 25

d: 50

mm

OK | Review | Cancel

Define the Beam Orientation: click **More** > **New**, and make sure that DY=0 and DZ=0 to complete the process of defining the beam section > **OK**.

Beam Orientation Definition

Name
BeamOrient1

Description

Orientation Angle
0   degree

Offsets
Location of:
⦿ Shape Origin ◯ Shear Center
with respect to the beam action coord. sys.

DX: 0   mm
DY: 0
DZ: 0

OK | Cancel

WCSCSYS_DEF

Define a simple supported constraint on the right end of the beam.
Constraints > Point(s) > pick **PNT1** > set the rotation about z-axis to free and fixed the others.
Constraints > Point(s) > pick **PNT2** > set the translation along x-axis to free and the rotation about z-axis to free.

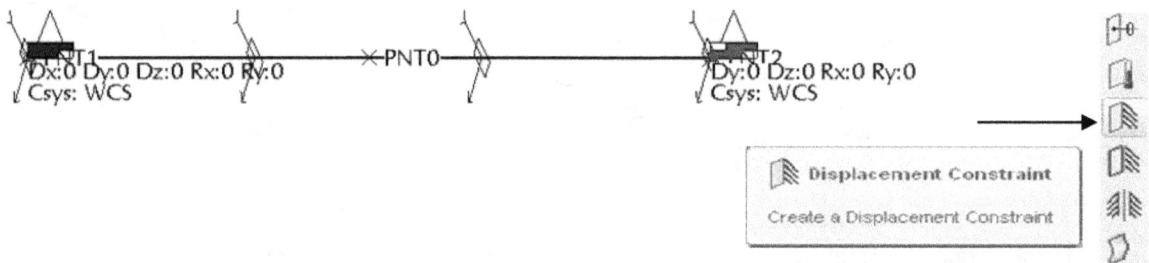

PNT1
Dx:0 Dy:0 Dz:0 Rx:0 Ry:0
Csys: WCS

PNT0

PNT2
Dy:0 Dz:0 Rx:0 Ry:0
Csys: WCS

Displacement Constraint
Create a Displacement Constraint

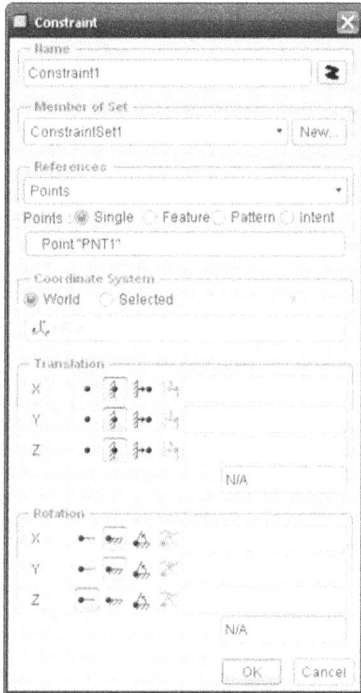

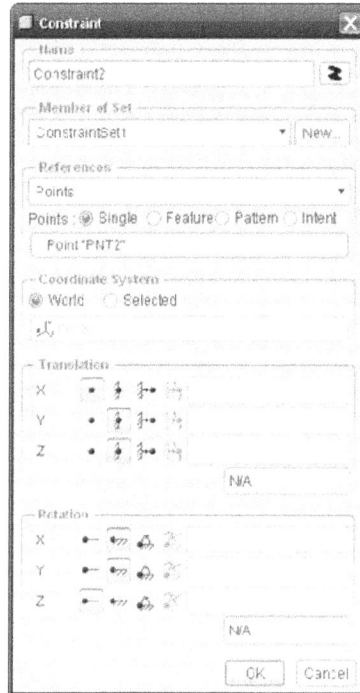

Select the icon of **Force/Moment > Point(s) >** pick PNT0.
Total Load > Uniform > type –100 in the force component **Y > OK.**

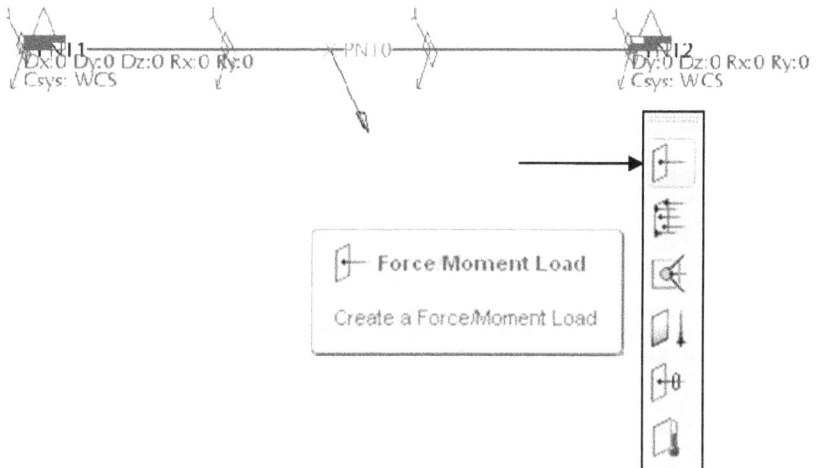

Step 3: Perform a static analysis
From the top menu, click the icon of **Run a Design Study > File > New Static.**

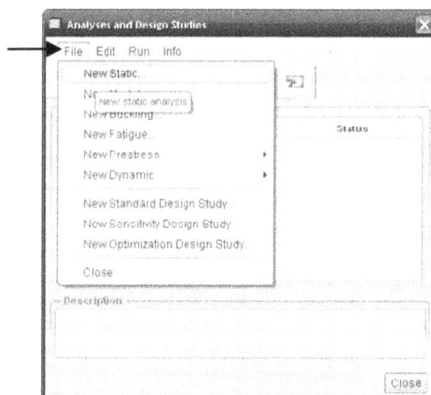

Type *simple_support_structure* as the file name > **OK**. Run the static analysis.

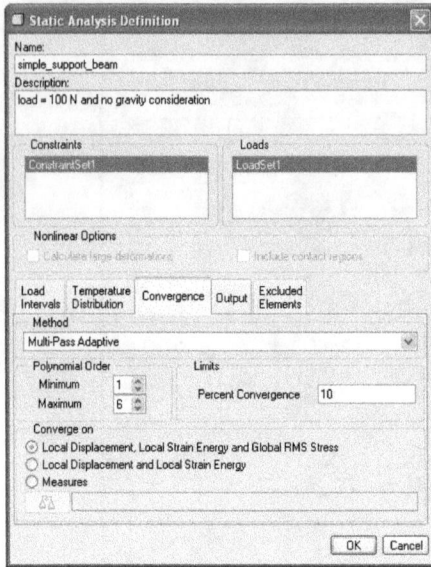

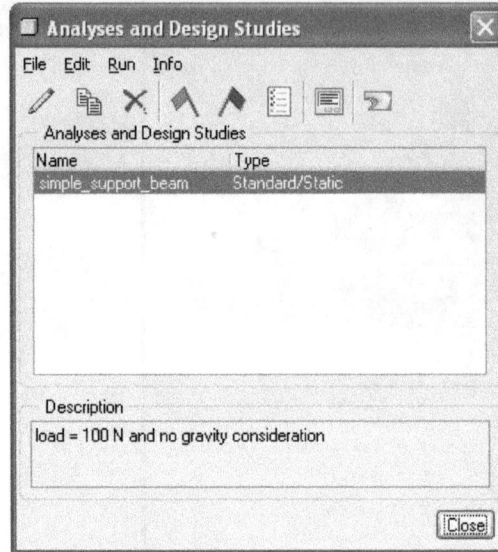

Note: The Percent Convergence is set at 10% (default value).

There is always a message displayed on screen. The message is asking the user "Do you want to run interactive diagnostics?" Click **Yes**. In this way a warning message will appear if an abnormal operation occurs.

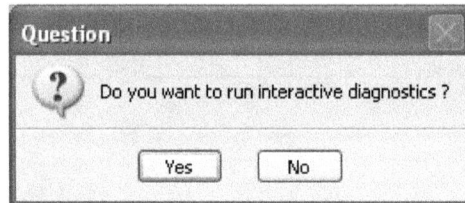

Step 4: Show the results obtained from FEA.

To study the obtained result, say the reaction forces at the two support locations, click the icon of **Review results > Model > Reaction at Point Constraint** > select **Y** > **OK and Show.** The magnitude is 50 N at each of the two supports, as shown.

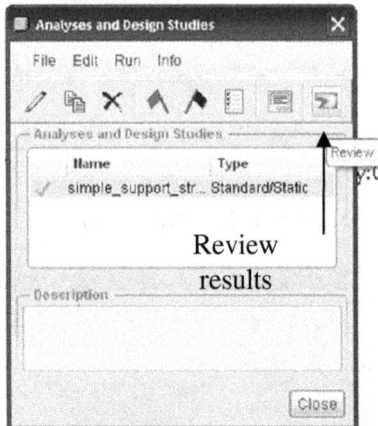

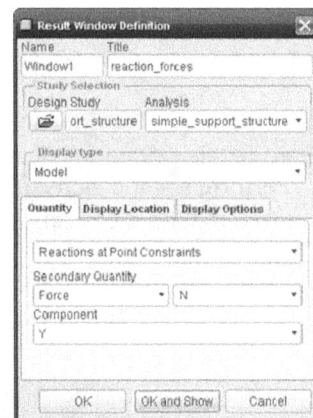

To plot the shear diagram and moment diagram, click the icon of **Review results > Graph > Shear & Moment** > select **Vy** and **Mz**. Click **Relative To** and pick the beam > press the middle button of mouse to accept the selection > click **OK** to accept the starting point set by default > **OK and Show.**

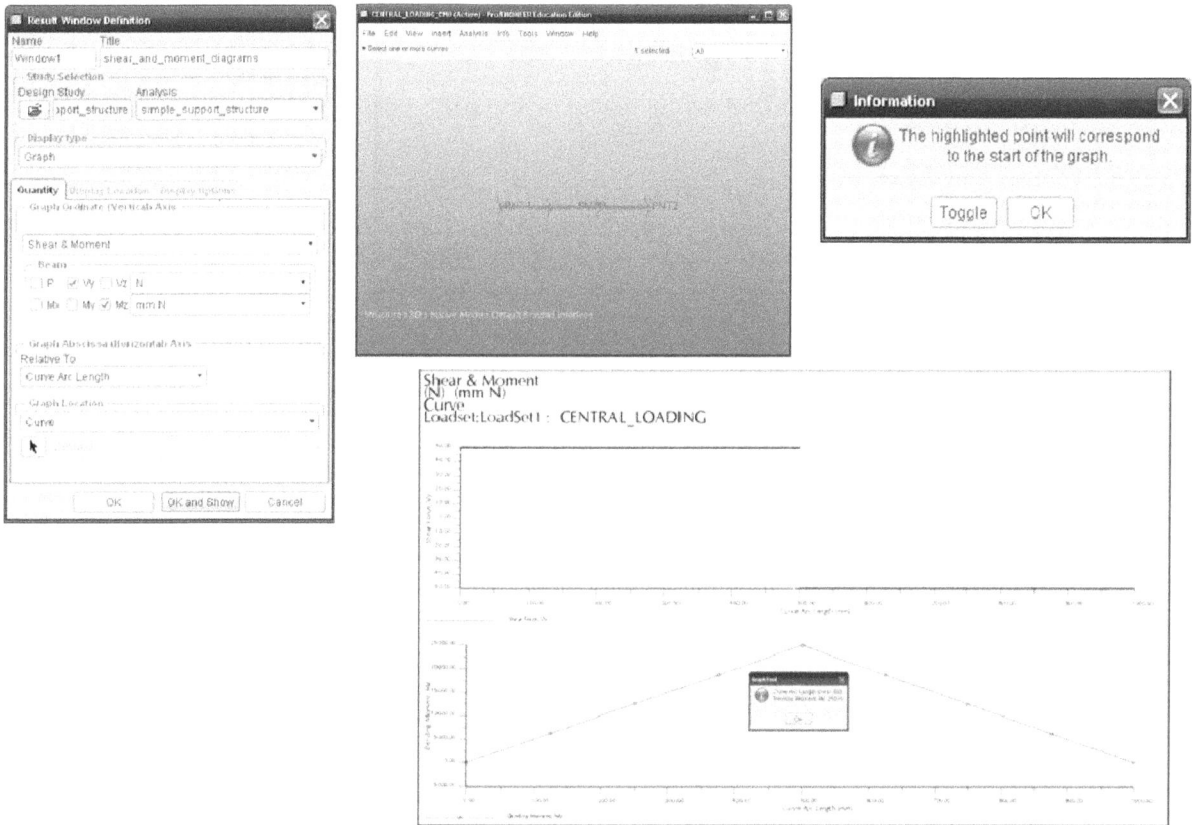

To show the distribution of displacement, click the icon of **Review Results** > type *deflection subjected to 100 N load only* as the title > select the folder called simple_support_beam > **Graph > Displacement** > select **Y** > **OK and Show**. The maximum deflection value at the middle section is 0.040315 mm.

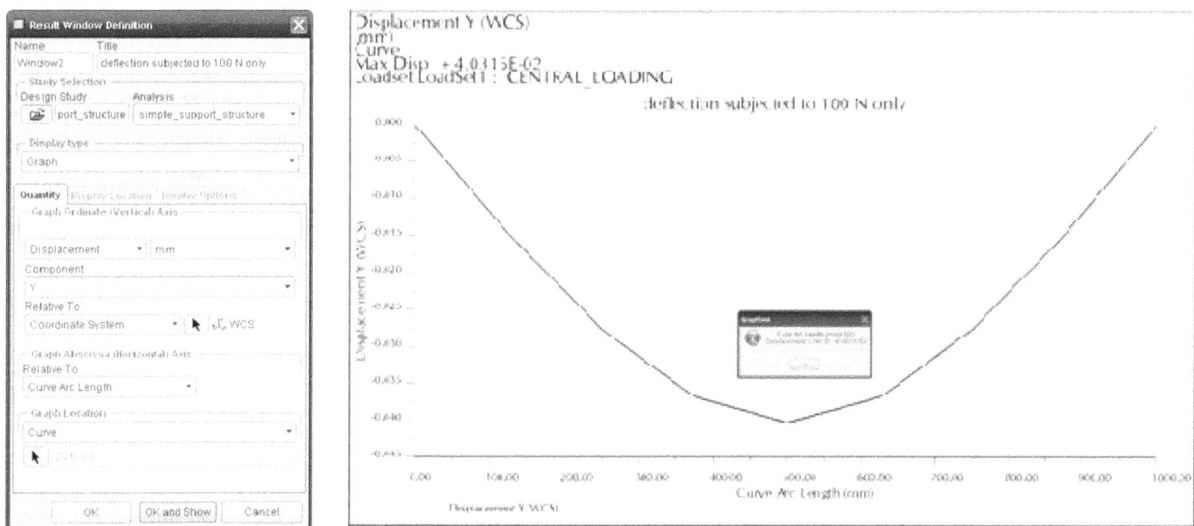

Step 5: Perform a static analysis for the gravity force only.

To run the FEA with the gravity force only, we need to define a second load set.

Click the icon of **Gravity Load** and select **New** again > type in –9810 > **Preview** > **OK.**

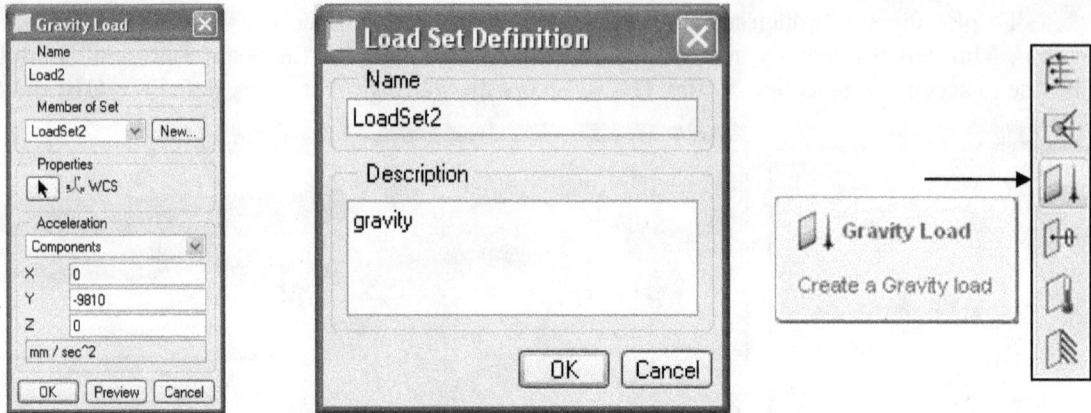

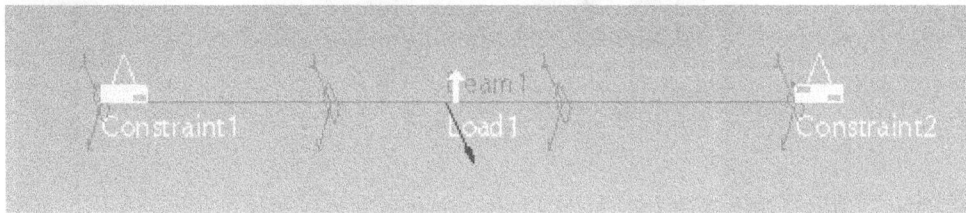

Click the icon of **Run a Design Study** from the main toolbar > **File** > Select **New Static**. Type *gravity_only* as the file name > **OK**. Run the static analysis.

To show the distribution of displacement, click the icon of **Review Results** > type *deflection_gravity_only* as the title > select the folder called simple_support_beam > **Graph** > Displacement > select **Y** > **OK and Show**. The maximum deflection value at the middle section is 0.024147 mm.

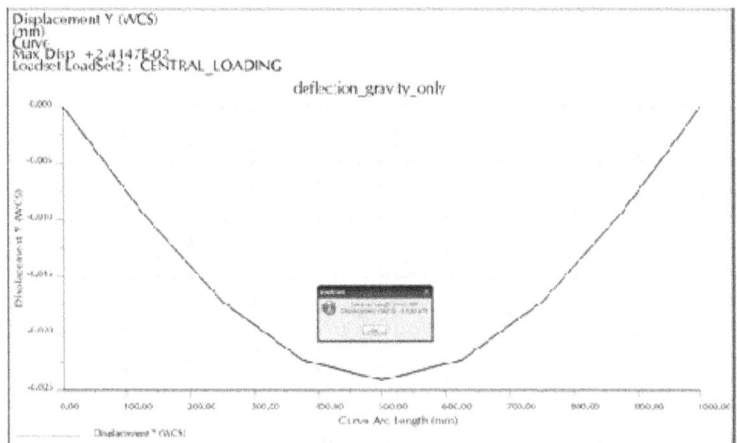

Step 5: Perform a static analysis for the presence of both the load of 100 N and the gravity force.

Click the icon of **Run a Design Study** from the main toolbar > **File** > Select **New Static**. Type *load_plus_gravity* as the file name > **OK**. Run the static analysis. Make sure that ConstraintSet1, LoadSet 1 and LoadSet2 are highlighted > click Sum load sets > **Multi-Pass Adaptive** > **OK** > **Run.**

Click the icon of **Review Results** > type *distribution_load_plus_gravity* as the title > select the folder called load_plus_gravity > **Graph** > **Displacement** > select **Y** > select the sketched curve > press the middle button of mouse > **OK** > **OK and Show**. The maximum deflection value at the middle section is 0.0644627 mm.

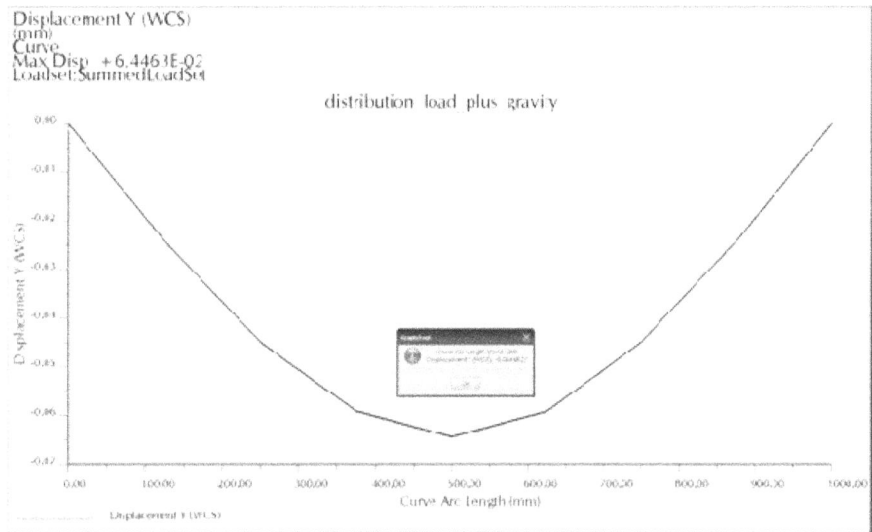

Now let us list these three values obtained from the FEA, and put the three graphs together to gain a clear picture on the gravity force effect on the total deflection of the beam.

	Load 100 N	Gravity Force	Deflection (mm)
Case 1	X		0.040315
Case 2		X	0.024147
Case 3	X	X	0.064462

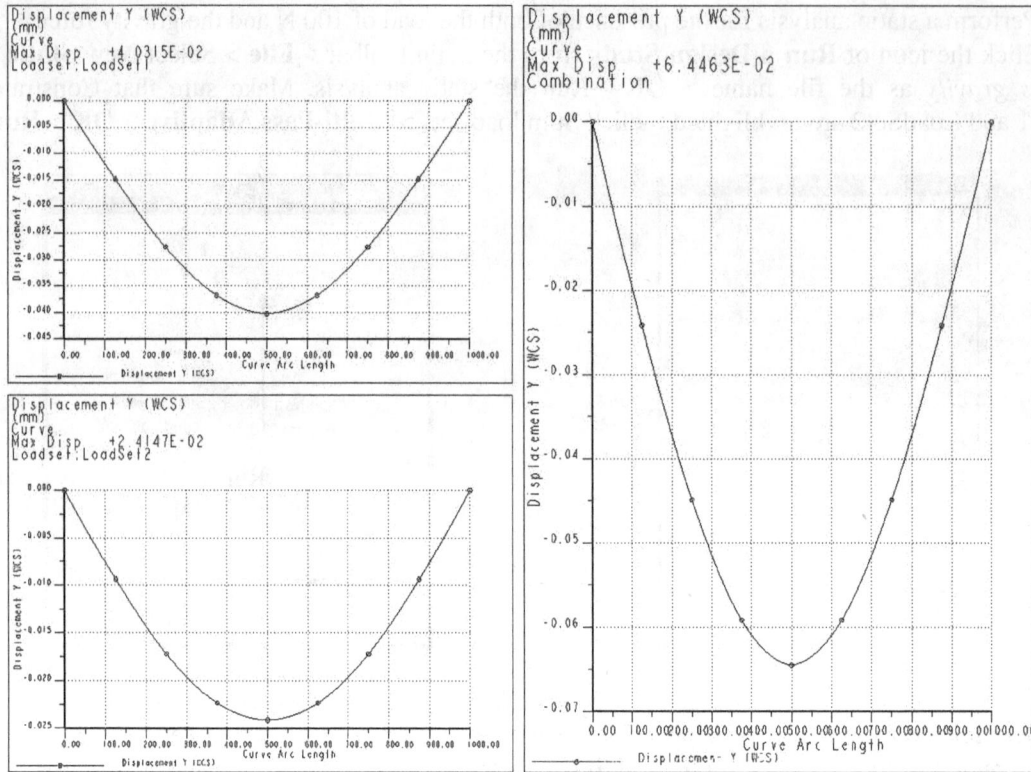

To verify the results obtained from the FEA, let us calculate the maximum deflection of this simple supported beam using an analytical formula. The formula shown below is used by most textbooks published in mechanics of materials,

$$\delta_{max} = \frac{PL^3}{48EI}$$

where E = 200 GPa and the moment inertia of I is given by

$$I = \frac{1}{12}bh^3 = \frac{1}{12}(25)(50)^3 \qquad (mm^4)$$

By substituting the load value of P and the span distanced of the beam structure of L, we have

$$\delta_{max} = \frac{PL^3}{48EI} = \frac{(100N)(1000mm)^3}{48(200\frac{N}{m^2})(\frac{1}{12})(25mm)(50mm)^3} = 0.04 \quad (mm)$$

This calculated value of 0.04 mm is very closed to the result obtained in case study 1, which was 0.040315. Note that the gravity force was not considered in the study of case 1. Therefore, the calculated value based on the analytical formula does not consider the effect of the gravity force on the deflection of the beam structure. Although the effect of the gravity force on the gravity force was small, which was 0.024147 mm; it should be considered because it accounts for almost 50% of the deflection value due to the external force. As a result, under circumstances where accuracy of the deflection of a beam structure is set to be high, the gravity force should be considered.

In the following, we modify the model used in the previous example. Instead of having a point load, we have a uniform load, as shown below.

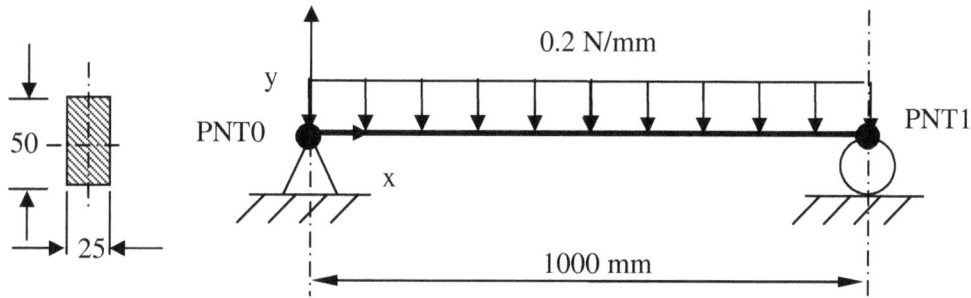

Step 1: Create a 2D model for the simple supported beam with a pin connection at the left side and with a simple support at the right side with a distributed load.

From the mail toolbar, select the icon of **Create** a new object. Make sure the **Part** mode is selected. Type *distributed_load_structure* and clear the box of **use default template > OK**

Select the unit system to be **mmns_part_solid** > type *distributed load structure* in **DESCRIPTION** and *student* in **MODELED_BY > OK**.

Define a datum curve, which is along the x-axis. The length of the datum curve is 1000 mm and it is symmetric about the y-axis.

Select the icon of **Sketch Tool** from the toolbar of datum creation.

Pick the **FRONT** datum plane as the sketch plane, and accept the default orientation setting, which uses the **RIGHT** datum plane and position it at the right side > **Sketch**.

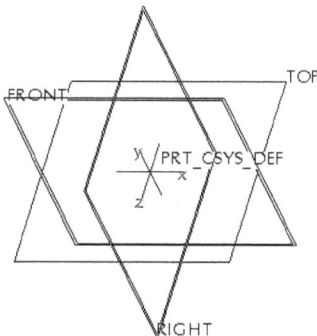

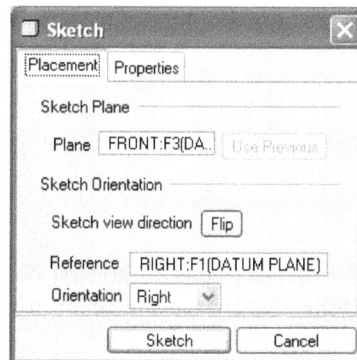

Click the icon of **Centerline** and sketch a centerline along the y-axis. This vertical centerline is used for symmetry.

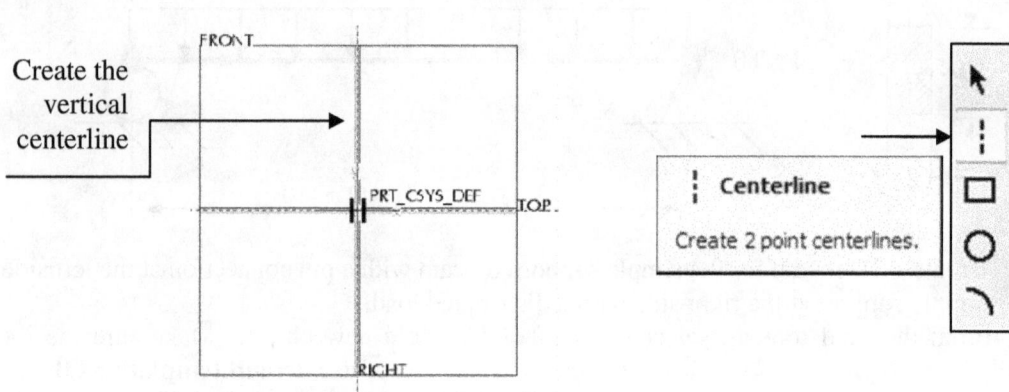

Create the vertical centerline

Click the icon of **Line** and sketch a line along the x-axis. The sketched line is symmetric about the vertical centerline. The dimension is 1000. Click the icon of **Done**.

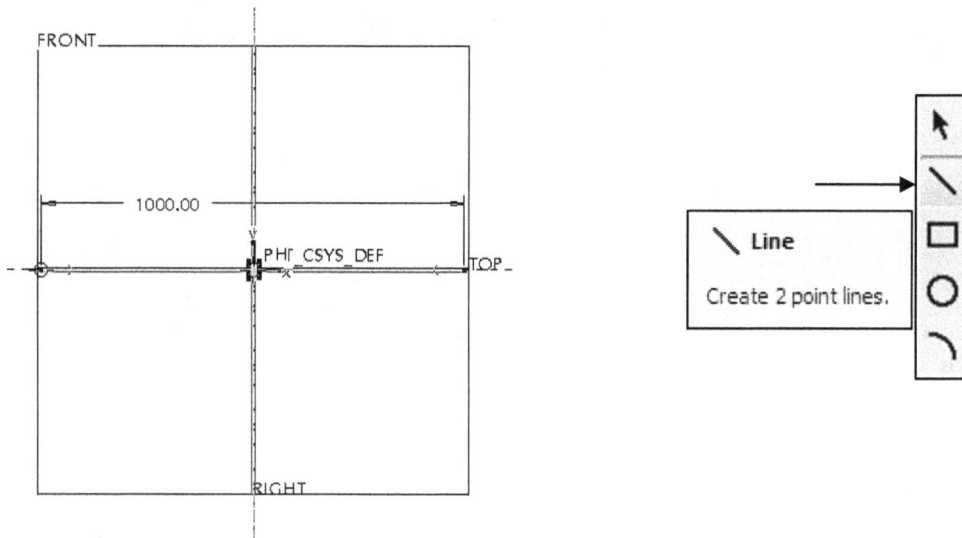

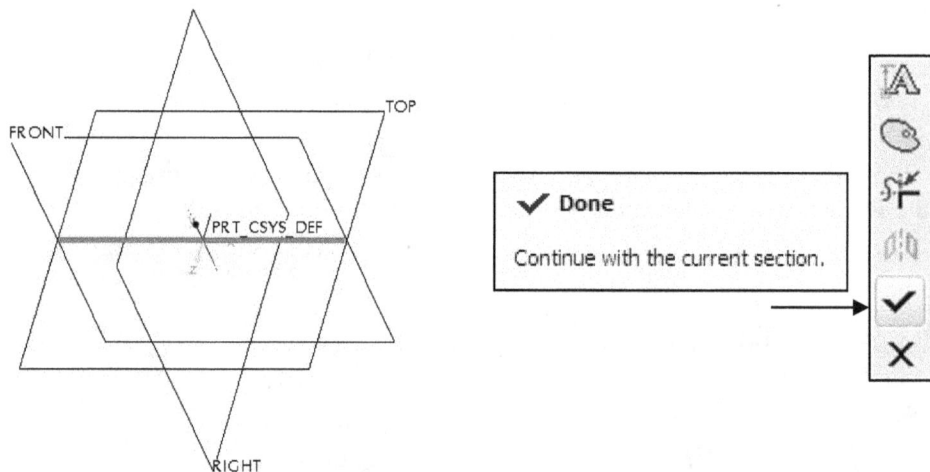

1000.00

Create two datum points. The first datum point is at the left end of the datum curve. Select the icon of **Datum Point Tool**. Pick the point at the left end of the datum curve on display. Make sure the offset value is set at 0.00.

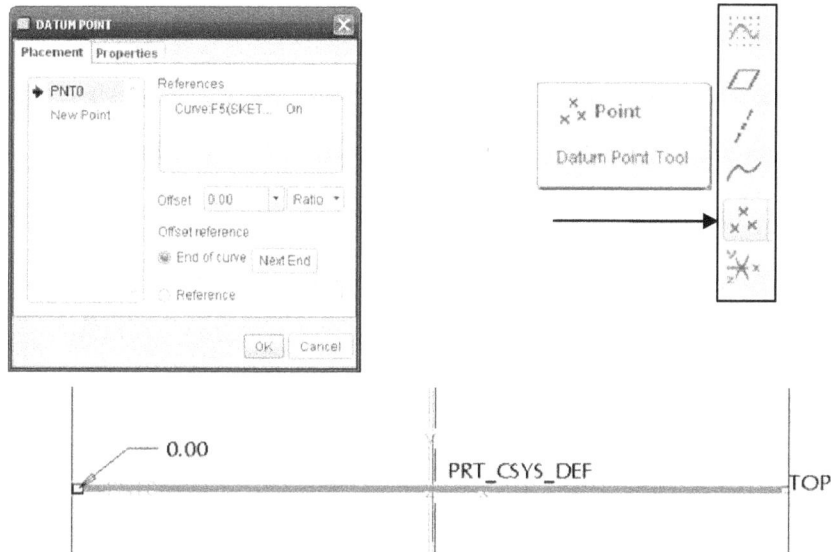

To define the second datum point, select the icon of **Datum Point Tool**. Pick the point at the right end of the datum curve. Make sure the offset value is set at 0.00.

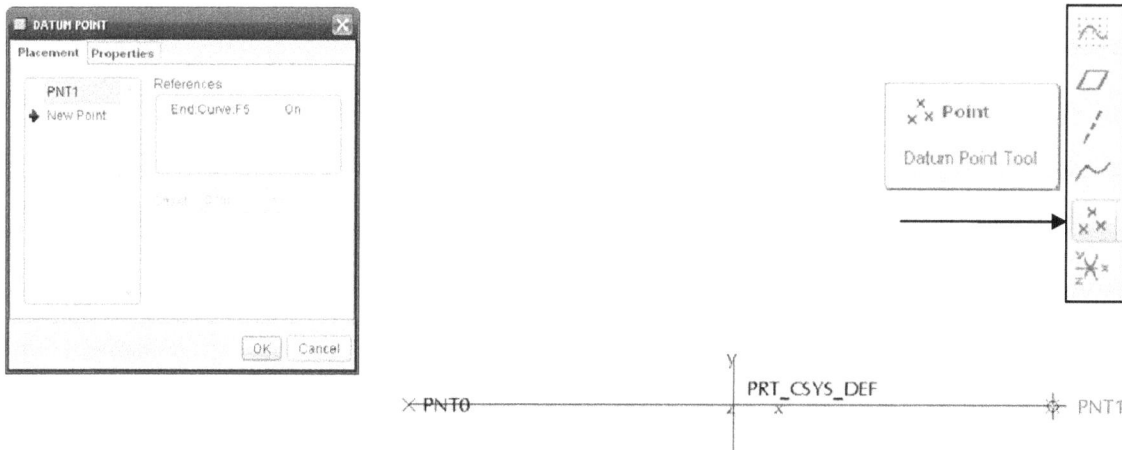

Step 2: Perform FEA under **Pro/Mechanica**

From the main toolbar or the menu toolbar, select **Applications > Mechanica > Structure > OK**.

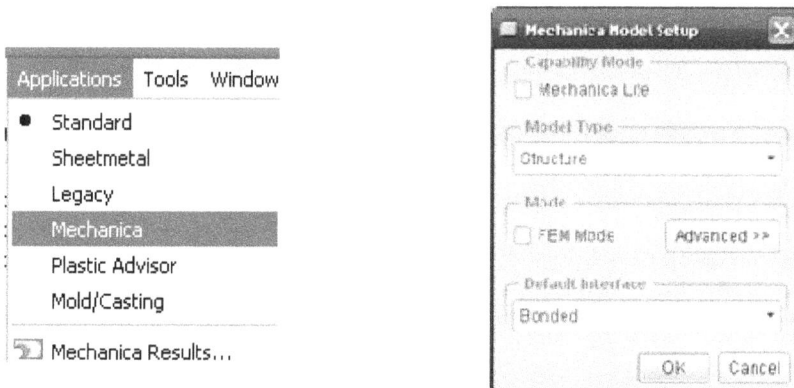

Pick the icon of **Beam**. Accept Beam1 as the name.

Accept the name: Beam1

Select Edge/Curve

Select Steel

Y vector

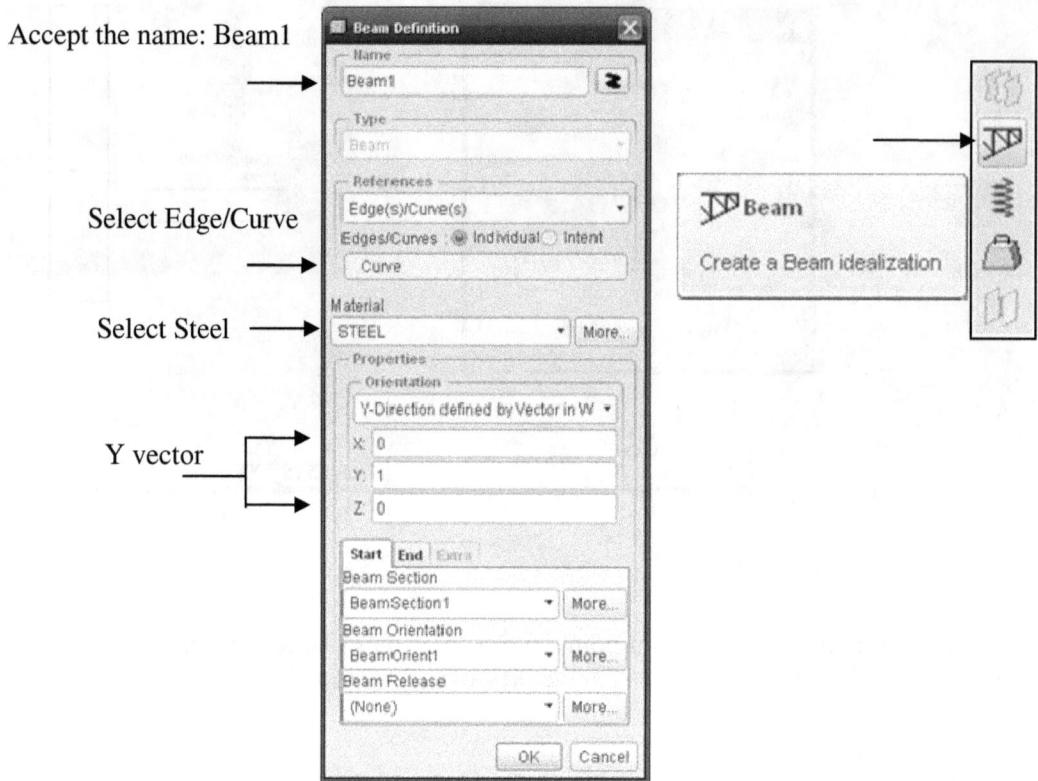

On the appeared **Beam Definition** window, do the followings:
Select **Edge/Curve** > pick the datum curve.

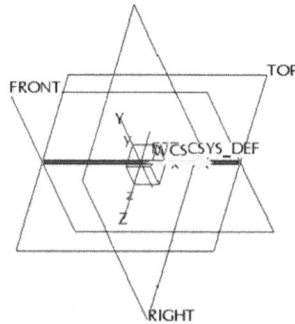

Select More from **Material** > Select **STEEL** > **Close**

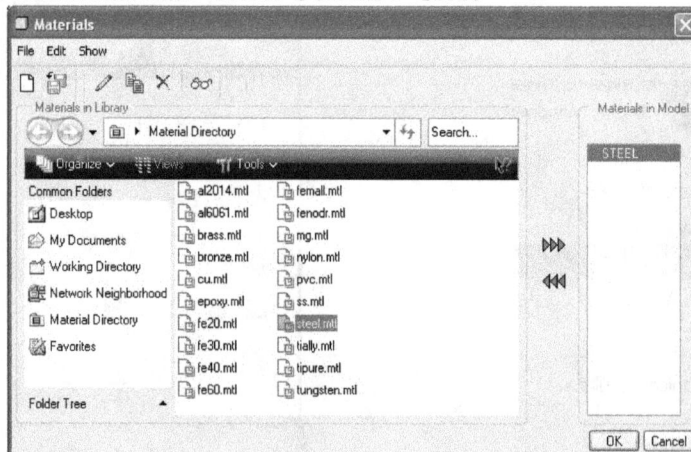

In the Properties field, set x = 0, y = 1 and z = 0 as the normal vector for orientation.

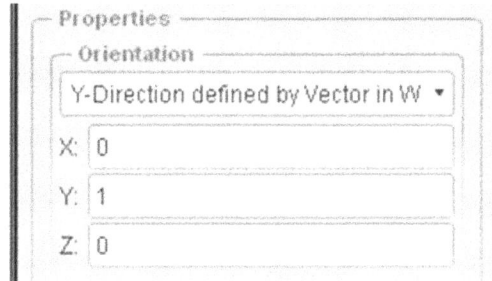

Select More from **Section** > **New** > select **Rectangle** > type *25* and *50* as b and d, respectively > **OK**.

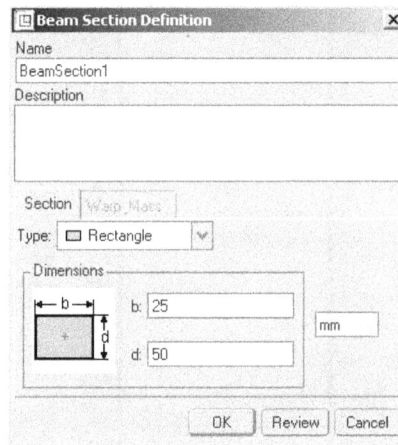

Define the Beam Orientation: click **More** > **New**, and make sure that DY=0 and DZ=0 to complete the process of defining the beam section > **OK**.

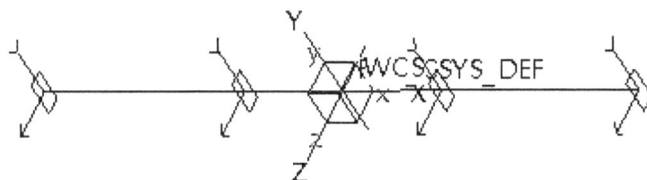

Define a simple supported constraint on the right end of the beam.
Constraints > Point(s) > pick **PNT0** > set the rotation about z-axis to free and fixed the others.
Constraints > Point(s) > pick **PNT1** > set the translation along x-axis to free and the rotation about z-axis to free.

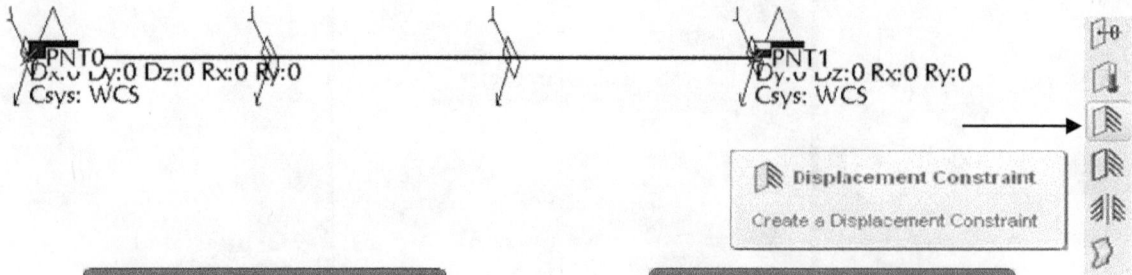

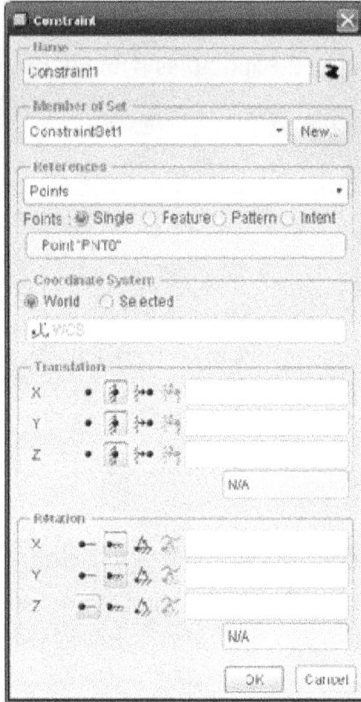

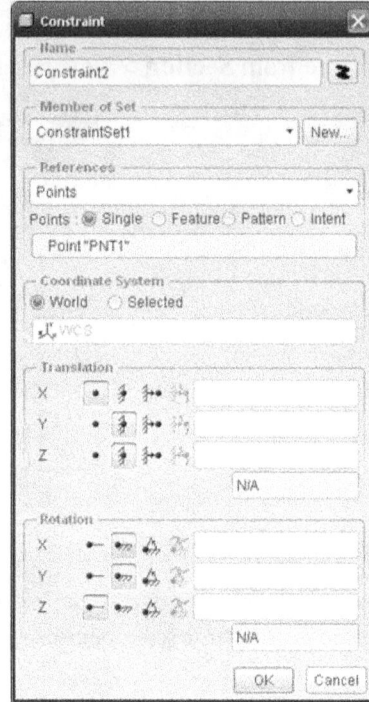

Select the icon of **Force/Moment > Point(s) >** pick PNT0.
Total Load > Uniform > type –200 in the force component **Y > OK.**

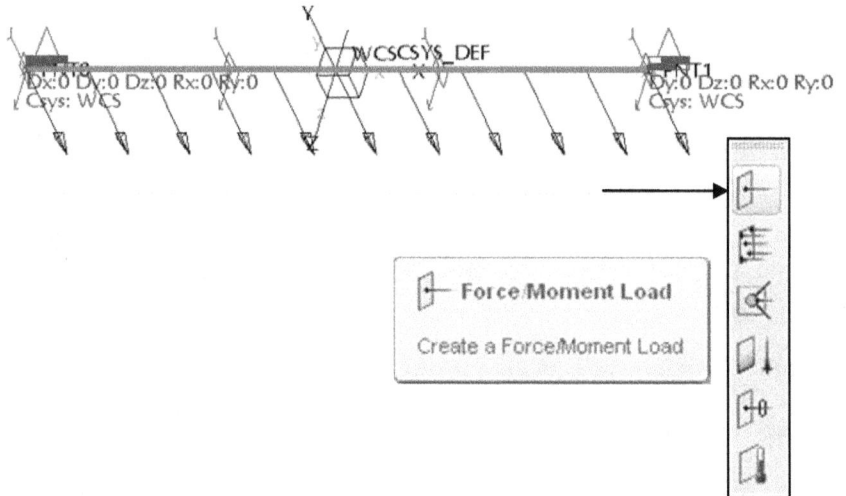

Step 3: Perform a static analysis
From the top menu, click the icon of **Run a Design Study > File > New Static.**

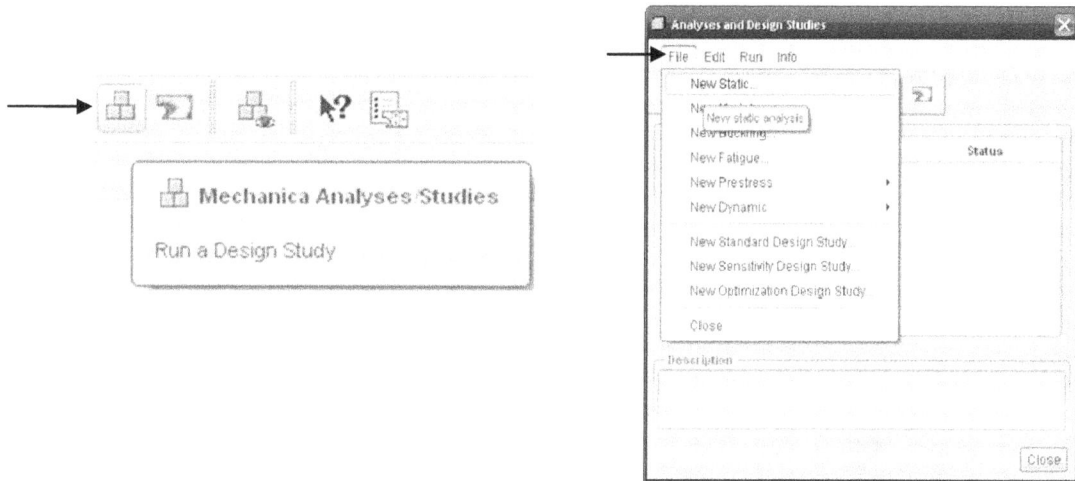

Type *distributed_load_structure* as the file name > **OK**. Run the static analysis.

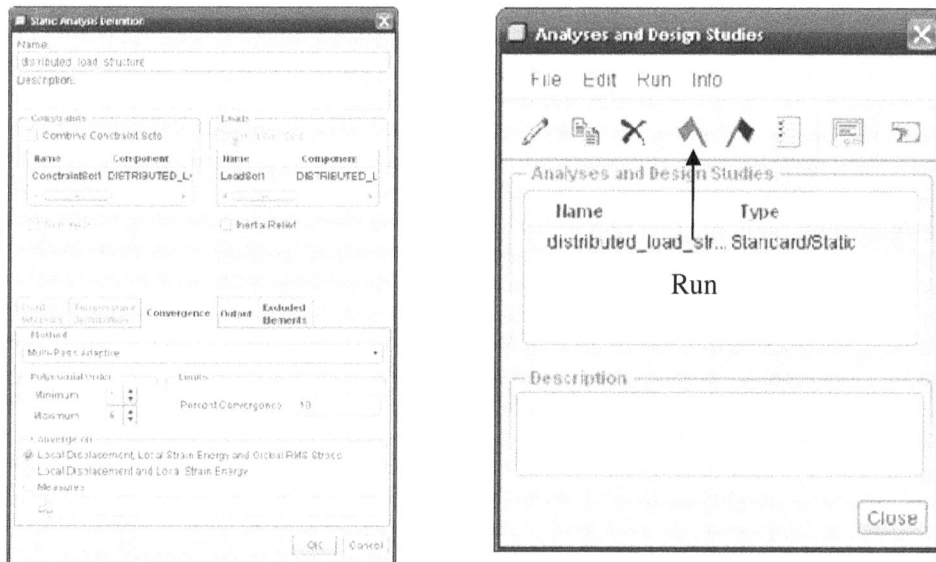

Note: The Percent Convergence is set at 10% (default value).

There is always a message displayed on screen. The message is asking the user "Do you want to run interactive diagnostics?" Click **Yes**. In this way a warning message will appear if an abnormal operation occurs.

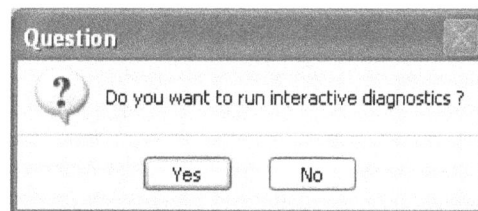

Step 4: Show the results obtained from FEA by plotting the distribution of deflection.

To study the obtained result, say the reaction forces at the two support locations, click the icon of **Review results > Model > Reaction at Point Constraint** > select **Y** > **OK and Show.** The magnitude is 100 N at each of the two supports, as shown.

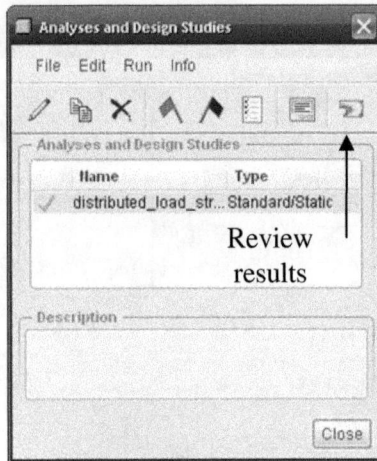

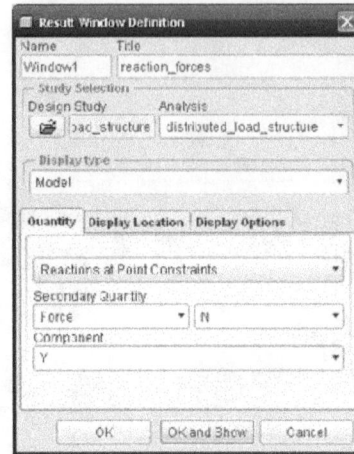

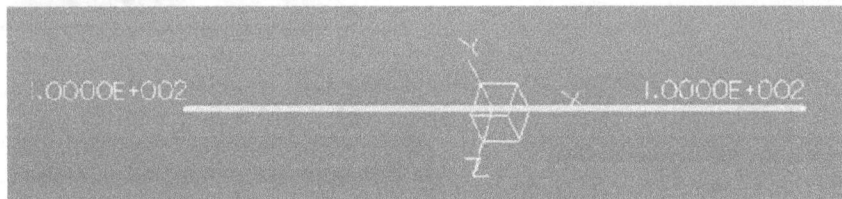

To plot the shear diagram and moment diagram, click the icon of **Review results** > **Graph** > **Shear & Moment** > select **Vy** and **Mz**. Click **Relative To** and pick the beam > press the middle buttom of mouse to accept the selection > click Toggle > **OK** to accept the starting point set by default > **OK and Show.**

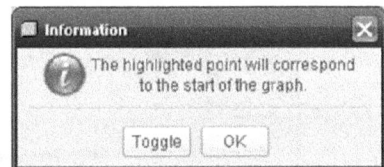

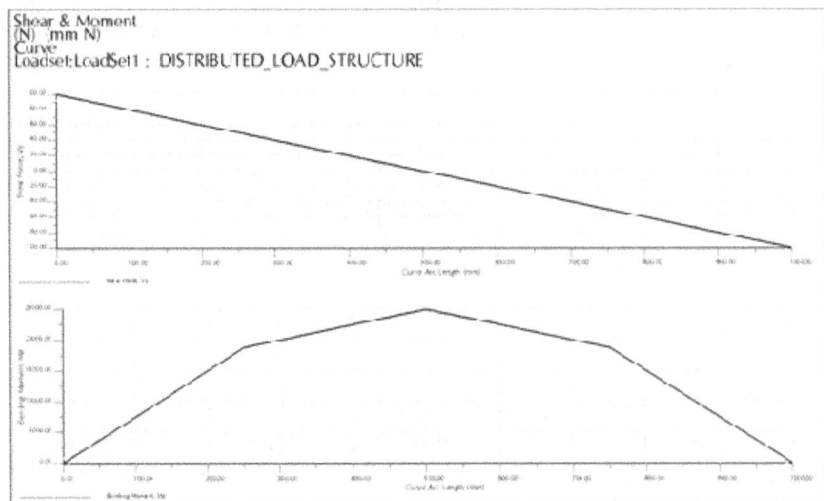

To show the distribution of displacement, click the icon of **Review Results** > type *distribution_of_displacement* as the title > select the folder called distributed_load_structure > **Graph > Displacement** > select **Y** > **OK and Show**. The maximum deflection value at the middle section is 0.050318 mm.

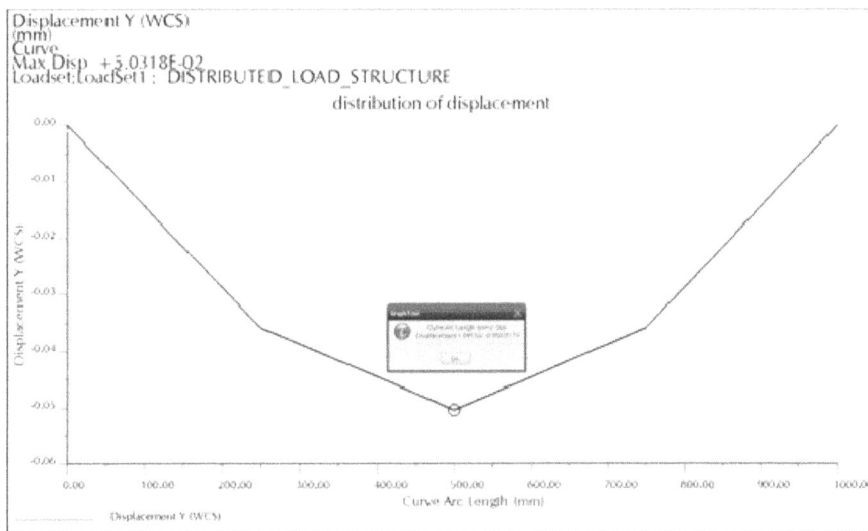

In the following, we modify the model used in the previous example. Instead of having a uniformly distributed load, we have a point load and a uniformly distributed load as well, as shown below.

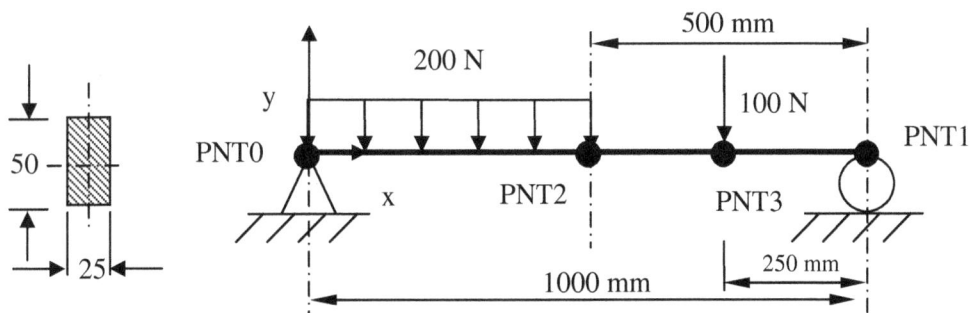

Step 1: Create a 2D model for the simple supported beam with a pin connection at the left side and with a simple support at the right side with a distributed load.

From the mail toolbar, select the icon of **Create** a new object. Make sure the **Part** mode is selected. Type *combined_load_structure* and clear the box of **use default template > OK**

Select the unit system to be **mmns_part_solid** > type *combined load structure* in **DESCRIPTION** and *student* in **MODELED_BY > OK**.

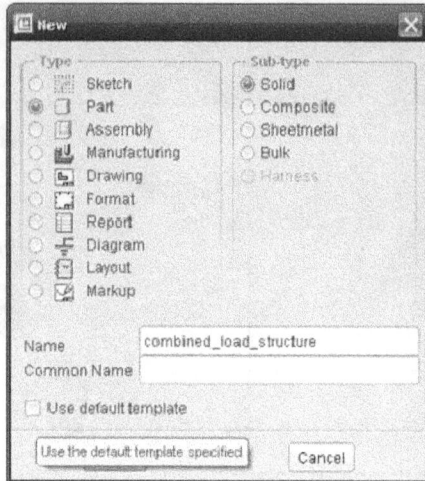

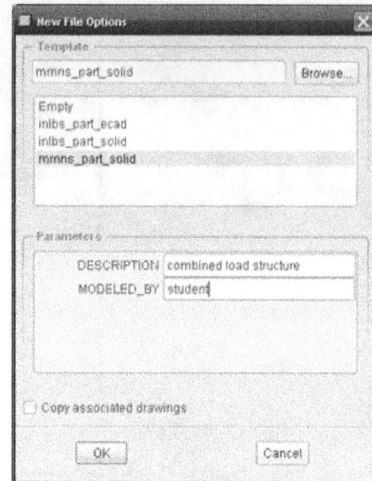

In this example, we need to use 2 sketches to create this beam structure. Let us work on the first sketch, which is a line along the x-axis. The length of the datum curve is 500 mm.

Select the icon of **Sketch Tool** from the toolbar of datum creation.

Pick the **FRONT** datum plane as the sketch plane, and accept the default orientation setting, which uses the **RIGHT** datum plane and position it at the right side > **Sketch**.

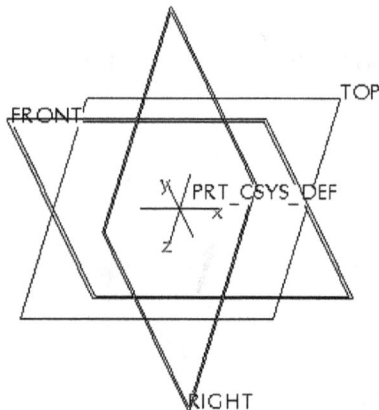

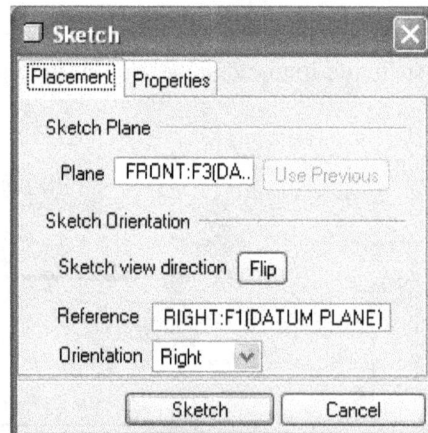

Click the icon of **Line** and sketch a line along the x-axis. The sketched line is on the left side, as shown. The dimension is 500. Click the icon of **Done**. Special attention is paid to the starting point, ending point and sketching direction, as shown.

500.00

PRT_CSYS_DEF

Line

Create 2 point lines.

Starting point   Sketching direction   Ending point

FRONT   TOP

PRT_CSYS_DEF

RIGHT

Done

Continue with the current section.

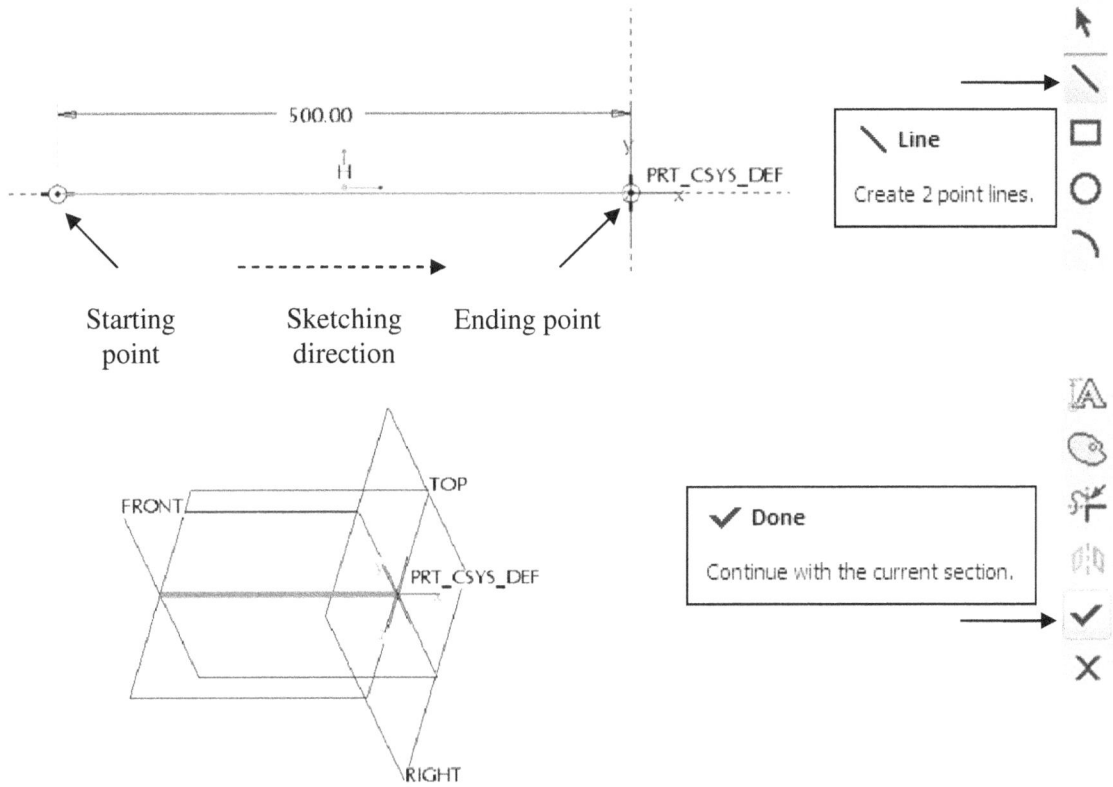

Now let us prepare the second sketch. Select the icon of **Sketch Tool** from the toolbar of datum creation, and pick **Use Previous**.

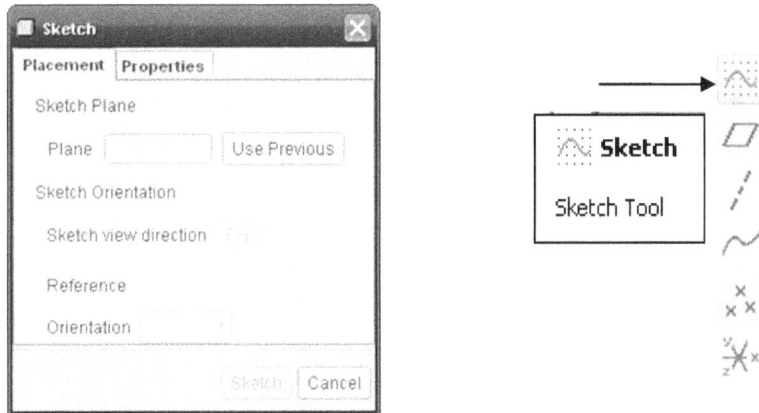

Sketch

Placement   Properties

Sketch Plane

Plane   Use Previous

Sketch Orientation

Sketch view direction

Reference

Orientation

Sketch   Cancel

Sketch

Sketch Tool

Click the icon of **Line** and sketch a line along the x-axis. The starting point of this line is the ending point of the line previously sketched. The dimension of this line is 500, as shown. Click the icon of **Done**.

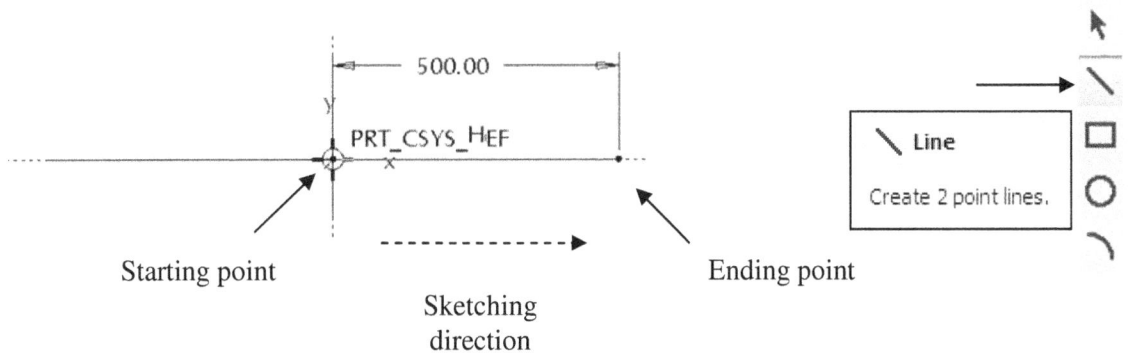

500.00

PRT_CSYS_HEF

Line

Create 2 point lines.

Starting point   Ending point

Sketching direction

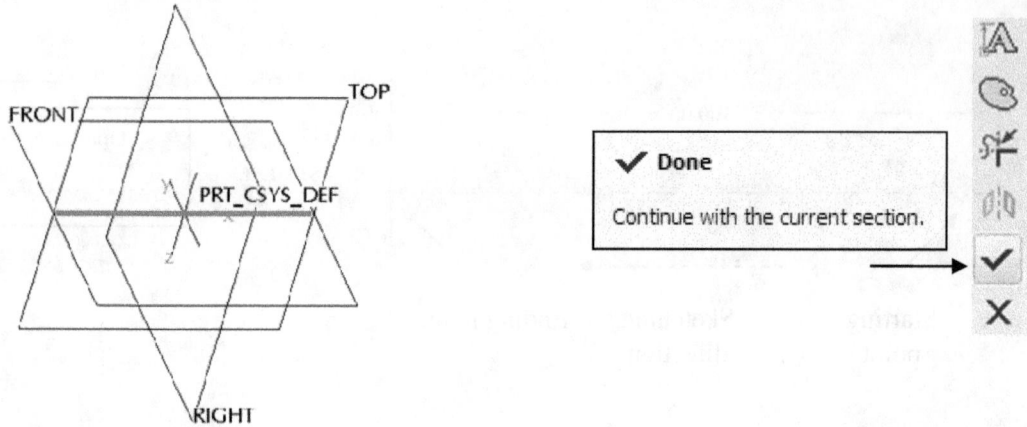

Create three datum points. The first datum point is at the central location of the datum curve. Select the icon of **Datum Point Tool**.

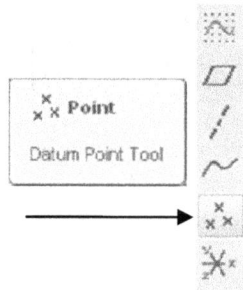

Pick the point at the central location of the entire datum curve on display. Make sure the offset value is set at 0.00.

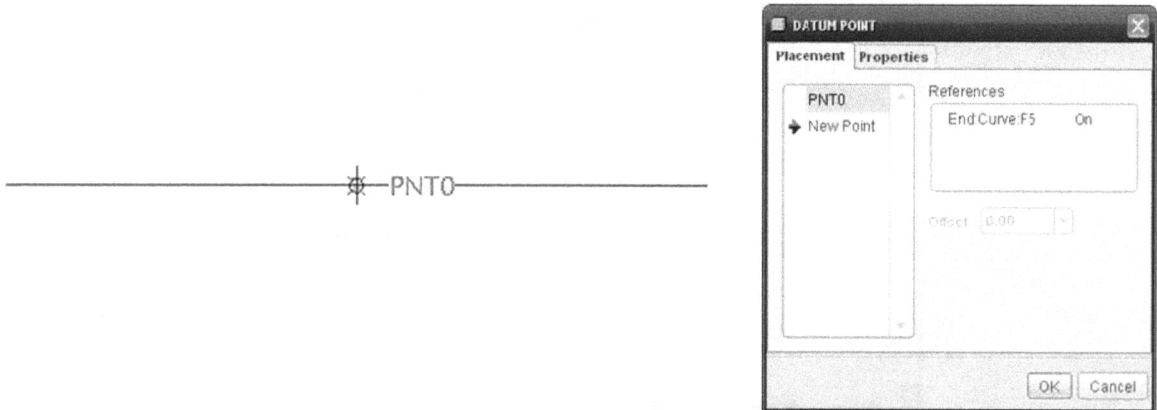

To define the second datum point, select the icon of **Datum Point Tool**. Pick the point at the left end of the datum curve on display. Make sure the offset value is set at 0.00.

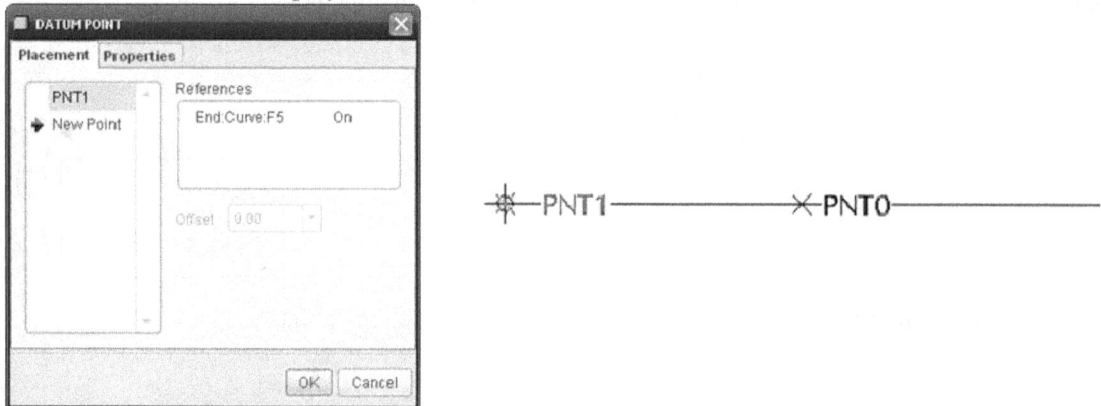

To define the third datum point, select the icon of **Datum Point Tool**. Pick the point at the left end of the datum curve. Make sure the offset value is set at 0.00.

Step 2: Perform FEA under **Pro/Mechanica**

From the main toolbar or the menu toolbar, select **Applications > Mechanica > Structure > OK.**

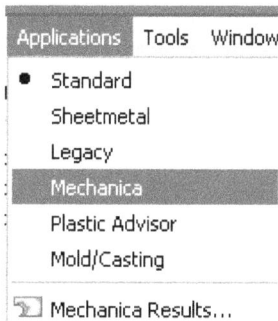

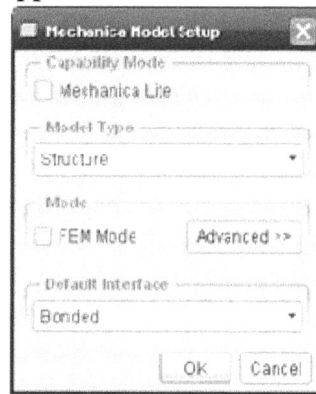

Pick the icon of **Beam**. Accept Beam1 as the name.

Accept the name: Beam1

Select Edge/Curve

Select Steel

Y vector

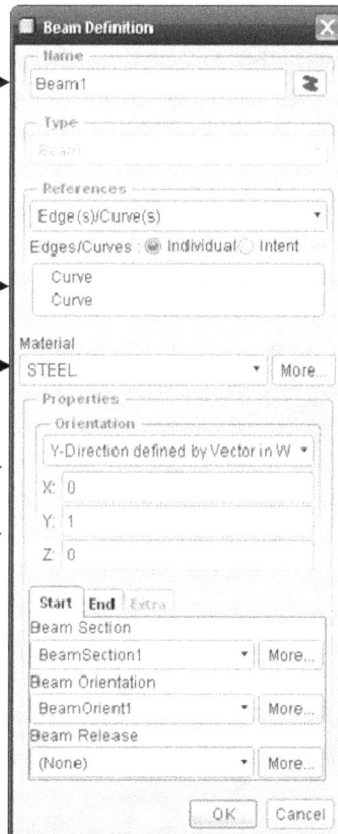

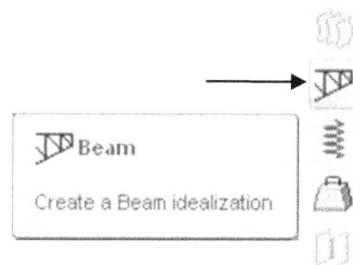

On the appeared **Beam Definition** window, do the followings:
Select **Edge/Curve** > pick the 2 datum curves while holding down the **Ctrl** key.

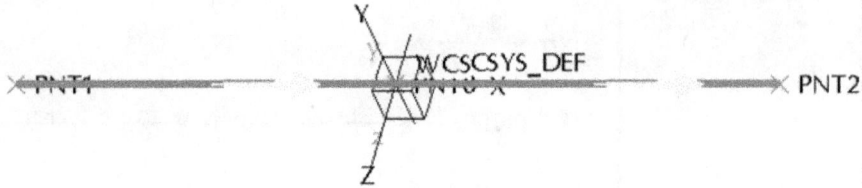

Select More from **Material** > Select **STEEL** > **Close**

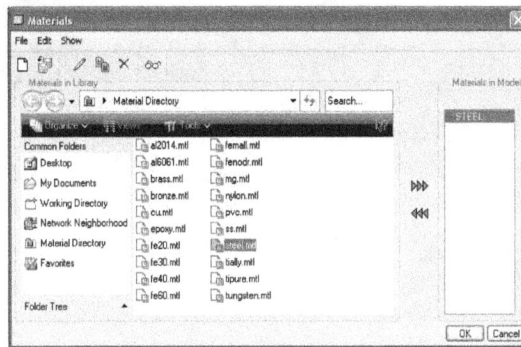

In the Properties field, set x = 0, y = 1 and z = 0 as the normal vector for orientation.

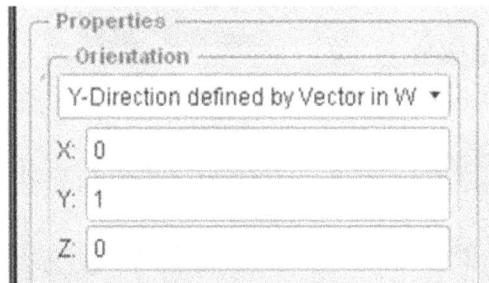

Select More from **Section** > **New** > select **Rectangle** > type *25* and *50* as b and d, respectively > **OK**.

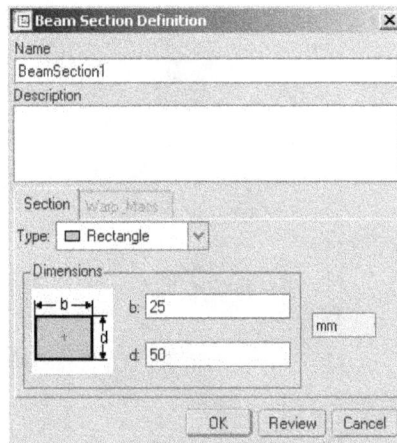

Define the Beam Orientation: click **More > New**, and make sure that DY=0 and DZ=0 to complete the process of defining the beam section > **OK**.

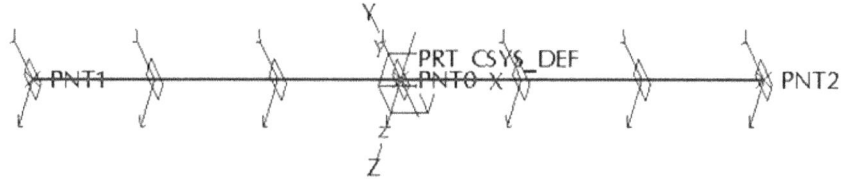

Define a simple supported constraint on the right end of the beam.

Constraints > Point(s) > pick **PNT1** > set the rotation about z-axis to free and fixed the others.

Constraints > Point(s) > pick **PNT2** > set the translation along x-axis to free and the rotation about z-axis to free.

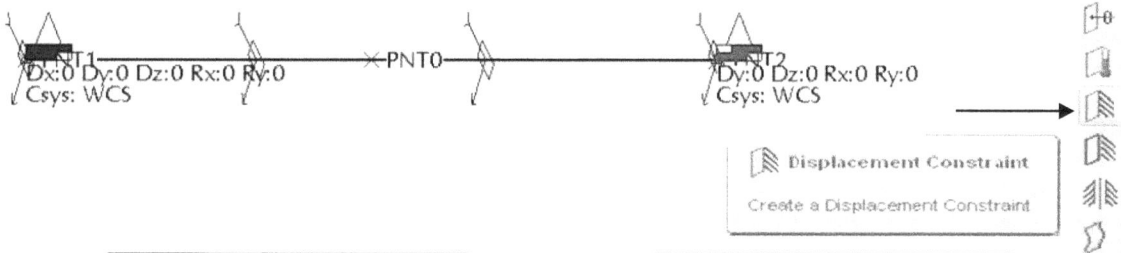

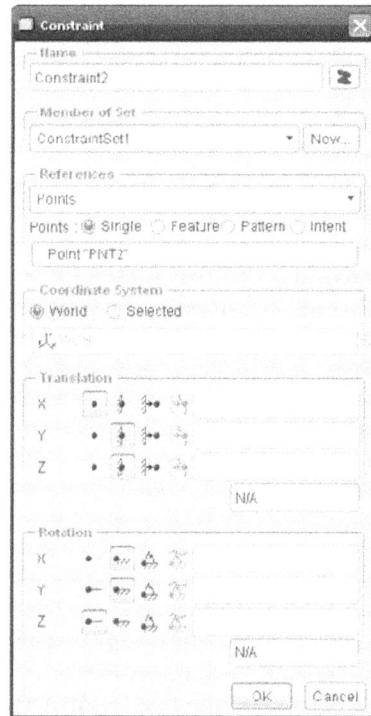

Select the icon of **Force/Moment > Edges/Curves)** > pick the line segment on the left.
Total Load > Uniform > type –200 in the force component **Y > OK.**

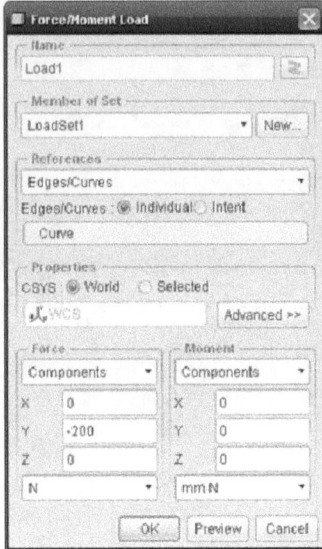

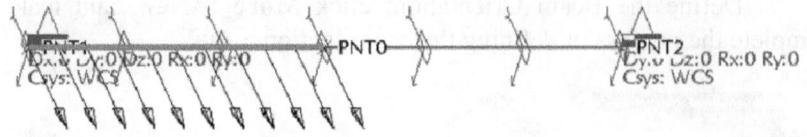

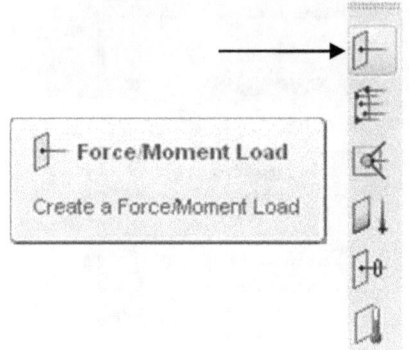

In order to define the point load at the middle of the line segment on the right side, we need to define a datum point. Can we define such a datum point under Pro/Mechanica? The answer is YES. Click the icon of **Datum Point Tool**. Pick a location on the line segment on the right side. Specify 0.5 as the offset value > **OK**.

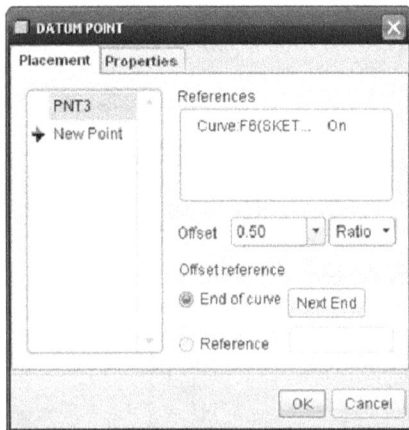

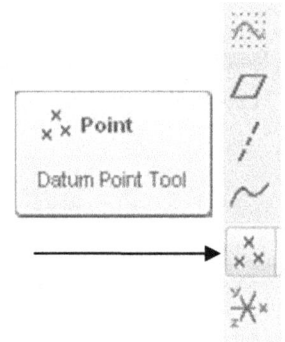

Select the icon of **Force/Moment > Point(s) >** pick PNT3 just created.
Total Load > Uniform > type –100 in the force component **Y > OK.**

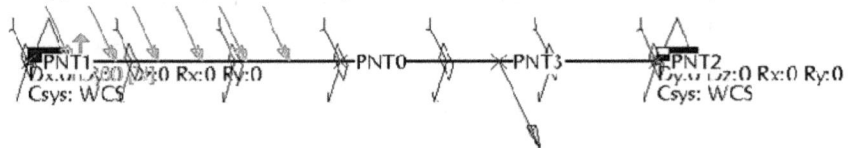

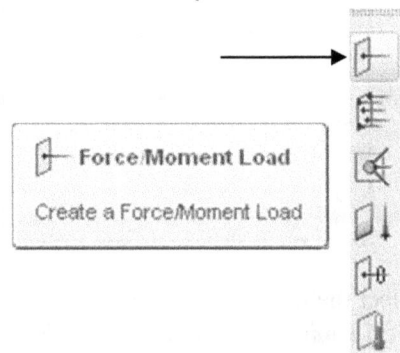

Step 3: Perform a static analysis
From the top menu, click the icon of **Run a Design Study > File > New Static.**

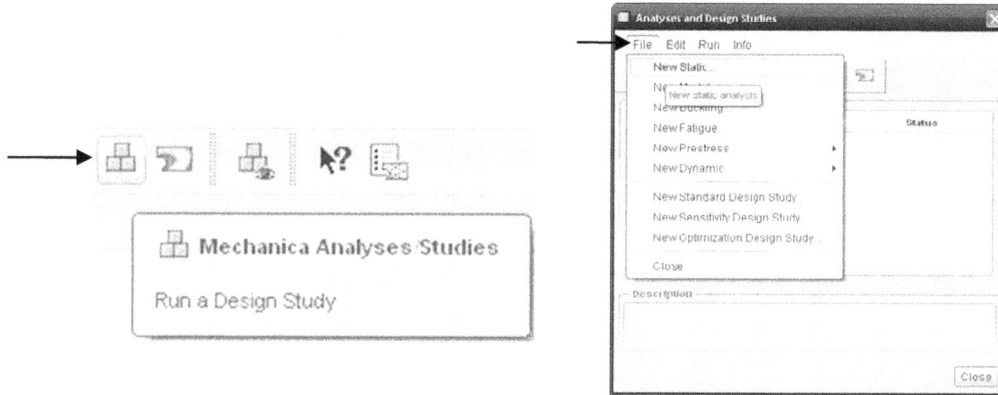

Type *combined_load_structure* as the file name > **OK**. Run the static analysis.

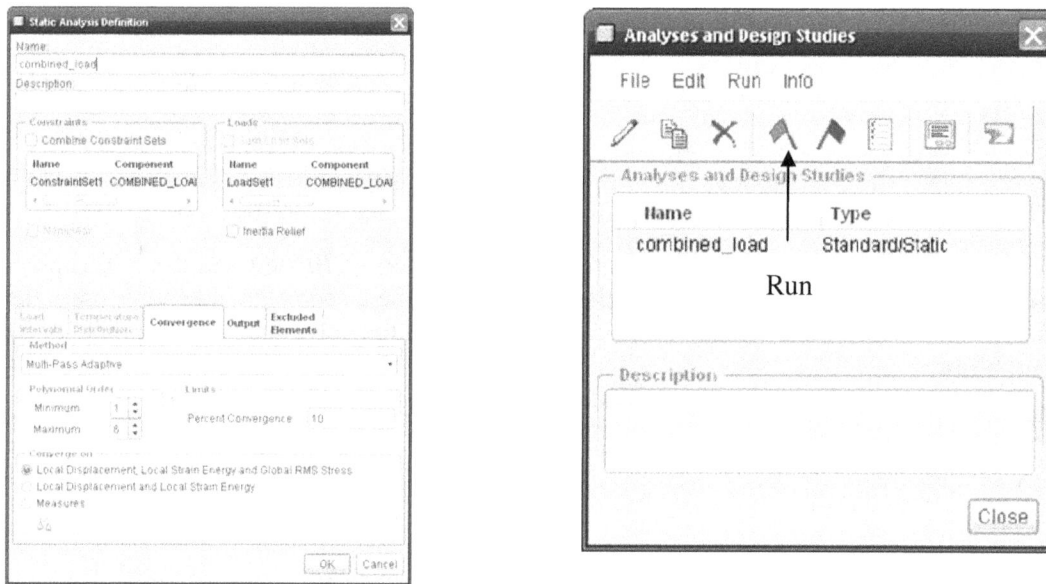

Note: The Percent Convergence is set at 10% (default value).
There is always a message displayed on screen. The message is asking the user "Do you want to run interactive diagnostics?" Click **Yes**. In this way a warning message will appear if an abnormal operation occurs.

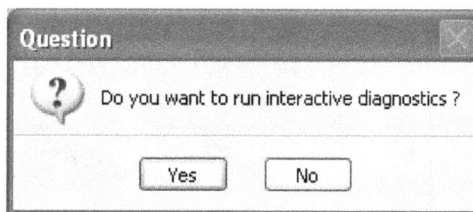

Step 4: Show the results obtained from FEA by plotting the reaction forces, share diagram and moment diagram.

To study the obtained result, say the reaction forces at the two support locations, click the icon of **Review results > Model > Reaction at Point Constraint > select Y > OK and Show.** The magnitude is 175 N at the left and is 125 at right, as shown.

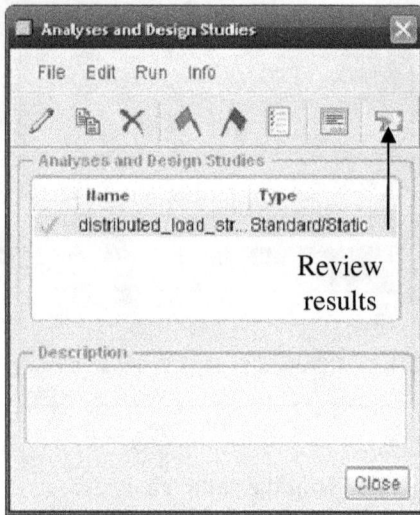

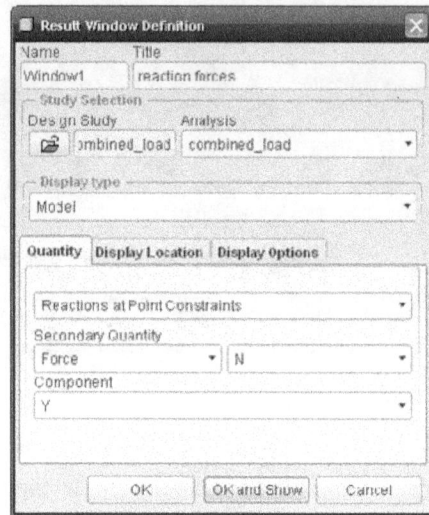

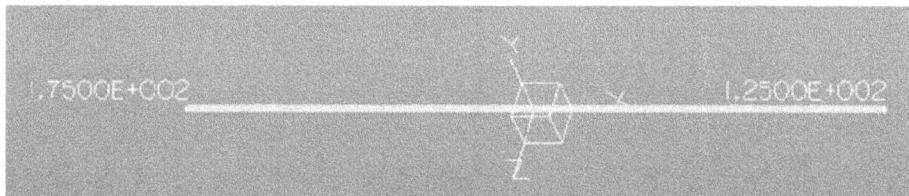

To plot the shear diagram and moment diagram, click the icon of **Review results > Graph > Shear & Moment > select Vy and Mz.** Click **Relative To** and pick the 2 segments of beam > press the middle button of mouse to accept the selection > click Toggle > **OK** to accept the starting point set by default > **OK and Show.**

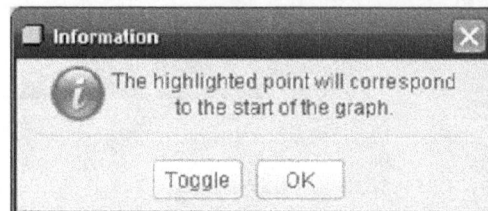

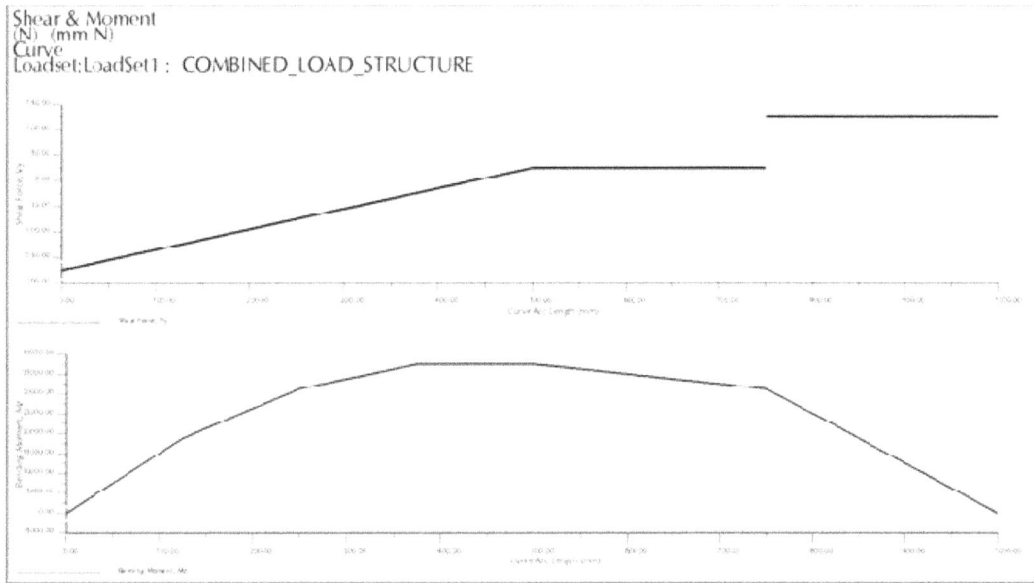

To show the distribution of displacement, click the icon of **Review Results** > type *distribution_of_displacement* as the title > select the folder called combomed_load > **Graph** > **Displacement** > select **Y** > **OK and Show**. The maximum deflection value at the middle section is 0.07798 mm.

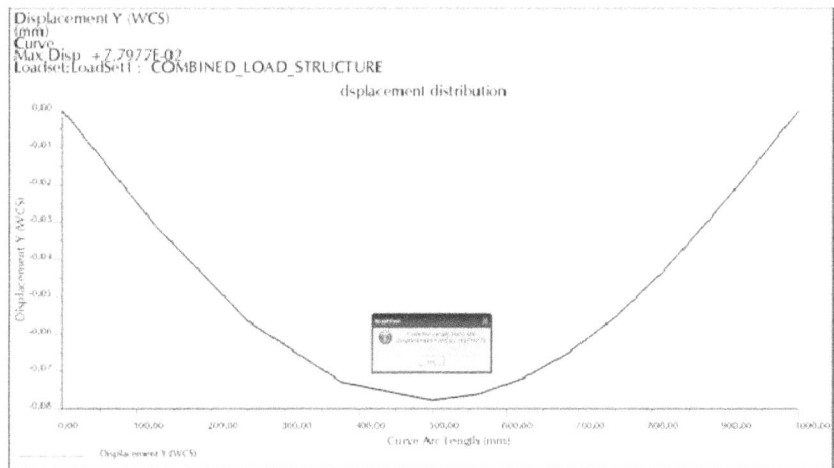

In the following, we further modify the model used in the previous examples. Instead of having a point load and a uniformly distributed load, we have a point load and non-uniformly distributed load.

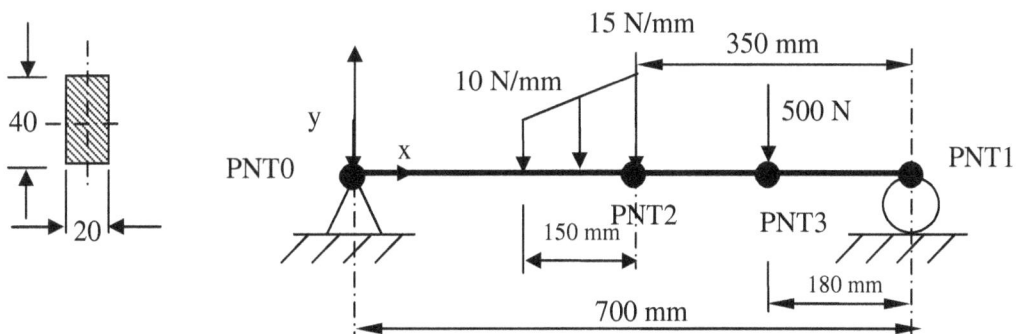

Step 1: Create a 2D model for the simple supported beam with a pin connection at the left side and with a simple support at the right side with a distributed load.

From the mail toolbar, select the icon of **Create** a new object. Make sure the **Part** mode is selected. Type *nonuniform_load* and clear the box of **use default template** > **OK**

Select the unit system to be **mmns_part_solid** > type *non-uniform load* in **DESCRIPTION** and *student* in **MODELED_BY** > **OK**.

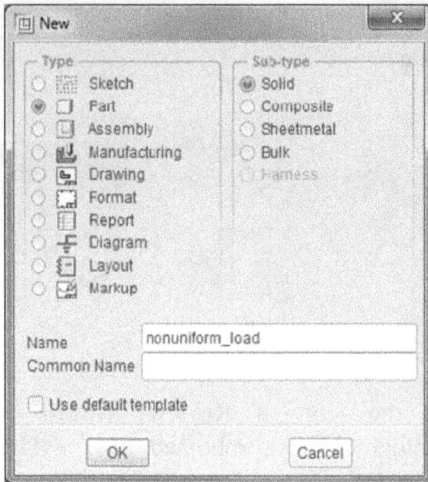

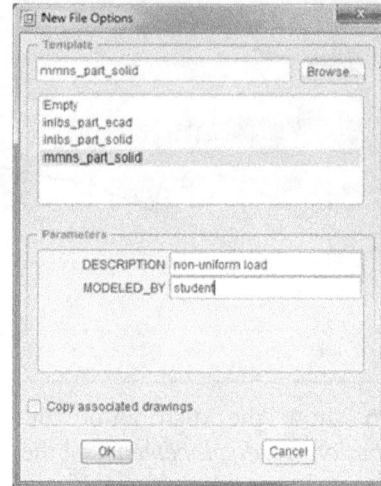

In this example, we need to use 3 sketches to create this beam structure. Let us work on the first sketch, which is a line along the x-axis. The length of the datum curve is 350 mm.

Select the icon of **Sketch Tool.** Pick the **FRONT** datum plane as the sketch plane, and accept the default orientation setting, which uses the **RIGHT** datum plane and position it at the right side > **Sketch**.

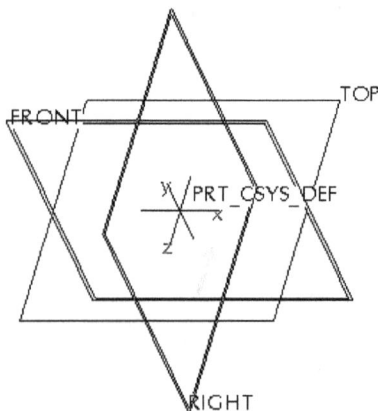

Click the icon of **Line** and sketch a line along the x-axis. The sketched line is on the right side, as shown. The dimension is 200. Click the icon of **Done**. Special attention is paid to the starting point, ending point and sketching direction, as shown.

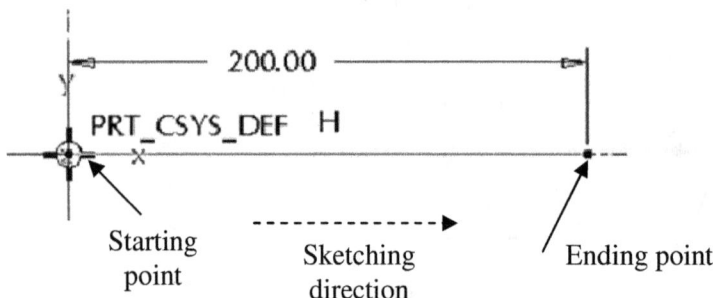

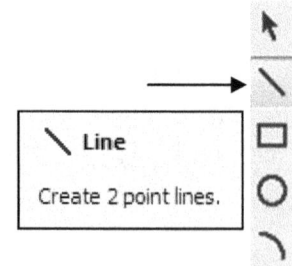

To prepare the second sketch, click the icon of **Sketch Tool > Use Previous**.

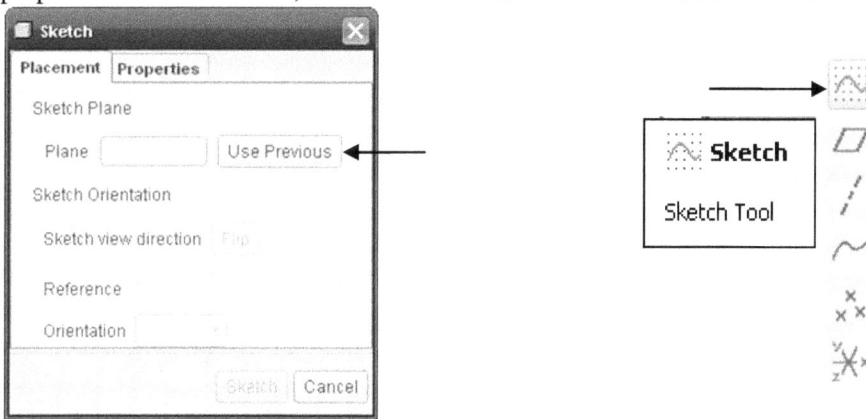

From the top menu, click **Sketch > References**. Select the sketched line as a new reference.

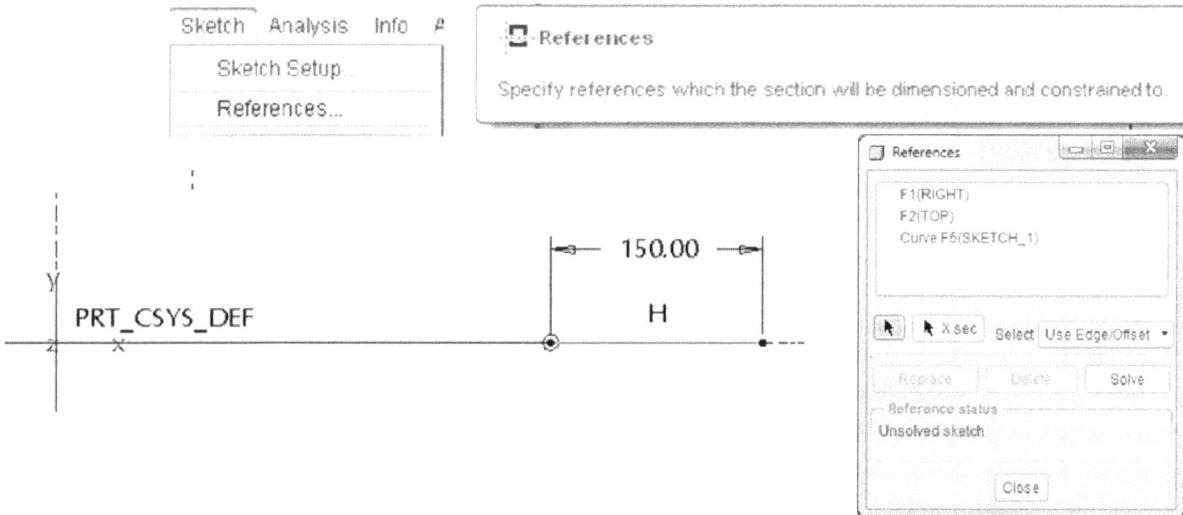

150.00

H

PRT_CSYS_DEF

Click the icon of **Line** and sketch a new line. Use the end point of the sketched line as the starting point of the new line. Specify the length equal to 150 > click the icon of **Done**.

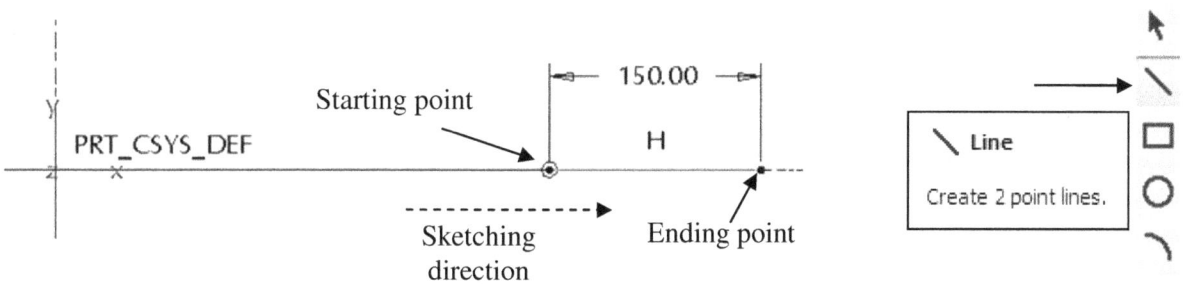

Starting point

150.00

H

PRT_CSYS_DEF

Sketching direction

Ending point

Line

Create 2 point lines.

To prepare the third sketch, click the icon of **Sketch Tool > Use Previous**.

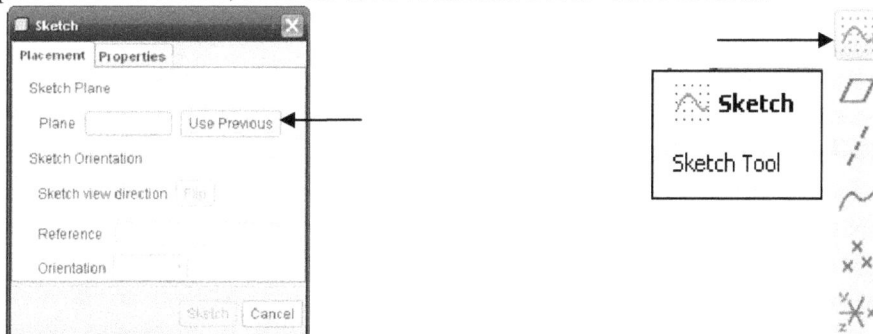

From the top menu, click **Sketch > References**. Select the sketched line as a new reference.

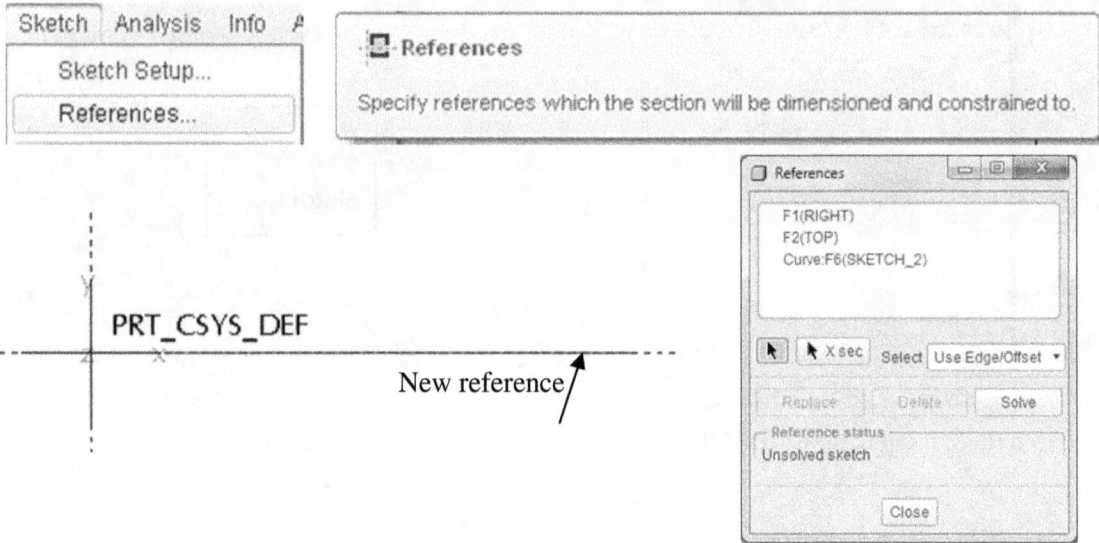

Click the icon of **Line** and sketch a new line. Use the end point of the sketched line as the starting point of the new line. Specify the length equal to 150 > click the icon of **Done**.

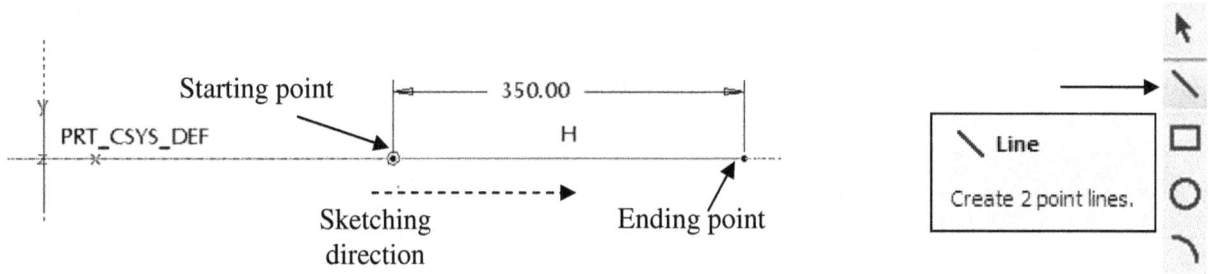

Create three datum points. The first datum point is at the central location of the datum curve. Select the icon of **Datum Point Tool**.

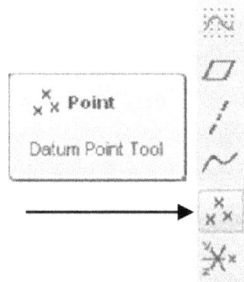

Pick the point at the central location of the entire datum curve on display. Make sure the offset value is set at 0.00.

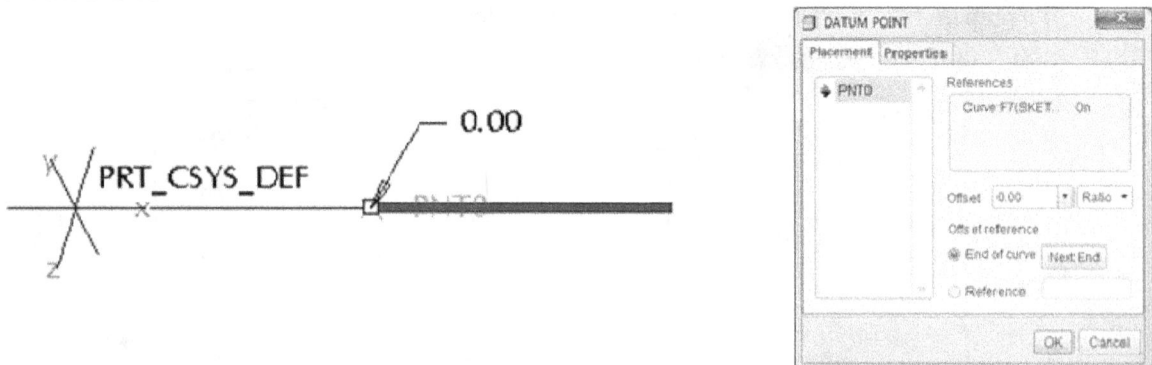

To define the second datum point, select the icon of **Datum Point Tool**. Pick the point at the left end of the datum curve on display. Make sure the offset value is set at 0.00.

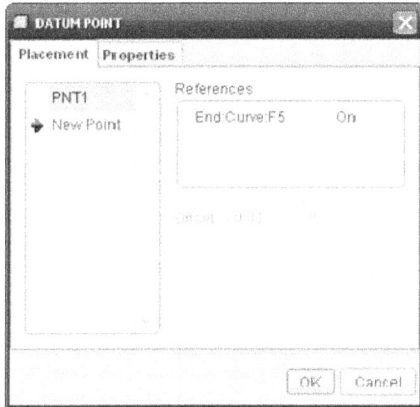

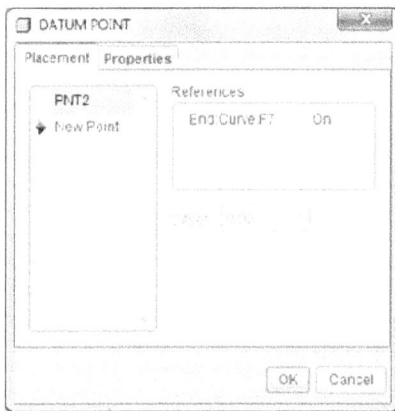

To define the third datum point, select the icon of **Datum Point Tool**. Pick the point at the left end of the datum curve. Make sure the offset value is set at 0.00.

To define the third datum point, select the icon of **Datum Point Tool**. Pick the point at the middle of the datum curve on the right. Make sure the offset value is set at 0.50.

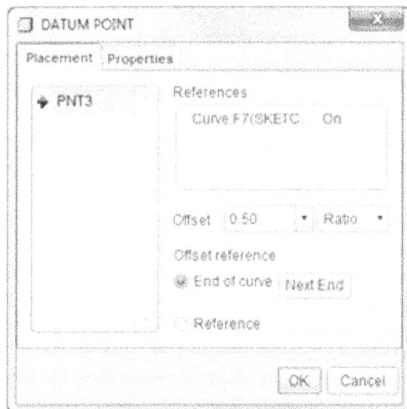

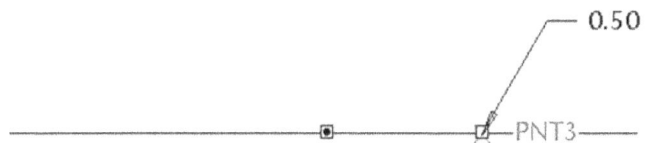

Step 2: Perform FEA under **Pro/Mechanica**

From the main toolbar or the menu toolbar, select **Applications > Mechanica > Structure > OK.** Pick the icon of **Beam**. Accept Beam1 as the name.

Accept the name: Beam1

Select Edge/Curve

Select Steel

Y vector

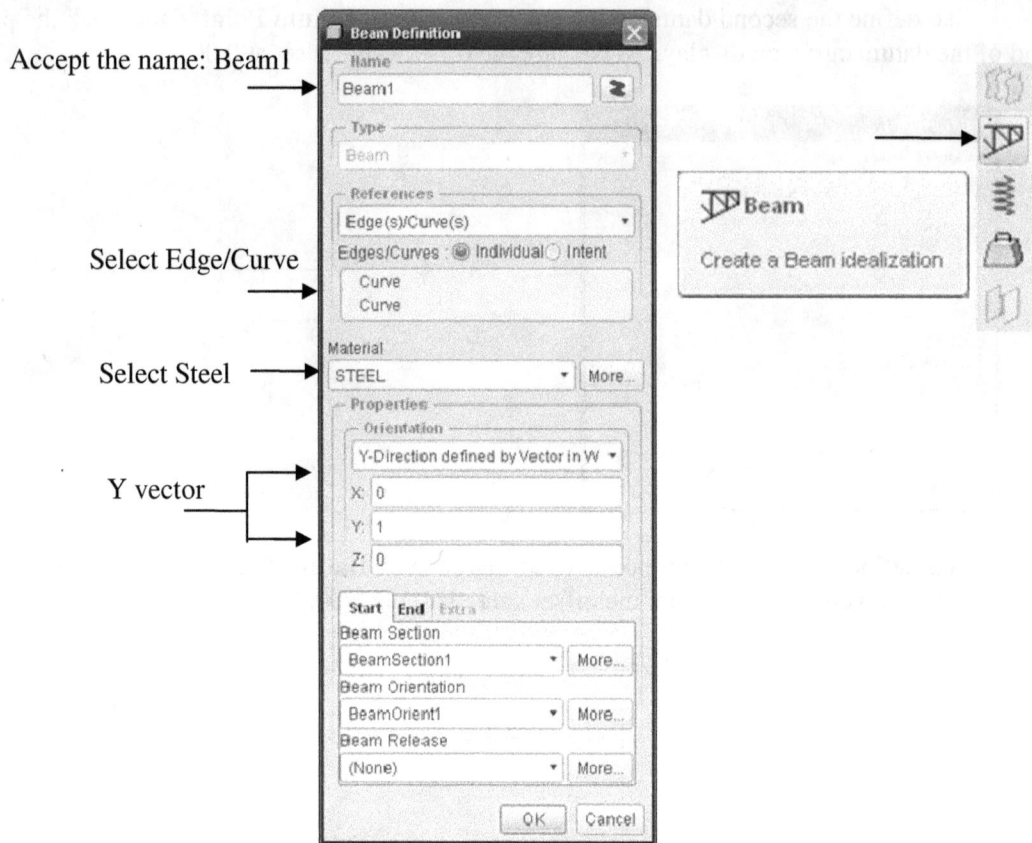

On the appeared **Beam Definition** window, do the followings:
Select **Edge/Curve** > pick the 3 datum curves while holding down the **Ctrl** key.

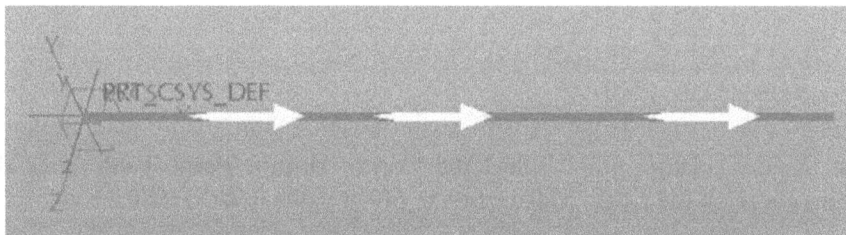

Select More from **Material** > Select **STEEL** > **Close**

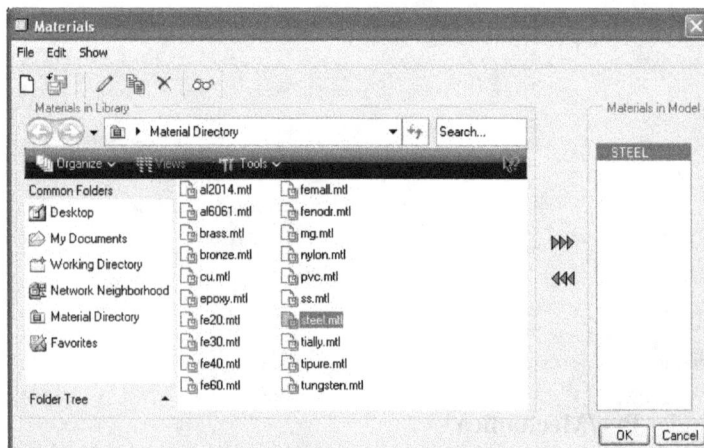

In the Properties field, set x = 0, y = 1 and z = 0 as the normal vector for orientation.

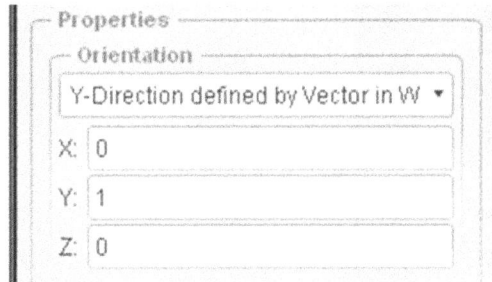

Select More from **Section** > **New** > select **Rectangle** > type *20* and *40* as b and d, respectively > **OK**.

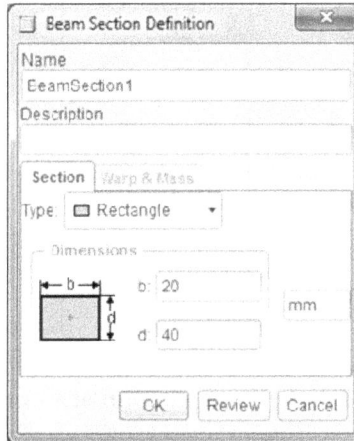

Define the Beam Orientation: click **More** > **New**, and make sure that DY=0 and DZ=0 to complete the process of defining the beam section > **OK**.

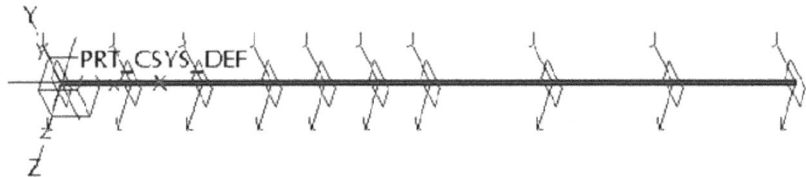

Define a simple supported constraint on the right end of the beam.

Constraints > **Point(s)** > pick **PNT1** > set the rotation about z-axis to free and fixed the others.

Constraints > **Point(s)** > pick **PNT2** > set the translation along x-axis to free and the rotation about z-axis to free.

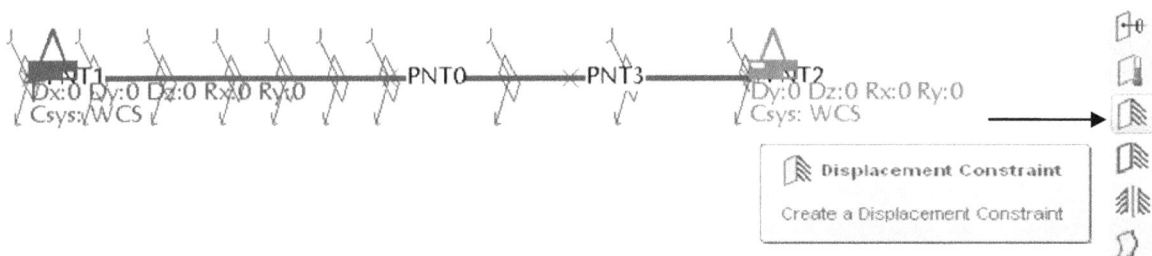

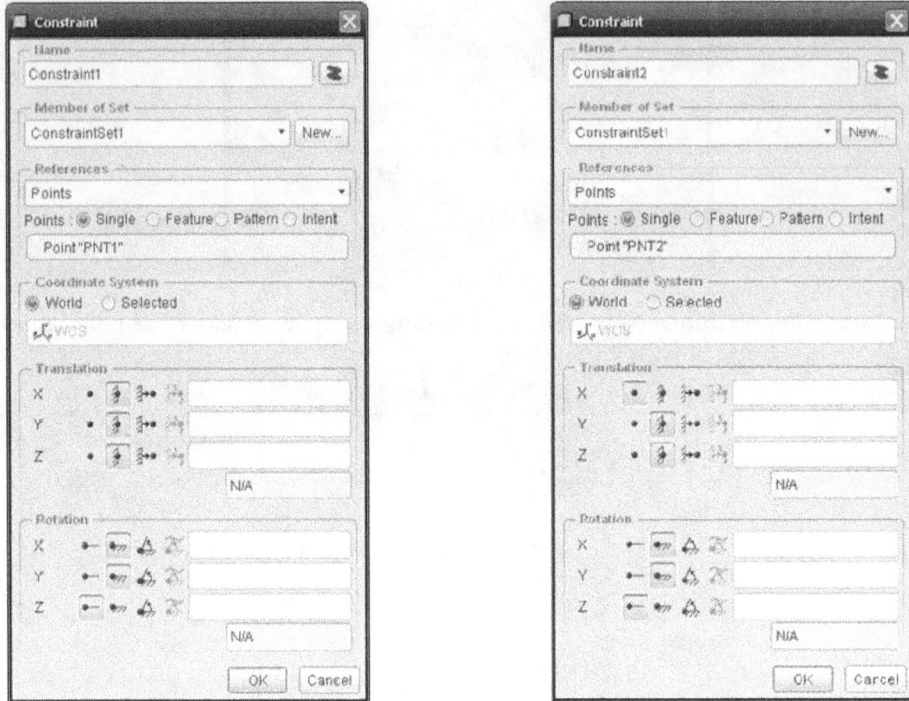

Select the icon of **Force/Moment > Edges/Curves) >** pick the line segment on the left. **Advanced > Force Per Unit Length > Function of Coordinates >** click *f(x)* to define the function.

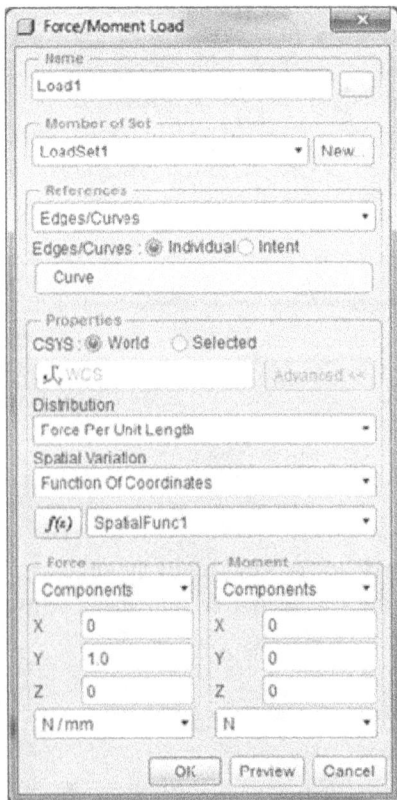

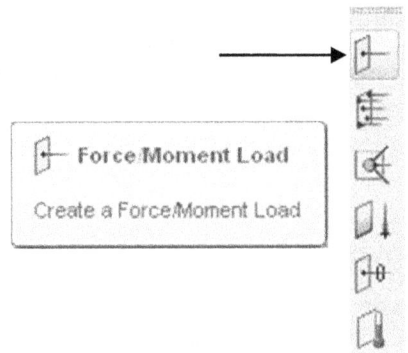

New > Table > Add and specify *2* as the number of rows. Type *200* and *350* as the x coordinates and *-10* and *-15* as magnitude value > OK. In the window of **Force/Moment Load**, specify **1** in the Y > **OK**.

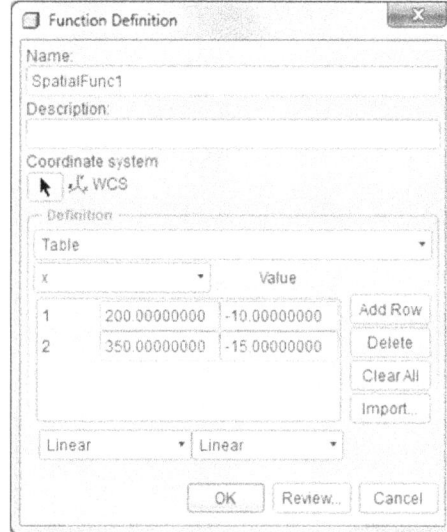

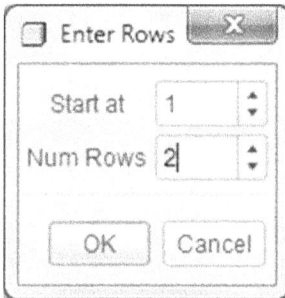

To define the point load, click icon of **Force/Moment > Point(s) >** pick PNT3 just created.
Total Load > Uniform > type *–500* in the force component **Y > OK.**

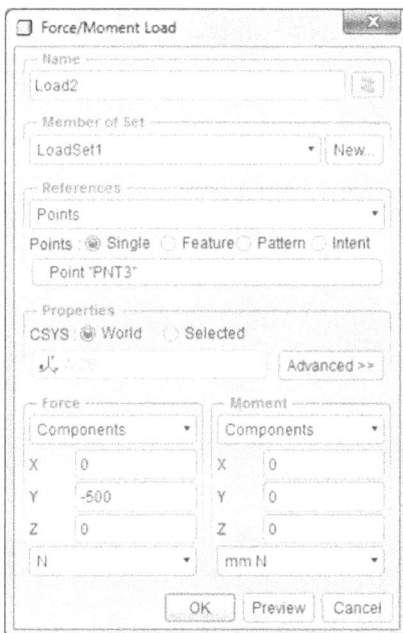

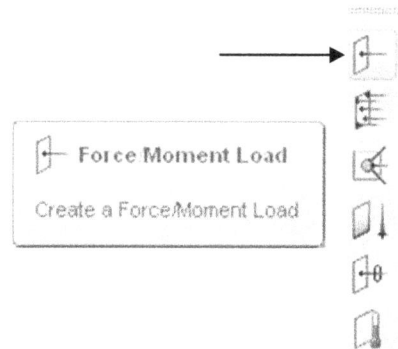

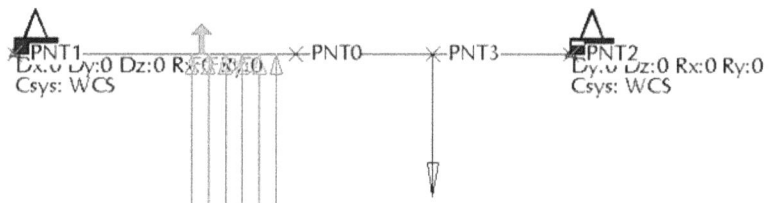

Step 3: Perform a static analysis

From the top menu, click the icon of **Run a Design Study > File > New Static.**

Type *nonuniform_load* as the file name **> OK**. Run the static analysis.

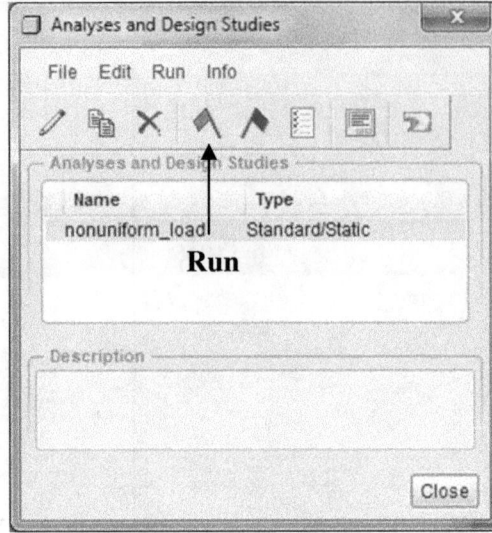

Step 4: Show the results obtained from FEA by plotting the reaction forces, share diagram and moment diagram.

To study the obtained result, say the reaction forces at the two support locations, click the icon of **Review results > Model > Reaction at Point Constraint** > select **Y** > **OK and Show.** The magnitude is 175 N at the left and is 125 at right, as shown.

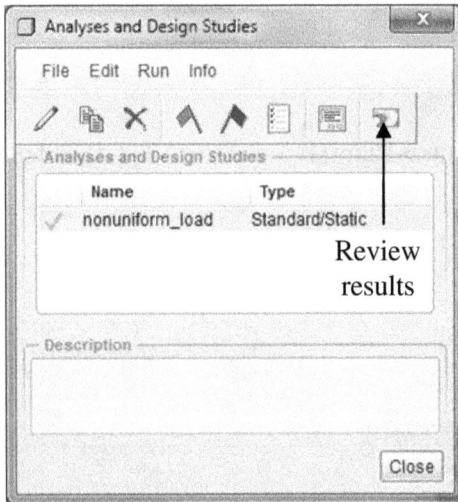

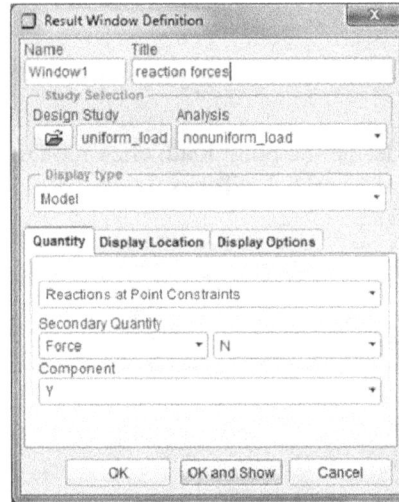

To plot the shear diagram and moment diagram, click the icon of **Review results > Graph > Shear & Moment** > select **Vy** and **Mz**. Click **Relative To** and pick the 3 segments of beam > press the middle button of mouse to accept the selection > click **OK** in the **Toggle** window to accept the starting point set by default > **OK and Show.**

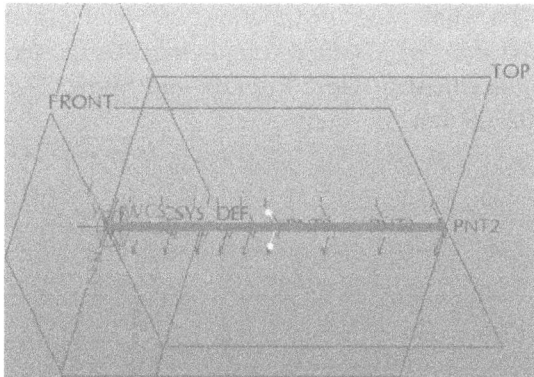

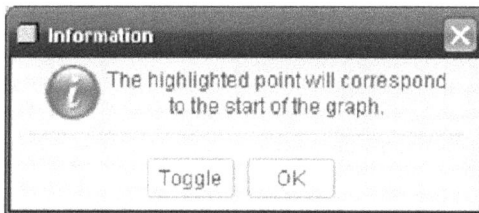

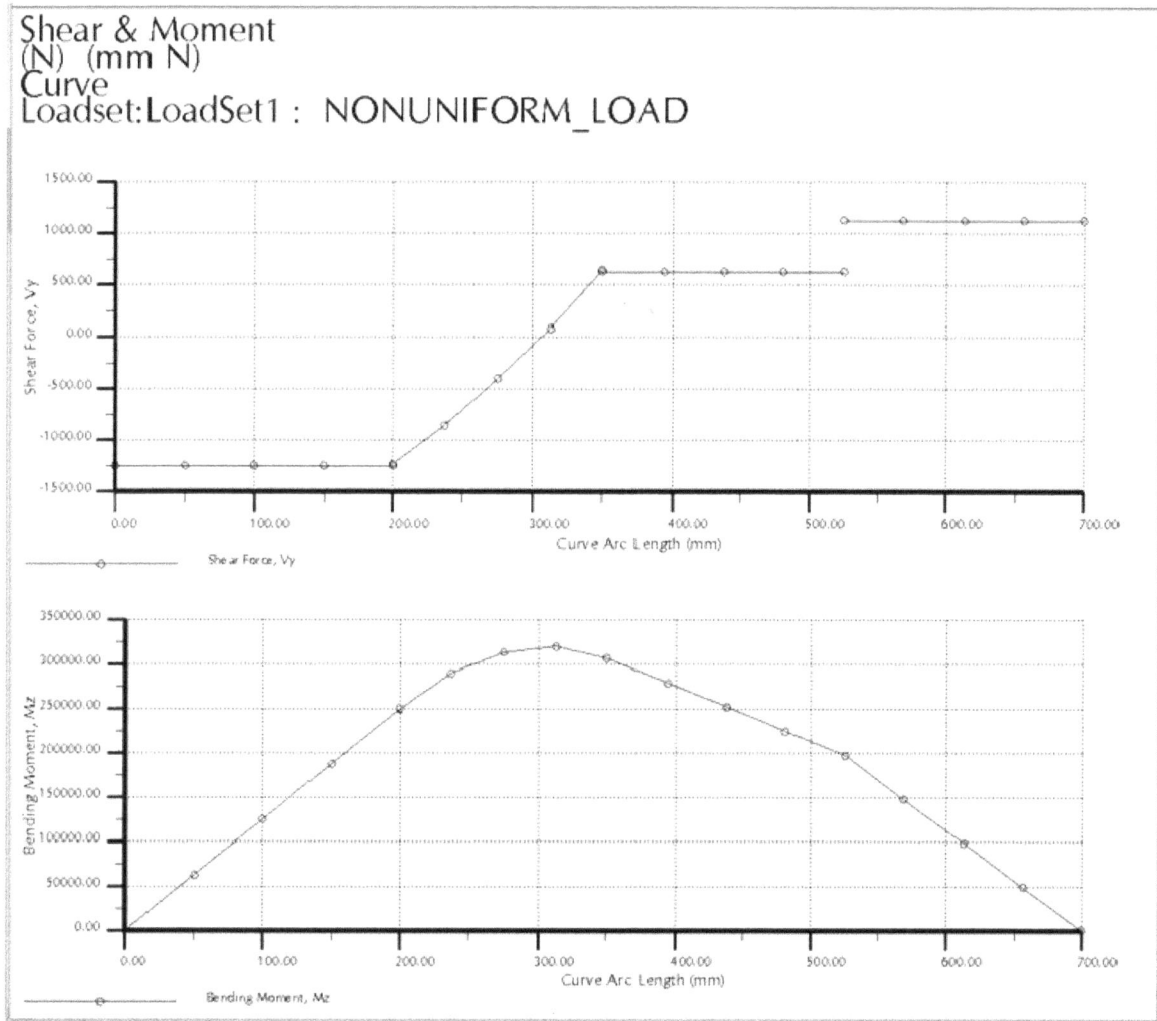

To show the distribution of displacement, click the icon of **Review Results** > type *distribution_of_displacement* as the title > select the folder called nonuniform_load > **Graph** > **Displacement** > select **Y** > **OK and Show**. The maximum deflection value at the middle section is 0.702739 mm.

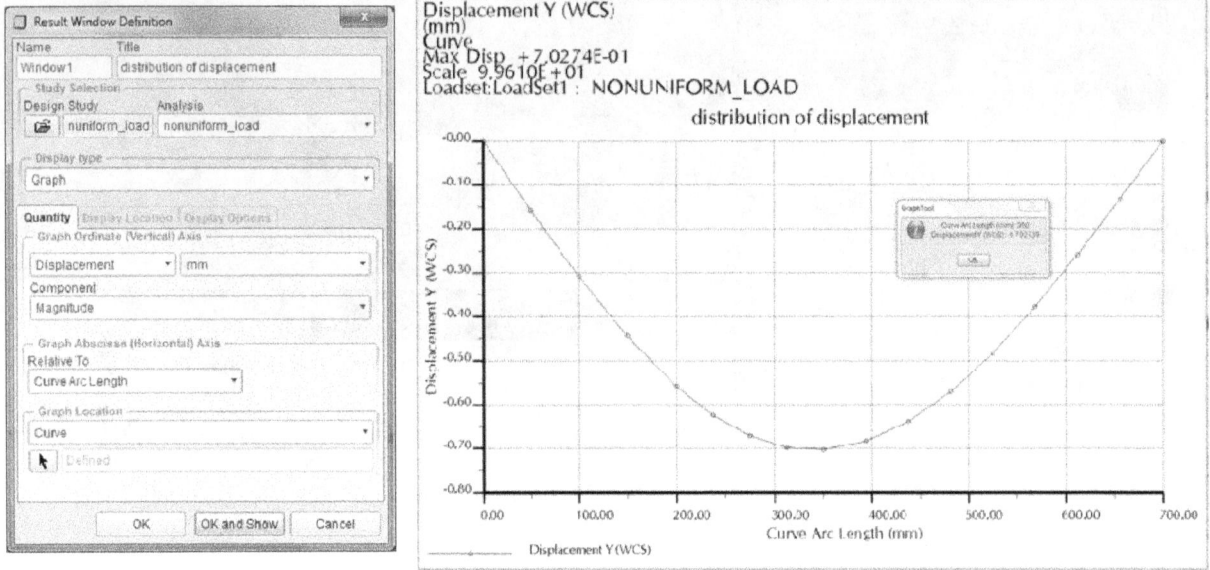

9.4 Case Study: FEA with a Roof Structure

The geometry of a frame structure is shown. Construct a 3D solid model of the frame structure. Perform FEA to evaluate the displacement pattern of the frame structure, indicating the magnitude and the location of the maximum displacement, and the magnitude and the location of the maximum value of the maximum principal stress. Fix the bottom surface of the frame structure to the ground. Assume the magnitude of the total load acting the top surface of the frame is 10,000 Newton.

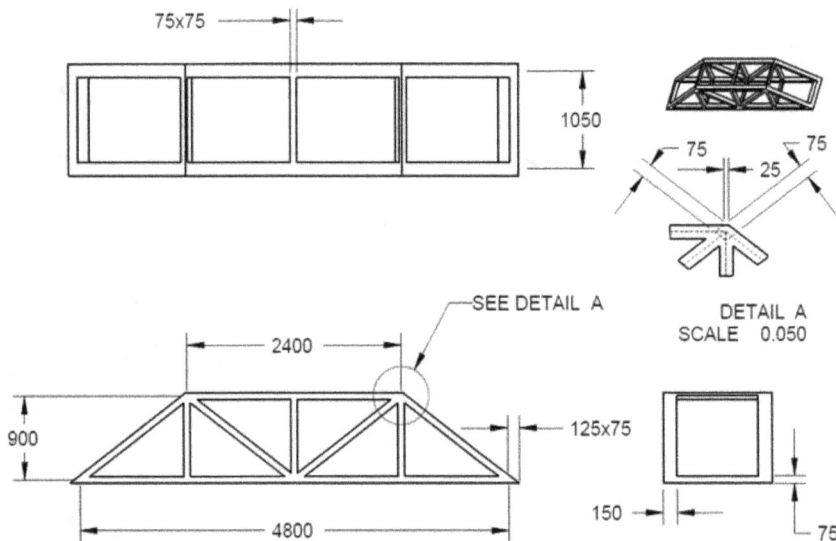

Step 1: Model the truss structure

From the mail toolbar, select the icon of **Create** a new object. Make sure the **Part** mode is selected. Type *truss_structure* and clear the box of **use default template** > **OK**

Select the unit system to be **mmns_part_solid** > type *truss structure* in **DESCRIPTION** and *student* in **MODELED_BY > OK**.

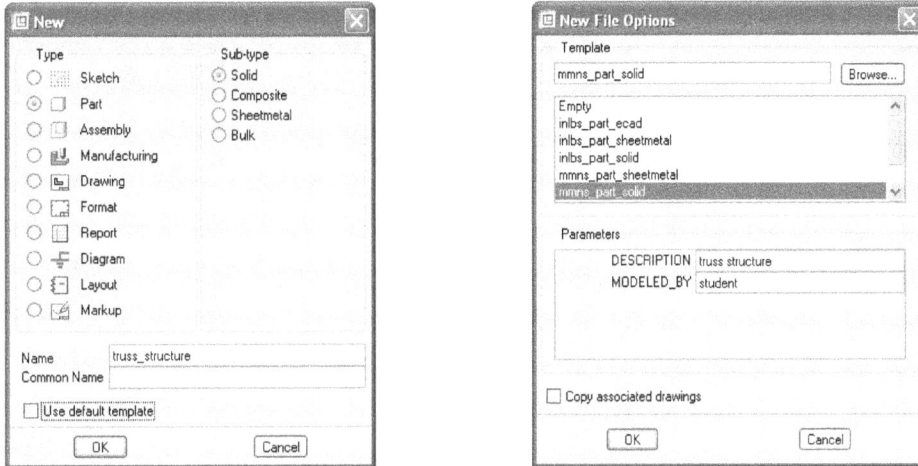

Select the icon of **Sketch Tool.** Pick the **FRONT** datum plane as the sketch plane, and accept the default orientation setting, which uses the **RIGHT** datum plane and position it at the right side > **Sketch.**

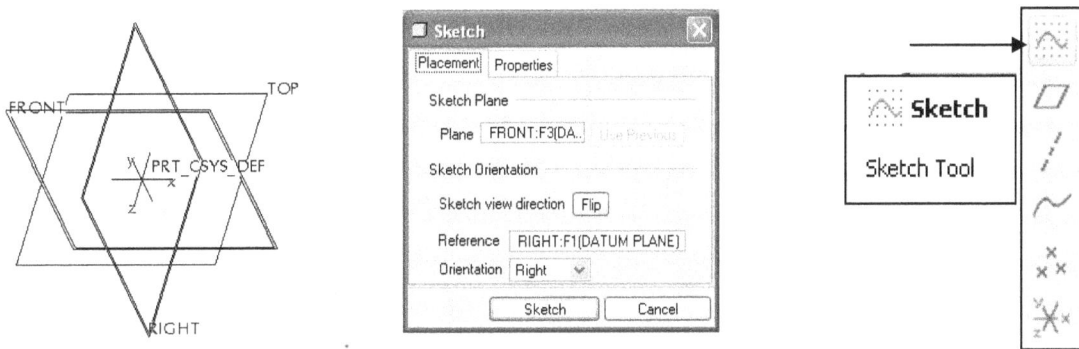

We first sketch a vertical centerline passing through the origin of the coordinate system, as shown. Right-click and hold, select **Centerline**. Users may select the icon of **Centerline** to sketch a vertical centerline.

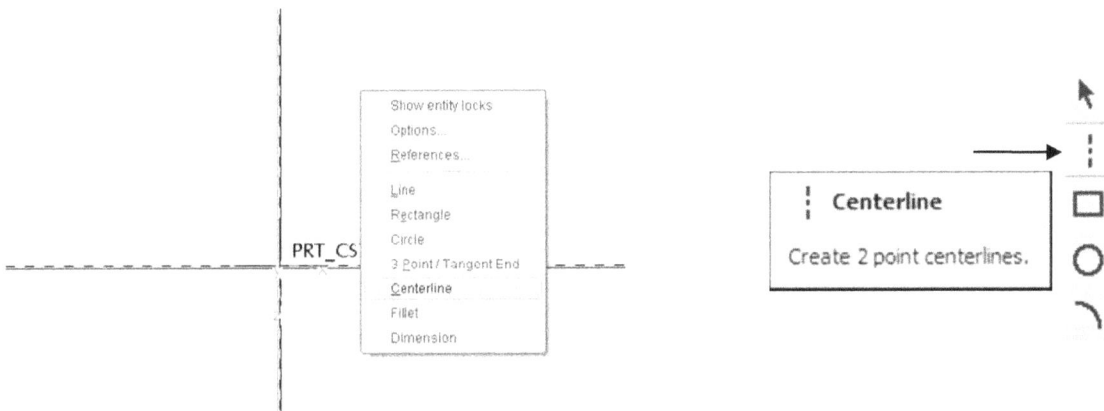

Pick the icon of **Line**, and sketch a horizontal line along the x-axis, which is symmetric about the vertical centerline with the dimension equal to 4800.

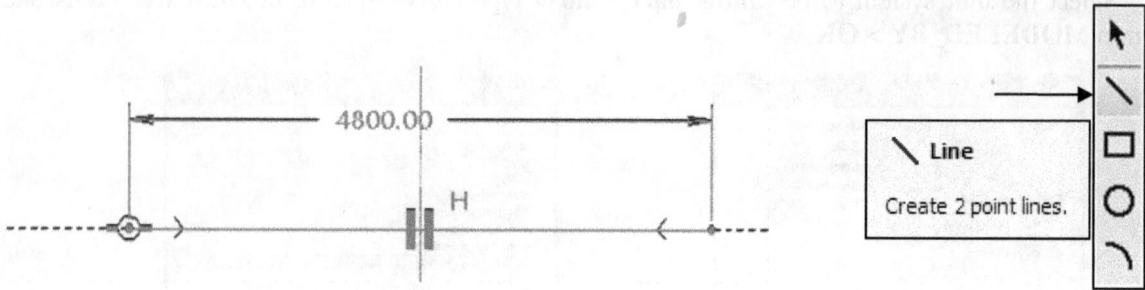

Sketch another horizontal line symmetric about the vertical centerline. The two dimensions are 2400 and 900, respectively.

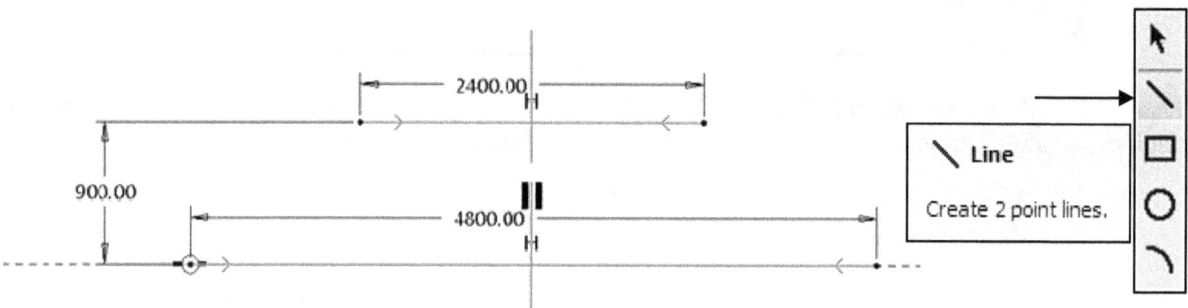

Sketch an inclined line by connecting 2 points, as shown.

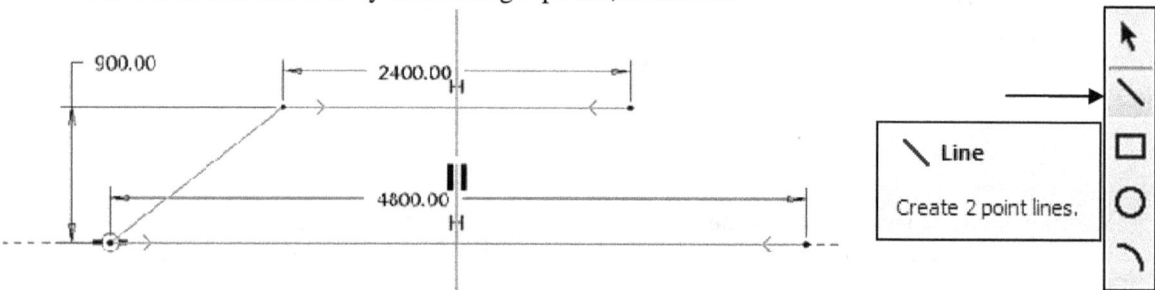

Repeat the previous procedure to sketch another inclined line by connecting another set of 2 points, as shown below:

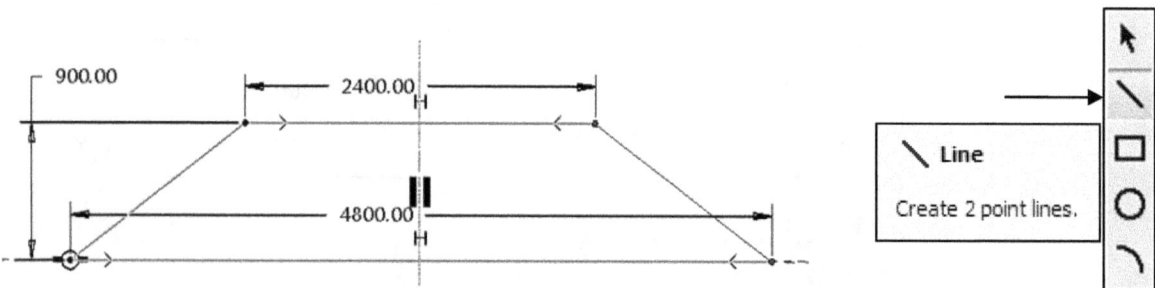

Upon completion, select the icon of **Done** from the toolbar of sketcher. The sketched 4 datum curves (actually, they are straight lines) represent the 4 bars of the truss structure.

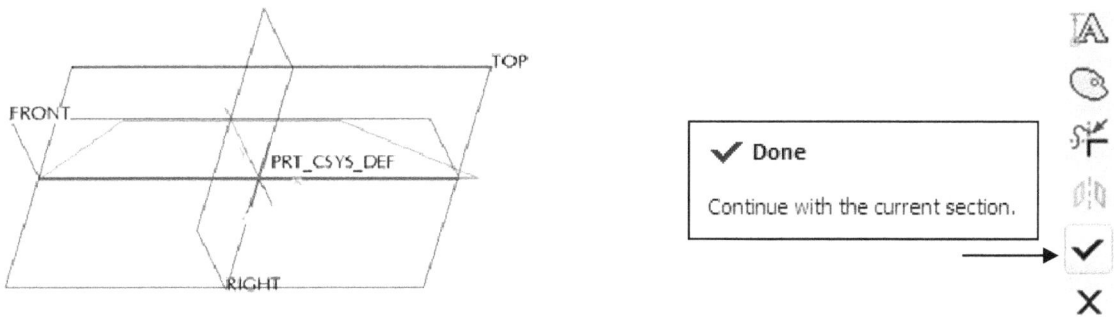

Now let us create six (6) datum points. The first datum point is at the end of the datum curve located at the very left side. Select the icon of **Datum Point Tool**. Make sure the offset value is set at zero > **OK**. PNT0 is on display.

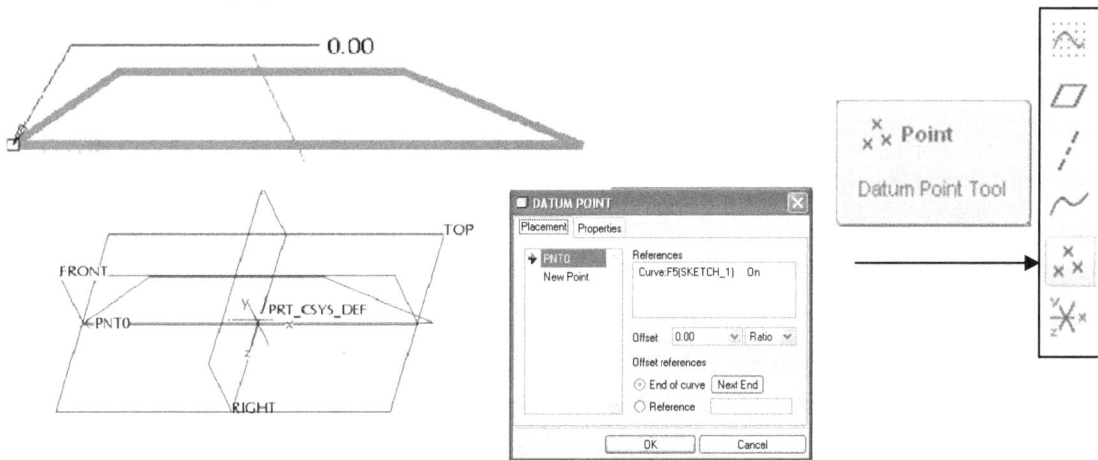

To define the second datum point, select the icon of **Datum Point Tool**. Pick the point at the very right corner of the datum curve shown below. Make sure the offset value is set at 0.00.

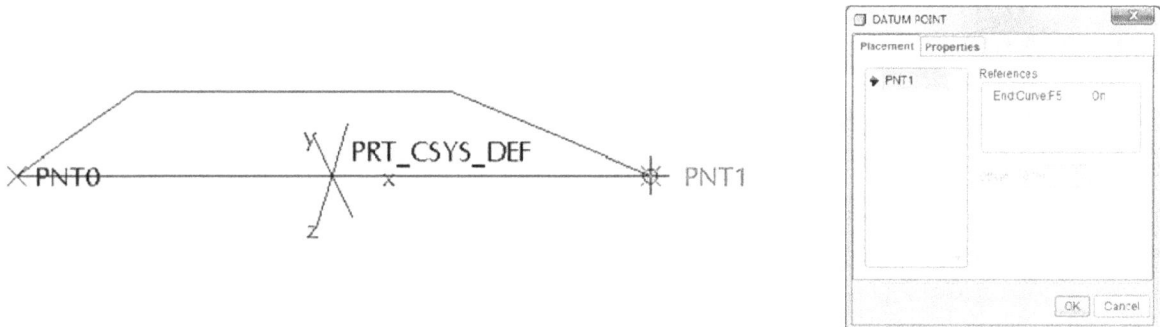

To define the third datum point, select the icon of **Datum Point Tool**. Pick the point at the left-upper corner of the datum curve on display. Make sure the offset value is set at 0.00.

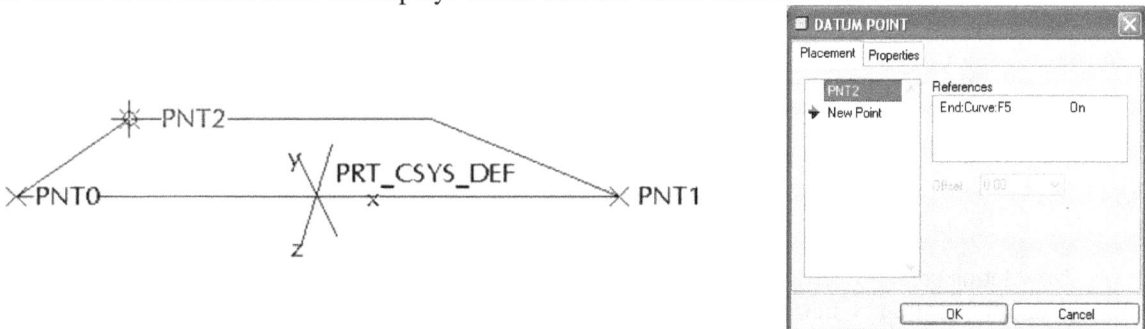

To define the 4th datum point, select the icon of Datum Point Tool. Pick a point on the right-upper corner of the datum curve, and set the offset value is set at 0.00, as shown below.

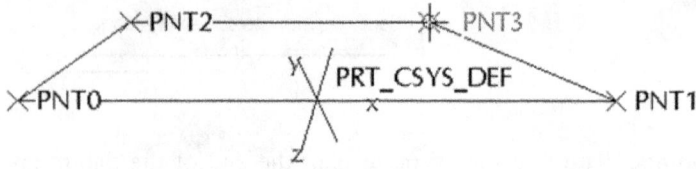

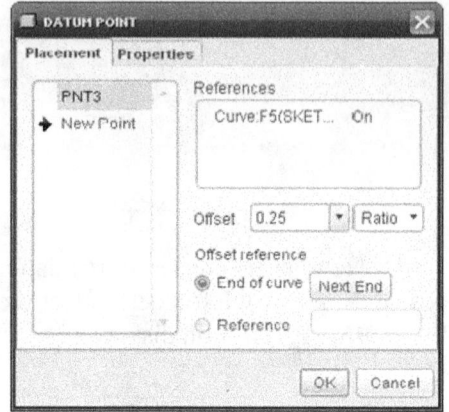

To define the 5th datum point, select the icon of Datum Point Tool. Pick a point on the horizontal line, and set the offset value is set at 0.75, as shown below.

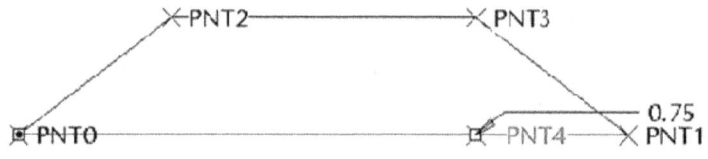

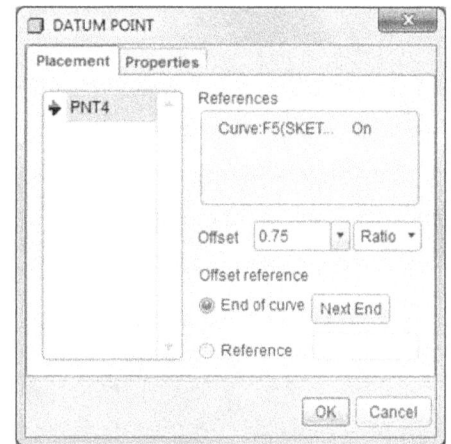

To define the 6th datum point, select the icon of Datum Point Tool. Pick a point on the horizontal line, and set the offset value is set at 0.25, as shown below.

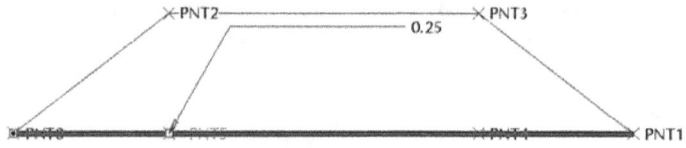

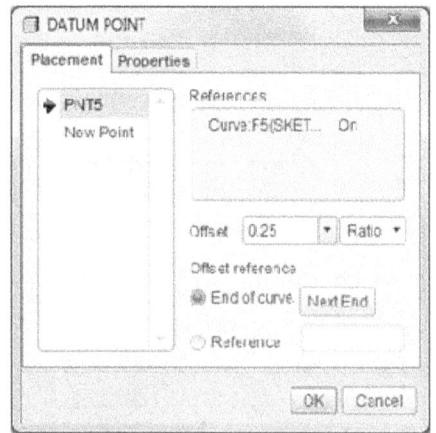

Now let us create 3 vertical datum curves (or lines). Select the icon of **Sketch Tool** from the toolbar of datum creation. Click **Use Previous**. We have selected **FRONT** as our sketch plane.

Click **Sketch** from the top menu and click **References**. Select 3 lines (upper horizontal line and the 2 inclined lines) and 2 points (PNT4 and PNT5) as new references, as shown.

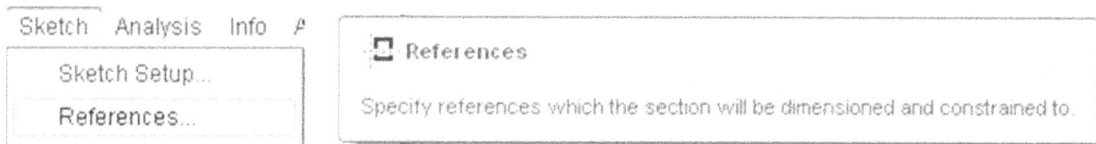

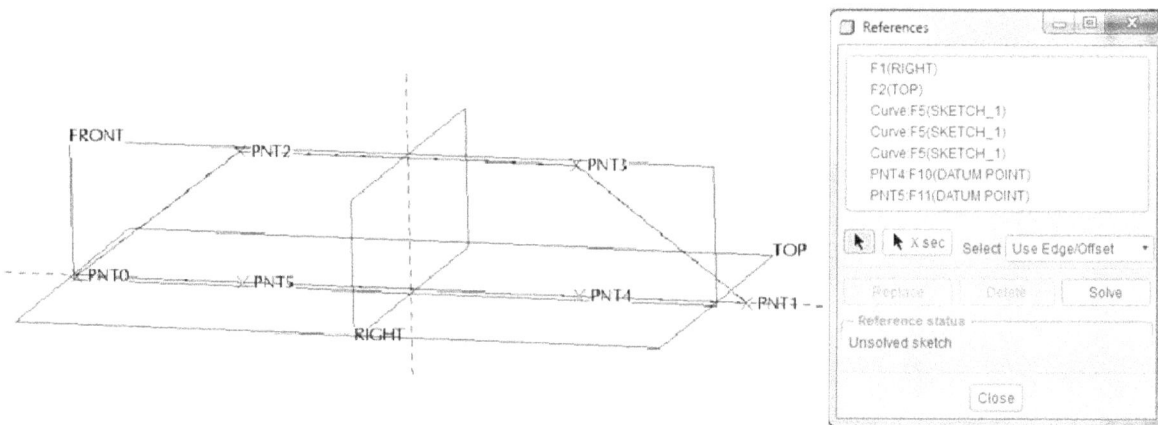

Pick the icon of **Line**, and sketch a vertical line by connecting **PNT3** and **PNT4**. Sketch a vertical line by connecting **PNT2** and **PNT5**. Sketch a vertical line by connecting the middle point of the upper horizontal line to the middle point of the lower horizontal line. Click the icon of **Done** or the check mark when completing the sketch of the 3 vertical lines.

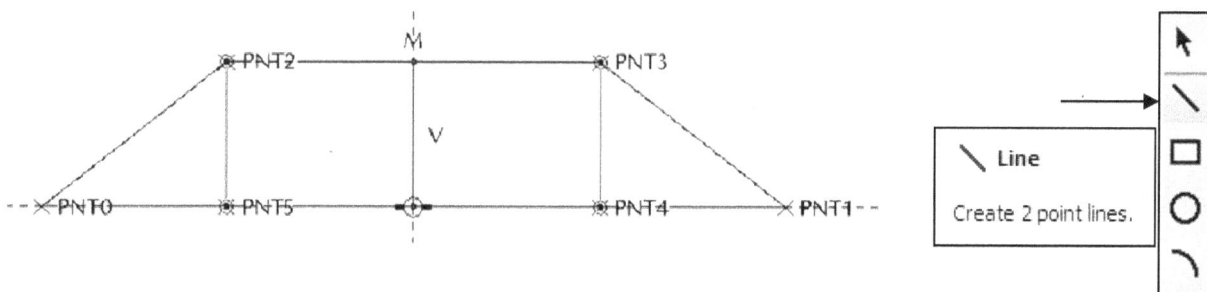

Now let us create 2 more datum curves (or lines). Select the icon of **Sketch Tool** from the toolbar of datum creation. Click **Use Previous**. We have selected **FRONT** as our sketch plane.

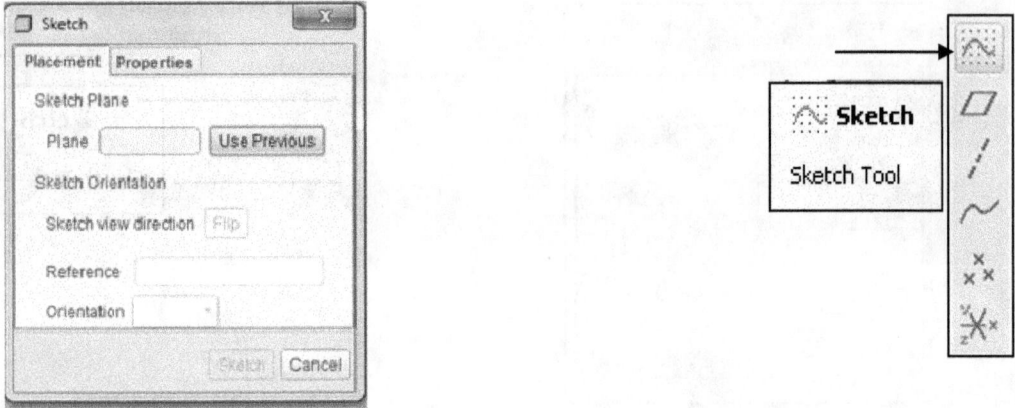

Click **Sketch** from the top menu and click **References**. Select 2 points (PNT2 and PNT3) as new references, as shown.

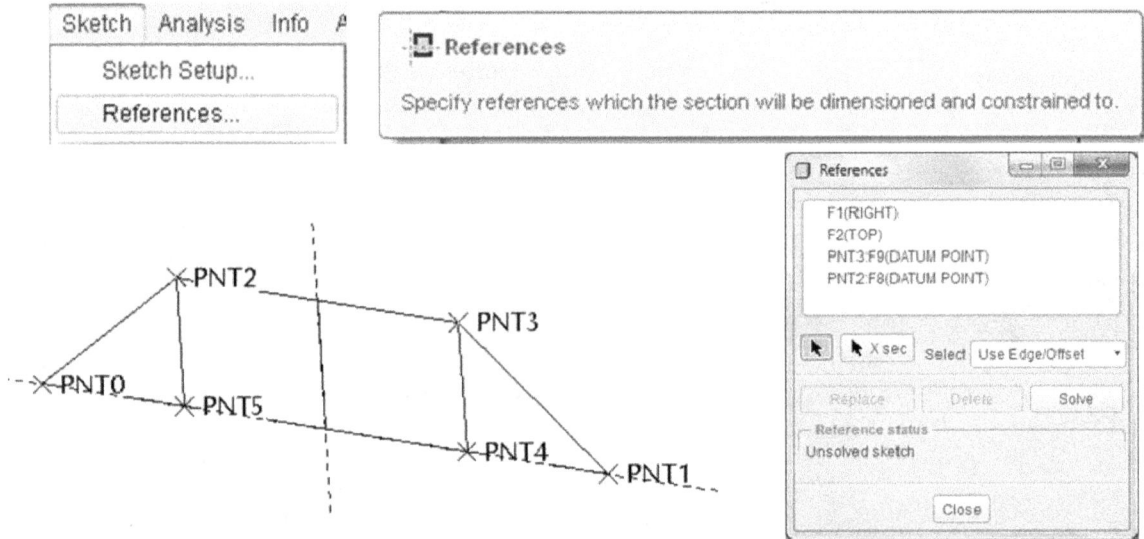

Pick the icon of **Line**, and sketch an inclined line by connecting **PNT3** and the middle point of the lower horizontal line. Sketch an inclined line by connecting **PNT2** and the middle point of the lower horizontal line. Click the icon of **Done** or the check mark when completing the sketch of the 2 inclined lines.

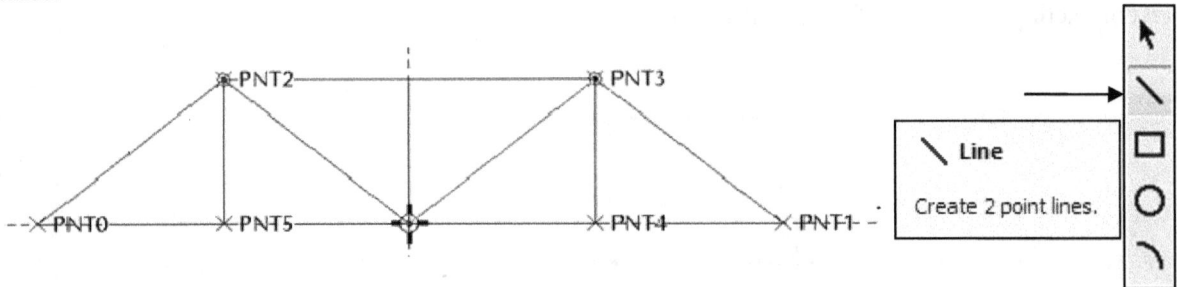

Now let us use **SWEEP** to create the first part of the truss structure with a rectangle section. The size is 150 x 75 mm.

From the main toolbar, **Insert > Sweep > Protrusion**, the window of Protrusion: Sweep appears.

Pick **Select Traj > One By One** and pick the 2 horizontal lines and the 2 inclined lines, as shown while holding down the **Ctrl** key > **OK** > **Done**

Insert Analysis Info Applications Tools Window He

/disable/enable the constraint L

Extrude...

Revolve...

Sweep ▶ Protrusion...

Menu Manager
▼ SWEEP TRAJ
Sketch Traj
Select Traj

Menu Manager
▼ CHAIN
One By One
Tangnt Chain
Curve Chain
Bndry Chain
Surf Chain
Intent Chain
Select
Done
Quit

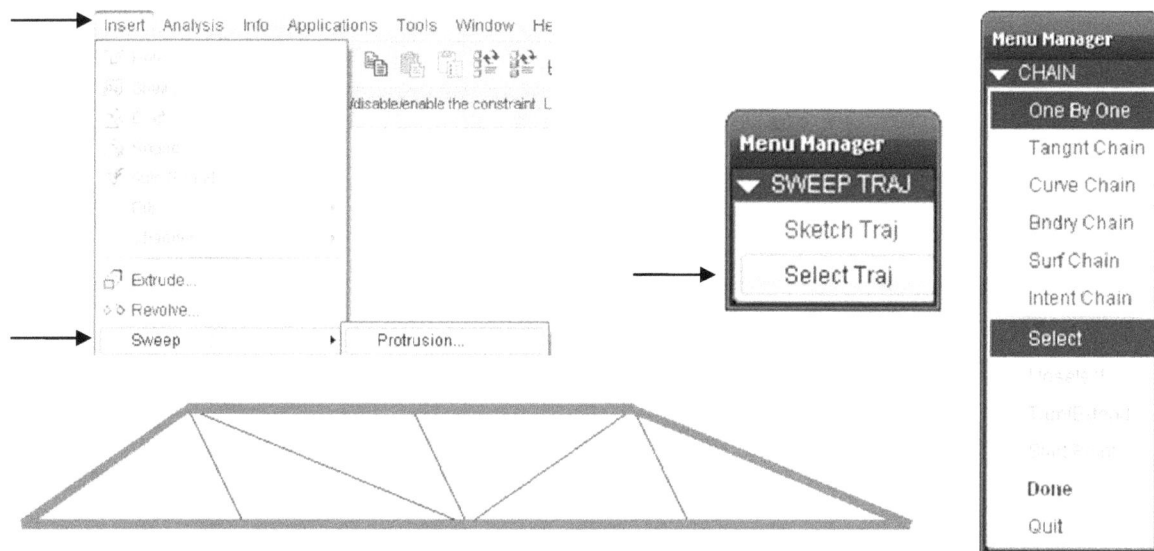

In the pop up window, select **No Inn Fcs > Done**.

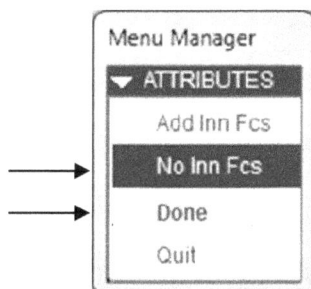

Menu Manager
▼ ATTRIBUTES
Add Inn Fcs
No Inn Fcs
Done
Quit

Pick the icon of **Rectangle**, and sketch a rectangle with dimensions: 150 x 75, as shown below:

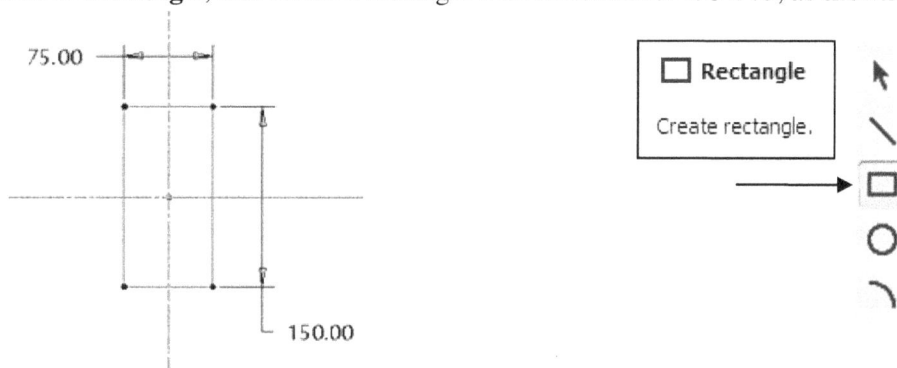

75.00

150.00

☐ **Rectangle**
Create rectangle.

Upon completing the sketch, pick the icon of **Done** from the toolbar of sketcher, and click **OK**.

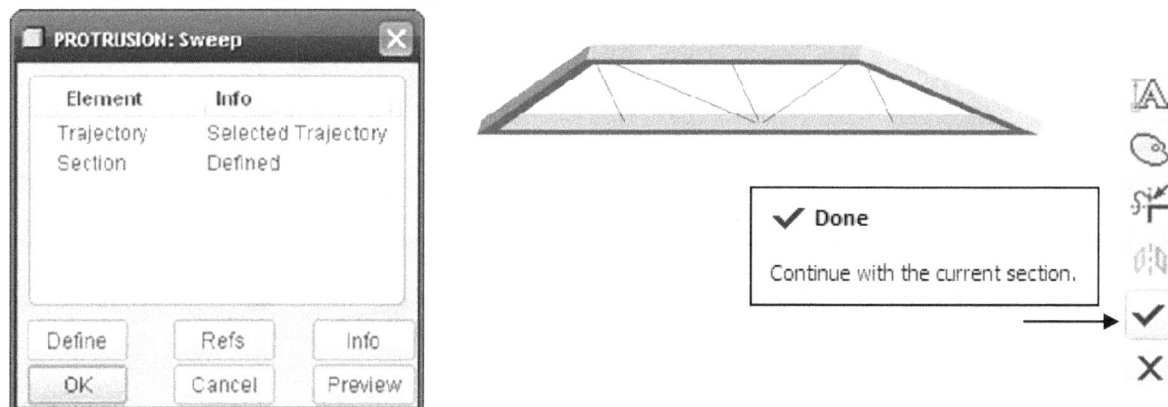

PROTRUSION: Sweep

Element	Info
Trajectory	Selected Trajectory
Section	Defined

Define	Refs	Info
OK	Cancel	Preview

✔ **Done**
Continue with the current section.

Now let us use **SWEEP** to create the second part of the truss structure with the rectangle section. From the menu bar, **Insert > Sweep > Protrusion**, the window of Protrusion: Sweep appears. Pick **Select Traj > One By One** and pick one of the 3 vertical lines > **OK > Done.** From the window of **Attributes** > select **Merge Ends** (because both ends are not free) > **Done.**

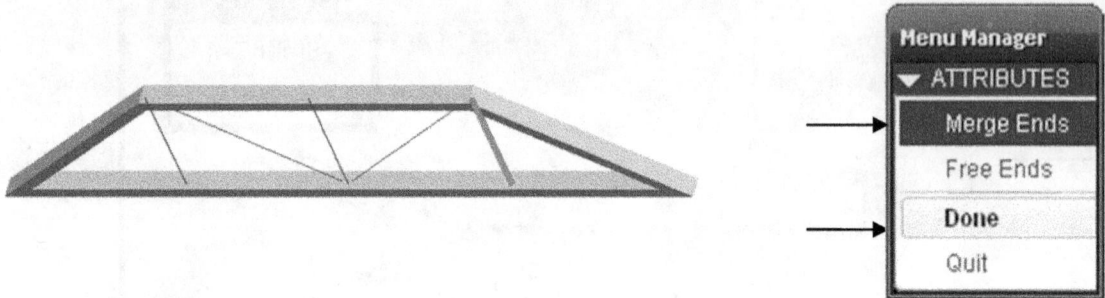

Pick the icon of **Rectangle**, and sketch a rectangle with dimensions: 150 x 75, as shown below:

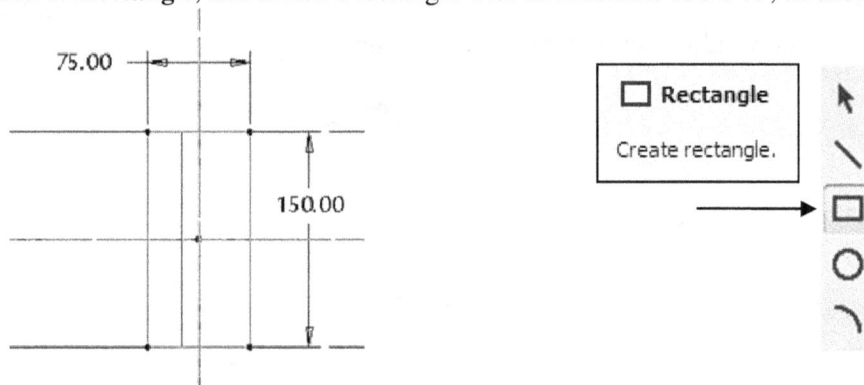

Upon completing the sketch, pick the icon of **Done** from the toolbar of sketcher, and click **OK**.

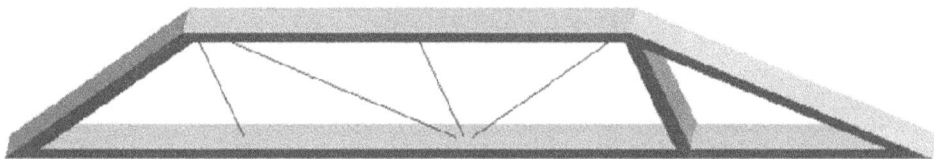

Repeat this sweep operation twice to create the other 2 vertical beams.

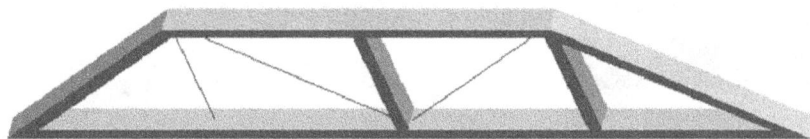

Now let us use **SWEEP** to create the third part of the truss structure with the rectangle section.
From the menu bar, **Insert > Sweep > Protrusion**, the window of Protrusion: Sweep appears.
Pick **Select Traj > One By One** and pick the 2 inclined lines > **OK > Done.**
From the window of **Attributes** > select **Merge Ends** (because all ends are not free) > **Done.**

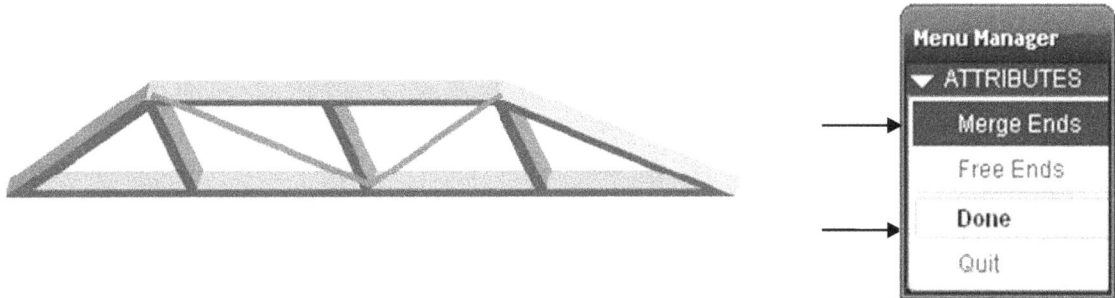

Pick the icon of **Rectangle**, and sketch a rectangle with dimensions: 150 x 75, as shown below:

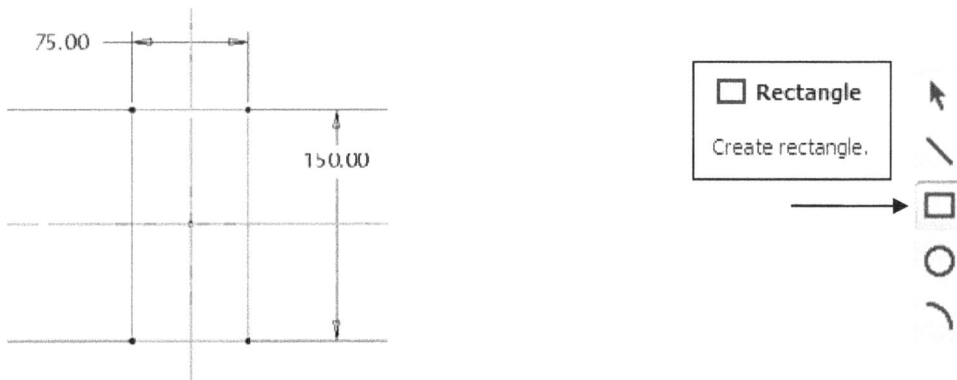

Upon completing the sketch, pick the icon of **Done** from the toolbar of sketcher, and click **OK**.

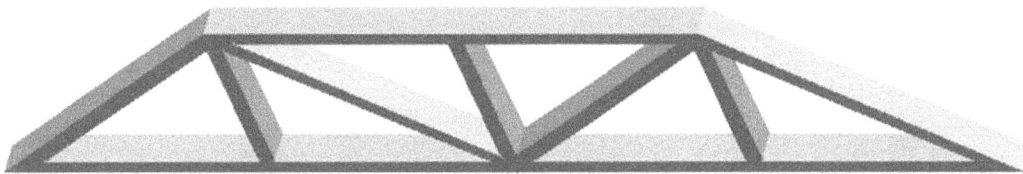

Create a datum plane. Click the icon of Datum Plane Tool. Specify 525 as the offset value and click **OK**.

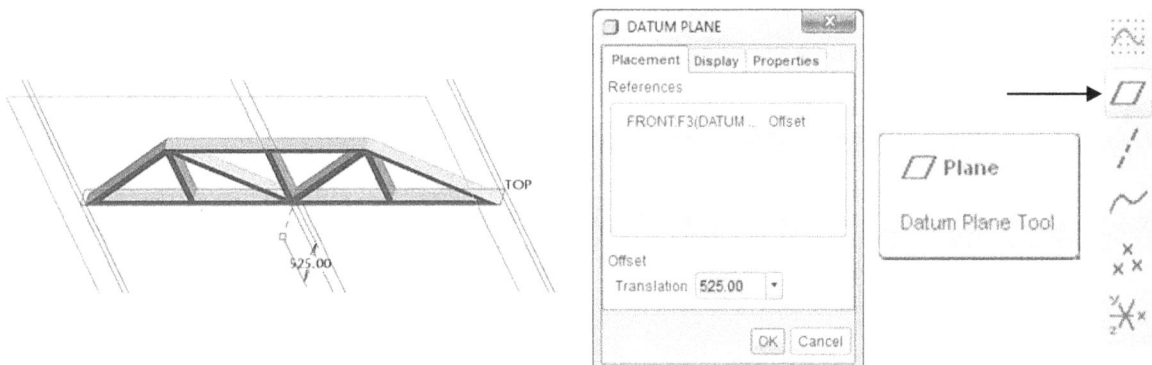

Use the mirror operation to get another set of the features just created. First select those features listed in the model tree while holding down the **Ctrl** key. Click the icon of **Mirror**. Select **DTM1** and click the icon of **Apply and Save** to complete the mirror operation.

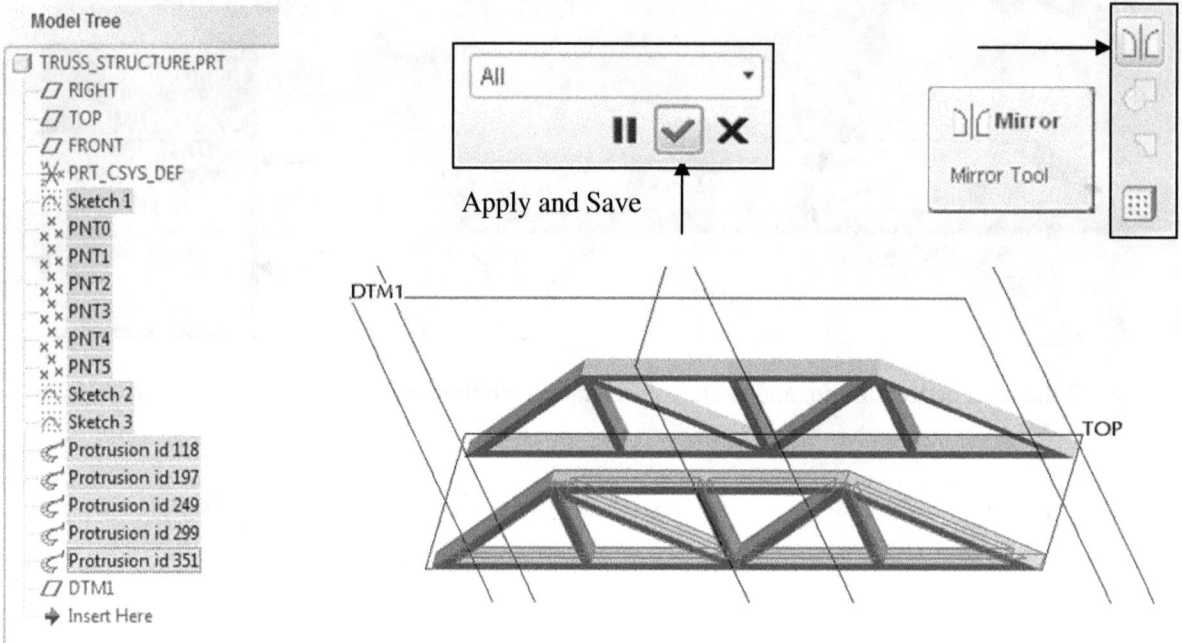

Now we need to create those beams to connect the constructed features, which are separated due to the mirror operation.

Pick up the icon of **Extrude**. Use **Up to Next** as the choice for defining the depth value. Activate **Placement > Define**.

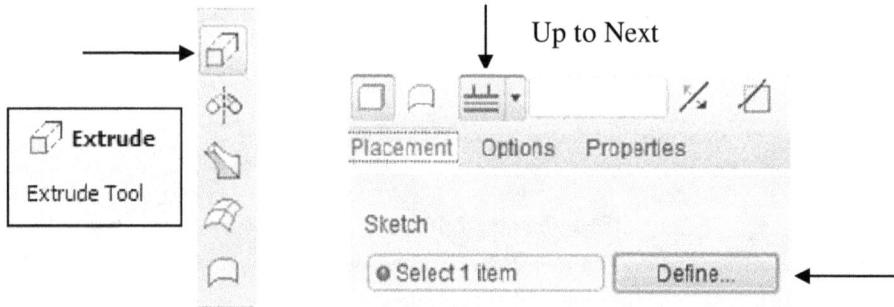

Select the surface, as shown below, as the sketch plane and click the box of **Sketch** to accept the **RIGHT** datum plane as the default reference to orient the sketch plane, as illustrated below:

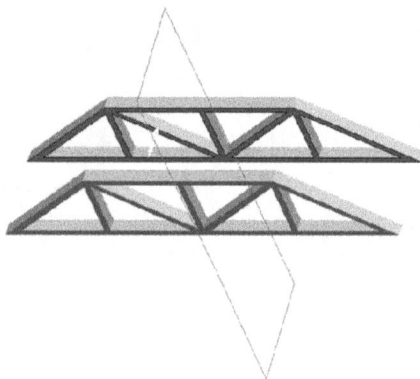

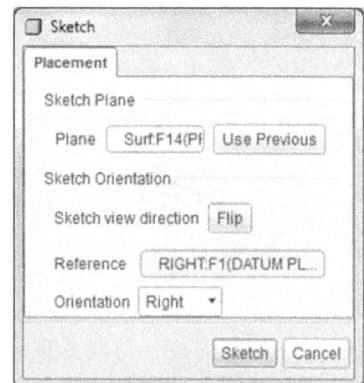

Click **Sketch** from the top menu and click **References**. Select the 4 lines as new references, as shown.

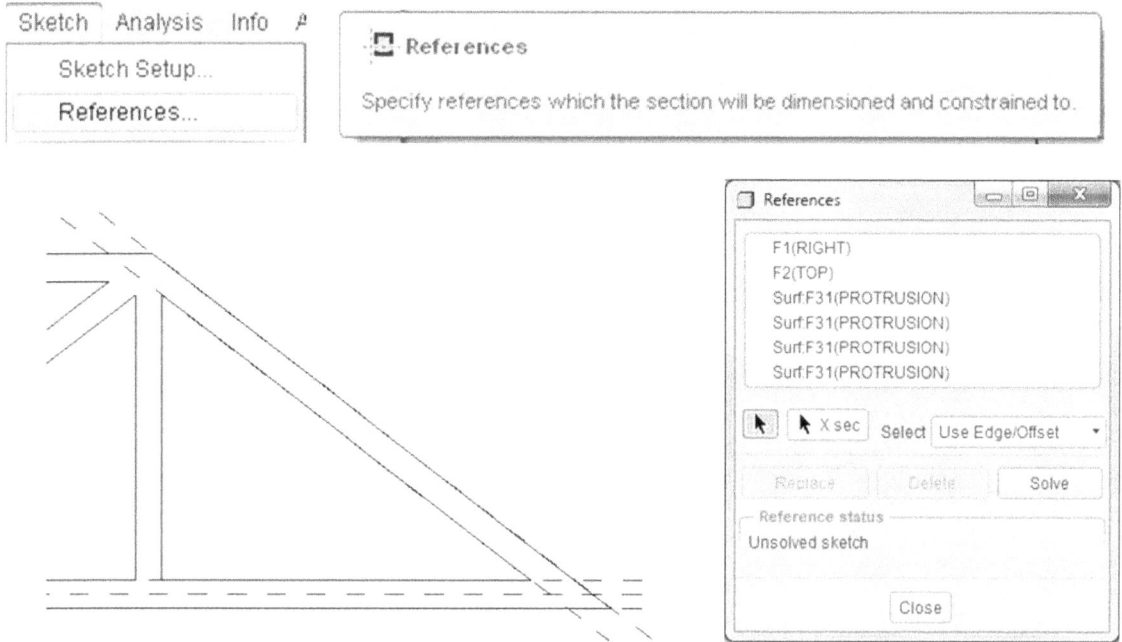

Pick the icon of **Line**, and sketch a parallel shape as shown below. Click the icon of **Done** or the check mark when completing the sketch. Click the icon of **Apply and Save**.

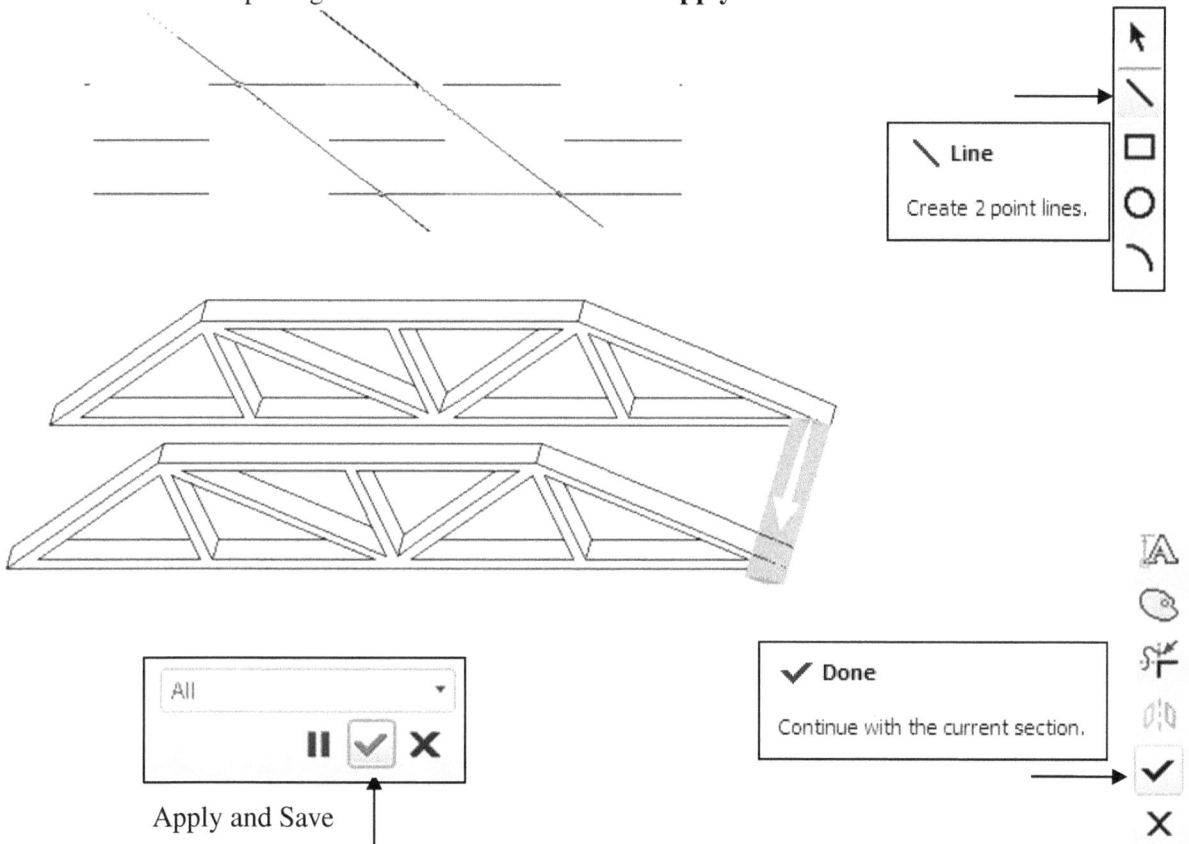

Apply and Save

Use the mirror operation to get the beam on the other side. First highlight the beam just created. Click the icon of **Mirror**. Select **RIGHT** and click the icon of **Apply and Save** to complete the mirror operation.

Protrusion id 1077
Protrusion id 1104
Protrusion id 1131
Protrusion id 1158
Protrusion id 1185
Extrude 1
Insert Here

Mirror
Mirror Tool

All

Apply and Save

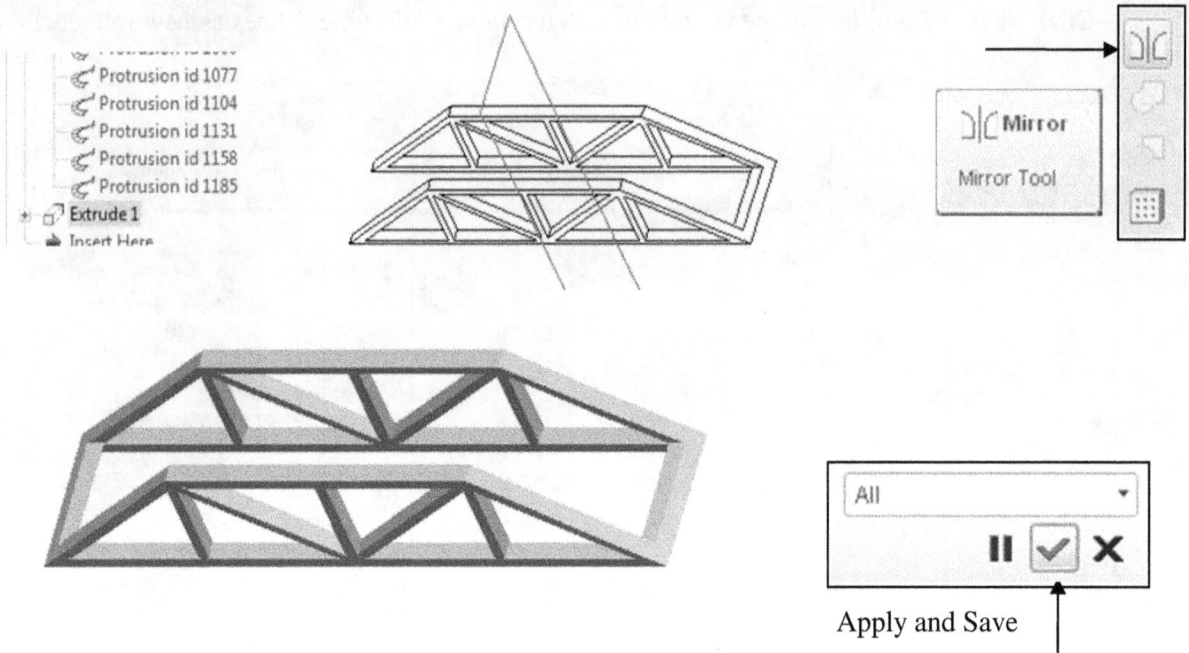

Pick up the icon of **Extrude**. Use **Up to Next** as the choice for defining the depth value. Activate **Placement > Define.**

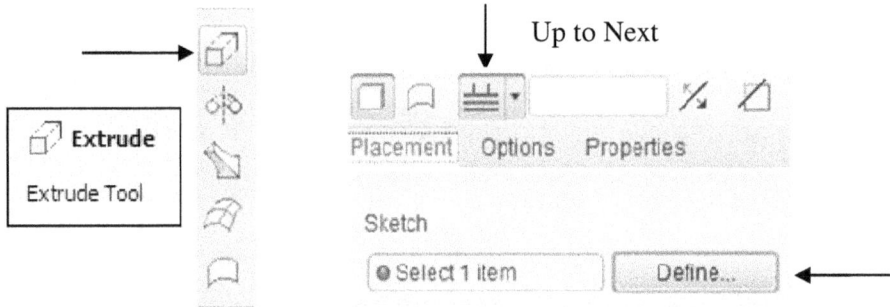

Up to Next

Extrude
Extrude Tool

Placement   Options   Properties

Sketch

Select 1 item   Define...

Select the surface, as shown below, as the sketch plane and click the box of **Sketch** to accept the **RIGHT** datum plane as the default reference to orient the sketch plane, as illustrated below:

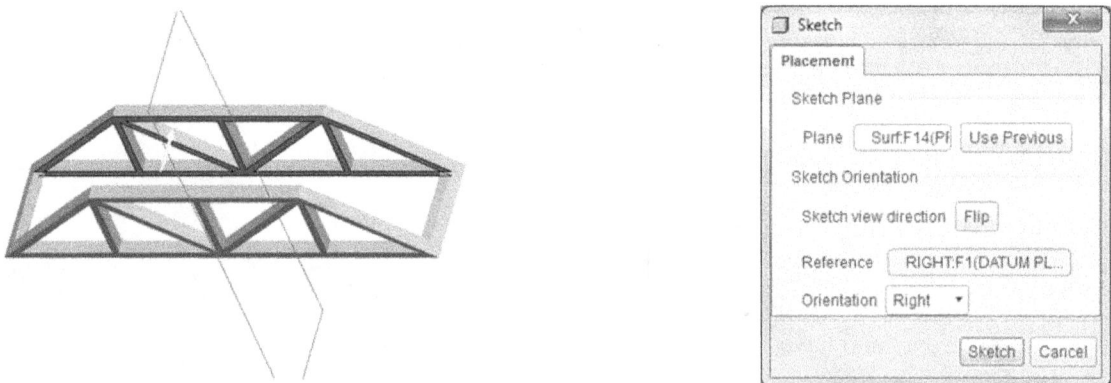

Sketch

Placement

Sketch Plane

Plane   Surf:F14(Pl   Use Previous

Sketch Orientation

Sketch view direction   Flip

Reference   RIGHT:F1(DATUM PL...

Orientation   Right

Sketch   Cancel

Click **Sketch** from the top menu and click **References**. Select the 6 lines as new references, as shown.

Sketch   Analysis   Info   A

Sketch Setup...

References...

References

Specify references which the section will be dimensioned and constrained to.

References

F1(RIGHT)
F2(TOP)
Surf:F33(PROTRUSION)
Surf:F33(PROTRUSION)
Surf:F31(PROTRUSION)
Surf:F31(PROTRUSION)
Surf:F31(PROTRUSION)
Surf:F14(PROTRUSION)

Select   Use Edge/Offset

Replace   Delete   Solve

Reference status
Unsolved sketch

Close

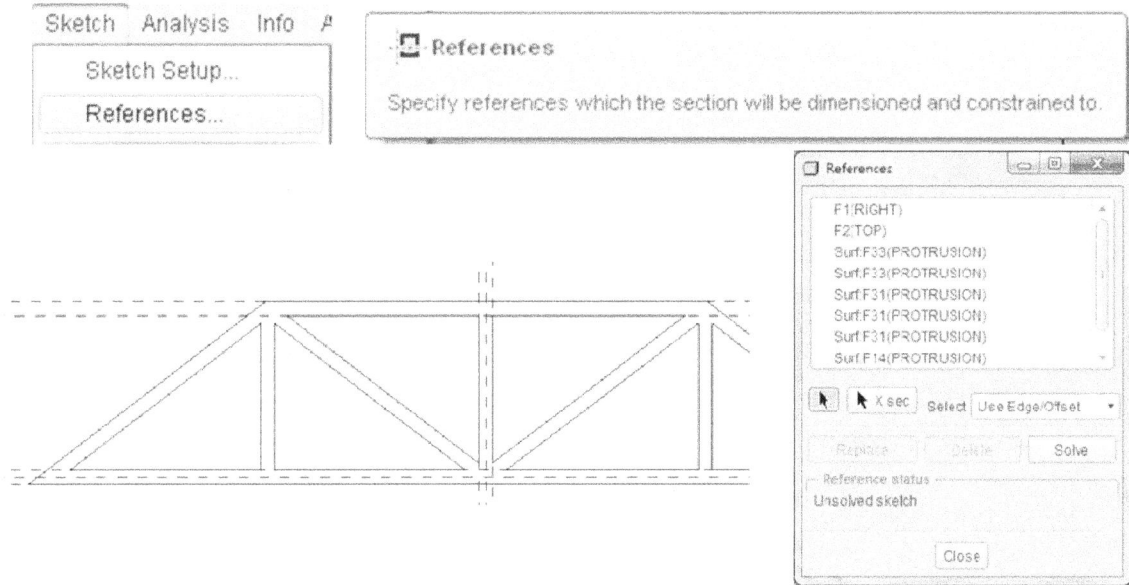

Pick the icon of **Rectangle**, and sketch 2 rectangles as shown below. Click the icon of **Done** or the check mark when completing the sketch. Click the icon of **Apply and Save**.

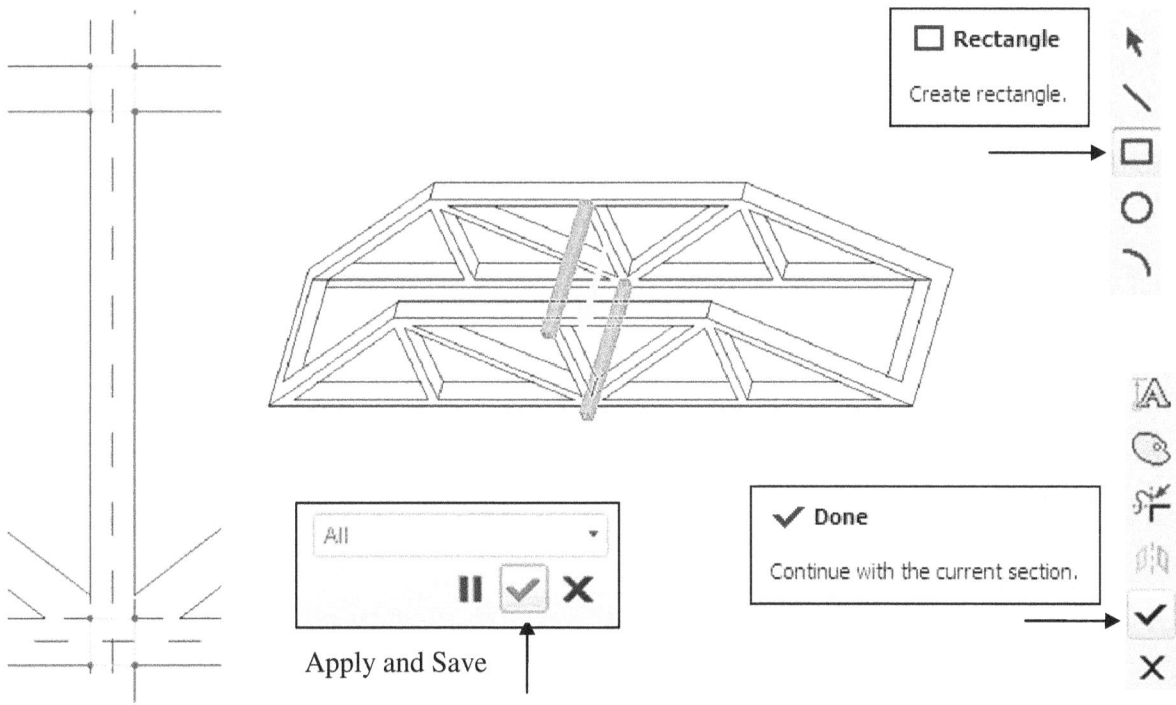

☐ **Rectangle**

Create rectangle.

All

Apply and Save

✓ **Done**

Continue with the current section.

Pick up the icon of **Extrude** displayed on the toolbar of feature creation. Use **Up to Next** as the choice for defining the depth value. Activate **Placement > Define.**

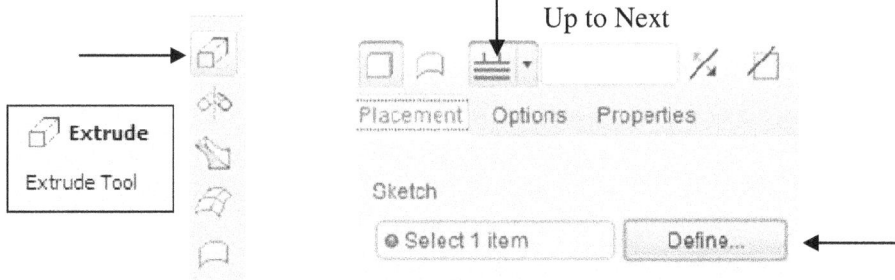

Up to Next

Placement   Options   Properties

☐ **Extrude**

Extrude Tool

Sketch

⊙ Select 1 item   Define...

Select the surface, as shown below, as the sketch plane and click the box of **Sketch** to accept the **RIGHT** datum plane as the default reference to orient the sketch plane, as illustrated below:

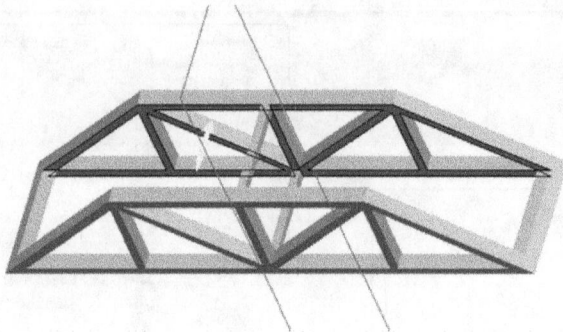

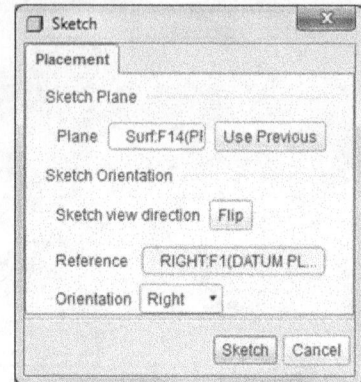

Click **Sketch** from the top menu and click **References**. Select the 8 lines as new references, as shown.

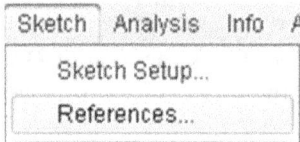

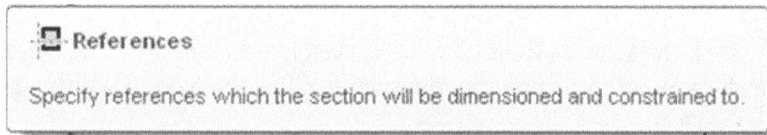

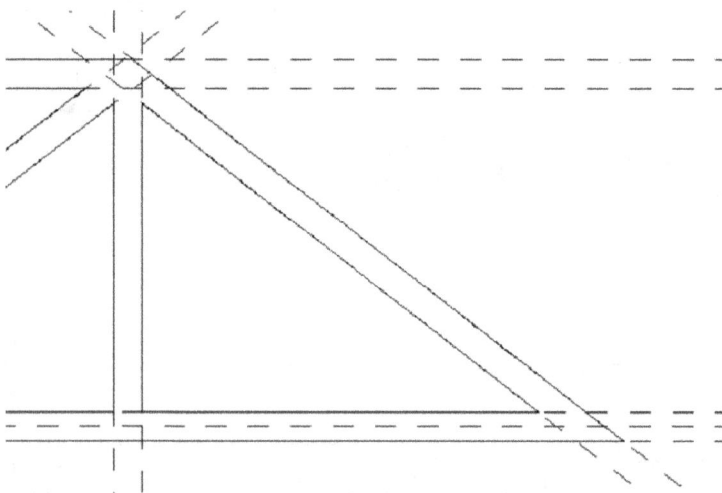

Pick the icon of **Rectangle**, and sketch 1 rectangle as shown below.

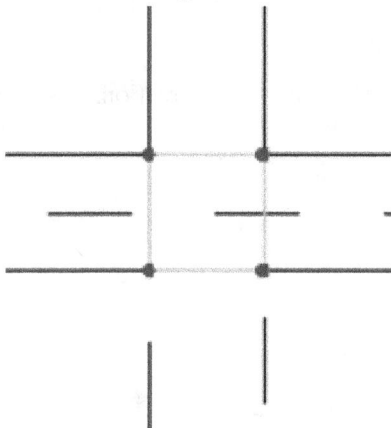

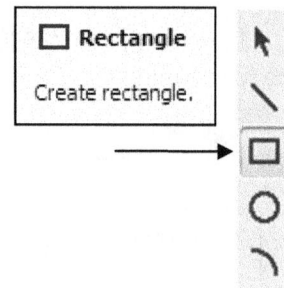

Pick the icon of **Line**, and sketch a polygon shape, as shown below. Click the icon of **Done** or the check mark when completing the sketch. Click the icon of **Apply and Save**.

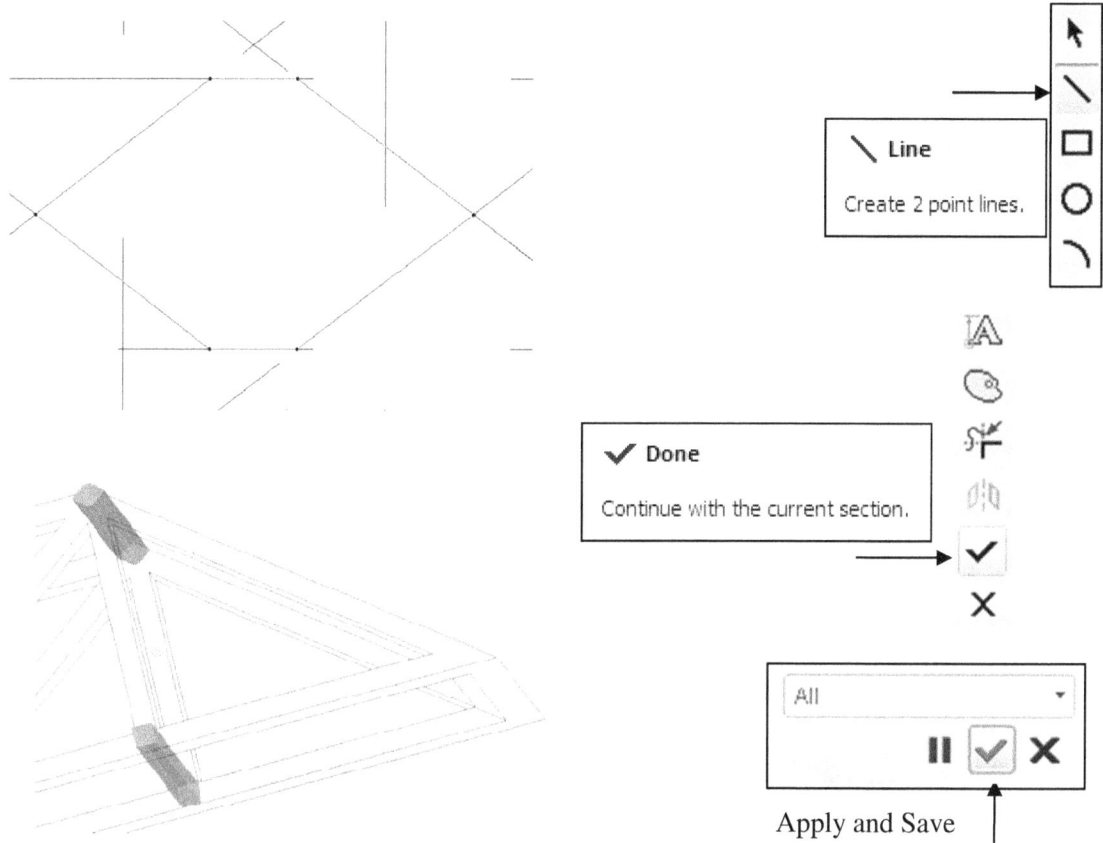

\ **Line**

Create 2 point lines.

✓ **Done**

Continue with the current section.

All

Apply and Save

Use the mirror operation to get the beam on the other side. First highlight the beam just created. Click the icon of **Mirror**. Select **RIGHT** and click the icon of **Apply and Save** to complete the mirror operation.

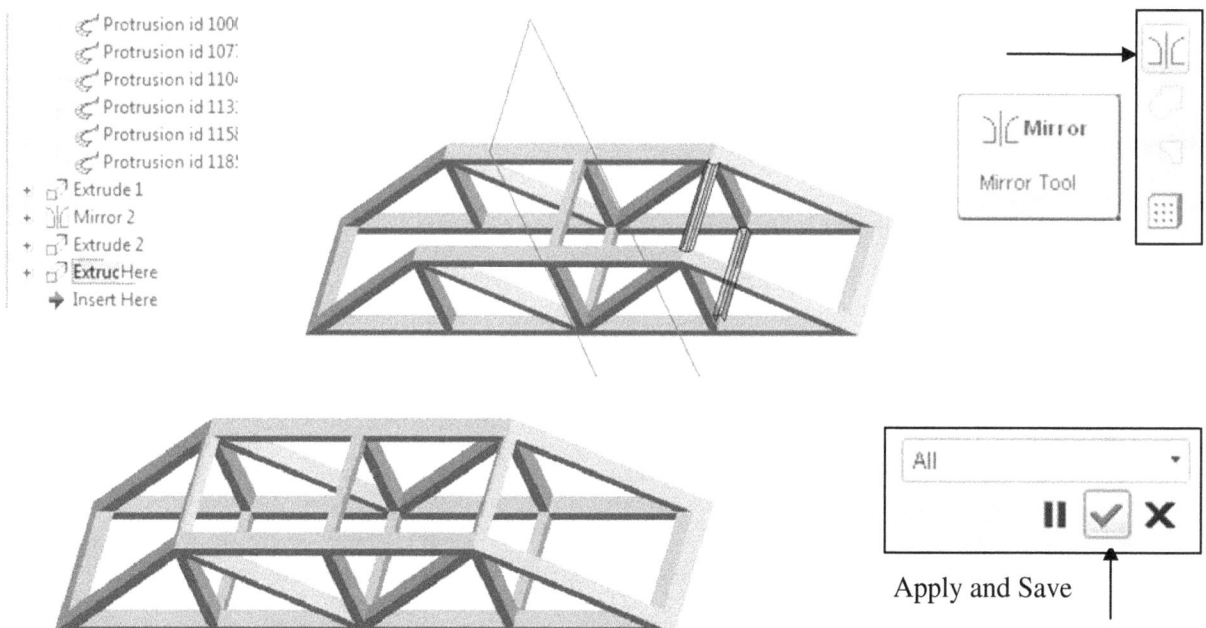

Protrusion id 100(
Protrusion id 107:
Protrusion id 110·
Protrusion id 113:
Protrusion id 115:
Protrusion id 118!
+ Extrude 1
+ Mirror 2
+ Extrude 2
+ **Extruc**Here
Insert Here

)|(**Mirror**

Mirror Tool

All

Apply and Save

Step 2: under the Pro/Mechanica environment, perform FEA.
From the top menu, click **Applications > Mechanica > Continue > Structure > OK.**

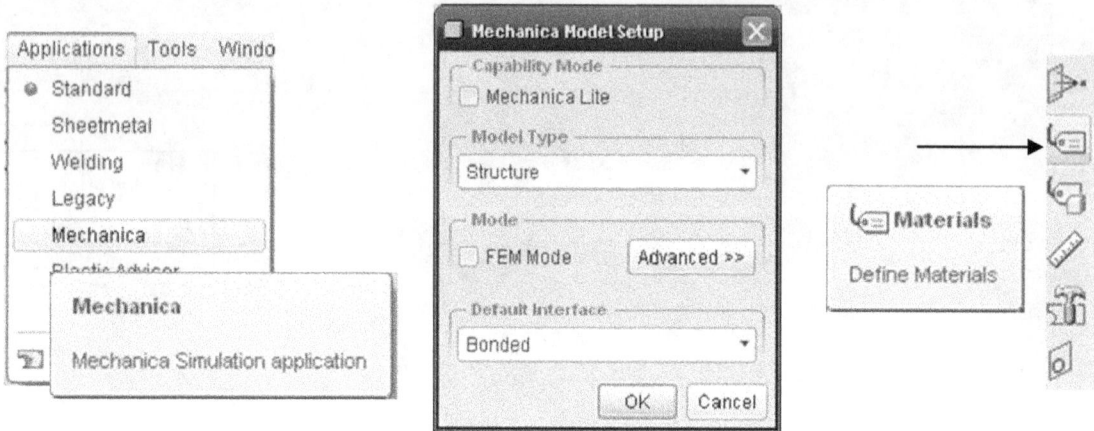

From the toolbar of functions, select the icon of **Materials >** select **PVC** from the left side called Materials in Library > click the directional arrow so that the selected material type goes to the right side called Materials in Model.

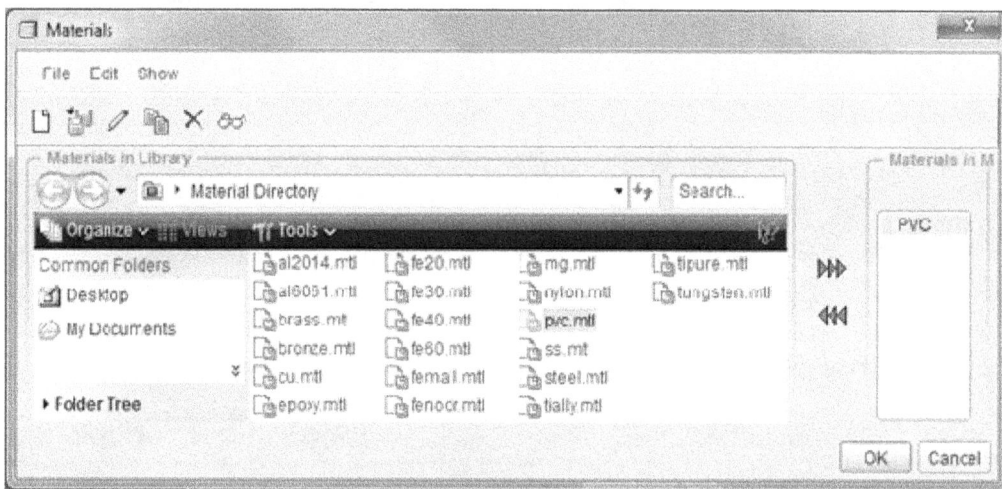

Click the icon of **Material Assignment**, the pop-up window indicates there is one component and the material type is PVC. Just click **OK.**

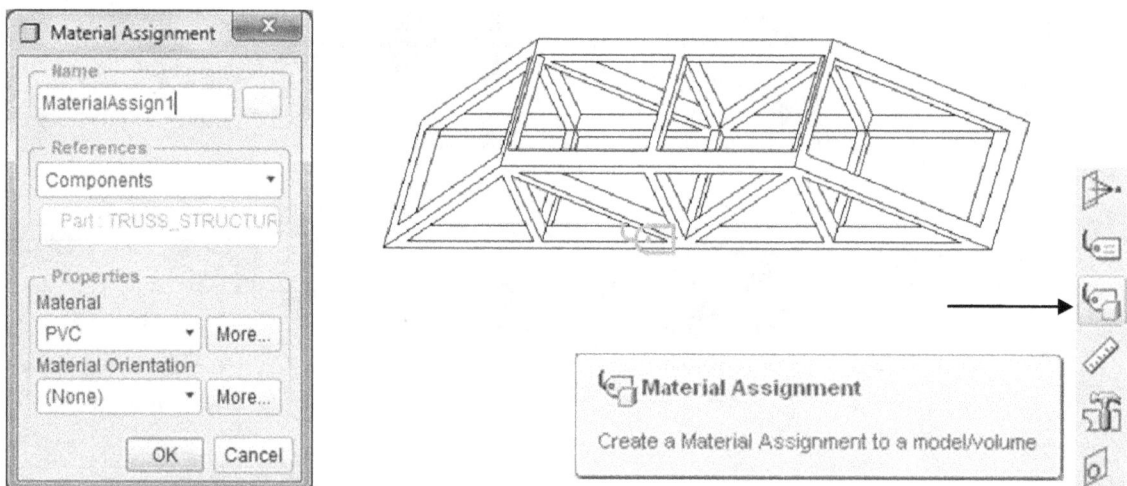

To define the fixed end constraint on the bottom surface, click the icon of **Displacement Constraint > Surface(s) >** pick the bottom surface > OK to set all 6 degrees of freedom equal to zero.

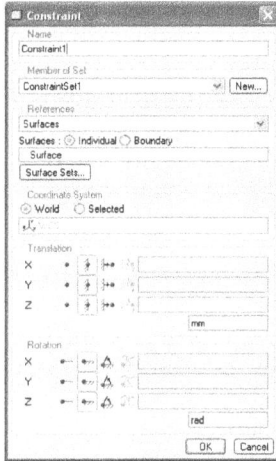

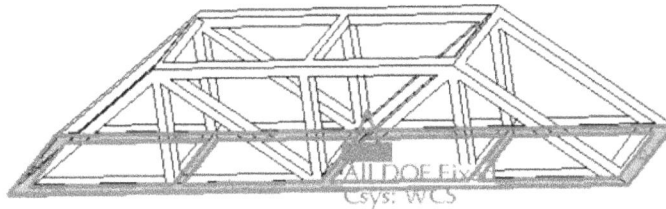

To define the load condition, click.the icon of **New Force/Moment > Surface(s) >** pick the top surface. Specify the force as 10000 N along the negative Y direction > **OK**.

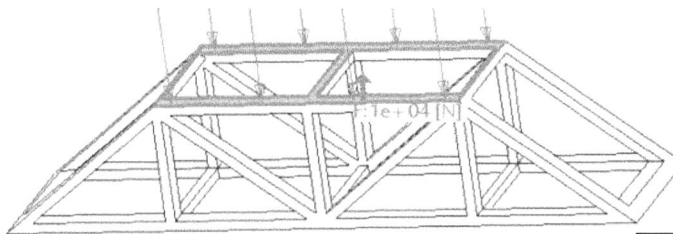

Select the icon of **Mechanica Analyses/Studies** from the main toolbar > **File > New Static >** type *truss_structure* as the name of the analysis folder > **OK.** Click the box of **Run.**

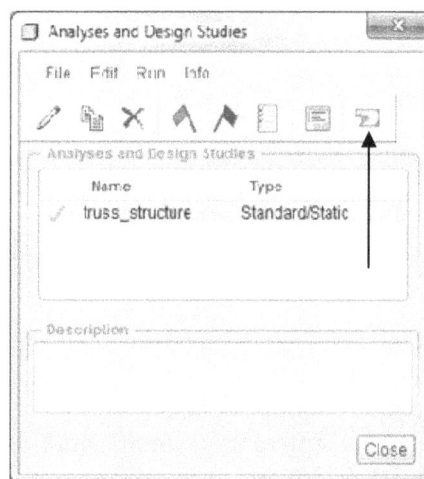

To visualize the displacement distribution, click the icon of **Review results > Displacement > Magnitude > OK and Show.** The maximum value is 0.67 mm.

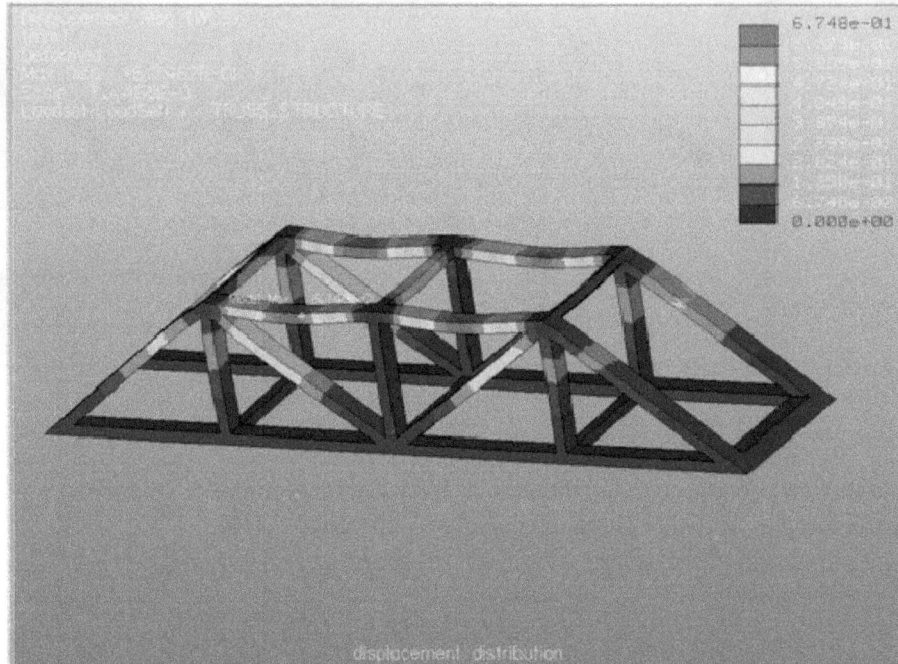

To visualize the distribution of the maximum principal stress, click the icon of **Review results > Stress > Max Principal > OK and Show.** The maximum value is 1.528 MPa.

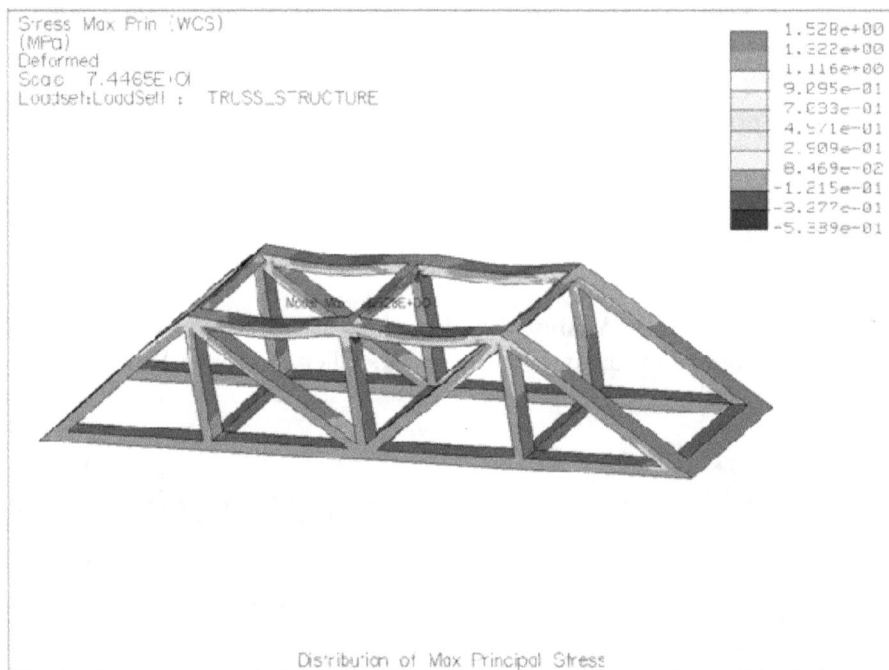

In the following, we use the beam elements to re-visit the roof structure. From the mail toolbar, select the icon of **Create** a new object. Make sure the **Part** mode is selected. Type *truss_beam* and clear the box of **use default template > OK**

Select the unit system to be **mmns_part_solid** > type *truss beam* in **DESCRIPTION** and *student* in **MODELED_BY > OK**.

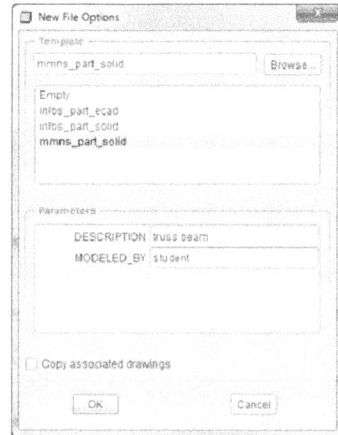

Select the icon of **Sketch Tool.** Pick the **FRONT** datum plane as the sketch plane, and accept the default orientation setting, which uses the **RIGHT** datum plane and position it at the right side > **Sketch.**

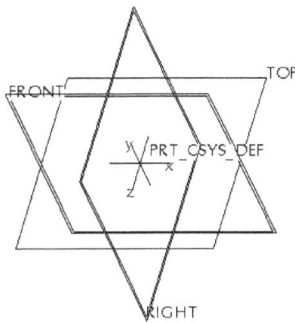

We first sketch a vertical centerline passing through the origin of the coordinate system, as shown.

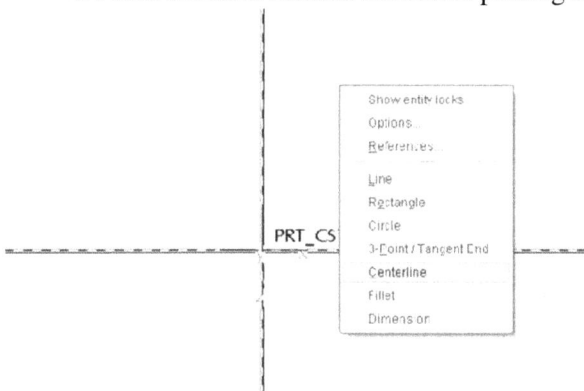

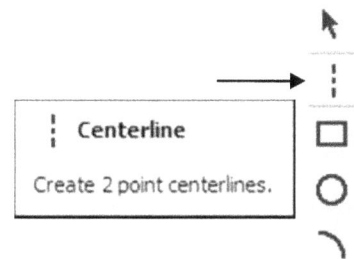

Pick the icon of **Line**, and sketch a horizontal line along the x-axis, which is symmetric about the vertical centerline with the dimension equal to 4800.

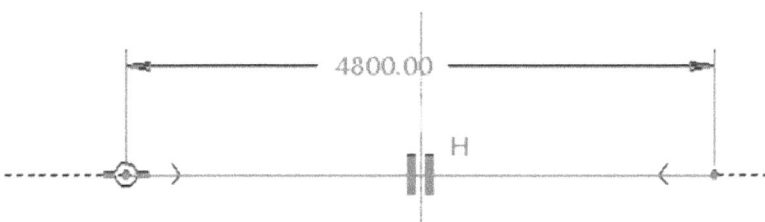

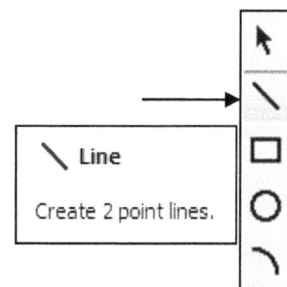

Sketch 4 more lines, as shown. Make sure the 4 lines are symmetric about the vertical centerline. The two dimensions are 2400 and 900, respectively.

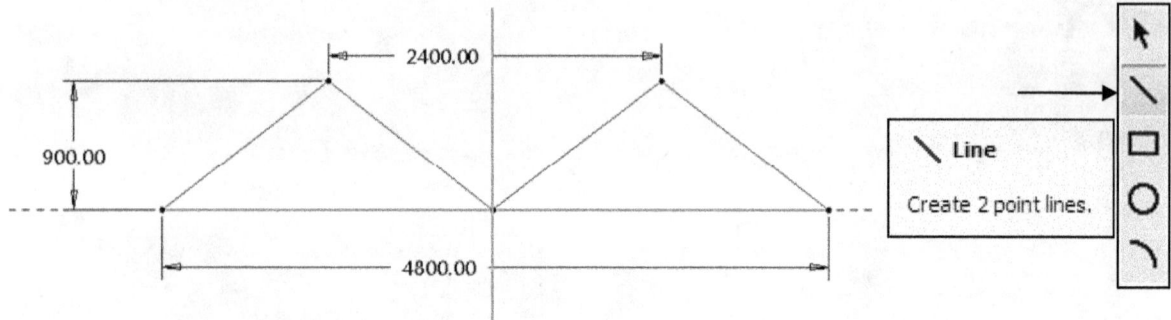

Upon completion, select the icon of **Done.** The sketched datum curves (actually, they are straight lines) represent the 5 beams of the truss structure.

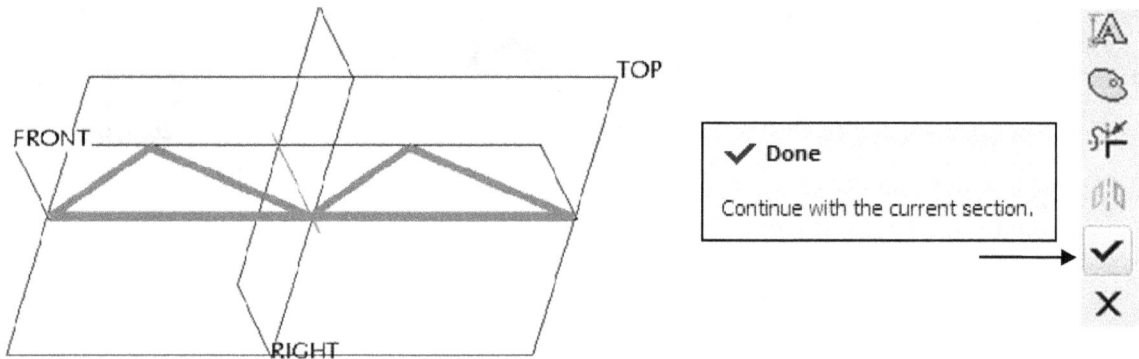

Now let us create one more datum curve. Select the icon of **Sketch Tool.** Pick **Use Previous.**

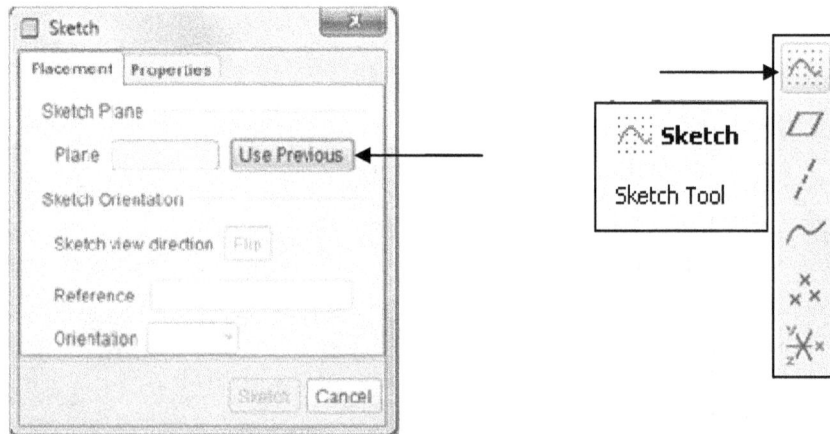

Click **Sketch** from the top menu and click **References**.

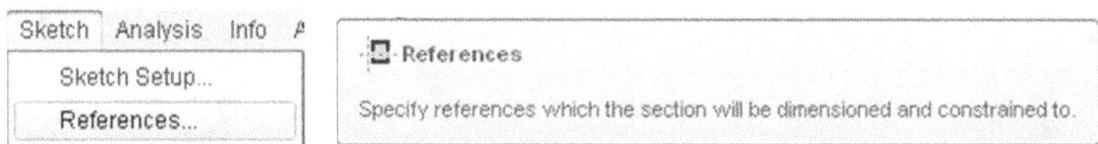

Pick the 4 inclined segments as 4 new references. Click the icon of **Line**, and sketch a horizontal line connecting the two corners. Click the icon of **Done**.

Add those 4 new references

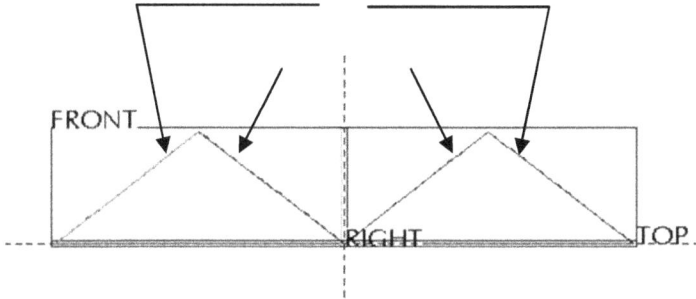

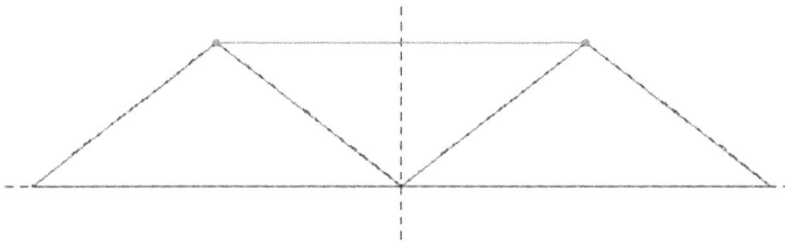

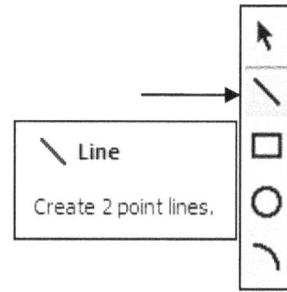

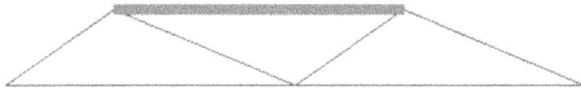

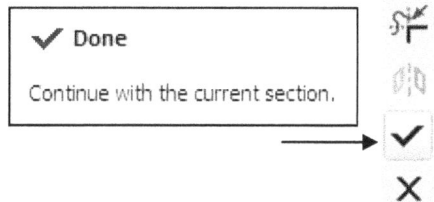

References

F1(RIGHT)
F2(TOP)
Curve:F5(SKETCH_1)
Curve:F5(SKETCH_1)
Curve:F5(SKETCH_1)
Curve:F5(SKETCH_1)

X sec   Select   Use Edge/Offset

Replace   Delete   Solve

Reference status
Unsolved sketch

Close

FRONT

RIGHT   TOP

Line
Create 2 point lines.

Done
Continue with the current section.

Create a datum plane. Click the icon of **Datum Plane Tool**. Pick the **FRONT** datum plane. Specify 525 as the offset value and click **OK**.

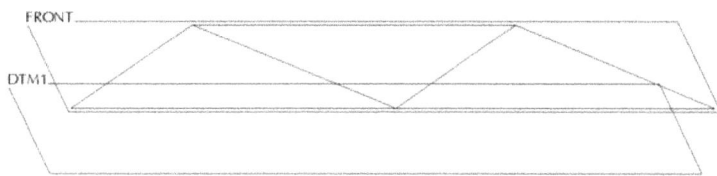

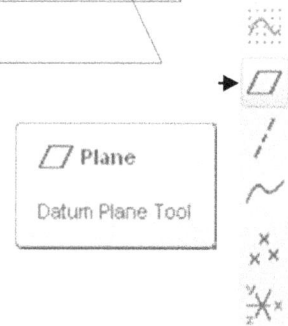

DATUM PLANE

Placement   Display   Properties

References

FRONT:F3(DATUM ...   Offset

Offset
Translation   525.00

OK   Cancel

FRONT

DTM1

Plane
Datum Plane Tool

Highlight Sketch 1 and Sketch 2 listed in the model tree while holding down the **Ctrl** key. Click the icon of **Mirror**. Select **DTM1** and click the icon of **Apply and Save** to complete the mirror operation.

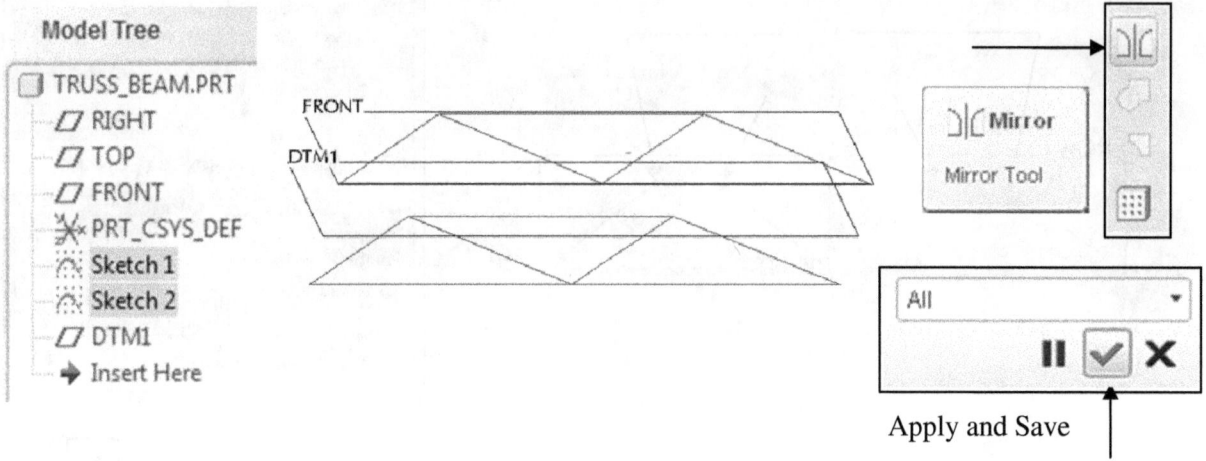

Now let us create 5 datum curves on the **TOP** datum plane. Select the icon of **Sketch Tool**. Pick the **TOP** datum plane > Sketch to accept the default setting for orientation.

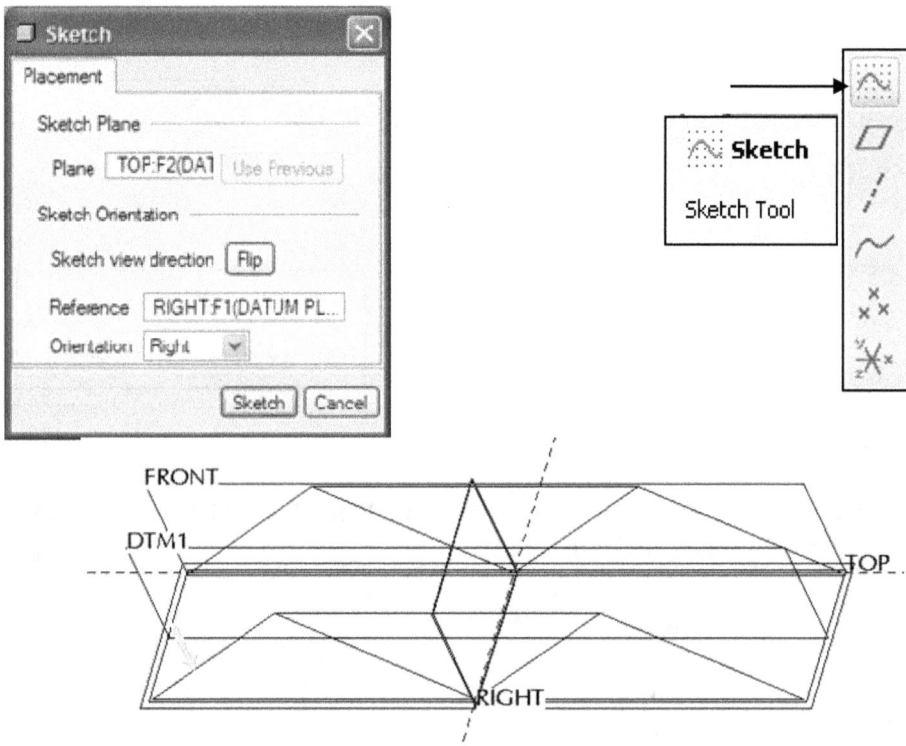

Click **Sketch** from the top menu and click **References**.

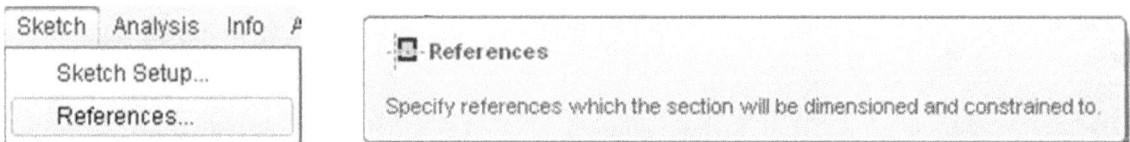

Pick the 6 segments as 6 new references Click the icon of **Line**, and sketch 5 vertical lines for connections. Click the icon of **Done**.

Add those 6 new references

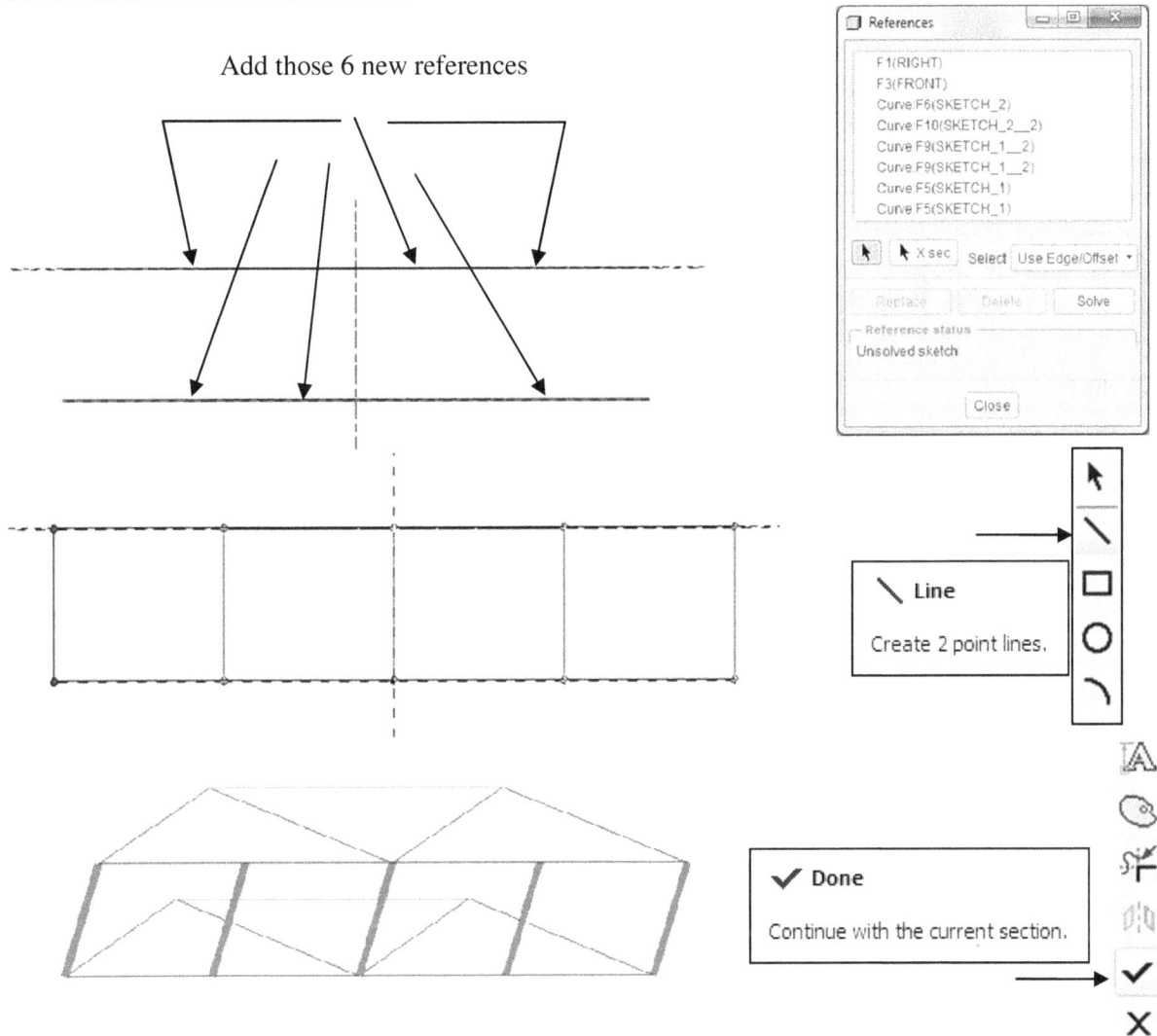

Create a datum plane. Click the icon of **Datum Plane Tool**. Pick the TOP datum plane from the model tree. Change Offset to Parallel. While holding down the Ctrl key. Pick a corner as shown> **OK**.

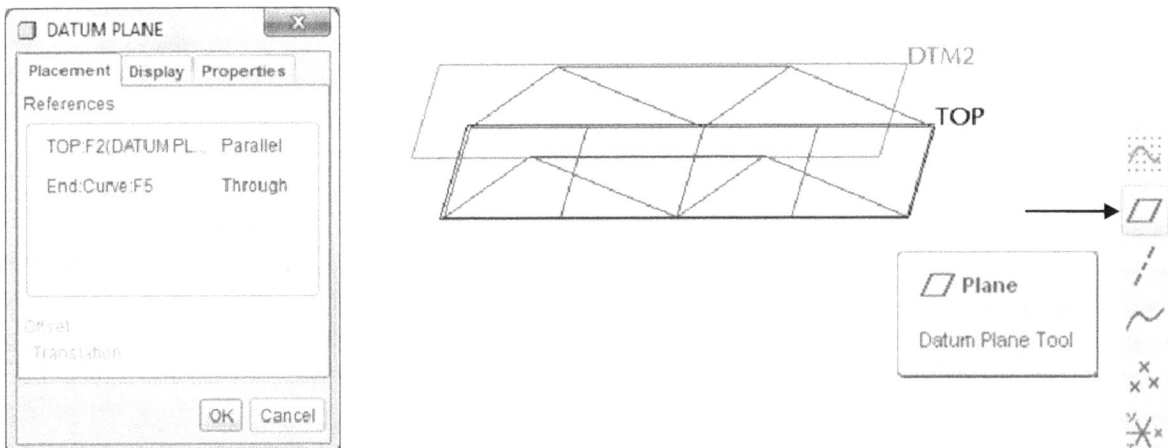

Now let us create 3 datum curves on the created datum plane. Select the icon of **Sketch Tool**. Pick DTM2 > Sketch to accept the default setting for orientation.

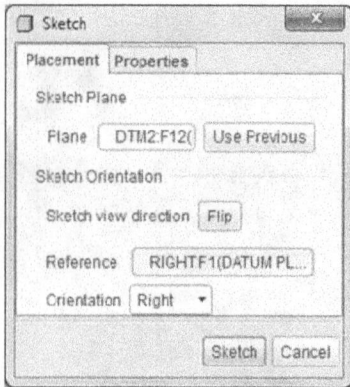

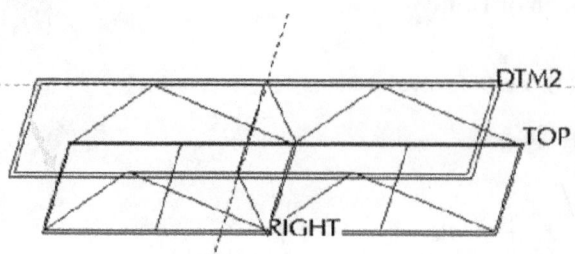

Click **Sketch** from the top menu and click **References**. Pick the 6 segments as 6 new references Click the icon of **Line**, and sketch 5 vertical lines for connections. Click the icon of **Done**.

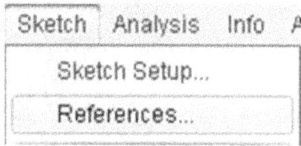

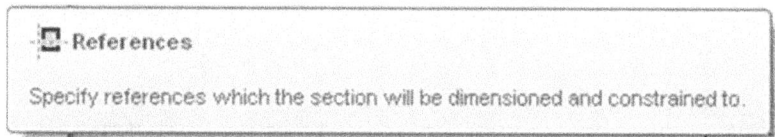

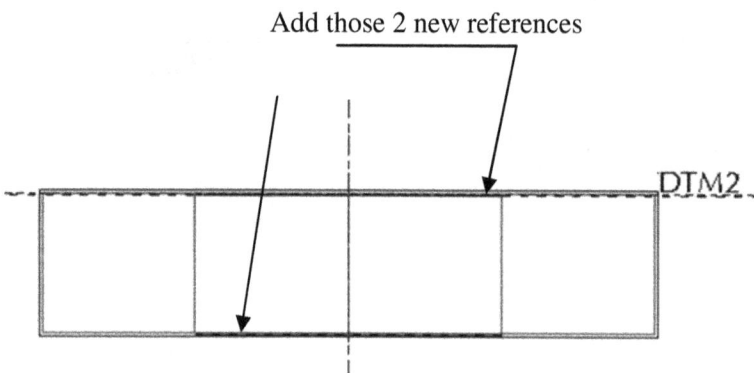

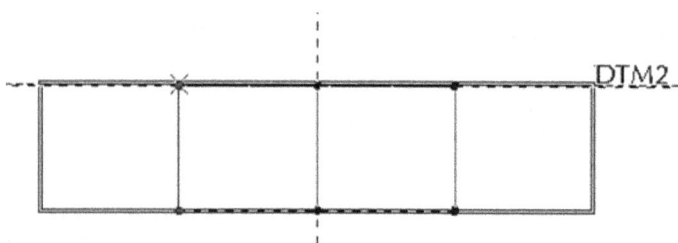

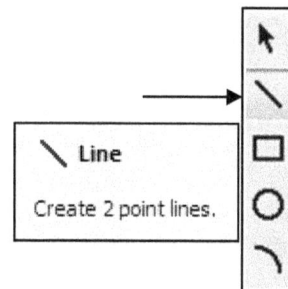

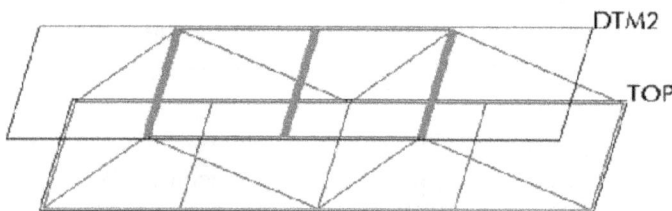

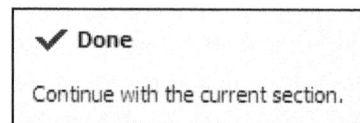

Now let us create 3 datum curves on the **FRONT** datum plane. Select the icon of **Sketch Tool.** Pick **FRONT > Sketch** to accept the default setting for orientation**.**

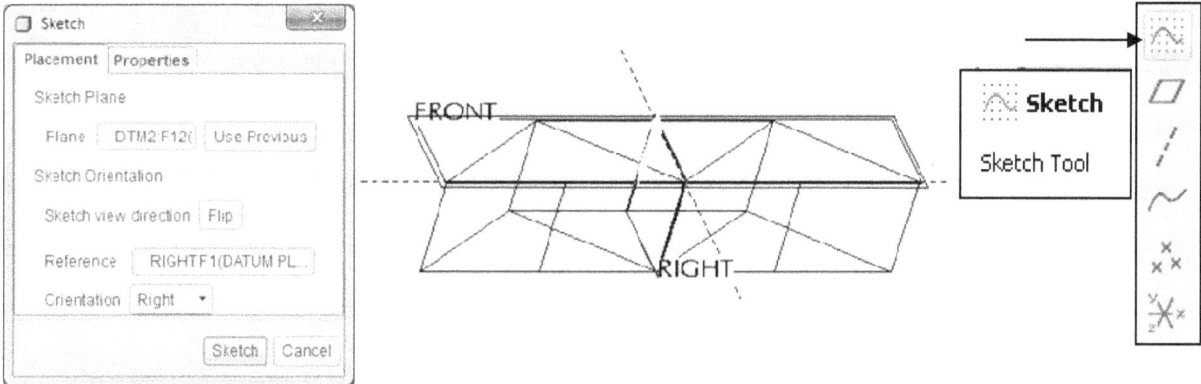

Click **Sketch** from the top menu and click **References**. Pick the 2 segments as 2 new references. Click the icon of **Line**, and sketch 3 vertical lines for connections. Click the icon of **Done.**

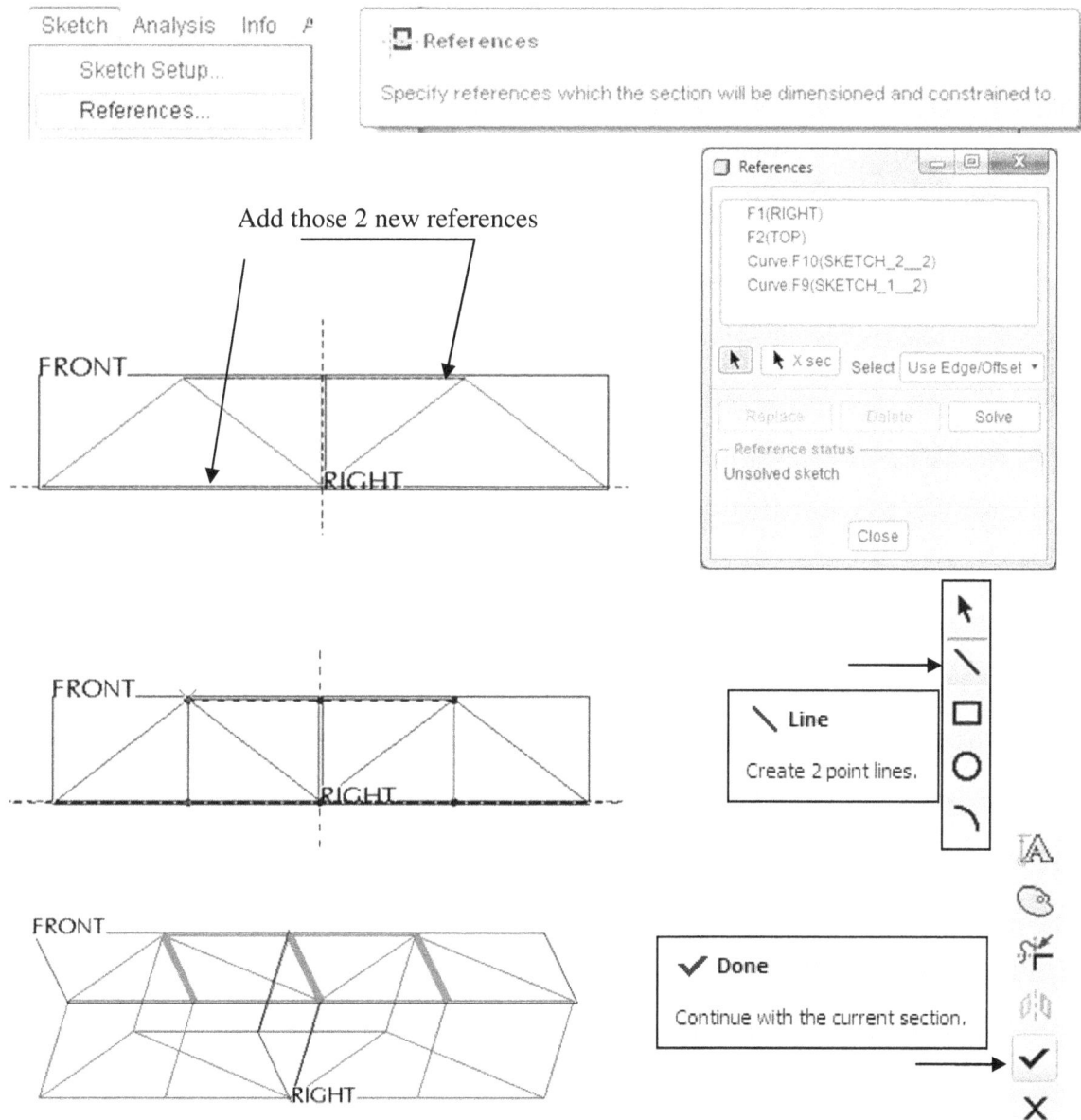

Add those 2 new references

Create a datum plane. Click the icon of **Datum Plane Tool**. Pick the FRONT datum plane from the model tree. Change **Offset** to **Parallel**. While holding down the **Ctrl** key. Pick a corner as shown> **OK**.

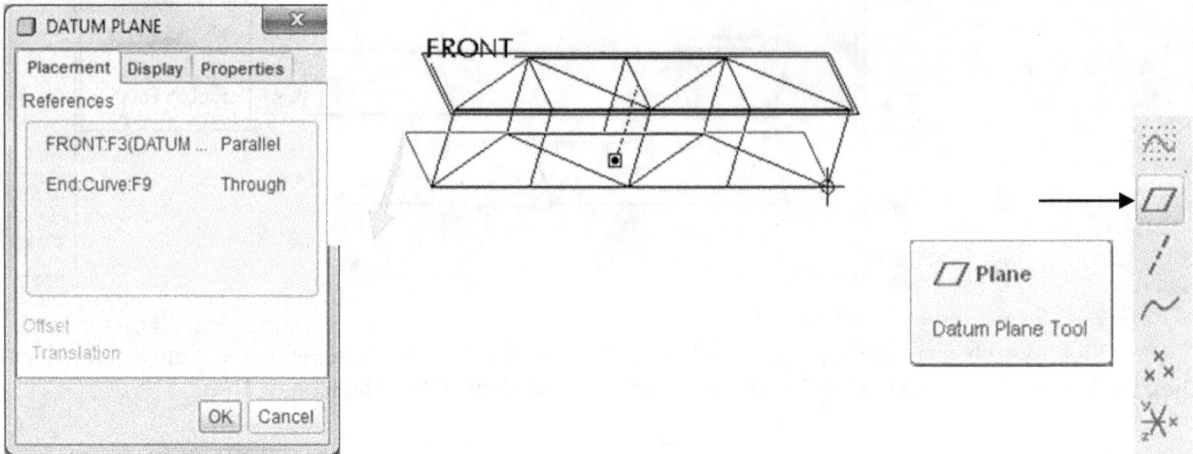

Now let us create another set of 3 datum curves on the **DTM3** datum plane. Select the icon of **Sketch Tool.** Pick DTM3 > **Sketch** to accept the default setting for orientation**.**

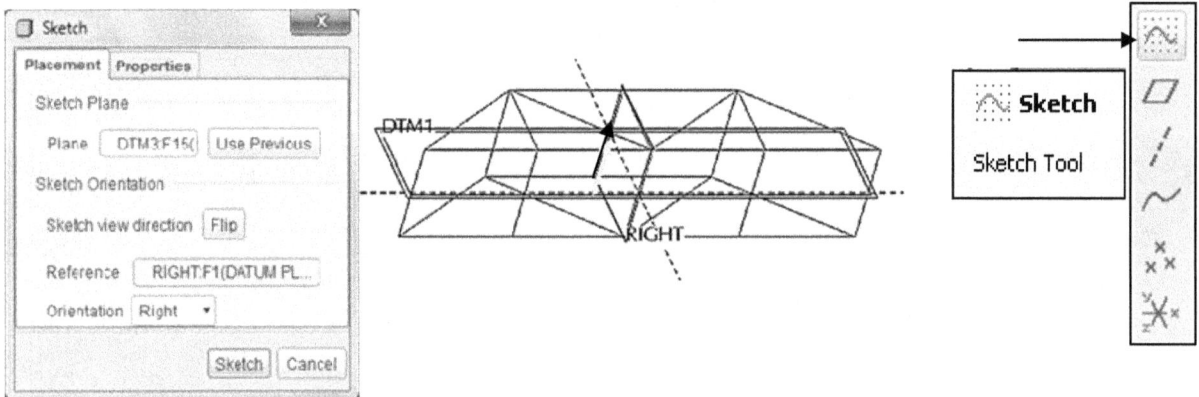

Click **Sketch** from the top menu and click **References**. Pick the 2 segments as 2 new references.

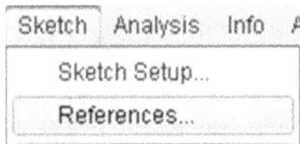

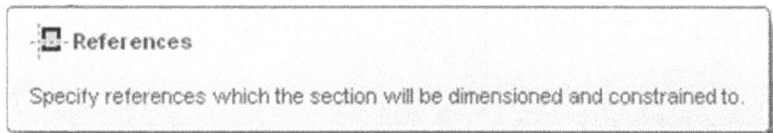

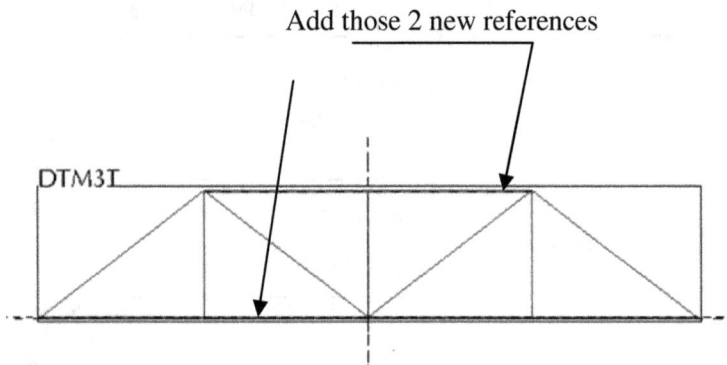

Add those 2 new references

Click the icon of **Line**, and sketch 3 vertical lines for connections. Click the icon of **Done**.

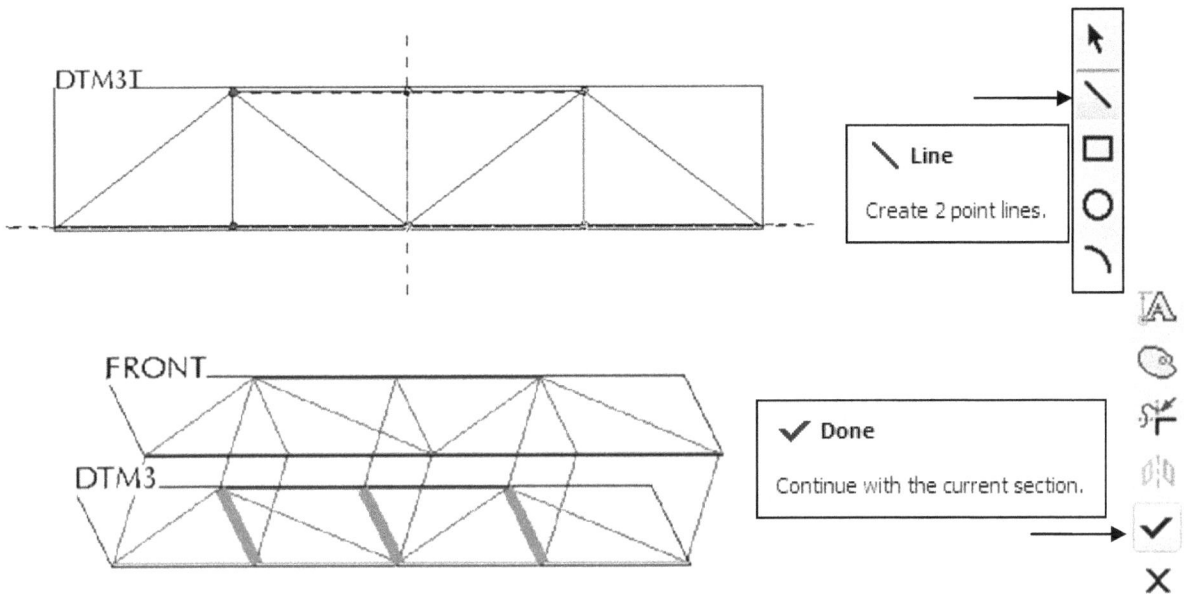

Step 2: Perform FEA under **Pro/Mechanica**
From the main toolbar or the menu toolbar, select **Applications > Mechanica > Structure > OK.**
Pick the icon of **Beam**. Accept Beam1 as the name.

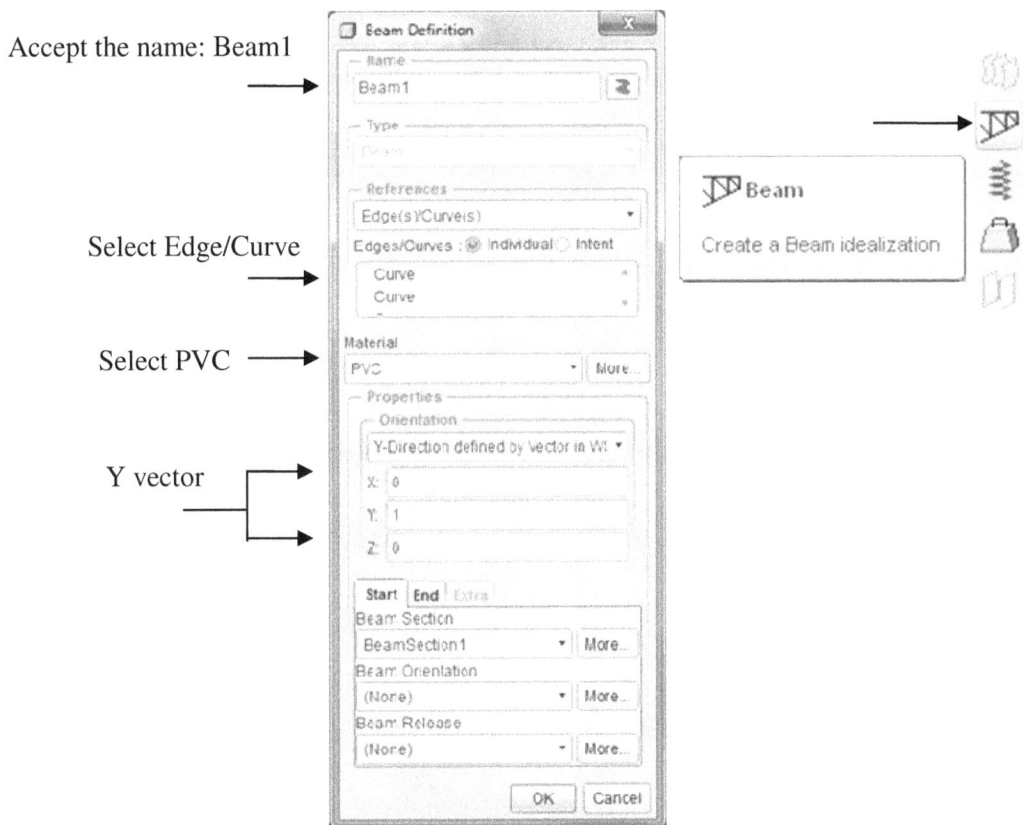

On the appeared **Beam Definition** window, do the followings:
Select **Edge/Curve** > pick the all datum curves while holding down the **Ctrl** key, except the 6 vertical straight lines (curves).

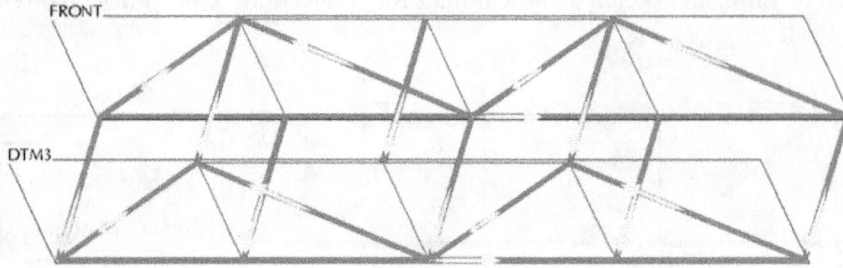

Select More from **Material** > Select **PVC** > **Close**

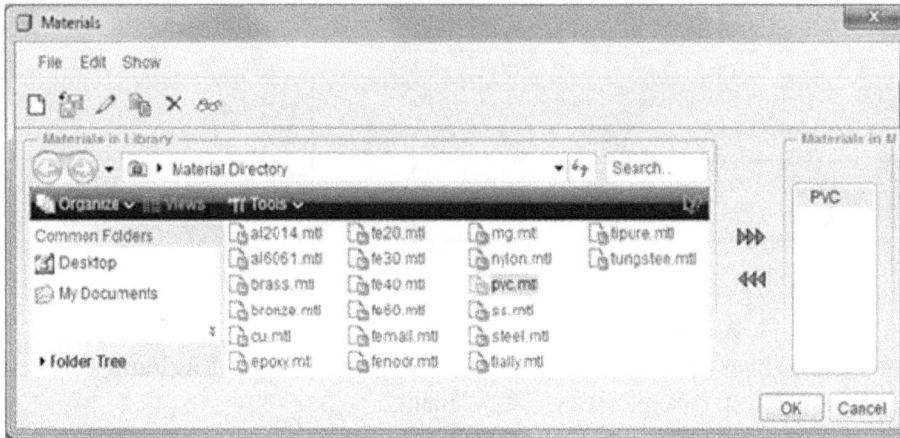

In the Properties field, set x = 0, y = 1 and z = 0 as the normal vector for orientation.

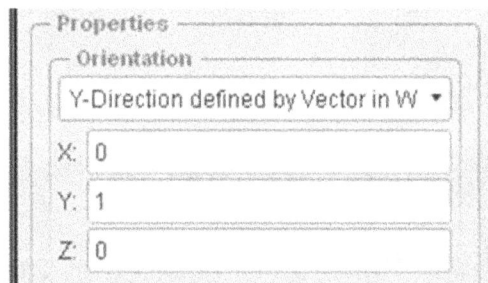

Select More from **Section** > **New** > select **Rectangle** > type *150* and *75* as b and d, respectively > **OK**.

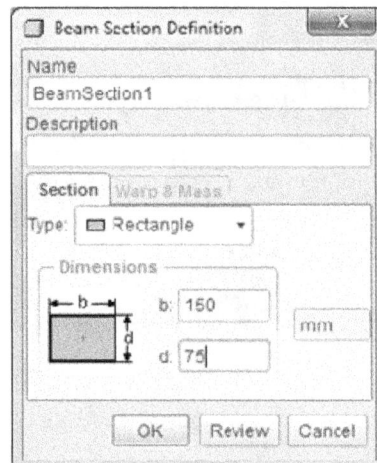

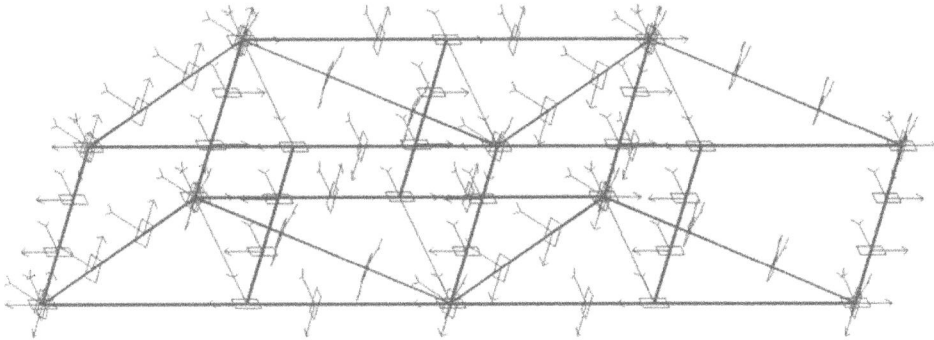

Pick the icon of **Beam**. Accept Beam2 as the name.

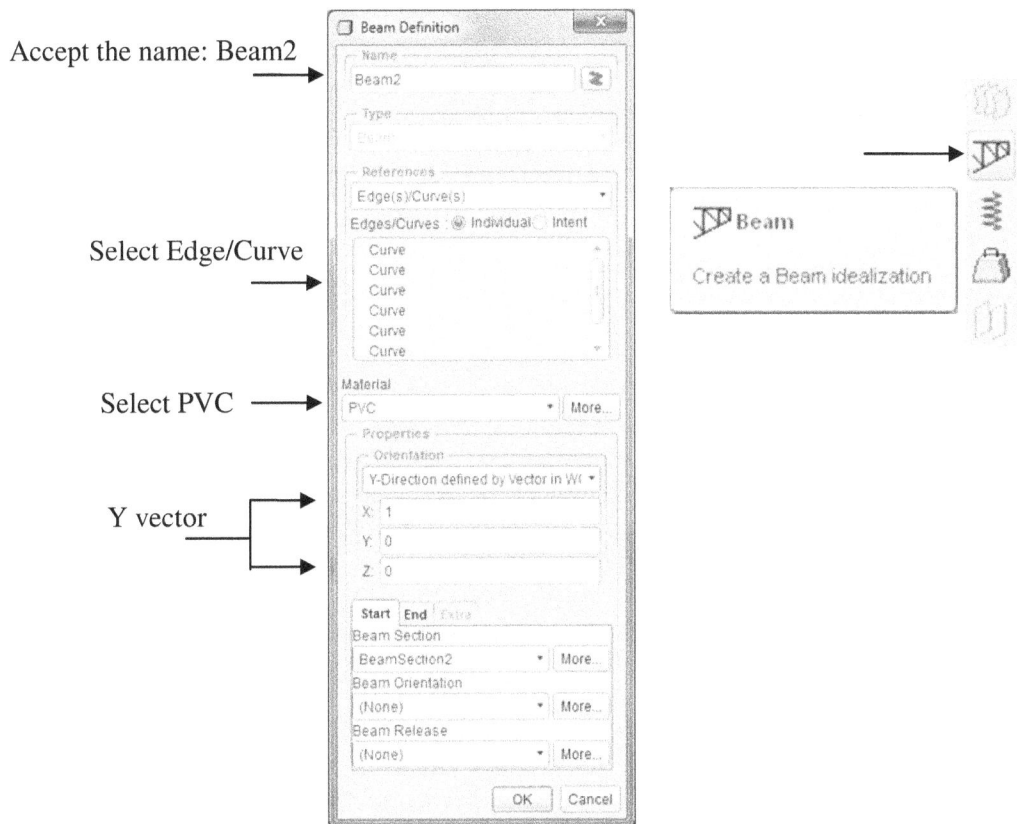

On the appeared **Beam Definition** window, do the followings:

Select **Edge/Curve**>pick the 6 vertical datum curves while holding down the **Ctrl** key.

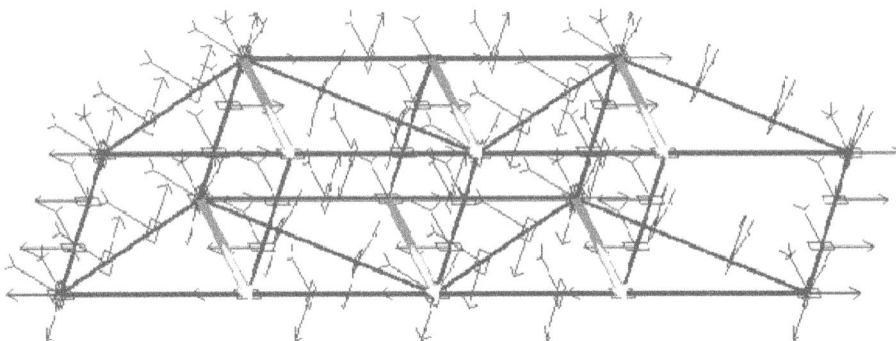

Specify STEEL and rectangular section with 150 and 75.

In the Properties field, set x = 1, y = 0 and z = 0 as the normal vector for orientation. Click **OK** to complete the process of defining BEAM 2.

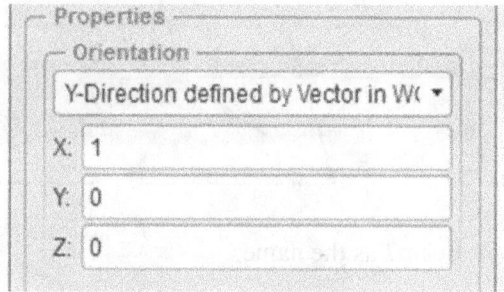

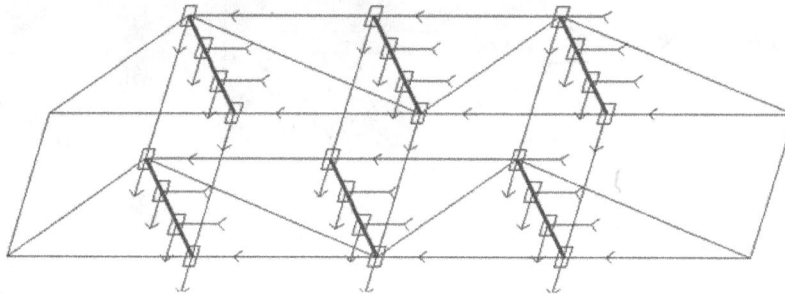

A special note on the reason why the beam elements associated with the 6 vertical lines are treated different from the others. There are two coordinate systems used by Pro/Mechanica. The first system is called BACS (Beam Action Coordinate System). BACS defines the X-axis of the beam. All the straight lines we have sketched represent the X-axis direction. The second system is called BSCS (Beam Shape Coordinate System). The beam cross sectional shape is defined by BSCS. When running FEA, it is required that the X-axis of BACS should be parallel to the X-axis of BSCS. The online help menu provides more information on BACS and BSCS.

Define fixed end constraint on the 10 vertices, as shown.

Constraints > Point(s) > pick the 10 points > **OK**.

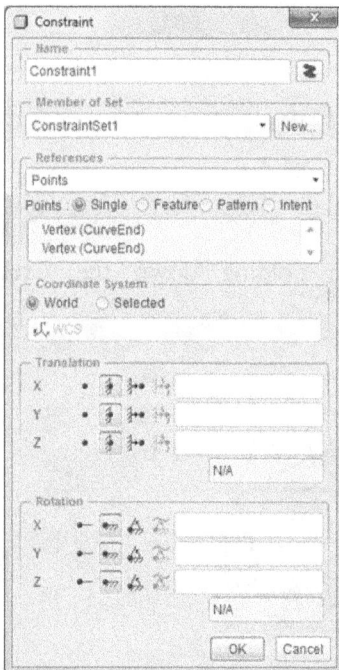

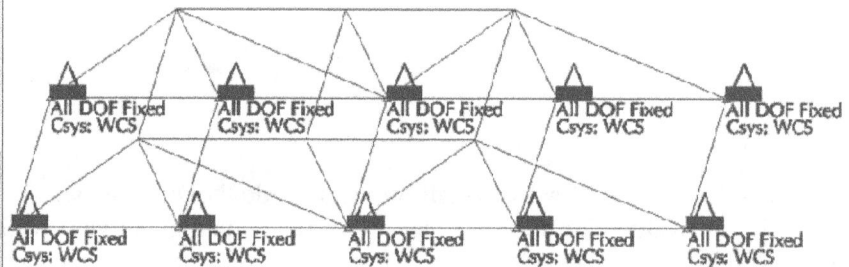

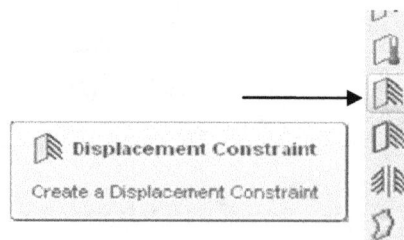

Define the load condition. Click the icon of **Force/Moment > Edges/Curves) >** pick the 5 lines on the top, as shown. Specify 10000 in the –Y direction > **OK**.

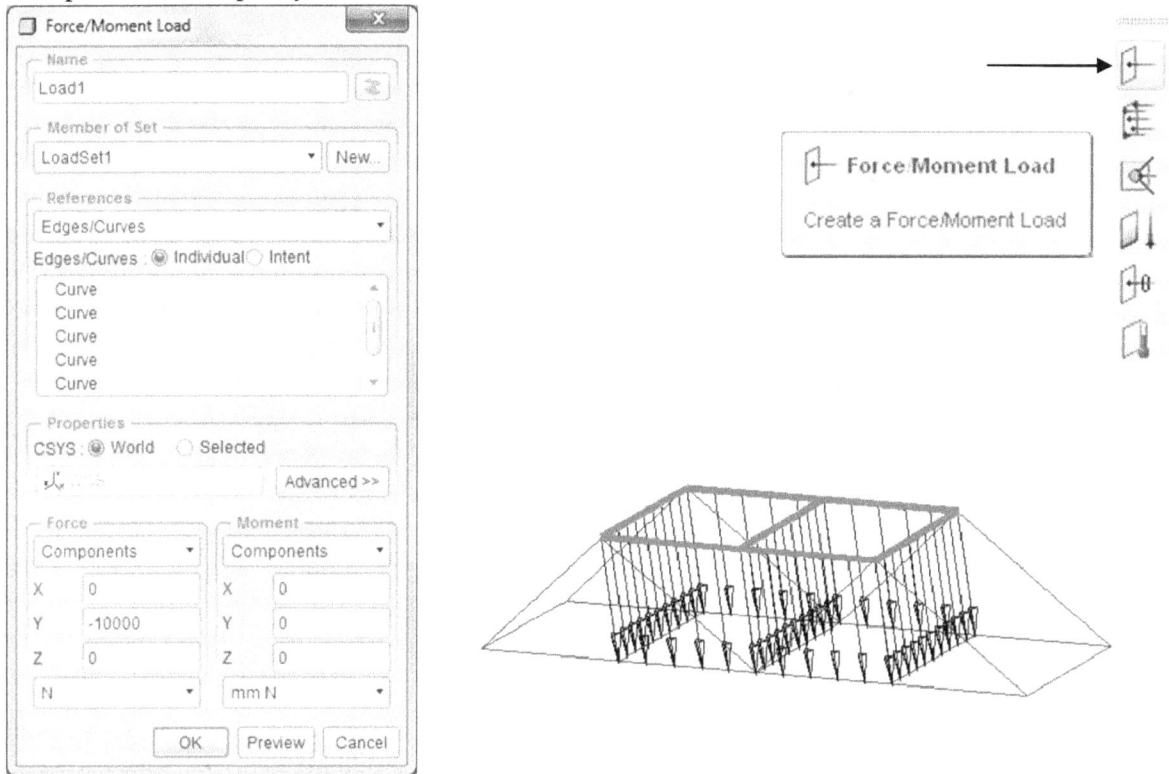

Step 3: Perform a static analysis
From the top menu, click the icon of **Run a Design Study > File > New Static**.
Type *nonuniform_load* as the file name > **OK**. Run the static analysis.

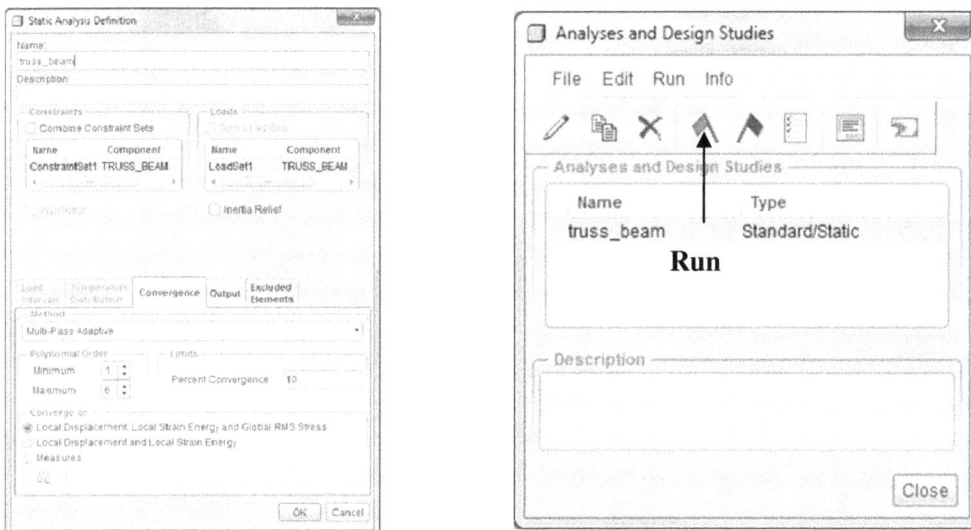

Step 4: Show the results obtained from FEA by plotting the reaction forces, share diagram and moment diagram.

To study the obtained result, say the reaction forces at the two support locations, click the icon of **Review results > Model > Reaction at Point Constraint >** select **Y > OK and Show.** The magnitude is 175 N at the left and is 125 at right, as shown.

To visualize the displacement distribution, click the icon of **Review results > Displacement > Magnitude > OK and Show.** The maximum value is 0.607 mm.

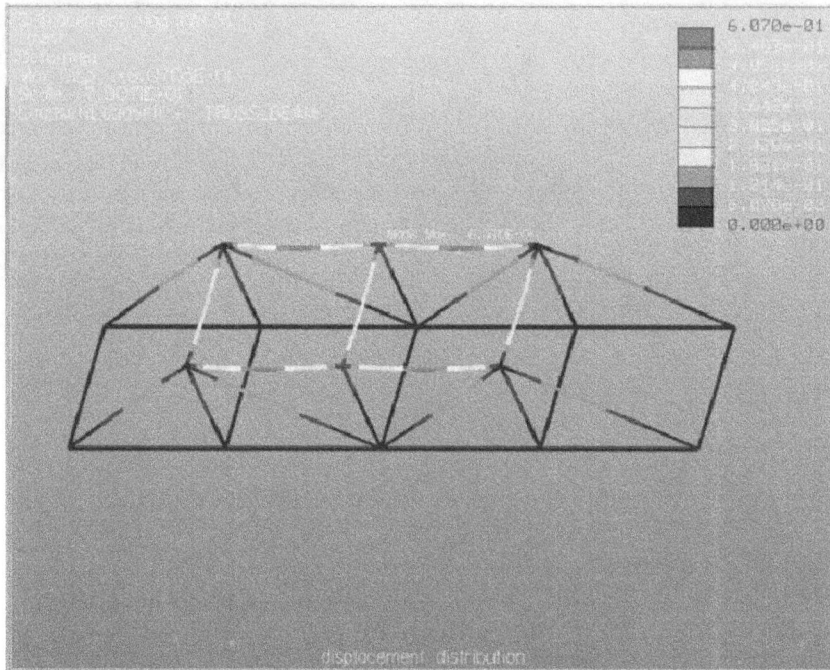

To visualize the distribution of the maximum principal stress, click the icon of **Review results > Stress > Max Principal > OK and Show.** The maximum value is 1.528 MPa.

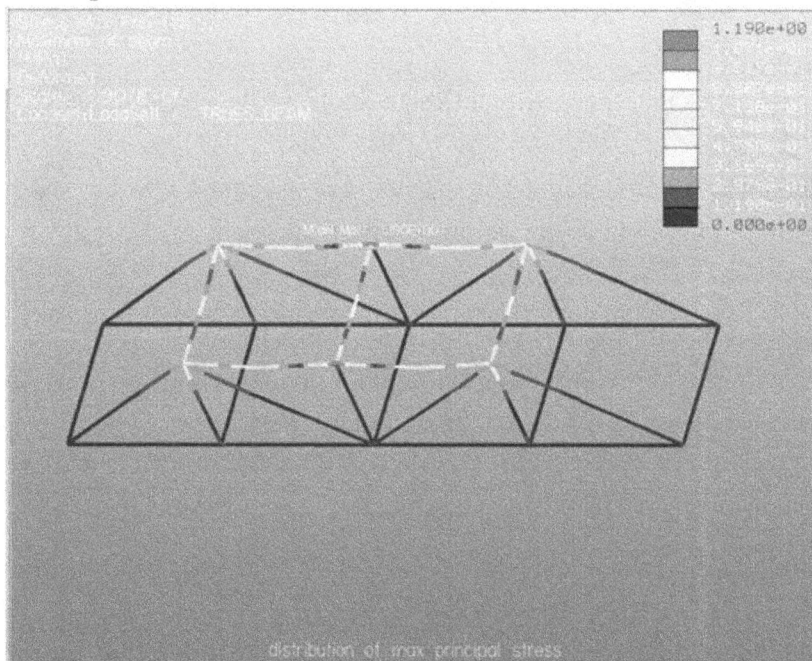

9.5   References

1. Algor Processor Reference Manual, Algor Interactive Systems, 150 Alpha Drive, Pittsburgh, PA 15238
2. O. Axelsson, Iterative Solution Methods, Cambridge University Press, New York, 1994.
3. K. J. Bathe, Finite Element Procedure, Prentice Hall, Upper Saddle River, New Jersey, 1996.

4. J. W. Dally and W. F. Riley, Experimental Stress Analysis, McGraw Hill, 1965.

5. J. W. Dally, Design Analysis of Structural Elements, Goodington, New York, NY, 2003.

6. Y. Y. Hsieh, Elementary Theory of Structure, 2nd edition, Prentice-Hall, Englewood Cliffs, NJ, 1982.

7. M. A. N. Hendriks, H Jongedijk, J. G. Rots and W. J. E. van Spange (eds), Finite Elements in Engineering and Science, DIANA Computational Mechanics, 1997.

8. M. Petyl, Introduction to Finite Element Vibration Analysis, Cambridge University Press, 1990.

9. T. H. H. Pian and P. Tong, Basis of Finite Element Methods for Solid Continua, Int. J. Numer. Methods Eng., Vol. 2, No. 7, April 1964, pp. 1333-1336.

10. P. G. Plockner, Symmetry in Structural Mechanics, J. of the Structural Division, American Society of Civil Engineers, Vol. 99, No. ST1, pp. 71-89, 1973.

11. J. N. Reddy, An Introduction to the Finite Element Method, McGraw-Hill Publishing Company, New York, NY, 1984.

12. T. Ross, T. Johns and Emile, Dynamic Buckling of Thin Walled Domes under External Water Pressure, Applied Solid Mechanics 2nd Conference, Ed. Tooth & Spence, Elsevier, pp. 211-224, 1987.

13. T. Ross, Finite Element Program in Structural Engineering and Continuum Mechanics, Horwood Publishing Limited, Chichester, England, 1996.

14. O. C. Zienkiewicz, The Finite Element Method, third edition, McGraw-Hill, New York, NY, 1977.

15. O. C. Zienkiewicz and R. L. Taylor, The Finite Element Method, 4th edition, Vol. 1, McGraw-Hill (UK),London, 1989.

9.6 Exercises

1. Use Pro/Mechanica to evaluate the displacement and von Mises stress for the beam shown below. You are asked to evaluate them for 3 sections, as shown. Fill the information in the attached table. Answer the following questions: (material type: steel)

   (1) What is the reaction force at the pin on the left side? What is the reaction force at the support on the right side?

   (2) Assume the design criterion is to minimize the maximum value of the displacement. Which type of beam would fit this requirement among the 3 types?

   (3) Assume the design criterion is to minimize the maximum value of the von Mises stress. Which type of beam would fit this requirement among the 3 types?

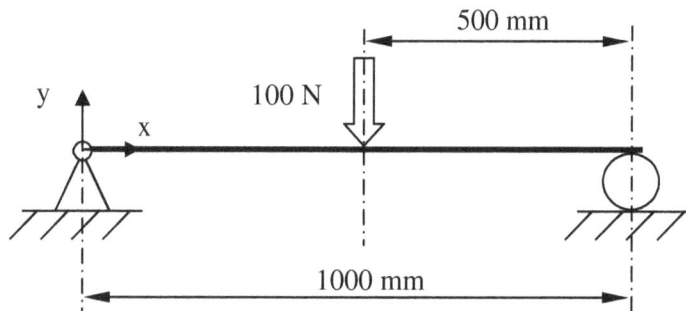

Beam Type	Key Dimensions (mm)	Section Area (mm^2)	δ_{max} (mm)	ovm_max (MPa)
I-beam	86x70x8x6			
Rectangular beam	70x22			
Hollow Circle beam	R30xR20			

2. Use Pro/Mechanica to evaluate the reaction forces at both ends. Determine the rotation angles at both ends as well. As a proximate approach, use PNT2 and PNT3, respectively. To evaluate the rotation angle at a support, the displacement in the Y or vertical direction should be evaluated first, as illustrated. (material type: steel)

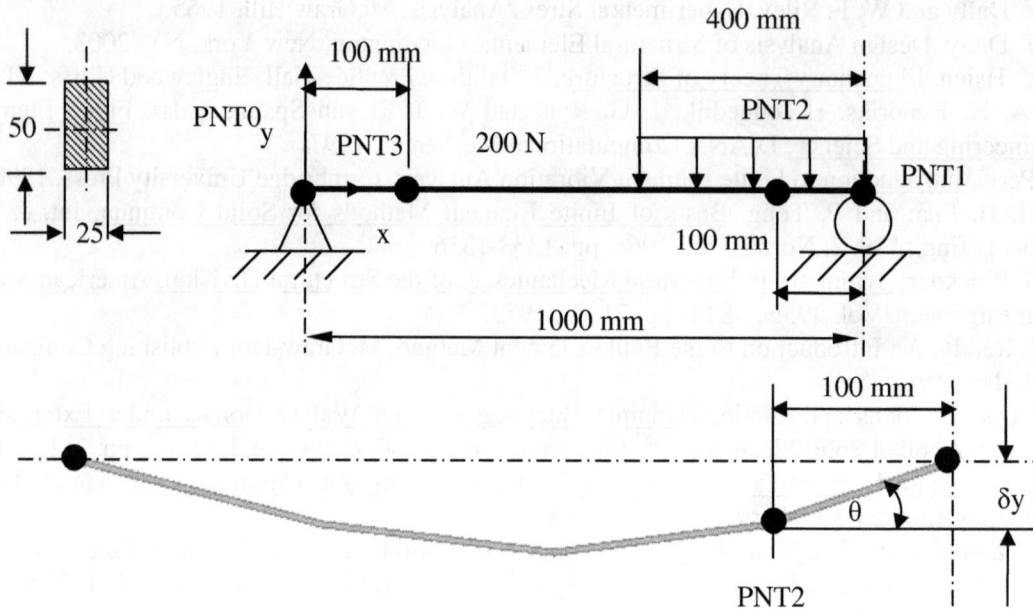

3.   Use Pro/Mechanica to evaluate the maximum displacement of the beam under the shown load conditions. Specify the location. Assume that the material type is STEEL. Neglect the effect of the gravity force. Also evaluate the maximum value of the von Mises stress and specify its location. Determine the reaction forces at both supports. Construct the shear and moment diagrams. If time permits, repeat the above calculations considering the presence of the gravity force.

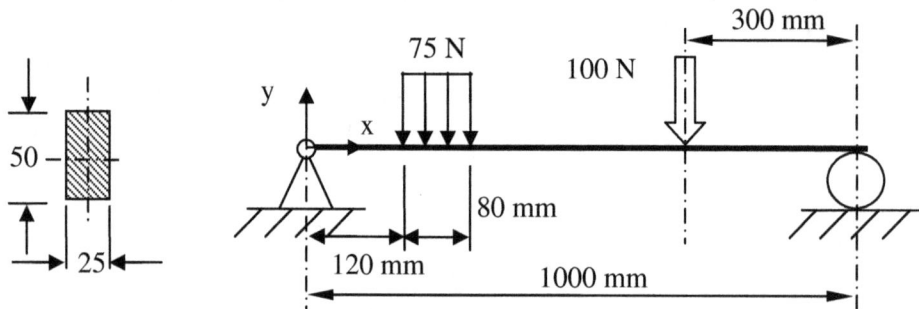

4.   Use Pro/Mechanica to evaluate the maximum displacement of the beam under the shown load conditions. Specify the location. Assume that the material type is STEEL. Neglect the effect of the gravity force. Also evaluate the maximum value of the von Mises stress and specify its location. Determine the reaction forces at both supports. Construct the shear and moment diagrams.

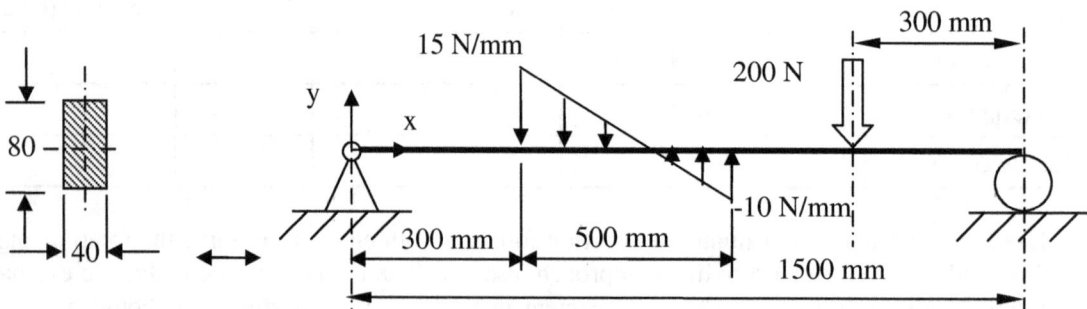

5.   A transmission tower is shown below. You are asked to create a 2D model and use the BEAM idealization to perform FEA. The material type is STEEL. The cross section is a hollow circle. The outer diameter is 80 mm and the inner diameter is 70 mm.

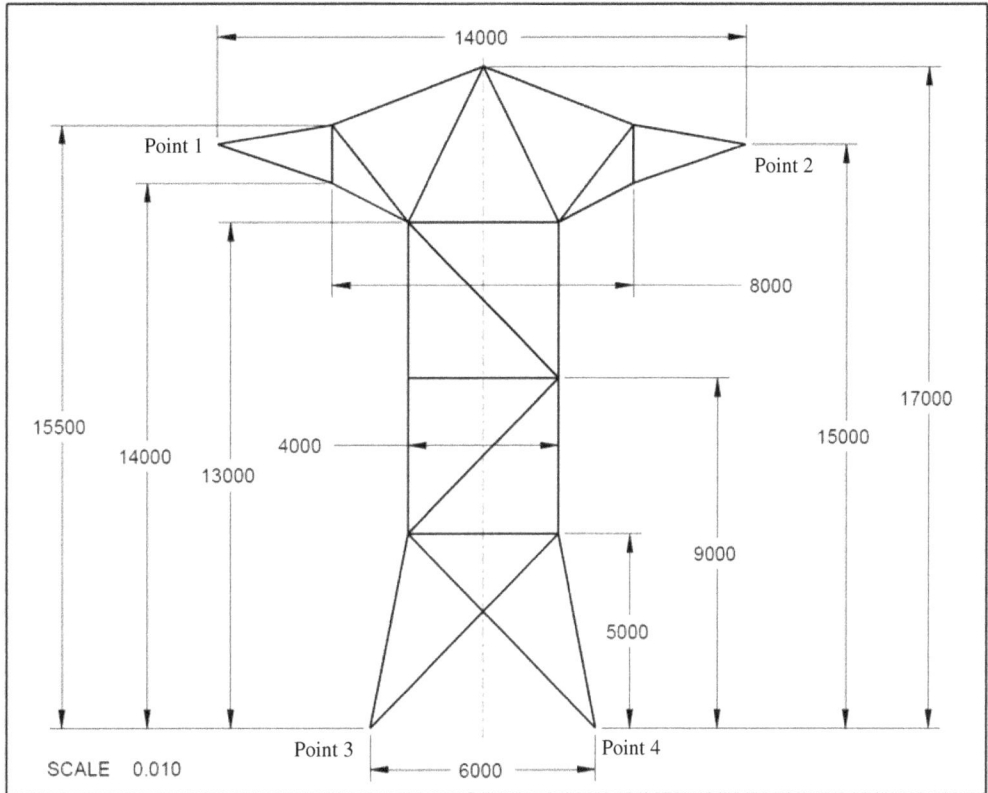

The 2 loads are acting at Point 1 and Point 2. The magnitude is 20,000 Newton each (Case 1: Downward and Case 2: 45 degrees and downward). For case 3 and case 4, add the gravity force into consideration. Point 3 and Point 4 are fixed to the ground. Show the distribution of displacement under loading and the maximum von Mises stress. Identify their locations. Is the design of this tower structure a good design? How to improve the design? Try yours and make comparison.

6.   Use Pro/Mechanica to evaluate the displacement of node 2, the reaction force components at Node 1 and the reaction force components at Node 3. You may pick up a material type for those beams.

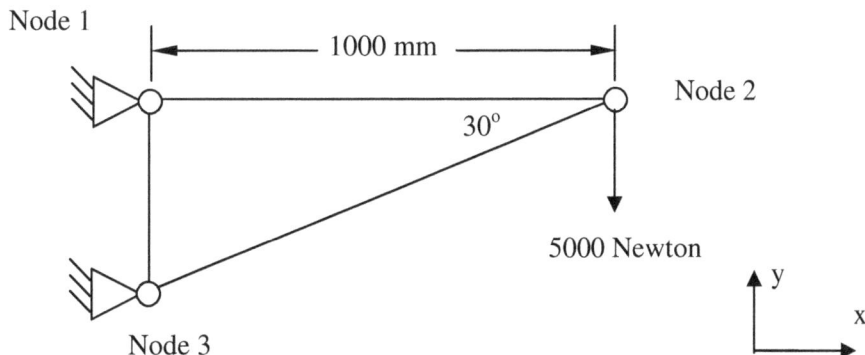

7.   Use Pro/Mechanica to evaluate the displacement of node 2, the displacement of node 3, the reaction force components at node 1 and the reaction force components at node 4. Assume the material type is STEEL. The cross section area is rectangular with 2 dimensions equal to 20 and 40 mm.

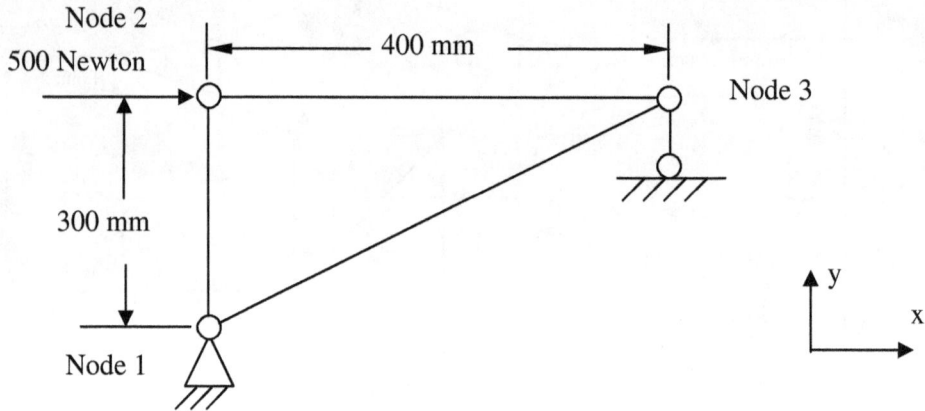

8.   The following figure illustrates a truss structure, which consists of 3 beams and they are rigidly connected. The dimension of the section area is 100 mm^2 (squared section). The geometrical information of the truss structure is shown in the figure. The material type is STEEL. Use Pro/Mechanica to evaluate the displacement components at Node 2, the displacement components at Node 3, the reaction force components at Node 1 and the reaction force components at Node 4.

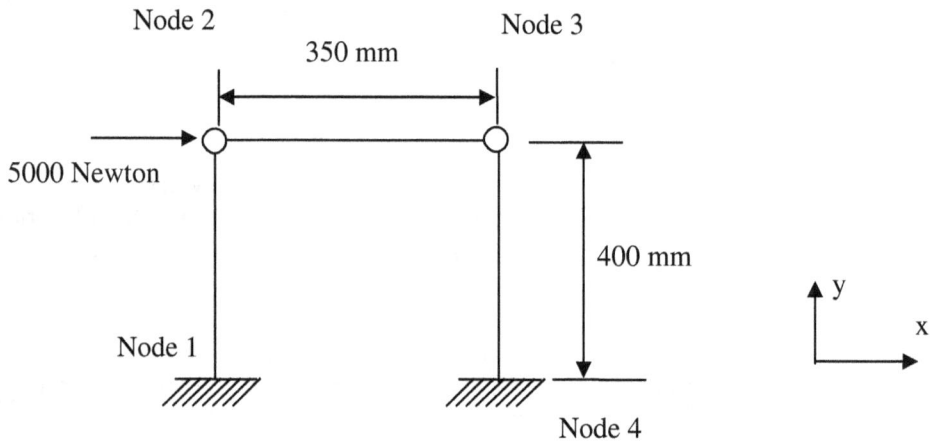

CHAPTER 10

FEA USING ANSYS

10.1 Introduction

In the previous 9 chapters, we have learned Pro/Mechanica. In this chapter, we will use **ANSYS** to perform FEA. As discussed in Chapter 1, ANSYS is a simulation software system designed to optimize product development processes. **ANSYS** has been widely used by engineers and designers in industry and academy. The software system focuses on the development of open and flexible solutions that enable users to analyze designs so as to provide a common platform for fast, efficient and cost- conscious product development, from design concept to final-stage testing and validation. The newest release is **ANSYS 13.0**.

Section 10.2 presents a shell model under the **ANSYS** design environment because of its simplicity of creating a 2D model. The procedure to perform FEA using ANSYS is demonstrated step by step. Section 10.3 presents an example with a 3D solid model for component. Section 10.4 presents a 3D solid model for an assembly, which consists of two components. Section 10.5 presents a detailed discussion on the issue of using Working Plane, which is subjected related to the global coordinate system and local coordinate systems. Sections 10.6, 10.7, and 10.8 demonstrate the use of different unit systems including CGS, BIN and BFT unit systems. Examples related to thermal analysis are presented in Section 10.8 and 10.9. Section 10.10 presents an example related to fluid dynamic study. The creation and execution of ANSYS command files are presented in Section 10.12.

10.2 FEA with a 2D Model

The following figure illustrates a plate structure. The dimensions of length and width are 800 and 100 mm, respectively. The thickness of the plate is 10 mm. The left end of the plate is fixed to the wall. The load of 100 N acts on the edge at the right end, which is set free to move. In terms of the material properties, we assume that the Young's modulus is 30 GPa, and the Poisson's ratio is 0.26. The objective of this study is to obtain the information on the maximum displacement and the maximum principal stress developed under the load condition.

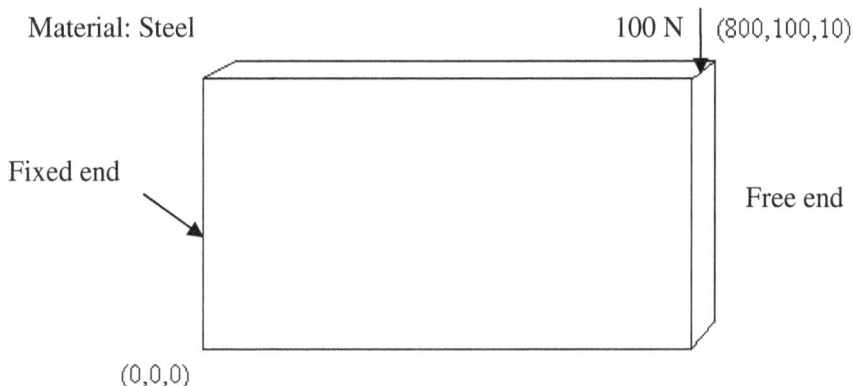

Step 1: Launch the **ANSYS** system.
Click **All Programs > ANSYS**.

Step 2: Define the Structural Analysis first.
From the main menu, select **Preferences**. Make sure that **Structural** is checked, and h-method is also selected > **OK**.

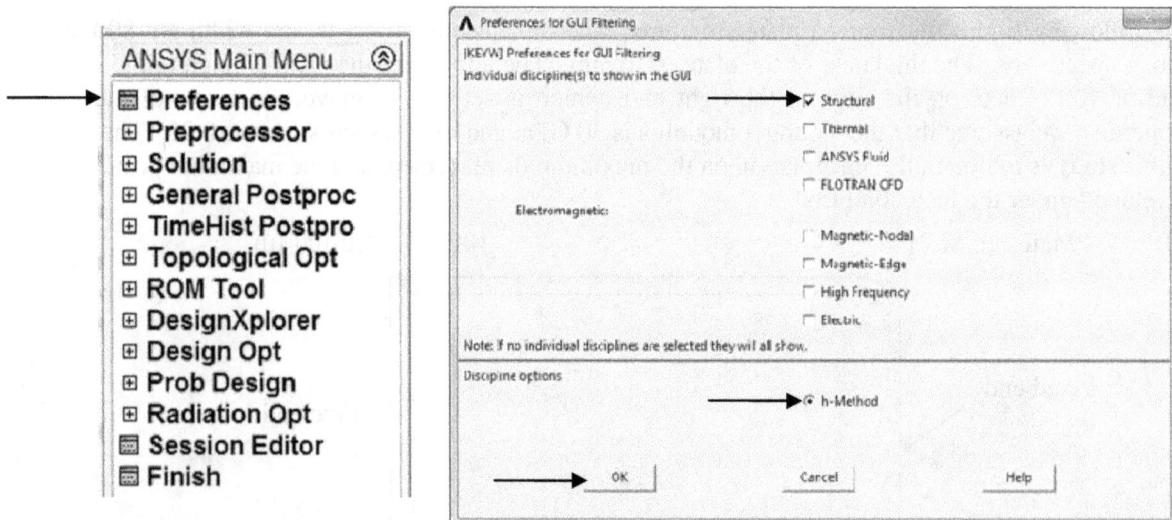

From the main menu, select **Preprocessor** and expand it. Select **Modeling > Create > Areas > Rectangle > By Dimensions**.

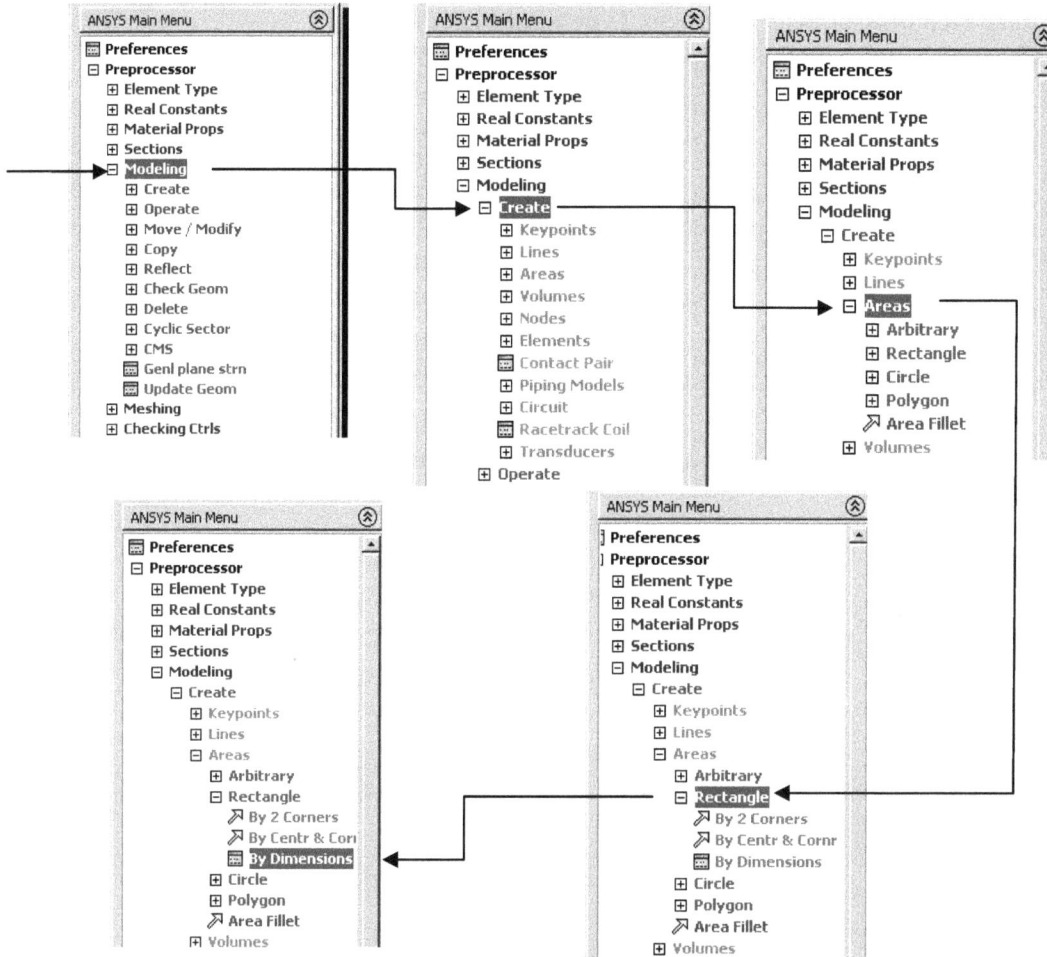

In the window for creating the rectangle by dimensions, type 0 and 0.8 for X1 and X2, and 0 and 0.1 for Y1 and Y2. Click **OK**.

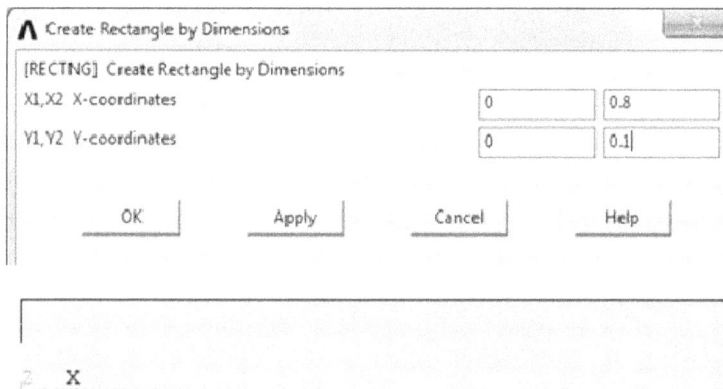

Click the icon of **Isometric View**.

Under the Preprocessor, select **Element Type > Add/Edit/Delete.**

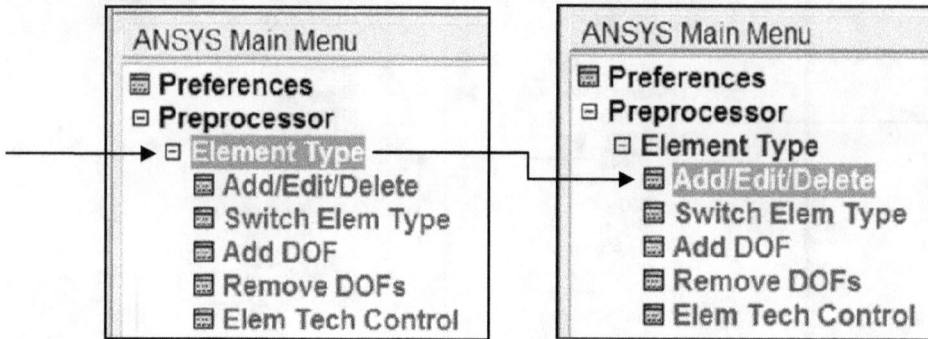

A window called **Element Types** appears. Click **Add > pick Solid > Quad 4node 182 > OK > Close**.

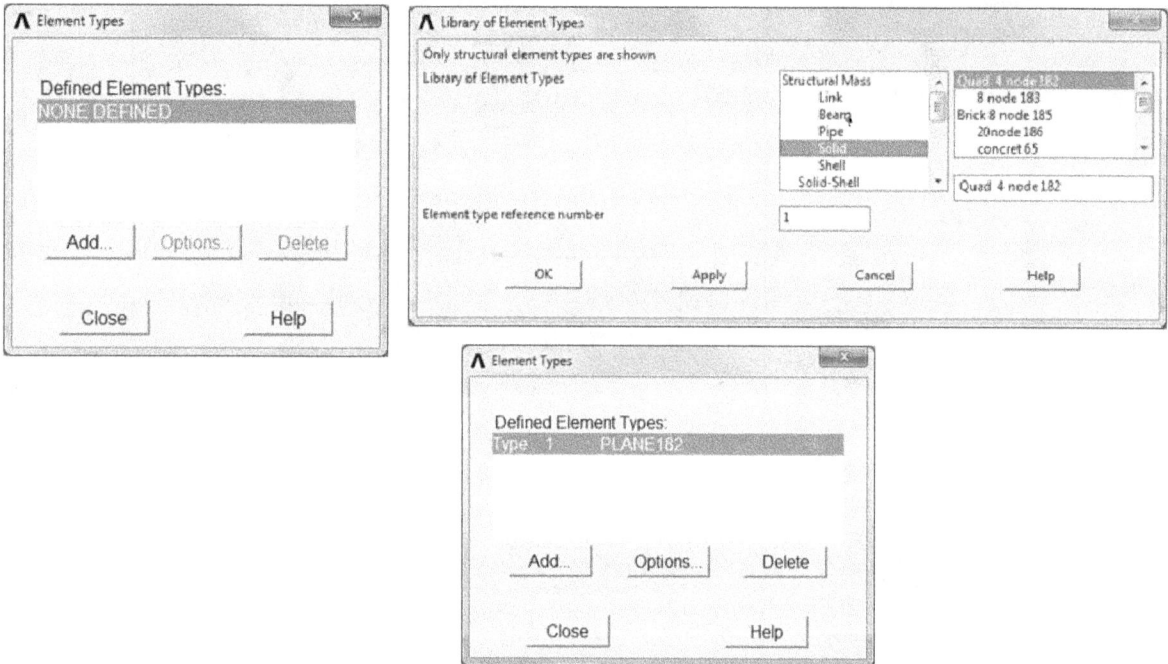

Click **Options > Plane stress w/Thk > OK > Close.**

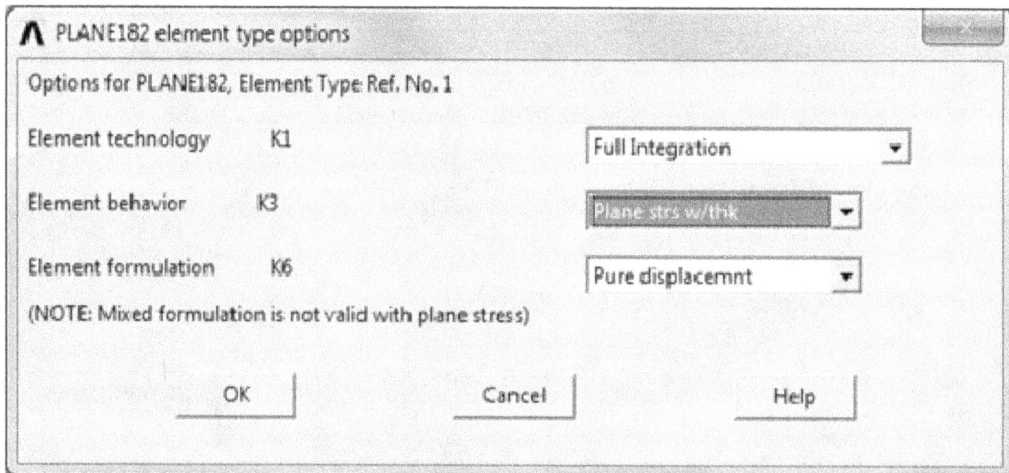

From the main menu, expand **Real Constants > Add/Edit/Delete > Add > OK**.

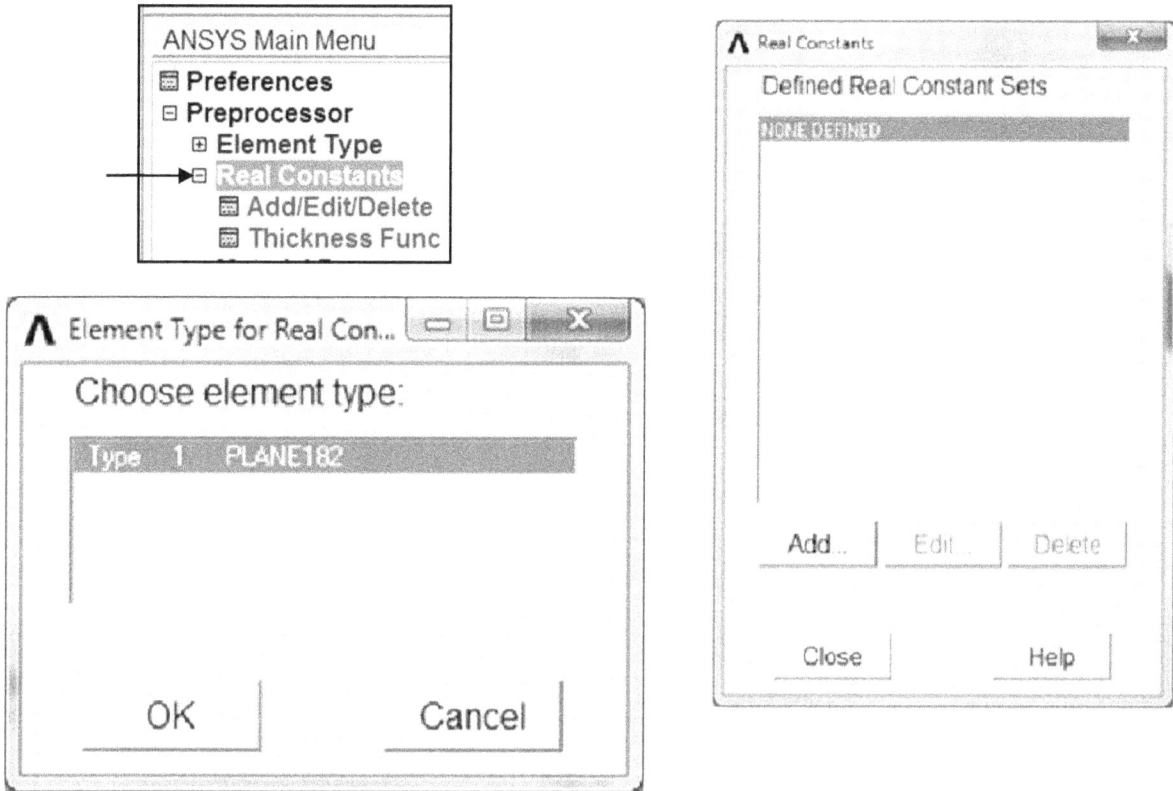

In the pop up window, set the thickness value to 0.01 > **OK > Close**.

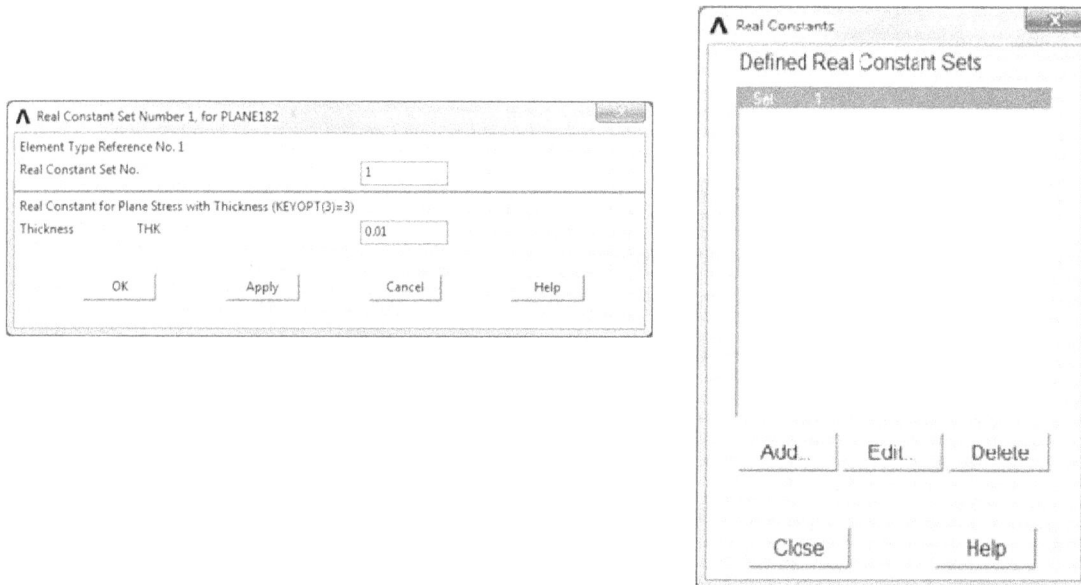

Step 3: Define the Material Properties

To select the material for the 2D model, select **Material Properties** from the Preprocessor menu. Select **Material Model**, and then select **Structural, Linear, Elastic** and **Isotropic** by left clicking on each one.

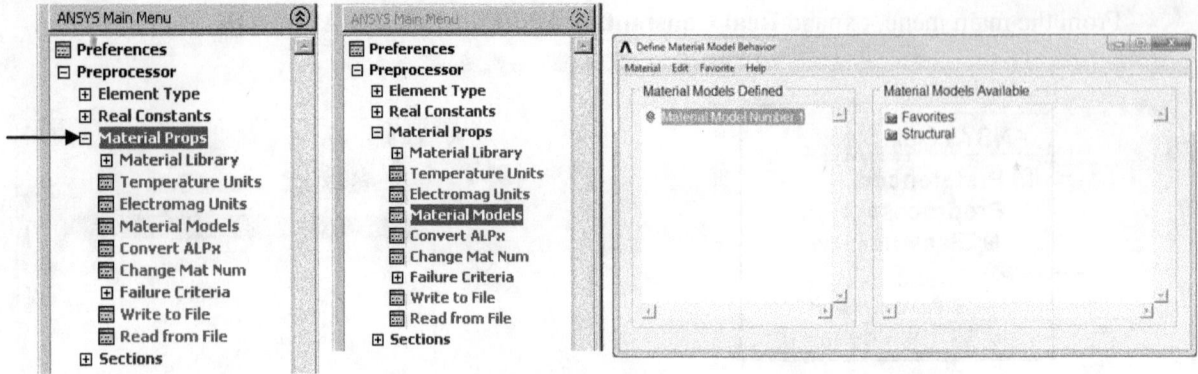

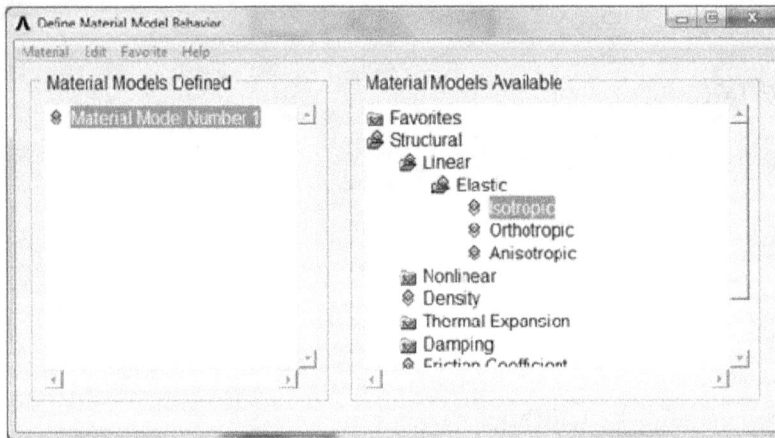

Type 3e10 and 0.26 as the values for Young's Modulus and Poisson's ratio, respectively. Then select **OK**. Click **Material > Exit**.

To check the unit system, select **Material Library** under the material properties menu. Then select **Units** and make sure that the **SI** unit system is highlighted. Click **OK**.

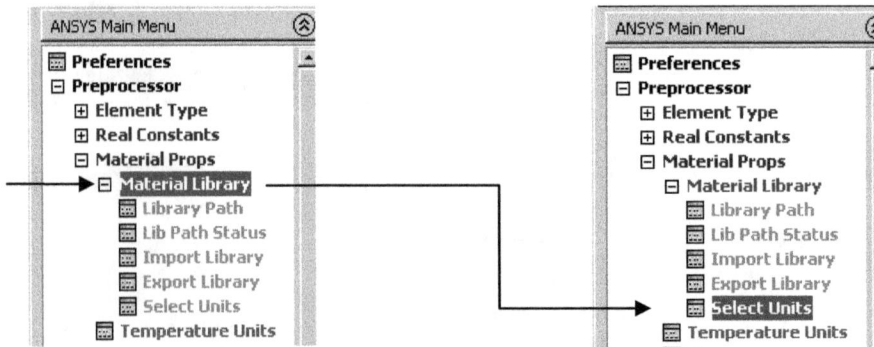

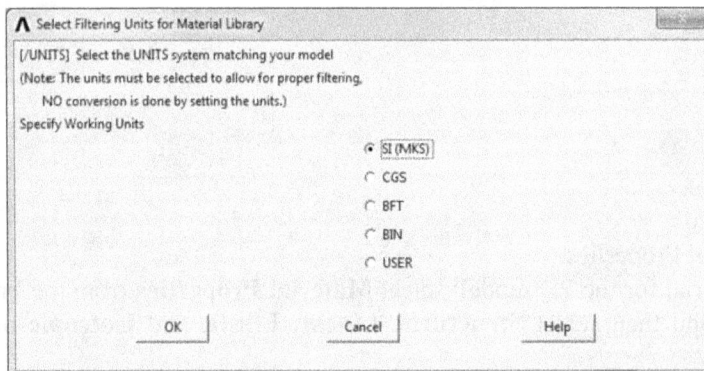

Step 4: Mesh Generation

To create the mesh for the shell model, select **MeshTool** from the Preprocessor menu. Then choose **Smart Size** and click **Mesh**.

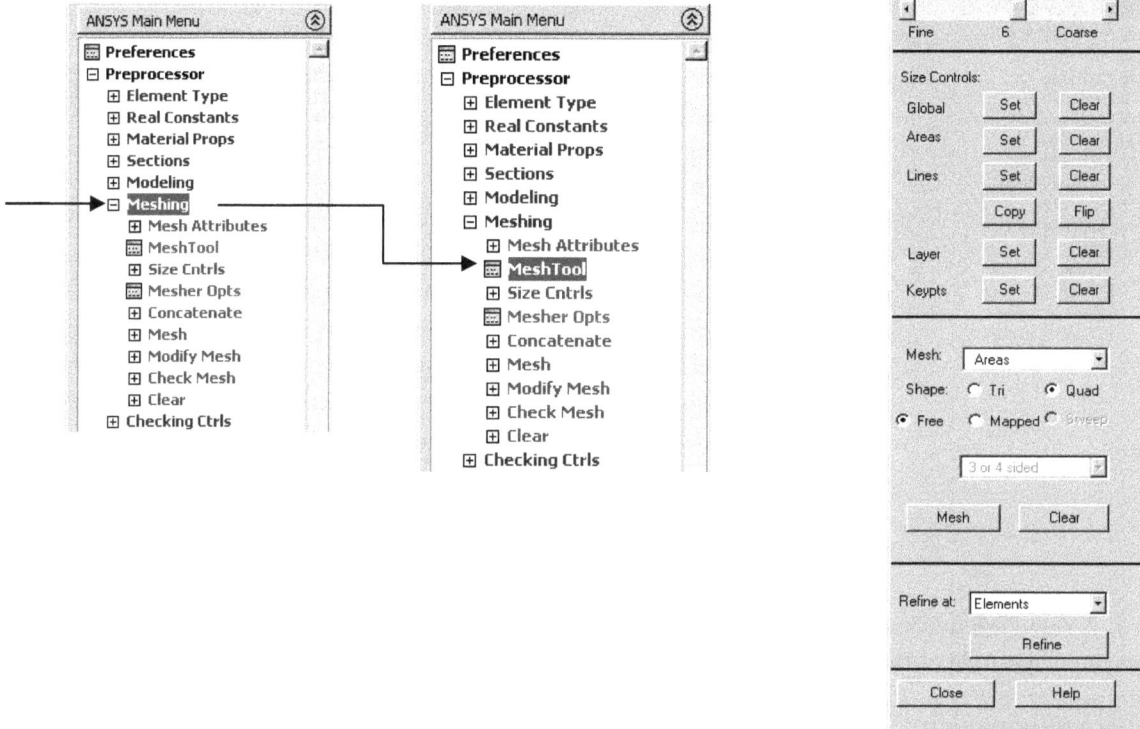

When the selection box comes up, select the area from the display window, and then click **OK**. Make sure that you close the **MeshTool** window.

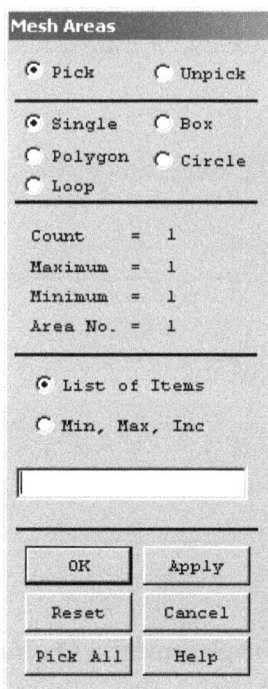

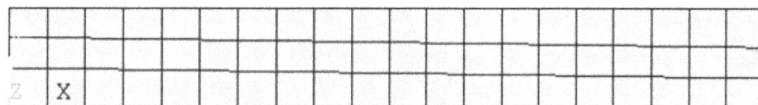

Step 5: Apply the load condition.
    To apply the load, on the Preprocessor menu, click **Load > Define Loads > Apply > Structural > Force/Moment > On Nodes**. Select the nodes at the upper right corner, and click **OK**.

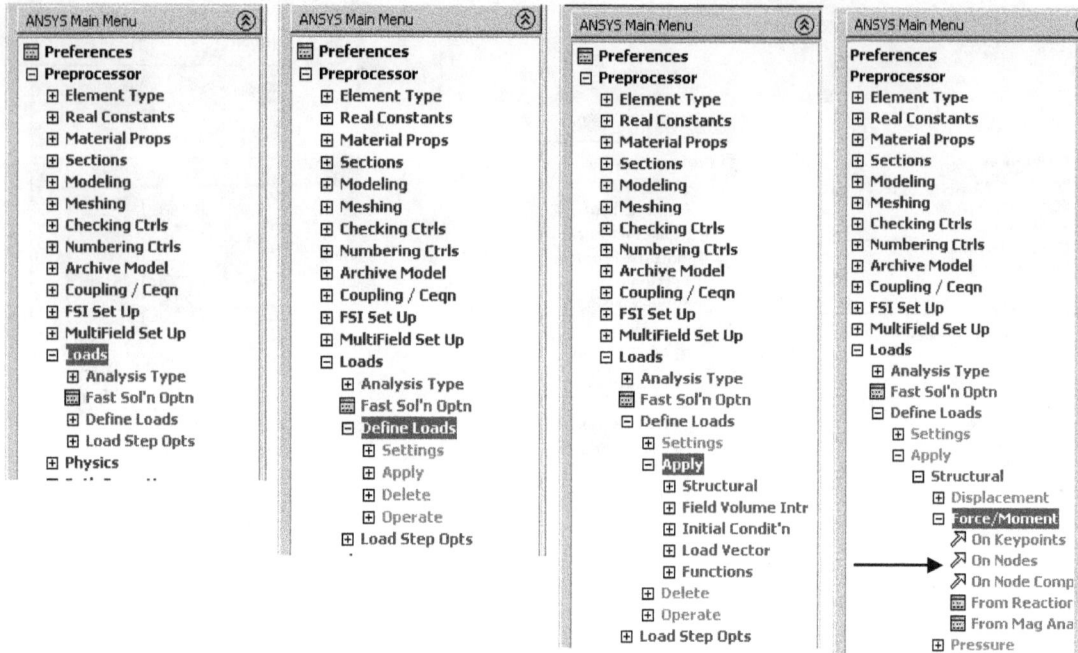

    A selection window appears. Pick the node at the upper right corner **> OK**. Select **FY** and type – 100 as the magnitude of the load.

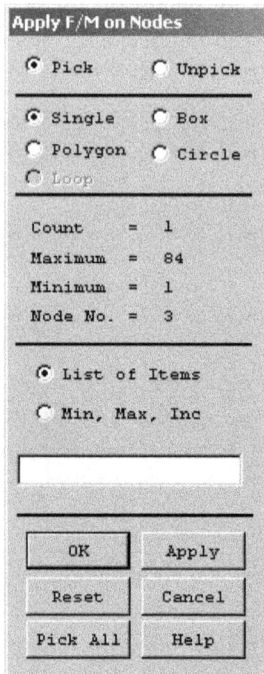

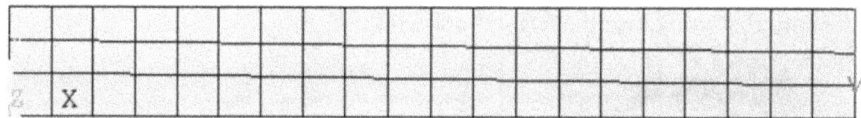

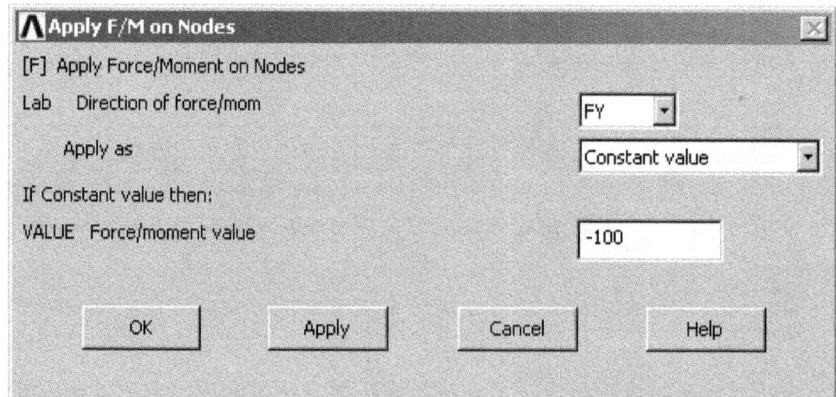

Step 6: Defining the Constraint Condition
    To define the constraint conditions, under the Structural menu, select **Displacement > On Nodes**.

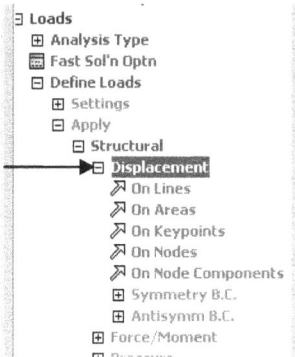

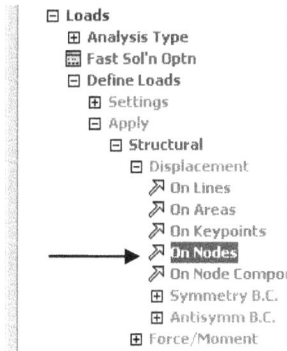

A selection window appears. Select Box as the method to pick a line. Holding down the right bottom of the mouse, sketch a box on the left side of the rectangle, as shown > **OK**.

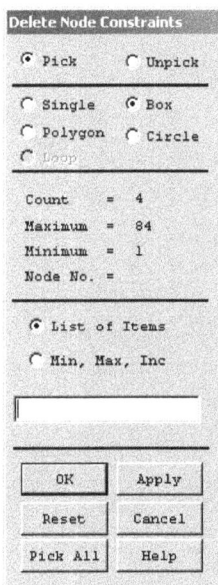

Select **All DOFS**, and type 0 as the value. Then click **OK**.

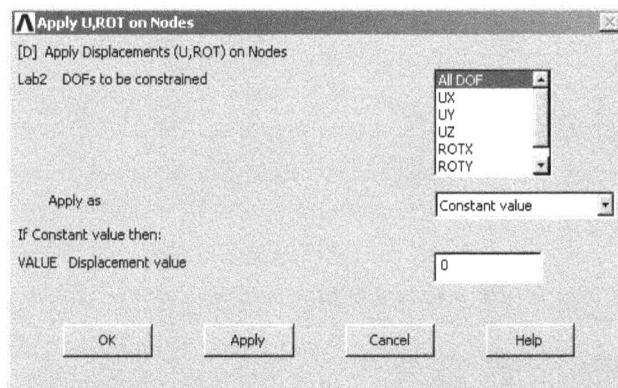

Step 7: Perform the analysis and obtain the solution.

To find the solution, run the program by selecting **Solution** from main menu > **Analysis Type** > **New Analysis** > **Static** > **OK**.

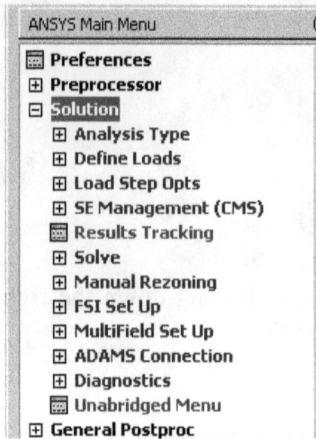

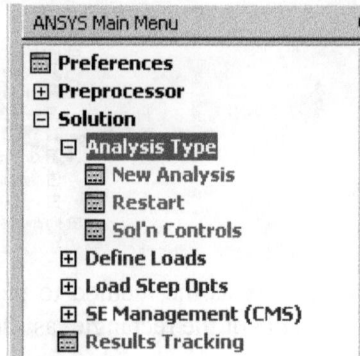

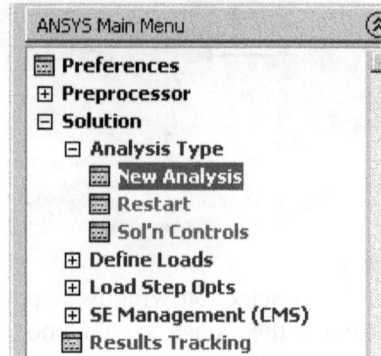

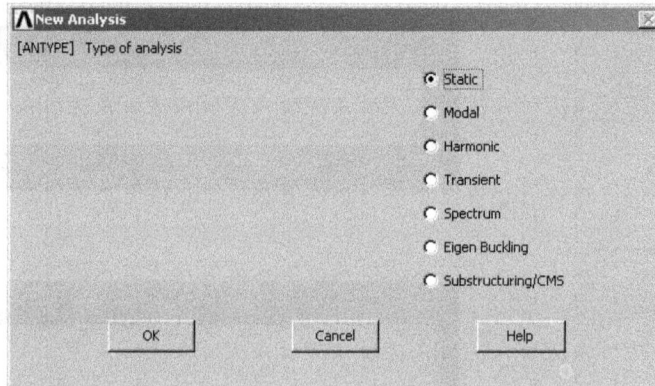

Choose **Current LS** > after reviewing the information, click **OK** to run the program.

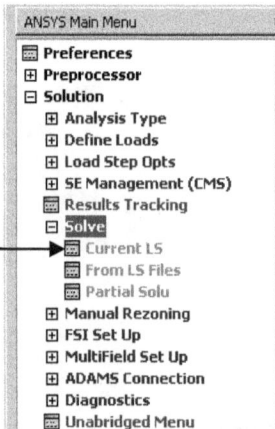

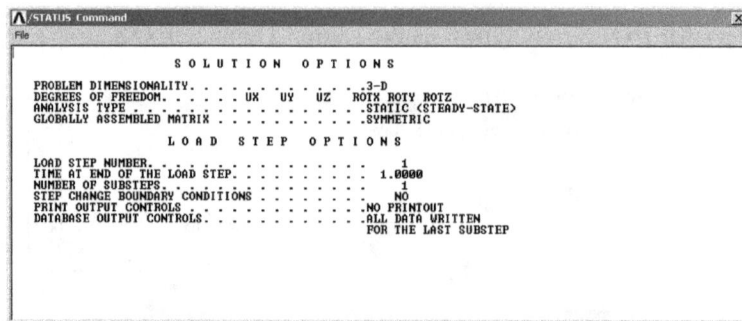

A window called Note appears when the solution is done > **Close**.

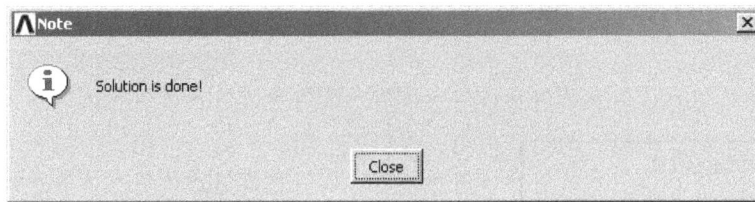

Step 8: Displaying Results

To review the results, select **General Postprocessing** from the main menu. Then click **Plot Results**, > **Deformed Shape** > select **Def + undeformed** > **OK**.

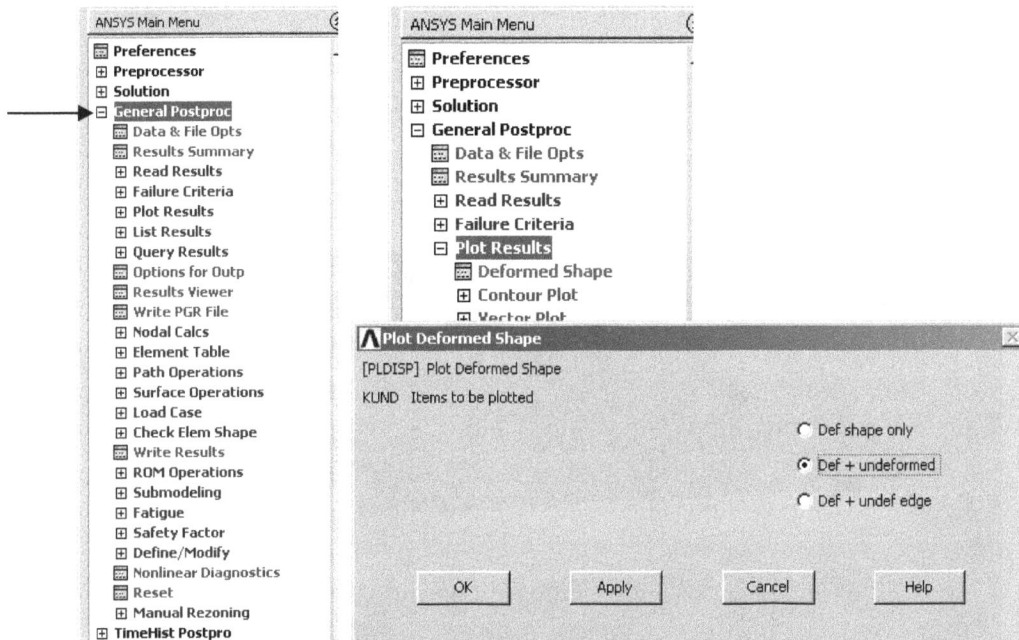

The following plot depicts the deformed shape. As indicated, the maximum deflection is 0.645 mm. Note DMX stands for Maximum Displacement.

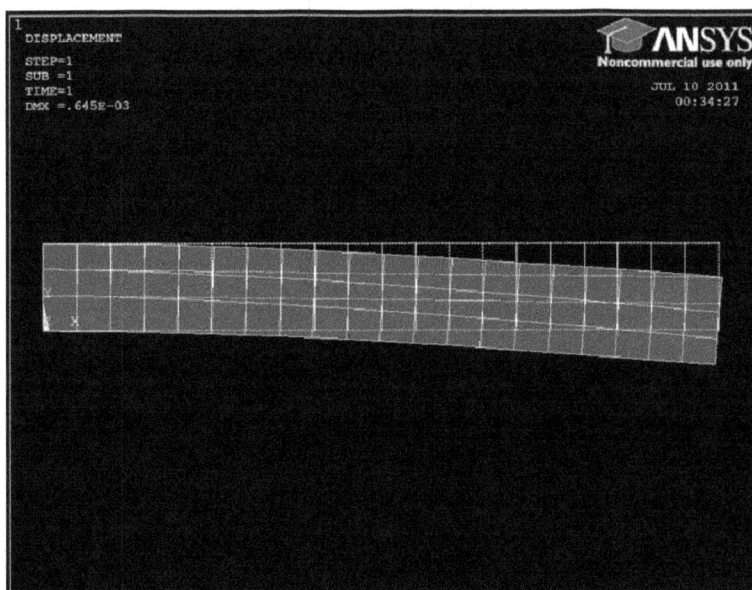

To plot a stress distribution, **Contour Plot > Element Solution** > pick a specific type of stress of interest, say the 1st Principal stress > **OK**.

When examining the information shown on the plot, the maximum stress value (tensile stress) is 4.88 MPa (SMX stands for Maximum Stress).

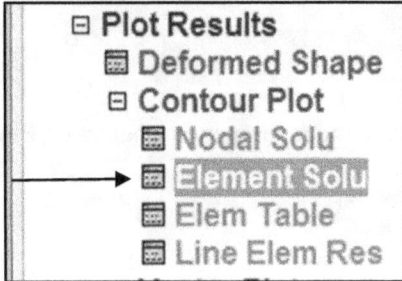

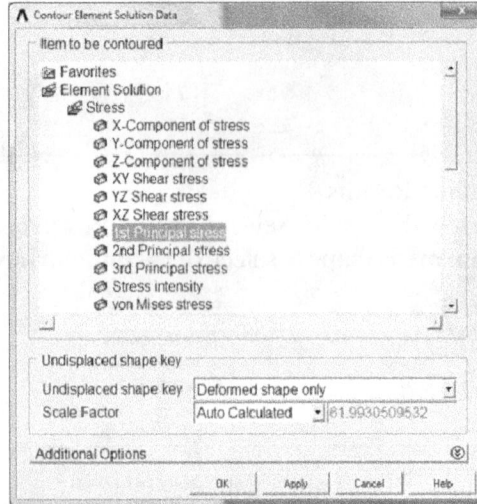

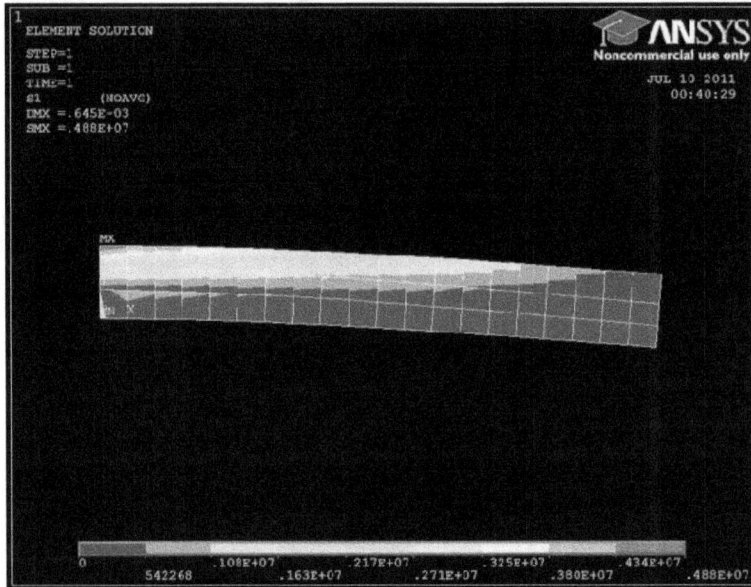

If desired, the results can also be animated. To do this, select **Plot Ctrls** from the upper menu, and click **Animation.** Then select **Deformed Shape** and accept the default settings > **OK**.

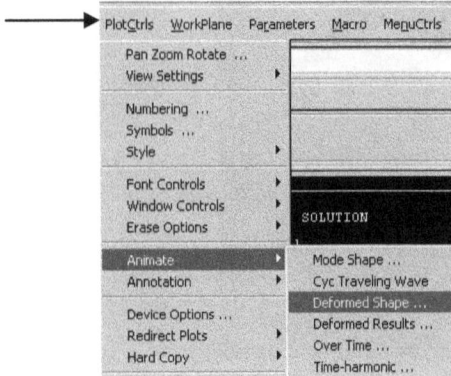

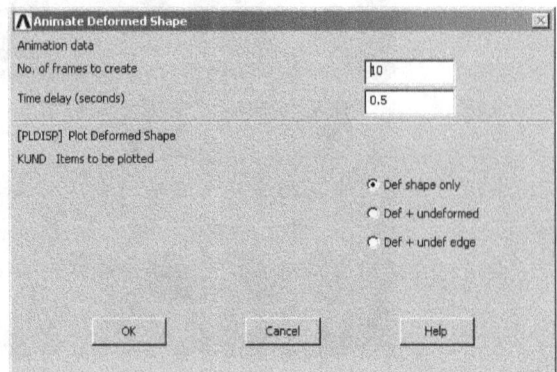

The following plot illustrates the animation in progress. To terminate the animation process, click **Close**.

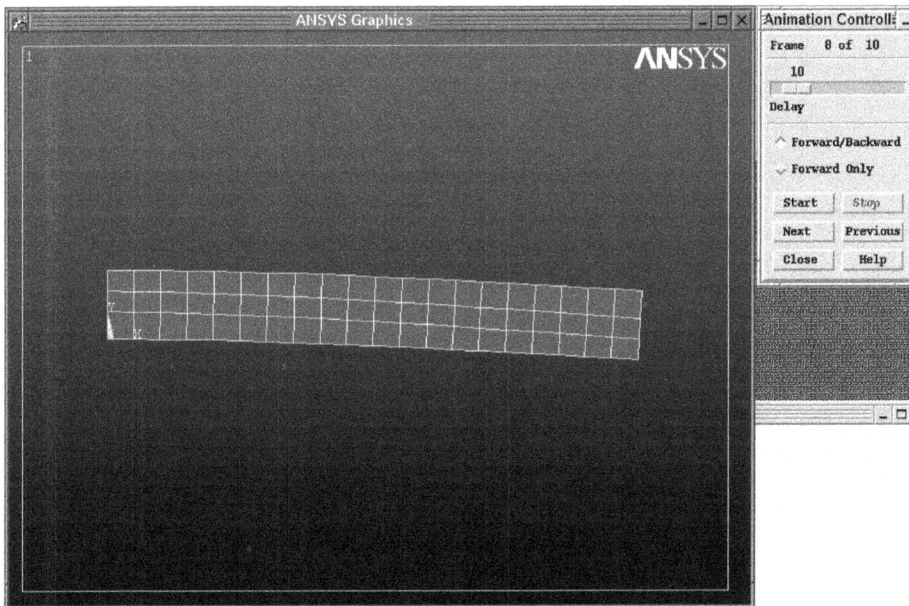

To print the results to an image file, select **PlotCtrls** from the top menu. Select **Hard Copies**, and then choose **Graphics Window**. Then save the file in a .jpg format.

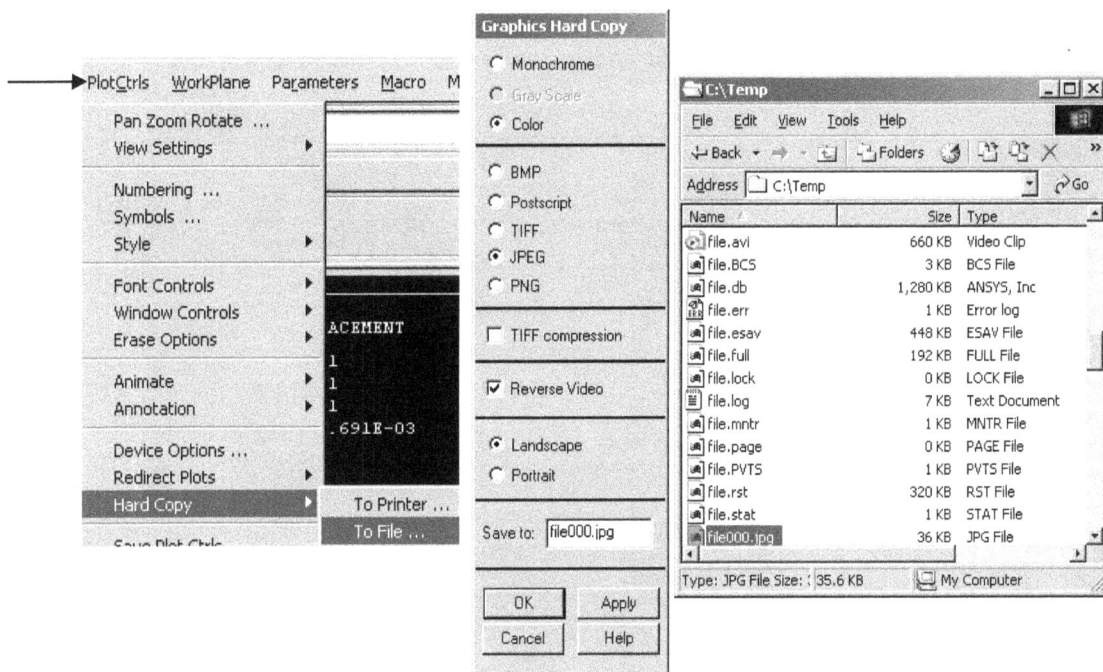

Also, the image can be saved by pressing the "**Print Screen**" key on the keyboard and pasting the image into an image-editing program, such as MS Paint or directly to a Word document.

10.3 FEA with a 3D Solid Model

The following figure illustrates a beam structure. The 3 dimensions of the beam structure are 800 x 100 x 50 mm. The left end of the plate is fixed to the wall. The load acts on part of the top surface, as shown. The right end is set free to move. In terms of the material properties, we assume that the Young's modulus is 30 GPa, and the Poisson's ratio is 0.26. The objective of this study is to obtain the information on the maximum principal stress developed under the load condition.

Uniform load: 100 Newton

(800, 100, 50)

(0,0,0)

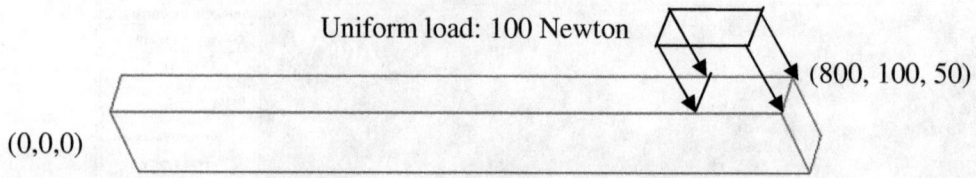

Step 1: Create a 3-D model for the cantilever beam.

From the **ANSYS** main menu, select **Preferences**. Make sure that **Structural** is selected and **h-method** is also selected > **OK**.

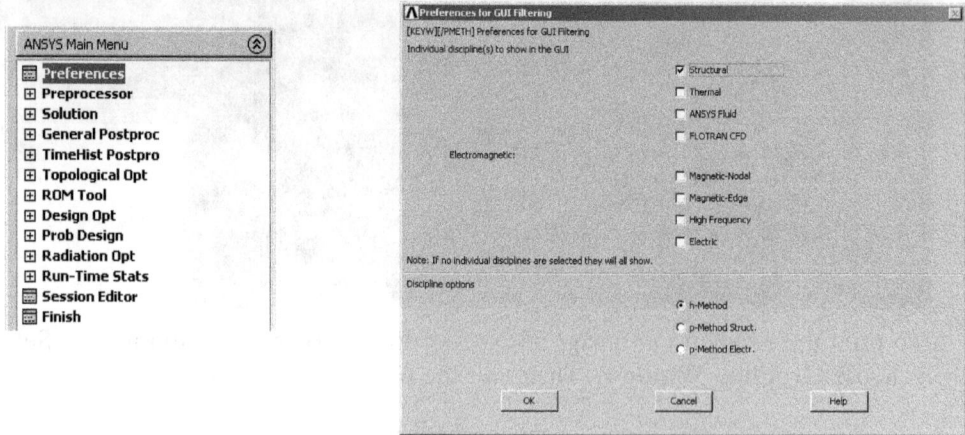

From the main menu, select **Preprocessor** and expand it. Select **Modeling > Create > Volumes > Block > By Dimensions**.

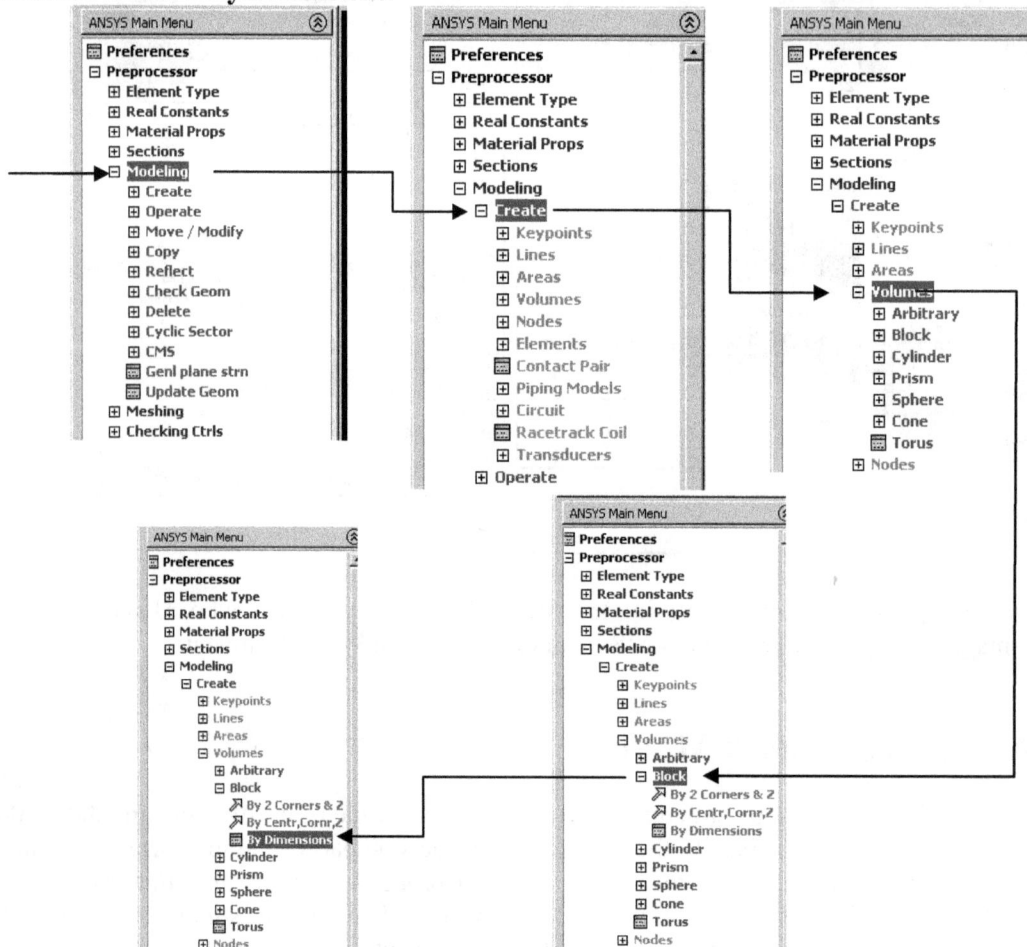

In the window for creating the rectangle by dimensions, type 0 and 0.8 for X1 and X2, type 0 and 0.1 for Y1 and Y2, and type 0 and 0.05 for Z1 and Z2 > **OK**.

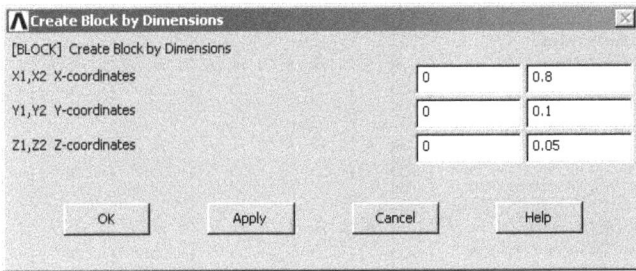

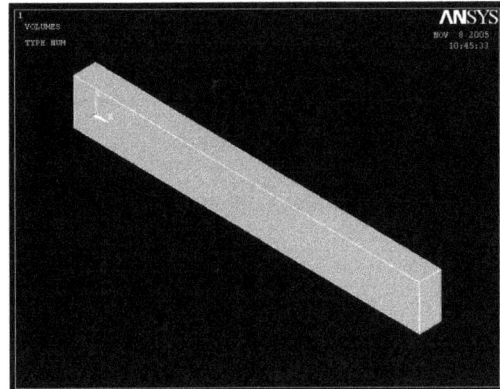

Select **Element Type > Add/Edit/Delete.**

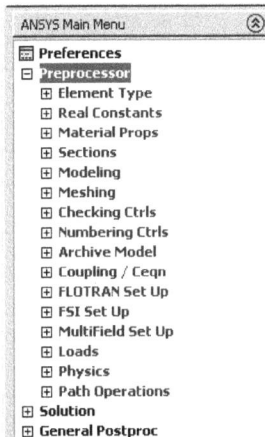

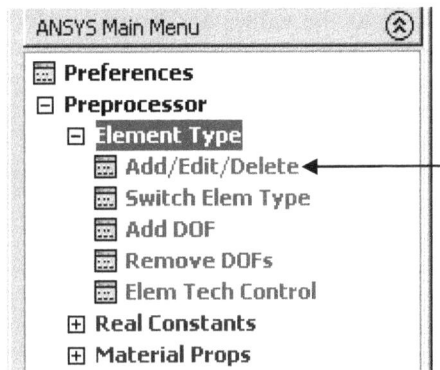

A window called **Element Types** appears. Click **Add > pick Solid > Brick 8node 185 > OK > Close**.

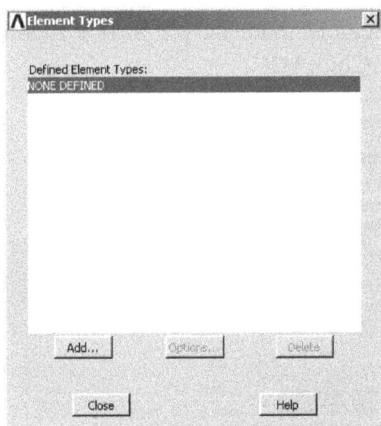

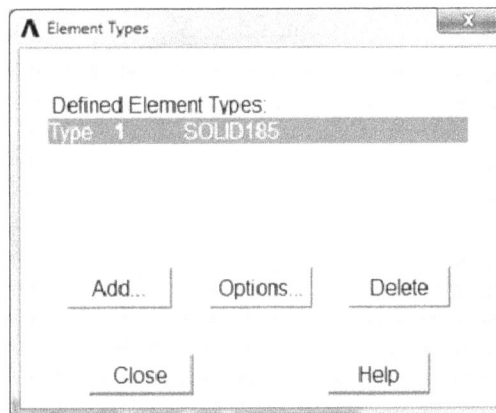

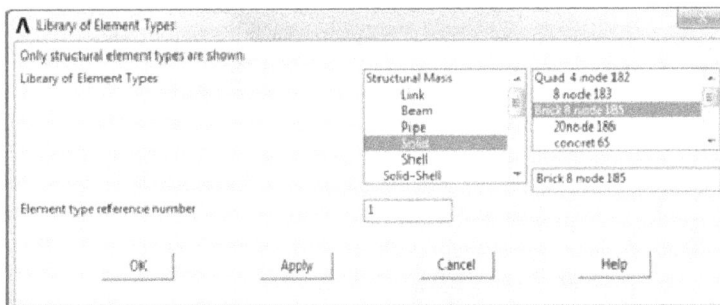

Step 3: Define the Material Properties

To select the material for the shell model, select **Material Properties** from the Preprocessor menu. Select **Material Model**, and then select **Structural, Linear, Elastic** and **Isotropic** by left clicking on each one.

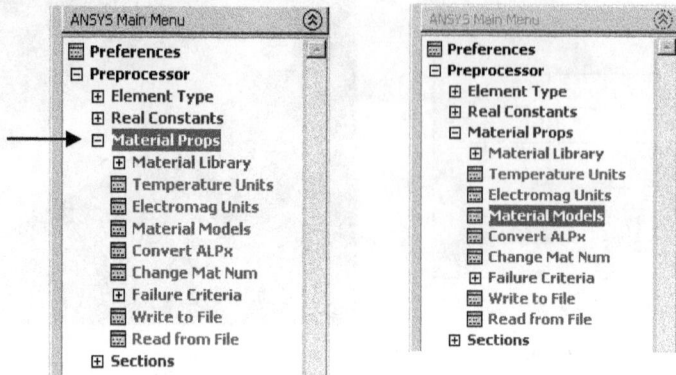

Type 3e10 and 0.26 as the values for Young's Modulus and Poisson's ratio, respectively. Then select **OK**. Click **Material > Exit**.

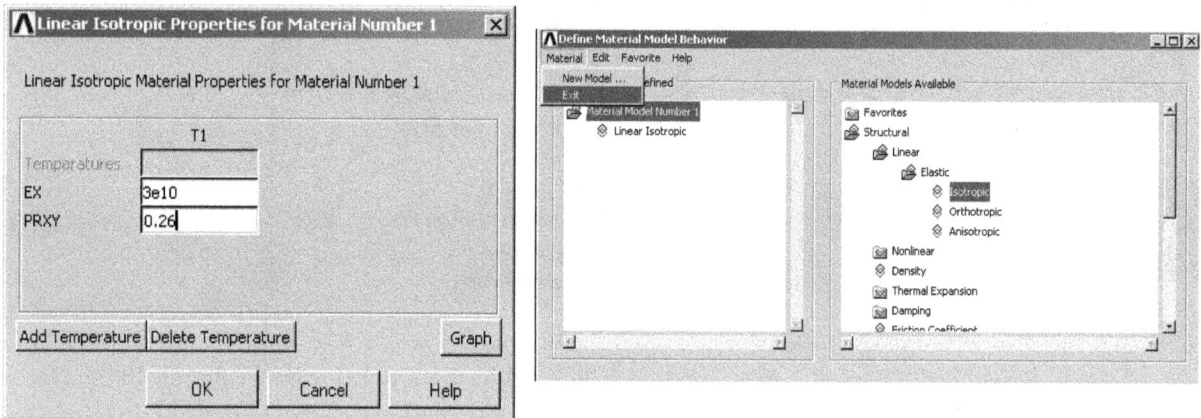

To check the unit system, select **Material Library** under the material properties menu. Then select **Units** and make sure that the **SI** unit system is highlighted. Click **OK**.

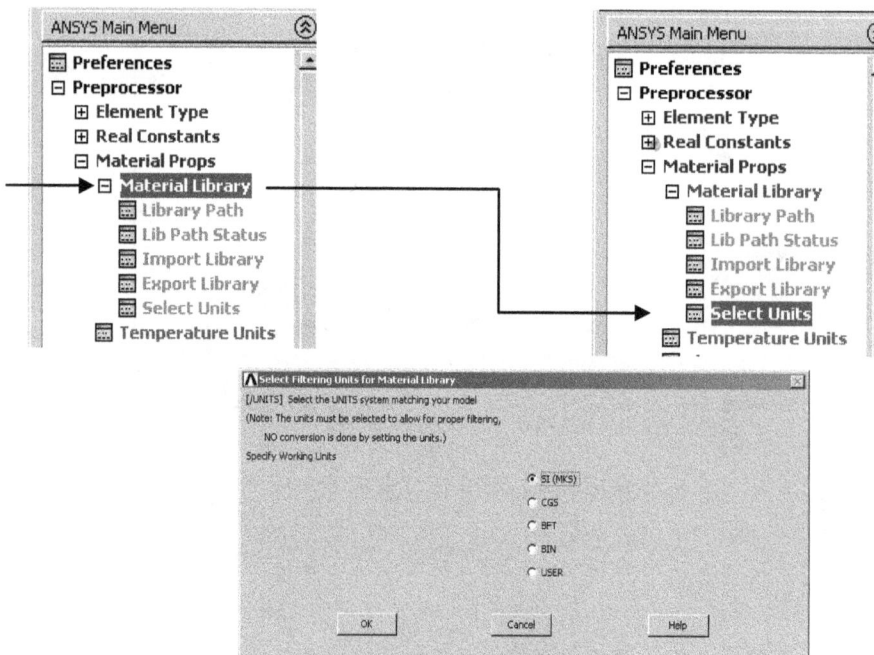

Step 4: Mesh Generation

To create the mesh for the shell model, select **MeshTool** from the Preprocessor menu. Then select **Hex** and **Mapped**, and click **Mesh**.

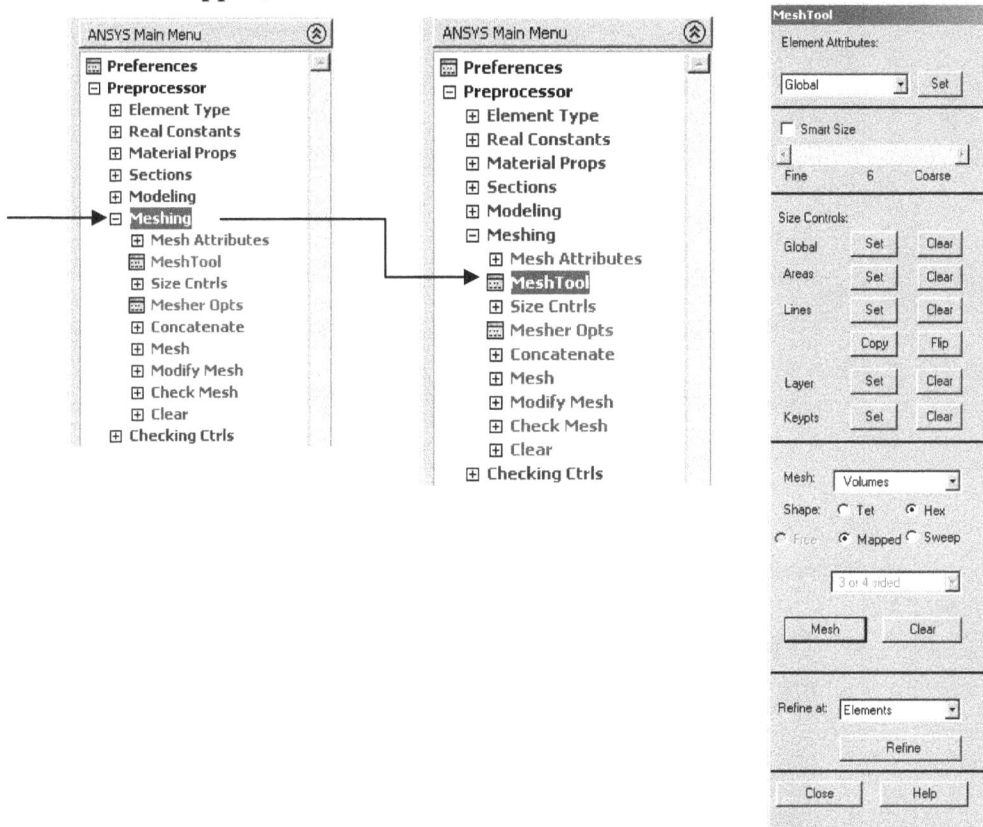

When the selection box comes up, select the volume from the display window, and then click **OK**. Make sure that you close the **MeshTool** window.

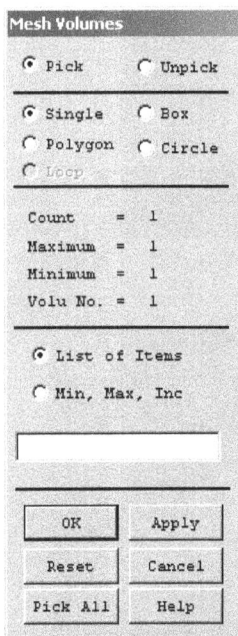

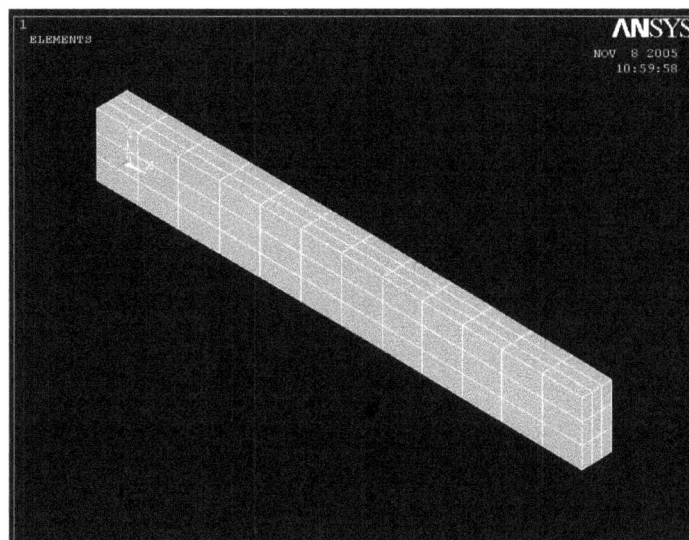

Step 5: Apply the load condition.

To apply the load, on the Preprocessor menu, click **Load > Define Loads > Apply > Structural > Force/Moment > On Nodes**.

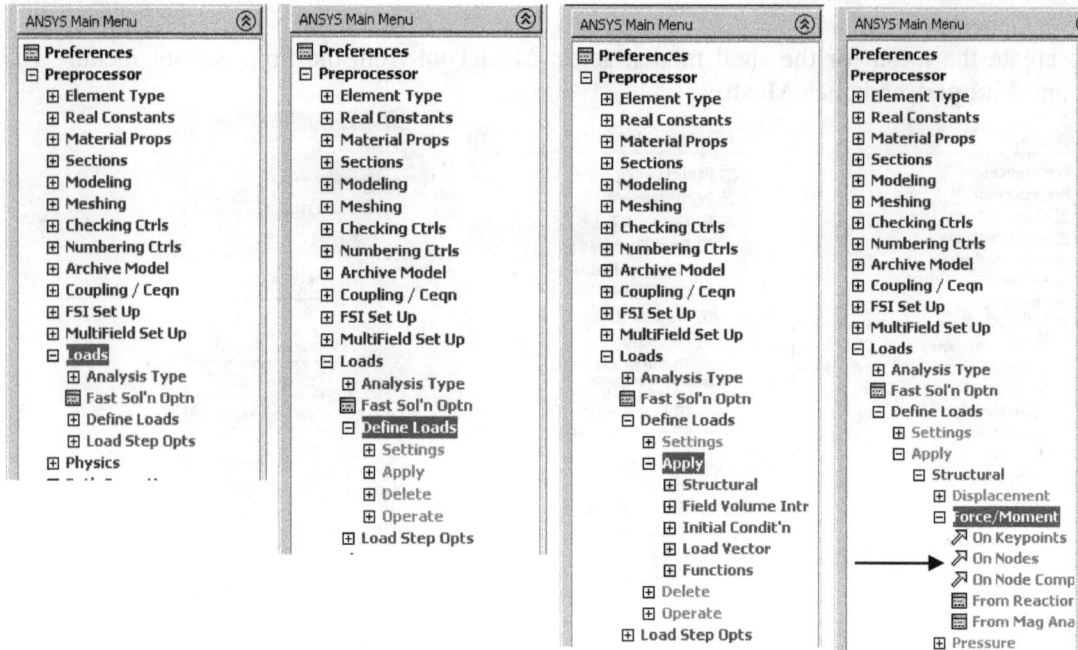

A selection window appears. Pick the 16 nodes from the top surface, as shown > **OK**.

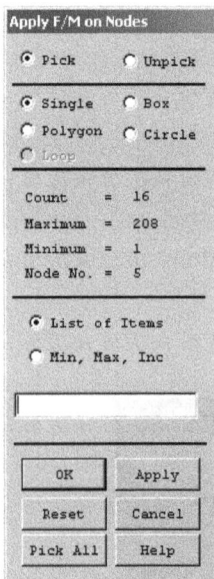

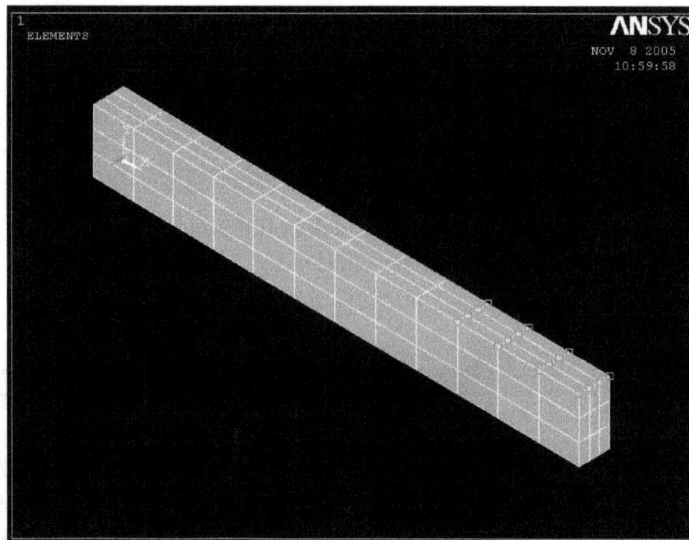

Select **FY** and type −100 as the magnitude of the load.

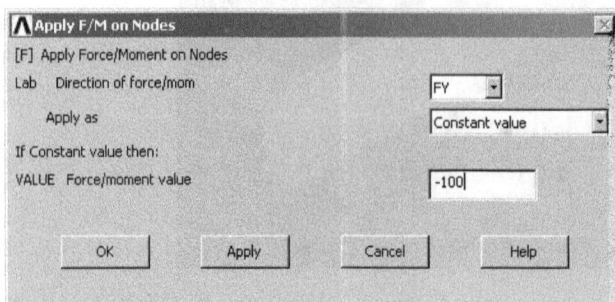

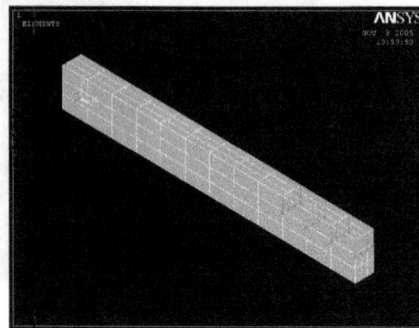

Step 6: Defining the Constraint Condition

To define the constraint conditions, under the Structural menu, select **Displacement > On Nodes**.

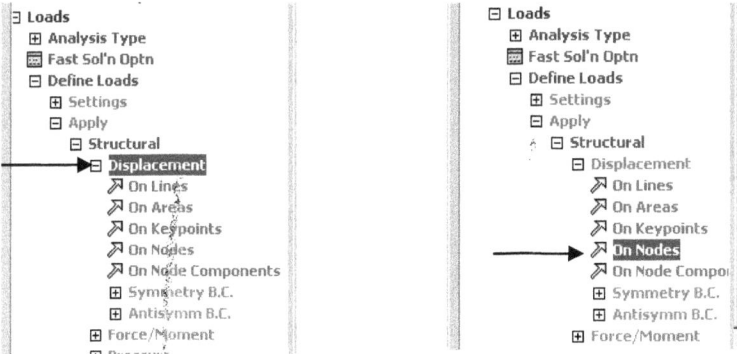

Before selecting the nodes, let us orient the cantilever beam to the FRONT view by clicking the icon of Front View.

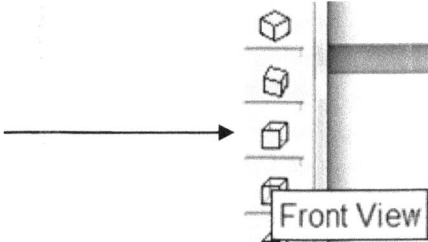

At this moment, a selection window appears. Select Box as the method to pick the nodes. Holding down the right bottom of the mouse, sketch a box on the left side of the cantilever beam, as shown > **OK**.

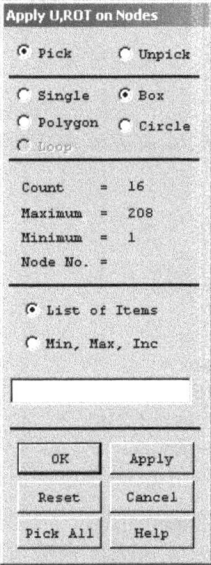

Select **All DOFS**, and type 0 as the value. Then click **OK**.

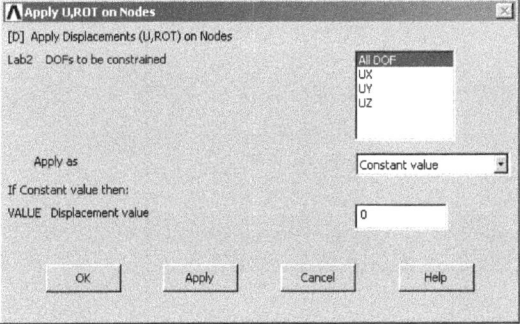

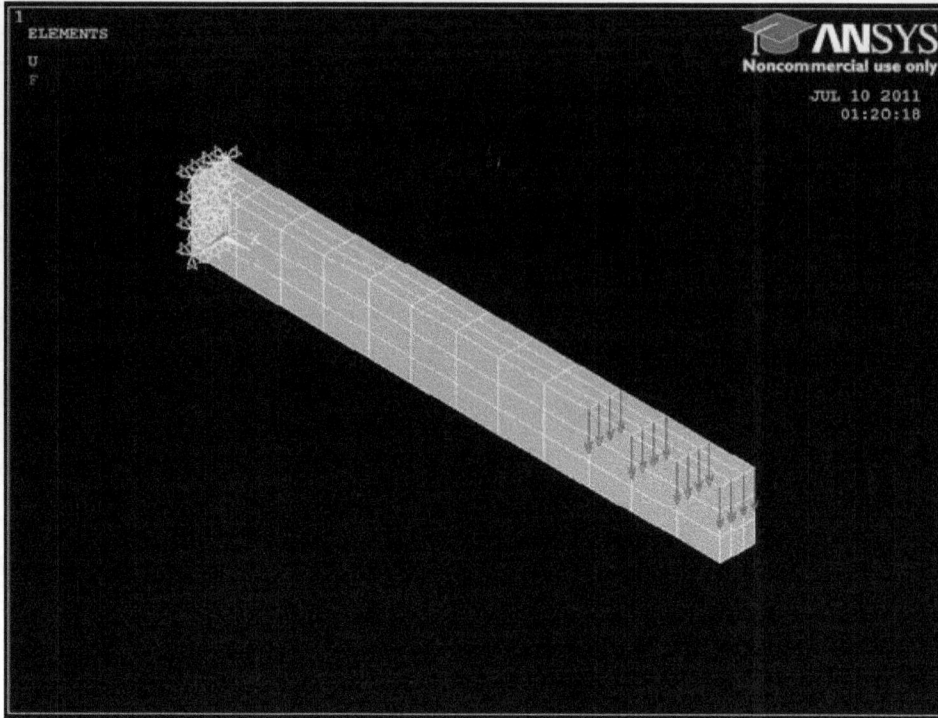

Step 7: Perform the analysis and obtain the solution.

To find the solution, run the program by selecting **Solution** from main menu > **Analysis Type** > **New Analysis** > **Static** > **OK**.

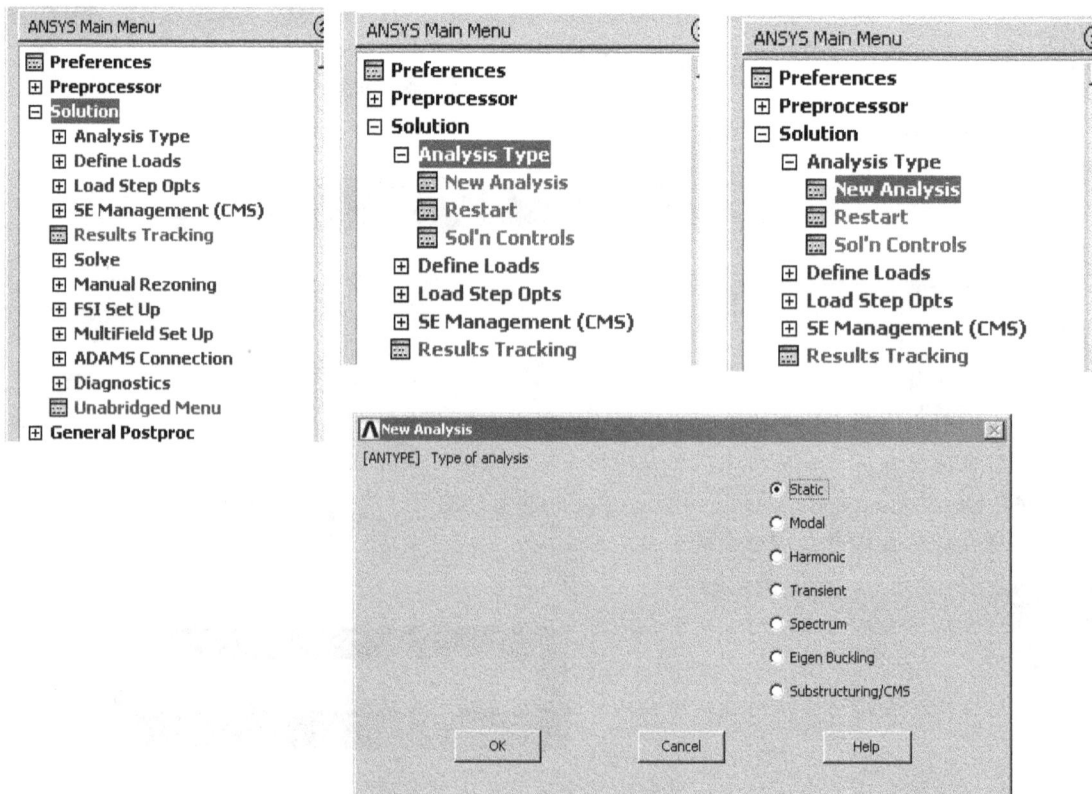

Choose **Current LS** > after reviewing the information, click **OK** to run the program.

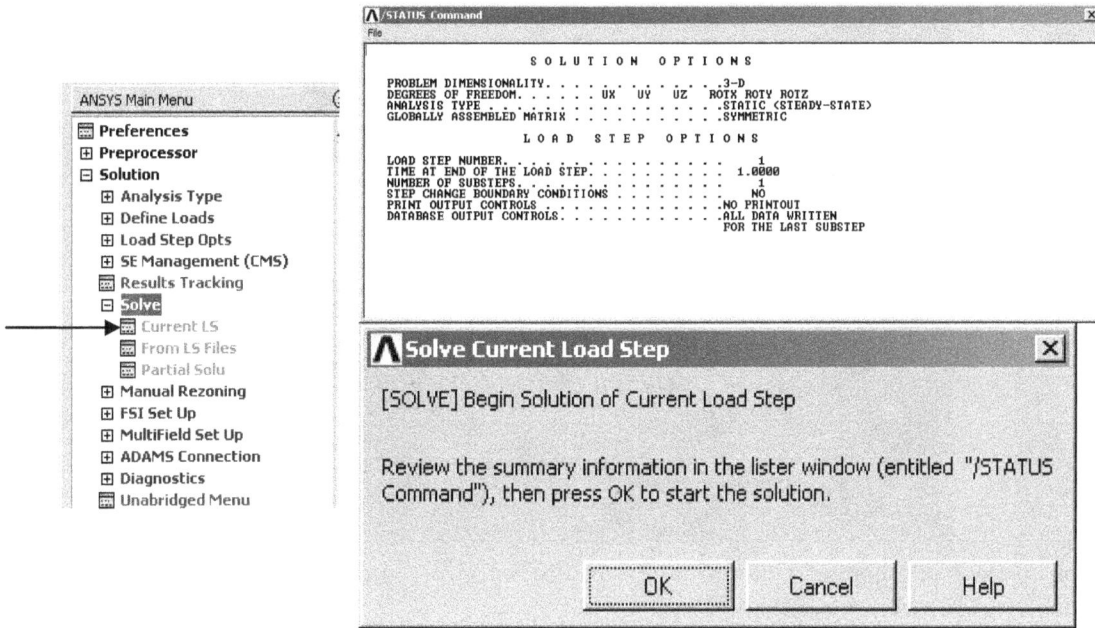

A window called Note appears when the solution is done > **Close**.

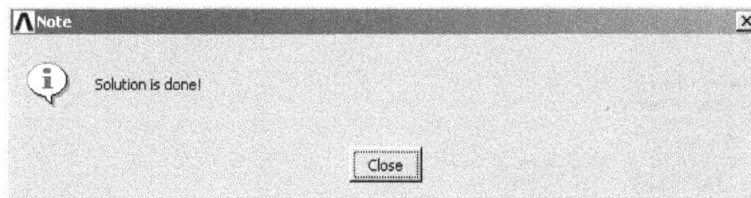

Step 8: Displaying Results

To review the results, select **General Postprocessing** from the main menu. Then click **Plot Results**, > **Deformed Shape** > select **Def + undeformed** > **OK**.

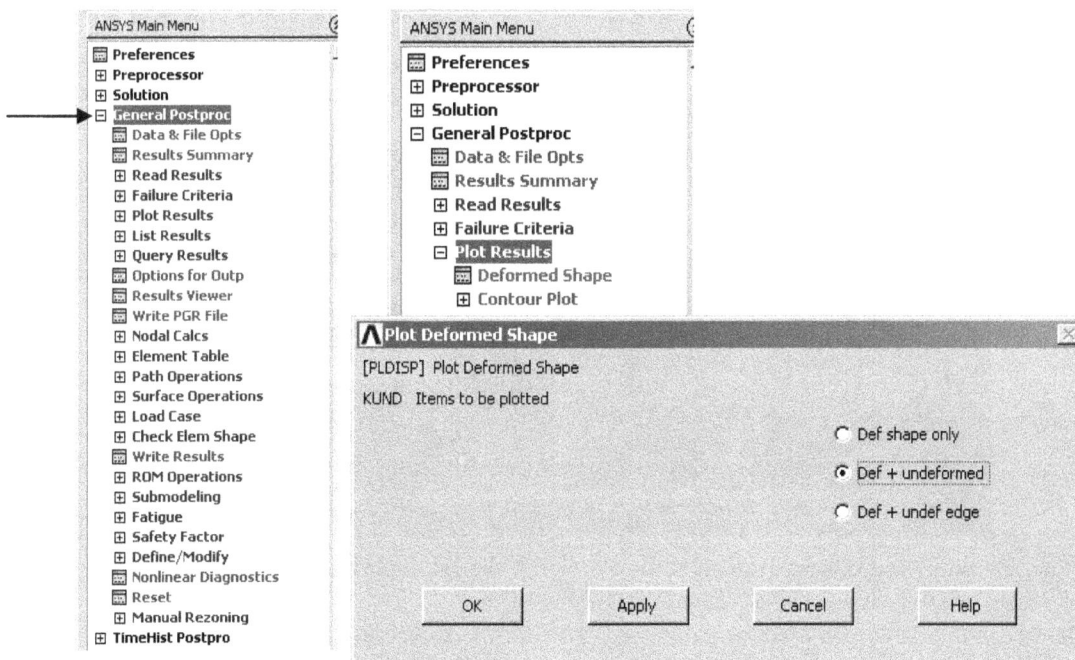

The following plot depicts the deformed shape, showing the maximum deflection is 1.576 mm.

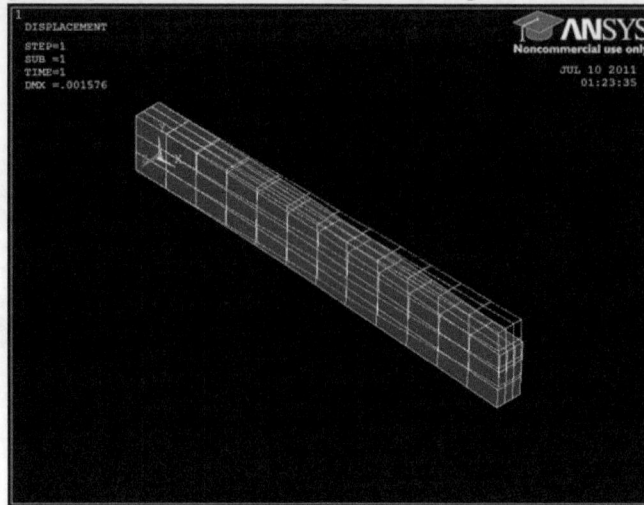

To plot a stress distribution, **Contour Plot > Nodal Solution** > pick a specific type of stress of interest, say the von Mises stress > **OK**.

When examining the information shown on the plot, the maximum stress value (tensile stress) is 3.33 MPa and the minimum stress value (compressive stress) is 0.3.33 MPa.

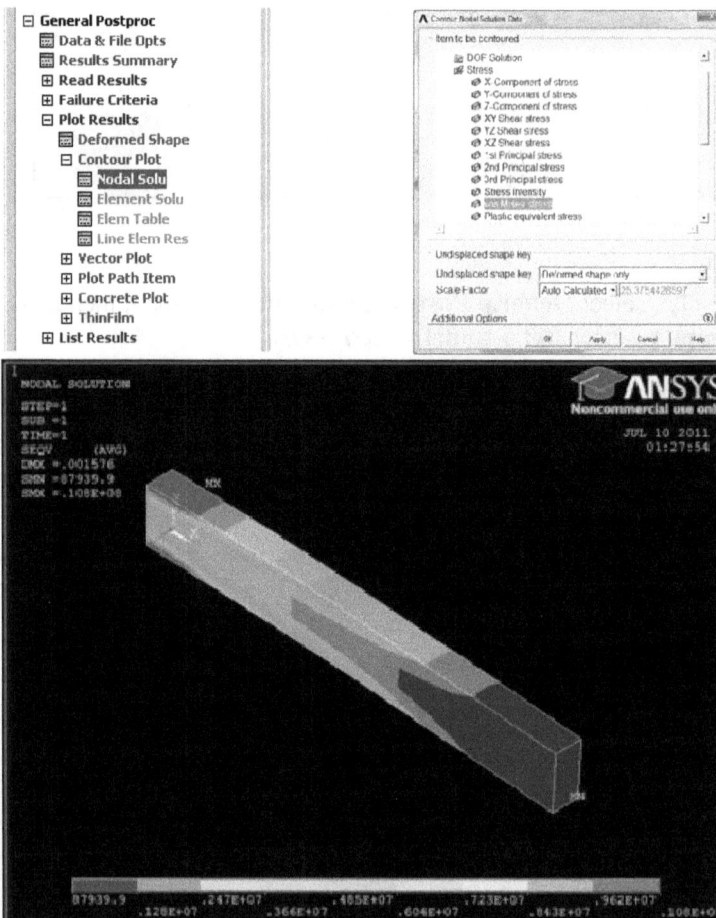

When examining the information shown on the plot, the maximum von Mises stress value is 10.8 MPa and the minimum von Mises stress value is 0.087939 MPa.

If desired, the results can also be animated. To do this, select **Plot Ctrls** from the upper menu, and click **Animation.** Then select **Deformed Shape** and accept the default settings > **OK**.

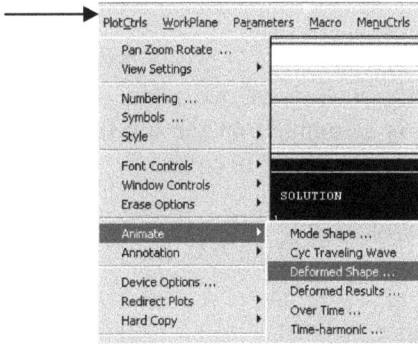

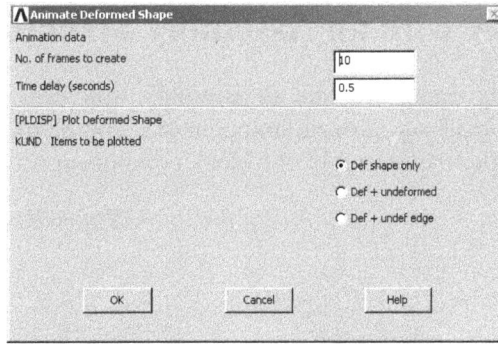

The following plot illustrates the animation in progress. To terminate the animation process, click **Close**.

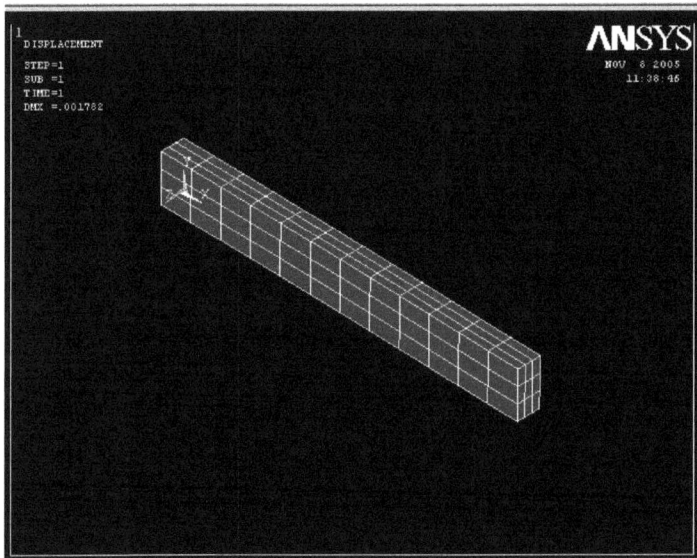

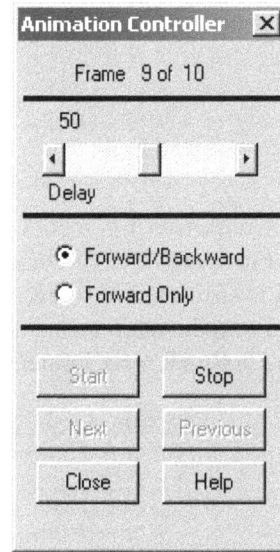

To print the results to an image file, select **PlotCtrls** from the top menu. Select **Hard Copies**, and then choose **Graphics Window**. Then save the file in a .jpg format.

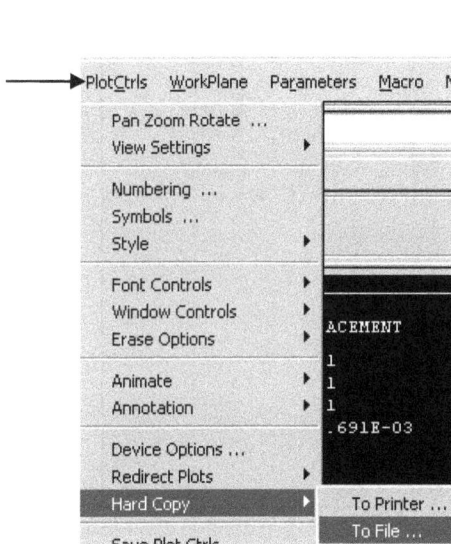

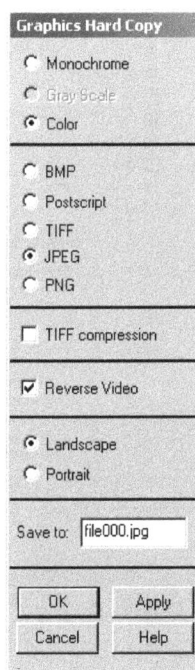

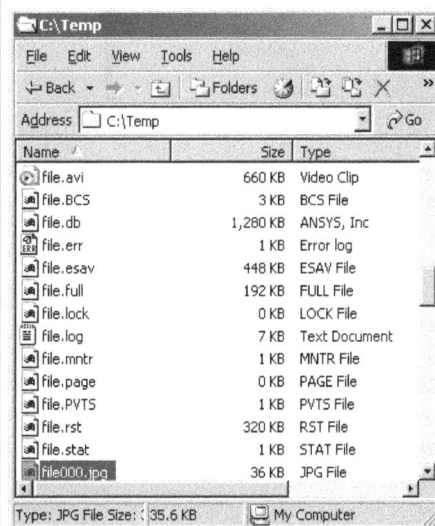

10.4 FEA with an Assembly Structure

The following figure presents an assembly structure. The assembly consists of two parts. They are a block component and a beam component. The block component is assembled on the top of the beam structure. After the assembly, the block component and the beam component are in contact.

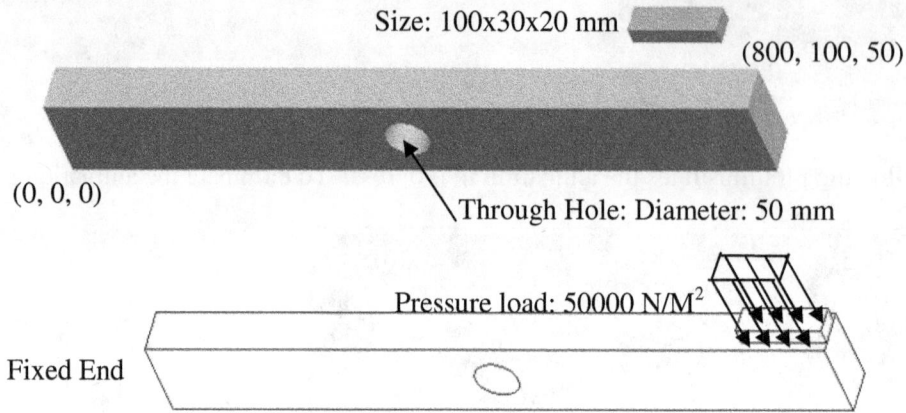

Size: 100x30x20 mm

(800, 100, 50)

(0, 0, 0)

Through Hole: Diameter: 50 mm

Pressure load: 50000 N/M^2

Fixed End

Step 1: Create two 3-D models for the cantilever beam and small block components.

From the ANSYS main menu, select **Preferences**. Make sure that **Structural** is selected and **h-method** is also selected > **OK**.

From the main menu, select **Preprocessor** and expand it. Select **Modeling > Create > Volumes > Block > By Dimensions**.

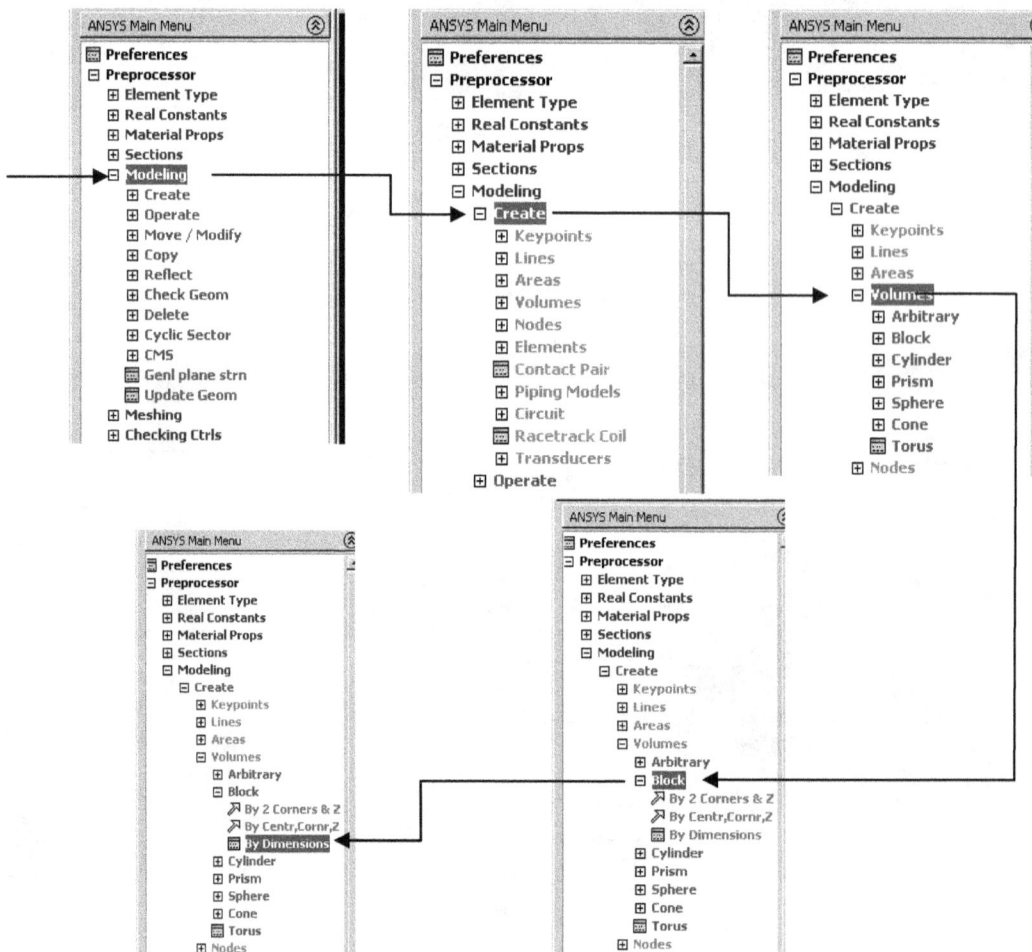

In the window for creating the rectangle by dimensions, type 0 and 0.8 for X1 and X2, type 0 and 0.1 for Y1 and Y2, and type 0 and 0.05 for Z1 and Z2 > **Apply**.

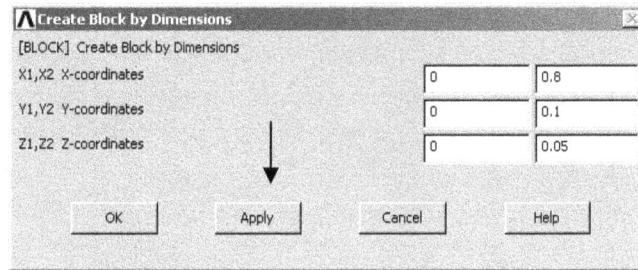

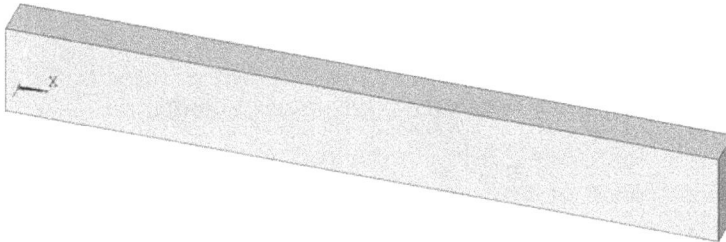

Note that "**Apply**" is used, not "**OK**" so that we are able to continue defining the small block. Type 700 and 800 for X1 and X2, 100 and 120 for Y1 and Y2, and 10 and 40 for Z1 and Z2 > **OK**.

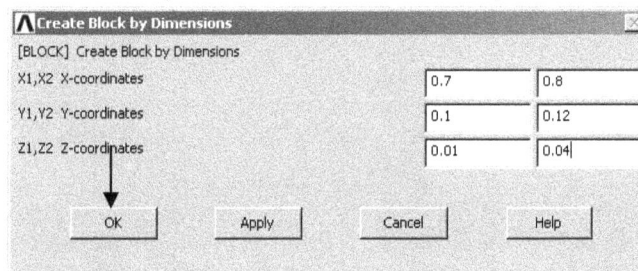

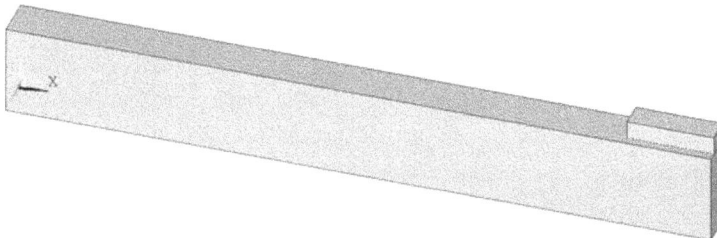

To create a hole at the central location, select **Create > Cylinder > Solid Cylinder**. Type the following values: the Center Location (X=0.4, Y=0.05), Radius (0.03), and Depth (0.05) > **OK**.

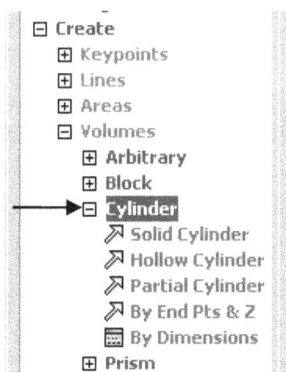

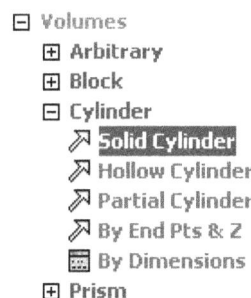

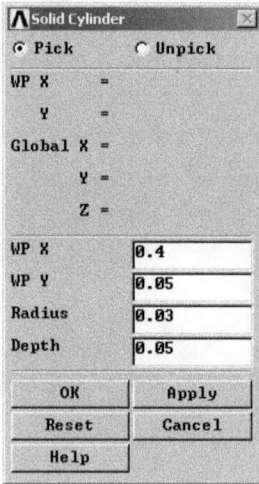

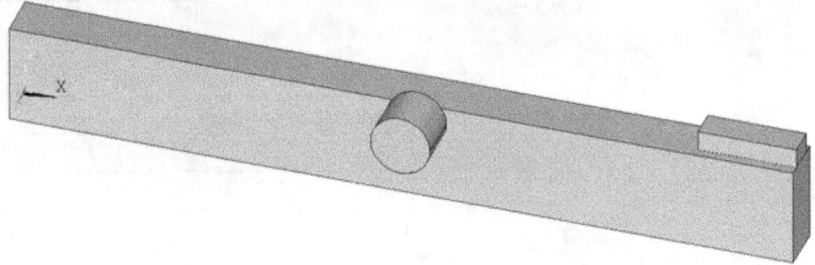

Under **Modeling**, select **Operate > Booleans > Subtract > Volume**.

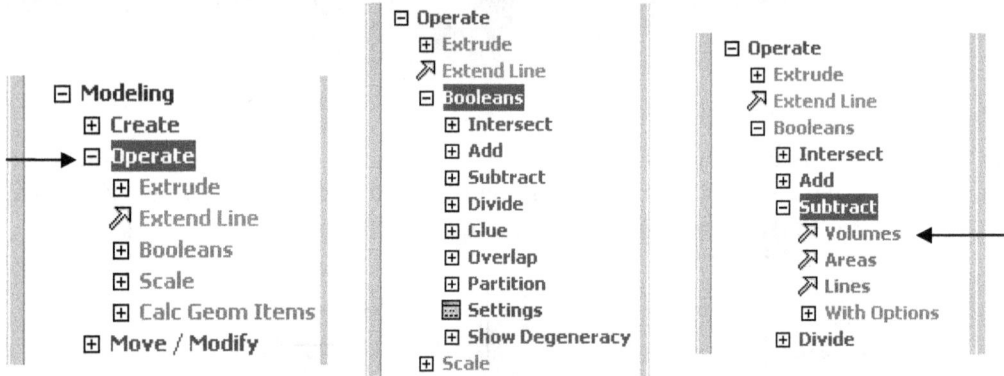

A window called **Subtract Volumes** appears, asking you to pick the base volume from which to subtract. Let us pick the big block or the cantilever beam first. A pop up window appears, indicating that the volume you picked is Volume 1. Because your pickup is correct, click **OK**. Afterwards, in the Subtract Volumes window, click **OK,** completing the process of picking up the base volume.

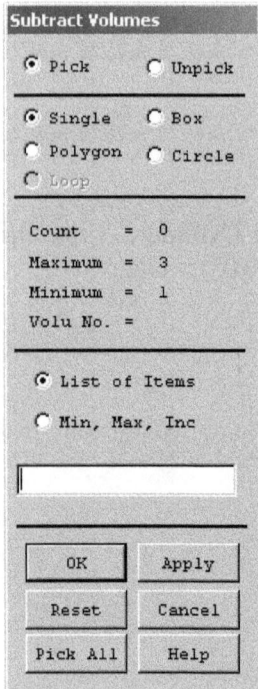

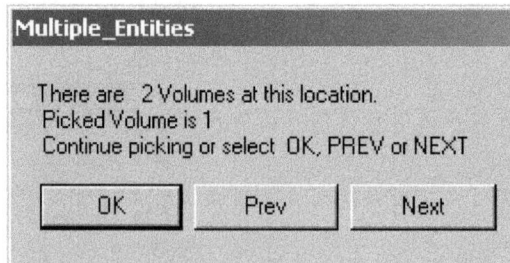

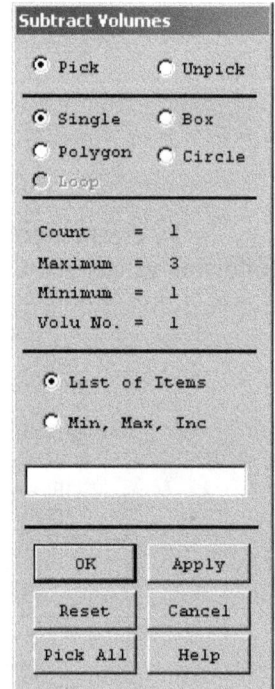

At this moment, the Subtract Volumes window is updated. The software system is asking the user to pick the volume to be subtracted. Pick the cylindrical volume. A pop up window appears, indicates that the volume you picked is Volume 3. Because your pickup is correct, click **OK**. Afterwards, in the Subtract Volumes window, click **OK**, completing the Booleans operation of subtraction.

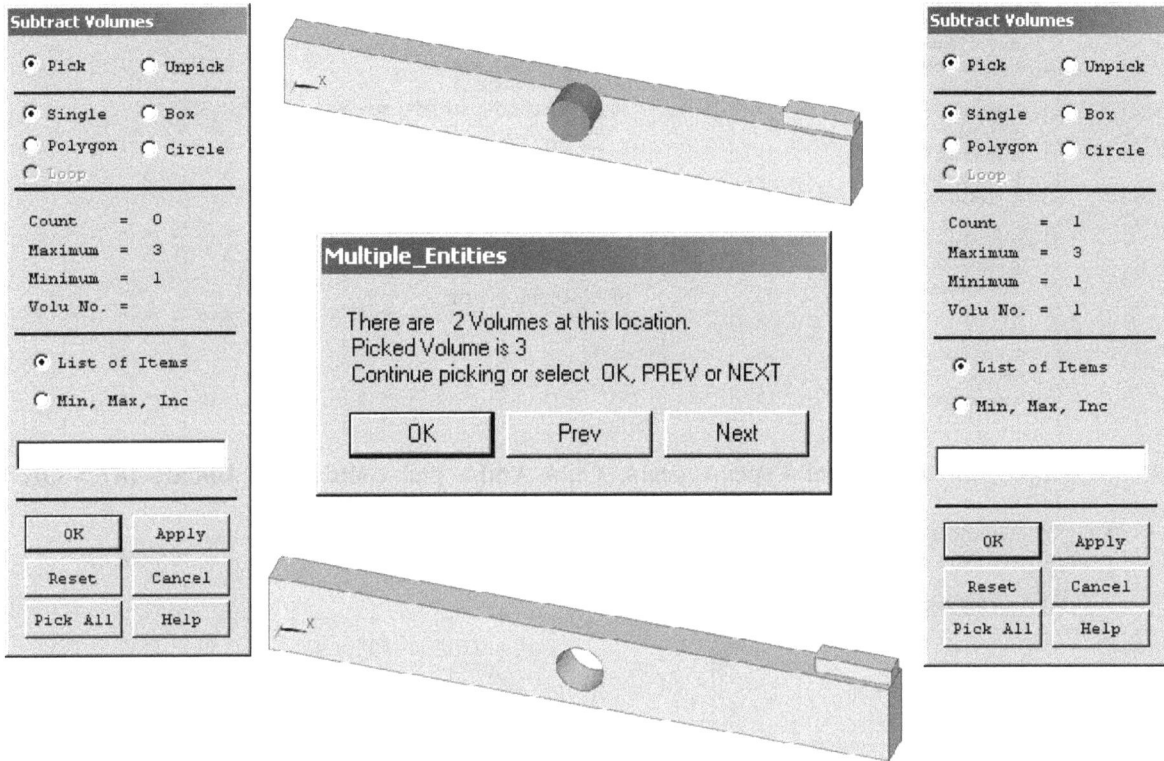

Step 2: Assemble the 2 Components together.

Now let us assemble the small block to the big block. **Operate > Booleans > Glue >** pick both blocks or just click the box called Pick All because there are only 2 volumes > **OK.**

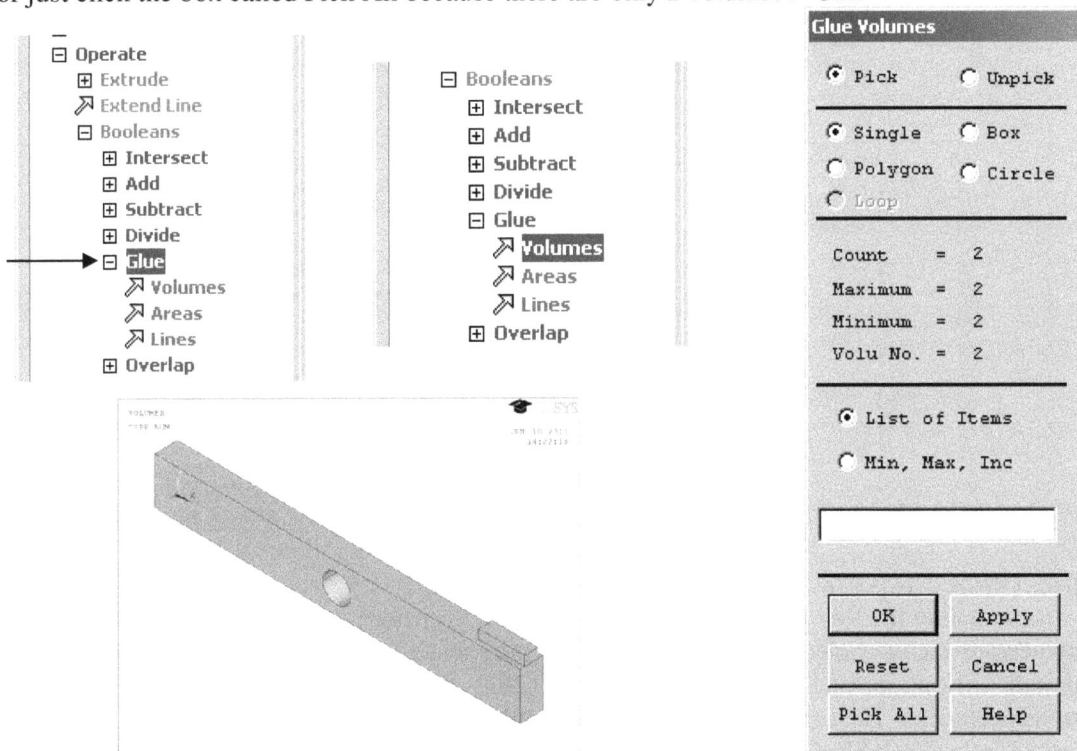

Step 3: Select Element Type
Select **Element Type > Add/Edit/Delete.**

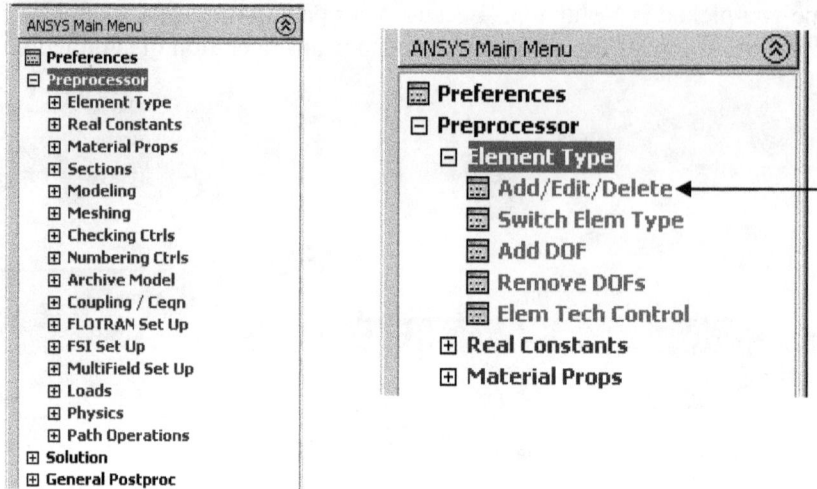

A window called **Element Types** appears. Click **Add > pick Solid > Tet 10node 187 > OK > Close**.

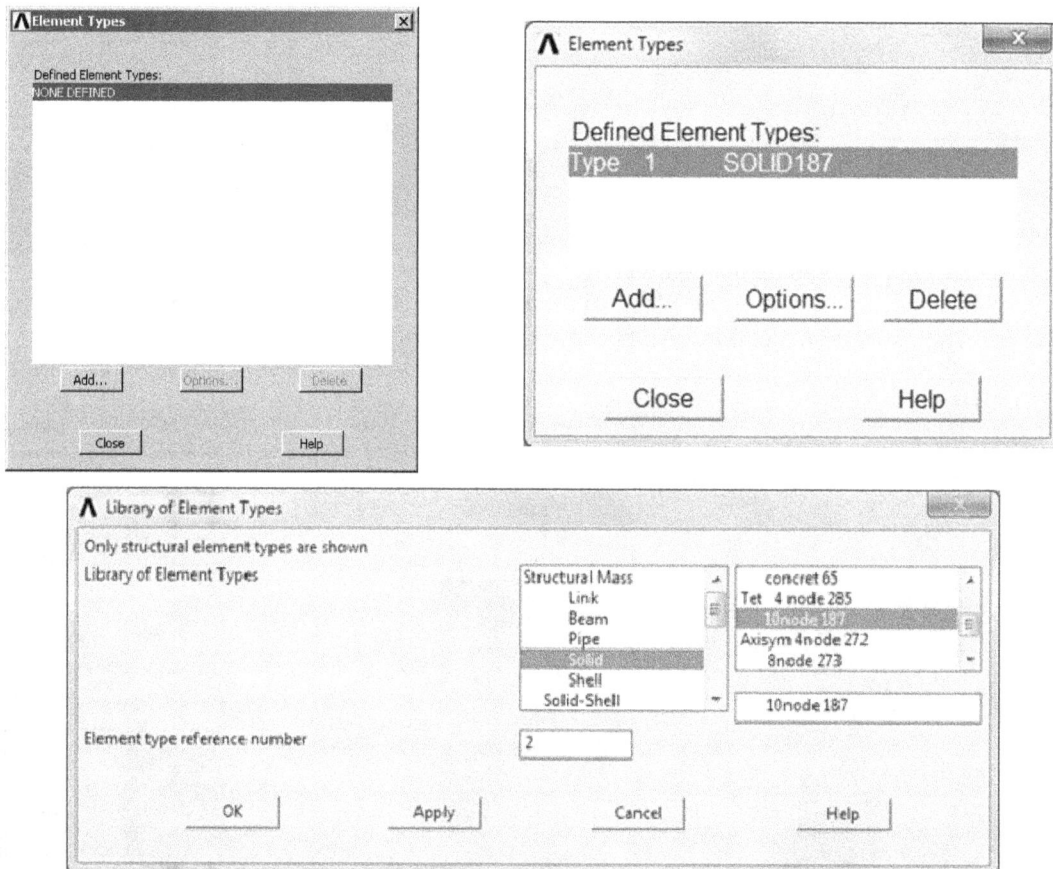

Step 4: Define 2 Types of Material for the Assembly System
To define the first type of material for the cantilever beam, select **Material Properties** from the **Preprocessor** menu. Select **Material Model**, and then select **Structural, Linear, Elastic** and **Isotropic** by double clicking on each one. Type 3e10 and 0.26 as the values for Young's Modulus and Poisson's ratio, respectively. Then select **OK**. Click **Material > New Model > OK** to define the second type of material for the small block > type in 3e9 and 0.40 > **OK > Material > Exit**

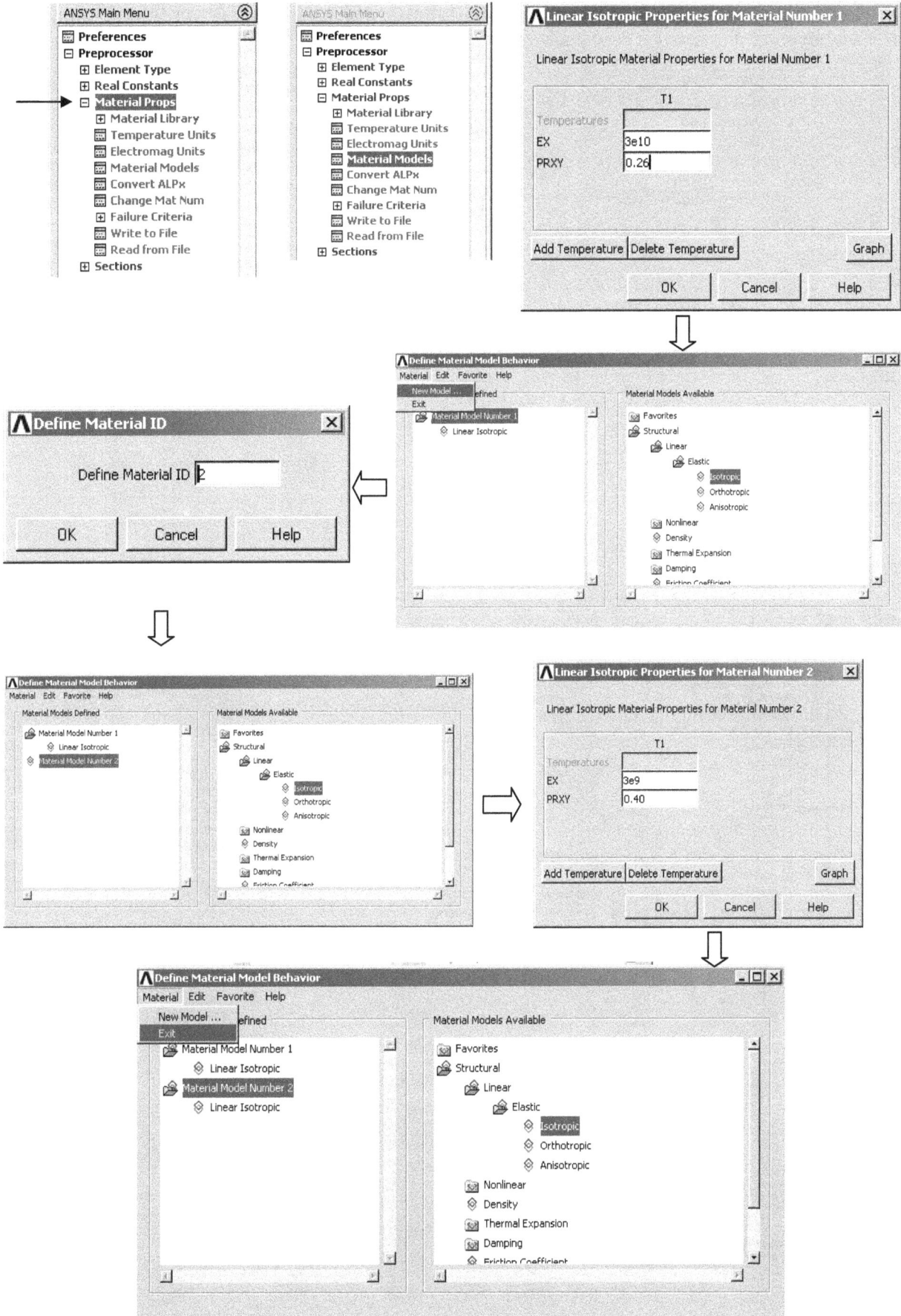

To check the unit system, select **Material Library** under the material properties menu. Then select **Units** and make sure that the **SI** unit system is highlighted. Click **OK**.

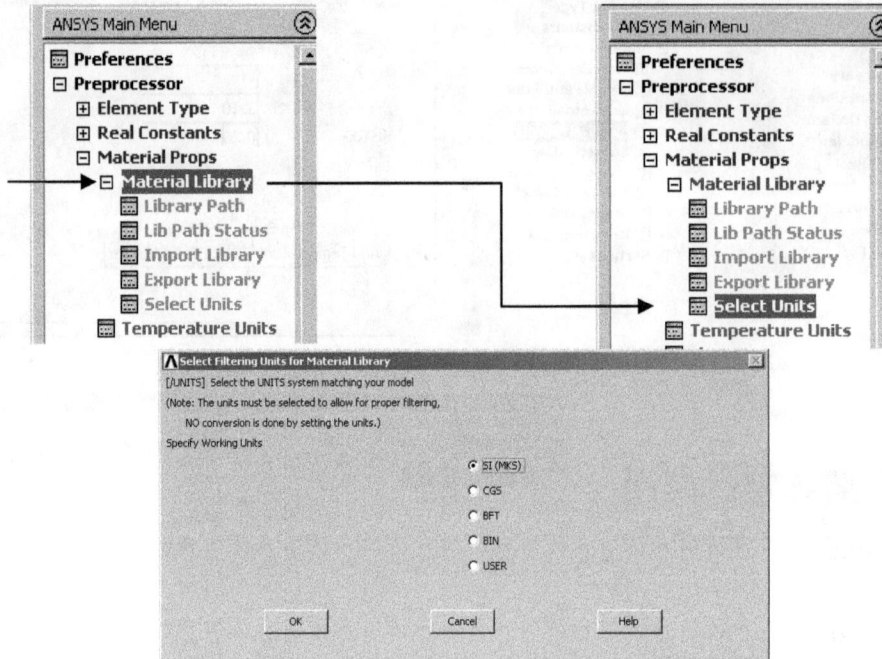

Step 5: Mesh Generation

To create the mesh for the shell model, select **MeshTool** from the Preprocessor menu. Change Global to **Volumes > Set >** pick the cantilever beam **>** OK. Make sure that Material Type 1 is the material type associated with the cantilever beam.

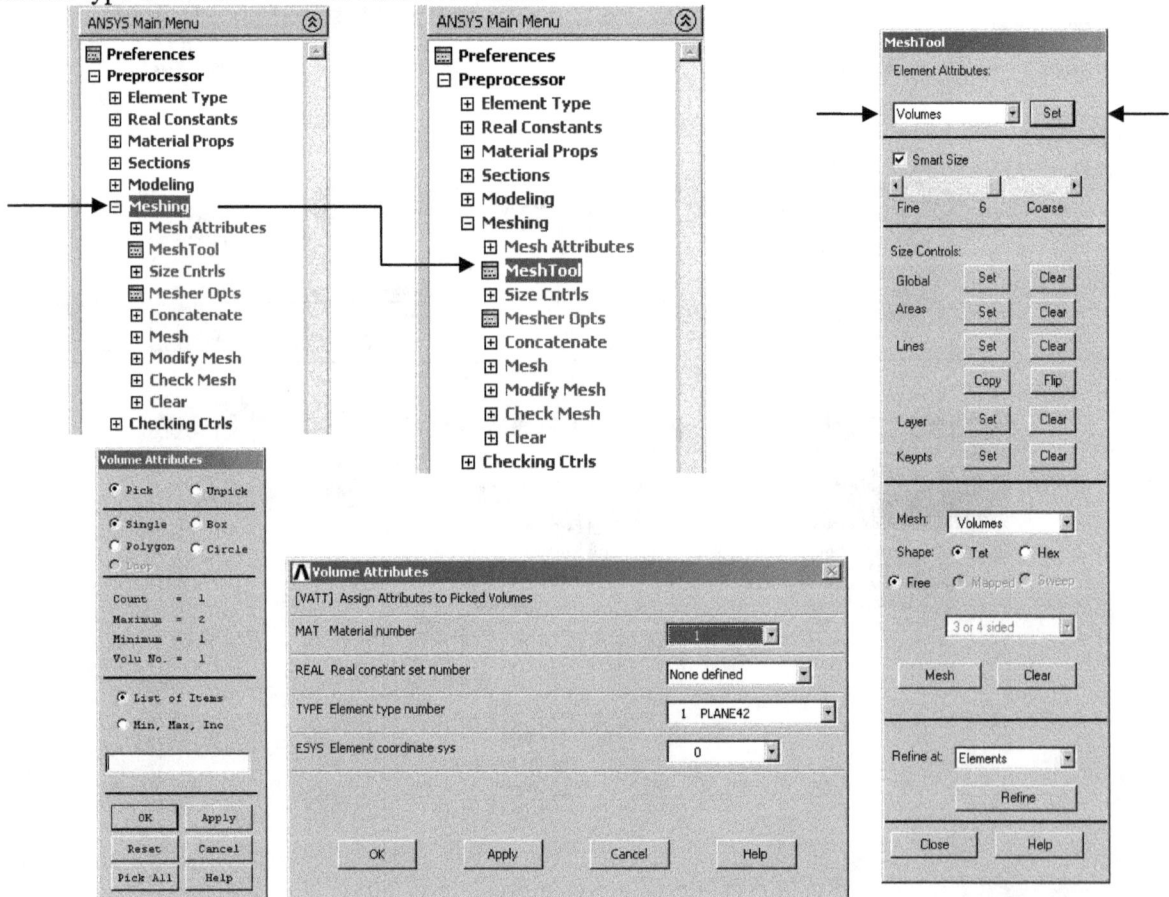

Again, click **Set** > pick the small block > **OK**. Make sure that Material Type 2 is the material type associated with the small block.

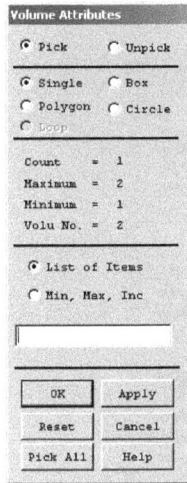

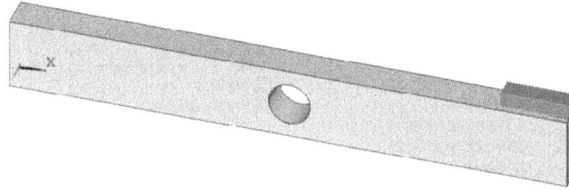

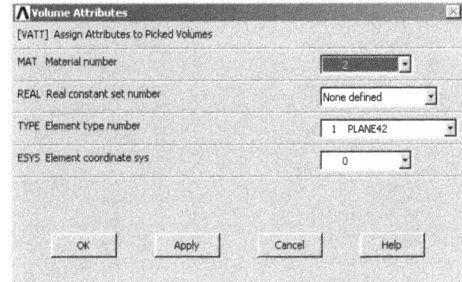

Select **Smart Size > Volume > Tet > Free > Mesh**. When the selection box comes up, click **Pick All > OK**. Click Close to close the **Mesh Tool** window.

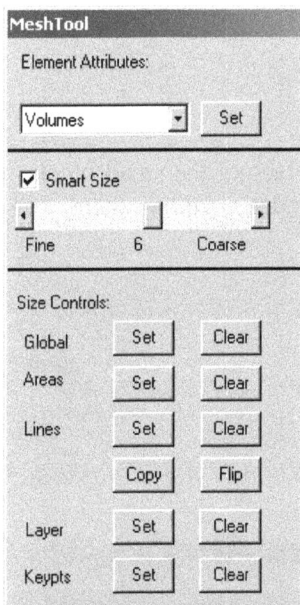

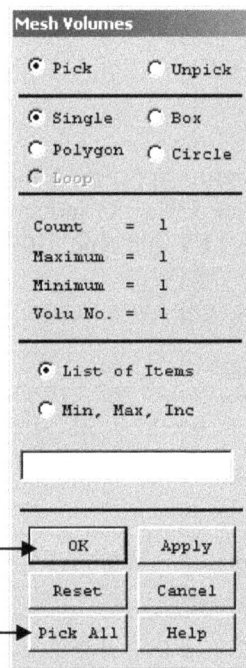

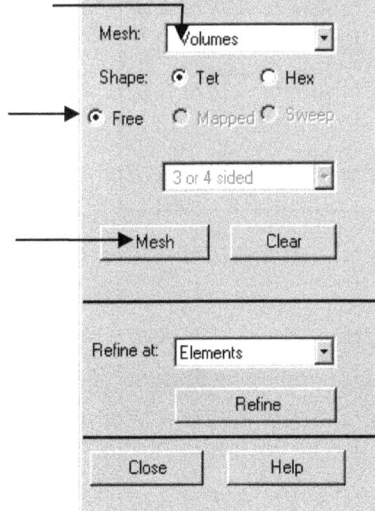

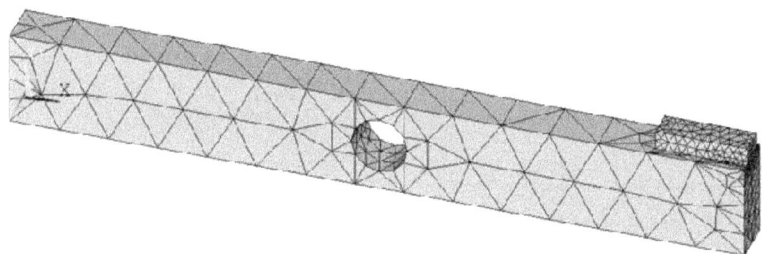

Step 5: Apply the load condition.

To apply the load, on the Preprocessor menu, click **Load > Define Loads > Apply > Structural > Force/Moment > On Nodes**.

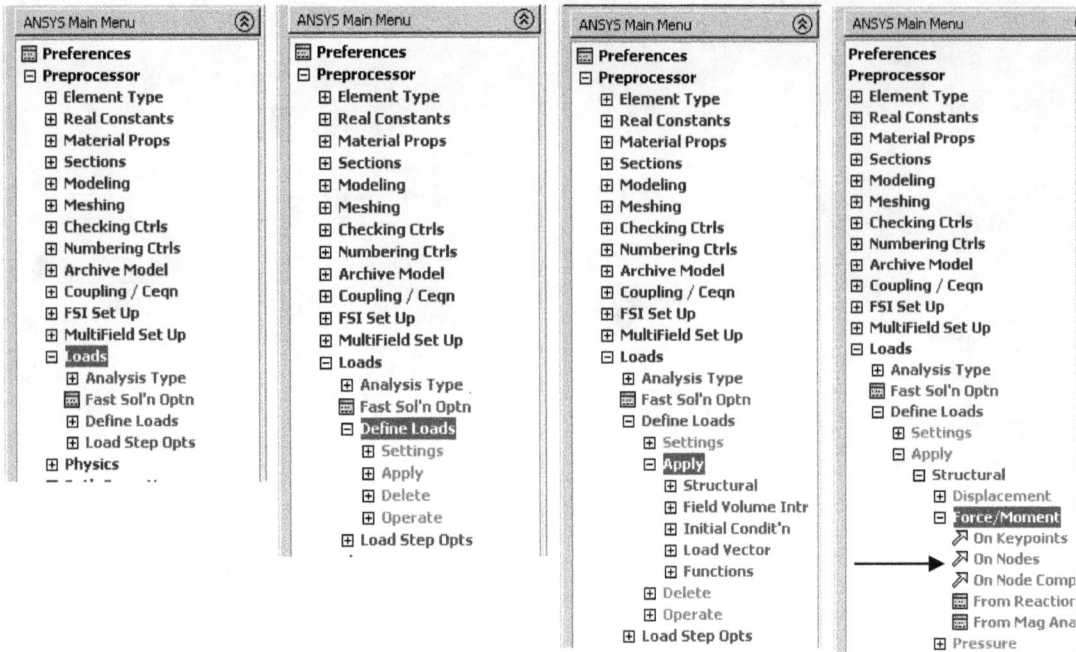

A selection window appears. Pick all the elements on the top surface of the small block (use the choice of **Box**) and click **OK**. As indicated, there are a total of 3213 nodes. Among them, 163 nodes have been picked. Select **FY** and type −100 as the magnitude of the load.

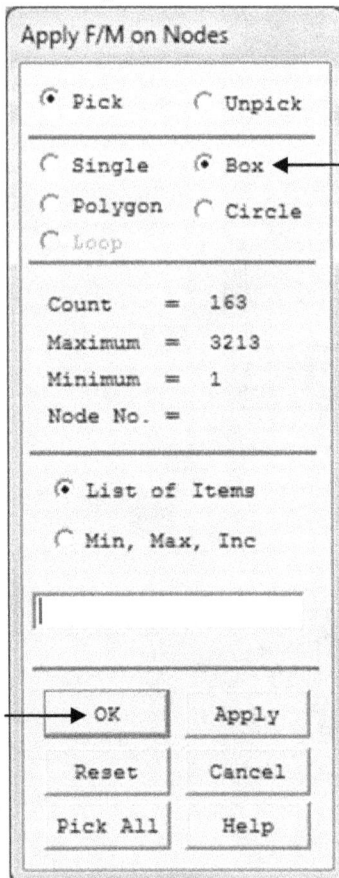

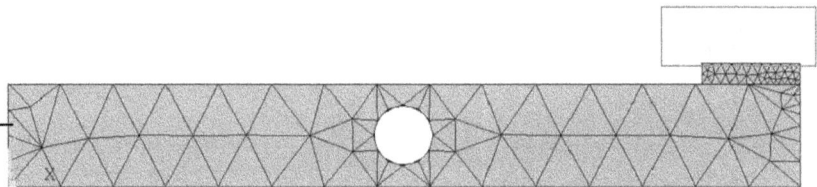

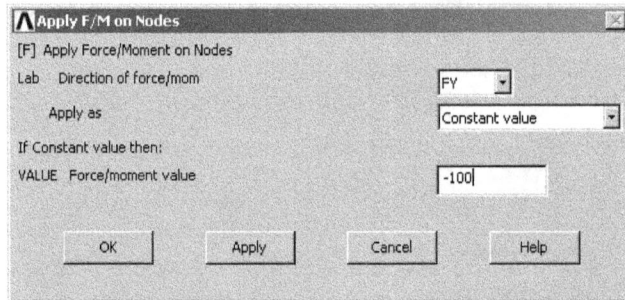

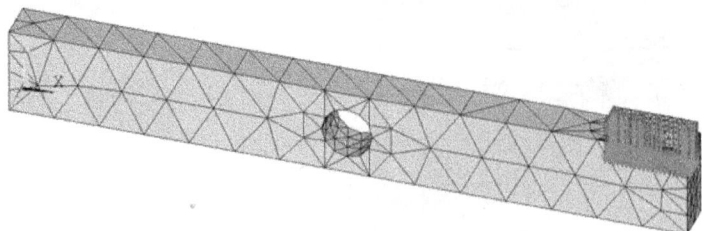

Step 6: Defining the Constraint Condition
To define the constraint conditions, under the Structural menu, select **Displacement > On Nodes**.

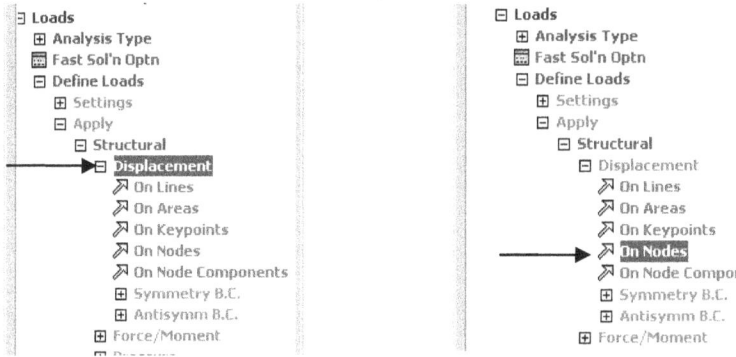

Before selecting the nodes, let us orient the cantilever beam to the FRONT view by clicking the icon of Front View.

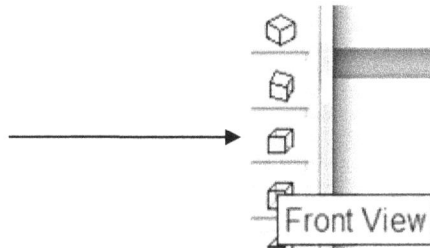

At this moment, a selection window appears. Select **Box** as the method to pick the nodes. Holding down the right bottom of the mouse, sketch a box on the left side of the cantilever beam, as shown > **OK**. As indicated, there are a total of 3213 nodes. Among them, 57 nodes have been picked. Select **All DOFS**, and type 0 as the value. Then click **OK.**

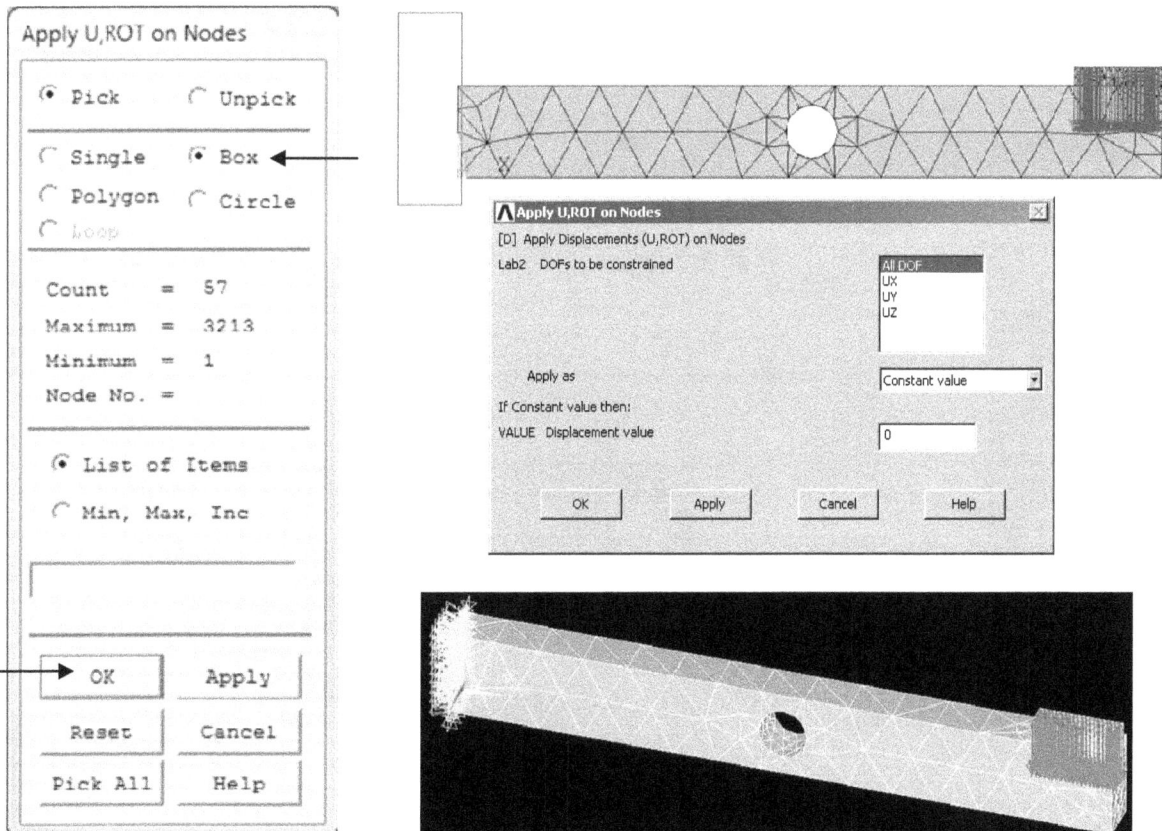

Step 7: Perform the analysis and obtain the solution.

To find the solution, run the program by selecting **Solution** from main menu > **Analysis Type** > **New Analysis** > **Static** > **OK**.

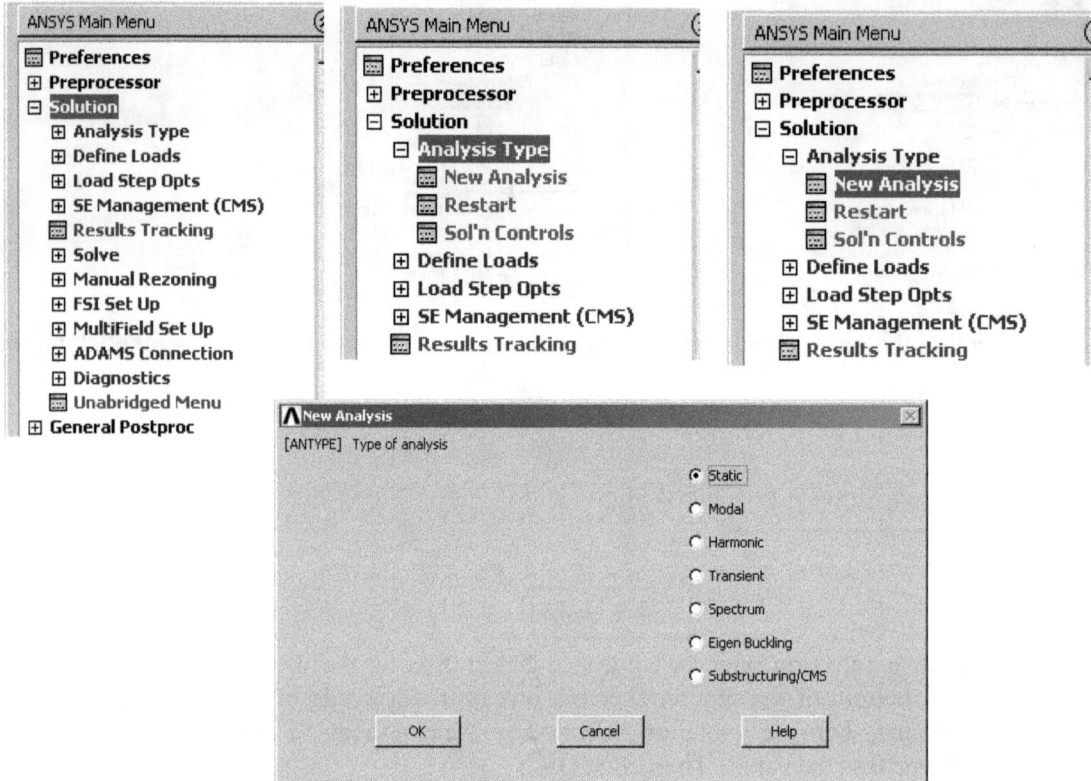

Choose **Current LS** > after reviewing the information, click **OK** to run the program.

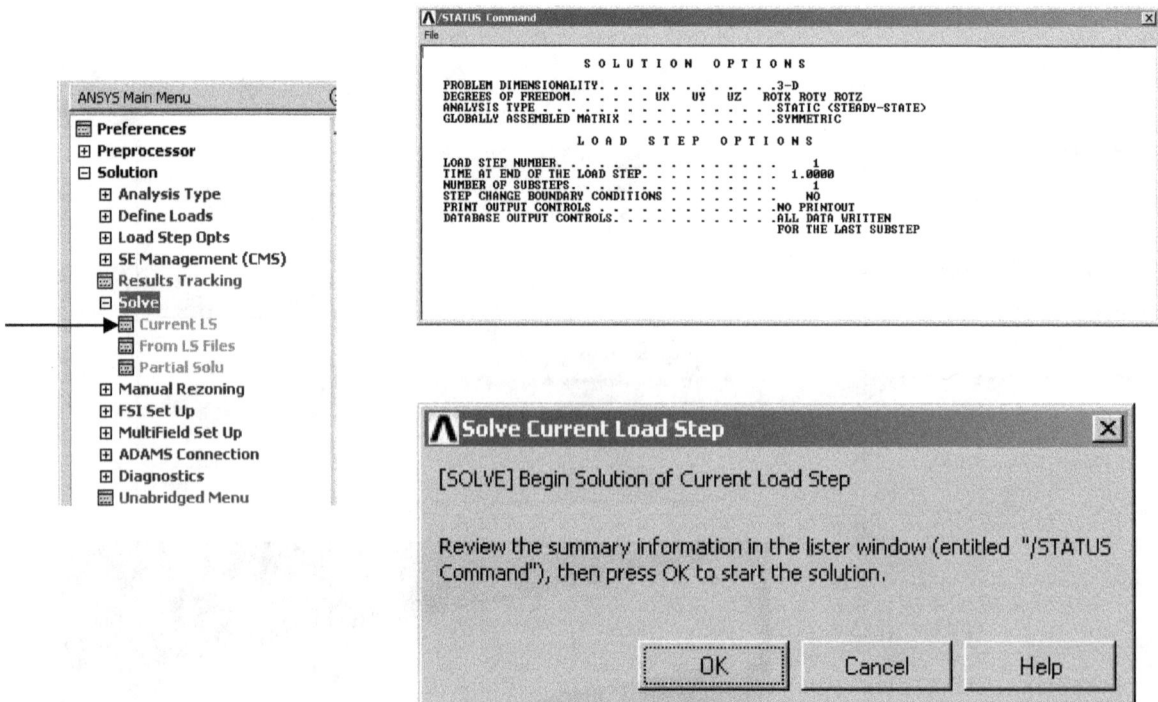

A window called Note appears when the solution is done > **Close**.

Step 8: Displaying Results

Under **General Postprocessing**, select **Plot Results > Contour Plot > Nodal Solution**. Select **Stress** from the list, say the von Mises stress > **OK**.

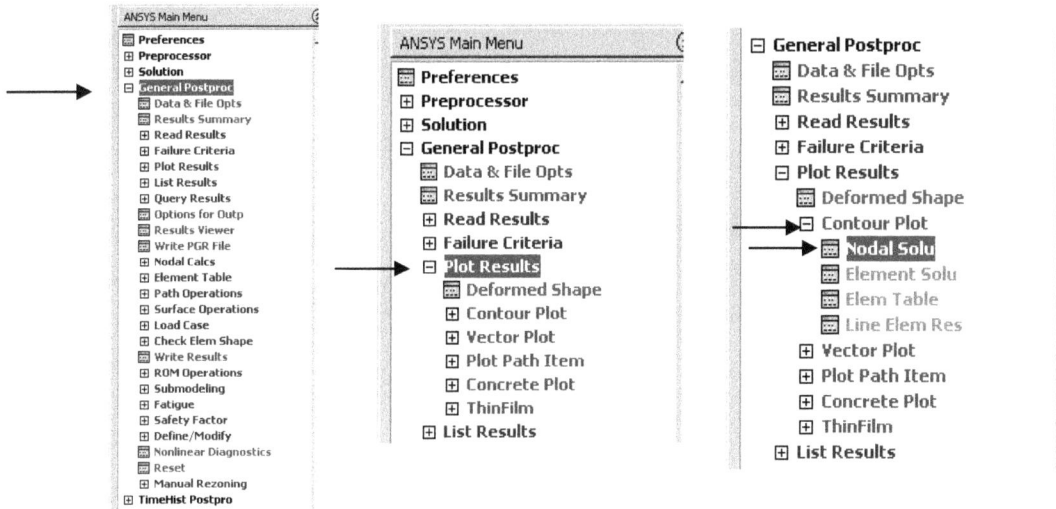

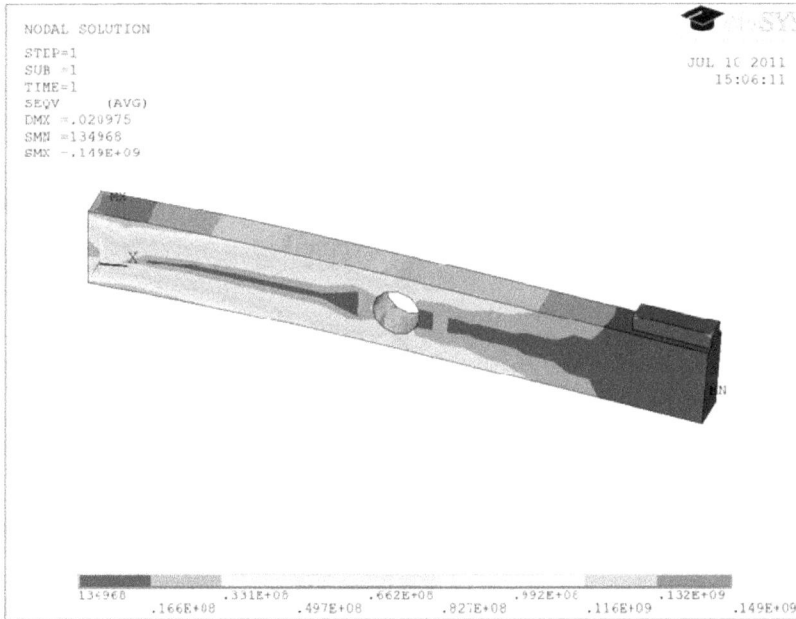

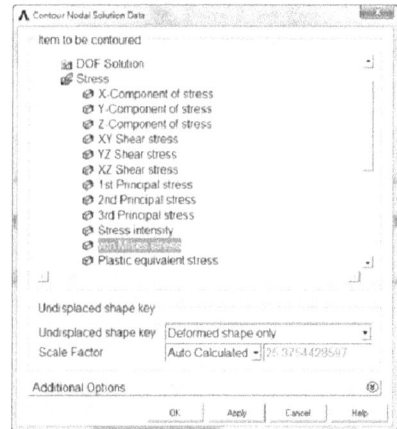

Again, select **Element Solution > Stress > Ist Principal Stress > OK**.

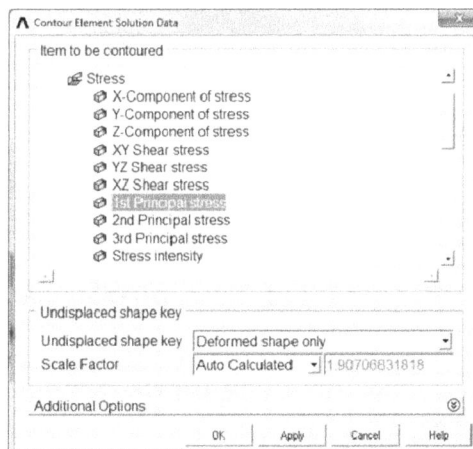

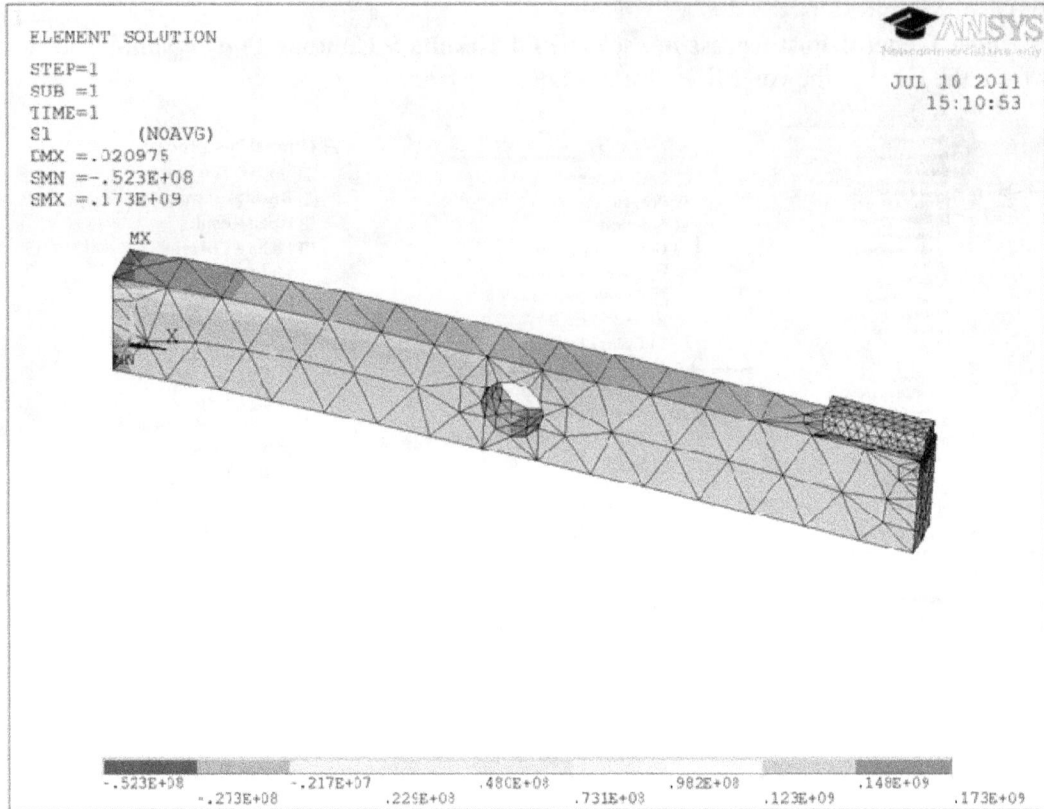

The above picture is plotted using a function called **Reverse Video**. To get to the function of Reverse Video, from the top menu, click **PlotCtrls > Style > Colors > Reverse Video**.

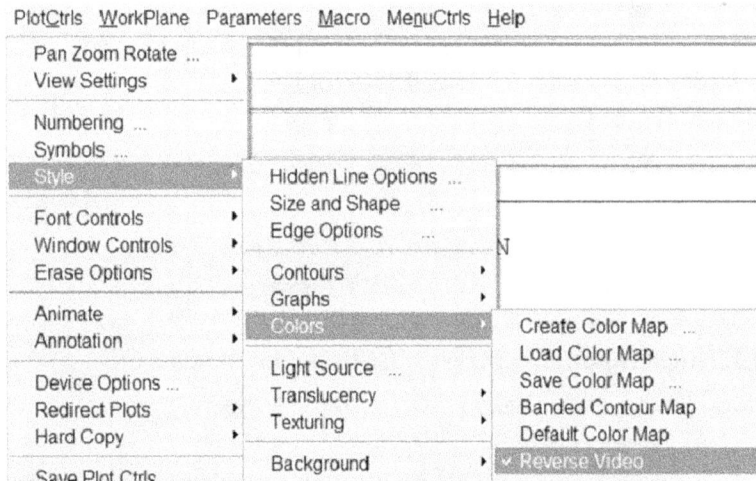

To print the results to an image file, select **PlotCtrls** from the top menu. Select **Hard Copies**, and then choose **Graphics Window**. Then save the file in a .jpg format.

Also, the image can be saved by pressing the "Print Screen" key on the keyboard and pasting the image into an image-editing program, such as MS Paint or directly to a Word document.

10.5 Working Plane (Global and Local Coordinate Systems)

The following figure illustrates the geometry of the component we are going to model and perform FEA. Note that the unit is SI unit. Assume that the hole on the left side is fixed to the ground and a load is acting on the hole at the right side, which is a bearing load.

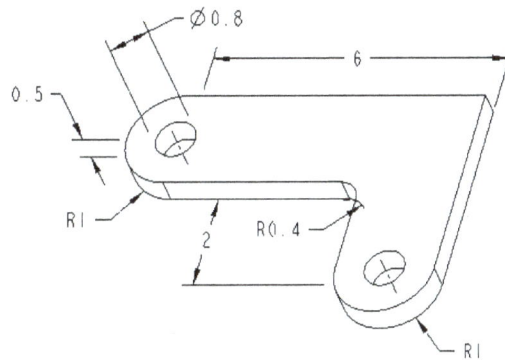

Step 1: Launch **ANSYS13.0**

To define the Structural Analysis, from the main menu or the model tree, select **Preferences**. > make sure that **Structural** is highlighted, and h-method is also selected > **OK**.

Step 2: Define a 2D Geometry.

From the main menu, select **Preprocessor** and expand it. Select **Modeling > Create > Areas > Rectangle > By Dimensions**.

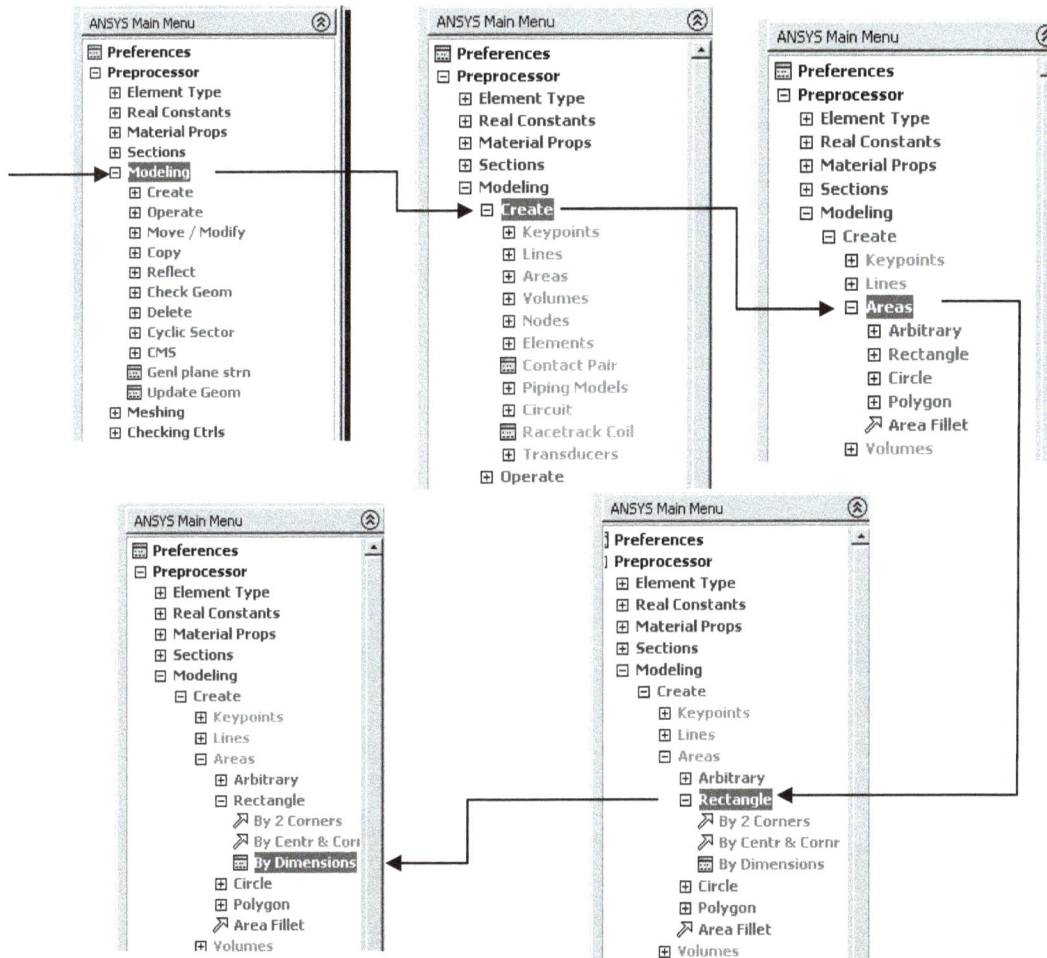

In the window for creating the rectangle by dimensions, type 0 and 6 for X1 and X2, and -1 and 1 for Y1 and Y2 > **Apply**.

Again, in the window for creating the rectangle by dimensions, type 4 and 6 for X1 and X2, and -1 and -3 for Y1 and Y2 > OK.

Create Rectangle by Dimensions

[RECTNG] Create Rectangle by Dimensions

X1,X2 X-coordinates | 0 | 6

Y1,Y2 Y-coordinates | -1 | 1

OK | Apply | Cancel | Help

Create Rectangle by Dimensions

[RECTNG] Create Rectangle by Dimensions

X1,X2 X-coordinates | 4 | 6

Y1,Y2 Y-coordinates | -1 | -3

OK | Apply | Cancel | Help

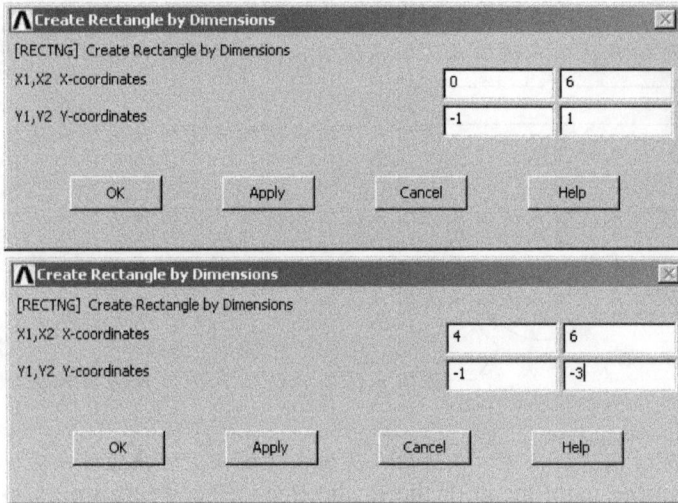

From the main menu, select **ProtCtrols > Numbering** > Turn on **Area Numbering > OK**. On the screen, A1 and A2 are on display.

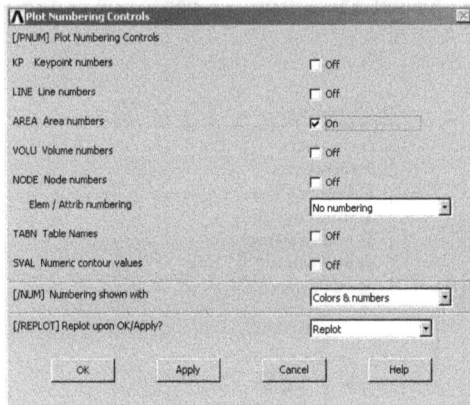

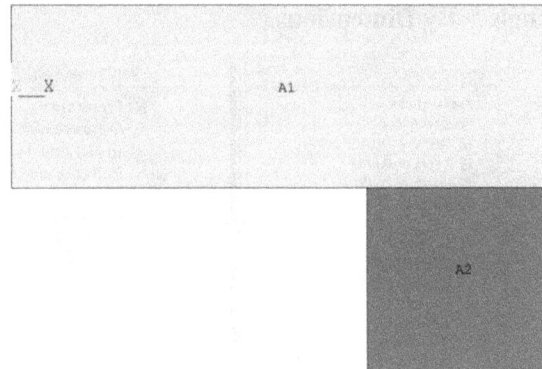

Plot Numbering Controls

[/PNUM] Plot Numbering Controls

KP   Keypoint numbers — Off
LINE Line numbers — Off
AREA Area numbers — On
VOLU Volume numbers — Off
NODE Node numbers — Off
   Elem / Attrib numbering — No numbering
TABN Table Names — Off
SVAL Numeric contour values — Off
[/NUM] Numbering shown with — Colors & numbers
[/REPLOT] Replot upon OK/Apply? — Replot

OK | Apply | Cancel | Help

From the main menu, select **WorkPlane > WP Settings > Polar > Grid and Triad** > enter *0.1* for snap increment > **OK**.

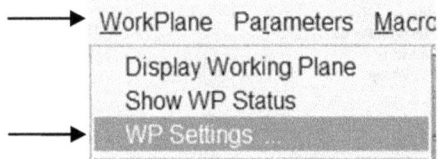

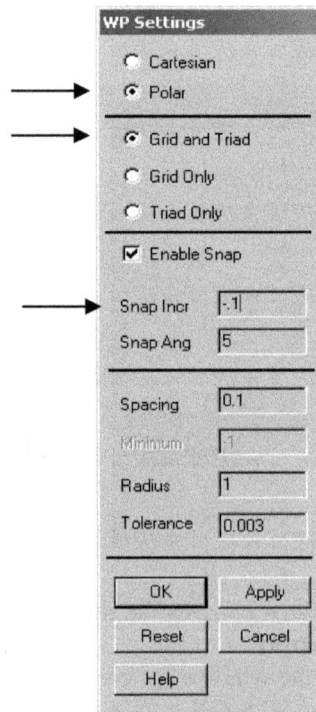

WorkPlane Parameters Macro
Display Working Plane
Show WP Status
WP Settings ...

WP Settings

C Cartesian
⊙ Polar

⊙ Grid and Triad
C Grid Only
C Triad Only

☑ Enable Snap

Snap Incr | .1
Snap Ang | 5

Spacing | 0.1
Minimum | -1
Radius | 1
Tolerance | 0.003

OK | Apply
Reset | Cancel
Help

To create the third area, which is a circle, **Modeling > Create > Areas > Circle > Solid Circle.**
Pick the point (x=0 and y=0) as the center point > move the mouse to radius of 1 and click the left button >OK.

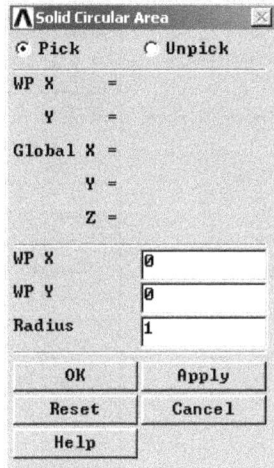

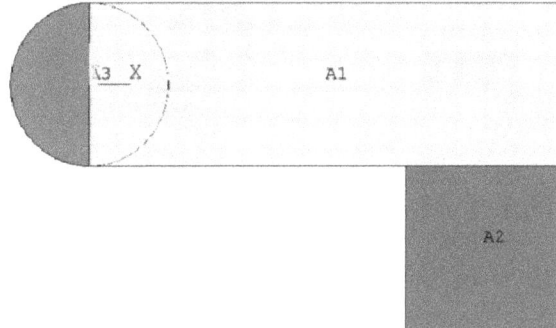

To create the circle at the other end of this component, we need to move the working plane first.
WorkPlane > Offset WP to > Keypoints > pick the lower left corner of the rectangle as the first key point and pick the lower right corner of the rectangle as the second key point > **OK**.

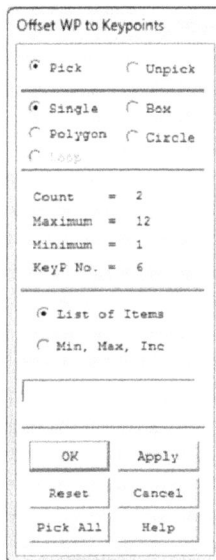

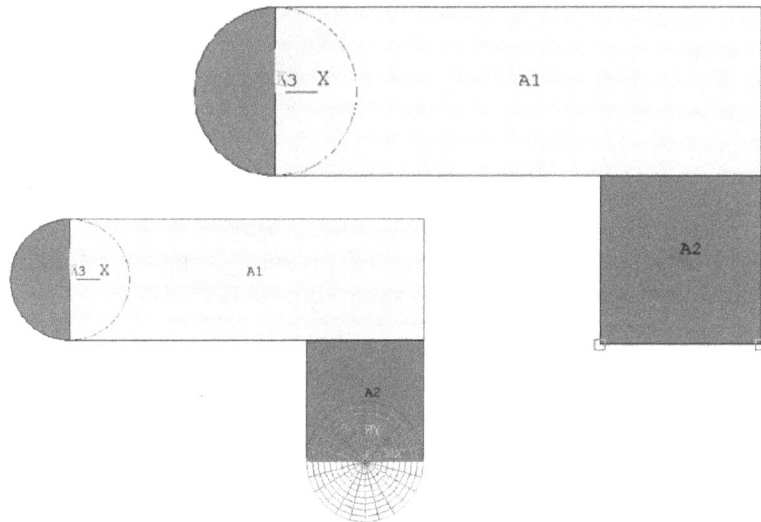

To create a new circle, **Modeling > Create > Areas > Circle.** Pick the point (x=0 and y=0) as the center of the circle and move the mouse to radius of 1 and click the left button > OK.

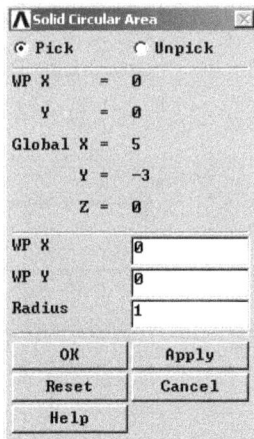

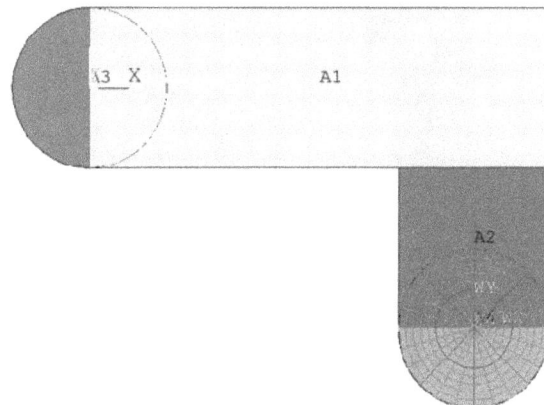

Now that the appropriate pieces of the model are defined, we need to add them together so the model becomes one continuous piece. **Modeling > Operate > Booleans > Add > Areas > Pick All > OK**.

Now let us add an inner fillet at the corner, as shown. **Modeling > Create > Lines > Line Fillet.** Pick the 2 lines > specify the radius value equal to 0.4 > **OK**.

To create an area for the inner fillet, **Modeling > Create > Areas > Arbitrary > By Lines.** Pick the 3 lines, as shown > **OK**.

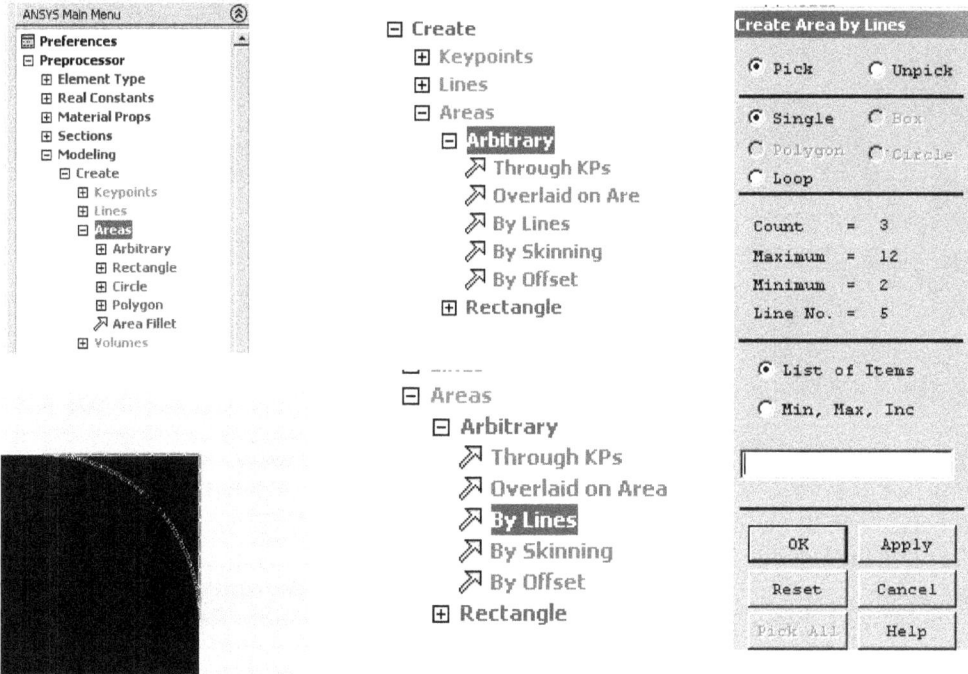

Now let us add this fillet area to the model. **Modeling > Operate > Booleans > Add > Areas > Pick All > OK**.

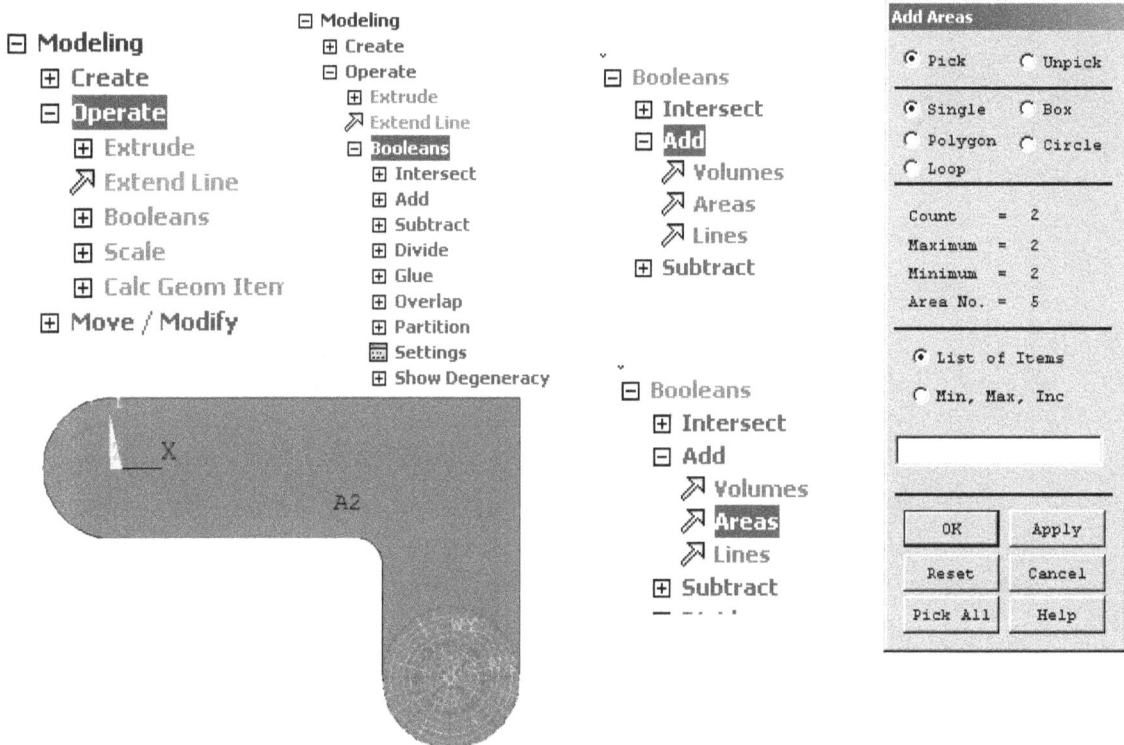

Now let us create the hole at the lower left end. Where is the Working Plane? From the top menu, click **WorkPlane > Display Working Plane** (toggle on). **Modeling > Create > Areas > Circle > Solid Circle.**

Pick the point (x=0 and y=0) as the center point > move the mouse to radius of 0.4 and click the left button >**OK**.

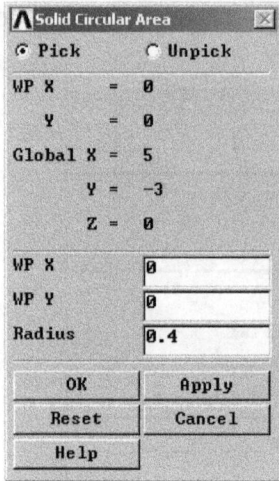

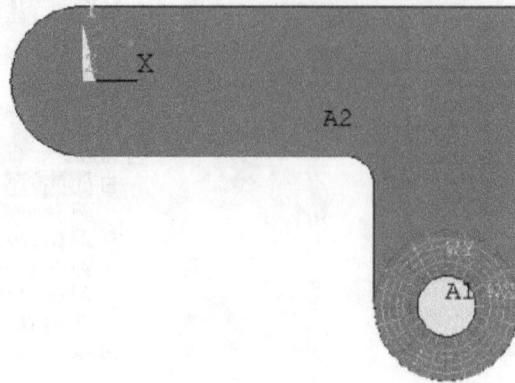

To create the circle at the other end of this component, we need to move the working plane first. **WorkPlane > Offset WP to > Global Origin**. To create a new circle, **Modeling > Create > Areas > Circle > Solid Circle.** Pick the point (x=0 and y=0) as the center of the circle and move the mouse to radius of 0.4 and click the left button > OK.

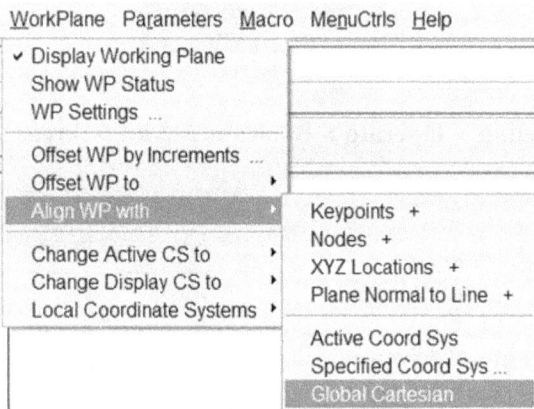

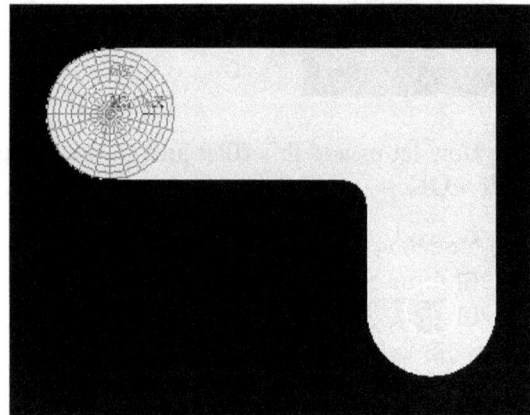

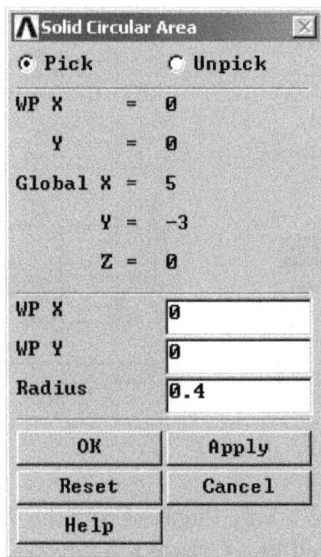

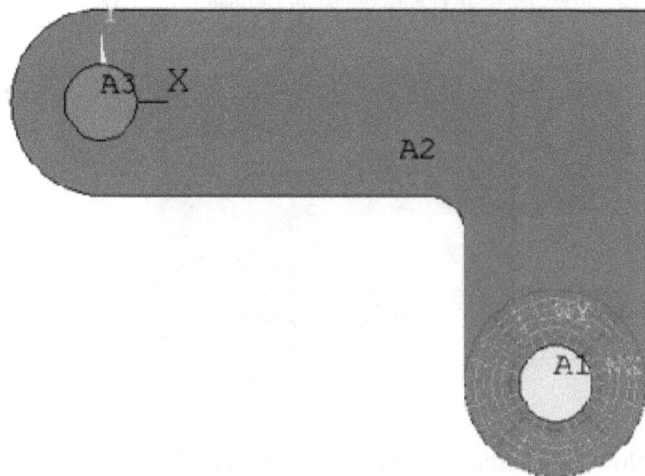

Now let us subtract these 2 circular areas from the 2D model. **Modeling > Operate > Booleans > Subtract > Areas > Pick** the area of the 2D model > **OK**.

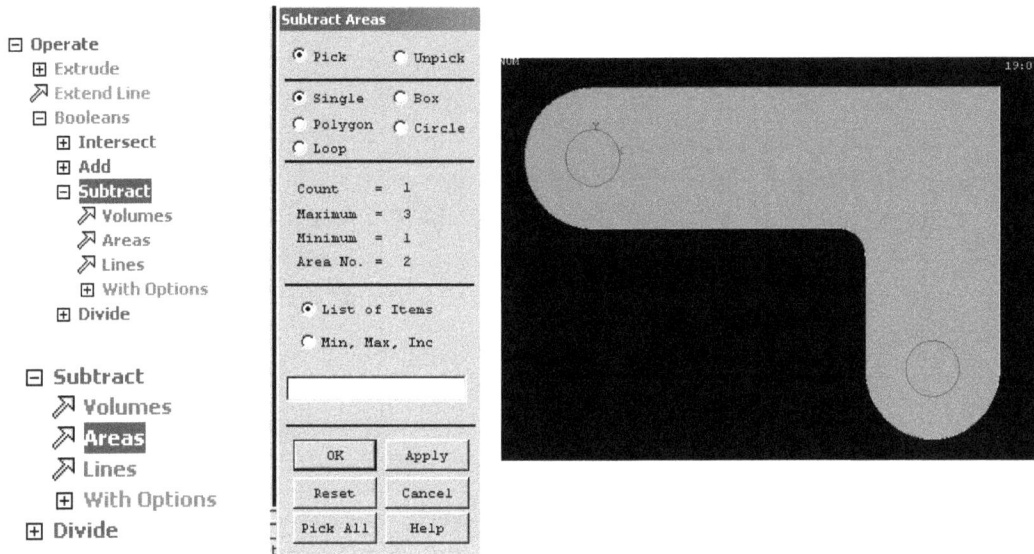

Afterwards, pick the 2 circular areas > **OK** > **OK**.

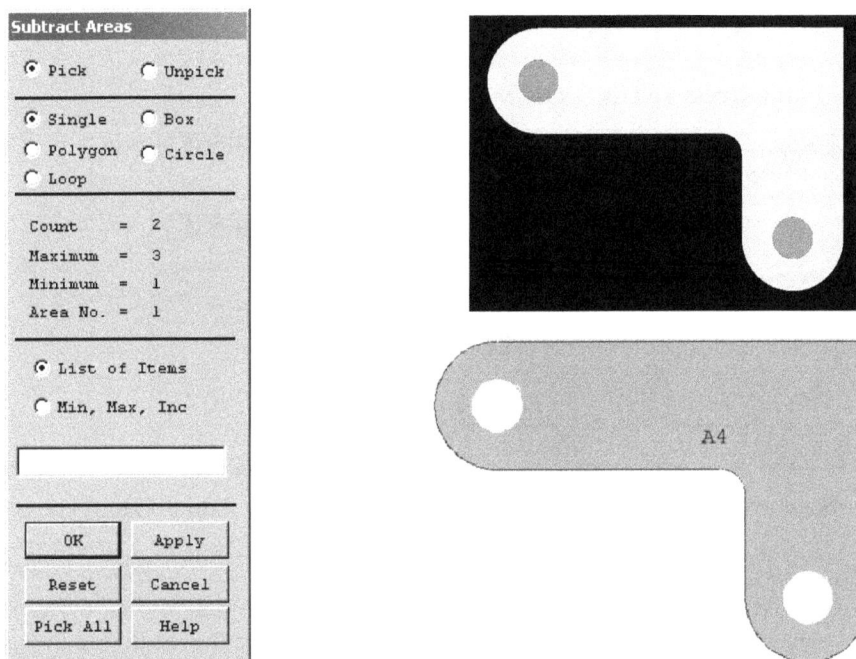

Step 3: Select the element type.
Select **Element Type > Add/Edit/Delete.**

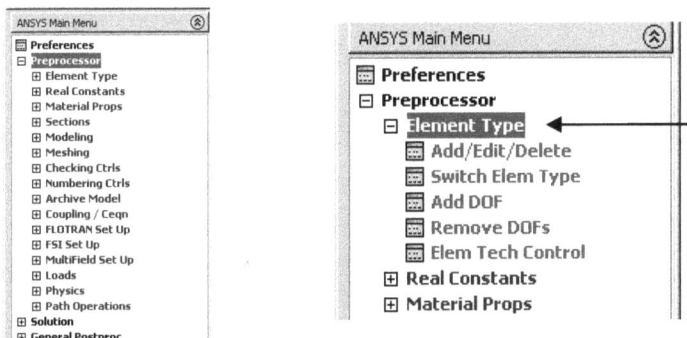

A window called **Element Types** appears. Click **Add > pick Solid > Quard 8node 183 > OK.**

In the **Element Types** window, click **Add > pick Solid > Quad 8 node 183 > OK > Close**.

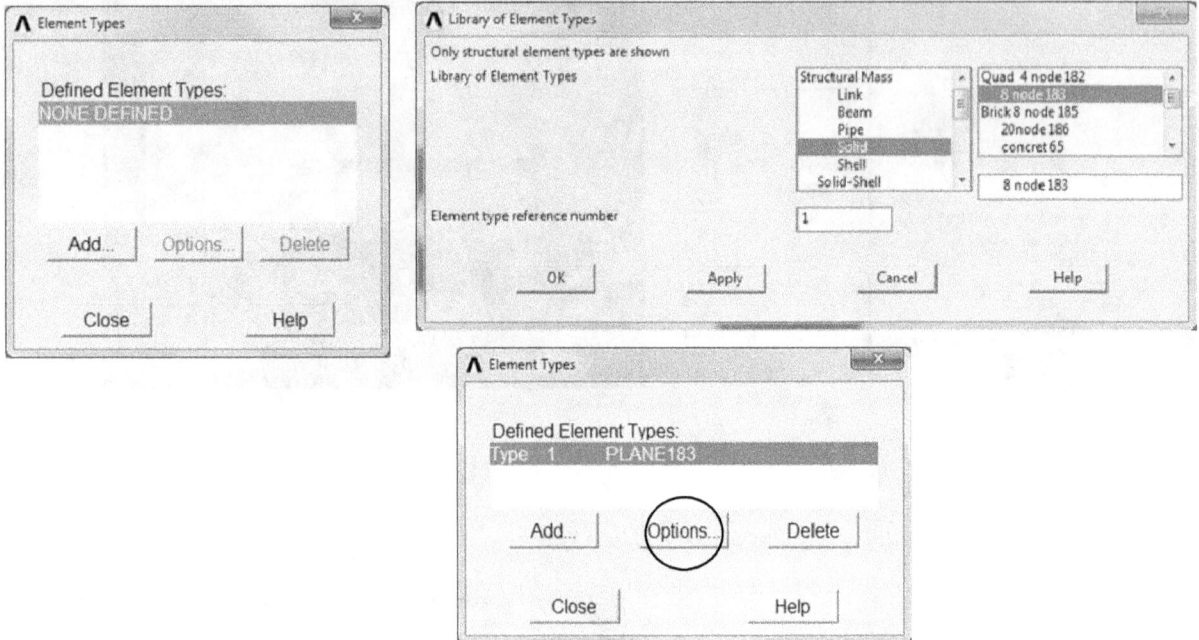

Click **Options > Plane stress w/Thk > OK > Close**.

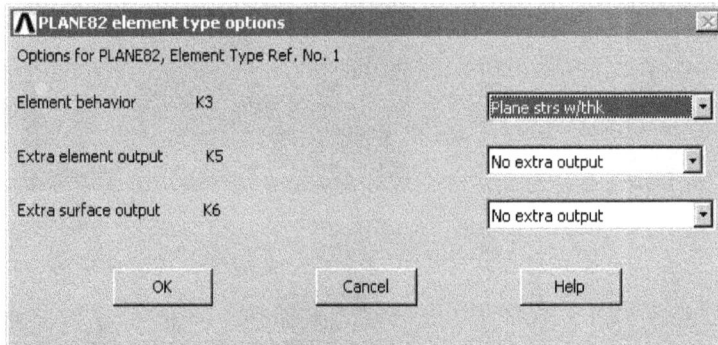

From the main menu, expand **Real Constants > Add/Edit/Delete > Add > OK**. Type *0.4* as the thickness value **> OK**.

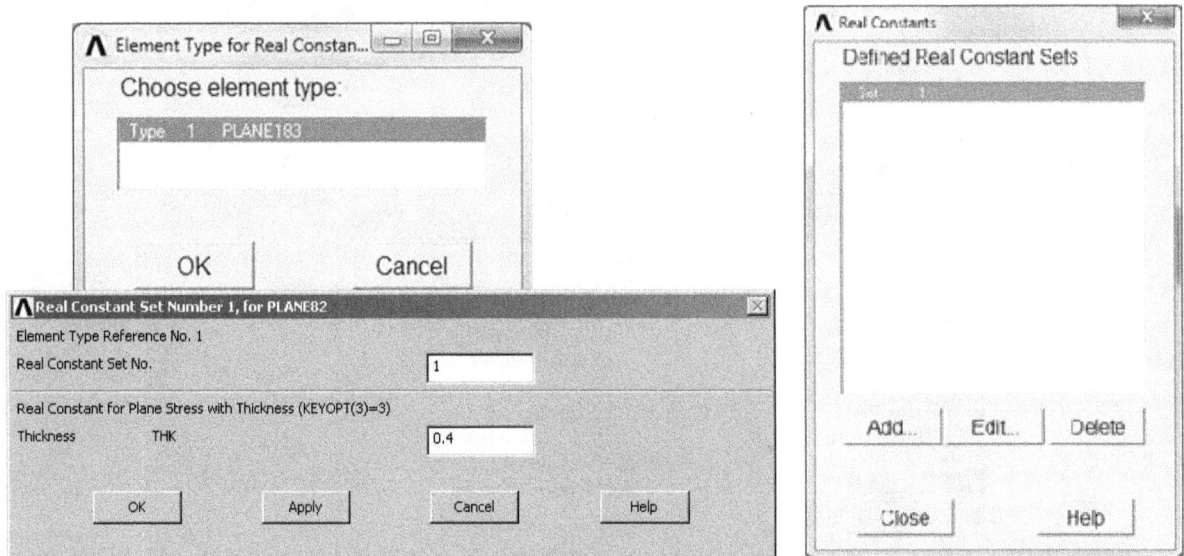

Step 4: Choosing a Material Type

To select the material for the shell model, select **Material Properties** from the Preprocessor menu. Select **Material Model**, and then select **Structural, Linear, Elastic** and **Isotropic** by double clicking on each one.

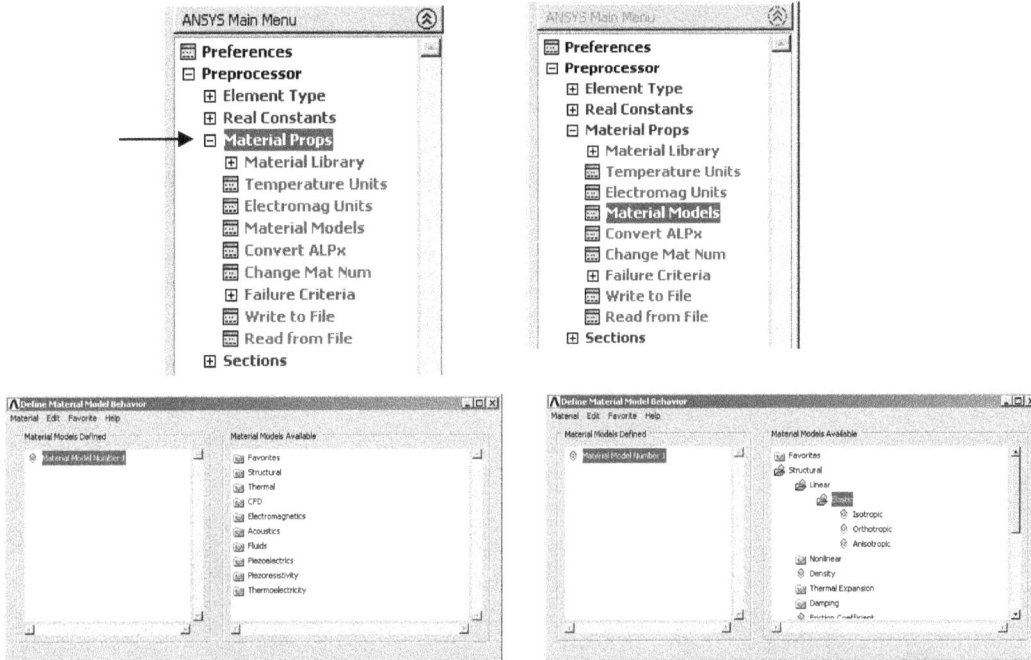

Type 3e10 and 0.26 as the values for Young's Modulus and Poisson's ratio, respectively. Then select OK. Click **Material > Exit**.

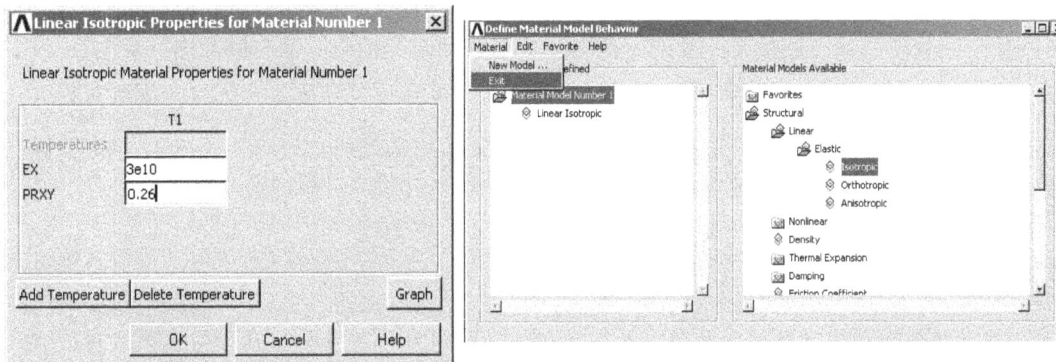

To check the unit system, select **Material Library** under the material properties menu. Then select **Units** and make sure that the **SI** unit system is highlighted. Click **OK**.

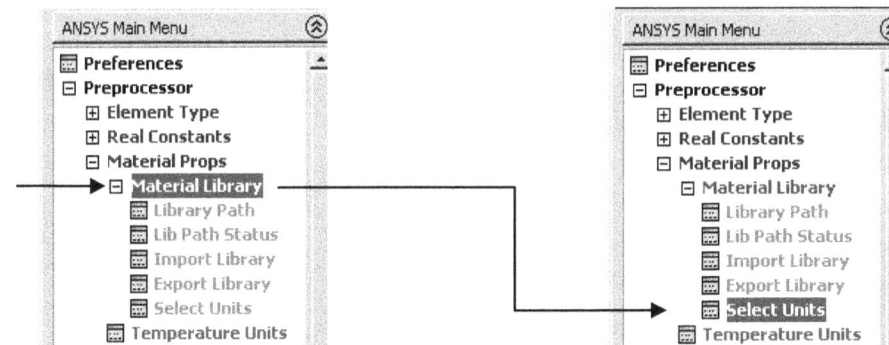

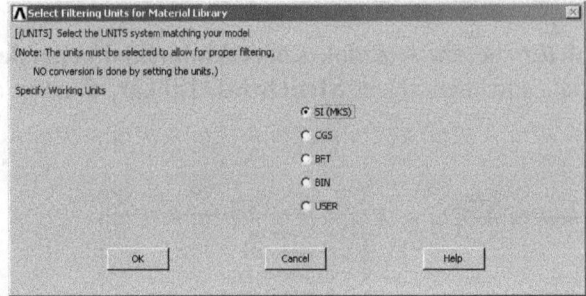

Step 5: Mesh Generation

To create the mesh for the shell model, select **MeshTool** from the Preprocessor menu. Then choose **Smart Size** and click **Mesh**.

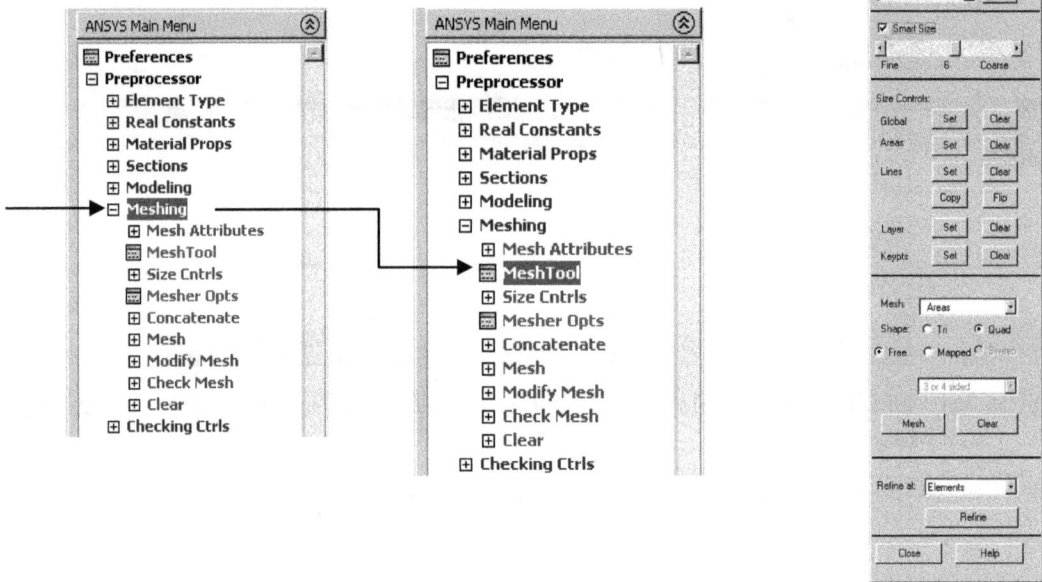

When the selection box comes up, select the area from the display window, and then click **OK**. Make sure that you close the **MeshTool** window when the meshing process is completed.

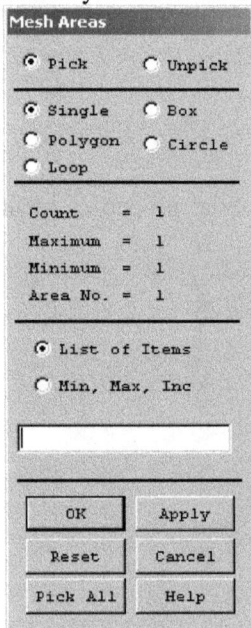

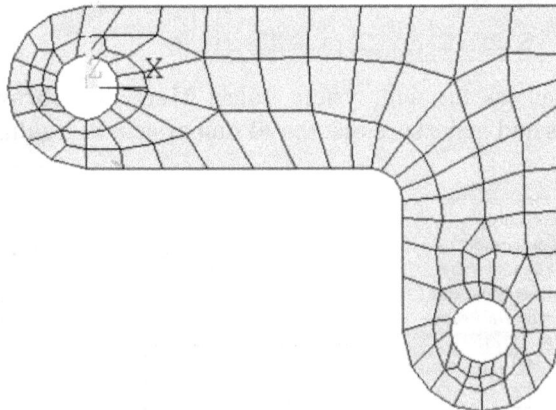

Step 6: Apply the load condition.

To apply the load, on the Preprocessor menu, click **Load > Define Loads > Apply > Structural > Pressure > On Lines**.

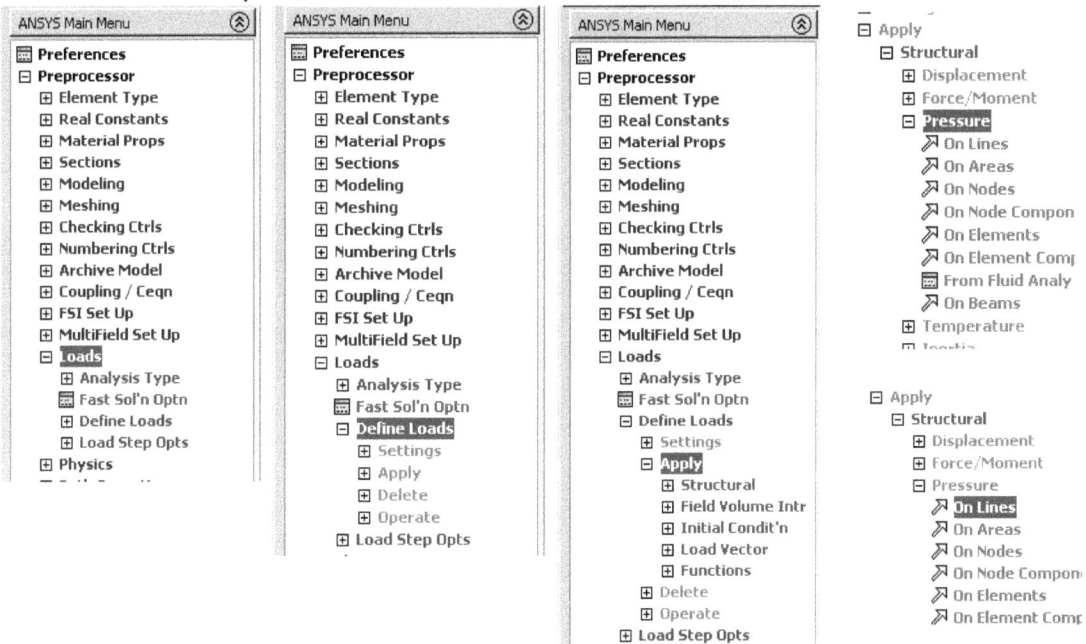

A selection window appears. Pick the arc at the bottom left > **Apply**. Enter 50 in the upper box and enter 500 in the lower box, defining a bearing load > OK.

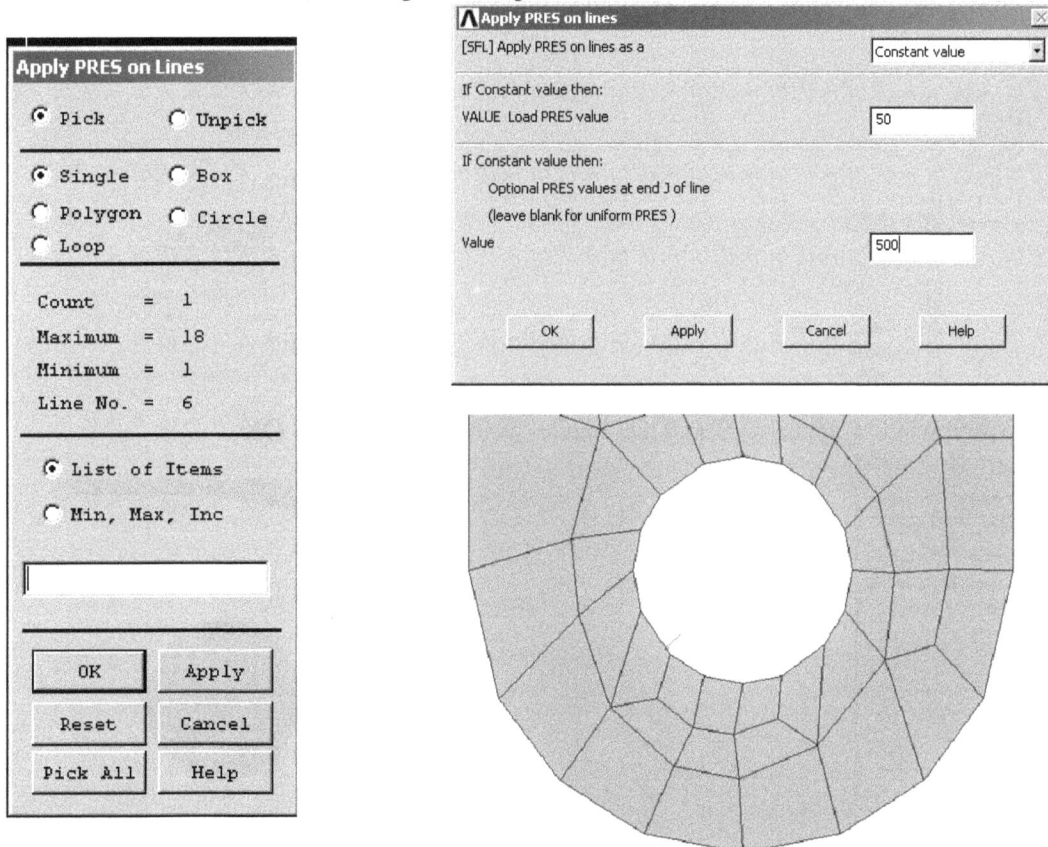

Pick the arc at the bottom right > **Apply**. Enter 500 in the upper box and enter 50 in the lower box, defining a bearing load > OK.

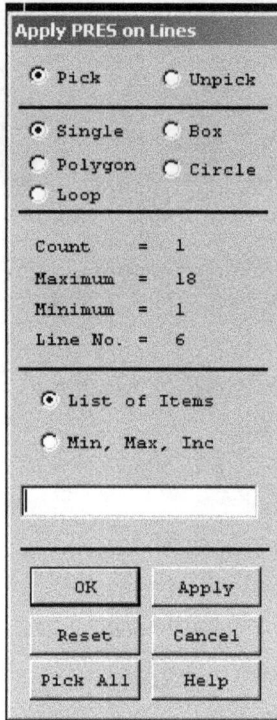

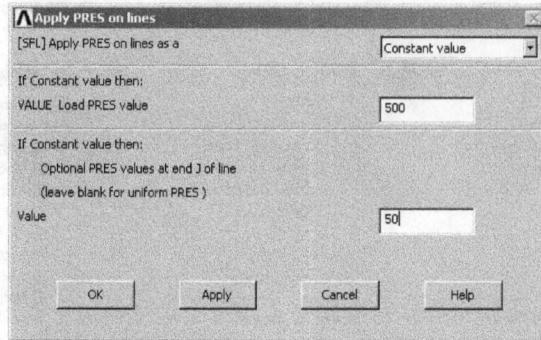

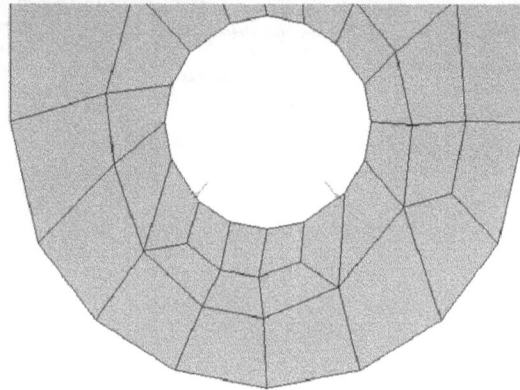

Step 7: Defining the Constraint Condition

To define the constraint conditions, under the Structural menu, select **Displacement > On Keypoints.**

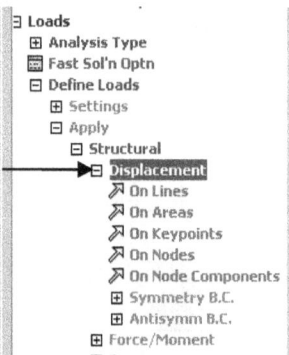

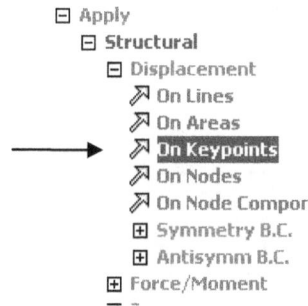

A selection window appears. Select the 4 nodes as the 4 keypoints > **OK**.

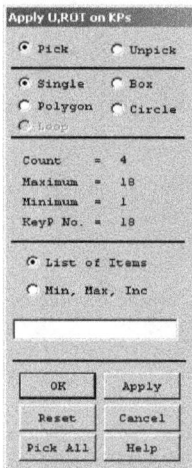

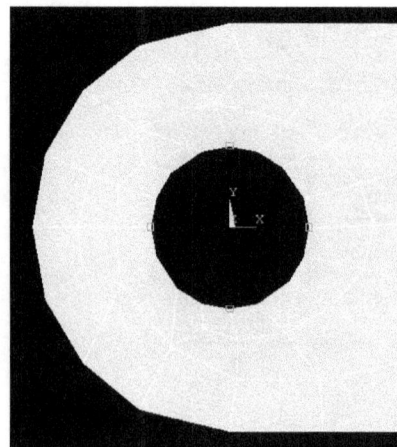

Select **All DOFS**, and type 0 as the value. Click **Yes** to expand displacement constraints to those nodes between the 4 keypoints. Afterwards, click **OK**.

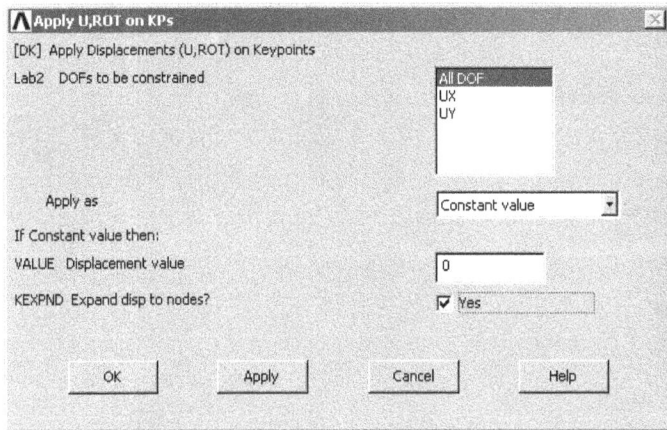

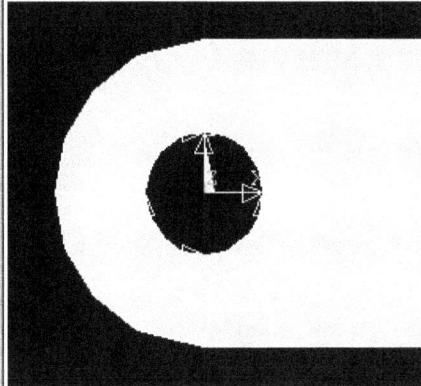

Step 7: Perform the analysis and obtain the solution.

To find the solution, run the program by selecting **Solution** from main menu > **Analysis Type** > **New Analysis > Static > OK**.

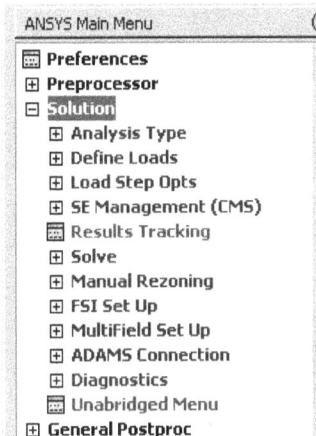

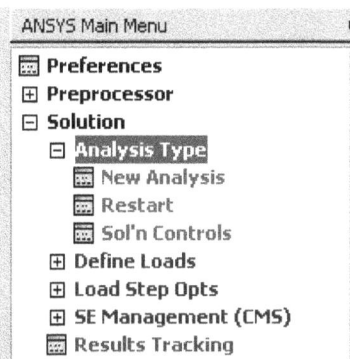

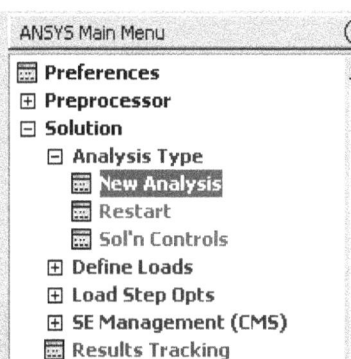

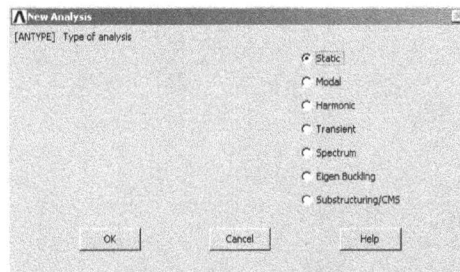

Choose **Current LS** > after reviewing the information, click **OK** to run the program.

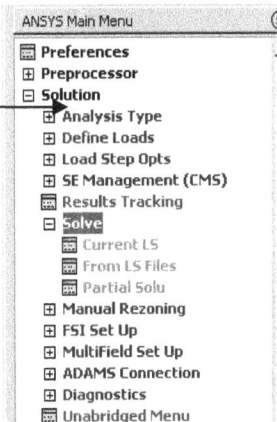

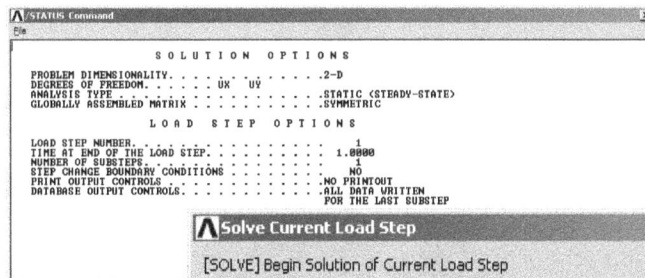

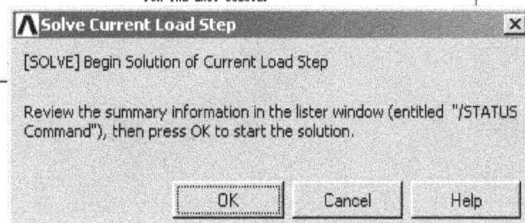

A window called Note appears when the solution is done > **Close**.

Step 8: Displaying Results

To review the results, select **General Postprocessing** from the main menu. Then click **Plot Results, > Deformed Shape** > select **Def + undeformed** > **OK**.

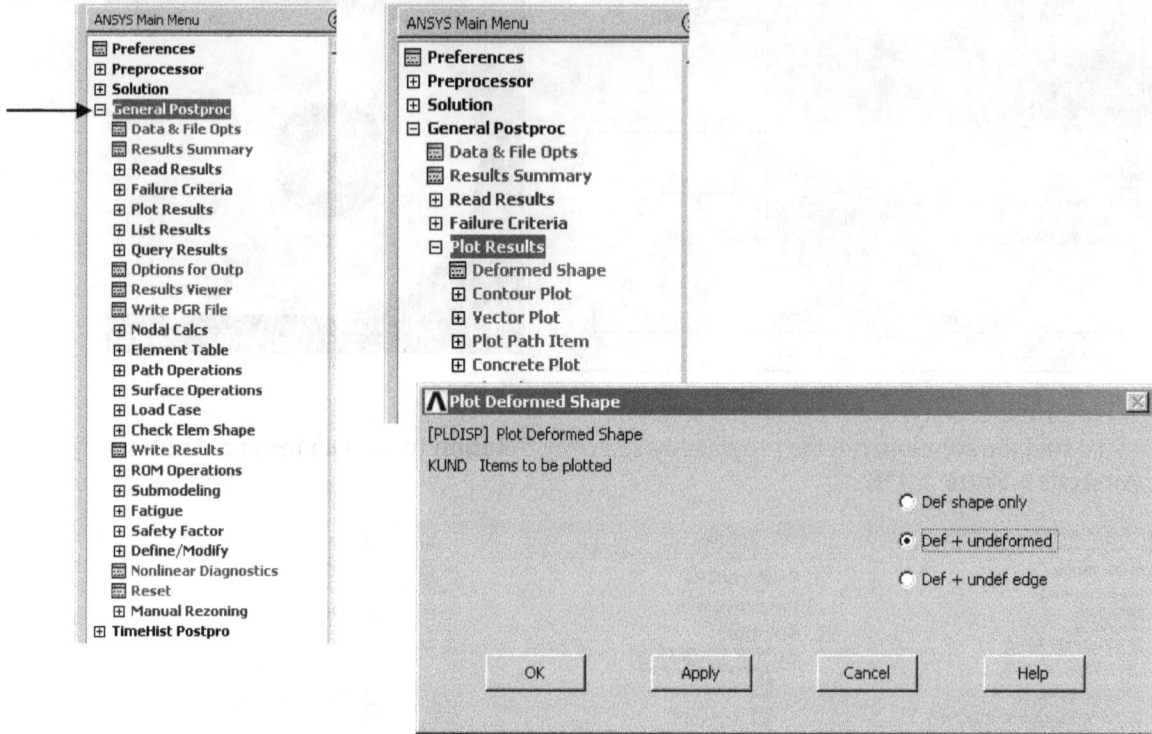

The following plot depicts the deformed shape.

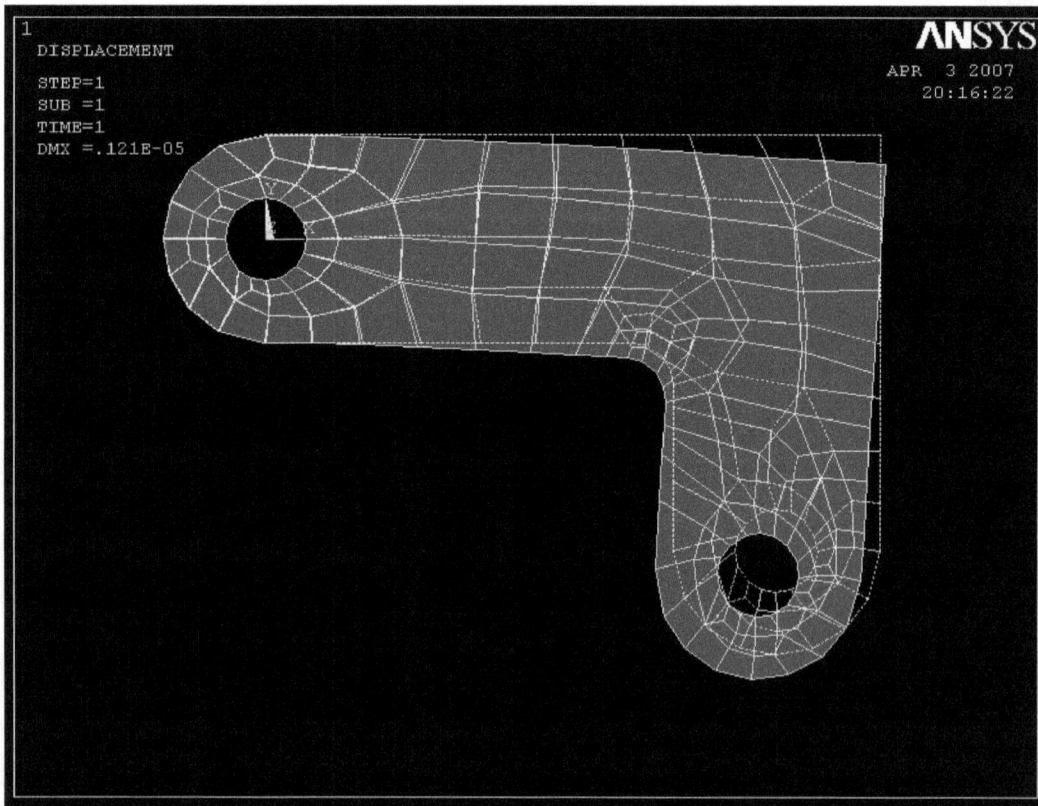

To plot a stress distribution, **Contour Plot > Nodal Solution** > pick a specific type of stress of interest, say von Mises stress.

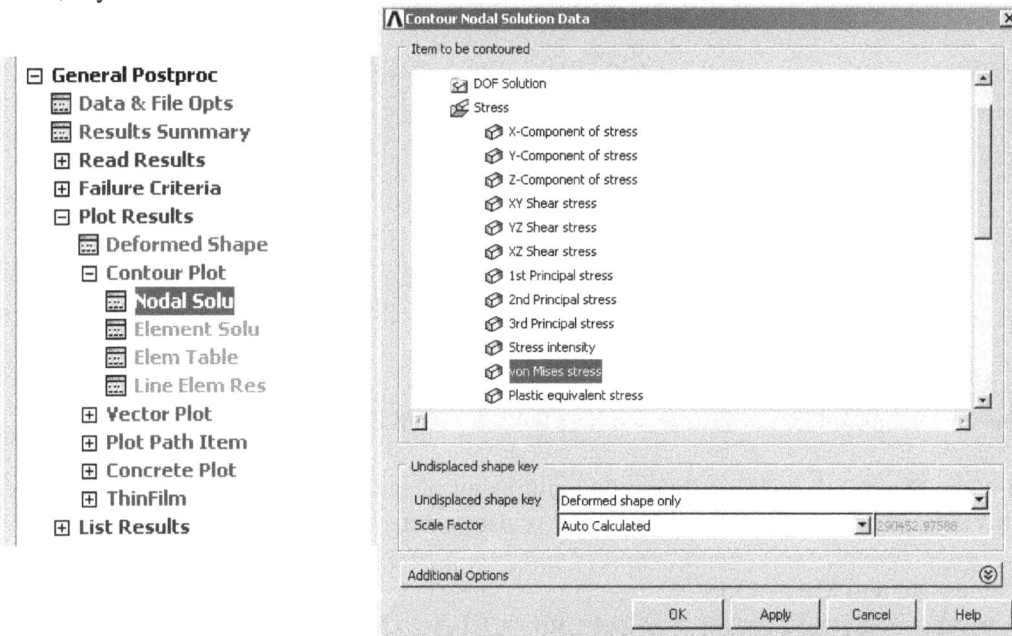

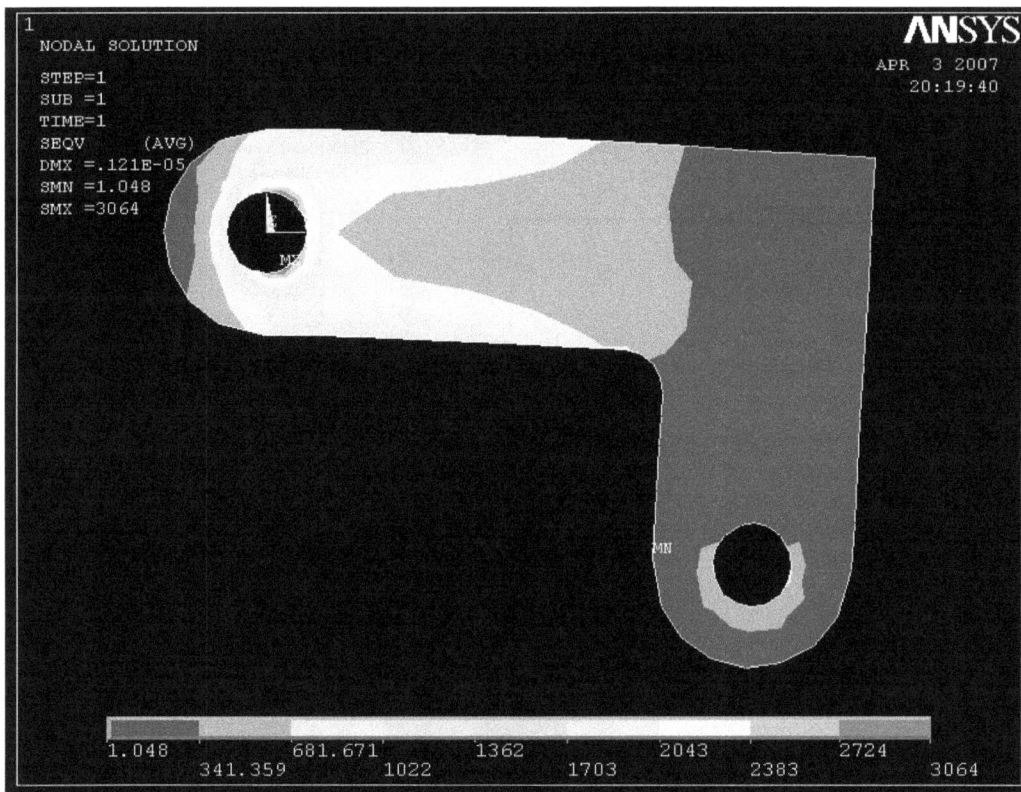

10.6 Working with CGS Unit System

The following figure illustrates the geometry of a wrench. Note the unit of dimensions shown in the drawing is *centimeter* or *cm*. When creating the geometrical model in **ANSYS**, use Metric CGS system (*Centimeter Gram Second*).

Unit: cm

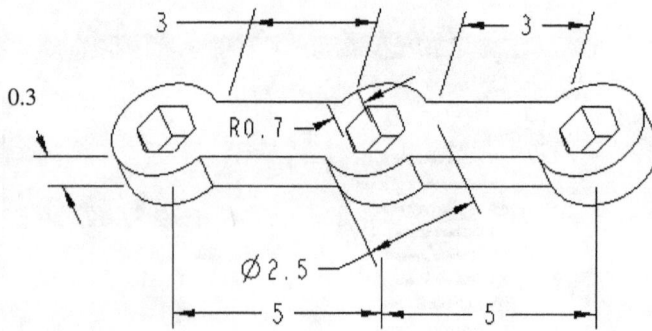

Step 1: Launch **ANSYS 13.0**

From the start menu, select **Programs > ANSYS 13.0 > ANSYS**, or directly click the icon of **ANSYS**.

To define the Structural Analysis, from the main menu or the model tree, select **Preferences**. > make sure that **Structural** is highlighted, and h-method is also selected > **OK**.

Step 2: Set the unit system to Metric CGS

Preprocessor > Material Library > Select Units > CGS > OK.

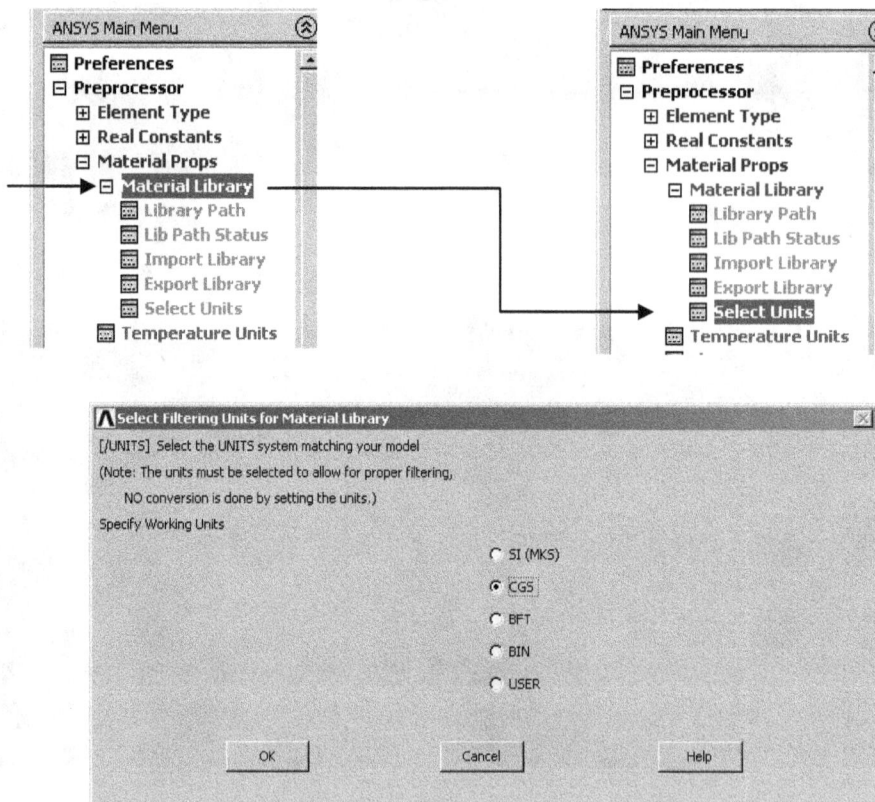

As illustrated, the basic quantities used in the Metric CGS system are

Length: centimeter (cm)
Mass: gram (g)
Time: second (s)
Temperature: Celsius (deg C)
Force: dyne (g*cm/s^2)

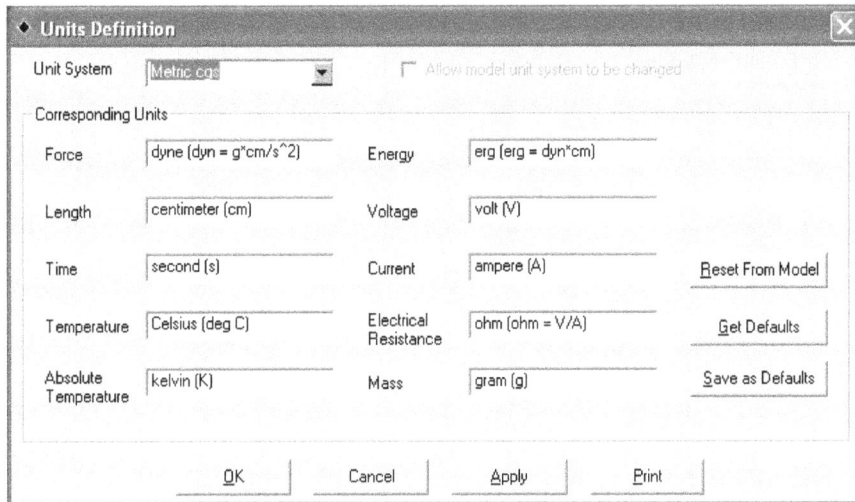

Step 3: Define a 2D Geometry.

From the main menu, select **Preprocessor** and expand it. Select **Modeling > Create > Areas > Rectangle > By Dimensions**.

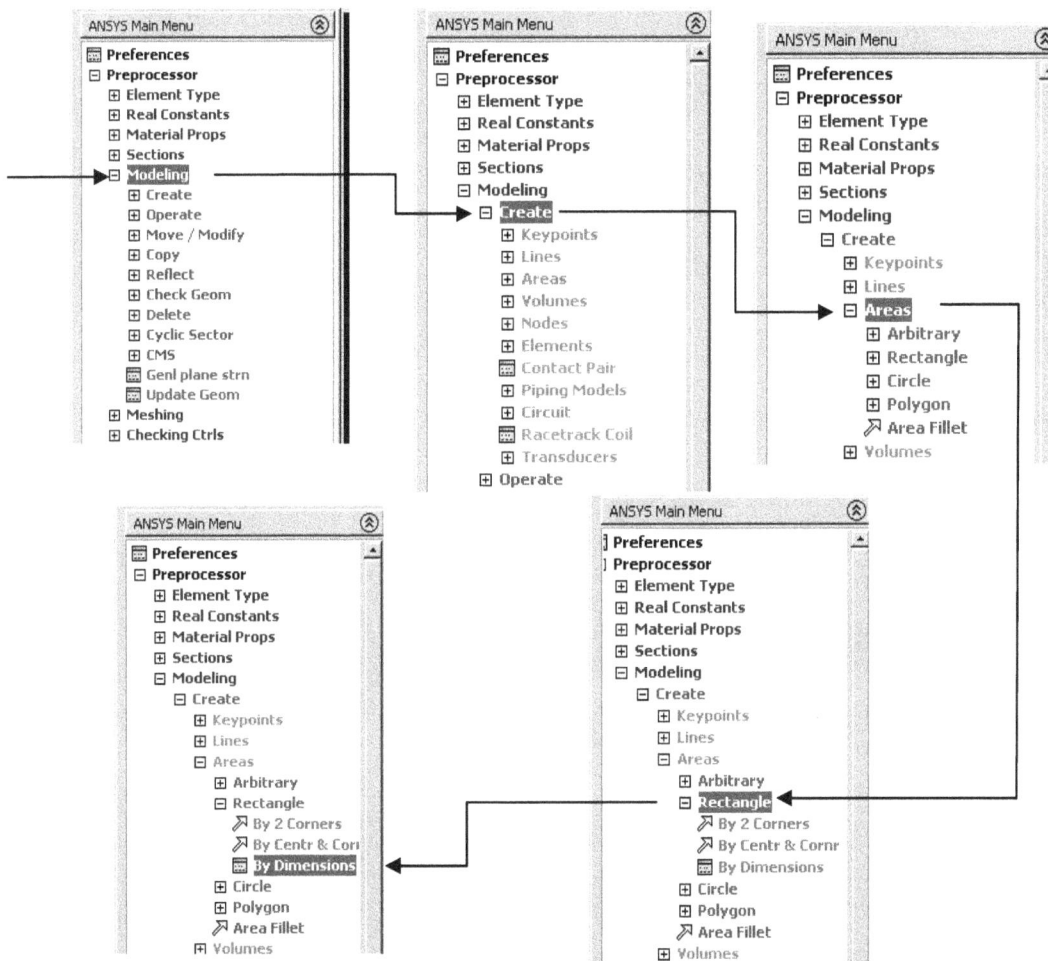

In the window for creating the rectangle by dimensions, type 2.25 and 5.25 for X1 and X2, and 0.5 and 2 for Y1 and Y2 > **Apply**.

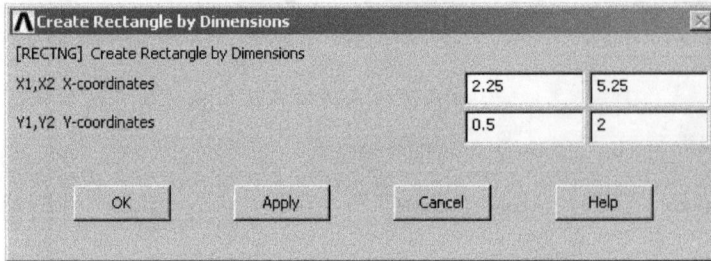

Again, in the window for creating the rectangle by dimensions, type 7.25 and 10.25 for X1 and X2, and -1 and 0.5 and 2 for Y1 and Y2 > **OK**.

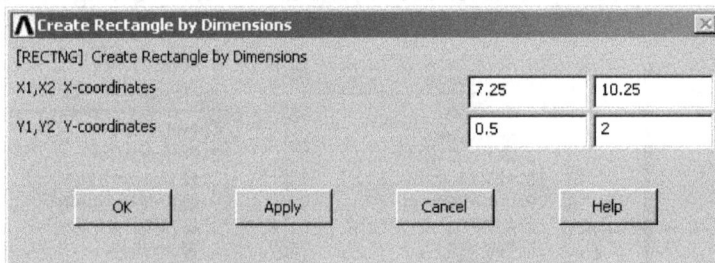

From the main menu, select **ProtCtrols > Numbering** > Turn on Area Numbering > **OK**. On the screen, A1 and A2 are on display.

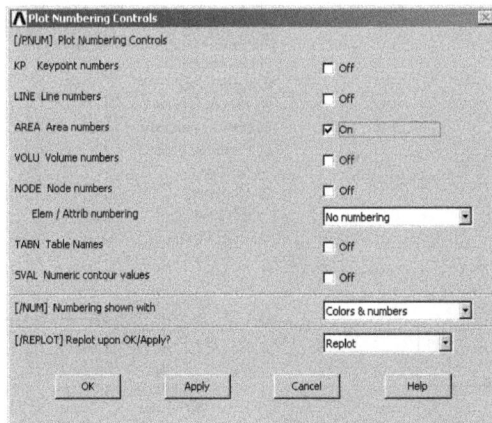

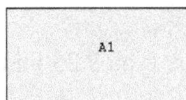

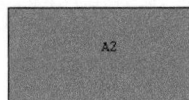

To create 3 circular areas, select **Modeling > Create > Areas > Circle > Solid Circle.**
Specify x = 1.25 and y = 1.25 as the center and specify 1.25 as the radius value > **Apply**.

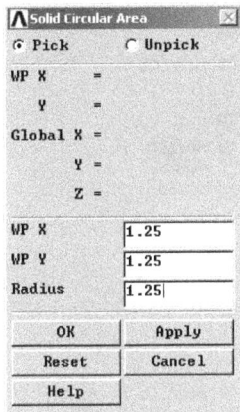

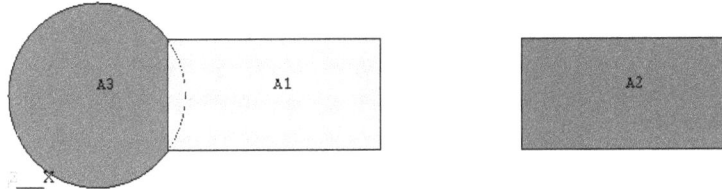

To create the second circle, specify x = 6.25 and y = 1.25 as the center and specify 1.25 as the radius value > **Apply**.

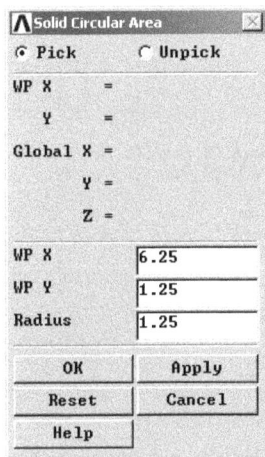

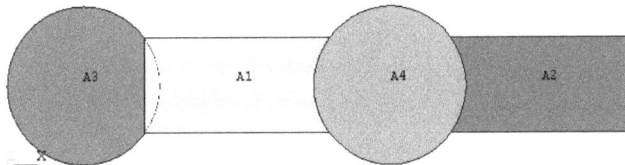

To create the second circle, specify x = 11.25 and y = 1.25 as the center and specify 1.25 as the radius value > **OK**.

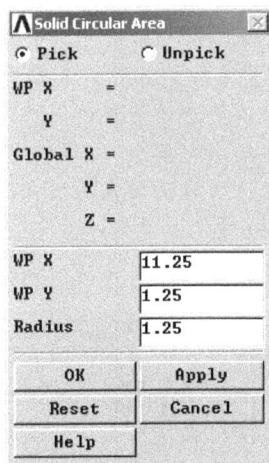

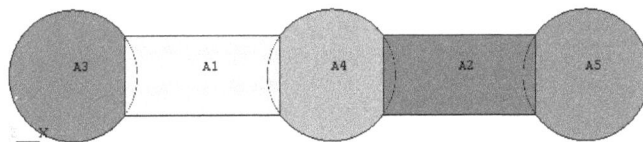

Now that the appropriate pieces of the model are defined, we need to add them together so the model becomes one continuous piece. **Modeling > Operate > Booleans > Add > Areas > Pick All > OK**.

⊟ **Modeling**
 ⊞ **Create**
 ⊟ **Operate**
  ⊞ **Extrude**
  ⬈ **Extend Line**
  ⊞ **Booleans**
  ⊞ **Scale**
  ⊞ **Calc Geom Item**
 ⊞ **Move / Modify**

⊟ Modeling
 ⊞ Create
 ⊟ Operate
  ⊞ Extrude
  ⬈ Extend Line
  ⊟ **Booleans**
   ⊞ Intersect
   ⊞ Add
   ⊞ Subtract
   ⊞ Divide
   ⊞ Glue
   ⊞ Overlap
   ⊞ Partition
   ▦ Settings
   ⊞ Show Degeneracy

⊟ **Booleans**
 ⊞ **Intersect**
 ⊟ **Add**
  ⬈ **Volumes**
  ⬈ **Areas**
  ⬈ **Lines**
 ⊞ **Subtract**

⊟ **Booleans**
 ⊞ **Intersect**
 ⊟ **Add**
  ⬈ **Volumes**
  ⬈ **Areas**
  ⬈ **Lines**
 ⊞ **Subtract**

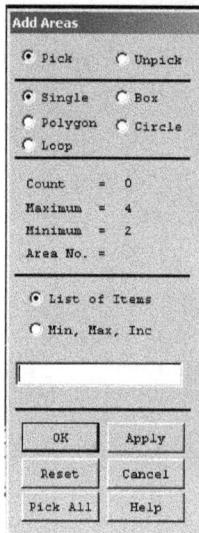

Now let us create the 3 hexagons and subtract them from Area 6. **Modeling > Create > Areas > Polygon > Hexagon.**

Specify x = 1.25 and y = 1.25 as the center, specify 0.7 as the radius value and specify 120 as the angle value > **Apply.**

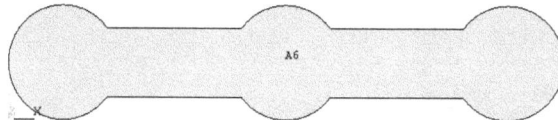

To create the second hexagon, specify x = 6.25 and y = 1.25 as the center, specify 0.7 as the radius value and specify 120 as the angle value > **Apply.**

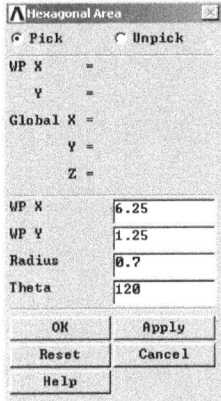

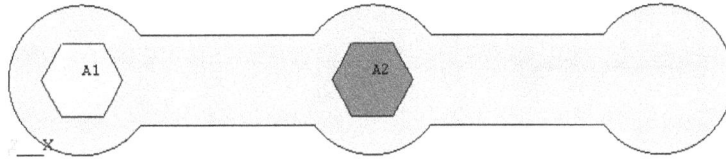

To create the third hexagon, specify x = 11.25 and y = 1.25 as the center, specify 0.7 as the radius value and specify 120 as the angle value > **OK.**

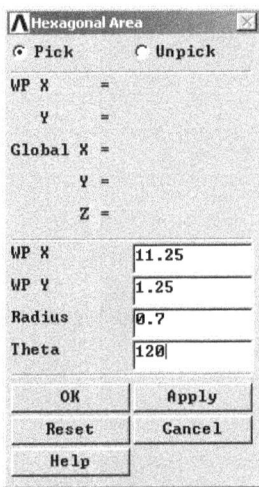

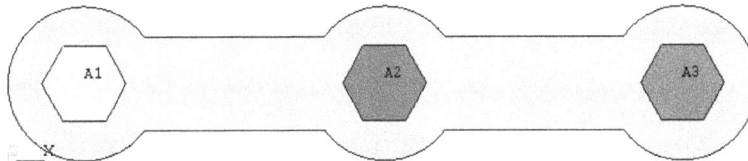

Now let us subtract these 3 areas from Area 6. **Modeling > Operate > Booleans > Subtract > Areas.**

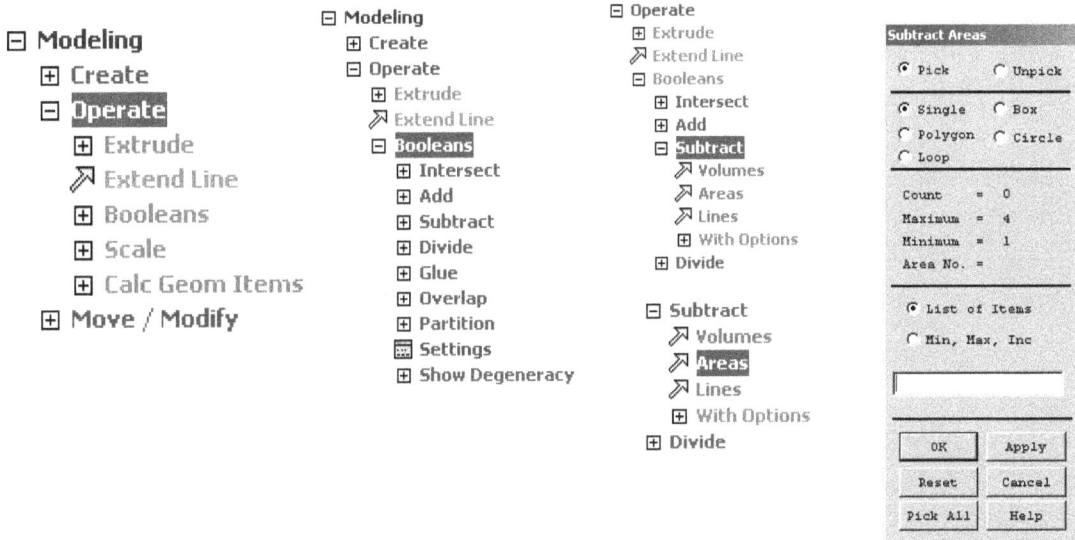

Pick Area 6 > **OK > Apply.**

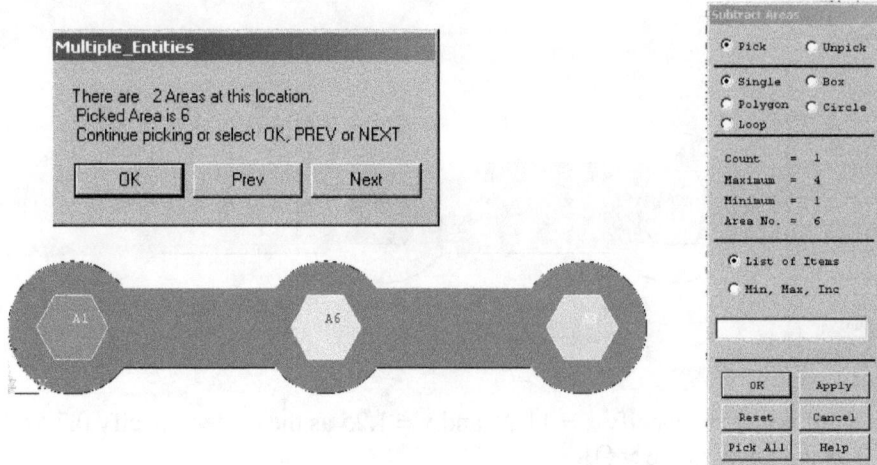

Pick the left hexagon area (A1), pick the center hexagon are (A2) and pick the right hexagon area (A3) > Apply > **OK**.

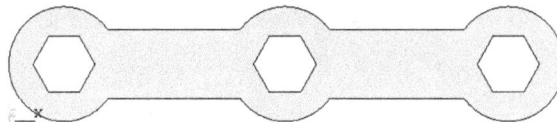

Step 4: Select the element type.

Select **Element Type > Add/Edit/Delete**. In the **Element Types** window, click **Add > pick Solid > Quad 8 node 183 > OK > Close**.

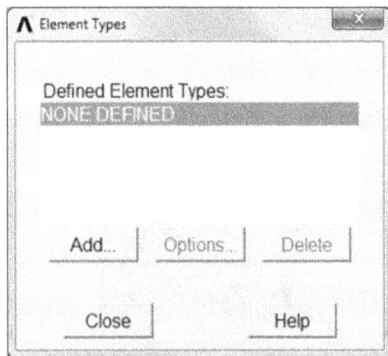

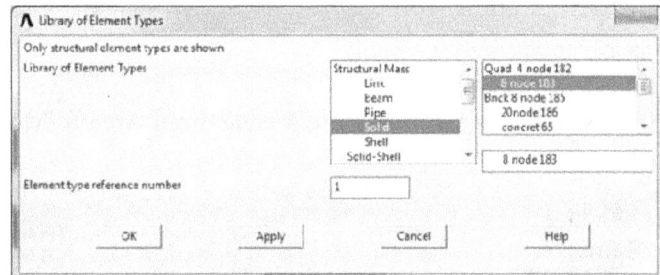

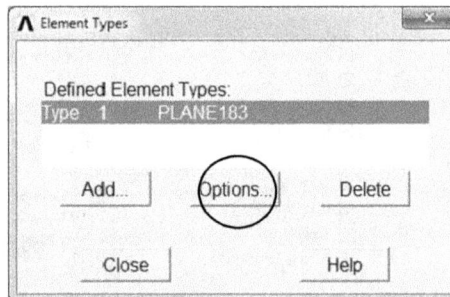

Click **Options > Plane stress w/Thk > OK > Close**.

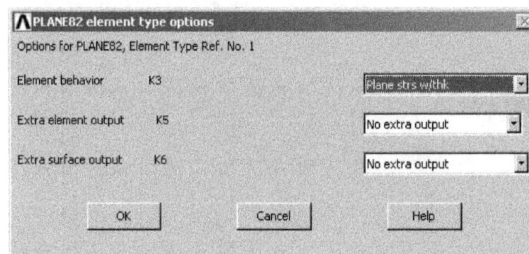

From the main menu, expand **Real Constants > Add/Edit/Delete > Add > OK**. Type *0.3* as the thickness value > **OK**.

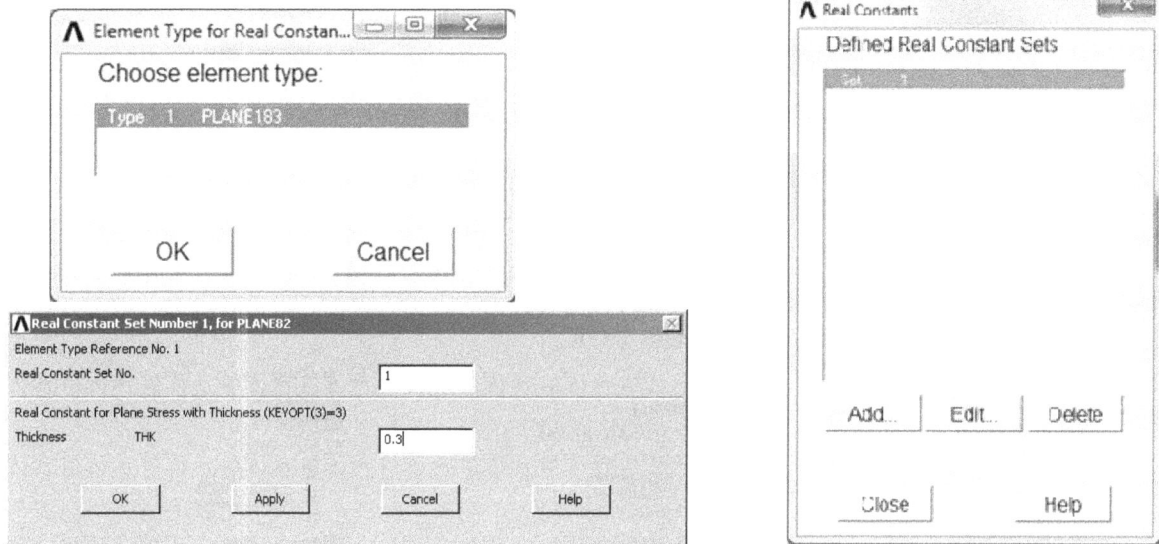

Step 5: Choosing a Material Type

To select the material for the shell model, select **Material Properties** from the Preprocessor menu. Select **Material Model**, and then select **Structural, Linear, Elastic** and **Isotropic** by double clicking on each one.

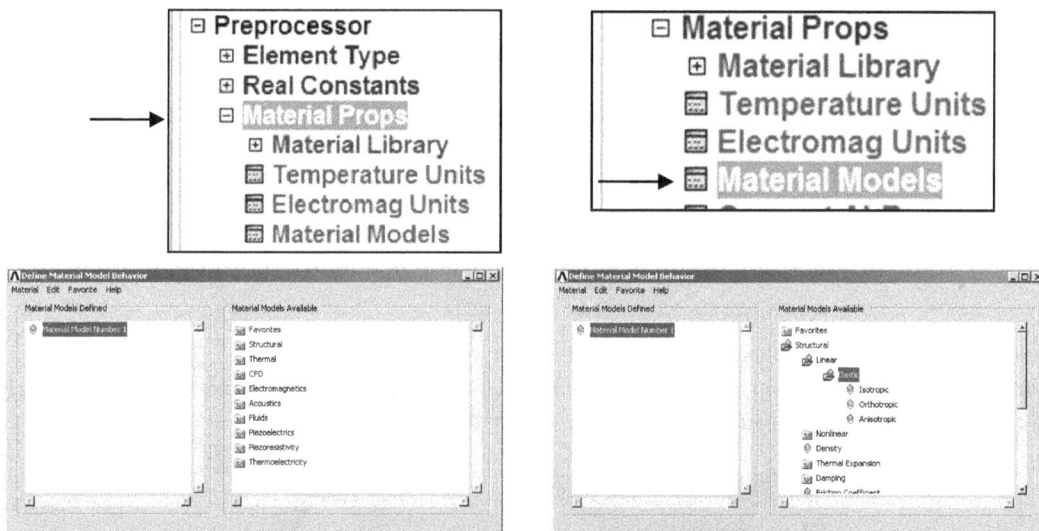

Type 3e11 and 0.26 as the values for Young's Modulus and Poisson's ratio, respectively. Then select OK. Click **Material > Exit**.

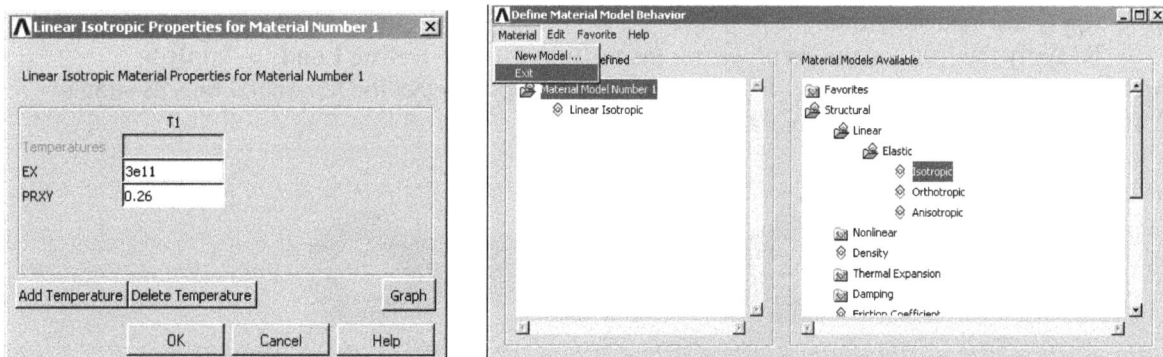

To check the unit system, select **Material Library** under the material properties menu. Then select **Units** and make sure that the **CGS** unit system is highlighted. Click **OK**.

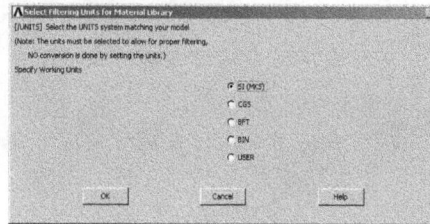

Step 6: Mesh Generation

To create the mesh for the shell model, select **MeshTool** from the Preprocessor menu. **Meshing > Size Cntrls > Manual Size > Global Size > Size**.

Specify 0.1 as the Element edge length > **OK**.

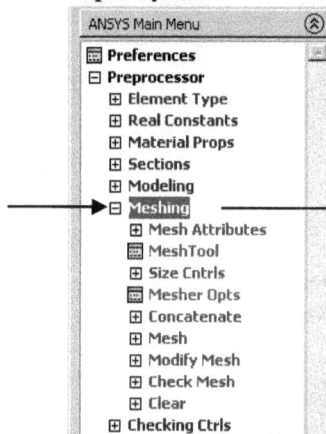

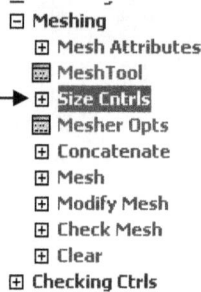

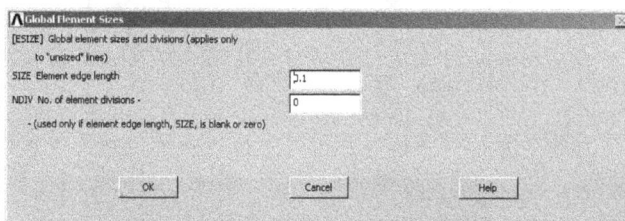

Meshing > Meshing Tool > Area > Free > Mesh.

When the selection box comes up, select **Pick All**, and then click **OK**. Make sure that you close the **MeshTool** window when the meshing process is completed.

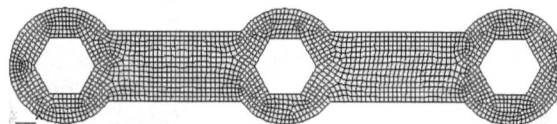

Step 7: Apply the load condition.

To apply the load, on the Preprocessor menu, click **Load > Define Loads > Apply > Structural > Pressure > On Lines**. In the Apply Pres window, Select the horizontal line, as shown > OK. Enter 8800000 in the upper box > **OK**.

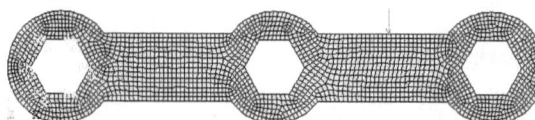

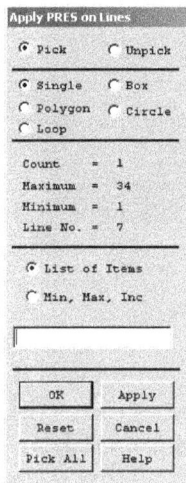

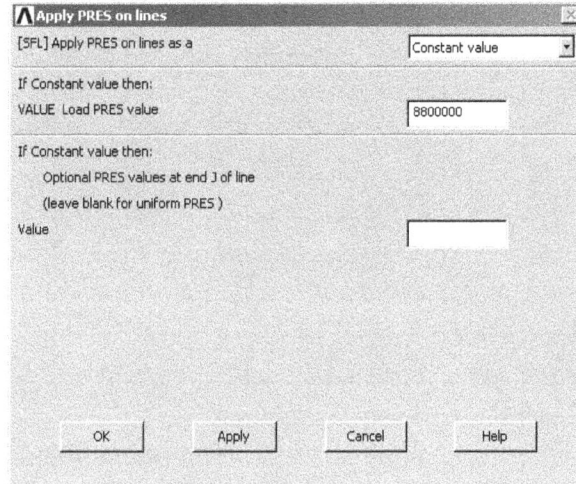

Step 8: Defining the Constraint Condition

To define the constraint conditions, under the Structural menu, select **Displacement > On Keypoints.**

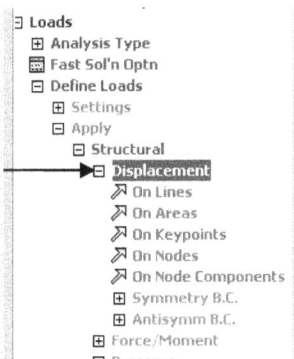

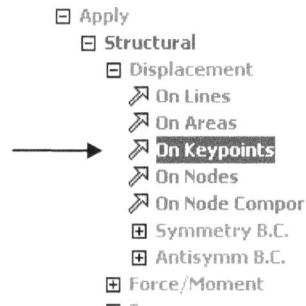

A selection window appears. Select the 6 corner keypoints > **OK**.

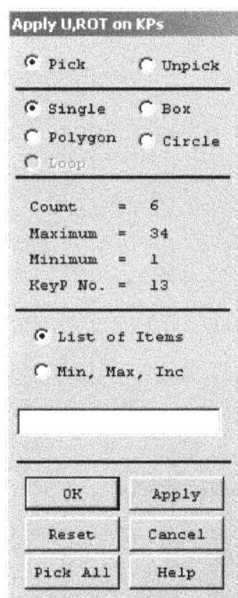

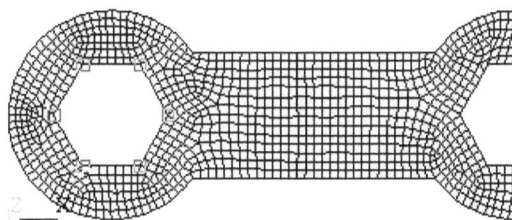

Select **All DOFS**, and type 0 as the value. Click **Yes** to expand displacement constraints to those nodes between the 6 keypoints. Afterwards, click **OK**.

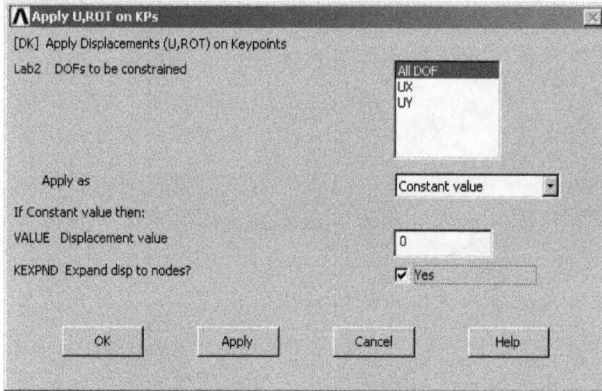

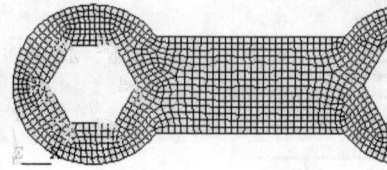

Step 7: Perform the analysis and obtain the solution.

To find the solution, run the program by selecting **Solution** from main menu > **Analysis Type** > **New Analysis** > **Static** > **OK**.

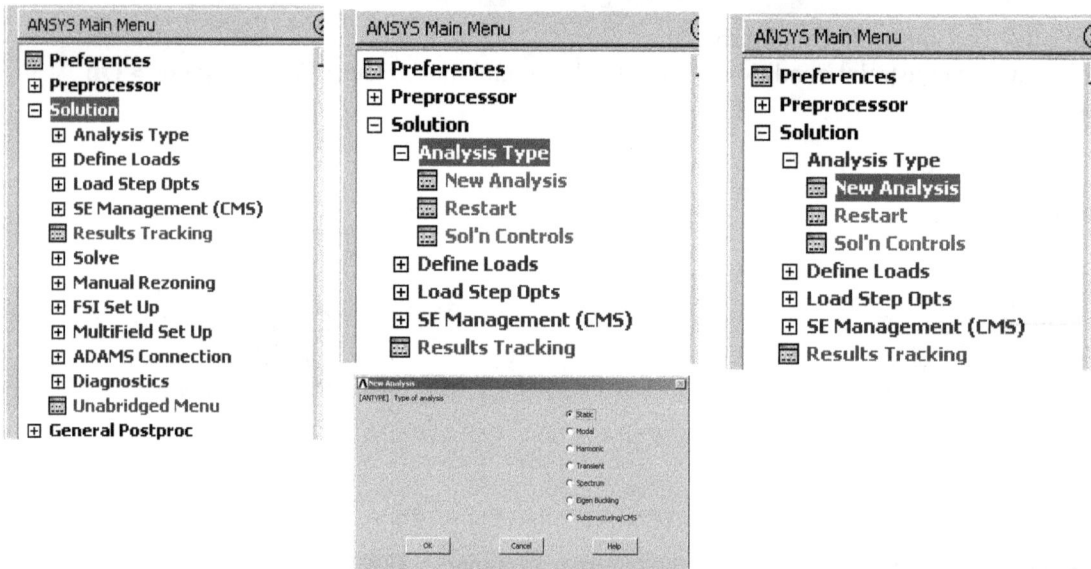

Choose **Current LS** > after reviewing the information, click **OK** to run the program.

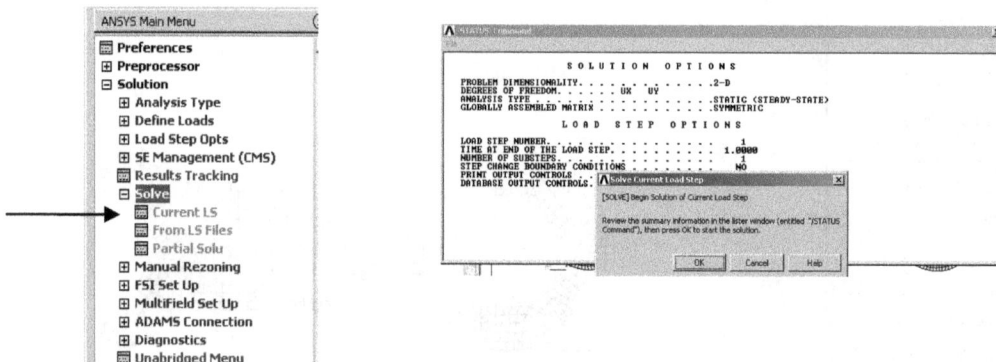

A window called Note appears when the solution is done > **Close**.

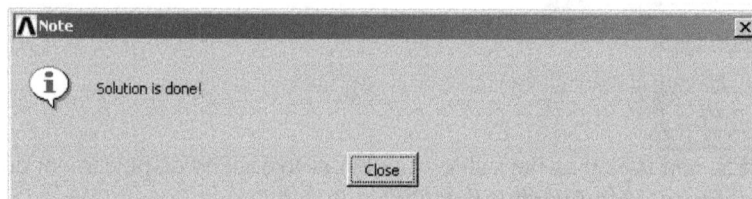

Step 8: Displaying Results

To review the results, select **General Postprocessing** from the main menu. Then click **Plot Results**, > **Deformed Shape** > select **Def + undeformed** > **OK**.

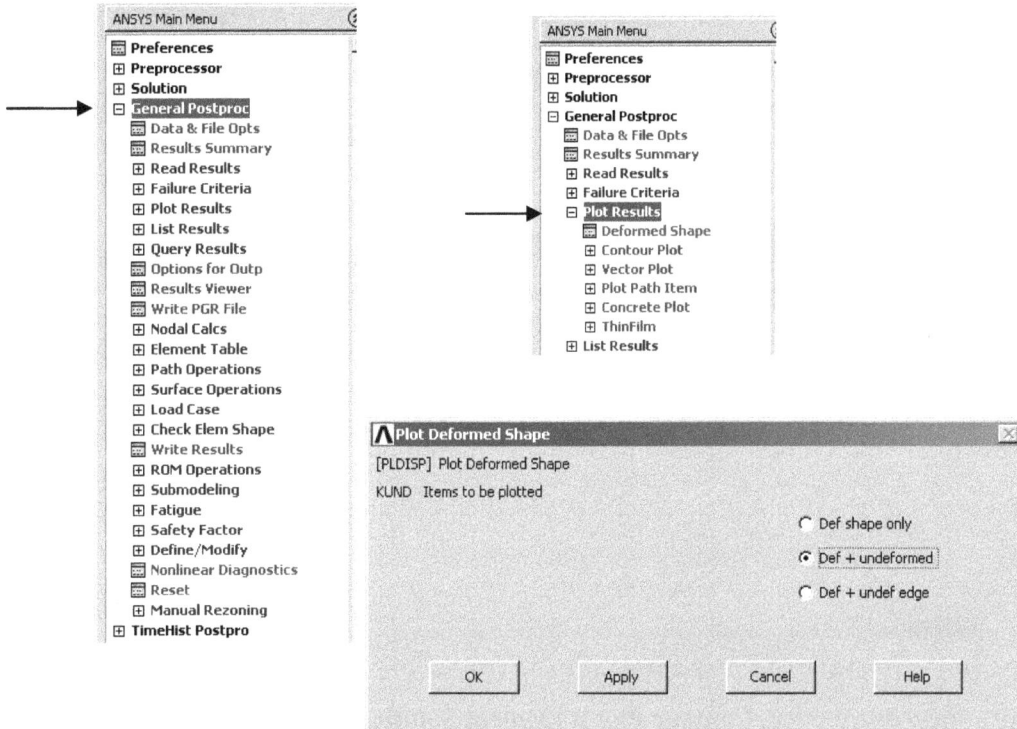

The following plot depicts the deformed shape.

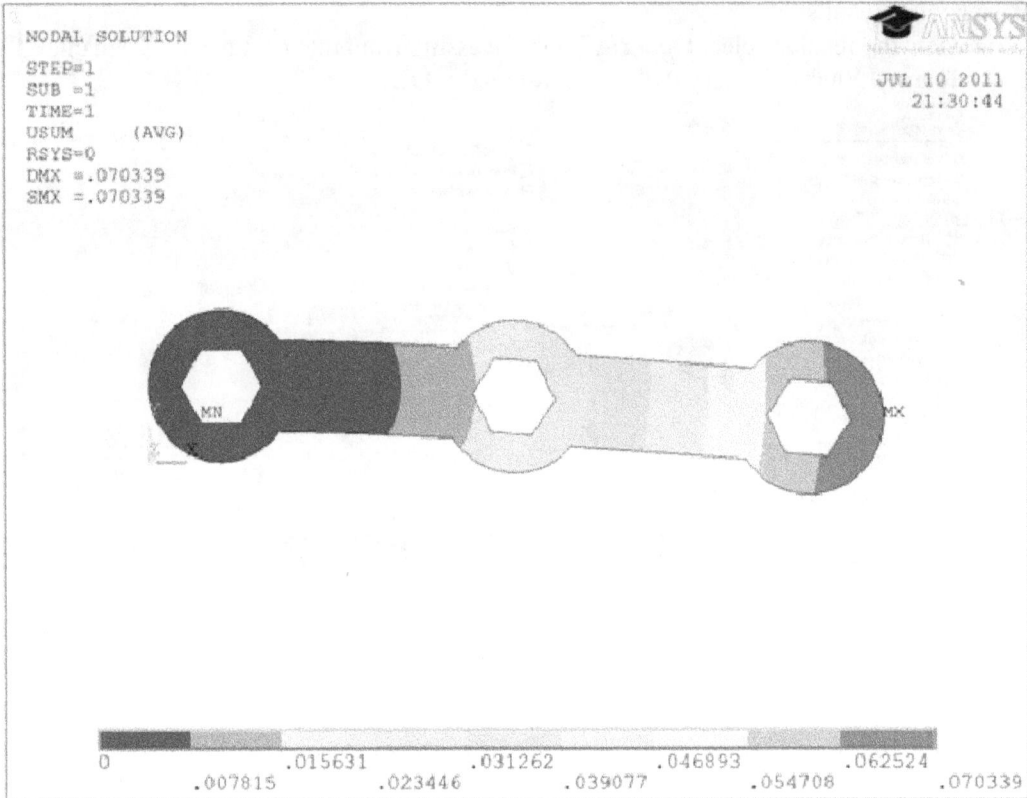

To plot a stress distribution, **Contour Plot > Element Solution >** pick a specific type of stress of interest, say von Mises stress.

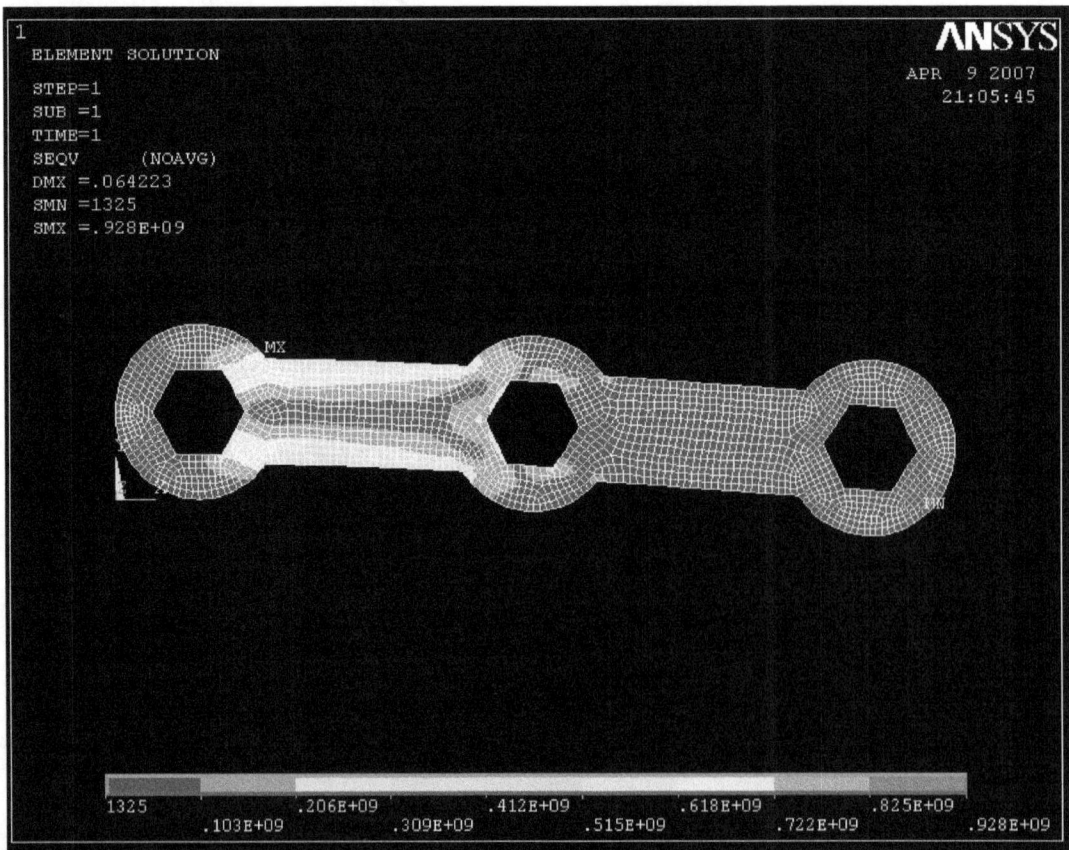

10.7 Working with INCH Unit System

The following figure illustrates a weld structure consisting of 3 plates. Under a static equilibrium condition, the 2 holes are fixed to the ground. A uniformly distributed load acts on the top surface of the horizontal surface. All the dimensions shown are inches. Evaluate the distribution of the first principal stress and the deformation patter in the vertical direction.

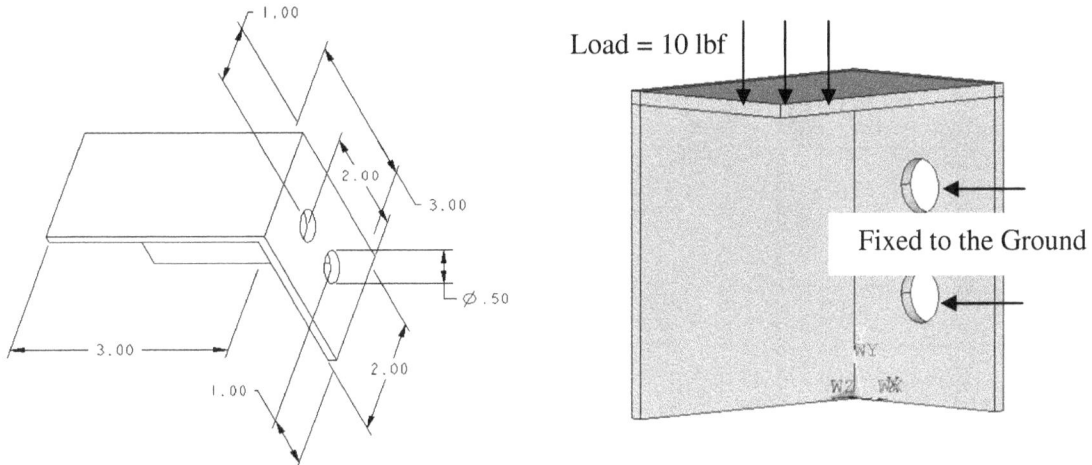

Step 1: Launch **ANSYS 13.0**

From the start menu, select **Programs > ANSYS 13.0 > ANSYS**, or directly click the icon of **ANSYS**.

To define the Structural Analysis, from the main menu or the model tree, select **Preferences**. > make sure that **Structural** is highlighted, and h-method is also selected > **OK**.

Step 2: Set the unit system to English (in) or BIN.

Preprocessor > Material Library > Select Units > BIN > OK.

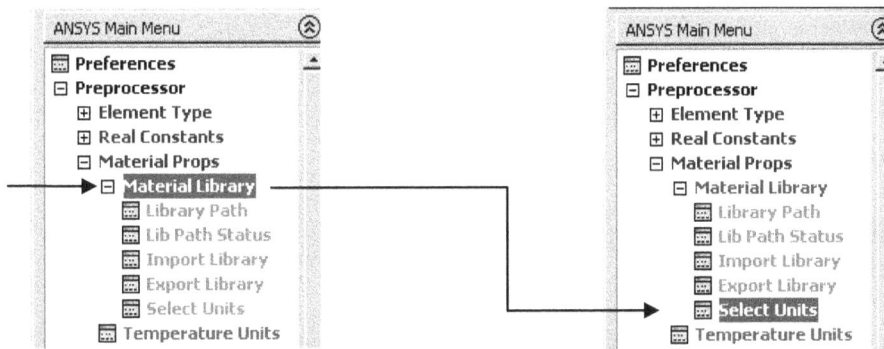

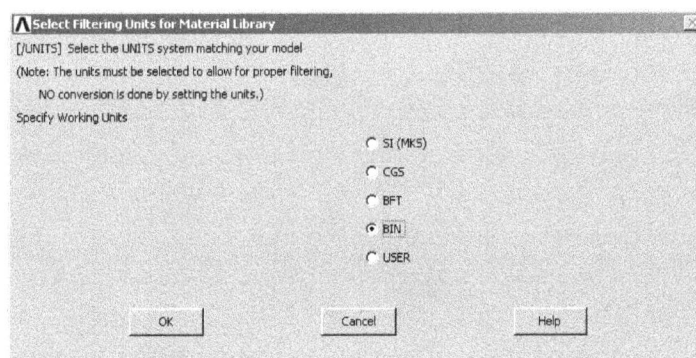

As illustrated, the basic quantities used in English (in) system are

Length:	inch (in)
Mass:	lbf*s^2/in
Force:	pounds (lbf)
Time:	second (s)
Temperature:	Fahrenheit (deg F)

Step 3: Create a 3D solid model

From the main menu, select **Preprocessor** and expand it. Select **Modeling > Create > Volumes > Block > By Dimensions**.

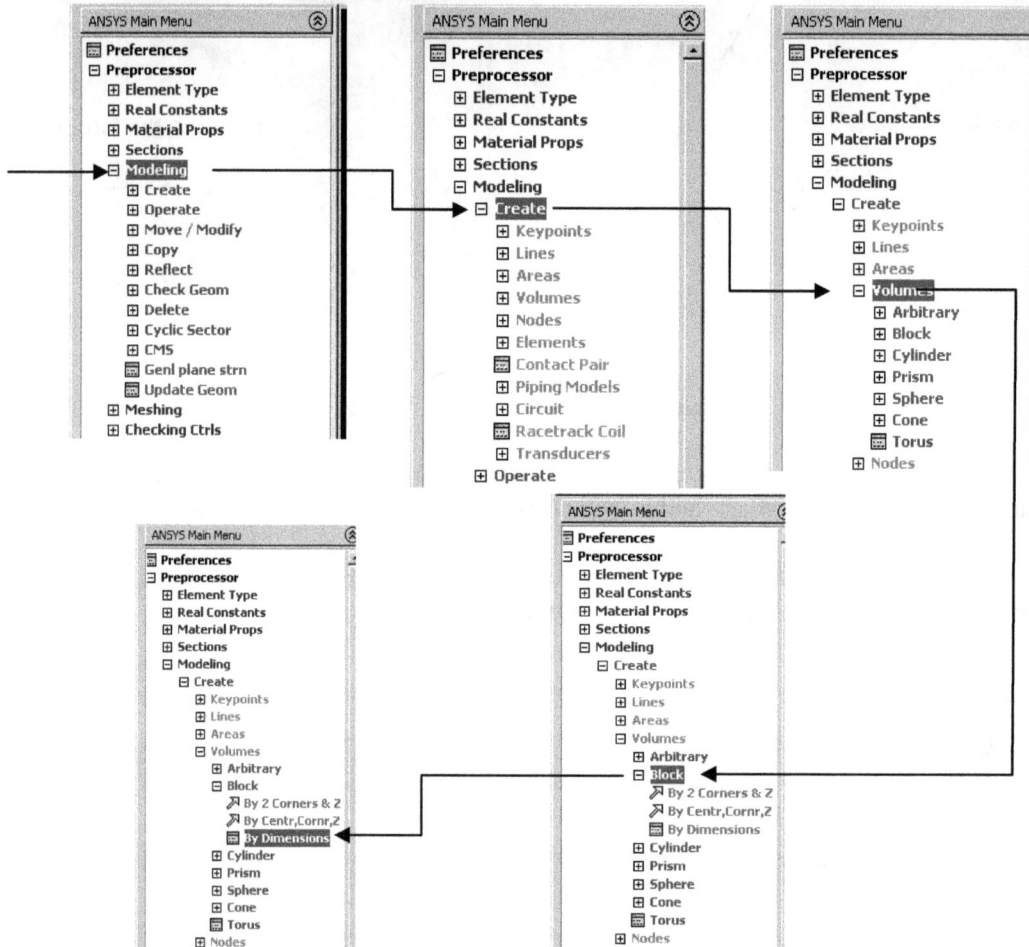

In the window for creating the rectangle by dimensions, type (0, 0, 0) for X1, Y1 and Z1. Type (2, 3, 0.125) for X2, Y2 and Z2 > **OK**.

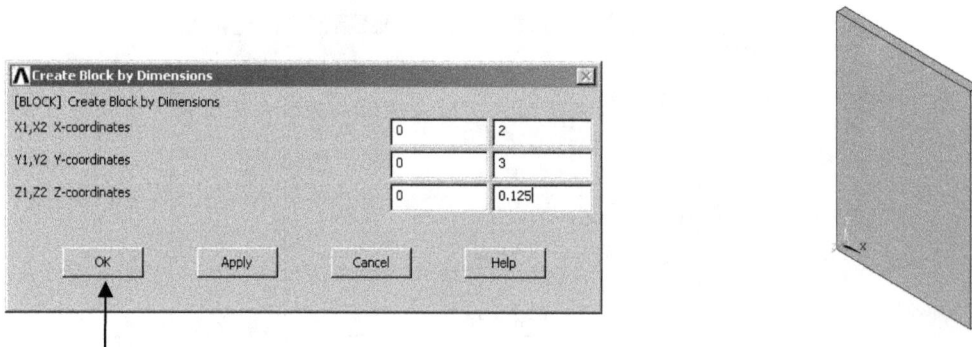

To create 2 cylinders, **Modeling > Create > Volumes > Cylinder > Solid Cylinder.** Type (1, 1) for the center location. Type 0.25 as the radius value and type 0.125 as the depth of the cylinder > **Apply**. The first cylinder is created, as shown.

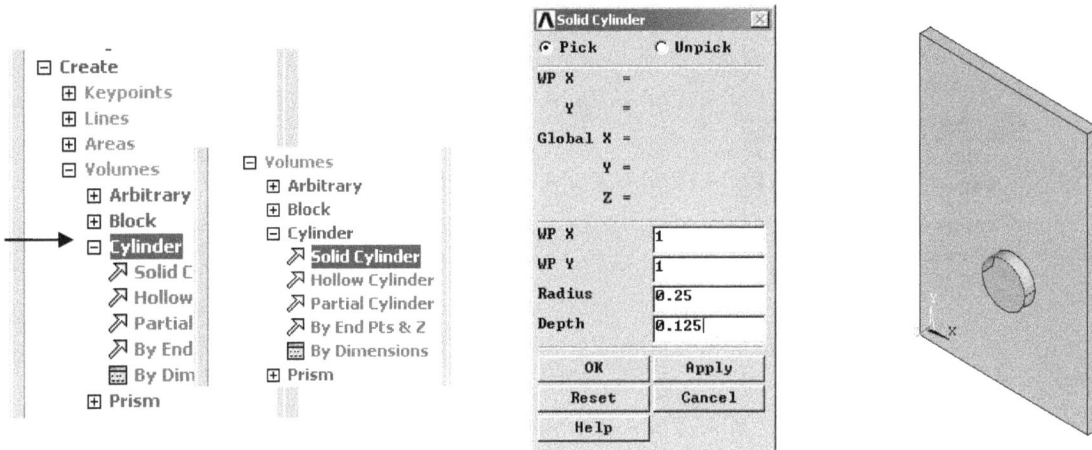

To create 2 cylinders, **Modeling > Create > Volumes > Cylinder > Solid Cylinder.** Type (1, 2) for the center location. Type 0.25 as the radius value and type 0.125 as the depth of the cylinder > **OK**. The second cylinder is created, as shown.

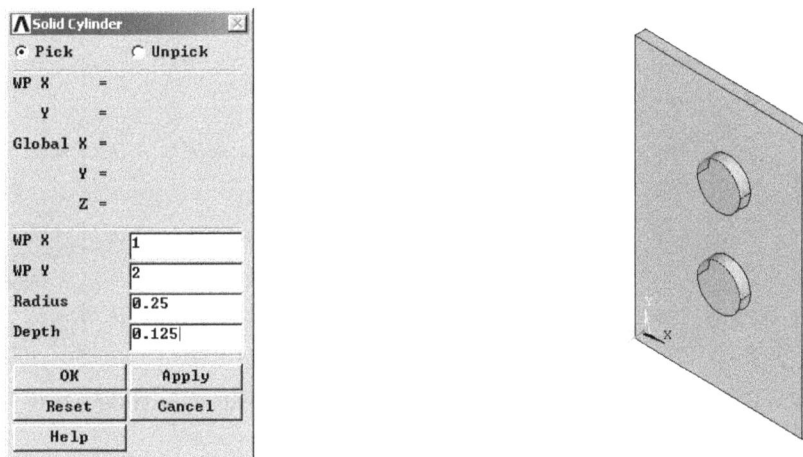

Under **Modeling**, select **Operate > Booleans > Subtract > Volume**. Pick the plate first and click **OK**, then pick the 2 cylinders and click **OK**.

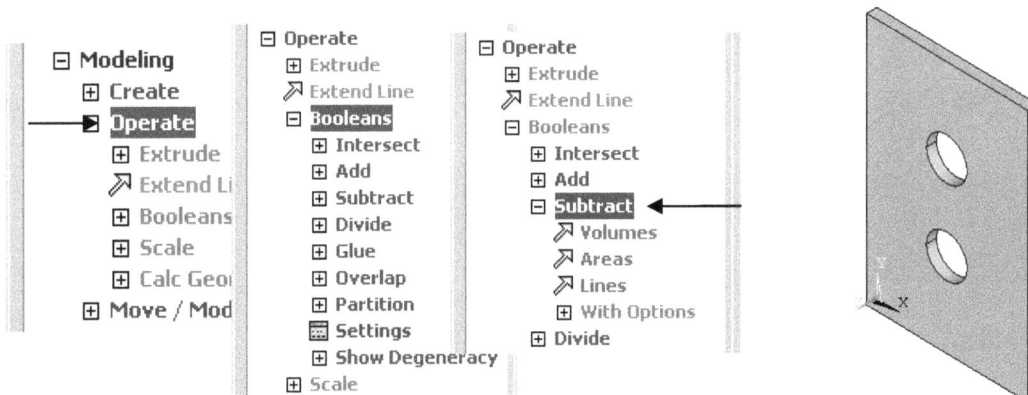

To create the top plate, let us move the workplane. From the main menu, click **Workplane > Offset WP by Increments**. A window called Offset WP appears. Type 0, 3, 0, as shown > **OK**.

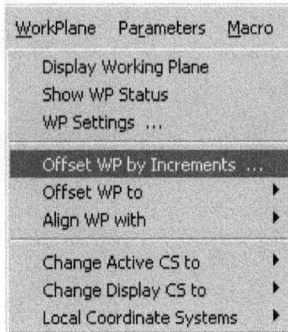

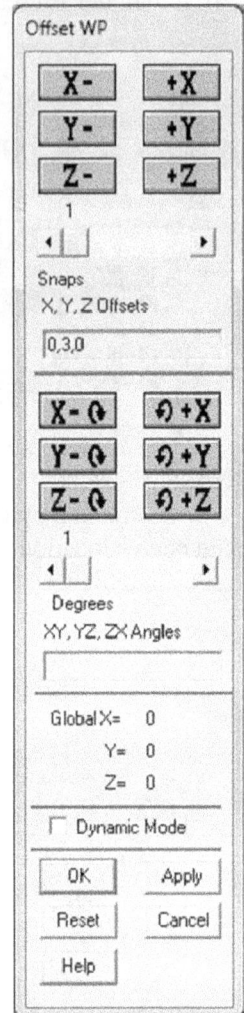

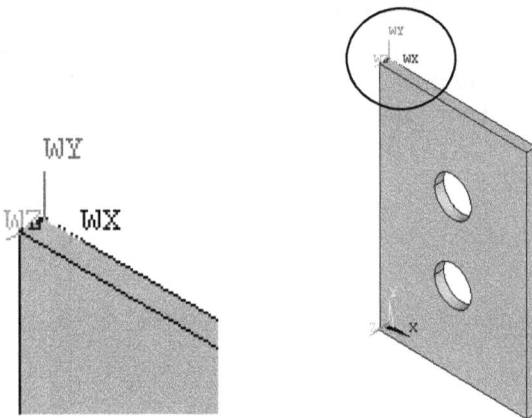

To rotate the WP, click Offset WP by Increment, move the Degrees Slide bar to 90 and press **+X rotation** button > **OK**.

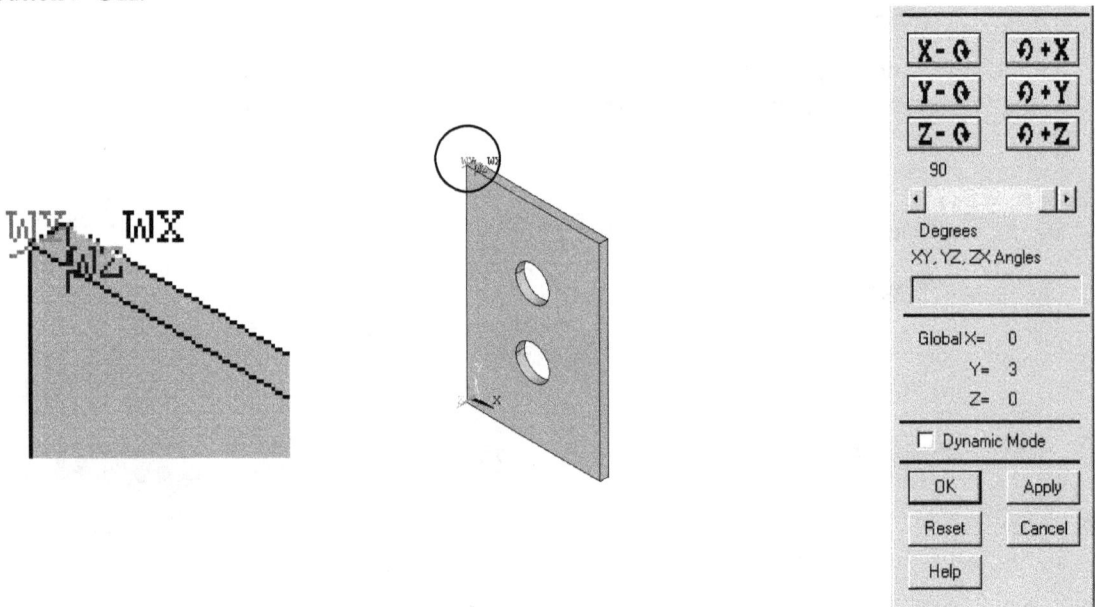

To create the second plate on the horizontal level, select **Modeling > Create > Volumes > Block > By Dimensions**. In the window for creating the rectangle by dimensions, type (0, 0, 0) for X1, Y1 and Z1. Type (2, 3, 0.125) for X2, Y2 and Z2 > **OK**.

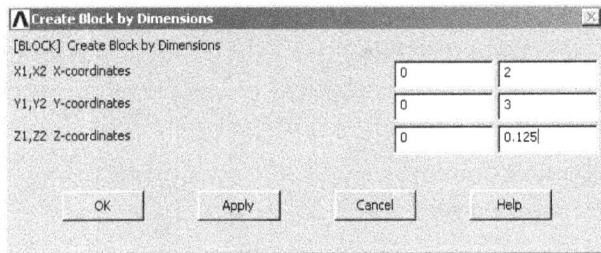

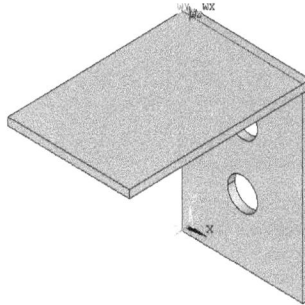

To create the third plate on the vertical level, let us move the workplane to its original position (global position and orientation)
    WorkPlane > Align WR with > Global Cartesian. To clearly display the workplane, **Plot > Volumes**

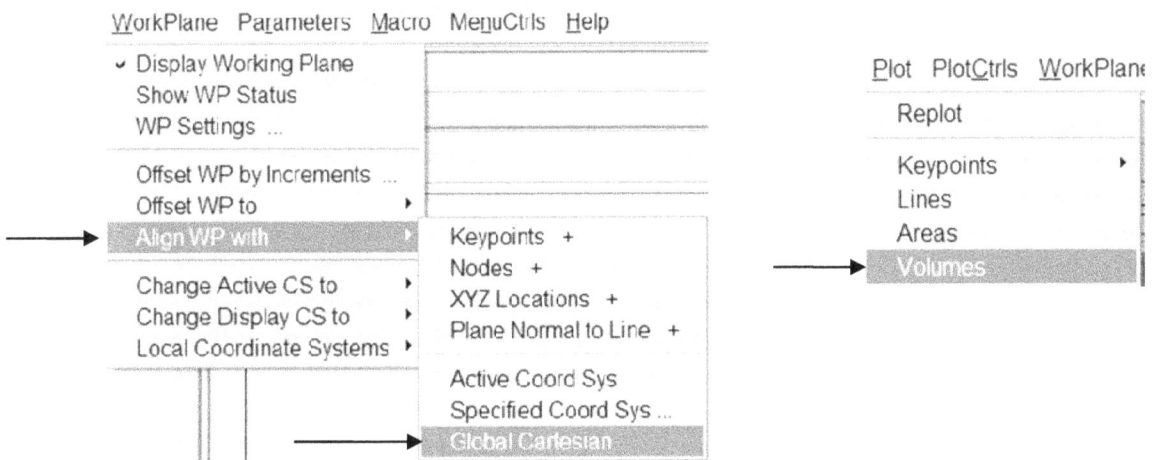

Select **Modeling > Create > Volumes > Block > By Dimensions**. . In the window for creating the rectangle by dimensions, type (0, 0, 0) for X1, Y1 and Z1. Type (0.125, 3, 3) for X2, Y2 and Z2 > **OK**.

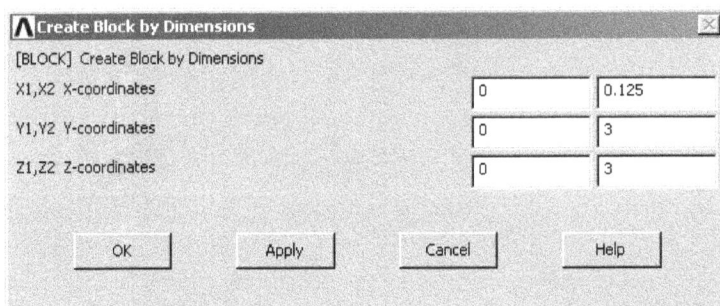

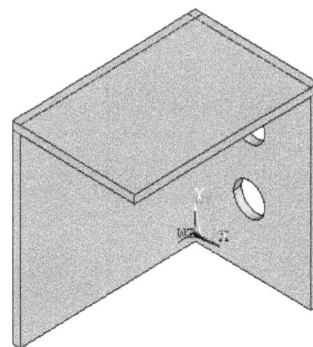

Now let us add these 3 volumes together. **Operate > Booleans > Add > Volumes > Pick All.**

Step 4: Select Element Type
       Select **Element Type > Add/Edit/Delete.**

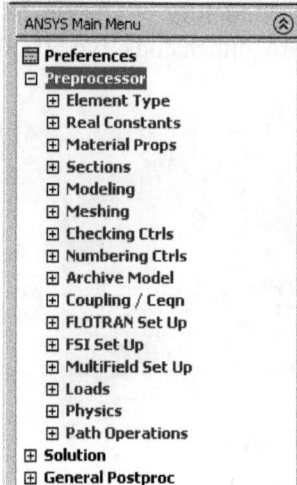

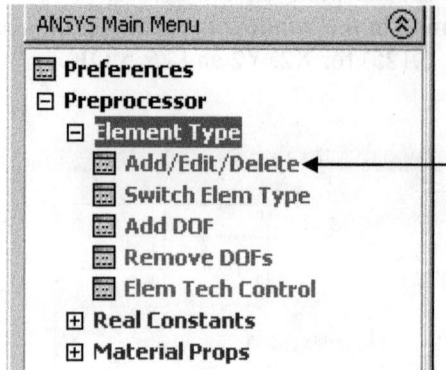

In the **Element Types** appears. Click **Add >** pick **Solid > Tet 10 node 187 > OK > Close**.

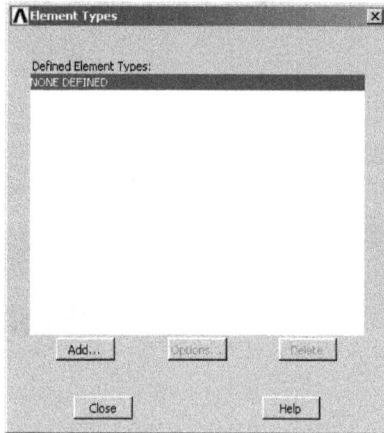

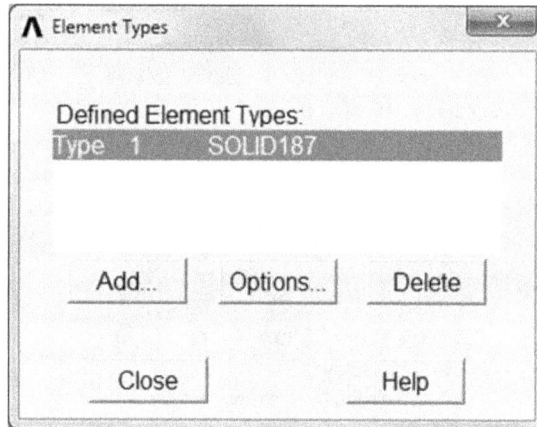

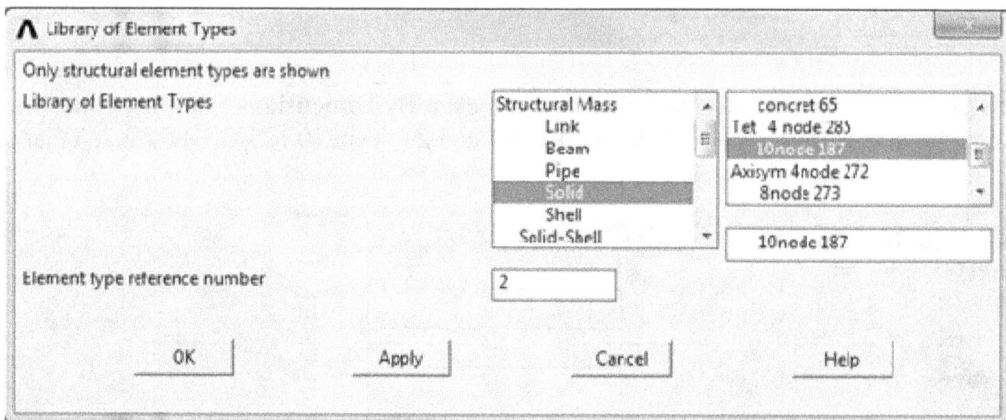

Step 5: Define the material type

Select **Material Properties** from the **Preprocessor** menu. Select **Material Model**, and then select **Structural, Linear, Elastic** and **Isotropic** by double clicking on each one. Type 29e6 and 0.26 as the values for Young's Modulus and Poisson's ratio, respectively. Then select **OK**. Click **Material > Exit**

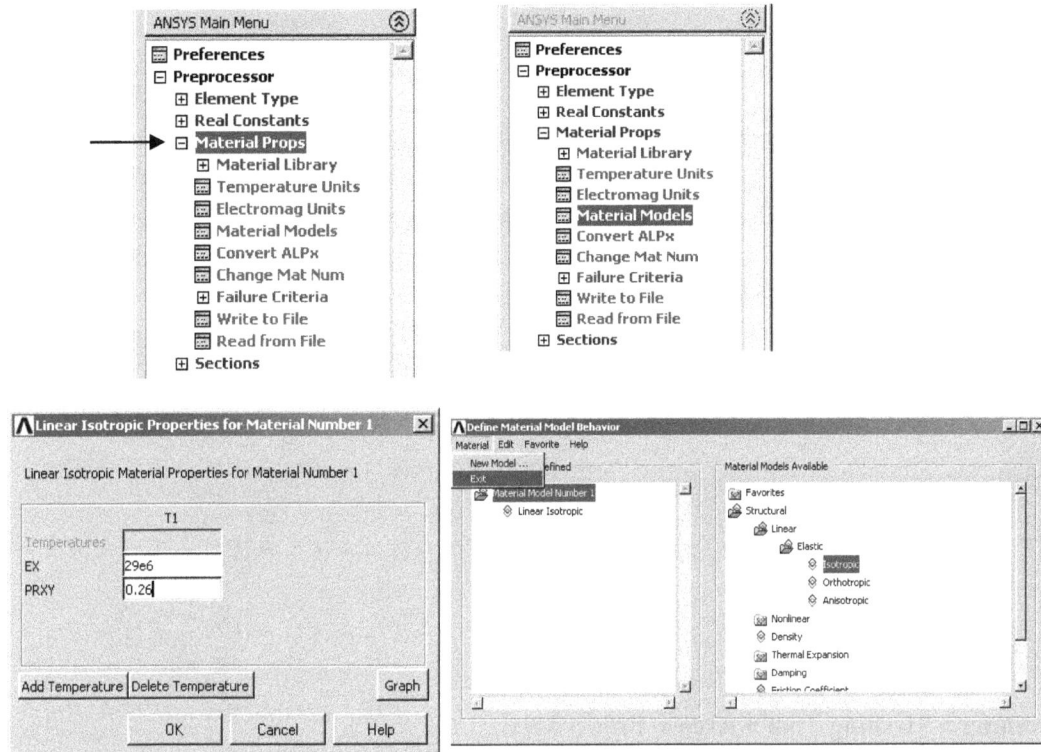

To check the unit system, select **Material Library** under the material properties menu. Then select **Units** and make sure that the **BIN** unit system is highlighted. Click **OK**.

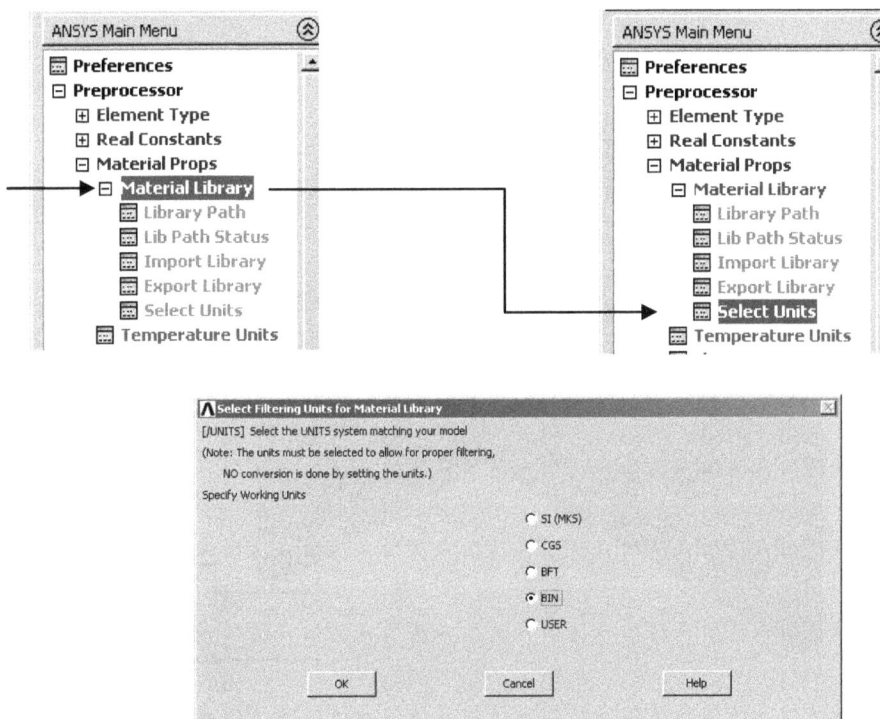

Step 6: Mesh Generation

To create the mesh for the shell model, select **MeshTool** from the Preprocessor menu.
Meshing > Size Cntrls > Manual Size > Global Size > Size.

Specify 0.1 as the Element edge length > **OK**.

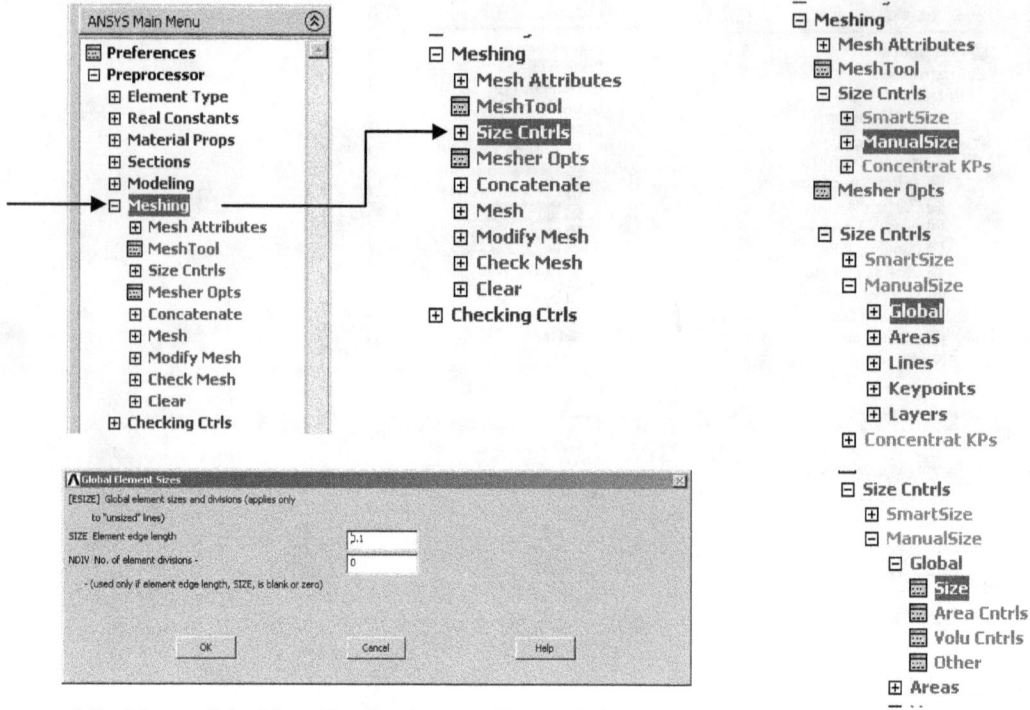

Meshing > Meshing Tool > Area > Free > Mesh.

When the selection box comes up, select **Pick All**, and then click **OK**. Make sure that you close the **MeshTool** window when the meshing process is completed.

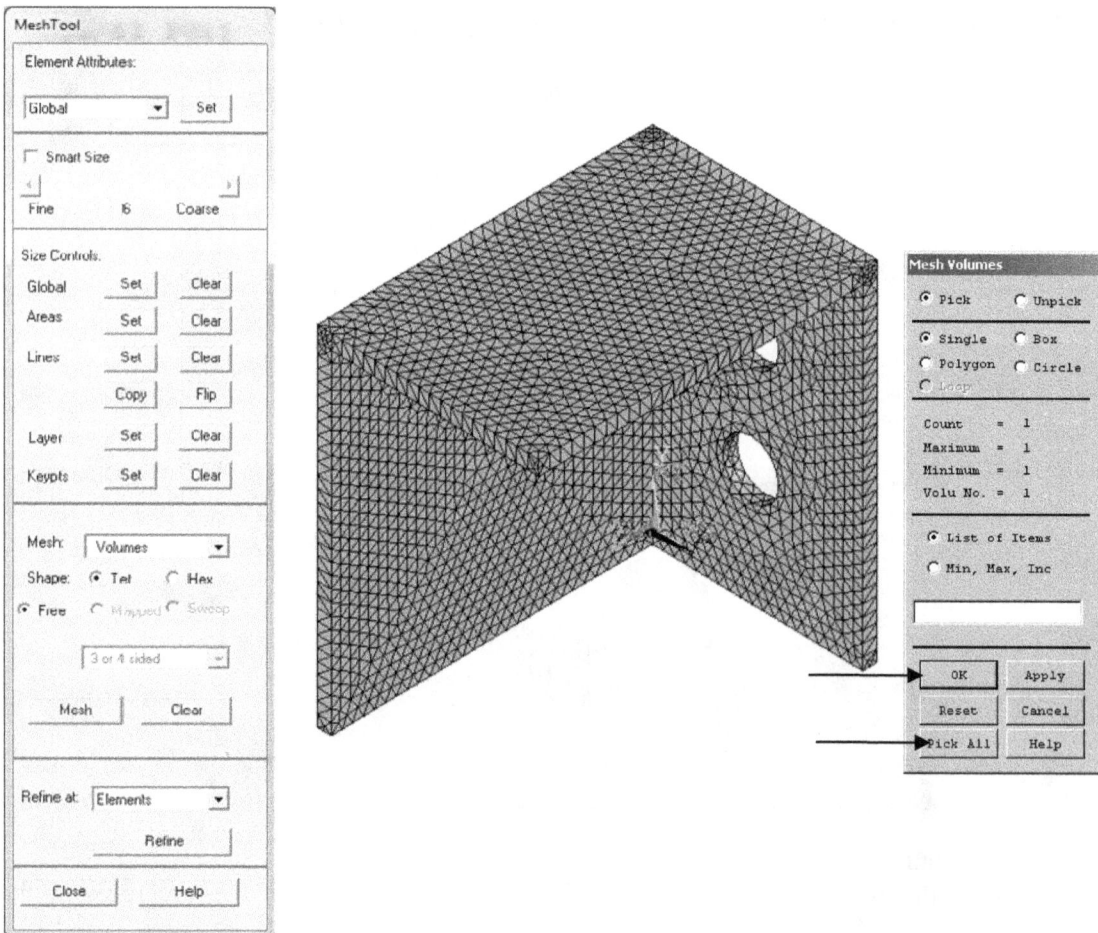

Step 7: Apply the load condition.

Assume that a pressure load is acting on the top surface of the horizontal plate.

To apply the load, on the Preprocessor menu, click **Load > Define Loads > Apply > Structural > Force/Moment> On Nodes**.

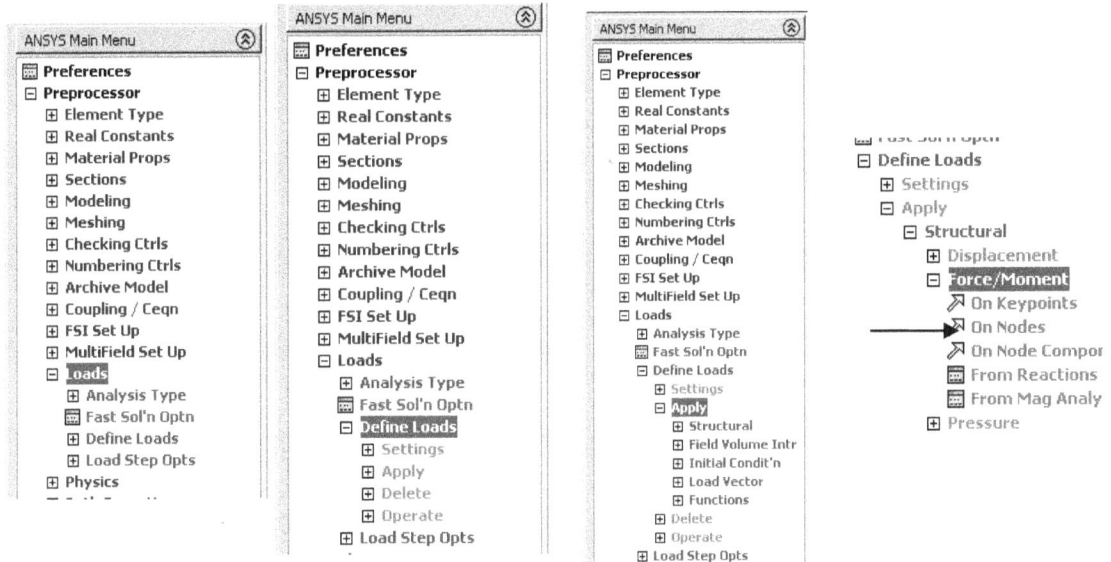

Pick the 9 nodes along the 2 dges, as shown.

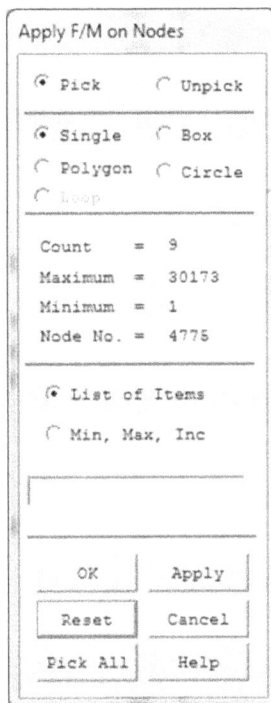

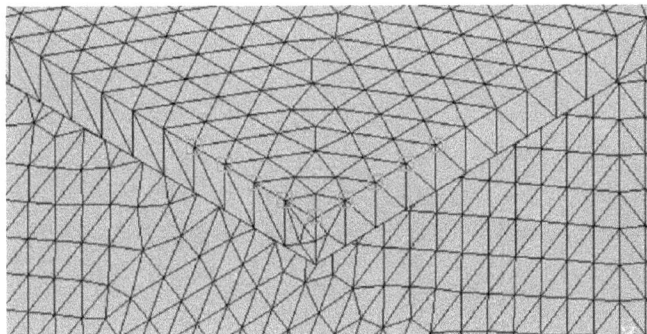

Specify -10 as the value of the total magnitude in the Y direction or FY.

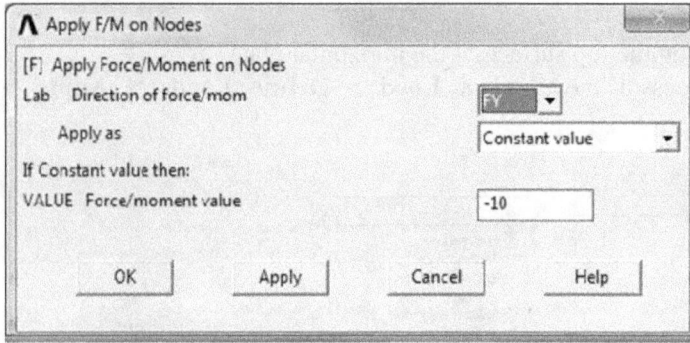

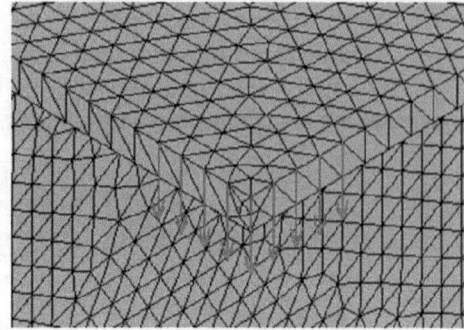

Step 8: Defining the Constraint Condition
To define the constraint conditions, under the Structural menu, select **Displacement > On Nodes**.

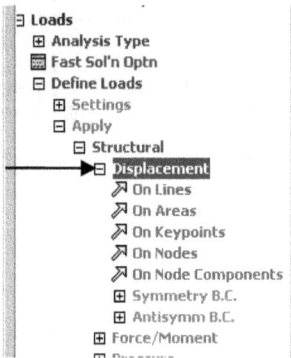

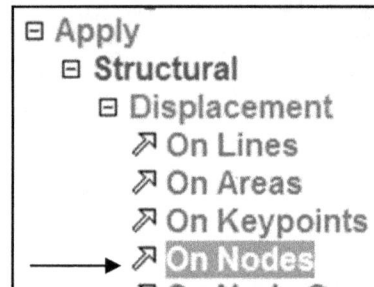

Change the picking mode to circle. Starting at the center of the holes and sketch a circle so that all keypoints are enclosed. There are 8 keypoints associated with a circle. Therefore, a total of 368 nodes are selected.

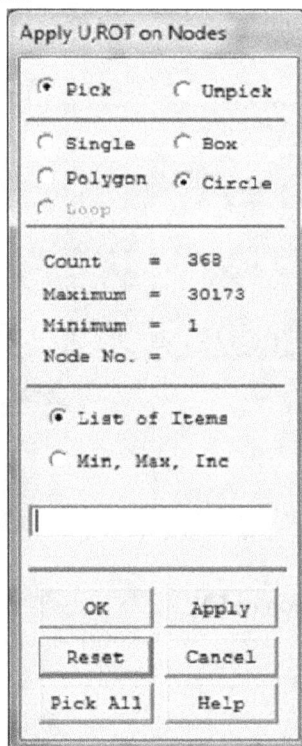

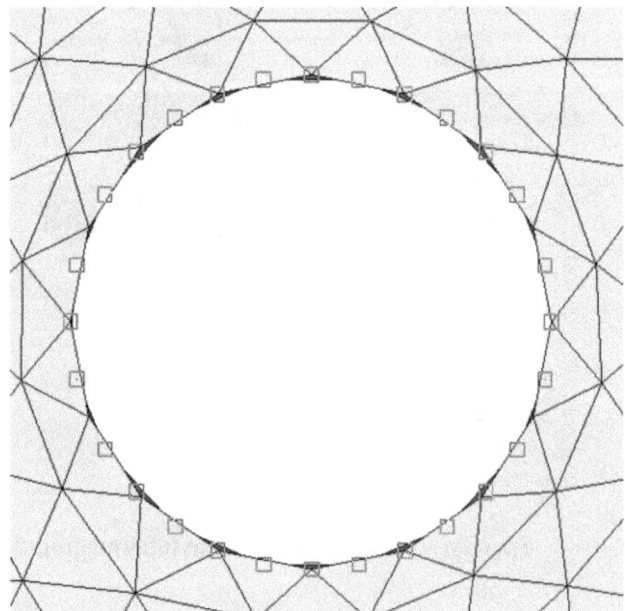

Select **All DOFS**, and type 0 as the value. Then click **OK**.

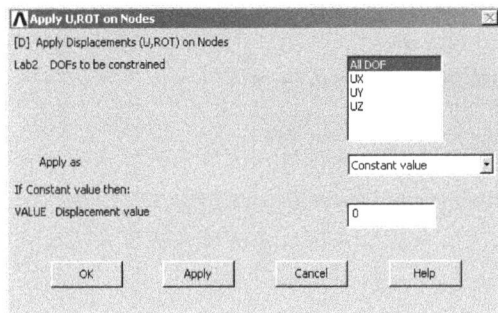

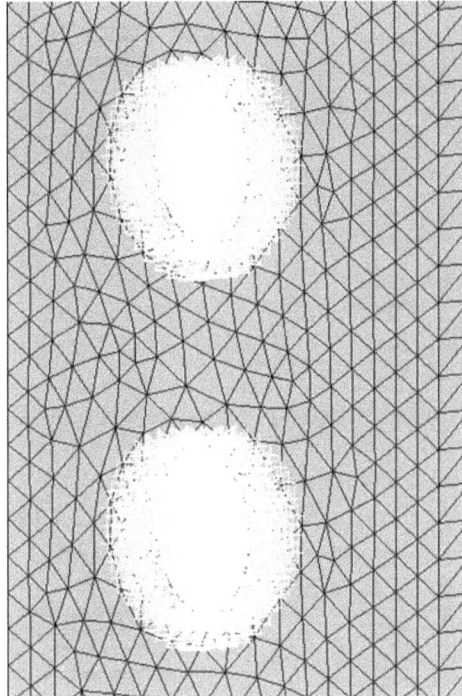

Step 9: Perform the analysis and obtain the solution.

To find the solution, run the program by selecting **Solution** from main menu > **Analysis Type** > **New Analysis** > **Static** > **OK**.

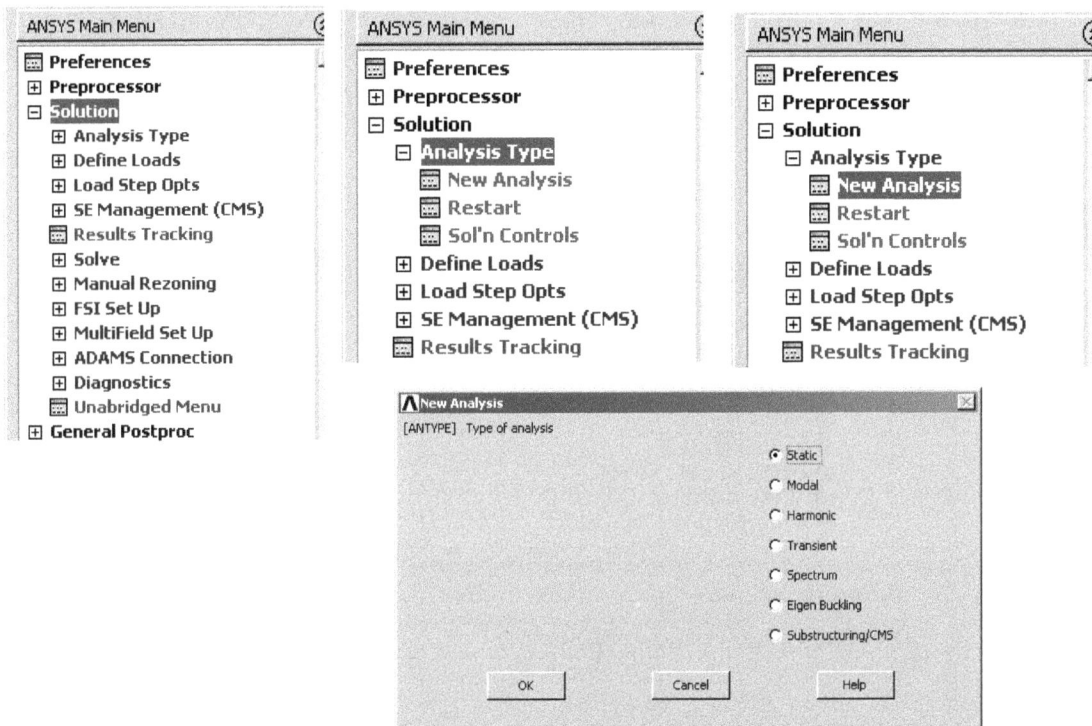

Choose **Current LS** > after reviewing the information, click **OK** to run the program.

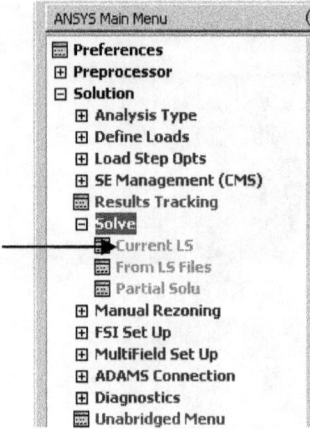

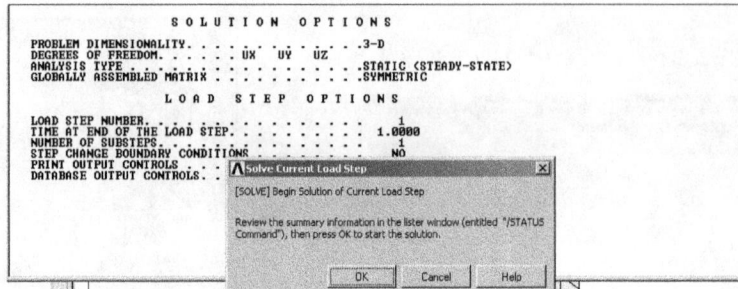

A window called Note appears when the solution is done > **Close**.

Step 10: Displaying Results

Under **General Postprocessing**, select **Plot Results** > **Contour Plot** > **Nodal Solution**. Select **Stress** from the list, say von Mises stress > **OK**.

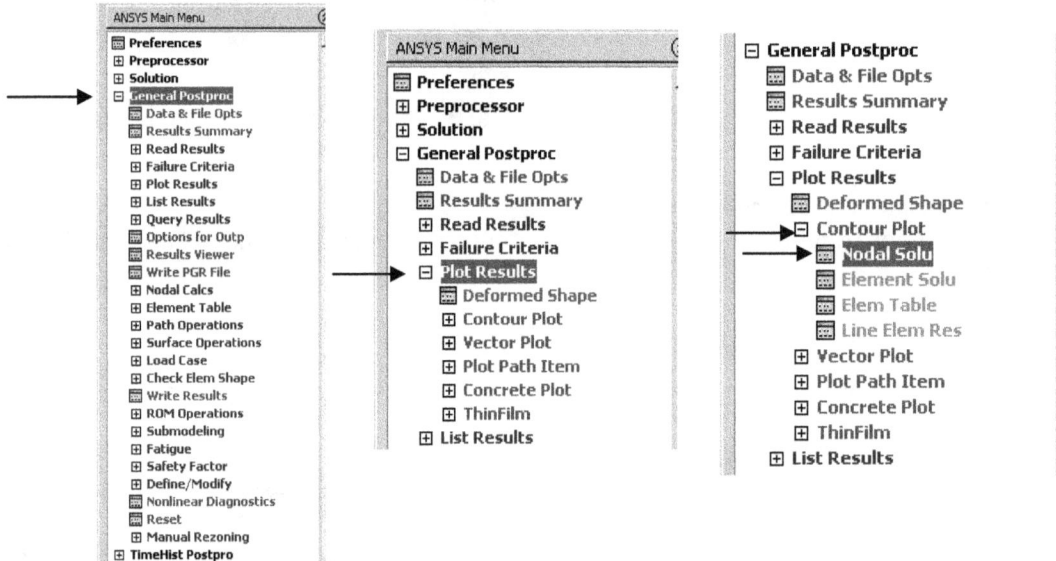

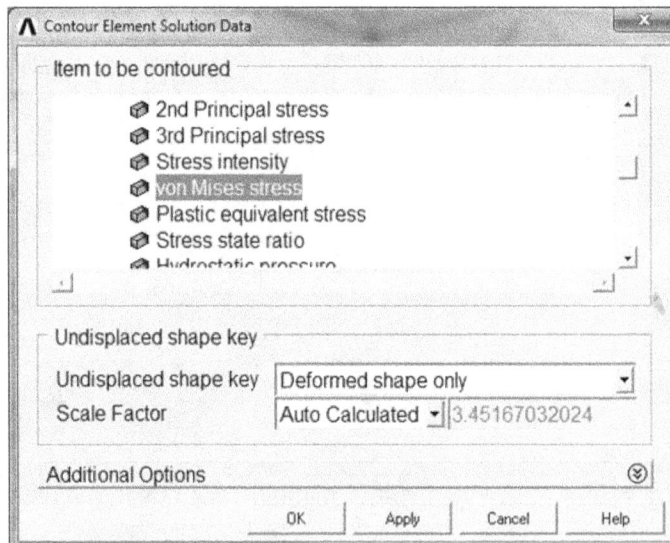

The following plot depicts the distribution of von Mises stress.

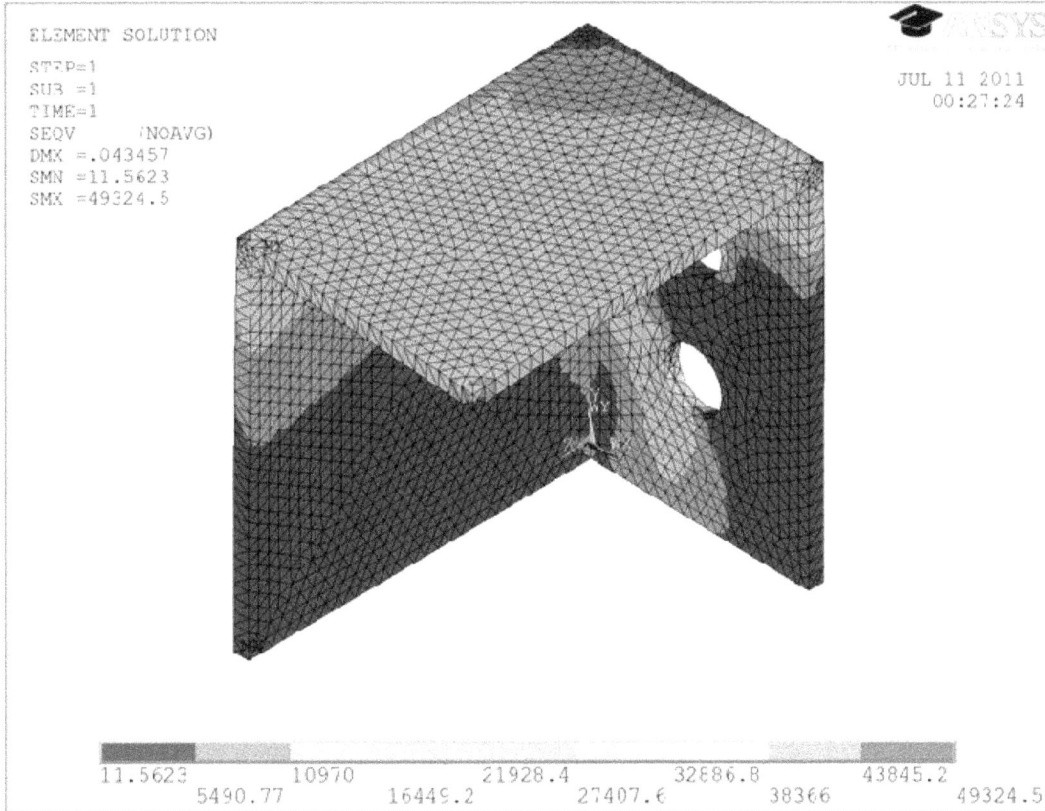

Again, select **Element Solution > Stress > Ist Principal Stress > OK**.

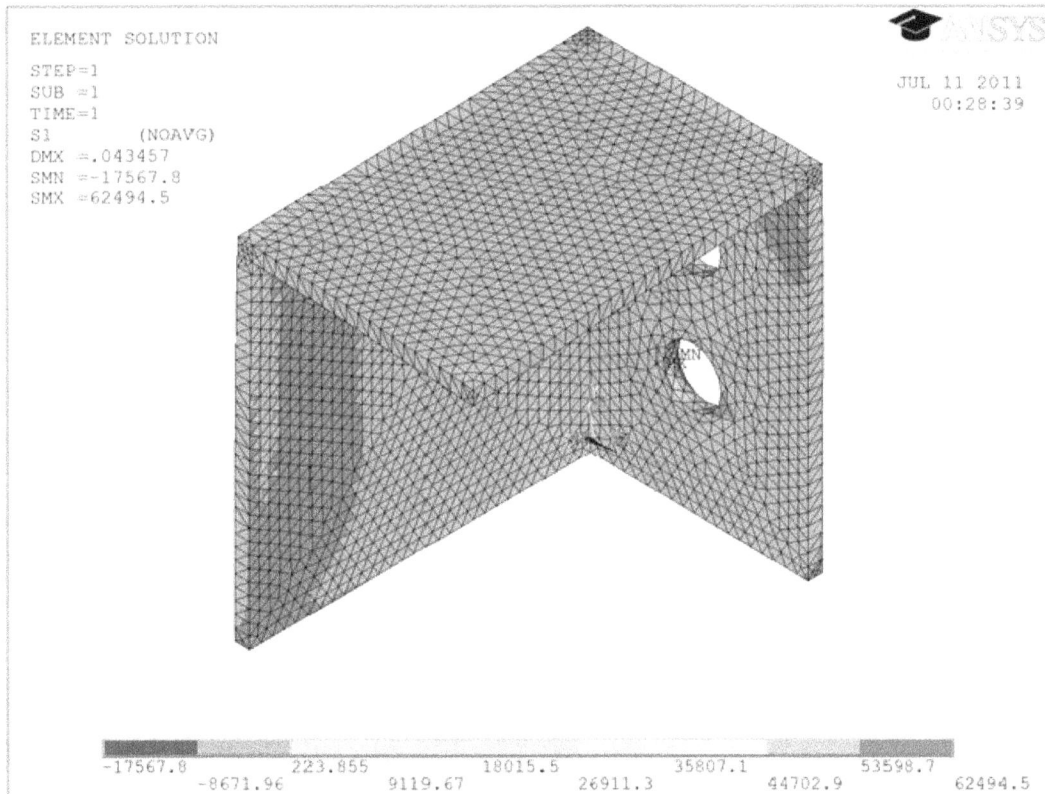

The following plot depicts the distribution of displacement. Note the scale is set by the user. As indicated, the scale is 10.

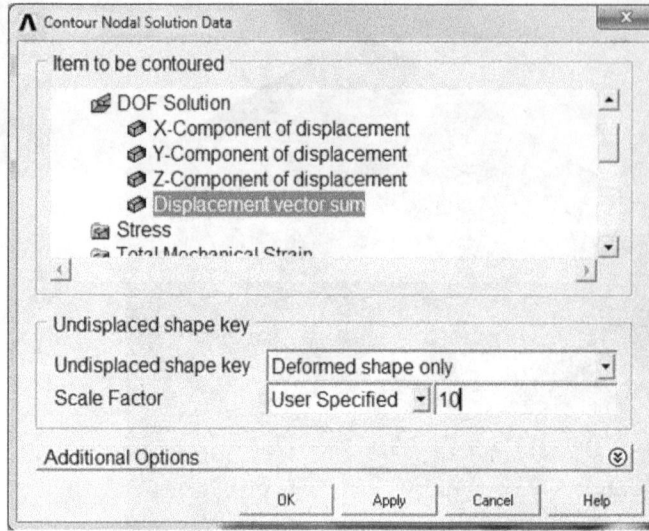

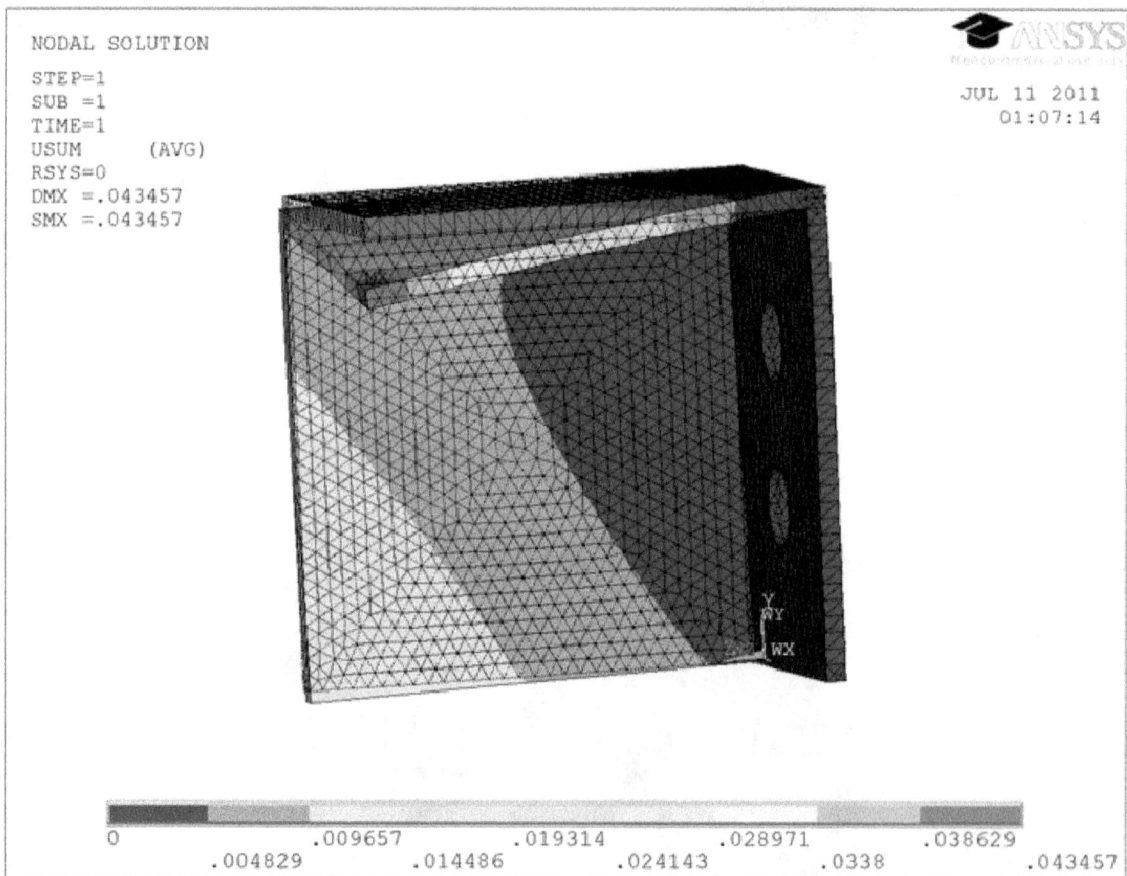

10.8 Thermal Analysis with 3D Solid Model (BFT Unit System)

The following figure illustrates a window structure consisting of 2 components, namely, the wall component and the plastic window component. The unit of the dimensions shown is feet. Perform a thermal analysis.

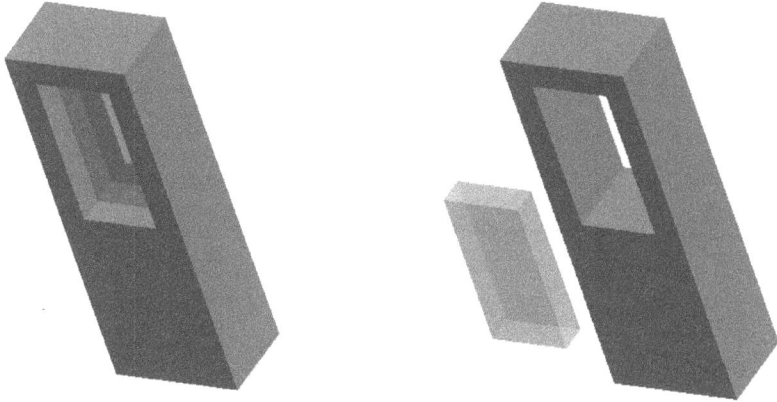

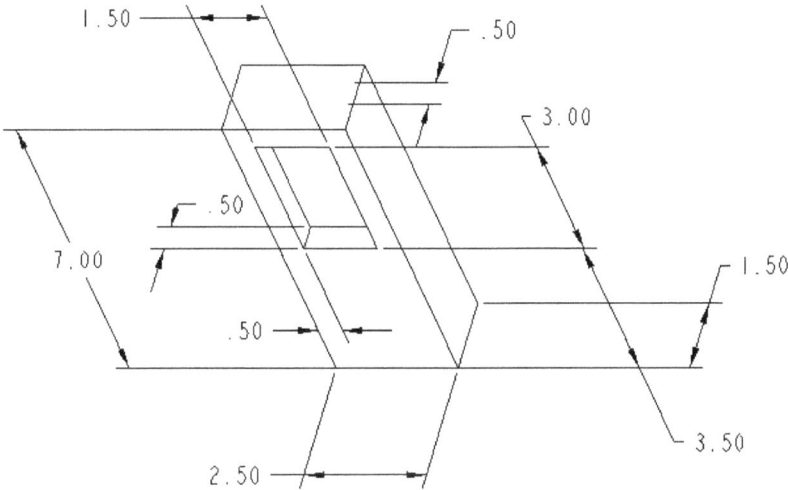

Step 1: Launch **ANSYS 13.0**

From the start menu, select **Programs > ANSYS 13.0 > ANSYS,** or directly click the icon of **ANSYS**.

To define a thermal analysis, from the main menu or the model tree, select **Preferences**. > make sure that **Thermal** is highlighted, and h-method is also selected > **OK**.

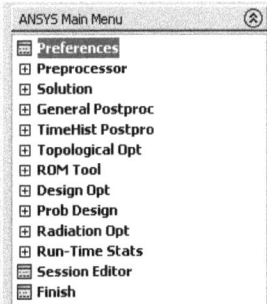

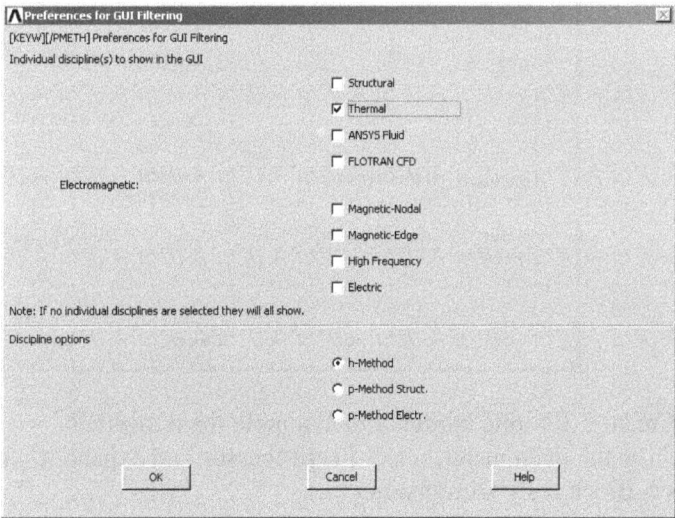

Step 2: Set the unit system to BFT
Preprocessor > Material Library > Select Units > BFT > OK.

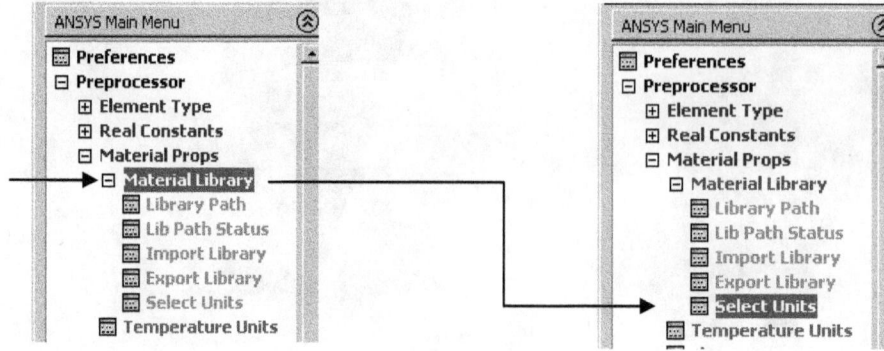

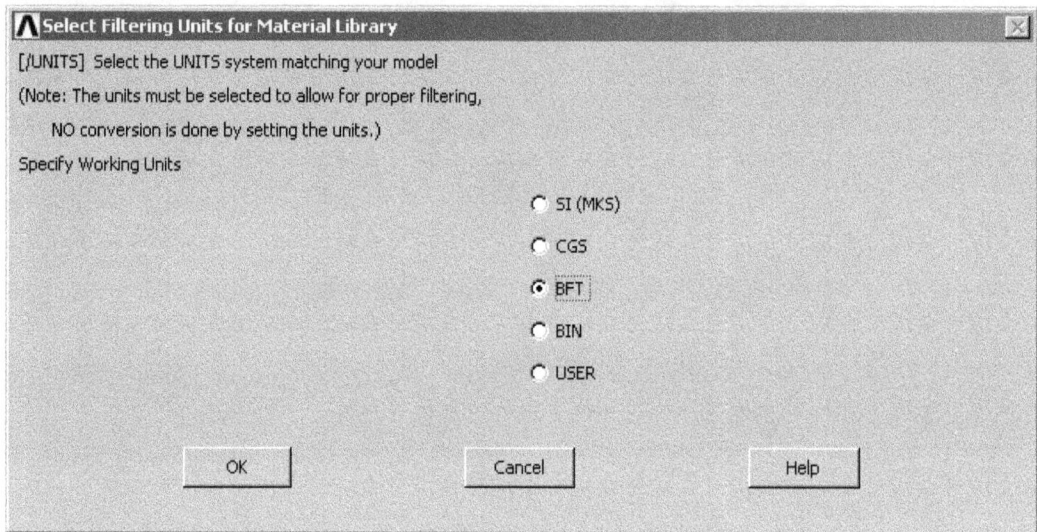

As illustrated, the basic quantities used in English (ft) system are Length: feet (ft), Mass: slug, Force: pounds (lbf). Time: second (s) and Temperature: Fahrenheit (deg F).

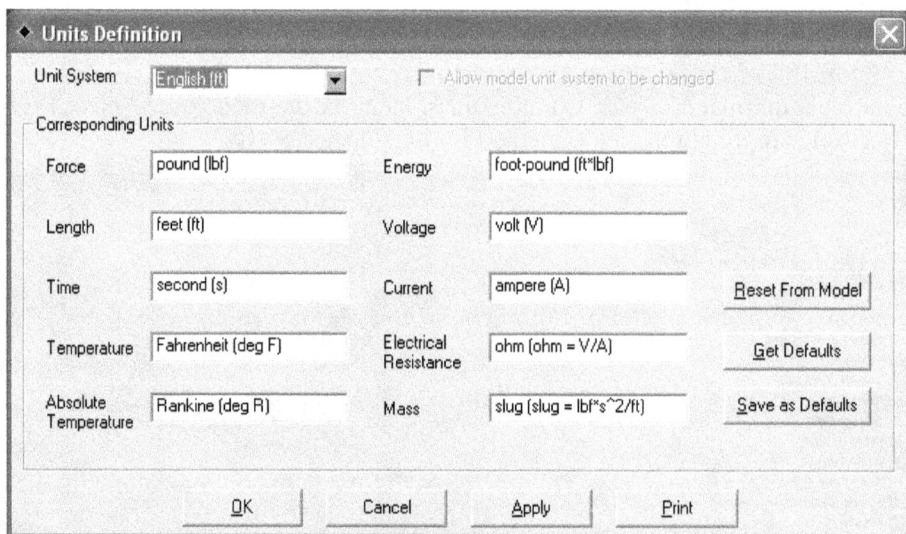

Step 3: Create 2 3D solid models and assemble them together.
From the main menu, select **Preprocessor** and expand it. Select **Modeling > Create > Volumes > Block > By Dimensions**.

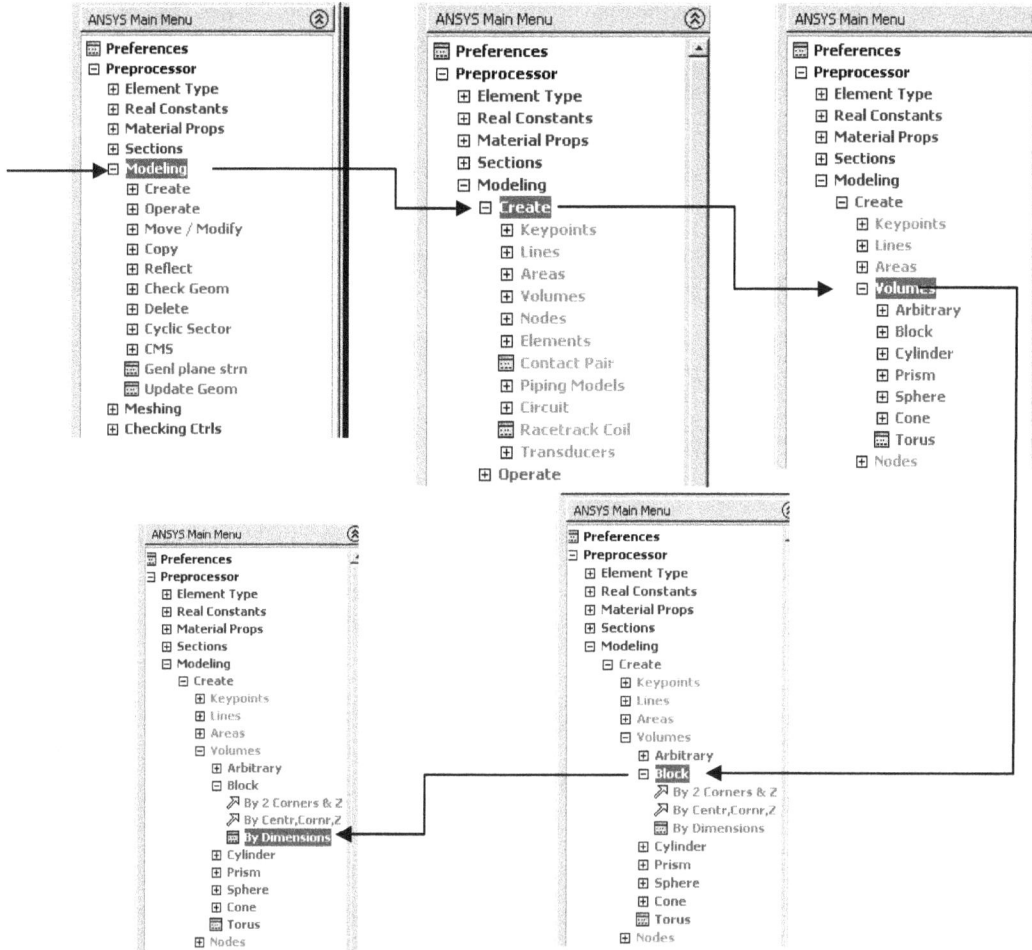

In the window for creating the rectangle by dimensions, type (0, 0, 0) for X1, Y1 and Z1. Type (2.5, 7, -1.5) for X2, Y2 and Z2 > **Apply**.

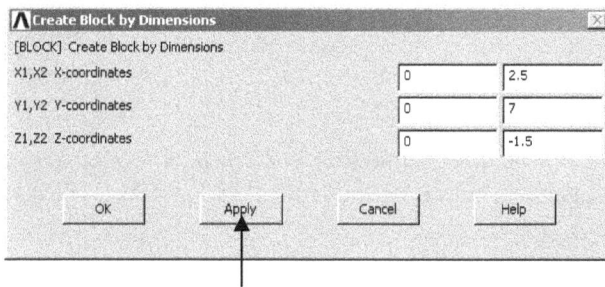

To create a second block, **Modeling > Create > Volumes > Block > By Dimensions.** Type (0.5, 3.5, 0) for the lower left corner in the front plane. Type (2, 6.5, -1.5) for the upper right corner in the back plane > **OK**.

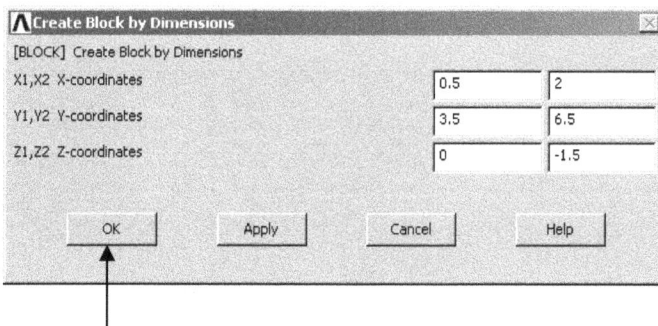

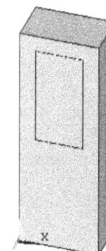

To remove the second volume from the first volume, under **Modeling**, select **Operate > Booleans > Subtract > Volume**. Pick Volume 1 first and click **OK**, then pick Volume 2, and click **OK**.

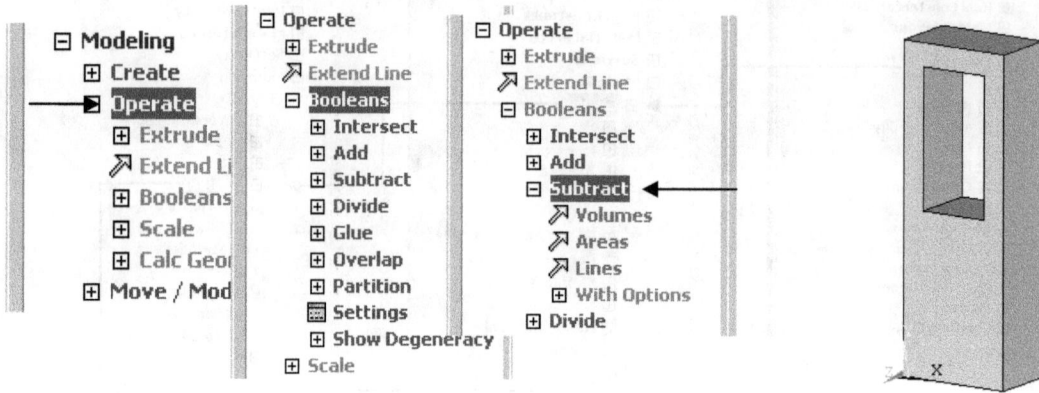

To create the plastic volume as the window, let us move the workplane. From the main menu, click **Workplane > Offset WP by Increments**. In the **Offset WP** window. Type *0, 0, -0.5*, as shown > **OK**.

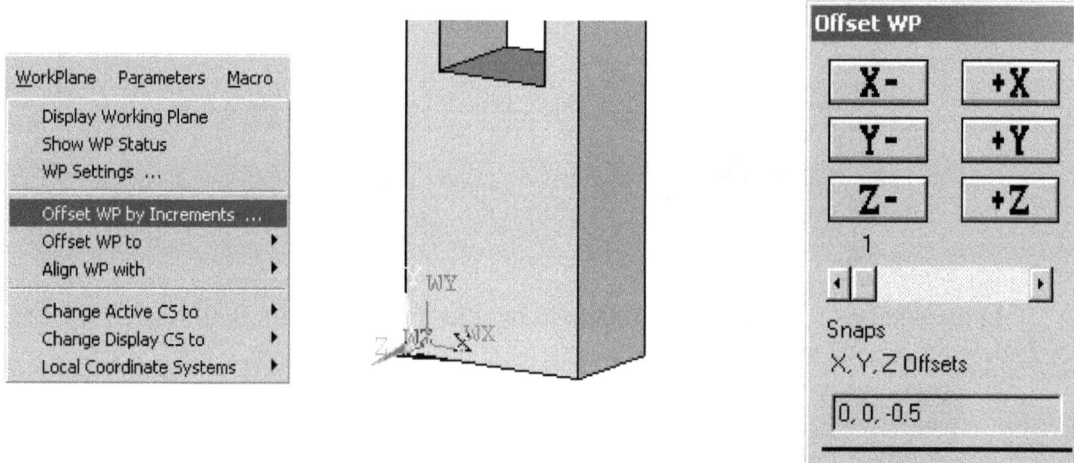

To create a third block for the plastic window, **Modeling > Create > Volumes > Block > By Dimensions**. Type (0.5, 3.5, 0) for the lower left corner as the front. Type (2, 6.5, -0.5) for the upper right corner as the back > **OK**.

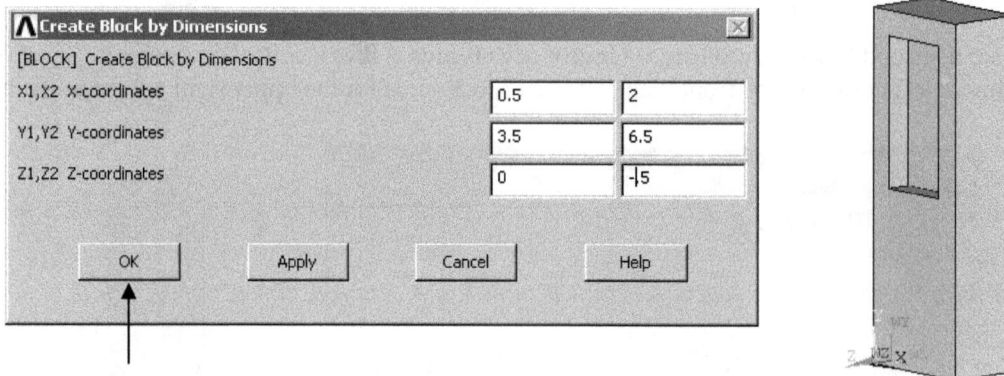

Now let us assemble the wall component and window component together. **Operate > Booleans > Glue > Volumes > Pick All.**

Step 4: Select Element Type
Select **Element Type > Add/Edit/Delete.**

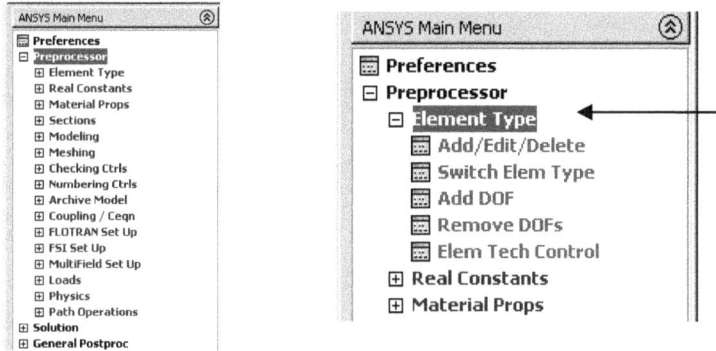

A window called **Element Types** appears. Click **Add** > pick **Solid** > **Tet 10node 87** > **OK** >
Close.

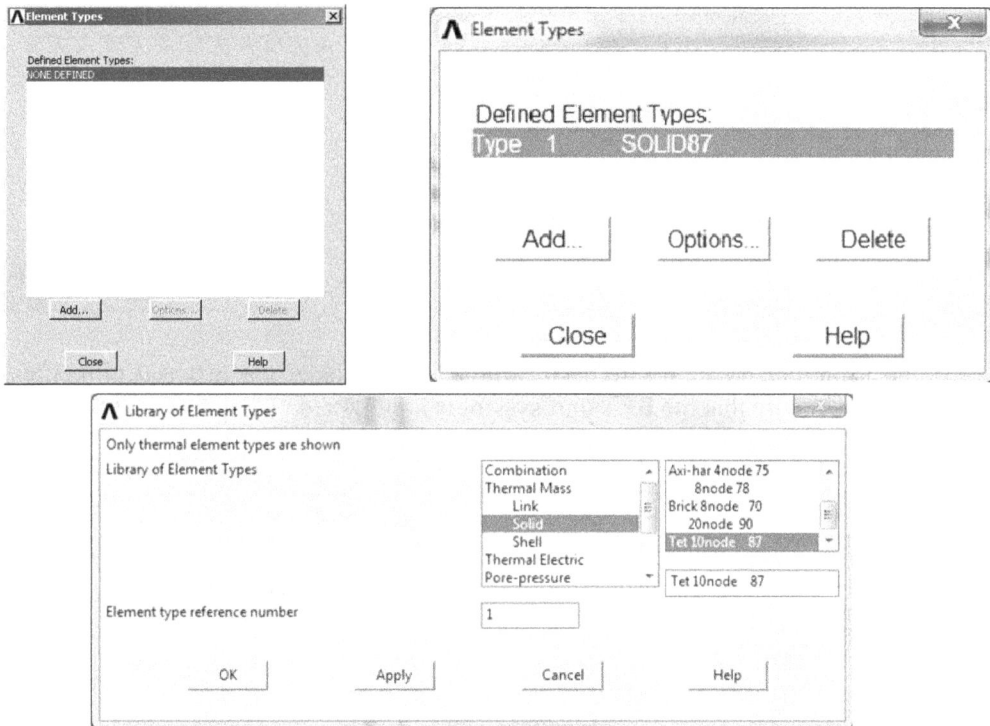

Step 4: Define the material type for each of the 2 components.
Select **Material Properties** from the **Preprocessor** menu. Select **Material Model**, and then
select **Thermal, Conductivity > Isotropic** by double clicking on each item. For material type 1 for the
wall component, specify 0.81 > **OK**.

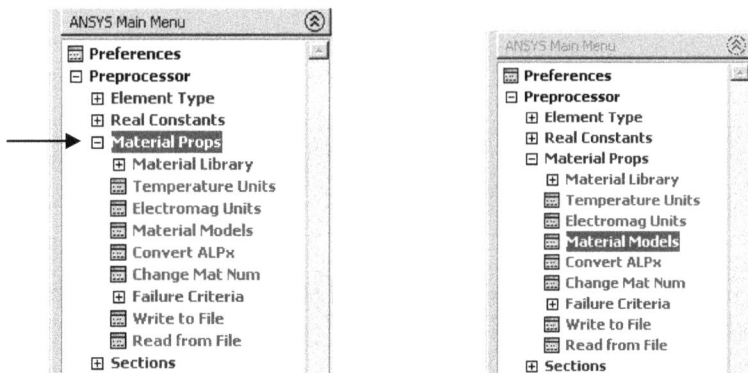

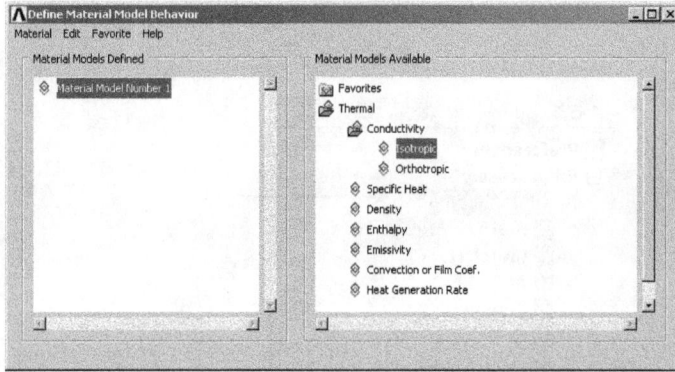

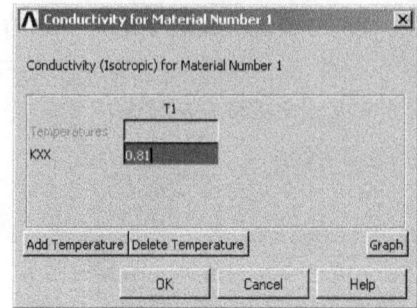

To define material type 2 for the plastic window component, Material > New Model > OK to accept 2 as Material ID > double click Isotropic and specify 0.195 > **OK**. **Material > Exit**.

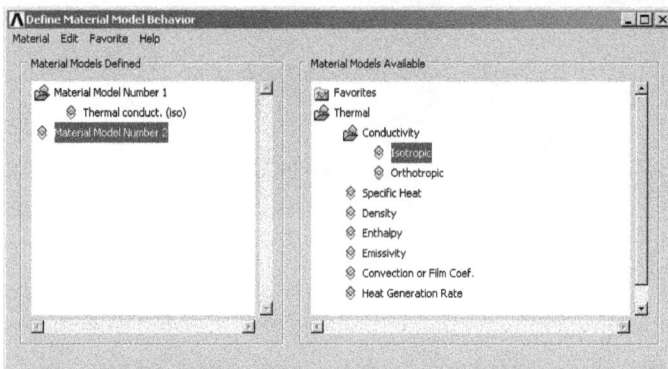

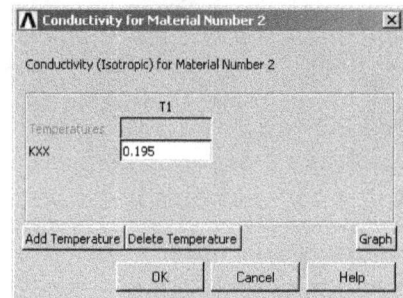

To check the unit system, select **Material Library** under the material properties menu. Then select **Units** and make sure that the **BFT** unit system is highlighted. Click **OK**.

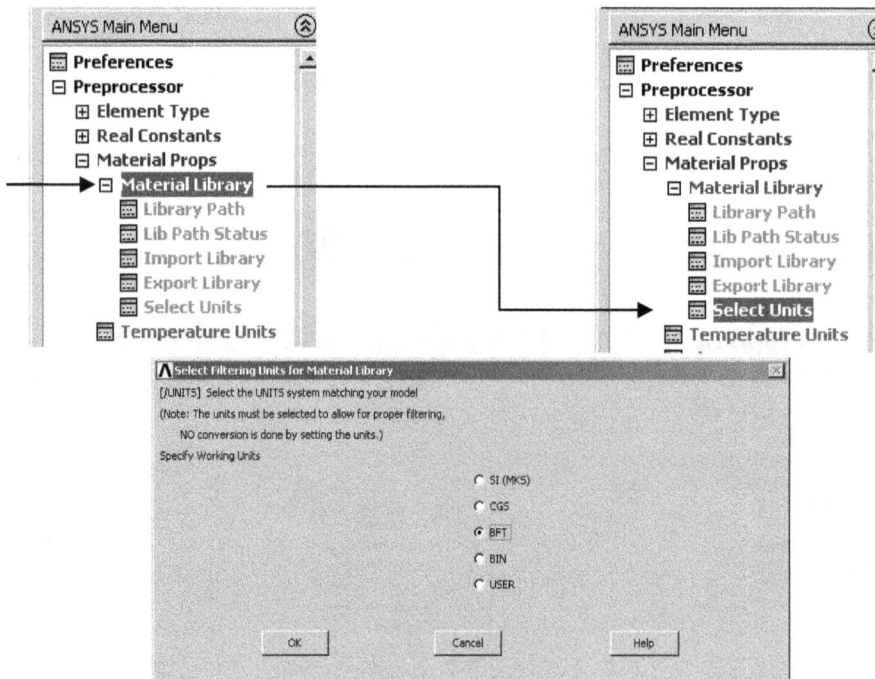

Step 5: Mesh Generation

To create the mesh for this assembly system, select **MeshTool** from the **Preprocessor** menu. **Meshing > Size Cntrls > Manual Size > Global Size > Size**.

Specify 0.25 as the Element edge length > **OK**.

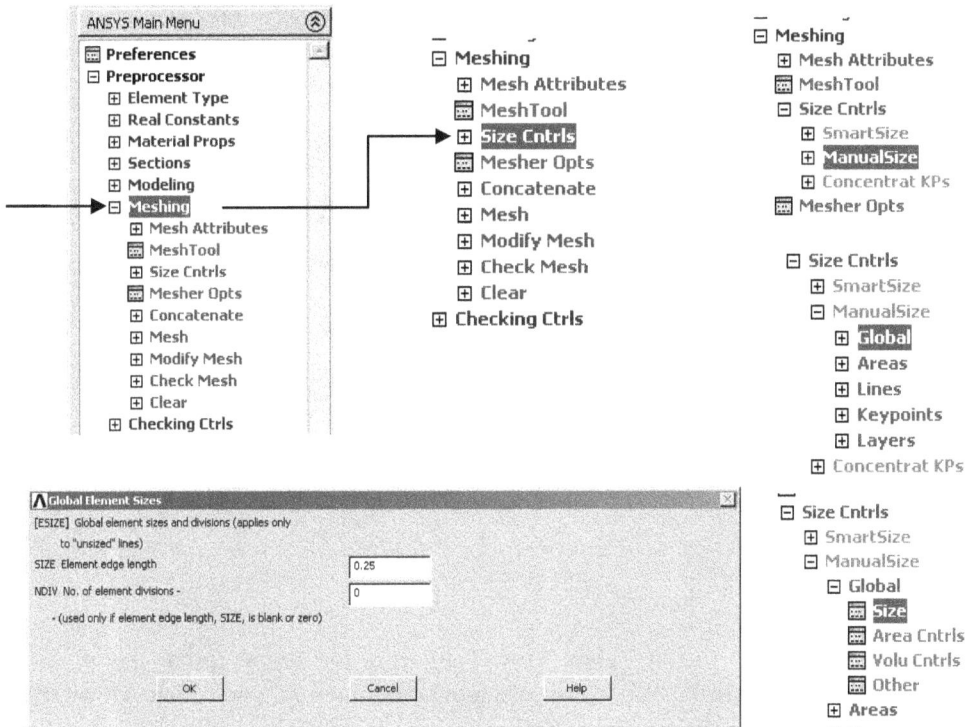

To assign a material type to a component in an assembly, select **MeshTool** and change **Global** to **Volumes > Set >** pick the wall component **> OK**. Make sure that Material Type 1 is the material type associated with the wall component.

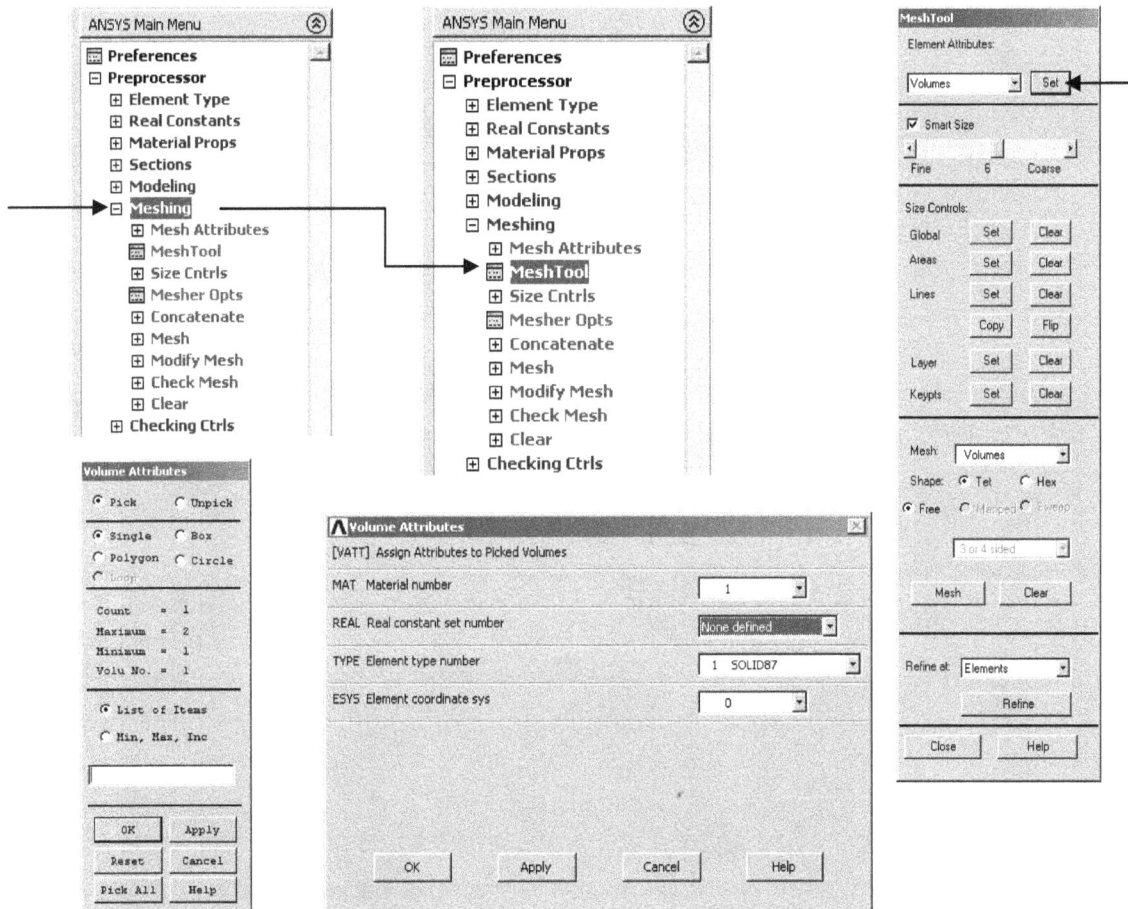

Again, click **Set** > pick the window component > **OK**. Make sure that Material Type 2 is the material type associated with the window component.

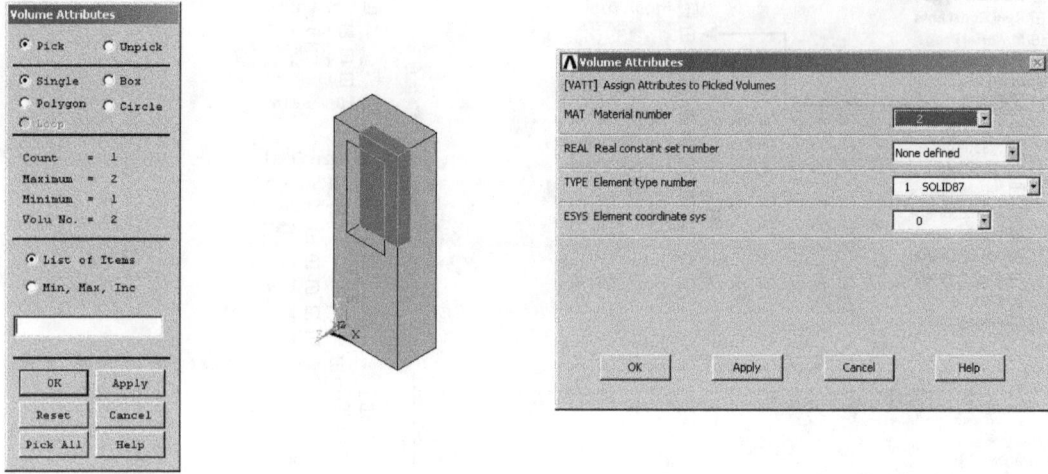

To create the mesh for this 3D model, select **MeshTool** from the Preprocessor menu. Select **Smart Size > Volume > Tet > Free > Mesh**. When the selection box comes up, click **Pick All > OK**.

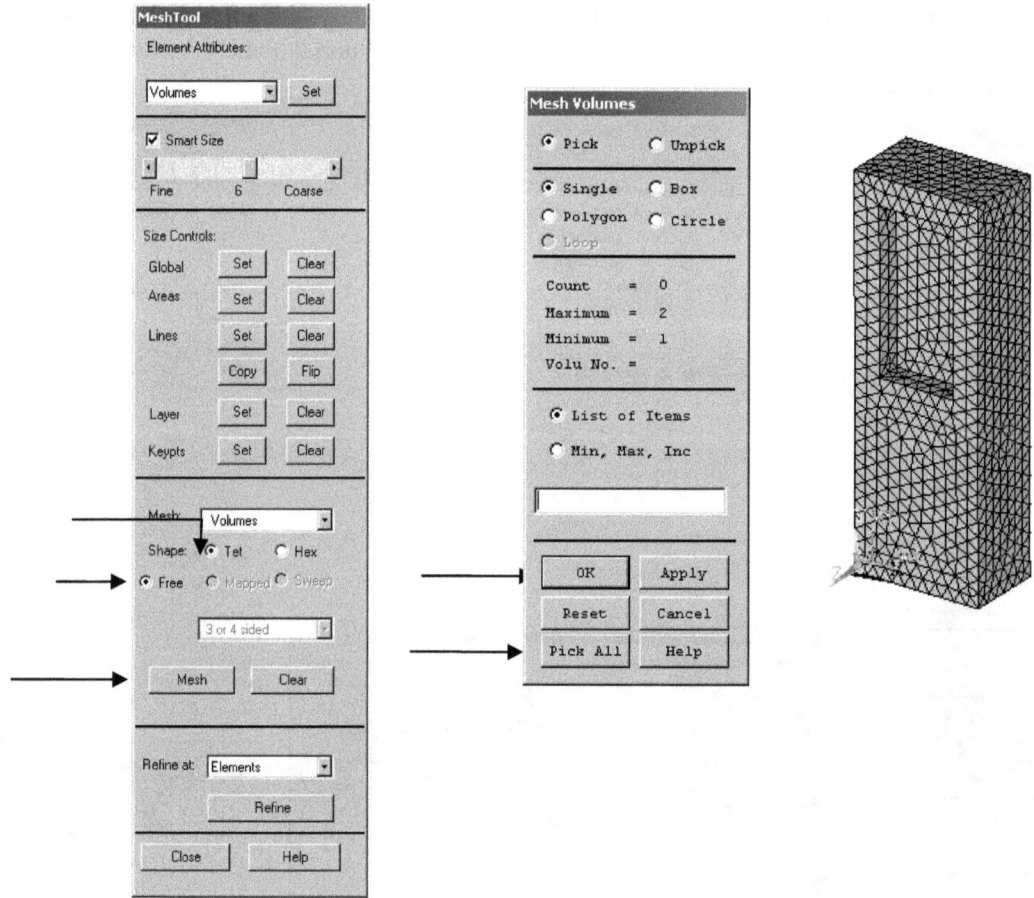

Step 6: Apply the convection boundary conditions.

From the main menu, Select > **Entities**. Specify **Areas > By Location > Z coordinate** > specify 0, -0.5 as the 2 limiting values (Min, Max) > **From Full > OK**.

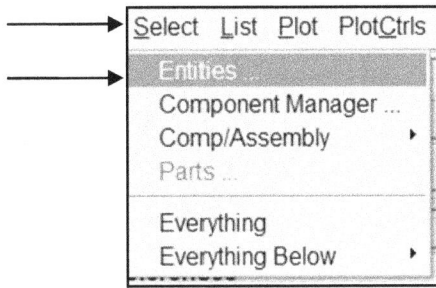

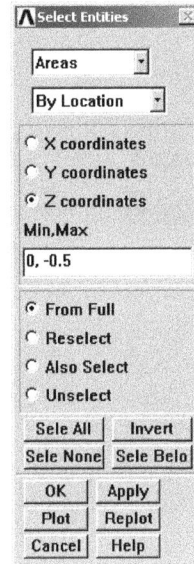

From the main menu, **Plot > Areas**.

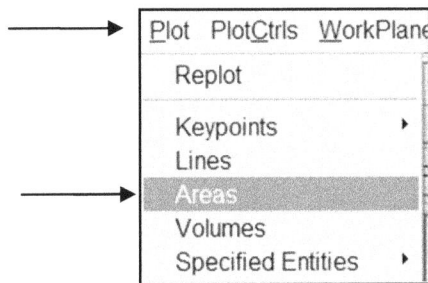

To apply the load condition, on the Preprocessor menu, click **Loads > Apply > Thermal > Convection > On Areas > Pick All**. Specify 1.46 as the coefficient of convection and 70 as the bulk temperature > **OK**.

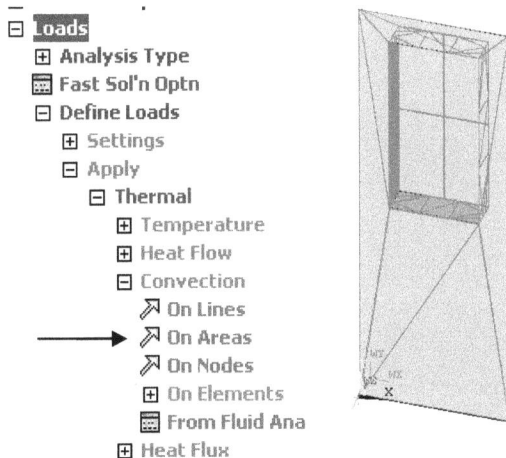

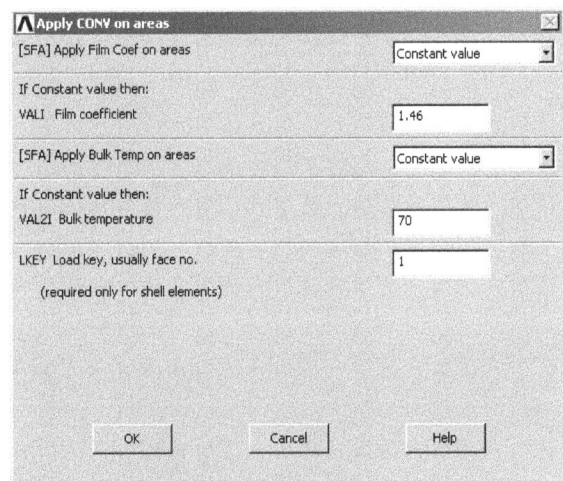

Select > Everything.

Select > Entities. Specify **Areas > By Location > Z coordinate** > specify -1.0, -1.5 as the 2 limiting values (Min, Max) > **From Full > OK**.

Plot > Areas.

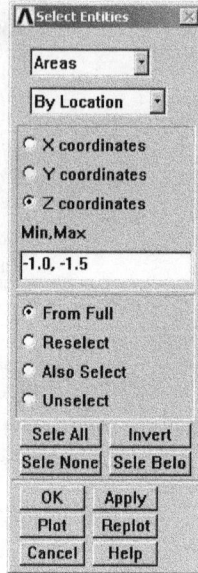

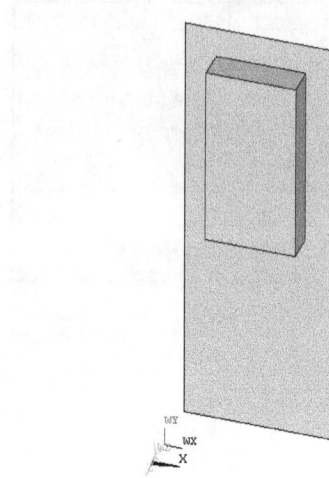

Click **Loads > Apply > Thermal > Convection > On Areas > Pick All**. Specify 10.5 as the coefficient of convection and 50 as the bulk temperature > **OK**.

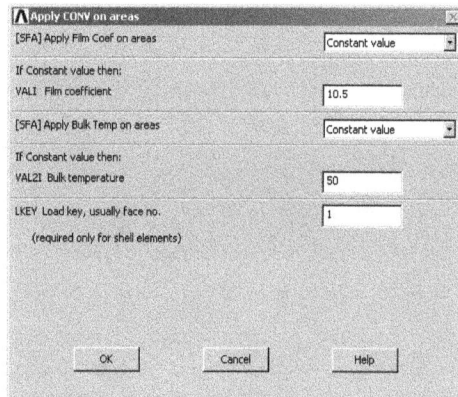

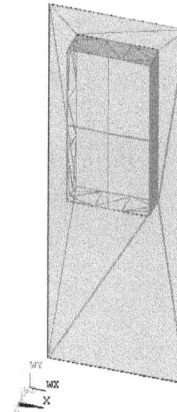

To view the applied boundary condition, from the main menu, **PlotCntrls > Symbols**. Check **All Applied BCs**

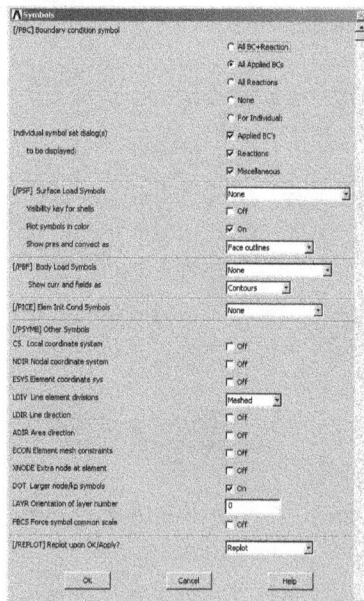

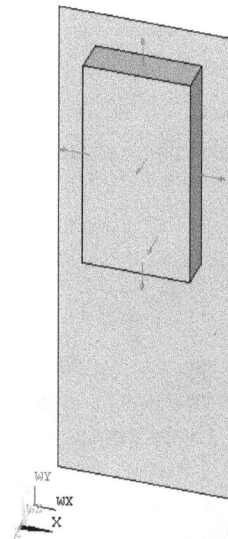

Select > Everything.
Plot > Areas.

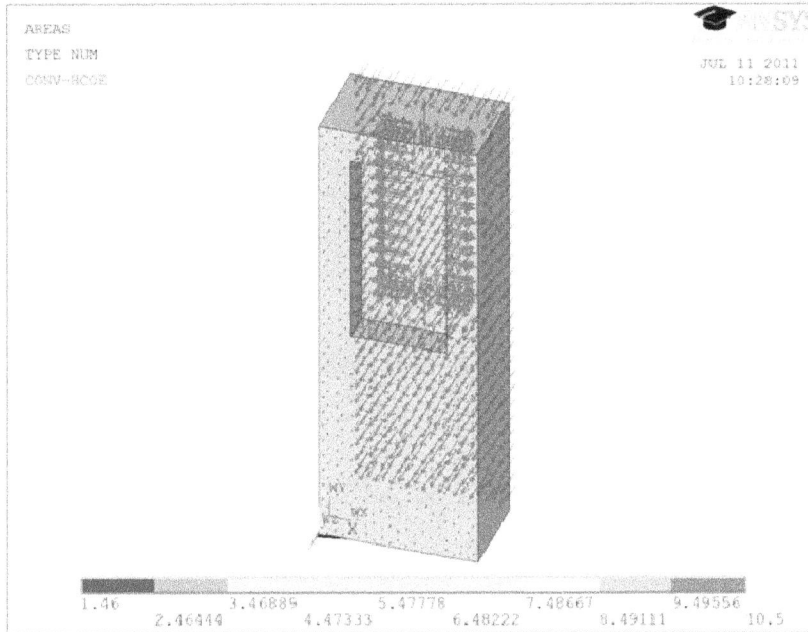

Step 7: Perform the analysis and obtain the solution.
To find the solution, run the program by selecting **Solution** from main menu > **Analysis Type** > **New Analysis > Steady-State > OK**.

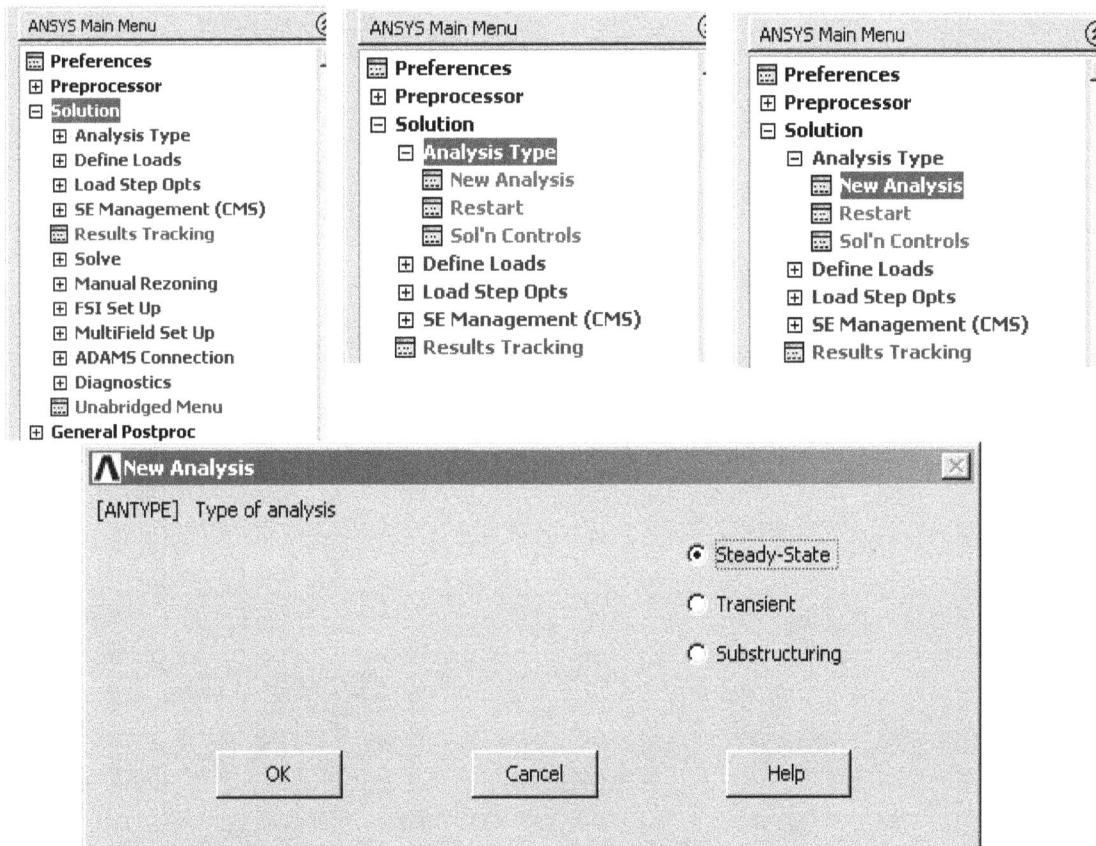

Choose **Current LS** > after reviewing the information, click **OK** to run the program.

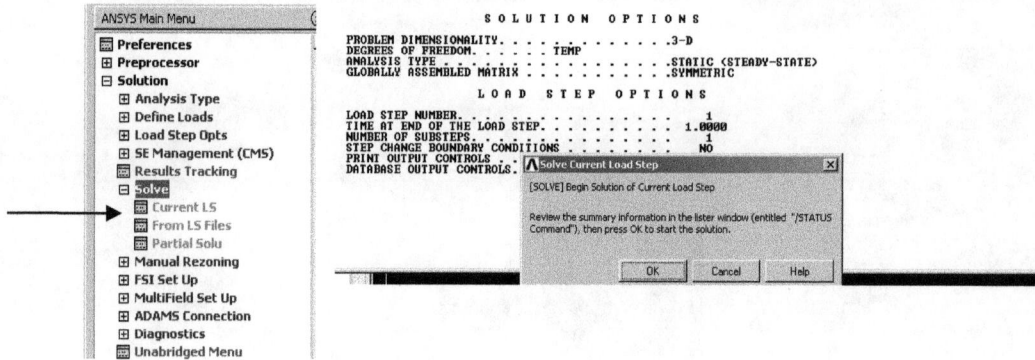

A window called Note appears when the solution is done > **Close**.

Step 8: Displaying Results

Under **General Postprocessing**, select **Plot Results** > **Contour Plot** > **Nodal Solution**. Select **DOF Solution** from the list > **Temperature** > **OK**.

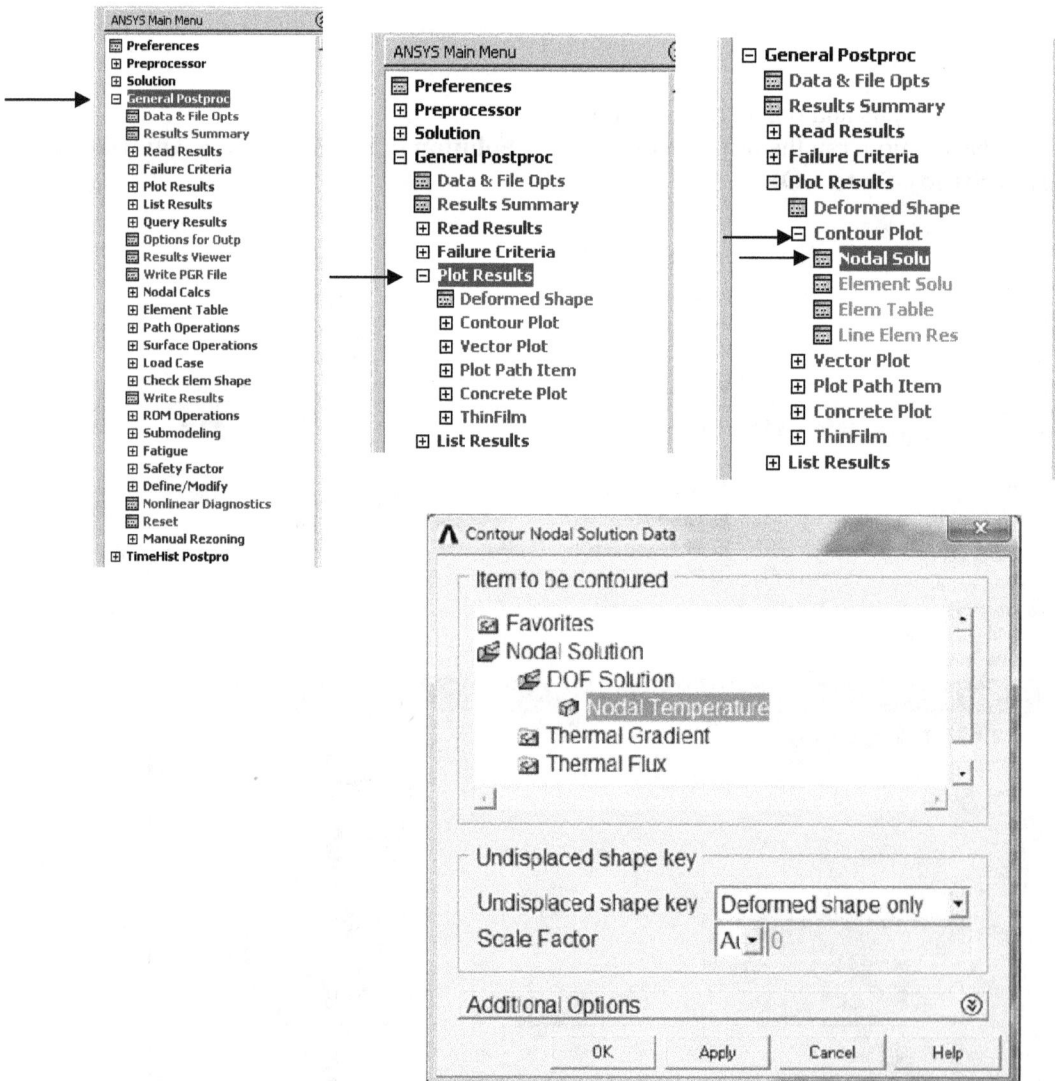

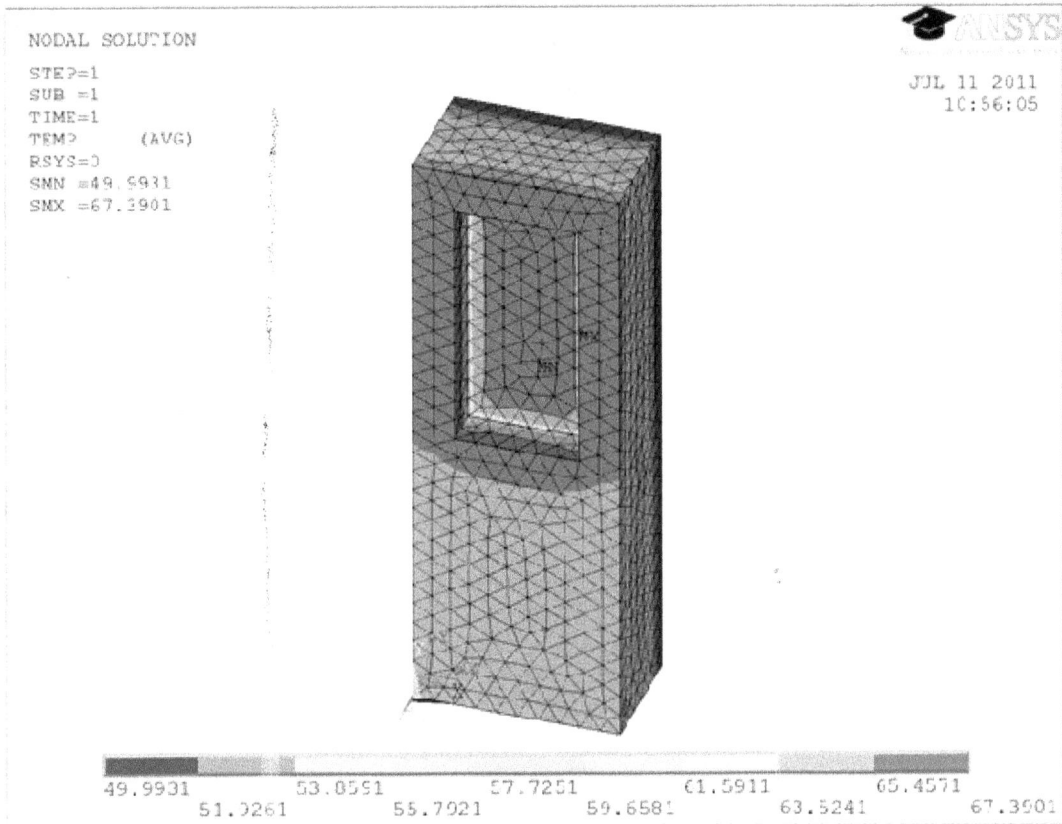

```
NODAL SOLUTION
STEP=1
SUB =1
TIME=1
TEMP    (AVG)
RSYS=0
SMN =49.9931
SMX =67.3901

JUL 11 2011
10:56:05
```

```
49.9931        53.8591        57.7251        61.5911        65.4571
      51.9261        55.7921        59.6581        63.5241        67.3901
```

To plot the heat flow vector, **Plot Results > Vector Plot > Predefined**. Select **Flux & gradient > Thermal flux TF > OK.**

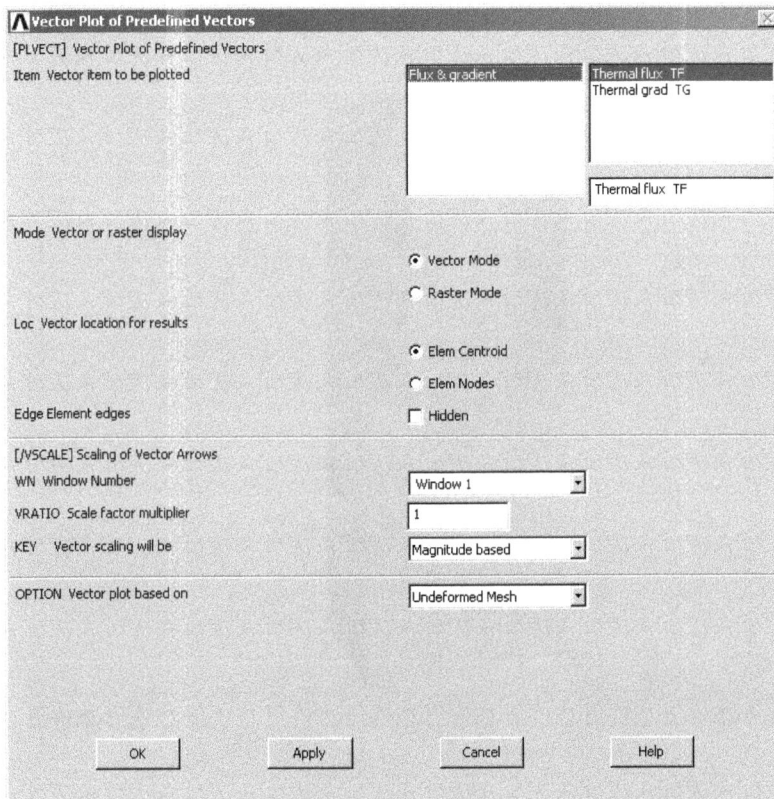

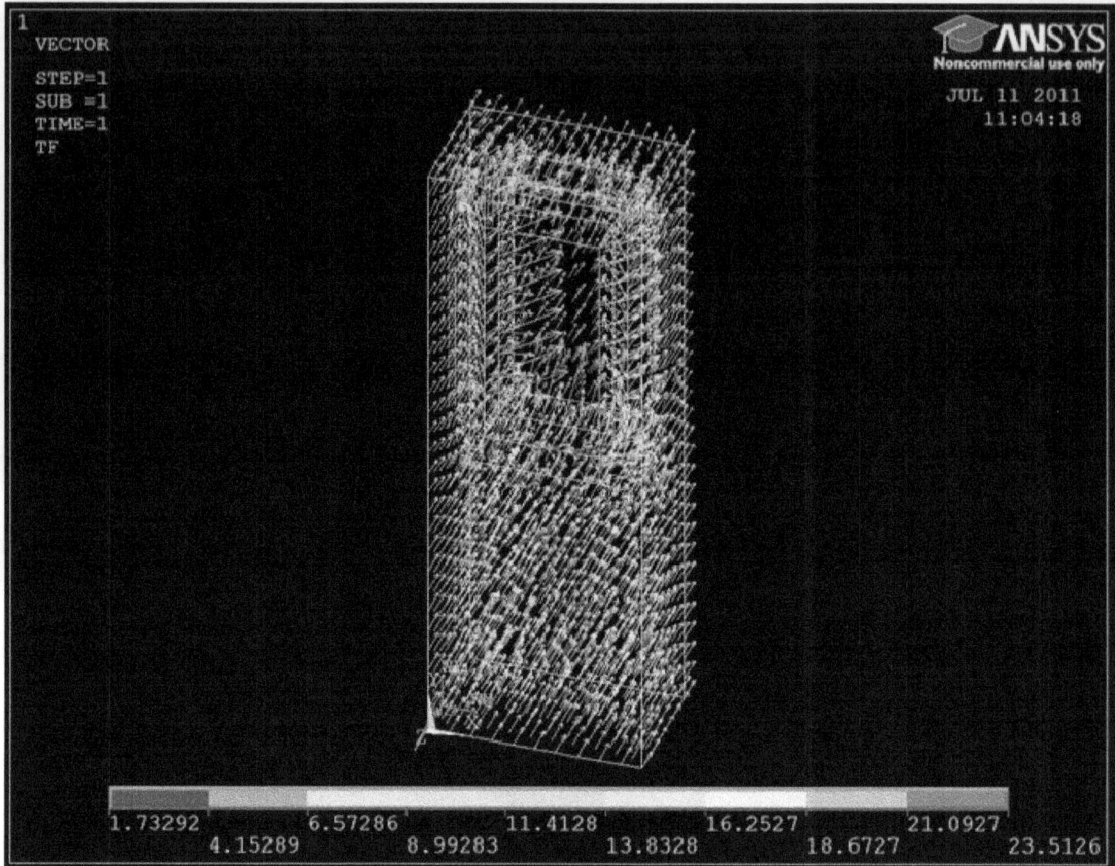

The following plot depicts the distribution of thermal gradient.

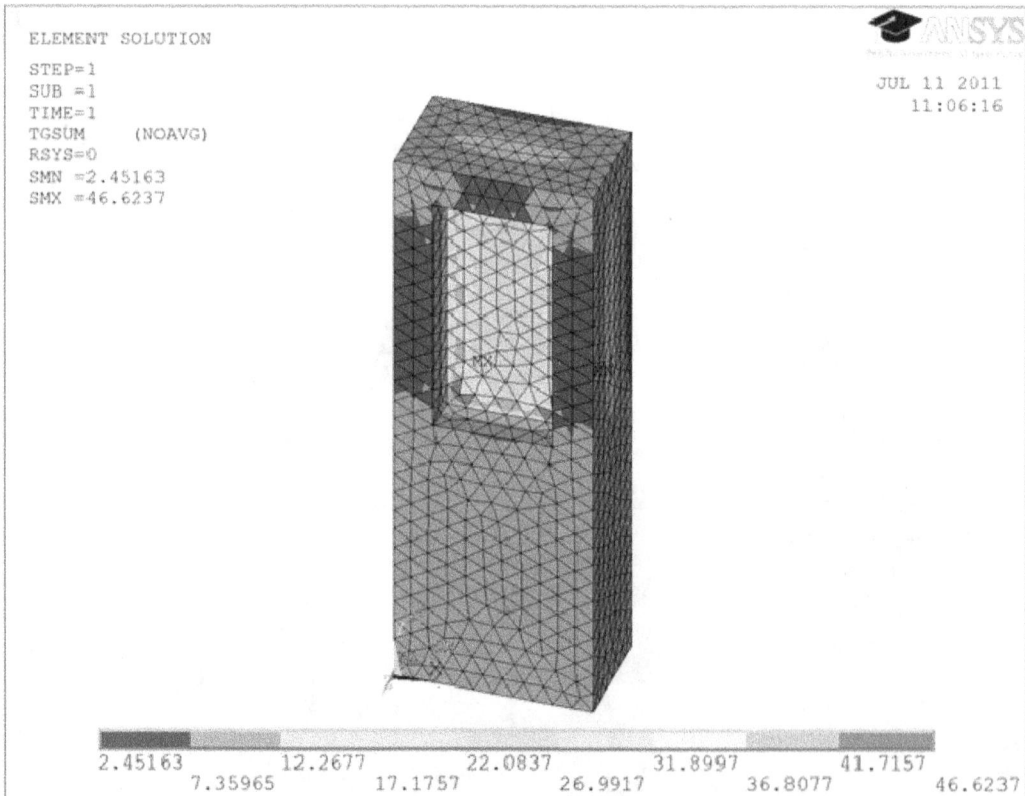

10.9 Thermal Analysis with 2D Model (BFT Unit System)

A small chimney structure is shown below. The unit of the dimensions shown is inch. The inner layer is constructed from concrete (thermal conductivity = 0.07 Btu/hr.in $^{\circ}$F). The outer layer is constructed from bricks (thermal conductivity = 0.004 Btu/hr.in $^{\circ}$F). Assume that the temperature of the hot airflow on the inside surface of the concrete layer is 140°F with the coefficient of convection equal to 0.037 Btu/hr.in^2 $^{\circ}$F The outside surface is exposed to the surrounding air, which is at 10oF with the coefficient of convection equal to 0.012 Btu/hr.in^2 $^{\circ}$F. Determine the temperature distribution under a steady-state condition, and plot the heat fluxes through these 2 layers.

Unit: inch

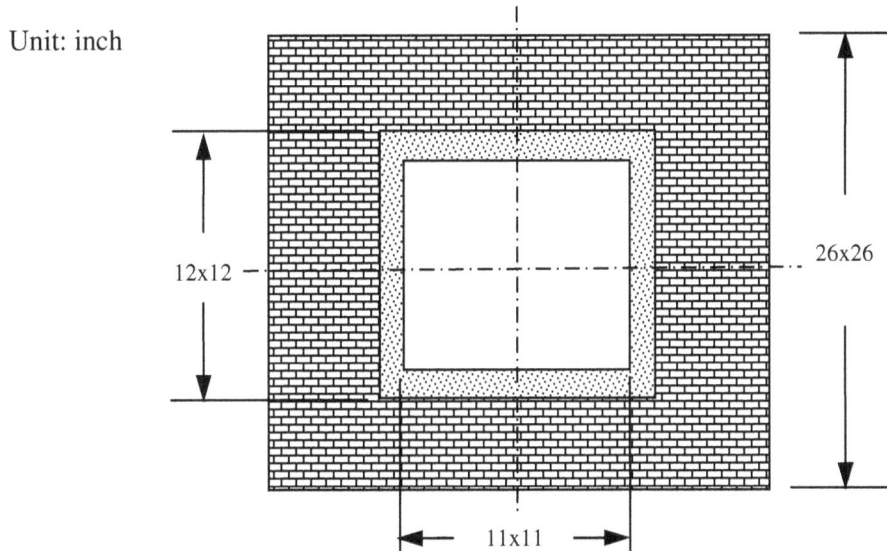

12x12

26x26

11x11

Step 1: Launch **ANSYS 13.0**

From the start menu, select **Programs > ANSYS 13.0 > ANSYS**, or directly click the icon of **ANSYS**.

To define the Structural Analysis, from the main menu or the model tree, select **Preferences**. > make sure that **Thermal** is highlighted, and h-method is also selected > **OK**.

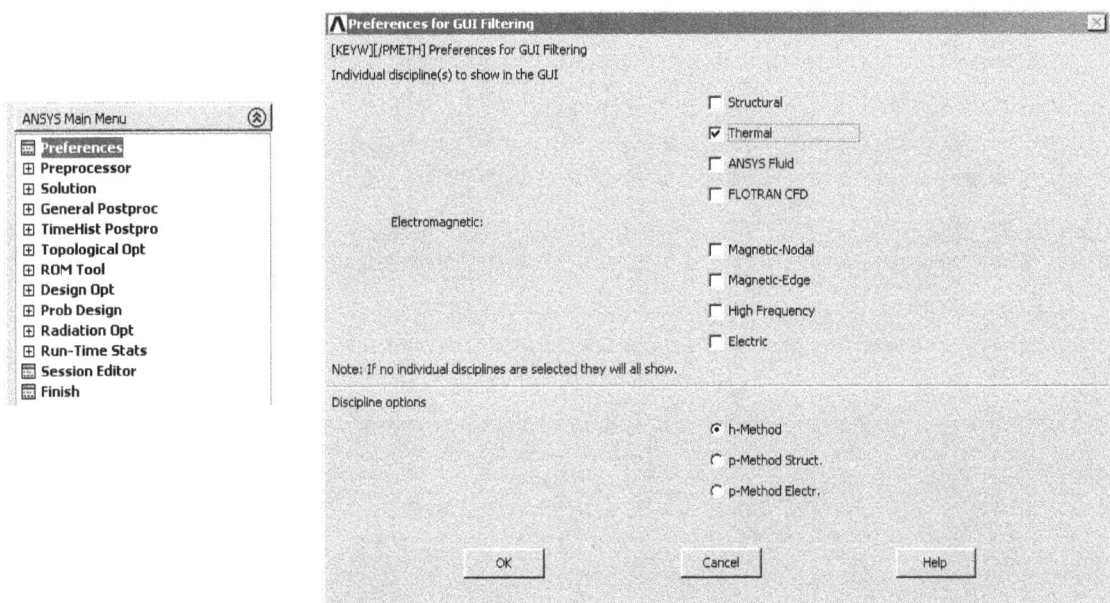

Step 2: Set the unit system to English (in), or **BIN**.

Preprocessor > Material Library > Select Units > BIN > OK.

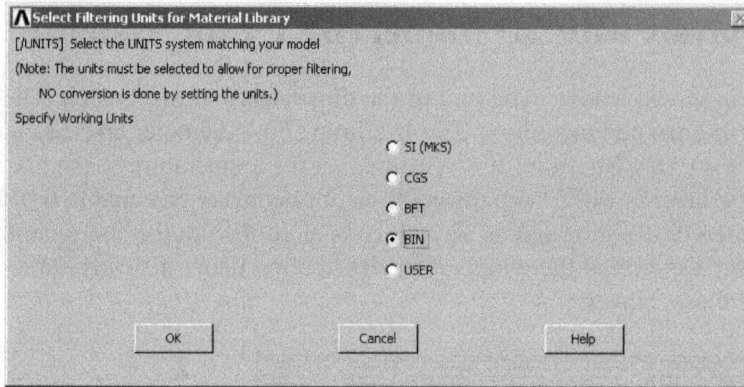

As illustrated, the basic quantities used in English (in) system are

Length:	inch (in)
Mass:	lbf*s^2/in
Force:	pounds (lbf)
Time:	second (s)
Temperature:	Fahrenheit (deg F)

Step 3: Define a 2D Geometry.

We start with the creation of a 2D model for the outer layer of bricks. From the main menu, select **Preprocessor** and expand it. Select **Modeling > Create > Areas > Rectangle > By Dimensions**.

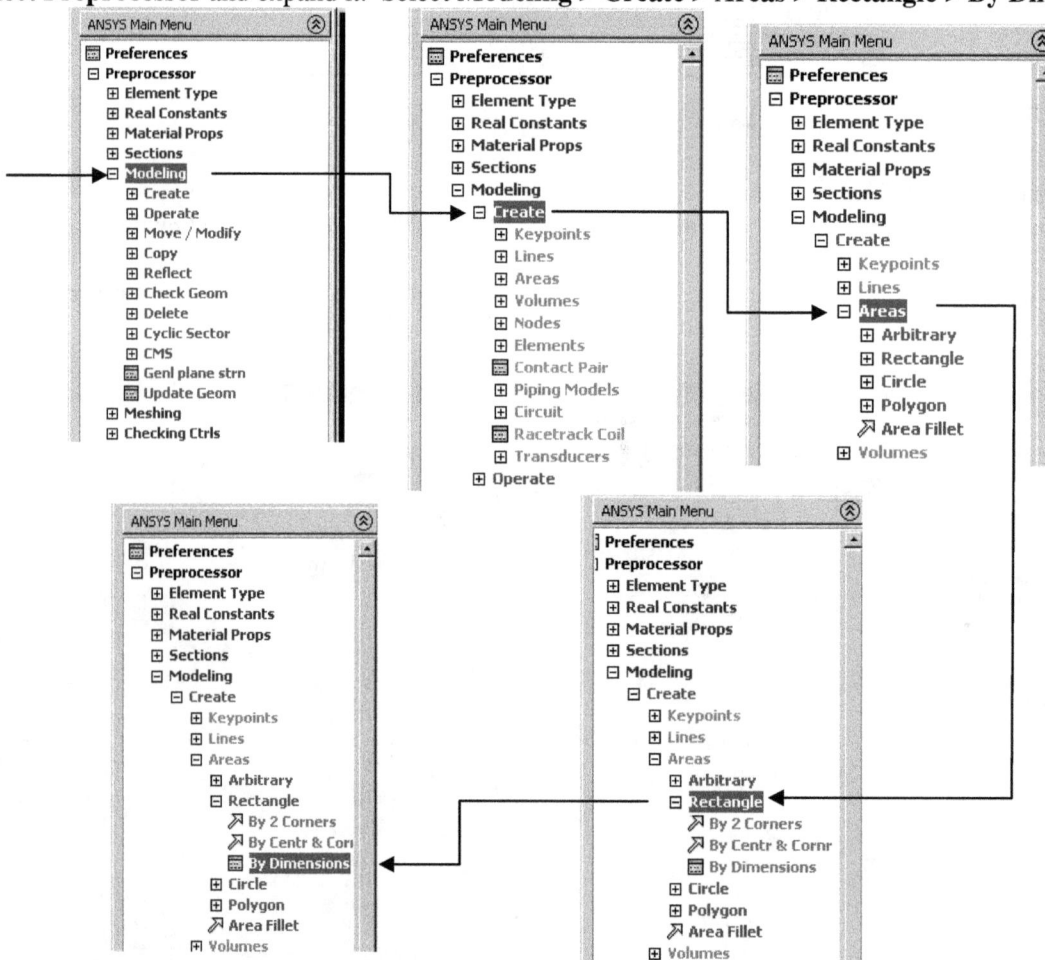

In the window for creating rectangle by dimensions, type 0 and 26 for X1 and X2, and 0 and 26 for Y1 and Y2 > **Apply**.

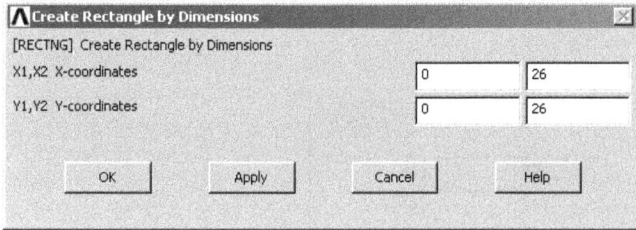

Again, in the window for creating the rectangle by dimensions, type 6 and 20 for X1 and X2, and 6 and 20 for Y1 and Y2 > **OK**.

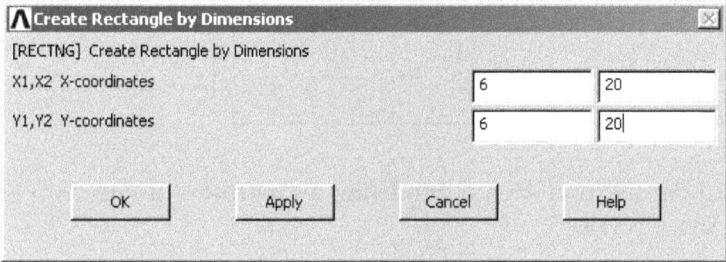

Now let us subtract Area 2 from Area 1. **Modeling > Operate > Booleans > Subtract > Areas.** Pick Area 1 > **OK**. Pick Area 2 > **OK**.

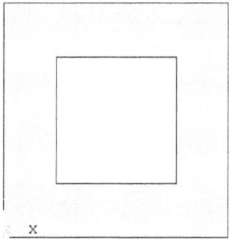

Now let us create a 2D model for the inner layer of concrete.

In the window for creating rectangle by dimensions, type 6 and 20 for X1 and X2, and 6 and 20 for Y1 and Y2 > **Apply**.

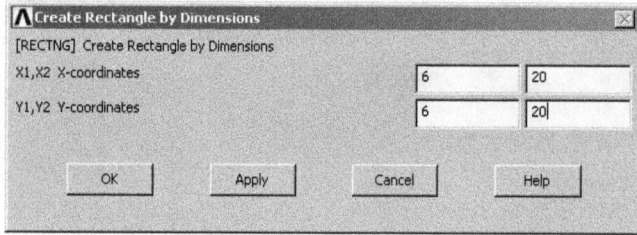 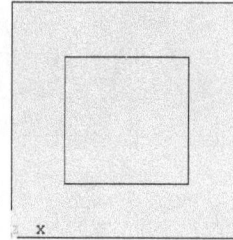

Again, in the window for creating the rectangle by dimensions, type 7 and 19 for X1 and X2, and 7 and 19 for Y1 and Y2 > **OK**.

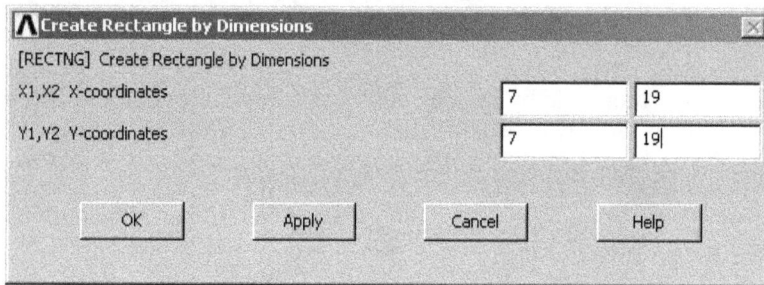 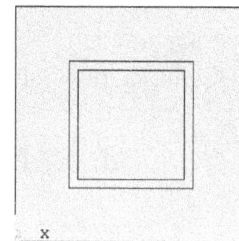

Now let us subtract Area 2 from Area 1. **Modeling > Operate > Booleans > Subtract > Areas.** Pick Area 2 > **OK**. Pick Area 1 > **OK**.

Now let us assemble the outer layer and inner layer components together. **Operate > Booleans > Glue > Areas > Pick All.**

Step 4: Select the element type.
        Select **Element Type > Add/Edit/Delete.**

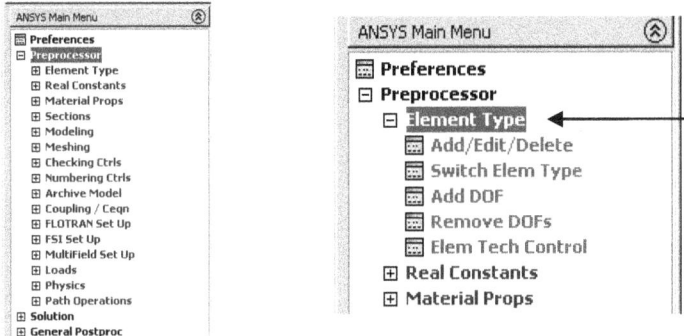

A window called **Element Types** appears. Click **Add** > pick **Solid** > **Quard 4node 55** > **OK.**

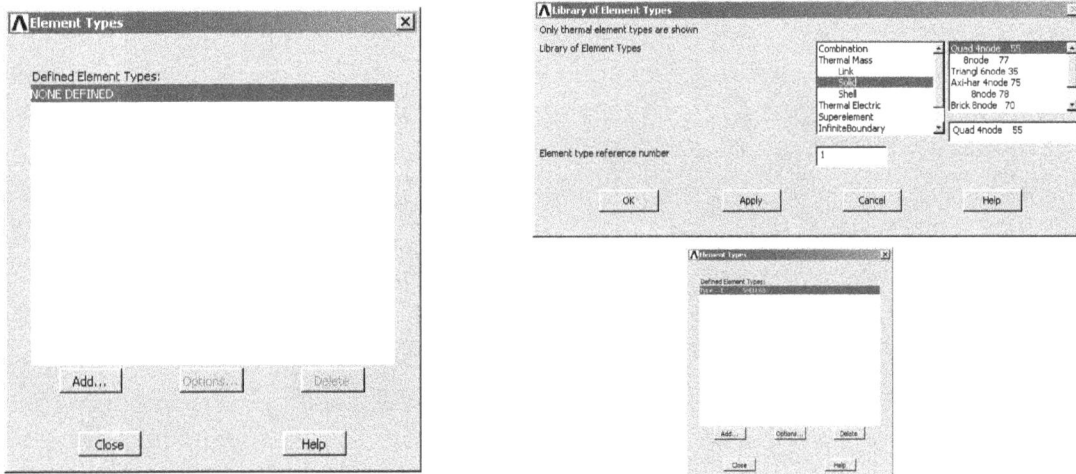

Step 4: Define the material type for each of the 2 components.
        Select **Material Properties** from the **Preprocessor** menu. Select **Material Model**, and then select **Thermal, Conductivity > Isotropic** by double clicking on each item. For material type 1 for the concrete component, specify 0.07 > **OK**.

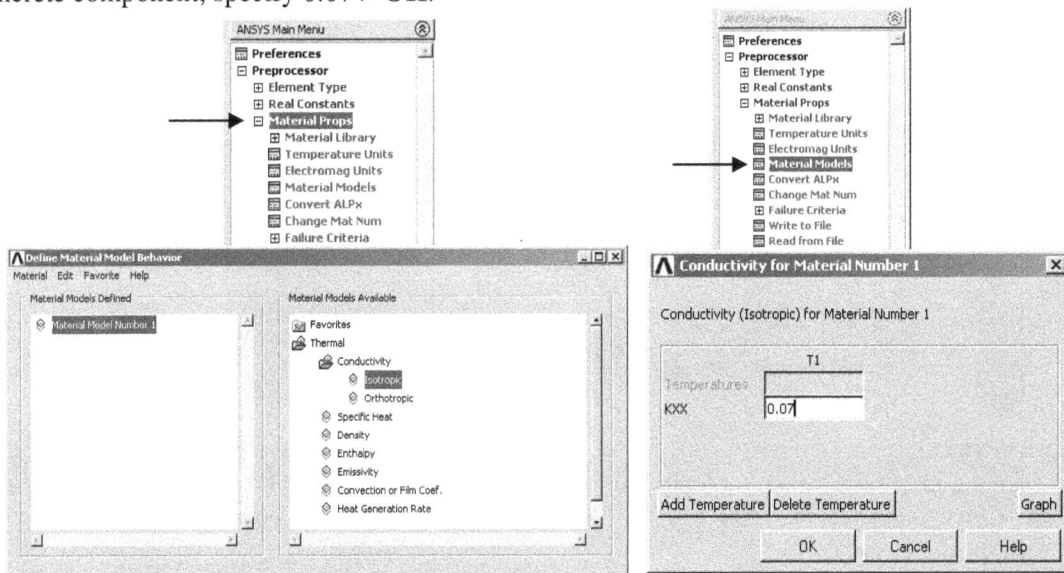

To define material type 2 for the brick outer layer, **Material > New Model > OK** to accept 2 as **Material ID** > double click Isotropic and specify 0.04 > **OK**. **Material > Exit**.

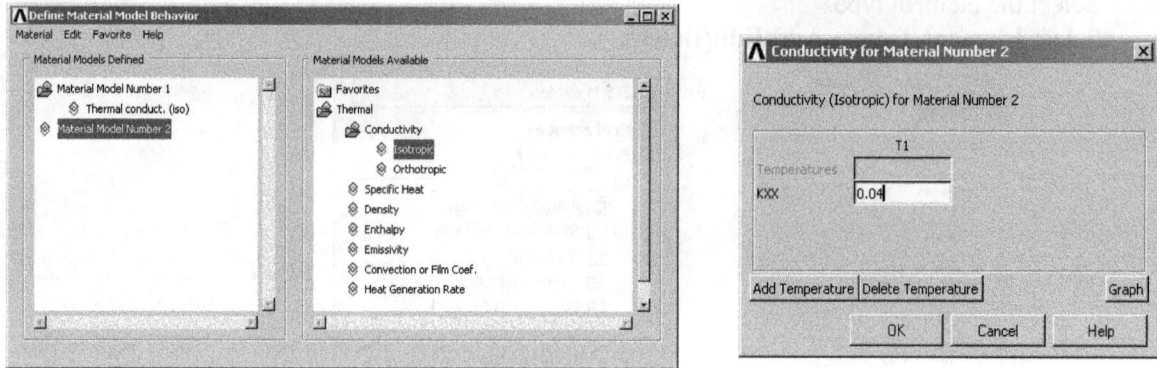

To check the unit system, select **Material Library** under the material properties menu. Then select **Units** and make sure that the **BIN** unit system is highlighted. Click **OK**.

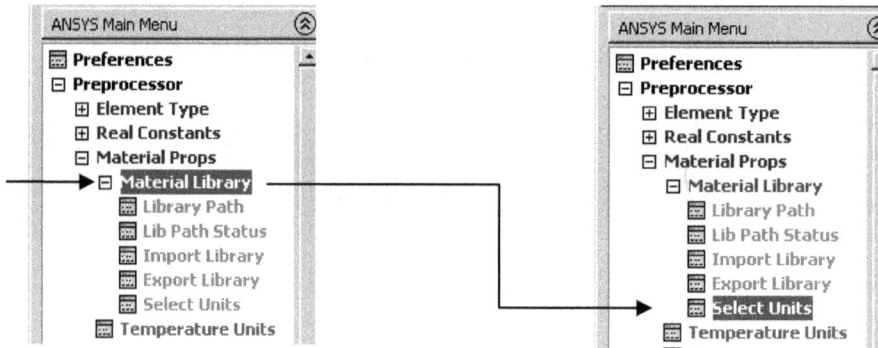

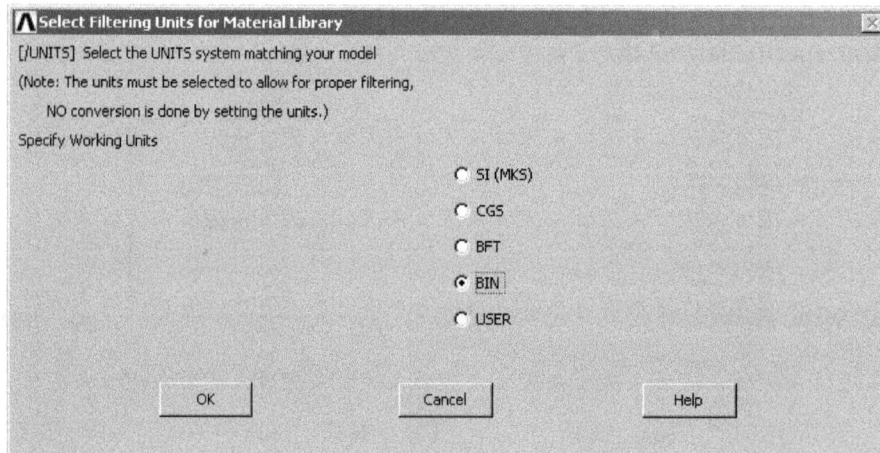

Step 5: Mesh Generation

To assign a material type to a component in an assembly, select **MeshTool** and change **Global** to **Areas > Set >** pick the concrete layer > **OK**. Make sure that Material Type 1 is the material type associated with the concrete layer.

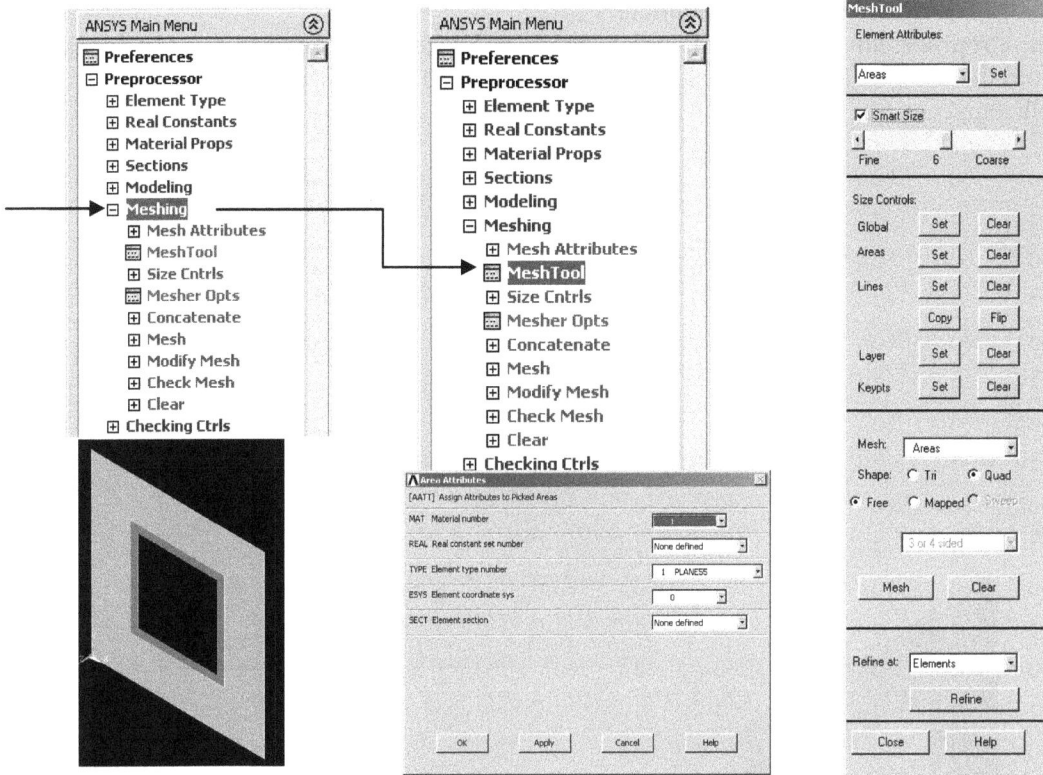

Again, click **Set** > pick the brick outer layer > **OK**. Make sure that Material Type 2 is the material type associated with the brick outer layer.

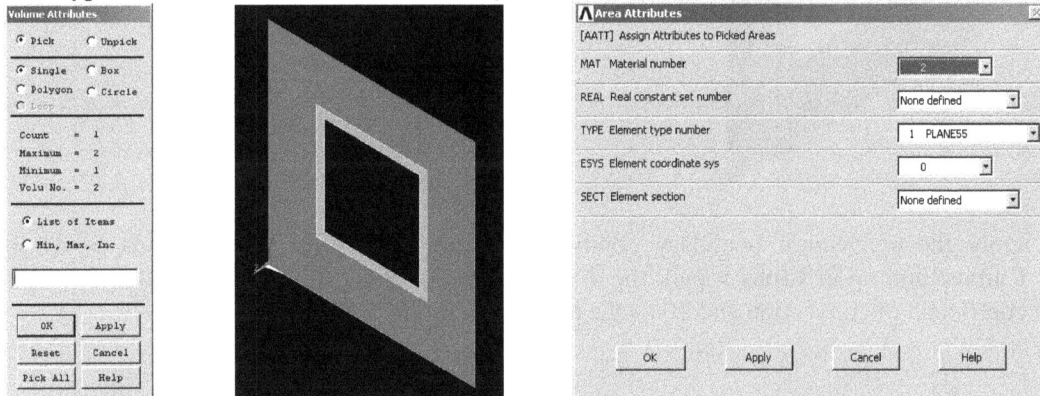

Meshing > Meshing Tool > Area > check the **Smart Size > Quad > Free > Mesh**.

When the selection box comes up, select **Pick All**, and then click **OK**. Make sure that you close the **MeshTool** window when the meshing process is completed.

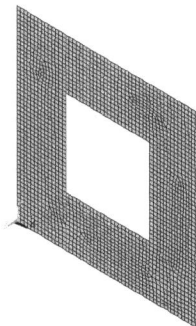

Step 6: Apply the convection boundary conditions.

On the **Preprocessor** menu, click **Loads > Apply > Thermal > Convection > On Lines >** pick the 4 lines from the inner side of the concrete layer. Specify 0.037 as the coefficient of convection and 140 as the bulk temperature > **OK**.

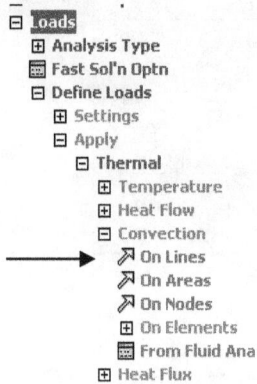

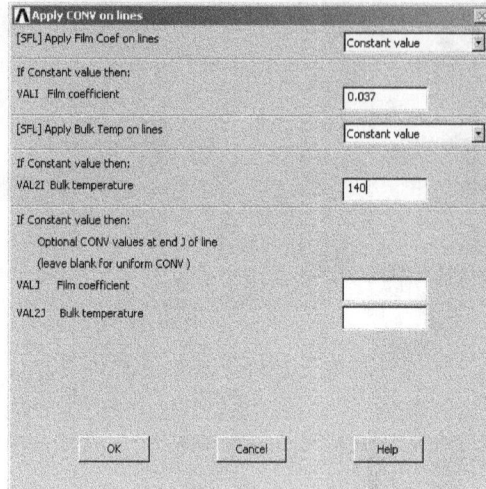

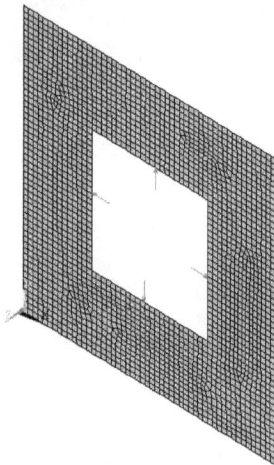

To apply the convection boundary condition on the outer layer, click **Loads > Apply > Thermal > Convection > On Lines >** pick the 4 lines from the outer side of the brick layer. Specify 0.012 as the coefficient of convection and 10 as the bulk temperature > **OK.**

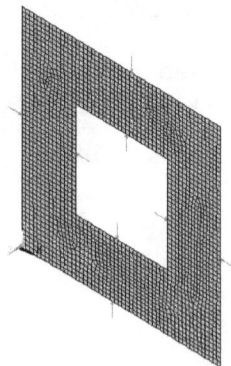

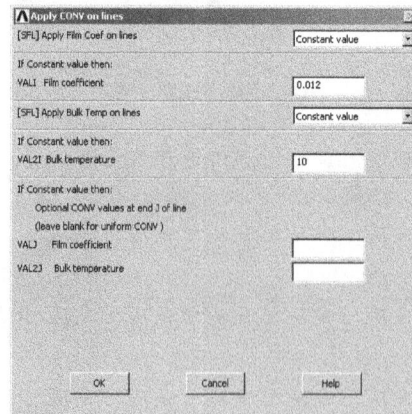

Step 7: Perform the analysis and obtain the solution.

To obtain the solution, run the program by selecting **Solution** from main menu > **Analysis Type > New Analysis > Steady-State > OK**.

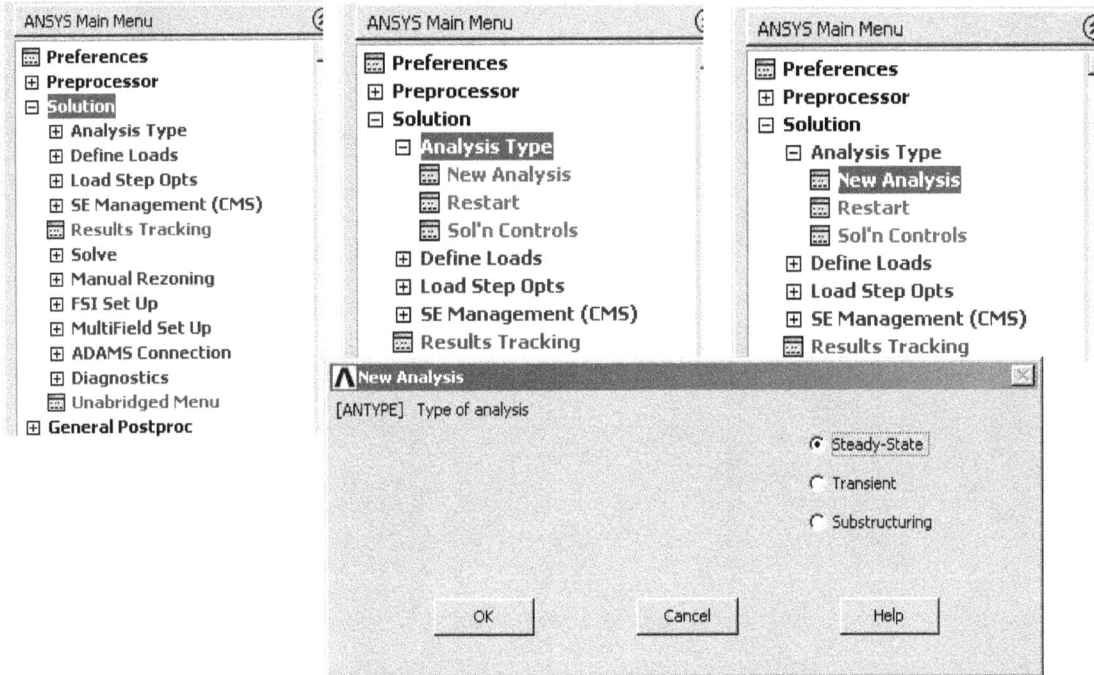

Choose **Current LS** > after reviewing the information, click **OK** to run the program.

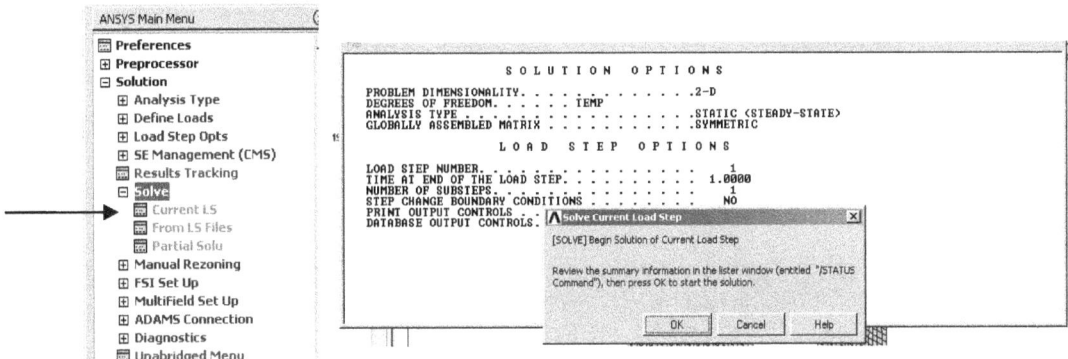

A window called Note appears when the solution is done > **Close**.

Step 8: Displaying Results

Under **General Postprocessing**, select **Plot Results** > **Contour Plot** > **Nodal Solution**. Select **DOF Solution** from the list > **Temperature** > **OK**.

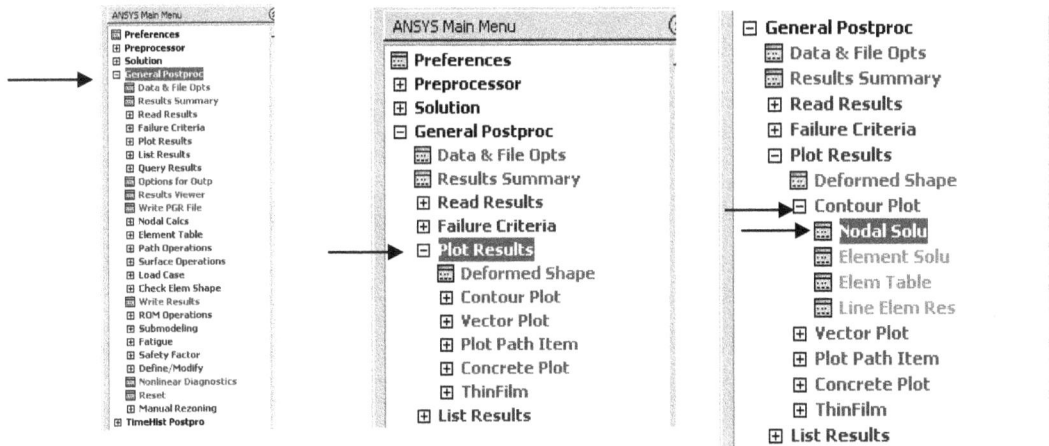

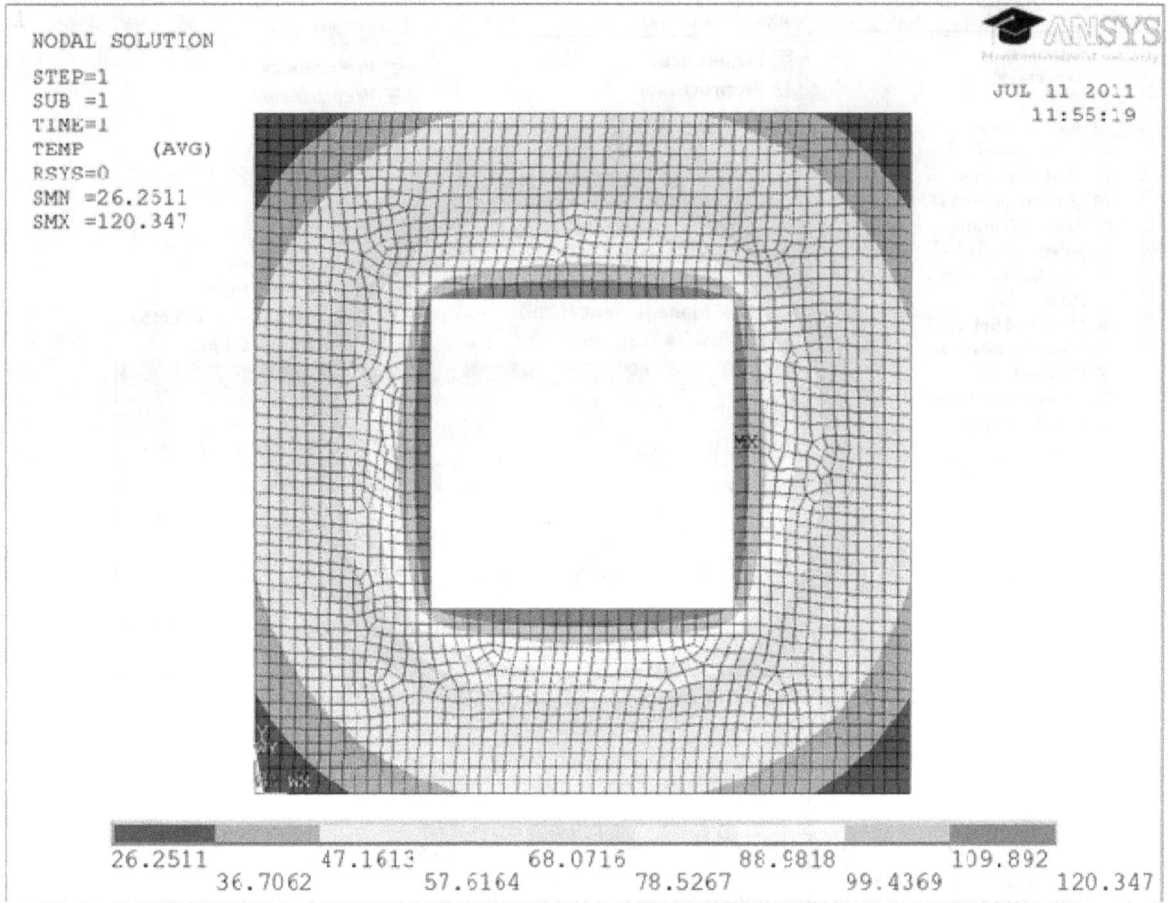

To plot the heat flow vector, **Plot Results > Vector Plot > Predefined**. Select **Flux & gradient > Thermal flux TF > OK.**

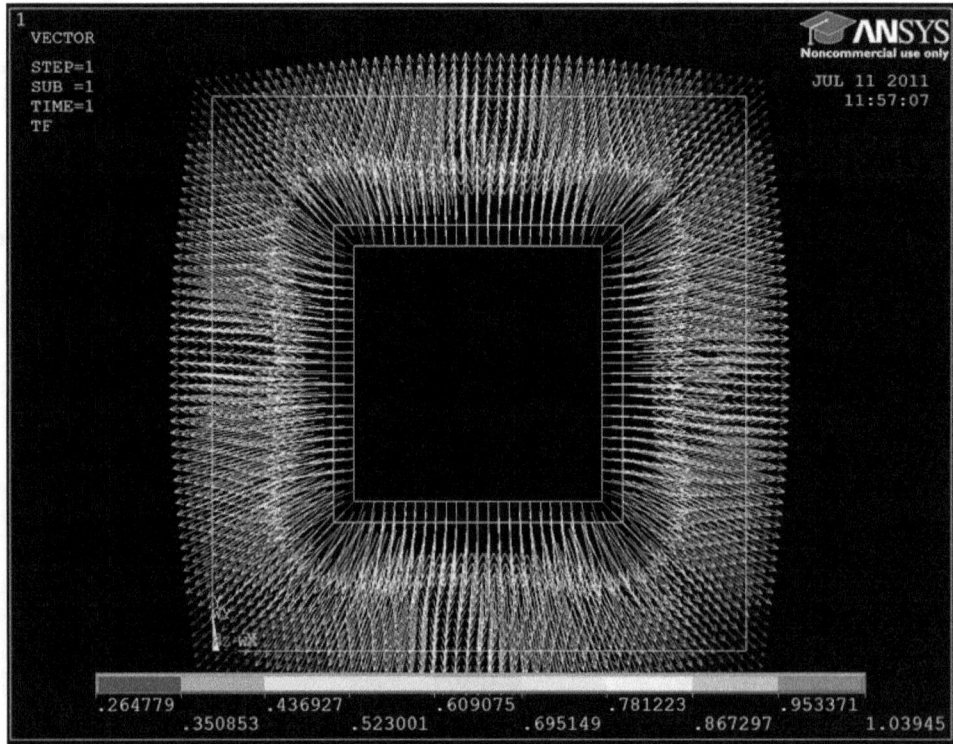

To plot the flux gradient along a specific path, we need to define the path first. To define a path, General **Postproc > Path Operations > Define Path > On Working Plane > Arbitrary > OK**. Make sure that the icon of the Working Plane is checked, not the Global > pick 2 points, as shown > specify 101 as the name of the defined path > **OK**.

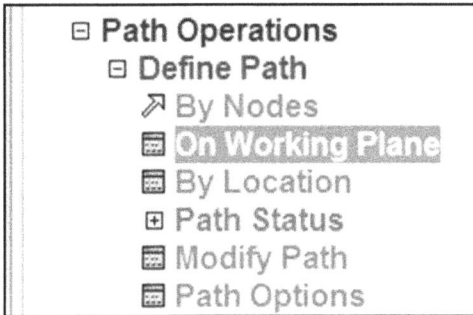

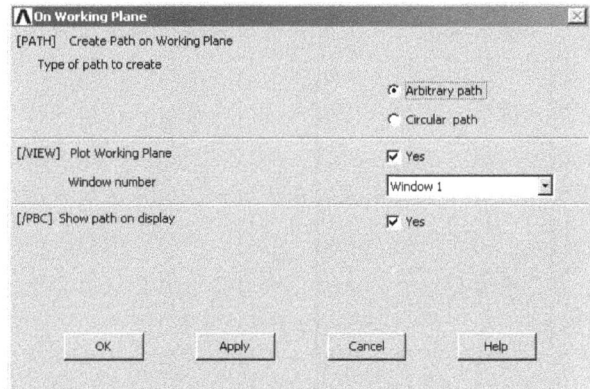

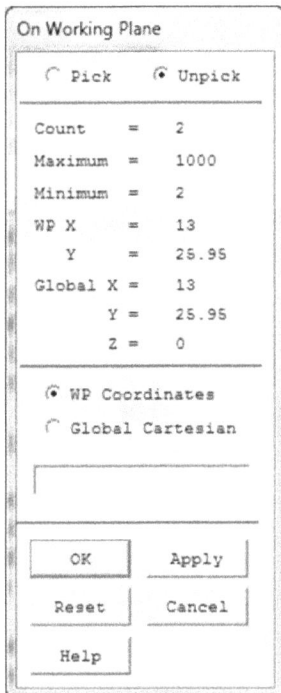

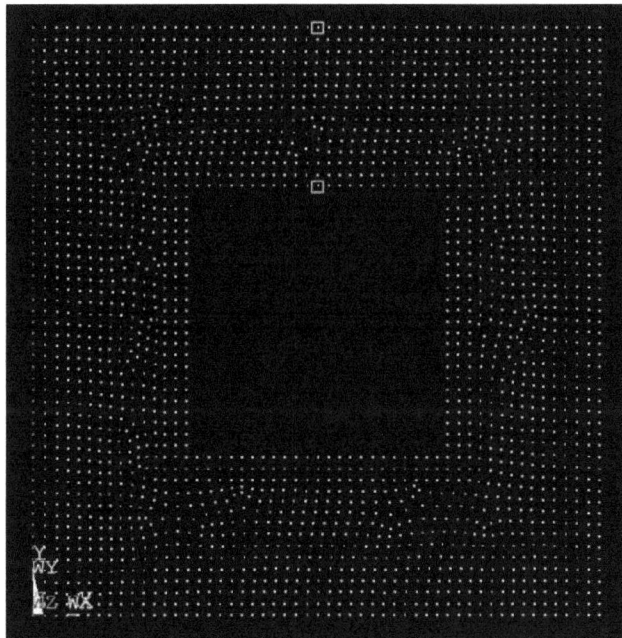

General Postproc > Path Operations > Map onto Path > specify dTdx, dTdy and TotalGra.

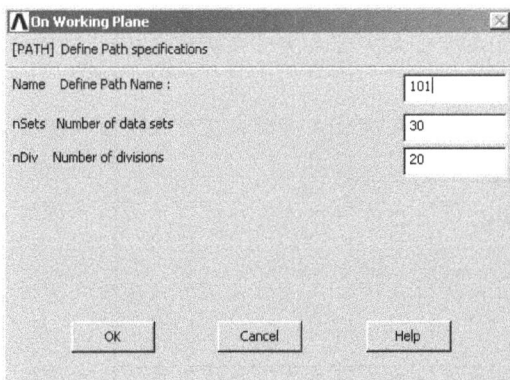

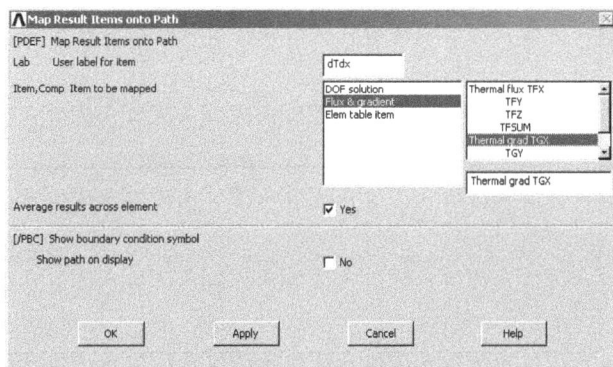

General Postproc > Path Operations > Plot Path Item > On Graph.

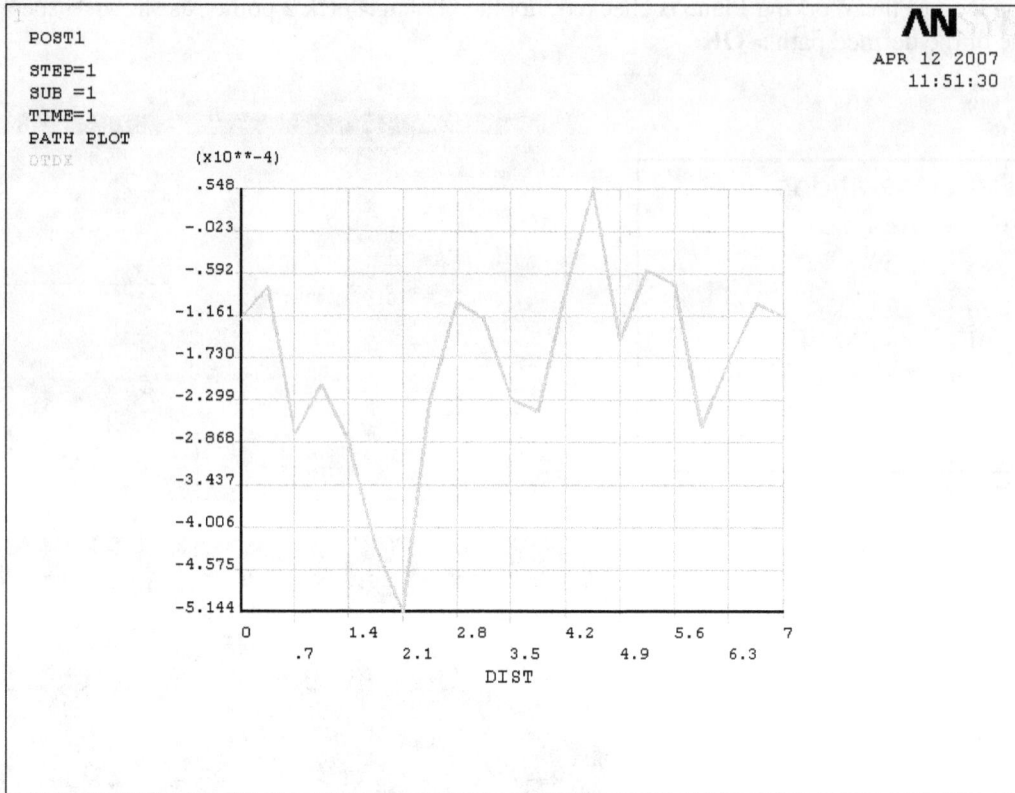

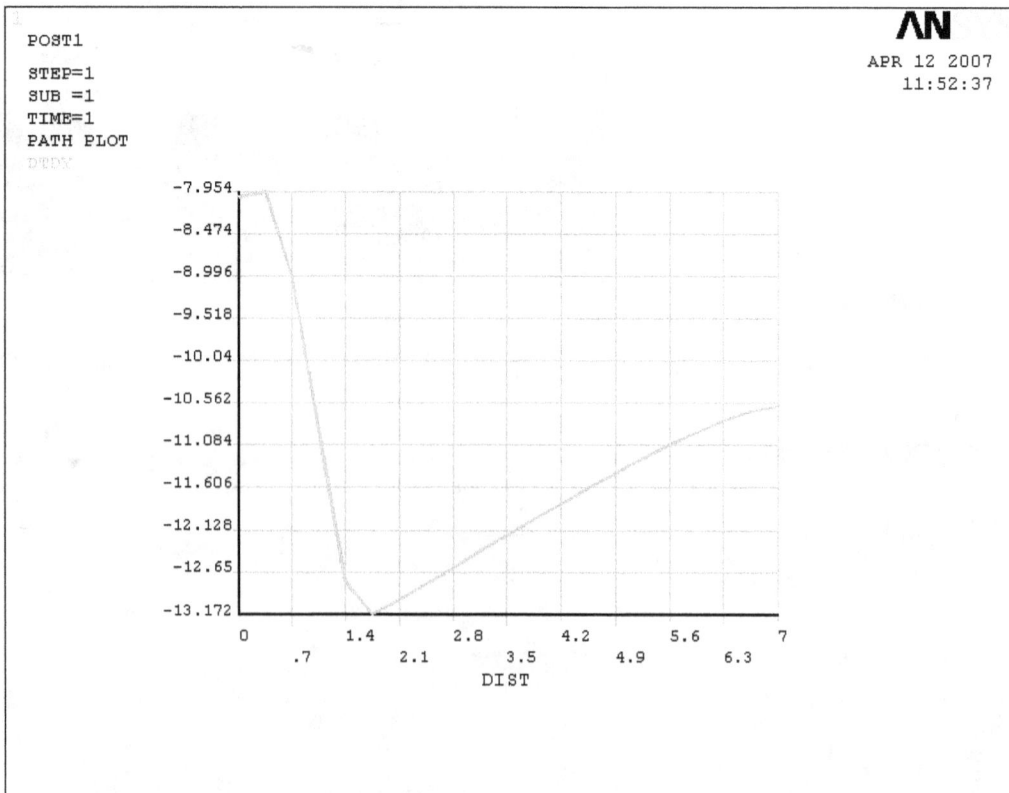

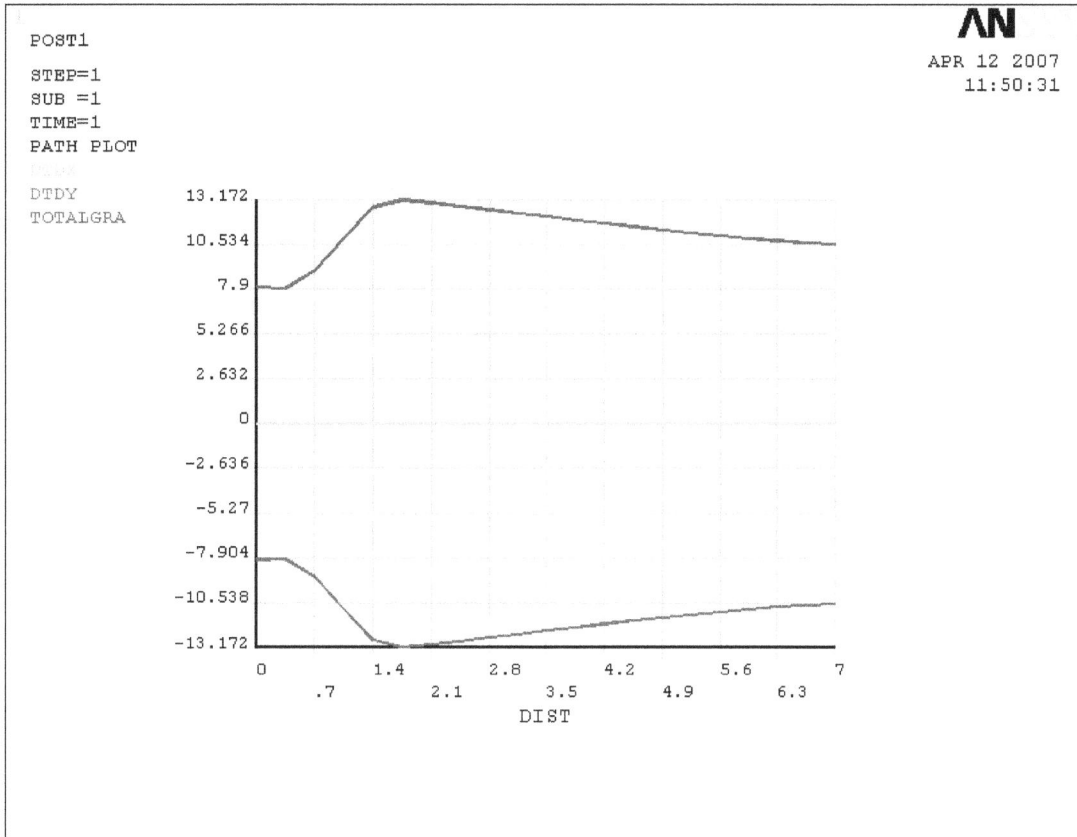

10.10 Fluid Mechanics Analysis with a 2D Model

To introduce the analysis of fluid mechanics, consider an ideal flow of air around a cylinder. As shown in the following figure, the velocity of air flow is 10 cm/s. The radius of the cylinder is 5 cm. Let us employ ANSYS to determine the velocity distribution around the cylinder. We assume that the free-stream velocity remains constant at a distance of 25 cm downstream and upstream of the cylinder.

Unit: cm

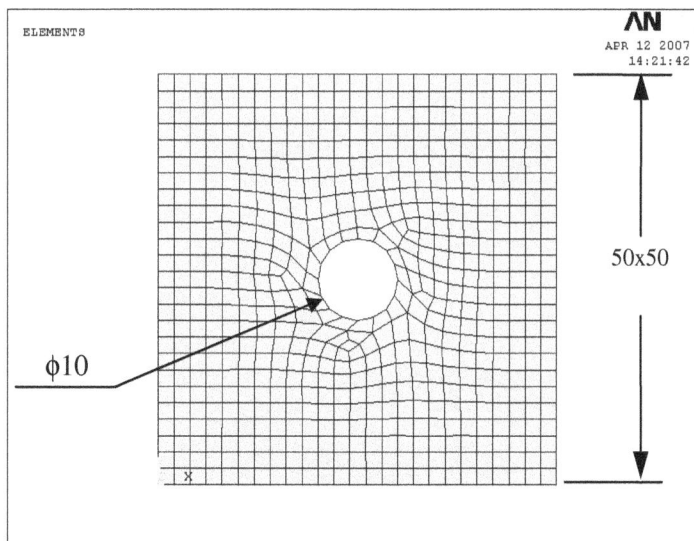

Step 1: Launch **ANSYS 13.0**

From the start menu, select **Programs > ANSYS 13.0 > ANSYS**, or directly click the icon of **ANSYS**.

To define the Structural Analysis, from the main menu or the model tree, select **Preferences**. > make sure that **Thermal** is highlighted, and h-method is also selected > **OK**.

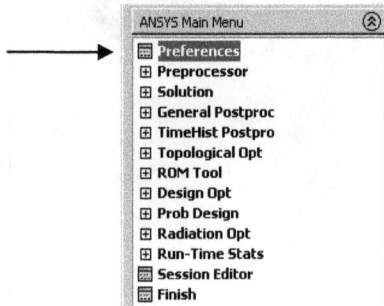

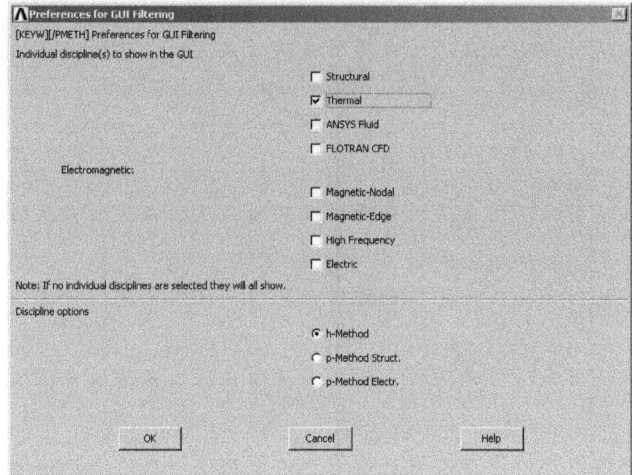

Step 2: Set the unit system to Metric (cm), or **CGS**.

Preprocessor > Material Library > Select Units > CGS > OK.

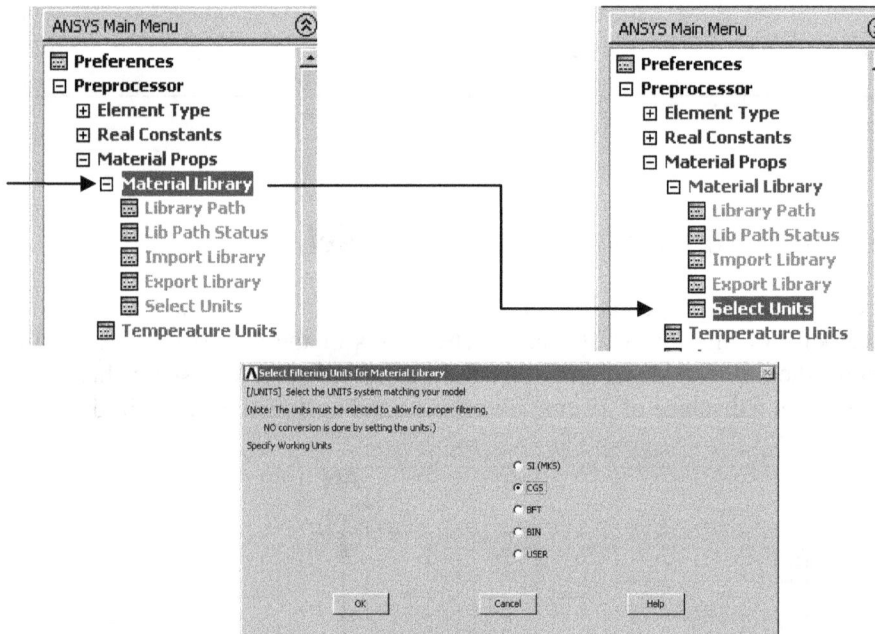

As illustrated, the basic quantities used in CGS system are

    Length: centimeter (cm)
    Mass: gram (g)
    Time: second (s)
    Temperature: Celsius (deg C)
    Force: dyne (g*cm/s^2)

Step 3: Define a 2D Geometry.

We start with the creation of a 2D model for the outer layer of bricks. From the main menu, select **Preprocessor** and expand it. Select **Modeling > Create > Areas > Rectangle > By Dimensions**.

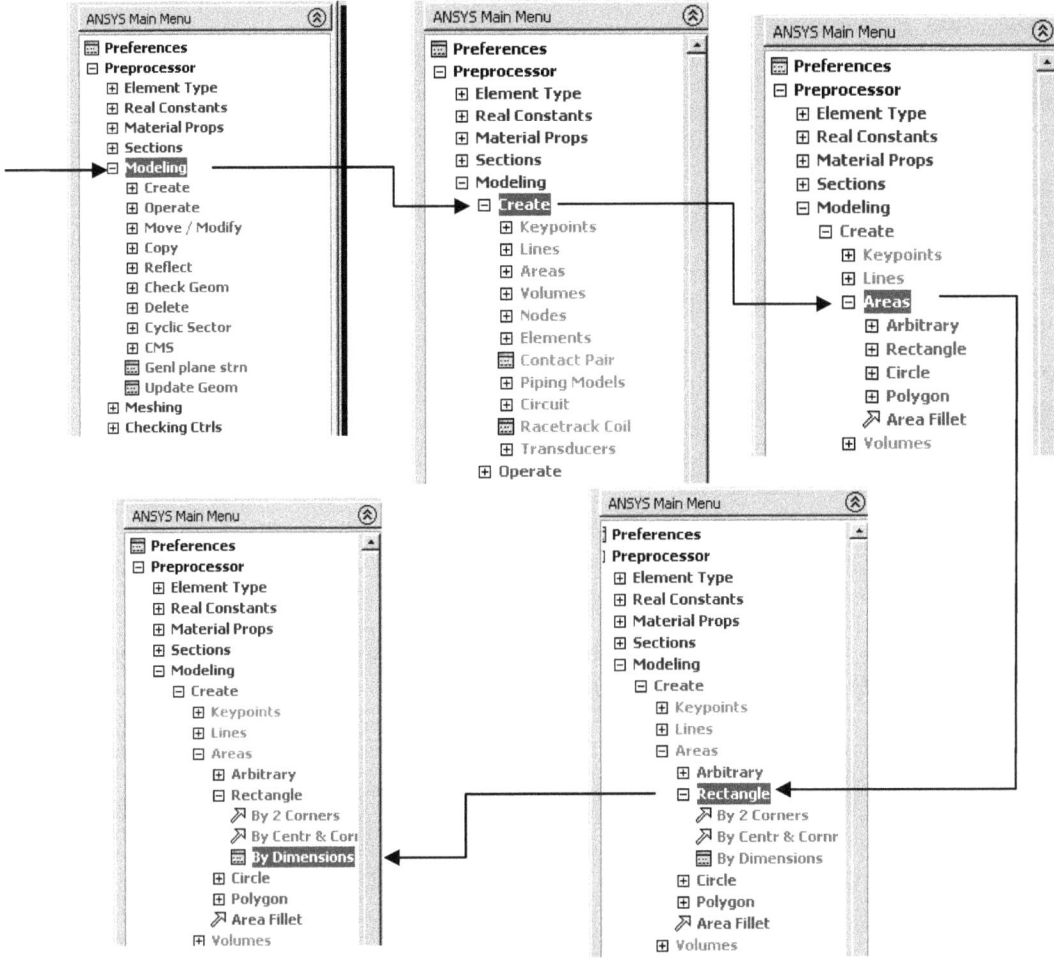

In the window for creating rectangle by dimensions, type 0 and 50 for X1 and X2, and 0 and 50 for Y1 and Y2 > **OK**.

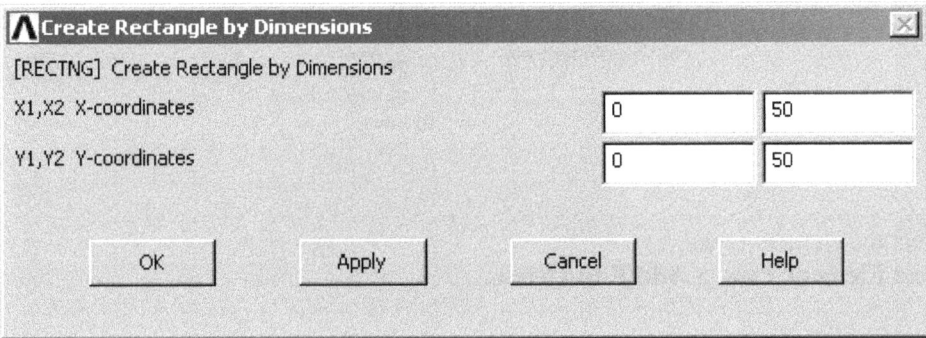

To create a circular area, **Circle > Solid Circle**. Specify 25, 25 as the center location and 5 as the radius value > **OK**.

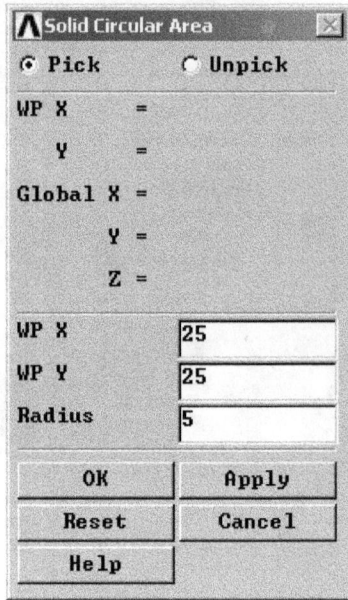

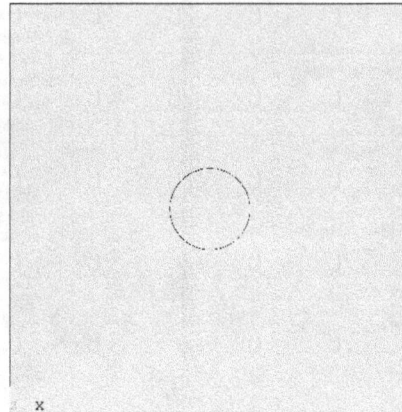

Now let us subtract Area 2 from Area 1. **Modeling > Operate > Booleans > Subtract > Areas.** Pick the rectangular area > **OK**. Pick the circular area > **OK**.

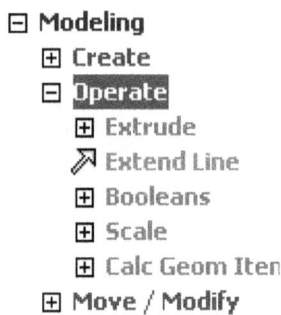

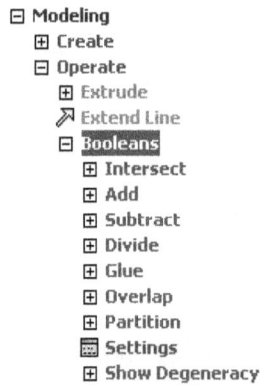

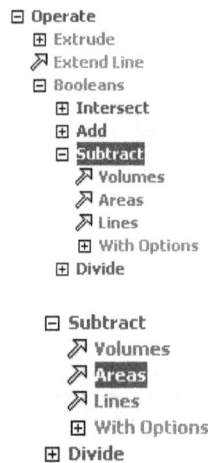

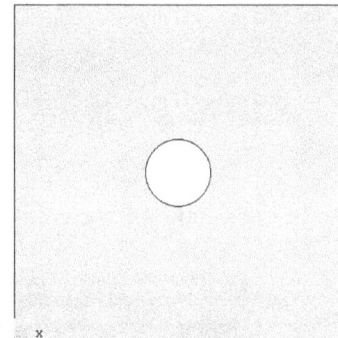

Step 4: Select the element type.
Select **Element Type > Add/Edit/Delete.**

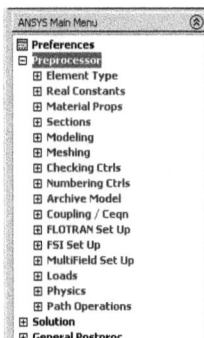

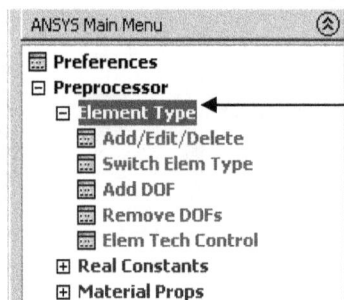

A window called **Element Types** appears. Click **Add** > pick **Solid > Quad 4nodes 55 > OK.**

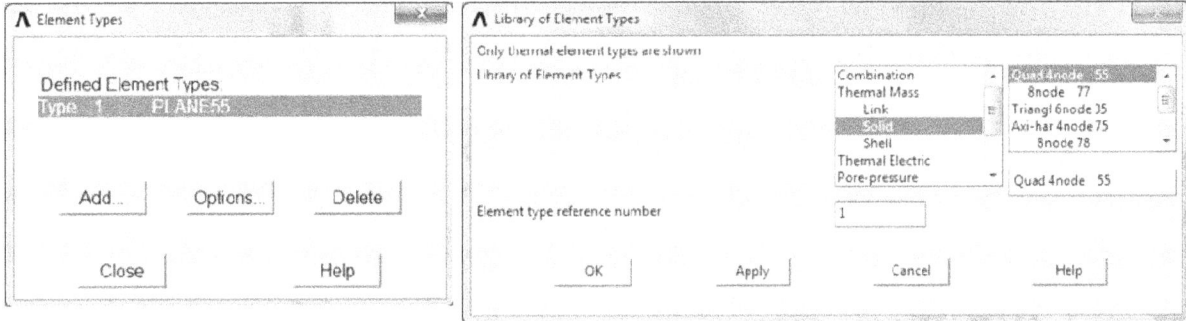

Step 4: Define the material type for the fluid media.

   Select **Material Properties** from the **Preprocessor** menu. Select **Material Model**, and then select **Thermal, Conductivity > Isotropic**. For material type 1 for the fluid media, specify 1 as the value of K_{xx} > **OK**.

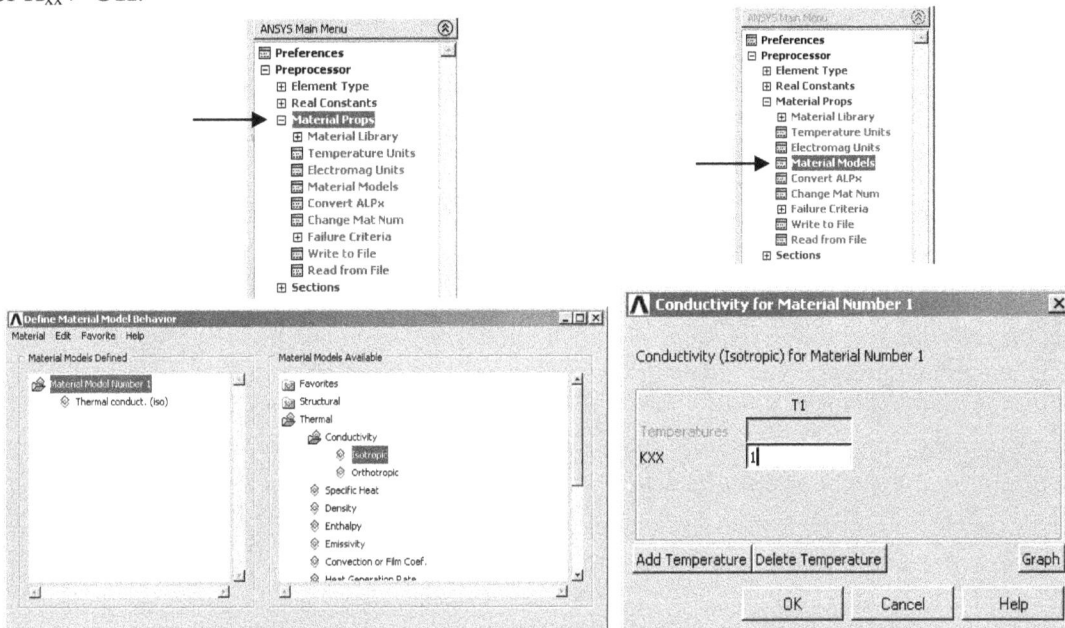

   To check the unit system, select **Material Library** under the material properties menu. Then select **Units** and make sure that the **CGS** unit system is highlighted. Click **OK**.

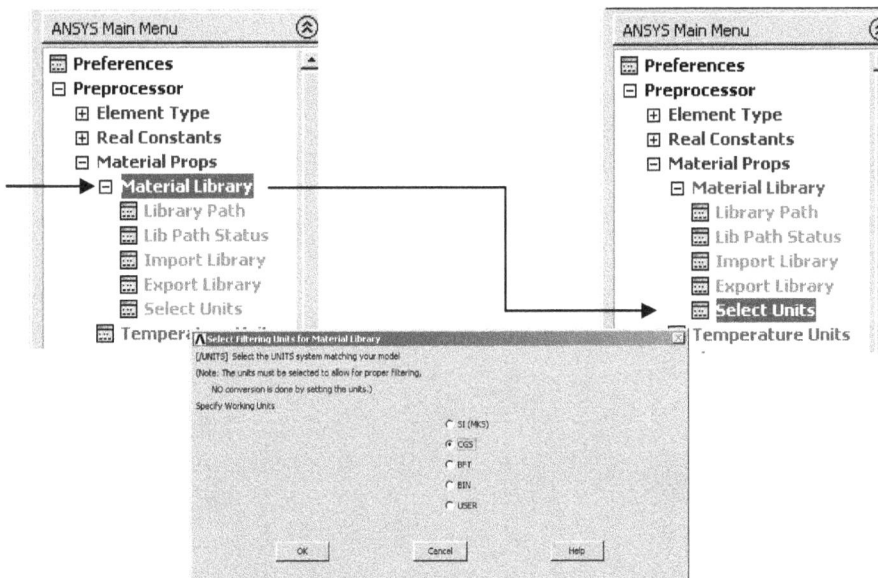

Step 5: Mesh Generation
To create the mesh for the shell model, select **MeshTool** from the Preprocessor menu. **Meshing > Size Cntrls > Manual Size > Global > Size**.
Specify 2 as the Element edge length > **OK**.

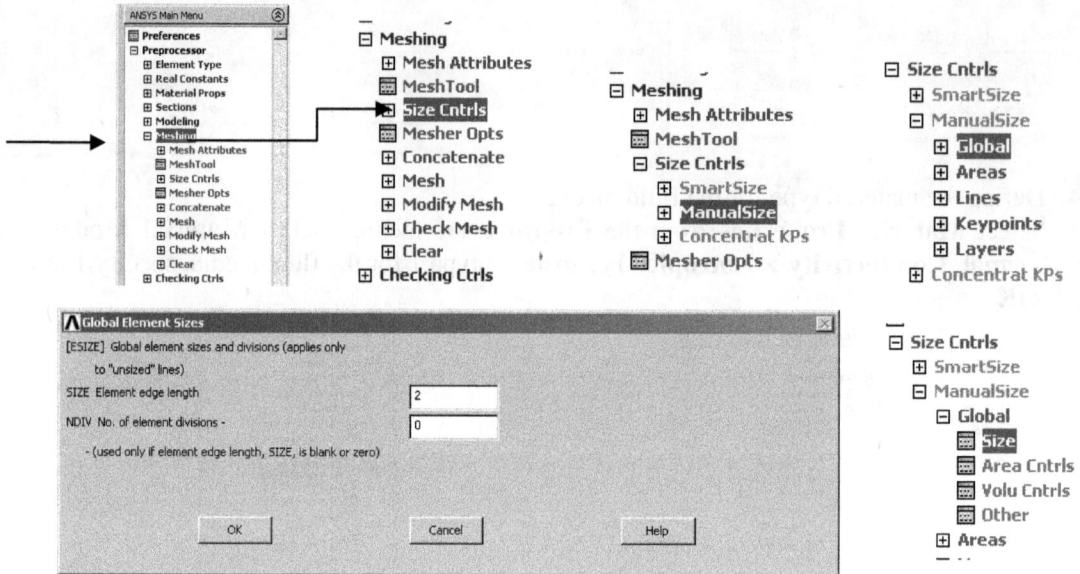

Meshing > Meshing Tool > Area > Quad > Free > Mesh.
When the selection box comes up, select **Pick All**. Make sure that you close the **MeshTool** window when the meshing process is completed.

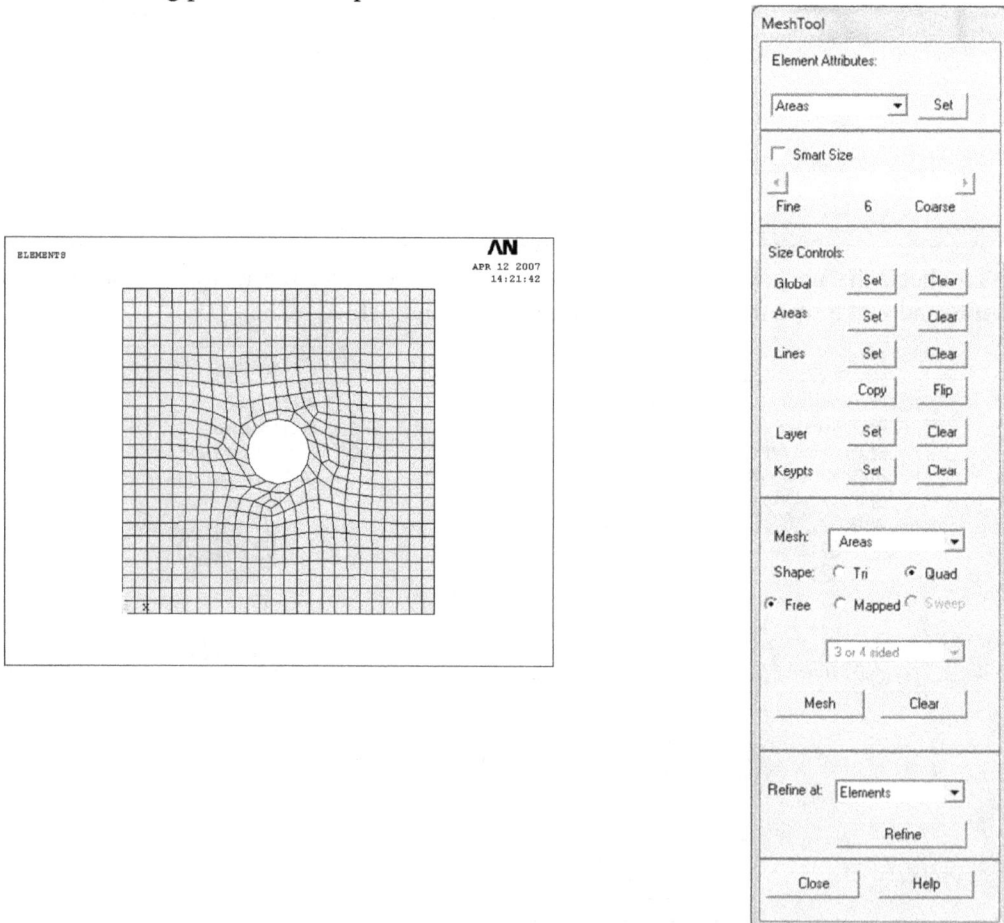

Step 6: Apply the convection boundary conditions.

On the **Preprocessor** menu, click **Loads > Define Loads > Apply > Thermal > Heat Flux > On Lines** > pick the edge on the left side > **OK**. Specify 10 > **OK**.

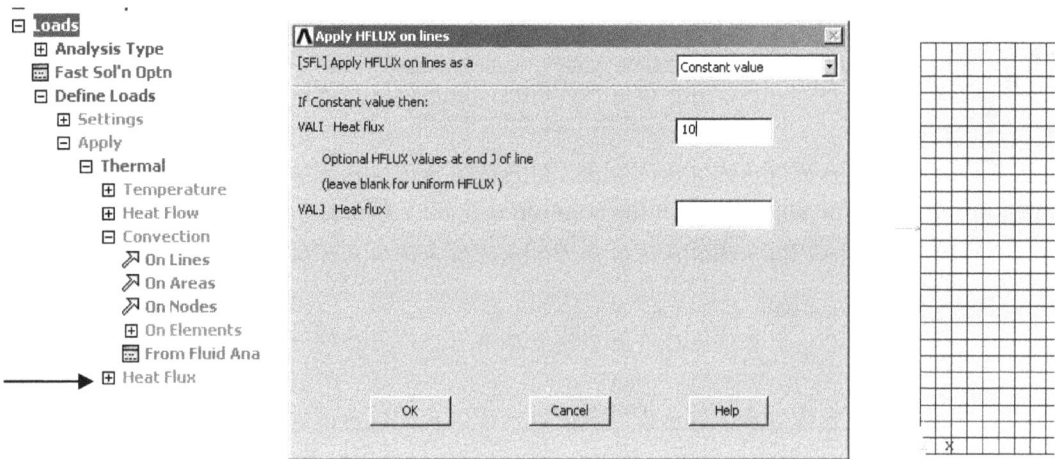

Loads > Define Loads > Apply > Thermal > Heat Flux > On Lines > pick the edge on the right side > **OK**. Specify -10 > **OK.**

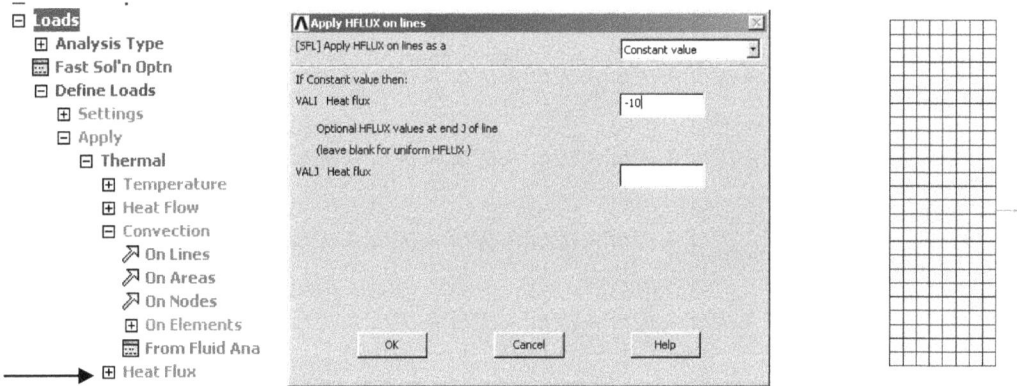

Step 7: Perform the analysis and obtain the solution.

To obtain the solution, run the program by selecting **Solution** from main menu > **Analysis Type > New Analysis > Steady-State > OK**.

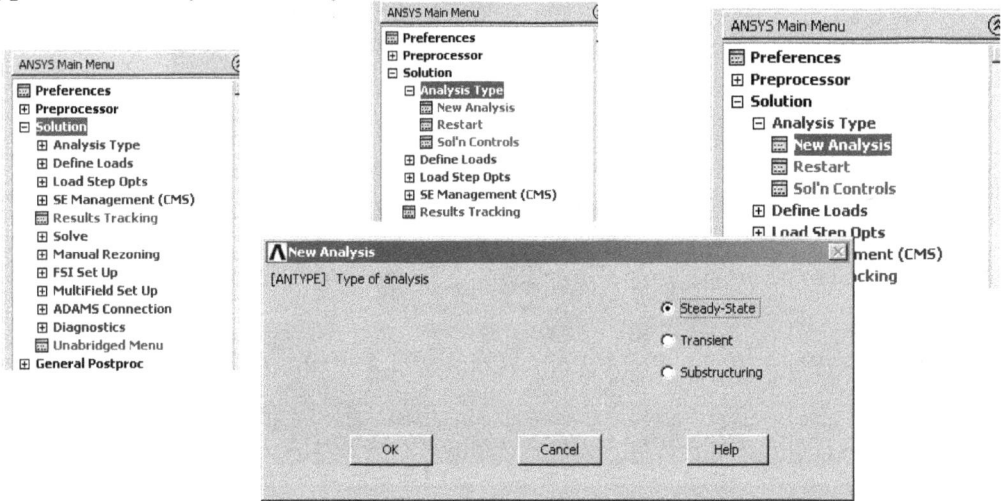

Solve > Current LS > after reviewing the information, click **OK** to run the program.

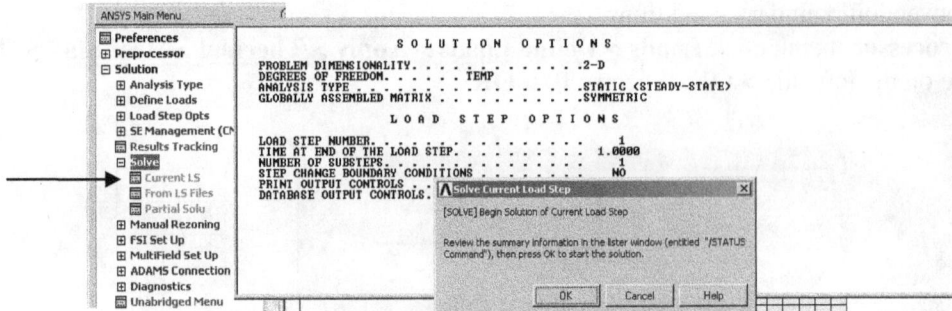

A window called Note appears when the solution is done > **Close**.

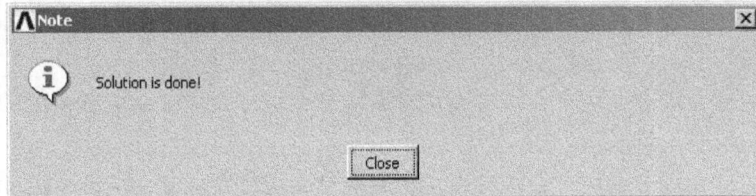

Step 8: Displaying Results

To plot the heat flow vector, **Plot Results > Vector Plot > Predefined**. Select **Flux & gradient > Thermal flux TF > OK**.

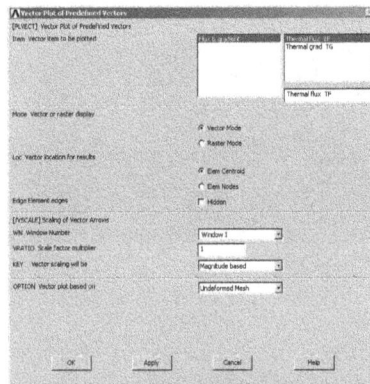

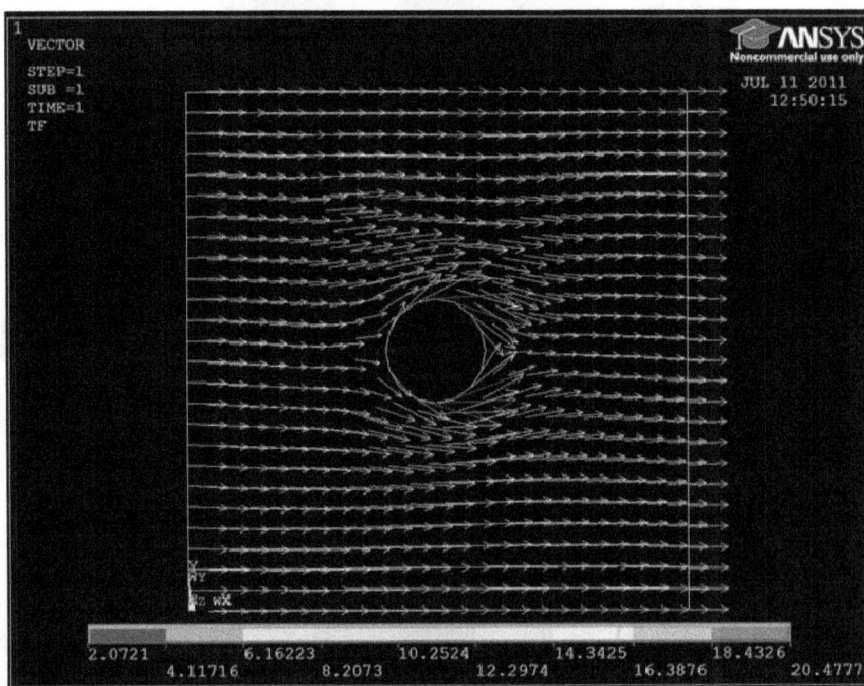

To plot the flux gradient along a specific path, we need to define the path first. To define a path, **General Postproc > Path Operations > Define Path > On Working Plane > Arbitrary > OK**.

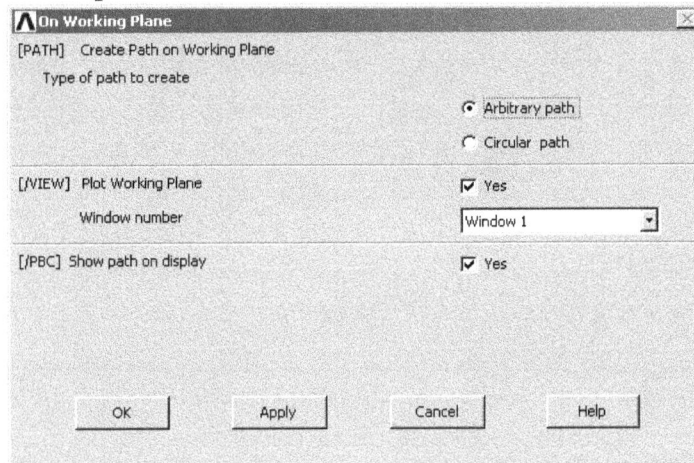

In the window, make sure that the icon of the **WP coordinates** is checked, not the **Global** > pick 2 points, as shown > specify 101 as the name of the defined path > **OK**.

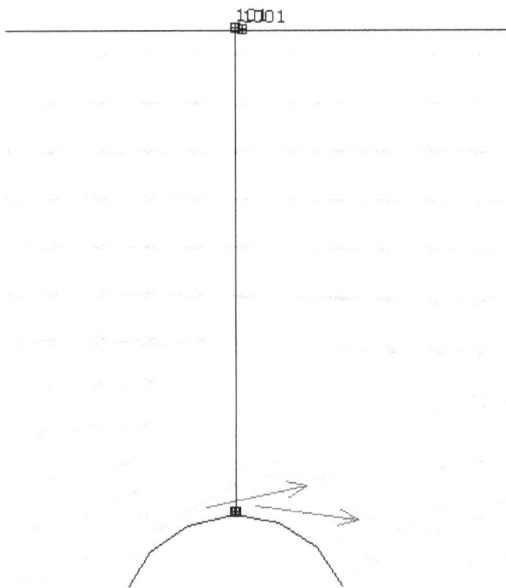

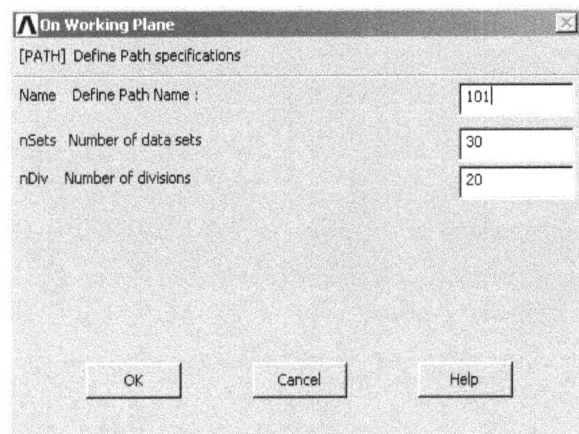

General Postproc > Path Operations > Map onto Path > specify *velocity* > pick **Flux & gradient > TFSUM > OK**.

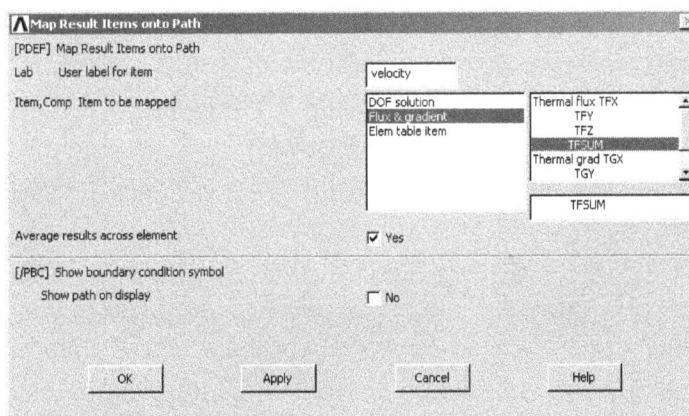

To visualize the obtained result, **General Postproc > Path Operations > Plot Path Item > On Graph**.

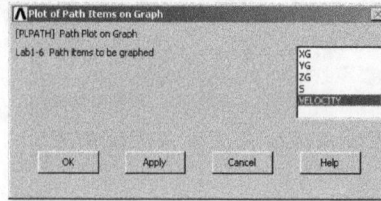

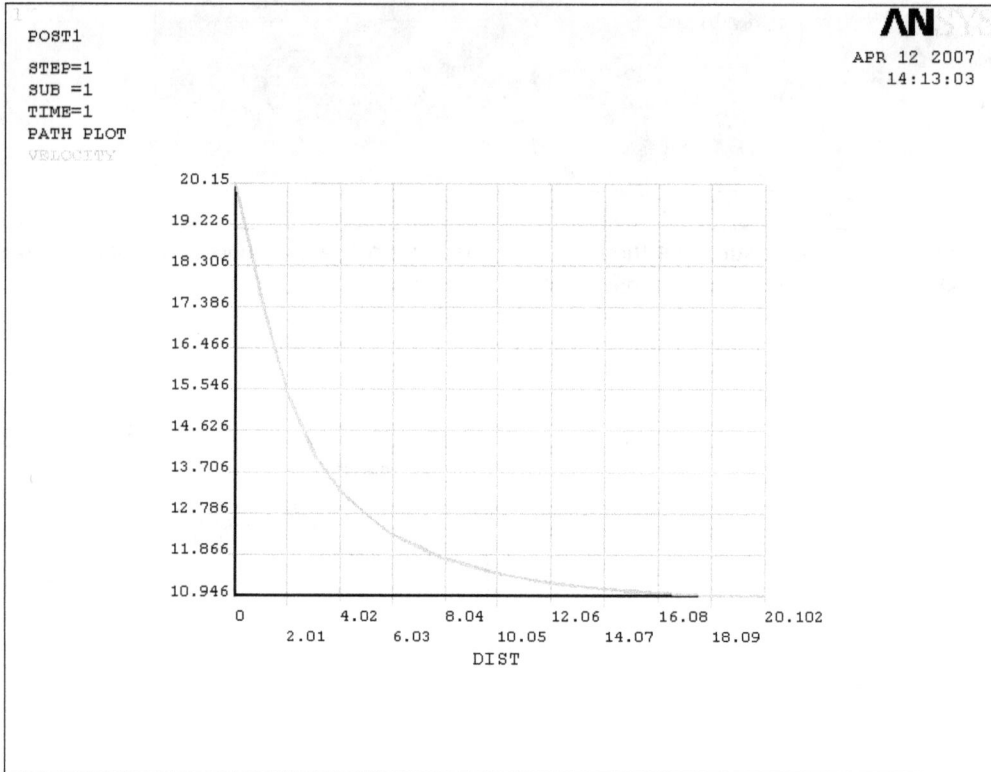

Repeat the above procedure by defining a new path and plot the Flux & gradient in the horizontal direction.

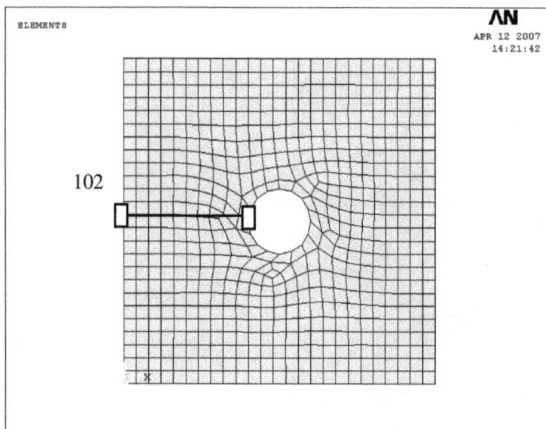

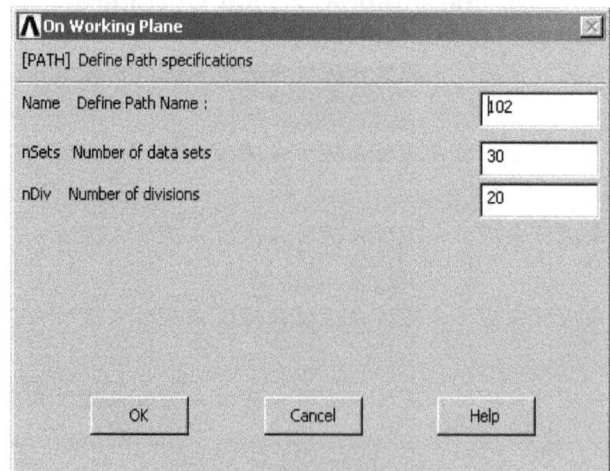

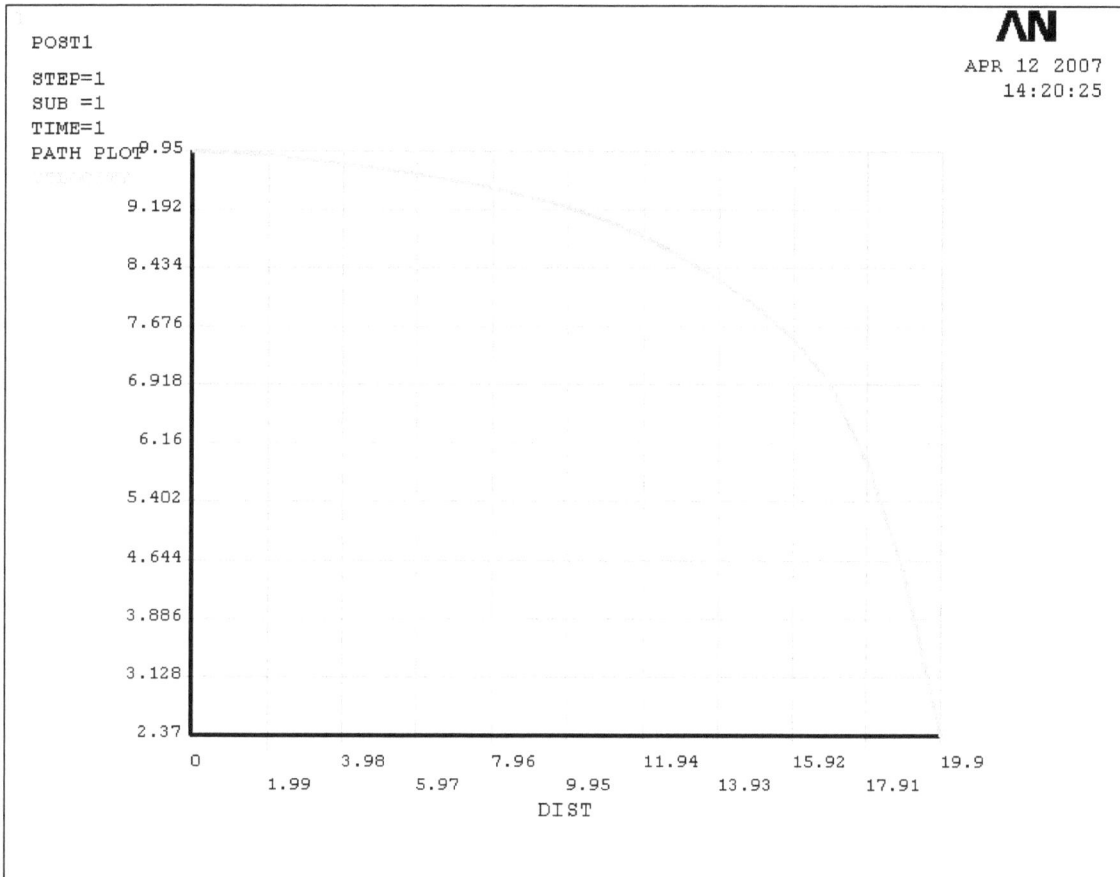

10.11 ANSYS Command File Creation and Execution

At this moment, readers are familiar with the procedure to perform FEA using ANSYS. It is the time for readers to practice the generation of commend files, a traditional method before the Graphic User Interface or the use of graphic icons and a pointing device, became available to control and execute an ANSYS program file. An example of a command file is shown below.

```
/filename, beam_hole_pressure_load
!-----------
!beam properties and dimension
!---------------
bwidth=8e-4                      !beam width (m)
bthick=1e-4                      !beam thickness(m)
bex=19e10                        !beam young's modulus (pa)
bNUXY= 0.27                      !beam poisson's ratio
bsmart = 3                       !smart size for beam meshing
!---------------
!cylinder
!-----------
cx=bwidth/2
cy=bthick/2
cr=4e-5
cd=bthick
cmart=2
!-------------
```

```
!boundary condition
!-------------
pressu=8e7                      !pressure force on the beam (pa)
!-------------
!begin model generation
!-------------
/prep7                          !entry with preprocessor command
!-------------
!Element Type and Material Properties
!-------------
ET,1,solid92                    !apply element type brick 8node 45
!*define the material properties
mp, ex,1,bex                    !young's modulus
mp, NUXY,1,bNUXY                !poisson's ratio
R,1                             !real constant related to material 1
!-------------
!Modeling
!-------------
blc4,,,bwidth,bthick,bthick     !create the beam volume
cyl4,cx,cy,cr,,,,cd             !create the circle volume
vsel,s,volu,,1                  !first select the beam volume
vsel,a,volu,,2                  !second select the circle volume
vsbv,1,2,,,delete               !boolean function subtract the circle volume from the beam volume
!-------------
!meshing 1
!-------------
Vsel, s, volu,,all              !select the volume after subtraction
Vatt,1,1,1                      !assign material, real constant relative material type, element type
Smartsize,bsmart                !generate mesh by using smart size
vmesh,all                       !generate mesh for the whole beam volume
!-------------
!meshing 2
!-------------
nsel,s,loc,x,cx+cr,cx-cr        !select the nodes around the circle
nrefine,all,,cmart              !refine mesh size at the circle by the nodes
!-------------
!applying boundary condition
!-------------
/psf,pres,,0                    !display load boundary condition
asel,s,loc,x,0                  !select the left constrait surface
nsla,s,1                        !select all the nodes on the selected surface
d,all,all                       !constraint DOF of the all nodes =0
asel,u,loc,x,0                  !unpick the selected left surface
asel,S,loc,y,bthick             !select the top surface
nsla,s,1                        !pick all the nodes on the surface
sf,all,pres,pressu              !apply a pressure load on the top surface
finish                          !finish the /prep7 command
allsel,all                      !select all commands
/pbc,all,,1                     !display boundary condition
/solu                           !enter solution command
solve                           !solve current LS
finish                          !finish the /solu command
```

There are several methods available to execute such a command file to run FEA. For example, from the File menu, select Read Input from ...An alternative method would be reading the command file right from the ANSYS command line. The most efficient method to execute a command file is "Copy the text file, paste it in the location called ANSYS Command Prompt, and press the Enter key."

Paste the copied file in this location, and press the **Enter** key

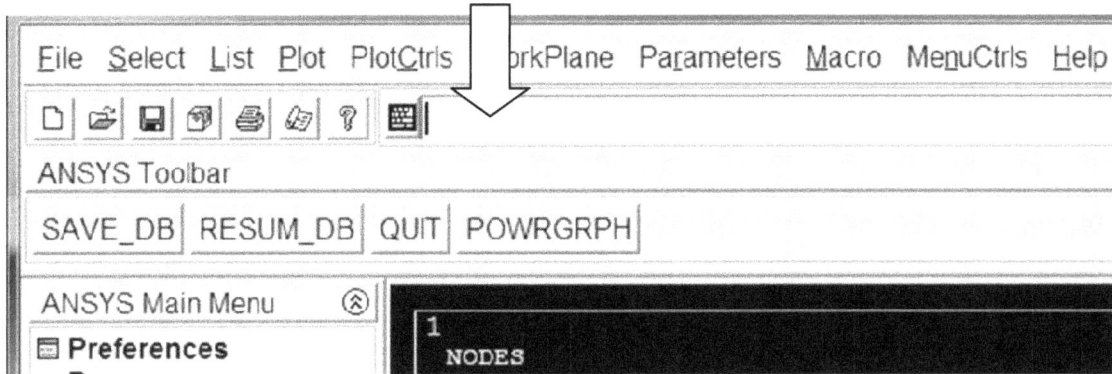

Readers may try this copy-paste-enter procedure to run FEA using the command file listed above. The result will be what is shown below.

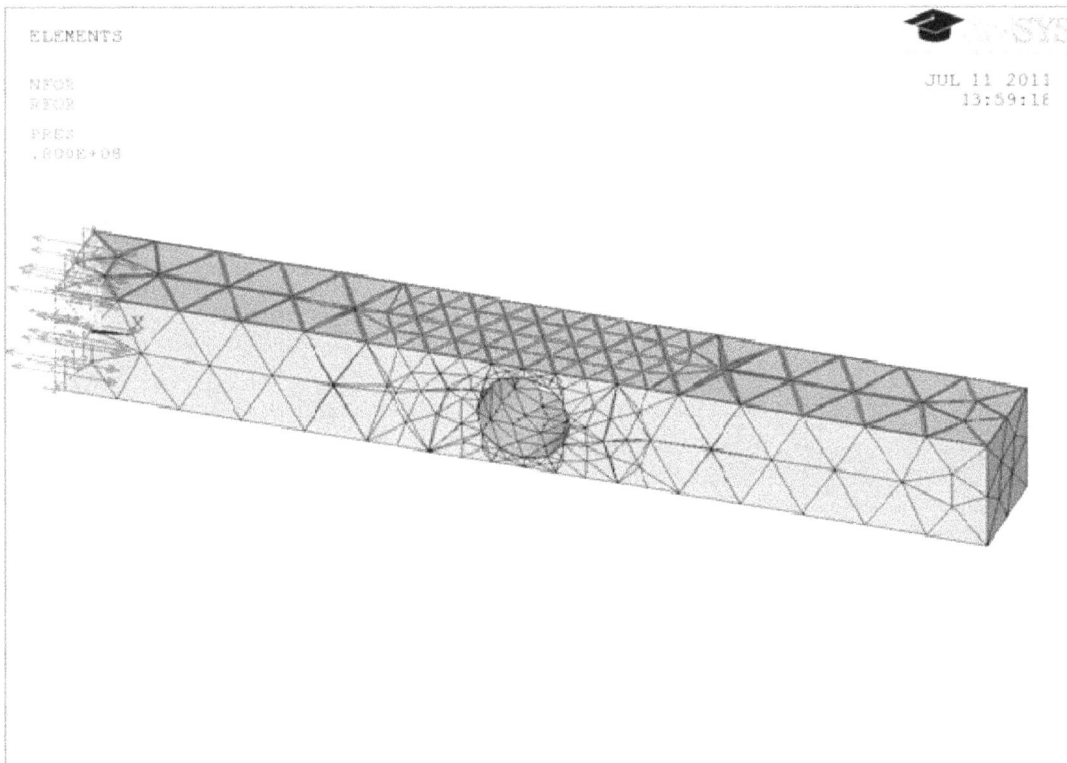

For readers to practice this copy-paste-enter procedure to run FEA using a command file, two additional files are listed below. The first file shows two holes associated with the beam. The second file demonstrates the procedure to relocate the 2 holes.

```
!------------------------
!beam with two holes
!------------------------
```

```
bw=8                        !Length of The beam
bt=1                        !Height of the beam
bd=0.5                      !Depth of the beam
bex=8e10                    !Youngs Modulus
bnuxy=0.26                  !Poisson ratio
force=-10                   !Force applied to the surface
sbeam=4                     !Mesh Size of the beam
!----------------
!cylinder1
!--------------
r=0.4                       !radius of the cylinder
cx1=bw/2                    !center location of the cylinder in x
cy1=bt/2                    !center location of the cylinder in y
ct1=bd                      !depth of the cylinder
!--------------
!cylinder2
!--------------
r2=r                        !radius of the cylinder
cx2=3*bw/4                  !center location of the cylinder in x
cy2=cy1                     !center location of the cylinder in y
ct2=bd                      !depth of the cylinder
!modeling
/prep7                      !Entry of creating the model
!--------------
!element type
!--------------
et,1,solid187               !et = element type, 1 = element type 1, solid187
mp,ex,1,bex                 !mp = material properties, ex = Youngs modulus, 1 = material type 1,
                            bex = value of Youngs modulus
mp,nuxy,1,bnuxy             !mp = material properties, nuxy= poisson ratio, 1 = material type 1,
                            bnuxy = value of the poisson ratio
R,1                         !Real constant
blc4,,,bw,bt,bd             !(1) Creating the beam from the origin with the given dimensions at the
begining of the code
cyl4,cx1,cy1,r,,,,ct1       !(2) create a cylinder1 x,y,r1,theta 1,r2,theta 2, depth
cyl4,cx2,cy2,r2,,,,ct2      !(3) "  " cylinder2 "      "
vsel,s,volu,,1              !Select the beam (1) volume
vsel,a,volu,,2              !also select First cylinder (2) volume
vsbv,1,2,,,delete           !subtract the first cylinder (2) volume from the beam (1)
vsel,s,volu,,4              !select the beam with hole (4) volume
vsel,a,volu,,3              !also select the second cylinder (3) volume
vsbv,4,3,,,delete           !subtract the second cylinder (3) volume from the beam with 1 hole (4)
                            volume

!--------------
!meshing
!--------------
vsel,s,volu,,all            !select the beam with the 2 holes
vatt,1,1,1                  !vatt = assign the material properties, element type, and real constant to
                            the whole volume
smartsize, sbeam            !define smart mesh
vmesh,all                   !mesh the whole volume
!--------------
!meshing 2
```

```
!--------------
nsel,s,loc,x,cx1-cr,cx1+cr              !nsel = node select, s = select, loc= from location, x = x direction, cx1-cr
                                         cx1+cr = from left of the circle to right
nsel,a,loc,x,cx2+cr2,cx2-cr2            !also select second circle
nrefine,all,2                          !refine all nodes meshing size to 2 (creating a fine mesh)
!--------------
!boundary
!--------------
nsel,s,loc,x,0                         !select nodes at the x=0 direction
d,all,all                              !degree of freedom of all the nodes on at the x=0 surface are constraint of
                                        all 6 degree of freedom.
asel,s,loc,y,bt                        !select area at the top surface of the beam
nsla,s,1                               !nsla= select all nodes related to top surface
f,all,fy,force                         !force applied in y direction (f can be only used for nodes)
allsel,all                             !allsel = select all of the codes
finish                                 !finish the prep7
/pbc,all,,1                            !display all the boundary conditions
/solu                                  !open the solve tab
solve                                  !solve the current setup
finish                                 !finish the code
```

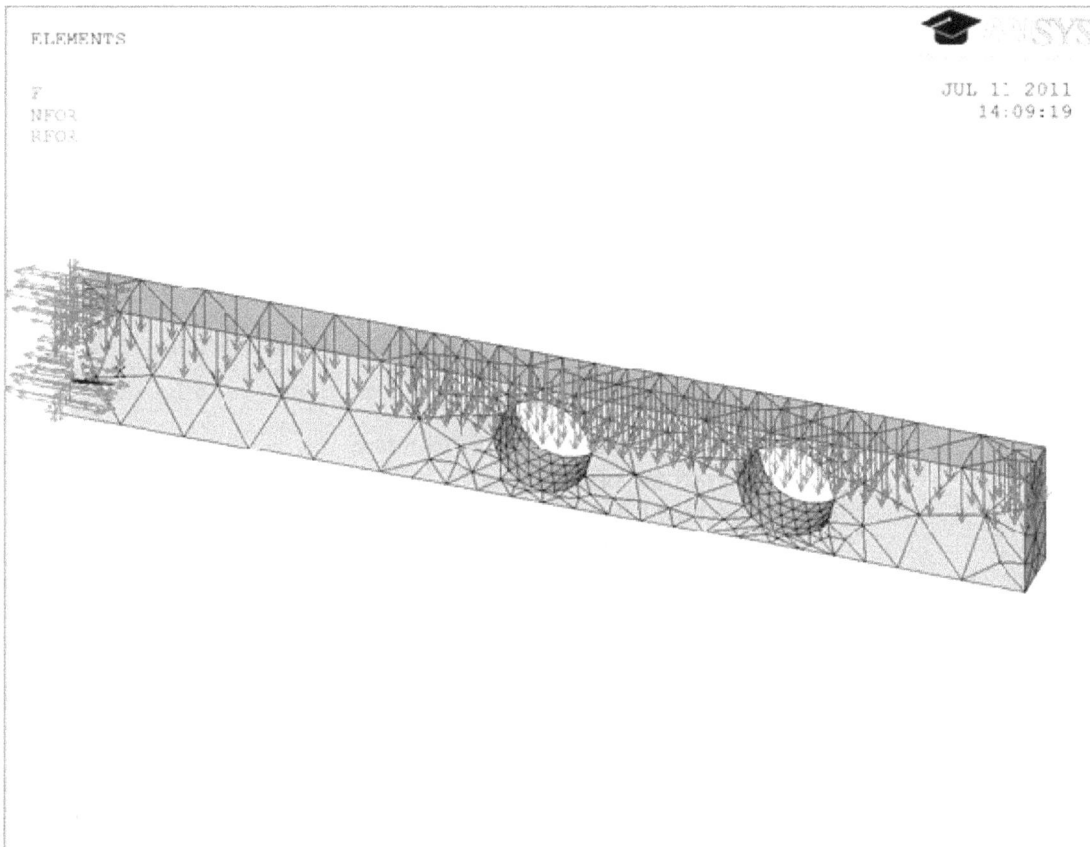

```
!-----------------------------------------
!beam with two holes relocated
!-----------------------------------------
bw=8                                   !Length of The beam
bt=1                                   !Height of the beam
bd=0.5                                 !Depth of the beam
bex=8e10                               !Youngs Modulus
```

```
bnuxy=0.26                      !Poisson ratio
force=-10                       !Force applied to the surface
sbeam=4                         !Mesh Size of the beam
!----------------
!cylinder1
!----------------
r=0.4                           !radius of the cylinder
cx1=bw/4                        !center location of the cylinder in x
cy1=bt/2                        !center location of the cylinder in y
ct1=bd                          !depth of the cylinder
!--------------
!cylinder2
!--------------
r2=r                            !radius of the cylinder
cx2=3*bw/4                      !center location of the cylinder in x
cy2=cy1                         !center location of the cylinder in y
ct2=bd                          !depth of the cylinder
!modeling
/prep7                          !Entry of creating the model
!--------------
!element type
!--------------
et,1,solid187                   !et = element type, 1 = element type 1, solid187
mp,ex,1,bex                     !mp = material properties, ex = Youngs modulus, 1 = material type 1,
                                bex = value of Youngs modulus
mp,nuxy,1,bnuxy                 !mp = material properties, nuxy= poisson ratio, 1 = material type 1,
                                bnuxy = value of the poisson ratio
R,1                             !Real constant
blc4,,,bw,bt,bd                 !(1) Creating the beam from the origin with the given dimensions at the
                                begining of the code
cyl4,cx1,cy1,r,,,,ct1           !(2) create a cylinder1 x,y,r1,theta 1,r2,theta 2, depth
cyl4,cx2,cy2,r2,,,,ct2          !(3) "  " cylinder2 "      "
vsel,s,volu,,1                  !Select the beam (1) volume
vsel,a,volu,,2                  !also select First cylinder (2) volume
vsbv,1,2,,,delete               !subtract the first cylinder (2) volume from the beam (1)
vsel,s,volu,,4                  !select the beam with hole (4) volume
vsel,a,volu,,3                  !also select the second cylinder (3) volume
vsbv,4,3,,,delete               !subtract the second cylinder (3) volume from the beam with 1 hole (4)
volume
!--------------
!meshing
!--------------
vsel,s,volu,,all                !select the beam with the 2 holes
vatt,1,1,1                      !vatt = assign the material properties, element type, and real constant to
                                the whole volume
smartsize, sbeam                !define smart mesh
vmesh,all                       !mesh the whole volume
!--------------
!meshing 2
!--------------
nsel,s,loc,x,cx1-cr,cx1+cr      !nsel = node select, s = select, loc= from location, x = x direction, cx1-cr
                                cx1+cr = from left of the circle to right
nsel,a,loc,x,cx2+cr2,cx2-cr2    !also select second circle
```

```
nrefine,all,2                   !refine all nodes meshing size to 2 (creating a fine mesh)
!---------------
!boundary
!---------------
nsel,s,loc,x,0                  !select nodes at the x=0 direction
d,all,all                       !degree of freedom of all the nodes on at the x=0 surface are constraint of
                                 all 6 degree of freedom.
asel,s,loc,y,bt                 !select area at the top surface of the beam
nsla,s,1                        !nsla= select all nodes related to top surface
f,all,fy,force                  !force applied in y direction (f can be only used for nodes)
allsel,all                      !allsel = select all of the codes
finish                          !finish the prep7
/pbc,all,,1                     !display all the boundary conditions
/solu                           !open the solve tab
solve                           !solve the current setup
finish                          !finish the code
```

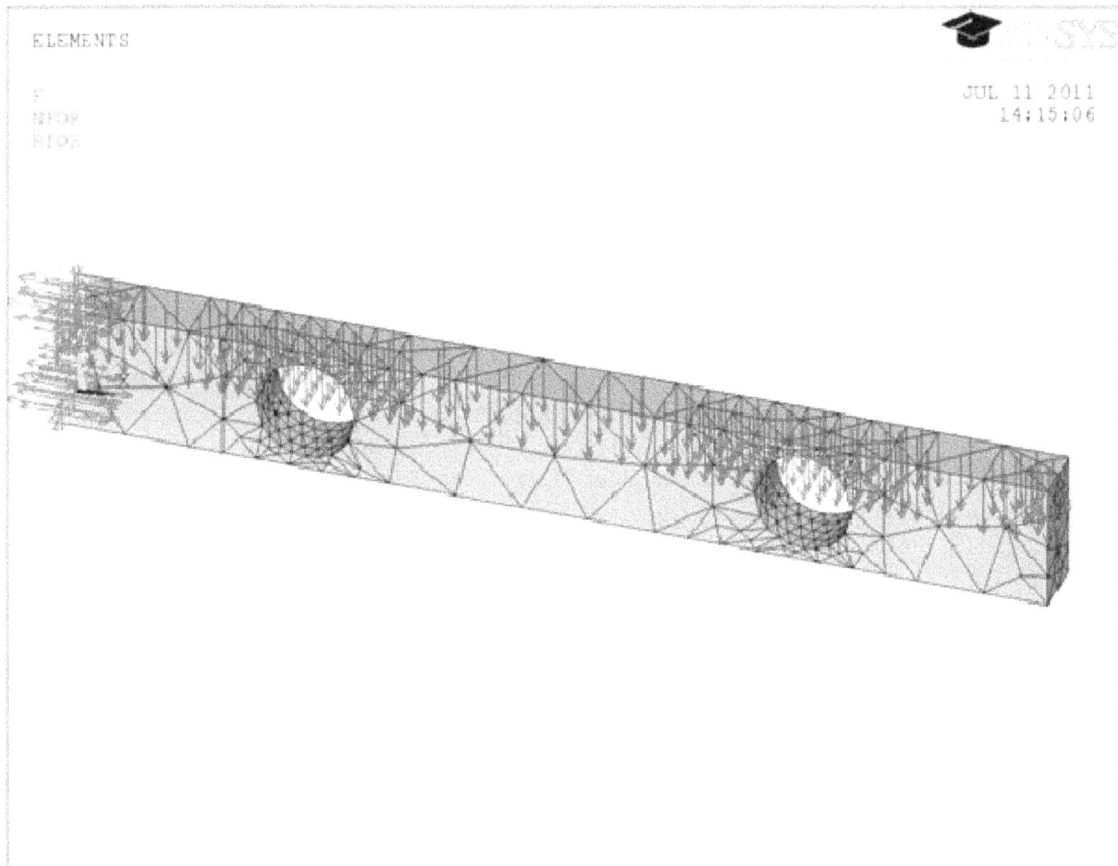

Writing a command file requires the knowledge of ANSYS commands. ANSYS contains hundreds of commands for generating geometry, applying loads and constraints, setting up different analysis types and post-processing. Readers may find useful information online. The address of the online ANSYS Command Menu is listed below.

http://www1.ansys.com/customer/content/documentation/100/ansys/acmd100.pdf

10.12 References

1. Algor Processor Reference Manual, Algor Interactive Systems, 150 Alpha Drive, Pittsburgh, PA 15238
2. O. Axelsson, Iterative Solution Methods, Cambridge University Press, New York, 1994.
3. K. J. Bathe, Finite Element Procedure, Prentice Hall, Upper Saddle River, New Jersey, 1996.
4. K. Bell, A Refined Triangular Plate Bending Finite Element, Int. J. Numer. Methods Eng., Vol. 1, No. 1 January 1969, pp. 101-122.
5. B. Carnahan, H. A. Luther, and J. O. Wikes, Applied Numerical Methods, Wiley New York, NY, 1969.
6. R. W. Clough, The Finite Element in Plane Stress Analysis, Proceedings of 2nd ASCE Conference on Electronic Computation, Pittsburgh, PA. September 1960.
7. R. D. Cook, D. S. Maikus and M. E. Plesha, Concepts and Applications of Finite Element Analysis, 3rd edition, Wiley, New York, NY, 1989.
8. J. W. Dally and W. F. Riley, Experimental Stress Analysis, McGraw Hill, 1965.
9. J. W. Dally, Design Analysis of Structural Elements, Goodington, New York, NY, 2003.
10. C. S. Desal and J. F. Abel, Introduction to the Finite Element Method, Van Nostrand Reinhold, New York, NY, 1972.
11. J. H. Earle, Graphics for Engineers, AutoCAD Release 13, Addision-Wesley, Reading, Massachusetts, 1996.
12. R. H. Gallagher, Finite Element Analysis Fundamentals, Prentice-Hall, Englewoood Cliffs, NJ, 1975.
13. J. M. Gere and S. P. Timoshenko, Mechanics of Materials, 3rd edition, PWS-KENT Publishing Company, Boston, 1990.
14. W. W. Hager, Applied Numerical Linear Algebra, Prentice-Hall, Englewood Cliffs, NJ, 1988.
15. Holland, The Finite Element Method in Plane Stress Analysis, in the Finite Element Method in Stress Analysis, I. Holland and K. Bell (eds), Tapir Press, Trondheim, Norway, 1969, Chapter 2.
16. Y. Y. Hsieh, Elementary Theory of Structure, 2nd edition, Prentice-Hall, Englewood Cliffs, NJ, 1982.
17. M. A. N. Hendriks, H Jongedijk, J. G. Rots and W. J. E. van Spange (eds), Finite Elements in Engineering and Science, DIANA Computational Mechanics, 1997.
18. K. H. Huebner, D. L. Dewhirst, D. E. Smith, and T. G. Byrom, The Finite Element Method for Engineers, 4th edition, John Wiley & Sons, Inc., 2001.
19. C.S. Krishnamoorthy, Finite Element Analysis, Theory and Programming, 2nd Ed., 1995.
20. D. L. Logan, A First Course in the Finite Element Method Using ALGOR, PWS Publishing, Boston, 1997.
21. H. C. Martin, Introduction to Matrix Methods of Structural Analysis, McGraw-Hill, New York, NY, 1966.
22. M. Ortiz, Y. Leroy and A. Needleman, A Finite Element Method for Localized Failure Analysis, Comp. Meth. Applied Mech. Enging, 61, pp. 189-214, 1987.
23. M. Petyl, Introduction to Finite Element Vibration Analysis, Cambridge University Press, 1990.
24. T. H. H. Pian and P. Tong, Basis of Finite Element Methods for Solid Continua, Int. J. Numer. Methods Eng., Vol. 2, No. 7, April 1964, pp. 1333-1336.
25. P. G. Plockner, Symmetry in Structural Mechanics, J. of the Structural Division, American Society of Civil Engineers, Vol. 99, No. ST1, pp. 71-89, 1973.
26. J. N. Reddy, An Introduction to the Finite Element Method, McGraw-Hill Publishing Company, New York, NY, 1984.
27. T. Ross, Finite Element Program in Structural Engineering and Continuum Mechanics, Horwood Publishing Limited, Chichester, England, 1996.
28. T. Ross, Advanced Applied Finite Element Methods, Horwood Publishing Limited, Chichester, England, 1998.
29. L. Stasa, Applied Finite Element Analysis for Engineers, Holt, Rinehart and Winston, New York, NY, 1985, pp. 101-157.

30. Y. Tada and G. C. Lee, Finite Element Solution to an Elastic Problem of Beams, Int. J. Numer. Methods Eng., Vol. 2, No. 2, April 1970, pp. 229-241.
31. R. L. Taylor, On Completeness of Shape Functions for Finite Element Analysis, Int. J. Numer., Methods Eng., Vol. 4., No. 1, 1972, pp. 17-22.
32. S. P. Timoshenko and D. H. Young, Theory of Structure, McGraw-Hill, Kogaksha Limited, 1965.
33. M. J. Turner, R. W. Clough, H. C. Martin, and L. J. Topp, Stiffness and Deflection Analysis of Complex Structures, J. of the Aeronautical Sciences, Vol. 23, No. 9, pp. 805-824, Sept. 1956.
34. O. C. Zienkiewicz, The Finite Element Method, third edition, McGraw-Hill, New York, NY, 1977.
35. O. C. Zienkiewicz and R. L. Taylor, The Finite Element Method, 4th edition, Vol. 1, McGraw-Hill (UK),London, 1989.

10.13 Exercises

1. The following figure illustrates the geometrical shape of the beam structure under consideration. The left end of the beam structure is a pin connection with the ground. The right end of the beam structure is a roller support. The total length of the beam structure is 2000 mm. The section area is a rectangle. The two dimensions are 25 mm and 40 mm, respectively. There are 3 loads acting on the beam. The first load acts on the right node. The magnitude is 4000 Newton and its direction is towards the right, or along the positive x axis direction. The second load is uniformly distributed load acting on the entire over length of the beam structure with the magnitude equal to 2 N/mm. The third load is a concentration load acting on a distance equal to 500 mm with reference to the right end of the beam structure. Pay attention to the assignment of the material properties. We assume that the Young's modulus is 30 MPa and the Poisson's ratio is 0.27. Therefore, the material is easy to be bent in the Y direction. Our objective of this study is to show the deflection pattern and to know the maximum deflection value and the location of the maximum deflection.

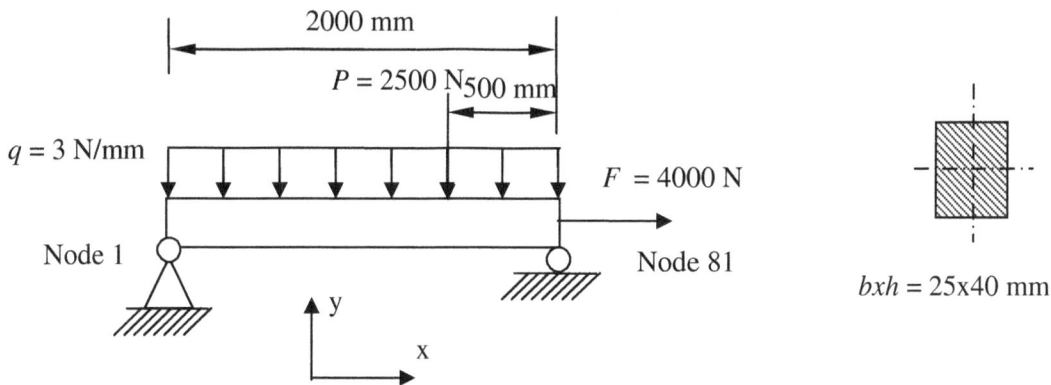

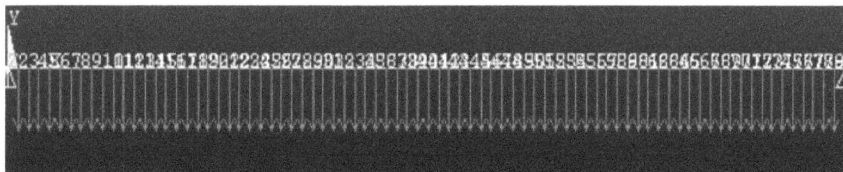

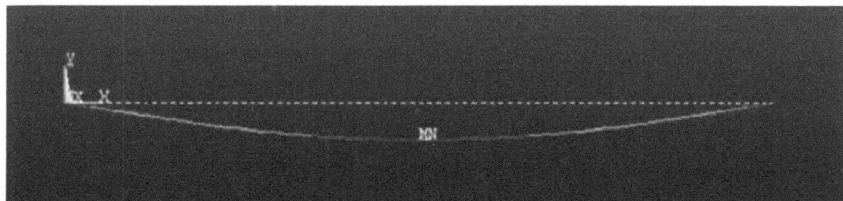

2. The following figure illustrates a truss structure, which consists of 3 bars. The dimension of the section area is 1000 mm^2. The geometrical information of the truss structure is shown in the figure. The material type is STEEL. Use **ANSYS** to construct the shear and moment diagrams.

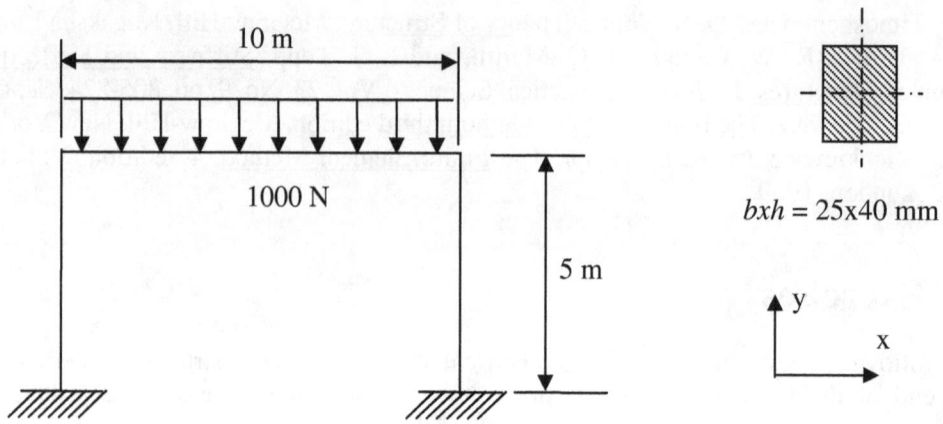

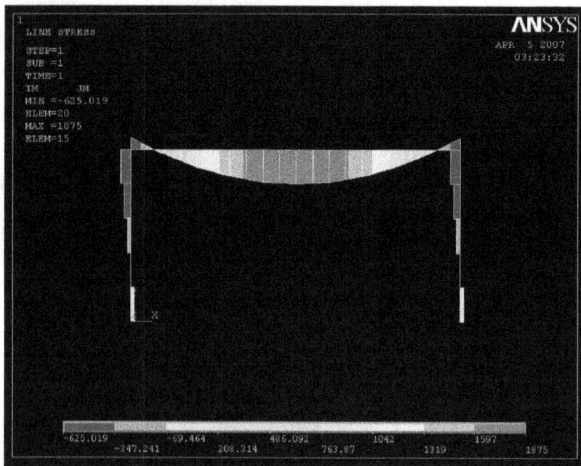

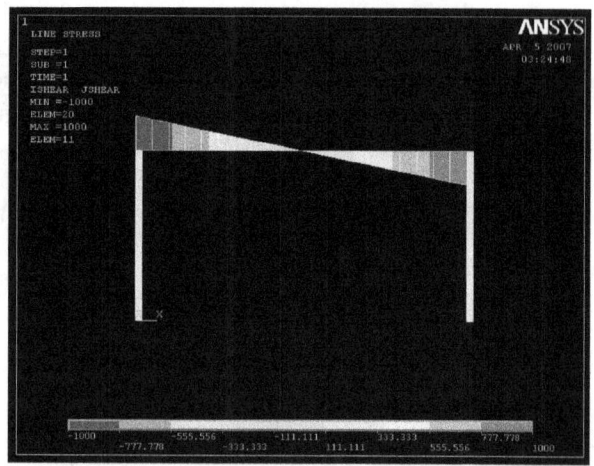

INDEX

www.ingramcontent.com/pod-product-compliance
Lightning Source LLC
Chambersburg PA
CBHW081238220326
41597CB00023BA/4024